AQA GCSE

MATHEMATICS
for Middle sets

Series editor: **Glyn Payne**

Authors:
Gwen Burns
Greg Byrd
Lynn Byrd
Crawford Craig
Janet Crawshaw
Fiona Mapp
Avnee Morjaria
Catherine Murphy
Katherine Pate
Glyn Payne
Ian Robinson
Harry Smith

www.pearsonschools.co.uk
✓ Free online support
✓ Useful weblinks
✓ 24 hour online ordering
0845 630 33 33

Part of Pearson

Longman is an imprint of Pearson Education Limited, a company incorporated in England and Wales, having its registered office at Edinburgh Gate, Harlow, Essex, CM20 2JE. Registered company number: 872828

www.pearsonschoolsandfecolleges.co.uk

Longman is a registered trademark of Pearson Education Limited

First published 2010
14 13 12 11 10
10 9 8 7 6 5 4 3 2 1 0

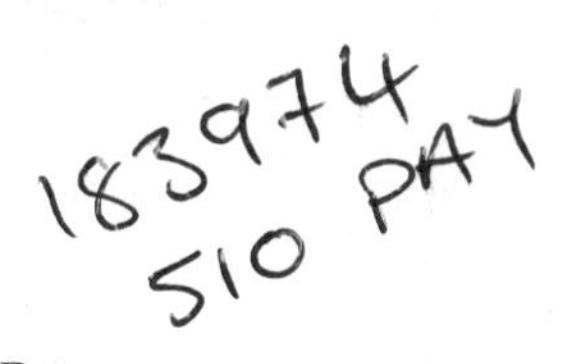

British Library Cataloguing in Publication Data
A catalogue record for this book is available from the British Library.
ISBN 978 1 408 23282 8

Edited by Gwen Burns, Nicola Morgan, Jim Newall, Maggie Rumble, Laurice Suess and Christine Vaughan
Designed by Pearson Education Limited
Typeset by Tech-Set Ltd, Gateshead
Original illustrations © Pearson Education Ltd 2010
Illustrated by Tech-Set Ltd
Cover design by Wooden Ark
Picture research by Chrissie Martin
Cover photo/illustration © iStockPhoto / Stuart Berman
Printed in Italy by Rotolito

Acknowledgements

The author and publisher would like to thank the following individuals and organisations for permission to reproduce photographs.

Alamy / Alex Segre p263; Alamy / Anthony Collins Cycling p472; Alamy / Bruce McGowan p368; Alamy / Carmen Sedano p390; Alamy / Charles Best p147; Alamy / John Glover p448; Alamy / Kevin Wheal p259; Alamy / Malcolm Case-Green p416; Alamy / Mark Sunderland p486; Alamy / Martin Pick p404; Alamy / Robert Convery p113; Alamy / Sinibomb p430; Alamy / Yiap Views p212; Corbis / Arcaid / Clive Nichols p458; Corbis / Crush p322; Corbis / Darrell Gulin p196; Corbis / Frank Lukasseck p276; Corbis / Herbert Pfarrhofer p95; Digital Vision p35; Getty Images / AFP / Martin Bureau p340; Getty Images / Iconica p158; Getty Images / PhotoDisc pp13, 159, 202, 223, 411, 536; Getty Images / Photographers Choice p86; Getty Images / Riser p460; Getty Images / Stone p1; ImageState / John Foxx Imagery p422; Masterfile / Allan Davey p132; Masterfile / Jeremy Maude p533; NASA / Goddard Space Flight Center p278; Pearson Education Ltd p297; Pearson Education Ltd / Gareth Boden p188; Pearson Education Ltd / Jules Selmes p548; Photolibrary / Digital Light Source p503; Photolibrary / Imagestate Media p242; Rex Features p330; Science Photo Library / Antonia Reeve p70; Science Photo Library / Chris Priest p208; Science Photo Library / D. van Ravenswaay p82; Science Photo Library / Detlev van Ravenswaay p357; Science Photo Library / Friedrich Saurer p138; Science Photo Library / Susumu Nishinaga p474; Science Photo Library / TRL Ltd p53; Shutterstock / 0833379753 p444; Shutterstock / Afaizal p198; Shutterstock / Alexander Raths p281; Shutterstock / Alexei Novikov p154; Shutterstock / Benis Arapovic p343; Shutterstock / BlueOrange Studio p130; Shutterstock / CBPix p488; Shutterstock / Elena Elisseva p41; Shutterstock / Erhan Daly p297; Shutterstock / Eric Isselee p54; Shutterstock / Freebird p124; Shutterstock / Galyna Andrushko p335; Shutterstock / IcemanJ p378; Shutterstock / Ioana Drutu p286; Shutterstock / Jan van der Hoeven p61; Shutterstock / Jenny T p311; Shutterstock / Jiri Juru p224; Shutterstock / Johann Helgason p394; Shutterstock / Kapu p443; Shutterstock / Liew Weng Keong p252; Shutterstock / Mary Katherine Donovan p76; Shutterstock / Meerok p374; Shutterstock / Michael Svoboda p520; Shutterstock / Mircea Maieru p78; Shutterstock / Monika 23 p215; Shutterstock / Monkey Business Images p24; Shutterstock / Monkey Business Images p510; Shutterstock / New Photo Service p426; Shutterstock / Olga Lyubkina p109; Shutterstock / Otmar Smit p405; Shutterstock / Prism68 p526; Shutterstock / Rachelle Burnside p396; Shutterstock / Rob Byron p284; Shutterstock / Rob Digphot p358; Shutterstock / Russ Witherington p463; Shutterstock / Sas Partout p73; Shutterstock / Tompet p179; Shutterstock / Vasilly Koval p349; Shutterstock / Vladimir Melnik p461; Shutterstock /Teresa Kasprzycka p19; Superstock / Greer & Associates p547.

The map on page 363 is reproduced by permission of Ordnance Survey on behalf of HMSO. © Crown copyright. All rights reserved. Licence number 100030901

Every effort has been made to contact copyright holders of material reproduced in this book. Any omissions will be rectified in subsequent printings if notice is given to the publishers.

Quick contents guide

About this book viii
About assessment objectives x
About functional maths xi

Blue type shows content that will only be examined at Higher Tier GCSE.

UNIT 1 Statistics and Number

1 Data collection	1
2 Fractions, decimals and percentages	19
3 Interpreting and representing data	35
4 Range and averages	53
Problem-solving: Statistics	69
Functional maths	70
Problem-solving: Number	72
5 Probability	73
6 Cumulative frequency	95
7 Ratio and proportion	109
Functional maths	130
8 Complex calculations	132
Problem-solving practice	143

UNIT 2 Number and Algebra Non-calculator

9 Number skills	147
10 Factors, powers and standard form	158
11 Basic rules of algebra	179
Functional maths	196
12 Fractions	198
13 Decimals	212
14 Equations and inequalities	223
15 Indices and formulae	242
16 Percentages	259
Problem-solving: Algebra	275
Functional maths	276
17 Sequences and proof	278
18 Linear graphs	297
19 Quadratic equations	322
Problem-solving practice	333

UNIT 3 Geometry and Algebra

20 Number skills revisited	335
21 Angles	340
22 Measurement 1	357
23 Triangles and constructions	368
24 Equations, formulae and proof	378
25 Quadrilaterals and other polygons	390
26 Perimeter, area and volume	404
27 3-D objects	416
28 Reflection, translation and rotation	422
29 Circles and cylinders	443
Problem-solving: Geometry	457
Functional maths	458
30 Measurement 2	460
Functional maths	472
31 Enlargement and similarity	474
32 Non-linear graphs	486
33 Constructions and loci	503
34 Pythagoras' theorem	520
35 Trigonometry	533
36 Circle theorems	547
Problem-solving practice	558

Answers 560
Index 611

Contents

About this book *viii*
About assessment objectives *x*
About functional maths *xi*

Blue type shows content that will only be examined at Higher Tier GCSE.

UNIT 1 Statistics and Number

1	**Data collection**	**1**
1.1	The data handling cycle	2
1.2	Gathering information	3
1.3	Types of data	5
1.4	Data collection	6
1.5	Grouped data	8
1.6	Two-way tables	10
1.7	Questionnaires	13
1.8	Sampling	15
	Review exercise	17
2	**Fractions, decimals and percentages**	**19**
2.1	Fraction of an amount	20
2.2	One quantity as a fraction of another	21
2.3	Calculating with fractions	22
2.4	Percentages of amounts	24
2.5	One quantity as a percentage of another	26
2.6	Percentage increase and decrease	28
2.7	Index numbers	31
	Review exercise	33
3	**Interpreting and representing data**	**35**
3.1	Drawing pie charts	36
3.2	Stem-and-leaf diagrams	38
3.3	Scatter diagrams	41
3.4	Lines of best fit and correlation	43
3.5	Frequency diagrams for continuous data	46
3.6	Frequency polygons	48
	Review exercise	50
4	**Range and averages**	**53**
4.1	Averages and range	54
4.2	Calculating the range, mode and median from a frequency table	57
4.3	Calculating the mean from an ungrouped frequency table	59
4.4	Finding the range, median and mode from a grouped frequency table	61
4.5	Estimating the mean from a grouped frequency table	63
	Review exercise	66
	Problem-solving: Statistics AOk	**69**
	Functional maths: Missed appointments	**70**
	Problem-solving: Number AOk	**72**
5	**Probability**	**73**
5.1	Probability that an event does not happen	74
5.2	Mutually exclusive events	76
5.3	Two-way tables	78
5.4	Expectation	81
5.5	Relative frequency	82
5.6	Independent events	86
5.7	Tree diagrams	88
	Review exercise	92
6	**Cumulative frequency**	**95**
6.1	Cumulative frequency	96
6.2	Box plots	101
6.3	Comparing data sets and drawing conclusions	104
	Review exercise	107
7	**Ratio and proportion**	**109**
7.1	Ratio	110
7.2	Ratios and fractions	113
7.3	Ratios in the form 1 : n or n : 1	115
7.4	Dividing quantities in a given ratio	116
7.5	More ratios	118
7.6	Proportion	120
7.7	Best buys	122
7.8	More proportion problems	124
	Review exercise	127
	Functional maths: Come dine with Brian	**130**

8 **Complex calculations** **132**
8.1 Repeated percentage change 133
8.2 Reverse percentages 135
8.3 Standard form 138
Review exercise 141

Problem-solving practice: Statistics **143**
Problem-solving practice: Number **145**

UNIT 2 Number and Algebra

9 **Number skills** **147**
Number skills: adding and subtracting 148
Number skills: multiplying and dividing 149
9.1 Multiplying and dividing whole numbers 150
9.2 Estimation 152
9.3 Negative numbers 154
Review exercise 156

10 **Factors, powers and standard form** **158**
10.1 Multiples 159
10.2 Factors 161
10.3 Squares, cubes and roots 164
10.4 Indices 166
10.5 Prime factors 168
10.6 Laws of indices and standard form 171
Review exercise 177

11 **Basic rules of algebra** **179**
Algebra skills: writing and simplifying expressions 180
11.1 Collecting like terms 181
11.2 Multiplying in algebra 183
11.3 Expanding brackets 184
11.4 Simplifying expressions with brackets 187
11.5 Factorising algebraic expressions 188
11.6 Expanding two brackets 191
Review exercise 194

Functional maths: eBay business **196**

12 **Fractions** **198**
12.1 Comparing fractions 199
12.2 Adding and subtracting fractions 200
12.3 Adding and subtracting mixed numbers 202
12.4 Multiplying fractions 203
12.5 Multiplying mixed numbers 205
12.6 Reciprocals 207
12.7 Dividing fractions 208
Review exercise 210

13 **Decimals** **212**
13.1 Adding and subtracting decimals 213
13.2 Converting decimals to fractions 215
13.3 Multiplying and dividing decimals 216
13.4 Converting fractions to decimals 219
Review exercise 221

14 **Equations and inequalities** **223**
14.1 Solving two-step equations 224
14.2 Writing and solving equations 227
14.3 Equations with brackets 229
14.4 Equations with an unknown on both sides 230
14.5 Equations with fractions 233
14.6 Inequalities 235
14.7 Simultaneous equations 237
Review exercise 240

15 **Indices and formulae** **242**
15.1 Using index notation 243
15.2 Laws of indices (index laws) 245
15.3 Writing your own formulae 248
15.4 Substituting into expressions 249
15.5 Substituting into formulae 252
15.6 Changing the subject of a formula 255
Review exercise 257

16 **Percentages** **259**
Number skills: fractions, decimals and percentages 260
Number skills: using percentages in calculations 260
16.1 Percentage increase and decrease 261
16.2 Calculations with money 263
16.3 Percentage profit or loss 267

16.4 Repeated percentage change 269
16.5 Reverse percentages 271
Review exercise 273

Problem-solving: Algebra AOk 275
Functional maths: Camping with wolves 276

17 Sequences and proof 278
17.1 Number patterns 279
17.2 Rules for sequences 281
17.3 Using the *n*th term 284
17.4 Sequences of patterns 286
17.5 Quadratic sequences 288
17.6 The *n*th term of a simple quadratic sequence 290
17.7 Proof 291
17.8 Using counter-examples 293
Review exercise 295

18 Linear graphs 297
18.1 Mid-point of a line segment 298
18.2 Plotting straight-line graphs 300
18.3 Equations of straight-line graphs 303
18.4 Conversion graphs 309
18.5 Real-life graphs 311
18.6 Using graphs to solve simultaneous equations 315
18.7 Using graphs to solve inequalities 317
Review exercise 319

19 Quadratic equations 322
19.1 Factorising the difference of two squares 323
19.2 Factorising quadratics of the form $x^2 + bx + c$ 324
19.3 Solving quadratic equations 327
19.4 Writing and solving quadratic equations 330
Review exercise 332

Problem-solving practice: Algebra 333

UNIT 3 Geometry and Algebra

20 Number skills revisited 335
Number skills: revision exercise 1 336
Number skills: revision exercise 2 338

21 Angles 340
Geometry skills: angles 341
21.1 Angle facts 343
21.2 Angles in parallel lines 345
21.3 Bearings 349
Review exercise 355

22 Measurement 1 357
22.1 Time and timetables 358
22.2 Converting between metric and imperial units 360
22.3 Maps and scale drawings 361
22.4 Accuracy of measurements 364
Review exercise 366

23 Triangles and constructions 368
23.1 Interior and exterior angles 369
23.2 Constructions 371
23.3 Congruent triangles 374
Review exercise 376

24 Equations, formulae and proof 378
Algebra skills: expressions 379
Algebra skills: brackets 379
Algebra skills: solving equations 380
Algebra skills: formulae 380
24.1 Equations and formulae 381
24.2 Proof 384
24.3 Trial and improvement 386
Review exercise 388

25 Quadrilaterals and other polygons 390
25.1 Quadrilaterals and algebra 391
25.2 Special quadrilaterals 394
25.3 Polygons 396
25.4 Coordinates 400
Review exercise 402

26 **Perimeter, area and volume** **404**
26.1 Perimeter and area of simple shapes 405
26.2 Perimeter and area of compound shapes 408
26.3 Volume and surface area of prisms 411
Review exercise 414

27 **3-D objects** **416**
27.1 Drawing 3-D objects 417
Review exercise 421

28 **Reflection, translation and rotation** **422**
28.1 Reflection on a coordinate grid 423
28.2 Translation 427
28.3 Rotation 430
28.4 Combined transformations 436
Review exercise 439

29 **Circles and cylinders** **443**
29.1 Circumference of a circle 444
29.2 Area of a circle 448
29.3 Cylinders 451
Review exercise 455

Problem-solving: Geometry AOk **457**
Functional maths: Roof garden **458**

30 **Measurement 2** **460**
30.1 Converting areas and volumes 461
30.2 Speed 463
30.3 Density 465
30.4 Dimension theory 467
Review exercise 470

Functional maths: Pembrokeshire cycle ride **472**

31 **Enlargement and similarity** **474**
31.1 Enlargement 475
31.2 Enlargements with fractional scale factors 479
31.3 Similarity 481
Review exercise 484

32 **Non-linear graphs** **486**
Algebra skills: quadratic expressions 487
32.1 Graphs of quadratic functions 488
32.2 Solving quadratic equations graphically 495
32.3 Graphs of cubic functions 498
Review exercise 500

33 **Constructions and loci** **503**
33.1 Constructions 504
33.2 Locus 510
Review exercise 518

34 **Pythagoras' theorem** **520**
34.1 Pythagoras' theorem 521
34.2 Finding the hypotenuse 522
34.3 Finding a shorter side 526
34.4 Calculating the length of a line segment 529
Review exercise 531

35 **Trigonometry** **533**
35.1 Trigonometry – the ratios of sine, cosine and tangent 534
35.2 Finding lengths using trigonometry 536
35.3 Finding angles using trigonometry 538
35.4 Applications of trigonometry and Pythagoras' theorem 540
35.5 Angles of elevation and depression 543
Review exercise 544

36 **Circle theorems** **547**
36.1 Circle properties 548
36.2 Circle theorems 551
Review exercise 555

Problem-solving practice: Geometry **558**

Answers *560*
Index *611*

All set to make the grade!

AQA GCSE Mathematics for Middle sets is specially written to help you get your best grade in the exams.

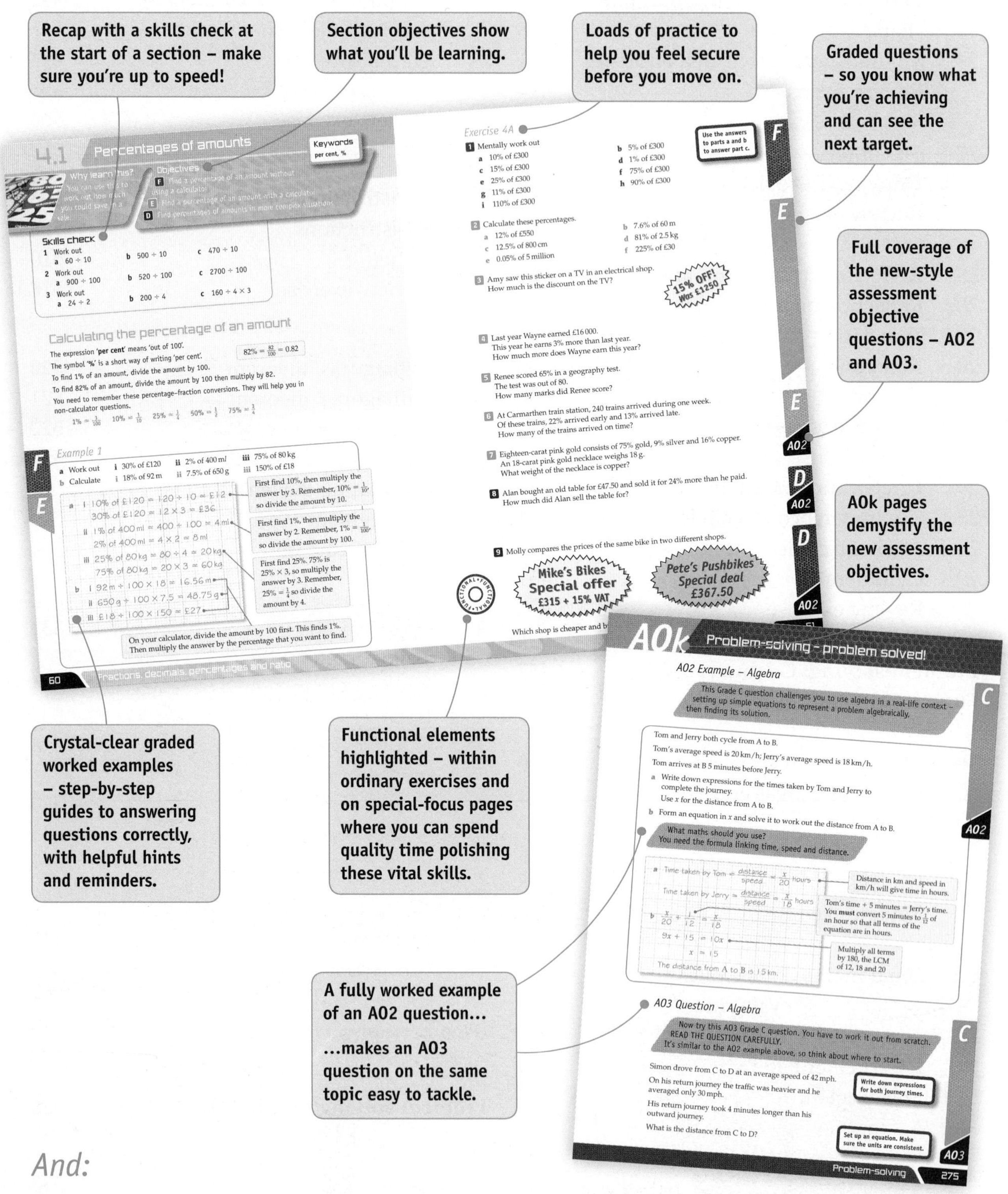

And:

- A pre-check at the start of each chapter helps you recall what you know!
- End-of-chapter graded review exercises consolidate your learning and include exam-style questions.

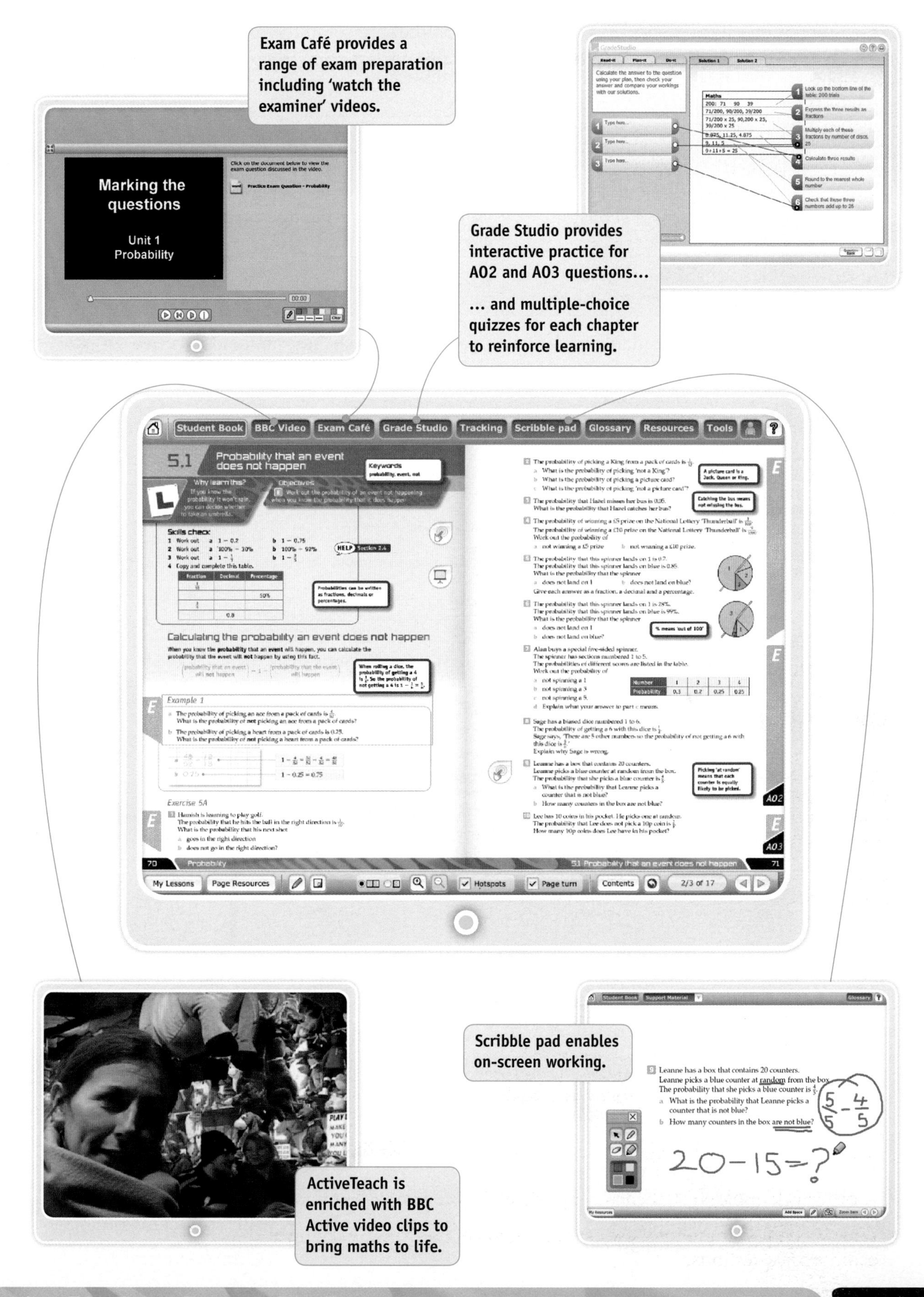
Exam Café provides a range of exam preparation including 'watch the examiner' videos.
Marking the questions
Unit 1 Probability
Grade Studio provides interactive practice for AO2 and AO3 questions...
... and multiple-choice quizzes for each chapter to reinforce learning.
Student Book
BBC Video
Exam Café
Grade Studio
Tracking
Scribble pad
Glossary
Resources
Tools
5.1 Probability that an event does not happen
Calculating the probability an event does not happen
My Lessons
Page Resources
Hotspots
Page turn
Contents
2/3 of 17
Scribble pad enables on-screen working.
20 − 15 = ?
ActiveTeach is enriched with BBC Active video clips to bring maths to life.

'Assessment Objectives' define the types of question that are set in the exam:

Assessment Objective	What it is	What this means	Approx % of marks in the exam
A01	Recall and use knowledge of the prescribed content.	Ordinary questions testing your knowledge of each topic.	50
A02	Select and apply mathematical methods in a range of contexts.	Problem-solving: find the ordinary maths you need to get to the correct answer.	30
A03	Interpret and analyse problems and generate strategies to solve them.	A step up from A02. There could be more than one way to tackle these.	20

The proportion of marks available in the exam varies with each Assessment Objective.

So it's worth making sure you know how to do AO2 and AO3 questions!

What does an AO2 question look like?

4 The diagram shows a shaded triangular shape made up from two right-angled triangles, one inside the other.
The larger triangle has a base of 15.6 cm and a height of 12 cm.
The smaller triangle has a base of 6.5 cm and a height of 5 cm.
Calculate the shaded area.

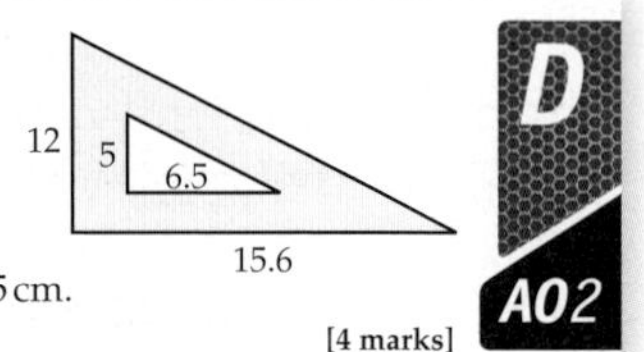

[4 marks]

D

AO2

This just needs you to (a) read and understand the question and (b) recall and apply the geometric formula for the area of a triangle. Simple!

We give you special help with AO2s on pages 69, 72, 143, 145, 275, 333, 457 and 558.

What does an AO3 question look like?

11 The results of a survey about the number of eggs in a sample of birds' nests are shown in a frequency table.

The frequency in the last row has been accidentally torn off.

The median number of eggs per nest is 2.5.

Calculate the mean number of eggs per nest.
Give your answer to two decimal places.

Number of eggs	Frequency
0	6
1	3
2	8
3	7
4	

[4 marks]

Here you need to read and analyse the question. Then use your statistics knowledge to solve this problem.

We give you special help with AO3s on pages 69, 72, 144, 146, 275, 334, 457 and 559.

Quality of written communication

There are a few extra marks in the exam if you take care to write your working 'properly'.

- Write legibly.
- Use the correct mathematical notation and vocabulary, showing that you can communicate effectively.
- Write logically in an organised sequence that the examiner can follow.

In the exam paper, such questions will be marked with a star (☆) – see the problem-solving practice pages 143–6, 333–4 and 558–9.

About functional elements

Functional maths means using maths effectively in a wide range of real-life contexts.

There are three 'key processes' for maths:

Key process	What it is	What this means
Representing	Understanding real-life problems and selecting the mathematics to solve them.	• Understanding the information in the question. • Working out what maths you need to use. • Planning the best order in which to do your working.
Analysing	Applying a range of mathematics within realistic contexts.	• Being organised and following your plan. • Using appropriate maths to work out the answer. • Checking your calculations. • Explaining your plan.
Interpreting	Communicating and justifying solutions and linking solutions back to the original context of the problem.	• Explaining what you've worked out. • Explaining how it relates to the question.

The proportion of functional maths marks in the GCSE exam depends on which tier you are taking:

GCSE tier	Approx % of marks in the exam
Foundation	30 to 40
Higher	20 to 30

So it's worth making sure you know how to do functional maths questions!

What does a question with functional maths look like?

C

4 The table shows the adult : child ratios for different ages of children in daycare.

Age of children	Adult : child ratio
Less than 2 years	1 : 3
2 years	1 : 4
3–7 years	1 : 8

These are the ages of children registered at a daycare centre in the school holidays.

Babies up to one year old	4
Toddlers over 1 year, but less than 2	7
Two-year-olds	8
Three-year-olds	6
Four-year-olds	12
Five-year-olds	9
Six-year-olds	5

Work out the number of adults needed to care for these children.

AO2

FUNCTIONAL

In the context of childcare, you need to read and understand the question.

Think what maths you need and plan the order in which you'll work.

Follow your plan. Check your calculations. Job done!

Don't miss the fun on our special functional maths pages: 70, 130, 196, 276, 458 and 472! These are not like the functional questions you'll get in GCSE but they'll give you more practice with the key processes.

Camping with wolves

1

Data collection

BBC Video

This chapter is about collecting information.

Do more people attend rock concerts now than 5 years ago? Does it vary by age group?

Objectives

This chapter will show you how to

- collect information **E**
- understand the data handling cycle **D**
- state a hypothesis **D**
- display information **D**
- design a questionnaire **C**
- look at sampling techniques **C**

Before you start this chapter

1 How will you know if rock concerts are better attended now than 5 years ago?

2 What kind of questions do you need to ask?

3 Where can you find information about the age of those who attend?

4 How can you obtain a representative sample of the views of those who attend rock concerts?

1.1 The data handling cycle

Keywords
data, hypothesis

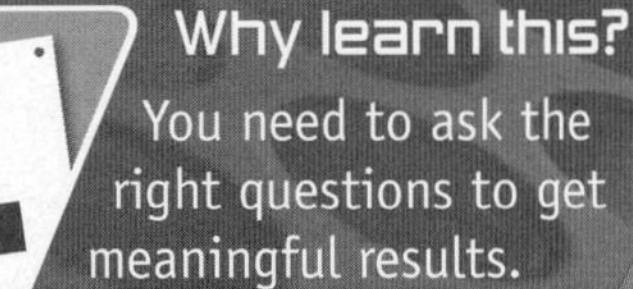

Why learn this?
You need to ask the right questions to get meaningful results.

Objectives
- **D** Learn about the data handling cycle
- **D** Know how to write a hypothesis

Skills check

1 How can you find out the following information?
- **a** The average amount of savings for your classmates.
- **b** What times trains leave your local station to go to Manchester.
- **c** The number of votes cast for each party in the last local council elections.

The data handling cycle

When you carry out a statistical investigation, you deal with lots of factual information, called **data**.

The investigation follows the data handling cycle:

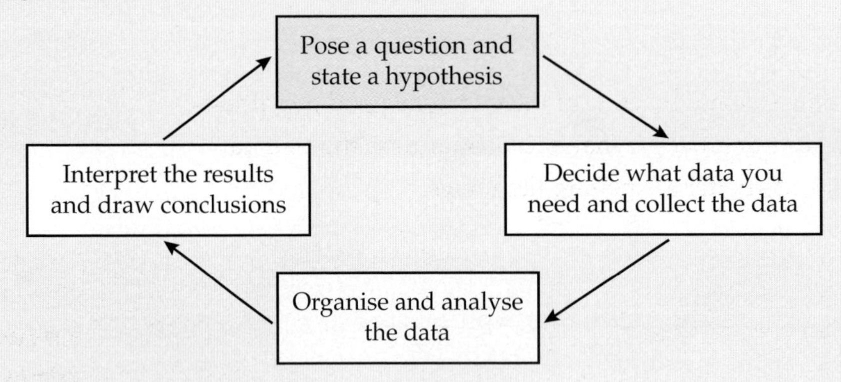

Stating a hypothesis

A **hypothesis** is a statement that can be tested to answer a question.
A hypothesis must be written so that it is either 'true' or 'false'.

When you write a hypothesis, use words with clear meaning.
Always make statements about things that can be measured.

Example 1

You are playing a game that involves rolling a dice. You think that a score of 6 occurs fewer times than it ought to.
State a hypothesis to investigate this.

A score of 6 happens fewer times than any other score.

You can easily test whether your hypothesis is true or false by carrying out a large number of trials (more than 100) and recording the number of times each score occurs.

Exercise 1A

1 Write a hypothesis to investigate each question.

a Who can run faster, boys in Year 10 or girls in Year 10?

b Do people do most of their shopping at the supermarket or at their local shop?

D

Example 2

D

Ben writes this hypothesis 'People who smoke die young.'

Give a reason why this is not a good hypothesis.

The statement is too vague.

'People who smoke' may smoke 1 cigarette a day, 20 a day or more.

'Young' means different things to different people.

Is under 20 'young'? Under 50?

What does 'People who smoke' mean?

What does 'young' mean?

Exercise 1B

1 Give one reason why each of these is not a good hypothesis.

a More young people than old people go to the cinema.

b Girls are better at spelling than boys.

D

1.2 Gathering information

Keywords
primary, secondary

Why learn this?

You need to find out some facts if you are going to test your hypothesis.

Objectives

D Know where to look for information

Skills check

1 Where do you go to find out the times of trains to Manchester?

2 Where and how do you find out the voting patterns in the last local council elections?

Data sources

After writing a hypothesis you need to think about how you are going to test it.

- What information do you need?
- Does the information exist already?
- How easy is it to get the information if it doesn't exist already?
- Where do you try to find the information if it does exist?

There are two types of data source: **primary** and **secondary**.

Primary data is data you collect yourself. You can ask people questions or carry out an experiment.

Secondary data is data that has already been collected. You can look at newspapers, magazines, the internet and many other sources.

D

Example 3

To test each hypothesis state

- whether you need primary data or secondary data
- how you would find or collect the data
- how you would use the data.

a The yearly total rainfall in Plymouth is greater than in Norwich.

b Girls in Year 10 at my school prefer English to maths.

a Secondary data

Find records of rainfall for Plymouth and Norwich on the internet.

It will then be easy to see which of them has more rainfall.

You can't record the data yourself so you need data collected by someone else.

b Primary data

Carry out a survey of the girls in Year 10.

Count the votes for each subject to find out which one is preferred.

This information needs to be collected.
A survey could be carried out during registration time by putting a note in each register asking for the two totals from each tutor group.

Exercise 1C

D

1 To test each hypothesis state

- whether you need primary data or secondary data
- how you would find or collect the data
- how you would use the data.

a In 2008, in England, more cars with petrol engines were bought than cars with diesel engines.

b Tenby has more hours of sunshine in June than Southend.

c People living in your street prefer Chinese takeaway to Indian.

d More people aged between 40 and 50 voted in the last general election than people aged between 20 and 30.

e There are more students in Year 11 who would prefer to go to London on an end-of-term trip than to a theme park or to Blackpool.

f Attendance at the local cinema has fallen steadily over the last 12 months.

1.3 Types of data

Keywords
qualitative, quantitative, discrete, continuous

Why learn this?
You need to know the correct terms to describe the data you collect.

Objectives
D Be able to identify different types of data

Skills check

1. How will the times of trains to Manchester be displayed?
2. How can you find out how many people go to Spain for their holidays?

Types of data

There are two types of data: **qualitative** and **quantitative**.

Qualitative data can only be described in words. It is usually organised into categories, such as colour (red, green, ...) or breed of dog (labrador, greyhound, ...).

Quantitative data can be given numerical values. Examples include shoe size or temperature.

Example 4

D

Is this data qualitative or quantitative?

a The number of tomatoes on a tomato plant.
b The taste of an orange.
c The weight of a car.

a Quantitative — The number is counted: 1, 2, 3, …
b Qualitative — The taste is described in words: sweet, bitter, juicy, …
c Quantitative — The weight is measured: 1485 kg, 1740 kg, 1520 kg, …

All quantitative data is either **discrete** or **continuous**.

Discrete data can only take certain values. It is usually whole numbers (the number of goals scored in a match) but may include fractions (shoe sizes).

Continuous data can take any value in a range and can be measured (tree heights).

Example 5

D

For each part, write whether the data is qualitative or quantitative.
If it is quantitative, is it discrete or continuous?

a The names of Formula 1 car racing teams.
b The sizes of women's clothes in a boutique.
c The names of members of a hockey team.
d The time taken to run 1500 metres.

Qualitative data is described in words. Quantitative data is numerical and can be discrete or continuous.

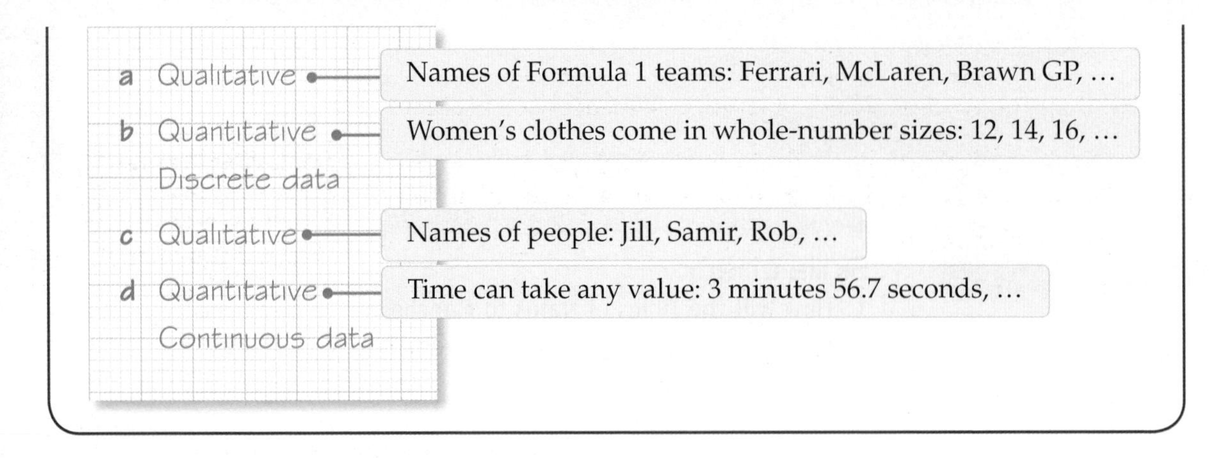

Exercise 1D

D

1 For each part, write whether the data is qualitative or quantitative.
If it is quantitative, is it discrete or continuous?
For each one give an example of a typical item of data.

a The hair colours of students in a class.

b The score when you throw three darts at a dartboard.

c The weight of coffee in a jar.

d The number of chairs in a classroom.

e The brands of soap on sale in a supermarket.

f The waist sizes of jeans on sale in a high street store.

g The makes of cars in the school car park.

h The heights of trees in the local park.

1.4 Data collection

Keywords
frequency table, tally mark, frequency, frequency distribution

Why learn this?

You need to be able to record large amounts of data in a way that is easy to understand.

Objectives

E Work out methods for gathering data efficiently

Skills check

1 Write the number given by each set of tally marks.

a |||| **b** ~~||||~~ ||| **c** ~~||||~~ ~~||||~~ ||||

2 Use tally marks to record each number.

a 5 **b** 9 **c** 17

Data collection

When you collect data you need to organise it in such a way that it is easy to read and understand.

A data collection table or **frequency table** has three columns: one for listing the items you are going to count, one for **tally marks** and one to record the **frequency** of each item.

Example 6

Here are the vowels from the first three sentences of a book.

e a e u i a o e e i a i a e a u i a o e
a e i o e a i i e u a e o a i e a u e a
o a e i o u i e a e e o u i e a i o i e

It is difficult to see how many times each one occurs.

Put the results into a frequency table (sometimes called a **frequency distribution**).

Vowel	Tally	Frequency
a	卌 卌 卌	15
e	卌 卌 卌 |||	18
i	卌 卌 |||	13
o	卌 |||	8
u	卌 |	6
	Total	60

The frequency is the total for each vowel.

Group the tally marks in 5s to make them easier to count.

You can easily see that e occurs most often.

Add the frequencies to get the total number of vowels.

Exercise 1E

E

1 A spinner with the colours red (R), blue (B), pink (P) and white (W) was spun. These are the results.

R R W P B P P B P R W R P P P
B R P W P W B P W B R R P W P
W B B R P R W R W B P B R W P

a Make a frequency table to show this information.

b Which colour occurred most often?

c How many times was the spinner spun?

E

2 Members of a class were asked, 'How many children are there in your family?' Their answers are given below.

3 4 1 2 1 4 5 3 6 2 3 1 3 2
3 2 4 1 2 3 1 4 2 1 2 5 2 4

a Draw a frequency table to show this information.

b What was the most frequent answer?

c How many children are in the class?

d Can you see a quick way to work out the total number of children in all the families?

Look closely at your frequency table. How many times does 3 occur? What does this mean?

3 Members of a class were asked how many pets they had. These are their answers.

1 4 1 1 2 3 1 6 3 2 1 4 1 2 3
2 4 4 1 5 2 3 1 2 1 4 3 2 1 2

a Draw a frequency table to show this information.

b What was the most common answer?

c How can you use the frequency table to work out the total number of pets?

AO2

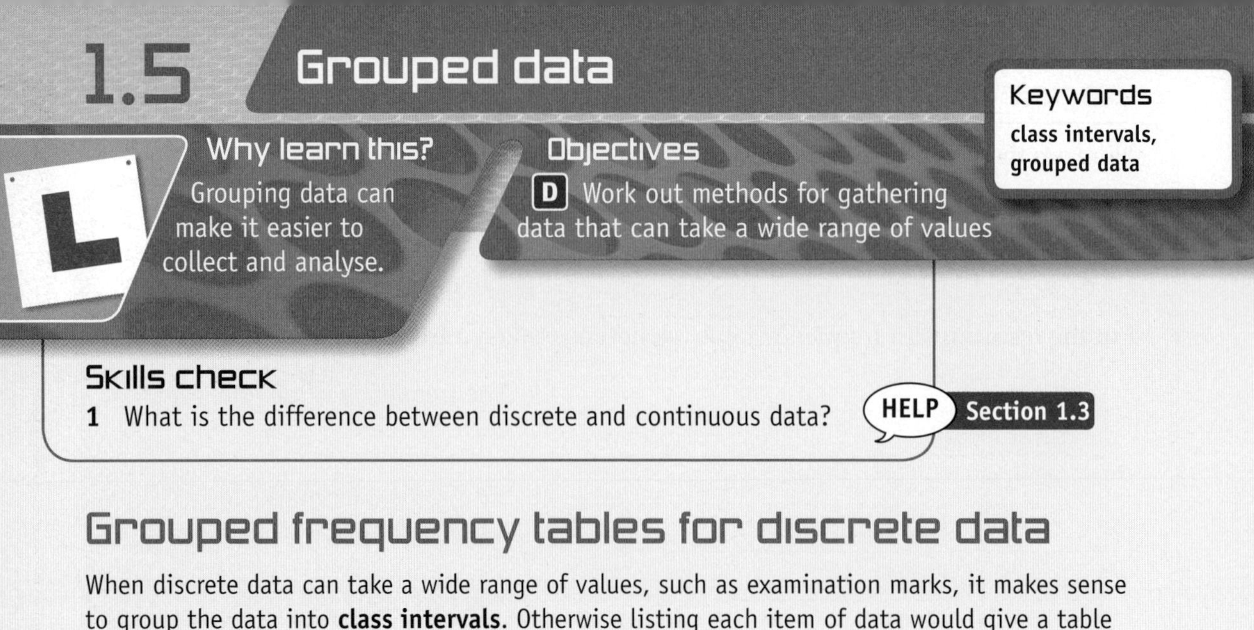

Keywords
class intervals, grouped data

Grouped frequency tables for discrete data

When discrete data can take a wide range of values, such as examination marks, it makes sense to group the data into **class intervals**. Otherwise listing each item of data would give a table that was too big and many items would occur only once.

Class intervals for **grouped data** should be equal sizes.

D

Example 7

In a quiz, there are 40 questions each worth 1 mark.
These are the scores of 30 people who entered the quiz.

23 14 17 36 25 31 20 38 33 28 29 25 19 22 36
30 34 35 36 28 19 21 26 30 32 35 28 31 27 25

Design a frequency table to show this information.

The groups must not overlap. 11–15, 16–20, etc. are called class intervals.

Score	Tally	Frequency
11–15	I	1
16–20	IIII	4
21–25	卌 I	6
26–30	卌 III	8
31–35	卌 II	7
36–40	IIII	4
	Total	30

The scores are discrete (see Section 1.3) and can be grouped. In this case it's sensible to put them into groups of 5 marks.

After the data has been put into the frequency table you can't identify individual scores unless you look back at the original data.

Grouped frequency tables for continuous data

Continuous data can take values anywhere in a range. The range can be wide. Height, weight and time are examples of continuous data.

To group continuous data, the class intervals must use inequality symbols, $\leqslant$ and $<$.
'$160 \leqslant h < 170$' means a height from 160 cm up to *but not including* a height of 170 cm.
A person of height 170 cm would belong in the next class interval, $170 \leqslant h < 180$.

Example 8

D

Here are the heights, to the nearest centimetre, of 25 basketball players.

215 220 211 212 198 190 210 208 206 212
208 218 210 199 204 206 207 188 209 207
210 200 203 205 222

Put these heights into a grouped frequency table.
Use the class intervals $180 \leqslant h < 190$, $190 \leqslant h < 200$, etc.

Be careful where you put heights that are on the boundary of a class interval.

Height, h (cm)	Tally	Frequency
$180 \leqslant h < 190$	|	1
$190 \leqslant h < 200$	|||	3
$200 \leqslant h < 210$	~~||||~~ ~~||||~~ |	11
$210 \leqslant h < 220$	~~||||~~ |||	8
$220 \leqslant h < 230$	||	2
	Total	25

210 goes in this class interval.

220 goes in this one.

Check that the total of your frequencies is 25.

Exercise 1F

D

1 Mrs Fisher gave her class a mental arithmetic test.
There were 20 questions.
Here are the students' marks.

7 13 18 9 5 12 14 11 16 8 19 11 6 16
17 15 18 10 11 15 4 10 7 15 6 14 12 9

a Put these marks into a grouped frequency table using the class intervals 1–5, 6–10, 11–15, 16–20.

b Three people were absent on the day of the test.
How many students are in Mrs Fisher's class?

c In which class interval do most marks occur?

d Did more students score over 50% or under 50%?

2 Here is a list of the pocket money (in £s) received in one particular week by 27 students.

18.40 9.50 12.00 12.25 14.40 5.50 7.80
11.45 16.30 12.45 9.60 13.60 10.00 16.75
13.20 12.80 9.80 10.20 7.75 15.50 8.70
13.50 12.00 8.80 6.65 11.25 5.60

Design a grouped frequency table to illustrate this data.
Choose suitable class intervals.

**Look for the smallest and largest amounts and choose equal class intervals.
Four or five equal class intervals ought to be enough.**

D

3 This grouped frequency table shows the heights of some plants.

Height, h (cm)	Frequency
$10 \leq h < 16$	9
$16 \leq h < 22$	11
$22 \leq h < 28$	15
$28 \leq h < 34$	28
$34 \leq h < 40$	17

a How many plants are in this survey?

b How many plants are less than 22 cm high?

c How many plants are at least 28 cm high?

A02

d Imagine all the plants lined up in a row from the smallest to the tallest. In which class interval would the plant in the middle of the row lie?

D

4 The weights of 30 people attending a fitness class were recorded to the nearest kilogram.
Here are the results.

46 57 49 66 82 64 55 61 69 68
75 53 94 60 89 64 80 55 83 85
74 53 65 59 54 75 70 78 90 72

a Put the results into a grouped frequency table using the class intervals $45 \leq w < 55$, $55 \leq w < 65$, etc.

b How many people weighed less than 75 kilograms?

c How many people weighed at least 65 kg?

d In which class interval are there the most people?

1.6 Two-way tables

Keywords
two-way table

Why learn this?
You need to be able to record two sets of related data in a clear way.

Objectives
D Work out methods for recording related data

Skills check

1 In this table, each row has the same total and each column has the same total. Work out the numbers represented by A, B, C and D.

4	9	8	3	16
2	9	A	11	B
18	C	4	D	2

2 In a class of 28 students, 23 own at least one pet.
How many do not own a pet?

Recording data in a two-way table

Data such as the eye colour of boys and girls in a particular class or year group needs to be presented in a way that makes it easy to answer simple questions.

A **two-way table** helps you to do this.

It shows how eye colour and gender are related.

	Blue eyes	Brown eyes	Other eye colour
Boys			
Girls			

Boys with blue eyes.

Girls whose eyes are neither blue nor brown.

The two-way table can be extended to show the totals for each row and column.

	Blue eyes	Brown eyes	Other eye colour	Total
Boys				
Girls				
Total				

The total number of boys.

The total number of students.

The total number of brown-eyed students.

Example 9

This two-way table shows the eye colour and gender of students in Mr Jamir's tutor group.

	Blue eyes	Brown eyes	Other eye colour	Total
Boys	3	8	2	13
Girls	7	6	5	18
Total	10	14	7	31

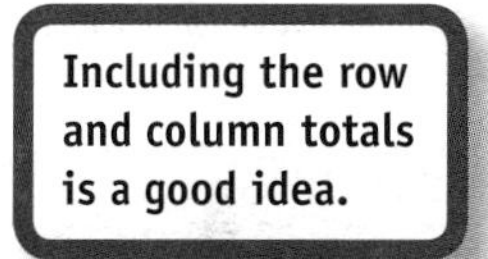

a How many girls have blue eyes?

b How many boys have neither blue nor brown eyes?

c How many students have brown eyes?

d How many students are in the tutor group?

a 7 girls have blue eyes.

Look at 'Girls'/'Blue eyes'.

b 2 boys have neither blue nor brown eyes.

Look at 'Boys'/'Other eye colour'.

c 14 students have brown eyes.

The total of the numbers in the 'Brown eyes' column: $8 + 6 = 14$

d There are 31 students in the tutor group.

This can come from the column or row totals: $10 + 14 + 7 = 31$ or $13 + 18 = 31$. The totals across and down must be the same.

D

Exercise 1G

D

1 The two-way table shows the GCSE maths grades of a group of students.

	A*	A	B	C	D	E	F	G	Total
Boys	10	8	12	8	3	9	5	1	
Girls	13	9	10	11	6	8	4	3	
Total	23				9				

a Copy the table and complete it by filling in all the missing numbers.

b How many boys obtained a grade C or better?

c How many girls got a result lower than grade D?

d Looking only at grades A*, A and B, who got more – the boys or the girls?

e How many students are represented by this two-way table?

2 The two-way table shows the number of adults per house and the number of televisions per house in 54 houses in a street.

	No TV	1 TV	2 TVs	3 TVs
1 adult	1	3	0	0
2 adults	1	12	8	2
3 adults	0	4	9	5
4 adults	0	1	5	3

a How many houses have exactly three adults and three TVs?

b How many houses have two TVs?

c How many houses have two adults living in the house?

d What percentage of the houses with one TV have three adults living in the house?

D

3 In a survey of 50 teachers about how they travel to work:

22 men came by car, 1 man came on the bus and 5 men walked to school.
There were 16 women altogether.
10 women came by car, 1 cycled and 3 came by bus.

a Design a two-way table to show this information.

b Complete the table, showing the totals for men and women and for each method of travel.

AO2

c How many teachers cycled to school?

d What percentage of the teachers walked to school?

D

4 The police carried out a spot check on vehicles passing through a town.
They inspected the lights and the tyres on each vehicle.
The two-way table shows the results of these checks.

	Satisfactory lights	Defective lights	Satisfactory tyres	Defective tyres
Motorbikes	13	7	15	5
Cars	35	11	42	4
Vans	19	3	16	6

a How many vans had satisfactory tyres?

b How many vehicles had defective lights?

c How many vehicles had satisfactory tyres?

d How many vehicles were stopped and checked?

1.7 Questionnaires

Keywords
survey, questionnaire, response, leading question

Why learn this?

The best surveys start with a good questionnaire.

Objectives

C Learn how to write good questions to find out information

Skills check

1 Describe a good method for recording data on a data collection sheet.

Questionnaires

To find out information, you might want to do a **survey**.

A survey collects primary data. One way to collect this data is to use a **questionnaire**.

A questionnaire is a form that people fill in. On the form are a number of questions.

Asking the right questions is important to find out what you want to know.

Key points

- Use simple language. Ask short questions that can be answered easily. Ask questions where the **response** is 'Yes' or 'No'.

Do you have breakfast every day?

Yes ☐ No ☐

- Always give a choice of answers with tick boxes.

The answer options provided must cover all possibilities.

How many sisters do you have?

0 ☐ 1 ☐ 2 ☐ 3 ☐ 4 ☐ More than 4 ☐

- Make sure the responses do not overlap and do not give too many choices. These choices are unsuitable for the previous question because they overlap:

0 ☐ 1–2 ☐ 2–3 ☐ 3–4 ☐ 4 or more ☐

Which box would you tick if you have 2 or 3 or 4 sisters?

- Ask a specific question. Make it clear and easy to answer.

How often do you use the internet?

Sometimes ☐ Occasionally ☐ Often ☐

'Sometimes', 'occasionally' and 'often' mean different things to different people, so don't use them.

This response choice is much better.

Never ☐ 1–2 times a week ☐ 3–6 times a week ☐ Every day ☐

- Never ask a personal question.

How old are you? ☐ years

Many people will refuse to answer this question, and some may give a false age.

This is a much better question.

How old are you?

Under 18 years ☐ 18 to 30 years ☐ Over 30 years ☐

Never ask people to put their names on the questionnaire. Some people may give a false name!

- Never ask a **leading question**.

Watching too much TV is bad for you. Don't you agree? Yes ☐ No ☐

A leading question encourages people to give a particular answer.

- Don't ask too many questions. If your questionnaire is too long, people won't want to answer it.

Recording your results

You can also use a two-way table to gather information. It is useful for recording responses to two related questions.

C

Example 10

Four athletics clubs, A, B, C and D, enter runners into a half-marathon.

Design a data collection sheet to show the finishing times of runners from all four clubs.

A data collection sheet is sometimes called an observation sheet.

Time (min)	A	B	C	D
$t < 70$		|		
$70 \leqslant t < 80$				|
$80 \leqslant t < 90$	|	||	|	|
$90 \leqslant t$	|||	||	卌	|||

In an exam question you may be asked to make up data, say 20 responses, to put into your two-way table.

Exercise 1H

1 Ann wants to find out what people think of their local health centre.
She includes these three questions in her questionnaire.

i What is your date of birth?

ii Don't you agree that it takes too long to get an appointment to see the doctor?

iii How many times did you visit the doctor last year?

Less than 3 times ☐ 3–7 times ☐ 7–10 times ☐ More than 10 times ☐

a Say why each question is unsuitable.

b Rewrite each question to make it suitable for a questionnaire.

2 These are questions about diet.

Eating plenty of vegetables each day is good for you. Don't you agree?

Strongly agree ☐ Agree ☐ Don't know ☐

a Give two reasons why this question is unsuitable.

Do you eat vegetables? Yes ☐ No ☐

If 'yes', how many times a week, on average, do you eat vegetables?

Once or less ☐ 2 or 3 times ☐ 4–7 times ☐ More than 7 times ☐

b Give two reasons why this is a good question.

3 Many of the workers at a factory use a car to travel to work.
The manager of the factory wants to find out if they are willing to car-share.
Write a suitable question with a response section to find out which days, from Monday to Friday, the workers are willing to car-share.

4 Julie is carrying out a survey about how much exercise students in her school do.
One of her questions is, 'How many days a week do you exercise for 30 minutes or more?'

a Design a response section for Julie's question.

b Write a question that she can use to find out what activities the students take part in.

5 a Design an observation sheet to show how far students travel to school.
It must show data for year groups 7, 8, 9, 10 and 11.

b Make up data for 20 students. Show their responses on your observation sheet.

C

C

AO2

1.8 Sampling

Keywords
population, sample, representative, bias, random

Why learn this?
You want a sample to give a fair and balanced range of views.

Objectives
C Know the techniques to use to get a reliable sample

Skills check

1 It takes 30 seconds to get a response to a question from one person. How long will it take to ask the question to every student in your school?

Sampling techniques

When you carry out a survey it would be too time consuming to ask everyone.

The total number of people you *could* ask is called the **population**.
It might be 952 students in your school or 1478 people who live in your local area.

Instead of asking everyone, you ask some of them.
This smaller group of people you *do* ask is called a **sample**.

You need to make sure that you obtain a **representative** sample.
A representative sample is a sample that will give you a fair and balanced range of people's opinions.

> **A sample that is not representative of the population will be biased.**

Random sampling allows every member of the population an equal chance of being selected. You could choose people by picking names out of a hat.

C

Example 11

Some students at the local school hear that the local council wants to make the centre of the town a pedestrian-only area.
They decide to find out how much support there is for this plan.

a Why will they have to use a sample?

b The students decide to ask people their views. They make a questionnaire to do this.
They carry out their survey in the town centre between 10 am and midday.
Give reasons why this sample will not be representative.

AO2

a It is impossible or too time consuming to ask all the people who live in the town.

b Many people are at work during the day so they will not go to the town centre at this time. These people will not be able to give their opinion.
Also, the people who are in the town centre are probably doing their shopping. They will be likely to support a traffic-free area because it will make their shopping experience more pleasant.

> The opinions of the people who will be in the town centre in the late morning might not be the same as the people who are there at other times.
> The sample is unlikely to include a representative number of motorists.
> The sample is therefore likely to be biased in favour of the pedestrian-only area.

Exercise 1I

C

1 To find out what people think of the new refuse-collection arrangements, a telephone survey was carried out between the hours of 9 am and 5 pm.
Give reasons why the sample might not be representative.

> **Who might not be contactable by phone at these times?**

2 A survey to find out about car ownership was carried out at an out-of-town shopping centre.
Is this likely to give a representative sample? Give reasons for your answer.

AO2

3 A survey into how much exercise young people take was carried out at the local sports centre.
Why is this sample likely to be unrepresentative?

4 Ying wants to know how people travel to work.
He interviews people by waiting just outside the bus station.
Is this likely to give a representative sample? Give reasons for your answer.

5 Gwen wants to know how much people spend on entertainment each week.
She carries out a survey by interviewing people in the town centre in the evening.
Do you think this will be a representative sample? Give reasons for your answer.

C

AO2

Review exercise

E

1 Some drivers were asked what make of car they drive.
Their responses were the makes Ford (F), Vauxhall (V), BMW (B), Audi (A) and Kia (K).
Here are the results.

F	B	A	B	F	K	V	F	V	B	A	F	V	B
V	A	K	B	A	K	F	B	F	V	F	B	A	K
F	V	F	V	F	V	B	F	B	F	A	A	F	B

a Design a frequency table to show this information. **[3 marks]**

b Which make of car is driven by the most people? **[1 mark]**

c How many drivers took part in the survey? **[1 mark]**

D

2 Write a hypothesis to investigate whether more people go to Spain or America for their holiday. **[1 mark]**

3 Look at this hypothesis.

'More teenagers read *Hi* magazine than read *Rumours* magazine.'

To test this hypothesis state
- whether you need primary data or secondary data
- how you would find or collect the data
- how you would use the data. **[2 marks]**

4 For each part, write whether the data is qualitative or quantitative.
If it is quantitative, is it discrete or continuous?
For each one give an example of a typical item of data.

a The countries in Europe. **[2 marks]**

b The colour of people's front doors. **[2 marks]**

c The number of people who attend Glastonbury Festival. **[2 marks]**

d The rainfall in Windermere on each day in April. **[2 marks]**

5 These are the sizes of shoes sold in a shop during one busy Saturday.

3	5	7	5	9	12	11	5	7	4	8	8	13	6	14	6	8	9	5	8
10	6	7	4	12	11	3	7	9	10	6	10	7	12	10	9	8	7	4	10
7	5	9	10	5	11	8	9	6	10	4	10	7	11	14	7	6	8	10	9

a Design a grouped frequency table to illustrate this data.
Use the class intervals 3–5, 6–8, 9–11 and 12–14. **[3 marks]**

b Which class interval has the most shoes? **[1 mark]**

c How can you tell which size of shoe was the most common? **[1 mark]**

6 These are the heights (h) of some plants at a garden centre, measured to the nearest centimetre.

12 15 21 22 14 31 34 17 25 30 24 13
26 17 20 10 15 38 29 35 19 20 33 23
30 14 33 26 31 17 39 22 16 32 11 14

a Put this information into a grouped frequency table. **[3 marks]**
Use the class intervals $10 \leqslant h < 15$, $15 \leqslant h < 20$, etc.

b Which class interval contained the most plants? **[1 mark]**

7 People who had been on holiday during the summer months were asked what accommodation they had stayed in.
The information was put into this two-way table.

	Caravan	B&B	Apartment	Hotel	Total
June		4	5	14	
July	8		12		37
August	11	10	15		46
September	6	7		13	30
Total	28		36	48	

a Copy the table and complete it by filling in all the missing numbers. **[5 marks]**

b How many people had stayed in a B&B? **[1 mark]**

c How many people stayed in a hotel in August? **[1 mark]**

d How many people stayed in an apartment in September? **[1 mark]**

e How many people took part in this survey? **[1 mark]**

C

8 Write a question Ian could use to find out how much time students spend doing homework during the week. Include a response section. **[3 marks]**

A02

9 Simon wants to find out which sport is most popular amongst 14- to 16-year-olds in his school.
He asks a group of Year 11 boys.
Give two reasons why this sample is not representative. **[2 marks]**

Chapter summary

In this chapter you have learned how to

- gather data efficiently E
- understand the data handling cycle D
- write a hypothesis D
- look for information D
- identify different types of data D
- gather data that can take a wide range of values D
- record related data D
- write good questions to find out information C
- get a reliable sample C

2

Fractions, decimals and percentages

This chapter is about working with fractions, decimals and percentages.

If you cook a batch of 40 pancakes, you need these skills to help share them equally with 5 friends.

Objectives

This chapter will show you how to

- **find a fraction of an amount using a calculator** E D
- **calculate with fractions using a calculator** E D
- **calculate the percentage of an amount** E D
- **write one quantity as a fraction of another** D
- **calculate a percentage increase or decrease** D
- **write one quantity as a percentage of another** D C
- **understand and use a retail prices index (RPI)** D C

Before you start this chapter

1 Write down how many
 a m are in 1 km b mg are in 1 gram
 c mm are in 1 cm d m*l* are in 1 litre

2 Write down the value of the 8 in each of these numbers.
 a 186.79 b 0.872 c 3.228

3 a Write down the common factors of 18 and 27.
 b Write down the highest common factor of 18 and 27.

4 Copy and complete these conversions.
 a 7 km = □ m
 b 32 mm = □ cm
 c 0.6 km = □ m = □ cm

5 Work out
 a 10% of £50 b 5% of £50 c 15% of £50

6 Anna needs £250 to buy a new bicycle. She saves £18 each week.
 For how many weeks must she save so that she has enough money?

7 a Convert 16% to a decimal.
 b Convert 116% to a decimal.

8 Calculate $60 \times \frac{40}{100}$

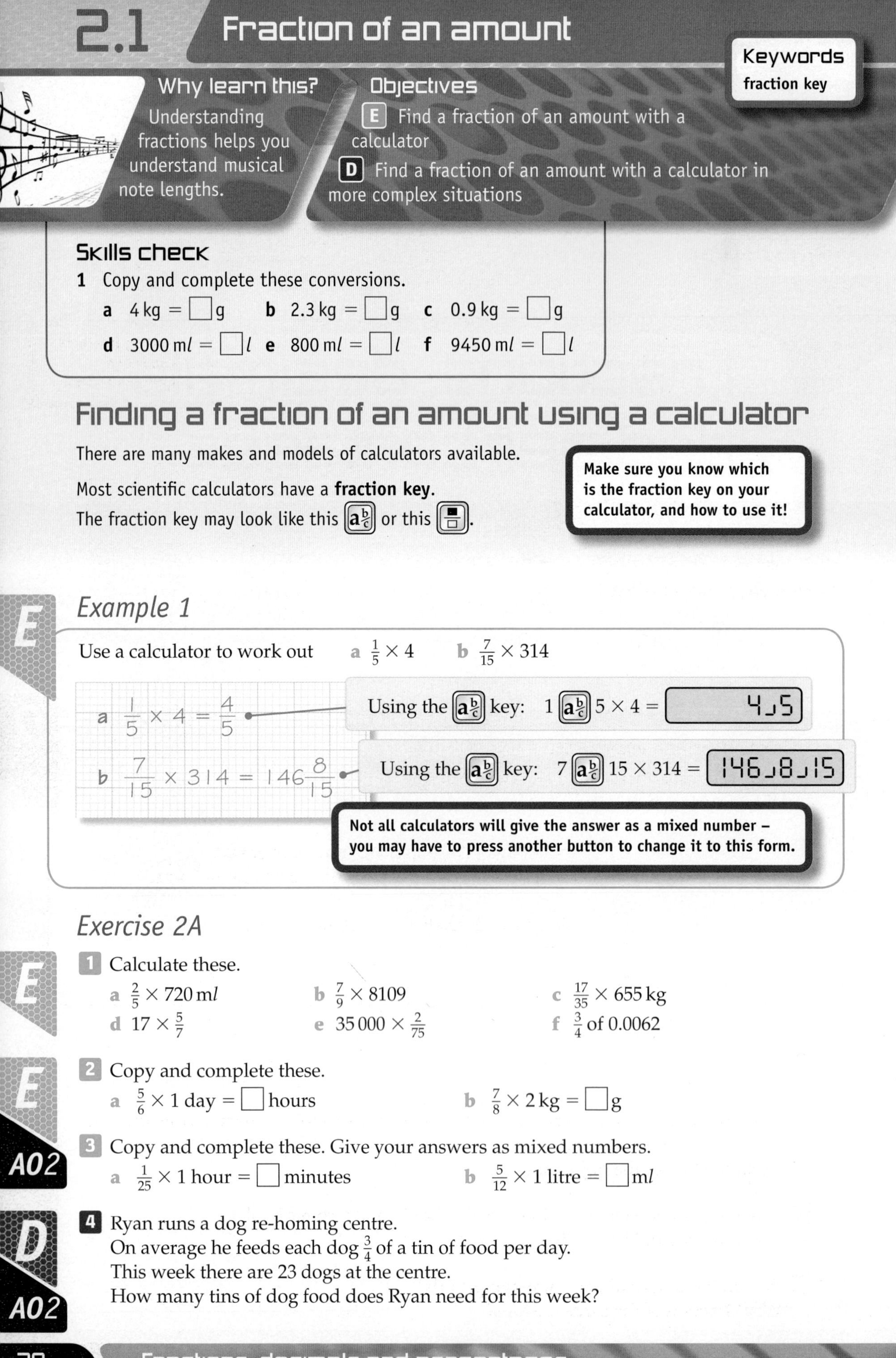

2.1 Fraction of an amount

Keywords
fraction key

Why learn this?

Understanding fractions helps you understand musical note lengths.

Objectives

E Find a fraction of an amount with a calculator

D Find a fraction of an amount with a calculator in more complex situations

Skills check

1 Copy and complete these conversions.

a 4 kg = □ g **b** 2.3 kg = □ g **c** 0.9 kg = □ g

d 3000 ml = □ l **e** 800 ml = □ l **f** 9450 ml = □ l

Finding a fraction of an amount using a calculator

There are many makes and models of calculators available.

Most scientific calculators have a **fraction key**.

The fraction key may look like this $\boxed{a\frac{b}{c}}$ or this $\boxed{\frac{\blacksquare}{\square}}$.

Make sure you know which is the fraction key on your calculator, and how to use it!

Example 1

Use a calculator to work out **a** $\frac{1}{5} \times 4$ **b** $\frac{7}{15} \times 314$

a $\frac{1}{5} \times 4 = \frac{4}{5}$

Using the $\boxed{a\frac{b}{c}}$ key: 1 $\boxed{a\frac{b}{c}}$ 5 × 4 = 4⌟5

b $\frac{7}{15} \times 314 = 146\frac{8}{15}$

Using the $\boxed{a\frac{b}{c}}$ key: 7 $\boxed{a\frac{b}{c}}$ 15 × 314 = 146⌟8⌟15

Not all calculators will give the answer as a mixed number – you may have to press another button to change it to this form.

Exercise 2A

E

1 Calculate these.

a $\frac{2}{5} \times 720$ ml **b** $\frac{7}{9} \times 8109$ **c** $\frac{17}{35} \times 655$ kg

d $17 \times \frac{5}{7}$ **e** $35\,000 \times \frac{2}{75}$ **f** $\frac{3}{4}$ of 0.0062

E

2 Copy and complete these.

a $\frac{5}{6} \times 1$ day = □ hours **b** $\frac{7}{8} \times 2$ kg = □ g

AO2

3 Copy and complete these. Give your answers as mixed numbers.

a $\frac{1}{25} \times 1$ hour = □ minutes **b** $\frac{5}{12} \times 1$ litre = □ ml

D AO2

4 Ryan runs a dog re-homing centre.
On average he feeds each dog $\frac{3}{4}$ of a tin of food per day.
This week there are 23 dogs at the centre.
How many tins of dog food does Ryan need for this week?

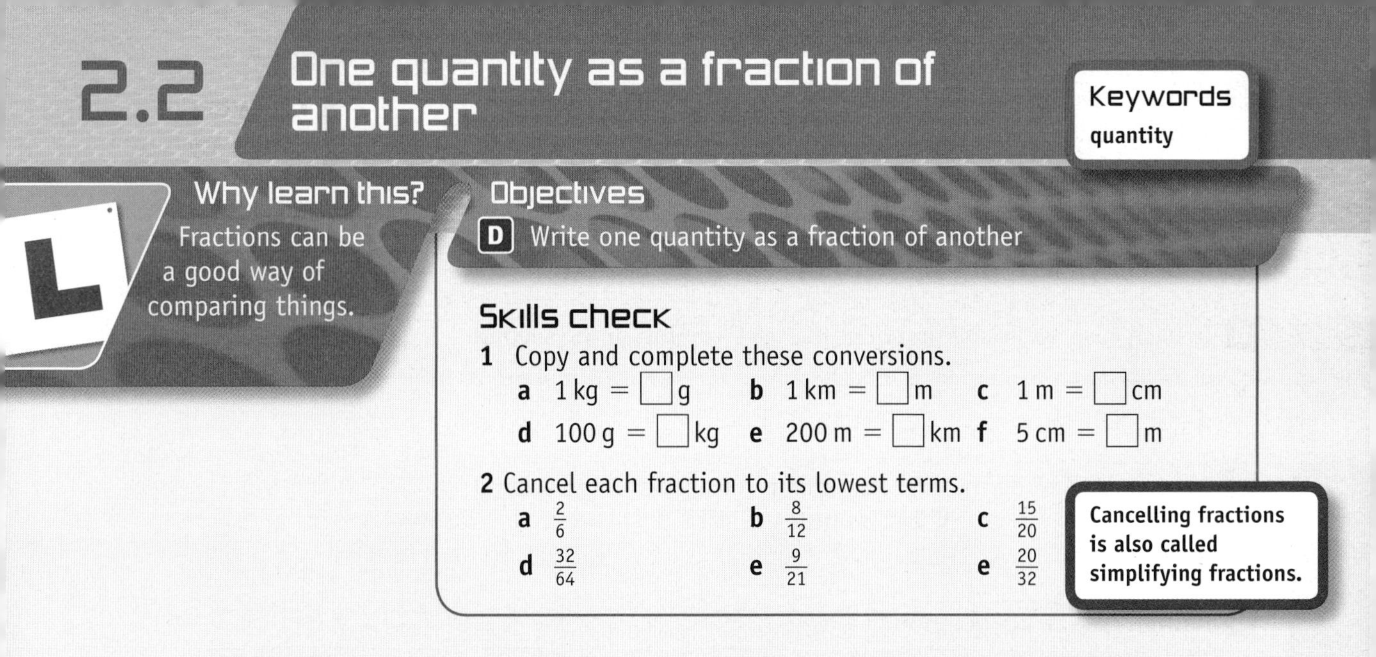

2.2 One quantity as a fraction of another

Keywords
quantity

Why learn this?
Fractions can be a good way of comparing things.

Objectives

D Write one quantity as a fraction of another

Skills check

1 Copy and complete these conversions.
   a 1 kg = ☐ g   b 1 km = ☐ m   c 1 m = ☐ cm
   d 100 g = ☐ kg   e 200 m = ☐ km   f 5 cm = ☐ m

2 Cancel each fraction to its lowest terms.
   a $\frac{2}{6}$ b $\frac{8}{12}$ c $\frac{15}{20}$
   d $\frac{32}{64}$ e $\frac{9}{21}$ e $\frac{20}{32}$

Cancelling fractions is also called simplifying fractions.

Writing one quantity as a fraction of another

To write one **quantity** as a fraction of another

- first write both quantities in the same units
- then write the first quantity over the second quantity
- finally, simplify the fraction.

Example 2

D

a Write 40p as a fraction of £6.

b In Diane's class there are 13 boys and 14 girls.
What fraction of the class are boys?

a $\frac{40}{600} = \frac{1}{15}$ — Write £6 as 600p and then write 40 over 600. Remember to cancel the fraction to its lowest terms.

b $\frac{13}{27}$ — 13 + 14 = 27 students altogether. 13 out of the 27 are boys.

Exercise 2B

D

1 In Polly's dog training class, three of the dogs pass their elementary certificate. The other four do not pass.
What fraction of the dogs pass?

2 Davin has 8 fiction books, 14 non-fiction books and 6 comic books.
What fraction of Davin's books are fiction?

3 Rowan has a normal pack of playing cards.
What fraction of the cards are aces?

D

4 During April it was sunny for 14 days.
What fraction of the days in April were not sunny?

A02

5 Donna said, 'I've got 5 red sweets and 9 blue sweets so the fraction of my sweets that are red is $\frac{5}{9}$.'
Is Donna correct? Explain your answer.

D

6 In each case, write the first quantity as a fraction of the second.

a £2, £7 **b** £2, £8 **c** 4 m*l*, 20 m*l*

d 8 kg, 20 kg **e** 1 hour, 24 hours **f** 6 weeks, 8 weeks

D

7 In each case, write the first quantity as a fraction of the second.

a 20p, £1 **b** 5 minutes, 1 hour **c** 5 days, 2 weeks

d 30 cm, 1 m **e** 750 g, 1 kg **f** 12 m, 1 km

8 Lauren gets £5 pocket money a week.
Each week she buys a magazine for £1.75 and she saves £1.50.
The rest is left over for her to spend on other things.
What fraction of her pocket money does Lauren

a spend on a magazine

b save

c have left over?

A02

2.3 Calculating with fractions

Why learn this?

By learning how to enter fractions into a calculator you can save a lot of time.

Objectives

E Use the fraction key on a calculator

D Use the fraction key on a calculator with mixed numbers

Skills check

1 Use the fraction key on a calculator to cancel these fractions.

a $\frac{24}{36}$ **b** $\frac{55}{75}$ **c** $\frac{52}{65}$

2 Use the fraction key on a calculator to convert these mixed numbers to improper fractions.

a $3\frac{8}{9}$ **b** $4\frac{12}{17}$ **c** $16\frac{18}{25}$

3 Use the fraction key on a calculator to convert these improper fractions to mixed numbers.

a $\frac{130}{7}$ **b** $\frac{196}{15}$ **c** $\frac{223}{32}$

Calculating with fractions using a calculator

To carry out any calculations involving fractions, enter the fractions on your calculator using the fraction key.

Make sure you practise using your calculator so you become skilled.

Example 3

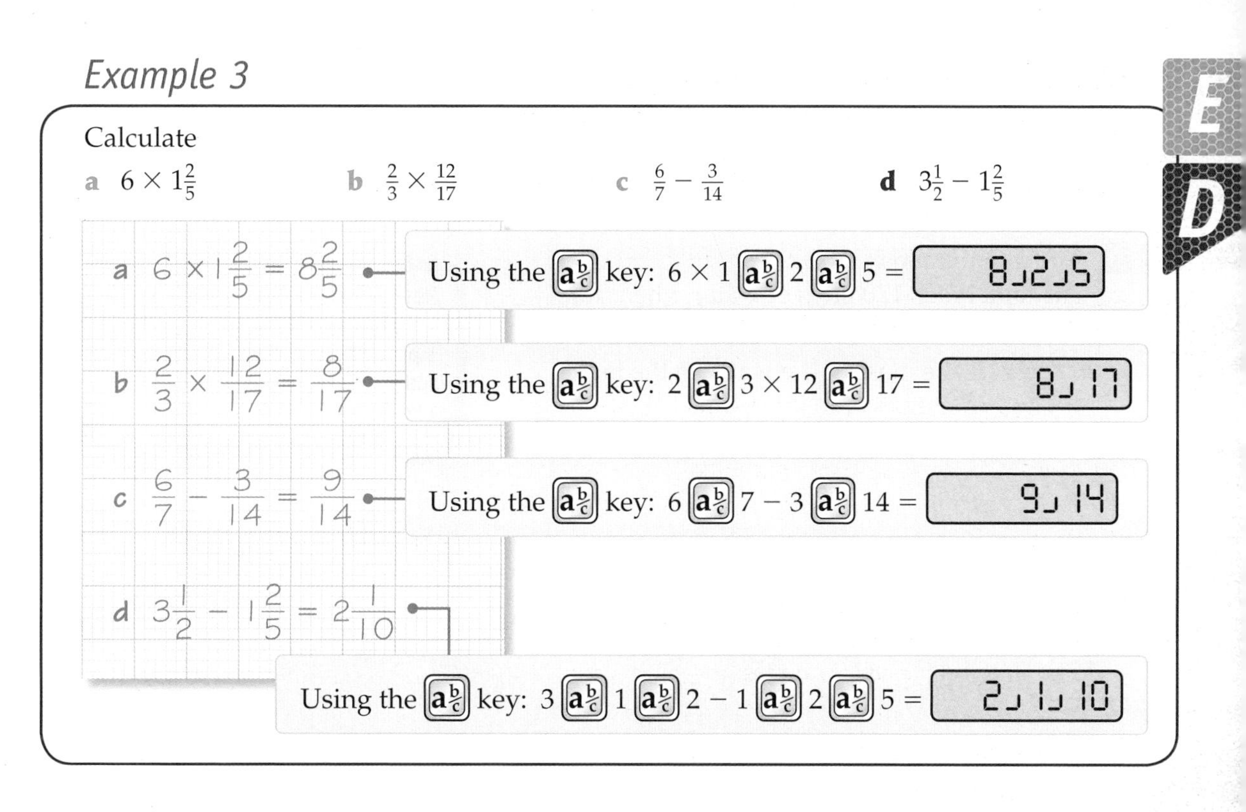

Exercise 2C

E

1 Work out these multiplications.
Give your answers as whole numbers.

a $\frac{1}{7} \times 63$ b $\frac{2}{9} \times 729$ c $405 \times \frac{2}{3}$

2 Work out these multiplications.
Give your answers as fractions in their simplest form.

a $\frac{2}{7} \times \frac{4}{5}$ b $\frac{3}{13} \times \frac{8}{25}$ c $\frac{4}{15} \times \frac{3}{20}$

3 Work out these calculations.
Give your answers as fractions or mixed numbers.

a $\frac{1}{2} + \frac{8}{9}$ b $\frac{3}{4} - \frac{7}{11}$ c $\frac{13}{14} + \frac{4}{5} - \frac{1}{12}$

D

4 Work out these calculations.
Give your answers as mixed numbers.

a $3\frac{1}{4} + 6\frac{2}{9}$ **b** $5\frac{1}{4} - 2\frac{3}{8}$ **c** $8\frac{2}{15} + 9\frac{11}{12}$

D

5 A football stadium has 72 000 seats. Each seat is $\frac{4}{5}$ m wide.
If all 72 000 seats were placed next to each other in a line, how long would the line be?
Give your answer in

a metres   **b** kilometres   **c** miles.

1 km = $\frac{5}{8}$ mile

A02

6 Fred added four identical mixed numbers.
He got the answer $8\frac{3}{4}$.
What were the mixed numbers that Fred added?

D

A03

7 A water-butt contains $123\frac{3}{4}$ litres of water.
The tap is opened so that $4\frac{1}{2}$ litres of water pour out every minute.
At this rate, how long will the water-butt take to empty?
Give your answer in minutes and seconds.

D

8 Which calculation gives the largest answer?

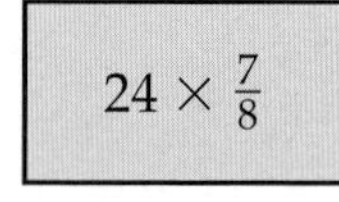

$24 \times \frac{7}{8}$

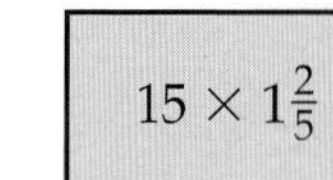

$15 \times 1\frac{2}{5}$

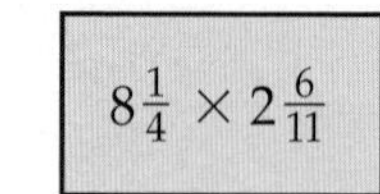

$8\frac{1}{4} \times 2\frac{6}{11}$

D
AO2

9 Anders buys a fish tank.
The tank holds 120 litres of water when full.
Anders fills the tank with water.

a What is the mass of the water in the tank?
Give your answer in kilograms.

b What do you notice about your answer?

> ***Metric–Imperial conversions***
> **1 litre is approximately $1\frac{3}{4}$ pints.**
> **1 pint of water weighs approximately $1\frac{1}{4}$ pounds (lb).**
> **1 kg is approximately $2\frac{1}{5}$ lb.**

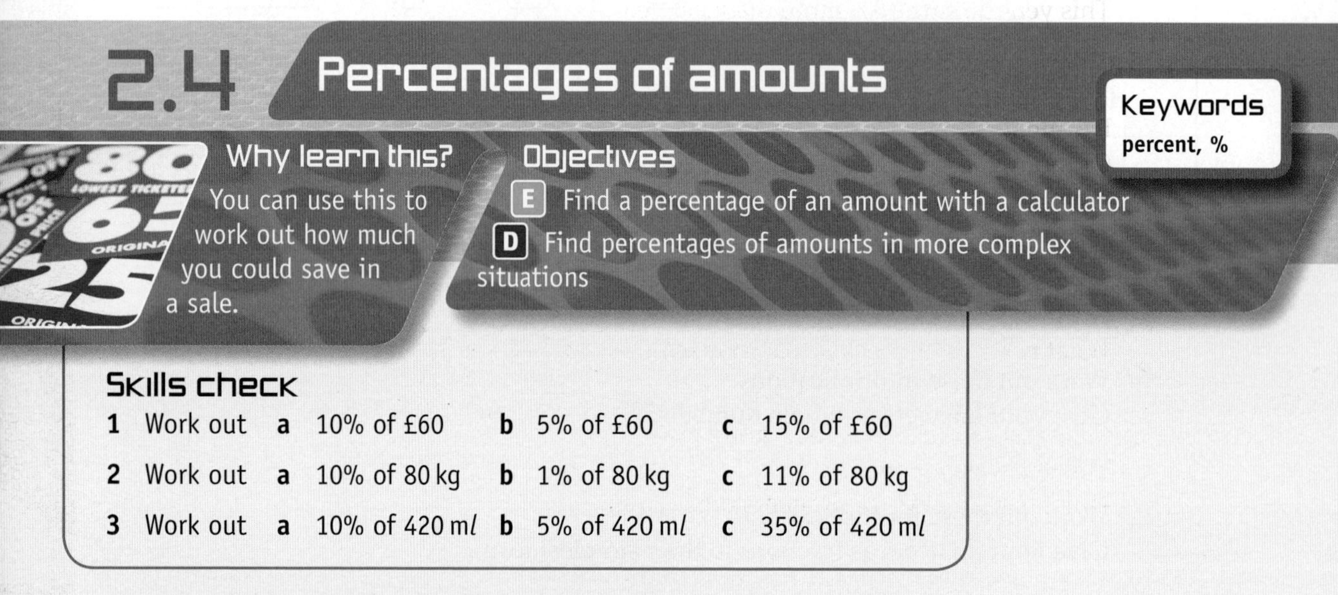

2.4 Percentages of amounts

Keywords
percent, %

Why learn this?
You can use this to work out how much you could save in a sale.

Objectives

E Find a percentage of an amount with a calculator

D Find percentages of amounts in more complex situations

Skills check

1 Work out a 10% of £60 b 5% of £60 c 15% of £60

2 Work out a 10% of 80 kg b 1% of 80 kg c 11% of 80 kg

3 Work out a 10% of 420 m*l* b 5% of 420 m*l* c 35% of 420 m*l*

Calculating the percentage of an amount

The word '**percent**' means 'out of 100'.

The symbol '**%**' is a short way of writing 'percent'.

$82\% = \frac{82}{100} = 0.82$

To find 1% of an amount, divide the amount by 100.

To find 82% of an amount, divide the amount by 100 and then multiply by 82.

E

Example 4

Calculate

a 18% of 92 m b 7.5% of 650 g c 150% of £18

a $92 \div 100 \times 18 = 16.56$ m

b $650 \div 100 \times 7.5 = 48.75$ g

c $18 \div 100 \times 150 = £27$

> On your calculator, divide the amount by 100 first.
> This finds 1%.
> Then multiply the result by the percentage that you want to find.

Exercise 2D

E

1 Calculate these percentages. Write down your workings.

a 12% of £550
b 7.6% of 60 m
c 12.5% of 800 cm
d 81% of 2.5 kg
e 0.05% of 5 million
f 225% of £30

2 Amy saw this sticker on a TV in an electrical shop.
How much is the discount on the TV?

3 Last year Wayne earned £16 000.
This year he earns 3% more than last year.
How much more does Wayne earn this year?

E

4 During one week, 240 trains arrived at Carmarthen train station.
Of these trains, 22% arrived early and 13% arrived late.
How many of the trains arrived on time?

AO2

5 Eighteen-carat pink gold consists of 75% gold, 9% silver and 16% copper.
An 18-carat pink gold necklace weighs 18 g.
What mass in the necklace is copper?

E

6 Renee scored 65% in a geography test.
The test was out of 80.
How many marks did Renee score?

D AO2

7 Alan bought an old table for £47.50 and sold it for 24% more than he paid.
How much did Alan sell the table for?

D AO3

FUNCTIONAL

8 Molly compares the prices of the same bike in two different shops.

Which shop is the cheaper and by how much?

2.5 One quantity as a percentage of another

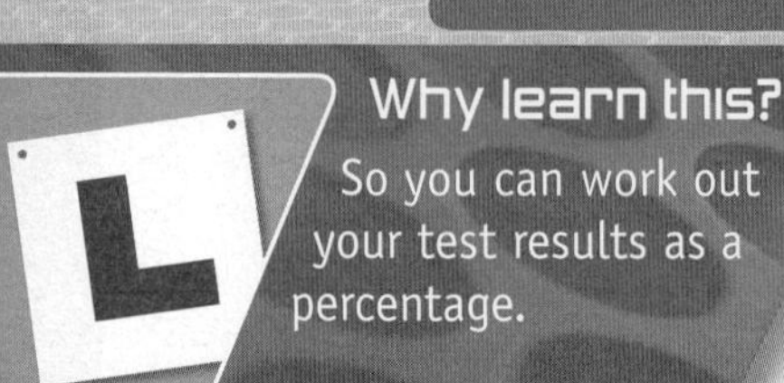

Why learn this?
So you can work out your test results as a percentage.

Objectives
D Write one quantity as a percentage of another
C Write one quantity as a percentage of another in more complex situations

Skills check

1 Write down how many
 a cm are in 1 m **b** m*l* are in 1 litre
 c c*l* are in 1 litre **d** g are in 1 kg
 e hours are in a day **f** months are in a year.

2 Copy and complete these divisions.
 a $\frac{3}{5} = 3 \div 5 = \square$ **b** $\frac{7}{28} = 7 \div \square = \square$ **c** $\frac{25}{40} = \square \div \square = \square$

Writing one quantity as a percentage of another

To write one quantity as a percentage of another
- write the first quantity as a fraction of the second.
- multiply the fraction by 100 to convert it to a percentage.

D C

Example 5

a Express £5 as a percentage of £25.
b Express 8 mm as a percentage of 2 cm.
c Daren bought a car for £8000 and sold it for £6500.
What percentage of the price that he paid has he lost?

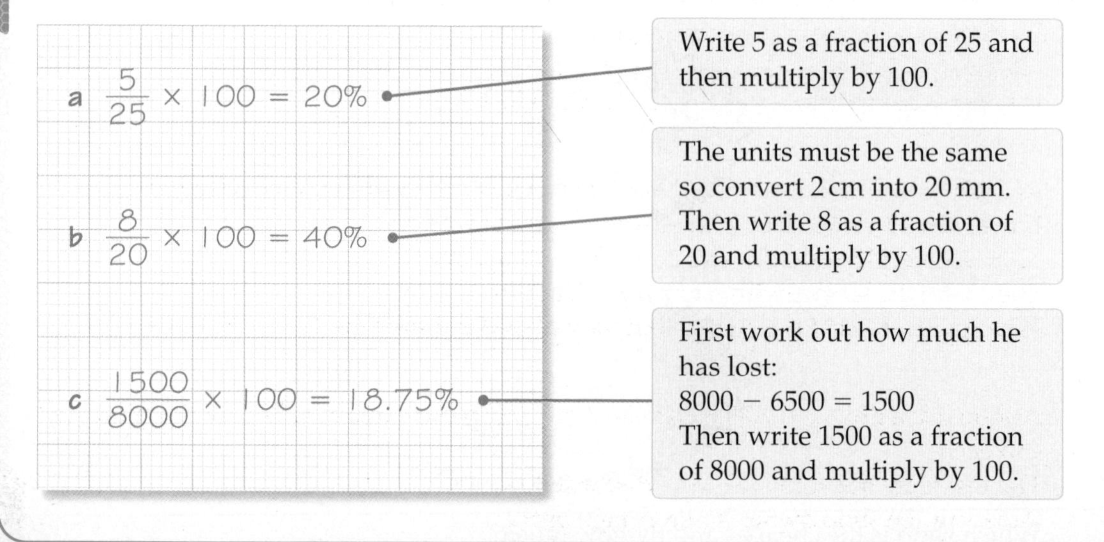

Exercise 2E

D

1 In each case, express the first quantity as a percentage of the second.

a £5, £50 **b** £5, £80 **c** £25, £75

d 4 hours, 1 day **e** 36 minutes, 1 hour **f** 125 g, 1 kg

g 13 weeks, 1 year **h** 14.5 cm, 1 m **i** 275 m*l*, 1 litre

Make sure the units of both quantities are the same.

2 Baz goes on a diet.
His starting weight is 93 kg.
Over 2 months his weight goes down by 5 kg to 88 kg.
What percentage of his starting weight has Baz lost?

3 Fathe goes to the shops with £12.
He buys a book for £4.50.
What percentage of his money has he got left?

4 Hari scores 26 out of 30 in an English test.
What percentage of the test does Hari get wrong?

5 A badminton club has 63 members.
The table shows the membership numbers.
What percentage of the members are

Men	Women	Girls	Boys
17	21	12	13

a men **b** female?

C

6 Billy weighed 102 kg at the start of his diet.
He now weighs 82 kg.
What percentage of his starting weight has he lost?
Give your answer to one decimal place.

7 Last year Llanreath Divers took £1800 in membership fees.
This year they took £2100 in membership fees.
What is the percentage increase in the amount taken in membership fees?

C

8 Jon Brower Minnoch was the heaviest man recorded in history.
In 16 months he lost 419 kg in weight.
His final weight was 216 kg.
What percentage of his starting weight did he lose?

9 In 2007 Little Haven won the South Pembrokeshire short-mat bowls league.
Out of the 20 games they played, they won 11 and drew 2.
They scored 813 shots for and had 515 shots against.
They won the league with a total of 102 points.
What percentage of the games that they played did they lose?

10 A shop owner buys in 2000 chocolate bars at 37p each and sells 1777 of them for 45p each.
The rest were not sold and were discarded.
Work out the shop owner's percentage profit.

AO2

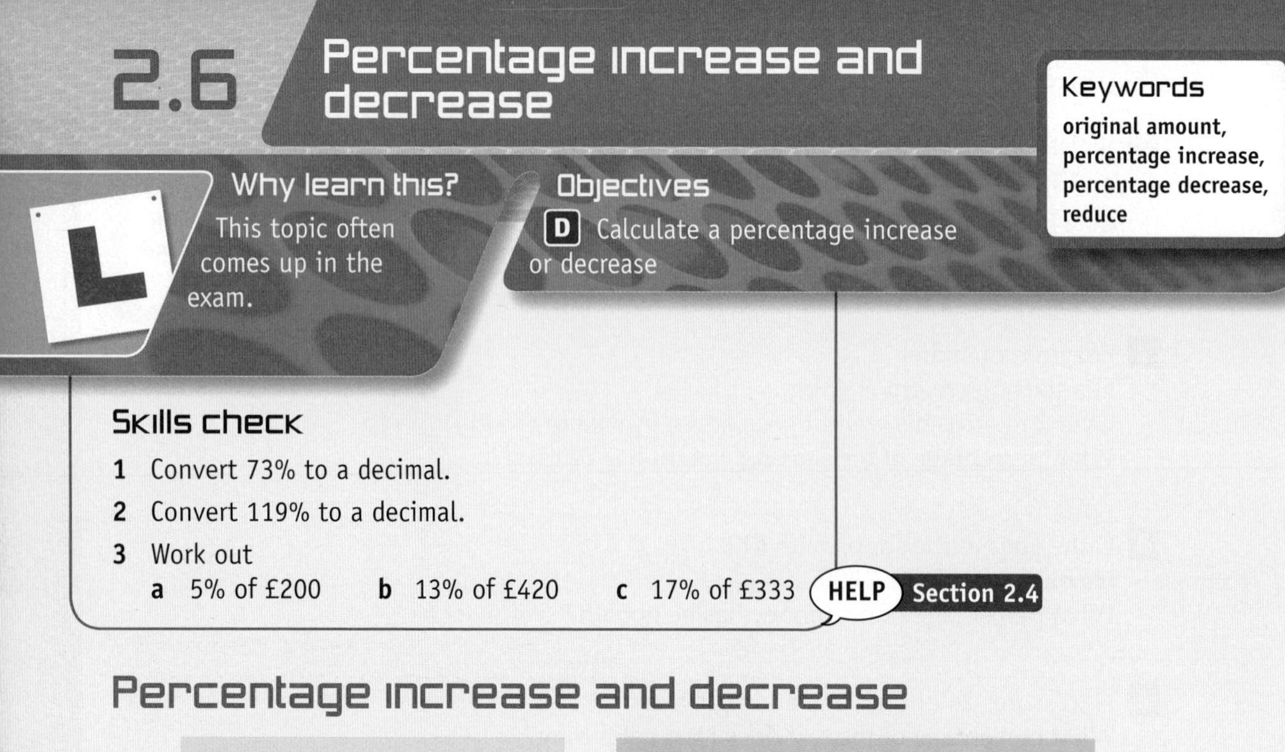

2.6 Percentage increase and decrease

Keywords
original amount, percentage increase, percentage decrease, reduce

Why learn this?
This topic often comes up in the exam.

Objectives
D Calculate a percentage increase or decrease

Skills check

1. Convert 73% to a decimal.
2. Convert 119% to a decimal.
3. Work out
   a 5% of £200 **b** 13% of £420 **c** 17% of £333

HELP Section 2.4

Percentage increase and decrease

Method A
1. Work out the actual increase (or decrease).
2. Add to (or subtract from) the **original amount**.

This method is most commonly used when working without a calculator.

Method B
1. Add the **percentage increase** to 100% (or subtract the **percentage decrease** from 100%).
2. Convert this percentage to a decimal.
3. Multiply it by the original amount.

This method is especially useful when using a calculator.

D

Example 6

John used to earn £320 a week. He has had an 8% pay rise.
What does he earn now?

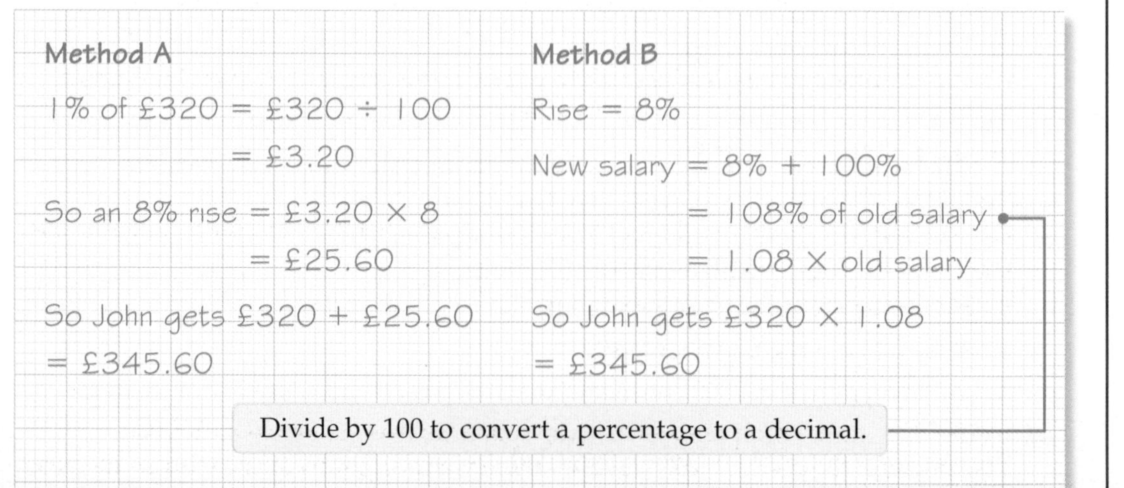

Example 7

The price of a bike is **reduced** by 20% in a sale. The original price was £150.
What is the sale price?

Method A

Decrease = £150 ÷ 5
= £30

Sale price is £150 − £30
= £120

20% is $\frac{1}{5}$, so divide by 5.

Method B

Decrease = 20%

New price = 100% − 20%
= 80% of old price
= 0.8 × old price

Sale price is £150 × 0.8
= £120

D

Exercise 2F

1 Work out the following.

a Increase 20 by 5%

b Increase 56 by 12%

c Decrease 3400 m*l* by 17%

d Decrease 480 by 32%

e Increase £135 by $7\frac{1}{2}$%

f Decrease 890 m*l* by 8.4%

2 Firefighters are given a 4% pay rise. If John earned £420 per week, how much will he earn after the pay rise?

3 There has been a 6% decrease in the number of reported thefts in Walton this past year. There were 350 reported thefts last year. How many were there this year?

4 A new car costs £8450. After two years the value of the car will have decreased by 43%. How much will the car be worth?

5 Sales of a magazine, costing £1.95, decreased by 11% this week. They sold 4300 copies last week. How much money have they lost on their sales this week?

6 Jane starts a new job earning £12 000 per year, increasing by 3% after 3 months.
Milly starts on £11 500 per year, increasing by 5% after 3 months.
Who has the greater salary after 3 months?

D

7 Ali weighed 84 kg. He lost 4% of his body weight when he started running but then put 2% of his new weight back on.

a How much was Ali's lowest weight?

b How much does Ali weigh now?

8 In 2007 the number of pairs of breeding sparrows was estimated as 200 000.
This decreased by 38% in 2008. In 2009 there was a slight recovery with an increase of 4%. How many breeding pairs were estimated at the end of 2009?

D

A02

VAT

VAT stands for value added tax. It is a tax that is added to the price of most items in shops and to many other services.

VAT is calculated as a percentage. Generally it is 17.5% in the UK.

VAT at 17.5% can be worked out by finding 10% + 5% + 2.5%.

Some goods have no VAT, for example books, most leisure activities and most food. A few items pay a reduced VAT rate (5%), for example children's car seats, domestic electricity and fuel.

D

Example 8

A digital camera is advertised for sale at £240 (excluding VAT).

How much will you have to pay?

'Excluding VAT' means that VAT has not been added on to the price.

Method A

VAT $= 17\frac{1}{2}\%$ of £240

$= \frac{17.5}{100} \times 240$

$=$ £42

Cost of digital camera

$=$ £240 + £42

$=$ £282

Method B

Increase $= 17\frac{1}{2}\%$

New cost $= 100\% + 17\frac{1}{2}\%$

$= 117\frac{1}{2}\%$ of old cost

$= 1.175 \times$ old cost

Cost of digital camera

$= 1.175 \times$ £240

$=$ £282

$117\frac{1}{2} \div 100 = 117.5 \div 100 = 1.175$
This is your multiplier to work out the new cost.

Exercise 2G

D

For Q1–5, assume the rate of VAT is 17.5%.

1 A DVD player costs £130 (excluding VAT).
A phone costs £40 (excluding VAT).
Work out **a** the VAT **b** the total cost of each item.

2 What is the total cost of a TV sold for £126 + VAT?

3 One car is sold at £4000 (including VAT) and another is sold at £3800 (excluding VAT).
Which car is the more expensive?

4 A meal for four costs £73.60 plus VAT.
If the four people decide to share the bill equally, how much will each pay?

D

AO2

5 The cost of a scooter is £376 + VAT. Jas decides to buy on credit.
He pays an initial deposit of 15% and then 12 monthly payments of £35.
How much more does Jas pay by buying on credit?

2.7 Index numbers

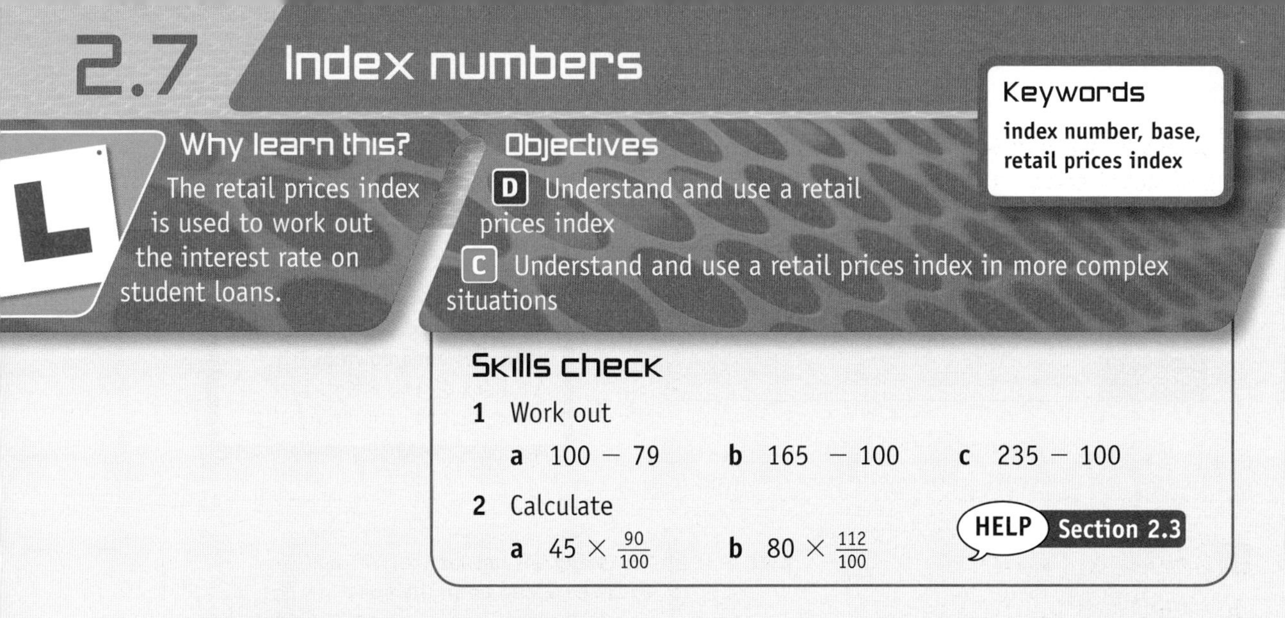

Objectives

D Understand and use a retail prices index

C Understand and use a retail prices index in more complex situations

Keywords
index number, base, retail prices index

Skills check

1 Work out

a $100 - 79$ b $165 - 100$ c $235 - 100$

2 Calculate

a $45 \times \frac{90}{100}$ b $80 \times \frac{112}{100}$

HELP Section 2.3

Retail prices index

An **index number** compares one number, usually a price, with another.
The figure that the numbers are compared with is called the **base**.
The index number is a percentage of the base, but the percentage sign is left out.
The base usually starts at 100.
The UK **retail prices index** started at base 100 in 1987.
In May 2009 the UK retail prices index was 211.3.
This means that average retail prices increased by 111.3% between 1987 and 2009.

Example 9

In 2000 the price of a litre of petrol was 69p.
The table shows the price index of petrol for 1999, 2001, 2002 and 2003, using 2000 as the base year.

Year	1999	2000	2001	2002	2003
Index	92	100	103	107	110
Price		69p			

Work out the price of petrol in 1999 and in the years 2001 to 2003.
Give your answers to one decimal place.

1999: $69\text{p} \times \frac{92}{100} = 63.5\text{p}$

In 1999 the index was less than the base, so the price of a litre of petrol must be less than 69p.

2001: $69\text{p} \times \frac{103}{100} = 71.1\text{p}$

In the years 2001 to 2003 the index was more than the base, so the price must be more than 69p.

2002: $69\text{p} \times \frac{107}{100} = 73.8\text{p}$

2003: $69\text{p} \times \frac{110}{100} = 75.9\text{p}$

Notice that you always use 69p for each year's calculation, as this is the price of petrol in the base year. So in the calculation for 2002, you don't use the 71.1p from 2001.

D

Exercise 2H

D

1 In 2006 the cost of a litre of petrol was 95p.
The table gives the price index of petrol for the next three years, using 2006 as the base year.

Year	2006	2007	2008	2009
Index	100	95	100	110
Price	95p			

An index of 95 means that the value has gone down by 5%. An index of 110 means that the value has gone up by 10%.

Work out the price of petrol from 2007 to 2009.
Give your answers in pence to one decimal place.

2 Compared with 2005 as the base, the index for the price of laptop computers is now 74.

a Has the price of laptop computers gone up or down?

b By what percentage has the cost of laptop computers changed?

3 This year the cost of a 4Gb memory stick is 40% lower than last year.
What is the index for the cost of a 4Gb memory stick this year, using last year as base?

4 In 1990 an average box of tissues cost 30p.
Taking 1990 as the base year of 100, work out how much an average box of tissues costs today, when the price index is 240.

D AO2

5 In 1990 the cost of 1 kg of bananas was £1.14.
Using 1990 as the base year, the price index of 1 kg of bananas was 75 in 2008.
Peter says, 'The price of bananas in 2008 is ¼ of the price they were in 1990.'
Is Peter correct? Explain your answer.

C AO2

6 A toy factory produced 455 000 toys in November 2008.
This represented an index of 78, using November 2002 as the base year with an index of 100.
How many toys did this factory produce in November 2002?

C AO3

7 The graph shows the exchange rates for the euro (€) and the pound (£) in 2008.

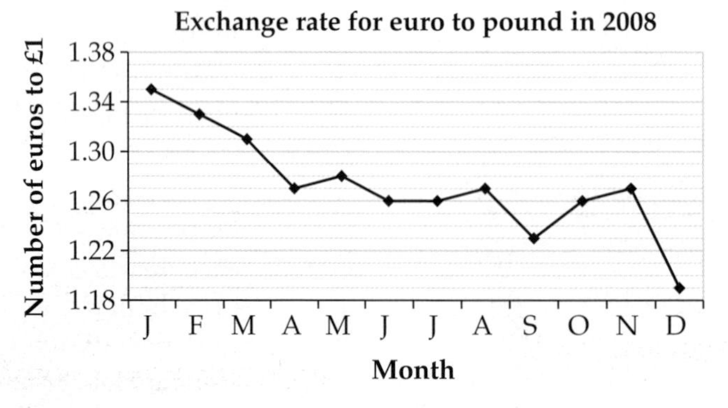

a What was the exchange rate in January 2008?

b Using January 2008 as the base of 100, work out the index for December 2008.

Review exercise

1 Ruth bought a house for £90 000.
Since she bought the house it has gone up in value by 6.5%.
How much has the house gone up in value? **[2 marks]**

E

2 Li-Mei saw this sticker on a sofa in a furniture shop.

Old price £1200
Now 22% off!!!

How much is the discount on the sofa? **[2 marks]**

3 Rohan scored 85% in a maths test.
The test was out of 60.
How many marks did Rohan score? **[2 marks]**

4 Tim added four identical mixed numbers.
He got the answer $4\frac{8}{23}$.
What were the mixed numbers that Tim added? **[2 marks]**

D AO2

5 Nigel is given £40 for his birthday. He spends £16.40 on clothes.
What percentage of his money has he got left? **[3 marks]**

D

6 All train fares are to increase by $7\frac{1}{2}$% next year.
At the moment, Jane pays £28 for her ticket and Peter pays £17.60.
How much will they have to pay next year? **[3 marks]**

7 In 1978 1 kg of mushrooms cost £1.53.
Taking 1978 as the base year of 100, work out how much 1 kg of mushrooms cost in 2008, when the price index was 172. **[2 marks]**

8 Tom invested £250 last year. Tom's investment has decreased in value by $2\frac{1}{2}$%.
How much is Tom's investment worth now? **[2 marks]**

9 Two posters advertise the same TV.

D

TVs R US
£240
+ VAT ($17\frac{1}{2}$%)
Buy now!

A.A. Electricals
Deposit: 22% of £280
plus
12 monthly payments of
£19.75

Which method is cheaper? Find how much can be saved using the cheaper method. **[6 marks]**

AO2

D
AO2

10 Heather sleeps for 8 hours every day.
What percentage of a week is Heather awake? **[3 marks]**

C

11 Last year a summer fête raised £1420 for charity.
This year it raised £1650.
What is the percentage increase in the amount raised for charity?
Give your answer to the nearest whole number. **[3 marks]**

C

12 Last year Amy bought a laptop computer.
This year the cost of the same laptop computer has fallen by £80 to £370.
By what percentage of last year's price has the cost of the laptop computer fallen?
Give your answer to one decimal place. **[3 marks]**

AO2

13 A specialist car manufacturer hand-built 20 cars in September 2007.
This represented an index of 80, using September 2000 as the base year with an index of 100.
How many cars were produced in this factory in September 2000? **[2 marks]**

Chapter summary

In this chapter you have learned how to

- use the fraction key on a calculator **E**
- find a fraction of an amount with a calculator **E**
- find a percentage of an amount with a calculator **E**
- find a fraction of an amount with a calculator in more complex situations **D**
- find percentages of amounts in more complex situations **D**
- use the fraction key on a calculator with mixed numbers **D**
- write one quantity as a fraction of another **D**
- write one quantity as a percentage of another **D**
- calculate a percentage increase or decrease **D**
- understand and use a retail prices index **D**
- write one quantity as a percentage of another in more complex situations **C**
- understand and use a retail prices index in more complex situations **C**

3 Interpreting and representing data

This chapter is about interpreting and representing data.

Is there a relationship between global temperature and the sizes of penguin colonies?

Objectives

This chapter will show you how to

- draw pie charts for various types of data E
- draw stem-and-leaf diagrams D
- draw frequency diagrams for continuous data with equal class intervals D
- draw and interpret scatter graphs D C
- draw frequency polygons for grouped data C

Before you start this chapter

1 Work out

a $\frac{1}{4}$ of 28 b $\frac{1}{3}$ of 120 c $360 - 47 - 116$ d $360 - 321 - 12$

2 How many degrees are there in a full circle?

3 What value is half way between the following pairs of numbers.

a 10 and 12 b 14 and 20 c 18 and 25 d 20 and 36

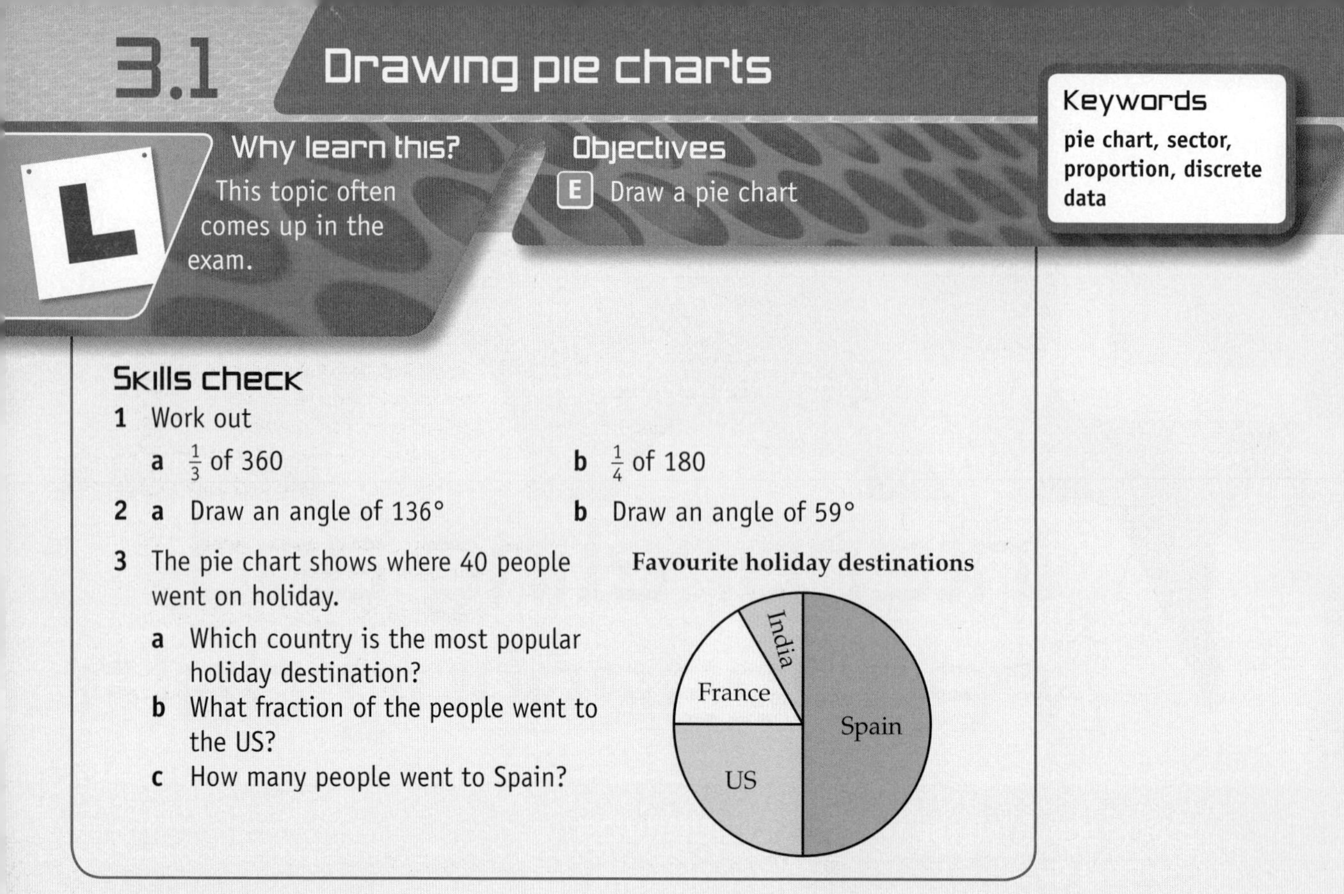

3.1 Drawing pie charts

Why learn this?
This topic often comes up in the exam.

Objectives
E Draw a pie chart

Keywords
pie chart, sector, proportion, discrete data

Skills check

1 Work out

a $\frac{1}{3}$ of 360 b $\frac{1}{4}$ of 180

2 a Draw an angle of 136° b Draw an angle of 59°

3 The pie chart shows where 40 people went on holiday.

a Which country is the most popular holiday destination?

b What fraction of the people went to the US?

c How many people went to Spain?

Drawing pie charts

A **pie chart** is a circle that is split up into sections (**sectors**). The different sectors represent the **proportions** of **discrete data**.

The angles in the sectors add up to 360°. The pie chart must be labelled and the angles accurately drawn.

Pie charts are a useful way of turning raw data into usable information, which is a key part of the data handling cycle.

A pie chart can be drawn from a frequency table.

E

Example 1

Jessica recorded the colour of cars parked at her school yesterday.
The table shows her results.
There are 24 cars in total.
Draw a pie chart to show this information.

Colour	Frequency
silver	14
red	3
blue	2
black	5

Method 1

$360° \div 24 = 15°$

There are 360° in a full circle. There are 24 cars. The size of the angle that represents 1 car is $\frac{360°}{\text{total frequency}}$.

15° represents 1 car.

Silver: $14 \times 15° = 210°$

Multiply the frequency of each colour by 15°.

Red: $3 \times 15° = 45°$

Blue: $2 \times 15° = 30°$

Black: $5 \times 15° = 75°$

Check that the angles add up to 360°.

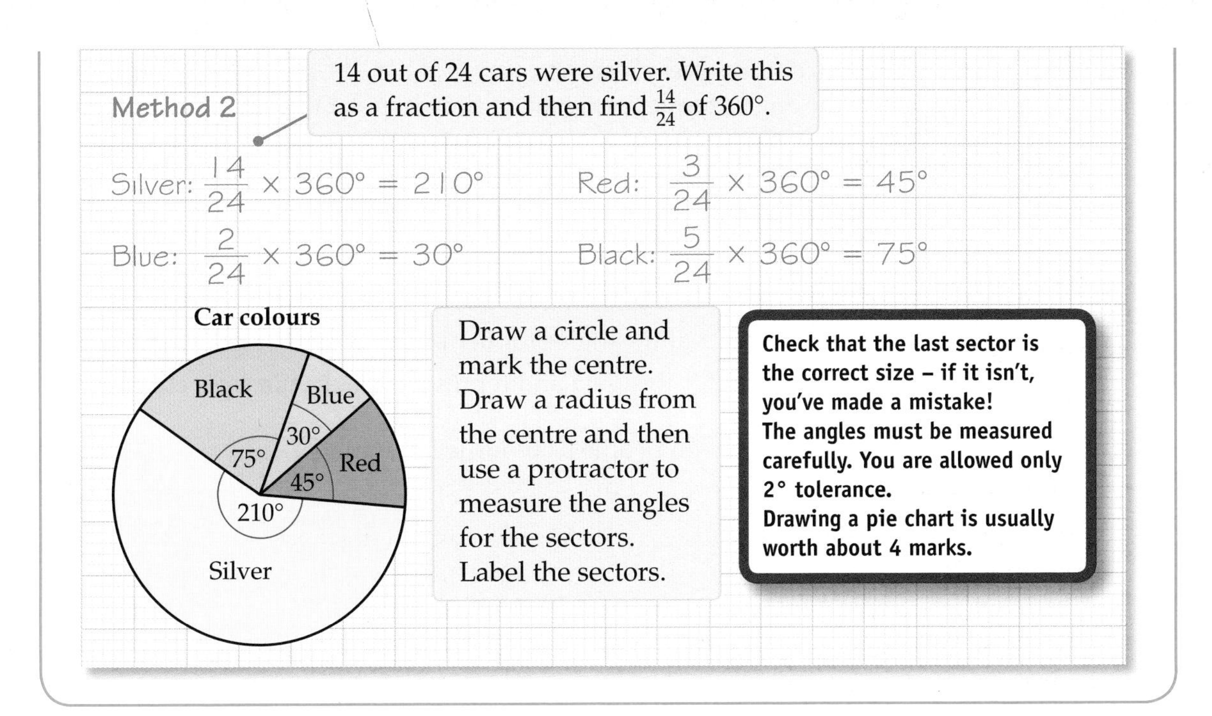

Exercise 3A

E

1 The frequency table shows the food bought in the school canteen.

Food	Frequency
pizza	20
sandwich	4
salad	3
pasta	9

Represent this data in a pie chart.

2 Asif carried out a survey on students' favourite pets.
His results are shown in the table.

Favourite pet	Frequency
cat	8
dog	2
fish	5
bird	3

Represent this data in a pie chart.

3 Reece owns a garden centre.
One day he records the numbers of electrical goods he sells.
His results are shown in the table.

Electrical goods	Number sold
lawnmower	9
strimmer	3
power saw	4
hedge trimmer	8

Represent this data in a pie chart.

E

4 Katie has collected this data on eye colour.
She has worked out the angles for her pie chart as shown in the table.

Eye colour	Frequency	Angle
blue	18	220°
brown	7	75°
green	5	60°

AO2

a Explain how you know that Katie's angles for her pie chart cannot be correct.
b Work out the correct angles for this data and represent this data in a pie chart.

E

5 Zi Ying has carried out a survey on people's favourite TV stations.

TV station	Frequency
BBC 1	21
BBC 2	8
ITV	12
Channel 4	4
Channel 5	10
Satellite	17

Draw a pie chart to illustrate this data.

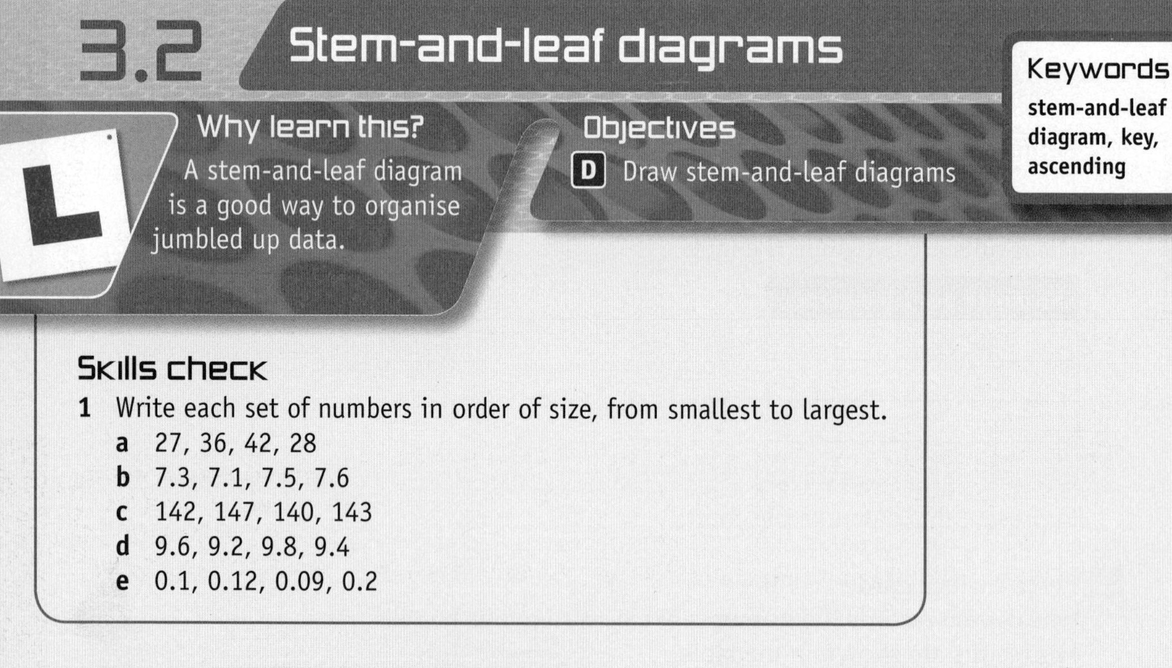

3.2 Stem-and-leaf diagrams

Keywords
stem-and-leaf diagram, key, ascending

Why learn this?
A stem-and-leaf diagram is a good way to organise jumbled up data.

Objectives
D Draw stem-and-leaf diagrams

Skills check

1 Write each set of numbers in order of size, from smallest to largest.
a 27, 36, 42, 28
b 7.3, 7.1, 7.5, 7.6
c 142, 147, 140, 143
d 9.6, 9.2, 9.8, 9.4
e 0.1, 0.12, 0.09, 0.2

Stem-and-leaf diagrams

Stem-and-leaf diagrams can be used to organise discrete data, so that analysis is easier.
The data is grouped and ordered according to size, from smallest to largest.
A **key** is needed to explain the numbers in the diagram.

Example 2

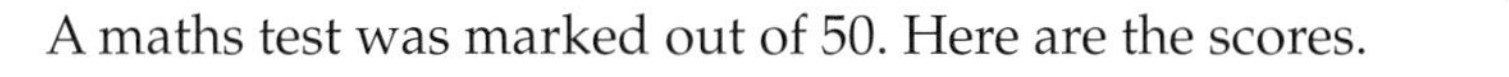

A maths test was marked out of 50. Here are the scores.

27 28 36 42 50 18 25 31 39 25 49 31
33 27 37 25 47 40 7 31 26 36 9 42

a Draw an ordered stem-and-leaf diagram to represent this data.
b How many students had a maths score of less than 20?

Write all the scores on a diagram. This data can be written with the 'tens' digit as the stem and the 'units' digit as the leaf.

a Unordered stem-and-leaf diagram:

```
0 | 7 9
1 | 8
2 | 7 8 5 5 7 5 6
3 | 6 1 9 1 3 7 1 6
4 | 2 9 7 0 2
5 | 0
```

Stem — Leaf

0 | 7 represents 7
1 | 8 represents 18

Key
3 | 6 means 36

The key shows how to read the values.

Ordered stem-and-leaf diagram:

Now rewrite the diagram with the leaves in **ascending** order.

```
0 | 7 9
1 | 8
2 | 5 5 5 6 7 7 8
3 | 1 1 1 3 6 6 7 9
4 | 0 2 2 7 9
5 | 0
```

Key
3 | 6 means 36

b 3 students

The scores 7, 9 and 18 are the only ones less than 20.

Check that the total number of values in the stem-and-leaf diagram is the same as the number of values in the original list.

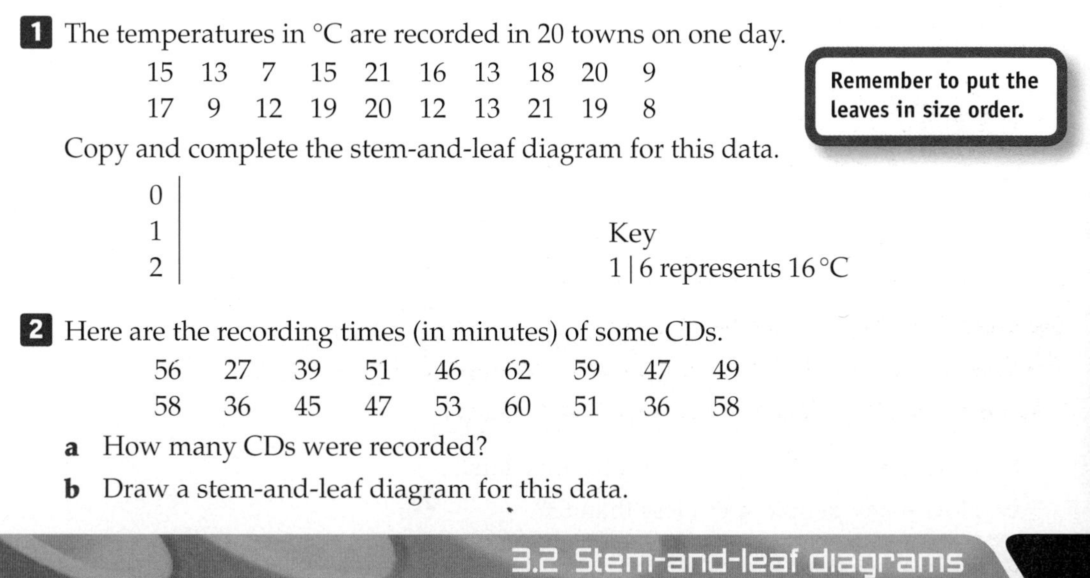

Exercise 3B

1 The temperatures in °C are recorded in 20 towns on one day.

15 13 7 15 21 16 13 18 20 9
17 9 12 19 20 12 13 21 19 8

Copy and complete the stem-and-leaf diagram for this data.

```
0 |
1 |
2 |
```

Key
1 | 6 represents 16 °C

Remember to put the leaves in size order.

2 Here are the recording times (in minutes) of some CDs.

56 27 39 51 46 62 59 47 49
58 36 45 47 53 60 51 36 58

a How many CDs were recorded?
b Draw a stem-and-leaf diagram for this data.

Decimal stem-and-leaf diagrams

Sometimes it may be necessary to have two digits for each of the leaves.

D

Example 3

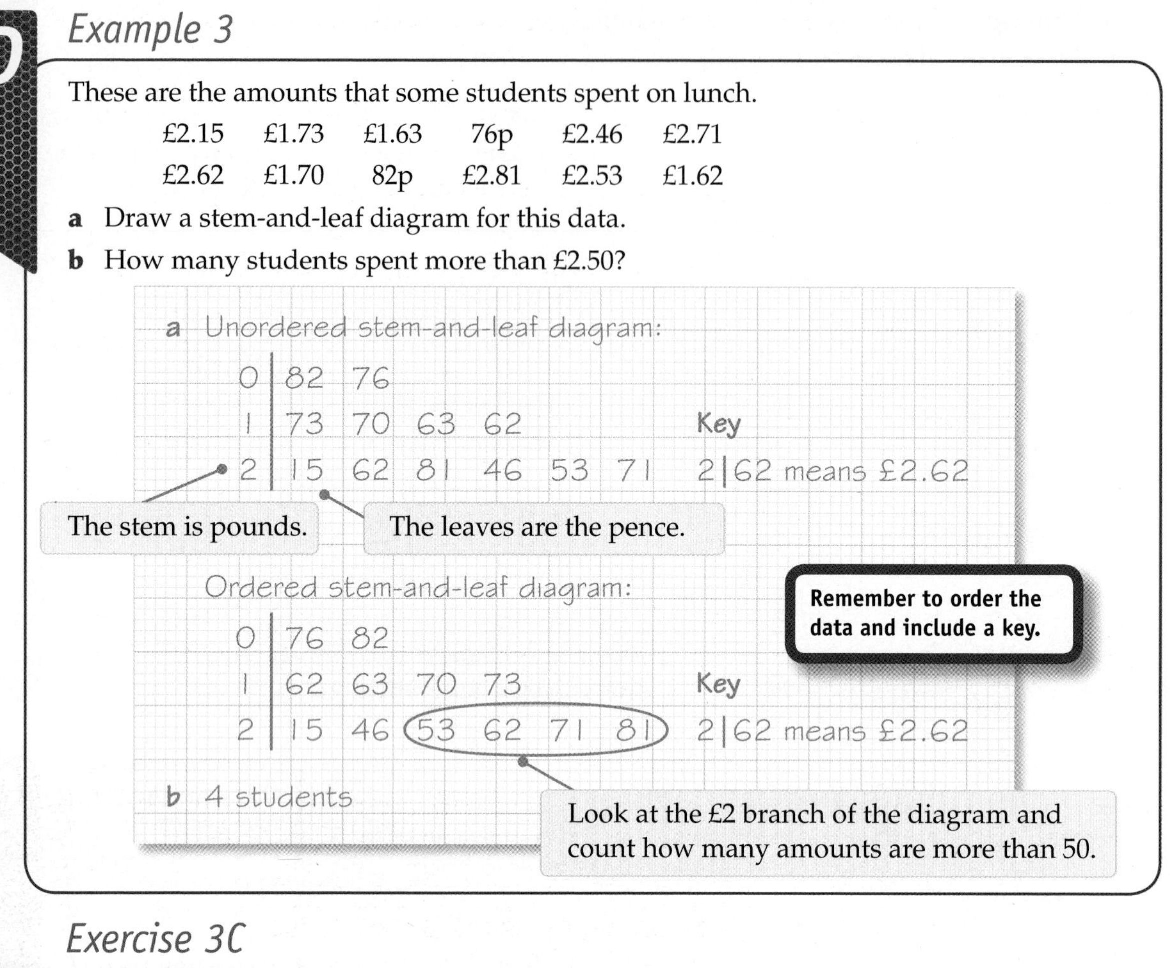

These are the amounts that some students spent on lunch.

£2.15 £1.73 £1.63 76p £2.46 £2.71
£2.62 £1.70 82p £2.81 £2.53 £1.62

a Draw a stem-and-leaf diagram for this data.

b How many students spent more than £2.50?

Exercise 3C

D

1 Twenty girls were timed over a 10-metre sprint.
Here are their times to the nearest tenth of a second.

2.8 4.6 3.7 3.1 4.7 2.9 3.2 4.0
4.1 2.9 3.6 4.3 3.9 2.8 3.3 3.9

a Draw a stem-and-leaf diagram for this data.

b How many girls had a time greater than 3.3 seconds?

2 These are the lengths of some tomato seedlings, to the nearest tenth of a centimetre.

5.6 4.2 2.9 3.1 3.5 4.3 0.7 2.1
4.3 5.7 2.7 3.6 3.2 1.9 0.9 2.3

a Draw a stem-and-leaf diagram for this data.

b How many seedlings were taller than 2.5 cm?

3 The customers in a coffee shop spent the following amounts.

£4.63 £4.91 £3.62 £5.25 £2.61 £4.86
£5.27 £3.75 £4.70 £2.93 £3.81 £5.23

a Draw a stem-and-leaf diagram for this data.

b How many people spent less than £3?

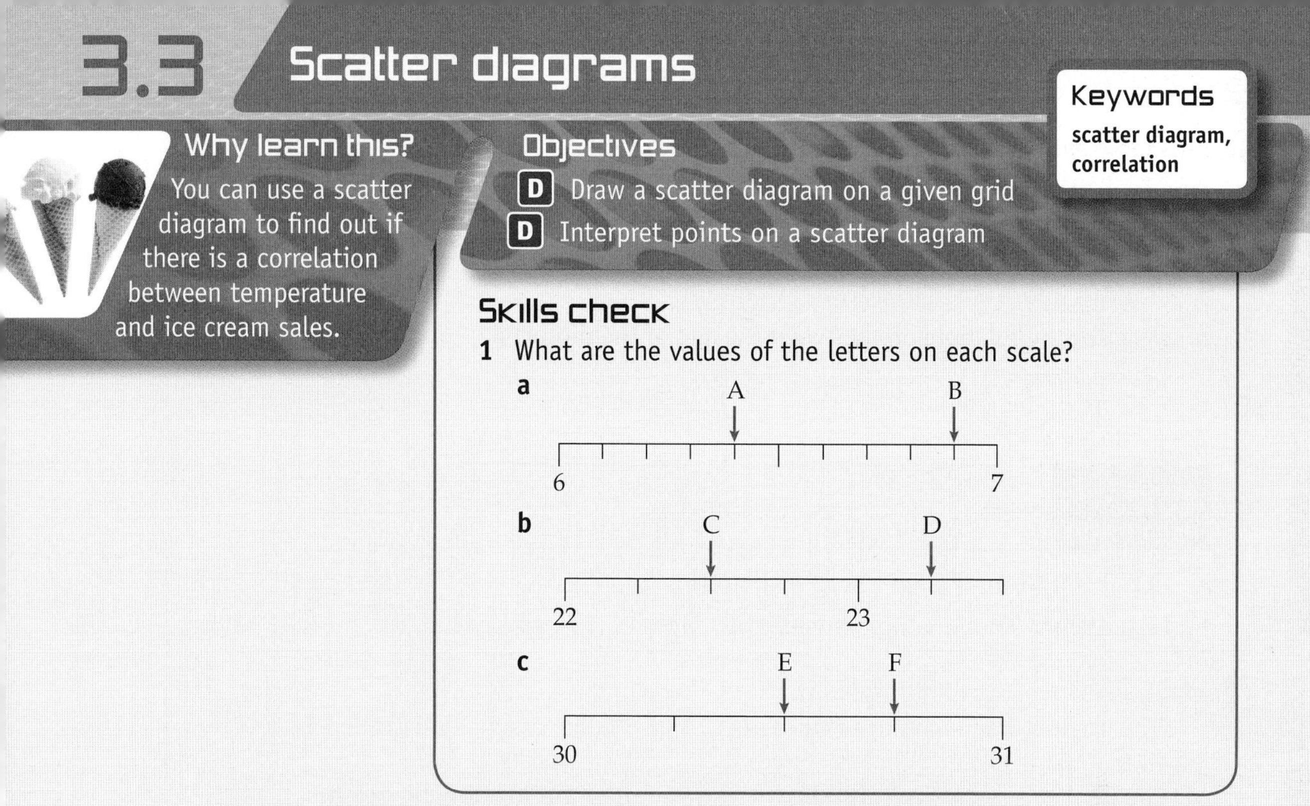

3.3 Scatter diagrams

Keywords
scatter diagram, correlation

Why learn this?

You can use a scatter diagram to find out if there is a correlation between temperature and ice cream sales.

Objectives

D Draw a scatter diagram on a given grid

D Interpret points on a scatter diagram

Skills check

1 What are the values of the letters on each scale?

a A, B (scale from 6 to 7)

b C, D (scale from 22 to 23)

c E, F (scale from 30 to 31)

Using and plotting scatter diagrams

Scatter diagrams are used to compare two sets of data. They show if there is a connection or relationship, called **correlation**, between the two quantities plotted.

Example 4

The table below shows the exam results in maths and science for 15 students.

Maths mark	40	50	56	62	67	43	74	57	75	48	83	50	64	80	70
Science mark	29	34	40	43	48	33	50	44	55	37	62	39	46	57	52

a Draw a scatter diagram for this data.
Plot the maths mark along the horizontal (x) axis.
Plot the science mark on the vertical (y) axis.

b Describe the relationship between the maths and science marks.

Tick off the pairs of data as you plot the points.

a

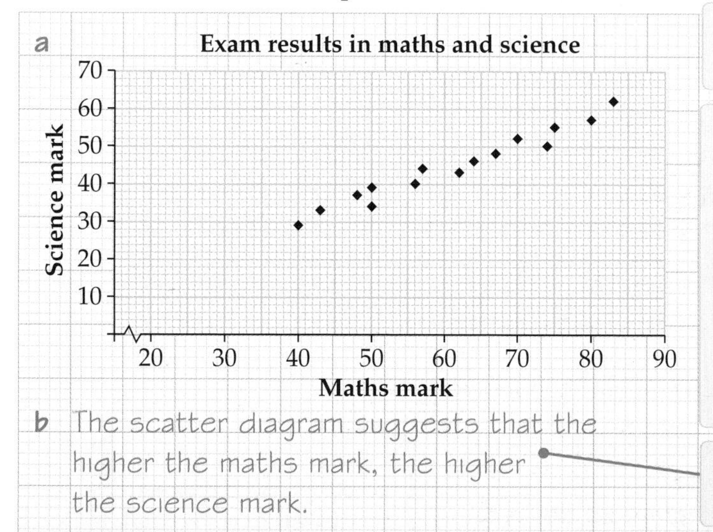

Draw the axes and mark a scale on each one.

For the first pair of values (maths 40, science 29), find 40 on the horizontal (maths) axis and then move up the graph until you reach 29 on the vertical (science) axis. Mark a cross at that point. Repeat for all the pairs of data in the table.

b The scatter diagram suggests that the higher the maths mark, the higher the science mark.

This is only a tendency and may not always be true.

Exercise 3D

D

1 This table shows information about the age and price of some motorbikes.

Age (years)	2	6	2	3	4	5	4	7	9	8
Price (£)	1300	1000	1800	1600	1200	1000	1400	600	200	400

a Plot this information as a scatter diagram.

b Describe the relationship between the age of the motorbikes and their price.

2 Erin recorded the weight (in kg) and the height (in cm) of each of ten children. The table shows her results.

Weight (kg)	39	46	42	50	49	52	39	53	50	44
Height (cm)	145	153	149	161	159	161	149	164	156	154

a Plot this information on a scatter diagram.

b Describe the relationship between the height and weight of the ten children.

D

3 Mr Gray is investigating this claim.

The greater the percentage attendance at his maths lessons, the higher the mark in the maths test.

Mr Gray collects some data and draws a scatter diagram to show his results.

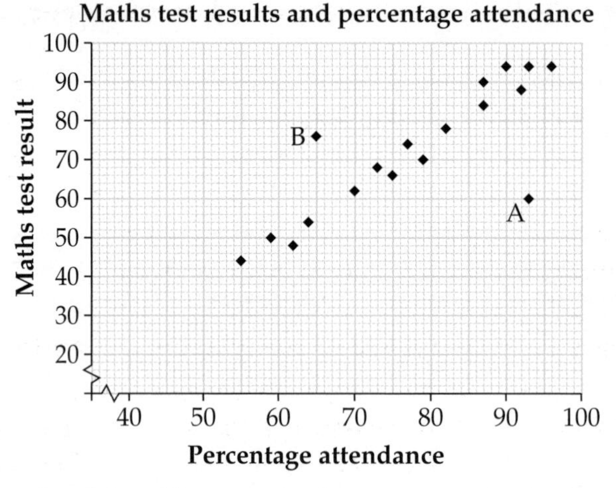

a Decide whether the claim, 'The greater the percentage attendance at his maths lessons, the higher the mark in the maths test', is correct.

AO2

b Students A and B do not fit the general trend. What can you say about
i student A **ii** student B?

D

4 Kushal is investigating this hypothesis.

The greater your hand span, the higher your maths test result.

He collects the following data.

Maths test result	30	30	34	35	38	42	45	50	50	57	57	58	65	65	70	75	75	80	80
Hand span	17	23	21	13	18	21	19	14	18	17	22	14	16	21	13	17	19	13	23

Decide whether Kushal's hypothesis is correct.
Give reasons for your answer.

A key part of the data handling cycle is drawing conclusions from the data.

AO3

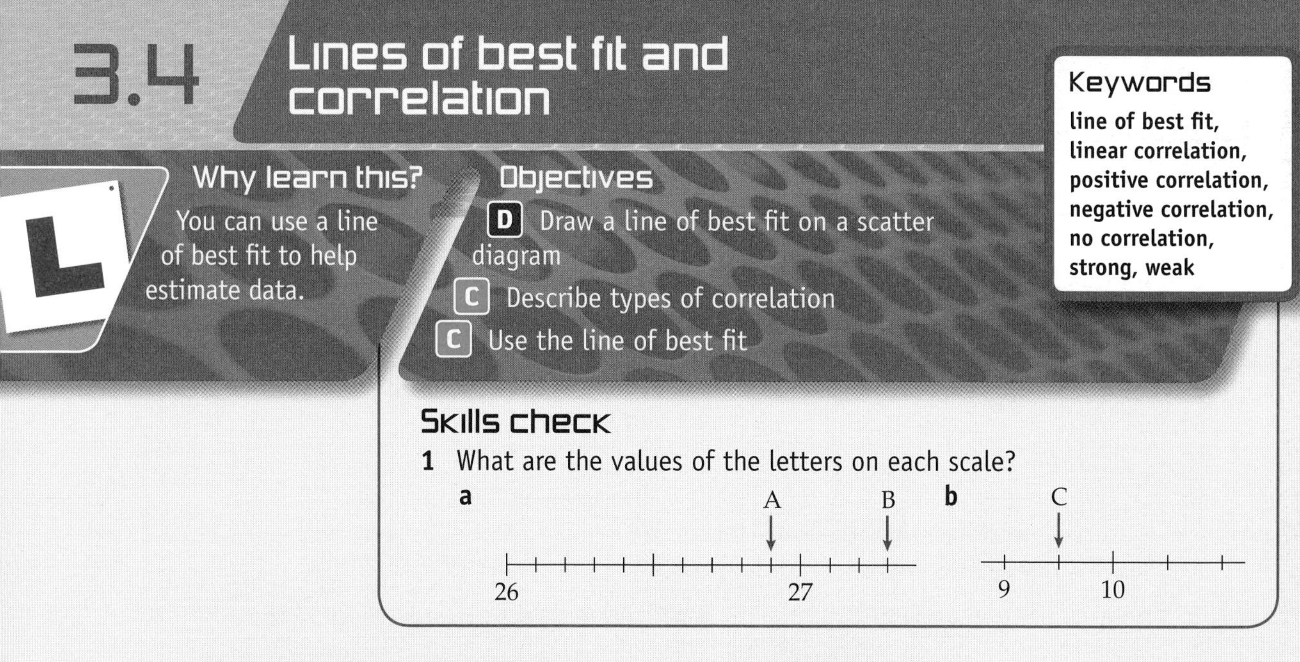

3.4 Lines of best fit and correlation

Keywords

line of best fit, linear correlation, positive correlation, negative correlation, no correlation, strong, weak

Why learn this?

You can use a line of best fit to help estimate data.

Objectives

D Draw a line of best fit on a scatter diagram

C Describe types of correlation

C Use the line of best fit

Skills check

1 What are the values of the letters on each scale?

a

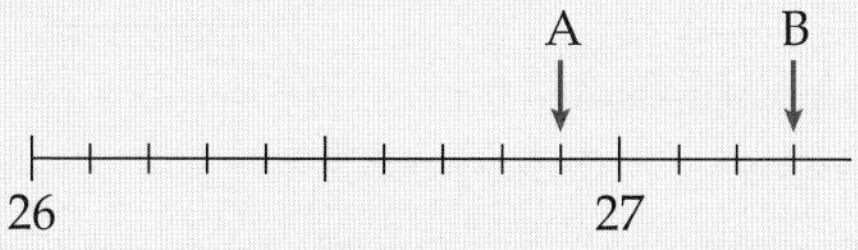

b

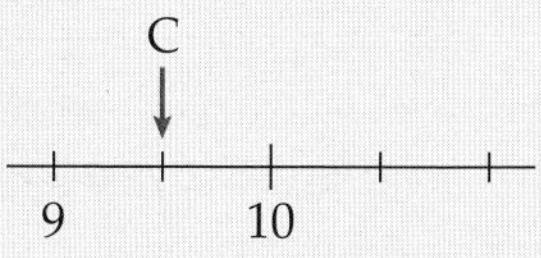

Lines of best fit and correlation

On a scatter diagram a **line of best fit** is a straight line that passes through the data, with an approximately equal number of points on either side of the line.

If a line of best fit can be drawn then there is some form of **linear correlation** between the two sets of data.

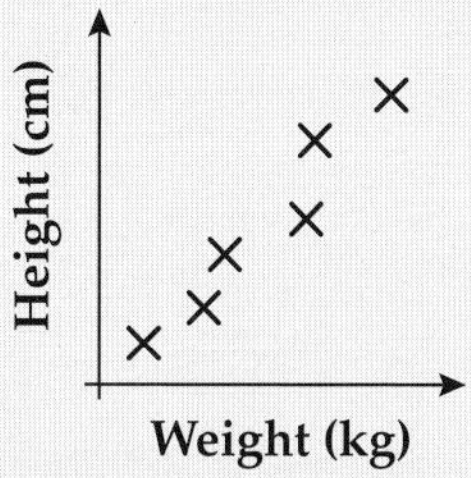

Positive correlation

As one quantity increases, the other one tends to increase.

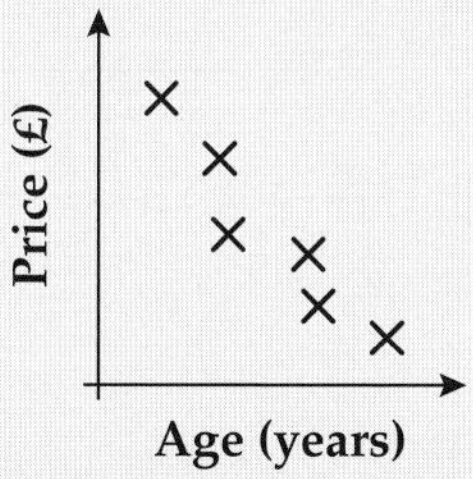

Negative correlation

As one quantity increases, the other one tends to decrease.

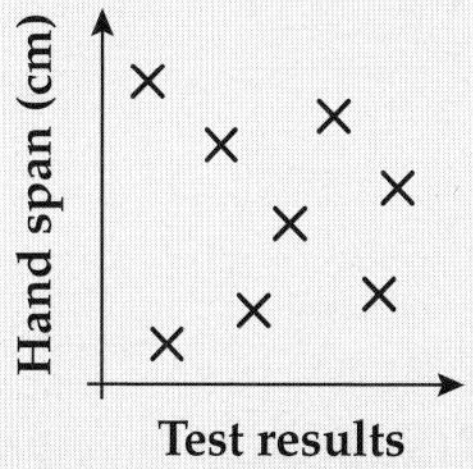

No (linear) correlation

These points are scattered randomly across the diagram. This is sometimes known as zero correlation.

When the points are close to the line of best fit, there is a **strong** linear correlation.

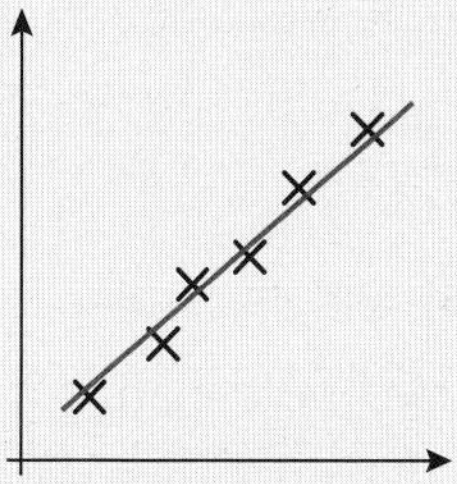

When the points are not all close to the line of best fit, there is a **weak** linear correlation.

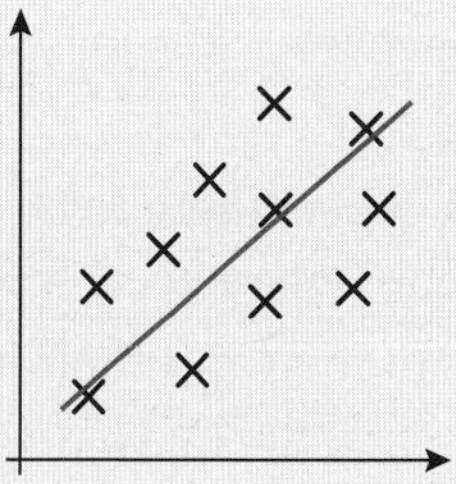

A line of best fit can be used to estimate the value of one quantity if the other value is known and there is a correlation.

C

Example 5

The scatter diagram shows students' marks in the end-of-term maths and science tests.
Robert scores 65 marks in his maths test.
Use a line of best fit to estimate his science mark.

Draw vertically up from 65 to the line of best fit. Draw a horizontal line across and read off the science mark, 47.

The scatter diagram shows that there is a positive correlation between the maths and science marks. Therefore a line of best fit can be used to estimate one mark when the other mark is known.

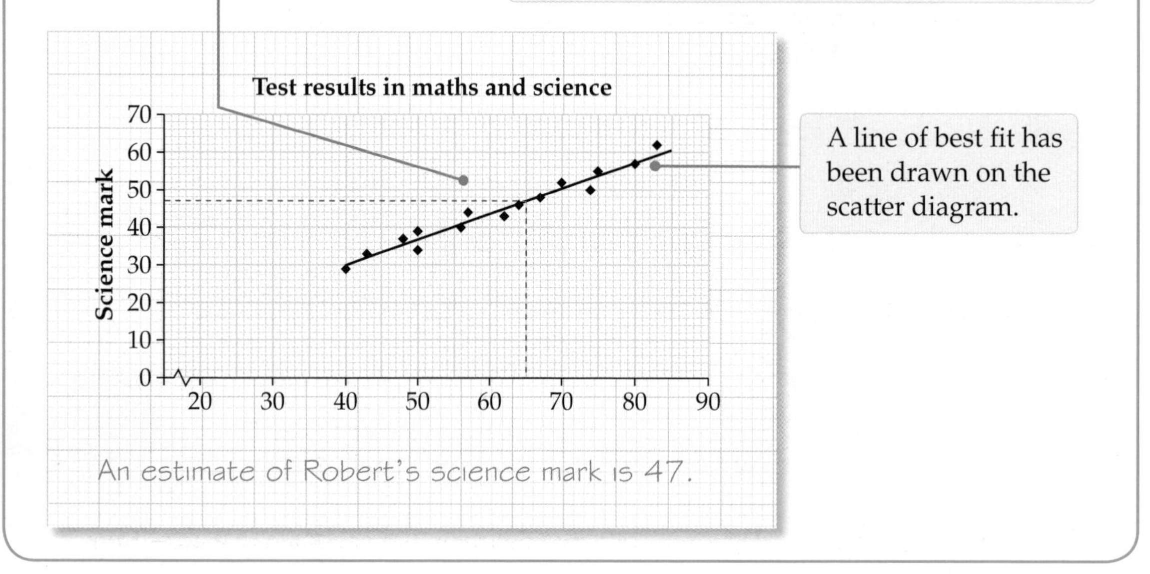

An estimate of Robert's science mark is 47.

Exercise 3E

C

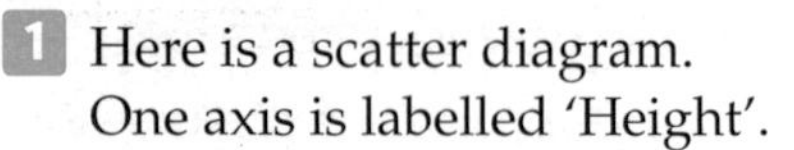

1 Here is a scatter diagram.
One axis is labelled 'Height'.

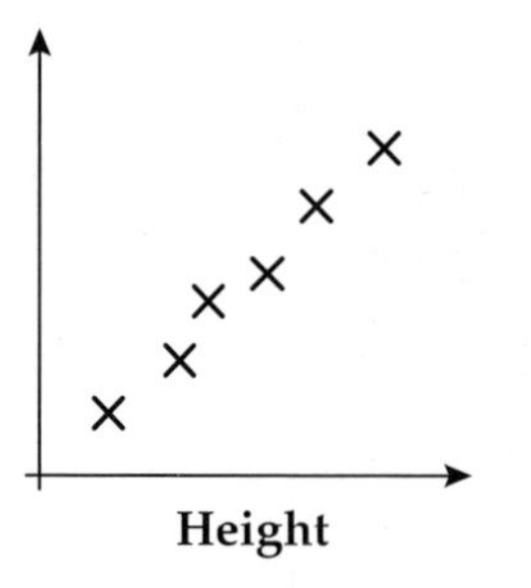

a What type of correlation does the scatter diagram show?

b Choose the most appropriate of these labels for the other axis.

A Maths mark **B** Arm span **C** Number of brothers **D** Length of hair

2 Would you expect there to be a positive correlation, a negative correlation or no correlation between each of these pairs of quantities?

a Heights of men and their IQ.

b The hat sizes of children and their height.

c The outside temperature and the number of cups of tea sold in a café.

d A car's fuel consumption and its speed.

3 The table shows the marks scored by students in their two maths exams.

Paper 1	10	68	80	46	24	84	60	16	90	32	26	94	56	76	36
Paper 2	8	66	78	44	22	74	56	18	86	24	30	92	48	72	34

a Copy the axes on to graph paper and draw a scatter diagram to show the information in the table.

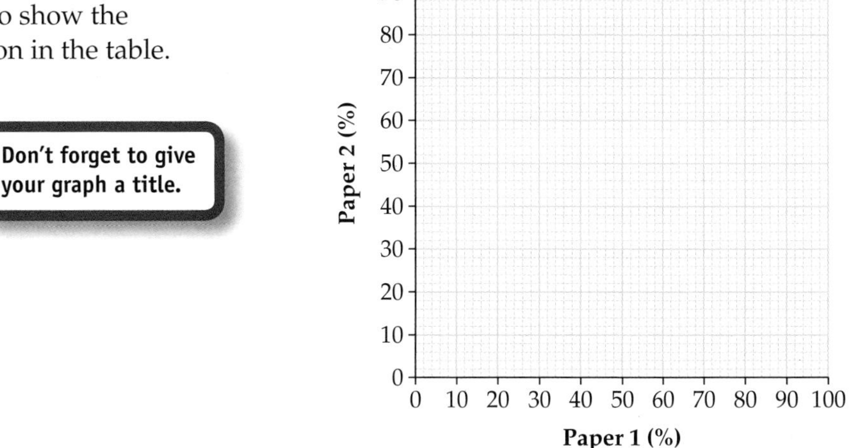

b Describe the correlation between the marks scored in the two exams.

c Draw a line of best fit on your scatter diagram.

d Use your line of best fit to estimate

i the Paper 2 score of a student whose score on Paper 1 is 64

ii the Paper 1 score of a student whose score on Paper 2 is 84.

4 The table shows the average price of a two-bedroom flat at certain distances from the mainline train station.

Distance from train station (km)	0	6	2	7	1	8	5	2	10	8	3	9	11	12
Average price (£000s)	213	198	207	197	211	192	203	210	186	194	207	190	184	182

a Copy the axes on to graph paper and draw a scatter diagram to show the information in the table.

Average price (£)
220 000
210 000
200 000
190 000
180 000
170 000
0 1 2 3 4 5 6 7 8 9 10 11 12
Distance from train station (km)

b Use your scatter diagram to estimate the price of a two-bedroom flat that is 4 km from the mainline train station.

c Describe the correlation between the distance from the mainline train station and the average price of a two-bedroom flat.

d Lucy needs to catch the train to work.
She has £195 000 to spend on a two-bedroom flat.
Approximately how far will her flat be from the station?

C

AO2

C

5 The scatter diagram shows the age and price of some used cars.
A line of best fit has been drawn.

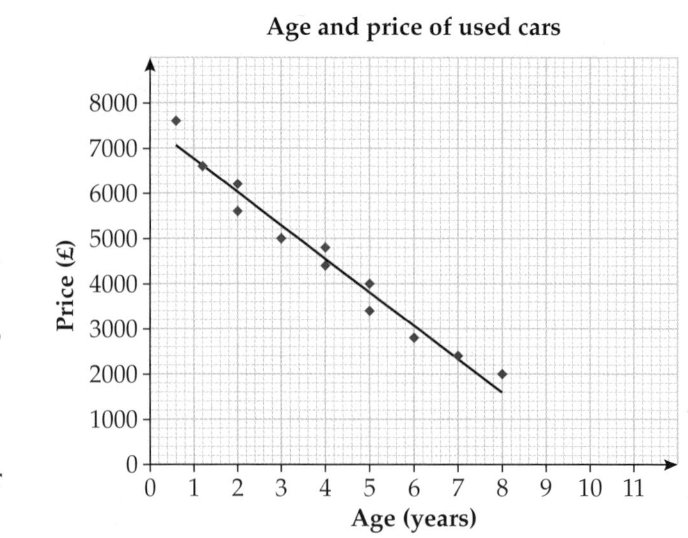

a Describe the correlation between the age of used cars and their price.

b Use the line of best fit to find

i the price of a 3-year-old used car

ii the age of a used car that costs £2200.

AO2

c Why would it not be useful to use the line of best fit to estimate the price of a 12-year-old car?

C

6 The table shows the number of female competitors in each Olympic Games from 1948 to 1984.

Female competitors	685	518	384	610	683	781	1070	1251	1088	1620
Year	1948	1952	1956	1960	1964	1968	1972	1976	1980	1984

a Draw a scatter graph for this data.

b Draw a line of best fit.

c State what type of correlation you see.

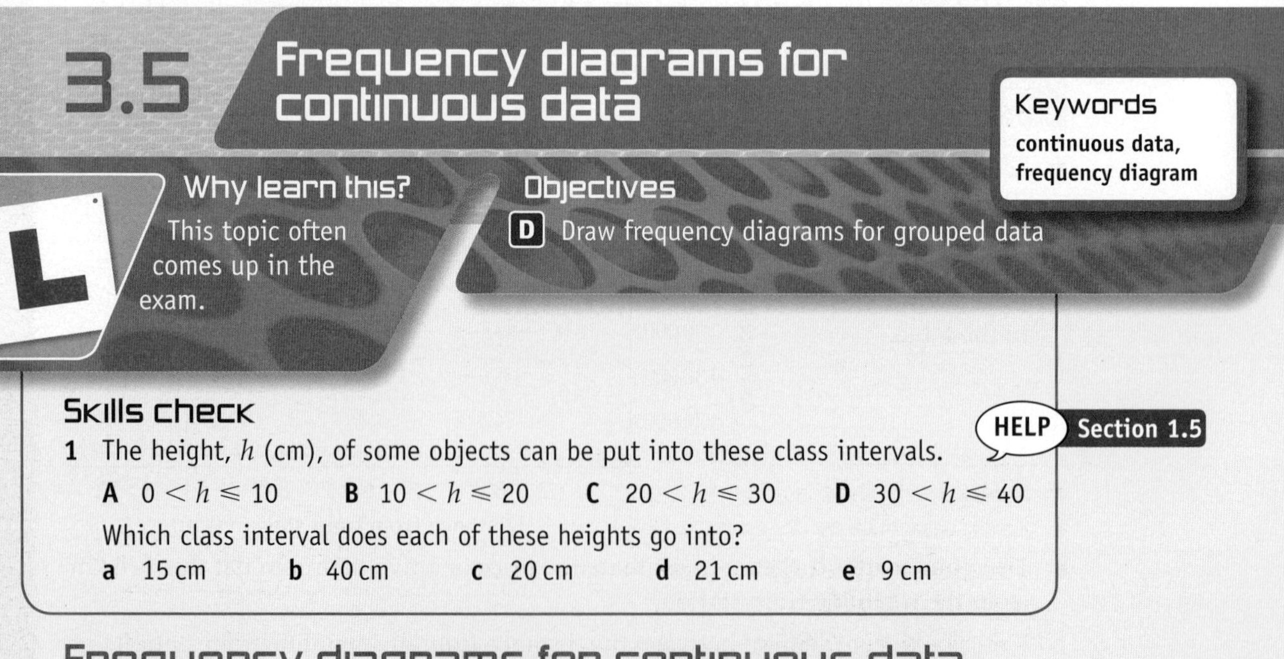

3.5 Frequency diagrams for continuous data

Keywords
continuous data, frequency diagram

Why learn this?
This topic often comes up in the exam.

Objectives
D Draw frequency diagrams for grouped data

Skills check

HELP Section 1.5

1 The height, h (cm), of some objects can be put into these class intervals.

A $0 < h \leqslant 10$ **B** $10 < h \leqslant 20$ **C** $20 < h \leqslant 30$ **D** $30 < h \leqslant 40$

Which class interval does each of these heights go into?

a 15 cm **b** 40 cm **c** 20 cm **d** 21 cm **e** 9 cm

Frequency diagrams for continuous data

Continuous data can be represented by a **frequency diagram**. A frequency diagram is similar to a bar chart except that it has no gaps between the bars.

Example 6

The heights of some swimmers were measured and recorded in the table.

Draw a frequency diagram to show this data.

Height, h (cm)	Frequency
$150 \leqslant h < 155$	5
$155 \leqslant h < 160$	8
$160 \leqslant h < 165$	11
$165 \leqslant h < 170$	7
$170 \leqslant h < 175$	3

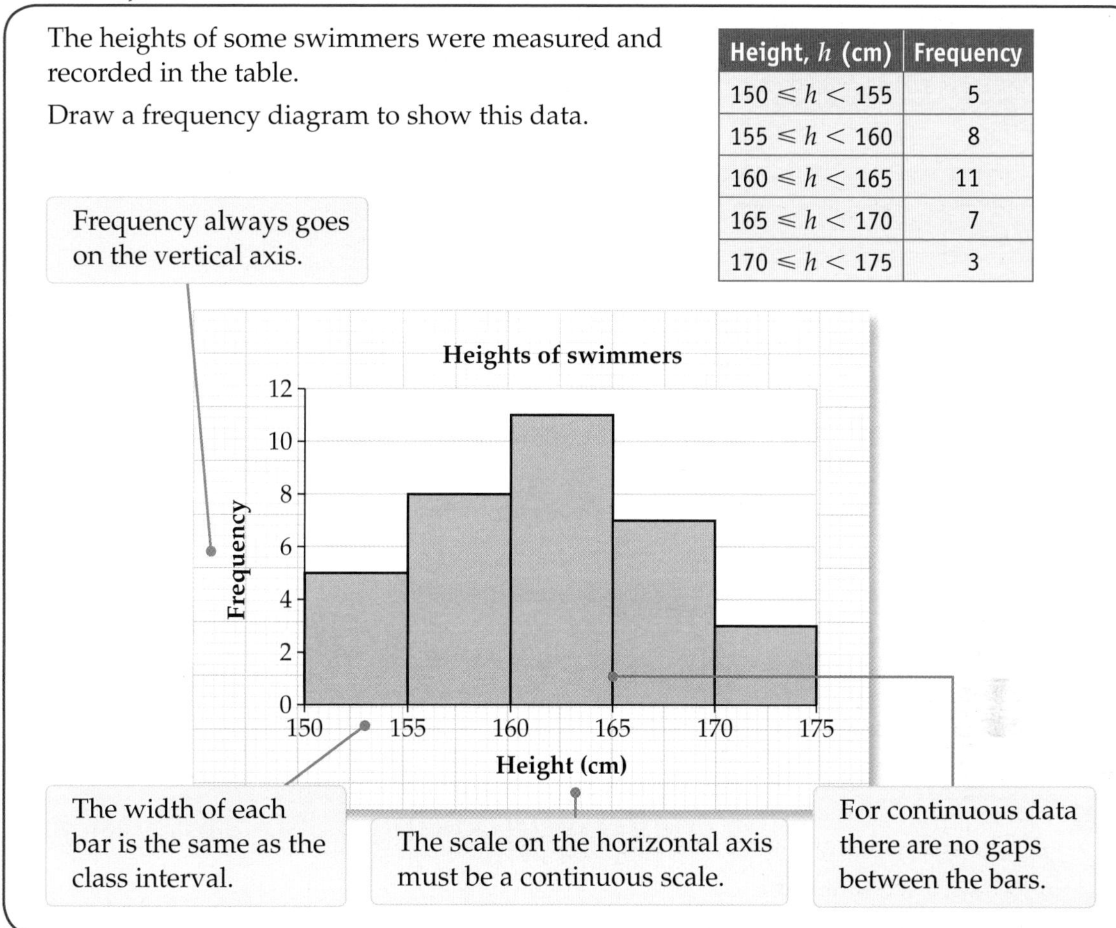

D

Exercise 3F

1 The weights of some potatoes were recorded to the nearest gram.
The table gives the results.
Draw a frequency diagram to show this information.

Weight, w (g)	Frequency
$0 \leqslant w < 20$	7
$20 \leqslant w < 40$	25
$40 \leqslant w < 60$	30
$60 \leqslant w < 80$	36
$80 \leqslant w < 100$	21
$100 \leqslant w < 120$	15
$120 \leqslant w < 140$	2

D

2 This table gives information about the ages of members at a leisure centre.

a Draw a frequency diagram for this data.

b The members are offered a special discount on night-club tickets.
Is this offer likely to be very popular?
Give reasons for your answer.

Age, y (years)	Frequency
$0 \leqslant y < 10$	23
$10 \leqslant y < 20$	45
$20 \leqslant y < 30$	56
$30 \leqslant y < 40$	36
$40 \leqslant y < 50$	49
$50 \leqslant y < 60$	32
$60 \leqslant y < 70$	16

It is important that you can interpret the data and draw conclusions from it as part of the data handling cycle.

D

3 A magazine carried out a survey of the ages of its readers.
The results of the survey are shown in the table.

Age, x (years)	Frequency
$20 \leqslant x < 30$	5
$30 \leqslant x < 40$	38
$40 \leqslant x < 50$	30
$50 \leqslant x < 60$	15
$60 \leqslant x < 70$	12

a Draw a frequency diagram to show the information.

AO2

b The magazine's editor is thinking of doing an article on the film *High School Musical*.

Do you think this is a good idea? Give reasons for your answer.

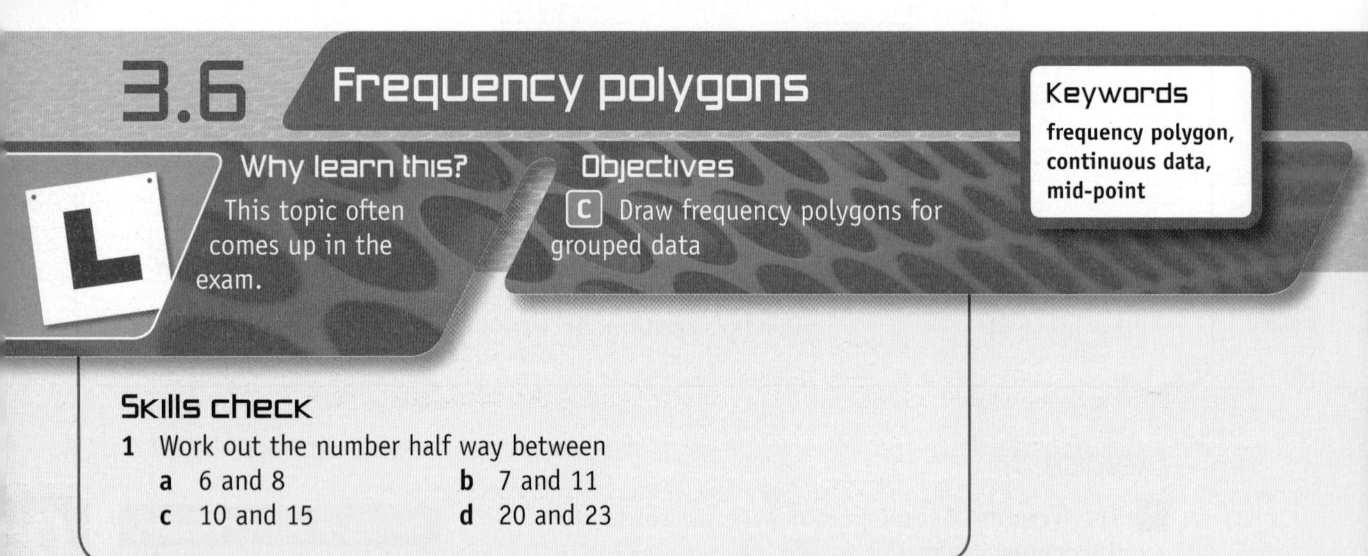

3.6 Frequency polygons

Keywords
frequency polygon, continuous data, mid-point

Why learn this?
This topic often comes up in the exam.

Objectives
C Draw frequency polygons for grouped data

Skills check

1 Work out the number half way between
 a 6 and 8   b 7 and 11
 c 10 and 15   d 20 and 23

Frequency polygons for grouped data

A **frequency polygon** shows patterns or trends in the data.

When drawing a frequency polygon for grouped or **continuous data**, the **mid-point** of each class interval is plotted against the frequency.

C

Example 7

The frequency table shows the times taken by a sample of students to solve a maths problem.
Draw a frequency polygon for this data.

Time, t (minutes)	Frequency
$0 \leqslant t < 5$	1
$5 \leqslant t < 10$	4
$10 \leqslant t < 15$	10
$15 \leqslant t < 20$	7
$20 \leqslant t < 25$	2

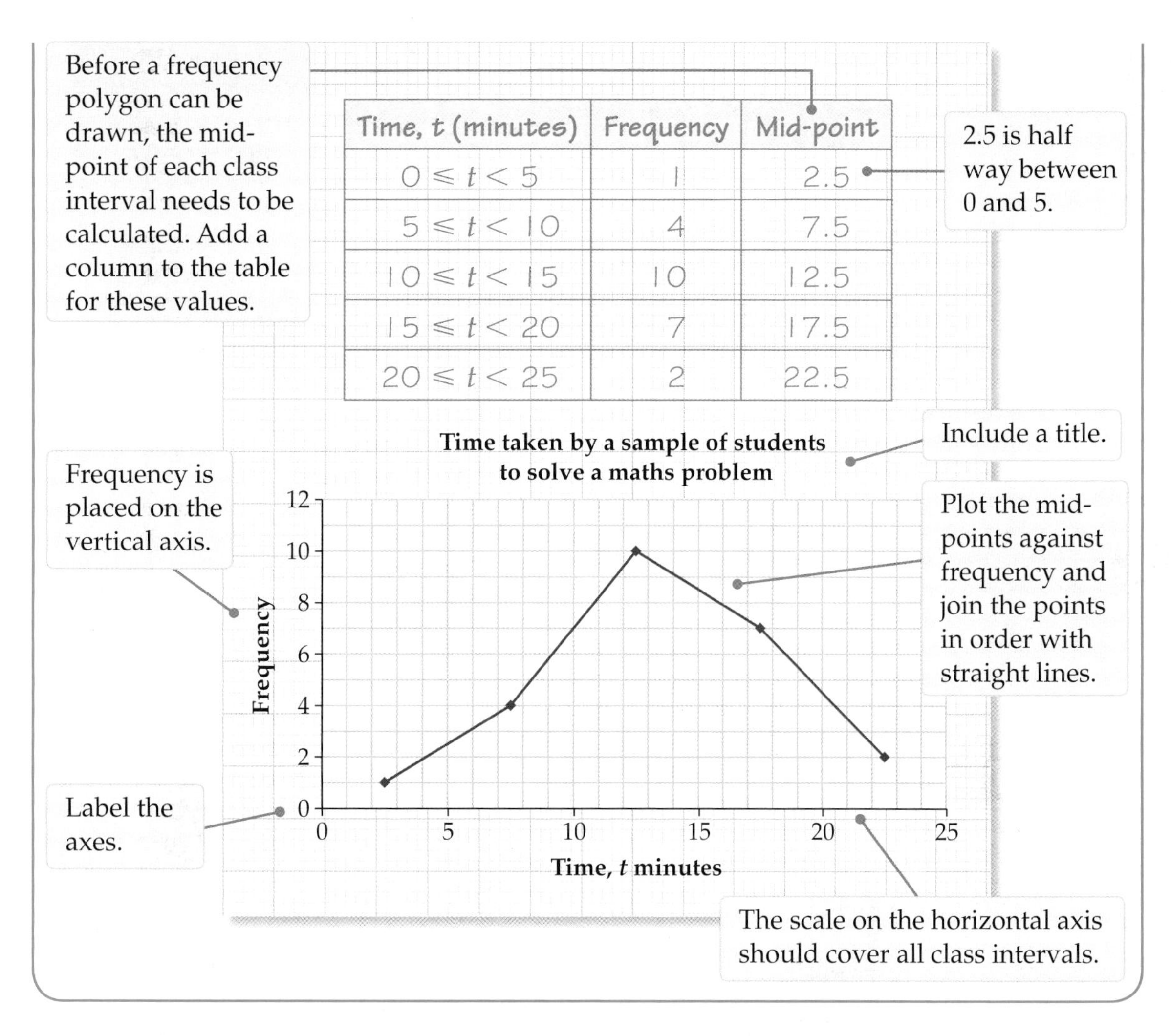

Exercise 3G

1 The heights of some seedlings are shown in the table.

Height, h (cm)	Number of seedlings	Mid-point
$5 \leqslant h < 10$	6	7.5
$10 \leqslant h < 15$	10	
$15 \leqslant h < 20$	12	
$20 \leqslant h < 25$	9	
$25 \leqslant h < 30$	3	

a Copy and complete the table.

b Draw a frequency polygon for this data.

2 The frequency table shows the weights of some Year 9 students.

a Copy and complete the table.

b Draw a frequency polygon for this data.

Weight, w (kg)	Frequency	Mid-point
$30 \leqslant w < 40$	5	
$40 \leqslant w < 50$	12	
$50 \leqslant w < 60$	20	
$60 \leqslant w < 70$	14	
$70 \leqslant w < 80$	6	

C

C

3 The table shows the times (in minutes) that some patients waited in a doctors' surgery.
Draw a frequency polygon for this data.

Time, t (minutes)	Frequency
$0 \leqslant t < 10$	10
$10 \leqslant t < 20$	5
$20 \leqslant t < 30$	4
$30 \leqslant t < 40$	1

4 A magazine carried out a survey of the ages of its readers.
The results of the survey are shown in the table.
Draw a frequency polygon to show this data.

Age group, x (years)	Frequency
$25 \leqslant x < 30$	25
$30 \leqslant x < 35$	38
$35 \leqslant x < 40$	17
$40 \leqslant x < 45$	12
$45 \leqslant x < 50$	12
$50 \leqslant x < 55$	6

C

5 The two frequency polygons show the heights of a group of Year 7 girls and boys.

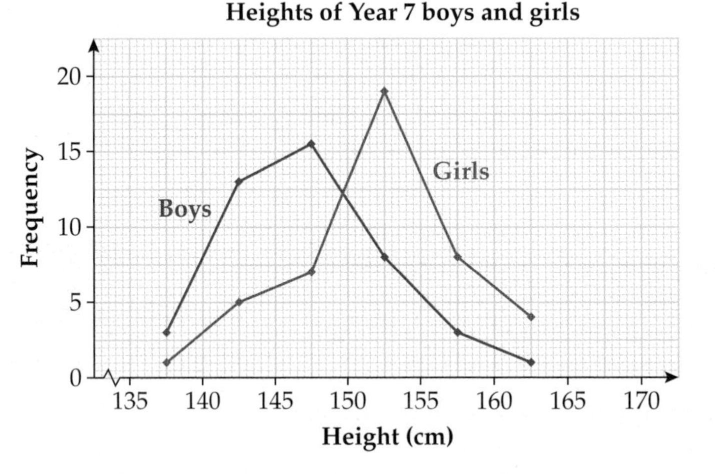

AO2

Compare the heights of the two groups. Give a reason for your answer.

Review exercise

E

1 Sophie went to a theme park.
The table shows the amounts she spent.
Draw and label a pie chart to represent these costs. **[4 marks]**

Item	Cost (£)
food	6
drinks	4
souvenirs	9
sunhat	5

D

2 Here are the times (in minutes) taken to walk around a park.

36	27	35	47	45	47	29
42	21	43	49	50	41	37
51	37	56	54	56	40	35

Draw a stem-and-leaf diagram to show this data. **[3 marks]**

D

3 The headteacher of a school sent out a questionnaire to a sample of Year 11 students.
One of the questions was:

How many hours do you spend on homework each week?

The results are shown in the table.

Time spent on homework each week, t (hours)	Number of students
$0 \leqslant t < 2$	1
$2 \leqslant t < 4$	4
$4 \leqslant t < 6$	10
$6 \leqslant t < 8$	25
$8 \leqslant t < 10$	30
$10 \leqslant t < 12$	22
$12 \leqslant t < 14$	13
$14 \leqslant t < 16$	4

a How many students said that they spend 12 or more hours doing homework each week? **[1 mark]**

b Draw a frequency diagram to represent this data. **[3 marks]**

C

4 The table shows the times taken by a group of students to complete a challenge, called Challenge 1.

Time, t (minutes)	Number of students
$10 \leqslant t < 20$	6
$20 \leqslant t < 30$	10
$30 \leqslant t < 40$	11
$40 \leqslant t < 50$	17
$50 \leqslant t < 60$	8
$60 \leqslant t < 70$	2

a Draw a frequency polygon for this data. **[2 marks]**

b The table shows the time taken for the same group of students to complete a different challenge, called Challenge 2.
Draw a frequency polygon for this data on the same grid as part **a**. **[2 marks]**

Time, t (minutes)	Number of students
$0 \leqslant t < 10$	4
$10 \leqslant t < 20$	14
$20 \leqslant t < 30$	20
$30 \leqslant t < 40$	10
$40 \leqslant t < 50$	6

c Compare the times taken for the students to complete the two challenges.
Give reasons for your answer. **[2 marks]**

5 Here is a scatter diagram. One axis is labelled 'Weight'.

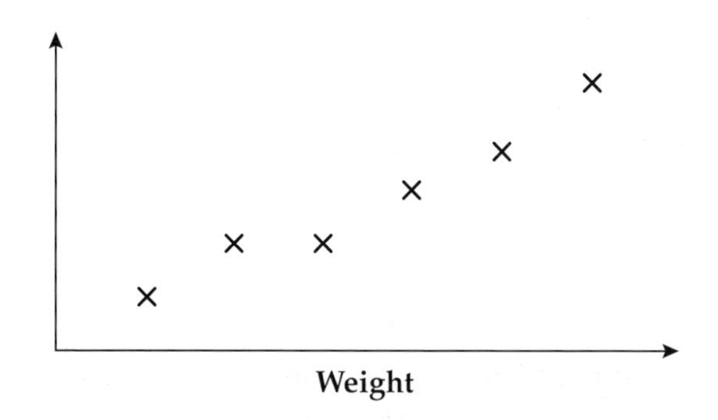

a For this graph, state the type of correlation.

b Choose the most appropriate of these labels for the other axis.

A Maths score **B** Length of hair
C Waist measurement **D** Hat size **[2 marks]**

AO2

C

6 The table shows the midday temperatures of some cities and the number of hours of sunshine one day in July.

City	Hours of sunshine	Midday temperature (°C)
London	6	18
Manchester	7	19
Sydney	3	12
Paris	9	22
Berlin	5	16
New York	11	24
Montreal	6	17
Abu Dhabi	12	26
Rio de Janeiro	8	20
Wellington	2	11

a Plot the data on a scatter diagram on a grid similar to the one below.

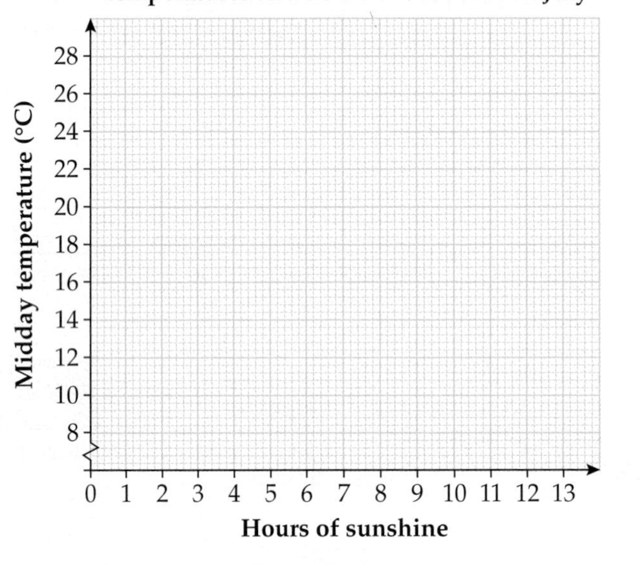

[2 marks]

b Describe the strength and type of correlation. **[1 mark]**

c Draw a line of best fit on your scatter diagram. **[1 mark]**

d Use your line of best fit to estimate the number of hours of sunshine at a place with an average temperature of 21 °C. **[1 mark]**

AO2

e It is not sensible to use your line of best fit to estimate the number of hours of sunshine in the Sahara Desert, where the mid-day temperature was 36 °C. Explain why. **[1 mark]**

Chapter summary

In this chapter you have learned how to

- draw a pie chart **E**
- draw a stem-and-leaf diagram **D**
- draw a scatter diagram on a given grid **D**
- interpret points on a scatter diagram **D**
- draw a frequency diagram for grouped data **D**
- draw a line of best fit on a scatter diagram **D**
- use the line of best fit **C**
- describe types of correlation **C**
- draw a frequency polygon for grouped data **C**

4

Range and averages

BBC Video

This chapter is about finding averages.

Car manufacturers use average-sized dummies to test the safety features of their new designs.

Objectives

This chapter will show you how to

- work out the mode, mean, median and range of a set of data E
- work out the median, mode and range from a frequency table or bar chart E
- calculate the mean from a frequency table D
- identify the modal class in a grouped frequency table D
- estimate the range and mean from a grouped frequency table C
- work out which class interval contains the median C

Before you start this chapter

1 Which number is half way between

a 8 and 10 b 24 and 32 c 100 and 101 d 0.8 and 1.1?

2 This frequency table shows the scores when a dice is rolled.

Score	1	2	3	4	5	6
Frequency	10	11	9	7	8	4

a How many times was the dice rolled?

b How many times was the score 5?

c How many more times was 2 scored than 6?

HELP Chapter 3

3 Copy and complete.

Write $<$, $>$ or $=$ between each pair of values.

a 4.6 cm ☐ 4.39 cm b 300 m ☐ 0.3 km c 15 cm ☐ 125 mm d 500 g ☐ 5 kg

4.1 Averages and range

Keywords
average, mean, median, mode, modal value, range

Why learn this?
The mean is the most commonly used average. For example, the mean lifespan of a chinchilla is 9 years

Objectives
E D Find the mean, median and range from a set of data, including data given in a stem-and-leaf diagram

Skills check

1 Work out the mean, median and range of each group of numbers.
 a 9 14 8 7 12
 b 0.5 1.0 0.8 1.7
 c 250 300 320 240 330

Averages and range

There are three different types of **average**.

- The **mean** is the most commonly used average.

$$\text{Mean} = \frac{\text{sum of all the data values}}{\text{number of data values}}$$

- The **median** is the middle value when the data is written in order.

  If there are an even number of pieces of data, then there are two middle values. The median is the value half way between them.

  With n data values, you can work out the median using the formula below.

$$\text{Median} = \left(\frac{n+1}{2}\right)\text{th value}$$

- The **mode** of a set of data is the number or item that occurs most often.
  The mode is the only average you can use for qualitative data.
  The mode is also called the **modal value**.

The **range** of a set of data is the difference between the largest value and the smallest value.

The range tells you how spread out the data is.

$$\text{Range} = \text{largest value} - \text{smallest value}$$

E

Example 1

This stem-and-leaf diagram shows the heights of six plant seedlings.

Stem	Leaf
6	2 3 3
7	2
8	2
9	0

Key
6|2 means 6.2 cm

a Calculate the mean height of these plant seedlings.
b Work out the median of this data.
c Work out the range of this data.

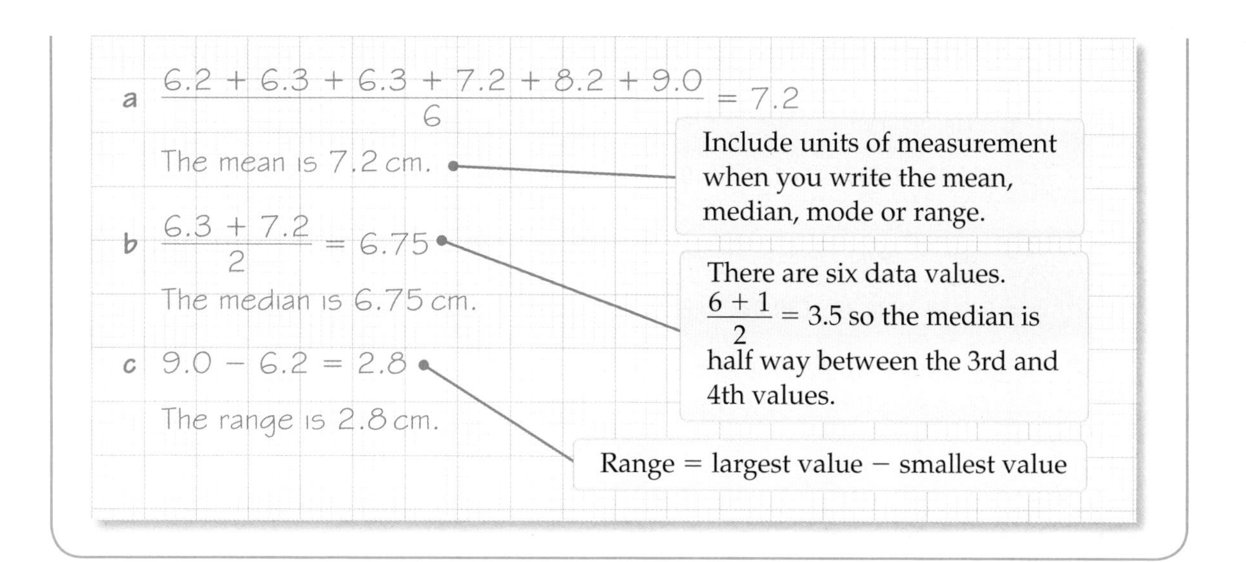

Exercise 4A

E

1 Imran recorded the prices of flight tickets from London to Mumbai in a stem-and-leaf diagram.

```
4 | 70 90
5 | 20 50 80
6 | 00
7 | 10 20
```

Key
4 | 70 means £470

a Work out the range of these prices.

b Calculate the mean cost per flight.

c Work out the median cost per flight.

2 This stem-and-leaf diagram shows the heights of some trees.

```
 7 | 5
 8 | 2 5 8
 9 | 3 8 9
10 | 0
```

Key
7 | 5 means 7.5 m

a Calculate the mean height.

b Work out the median of this data.

c Work out the range of this set of data.

3 This stem-and-leaf diagram shows the results from the men's 400 m final at the Sydney Olympic Games.

```
43 | 8
44 | 4 7
45 | 0 1 3 4
```

Key
43 | 8 means 43.8 seconds

a Work out the range for this data.

b Work out the median time.

c Calculate the mean time.
Round your answer to one decimal place.

4 This table shows the monthly gas bills for a family in 2009.

Month	Jan	Feb	Mar	Apr	May	Jun
Bill	£80.22	£91.50	£62.35	£45.05	£18.50	£22.24

Month	Jul	Aug	Sept	Oct	Nov	Dec
Bill	£19.69	£24.38	£30.03	£42.97	£88.30	£96.62

The family chooses to pay the same amount each month by direct debit.
Use the mean to estimate how much they should pay each month.

E
AO2

5 The mean of four numbers is 7.

a What is the sum of the four numbers?

b Three of the numbers are 10, 7 and 5.
Work out the fourth number.

E

6 The mean of four numbers is 2.5.
Three of the numbers are 1, 2 and 5.
Work out the fourth number.

AO3

7 Write down five different numbers with a mean of 7.

8 Write down four different numbers with a mean of 6.8.

E

9 Count the letters in each word of this sentence.

The median is the middle value.

What is the median word length in the sentence?

AO2

10 In a survey of 45 people, the median salary was £22 000 a year.
How many people in the survey had a salary greater than £22 000 a year?

E

11 Alice was measuring the lengths of some chips, but spilt tomato sauce on her paper.

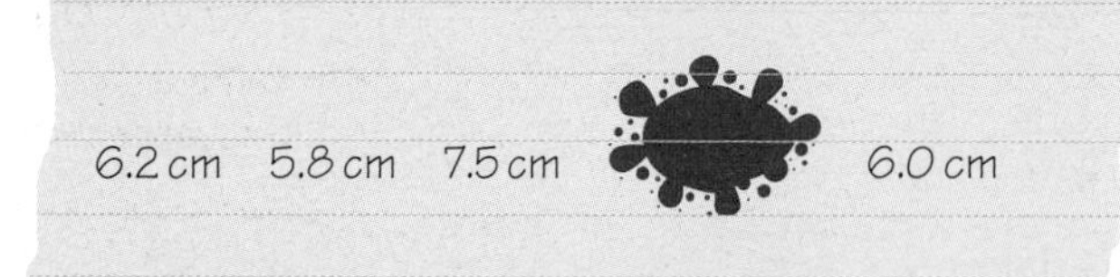

The range of her values was 2.5 cm.

AO3

a What is the smallest possible value for the missing measurement?

b What is the largest possible value for the missing measurement?

D

12 Kieran wrote down the test marks of 11 members of his class.

17 20 11 19 15 12 20 8 12 20 18

a Draw a stem-and-leaf diagram for this data.

b Work out the range of this data.

c Work out the median mark.

d Calculate the mean mark.
Give your answer to two significant figures.

13 These are the numbers of strawberries on nine plants.

13 22 19 7 21 10 15 20 19

a Draw a stem-and-leaf diagram for this data.

b Work out the median number of strawberries per plant.

c Calculate the mean of this data.

Give your answer to two significant figures.

D

14 Supraj has written down three numbers with a mean of 120 and a range of 60.
The first number is 110.
What are the other two numbers?

D

15 Will and Laura have started recording their scores in three different tests.

	Maths	English	Science
Will	65%	80%	50%
Laura			65%

Laura and Will had the same mean test score.
The range of Laura's results was twice the range of Will's results.
Laura's best result was in English.
Copy and complete the table using this information.

AO3

4.2 Calculating the range, mode and median from a frequency table

Keywords
ungrouped frequency table

Why learn this?

When you collect data from a survey or experiment, your results will often be in a frequency table.

Objectives

E Calculate the mode, median and range from an ungrouped frequency table

Skills check

This frequency table shows how many pets are owned by the members of a class.

Number of pets	0	1	2	3	4	5
Frequency	5	13	9	4	2	1

1 How many students are there in the class?

2 How many students had fewer than two pets?

3 What was the greatest number of pets owned?

HELP Section 1.4

Frequency table averages

You can calculate the range, mode and median when data is presented in an **ungrouped frequency table.**

The range is the difference between the largest value and the smallest value.

To work out the mode, look for the highest frequency.

Add up the frequencies to find the number of data values.

With n data values, you can work out the median using this formula.

$$\text{Median} = \left(\frac{n+1}{2}\right)\text{th value}$$

Then count through the frequencies to obtain the correct value.

E

Example 2

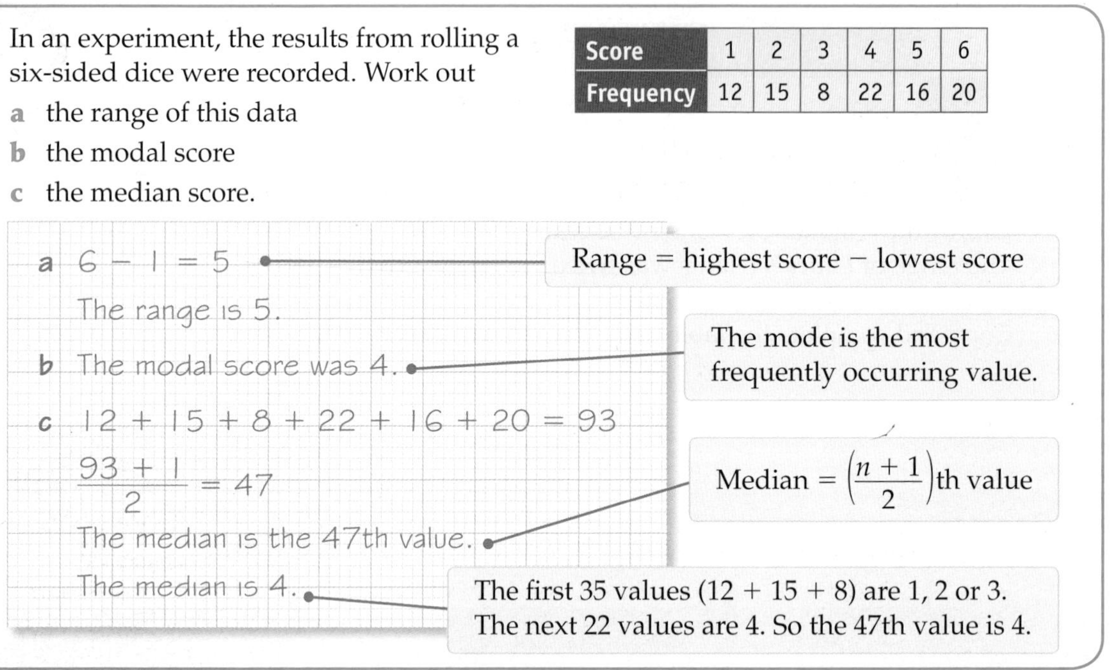

In an experiment, the results from rolling a six-sided dice were recorded. Work out

a the range of this data
b the modal score
c the median score.

Score	1	2	3	4	5	6
Frequency	12	15	8	22	16	20

a $6 - 1 = 5$
The range is 5.

b The modal score was 4.

c $12 + 15 + 8 + 22 + 16 + 20 = 93$
$\frac{93 + 1}{2} = 47$
The median is the 47th value.
The median is 4.

Range = highest score − lowest score

The mode is the most frequently occurring value.

Median $= \left(\frac{n+1}{2}\right)$th value

The first 35 values (12 + 15 + 8) are 1, 2 or 3. The next 22 values are 4. So the 47th value is 4.

Exercise 4B

E

1 A football team recorded the number of goals they scored in each match for one season.
Work out
a the range of this data
b the modal number of goals scored
c the median number of goals scored.

Number of goals scored	0	1	2	3	4
Frequency	11	14	10	5	3

First work out how many matches the team played in the whole season.

2 Amit asked a group of people to predict the results of five coin flips.
He recorded the number of correct predictions in a frequency table.
Work out
a the range of this data
b the modal number of correct predictions
c the median number of correct predictions.

Number of correct predictions	Frequency
0	1
1	2
2	10
3	7
4	5
5	0

3 James surveyed the lengths of time people had to wait for a bus, to the nearest minute.

Waiting time (minutes)	1	2	3	4	5	6	7	8	9
Frequency	0	15	20	17	8	9	6	0	0

a Work out the range of this data.

b What was the modal waiting time?

c Cally said, 'The median waiting time was 5 minutes.'
Show working to explain why Cally was wrong.

E

AO2

4 This table shows the number of hours of television watched each night by the members of a class.

Number of hours of television watched	0	1	2	3	4	More than 4
Frequency	3	8	11	5	4	2

a Is it possible to calculate the range of this data?
Give a reason for your answer.

b Jacob said, 'It is impossible to calculate the median of this data.'
Use working to show that he is wrong.

E

AO3

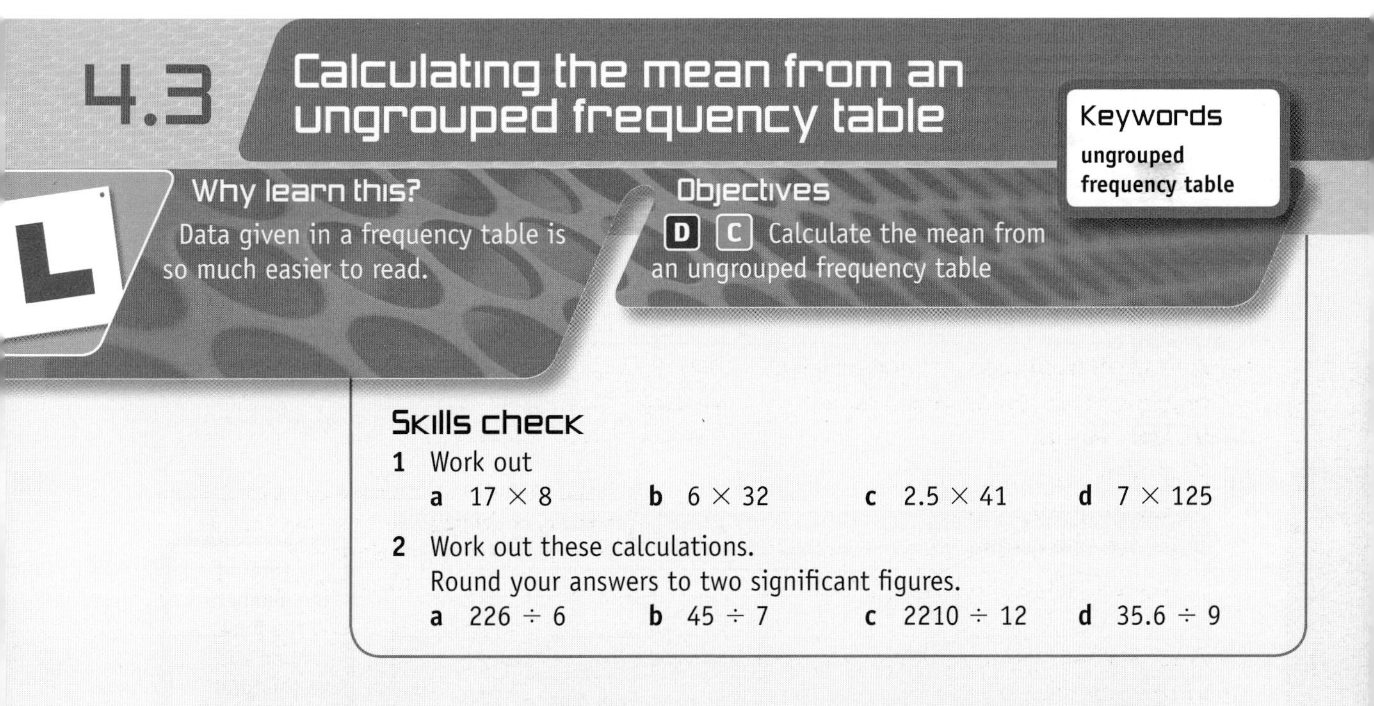

4.3 Calculating the mean from an ungrouped frequency table

Keywords
ungrouped frequency table

Why learn this?

Data given in a frequency table is so much easier to read.

Objectives

D C Calculate the mean from an ungrouped frequency table

Skills check

1 Work out

a 17 × 8 b 6 × 32 c 2.5 × 41 d 7 × 125

2 Work out these calculations.
Round your answers to two significant figures.

a 226 ÷ 6 b 45 ÷ 7 c 2210 ÷ 12 d 35.6 ÷ 9

Calculating the mean from a frequency table

To calculate the mean from an **ungrouped frequency table**, first work out the sum of all the data values and the total frequency. Then you can calculate the mean using this formula.

$$\text{Mean} = \frac{\text{sum of all the data values}}{\text{number of data values}}$$

D

Example 3

This frequency table shows the number of goals a football team scored in each match for one season.

a How many matches did the team play?

b Work out the total number of goals scored in the season.

c Calculate the mean number of goals scored per match.

Number of goals scored	Frequency
0	8
1	15
2	12
3	7
4	3

Add an extra column to your frequency table to show the total number of goals scored in each row.

a $8 + 15 + 12 + 7 + 3 = 45$

The team played 45 matches.

b

Number of goals scored	Frequency	Number of goals × frequency
0	8	$0 \times 8 = 0$
1	15	$1 \times 15 = 15$
2	12	$2 \times 12 = 24$
3	7	$3 \times 7 = 21$
4	3	$4 \times 3 = 12$
Total	45	72

The team scored 3 goals in a match 7 times. The total number of goals they scored in these matches is $3 \times 7 = 21$ goals.

The team scored 72 goals in the season.

c Mean $= \frac{72}{45} = 1.6$ goals per match.

$$\text{Mean} = \frac{\text{total number of goals scored}}{\text{number of matches played}}$$

Exercise 4C

D

1 Darren counted the people in each checkout queue at a supermarket.

Number of people in queue	Frequency	Number of people × frequency
0	4	
1	6	
2	13	
3	2	$3 \times 2 = 6$
4	0	
Total		

The total of the numbers in the final column will be the total number of people queuing.

a How many checkouts were there at the supermarket?

b Copy and complete the table to work out the total number of people queuing.

c Calculate the mean number of people per queue at the supermarket.

2 Alison counted the buses passing a road junction each minute for an hour.

Number of buses	Frequency	Number of buses × frequency
0	8	
1	17	
2	12	
3	14	
4	5	
5	4	
Total		

Copy and complete the table. Use it to calculate the mean number of buses per minute, correct to one decimal place.

D

3 Students at a school are awarded gold stars for good work. This table shows the number of gold stars awarded to the students in three different year groups in one month.

Number of gold stars	Year 7 frequency	Year 8 frequency	Year 9 frequency
0	12	2	22
1	15	20	36
2	40	28	18
3	22	31	11
4	6	15	9
5	2	14	5
6	13	6	8

a How many students are in Year 8?

b How many gold stars were awarded in total to students in Year 9?

c How many more students won six gold stars in Year 7 than in Year 8?

d Calculate the mean number of gold stars awarded per student for the students in Year 7. Give your answer correct to two decimal places.

e The year group with the highest mean number of gold stars awarded per student is awarded a trophy. Which year group won the trophy?

C

AO2

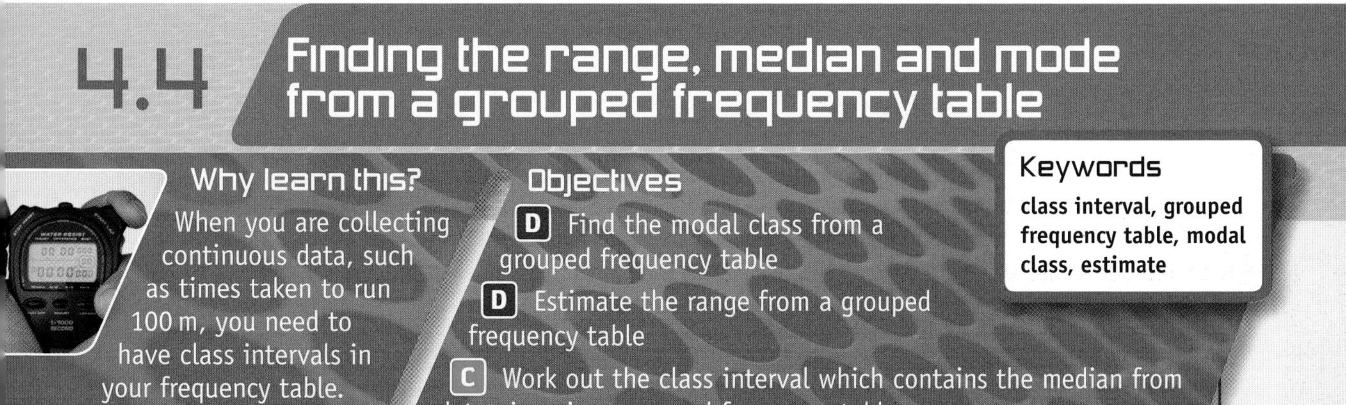

4.4 Finding the range, median and mode from a grouped frequency table

Why learn this?

When you are collecting continuous data, such as times taken to run 100 m, you need to have class intervals in your frequency table.

Objectives

D Find the modal class from a grouped frequency table

D Estimate the range from a grouped frequency table

C Work out the class interval which contains the median from data given in a grouped frequency table

Keywords
class interval, grouped frequency table, modal class, estimate

Skills check

HELP Section 1.3

1 Is this data discrete or continuous?

a The heights of some trees. b The number of items in a pencil case.

2 Write $>$, $<$ or $\leqslant$ between each pair of numbers.

a 4 ☐ 6 b 3.2 ☐ 3.25 d 0.1 ☐ 0.06

Grouped frequency table averages

Sometimes the data in a frequency table is grouped into different **class intervals.**

When the data is arranged in a **grouped frequency table**, you don't know the exact data values. This means you can't calculate the mode, median and range exactly.

The class interval with the highest frequency is called the **modal class.**

You can **estimate** the range using this formula.

Estimated range = highest value of highest class interval − lowest value of lowest class interval

With n data values, the median is the $\left(\frac{n+1}{2}\right)$th data value.

You can use this formula to work out which class interval contains the median.

D C

Example 4

This frequency table shows the heights of some plant seedlings.

a Write down the modal class.

b Estimate the range of this data.

c Which class interval contains the median?

Height, h (cm)	Frequency
$5 \leqslant h < 10$	14
$10 \leqslant h < 15$	11
$15 \leqslant h < 20$	8
$20 \leqslant h < 25$	2

a The modal class is $5 \leqslant h < 10$.

The class interval $5 \leqslant h < 10$ has the highest frequency.

b $25 - 5 = 20$

An estimate for the range is 20 cm.

Range = highest value − lowest value

c There are 35 data values.

Add up the frequencies to find the total number of data values.

$\frac{35+1}{2} = 18$

So the median is the 18th data value.

The median is in the class interval $10 \leqslant h < 15$.

The first 14 values are in the class interval $5 \leqslant h < 10$. The next 11 values are in the class interval $10 \leqslant h < 15$.

Exercise 4D

D C

1 Roselle weighed some eggs, and recorded her results in a frequency table.

a Write down the modal class.

b Estimate the range of this data.

c How many eggs did Roselle weigh in total?

d Which class interval contains the median?

Weight, w (g)	Frequency
$45 \leqslant w < 50$	3
$50 \leqslant w < 55$	8
$55 \leqslant w < 60$	11
$60 \leqslant w < 65$	7

2 In an experiment, Andrew used a computer to record the reaction times of a group of his friends. **D**

Reaction time, t (milliseconds)	$100 \leqslant t < 200$	$200 \leqslant t < 300$	$300 \leqslant t < 400$	$400 \leqslant t < 500$
Frequency	2	8	5	4

a Write down the modal class.

b Estimate the range of this data.

c Which class interval contains the median? **C**

3 This frequency table shows the number of hits on a website each day for eight weeks. **D**

Number of hits	0–999	1000–1999	2000–2999	3000–3999	4000–4999	5000–5999
Frequency	0	6	18	21	9	2

a Estimate the range of this data.

b Write down the modal class.

c Which class interval contains the median?

d In the next week, the website server was broken, and the website received no hits each day. When this data is added to the table what effect will it have on **C**

  i the modal class

  ii the estimated range

  iii the class interval containing the median?

e Aaron says, 'It is impossible to calculate the mean for the data in the table exactly.' Is he correct? Give a reason for your answer. **AO2**

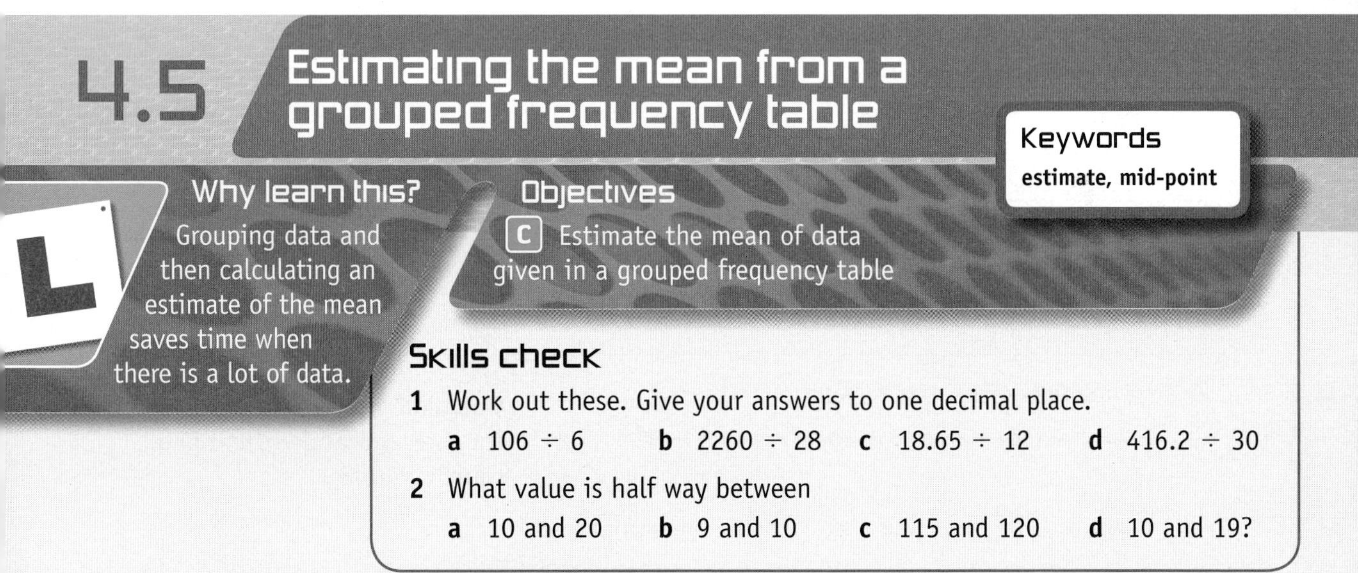

4.5 Estimating the mean from a grouped frequency table

Keywords
estimate, mid-point

Why learn this?
Grouping data and then calculating an estimate of the mean saves time when there is a lot of data.

Objectives
C Estimate the mean of data given in a grouped frequency table

Skills check

1 Work out these. Give your answers to one decimal place.

a $106 \div 6$ **b** $2260 \div 28$ **c** $18.65 \div 12$ **d** $416.2 \div 30$

2 What value is half way between

a 10 and 20 **b** 9 and 10 **c** 115 and 120 **d** 10 and 19?

Estimating the mean from a grouped frequency table

When data is arranged in a grouped frequency table, you don't know the exact data values. This means you can't calculate the mean exactly.

You can **estimate** the mean by assuming that every data value lies exactly in the middle of a class interval. You need to work out the **mid-point** of each class interval.

$$\text{Mid-point} = \frac{\text{maximum value of class interval} + \text{minimum value of class interval}}{2}$$

$$\text{Estimate of mean} = \frac{\text{total of 'mid-point} \times \text{frequency' column}}{\text{total frequency}}$$

Example 5

This frequency table shows the heights of the trees in a park.

Height, h (cm)	Frequency
$0 \leqslant h < 5$	10
$5 \leqslant h < 10$	18
$10 \leqslant h < 15$	6
$15 \leqslant h < 20$	2

a Work out the mid-point of the class interval $10 \leqslant h < 15$.

b Estimate the total height of all the trees in the park.

c Calculate an estimate for the mean height of the trees in the park.

Mid-point $= \dfrac{\text{maximum class interval value} + \text{minimum class interval value}}{2}$

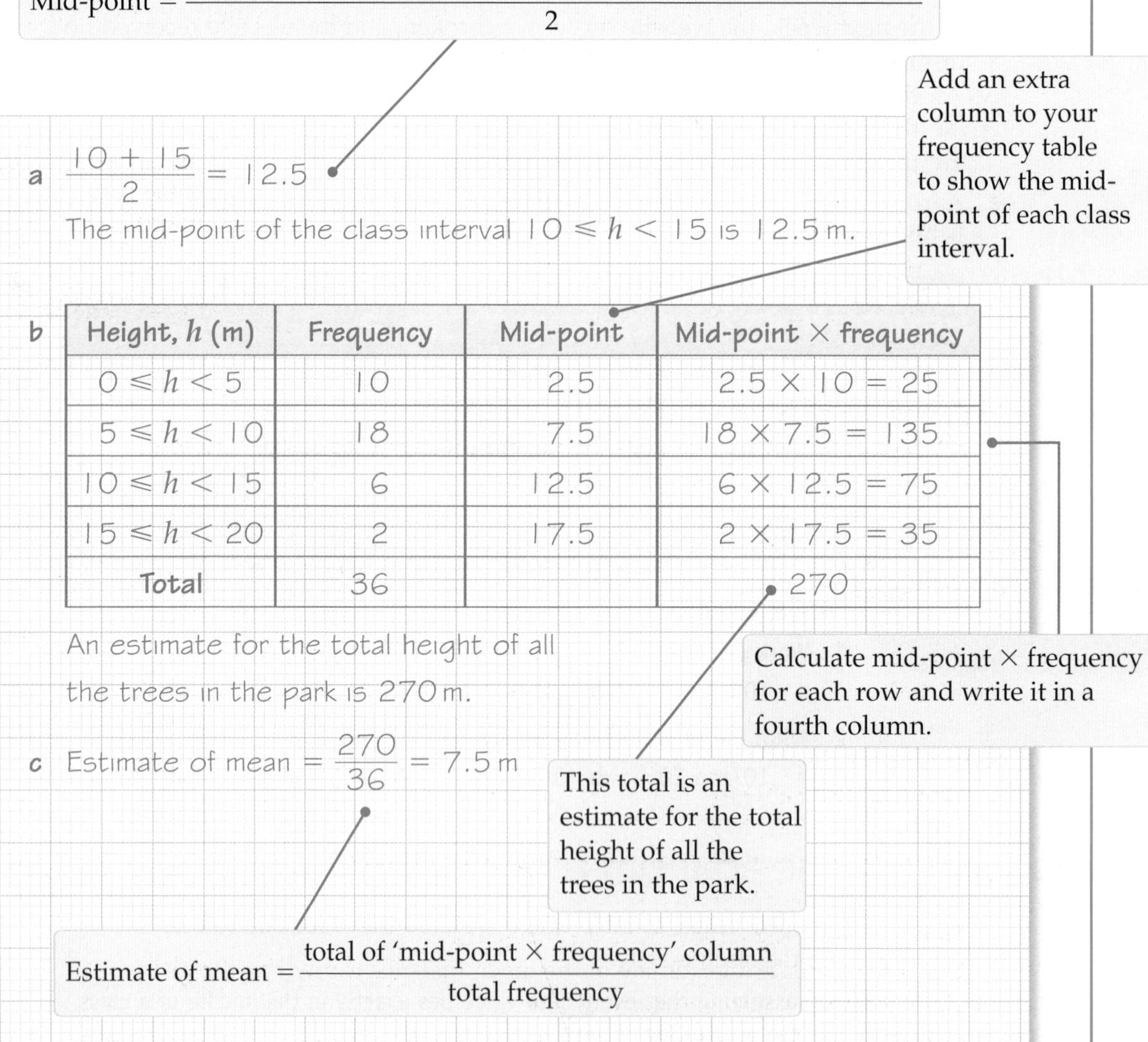

a $\dfrac{10 + 15}{2} = 12.5$

The mid-point of the class interval $10 \leqslant h < 15$ is 12.5 m.

Add an extra column to your frequency table to show the mid-point of each class interval.

b

Height, h (m)	Frequency	Mid-point	Mid-point × frequency
$0 \leqslant h < 5$	10	2.5	$2.5 \times 10 = 25$
$5 \leqslant h < 10$	18	7.5	$18 \times 7.5 = 135$
$10 \leqslant h < 15$	6	12.5	$6 \times 12.5 = 75$
$15 \leqslant h < 20$	2	17.5	$2 \times 17.5 = 35$
Total	36		270

Calculate mid-point × frequency for each row and write it in a fourth column.

An estimate for the total height of all the trees in the park is 270 m.

This total is an estimate for the total height of all the trees in the park.

c Estimate of mean $= \dfrac{270}{36} = 7.5$ m

Estimate of mean $= \dfrac{\text{total of 'mid-point} \times \text{frequency' column}}{\text{total frequency}}$

Exercise 4E

1 This frequency table shows the times taken by members of a class to solve a puzzle.

Time taken, t (minutes)	Frequency	Mid-point	Mid-point × frequency
$0 \leqslant t < 5$	3	2.5	$2.5 \times 3 = 7.5$
$5 \leqslant t < 10$	15		
$10 \leqslant t < 15$	8		
$15 \leqslant t < 20$	2		
$20 \leqslant t < 25$	5		
Total			

a Work out the mid-point of the class interval $5 \leqslant h < 10$.

b Copy and complete the table to work out an estimate for the total time taken by the whole class.

c Calculate an estimate for the mean time taken, correct to one decimal place.

C

2 Jaden used a frequency table to record the number of times each student in her year group logged in to the school's intranet in one week.

Number of log-ins	Frequency	Mid-point	Mid-point × frequency
0–4	22		
5–9	31	7	
10–14	17		
15–19	20		
20–24	6		
Total			

The mid-point of the class interval 5–9 is $\frac{5+9}{2} = 7$.

Copy and complete the table.
Use it to calculate an estimate for the mean number of log-ins per student.
Give your answer correct to one decimal place.

3 Archie carried out an experiment to find out how far the members of his tennis club could throw a tennis ball.
He used a tally chart to record his results.

Distance thrown, d (m)	Boys' tally	Girls' tally
$0 \leqslant d < 8$	|||	||
$8 \leqslant d < 16$	卌 |	卌 卌 ||
$16 \leqslant d < 24$	卌 卌 ||	卌
$24 \leqslant d < 32$	||||	|||
$32 \leqslant d < 40$	|	卌 ||
$40 \leqslant d < 48$	|	

a Will Archie be able to use his tally chart to calculate the exact mean distance thrown for the boys?
Give a reason for your answer.

b Draw a frequency table for the girls' results.
Use it to calculate an estimate of the mean distance thrown by the girls.
Give your answer to one decimal place.

c Calculate an estimate for the mean distance thrown by the whole tennis club.
Give your answer to one decimal place.

d In his conclusion Archie wrote, 'The girls' results were below average. This means that the boys could throw the tennis ball further.'
Do you agree with this statement? Give reasons for your answer.

C

AO2

Review exercise

E

1 The reaction times (in seconds) of seven students are given below.

0.17 0.2 0.31 0.19 0.25 0.33 0.4

a Work out the median reaction time. **[1 mark]**

b Work out the range of this data. **[1 mark]**

c Calculate the mean of this data.
Give your answer correct to two significant figures. **[2 marks]**

2 This stem-and-leaf diagram shows the heights of the members of a school basketball team.

14	3	6	6	7	8	9		
15	0	1	2	2	6	8	9	9
16	0	2	3	6	6			

Key
14 | 3 means 143 cm

a Work out the range of the heights. **[1 mark]**

b Work out the median height. **[1 mark]**

c Calculate the mean height of a member of the team to the nearest cm. **[2 marks]**

3 Write down five numbers with a mean of 7 and a range of 5. **[3 marks]**

D

4 In a survey, people were asked how many times they visited the doctor each month.

Number of visits to the doctor	Frequency	Number of visits × frequency
0	18	
1	17	
2	6	
3	3	
4	0	
Total		

a Work out the range of this data. **[1 mark]**

b Write down the modal number of visits per month. **[1 mark]**

c Calculate the median of this data. **[1 mark]**

d Copy and complete the frequency table.
Calculate the mean number of visits per month, correct to one decimal place. **[3 marks]**

D

5 This bar chart shows the number of players in different games of Monopoly.

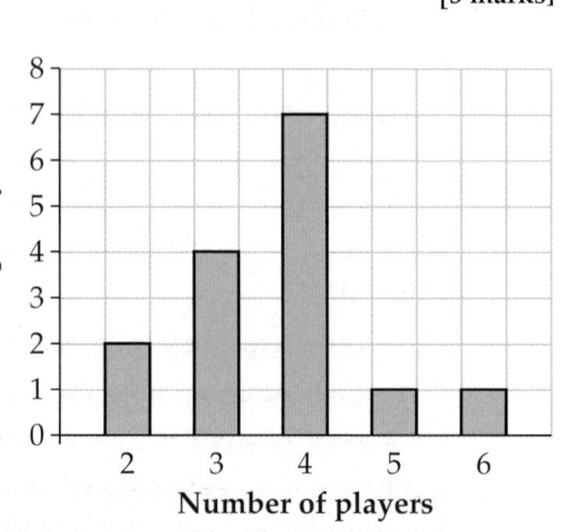

a Write down the modal number of players. **[1 mark]**

b Work out the range of this data. **[1 mark]**

c Work out the median number of players. **[1 mark]**

d Draw a suitable frequency table and use it to calculate the mean number of players in each game.
Give your answer correct to one decimal place. **[3 marks]**

AO2

6 Amit measured the lengths of earthworms for a biology project.

Length, l (cm)	Frequency	Mid-point	Mid-point × frequency
$3 \leqslant l < 4$	2		
$4 \leqslant l < 5$	6		
$5 \leqslant l < 6$	11		
$6 \leqslant l < 7$	12		
$7 \leqslant l < 8$	9		
Total			

a Estimate the range of this data. **[1 mark]**

b Write down the modal class interval. **[1 mark]**

c Which class interval contains the median? **[1 mark]**

d Copy and complete the table.
Calculate an estimate for the mean length of an earthworm. **[3 marks]**

D C

7 Zoe counted the tomatoes on each of the tomato plants in the school greenhouse.

Number of tomatoes	Frequency
0–4	3
5–9	11
10–14	8
15–19	4
20–24	1

a Estimate the range of this data **[1 mark]**

b Write down the modal class interval. **[1 mark]**

c Which class interval contains the median? **[1 mark]**

d Calculate an estimate for the mean number of tomatoes per plant. Give your answers to one decimal place. **[3 marks]**

e Explain why it is not possible to work out the exact mean number of tomatoes per plant. **[1 mark]**

D C AO2

8 The table shows the amounts spent on food by 25 students on a school trip to a theme park.

Amount spent, x (£s)	Frequency
$0 \leqslant x < 5$	12
$5 \leqslant x < 10$	8
$10 \leqslant x < 15$	3
$15 \leqslant x < 20$	2

a Which class interval contains the median? **[1 mark]**

b Use mid-points to calculate an estimate for the mean amount spent per student. **[3 marks]**

c Explain why it is not possible to calculate the exact mean. **[1 mark]**

C

9 Yaniv collected data about the fuel consumption (in miles per gallon) of 30 different cars.

65.5 19.6 18.0 42.6 49.8 61.1 58.3 52.4 56.0 60.8
44.8 46.1 42.3 55.9 51.0 52.5 59.6 48.1 14.3 62.4
45.3 53.8 51.6 52.9 48.3 47.2 52.2 56.2 49.0 51.9

a Choose suitable class intervals and draw a grouped frequency table for this data. **[2 marks]**

b Use your frequency table to estimate the mean fuel consumption. **[3 marks]**

c Calculate the exact mean fuel consumption from the original data.
Give your answer to two decimal places. **[2 marks]**

d Compare your estimate with the exact mean.
How will the choice of class intervals affect the accuracy of your estimate? **[1 mark]**

AO2

C

10 This frequency table shows the numbers of DVDs owned by the students in a class.

Number of DVDs owned	Frequency
0–9	11
10–19	14
20–29	6
30–39	4

a What is the mid-point of the class interval 10–19? **[1 mark]**

b Calculate an estimate for the mean number of DVDs owned per student. Give your answer to one decimal place. **[3 marks]**

C

AO3

11 The results of a survey about the number of eggs in a sample of birds' nests are shown in a frequency table.

The frequency in the last row has been accidentally torn off.

The median number of eggs per nest is 2.5.

Calculate the mean number of eggs per nest. Give your answer to two decimal places.

Number of eggs	Frequency
0	6
1	3
2	8
3	7
4	

[4 marks]

Chapter summary

In this chapter you have learned how to

- calculate the mode, median and range from an ungrouped frequency table **E**
- find the mean, median and range from a set of data, including data given in a stem-and-leaf diagram **E** **D**
- calculate the mean from an ungrouped frequency table **D**
- find the modal class from a grouped frequency table **D**
- estimate the range from a grouped frequency table **D**
- work out the class interval which contains the median from data given in a grouped frequency table **C**
- estimate the mean of data given in a grouped frequency table **C**

AO2 Example – Statistics

C

This Grade C question challenges you to use statistics in a real-life context – working with the data to solve a problem.

The frequency table shows the times that some motorists were stuck in traffic jams (to the nearest 10 minutes). One of the frequencies is missing (shown by n in the table).

The mean waiting time is 28 minutes.

Number of minutes (to the nearest 10)	Frequency
10	4
20	n
30	12
40	5
50	2

a How many motorists are in this survey?

b What is the total number of minutes of waiting time for those motorists whose data can be seen in the table?

c Write an expression for the total number of minutes of waiting time for those motorists who waited 20 minutes?

d Write an expression for the mean waiting time and use it to calculate the missing frequency.

AO2

What maths should you use?

a Total number of motorists $= 4 + n + 12 + 5 + 2 = 23 + n$ — Sum the frequencies.

b Total $= (4 \times 10) + (12 \times 30) + (5 \times 40) + (2 \times 50) = 700$

c Total $= 20n$

Don't forget to multiply the times by the frequencies.

d $\frac{700 + 20n}{23 + n} = 28$

$700 + 20n = 28(23 + n)$

$= 644 + 28n$

$700 - 644 = 28n - 20n$

$56 = 8n$

$n = 7$ so the missing frequency is 7.

- Set up an equation using the formula for the mean waiting time: $\frac{\text{total number of minutes waiting time}}{\text{total number of motorists}}$
- Solve the equation

AO3 Question – Statistics

C

Now try this AO3 Grade C question. You have to work it out from scratch.
READ THE QUESTION CAREFULLY.
It's similar to the AO2 example above, so think about where to start.

The frequency table shows the journey time to school by some students (to the nearest 5 minutes).

One of the frequencies is missing (shown by n in the table).

The mean journey time is 19 minutes.

Calculate the missing frequency.

You **must** show your working.

Number of minutes (to the nearest 5)	Frequency
10	12
15	3
20	11
25	n
30	5

Work out the total time by multiplying the times by the frequencies.

Use the formula for the mean journey time to set up an equation and then solve it.

AO3

Missed appointments

At a dental surgery, patients make an appointment to see either a dentist or a hygienist or both. If the patient misses their appointment they are fined according to the length of the appointment they missed. They are fined at the rate of £60 per hour.

Question bank

1 How long is a dentist appointment?

2 In 2009, in which month were the most dentist appointments missed?

3 How much is a patient fined for missing a hygienist appointment?

At the end of 2009, the practice manager writes a report on the number of appointments that patients missed. The practice manager writes, 'Altogether the total money due from fines comes to £28 620.'

4 Is the practice manager's total correct? Show working to justify your answer.

5 Calculate the mean, median and modal monthly number of dentist appointments missed in 2009.

In his report, the practice manager also writes, 'On average, more dentist appointments were missed per month than hygienist appointments.'

6 Is the practice manager correct? Explain clearly the reasons for making your decision. You could use some of your previous working to help you.

Information bank

Surgery facts

The surgery opens from 9 am to 6 pm Monday to Friday.

The surgery closes for lunch from 1 pm to 2 pm.

There are six dentists at the surgery.

A dentist appointment lasts 15 minutes.

There are three hygienists at the surgery.

A hygienist appointment lasts 10 minutes.

The total number of hygienist appointments missed in 2009 is 1008.

The median monthly number of hygienist appointments missed in 2009 is 81.

The modal monthly number of hygienist appointments missed in 2009 is 96.

There are two receptionists at the surgery.

Salary facts

A dentist at the surgery earns £86 000 per year.

A hygienist at the surgery earns £26 000 per year.

A receptionist at the surgery earns £15 000 per year.

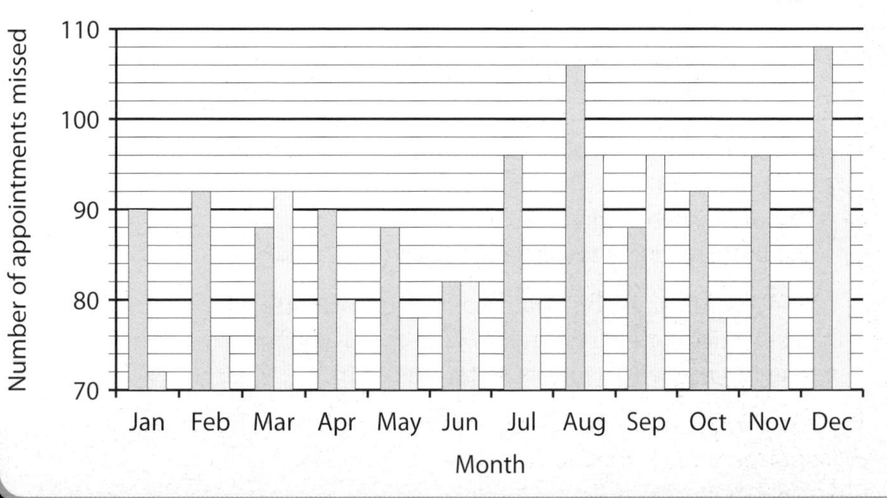

AO2 Example – Number

This Grade C question challenges you to use number in a real-life context – applying maths you know to solve a problem.

Peter supports a football team playing in the Champions League.
He wants to go to watch a match for which tickets cost £45.
He can only afford to pay £27.

a What percentage of the ticket price can Peter afford?

Peter plays football for a local junior team and his Dad says he will give him 5% of the ticket price for every 2 goals Peter scores for his local team.
So far Peter has scored 9 goals.

AO2

b How many more goals must he score so that he is able to buy the ticket?

What maths should you use?
You need a formula to work out percentages.

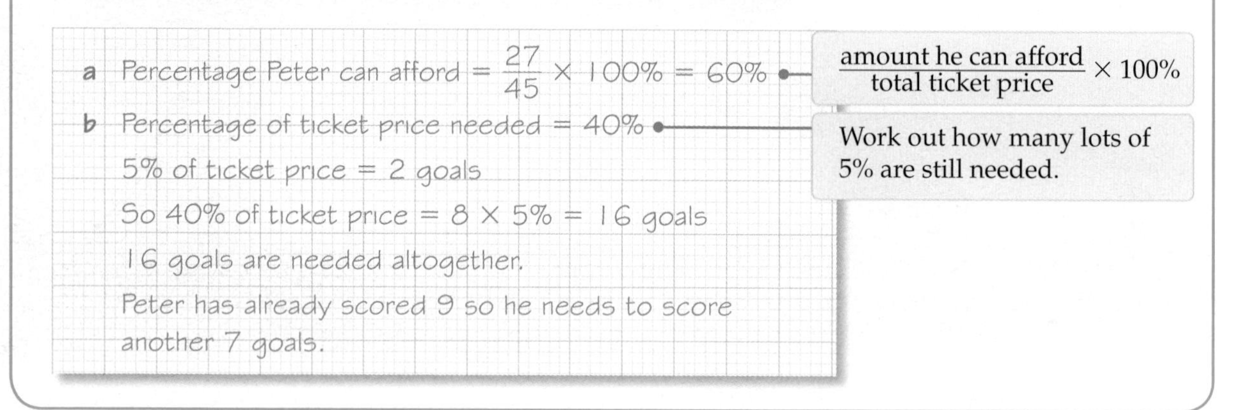

a Percentage Peter can afford $= \frac{27}{45} \times 100\% = 60\%$

b Percentage of ticket price needed = 40%

5% of ticket price = 2 goals

So 40% of ticket price = 8 × 5% = 16 goals

16 goals are needed altogether.

Peter has already scored 9 so he needs to score another 7 goals.

AO3 Question – Number

Now try this AO3 Grade C question. You have to work it out from scratch.
READ THE QUESTION CAREFULLY.
It's similar to the AO2 example above, so think about where to start.

Jane wants to go to a live performance of a TV talent show at the end of term.
The price of the ticket is £75.
She has saved £54 towards the cost of the ticket.

What percentage of the ticket cost still needs to be found?

Jane's Mum says she will help her and promises her 4% of the ticket price for every 5 merits Jane earns at school.
Jane has earned 19 merits so far this term.

AO3

How many more merits must Jane earn so that she will have enough money for the TV talent show ticket?

5 Probability

BBC Video

This chapter is about predicting the chance of things happening.

Jelly beans come in 60 different flavours! If there are 65 beans in a bag, what is the chance of picking your favourite?

Objectives

This chapter will show you how to

- work out the probability of an event not happening E
- identify mutually exclusive events D
- understand and use two-way tables E D
- predict the number of times an event is likely to happen D
- calculate relative frequencies and estimate probabilities C
- calculate the probability of two independent events happening at the same time C
- draw and use tree diagrams B

Before you start this chapter

1 This is the list of vegetables available in a school canteen.

Today's vegetables
Broccoli
Carrots
Peas
Sweetcorn

Bryoni chooses two different types of vegetables. List all the possible combinations of vegetables that Bryoni could choose.

2 Here is a probability scale with arrows.

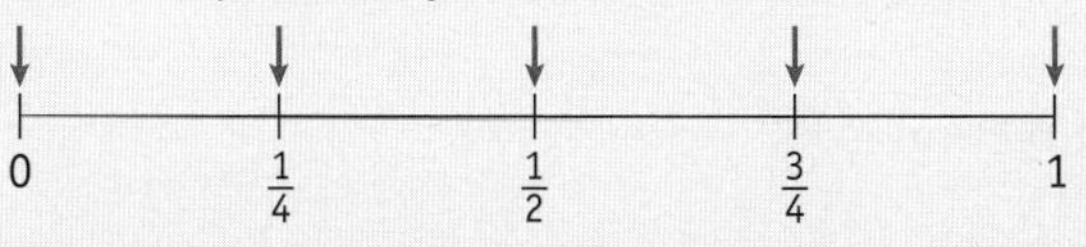

Work out the probability of each of these spinners landing on red.

Which arrow should be labelled with which spinner?

3 Sam rolls a fair six-sided dice. Work out the probability of

a rolling a 1

b rolling a number less than 7

c rolling an even number

d rolling a 12.

> **You can write probabilities using this notation:**
> **P(1) P(<7) P(even)**

4 One letter is chosen at random from the word

PROBABILITY

Work out the probability that the letter is

a the letter B b a vowel

c made up entirely of straight lines.

5.1 Probability that an event does not happen

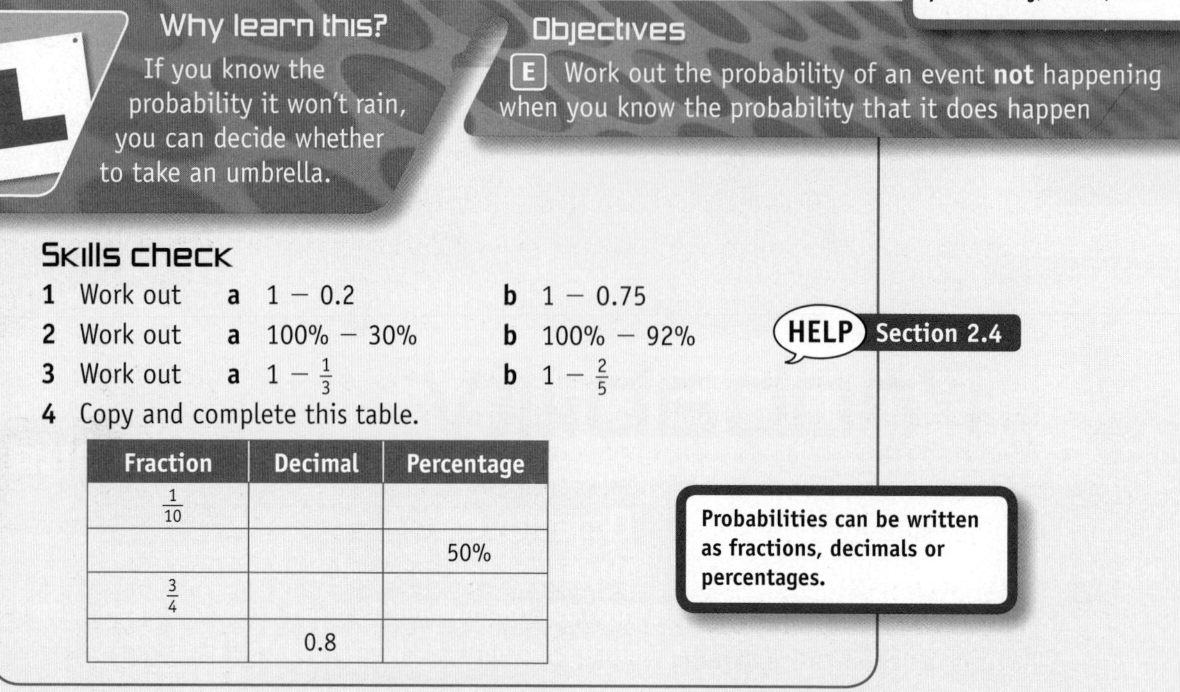

Keywords
probability, event, not

Why learn this?

If you know the probability it won't rain, you can decide whether to take an umbrella.

Objectives

E Work out the probability of an event **not** happening when you know the probability that it does happen

Skills check

1 Work out **a** $1 - 0.2$ **b** $1 - 0.75$

2 Work out **a** $100\% - 30\%$ **b** $100\% - 92\%$

HELP Section 2.4

3 Work out **a** $1 - \frac{1}{3}$ **b** $1 - \frac{2}{5}$

4 Copy and complete this table.

Fraction	Decimal	Percentage
$\frac{1}{10}$		
		50%
$\frac{3}{4}$		
	0.8	

Probabilities can be written as fractions, decimals or percentages.

Calculating the probability an event does **not** happen

When you know the **probability** that an **event** will happen, you can calculate the probability that the event will **not** happen by using this fact.

$$\left(\text{Probability that an event will \textbf{not} happen}\right) = 1 - \left(\text{probability that the event will happen}\right)$$

When rolling a dice, the probability of getting a 4 is $\frac{1}{6}$. So the probability of not getting a 4 is $1 - \frac{1}{6} = \frac{5}{6}$.

Example 1

E

a The probability of picking an ace from a pack of cards is $\frac{4}{52}$.
What is the probability of **not** picking an ace from a pack of cards?

b The probability of picking a heart from a pack of cards is 0.25.
What is the probability of **not** picking a heart from a pack of cards?

a $\frac{48}{52} = \frac{12}{13}$ — $1 - \frac{4}{52} = \frac{52}{52} - \frac{4}{52} = \frac{48}{52}$

b 0.75 — $1 - 0.25 = 0.75$

Exercise 5A

E

1 Hamish is learning to play golf.
The probability that he hits the ball in the right direction is $\frac{1}{10}$.
What is the probability that his next shot

a goes in the right direction

b does not go in the right direction?

2 The probability of picking a King from a pack of cards is $\frac{1}{13}$.

A picture card is a Jack, Queen or King.

a What is the probability of picking 'not a King'?

b What is the probability of picking a picture card?

c What is the probability of picking 'not a picture card'?

3 The probability that Hazel misses her bus is 0.05.
What is the probability that Hazel catches her bus?

Catching the bus means not missing the bus.

4 The probability of winning a £5 prize on the National Lottery 'Thunderball' is $\frac{3}{100}$.
The probability of winning a £10 prize on the National Lottery 'Thunderball' is $\frac{9}{1000}$.
Work out the probability of

a not winning a £5 prize   b not winning a £10 prize.

5 The probability that this spinner lands on 1 is 0.7.
The probability that this spinner lands on blue is 0.85.
What is the probability that the spinner

a does not land on 1   b does not land on blue?

Give each answer as a fraction, a decimal and a percentage.

6 The probability that this spinner lands on 1 is 28%.
The probability that this spinner lands on blue is 99%.
What is the probability that the spinner

a does not land on 1

b does not land on blue?

% means 'out of 100'

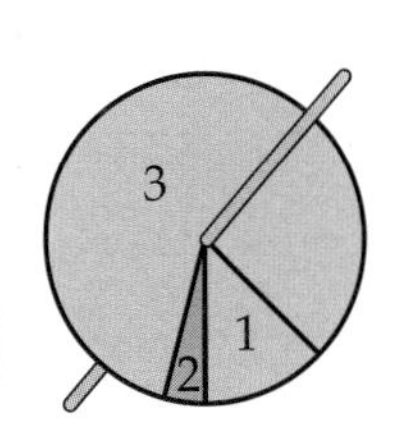

E

7 Alan buys a special spinner.
The spinner has sections numbered 1 to 5.
The probabilities of different scores are listed in the table.
Work out the probability of

a not spinning a 1

b not spinning a 3

c not spinning a 5.

d Explain what your answer to part c means.

Number	1	2	3	4
Probability	0.3	0.2	0.25	0.25

8 Sage has a biased dice numbered 1 to 6.
The probability of getting a 6 with this dice is $\frac{1}{3}$.
Sage says, 'There are 5 other numbers so the probability of not getting a 6 with this dice is $\frac{5}{6}$.'
Explain why Sage is wrong.

9 Leanne has a box that contains 20 counters.
Leanne picks a blue counter at random from the box.
The probability that she picks a blue counter is $\frac{4}{5}$.

a What is the probability that Leanne picks a counter that is not blue?

b How many counters in the box are not blue?

Picking 'at random' means that each counter is equally likely to be picked.

AO2

E

10 Lee has 10 coins in his pocket. He picks one at random.
The probability that Lee does not pick a 10p coin is $\frac{2}{5}$.
How many 10p coins does Lee have in his pocket?

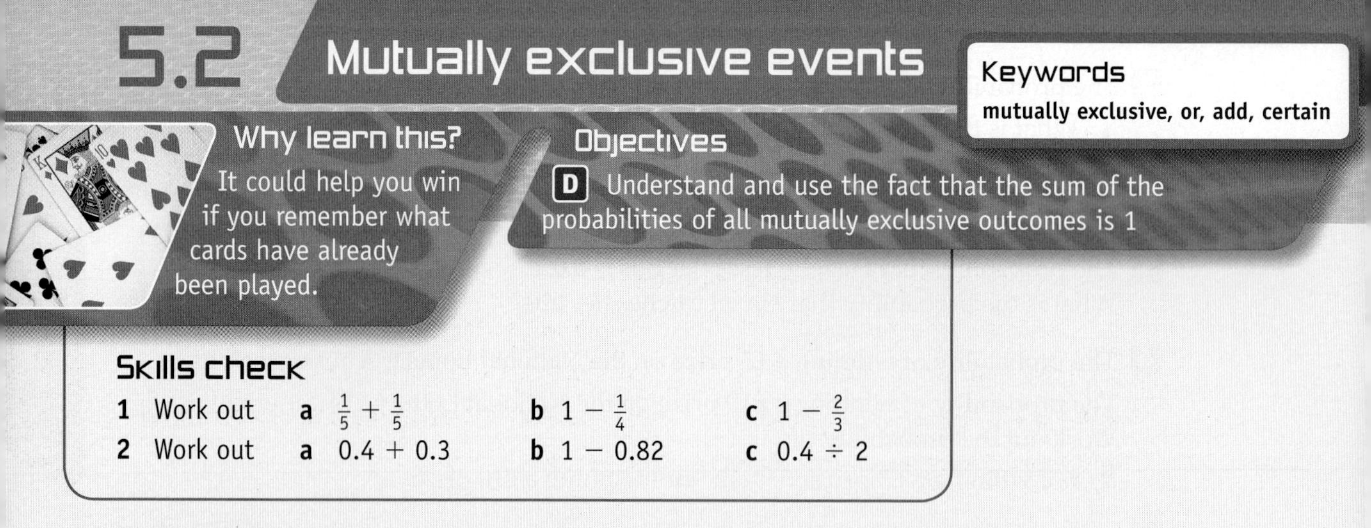

5.2 Mutually exclusive events

Keywords
mutually exclusive, or, add, certain

Why learn this?
It could help you win if you remember what cards have already been played.

Objectives

D Understand and use the fact that the sum of the probabilities of all mutually exclusive outcomes is 1

Skills check

1 Work out **a** $\frac{1}{5} + \frac{1}{5}$ **b** $1 - \frac{1}{4}$ **c** $1 - \frac{2}{3}$

2 Work out **a** $0.4 + 0.3$ **b** $1 - 0.82$ **c** $0.4 \div 2$

Mutually exclusive events

Mutually exclusive events cannot happen at the same time. When you roll a dice you cannot get a 1 and a 6 at the same time. When you flip a coin you can get either a head **or** a tail, but not both at the same time.

For any two events, A and B, which are mutually exclusive

$$P(A \text{ or } B) = P(A) + P(B)$$

For a fair dice, the probability of rolling a 2 is $\frac{1}{6}$.

You can write this as $P(2) = \frac{1}{6}$

Also, $P(1) = \frac{1}{6}$, $P(3) = \frac{1}{6}$, $P(4) = \frac{1}{6}$, $P(5) = \frac{1}{6}$ and $P(6) = \frac{1}{6}$

Add together the probabilities of all the possible outcomes.

$P(1) + P(2) + P(3) + P(4) + P(5) + P(6) = \frac{1}{6} + \frac{1}{6} + \frac{1}{6} + \frac{1}{6} + \frac{1}{6} + \frac{1}{6} = 1$

This is because you are **certain** to roll either 1 or 2 or 3 or 4 or 5 or 6.

P(A) means the probability of event A occurring.

Rolling a 2 and not rolling a 2 are mutually exclusive events.

The total sum of their probabilities is $\frac{1}{6} + \frac{5}{6} = 1$. This is because you are certain to get either '2' or 'not 2'.

D

Example 2

a Work out the probability of rolling a 5 **or** a 6 with an ordinary dice.

b This spinner has four sections numbered 5 to 8. The table shows the probability of the spinner landing on each number.

Number	5	6	7	8
Probability	0.2	0.2	0.2	?

What is the probability that the spinner lands on 8?

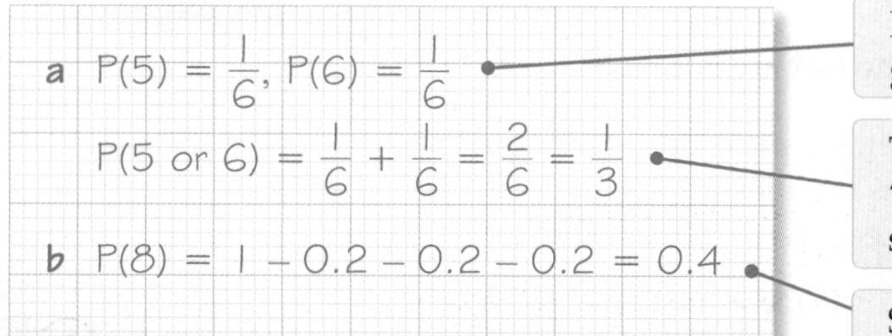

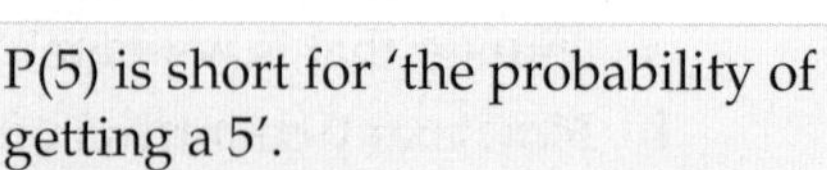
P(5) is short for 'the probability of getting a 5'.

The events 'landing on 5' and 'landing on 6' are mutually exclusive so **add** the probabilities together.

Subtract P(5), P(6) and P(7) from 1.

Exercise 5B

1 A box of chocolates contains 15 identical looking chocolates.
Six of the chocolates have toffee centres, four are solid chocolate, three have soft centres and two have nut centres.
One chocolate is taken from the box at random.
What is the probability that the chocolate

a doesn't have a nut centre
b doesn't have a toffee centre
c has a toffee or a chocolate centre
d has a toffee or a soft centre
e has a toffee or a nut centre
f doesn't have a toffee or a nut centre
g doesn't have a soft or a nut or a toffee centre?

E D

2 A tin contains biscuits.
One biscuit is taken from the tin at random.
The table shows the probabilities of taking each type of biscuit.

Biscuit	Probability
digestive	0.4
wafer	
cookie	0.15
ginger	0.25

a What is the probability that the biscuit is a digestive or a cookie?
b What is the probability that the biscuit is a wafer?

D

3 A bag contains cosmetics.
One cosmetic is taken from the bag at random.
The table shows the probabilities of taking each type of cosmetic.
There are three times as many eyeshadows as blushers.

Cosmetic	Probability
eyeliner	0.3
lipgloss	0.3
eyeshadow	
blusher	

What is the probability that the cosmetic is an eyeshadow?

D

4 A bag contains 36 marbles of three different colours, red (R), blue (B) and yellow (Y).

$$P(R) = \frac{5}{12} \qquad P(Y) = \frac{1}{4}$$

a Work out the probability of picking a blue marble.
b Work out the number of marbles of each colour in the bag.

AO2

5 David puts 15 CDs into a bag.
Elliot puts 9 computer games into the same bag.
Fern puts some DVDs into the bag.
The probability of taking a DVD from the bag at random is $\frac{1}{3}$.
How many DVDs did Fern put into the bag?

D AO3

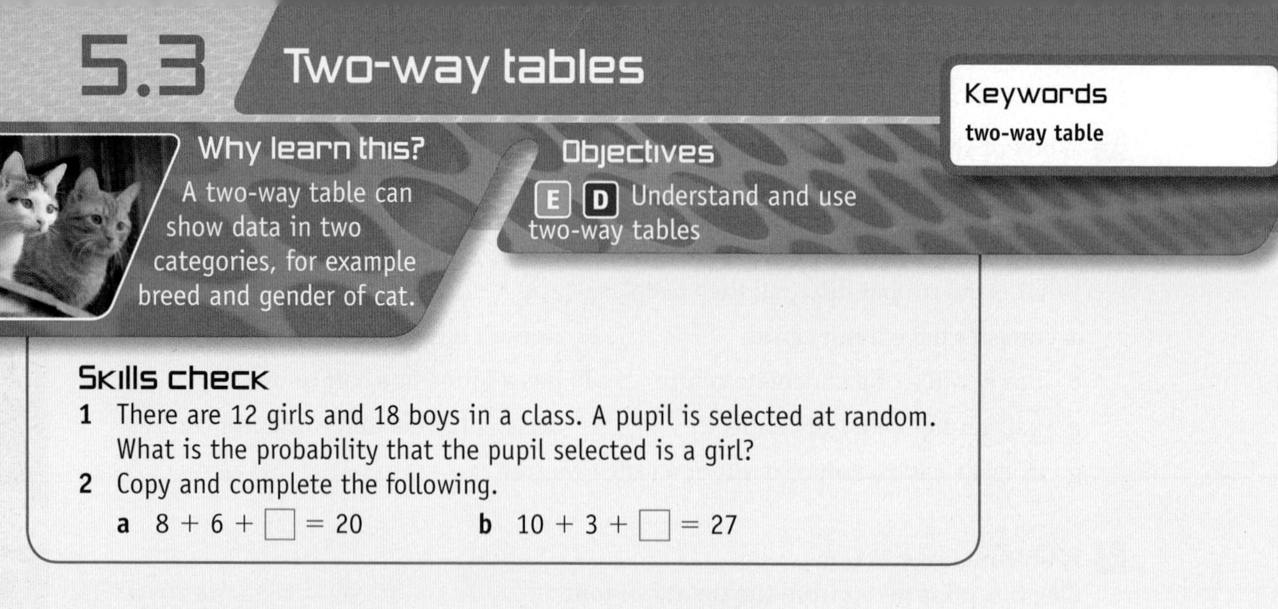

5.3 Two-way tables

Keywords
two-way table

Why learn this?

A two-way table can show data in two categories, for example breed and gender of cat.

Objectives

E **D** Understand and use two-way tables

Skills check

1. There are 12 girls and 18 boys in a class. A pupil is selected at random. What is the probability that the pupil selected is a girl?
2. Copy and complete the following.
   a $8 + 6 + \square = 20$ **b** $10 + 3 + \square = 27$

Probability from two-way tables

A **two-way table** shows two or more sets of data at the same time.

You can work out probabilities from a two-way table.

E

Example 3

The two-way table shows the number of doors and the number of windows in each office in a block.

		Number of doors			Total
		1	2	3	
Number of windows	1	3	0	0	3
	2	7	3	0	10
	3	5	5	3	13
	4	1	1	2	4
Total		16	9	5	30

Including the totals helps with checking your working.

a How many offices have three doors?

b One office is chosen at random. What is the probability that it has
 i three doors **ii** two doors and three windows?

c An office with two windows is selected. What is the probability that it will have two doors?

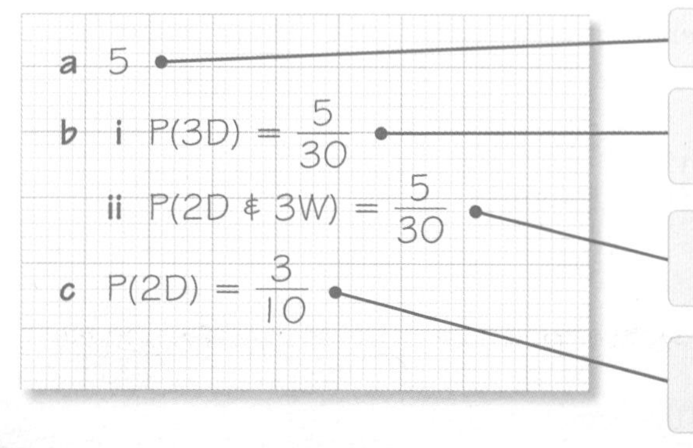

$0 + 0 + 3 + 2 = 5$

5 offices have 3 doors.
There are 30 offices in total.

5 offices have 2 doors and 3 windows.
There are 30 offices in total.

$7 + 3 + 0 = 10$, so 10 offices have 2 windows.
Out of these 10 offices, 3 have 2 doors.

Exercise 5C

1 The two-way table shows the number of doors and the number of windows in each room in a mansion.

		Number of doors				Total
		1	2	3	4	
Number of windows	1	6	1	0	0	7
	2	2	3	2	0	7
	3	2	4	0	1	7
	4	5	4	1	1	11
	5	0	2	3	3	8
Total		15	14	6	5	40

a How many rooms are there in the mansion?

b How many of the rooms have one door?

c How many of the rooms have one window?

d One room is chosen at random.
What is the probability that the room
- i has one door
- ii has one door and three windows
- iii has the same number of doors as windows?

2 The table shows the number of Year 10 students at Brightspark High School who do or don't have part-time jobs.
One student is chosen at random.
What is the probability that this student

a is a boy who has a part-time job

b is a girl who has a part-time job

c doesn't have a part-time job?

	Year 10 students		Total
	Job	No job	
Boys	62	41	103
Girls	58	39	97
Total	120	80	200

3 The table shows the age and sex of a sample of 50 pupils in a school.

	Age in years						Total
	11	12	13	14	15	16	
Boys	3	3	6	3	5	4	24
Girls	3	5	3	5	4	6	26
Total	6	8	9	8	9	10	50

A pupil is chosen at random.
What is the probability that this pupil

a is a 13-year-old boy

b is a 16-year-old girl

c is 14 years old

d is at least 14 years old

e is not 11 years old?

There are 1000 pupils in the school altogether. How many of these are likely to be

f 11 years old

g girls?

E

4 The table shows the number of walkers on a trek and whether they are wearing trainers or walking boots.

	Wearing trainers	Wearing walking boots	Total
Boys	3	21	24
Girls	17	9	26
Total	20	30	50

a A walker is chosen at random.
What is the probability that they are wearing walking boots?

b A boy is chosen at random.
What is the probability that he is wearing trainers?

c A walker wearing trainers is chosen at random.
What is the probability that it is a girl?

5 The table shows the history exam results of 50 students.
A student is chosen at random.
What is the probability that the student

a has passed the exam

b is male

c is a male who has passed the exam?

	Pass	Fail	Total
Male	13	15	28
Female	14	8	22
Total	27	23	50

D

6 Professor Newton has spilled tea on the table that shows the test results of his science class.
Use Professor Newton's notes below to work out the numbers underneath the tea stains.

	Pass	Fail
Male		
Female		

Altogether 60 students took the test.
The probability that a student failed is $\frac{29}{60}$.
The probability that the student is male is $\frac{1}{2}$.
The probability that the student failed and is male is $\frac{1}{4}$.

7 The tables show the driving test results of Ace Driving School and UCan Driving School.

Ace

	Pass	Fail	Total
Male	12	4	16
Female	10	4	14
Total	22	8	30

UCan

	Pass	Fail	Total
Male	3	3	6
Female	9	0	9
Total	12	3	15

a A learner driver is chosen at random from Ace Driving School.
What is the probability that they passed their driving test?

b A learner driver is chosen at random from UCan Driving School.
What is the probability that they passed their driving test?

c One learner from one of the driving schools is chosen at random to be interviewed by local radio. What is the probability that the learner is from Ace Driving School and failed their test?

A02

5.4 Expectation

Keywords
likely, estimate, trial

Why learn this?

Knowing the expected number of 6s in a number of rolls could help you work out if a dice is fair.

Objectives

D Predict the likely number of successful events given the probability of any outcome and the number of trials or experiments

Skills check

1 Alice rolls a fair six-sided dice.
What is the probability that she rolls
a a 2 **b** an odd number
c a number less than 3?

2 Work out
a $\frac{1}{2} \times 40$ **b** $\frac{1}{3} \times 15$ **c** $\frac{2}{5} \times 30$

The number of times an event is likely to happen

Sometimes you will want to know the number of times an event is **likely** to happen.

You can work out an **estimate** of the frequency using this formula.

Expected frequency = probability of the event happening × number of **trials**

Example 4

In a game a fair six-sided dice is rolled 30 times.

a How many 6s would you expect to get?

b How many even numbers would you expect to get?

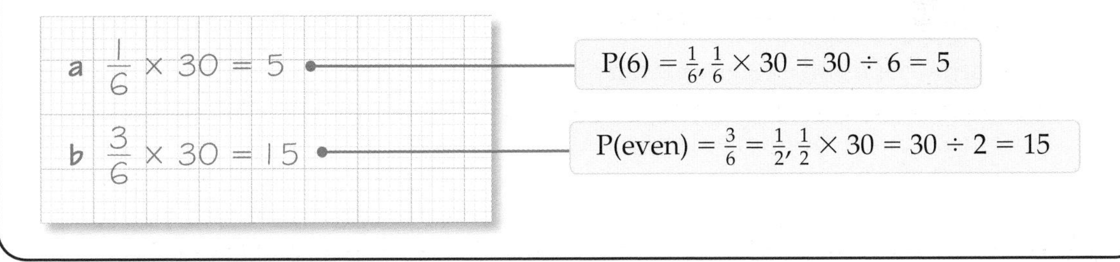

Exercise 5D

1 A fair coin is flipped 500 times.
How many times would you expect to get 'heads'?

2 A fair dice is rolled 60 times.
How many times would you expect it to land on 2?

3 In an experiment, a card is drawn at random from a normal pack of playing cards.
This is done 520 times.
How many times would you expect to get
a a red card **b** a heart
c a King **d** the King of hearts?

D

4 Asif buys 60 scratch-cards. Each scratch-card costs £1.
The probability of winning the £20 prize with each scratch-card is $\frac{1}{30}$.

a How many times is Asif likely to win?

b How much money is Asif likely to win?

c Overall, how much money is Asif likely to lose?

D

5 In a bag there are 15 red, 5 blue, 5 green and 5 orange counters.
Moira takes a counter at random from the bag, notes the colour, then puts the counter back in the bag. She does this 1500 times.
How many times would you expect her to take a blue counter from the bag?

6 The probability that a slot machine pays out its £10 jackpot is $\frac{1}{80}$.
The rest of the time it pays out nothing.
Jimmy plays the slot machine 400 times.

a How many times is Jimmy likely to win?

b How much money is Jimmy likely to win?

Each game costs 20p to play.

c How much does it cost Jimmy to play the 400 games?

d Is Jimmy likely to make a profit? Give a reason for your answer.

7 In a bag there are 20 counters. Ten of the counters are red, five are blue, four are green and one is gold. A counter is taken at random from the bag then replaced. This is done 300 times.
How many times would you expect to get

a a gold counter

b a red counter

c not a blue counter

d a white counter

e a blue or a green counter

f not a red or a blue counter?

A02

8 At a summer fête, Alun runs a charity 'Wheel of fortune' game.
He charges £1 to spin the wheel.
If the arrow lands on a square number he gives a prize of £2.
Altogether 200 people play the game.
How much money would you expect Alun to make for charity?

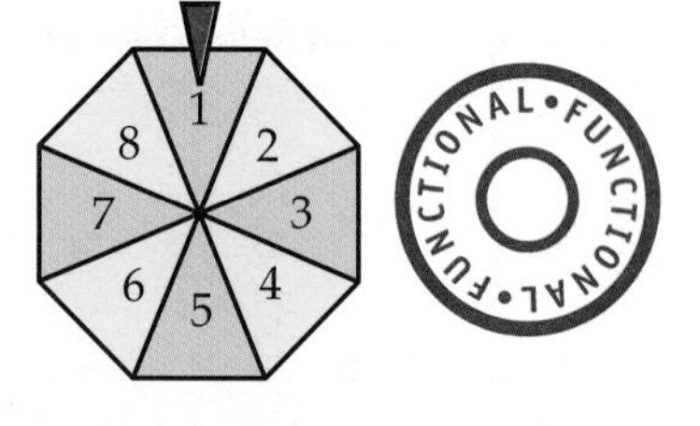

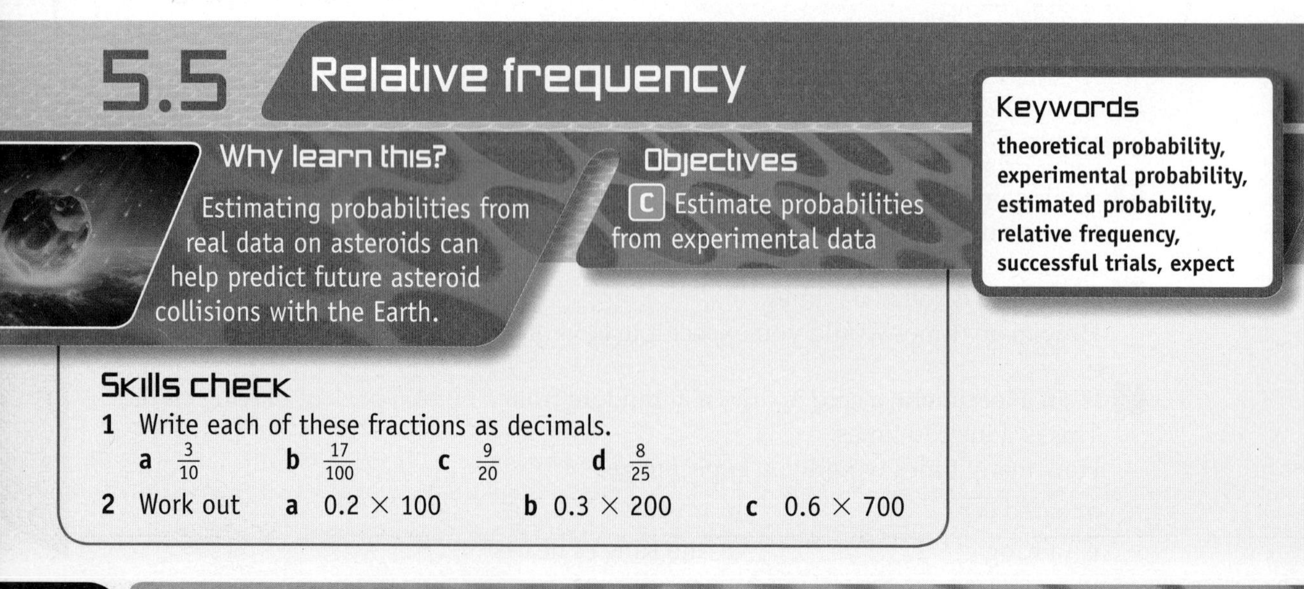

5.5 Relative frequency

Why learn this?

Estimating probabilities from real data on asteroids can help predict future asteroid collisions with the Earth.

Objectives

C Estimate probabilities from experimental data

Keywords

theoretical probability, experimental probability, estimated probability, relative frequency, successful trials, expect

Skills check

1 Write each of these fractions as decimals.

a $\frac{3}{10}$ **b** $\frac{17}{100}$ **c** $\frac{9}{20}$ **d** $\frac{8}{25}$

2 Work out **a** 0.2×100 **b** 0.3×200 **c** 0.6×700

Calculating relative frequency

For a fair dice, the **theoretical probability** of getting a 3 is $\frac{1}{6}$.

For some events, you don't know the theoretical probability. For example, when you drop a drawing pin, what is the probability that it lands 'point up'?

You could carry out an experiment. Drop the drawing pin many times and record the number of times it lands point up. Then work out the **experimental** or **estimated probability**. This estimated probability is called the **relative frequency**.

$$\text{Relative frequency} = \frac{\text{number of } \textbf{successful trials}}{\text{total number of trials}}$$

The theory (or idea) is that there are 6 possible outcomes and they are all equally likely, so each has probability $\frac{1}{6}$.

As the number of trials increases, the relative frequency approaches the theoretical probability.

Example 5

C

Anil carries out an experiment. He drops a drawing pin and records the number of times it lands 'point up'.

Here are his results at different stages of his 2000 trials.

Number of times pin is dropped	Number of times pin lands 'point up'	Relative frequency
100	82	
200	101	
500	326	
1000	586	
1500	882	
2000	1194	

a Calculate the relative frequency at each stage of the testing.

b Estimate the probability of this drawing pin landing 'point up'.

c 15 000 of these drawing pins are dropped. How many do you **expect** to land 'point up'?

d Draw a graph of number of trials against relative frequency to illustrate the results.

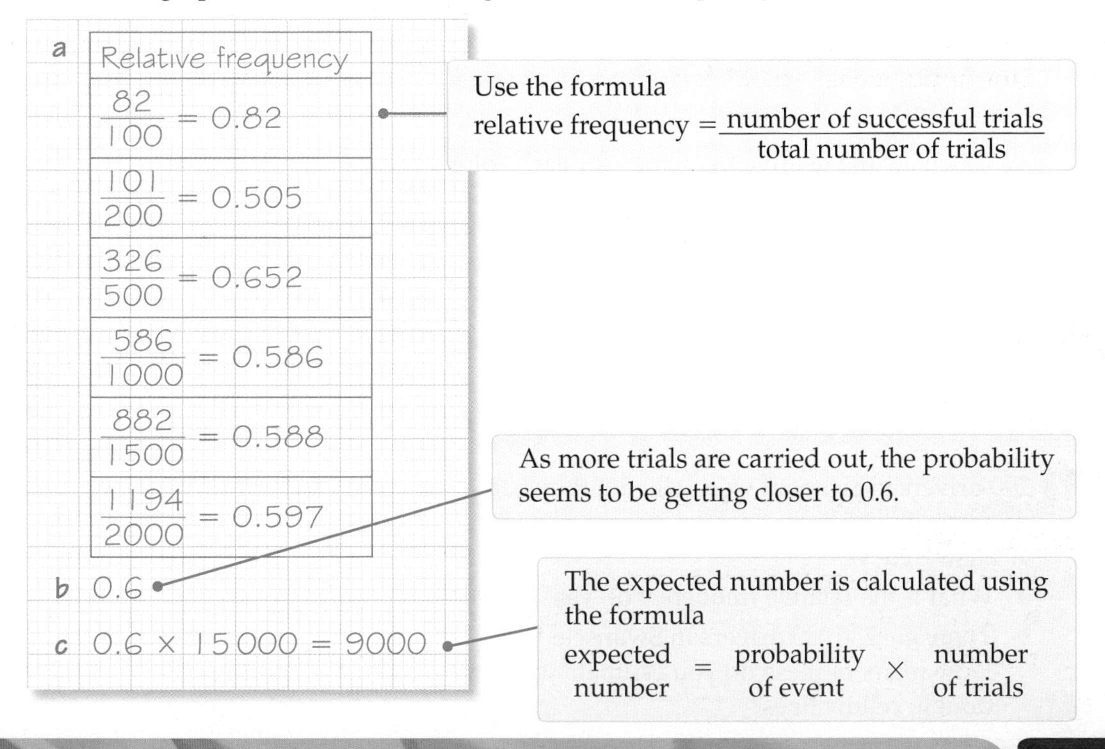

d

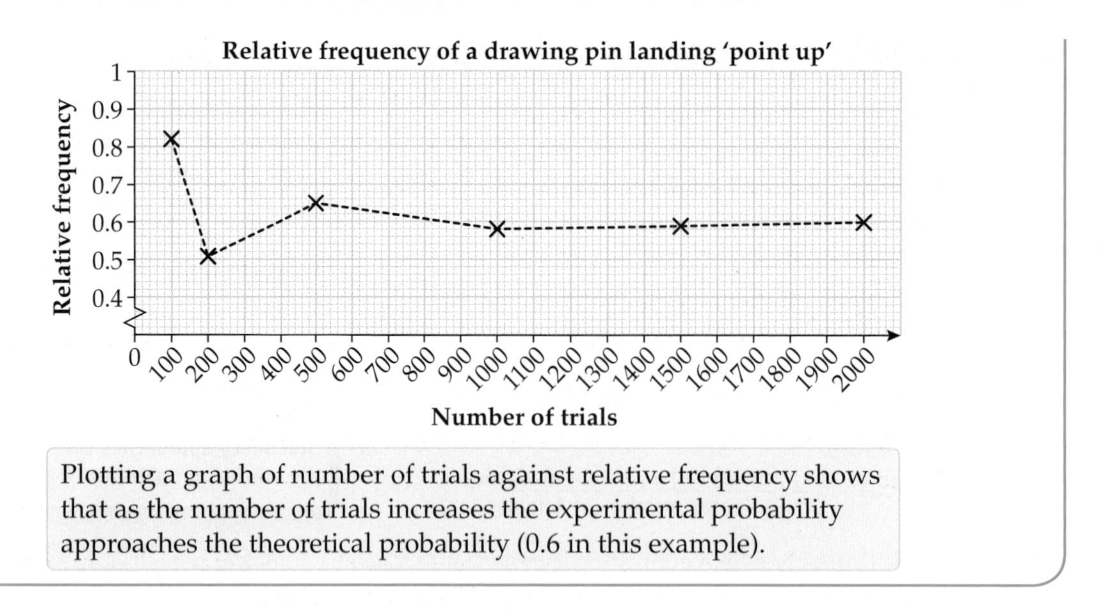

Plotting a graph of number of trials against relative frequency shows that as the number of trials increases the experimental probability approaches the theoretical probability (0.6 in this example).

Exercise 5E

C

1 A bag contains 50 coloured discs. The discs are either blue or red. Jake conducts an experiment to see if he can work out how many of each colour there are.
He takes out a disc, records its colour then replaces it in the bag.
He keeps a tally of how many of each colour there are after different numbers of trials.
The table shows his results.

Number of trials	Number of blue discs	Relative frequency (blue)	Number of red discs	Relative frequency (red)
20	15		5	
50	29		21	
100	56		44	
200	124		76	
500	341		159	
750	480		270	
1000	631		369	
1500	996		504	
2000	1316		684	

a Calculate the relative frequency for each colour at each stage of the experiment.

b Estimate the theoretical probability of obtaining
i a blue disc **ii** a red disc.

c Work out how many of each colour there are in the bag.

d Draw a graph of number of trials against relative frequency to illustrate your results.
Plot the graphs for blue and red discs on the same axes.

Use a horizontal scale as in Example 5 and a vertical scale of 1 cm for 0.1 with the vertical axis going from 0 to 1.

2 200 drivers in Swansea were asked if they had ever parked their car on double yellow lines.
47 answered 'yes'.

a What is the relative frequency of 'yes' answers?

b There are 230 000 drivers in Swansea.
How many of these do you estimate will have parked their car on double yellow lines?

C

3 Salib thinks his dice is biased as he never gets a 6 when he wants to. To test it, he rolls the dice and records the number of 6s he gets. The table shows his results.

Number of rolls	20	50	100	150	200	500
Number of sixes	1	11	14	24	32	84
Relative frequency						

a Calculate the relative frequency of scoring a 6 at each stage of Salib's experiment.

b What is the theoretical probability of rolling a 6 with a fair dice?

c Do you think that Salib's dice is biased? Explain your answer.

d Salib rolls the dice 1200 times. How many 6s do you expect him to get?

4 Maleek thinks her dice is biased. To test this theory she rolls the dice 200 times and records the scores. Her results are shown in the table below.

Score on the dice	1	2	3	4	5	6
Frequency	35	22	25	27	51	40
Relative frequency						

a Calculate the relative frequency of rolling each number.

b What is the theoretical probability of rolling each number on a fair dice?

c Do you think Maleek's dice is fair? Explain your answer.

AO2

C

5 George and Zoe each carry out an experiment with the same four-sided spinner. The tables show their results.

George's results

Number on spinner	1	2	3	4
Frequency	3	14	10	13

Zoe's results

Number on spinner	1	2	3	4
Frequency	45	53	48	54

George thinks the spinner is biased. Zoe thinks the spinner is fair. Who is correct? Explain your answer.

6 Peter wants to test if a spinner is biased.

The spinner has five equal sections labelled 1, 2, 3, 4, 5. Peter spins the spinner 20 times. Here are his results.

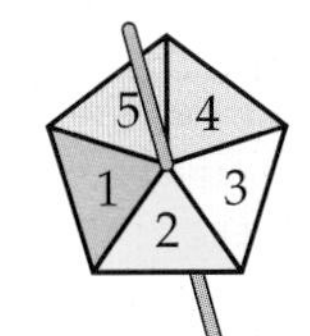

2 1 3 1 5 5 1 4 3 5
4 2 1 5 1 4 3 1 2 1

a Copy and complete the relative frequency table.

Number	1	2	3	4	5
Relative frequency					

b Peter thinks that the spinner is biased. Write down the number you think the spinner is biased towards. Explain your answer.

c What could Peter do to make sure his results are more reliable?

AO3

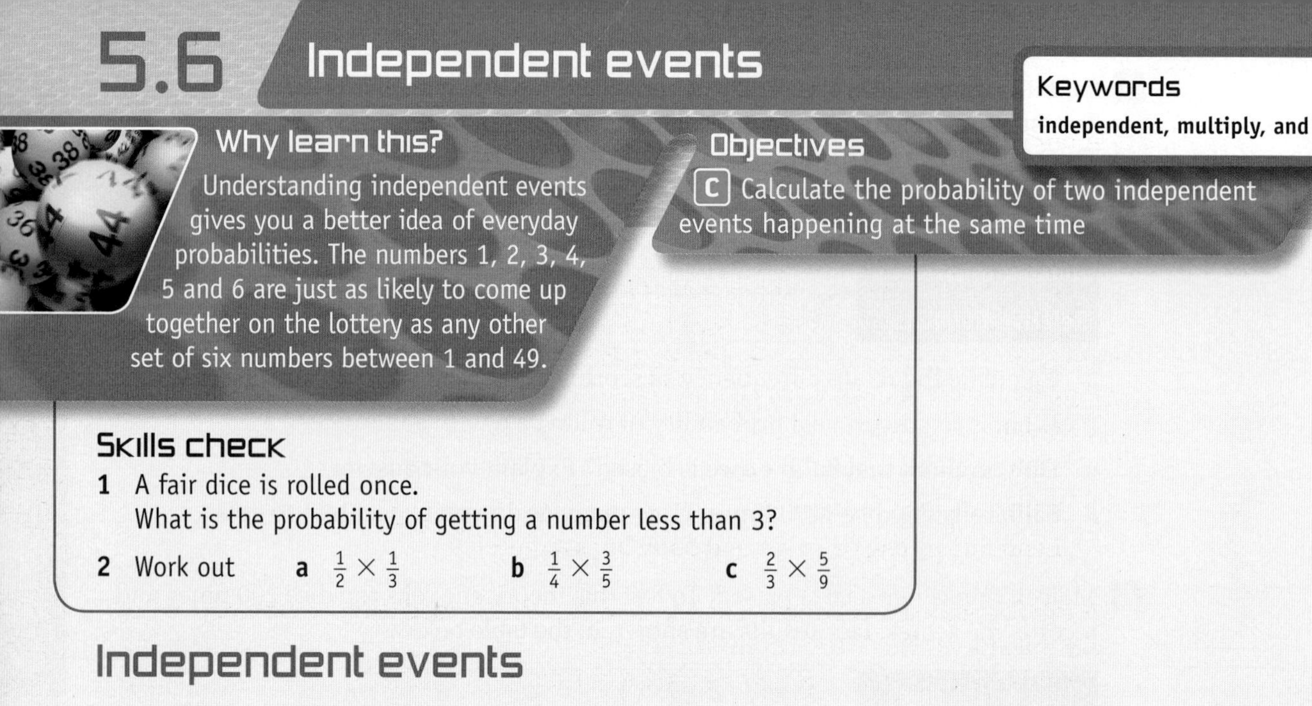

5.6 Independent events

Keywords
independent, multiply, and

Why learn this?

Understanding independent events gives you a better idea of everyday probabilities. The numbers 1, 2, 3, 4, 5 and 6 are just as likely to come up together on the lottery as any other set of six numbers between 1 and 49.

Objectives

C Calculate the probability of two independent events happening at the same time

Skills check

1 A fair dice is rolled once.
What is the probability of getting a number less than 3?

2 Work out **a** $\frac{1}{2} \times \frac{1}{3}$ **b** $\frac{1}{4} \times \frac{3}{5}$ **c** $\frac{2}{3} \times \frac{5}{9}$

Independent events

Two events are **independent** if the outcome of one does not affect the outcome of the other. When you roll two dice at the same time, the number you get on one dice does not affect the number you get on the other. To calculate the probability of two independent events happening at the same time, you **multiply** the individual probabilities.

When A and B are independent events P(A **and** B) = P(A) × P(B)

C

Example 6

a A fair dice is rolled twice.
What is the probability of getting two 6s?

b A coin and a dice are thrown together.
What is the probability of getting a head and a 1?

a $P(6) = \frac{1}{6}$ — The individual probability of getting a 6.

P(6 **and** another 6) — The combined probability. Here '**and**' means multiply.

$= \frac{1}{6} \times \frac{1}{6} = \frac{1}{36}$

This is a sample space diagram. It lists all the possible outcomes when a dice is rolled twice. There are 36 possible outcomes and only one with two 6s, which confirms that the probability $= \frac{1}{36}$.

	1	2	3	4	5	6
1	1,1	1,2	1,3	1,4	1,5	1,6
2	2,1	2,2	2,3	2,4	2,5	2,6
3	3,1	3,2	3,3	3,4	3,5	3,6
4	4,1	4,2	4,3	4,4	4,5	4,6
5	5,1	5,2	5,3	5,4	5,5	5,6
6	6,1	6,2	6,3	6,4	6,5	6,6

Total number of outcomes = total number of outcomes for event A × total number of outcomes for event B

A sample space diagram is one way of showing all the possible outcomes of an experiment.

b $P(H) = \frac{1}{2}$, $P(1) = \frac{1}{6}$ — The individual probabilities of getting a head and a 1.

P(H **and** 1) $= \frac{1}{2} \times \frac{1}{6} = \frac{1}{12}$ — The combined probability. Here '**and**' means multiply.

Exercise 5F

C

1 A fair six-sided dice is rolled twice.

a What is the probability of getting a 5 and then a 4?

b What is the probability of getting two 3s?

2 When Lynn and Sally go to the shop, the probability that Lynn buys a bag of crisps is $\frac{1}{3}$. The probability that she buys a muesli bar is $\frac{1}{4}$. The probability that Sally buys a bag of crisps is $\frac{1}{2}$. The probability that she buys a muesli bar is $\frac{1}{4}$. The girls choose independently of each other.
Calculate the probability that

a both girls buy a bag of crisps

b both girls buy a muesli bar

c Lynn buys a bag of crisps and Sally buys a muesli bar.

C

3 A fair four-sided dice, numbered 1 to 4, and a fair 10-sided dice, numbered 1 to 10, are rolled together. Their scores are added.
What is the probability of getting a total of 14?

4 In roulette, a ball is spun around a wheel. Players bet on where the ball will land. A roulette wheel has 37 sections numbered 0 to 36. The zero is coloured green and the other numbers are coloured red or black, as shown in the diagram.

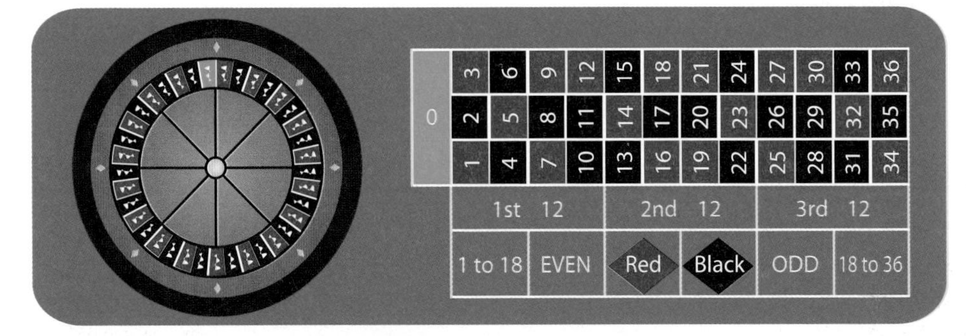

A £10 bet could win these amounts.

A £10 bet on	Possible win
a single number e.g. 17	£360
a pair of numbers e.g. 14 and 17	£180
a group of four numbers e.g. 14, 17, 15 and 18	£90
a group of 12 numbers e.g. 13 to 24	£30
a red or a black number	£20
an odd or an even number	£20

A player places two £10 bets. What is the probability that she wins

a £720 **b** £60?

5 Joe spins a spinner with five equal sectors numbered from 1 to 5.
Sarah rolls a fair dice numbered from 1 to 6.
Work out the probability that

a they both obtain a 3

b the total of their scores is 2

c the total of their scores is 5

d Sarah's score is twice Joe's score

e they both obtain an even number.

How will a total of 5 arise? You will need to add some probabilities.

AO2

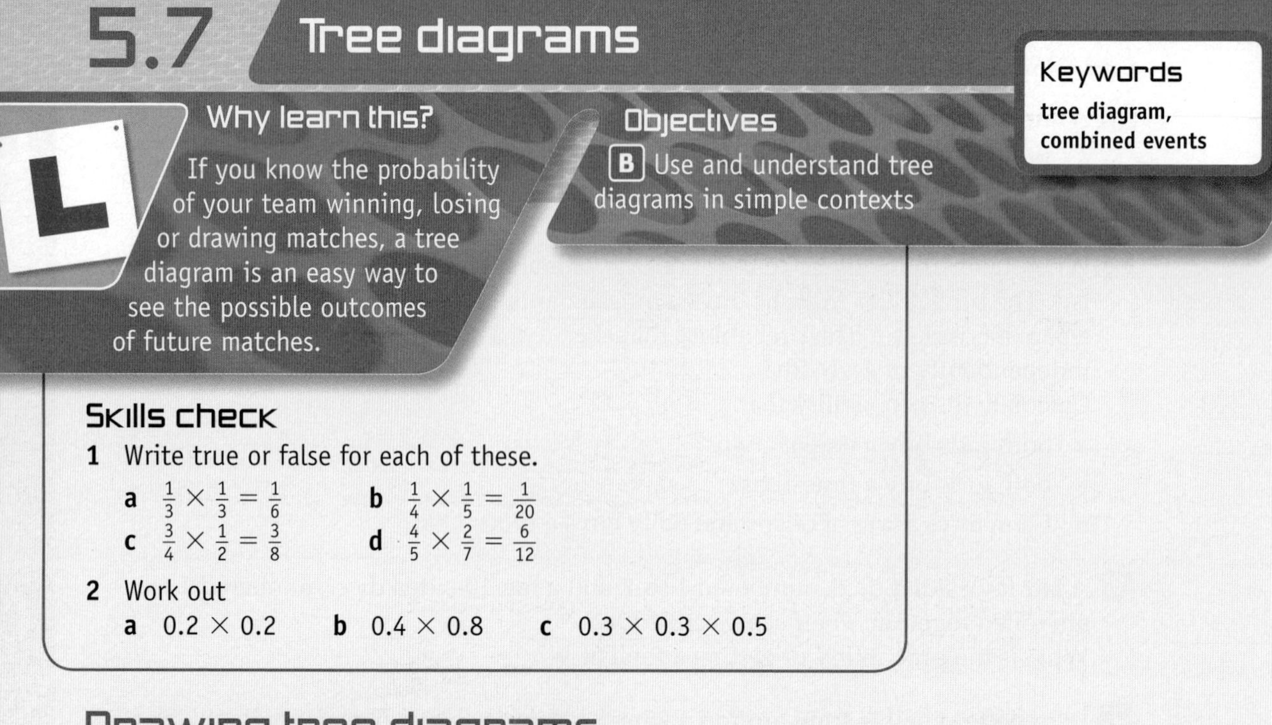

5.7 Tree diagrams

Keywords
tree diagram, combined events

Why learn this?

If you know the probability of your team winning, losing or drawing matches, a tree diagram is an easy way to see the possible outcomes of future matches.

Objectives

B Use and understand tree diagrams in simple contexts

Skills check

1 Write true or false for each of these.

a $\frac{1}{3} \times \frac{1}{3} = \frac{1}{6}$ **b** $\frac{1}{4} \times \frac{1}{5} = \frac{1}{20}$

c $\frac{3}{4} \times \frac{1}{2} = \frac{3}{8}$ **d** $\frac{4}{5} \times \frac{2}{7} = \frac{6}{12}$

2 Work out

a 0.2×0.2 **b** 0.4×0.8 **c** $0.3 \times 0.3 \times 0.5$

Drawing tree diagrams

A **tree diagram** can show all the possible outcomes of two or more **combined events**, and their probabilities.

Example 7

Imagine you have a bag of discs, some are red and some are blue.
You pick out a disc, record its colour then replace it in the bag.
Then you pick out a second disc.

You can show all the possible results by drawing a tree diagram.

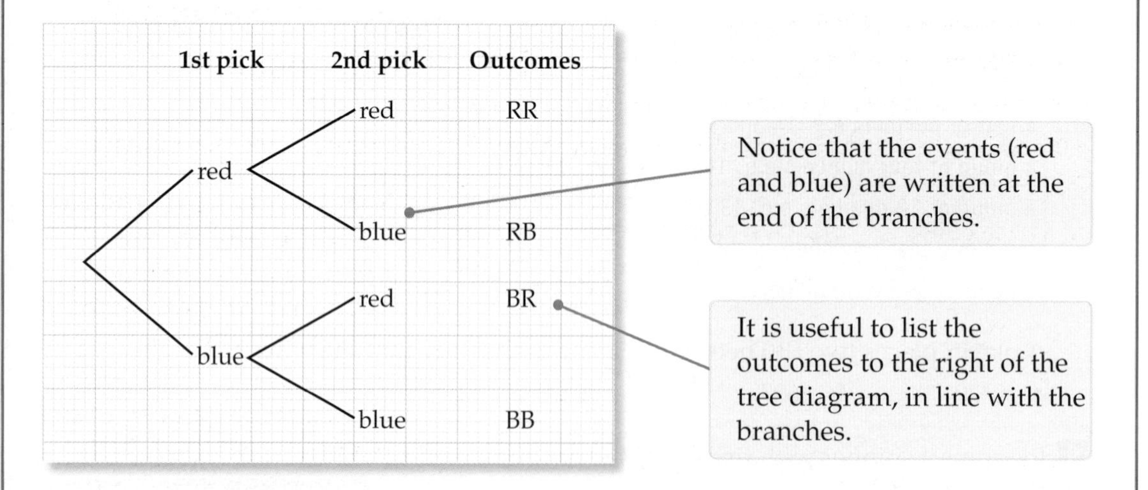

There is only one way of obtaining the outcomes 'red, red' or 'blue, blue' (usually written 'both red' or 'both blue').

There are two ways of obtaining the outcome 'one of each colour' because either 'red, blue' or 'blue, red' fit this description.

B

Exercise 5G

Draw a tree diagram to show all the possible outcomes in each question.
For each one, think about the possible outcomes if you carried out the experiment.
Label the branches with the appropriate outcomes.

1 A card is selected at random from a normal pack of 52 cards.
Its colour is recorded then it is replaced.
A second card is selected.

2 A 2p coin is flipped then a 10p coin is flipped.

3 On a certain day Tom takes his driving test. He either passes or fails.
The next day his friend Marcus takes his driving test. He either passes or fails.

4 A bag contains red, blue and yellow discs.
Tiffany picks out a disc, records its colour then replaces it.
She then picks out a second disc.

5 For breakfast today I have a choice of cereal, toast or porridge.
Tomorrow I can choose between the same three items.

6 A box contains black and white balls.
Anisha picks out a ball, records its colour then replaces it.
She picks out a second ball, records its colour then replaces it.
She picks out a third ball.

Calculations using tree diagrams

There are always 5 discs in the bag, 3 red and 2 blue.

Look again at Example 7.

The bag contained red and blue discs.

Suppose there are 3 red discs and 2 blue discs.

The probability of picking a red on the first pick is $\frac{3}{5}$ and picking a blue is $\frac{2}{5}$.

Since the discs are replaced, these probabilities always remain the same.

You can write the probabilities on the appropriate branches, so the completed tree diagram looks like this.

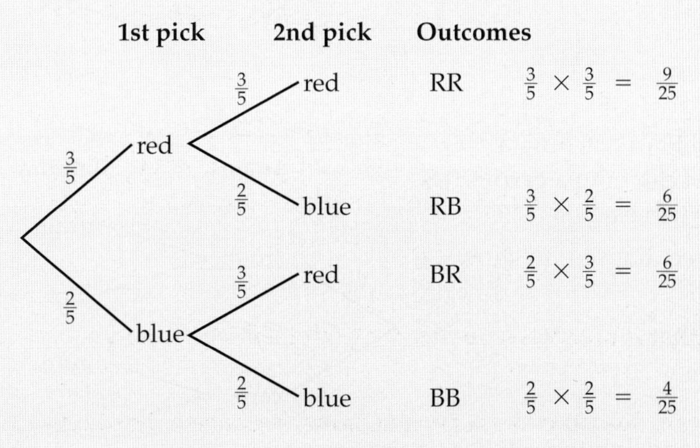

The events are independent so you can multiply the probabilities on the branches to calculate the probability of each final outcome.

The sum of the probabilities of the final outcomes is $\frac{9}{25} + \frac{6}{25} + \frac{6}{25} + \frac{4}{25} = \frac{25}{25} = 1$

This is always the case since the final outcomes represent everything that can possibly happen.

B

Example 8

Use the tree diagram above to answer these questions.

a What is the probability that the two discs are the same colour?

b What is the probability that the two discs are different colours?

c What is the probability of picking at least one red disc?

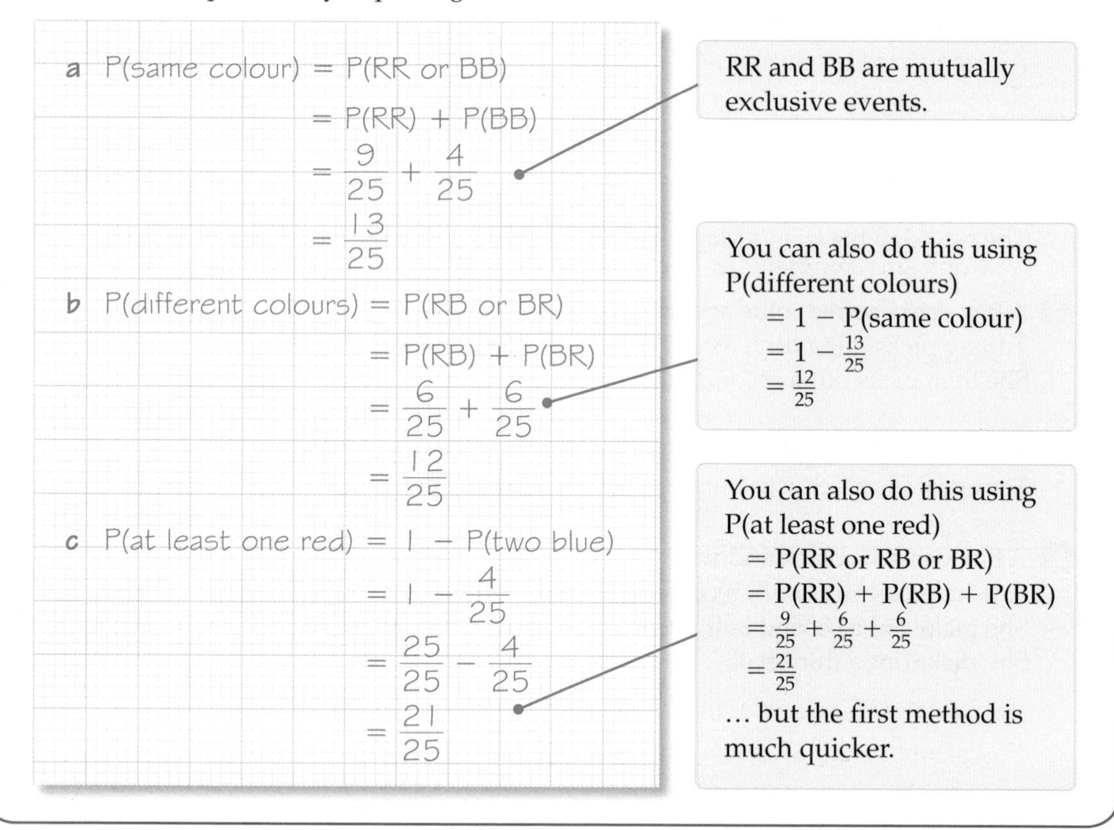

Exercise 5H

B

1 **a** David flips a coin twice.
Copy and complete the tree diagram to show all the possible outcomes and their probabilities.

b Work out the probability that David flips

i two heads

ii a head and then a tail

iii a tail and then a head

iv two tails.

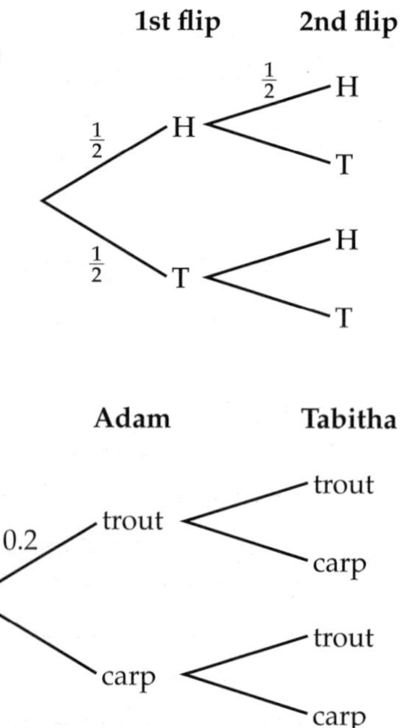

2 At a fishing lake the probability of catching a trout is 0.2. The probability of catching a carp is 0.8. Adam and Tabitha catch one fish each.

a Copy and complete the tree diagram to show all the possible outcomes.

b Work out the probability that

i they both catch a trout

ii they both catch a carp

iii Adam catches a trout and Tabitha catches a carp

iv one catches a trout and the other a carp.

3 Zina has a pack of cards. She shuffles the pack then takes a card at random. She replaces the card, shuffles the pack, then takes another card.

a What is the probability that the first card Zina takes is a Jack?

b What is the probability that the first card Zina takes isn't a Jack?

c Work out the probability that

i both cards are a Jack

ii neither card is a Jack

iii the first card isn't a Jack, but the second card is.

B

4 Aleksy has a 5p coin, a 10p coin and a 50p coin. He spins the coins at the same time and records whether they land heads up or tails up.

a Copy and complete the tree diagram to show all the possible outcomes.

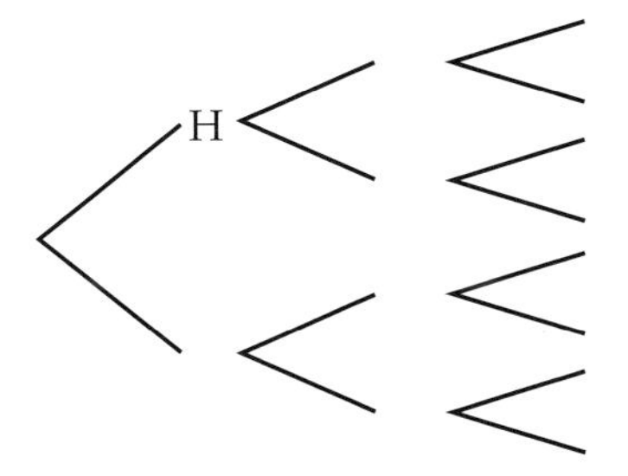

b Calculate the probability that the 5p and the 10p coins land heads up and the 50p coin lands tails up.

c Calculate the probability that all coins land heads up.

d Calculate the probability that one coin lands heads up and two coins land tails up.

e Calculate the probability that at least one coin lands tails up.

B

5 Siobhan has two maths tests next week. She estimates the probability of her passing the geometry test is 0.8, but the probability of her passing the statistics test is only 0.2.

pass

a Copy and complete the tree diagram to show all the possible outcomes.

b Work out the probability that Siobhan only passes the geometry test.

c Work out the probability that Siobhan passes neither test.

d Work out the probability that Siobhan passes only one test.

6 On Errol's way to work there are two sets of traffic lights.
The probability that the first set is green is $\frac{1}{4}$.
The probability that the second set is green is $\frac{1}{3}$.

green

a Copy and complete the tree diagram to show all possible outcomes.

b Work out the probability that Errol is stopped at both sets of lights.

c Work out the probability that he is stopped at exactly one set of lights.

d Work out the probability that he is not stopped at either set of lights.

AO2

B

7 A fair dice is rolled twice. You are only interested in whether a 6 is rolled.

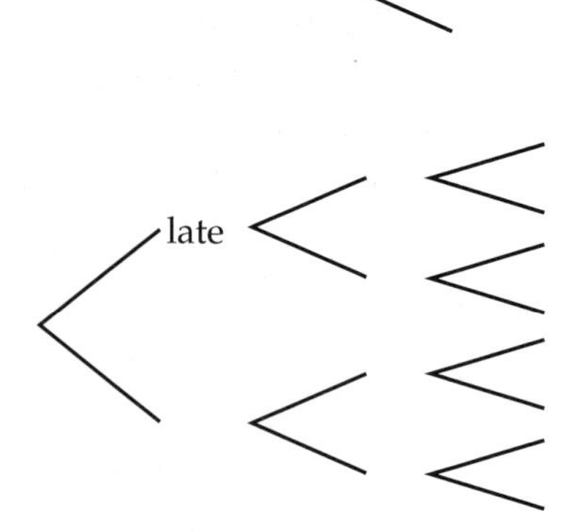

a Copy and complete the tree diagram to show all possible outcomes.

b Work out the probability of getting

i two 6s **ii** no 6s

iii a 6 followed by not a 6.

8 Luke, Matthew and Sophie all work for the same company.
The probability that Luke is late for work is 0.3.
The probability that Matthew is late for work is 0.4.
The probability that Sophie is late for work is 0.2.

a Copy and complete the tree diagram to show all possible outcomes.

b On any day, what is the probability that

i all three are late for work

ii all three are on time

iii Sophie is late but Luke and Matthew are on time

iv at least one of them is late?

9 Tabitha, Matilda and Oscar are taking an English exam.
The probability that Tabitha passes is 0.9.
The probability that Matilda passes is 0.7.
The probability that Oscar passes is 0.6.

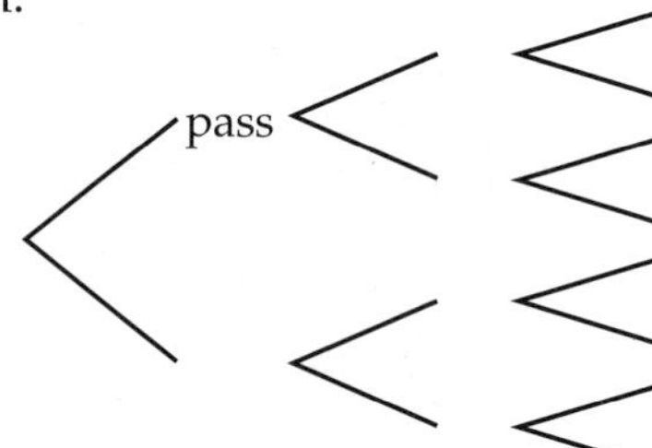

a Copy and complete the tree diagram to show all possible outcomes.

b What is the probability that

i all three pass

ii Tabitha and Oscar pass but Matilda fails

iii all three fail

iv at least one of them passes

v any two of them pass?

A02

Review exercise

E

1 Colleen has these two four-sided spinners.
She spins the spinners at the same time.
She works out the difference between the numbers that the spinners land on.

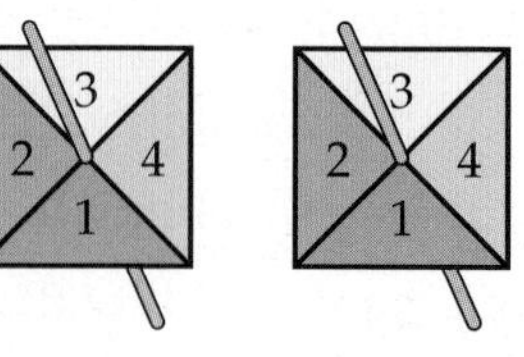

a Copy and complete the sample space diagram to show all the possible differences.

	1	2	3	4
1	0	1	2	
2	1			
3	2			
4				

[1 mark]

b What is the probability that the difference is 0? **[1 mark]**

c What is the probability that the difference is an odd number? **[1 mark]**

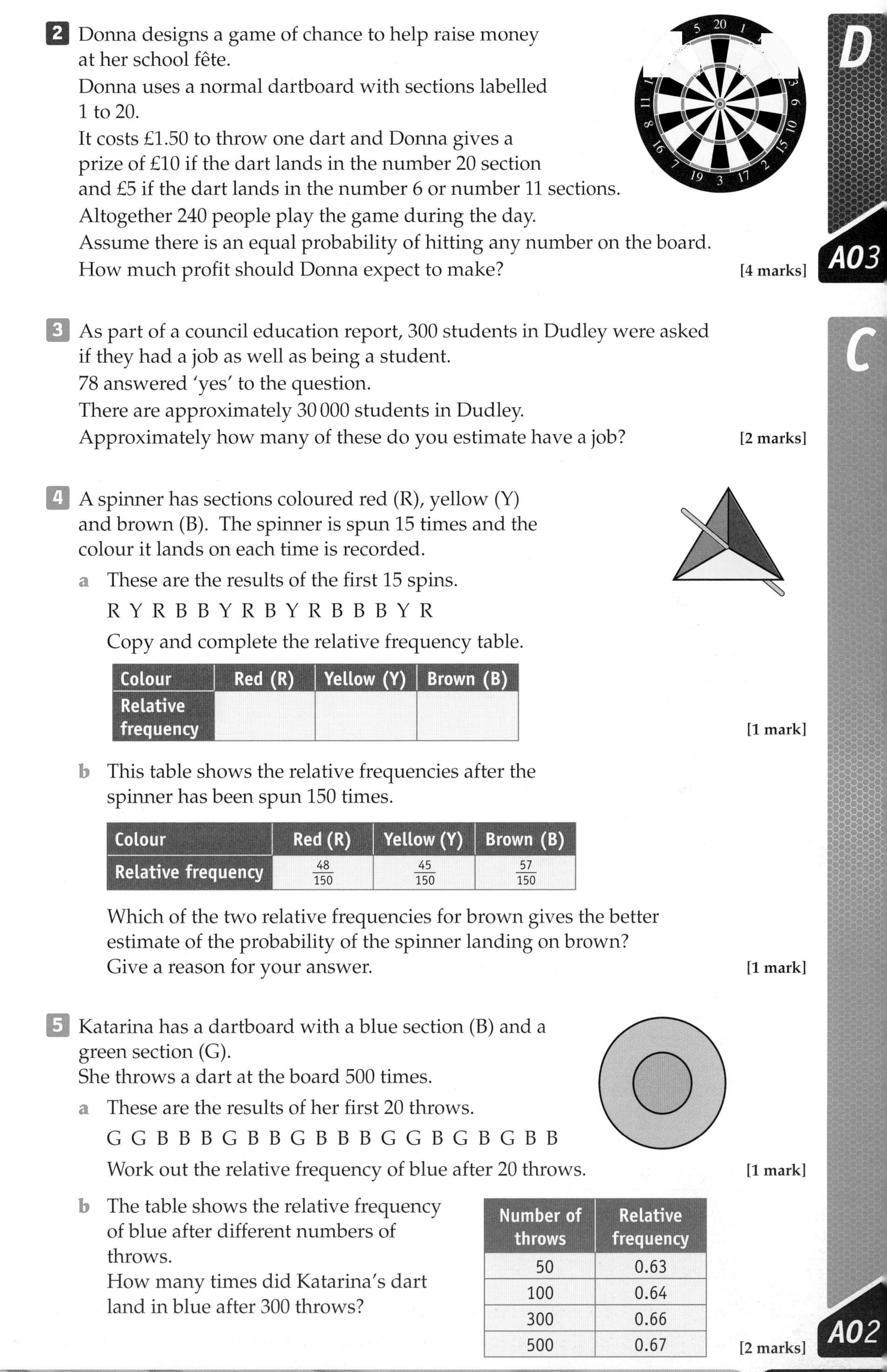

2 Donna designs a game of chance to help raise money at her school fête.
Donna uses a normal dartboard with sections labelled 1 to 20.
It costs £1.50 to throw one dart and Donna gives a prize of £10 if the dart lands in the number 20 section and £5 if the dart lands in the number 6 or number 11 sections.
Altogether 240 people play the game during the day.
Assume there is an equal probability of hitting any number on the board.
How much profit should Donna expect to make? **[4 marks]**

D

AO3

3 As part of a council education report, 300 students in Dudley were asked if they had a job as well as being a student.
78 answered 'yes' to the question.
There are approximately 30 000 students in Dudley.
Approximately how many of these do you estimate have a job? **[2 marks]**

C

4 A spinner has sections coloured red (R), yellow (Y) and brown (B). The spinner is spun 15 times and the colour it lands on each time is recorded.

a These are the results of the first 15 spins.

R Y R B B Y R B Y R B B B Y R

Copy and complete the relative frequency table.

Colour	Red (R)	Yellow (Y)	Brown (B)
Relative frequency			

[1 mark]

b This table shows the relative frequencies after the spinner has been spun 150 times.

Colour	Red (R)	Yellow (Y)	Brown (B)
Relative frequency	$\frac{48}{150}$	$\frac{45}{150}$	$\frac{57}{150}$

Which of the two relative frequencies for brown gives the better estimate of the probability of the spinner landing on brown?
Give a reason for your answer. **[1 mark]**

5 Katarina has a dartboard with a blue section (B) and a green section (G).
She throws a dart at the board 500 times.

a These are the results of her first 20 throws.

G G B B B G B B G B B B G G B G B G B B

Work out the relative frequency of blue after 20 throws. **[1 mark]**

b The table shows the relative frequency of blue after different numbers of throws.
How many times did Katarina's dart land in blue after 300 throws?

Number of throws	Relative frequency
50	0.63
100	0.64
300	0.66
500	0.67

[2 marks]

AO2

B

6 A catering company produces hot meals for parties.
It offers three main courses: chicken (C), beef (B) or vegetarian (V).
It offers two types of potatoes: roast (R) or new (N).
The company uses previous data to estimate the numbers of different types of meals that need to be cooked.
Previous data shows that the probability of a person choosing chicken is $\frac{1}{2}$, beef $\frac{1}{3}$ and vegetarian $\frac{1}{6}$.
The probability of a person choosing roast potatoes is $\frac{1}{4}$ and new potatoes is $\frac{3}{4}$.

a Copy and complete the tree diagram to show all the possible outcomes. **[2 marks]**

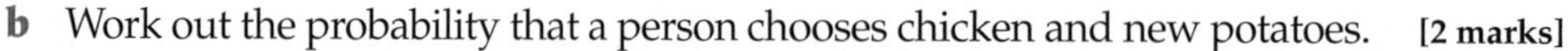

b Work out the probability that a person chooses chicken and new potatoes. **[2 marks]**

c Work out the probability that a person chooses meat and roast potatoes. **[3 marks]**

d At the next party, 120 guests are expected.
Estimate the number of 'vegetarian and new potato' meals the company needs to cook. **[2 marks]**

AO2

7 A fair six-sided dice, numbered from 1 to 6, is thrown three times.
What is the probability of getting

a three 6s

b two 6s

c at least two 6s

d no 6s

e at least one 6?

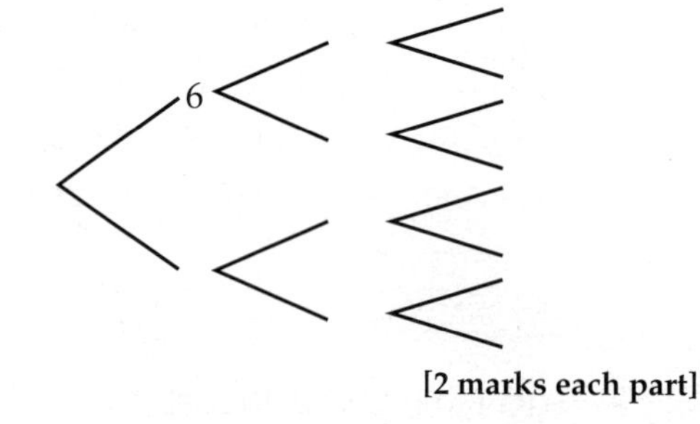

[2 marks each part]

Chapter summary

In this chapter you have learned how to

- work out the probability of an event **not** happening when you know the probability that it does happen **E**
- understand and use two-way tables **E** **D**
- understand and use the fact that the sum of the probabilities of all mutually exclusive outcomes is 1 **D**
- predict the likely numbers of successful events given the probability of any outcome and the number of trials or experiments **D**
- estimate probabilities from experimental data **C**
- calculate the probability of two independent events happening at the same time **C**
- use and understand tree diagrams in simple contexts **B**

6 Cumulative frequency

This chapter is about analysing information.

These runners are crossing the Reichsbruecke bridge in the Vienna City Marathon. Analysing data of the runners' times will show what percentage took between 2 hours 30 minutes and 3 hours 30 minutes.

Objectives

This chapter will show you how to

- compile a cumulative frequency table for continuous (grouped) data B
- draw cumulative frequency diagrams to find the median and quartiles of grouped data B
- use cumulative frequency diagrams to analyse other features of grouped data B
- draw box plots from a cumulative frequency diagram B
- use cumulative frequency diagrams and box plots to compare data and draw conclusions B

Before you start this chapter

1 a What must you do before you can find the median of these numbers?

8 31 4 20 49 12 6 34 5 16 22 9 43

b What is the median?

c The number 49 is removed. What happens to the median?

2 Write down six different numbers with a median of 7.5 and a range of 7.

3 This frequency table shows the scores of a group of students who took a mental maths test.

Test score	2	3	4	5	6	7	8	9	10
Frequency	4	2	3	2	5	8	6	5	1

a What is the range of these marks?

b Work out the median test score.

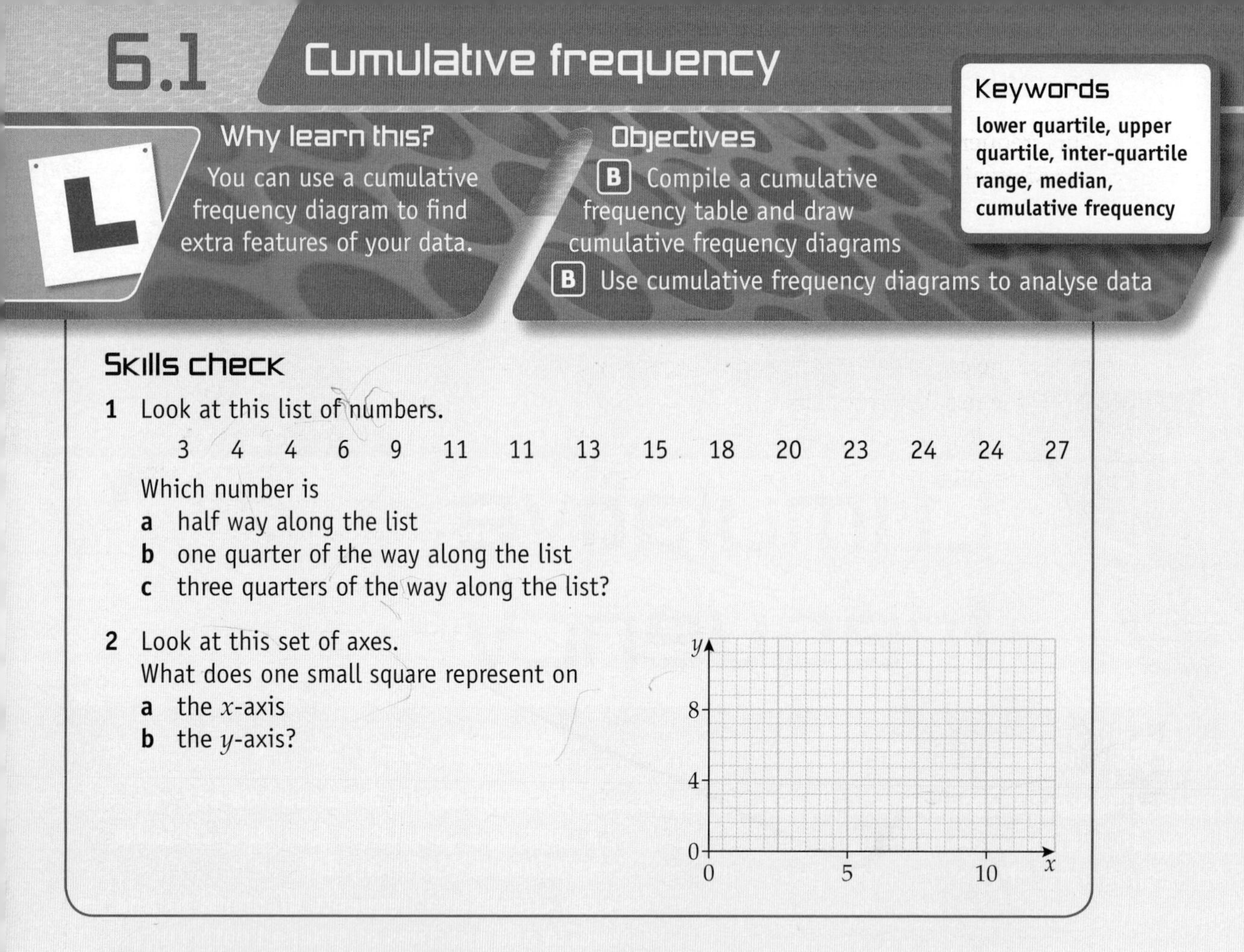

6.1 Cumulative frequency

Why learn this?
You can use a cumulative frequency diagram to find extra features of your data.

Objectives

B Compile a cumulative frequency table and draw cumulative frequency diagrams

B Use cumulative frequency diagrams to analyse data

Keywords
lower quartile, upper quartile, inter-quartile range, median, cumulative frequency

Skills check

1 Look at this list of numbers.

3 4 4 6 9 11 11 13 15 18 20 23 24 24 27

Which number is

a half way along the list
b one quarter of the way along the list
c three quarters of the way along the list?

2 Look at this set of axes.
What does one small square represent on

a the x-axis
b the y-axis?

Drawing and using a cumulative frequency diagram for grouped data

The **lower quartile** is the data item one quarter of the way along the list when the data is in ascending order.

The **upper quartile** is the data item three quarters of the way along the list when the data is in ascending order.

The **inter-quartile range** is the difference between the values of the upper and lower quartiles of the data.

Inter-quartile range = upper quartile − lower quartile

It is a single value and is often used instead of the range to indicate the spread of the data.

When data is grouped into class intervals you can work out which class interval contains the **median** or quartiles, but you cannot give the exact values. Continuous data is always grouped into class intervals in a frequency table (see Section 1.5).

Cumulative frequency is a running total of the frequencies.

You can use a cumulative frequency diagram to estimate the median, the lower quartile and the upper quartile of the data. They can only be estimates because when data is grouped some of the detail of the original data is lost.

In cumulative frequency questions there is usually a large amount of data.
If the number of data items, n, is large ($n \geqslant 25$) you can use these formulae for locating the positions of the median and quartiles.

Median = $\frac{n}{2}$th value Lower quartile = $\frac{n}{4}$th value Upper quartile = $\frac{3n}{4}$th value

Example 1

This frequency table shows the times taken by some students to work out the answer to a mathematical puzzle.

Time taken, t (minutes)	$0 < t \leqslant 1$	$1 < t \leqslant 2$	$2 < t \leqslant 3$	$3 < t \leqslant 4$	$4 < t \leqslant 5$	$5 < t \leqslant 6$	$6 < t \leqslant 7$
Frequency	3	4	7	11	9	4	2

a Draw a cumulative frequency diagram to illustrate this data.

b Use the cumulative frequency diagram to estimate

i the median time taken

ii the lower and upper quartiles

iii the inter-quartile range.

First create a cumulative frequency table. Add an extra row or column to the frequency table to calculate a running total of the frequencies (cumulative frequency).

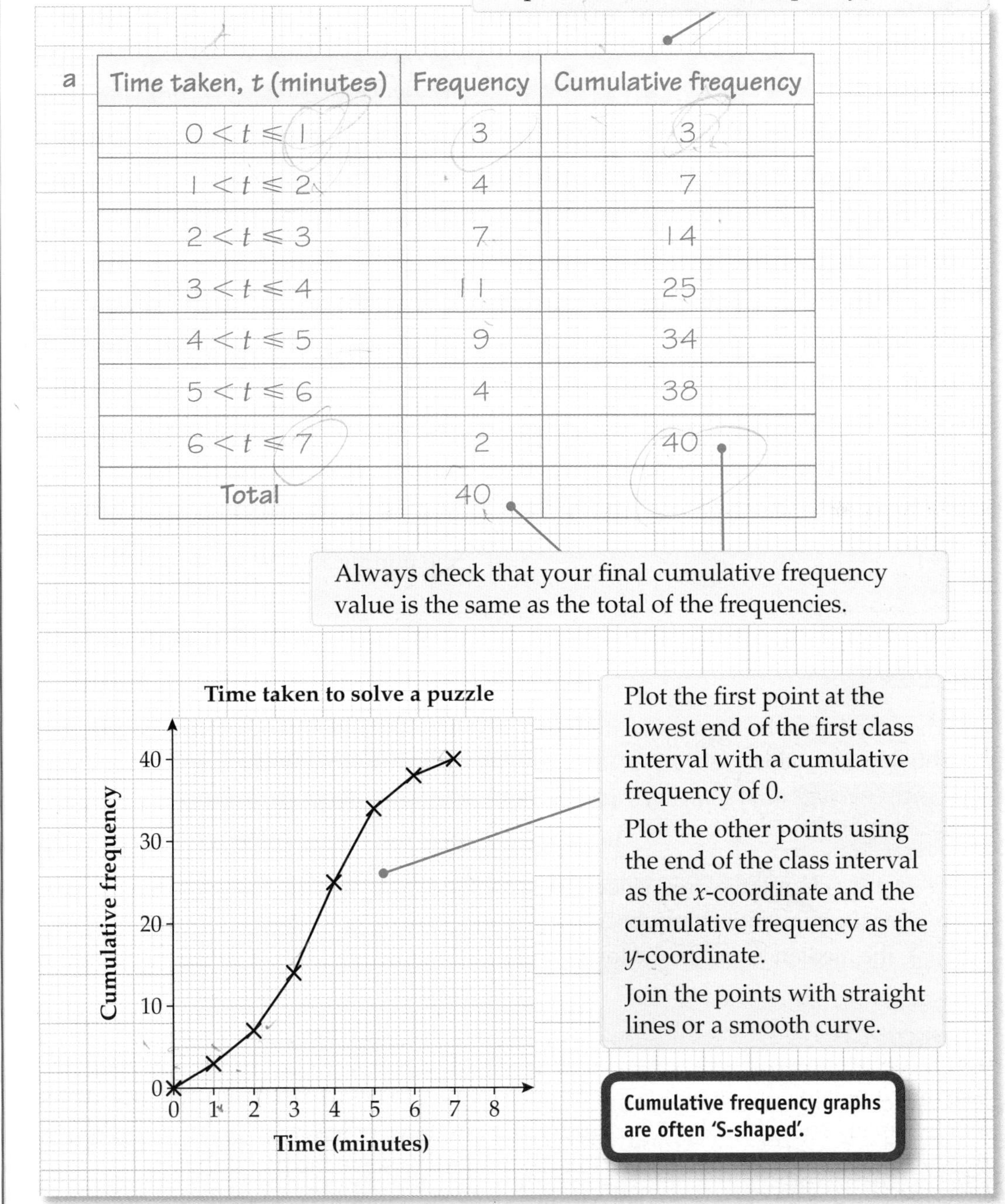

a

Time taken, t (minutes)	Frequency	Cumulative frequency
$0 < t \leqslant 1$	3	3
$1 < t \leqslant 2$	4	7
$2 < t \leqslant 3$	7	14
$3 < t \leqslant 4$	11	25
$4 < t \leqslant 5$	9	34
$5 < t \leqslant 6$	4	38
$6 < t \leqslant 7$	2	40
Total	40	

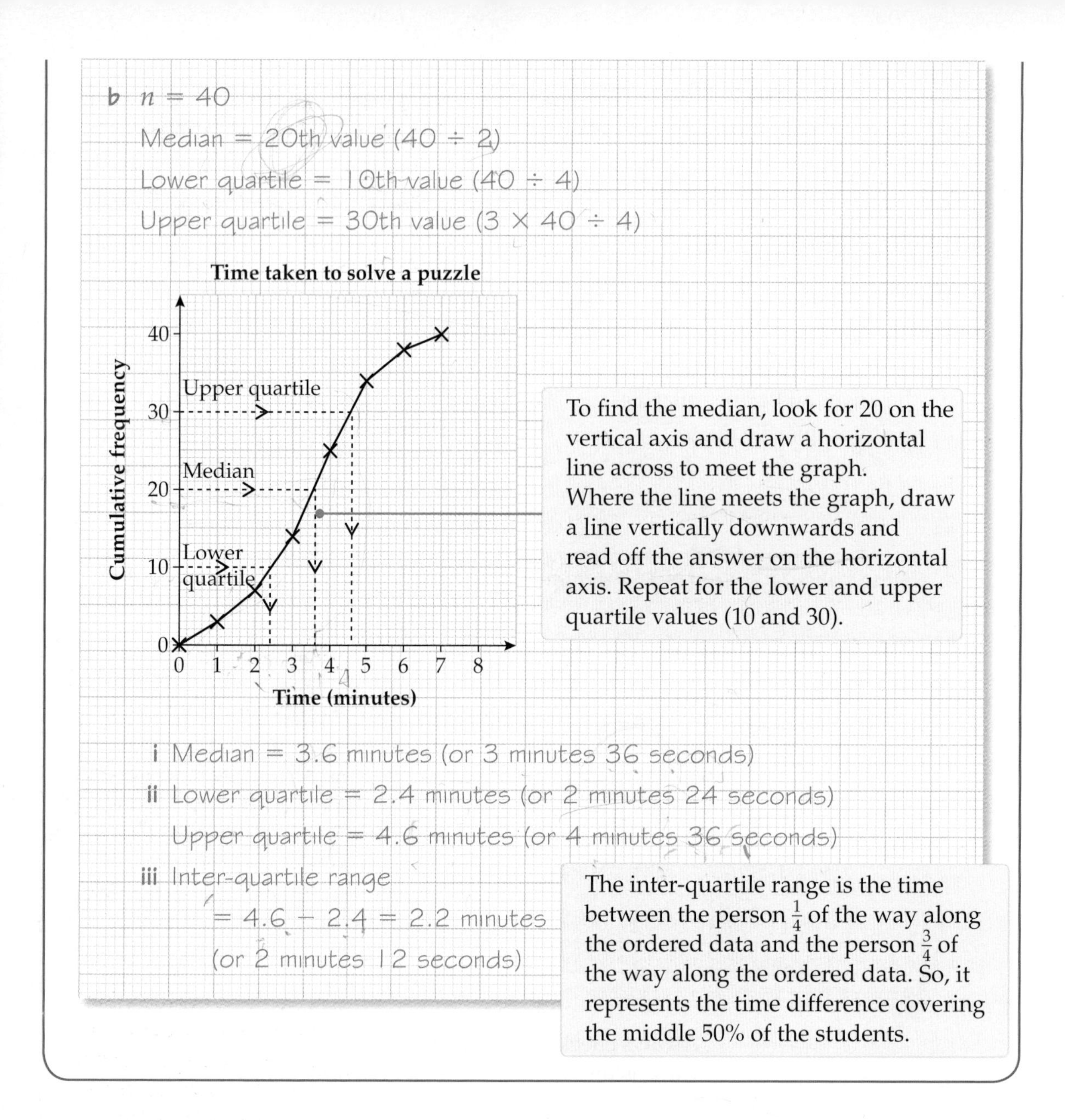

Exercise 6A

B

1 The times spent by a group of students using their mobile phones in one day are shown in the table.

Time, t (minutes)	$0 < t \leq 10$	$10 < t \leq 20$	$20 < t \leq 30$	$30 < t \leq 40$	$40 < t \leq 50$	$50 < t \leq 60$
Frequency	6	10	21	46	11	6

a Draw a cumulative frequency diagram to illustrate this data.

b Use the cumulative frequency diagram to estimate

i the median time **ii** the lower and upper quartiles **iii** the inter-quartile range.

B

Example 2

A02

Use the cumulative frequency graph from Example 1 to answer the following questions.

a How many students solved the puzzle in 2.4 minutes or less?

b How many students took longer than $5\frac{1}{2}$ minutes to solve the puzzle?

c What percentage of students solved the puzzle in 1 minute 30 seconds or less?

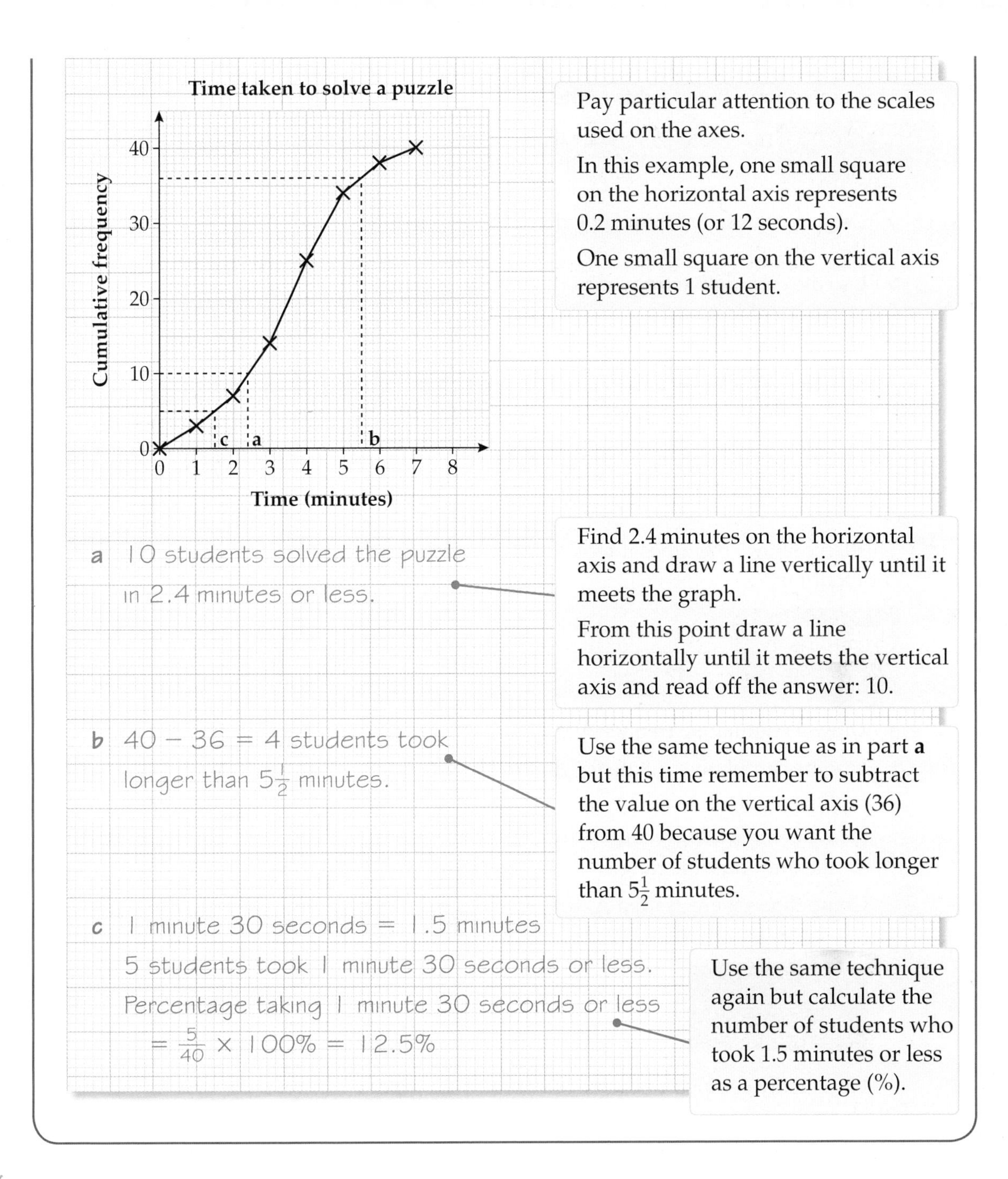

Exercise 6B

1 This table shows the times taken by some students to walk to school in the morning.

Time, t (minutes)	$0 < t \leqslant 5$	$5 < t \leqslant 10$	$10 < t \leqslant 15$	$15 < t \leqslant 20$	$20 < t \leqslant 25$	$25 < t \leqslant 30$
Frequency	8	8	31	2	7	4

a Draw a cumulative frequency diagram to illustrate this data.

b Use the cumulative frequency diagram to estimate

i the median time taken

ii the lower and upper quartiles

iii the inter-quartile range.

c How many students took 2 minutes or less to walk to school?

d How many students took between 13 and 22 minutes to walk to school?

e What percentage of students took over 24 minutes to walk to school?

B

AO2

B

2 Grace took some letters to the Post Office.
This table shows the weights of the letters.

Weight, g (grams)	$20 < g \leq 30$	$30 < g \leq 40$	$40 < g \leq 50$	$50 < g \leq 60$	$60 < g \leq 70$	$70 < g \leq 80$	$80 < g \leq 90$
Frequency	8	14	17	8	6	4	3

a Draw a cumulative frequency diagram to illustrate this data.

b Use the cumulative frequency diagram to estimate
  i the median weight
  ii the lower and upper quartiles
  iii the inter-quartile range.

c How many letters weighed more than 75 grams?

d How many letters weighed between 62 and 84 grams?

e What percentage of letters weighed 41 grams or less?

> **You can start the horizontal axis at 20 but the vertical axis must start at zero.**

3 This table shows the times it took competitors to complete a crossword in a competition.

Time, t (minutes)	$10 < t \leq 15$	$15 < t \leq 20$	$20 < t \leq 25$	$25 < t \leq 30$	$30 < t \leq 35$	$35 < t \leq 40$
Frequency	12	15	27	50	13	3

a Draw a cumulative frequency diagram to illustrate this data.

b Use the cumulative frequency diagram to estimate
  i the median time taken
  ii the lower and upper quartiles
  iii the inter-quartile range.

c Competitors were graded A, B, C or D according to the time they took to complete the crossword.

Grade	A	B	C	D
Time, t (minutes)	$t \leq 13$	$13 < t \leq 22$	$22 < t \leq 33$	$33 < t$

How many competitors achieved each of the grades?

4 In France, speed limits are given in kilometres per hour.
This table shows the speeds of 140 motorists on a French toll motorway.

Speed, v (km/h)	$65 < v \leq 75$	$75 < v \leq 85$	$85 < v \leq 95$	$95 < v \leq 105$
Frequency	14	18	10	16

Speed, v (km/h)	$105 < v \leq 115$	$115 < v \leq 125$	$125 < v \leq 135$	$135 < v \leq 145$
Frequency	38	26	12	6

a Draw a cumulative frequency diagram to illustrate this data.

b Use the cumulative frequency diagram to estimate
  i the median speed
  ii the lower and upper quartiles
  iii the inter-quartile range.

A02

c The speed limit on French toll motorways is 130 km/h in dry weather and 110 km/h in wet weather.

i Assume the weather is dry. Use your cumulative frequency diagram to find out how many drivers are breaking the speed limit.

ii How many more drivers would be breaking the speed limit if this diagram represented data recorded in wet weather?

d To convert miles per hour to kilometres per hour you multiply by 1.6.
In Britain, the speed limit on motorways is 70 mph.
How many of the French drivers were travelling at a speed greater than the speed limit on British motorways?

B

A02

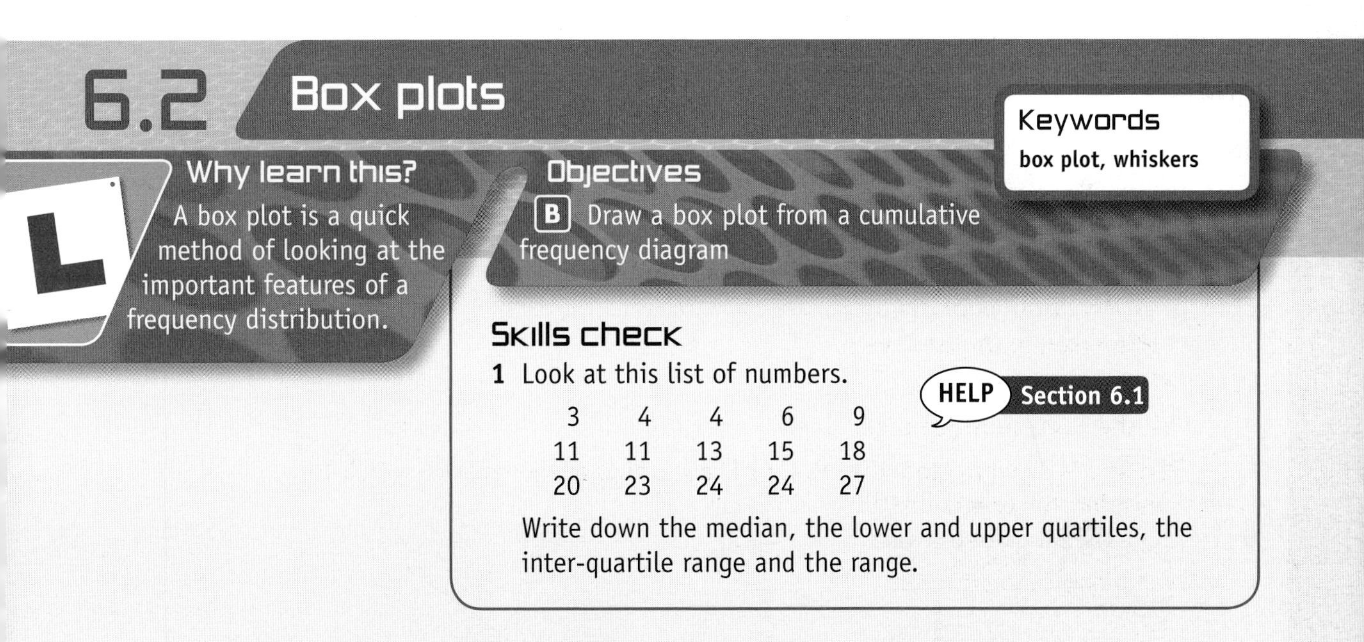

6.2 Box plots

Keywords
box plot, whiskers

Why learn this?

A box plot is a quick method of looking at the important features of a frequency distribution.

Objectives

B Draw a box plot from a cumulative frequency diagram

Skills check

1 Look at this list of numbers.

3	4	4	6	9
11	11	13	15	18
20	23	24	24	27

Write down the median, the lower and upper quartiles, the inter-quartile range and the range.

HELP Section 6.1

Drawing a box plot from a cumulative frequency diagram

A **box plot** (sometimes called a box-and-whisker diagram) shows the range, the lower and upper quartiles (and hence the inter-quartile range) and the median of a set of data.

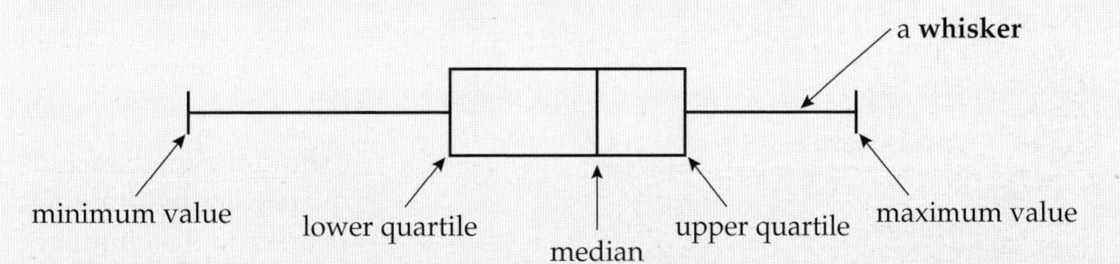

Box plots are usually drawn on a scale. Use the same scale as the cumulative frequency diagram from which the data comes.

A box plot has advantages and disadvantages. It is easy to read and provides information at a glance, but it gives no details about the number of data items in the distribution.

B

Example 3

Draw a box plot to illustrate the data from Examples 1 and 2.

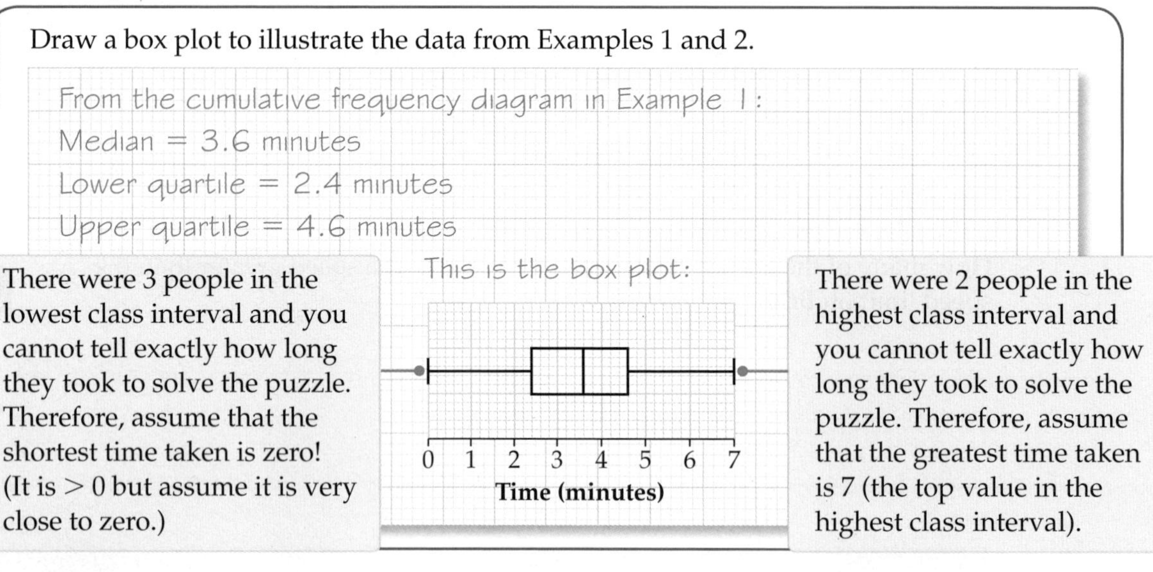

B

Exercise 6C

1 **a** to **e** Draw a box plot for each of the five questions in Exercises 6A and 6B. Remember to clearly label the scale on the x-axis.

B

Example 4

This cumulative frequency diagram shows the heights, in centimetres, of 60 plants.

a Find

i the median

ii the lower and upper quartiles.

b Estimate the range of the data.

c Draw a box plot to represent this data.

d Estimate the range of heights for the tallest 25% of plants.

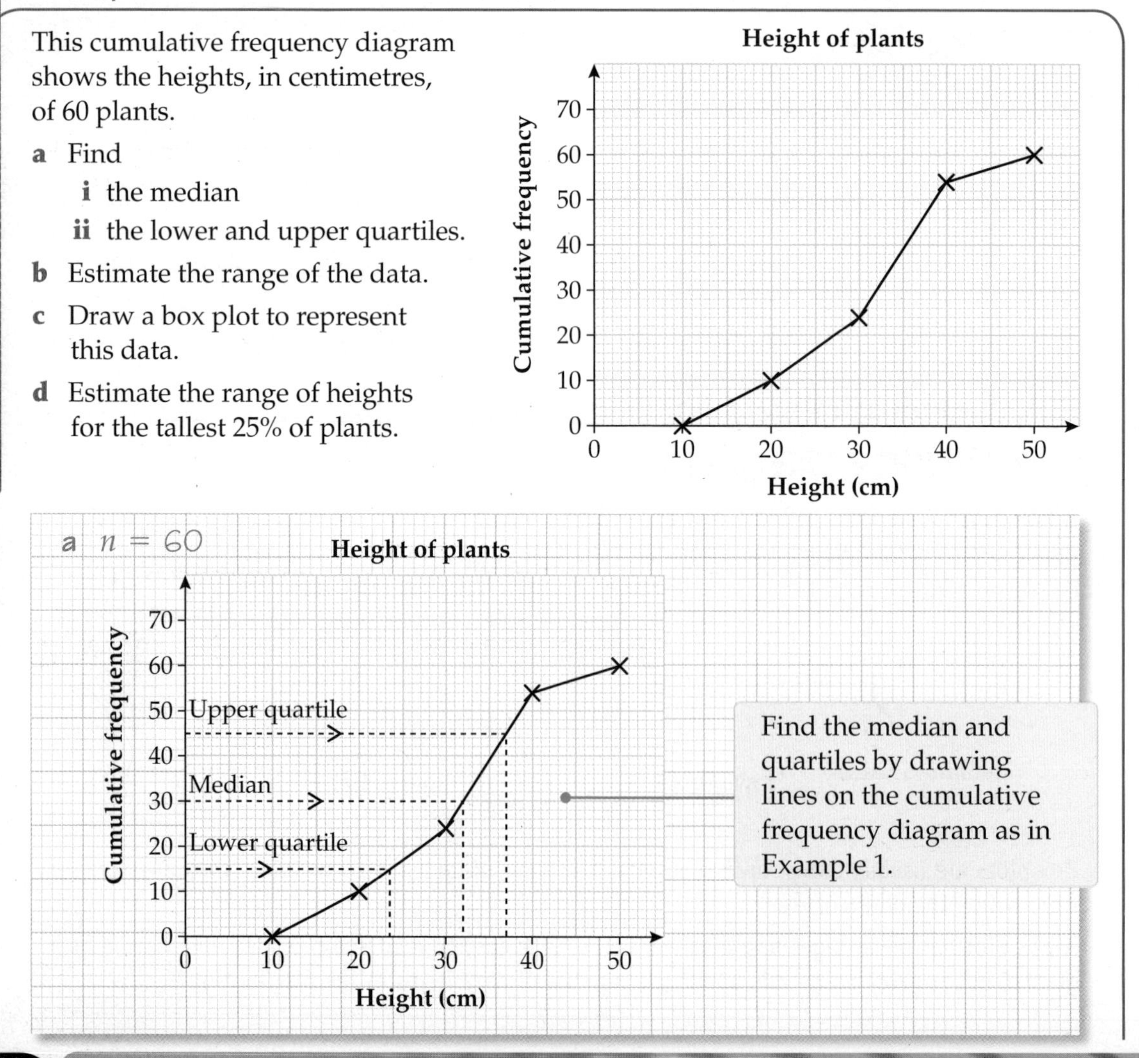

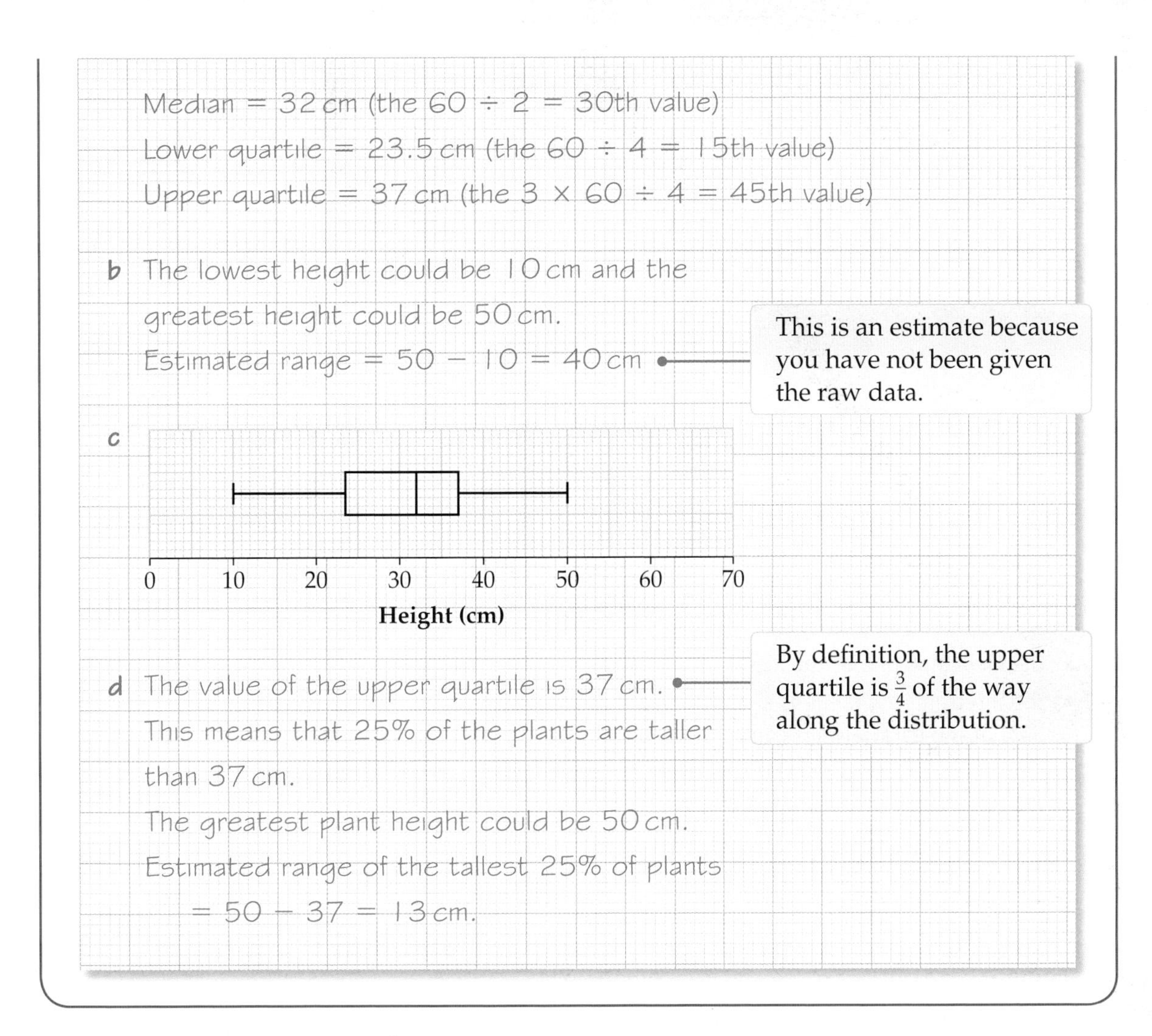

Median = 32 cm (the 60 ÷ 2 = 30th value)

Lower quartile = 23.5 cm (the 60 ÷ 4 = 15th value)

Upper quartile = 37 cm (the 3 × 60 ÷ 4 = 45th value)

b The lowest height could be 10 cm and the greatest height could be 50 cm.

Estimated range = 50 − 10 = 40 cm

This is an estimate because you have not been given the raw data.

c

d The value of the upper quartile is 37 cm.

This means that 25% of the plants are taller than 37 cm.

By definition, the upper quartile is $\frac{3}{4}$ of the way along the distribution.

The greatest plant height could be 50 cm.

Estimated range of the tallest 25% of plants
= 50 − 37 = 13 cm.

Exercise 6D

1 For each of these box plots, work out

i the median

ii the inter-quartile range

iii the range.

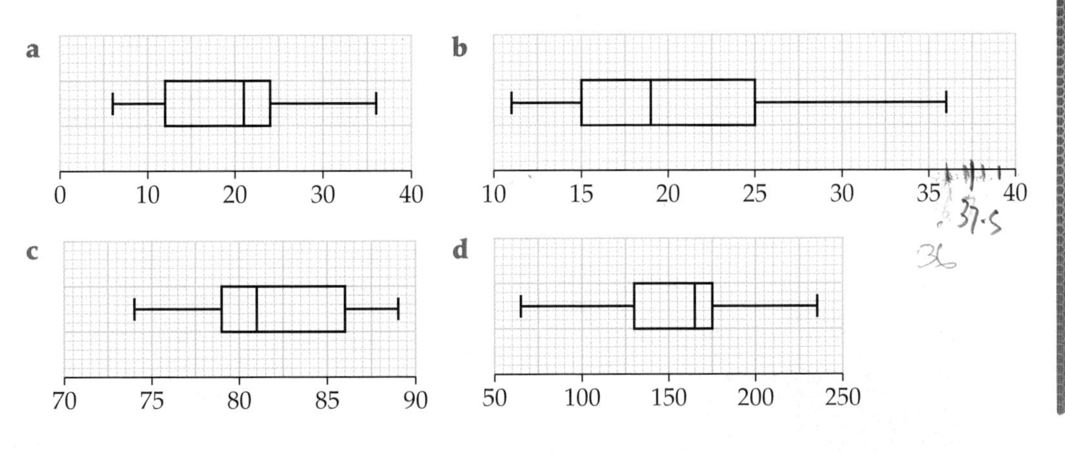

6.3 Comparing data sets and drawing conclusions

Why learn this?

This topic often comes up in the exam.

Objectives

B Use cumulative frequency diagrams and box plots to compare data and draw conclusions

Skills check

1 Look at the two lists of numbers.

HELP Section 6.1

A	3	7	9	13	15	20	21
B	7	8	11	12	13	17	20

a Which list has the higher median?
b Which list has the larger inter-quartile range?

Comparing data sets and drawing conclusions

When comparing frequency distributions remember that

- the median is one of the measures of 'average'. Always look to compare the medians.
- the inter-quartile range is a measure of consistency and tells you the spread of the middle 50% of the data. Always comment on the size of the inter-quartile range.

B

Example 5

Eighty of each of two types of battery, Superamp and Powerplus, were tested.

The minimum life of each type was 16 hours.
The maximum life of Superamp was 36 hours.
The maximum life of Powerplus was 40 hours.

This frequency table shows the test results for each type of battery.

Battery life, h (hours)	Superamp	Powerplus
$16 < h \leqslant 20$	7	14
$20 < h \leqslant 24$	11	16
$24 < h \leqslant 28$	16	20
$28 < h \leqslant 32$	40	16
$32 < h \leqslant 36$	6	8
$36 < h \leqslant 40$	0	6
Total	80	80

a Draw a cumulative frequency diagram for each type of battery.
b Estimate the median and the inter-quartile range for each type of battery.
c Draw a box plot for each type of battery.
d Which type of battery is the better buy? Explain your answer.

A03

a

Battery life, h (hours)	Superamp frequency	Superamp cum. freq.	Powerplus frequency	Powerplus cum. freq.
$16 < h \leqslant 20$	7	7	14	14
$20 < h \leqslant 24$	11	18	16	30
$24 < h \leqslant 28$	16	34	20	50
$28 < h \leqslant 32$	40	74	16	66
$32 < h \leqslant 36$	6	80	8	74
$36 < h \leqslant 40$	0	80	6	80
Total	80		80	

Both sets of cumulative frequencies can be written in the same table.

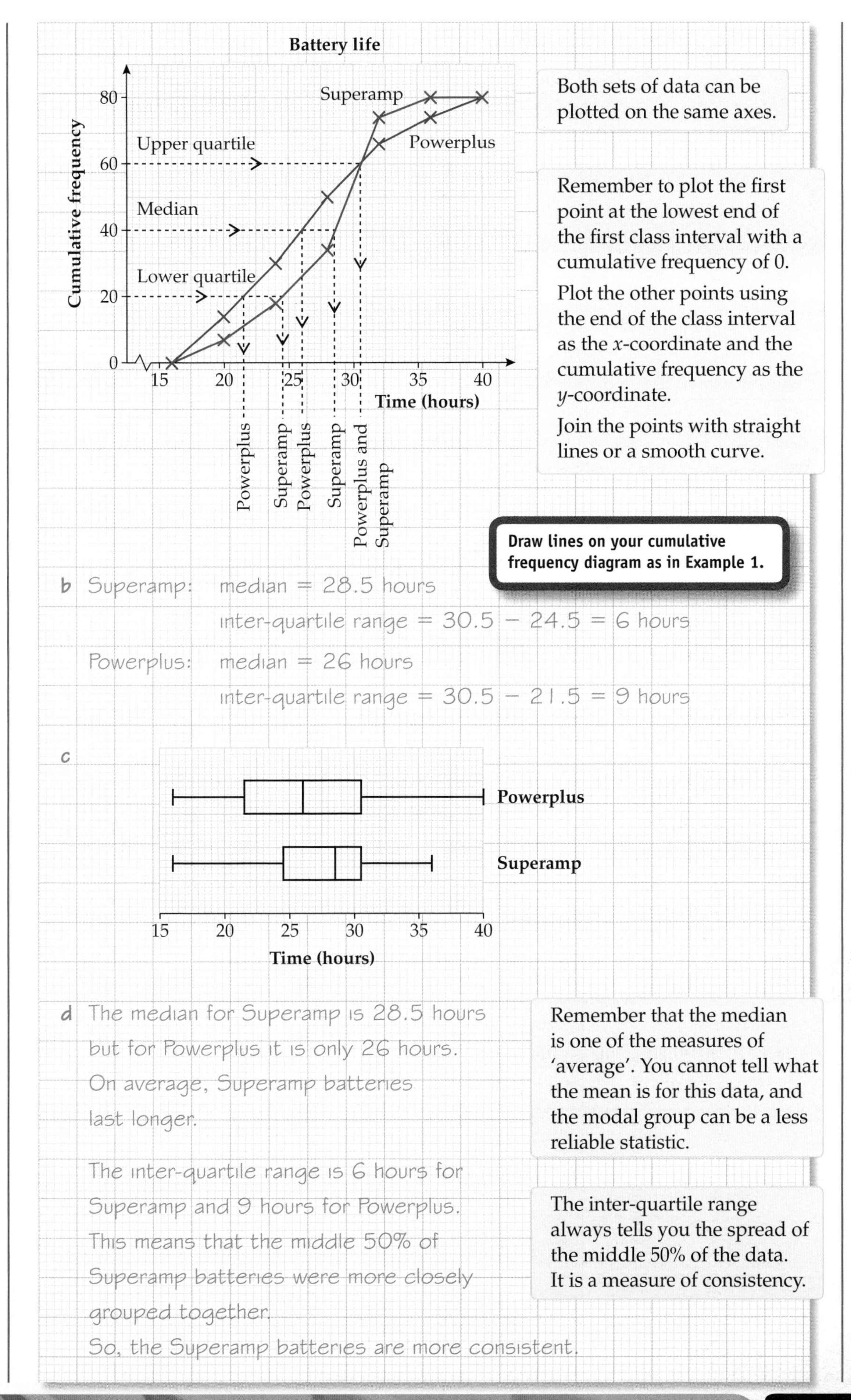

Both sets of data can be plotted on the same axes.

Remember to plot the first point at the lowest end of the first class interval with a cumulative frequency of 0.

Plot the other points using the end of the class interval as the x-coordinate and the cumulative frequency as the y-coordinate.

Join the points with straight lines or a smooth curve.

Draw lines on your cumulative frequency diagram as in Example 1.

b Superamp: median = 28.5 hours

inter-quartile range = 30.5 − 24.5 = 6 hours

Powerplus: median = 26 hours

inter-quartile range = 30.5 − 21.5 = 9 hours

c

d The median for Superamp is 28.5 hours but for Powerplus it is only 26 hours. On average, Superamp batteries last longer.

Remember that the median is one of the measures of 'average'. You cannot tell what the mean is for this data, and the modal group can be a less reliable statistic.

The inter-quartile range is 6 hours for Superamp and 9 hours for Powerplus. This means that the middle 50% of Superamp batteries were more closely grouped together.

So, the Superamp batteries are more consistent.

The inter-quartile range always tells you the spread of the middle 50% of the data. It is a measure of consistency.

Comparing the lower quartiles,
Superamp = 24.5 hours and
Powerplus = 21.5 hours.
This means that 60 out of 80 (75%) of the Superamp batteries lasted over 24.5 hours.
Using the cumulative frequency diagram you can see that the number of Powerplus batteries lasting over 24.5 hours is only about 48.

The lower quartile is the point 25% into the data, so 75% of the data lies above it.

Although some of the Powerplus batteries lasted more than 36 hours, overall they are less reliable.
The Superamp batteries are the better buy.

Exercise 6E

B

1 The box plots show the lengths jumped by 50 boys and 40 girls in the triple jump.

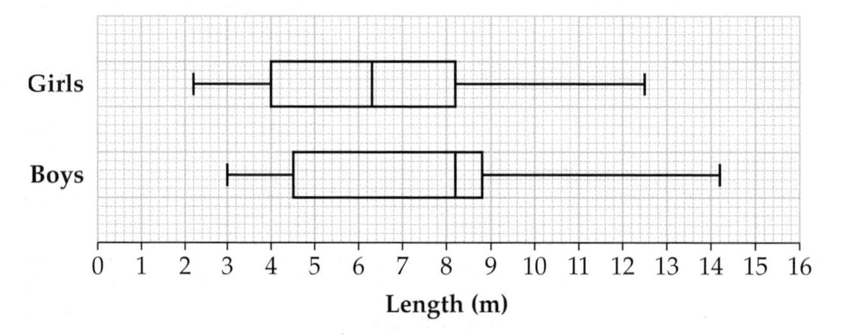

a What is the median length jumped by the girls?
b What was the longest jump made by a boy?
c Give **two** differences between the lengths jumped by the boys and the lengths jumped by the girls.
d How many students jumped more than 8.2 metres?

AO2

B

2 Potatoes of variety X and Y are weighed.
The frequency distributions of each variety are shown on the diagram.

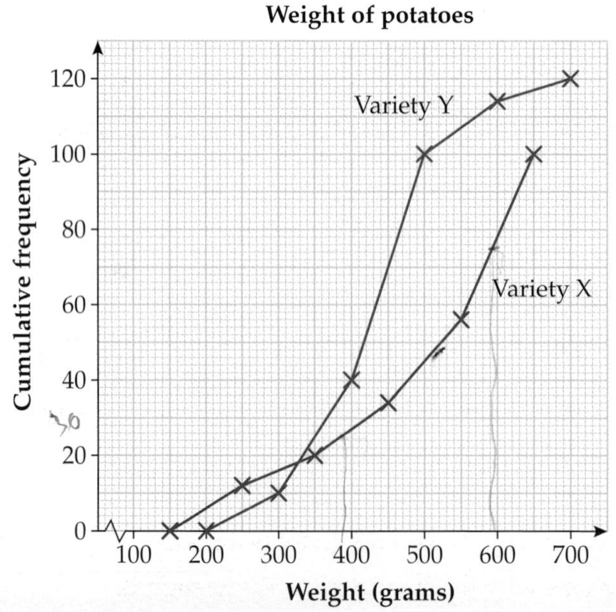

AO3

a How many potatoes of variety X were weighed?

b Estimate the minimum and maximum weight of potatoes from variety X.

c Work out the median and the inter-quartile range for variety X.

d How many potatoes of variety Y were weighed?

e Estimate the minimum and maximum weight of potatoes from variety Y.

f Work out the median and the inter-quartile range for variety Y.

g Draw box plots for both varieties, using a scale of 100 to 800 on the horizontal axis.
Draw one box plot directly beneath the other.

h Compare the two varieties, giving reasons for any statements you make.

B

AO3

Review exercise

1 A group of university physics students and a group of university history students were given 20 long multiplication calculations to work out.
The times taken (in minutes) by the physics students are shown in the box plot.

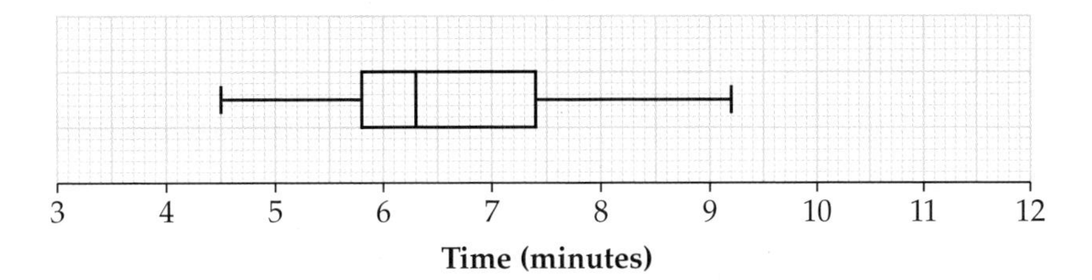

The history students had a shortest time of 4.2 minutes, a longest time of 10.5 minutes, a median of 7.2 minutes, a lower quartile of 6 minutes and an upper quartile of 8.5 minutes.

a Copy the box plot for the physics students and on the same scale draw the box plot for the history students. **[2 marks]**

b Comment on the differences between the two frequency distributions. **[2 marks]**

2 Joe is planning his summer holidays.
He looks at data about two holiday destinations, A and B.
The box plots show the temperatures from May to September.

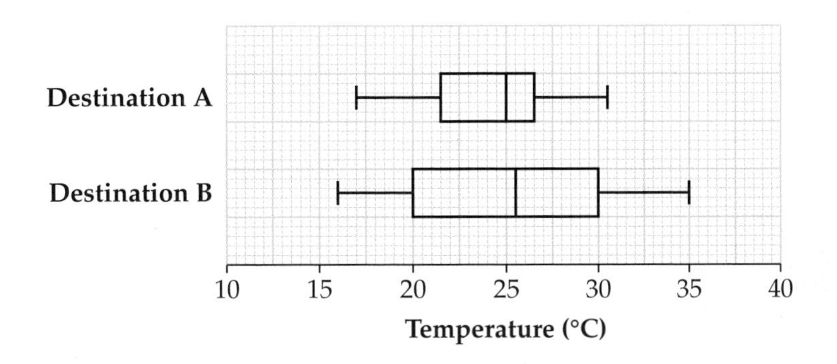

a Give **two** reasons why Joe should go to destination A. **[2 marks]**

b Give **two** reasons why Joe should go to destination B. **[2 marks]**

B

AO2

B

3 Two products to improve the growth of tomato plants are being tested, *Lottagrow* and *Supertom*.

The heights of plants are measured after treatment by each product.
These frequency tables show the results.

Lottagrow

Height, h (cm)	Frequency
$50 < h \leq 55$	4
$55 < h \leq 60$	20
$60 < h \leq 65$	30
$65 < h \leq 70$	13
$70 < h \leq 75$	15
$75 < h \leq 80$	10

Supertom

Height, h (cm)	Frequency
$50 < h \leq 55$	2
$55 < h \leq 60$	13
$60 < h \leq 65$	25
$65 < h \leq 70$	28
$70 < h \leq 75$	18
$75 < h \leq 80$	2

The *Lottagrow* tomatoes had a minimum height of 50 cm and a maximum height of 80 cm.

The *Supertom* tomatoes had a minimum height of 53 cm and a maximum height of 77 cm.

a Draw a cumulative frequency diagram for each product. **[6 marks]**

> **Use a scale of 2 cm for 5 units on the x-axis, with values from 50 to 80.**
> **Use a scale of 2 cm for 20 units on the y-axis with values from 0 to 100.**

b Draw a box plot for each product. **[4 marks]**

c Estimate the median and inter-quartile range for each product. **[6 marks]**

AO3

d Which product would you recommend?
Give at least **two** reasons for your choice. **[2 marks]**

Chapter summary

In this chapter you have learned how to

- compile a cumulative frequency table and draw cumulative frequency diagrams **B**
- use cumulative frequency diagrams to analyse data **B**
- draw a box plot from a cumulative frequency diagram **B**
- use cumulative frequency diagrams and box plots to compare data and draw conclusions **B**

7 Ratio and proportion

This chapter shows you how to use ratio and proportion to solve problems.

Whatever fruits, ice and yoghurt you use, they need to be in the correct ratio to make a delicious smoothie.

Objectives

- simplify ratios E
- use ratios E D
- use ratio notation, including reduction to its simplest form and its various links to fraction notation D
- write ratios as fractions D C
- write ratios in the form $1:n$ and $n:1$ C
- solve word problems involving ratio and proportion, including using informal strategies and the unitary method of solution D C
- divide a quantity in a given ratio D C
- solve problems involving quantities that vary in direct or inverse proportion D C B

Before you start this chapter

1 Which numbers are factors of (divide into)
a 4 and 12 **b** 9 and 15 **c** 36 and 48?

2 What is the highest common factor for each pair of numbers in Q1?

3 Convert
a 4 km to metres **b** 20 mm to cm **c** 3.5 kg to grams
d 2750 *ml* to litres **e** 240 minutes to hours.

4 Work out **a** $\frac{1}{9} + \frac{4}{9}$ **b** $1 + \frac{3}{4}$ **c** $1 - \frac{3}{4}$

7.1 Ratio

Keywords
ratio, simplify, maps, scale drawing

Why learn this?
You need ratios to adapt recipes for different numbers of people.

Objectives
E Simplify a ratio to its lowest terms
D Use a ratio in practical situations

Skills check

1 Copy and complete these length conversions.
 a 1 cm = □ mm b 1 m = □ cm c 1 km = □ m

2 Copy and complete these conversions.
 a 5 km = □ m b 47 mm = □ cm c 420 m = □ km
 d 360 m = □ cm e 2.5 km = □ m f 0.8 km = □ m = □ cm

Writing and simplifying ratios

A **ratio** compares two or more quantities.

In a cranberry and raspberry smoothie, you use 50 g of cranberries and 100 g of raspberries. You can write this using the ratio symbol.

cranberries : raspberries = 50 : 100

You can **simplify** ratios. Divide all the numbers in the ratio by the highest number possible.

cranberries : raspberries
50 : 100 (÷ 50, ÷ 50)
1 : 2

Divide by the highest common factor.

When you simplify ratios you must make sure that all the quantities are in the same units.

E

Example 1

Simplify these ratios.

a 2 m : 50 cm b $1\frac{1}{2}$ kg : $4\frac{1}{2}$ kg

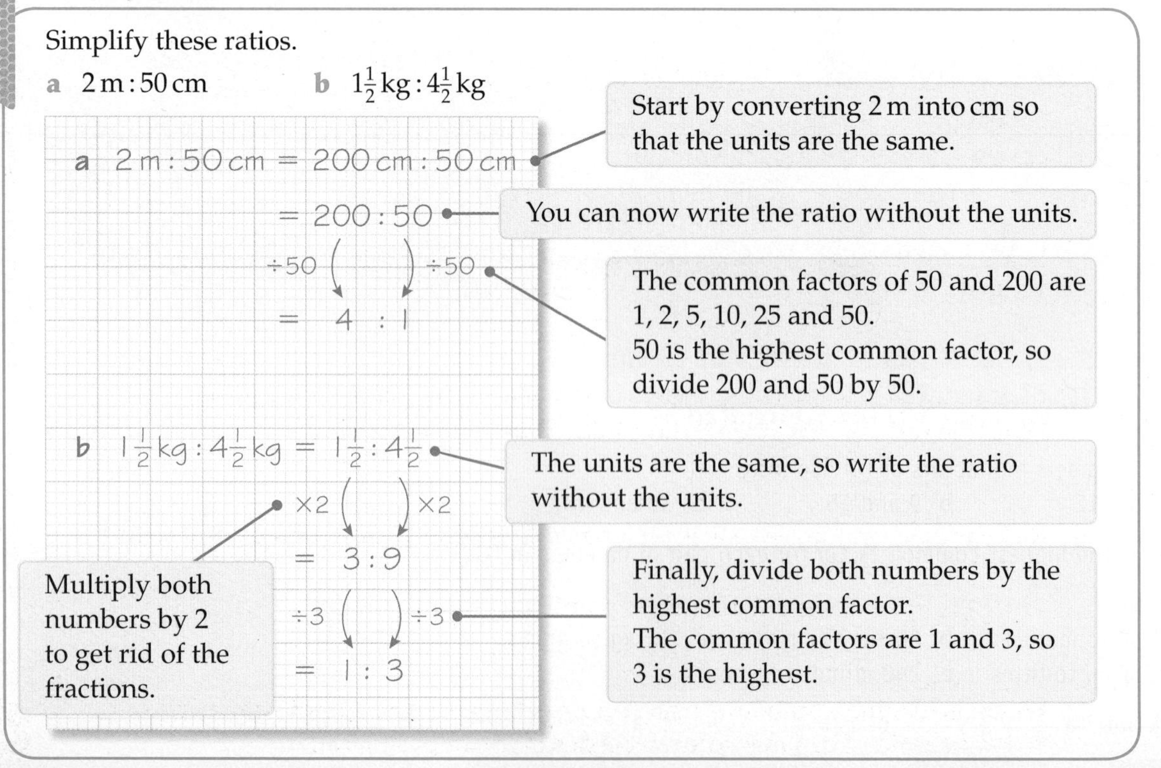

Exercise 7A

1 Simplify these ratios.

a $6:3$
b $5:15$
c $8:12$
d $25:15$
e $40:60$
f $75:100$

2 Simplify these ratios.

a 6 cm : 15 cm
b 20p : 40p
c 2 cm : 10 mm
d 500 m : 2000 cm
e 3 kg : 600 g
f 5 litres : 5 m*l*

Change the parts into the same units first.

3 Simplify these ratios.

a $1\frac{1}{2}:3$
b $2\frac{1}{2}:15$
c $2\frac{1}{4}:6$
d 0.8 mm : 1.2 mm
e 2.5 m : 3.5 m
f 1.6 kg : 0.4 kg

In parts d, e and f, start by multiplying both numbers by 10.

4 Which of these ratios are equivalent to $3:2$?

A $10:7$
B $15:10$
C $30:20$
D $18:15$

5 Which of these ratios are equivalent to $4:5$?

A $1:1.25$
B $20:35$
C $10:12.5$
D $6:7\frac{1}{2}$

E

6 Bethany said that the ratio $45:50$ simplifies to $10:9$.
Is Bethany correct?
Give a reason for your answer.

7 In a school there are 450 boys and 540 girls.
Write the ratio of boys to girls in its simplest form.

8 Moira makes a drink using $1\frac{1}{2}$ cups of lemonade and $\frac{1}{3}$ cup of lime juice.
Write the ratio of lemonade to lime juice in its simplest form.

9 Ricky is a decorator. He makes pink paint by mixing $1\frac{1}{3}$ litres of red paint with $2\frac{1}{2}$ litres of white paint.
Write the ratio of red paint to white paint in its simplest form.

10 Enitan is making a smoothie using 0.1 kg of bananas and 60 g of strawberries.
Write the ratio of bananas to strawberries in its simplest form.

AO2

Using ratios

Ratios are often used in real-life situations, for example when cooking and entertaining, and when describing the steepness of a hill.

If you know how much of each ingredient you need to make two smoothies, you can use ratios to work out how much of each ingredient you would need to make any number of smoothies.

D

Example 2

To make 2 berry smoothies you need 100 g raspberries and 50 g blackberries.
How much of each ingredient do you need to make

a 6 smoothies **b** 5 smoothies?

a For 6 smoothies:
$3 \times 100\text{ g} = 300\text{ g}$ raspberries
$3 \times 50\text{ g} = 150\text{ g}$ blackberries

The recipe makes 2 smoothies. 6 smoothies is 3 lots of 2. Multiply all the quantities by 3.

b For 1 smoothie:
$100\text{ g} \div 2 = 50\text{ g}$ raspberries
$50\text{ g} \div 2 = 25\text{ g}$ blackberries

First work out the quantities for 1 smoothie.

For 5 smoothies:
$50\text{ g} \times 5 = 250\text{ g}$ raspberries
$25\text{ g} \times 5 = 125\text{ g}$ blackberries

Then multiply the quantities for 1 smoothie by 5.

Exercise 7B

D

1 This recipe makes 10 pancakes.

500 g flour
2 eggs
300 m*l* milk

Work out the quantities for

a 30 pancakes **b** 5 pancakes **c** 15 pancakes.

2 A recipe for mushroom soup uses 100 g of mushrooms.
The recipe is for 4 people.
What weight of mushrooms is needed to make soup for

a 8 people **b** 2 people **c** 14 people?

3 It takes a photocopier 20 seconds to produce 15 copies.

a How long will it take the photocopier to produce 21 copies?

b How many copies could it produce in 44 seconds?

4 A builder makes concrete with 7 bags of sand and 3 bags of cement.

a If he has 28 bags of sand, how many bags of cement will he need to make concrete?

b The builder has 15 bags of cement. How many bags of sand should he order for the next day?

5 Cupro-nickel is used to make key rings. It is made by mixing copper and nickel in the ratio 5 : 2.

a How much copper would you need to mix with 6 kg nickel?

b How much nickel would you need to mix with 10 kg copper?

D

6 Here is a recipe for gooseberry fool.

D

Gooseberry fool

400 g gooseberries
60 g sugar
240 m*l* milk
1 tablespoon custard powder

I only have 300 g gooseberries.
I want to use all my gooseberries.
How much of each of the other ingredients should I use?

7 Here is a recipe for potato layer bake.
I have 1.75 kg of potatoes and I want to use them all.

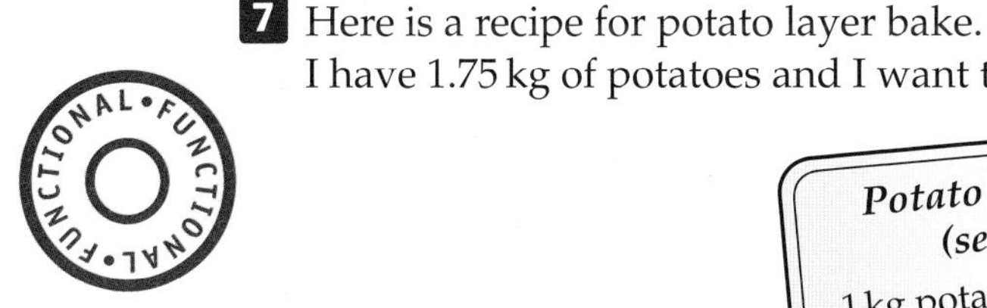

Potato layer bake (serves 4)

1 kg potatoes
100 g onion
50 g butter
180 g cheddar cheese
300 m*l* milk

a How much of each of the other ingredients should I use?

b How many people will my potato layer bake serve?

AO2

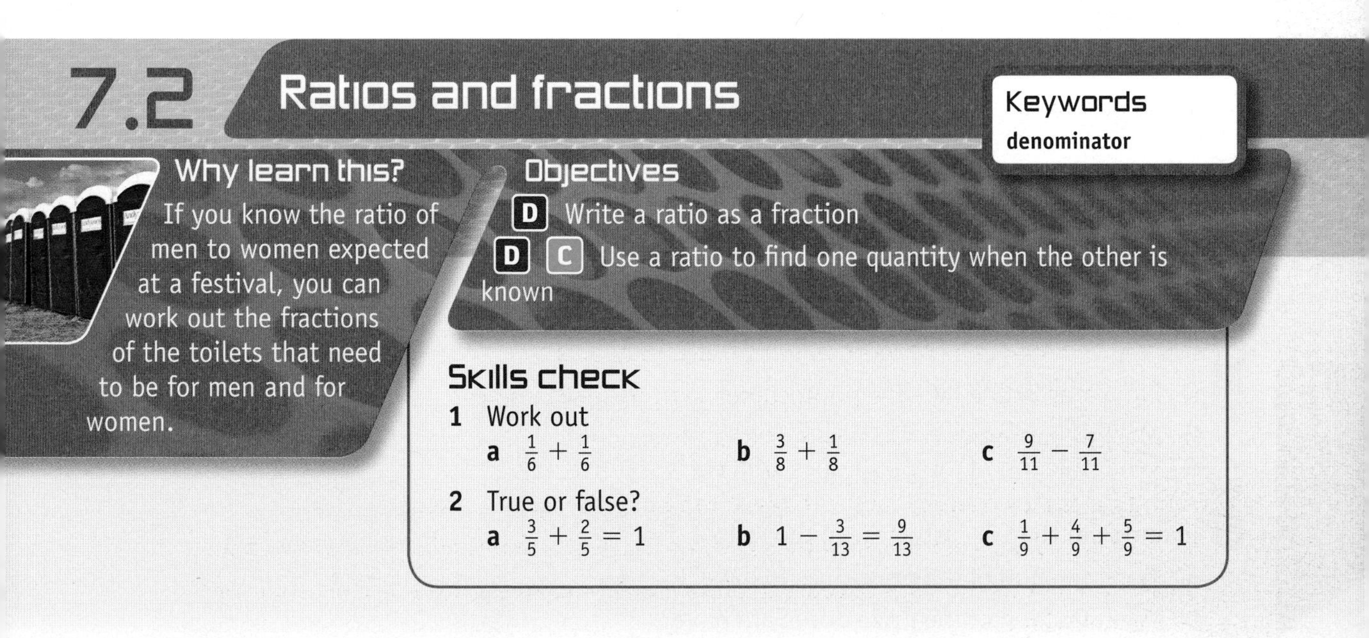

7.2 Ratios and fractions

Keywords
denominator

Why learn this?

If you know the ratio of men to women expected at a festival, you can work out the fractions of the toilets that need to be for men and for women.

Objectives

D Write a ratio as a fraction

D **C** Use a ratio to find one quantity when the other is known

Skills check

1 Work out

a $\frac{1}{6} + \frac{1}{6}$ **b** $\frac{3}{8} + \frac{1}{8}$ **c** $\frac{9}{11} - \frac{7}{11}$

2 True or false?

a $\frac{3}{5} + \frac{2}{5} = 1$ **b** $1 - \frac{3}{13} = \frac{9}{13}$ **c** $\frac{1}{9} + \frac{4}{9} + \frac{5}{9} = 1$

Writing a ratio as a fraction

You can write a ratio as a fraction by simply changing the whole numbers in the ratio into fractions with the same **denominator**.

In a class of students, if the ratio of boys to girls is 2 : 3, then $\frac{2}{5}$ of the class are boys and $\frac{3}{5}$ of the class are girls.

The denominator is found by adding the whole numbers in the ratio. In this case, 2 + 3 = 5.

Thinking of ratios as fractions can help when solving ratio problems.

D C

Example 3

a Trish and Del share a pie in the ratio 1 : 3.
What fraction of the pie does Trish eat?
What fraction of the pie does Del eat?

b A supermarket sells cheese pizzas and meat pizzas in the ratio 4 : 3.
One day 120 cheese pizzas are sold.
How many meat pizzas are sold?

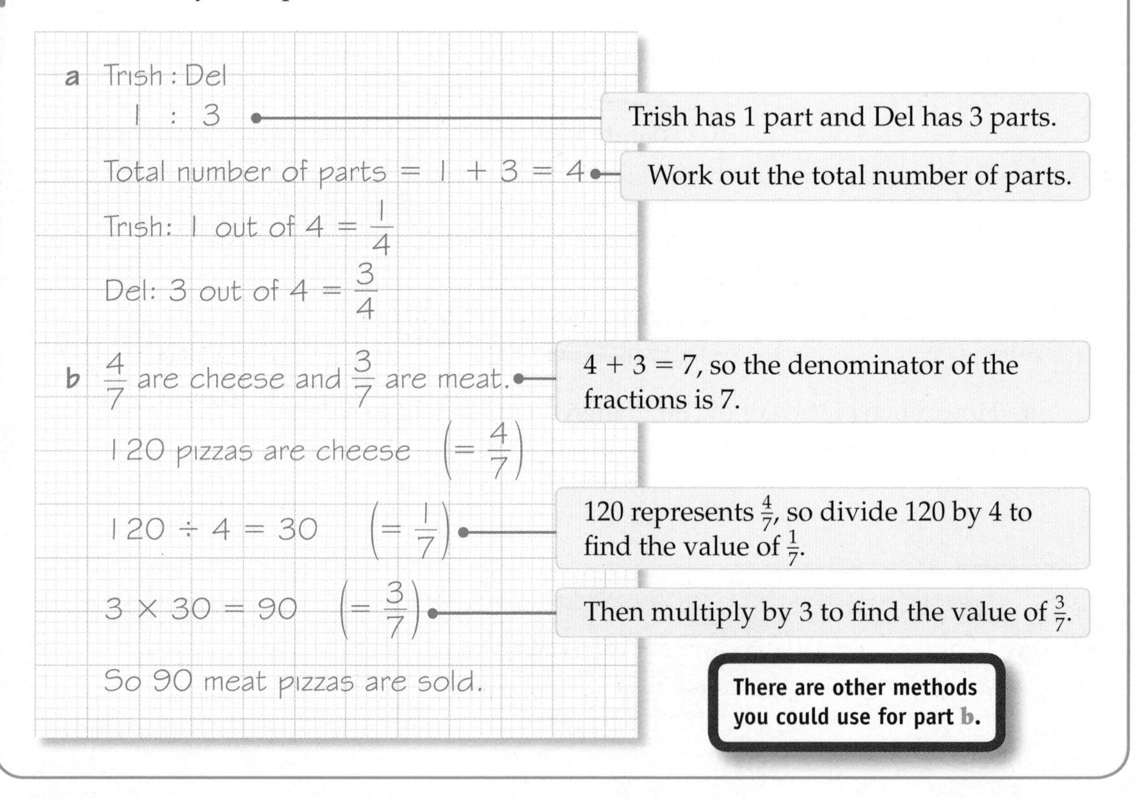

Exercise 7C

D

1 Kim and Tom share a pizza in the ratio 4 : 5.

a What fraction of the pizza does Kim eat?

b What fraction of the pizza does Tom eat?

2 Jo and Sam share a cash prize in the ratio 2 : 5.

a What fraction of the prize does Jo have?

b What fraction of the prize does Sam have?

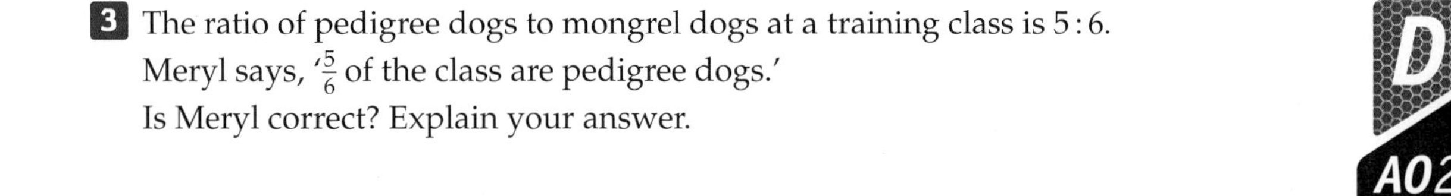

3 The ratio of pedigree dogs to mongrel dogs at a training class is 5 : 6.
Meryl says, '$\frac{5}{6}$ of the class are pedigree dogs.'
Is Meryl correct? Explain your answer.

C

4 Peter and David share an inheritance from their grandmother in the ratio 2 : 3.
Peter receives £600.
How much does David receive?

5 A recipe for fruit cake uses sultanas and raisins in the ratio 5 : 3.
Sharon uses 150 g of raisins.
What weight of sultanas does she use?

C AO2

6 In a bag of apples the ratio of bruised to not-bruised apples is 1 : 6.
There are 8 bruised apples in the bag.
How many apples are there in the bag altogether?

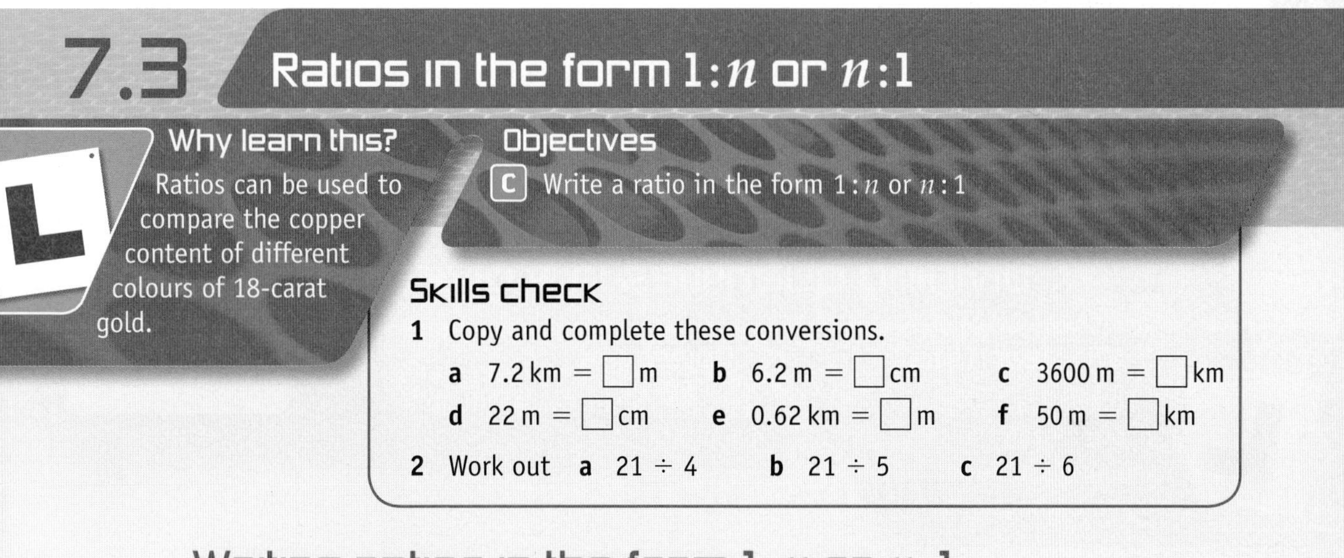

7.3 Ratios in the form 1 : n or n : 1

Why learn this?
Ratios can be used to compare the copper content of different colours of 18-carat gold.

Objectives
C Write a ratio in the form 1 : n or n : 1

Skills check

1 Copy and complete these conversions.
 a 7.2 km = ☐ m b 6.2 m = ☐ cm c 3600 m = ☐ km
 d 22 m = ☐ cm e 0.62 km = ☐ m f 50 m = ☐ km

2 Work out a 21 ÷ 4 b 21 ÷ 5 c 21 ÷ 6

Writing ratios in the form 1 : n or n : 1

To write a ratio in the form 1 : n or n : 1
- first decide which number in the ratio you want to be 1
- then divide all the numbers in the ratio by that number.

Example 4

C

a Write the ratio 15 : 10 in the form n : 1. **b** Write the ratio 5 : 3 in the form 1 : n.

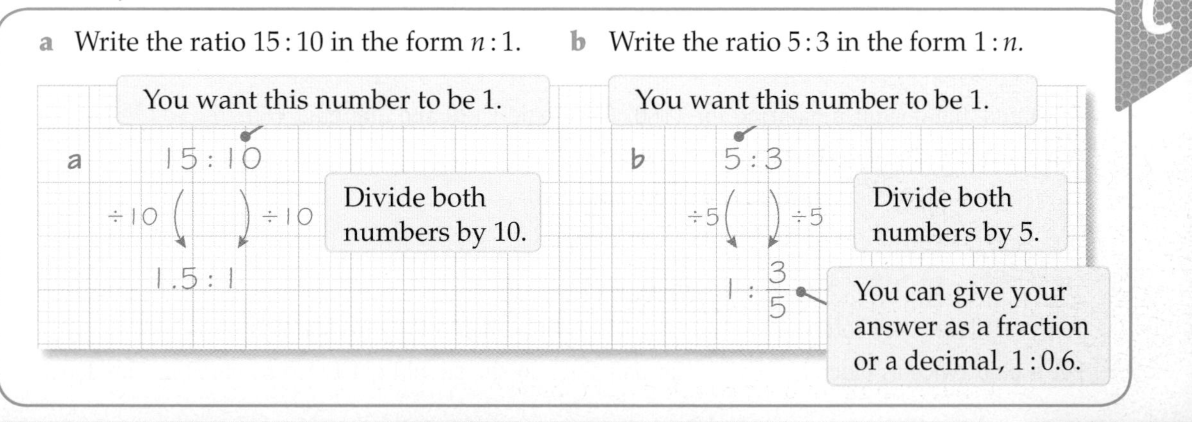

Exercise 7D

1. Write these ratios in the form $n:1$.

   a 6 : 2 b 12 : 6 c 21 : 3 d 5 : 4

2. Write these ratios in the form $1:n$.

   a 3 m : 12 m b 400 cm : 15 m c 700 g : 1.2 kg
   d 5 litres : 7500 m*l* e 2 years : 8 months f 2 days : 15 hours

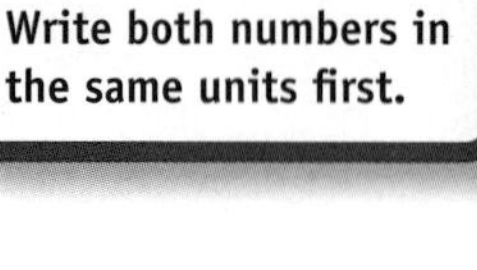

C AO2

3. a The ratio of gold to silver in necklace A is 229 : 21.
   Write this as a ratio in the form $n:1$.

   b The ratio of gold to silver in necklace B is 117 : 83.
   Write this as a ratio in the form $n:1$.

   c Compare your answers to parts **a** and **b**.
   Which necklace has the higher proportion of gold?
   Explain your answer.

C AO3

4. The ratio of gold to silver to copper in one type of gold bar is 75 : 3 : 22.
   The ratio of gold to silver to copper in a second type of gold bar is 15 : 1 : 4.
   Which gold bar, the first or the second, has the higher proportion of copper?
   Show workings to support your answer.

7.4 Dividing quantities in a given ratio

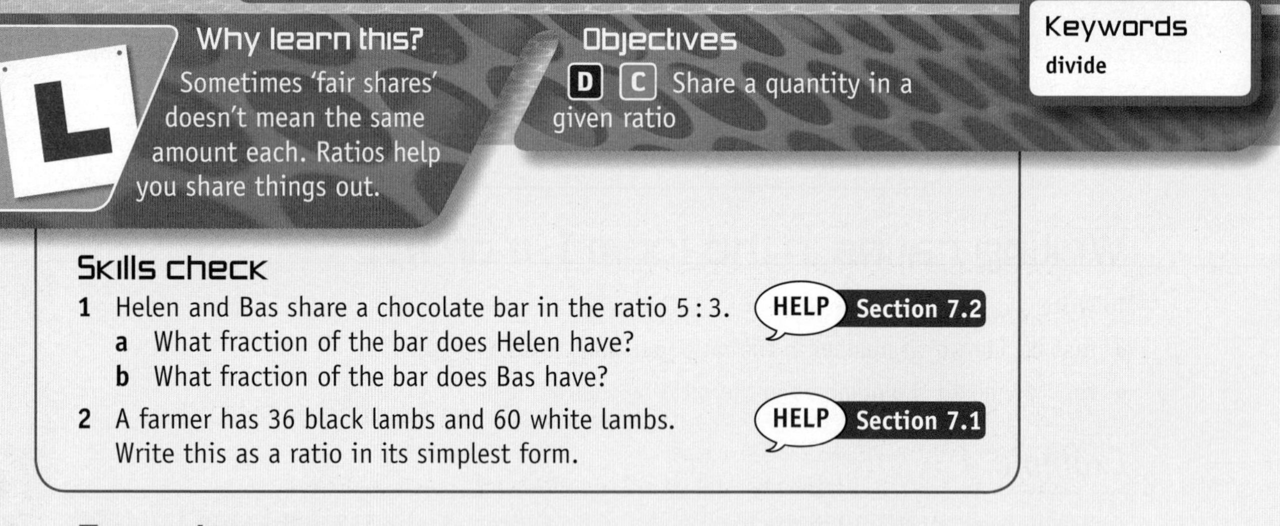

Why learn this?

Sometimes 'fair shares' doesn't mean the same amount each. Ratios help you share things out.

Objectives

D C Share a quantity in a given ratio

Keywords

divide

Skills check

1 Helen and Bas share a chocolate bar in the ratio 5 : 3. HELP Section 7.2
   a What fraction of the bar does Helen have?
   b What fraction of the bar does Bas have?

2 A farmer has 36 black lambs and 60 white lambs. HELP Section 7.1
   Write this as a ratio in its simplest form.

Fair shares

Petra and Jim bought a painting for £180. Petra paid £120 and Jim paid £60.

You can write the amounts they paid as a ratio.

Petra Jim
£120 : £60
2 : 1

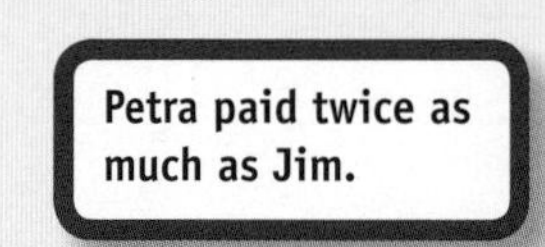

Ten years later they sold the painting for £540. How should they **divide** this fairly?

Petra paid twice as much as Jim for the painting, so she should get twice as much as Jim does.

Example 5

Share £540 between Petra and Jim in the ratio 2 : 1.

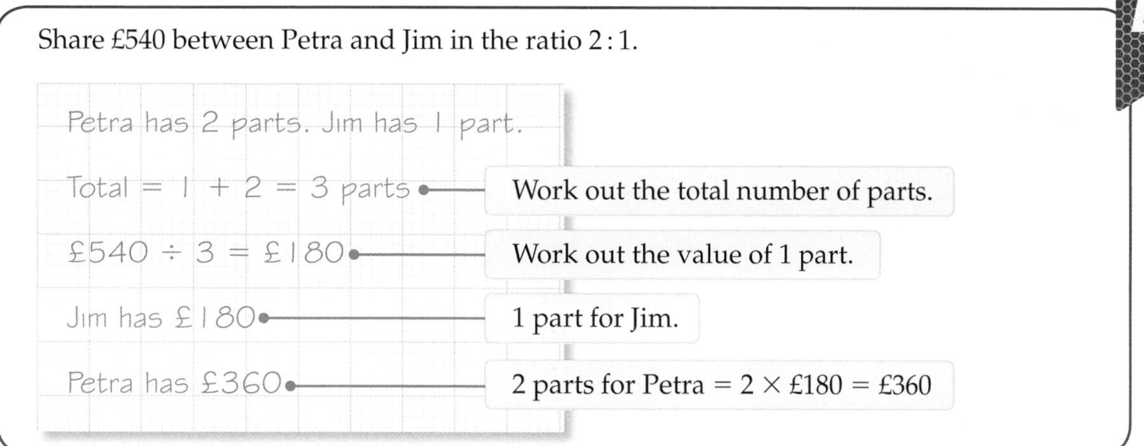

D

Sharing in a given ratio

To share a quantity in a given ratio

- work out the total number of parts to share into
- work out the value of 1 part
- work out the value of each share.

Exercise 7E

D

1 Share these amounts in the ratios given.

a £36 in the ratio 2 : 1 **b** £12 in the ratio 3 : 1

C

c 32 litres in the ratio 5 : 3 **d** 66 kg in the ratio 4 : 7

e £500 in the ratio 2 : 2 : 1 **f** 250 cm in the ratio 10 : 6 : 9

Work out the total number of parts first.

C

2 Dan and Tris buy a lottery ticket. Dan pays 30p and Tris pays 70p.

a Write the amounts they pay as a ratio.

Dan and Tris win £100 000.

b Work out how much prize money each should have.

3 Alix and Liberty buy a racehorse. Alix pays £4000 and Liberty pays £5000.
The racehorse wins a £27 000 prize.
Work out how much money Alix and Liberty should each receive.

4 Pip, Wilf and Anna share 300 g of fudge in the ratio 1 : 2 : 3.
How much fudge does each have?

5 Purple paint is made from red, blue and white paint in the ratio 2 : 3 : 4.
Meg makes 18 litres of paint.
How much of each colour does she use?

6 Fred, Sid and Ruby inherit £50 000.
They share it in the ratio of their ages.
Ruby is 10, Sid is 14 and Fred is 19.
How much does each receive?

Work out the amounts to the nearest pound.

7.5 More ratios

Why learn this?
The government decides safe adult to child ratios for childcare. For children under two, the ratio of adults to children must be 1 : 3.

Objectives
C Solve word problems involving ratio

Skills check

1 Share 400 kg in the ratio 3 : 5.

2 Share 24 litres in the ratio 1 : 2 : 3.

HELP Section 7.4

Word problems

For 3–7-year-olds on a playscheme, the law says that the ratio of adults to children must be at least 1 : 8.

This means that for every 8 children there must be at least 1 adult.

You can use ratios to work out the numbers of adults for different numbers of children.

Example 6

On a playscheme, the minimum ratio of adults to children is 1 : 8.

a There are 7 adults. How many children can they look after?

b How many adults do you need for 40 children?

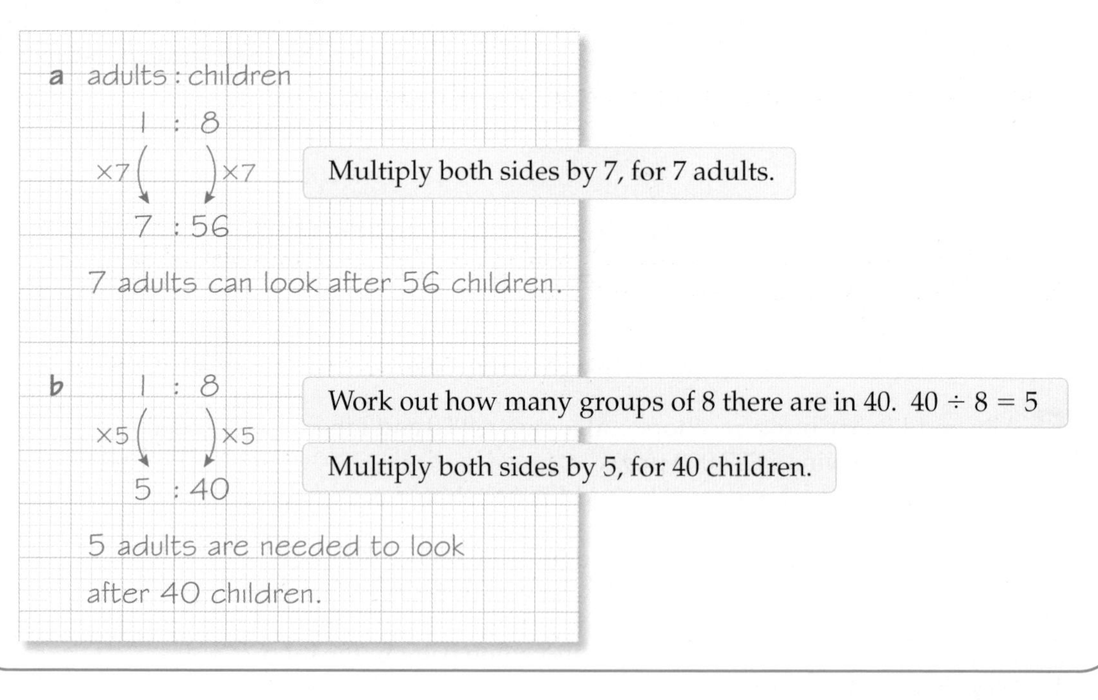

Exercise 7F

1 For children under two years old, the ratio of adults to children must be at least 1 : 3. How many under-two-year-olds can these numbers of adults care for?

a 4 adults **b** 6 adults **c** 9 adults

2 In a nursery group of two-year-olds, the ratio of adults to children must be at least 1 : 4.
How many adults do you need for

a 24 children b 32 children c 42 children?

C

3 Julie makes celebration cakes.
Her recipe uses 3 eggs for every 150 g of flour.

a Write the amounts of eggs and flour as a ratio.
b Simplify your ratio.
c How much flour does she need for 12 eggs?
d How many eggs does she need for 900 g of flour?

C

4 Bronze is made from 88% copper and 12% tin.

a Write the amounts of copper and tin as a ratio.
b Write your ratio in its simplest form.
c Tom has 55 kg of copper to make bronze. How much tin does he need?

5 Milly keeps hens. Some lay brown eggs, some lay white eggs.
The ratio of brown eggs to white eggs is 7 : 2.

a One week there are 56 brown eggs. How many white eggs are there?
b One month there are 84 white eggs. How many brown eggs are there?

AO2

6 To make strawberry jam, you need 3 kg sugar for every 3.5 kg of strawberries.
Penny has 10.5 kg of strawberries.
How much sugar does she need?

Putting the answer in context

Sometimes you need to round up or down to get a sensible answer.

Example 7

C

On a nursery school outing, the ratio of adults to children must be at least 3 : 8.
How many adults do you need for 50 children?

adults : children

3 : 8

×6.25 ×6.25

18.75 : 50

For 50 children, you need 19 adults.

Work out how many groups of 8 there are in 50.
$50 \div 8 = 6.25$

Multiply both sides by 6.25 to get 50 children.

You can't have 18.75 adults, so round up to 19.

Exercise 7G

1 On a playscheme, the ratio of adults to children must be at least 1 : 8.
Work out how many adults you need for 36 children.

Your answer must be a whole number of people.

C

C

2 On a school trip, the ratio of adults to children must be at least 4 : 10.
Work out how many adults you need for 35 children.

3 A fruit cocktail recipe uses 5 peaches for 200 g of cherries.
How many peaches do you need for 1 kg of cherries?

4 The table shows the adult : child ratios for different ages of children in daycare.

Age of children	Adult : child ratio
Less than 2 years	1 : 3
2 years	1 : 4
3–7 years	1 : 8

These are the ages of children registered at a daycare centre in the school holidays.

FUNCTIONAL

Babies up to one year old	4
Toddlers over 1 year, but less than 2	7
Two-year-olds	8
Three-year-olds	6
Four-year-olds	12
Five-year-olds	9
Six-year-olds	5

AO2

Work out the number of adults needed to care for these children.

7.6 Proportion

Keywords
direct proportion, unitary method

Why learn this?
Proportion calculations help you work out prices for different numbers of items.

Objectives
D Understand direct proportion
D Solve proportion problems, including using the unitary method

Skills check

1 Work out
 a 8 × 15p b 3 × 22p c 25p × 4 d 40p × 6

2 Work out
 a 48 ÷ 6 b 72 ÷ 12 c £1.80 ÷ 3

Direct proportion

Oranges cost 30p each. The number of oranges and the cost are in **direct proportion**.

When two values are in direct proportion:

- if one value is zero, so is the other

0 oranges cost 0 pence.

- if one value doubles, so does the other.

×2 **1 orange costs 30p** ×2
2 oranges cost 60p

When two quantities are in direct proportion, their ratios stays the same as they increase or decrease.

Unitary method

Four apples cost 64p.

You can work out the cost of one apple: 64p ÷ 4 = 16p

Then you can use this to work out the cost of seven apples: 7 × 16p = £1.12

This is called the **unitary method**, because you work out the value of one unit first.

Example 8

D

Jim works 12 hours a week for £78.
He increases his hours to 15 per week.
How much will he now be paid?

£78 ÷ 12 = £6.50 — Work out his pay for 1 hour.

£6.50 × 15 = £97.50 — Multiply his pay for 1 hour by 15.

Sometimes the unitary method is not the quickest way to solve a problem.
Look for patterns in the numbers.

Example 9

D

Six grapefruit cost £2.70.
Work out the cost of 24 grapefruit.

6 grapefruit cost £2.70

24 grapefruit cost £2.70 × 4 = £10.80 — 24 is 4 × 6, so multiply the cost by 4.

Exercise 7H

D

1 Eight pencils cost 96p.
Work out the cost of

a 16 pencils **b** 4 pencils **c** 20 pencils.

2 Five pantomime tickets cost £75.

a Work out the cost of 1 pantomime ticket.

b Work out the cost of 4 pantomime tickets.

D

3 Tanya is paid £40.95 for 7 hours' work.

a How much will she be paid for 10 hours' work?

b How many hours did she work to earn £93.60?

AO2

D

4 Four adult train tickets cost £84.
Work out the cost of 3 of these tickets.

5 Seven senior citizen cinema tickets cost £26.60.
Work out the cost of 5 of these tickets.

6 Wei Yen delivers leaflets.
She is paid £7.20 for delivering 300 leaflets.
How much will she get for delivering 450 leaflets?

7 16 sausages weigh 2 kg.
How much will 24 sausages weigh?

8 Three punnets of strawberries weigh 1.35 kg.
Work out the weight of 7 punnets of strawberries.

9 A catering firm charges £637.50 for a buffet for 75 people.
How much will it charge for 225 people?

10 A horse eats one bale of hay every 4 days.
How many bales is it likely to eat in November?

AO2

11 There are 24 students in a class.
The teacher buys each student a maths revision guide.
The total cost is £95.76.
Two more students join the class.
How much will it cost to buy them revision guides?

7.7 Best buys

Keywords
best buy

Why learn this?
You can use the unitary method to work out which product gives better value for money.

Objectives
D Work out which product is the better buy

Skills check

1 Work out
  a 125p ÷ 50
  b 360p ÷ 330
  c 142p ÷ 120
  Round your answers **b** and **c** to the nearest tenth of a penny.

Getting value for money

The **best buy** means the product that gives you the best value for money.

To compare two prices and sizes, work out the price for one unit for each size.

Example 10

Shampoo comes in two sizes.
Which is the better buy?

75 ml for £2.25 — 120 ml for £3.25

Small: £2.25 = 225p — Convert the price to pence.

225 ÷ 75 = 3p for 1 ml — Divide price by quantity to get the price for 1 ml.

Large: £3.25 = 325p

325 ÷ 120 = 2.7p for 1 ml

The larger bottle is the better buy. — Compare the prices for 1 ml and decide which is cheaper.

Sometimes it is easier to work out how much you get for 1p or £1.

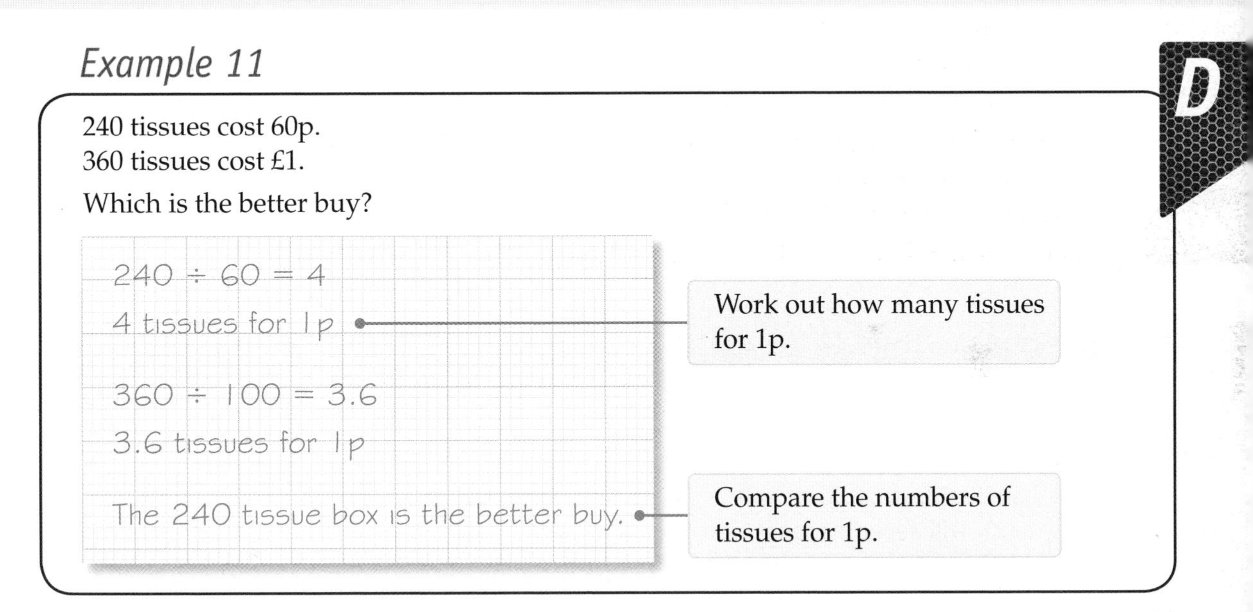

Example 11

240 tissues cost 60p.
360 tissues cost £1.
Which is the better buy?

240 ÷ 60 = 4
4 tissues for 1p — Work out how many tissues for 1p.

360 ÷ 100 = 3.6
3.6 tissues for 1p

The 240 tissue box is the better buy. — Compare the numbers of tissues for 1p.

Exercise 7I

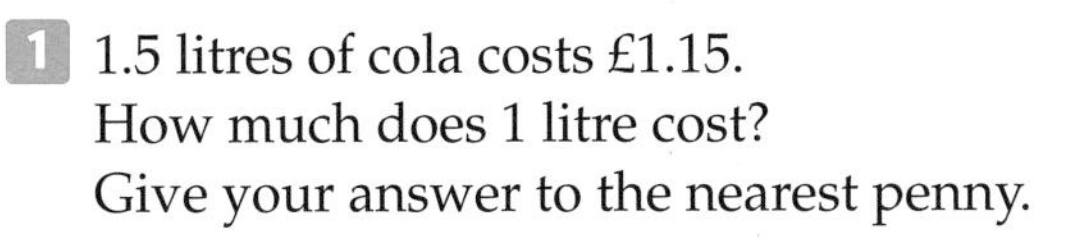

1 1.5 litres of cola costs £1.15.
How much does 1 litre cost?
Give your answer to the nearest penny.

2 5 litres of paint costs £20.
How much paint do you get for £1?

3 A large pack of cereal costs £2.34 for 500 g.

a Work out the cost of 1 g of cereal in this pack.

A small pack of the same cereal costs 98p for 200 g.

b Work out the cost of 1 g of cereal in this pack.

c Which pack is the better buy?
Explain your answer.

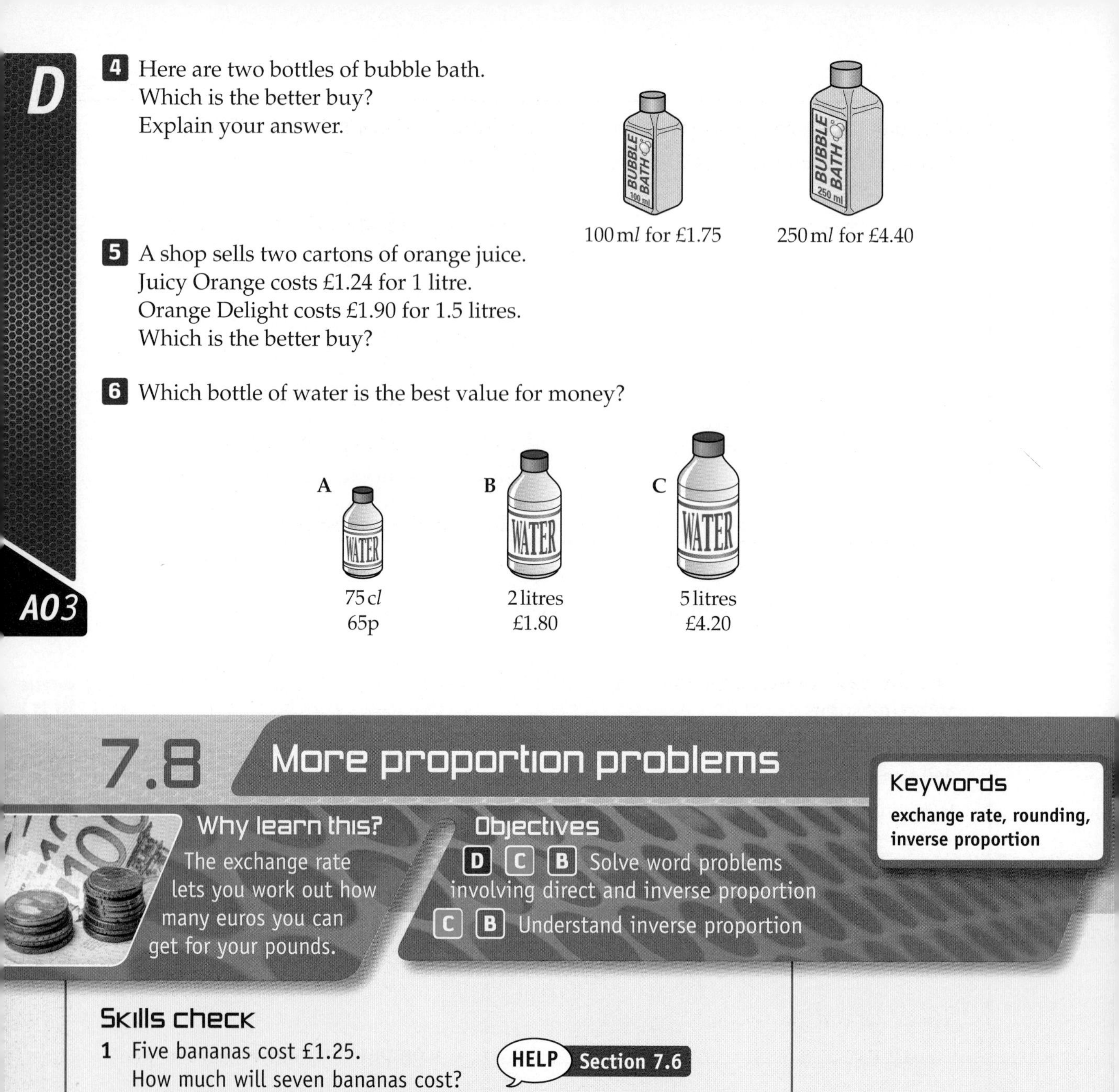

4 Here are two bottles of bubble bath.
Which is the better buy?
Explain your answer.

100 ml for £1.75   250 ml for £4.40

5 A shop sells two cartons of orange juice.
Juicy Orange costs £1.24 for 1 litre.
Orange Delight costs £1.90 for 1.5 litres.
Which is the better buy?

6 Which bottle of water is the best value for money?

A 75 cl 65p

B 2 litres £1.80

C 5 litres £4.20

7.8 More proportion problems

Keywords
exchange rate, rounding, inverse proportion

Why learn this?
The exchange rate lets you work out how many euros you can get for your pounds.

Objectives
D C B Solve word problems involving direct and inverse proportion
C B Understand inverse proportion

Skills check

1 Five bananas cost £1.25.
How much will seven bananas cost?

HELP Section 7.6

2 Round these amounts to **i** the nearest penny **ii** the nearest pound.
a £3.566   **b** £15.245   **c** £124.37824

Exchange rate

The **exchange rate** tells you how many euros (€), or dollars, or rupees (or any other currency) you can buy for £1. The rate varies from day to day.

One day the exchange rate for euros is

£1 = €1.2
So £2 = €2.4
£10 = €12

The two currencies are in direct proportion:
- **when one value is zero, so is the other**
- **when one value doubles, so does the other.**

You can use the unitary method to convert between currencies. First find the value of one unit of the currency.

Example 12

One day the exchange rate for pounds (£) to US dollars (US$) is £1 = US$1.6.

Convert **a** £10 to US$ **b** US$25 to pounds.

a £1 = US$1.6 — Multiply both values by 10.
so £10 = US$16

b £1 = US$1.6
so £1 ÷ 1.6 = US$1 — Divide both sides by 1.6 to find the value of US$1. Use a calculator for 1 ÷ 1.6 = 0.625.
£0.625 = US$1
25 × £0.625 = US$25 — Multiply by 25 to get the value of US$25.
£15.625 = US$25
US$25 = £15.63 — **Round** to the nearest penny.

D

Exercise 7J

Use this table of exchange rates for these questions.

Exchange rates one day in 2009	
£1	€1.2 (euros)
£1	US$1.6
£1	Aus$1.96 (Australian dollars)
£1	4.6 Polish zloty
£1	124 Pakistani rupees

D

1 Convert

a £25 to euros
b £40 to US dollars
c £30 to Australian dollars
d £15 to Polish zloty
e £50 to Pakistani rupees.

2 Convert these amounts to pounds.

a €35 **b** US$30 **c** 300 Pakistani rupees
d US$80 **e** 400 Polish zloty **f** €98
g 5000 Pakistani rupees **h** Aus$500 **i** 1000 Polish zloty

D AO2

3 Dale goes to Australia for his holiday. When he gets back he has Aus$48 left.
Tomas goes to Poland. When he gets back he has 123 Polish zloty left.

a Convert each amount to pounds.
b Who brings back more money, Dale or Tomas?

C

4 Which is worth more, 7000 Pakistani rupees or US$70?

5 Sadie goes to New York. She sees a games console in a shop for US$399.99.
Sadie calls her friend Jamila in London.
Jamila says the same console costs £279.99 in London.
Where should Sadie buy the console?
Explain why.

Inverse proportion

It takes 2 people 1 day to put up a fence.
Working at the same rate

- it would take 1 person 2 days to put up the same fence
- it would take 4 people $\frac{1}{2}$ a day to put up the same fence.

As the number of people goes down, the time taken goes up.
As the number of people goes up, the time taken goes down.

When two values are in **inverse proportion**, one increases at the same rate as the other decreases.

C

AO2

Example 13

It takes 4 people 8 hours to paint a wall along one side of a large garden.
There is an identical wall along the other side of the garden.

a How long will it take 3 people to paint this wall?

b How long will it take 5 people to paint this wall?

Give your answers to the nearest hour.

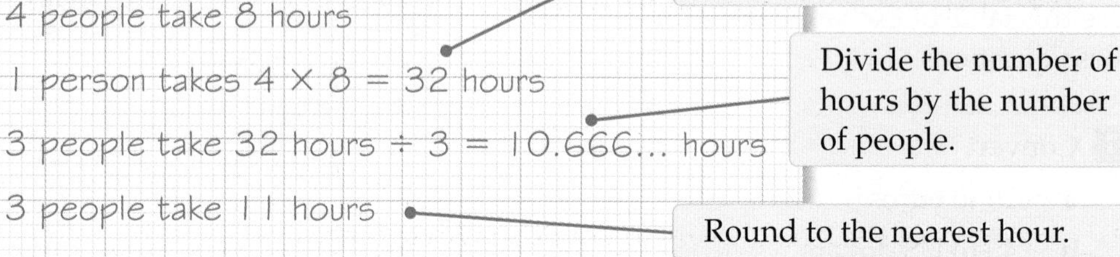

a 4 people take 8 hours

1 person takes $4 \times 8 = 32$ hours

3 people take 32 hours $\div 3 = 10.666...$ hours

3 people take 11 hours

b 5 people take $32 \div 5 = 6.4$ hours

$= 6$ hours

Work out the time for 1 person. 1 person will take 4 times as long as 4 people.

Divide the number of hours by the number of people.

Round to the nearest hour.

Divide the number of hours taken by 1 person by the number of people.

Round to the nearest hour.

Exercise 7K

C

AO2

1 It takes 5 children 45 minutes to build a sandcastle.
How long will it take 3 children to build an identical sandcastle?

2 It takes 2 electricians 3 days to rewire a house.
How long will it take 5 electricians to rewire an identical house?

3 In one day, 10 woodcutters chop 18 trees into logs.

a How many trees could 5 woodcutters chop into logs in one day?

b How many days will it take 5 woodcutters to chop 36 trees into logs?

4 Three women dig a vegetable plot in 4 hours.

a How long will it take 1 woman to dig the same size plot?

b How long will it take 5 women?
Give your answer in hours and minutes.

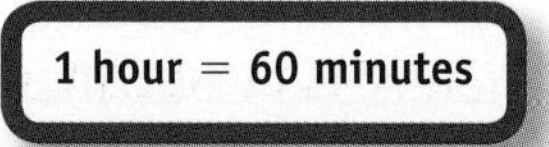

5 A farmer wants to plant 320 hawthorn bushes to make a hedge.
He works out that 1 person can plant 6 bushes in an hour.
He wants the whole hedge planted in 8 hours or less.
How many people does the farmer need for the job?

C AO3

6 It takes 60 workers 1 hour 20 minutes to erect a stage for a concert.
How long will it take 45 workers to erect the stage.
Give your answer in hours and minutes.

7 In a hotel, it takes one cleaner 90 minutes to clean 5 rooms.
The hotel has 48 rooms.
All the rooms have to be cleaned between 10 am and 2 pm.
How many cleaners does the hotel need?

B

8 Pen is retiling her bathroom.
There were 200 old tiles on the walls. Each tile covered 225 cm^2.
Each of the new tiles covers 300 cm^2.
How many of the new tiles does she need?

AO3

Review exercise

1 A model of St Paul's Cathedral in London is 70 cm long.
The model is built using a ratio of 1 : 250.
Work out the real-life length of the cathedral.
Give your answer in metres. **[2 marks]**

2 $\frac{9}{20}$ of the people at a pub quiz are men.
What is the ratio of men to women at the pub quiz? **[2 marks]**

D AO2

3 Washing powder is sold in two sizes.

The small packet contains 350 g and costs £2.25.
The large packet contains 800 g and costs £4.99.
Which is the better buy? **[3 marks]**

4 A car travels 500 km on 25 litres of petrol.
How far will it travel on 60 litres of petrol? **[2 marks]**

D

5 A printer can print 8 pages in one minute.
How many minutes will it take to print a 240-page document? **[2 marks]**

C

6 In a class of 27 pupils, the ratio of boys to girls is 4 : 5.
How many of the pupils are

a girls **[1 mark]**

b boys? **[1 mark]**

7 Two sisters share £45 in the ratio 2 : 3.
How much is the larger share? **[2 marks]**

8 Ana and Berwyn share a lottery win in the ratio 3 : 4.
Berwyn receives £2000.
How much does Ana receive? **[2 marks]**

9 Mr Jones leaves £10 000 to his grandchildren, Al, Deb and Cat, to be shared in the ratio 5 : 3 : 2.
Work out how much each receives. **[3 marks]**

10 A chemical reaction uses chemical A and chemical B in the ratio 3 : 7.
A chemist has 450 mg of chemical A.
How much of chemical B does she need? **[2 marks]**

C
AO3

11 A camera in England costs £50.
The same camera in France costs €75 (euros).
The exchange rate is £1 = €1.2.
In which country is the camera cheaper, and by how much?
You must show your working.
State the units of your answer. **[3 marks]**

C

12 1.5 litres of paint covers an area of 36 m^2.

a How many litres of paint do you need to cover 94 m^2? **[2 marks]**

b Paint comes in 1.5 litre tins.
How many 1.5 litre tins do you need to buy? **[1 mark]**

You may need to round up to get a sensible answer.

AO2

13 It takes 3 plumbers 7 hours to install a central heating system.
How long will it take 2 plumbers to do the same job? **[2 marks]**

B
AO3

14 Builders remove 4000 tiles from the school hall floor.
Each tile measures 50 cm by 50 cm.
The builders lay a new laminate floor in the hall.
Laminate comes in strips measuring 30 cm by 2 metres.
How many laminate strips will they need? **[4 marks]**

15 A factory makes jeans.
Four machinists produce 21 pairs of jeans in one day.
To fill a large order, the factory needs to produce 800 pairs of jeans in two days.
How many machinists do they need?

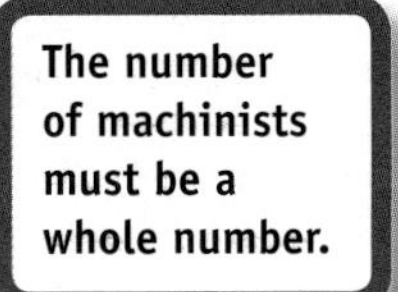

[4 marks]

Chapter summary

In this chapter you have learned how to

- simplify a ratio to its lowest terms E
- use a ratio in practical situations D
- write a ratio as a fraction D
- understand direct proportion D
- solve proportion problems, including using the unitary method D
- work out which product is the better buy D
- use ratio to find one quantity when the other is known D C
- share a quantity in a given ratio D C
- write a ratio in the form $1 : n$ or $n : 1$ C
- solve word problems involving ratio C
- solve word problems involving direct and inverse proportion D C B
- understand inverse proportion C B

Come dine with Brian

It's Brian's birthday, and he wants to have some fun. He decides to cook a meal for himself and seven friends.

Here is the food he plans to serve.

menu
stilton soup
leek and macaroni bake
rhubarb and orange crumble

Question bank

1 How much butter does Brian need for the Stilton soup?

2 What is the total amount of butter that Brian needs for the meal?

3 Write a list showing the total amounts of all the ingredients that Brian needs for the rhubarb and orange crumble.

Brian already has some of the ingredients he needs in his kitchen cupboard.

Brian says, 'I only need to buy seven ingredients as I already have the rest.'

4 Is Brian correct? Explain your answer.

Brian also says, 'It should only take about 3 hours to prepare and cook the whole meal.'

5 Is Brian correct? Show working to support your answer.

6 What other things do you think Brian needs to buy to go with the meal?

Information bank

Chef's top tips

- Soup can be prepared in advance and re-heated before serving.
- One vegetable stock cube will make 200 ml of vegetable stock.
- Leek and macaroni bake must be served straight from the oven.
- Organic leeks have a better flavour than normal leeks.
- Make your own breadcrumbs from slices of white bread.
- Tinned rhubarb can be used instead of fresh rhubarb.
- For an extra-sweet finish, sprinkle icing sugar on top of the crumble.

stilton soup [serves 4]

125 g Stilton cheese
50 g butter
40 g plain flour
900 ml vegetable stock
300 ml skimmed milk
60 ml double cream

45 ml white wine
1 onion

Cooking time: 45 minutes
Preparation time: 25 minutes

leek and macaroni bake [serves 6]

450 g leeks
300 g Cheddar cheese
240 g macaroni
75 g butter
60 g breadcrumbs
60 g plain flour

900 ml full-fat milk
30 g chives

Cooking time: 45 minutes
Preparation time: 20 minutes
Oven temperature: 190°C

rhubarb and orange crumble [serves 10]

1.4 kg rhubarb
225 g butter
150 g caster sugar
150 g self-raising flour
150 g plain flour
125 g granulated sugar

2 eggs
1 orange

Cooking time: 1 hour
Preparation time: 30 minutes
Oven temperature: 200°C

In his kitchen, Brian already has:

250 g butter	400 g Cheddar cheese
1 kg plain flour	$\frac{1}{2}$ dozen eggs
1 kg self-raising flour	8 vegetable stock cubes
1 *l* skimmed milk	500 g granulated sugar
1 *l* full-fat milk	$\frac{1}{2}$ bottle of red wine
250 m*l* single cream	$\frac{1}{2}$ loaf of white bread
250 m*l* double cream	1 onion
500 g spaghetti	3 oranges

8 Complex calculations

This chapter is about using calculators to solve problems.

The computers that got Apollo 11 to the Moon were *less* powerful than today's average scientific calculator!

Objectives

This chapter will show you how to

- **solve problems involving repeated percentage change** C
- **solve reverse percentage problems** B
- **handle very large and very small numbers using standard form** B

Before you start this chapter

1 Work out these without using a calculator.

a Increase £250 by 8%. b What is 266 ÷ 0.7? c Divide 45 000 000 by 0.005.

2 Write these numbers correct to one decimal place.

a 5.681 b 11.06 c 21.974

3 Write these numbers correct to two significant figures.

a 6842 b 153 945 c 0.000 382

8.1 Repeated percentage change

Keywords
compound interest, invest, multiplier

Why learn this?
Savings accounts are all about repeated percentage change.

Objectives
C Perform calculations involving repeated percentage changes

Skills check

1 Work out
   a 6% of £500 **b** 3% of £2000
2 Pete scored 85% in a history test.
   The test was out of 40.
   How many marks did Pete score?

Compound interest

Examples of repeated percentage change include **compound interest** and population changes.

When you **invest** money the interest you earn is usually calculated using compound interest.

In compound interest
- the rate of interest is fixed
- the amount of interest you receive each year is not the same
- each year's interest is calculated on the sum of money you invested in the first place *plus* any interest you have already received.

When you put money into a savings account, the interest you earn is a percentage of the amount in your account.

The amount you pay back on a loan is also calculated using compound interest.

Example 1

Venetia puts £800 into a savings account earning 6% per annum interest.
How much does she have after two years?

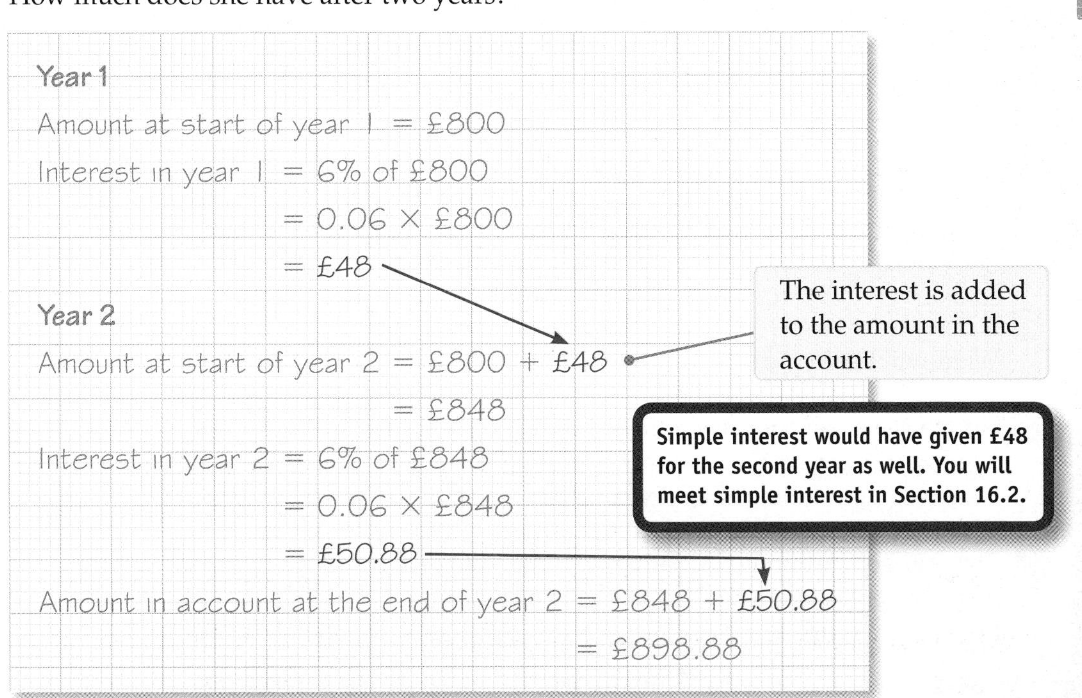

Repeated percentage change

When the same percentage change is repeated, as in Example 1, you can simplify the calculations.

1 Add the rate of interest to 100%.
2 Convert this percentage to a decimal to get the **multiplier**.
3 Multiply the original amount by the multiplier as many times as the number of years for which the money is invested.

Using this method for Example 1 ...

1 6% + 100% = 106%
2 106% = 1.06
3 £800 × 1.06 × 1.06 = £898.88

1.06 is called the multiplier.

Multiply by 1.06 *twice* because the money is invested for *two* years.

C

Example 2

There are approximately 5000 whales of a certain species, but scientists think that their numbers are reducing by 8% each year.

Estimate how many of these whales there will be in three years.

The multiplier is 100% − 8% = 92%
= 0.92

This time there is a loss each year so you must subtract the percentage from 100%.

Number in three years' time = 5000 × 0.92 × 0.92 × 0.92
= 3893.44

The estimated population in three years' time is 3893 whales.

Exercise 8A

C

1 Amy invests £135 for two years at a rate of 4% per annum compound interest. How much will she have at the end of the two years?

2 A new car costs £12 000. Each year the value of the car depreciates by 8%. What will the car be worth at the end of three years?

3 Sami has a choice of ways to invest the £250 legacy left by her grandfather. She can invest it at 4% per annum for 5 years or she can invest it for 2 years at 5% per annum followed by 3 years at 3% per annum. Which way is better?

C A02

4 The seal population in Scotland is estimated to decline at a rate of 16% each year owing to pollution. In 2007, 3000 seals were counted. How many years will it take for the seal population to fall below 1000?

C

5 A company leases a car at £260 per month. In the leasing agreement, the price is reduced by 2% each month after the first 6 months.

a Work out the difference between the amounts paid in the 6th and 7th months.

b How much will the car cost in total in the first 9 months?

6 Mamet earns £14 250 per year. His contract says this will increase each year in line with the annual rate of inflation. An estimate for the annual rate of inflation is 3%.

a Find his salary at the end of three years.

b Find his monthly pay at the end of five years.

c If the rate of inflation is 5% instead of 3%, how much more is his monthly pay at the end of five years?

Compare the monthly pay for the two rates of inflation.

7 How long will it take each of the following investments to reach £1 000 000?

a £200 000 invested at a rate of 15%.

b £150 000 invested at a rate of 19%.

AO2

8 If £7400 grows to £10 873 in five years, what is the rate of compound interest?

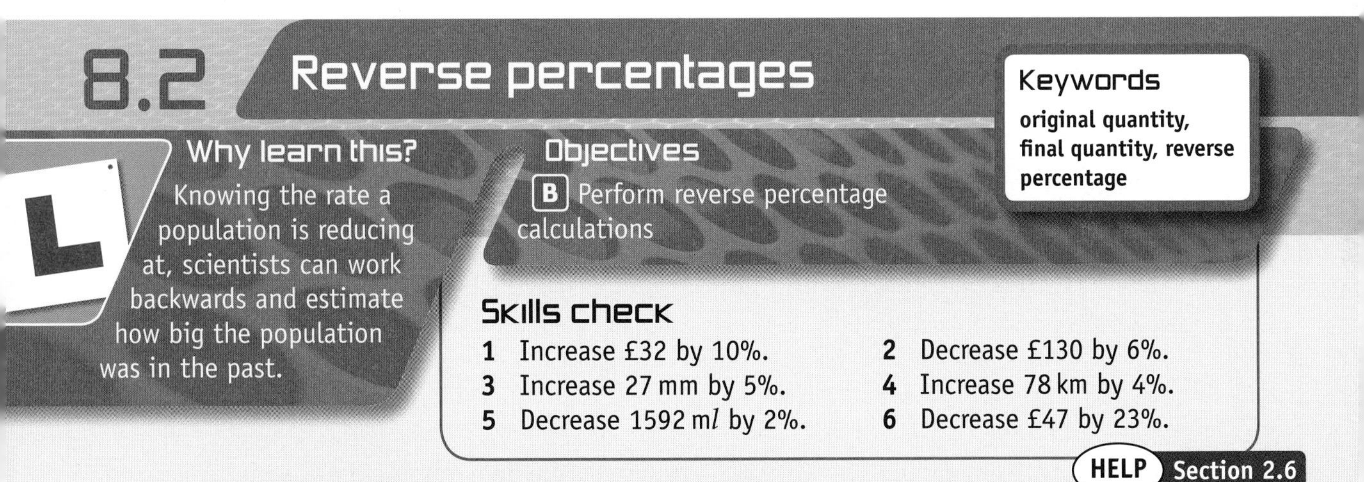

8.2 Reverse percentages

Why learn this?

Knowing the rate a population is reducing at, scientists can work backwards and estimate how big the population was in the past.

Objectives

B Perform reverse percentage calculations

Keywords

original quantity, final quantity, reverse percentage

Skills check

1 Increase £32 by 10%.
2 Decrease £130 by 6%.
3 Increase 27 mm by 5%.
4 Increase 78 km by 4%.
5 Decrease 1592 ml by 2%.
6 Decrease £47 by 23%.

HELP Section 2.6

Finding the original quantity 1

In some problems you are told the quantity after a percentage increase or decrease and you have to work out the **original quantity**.

Remember that the original quantity in any calculation is always taken to be 100% and the **final quantity** will be a given percentage above or below 100%, depending on whether there has been an increase or a decrease.

Consider this question.

A bike is reduced by 20% in a sale. The sale price is £135.
What was the original price?

You have not been told the original amount, the 100% figure and you need to work it out.

These questions are called **reverse percentage** problems.

There are two methods for answering these questions.

Method A	Method B
1 Work out what percentage the final quantity represents.	1 Work out what percentage the final quantity represents.
2 Divide by this percentage to find 1%.	2 Divide by 100 to get the multiplier.
3 Multiply by 100 to get the 100% figure.	3 Divide the final quantity by the multiplier.

B

Example 3

A magazine's circulation is 30% down on last month's figure.
This month they sold 490 copies of the magazine.
How many copies did they sell last month?

You do not have last month's figure, which is the original amount.

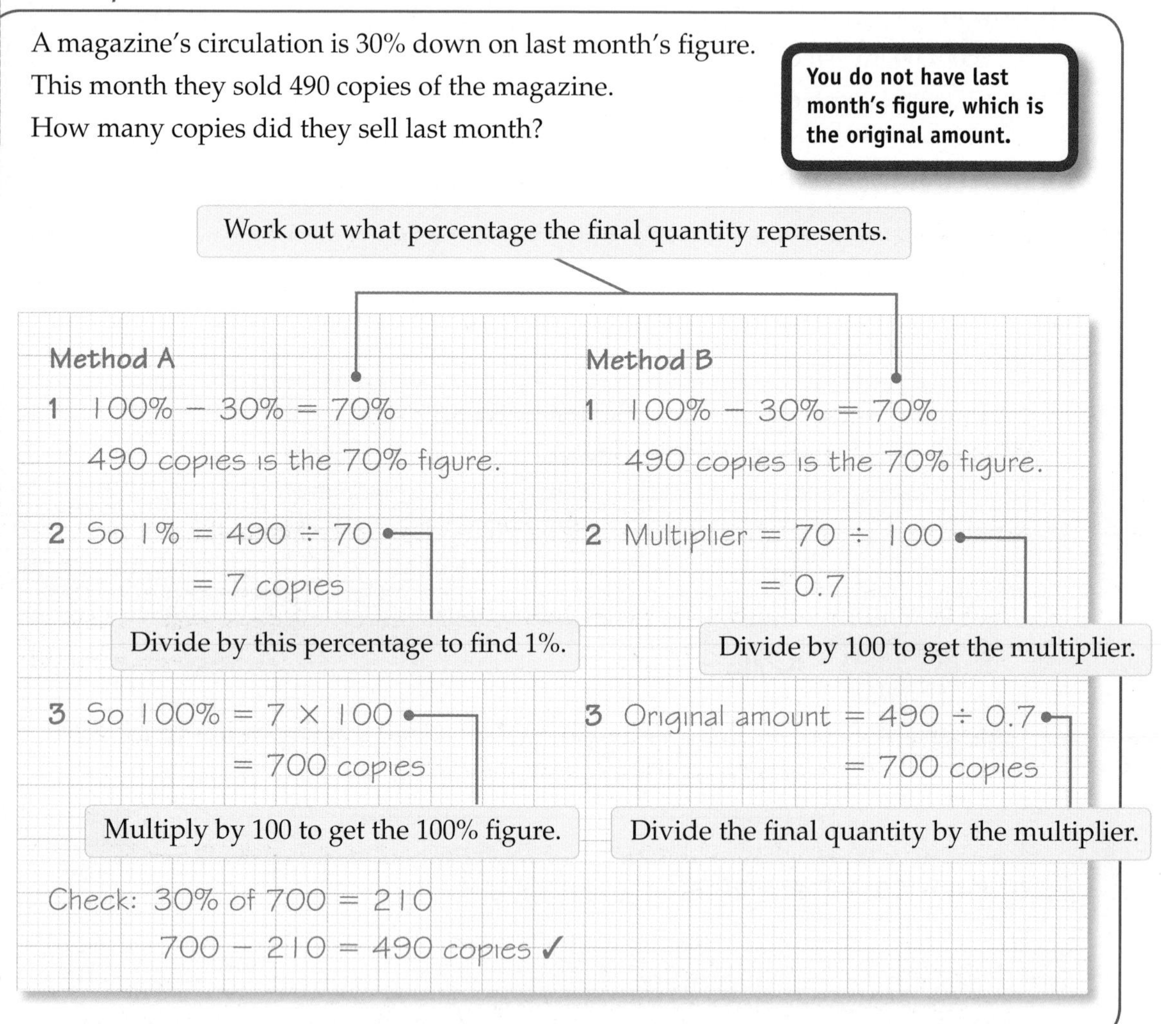

Finding the original quantity 2

You can use the same methods for a percentage increase.

Example 4

The number of students applying to Intelligentsia University has increased by 15% compared to last year.

This year the university had 9890 applications.

How many applications did it have last year?

Again you do not have the original amount, the number of applications before the rise.

Method A

100% + 15% = 115%

9890 applications is the 115% figure.

So 1% = 9890 ÷ 115
= 86 applications

So 100% = 86 × 100
= 8600 applications

Method B

100% + 15% = 115%

9890 applications is the 115% figure.

Multiplier = 115 ÷ 100
= 1.15

Original amount = 9890 ÷ 1.15
= 8600 applications

B

Exercise 8B

1 A laptop is reduced by 12% in a sale. It now costs £334.40. What was the original price of the laptop?

2 After a pay rise of 4%, Kate earns £14 560. What did she earn before the pay rise?

3 A mobile phone was priced at £135 including VAT. What would be the price without VAT?

Assume VAT = 17.5%.

4 A spring is stretched by 32% to a length of 22 cm. What length was it before it was stretched? Give your answer to the nearest millimetre.

5 A seal pup increases its body weight by $4\frac{1}{2}$% in one day. It now weighs 6.27 kg. What did it weigh yesterday?

6 A shop advertises, 'We will pay your VAT'. How much would I pay for a bike priced at £87.50 inclusive of VAT?

B

7 After paying a 12% deposit on a house, Ben needs a mortgage for £171 160. The estate agent will charge him $2\frac{1}{4}$% of the full cost of the house for a survey of the property. How much will he have to pay for the survey?

B

8 During heating, a metal rod expands by $1\frac{1}{4}$%.

a If the rod measures 40.5 cm when hot, what was its length originally?

b After 10 minutes the metal rod is still 0.5% longer than its original length. What is the length of the rod at this time?

AO2

8.3 Standard form

Keywords
standard form

Why learn this?

You can use standard form to write very large and very small numbers. The distance from the Earth to the Sun is approximately 1.5×10^{11} metres.

Objectives

B Interpret and use standard form

Skills check

1 Work out

a $325 \div 100\,000$ **b** $0.008\,57 \times 1000$ **c** $3.665 \times 10\,000$

2 Work out

a 1.5×2 **b** $3.6 \div 20$ **c** $3.2 \div 0.08$

Using and interpreting standard form

It is often convenient to write very large numbers or very small numbers in **standard form**.

All numbers can be expressed as $x \times 10^n$, where x is a number between 1 and 10 and n is an integer.

Numbers greater than 1

B

Example 5

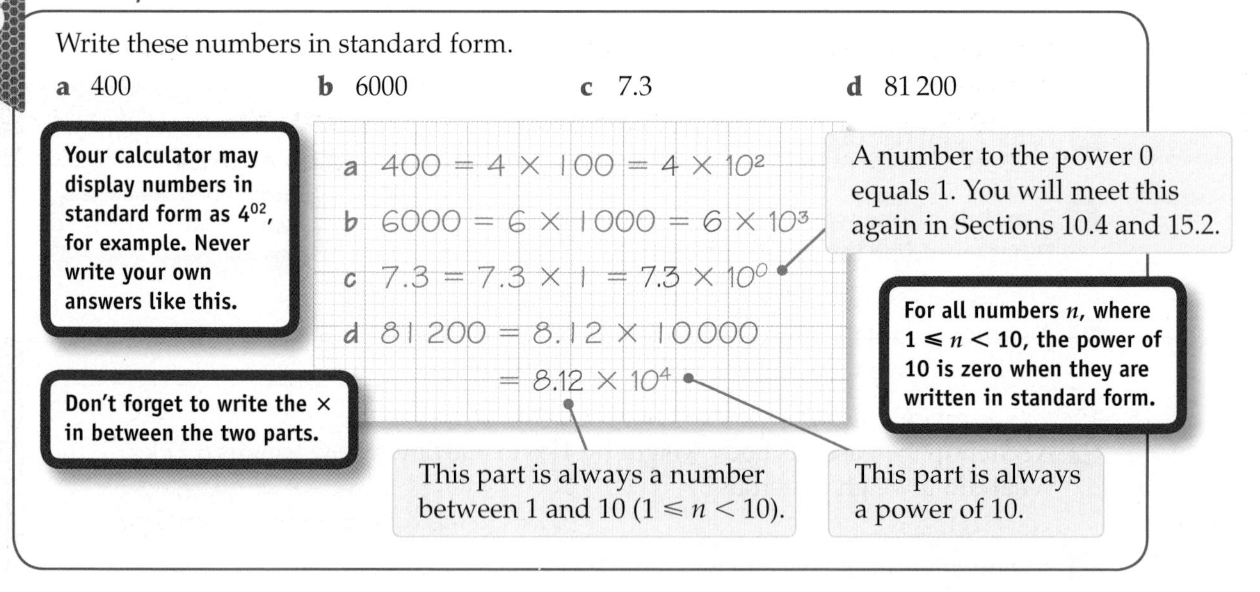

Write these numbers in standard form.

a 400 **b** 6000 **c** 7.3 **d** 81 200

Exercise 8C

B

1 Write these numbers in standard form.

a 3 000 000 **b** 7400 **c** 32 000 **d** 603 500 **e** 108

f 68 **g** 650.5 **h** 99.9 **i** 5 **j** 2.04

2 Write down the power of 10 for each of these numbers when they are written in standard form.

a 45.7 **b** 3.002 **c** 8293

d 94 000 **e** five million **f** one hundred thousand

Numbers less than 1

Example 6

Write these numbers in standard form.

a 0.6 **b** 0.03 **c** 0.0078 **d** 0.000 419

a $0.6 = 6 \times 0.1 = 6 \times 10^{-1}$

b $0.03 = 3 \times 0.01 = 3 \times 10^{-2}$

c $0.0078 = 7.8 \times 0.001 = 7.8 \times 10^{-3}$

d $0.000\,419 = 4.19 \times 0.0001 = 4.19 \times 10^{-4}$

A negative power is the reciprocal of the corresponding positive power. You will meet this again in Section 10.4. For example, $10^{-2} = \frac{1}{10^2} = \frac{1}{100} = 0.01$

Note that the power of 10 is always negative for numbers less than 1.

This part is always a number between 1 and 10 ($1 \leqslant n < 10$).

Exercise 8D

1 Write these numbers in standard form.

a 0.0005 **b** 0.006 **c** 0.4 **d** 0.000 12
e 0.0717 **f** 0.000 197 5 **g** 0.9009 **h** 0.001 000 3

2 Write down the power of 10 for each of these numbers when they are written in standard form.

a 0.0745 **b** 0.003 090 4 **c** 0.000 054 6
d 0.000 000 027 1 **e** one thousandth **f** three quarters

Changing between standard form and a decimal number

Example 7

Write these standard form numbers as decimal numbers.

a 3.709×10^5 **b** 4.19×10^{-4}

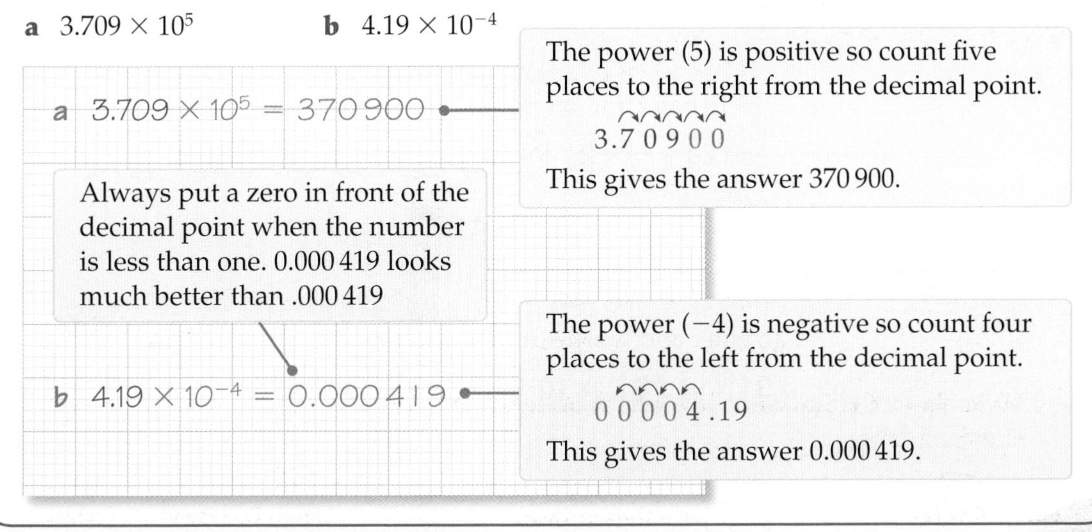

Exercise 8E

B

1 Write these as decimal numbers.

a 5×10^4 b 3.8×10^3 c 6×10^{-3} d 7.26×10^9

e 8.492×10^{-2} f 4.37×10^6 g 1.006×10^{-4} h 6.2387×10^3

2 Write these numbers given in standard form as decimal numbers.

a The distance from the Earth to the Sun is approximately 1.488×10^8 kilometres.

b The average width of an iris of an eye is 1×10^{-2} metres.

c The mass of a billion molecules of water is 3×10^{-14} grams.

3 These numbers are *not* in standard form. Rewrite them in standard form.

a 123×10^2 b 0.8×10^7 c 17×10^{-2}

d 0.25×10^{-4} e 18 million f $\frac{1}{8}$

g $36 \times 10^4 \times 0.006$ h $\sqrt{40 \times 10}$

Calculators and standard form

To enter a number given in standard form on your calculator you can use the button marked

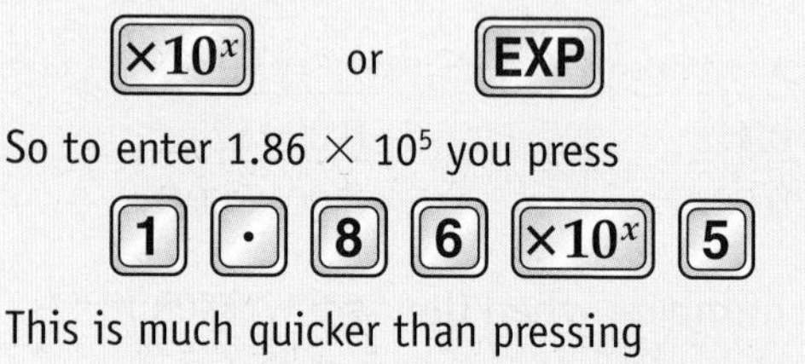

This is much quicker than pressing

1 · 8 6 × 1 0 $x^{■}$ 5

Never press × or 1 0 when using ×10^x or EXP.

B

Example 8

Work out

a $(2.2 \times 10^3) \times (4 \times 10^5)$ b $(3.3 \times 10^6) \div (3 \times 10^{-4})$

a $(2.2 \times 10^3) \times (4 \times 10^5)$

To enter and work out this, you press

2 · 2 ×10^x 3 × 4 ×10^x 5 =

The answer is 880 000 000 or 8.8×10^8.

b $(3.3 \times 10^6) \div (3 \times 10^{-4})$

To enter and work out this, you press

3 · 3 ×10^x 6 ÷ 3 ×10^x − 4 =

The answer is 11 000 000 000 or 1.1×10^{10}.

Exercise 8F

1 Work out these, giving your answers in standard form.

a $(4.2 \times 10^4) \times (5.3 \times 10^3)$ **b** $(8.4 \times 10^5) \div (2 \times 10^4)$

c $(5.6 \times 10^3) \div (4 \times 10^{-2})$ **d** $(1.6 \times 10^{-2}) \div (4 \times 10^{-4})$

2 This year it is estimated that, on average, each person in the United Kingdom will spend £170 making mobile phone calls. If the population is 6.1×10^7, what is the total amount spent on mobile phone calls? Give your answer in standard form.

3 The radius of the Earth is approximately 6.4×10^3 km.
Estimate the volume of the Earth if you assume it is a sphere.

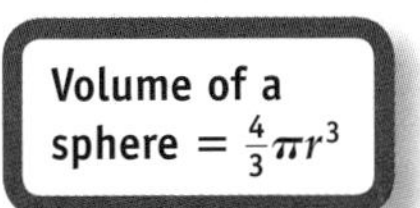

4 There are approximately 6×10^9 people in the world and they eat approximately 7×10^7 tonnes of fish per year. How much fish does each person eat on average?

5 The diameter of the Sun is approximately 1.392×10^6 km.
Find its surface area using the formula $4\pi r^2$.

6 A rectangular picture measures 1.4×10^2 cm by 2.7×10^3 cm. What is

a the area

b the perimeter of the picture?

7 After the Sun, the next nearest star to the Earth is Proxima Centauri. It takes about 4.24 years for light from this star to reach the Earth.
If light travels at 1.86×10^5 miles per *second*, estimate the distance of Proxima Centauri from Earth.

8 A teaspoon of oil (5 m*l*) is spilled onto a flat area of water.
It covers 0.4 hectares. What is the thickness of the oil?

1 hectare = 100 m × 100 m

Review exercise

1 Amy invests £1025 for four years at a rate of 4% per annum compound interest.
George invests £997 for four years at a rate of 5% per annum compound interest.
Who will have more money after four years?
Show your working. **[3 marks]**

2 The population of robins is decreasing in parts of Britain. One estimate is that a population of 2000 pairs is decreasing at a rate of 3% each year.
How many pairs of robins will there be at the end of five years? **[3 marks]**

3 Jane earns £21 000 a year. This increases by 6% each year.
How long will it take her to earn over £25 000 per year? **[3 marks]**

4 Kaps buys a guitar and then sells it on for £313.65. This is an increase of 23% on the price he paid for it. How much profit has Kaps made? **[3 marks]**

5 The value of a car depreciates by 17% in its first year.

a If it is worth £12 035 after one year, what was its original price? **[3 marks]**

b The original price included VAT.
How much was the car excluding VAT? **[3 marks]**

Assume VAT = 17.5%.

B

6 Hema pays a restaurant bill of £63.28 but later realises that she has been overcharged by 12%. How much should Hema have paid? **[3 marks]**

7 The total number of pupils in school this year is 1034. This is a 6% decrease on last year.

a How many pupils were there last year? **[3 marks]**

b Next year they expect an increase in pupils of 3% on this year. How many pupils will there be then? **[2 marks]**

8 If $a = 1.2 \times 10^7$ and $b = 4.2 \times 10^6$, find the following, giving your answers in standard form.

a $3b$ **[1 mark]**

b ab **[1 mark]**

c a^2 **[1 mark]**

d $a + b$ **[1 mark]**

e $b \div a$ **[1 mark]**

9 The number of krill (a type of plankton) is estimated to be 6×10^{15} and their mass is about 6.6×10^8 tonnes. What is the mass of one krill in grams? **[2 marks]**

1 tonne = 1000 kg

Chapter summary

In this chapter you have learned how to

- perform calculations involving repeated percentage change **C**
- perform reverse percentage calculations **B**
- interpret and use standard form **B**

Quality of written communication: Some questions on this page are marked with a star ☆. In the exam, this sort of question may earn you some extra marks if you

- use correct and accurate maths notation and vocabulary
- organise your work clearly, showing that you can communicate effectively.

E AO2

1 The mean of five numbers is 8.

a What is the sum of the five numbers? [1]

b Four of the numbers are 2, 3, 6 and 11. Work out the fifth number. [1]

D AO2

2 Isaiah used a computer program to record the reaction times of the students in his year group. He recorded his results in this grouped frequency table.

Reaction time, t (s)	$0.1 < t \leqslant 0.2$	$0.2 < t \leqslant 0.3$	$0.3 < t \leqslant 0.4$	$0.4 < t \leqslant 0.5$	$0.5 < t \leqslant 0.6$	$0.6 < t \leqslant 0.7$
Frequency	7	18	26	21	11	2

a Draw a cumulative frequency diagram to illustrate this data. [3]

b Estimate the number of students who had a reaction time of less than 0.25 seconds. [1]

c Draw a box plot for this data. [4]

C AO2

☆ **3** A health club wants to carry out a survey about its facilities. The manager surveys the first twenty people who visit the health club on a Monday morning. Is this likely to be a representative sample? Give a reason for your answer. [2]

☆ **4** Kirsten is investigating the effect post-16 education has on salaries. She asked 15 people about their starting salary in their first job.

Number of years of post-16 education	2	2	0	5	5	4	2	3	0	6	4	0	5	2	1
Starting salary in first job (£1000s)	17	21	12	24.5	23	20.5	18.5	18	10.5	25	19.5	13	20	14.5	15

a Draw a scatter graph for this data [3]

b Draw a line of best fit on your scatter graph [1]

c Use your line of best fit to estimate the number of years of post-16 education for someone with a starting salary of £17 000. [1]

d Helen has just finished a PhD. She has completed 10 years of post-16 education. Why would it not be sensible to use this data to estimate her starting salary in her first job? [1]

B AO2

5 The probability that it rains in the morning is 0.6. The probability that it rains in the afternoon is 0.8.

a Copy and complete this tree diagram to show all the possible outcomes. [3]

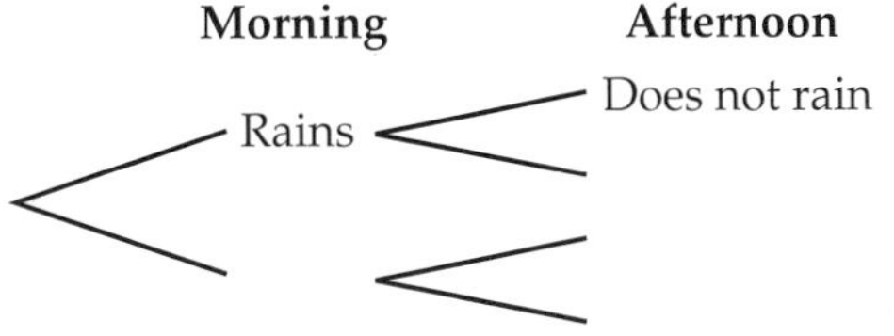

b Work out the probability that it rains at some point during the day. [2]

c Calculate the probability that it does not rain all day. [1]

Quality of written communication: Some questions on this page are marked with a star ☆. In the exam, this sort of question may earn you some extra marks if you

- use correct and accurate maths notation and vocabulary
- organise your work clearly, showing that you can communicate effectively.

E

1 Write down 5 whole numbers with a mean of 4.2. [2]

AO3

D

☆ **2** Olivia is conducting a survey for a project on road safety. She wants to know how the members of her class travel to school, how long it takes them and how many roads they have to cross.

a Write three questions that Olivia could include in her survey. [3]

b Design response boxes for each question. [3]

3 A bag contains black, blue and red marbles. There are 11 black marbles and 3 blue marbles in the bag. The probability of picking a red marble is $\frac{3}{5}$. How many marbles are in the bag in total? [2]

4 Gemma is running a spinner game at her school fete.

180 people play her game.
How much profit would you expect Gemma to make? [3]

AO3

C

☆ **5** Pippa and Nisha are playing a game. They each roll a fair 6-sided dice.
If the difference between the two numbers is odd then Pippa wins a point.
If the difference between the two numbers is even, Nisha wins a point.
If the two numbers are the same nobody wins a point.

a Is this game fair? Give a reason for your answer. [3]

b Calculate Pippa and Nisha's expected scores after 150 rolls. [2]

AO3

B

☆ **6** Mr Keane was comparing his class's scores on the mock exam with their scores on the actual exam. He recorded his data in a cumulative frequency graph.

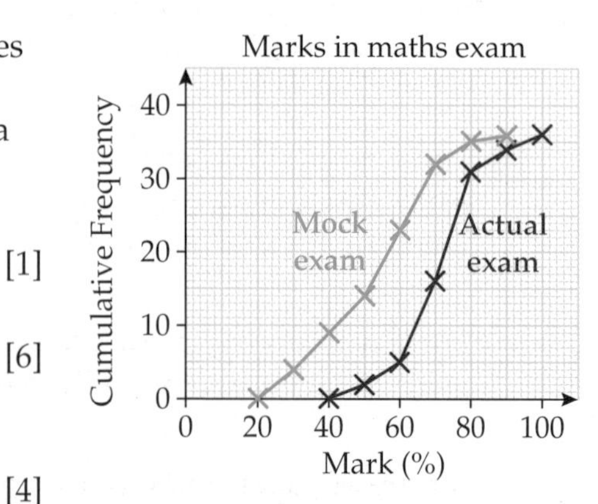

a How many students does Mr Keane have in his class? [1]

b Draw box plots for the mock exam results and actual exam results. [6]

c Compare the scores from the two exams. Give reasons for any statements that you make. [4]

d Do you think the more able students improved more or less than the less able students between the two tests. Give a reason for your answer. [2]

AO3

Problem-solving practice: Number

Quality of written communication: Some questions on this page are marked with a star ☆. In the exam, this sort of question may earn you some extra marks if you

- use correct and accurate maths notation and vocabulary
- organise your work clearly, showing that you can communicate effectively.

E

1 In a survey of 250 year 10 students, 78% said they ate school dinners.
How many students did not eat school dinners? [2]

☆ 2 Emma has bought a pack of 8 yoghurts. 3 of them are strawberry and the rest are vanilla. Emma says that the ratio of strawberry yoghurts to vanilla yoghurts is 3 : 8. Is Emma correct? Give a reason for your answer. [2]

A02

D

3 Jonah is making pancakes for a party.
He uses this recipe.
Jonah buys milk in 2 pint bottles.
How many bottles will he need to buy to make 150 pancakes?

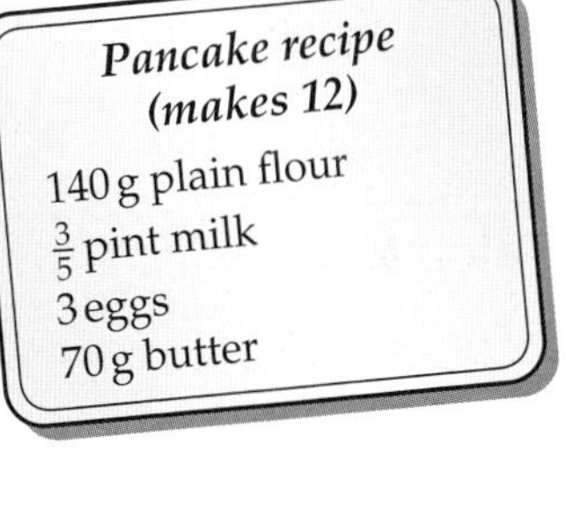

[2]

4 The diagram shows part of a supermarket receipt.
Calculate the cost of

5 × tin baked beans	£3.10
2 × farmhouse loaf (800 g)	£2.08

a 3 farmhouse loaves [2]

b 2 tins of baked beans. [2]

A02

C

5 A mining company calculates price indexes for iron ore using 2005 as the base year.

Year	2005	2006	2007	2008	2009
Price per tonne (pence)	60	72	90	140	126

a Calculate the price index for iron ore in 2006. [2]

b Calculate the price index for iron ore in 2009. [1]

c In 2010 the price index was 182.
Calculate the price per tonne of iron ore in 2010. [2]

6 Building mortar is made by mixing cement, sand and lime in the ratio 4 : 12 : 1.
Andy has 12 kg of cement and plenty of sand and lime.
Calculate the total amount of mortar Andy can make. [3]

7 Three workers can paint a room in 4 hours. How long will it take five workers to paint an identical room? Give your answer in hours and minutes. [3]

A02

B

8 A ream of paper contains 500 sheets. It is 4.2 cm thick. Calculate the thickness of one sheet of paper in metres. Give your answer in standard form. [2]

A02

Quality of written communication: Some questions on this page are marked with a star ☆. In the exam, this sort of question may earn you some extra marks if you
- use correct and accurate maths notation and vocabulary
- organise your work clearly, showing that you can communicate effectively.

E

1 A machine at a car factory produces 2200 engine components each day. 2.5% of all the components it produces are faulty. 60% of the non-faulty components are used immediately and the rest are sent to a warehouse.
How many components are sent to the warehouse each day? [3]

AO3

D

2 A petrol generator uses $3\frac{1}{3}$ litres of petrol each hour. The tank on the generator can hold $42\frac{1}{2}$ litres of petrol. How long can the generator run on a full tank of petrol? Give your answer in hours and minutes. [3]

3 Callum is comparing two mobile phone packages.

Business First	**Leisure One**
£18 per month + 17.5% VAT	One off yearly payment of £260
10% DISCOUNT FOR DIRECT DEBIT	20% DISCOUNT FOR DIRECT DEBIT

Which package is cheaper? [5]

☆ **4** A supermarket sells coffee in two different sizes.
Which packet offers the best value for money?
Give a reason for your answer. [3]

AO3

C

5 Jonathan mixes squash and water in the ratio 2 : 7 to make a drink.
Chloe mixes squash and water in the ratio 3 : 10.
Whose drink has the higher proportion of squash? [3]

6 Shilpa and Alex are taking part in Young Enterprise. They have set up a company which packages and sells pot pourri. They buy six 2.5 kg bags of pot pourri costing £14 each. They package up the pot pourri in 250 g boxes which sell for £2.99 each.
The packaging for each box costs 45p.
Shilpa and Alex sell all their pot pourri.
Calculate their profit as a percentage of the total amount they spent. [5]

7 Theo and Deborah are investing in a company. Theo invests £30 000 and Deborah invests £50 000. Between them they have a 15% share in the company. After five years the company is sold for £1.2 million.
Calculate the amount of money that they each receive. [3]

AO3

B

8 One fisherman can mend $2\,m^2$ of net in 45 minutes.
The captain of a boat wants to mend this rectangular fishing net in less than three hours.
How many fishermen will he need? [3]

5.1 m
10.2 m

AO3

9 Number skills

BBC Video

This chapter explores different written methods of calculation.

The Pontysyllte Aqueduct in North Wales was built more than 200 years ago, long before calculators were invented.

All the engineering calculations had to be done using pencil and paper.

Objectives

This chapter will show you how to

- use standard column procedures for multiplication of integers E
- give solutions in the context of the problem E
- round to one significant figure E
- multiply and divide by a negative number E
- multiply and divide integers E D
- check and estimate answers to problems E D
- estimate answers to problems involving decimals E D C

Before you start this chapter

Put your calculator away!

1 What is the value of the digit 7 in each of these numbers?

a 374 b 7250
c 2.67 d 17 900
e 13.027 f 89.70

2 Work out

a $4 \times 6 - 3$ b $8 + 6 \div 2$
c $12 - (3 + 2)$ d $24 \div (3 \times 4) + 5$

3 Arrange each set of numbers in order, from smallest to largest.

a 3, −4, 5, −7 b −6, 2, −8, 1.5
c 7, −9, 4.5, −3 d −2.5, 4, −1, 0.5

P Number skills: adding and subtracting

1 Work out

a $324 + 123$ b $1475 + 987$ c $876 - 449$ d $1140 - 576$

2 Find the total of £359, £78 and £265.

3 What is the difference between 1220 and 662?

4 From the numbers in the box, write down

a two numbers which have a sum of 100

b two numbers which have a difference of 50.

52 28 32 58 42 38 48 82 22 68

5 Work out

a $3.2 + 5.8$ b $3.62 + 2.24$ c $5.62 + 3.3$ d $2.4 + 5.16$

6 Work out

a $9.8 - 6.9$ b $4.2 - 1.9$ c $6.24 - 4.2$ d $8.7 - 6.94$

7 Steven pays for his shopping with a £10 note.
He receives £2.18 in change.
What was the cost of Steven's shopping?

8 Work out

a $3.02 + 2.9$ b $0.6 + 0.27$

c $11.59 + 2.4 + 7.27$ d $0.09 + 5.99 + 0.345$

9 Work out

a $3.09 - 1.7$ b $2.7 - 1.73$ c $9.91 - 7.84$ d $10 - 2.90$

10 The sum of two numbers is 312.6. One of the numbers is 156.2.
What is the other number?

11 A car in a showroom showed a total mileage of 8569 on its milometer.
On Monday the car was taken for a test run by a customer.
The customer did 52 miles on the test run.
On Tuesday the same car was taken for another test run.
The car was returned with 8720 on its milometer.
What distance was covered in the second test run?

12 Work out

a $6 - 4$ b $4 + (-8)$ c $-7 + (-5)$

d $-9 - (-5)$ e $12 + (-8)$ f $-11 - 6$

13 Copy and complete these calculations.

a $4 + \square = -7$ b $-4 + \square = 0$ c $-8 + \square = 2$

d $6 + \square = -2$ e $\square - (-4) = 7$ f $-3 + (-4) = \square$

14 Find the difference between

a -11 and -3 b -7 and 4 c 25 and -11

P Number skills: multiplying and dividing

1 Work out

a 23×10 b 45×100 c 82×1000 d 526×100 e 379×10

2 What must 725 be multiplied by to get 725 000?

3 Work out

a 2.7×10 b 73.9×100 c 4.5×1000 d 0.2×10 e 3.25×100

4 What is the value of the digit 5 in the answer to 4.57×100?

5 A theatre has 46 rows of 30 seats.
How many seats are there altogether?

6 A box contains 100 items each weighing 1.6 kg.
What is the total mass of the items in the box?

7 Work out

a 13×6 b 23×7 c 36×20 d 287×9 e 642×5

8 Work out the product of 189 and 4.

9 Work out

a $350 \div 10$ b $4600 \div 100$ c $5000 \div 1000$ d $35 \div 10$ e $460 \div 100$

10 Copy and complete these calculations.

a $\square \times 0.5 = 5$ b $36 \div \square = 0.36$ c $49 \div \square = 4.9$ d $2015 \div \square = 2.015$

11 Paperclips are packed into boxes of 100.
There are 230 000 paperclips.
How many boxes are there?

12 What is the value of the digit 8 in the answer to $208 \div 100$?

13 Work out

a $36 \div 3$ b $182 \div 7$ c $344 \div 4$ d $351 \div 9$

14 Share £296 between eight people.

15 An egg box holds six eggs. There are 278 eggs to put in boxes.

a How many egg boxes can be filled?

b How many eggs will be left over?

16 Use this fact to help you complete the calculations below.

$15 \times 5 = 75$

a $75 \div \square = 15$ b $15 \times 6 = \square$ c $15 \times 50 = \square$

17 Is each statement true or false?
Use an inverse operation to check each calculation.

a $36 + 27 = 63$ b $49 \times 3 = 174$ c $126 - 49 = 73$ d $132 \div 6 = 22$

9.1 Multiplying and dividing whole numbers

Why learn this?
Race engineers multiply the amount of fuel needed per lap by the number of laps to work out how much fuel is required for the whole race.

Objectives
E D Multiply whole numbers using written methods
E D Use repeated subtraction for division of whole numbers
E D Round up or down in context

Keywords
multiplication, repeated subtraction, division, divisor, remainder, round

Skills check

1 Work out
a 34×5 **b** 47×4 **c** 56×8 **d** 9×63

2 Work out
a 22×10 **b** 22×20 **c** 42×100 **d** 36×300

Methods of multiplication

You can use the grid method or the standard column method for written **multiplication**.

Grid method

Work out 24 × 23.

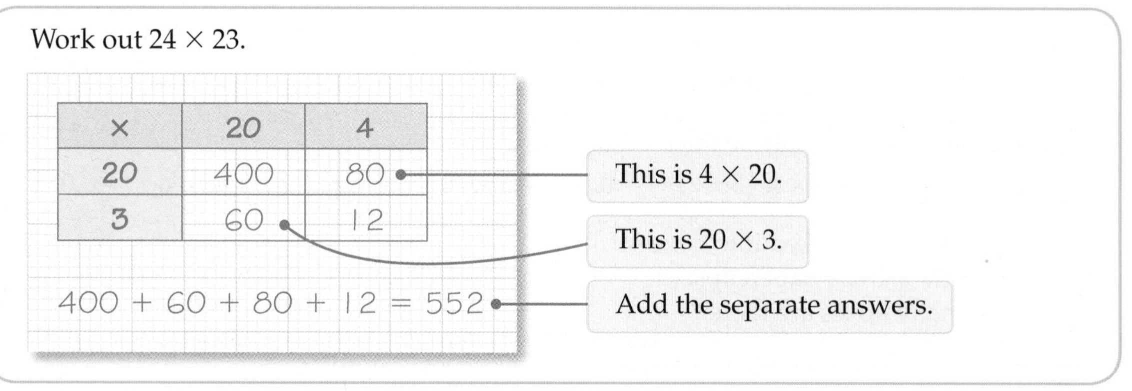

Standard method

Work out 24 × 23.

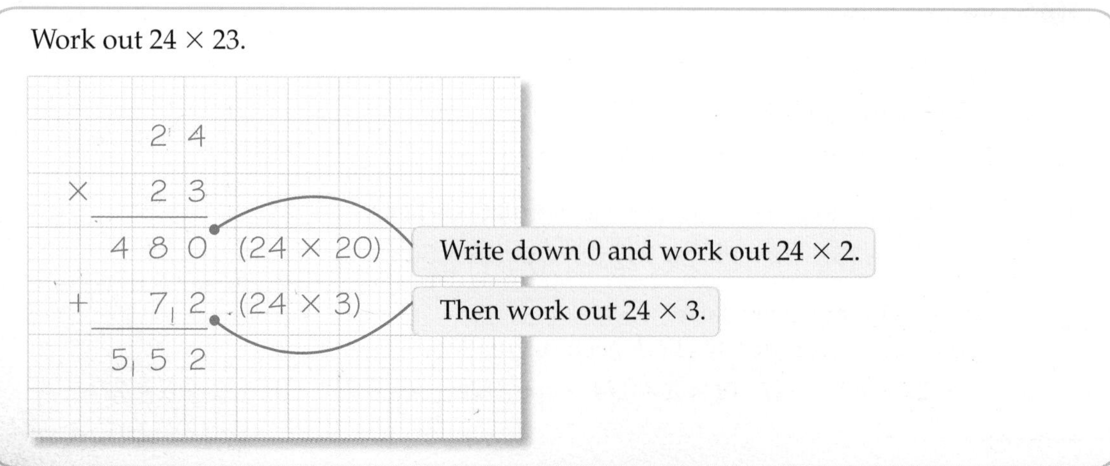

Exercise 9A

1 Jamie travels to Crewe 17 times a year to visit his family.
He travels by train and each return ticket costs £58.
He has saved £950 towards travel for the next year.

a What is the total cost of Jamie's travel for 1 year?

b Has Jamie saved enough money?
If not, how much more does he need to save?

> **You must show your working in these questions. This is very important on a non-calculator paper.**

2 Javier travels to Liverpool 24 times a year to support his football team.
He travels by train and each return ticket costs £126.
He has saved £2860 towards travel for the 2009–2010 season.
Has Javier saved enough money? If not, how much more does he need to save?

E

3 Yossi is a potato farmer. He is preparing an area of land for planting.
He needs to spread fertiliser over a rectangular piece of land which measures 195 m by 62 m.
Each bag of fertiliser covers an area of 54 m^2.
Yossi has 230 bags available.
Does Yossi have enough bags of fertiliser to complete the work?

> **Area = length × width**

AO3

Methods of division

You can use **repeated subtraction** for **division**.
The number you divide by is called the **divisor**.
Keep subtracting multiples of the divisor until you cannot subtract any more.
Then see how many lots of the divisor you have subtracted altogether.

Sometimes you will not be able to get to zero by subtracting multiples of the divisor.
The number you have left is called the **remainder**.

If you are answering a word problem, you may need to decide whether to **round** up or down.
For example, when working out the number of buses needed, the exact answer might be 3.6.
But you cannot have part of a bus, so you need to decide whether you need 3 buses or 4 buses.

Dividing using repeated subtraction

Work out 274 ÷ 12.

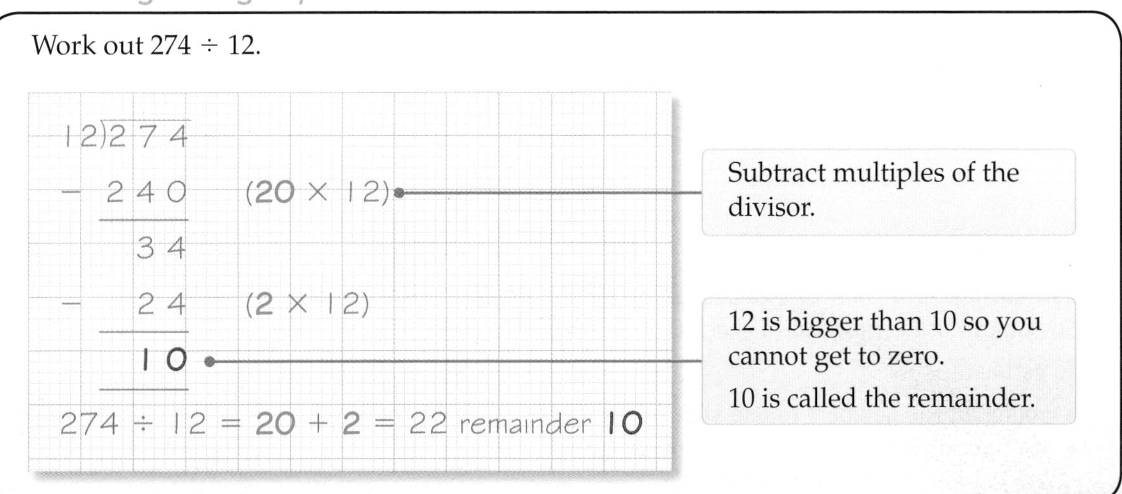

Exercise 9B

E

1 One tin of paint covers an area of 25 m².
The room to be painted has a wall area of 210 m².
How many tins of paint are needed for two coats?

You must show your working in these questions. This is very important on a non-calculator paper.

2 Nathan is laying laminate flooring in his office.
The area of the office is 600 m².
Each length of laminate covers an area of 3 m².
Each pack of laminate has 14 lengths.

a What is the total area covered by one pack of laminate?

b How many packs of laminate will Nathan need?

E AO3

3 Gwenno is laying laminate flooring in her office.
The area of the office is 504 m².
Each strip of laminate covers an area of 2 m².
Each laminate pack contains 16 strips.
How many packs of laminate flooring will Gwenno need?

D AO3

4 On Monday a restaurant receives a delivery of 28 trays of eggs.
Each tray holds 24 eggs.
One out of every 16 eggs is damaged and must be thrown away.
On average the restaurant uses 80 eggs per day.
How many days will this delivery of eggs last?

9.2 Estimation

Why learn this?

You can use estimation to check that you have enough materials to complete a DIY job.

Objectives

E D Check and estimate answers to problems

E D C Estimate answers to problems involving decimals

E D C Make estimates and approximations of calculations

Keywords

estimate, approximate, significant figure

Skills check

1 Work out

a $5 + 6 \times 2$ b 39×100 c $\frac{70 \times 9}{10}$

2 Round each of these numbers to one significant figure.

a 324   b 5500   c 2.8

Estimating

Estimating the answer to a calculation gives you an **approximate** answer.
So you can use estimation to check that an answer is about right.

To estimate

- round all the numbers to one **significant figure**
- do the calculation using these approximations.

Example 1

Use approximation to estimate the answer to each of these calculations.

a 48×23　　b 25.2×6.4

c $38.26 \div 4.58$　　d $\frac{119 \times 5.4}{46}$

a $48 \times 23 \approx 50 \times 20 = 1000$

48 rounded to 1 s.f. is 50.
23 rounded to 1 s.f. is 20.

'≈' means 'approximately equal to'.

b $25.2 \times 6.4 \approx 30 \times 6 = 180$

25.2 rounded to 1 s.f. is 30.
6.4 rounded to 1 s.f. is 6.

c $38.26 \div 4.58 \approx 40 \div 5 = 8$

38.26 rounded to 1 s.f. is 40.
4.58 rounded to 1 s.f. is 5.

d $\frac{119 \times 5.4}{46} \approx \frac{100 \times 5}{50} = \frac{500}{50} = 10$

119 rounded to 1 s.f. is 100.
5.4 rounded to 1 s.f. is 5.
46 rounded to 1 s.f. is 50.

Exercise 9C

E

1 By rounding to one significant figure, decide which is the best estimate for each calculation.

		Estimate A	Estimate B	Estimate C
a	3.7×4.9	12	20	16
b	28.6×2.2	40	90	60
c	$12.4 \div 3.5$	6	3	5
d	$24.9 \div 4.7$	7	12	4

2 Work out an approximate value for

a 8.4×9.5　　b $362 \div 2.4$　　c 5.2×4.4　　d $16.9 \div 3.5$

3 Use approximation to estimate the answer to each of these calculations.

a 42×37　　b 18×62　　c $352 \div 76$　　d $872 \div 25$

4 A pack of 500 sheets of printer paper costs £5.89.
Estimate the cost of 28 packs.

D

5 Estimate the answer to each of these calculations.

a $\frac{436 + 394}{109}$　　b $\frac{1248 - 560}{77}$　　c $\frac{27 \times 105}{55}$

d $\frac{40.26 \times 8.49}{16.4}$　　e $\frac{14.5 + 86.02}{1.2}$　　f $\frac{324 \times 7.63}{75.9}$

6 Estimate the answer to each of these calculations.

a $\frac{5.4 \times 19.8}{4.3 - 2.2}$　　b $\frac{12.7 \times 39.4}{(8.6 - 4.8)}$　　c $\frac{17.32 + 14.29}{4.08 - 1.79}$

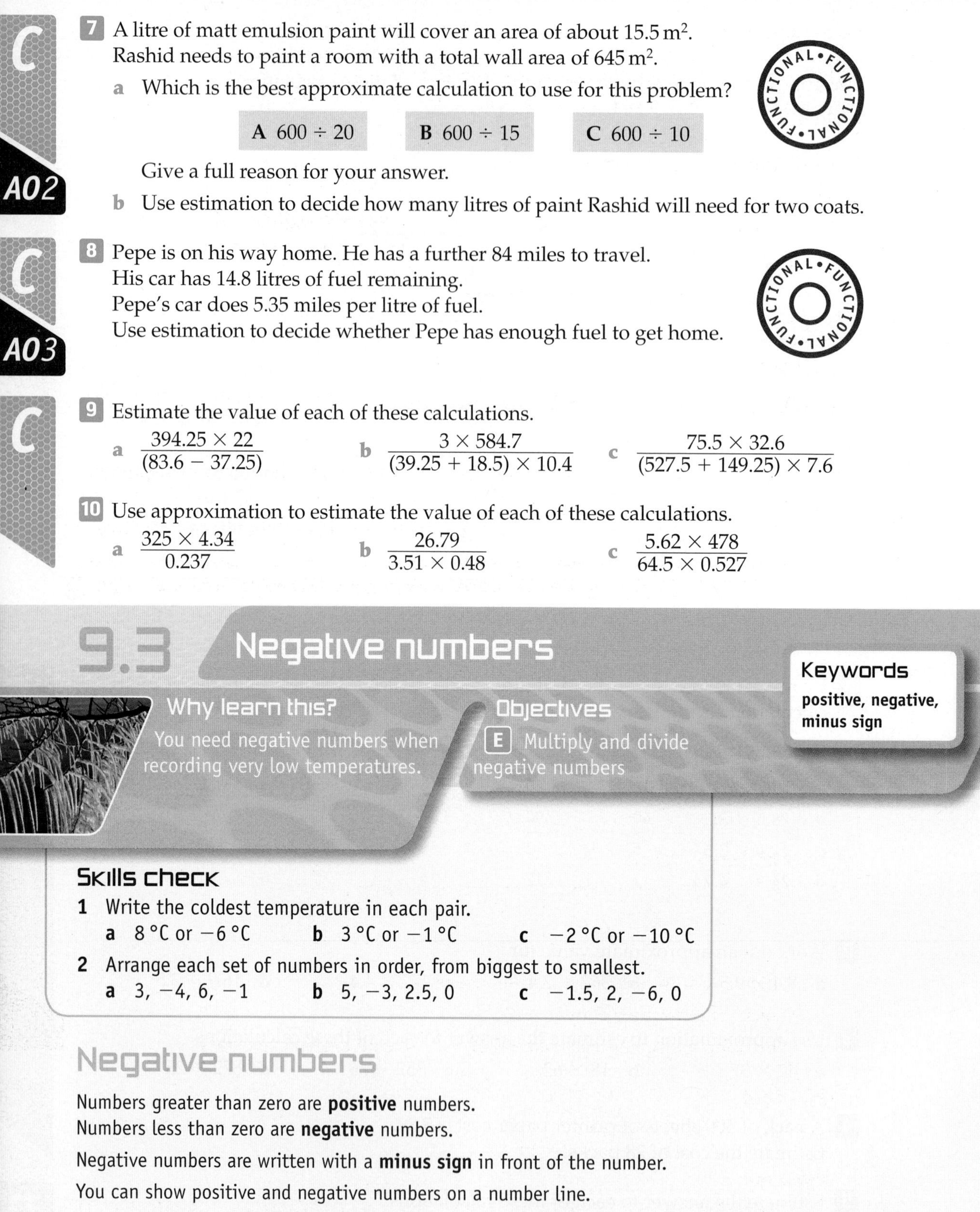

C

AO2

7 A litre of matt emulsion paint will cover an area of about 15.5 m². Rashid needs to paint a room with a total wall area of 645 m².

FUNCTIONAL

a Which is the best approximate calculation to use for this problem?

A $600 \div 20$ | **B** $600 \div 15$ | **C** $600 \div 10$

Give a full reason for your answer.

b Use estimation to decide how many litres of paint Rashid will need for two coats.

C

AO3

8 Pepe is on his way home. He has a further 84 miles to travel.
His car has 14.8 litres of fuel remaining.
Pepe's car does 5.35 miles per litre of fuel.
Use estimation to decide whether Pepe has enough fuel to get home.

FUNCTIONAL

C

9 Estimate the value of each of these calculations.

a $\dfrac{394.25 \times 22}{(83.6 - 37.25)}$ b $\dfrac{3 \times 584.7}{(39.25 + 18.5) \times 10.4}$ c $\dfrac{75.5 \times 32.6}{(527.5 + 149.25) \times 7.6}$

10 Use approximation to estimate the value of each of these calculations.

a $\dfrac{325 \times 4.34}{0.237}$ b $\dfrac{26.79}{3.51 \times 0.48}$ c $\dfrac{5.62 \times 478}{64.5 \times 0.527}$

9.3 Negative numbers

Why learn this?
You need negative numbers when recording very low temperatures.

Objectives
E Multiply and divide negative numbers

Keywords
positive, negative, minus sign

Skills check

1 Write the coldest temperature in each pair.
a 8 °C or −6 °C b 3 °C or −1 °C c −2 °C or −10 °C

2 Arrange each set of numbers in order, from biggest to smallest.
a 3, −4, 6, −1 b 5, −3, 2.5, 0 c −1.5, 2, −6, 0

Negative numbers

Numbers greater than zero are **positive** numbers.
Numbers less than zero are **negative** numbers.

Negative numbers are written with a **minus sign** in front of the number.

You can show positive and negative numbers on a number line.

−10 −9 −8 −7 −6 −5 −4 −3 −2 −1 0 1 2 3 4 5 6 7 8 9 10

If you add a negative number, the result is smaller than the number you started with.
Adding a negative number is the same as subtracting a positive number.

If you subtract a negative number, the result is bigger than the number you started with.
Subtracting a negative number is the same as adding a positive number.

When multiplying or dividing two numbers, you need to check the signs.

- If the signs are the same, the answer is positive. For example $3 \times 3 = 9$ and $-3 \times -3 = 9$

$- \times - = +$
$+ \times + = +$

- If the signs are different, the answer is negative. For example $3 \times -3 = -9$

$- \times + = -$
$+ \times - = -$

Example 2

Work out

a $(-2) \times (+3)$ b $(-4) \times (-5)$
c $(+18) \div (-3)$ d $(-25) \div (-5)$

a $(-2) \times (+3) = -6$ — The signs are different, so the answer is negative.

b $(-4) \times (-5) = 20$ — The signs are the same, so the answer is positive.

When the answer is positive, you don't need to write the + sign.

c $(+18) \div (-3) = -6$ — The signs are different, so the answer is negative.

d $(-25) \div (-5) = 5$ — The signs are the same, so the answer is positive.

E

Exercise 9D

1 Look at these numbers.

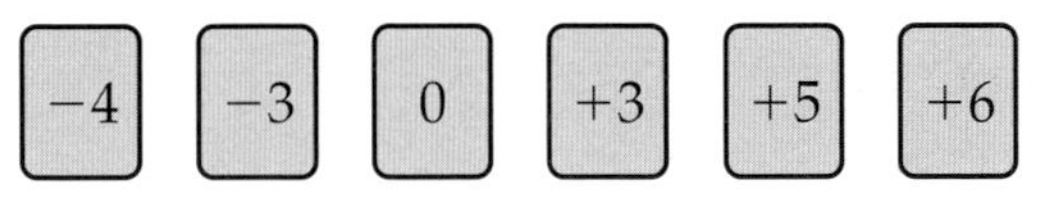

a Which two numbers have a difference of 7?
b Which two numbers sum to −1?
c Which three numbers have a total of −1? Give all possible answers.

E AO2

2 Work out

a $2 \times (-3)$ b $3 \times (-3)$ c $(-2) \times (+5)$ d $(-4) \times (+3)$ e $(-2) \times (-4)$

E

3 Work out

a $10 \div (-2)$ b $12 \div (-4)$ c $(-9) \div 3$ d $(-24) \div 6$ e $(-16) \div (-4)$

4 Work out the product of −8 and 2.

5 Work out

a $(-36) \div 6$ b $-36 \div (-6)$
c $\frac{-18}{2}$ d $\frac{-18}{-3}$

$\frac{-18}{2}$ means $-18 \div 2$.

E

6 Copy and complete these calculations.

a $6 \times \square = -18$ b $24 \div \square = -3$ c $\square \times (-7) = 21$

d $-40 \div \square = -10$ e $\square \div (-5) = -9$ f $\frac{-30}{\square} = 6$

7 What number gives an answer of 24 when multiplied by -4?

E AO2

8 Here is a sequence of numbers.

$1, -2, 4, -8, \ldots$

Work out the next three numbers in this sequence.

E AO3

9 Find two different pairs of numbers that have a difference of 7 and a product of -12.

10 The sum of two numbers is 8.
When they are divided, the answer is -3.
What are the two numbers?

> The first question to ask yourself is, 'Which two numbers when divided give an answer of -3?'
> The '$-$' sign tells you that one number is positive and the other is negative.

D

11 Work out

a $(-2) \times (-3) \times 3$ b $3 \times (-4) \times (-2)$ c $(-5) \times (-2) \times (-3)$

d $6 \div (-2) \div 1$ e $(-8) \div 4 \div (-2)$ f $(-12) \div (-1) \div (-3)$

12 Copy and complete these calculations.

a $(-3) \times (-4) \div 6 = \square$ b $(-16) \div 8 \times (-3) = \square$

c $(-4) \times \square \div (-2) = 12$ d $\square \div (-4) \times (-3) = -12$

D AO3

13 You are given that

$a + b + c = 1$
$a \times b \times c = 36$
$c \div b = -3$
$c \div a = -2$
a is a negative number.

What are the values of a, b and c?

Review exercise

E

1 Write the number that is

a 200 more than 1270 **[1 mark]**
b 10 less than 105 **[1 mark]**
c 20 multiplied by 40 **[1 mark]**
d 5 less than 0 **[1 mark]**
e 2 more than -6 **[1 mark]**
f -8 divided by -4 **[1 mark]**

2 Work out

a $4 - 6$ **[1 mark]**
b $5 \times (-3)$ **[1 mark]**
c $-18 \div (-6)$ **[1 mark]**

3 Here is a sequence of numbers.

$-24, 12, -6, \ldots$

Write the next two numbers in this sequence. **[2 marks]**

E AO2

4 Estimate the value of $\frac{385 \times 54}{1010}$. **[2 marks]**

D

5 Work out

a $-2 \times (-3) \times 3$ **[1 mark]**

b $-4 \div 2 \times (-5)$ **[1 mark]**

6 Hill View School has 36 teachers.
The student–teacher ratio is 22 to 1.

D AO2

a How many students are there at Hill View School? **[2 marks]**

b The students are organised in 32 classes.
31 classes all have an equal number of students.
What are the student class sizes at Hill View School? **[3 marks]**

7 A youth club is arranging a family day trip to a theme park.
236 adults and 460 children are going on the trip.
They are travelling by coach. Each coach seats 48 people and costs £370 to hire.

FUNCTIONAL • FUNCTIONAL •

C AO2

a How many coaches will be needed? **[2 marks]**

b Tickets to the theme park cost £20 for adults and £13 for children.
How much will the tickets cost the youth club? **[2 marks]**

c The youth club is charging £32 for adults and £20 for children.
Will the youth club make a profit from the trip? If yes, how much profit? **[2 marks]**

8 Here is a calculation.

$$\frac{34.96}{3.61 \times 0.54}$$

C

Pepe, Daniel and Xabi use calculators to work it out.
Pepe gets an answer of 179.3, Daniel gets 17.93 and Xabi gets 5.23.
Use approximation to decide who is most likely to be correct. **[3 marks]**

9 Use approximation to estimate the value of $\frac{278 \times 3.62}{0.248}$
Show your working. **[3 marks]**

Chapter summary

In this chapter you have learned how to

- multiply and divide negative numbers E
- multiply whole numbers using written methods E D
- use repeated subtraction for division of whole numbers E D
- round up or down in context E D
- check and estimate answers to problems E D
- estimate answers to problems involving decimals E D C
- make estimates and approximations of calculations E D C

10

Factors, powers and standard form

This chapter is about multiples, factors, powers, roots and standard form.

In the Chinese calendar two separate cycles interact. There are 10 heavenly stems and 12 zodiac animals. You can use lowest common multiples to work out when a certain year will come round again.

Objectives

This chapter will show you how to

- find multiples and factors **E**
- recognise and use prime numbers **E**
- recall the cubes of 2, 3, 4, 5 and 10 **E**
- use the terms positive and negative square root **D**
- calculate common factors, highest common factors and lowest common multiples **C**
- write a number as a product of prime factors **C**
- use index laws for multiplication and division of integer powers **C** **B**
- write numbers in standard index form **B**
- calculate using standard index form **B**

Before you start this chapter

Put your calculator away!

1 Which of the numbers in the cloud are divisible by

a 6

b 9

c 7

d 4

56 42 24 63 54 88 25 17 30

2 Work out

a 6^2

b $(-1)^2$

c $(-9)^2$

d 11^2

3 Work out

a 22×100

b 0.46×1000

c $22.8 \div 10$

d $17 \div 1000$

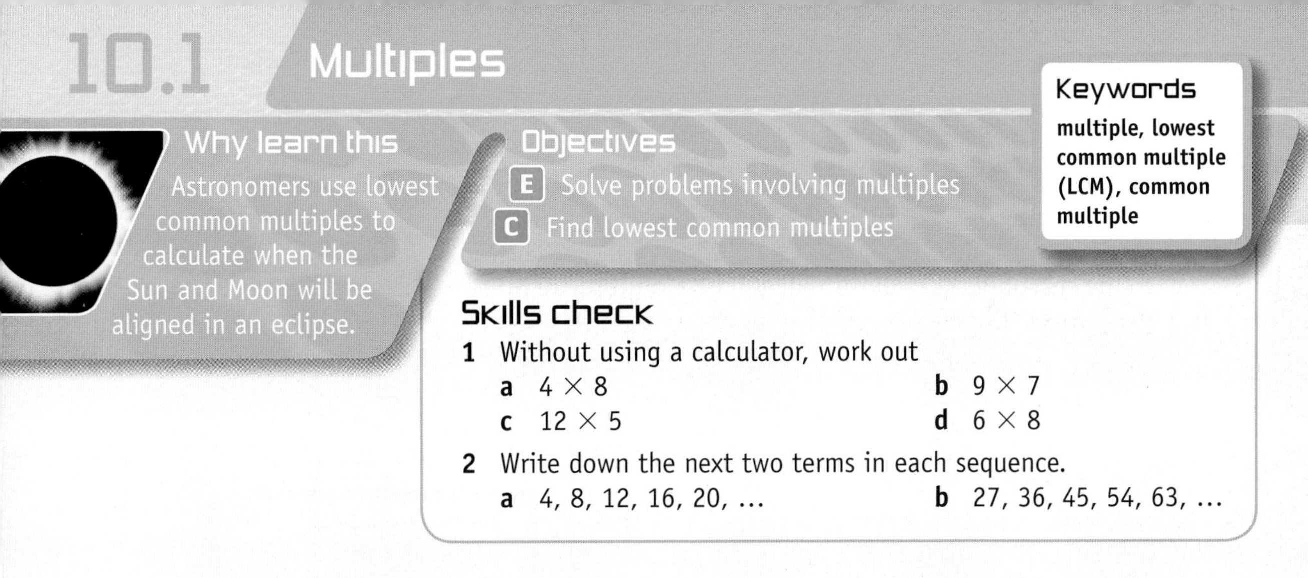

10.1 Multiples

Why learn this
Astronomers use lowest common multiples to calculate when the Sun and Moon will be aligned in an eclipse.

Objectives
E Solve problems involving multiples
C Find lowest common multiples

Keywords
multiple, lowest common multiple (LCM), common multiple

Skills check

1 Without using a calculator, work out
 a 4×8 **b** 9×7
 c 12×5 **d** 6×8

2 Write down the next two terms in each sequence.
 a 4, 8, 12, 16, 20, … **b** 27, 36, 45, 54, 63, …

Multiples

When a number is multiplied by a whole number the answer is a **multiple** of the first number. The multiples of a number are the answers in its times table.

The multiples of 5 are 5, 10, 15, 20, 25, …

All the multiples of a number are divisible by that number.

> **'Is divisible by …' means 'can be divided exactly by …'**

Example 1

E

a Write down the multiples of 7 between 50 and 60.
b Is 84 a multiple of 3? Give a reason for your answer.

a $7 \times 8 = 56$
56 is the only multiple of 7 between 50 and 60.

> Use trial and improvement.
> $7 \times 7 = 49$
> $7 \times 8 = 56$
> $7 \times 9 = 63$

b Yes. $84 \div 3 = 28$

> If a number is divisible by 3 then it is a multiple of 3.

Exercise 10A

E

1 Write down three multiples of 9 that are larger than 100.

2 Write down a multiple of 12 between 80 and 90.

3 Is 412 a multiple of 3? Give a reason for your answer.

4 Write down all the numbers from the cloud which are
 a multiples of 10
 b multiples of 7.

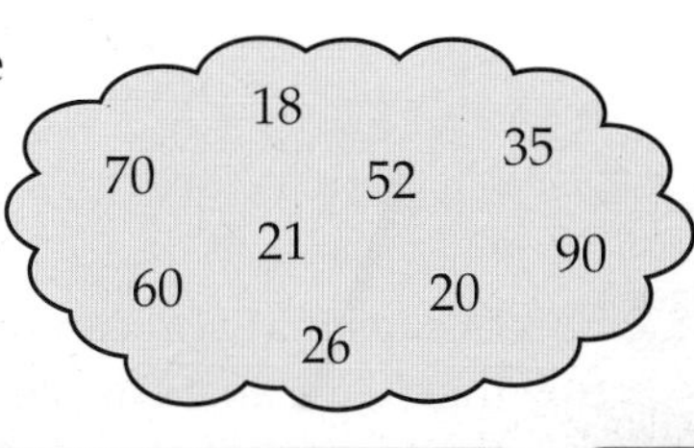

5 Mince pies come in boxes of 8. Beth has bought some boxes of mince pies for a party. Beth counts the mince pies and says there are 130. Tom counts them and says there are only 128. Who is correct? Give a reason for your answer.

6 Nisha has written down a multiple of 8. She says that if she adds a zero to the end of her number it will still be a multiple of 8. Is Nisha correct? Give a reason for your answer.

Lowest common multiples

The **lowest common multiple (LCM)** of two numbers is the smallest number that is a multiple of both numbers.

The lowest common multiple is also known as the least common multiple.

The multiples of 3 are 3, 6, 9, **12**, 15, 18, 21, **24**, ...

The multiples of 4 are 4, 8, **12**, 16, 20, **24**, ...

The **common multiples** of 3 and 4 are 12, 24, ...

So the lowest common multiple of 3 and 4 is 12.

Another way to find LCMs is covered in Section 10.5.

C

Example 2

What is the lowest common multiple of 6 and 8?

Write down the multiples of both numbers and circle the common multiples.

Multiples of 6: 6, 12, 18, (24), 30, 36, 42, (48), ...

Multiples of 8: 8, 16, (24), 32, 40, (48), ...

The common multiples are 24, 48, ...

The lowest common multiple is 24.

$6 \times 8 = 48$ is definitely a common multiple of 6 and 8 so you can stop at 48.

Exercise 10B

D

1 **a** Write down the first ten multiples of 6.
b Write down the first ten multiples of 9.
c What is the LCM of 6 and 9?

LCM stands for lowest common multiple.

2 Write down two common multiples of
a 2 and 3 **b** 4 and 5 **c** 2 and 10 **d** 9 and 3

C

3 Work out the lowest common multiple of
a 5 and 6 **b** 8 and 10 **c** 2 and 5
d 10 and 15 **e** 12 and 15 **f** 20 and 30

4 Shazia says that you can find the LCM of two numbers by multiplying them together. Give an example to show that Shazia is wrong.

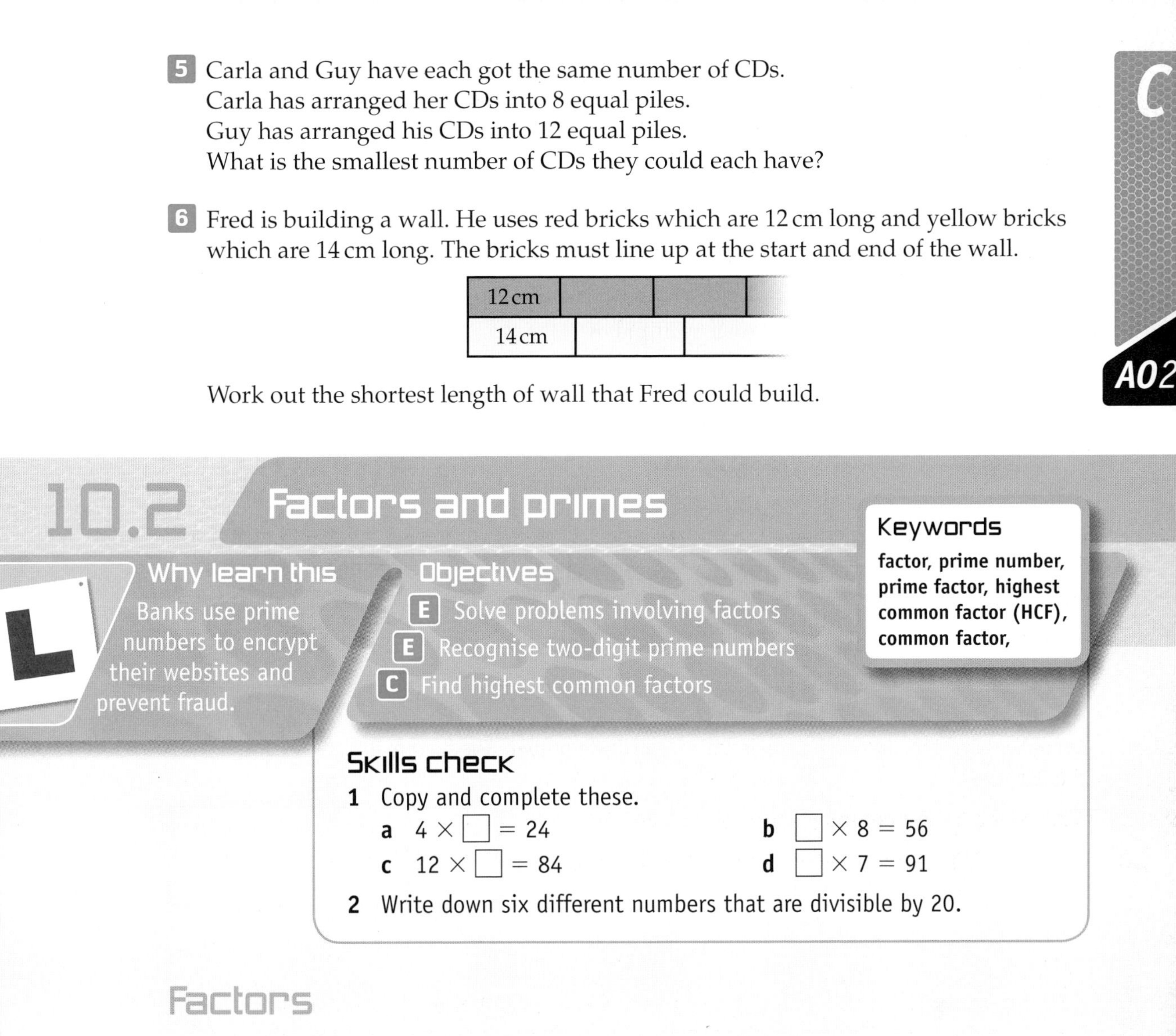

5 Carla and Guy have each got the same number of CDs.
Carla has arranged her CDs into 8 equal piles.
Guy has arranged his CDs into 12 equal piles.
What is the smallest number of CDs they could each have?

6 Fred is building a wall. He uses red bricks which are 12 cm long and yellow bricks which are 14 cm long. The bricks must line up at the start and end of the wall.

Work out the shortest length of wall that Fred could build.

C

AO2

10.2 Factors and primes

Why learn this

Banks use prime numbers to encrypt their websites and prevent fraud.

Objectives

E Solve problems involving factors

E Recognise two-digit prime numbers

C Find highest common factors

Keywords

factor, prime number, prime factor, highest common factor (HCF), common factor,

Skills check

1 Copy and complete these.

a $4 \times \square = 24$

b $\square \times 8 = 56$

c $12 \times \square = 84$

d $\square \times 7 = 91$

2 Write down six different numbers that are divisible by 20.

Factors

A **factor** of a number is a whole number that divides into it exactly.
The factors of a number always include 1 and the number itself.

The factors of 6 are 1, 2, 3 and 6.

Factors come in pairs. You can use factor pairs to help you find factors.

Factor pairs of 6:
$1 \times 6 = 6$
$2 \times 3 = 6$

Example 3

E

Write down all the factors of 20.

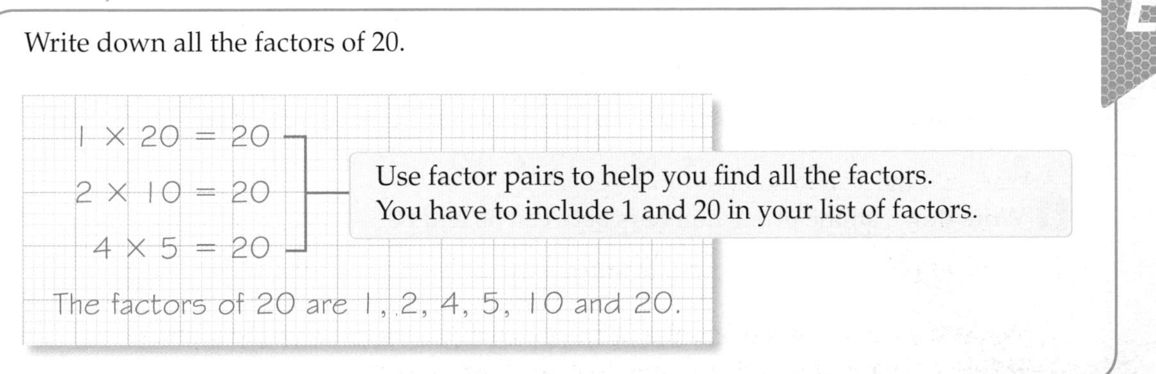

Exercise 10C

E

1 Write down all the factors of

a 10 b 18 c 24 d 30

2 John says that the factors of 12 are 2, 3, 4 and 6.
Is he correct? Give a reason for your answer.

3 Choose a number from the cloud which is

a a factor of 36 b a multiple of 7

c a factor of 40 *and* a multiple of 10

d a factor of 24 *and* a factor of 16.

E A02

4 Mike says that every number has an even number of factors.
Is he correct? Give a reason for your answer.

Prime numbers

A number with exactly two factors is called a **prime number.** The factors are always 1 and the number itself.

The first few prime numbers are 2, 3, 5, 7, 11, …

A prime number which is a factor of another number is called a **prime factor.**

1 is *not* a prime number. It only has one factor.
2 is the only even prime number.

E

Example 4

a Is 27 a prime number? Give a reason for your answer.

b Write down two prime numbers which add up to 18.

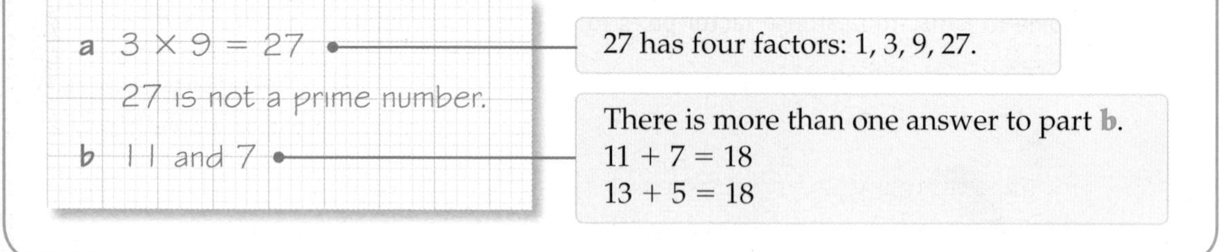

Exercise 10D

E

1 Write down a multiplication fact to show that each number is *not* a prime number.

a 15 b 21 c 63 d 121 e 33 f 91

2 Write down all the prime numbers between 30 and 50.

3 Write down two prime numbers which add up to 16.

4 Write down all the prime factors of

a 12 b 30 c 70 d 44

5 Show that 20 can be written as the sum of two prime numbers in two different ways.

E

AO2

Highest common factors

The **highest common factor (HCF)** of two numbers is the largest number that is a factor of both numbers.

The factors of 12 are **1**, **2**, 3, **4**, 6 and 12.

The factors of 16 are **1**, **2**, **4**, 8 and 16.

The **common factors** of 12 and 16 are 1, 2 and 4.

So the highest common factor of 12 and 16 is 4.

Another way to find HCFs is covered in Section 10.5.

Example 5

C

What is the highest common factor of 20 and 30?

Write down the factors of both numbers and circle the common factors.

Factors of 20: (1), (2), 4, (5), (10), 20

Factors of 30: (1), (2), 3, (5), 6, (10), 15, 30

The common factors are 1, 2, 5 and 10.

The highest common factor is 10.

Remember to include 1 and the number itself in your list of factors.

Exercise 10E

1 **a** Write down the factors of 12.

b Write down the factors of 8.

c What is the HCF of 12 and 8?

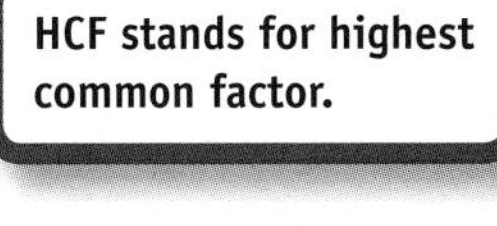

C

2 Work out the highest common factor of

a 15 and 25	**b** 14 and 12	**c** 21 and 15
d 24 and 20	**e** 8 and 10	**f** 8 and 16

3 Write down two numbers larger than 10 with an HCF of 8.

4 Lydia is making Christmas decorations. She needs to cut identical squares out of a rectangular sheet of paper.

a What is the largest square size Lydia can use without wasting any paper?

b How many of these squares will Lydia be able to cut from this sheet of paper?

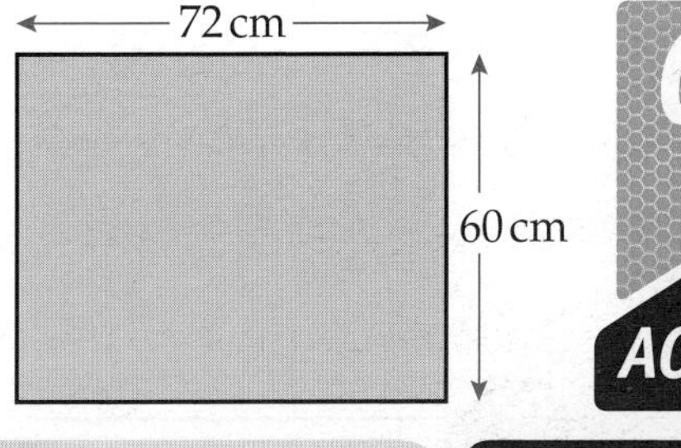

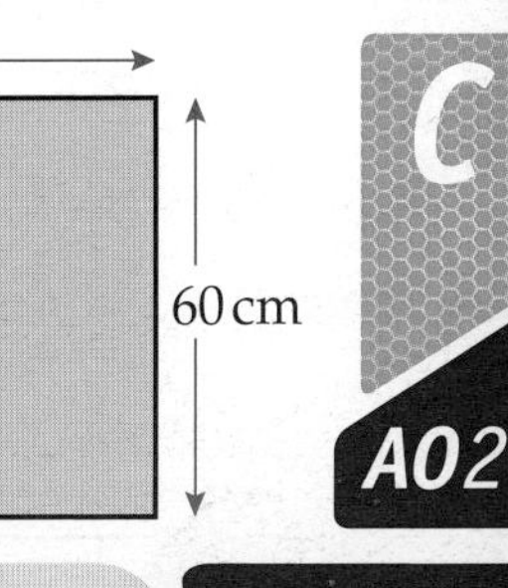

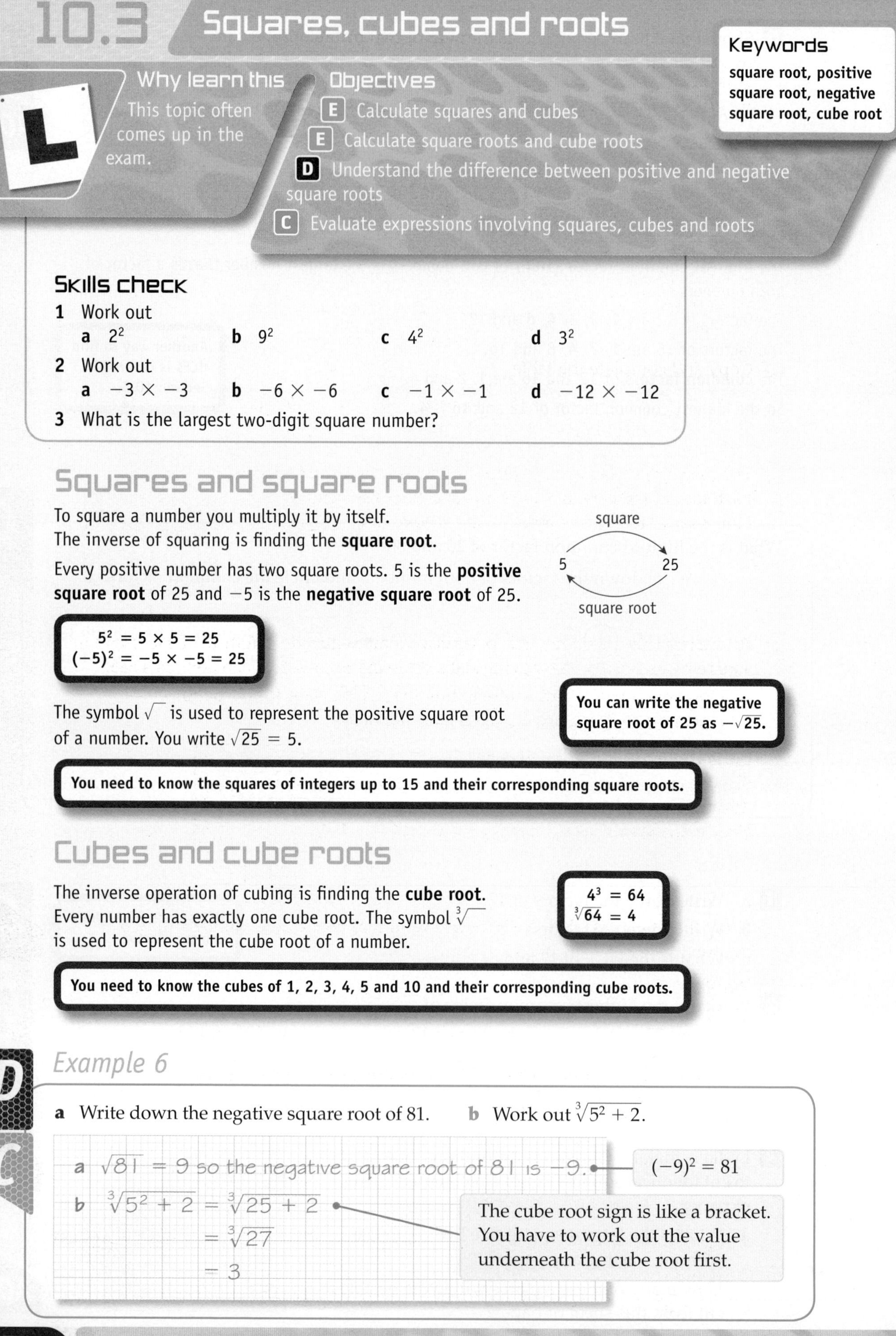

10.3 Squares, cubes and roots

Keywords
square root, positive square root, negative square root, cube root

Why learn this
This topic often comes up in the exam.

Objectives
- E Calculate squares and cubes
- E Calculate square roots and cube roots
- D Understand the difference between positive and negative square roots
- C Evaluate expressions involving squares, cubes and roots

Skills check

1 Work out
   a 2^2 b 9^2 c 4^2 d 3^2
2 Work out
   a -3×-3 b -6×-6 c -1×-1 d -12×-12
3 What is the largest two-digit square number?

Squares and square roots

To square a number you multiply it by itself.
The inverse of squaring is finding the **square root.**

Every positive number has two square roots. 5 is the **positive square root** of 25 and -5 is the **negative square root** of 25.

square
5 25
square root

$5^2 = 5 \times 5 = 25$
$(-5)^2 = -5 \times -5 = 25$

The symbol $\sqrt{\ }$ is used to represent the positive square root of a number. You write $\sqrt{25} = 5$.

You can write the negative square root of 25 as $-\sqrt{25}$.

You need to know the squares of integers up to 15 and their corresponding square roots.

Cubes and cube roots

The inverse operation of cubing is finding the **cube root.** Every number has exactly one cube root. The symbol $\sqrt[3]{\ }$ is used to represent the cube root of a number.

$4^3 = 64$
$\sqrt[3]{64} = 4$

You need to know the cubes of 1, 2, 3, 4, 5 and 10 and their corresponding cube roots.

D C

Example 6

a Write down the negative square root of 81. **b** Work out $\sqrt[3]{5^2 + 2}$.

a $\sqrt{81} = 9$ so the negative square root of 81 is -9. — $(-9)^2 = 81$

b $\sqrt[3]{5^2 + 2} = \sqrt[3]{25 + 2}$
$= \sqrt[3]{27}$
$= 3$

The cube root sign is like a bracket. You have to work out the value underneath the cube root first.

Exercise 10F

1 Work out

a nine squared

b five cubed

c the positive square root of 36

d the cube root of 8.

2 Work out

a 7^2 b 10^2 c 5^3 d 10^3

3 Work out

a $\sqrt{36}$ b $\sqrt{49}$ c $\sqrt{121}$ d $\sqrt{64}$

4 Work out

a $\sqrt[3]{125}$ b $\sqrt[3]{1}$ c $\sqrt[3]{1000}$ d $\sqrt[3]{27}$

5 Copy and complete the table.

x	1	2	3	4	5	6	7	8	9	10	11	12	13	14	15
x^2			9							100					

6 The diagram shows a cube decorated with star-shaped stickers.
Each face of the cube has the same number of stickers.
How many stickers are there on the whole cube?

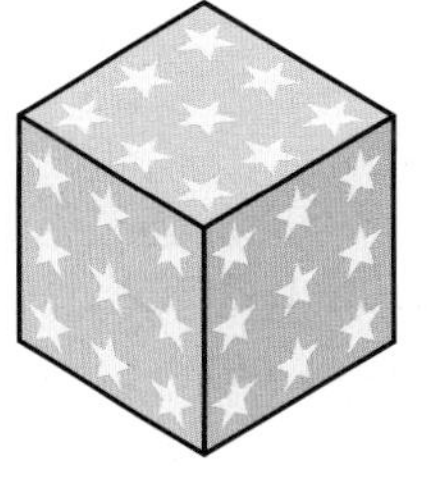

7 Andre is paving his patio. His patio is a square with side length 13 m. He uses 1 m square paving slabs. The paving slabs come in boxes of 20. Each box costs £45.

a How much will it cost Andre to buy the paving slabs for his patio?

b How many paving slabs will he have left over?

8 Carla has a bag of 100 small cubes.
She uses her cubes to make this larger cube.
How many small cubes does she have left?

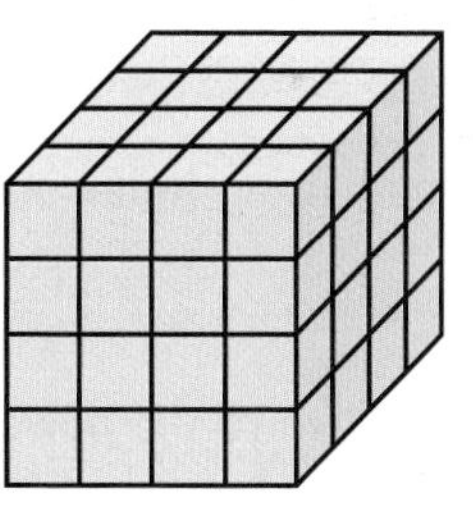

9 Show that it is possible to write 50 as the sum of two square numbers in two different ways.

10 Work out the negative square root of

a 49 b 25 c 4 d 9

11 Write down two possible values that would make this statement true.

$\square^2 + 9 = 45$

12 Work out

a $\sqrt{2^2 + 5}$ b $\sqrt{6^2 + 8^2}$ c $\sqrt{5^2 - 4^2}$ d $\sqrt{13^2 - 12^2}$

13 Work out

a $\sqrt{5^3 - 5^2}$ b $\sqrt{2^3 + 2^3}$ c $\sqrt[3]{11^2 + 2^2}$ d $\sqrt[3]{3^2 - 1}$

E E AO2 D D AO2 C

C

14 Estimate the answers to these calculations by rounding each value to the nearest whole number.

a $6.7^2 + 3.1^2$ **b** $5.04^3 - 3.28^3$ **c** $\sqrt{19.8 - 4.1}$ **d** $\sqrt[3]{2.1^2 + 1.7^2}$

C
AO3

15 Amber and Simon have each made a 10 cm square pattern using 1 cm square tiles.
Amber gives some tiles to Simon. They are both able to arrange their tiles exactly into square patterns.
How many tiles did Amber give to Simon?

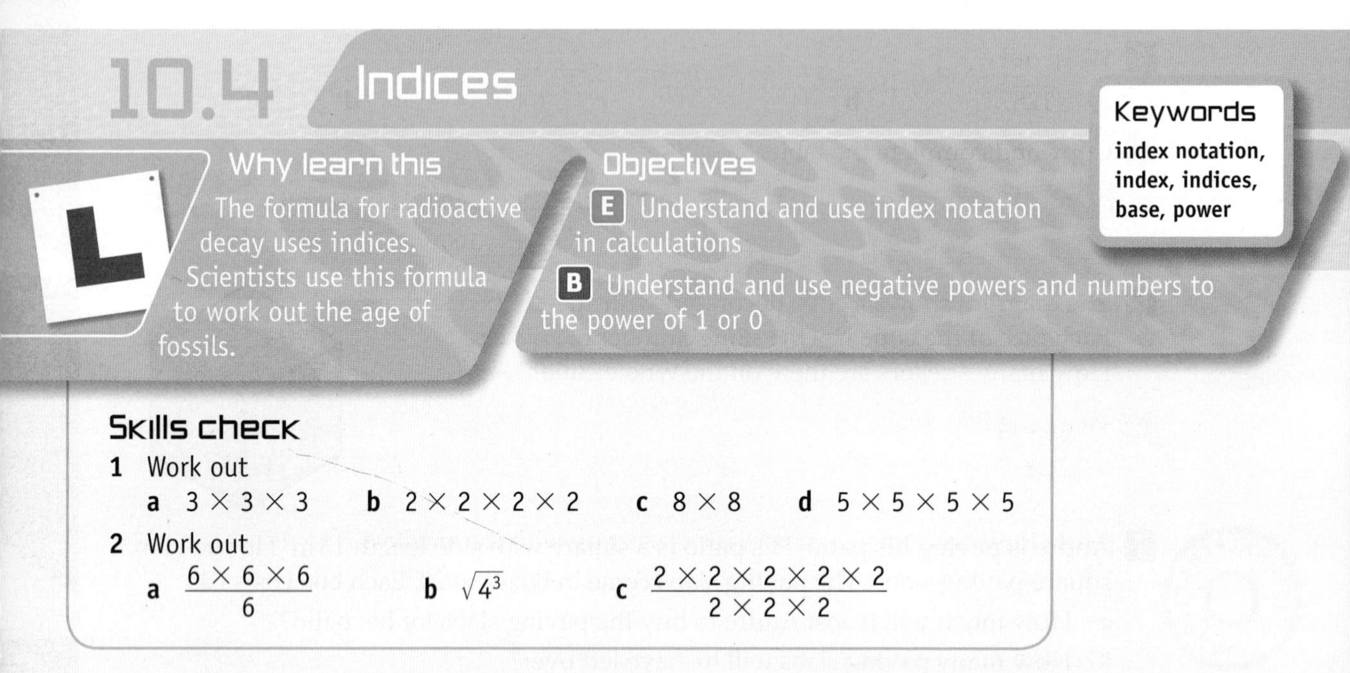

10.4 Indices

Keywords
index notation, index, indices, base, power

Why learn this

The formula for radioactive decay uses indices. Scientists use this formula to work out the age of fossils.

Objectives

E Understand and use index notation in calculations

B Understand and use negative powers and numbers to the power of 1 or 0

Skills check

1 Work out

a $3 \times 3 \times 3$ **b** $2 \times 2 \times 2 \times 2$ **c** 8×8 **d** $5 \times 5 \times 5 \times 5$

2 Work out

a $\frac{6 \times 6 \times 6}{6}$ **b** $\sqrt{4^3}$ **c** $\frac{2 \times 2 \times 2 \times 2 \times 2}{2 \times 2 \times 2}$

Positive and negative indices

You can write numbers using **index notation.**

$$2 \times 2 \times 2 \times 2 \times 2 \times 2 = 2^6$$

This number is called the **index**. The plural of index is **indices**.

This number is called the **base**.

You write 2^6. You say 'two to the **power** of six'.

This sequence of numbers halves each time.

$2^5 = 2 \times 2 \times 2 \times 2 \times 2 = 32$

$2^4 = 2 \times 2 \times 2 \times 2 = 16$

$2^3 = 2 \times 2 \times 2 = 8$

$2^2 = 2 \times 2 = 4$

The next power in this sequence is 2^1. You can continue the pattern by halving.

$2^1 = 2$

Any number to the power of 1 is the number itself.

$2^0 = 1$

Any number to the power of 0 is equal to 1.

$2^{-1} = \frac{1}{2}$

$2^{-2} = \frac{1}{4}$

$2^{-3} = \frac{1}{8}$

You can write 2^{-2} as $\frac{1}{2^2}$ and 2^{-3} as $\frac{1}{2^3}$.
If the index is negative, first change it to positive.
Then write the number as the denominator of a fraction.

Example 7

Work out

a 3^4 b 4^{-2}

a $3^4 = 3 \times 3 \times 3 \times 3$
$= 81$

Work out $3 \times 3 \times 3$ first, then multiply the answer by 3.

b $4^{-2} = \frac{1}{4^2}$
$= \frac{1}{16}$

You can give your answer as a fraction.

Exercise 10G

E

1 Work out

a 3^5 b 4^4 c 5^6 d 10^5 e 6^3 f 7^3

2 Work out

a $2^4 \times 5$ b $4^3 \times 3$ c $2^3 \times 3^3$ d $5^5 \div 5$ e $6^3 \div 3^2$ f $4^3 \div 2^5$

3 Write these numbers in order of size, starting with the smallest.

$2^2 \times 3^2$ 3^3 2^4 1^8 $\sqrt{225}$

E

4 This is a famous riddle.

> As I was going to St Ives I met a man with seven wives. Every wife had seven sacks, and every sack had seven cats. Every cat had seven kittens. Kittens, cats, sacks and wives, how many were going to St Ives?

a What is the answer to the riddle?

b Use index notation to write down the total number of kittens.

c Work out the total number of kittens and cats.

AO2

B

5 Work out the value of

a 8^0 b 6^1 c 100^0 d 17^1 e 29^1 f 83^0

6 Write each of these as a fraction in its lowest terms.

a 4^{-1} b 3^{-2} c 7^{-1} d 2^{-2} e 2×8^{-1} f 3×6^{-2}

E

FUNCTIONAL • FUNCTIONAL •

7 The formula $N = 80 \times 2^t$ is used to calculate the number of bacteria on a microscope slide after t hours.

a Use this formula to work out the number of bacteria on the microscope slide after

i 2 hours

ii 6 hours.

After 2 hours, $t = 2$.

b How many bacteria do you think were placed on the microscope slide at the beginning of the experiment? Give a reason for your answer.

AO2

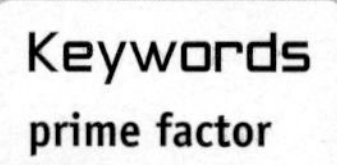

Why learn this

You can use prime factors to calculate lowest common multiples and highest common factors much more quickly.

Objectives

C Write a number as a product of prime factors using index notation

C Use prime factors to find HCFs and LCMs

Skills check

1 Write down all the prime numbers from the cloud.

HELP Section 10.2

2 Write down a multiplication fact to show that 111 is not a prime number.

Writing a number as a product of prime factors

You can write any number as a product of **prime factors**.

You can use a factor tree to write a number as a product of prime factors.

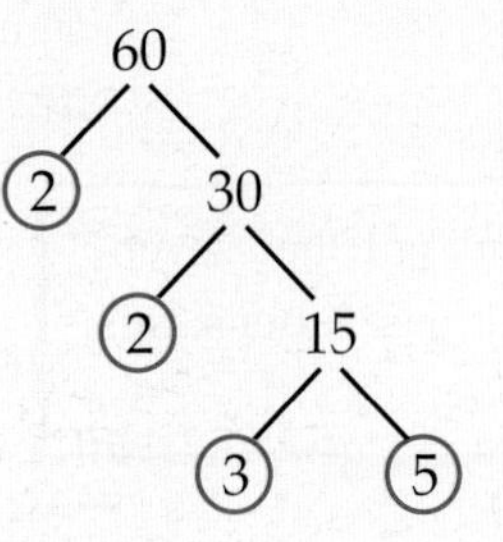

- Split each number up into factor pairs.
- When you reach a prime number, draw a circle around it. These are the ends of the branches.
- The answer is the product of the prime numbers on the branches.

$60 = 2 \times 2 \times 3 \times 5$

You can write this using index notation as $2^2 \times 3 \times 5$.

You can also use repeated division to write a number as a product of prime factors.

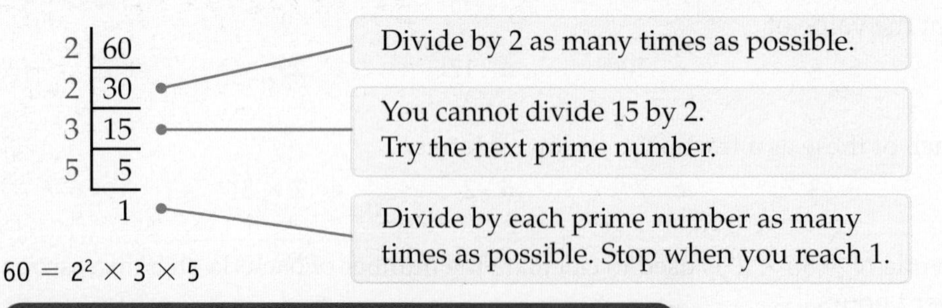

$60 = 2^2 \times 3 \times 5$

When writing a number as a product of prime factors, write the prime factors in order from smallest to largest.

Example 8

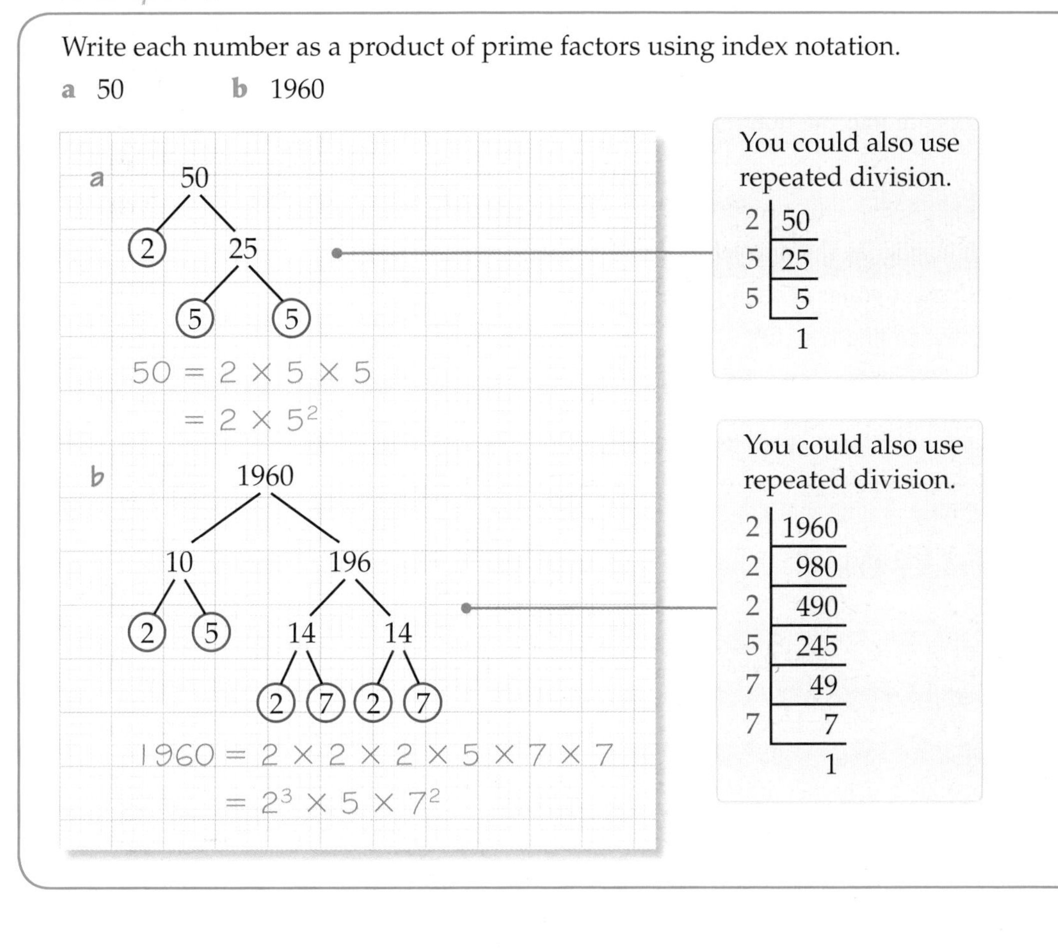

Write each number as a product of prime factors using index notation.

a 50 **b** 1960

a

$50 = 2 \times 5 \times 5$

$= 2 \times 5^2$

b

$1960 = 2 \times 2 \times 2 \times 5 \times 7 \times 7$

$= 2^3 \times 5 \times 7^2$

You could also use repeated division.

2	50
5	25
5	5
	1

You could also use repeated division.

2	1960
2	980
2	490
5	245
7	49
7	7
	1

Exercise 10H

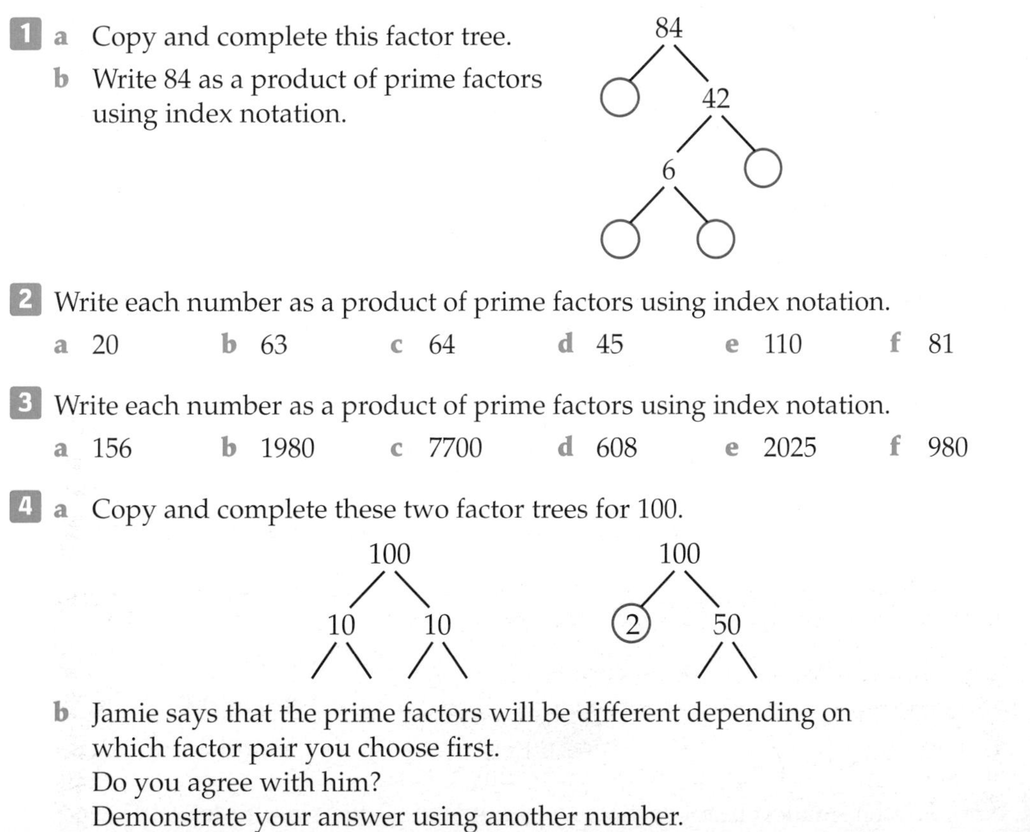

1 **a** Copy and complete this factor tree.

b Write 84 as a product of prime factors using index notation.

2 Write each number as a product of prime factors using index notation.

a 20 **b** 63 **c** 64 **d** 45 **e** 110 **f** 81

3 Write each number as a product of prime factors using index notation.

a 156 **b** 1980 **c** 7700 **d** 608 **e** 2025 **f** 980

4 **a** Copy and complete these two factor trees for 100.

b Jamie says that the prime factors will be different depending on which factor pair you choose first.
Do you agree with him?
Demonstrate your answer using another number.

C

AO2

Using prime factors to find the HCF

- Write each number as the product of prime factors.
- If a prime number is in *both* lists, circle the *lowest* power.
- Multiply these to find the HCF.

If a prime number is only in one of the lists you can't include it in your HCF.

Using prime factors to find the LCM

- Write each number as the product of prime factors.
- Circle the *highest* power of each prime number.
- Multiply these to find the LCM.

C

Example 9

Work out

a the HCF of 180 and 168 b the LCM of 24 and 60.

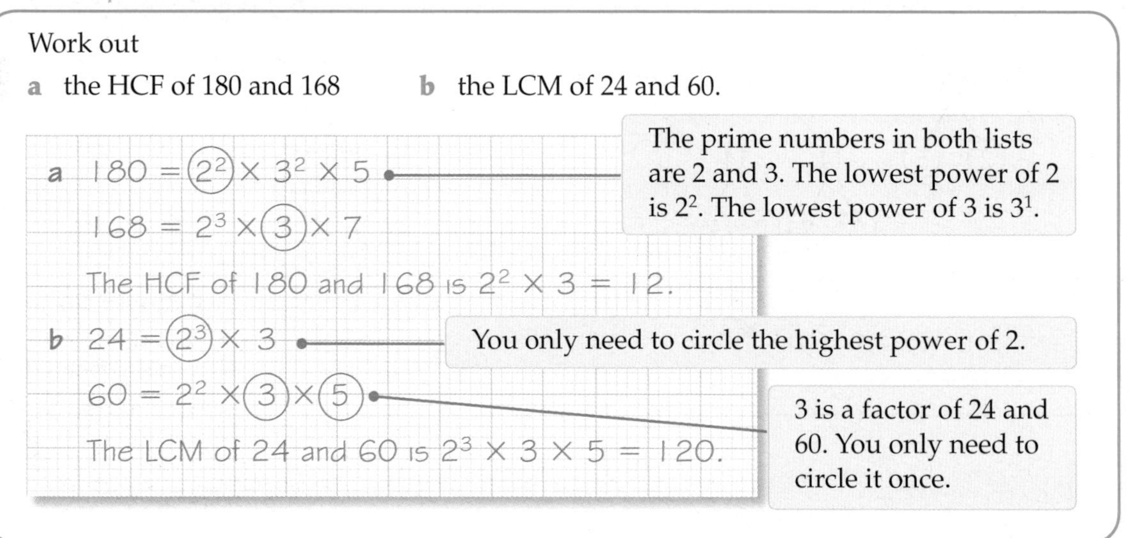

Exercise 10I

C

1 a Write 90 as a product of prime factors.
b Write 165 as a product of prime factors.
c Find the HCF of 90 and 165.

2 a Write 42 as a product of prime factors.
b Write 30 as a product of prime factors.
c Find the LCM of 42 and 30.

3 Work out the highest common factor of each pair of numbers.

a 32 and 56 b 80 and 72 c 27 and 45
d 100 and 75 e 48 and 64 f 60 and 160

4 Work out the largest whole number that will divide exactly into 264 and 150.

5 Work out the lowest common multiple of each pair of numbers.

a 18 and 20 b 6 and 32 c 27 and 15
d 9 and 75 e 60 and 80 f 14 and 21

6 Work out the smallest number that is a multiple of 90 *and* a multiple of 105.

7 Work out the highest common factor of 2016 and 1512.

8 Tarik and Archie have the same amount of money.
Tarik's money is all in 20p pieces and Archie's money is all in 50p pieces.
What is the smallest amount of money that they could each have?

9 Amy is investigating the relationship between the LCM and the HCF.

a Work out the HCF of 18 and 30.

b Amy says that she can find the LCM of 18 and 30 using the rule LCM $= \frac{18 \times 30}{\text{HCF}}$.
Show working to check that Amy's rule works.

c Show that Amy's rule will also work for 16 and 40.

C AO2

10 David has a pack of playing cards with some missing.
He arranges his playing cards into 15 rows of equal length. He then rearranges his playing cards into 9 rows of equal length.
How many cards are missing from David's pack?

A normal pack of playing cards contains 52 cards.

C AO3

10.6 Laws of indices and standard form

Why learn this

You can use standard form to write very large and very small numbers. The mass of a water molecule is about 3×10^{-26} kg.

Objectives

C B Use laws of indices to multiply and divide numbers written in index notation

B Carry out calculations with numbers given in standard form

Keywords
laws of indices, standard form

Skills check

1 Write each of these using index notation.

a $2 \times 2 \times 2$　　b $7 \times 7 \times 7 \times 7$

c $8 \times 8 \times 8$　　d $3 \times 3 \times 3 \times 3 \times 3$

2 Work out the value of

a 2^5　　b 3^4　　c 10^3　　d 2^7

Laws of indices

To *multiply* powers of the same number you *add* the indices.

$$4^2 \times 4^5 = (4 \times 4) \times (4 \times 4 \times 4 \times 4 \times 4) = 4^7$$

You can multiply in any order so the brackets don't matter.

Using the **laws of indices**:

$4^2 \times 4^5 = 4^{2+5} = 4^7$

To *divide* powers of the same number you *subtract* the indices.

$$6^5 \div 6^3 = \frac{6 \times 6 \times \cancel{6} \times \cancel{6} \times \cancel{6}}{\cancel{6} \times \cancel{6} \times \cancel{6}} = \frac{6 \times 6}{1} = 6^2$$

You can cancel 6 three times at the top and at the bottom of the fraction.

Using the laws of indices:

$6^5 \div 6^3 = 6^{5-3} = 6^2$

You can use the laws of indices to understand negative powers.

$$9^3 \div 9^7 = \frac{\not{9} \times \not{9} \times \not{9}}{9 \times 9 \times 9 \times 9 \times \not{9} \times \not{9} \times \not{9}}$$
$$= \frac{1}{9 \times 9 \times 9 \times 9}$$
$$= \frac{1}{9^4}$$

Using the laws of indices:

$9^3 \div 9^7 = 9^{3-7} = 9^{-4}$

This shows that $9^{-4} = \frac{1}{9^4}$.

You will meet laws of indices again in Section 15.2.

C B

Example 10

Simplify

a $5^8 \times 5^3$ b $2^3 \times 2^4 \times 2$ c $3^2 \div 3^6$

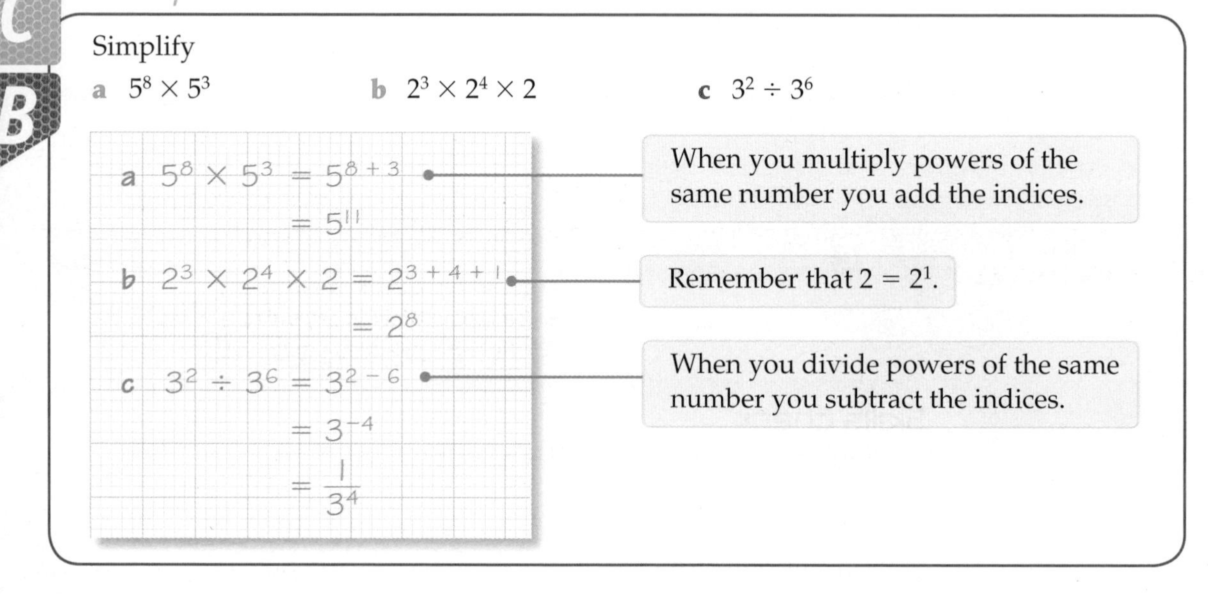

Exercise 10J

C

1 Write each of these expressions as a single power.

a $6^7 \times 6^3$ b $4^7 \times 4^3$ c 5×5^3 d $9^3 \times 9^3$

2 Write each of these expressions as a single power.

a $3^7 \div 3^3$ b $6^{10} \div 6^5$ c $4^{17} \div 4^{12}$ d $5^3 \div 5$

3 Write each of these expressions as a single power.

a $8^4 \times 8 \times 8^6$ b $2^2 \times 2^7 \times 2^4$ c $8^3 \times 8^5 \times 8$ d $9 \times 9^3 \times 9$

4 Write down the value of each of these expressions.

a $5^8 \div 5^5$ b $4^{13} \div 4^{11}$

c $2^{10} \div 2^6$ d $7^6 \div 7^3$

Give your answers as a whole number.

5 Alison writes that $7^{10} \div 7^2 = 7^5$.
Is Alison correct? Give a reason for your answer.

6 Write each of these expressions as a single power of 10.

a $(10^4)^2$ b $(10^2)^3$

C

AO2

B

7 Write each of these expressions as a single power.

a $5^2 \div 5^9$ b $\frac{3^2}{3^3}$ c $7 \div 7^5$ d $\frac{9^2}{9^6}$

8 Write each of these expressions as a fraction.

a $4^3 \div 4^5$ b $2 \div 2^4$ c $\frac{3^2 \times 3}{3^5}$ d $\frac{16}{4^4}$

9 Write down the value of

a $6^{-2} \times 6^4$ b $3^{-3} \times 3^6$ c $\frac{5^{-1} \times 5^6}{25}$ d $\frac{6^6 \times 6^{-1}}{6^4}$

Powers of different numbers

You can only use the laws of indices to multiply and divide powers of the same number.

To simplify powers of different numbers you need to look at each base separately.

$$(2^4 \times 3^2) \times (2 \times 3^7) = (2^4 \times 2) \times (3^2 \times 3^7)$$
$$= 2^5 \times 3^9$$

You can multiply in any order.

Example 11

B

Work out the value of

a $(5^3 \times 3^2) \times (5^{-1} \times 3)$ **b** $\frac{7^5 \times 10}{7^3 \times 10^2}$

a $(5^3 \times 3^2) \times (5^{-1} \times 3) = (5^3 \times 5^{-1}) \times (3^2 \times 3)$
$= 5^2 \times 3^3$
$= 25 \times 27$
$= 675$

Use long multiplication.

```
    2 5
×   2 7
  1 ₃7 5
 ₁5 0 0
  6 7 5
```

b $\frac{7^5 \times 10}{7^3 \times 10^2} = \frac{7^5}{7^3} \times \frac{10}{10^2}$
$= 7^2 \times 10^{-1}$
$= 49 \times \frac{1}{10}$
$= 4.9$

Multiplying by $\frac{1}{10}$ is the same as dividing by 10.

Exercise 10K

B

1 Work out the value of each of these.
Give your answer as a decimal number where appropriate.

a $(5^2 \times 2^2) \times (2^{-1} \times 5)$

b $(4^5 \times 10^2) \times (10 \times 4^{-2})$

c $(2^5 \times 6^3) \times (2^{-7} \times 6^{-1})$

d $(5 \times 10^2) \times (5^2 \times 10^{-3})$

e $(2 \times 3^2 \times 5^2) \times (2^2 \times 3)$

f $(5^2 \times 4^2 \times 10) \times (5^{-3} \times 4^{-1} \times 10^2)$

2 Work out the value of each of these.
Give your answer as a decimal number where appropriate.

a $\dfrac{3^2 \times 6^6}{3^3 \times 6^4}$

b $\dfrac{6^2 \times 3^5}{6^4 \times 3^3}$

c $\dfrac{10 \times 5^9}{10^4 \times 5^6}$

d $\dfrac{2 \times 10^7}{2^4 \times 10^4}$

B A03

3 Salma says that you can write $2^{11} \times 8^2$ as a single power of 2.
Show working to explain why Salma is correct.

Standard form

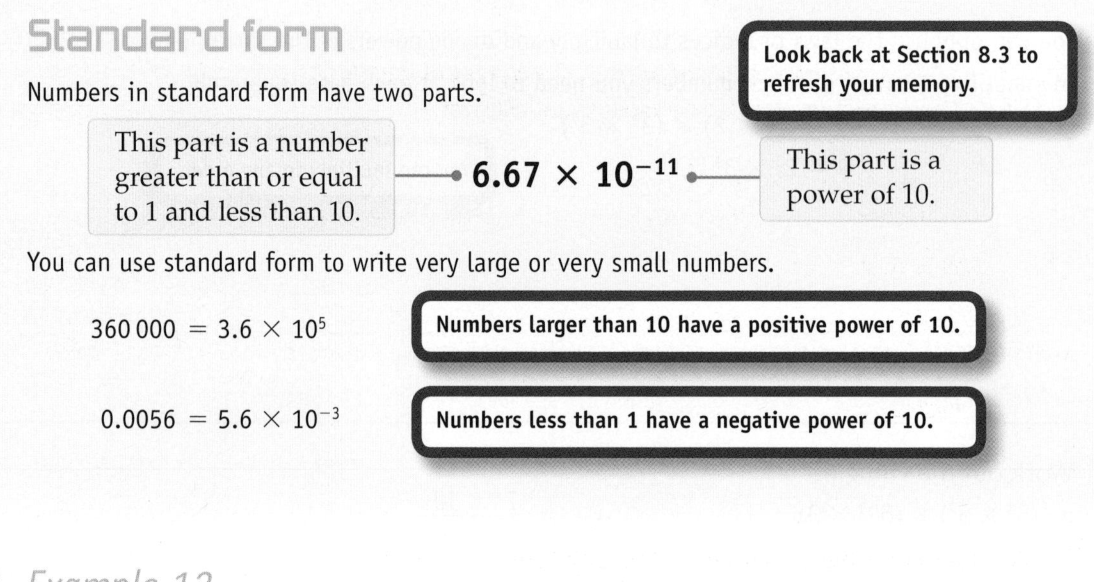

Look back at Section 8.3 to refresh your memory.

Numbers in standard form have two parts.

This part is a number greater than or equal to 1 and less than 10. —— 6.67×10^{-11} —— This part is a power of 10.

You can use standard form to write very large or very small numbers.

$360\,000 = 3.6 \times 10^5$ — **Numbers larger than 10 have a positive power of 10.**

$0.0056 = 5.6 \times 10^{-3}$ — **Numbers less than 1 have a negative power of 10.**

B

Example 12

Write these numbers in order of size, starting with the smallest.

6.4×10^3 $\quad 1 \times 10^5$ $\quad 7.4 \times 10^{-6}$ $\quad 5.2 \times 10^3$ $\quad 4.3 \times 10^5$

7.4×10^{-6} — Look at the indices first. 7.4×10^{-6} has the lowest index, so it is the smallest number.

5.2×10^3

6.4×10^3 — 5.2×10^3 and 6.4×10^3 have the same index. 5.2 is smaller than 6.4, so 5.2×10^3 is smaller.

1×10^5

4.3×10^5

Exercise 10L

1 Write each of these numbers in standard form.

a 34 000 b 2600 c 740
d 200 000 e 6 030 000 f 22.6

2 Write each of these numbers in standard form.

a 0.0006 b 0.0032 c 0.003 09
d 0.445 e 0.01 f 0.000 009 8

3 Write each of these as a decimal number.

a 2×10^4 b 6.2×10^3 c 4.54×10^{-3}
d 2.07×10^{-1} e 7.9×10^7 f 4.551×10^{-6}

4 Write these numbers in order of size, starting with the smallest.

8.04×10^3 4.8×10^4 9.31×10^{-1} 3.2×10^{-2} 8.66×10^{-1} 1×10^7

5 In the 2001 census the population of the UK was recorded as 58 789 194.
Round this number to 3 significant figures and write your answer in standard form.

6 The table shows the masses of atoms of different elements.
Write the elements in order of mass, smallest first.

Substance	Mass of atom (kg)
carbon	2.0×10^{-26}
plutonium	4.1×10^{-25}
gold	3.3×10^{-25}
iron	9.1×10^{-26}
hydrogen	1.7×10^{-27}
tungsten	3.1×10^{-25}

7 A human hair has a diameter of 0.002 54 cm.
Write this in metres using standard form.

Calculating with numbers in standard form

To multiply or divide numbers in standard form you multiply or divide each part separately. You can use the laws of indices to convert your answer back into standard form if necessary.

To add or subtract numbers in standard form you write them as decimal numbers first. Then you add or subtract. Finally, you write your answer in standard form.

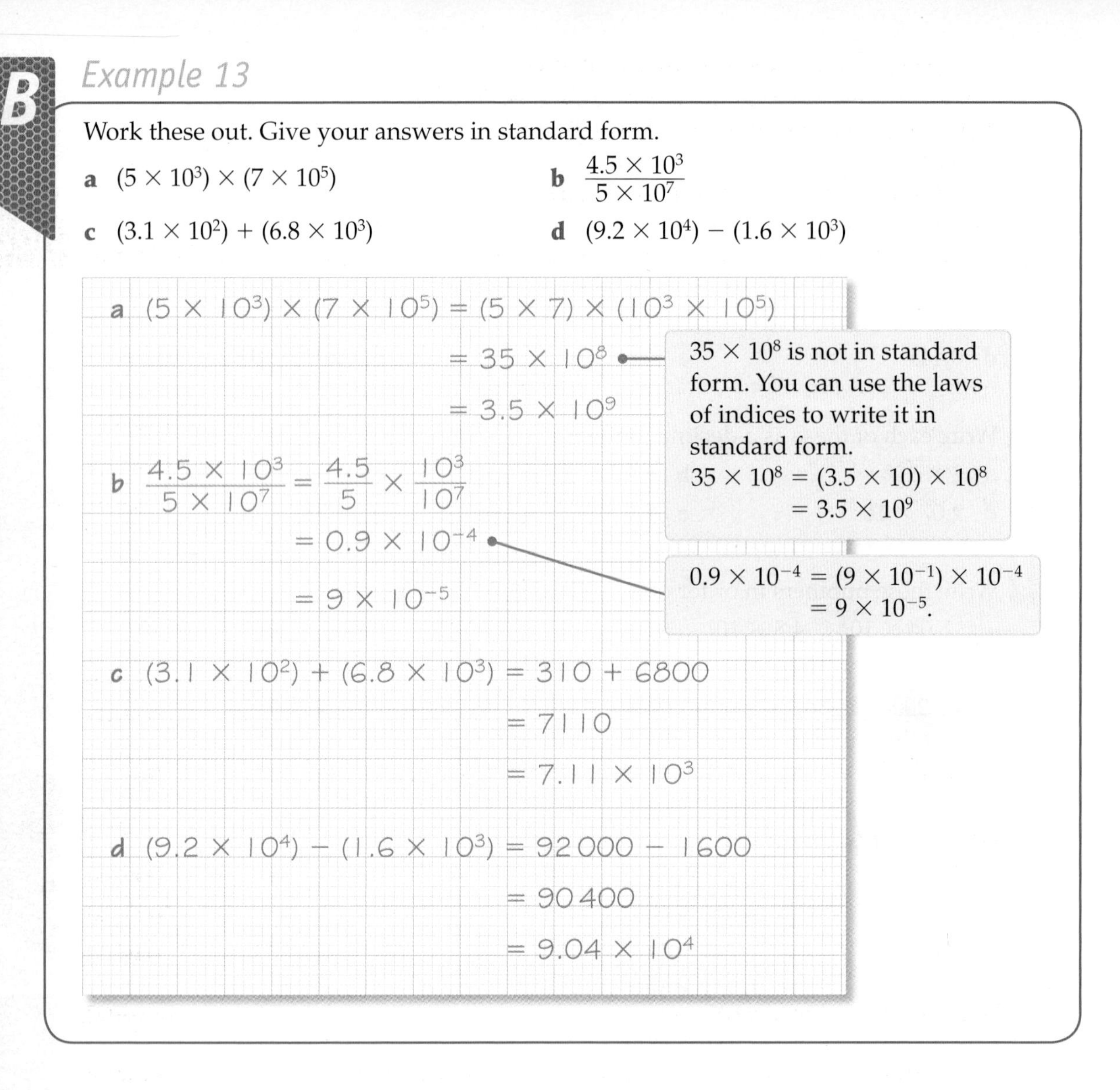

Example 13

Work these out. Give your answers in standard form.

a $(5 \times 10^3) \times (7 \times 10^5)$ **b** $\dfrac{4.5 \times 10^3}{5 \times 10^7}$

c $(3.1 \times 10^2) + (6.8 \times 10^3)$ **d** $(9.2 \times 10^4) - (1.6 \times 10^3)$

a $(5 \times 10^3) \times (7 \times 10^5) = (5 \times 7) \times (10^3 \times 10^5)$
$= 35 \times 10^8$
$= 3.5 \times 10^9$

35×10^8 is not in standard form. You can use the laws of indices to write it in standard form.
$35 \times 10^8 = (3.5 \times 10) \times 10^8$
$= 3.5 \times 10^9$

b $\dfrac{4.5 \times 10^3}{5 \times 10^7} = \dfrac{4.5}{5} \times \dfrac{10^3}{10^7}$
$= 0.9 \times 10^{-4}$
$= 9 \times 10^{-5}$

$0.9 \times 10^{-4} = (9 \times 10^{-1}) \times 10^{-4}$
$= 9 \times 10^{-5}$.

c $(3.1 \times 10^2) + (6.8 \times 10^3) = 310 + 6800$
$= 7110$
$= 7.11 \times 10^3$

d $(9.2 \times 10^4) - (1.6 \times 10^3) = 92\,000 - 1600$
$= 90\,400$
$= 9.04 \times 10^4$

Exercise 10M

1 Work these out. Give your answers in standard form.

a $(2 \times 10^6) \times (4 \times 10^2)$ **b** $(6 \times 10^4) \times (8 \times 10^5)$
c $(2.5 \times 10^2) \times (8 \times 10^7)$ **d** $(9 \times 10^{-6}) \times (1 \times 10^{11})$
e $(2.5 \times 10^4) \times (2 \times 10^{-9})$ **f** $(3 \times 10^{-1}) \times (6.2 \times 10^{-6})$

2 Work these out. Give your answers in standard form.

a $\dfrac{8 \times 10^{12}}{4 \times 10^7}$ **b** $\dfrac{8.4 \times 10^5}{2 \times 10^2}$
c $\dfrac{3 \times 10^8}{6 \times 10^3}$ **d** $(4.2 \times 10^4) \div (2 \times 10^9)$
e $(1 \times 10^3) \div (4 \times 10^{11})$ **f** $(3.5 \times 10^4) \div (5 \times 10^8)$

3 Work these out. Give your answers in standard form.

a $(7 \times 10^3) + (6 \times 10^4)$ **b** $(1.7 \times 10^5) + (3.5 \times 10^4)$
c $(2.2 \times 10^4) - (9 \times 10^3)$ **d** $(8.45 \times 10^5) - (2 \times 10^4)$
e $(8 \times 10^3) + (6.9 \times 10^5)$ **f** $(9.4 \times 10^4) - (3 \times 10^2)$

4 A water molecule has a mass of 3×10^{-26} kg.
A glass of water contains 1.2×10^{25} molecules of water.
Work out the mass of the water in the glass. Give your answer in grams.

5 An adult brain has a mass of 1.4×10^{3} grams.
It contains 7×10^{10} brain neurons.
Work out the average mass of a neuron. Give your answer in grams in standard form.

B
AO2

Review exercise

E

1 Write down the numbers from the cloud which are

a multiples of 3 [2 marks]

b prime numbers [2 marks]

c factors of 100. [2 marks]

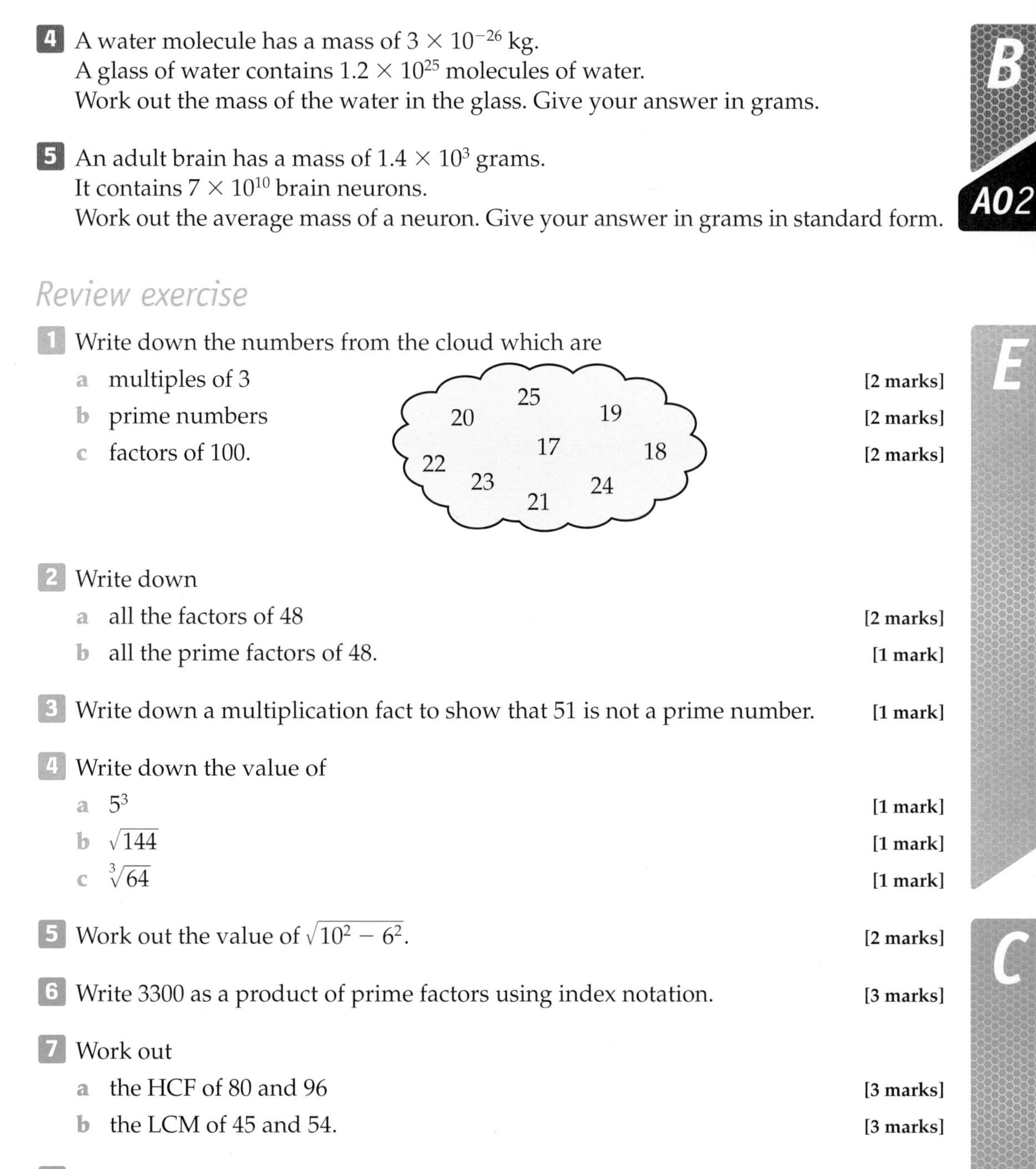

2 Write down

a all the factors of 48 [2 marks]

b all the prime factors of 48. [1 mark]

3 Write down a multiplication fact to show that 51 is not a prime number. [1 mark]

4 Write down the value of

a 5^3 [1 mark]

b $\sqrt{144}$ [1 mark]

c $\sqrt[3]{64}$ [1 mark]

C

5 Work out the value of $\sqrt{10^2 - 6^2}$. [2 marks]

6 Write 3300 as a product of prime factors using index notation. [3 marks]

7 Work out

a the HCF of 80 and 96 [3 marks]

b the LCM of 45 and 54. [3 marks]

8 Write each of these expressions as a single power.

a $4^4 \times 4^2$ [1 mark]

b $3^8 \times 3 \times 3^5$ [1 mark]

c $6^{12} \div 6^5$ [1 mark]

C

9 Daisy is tiling her bathroom floor.

260 cm

180 cm

She wants to use identical square tiles to completely cover the floor with no overlap.
Work out the largest size of square tile Daisy can use. [3 marks]

AO2

B

10 Work out the value of $(4^3 \times 10^{-1}) \times (4^{-4} \times 10^3)$ **[2 marks]**

11 Write each of these as a decimal number.

a 17^0 **[1 mark]**

b 2^{-1} **[1 mark]**

12 Work these out. Give your answers in standard form.

a $(6.2 \times 10^6) \times (3 \times 10^3)$ **[2 marks]**

b $(2.4 \times 10^3) \div (2 \times 10^{11})$ **[2 marks]**

B

AO2

13 An ant colony contains 5×10^5 ants. The total mass of all the ants in the colony is 1.5 kg.
Work out the average mass of an ant.
Give your answer in kg in standard form. **[3 marks]**

B

AO3

14 Anselm wants to take these files to a friend's house.

Filename	File size (kb)
project_ideas.doc	7.1×10^2
science1.mpg	8.34×10^5
blues_in_C.wav	9.42×10^4
me_bowling.jpg	2.41×10^3

FUNCTIONAL • FUNCTIONAL •

He has a memory stick which can hold 1 000 000 kb of data.
Will all of his files fit on the memory stick? Show all of your working. **[4 marks]**

Chapter summary

In this chapter you have learned how to

- solve problems involving multiples E
- solve problems involving factors E
- recognise two-digit prime numbers E
- calculate squares and cubes E
- calculate square roots and cube roots E
- understand and use index notation in calculations E
- understand the difference between positive and negative square roots D
- find lowest common multiples C
- find highest common factors C
- evaluate expressions involving squares, cubes and roots C
- write a number as a product of prime factors using index notation C
- use prime factors to find HCFs and LCMs C
- use laws of indices to multiply and divide numbers written in index notation C B
- understand and use negative powers and numbers to the power of 1 or 0 B
- carry out calculations with numbers given in standard form B

11 Basic rules of algebra

This chapter is about how to manipulate algebraic expressions.

Text messaging uses abbreviations to keep messages short. Mathematicians use algebra to communicate their ideas in a short form.

Objectives

This chapter will show you how to

- manipulate algebraic expressions by
  - collecting like terms **E**
  - multiplying a single term over a bracket **D**
  - taking out common factors **D** **C**
- expand the product of two linear expressions **C** **B**

Before you start this chapter

1 Write these as number sentences. Work out the answers.

a 3 less than 7   b 5 more than 3

c 2 less than 8   d 3 more than 10

7 − □ = □

2 Write these as multiplications. Work out the answers.

a $4 + 4 + 4$

b $5 + 5 + 5 + 5 + 5 + 5$

c $3 + 3 + 3 + 3$

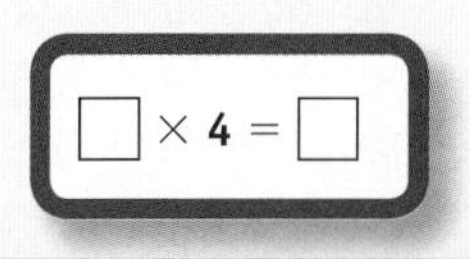

Writing simple expressions

a There are 5 boxes of strawberries. Each has n strawberries. How many strawberries are there altogether?

b x oranges are shared equally between three people. How many oranges does each person get?

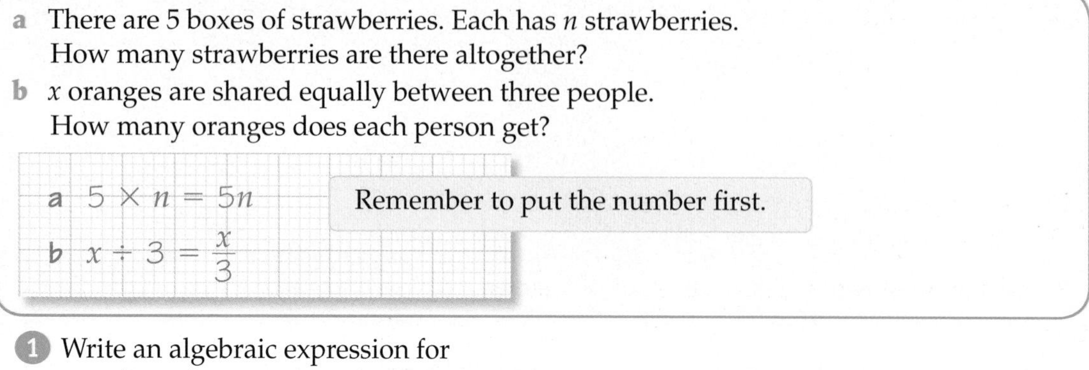

1 Write an algebraic expression for

a 3 lots of y **b** z divided by 3 **c** n multiplied by 10
d 12 divided by x **e** p shared between 4 **f** 8 lots of m

2 Use algebra to write expressions for these calculations.
Use x to represent the starting number.

a Starting number, add 2.
b Starting number, multiply it by 7.
c Starting number, take away 5.
d Starting number, double it.
e Starting number, halve it.
f Starting number, multiply it by 3, then subtract 2.
g Starting number, divide it by 4, then add 5.

3 Write an expression for the total cost (in £s) of

a 4 singles at £x each
b 6 albums at £y each
c d singles at £4 each
d t albums at £13 each.

Simplifying expressions

Simplify

a $6b + 5b - 7b$ **b** $6x - 2x + 8x - 10x$

a $6b + 5b - 7b = 11b - 7b$
$= 4b$

$6b + 5b = 11b$
$11 - 7 = 4$

b $6x - 2x + 8x - 10x = 6x + 8x - 2x - 10x$
$= 14x - 12x$
$= 2x$

Just as with numbers, you can add and subtract the terms in any order as long as each term keeps its own sign.
$6 - 2 + 8 - 10 = 2$

1 Simplify

a $a + a + a + a + a + a$ **b** $g + g$ **c** $5c + c$
d $4t + 5t + 3t$ **e** $x + 6x + 3x$ **f** $5l + l + 8l + 2l$

2 Simplify

a $10b - 7b$ **b** $6y - y$ **c** $12z - 8z$
d $8t - 2t$ **e** $6j + 5j - 7j$ **f** $5u + 3u - 6u$
g $h + h + h$ **h** $7t - 5t + 3t$ **i** $6x - x - 3x$

11.1 Collecting like terms

Keywords
term, like term

Why learn this?
Algebra is the language of maths. People across the world use it to communicate mathematical ideas.

Objectives

E Simplify algebraic expressions by collecting like terms

Skills check

1 Simplify

a $a + a + a$ **b** $p + p$ **c** $x + x + x + x + x$
d $2b + 3b + 4b$ **e** $5x + 3x + 2x$ **f** $2m - 5m + 9m$

Collecting like terms

The expression $3a + 6b + 5a$ has three **terms**.
The three terms are $3a$, $6b$ and $5a$.

Terms that use the same letter are called **like terms**.
So $3a$ and $5a$ are like terms.

You can simplify algebraic expressions by collecting like terms together.

$$3a + 6b + 5a = 3a + 5a + 6b$$
$$= 8a + 6b$$

Rearrange so that like terms are next to each other.
Keep the + or − sign with the term.

Example 1

Simplify these expressions by collecting like terms.

a $2a + 7b + 3a$ **b** $4p + 5 + 3p - 2$ **c** $6f + 5g - 4f + g$

a $2a + 7b + 3a = 2a + 3a + 7b$
$= 5a + 7b$

b $4p + 5 + 3p - 2 = 4p + 3p + 5 - 2$
$= 7p + 3$

Collect the terms in p, and collect the terms which are just numbers.

c $6f + 5g - 4f + g = 6f - 4f + 5g + g$
$= 2f + 6g$

Remember to keep the − sign with the $4f$.

E

Exercise 11A

1 Simplify these expressions by collecting like terms.

Remember, x means $1x$.

a $2c + 7d + 3c$ **b** $3m + 4r + 2m$ **c** $6x + 4y + x$
d $7p + 2 + 5p + 3$ **e** $5j + 4 + 6j + 1$ **f** $6w + 7 + 2w + 3$

E

2 Simplify these expressions by collecting like terms.

a $5x + 4y - 2x$

b $7a + 3b - 5a$

c $8k + 3m - 4k$

d $6p + 7 + 2p - 3$

e $5t + 2 - 3t + 1$

f $9z + 4 - 8z + 6$

3 Simplify these expressions.

a $6a + 5b + 4a + 2b$

b $5q + 8r + 6q + r$

c $6k + 5l - 4k + l$

d $7v + 2w - 5v - 3w$

e $4y + 4z - 3y - z$

f $9c - 7d - 8c - 6d$

> **Keep the – sign with the $4k$.**

4 Simplify these expressions.

a $5x + 3x + 2y + 2x + 4y$

b $7g + 3h - 4g + g - 2h$

c $4a + 6b - 2a + 3c + 2b - 4c$

d $5j + 6k + j + 4l - 2k + 3l$

e $8d + 5e + 6f - 3d + e - 4f$

f $6x - 2y - 5x + 8z - 3y - 4z$

5 Simplify by collecting like terms.

a $6xy + 5x^2 + 3xy$

b $4m^2 + 2m - 3m^2$

c $4ab + 6a + 3ab - 2a$

d $6x^2 + 5x + 4x + 2x^2$

e $7t^2 + 4 - 3t^2 + 1$

f $9xy + 5x - 6xy + 6x^2 + 2x$

g $4ab + 6a + 2ab - 5b - 3ab + a - 2b$

> **xy means $x \times y$**
> **x^2 means $x \times x$**
> **xy and x^2 are *not* like terms.**

Example 2

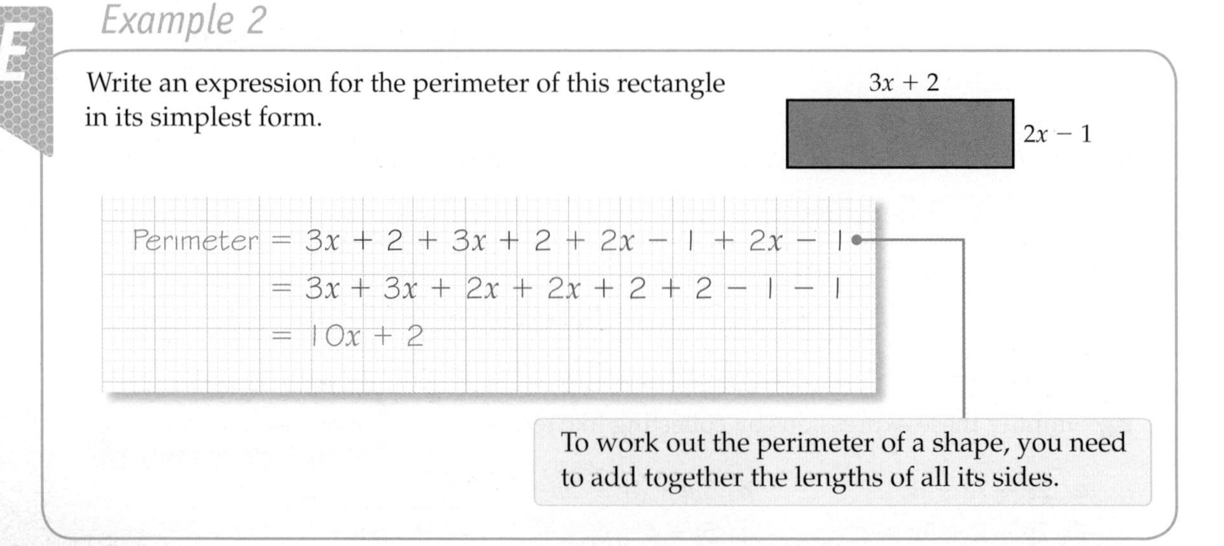
Write an expression for the perimeter of this rectangle in its simplest form.

Perimeter $= 3x + 2 + 3x + 2 + 2x - 1 + 2x - 1$

$= 3x + 3x + 2x + 2x + 2 + 2 - 1 - 1$

$= 10x + 2$

To work out the perimeter of a shape, you need to add together the lengths of all its sides.

Exercise 11B

1 Write an expression for the perimeter of each of these shapes.
Write each expression in its simplest form.

a A square, side $3a$.

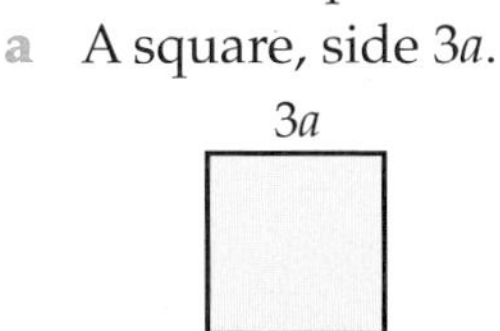

b A square, side $x + 1$.

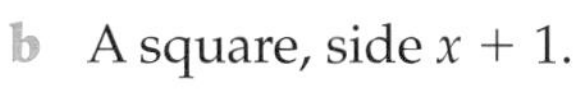

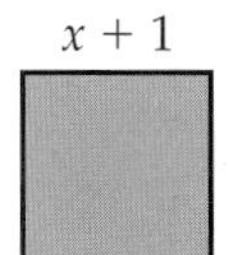

c A rectangle, width y and length $2x$.

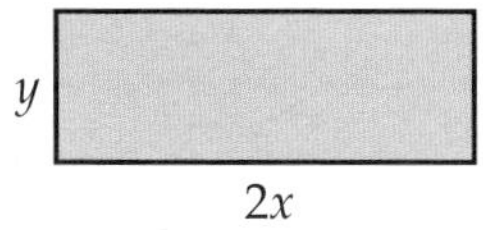

(Rectangle labelled y and $2x$)

d A rectangle, width $2x$ and length $3x + 1$.

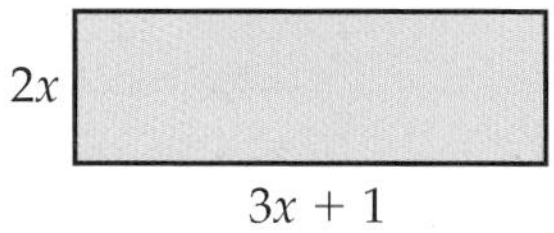

(Rectangle labelled $2x$ and $3x + 1$)

E
AO2

2 In a magic square, the expressions in each row, each column and in the two diagonals all add to the same total.
Copy and complete each of these magic squares.

a

$6a + 7b$		
$a + 6b$		
$8a + 2b$		$4a + 3b$

b

$3a + 2b$		$8a - 2b$
	$5a + b$	
$2a + 4b$		

c

$7a + b + 2c$		
$2a + 5b - 2c$	$10a - 2b + 5c$	$3a + 3b$

E
AO3

11.2 Multiplying in algebra

Keywords
algebraic

Why learn this?

You need to know how to use all four operations (+, −, ×, ÷) with algebra.

Objectives

E Multiply together two simple algebraic expressions

Skills check

1 Work out
a 3×4 **b** 6×7 **c** 3×8

2 Simplify
a $a + a + a + a$ **b** $b + b + b$ **c** 5 lots of n

Multiplying expressions

When multiplying **algebraic** expressions, there are some rules to follow.

Expression	You write
$3 \times x$ or $x \times 3$	$3x$
$d \div 4$	$\frac{d}{4}$
$g \times h$	gh
$x \times y$	xy
$y \times y$	y^2

Write the number first.

Leave out the × sign.

Write letters in alphabetical order.

y^2 is 'y squared'.

To multiply algebraic expressions

- multiply the numbers
- then multiply the letters.

For example

$$3f \times 4g = 3 \times f \times 4 \times g$$
$$= 3 \times 4 \times f \times g$$
$$= 12 \times fg$$
$$= 12fg$$

$f \times 4 = 4 \times f$

E

Example 3

Simplify

a $2 \times 5b$ **b** $3c \times 2d$

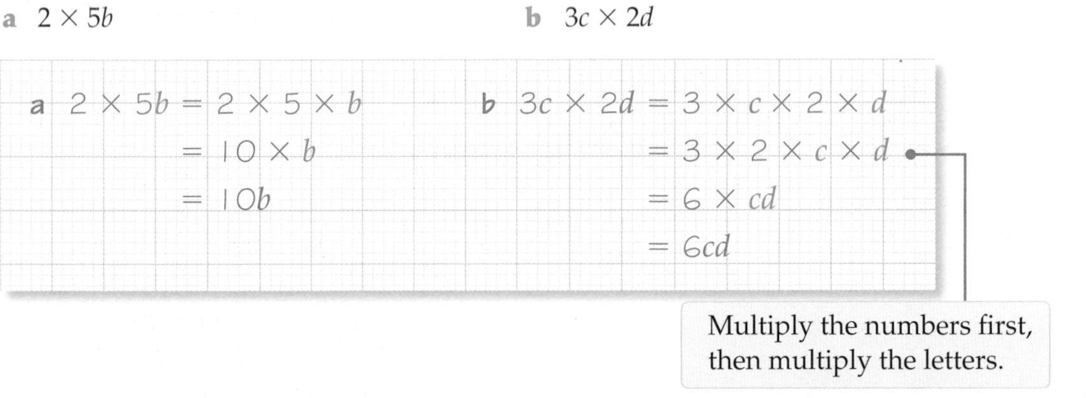

a $2 \times 5b = 2 \times 5 \times b$
$= 10 \times b$
$= 10b$

b $3c \times 2d = 3 \times c \times 2 \times d$
$= 3 \times 2 \times c \times d$
$= 6 \times cd$
$= 6cd$

Multiply the numbers first, then multiply the letters.

Exercise 11C

E

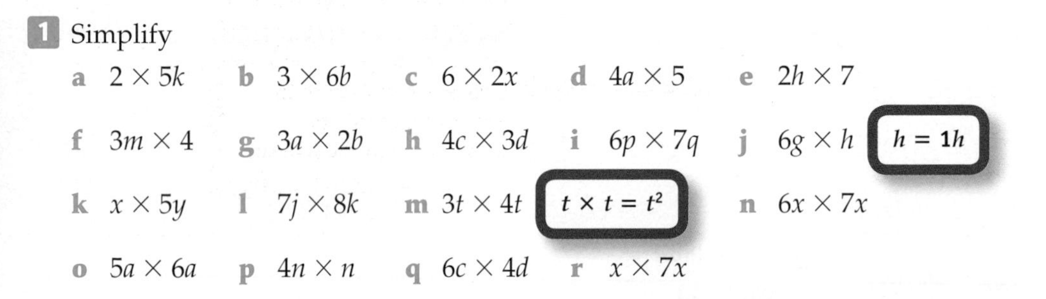

1 Simplify

a $2 \times 5k$ **b** $3 \times 6b$ **c** $6 \times 2x$ **d** $4a \times 5$ **e** $2h \times 7$

f $3m \times 4$ **g** $3a \times 2b$ **h** $4c \times 3d$ **i** $6p \times 7q$ **j** $6g \times h$

$h = 1h$

k $x \times 5y$ **l** $7j \times 8k$ **m** $3t \times 4t$ **n** $6x \times 7x$

$t \times t = t^2$

o $5a \times 6a$ **p** $4n \times n$ **q** $6c \times 4d$ **r** $x \times 7x$

11.3 Expanding brackets

Keywords
brackets, expand

Why learn this?

For every expression with brackets, there is an equivalent expression without brackets

Objectives

D Multiply terms in a bracket by a number outside the bracket

D Multiply terms in a bracket by a term that includes a letter

Skills check

1 Simplify

a $5 \times t$ **b** $7 \times 4x$ **c** $6 \times y^2$

HELP Section 9.3

2 Work out

a 2×-4 **b** 3×-6 **c** -2×-2

Expanding brackets

Some algebraic expressions have **brackets**.

$6(x + 4)$ means $6 \times (x + 4)$

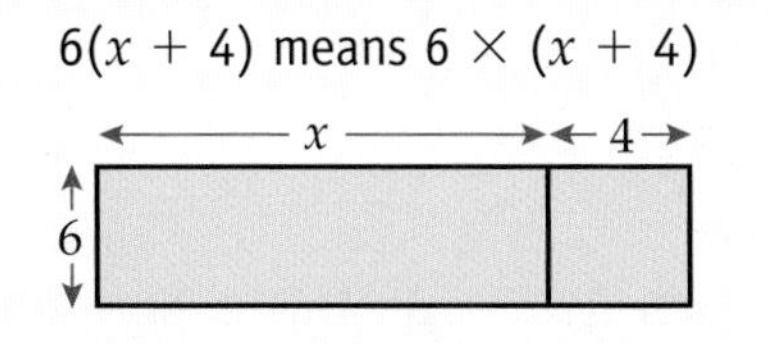

You usually write expressions like this without the multiplication sign.

You multiply each term inside the bracket by 6.

$$6(x + 4) = 6 \times x + 6 \times 4$$
$$= 6x + 24$$

To **expand** a bracket, multiply each term inside the bracket by the term outside the bracket.

Expanding brackets is sometimes called 'multiplying out the brackets'.

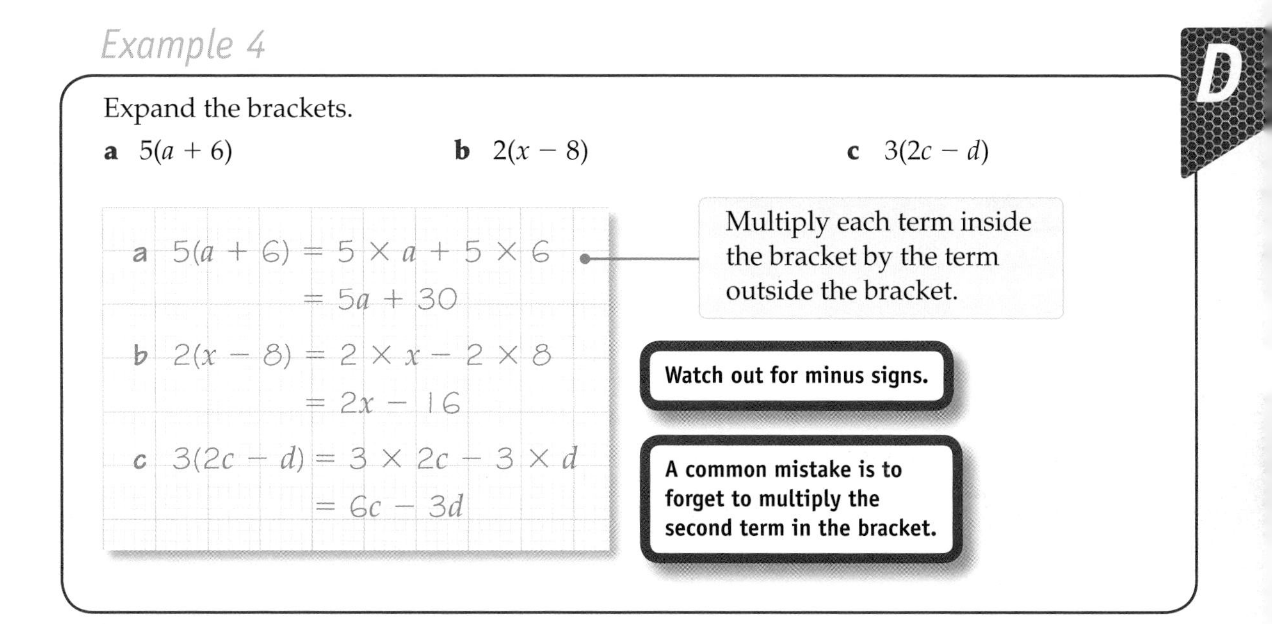

Example 4

D

Expand the brackets.

a $5(a + 6)$ **b** $2(x - 8)$ **c** $3(2c - d)$

a $5(a + 6) = 5 \times a + 5 \times 6$
$= 5a + 30$

Multiply each term inside the bracket by the term outside the bracket.

b $2(x - 8) = 2 \times x - 2 \times 8$
$= 2x - 16$

Watch out for minus signs.

c $3(2c - d) = 3 \times 2c - 3 \times d$
$= 6c - 3d$

A common mistake is to forget to multiply the second term in the bracket.

Exercise 11D

D

1 Multiply out the brackets.

a $5(p + 6)$ **b** $3(a + 5)$ **c** $7(k + 2)$ **d** $4(m + 9)$
e $5(7 + f)$ **f** $2(8 + q)$ **g** $2(a + b)$ **h** $5(x + y)$
i $8(g + h + i)$ **j** $4(u + v + w)$ **k** $2(a + b + c)$ **l** $9(d + e + 6)$

2 Expand the brackets.

a $2(y - 8)$ **b** $3(x - 5)$ **c** $6(b - 4)$ **d** $7(d - 8)$
e $2(7 - x)$ **f** $4(8 - n)$ **g** $5(a - b)$ **h** $2(x - y)$
i $7(4 + p - q)$ **j** $8(a - b + 6)$ **k** $5(c - 7 + d)$ **l** $3(u - v + w)$

3 Expand these expressions.

$-2 \times 3k$
$= -2 \times 3 \times k$
$= -6k$

a $-2(3k + 4)$ **b** $-3(2x + 6)$ **c** $-5(3n + 1)$
d $-4(3t + 5)$ **e** $-3(4p - 1)$ **f** $-2(3x - 7)$
g $-6(x - 3)$ **h** $-5(2x - 3)$ **i** $-8(3x - 4)$

D

4 Expand the brackets.

a $3(2c + 6)$ b $4(3m + 2)$ c $5(4t + 3)$ d $6(4y + 9)$
e $4(3e + f)$ f $2(5p + q)$ g $3(2a - b)$ h $6(3c - 2d)$
i $2(m - 4n)$ j $7(2x + y - 3)$ k $6(3a - 4b + c)$ l $4(2u - 5v - 3w)$

5 Multiply out the brackets.

a $2(x^2 + 3x + 2)$ b $3(x^2 + 5x - 6)$ c $2(a^2 - a + 2)$ d $4(y^2 - 3y - 10)$

D

6 Write an expression for the area of each shape. Then expand the brackets.

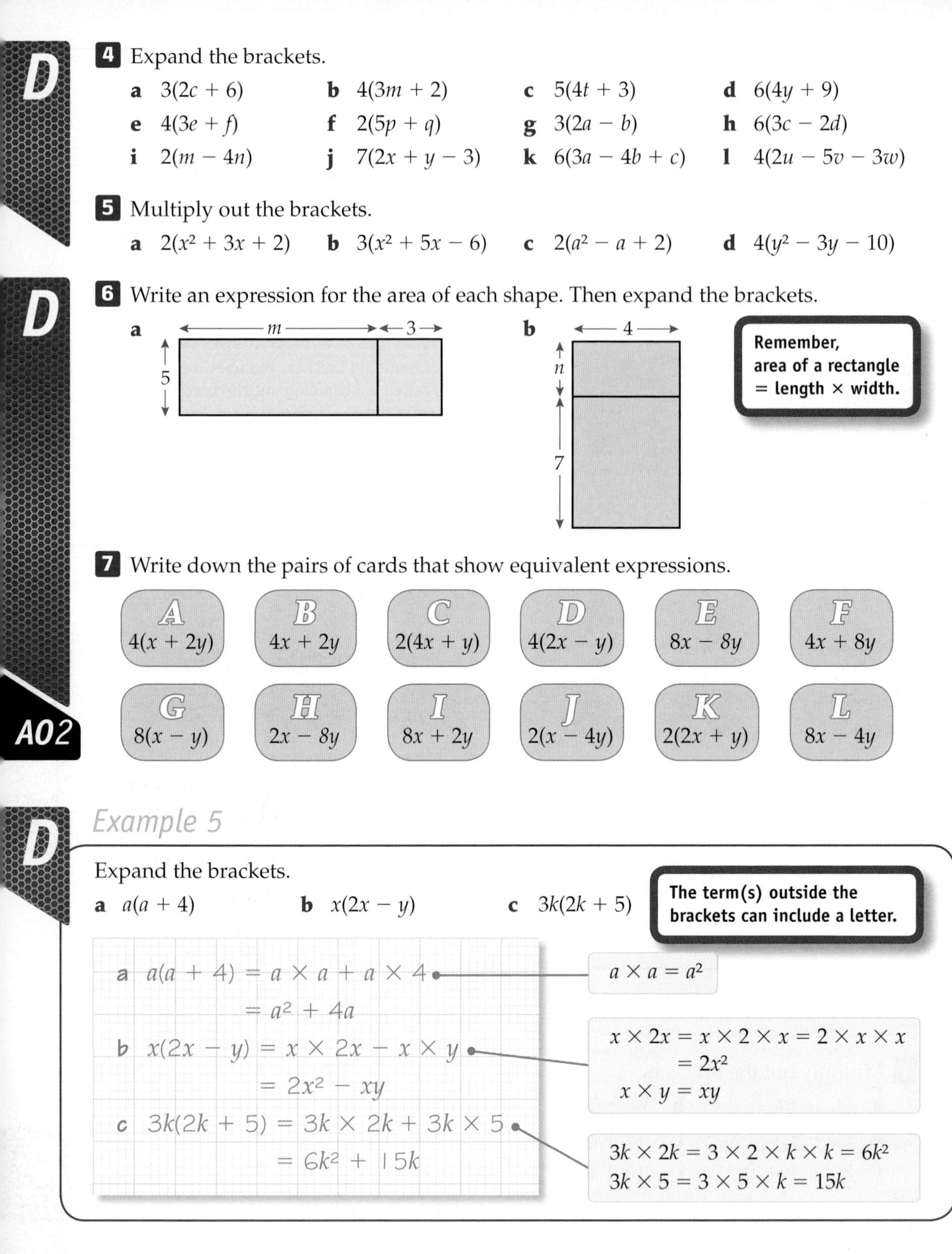

Remember, area of a rectangle = length × width.

AO2

7 Write down the pairs of cards that show equivalent expressions.

A	B	C	D	E	F
$4(x + 2y)$	$4x + 2y$	$2(4x + y)$	$4(2x - y)$	$8x - 8y$	$4x + 8y$

G	H	I	J	K	L
$8(x - y)$	$2x - 8y$	$8x + 2y$	$2(x - 4y)$	$2(2x + y)$	$8x - 4y$

D

Example 5

Expand the brackets.

a $a(a + 4)$ b $x(2x - y)$ c $3k(2k + 5)$

The term(s) outside the brackets can include a letter.

a $a(a + 4) = a \times a + a \times 4$
$= a^2 + 4a$

$a \times a = a^2$

b $x(2x - y) = x \times 2x - x \times y$
$= 2x^2 - xy$

$x \times 2x = x \times 2 \times x = 2 \times x \times x$
$= 2x^2$
$x \times y = xy$

c $3k(2k + 5) = 3k \times 2k + 3k \times 5$
$= 6k^2 + 15k$

$3k \times 2k = 3 \times 2 \times k \times k = 6k^2$
$3k \times 5 = 3 \times 5 \times k = 15k$

Exercise 11E

D

Expand the brackets.

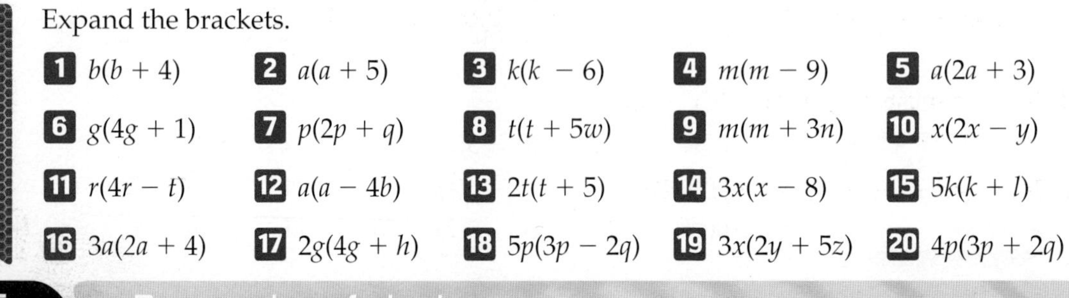

1 $b(b + 4)$ **2** $a(a + 5)$ **3** $k(k - 6)$ **4** $m(m - 9)$ **5** $a(2a + 3)$
6 $g(4g + 1)$ **7** $p(2p + q)$ **8** $t(t + 5w)$ **9** $m(m + 3n)$ **10** $x(2x - y)$
11 $r(4r - t)$ **12** $a(a - 4b)$ **13** $2t(t + 5)$ **14** $3x(x - 8)$ **15** $5k(k + l)$
16 $3a(2a + 4)$ **17** $2g(4g + h)$ **18** $5p(3p - 2q)$ **19** $3x(2y + 5z)$ **20** $4p(3p + 2q)$

Simplifying expressions with brackets

Why learn this?

Simplifying expressions makes them easier to deal with.

Objectives

D **C** Simplify expressions involving brackets

Skills check

1 Expand

a $3(p - 7)$ b $4(2x + 3)$

c $-2(x + 8)$ d $-3(m - 2)$

2 Simplify $8d + 4e - 3d - 6e$

Simplifying expressions with brackets

To add or subtract expressions with brackets

- expand the brackets
- then collect like terms to simplify your answer.

Example 6

Expand and simplify

a $3(2x + 5) + 2(x - 4)$ **b** $3(2t + 1) - 2(2t + 4)$

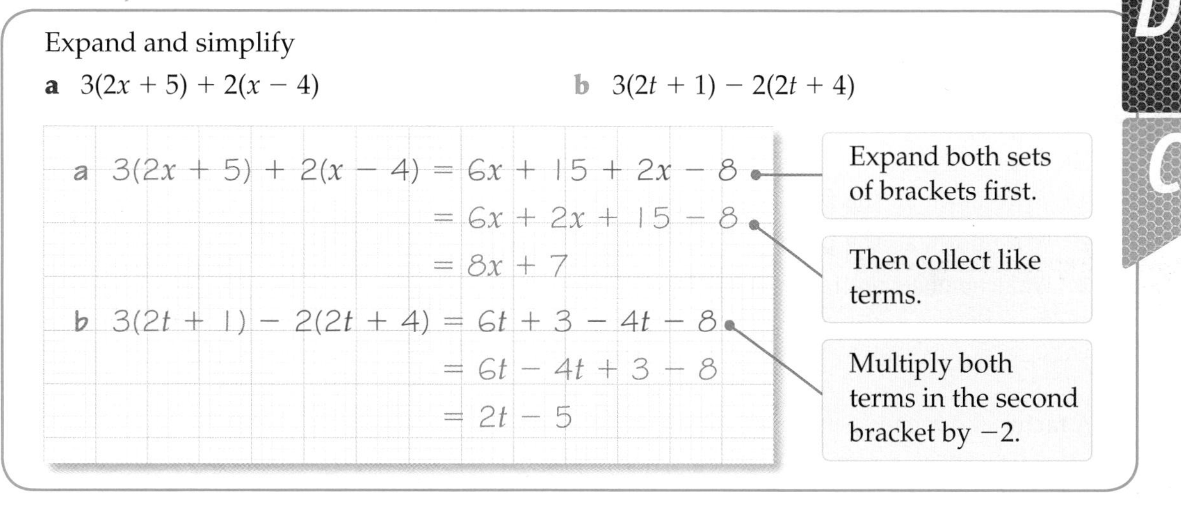

D C

Exercise 11F

1 Expand and simplify

a $3(y + 4) + 2y + 10$ **b** $2(k + 6) + 3k + 9$

c $4(a + 3) - 2a + 6$ **d** $3(t - 2) + 4t - 10$

e $4(x + 7) + 3(x + 4)$ **f** $3(x - 5) + 2(x - 3)$

g $3(2y + 3) - 2(y + 5)$ **h** $5(2x + 5) + 3(x - 4)$

i $2(4n + 5) - 5(n - 3)$ **j** $4(2x - 1) + 2(3x - 2)$

D C

2 Show that $2(x + 5) + 3x = 5(x + 2)$.

3 Show that $6(t - 5) + 6 = 6(t - 4)$.

4 Expand and simplify

a $3(2b + 1) - 2(2b + 4)$

b $4(2m + 3) - 2(2m + 5)$

c $5(2k + 2) - 4(2k + 6)$

d $2(4p + 1) - 4(p - 3)$

e $5(2g - 4) - 2(4g - 6)$

f $2(w - 4) - 3(2w - 1)$

C

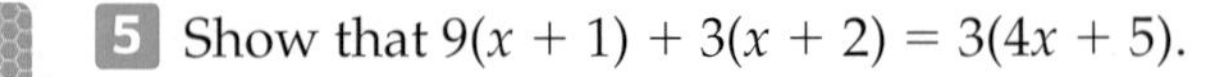

5 Show that $9(x + 1) + 3(x + 2) = 3(4x + 5)$.

AO2

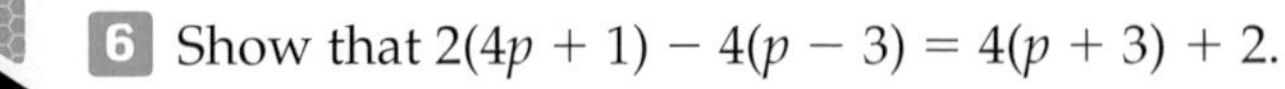

6 Show that $2(4p + 1) - 4(p - 3) = 4(p + 3) + 2$.

11.5 Factorising algebraic expressions

Keywords
factor, common factor, factorise

Why learn this?

It's useful to know how to write expressions in equivalent ways. Buying two dustpans and two brushes, $2d + 2b$, is the equivalent of buying two dustpan and brush sets, $2(d + b)$.

Objectives

D Recognise factors of algebraic terms

D Simplify algebraic expressions by taking out common factors

Skills check

HELP Section 10.2

1. Write down all the factors of 18.
2. Write down all the factors of 24.
3. What numbers are factors of 18 *and* 24?

Factors

A **factor** of a term is a number or letter that divides into it exactly.

2 is a factor of 4. 2 is a factor of $6x$.

A **common factor** of two terms is a factor of both of them.

2 is a common factor of 4 and $6x$.

D

Example 7

a Which of these terms have 2 as a factor?

$12 \quad 7 \quad 6t \quad 10x \quad 5y^2$

b Write each term from part **a** as $2 \times \square$

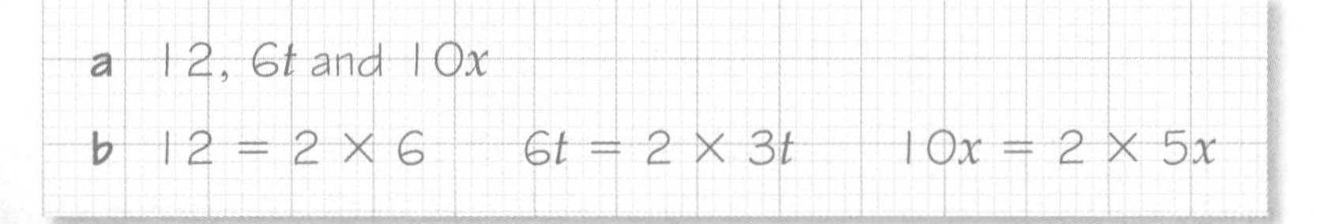

a 12, $6t$ and $10x$

b $12 = 2 \times 6$ $\quad 6t = 2 \times 3t$ $\quad 10x = 2 \times 5x$

Example 8

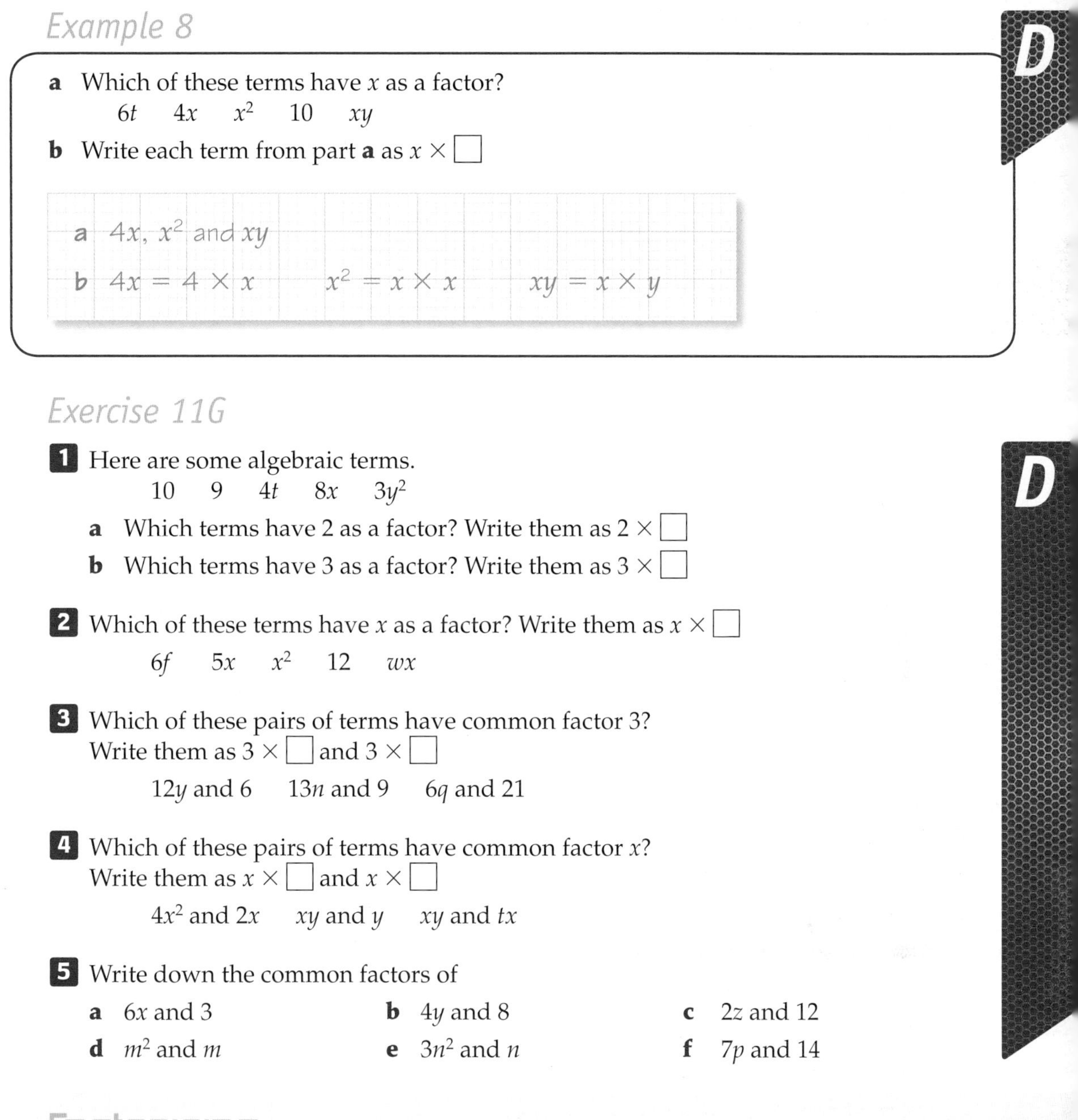

a Which of these terms have x as a factor?

$6t \quad 4x \quad x^2 \quad 10 \quad xy$

b Write each term from part **a** as $x \times \square$

a $4x$, x^2 and xy

b $4x = 4 \times x \qquad x^2 = x \times x \qquad xy = x \times y$

Exercise 11G

D

1 Here are some algebraic terms.

$10 \quad 9 \quad 4t \quad 8x \quad 3y^2$

a Which terms have 2 as a factor? Write them as $2 \times \square$

b Which terms have 3 as a factor? Write them as $3 \times \square$

2 Which of these terms have x as a factor? Write them as $x \times \square$

$6f \quad 5x \quad x^2 \quad 12 \quad wx$

3 Which of these pairs of terms have common factor 3?
Write them as $3 \times \square$ and $3 \times \square$

$12y$ and 6 $\quad 13n$ and 9 $\quad 6q$ and 21

4 Which of these pairs of terms have common factor x?
Write them as $x \times \square$ and $x \times \square$

$4x^2$ and $2x \quad xy$ and $y \quad xy$ and tx

5 Write down the common factors of

a $6x$ and 3 **b** $4y$ and 8 **c** $2z$ and 12

d m^2 and m **e** $3n^2$ and n **f** $7p$ and 14

Factorising

Factorising an algebraic expression is the opposite of expanding brackets.
Start by writing a common factor of both terms outside the bracket.
Then work out the terms inside the bracket.

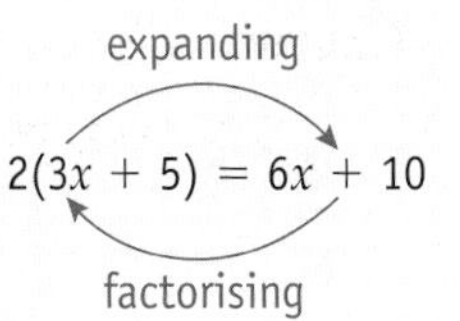

Example 9

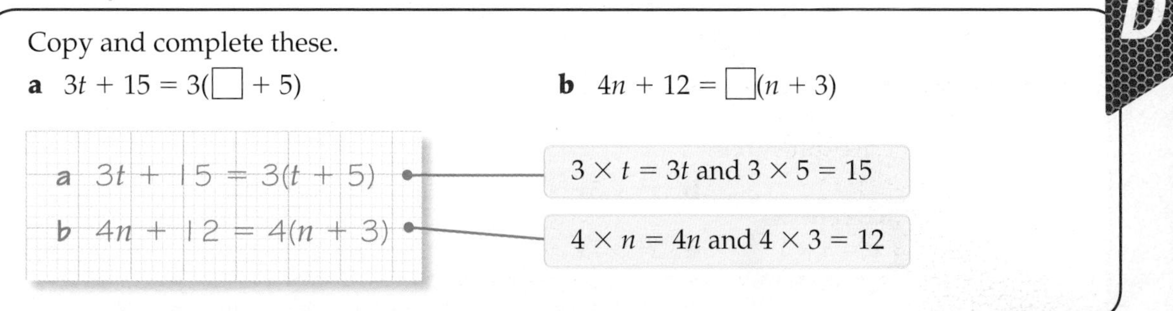

Copy and complete these.

a $3t + 15 = 3(\square + 5)$ **b** $4n + 12 = \square(n + 3)$

a $3t + 15 = 3(t + 5)$ — $3 \times t = 3t$ and $3 \times 5 = 15$

b $4n + 12 = 4(n + 3)$ — $4 \times n = 4n$ and $4 \times 3 = 12$

Exercise 11H

Copy and complete these. Check your answers by expanding the brackets.

1 $3x + 15 = 3(\square + 5)$ **2** $5a + 10 = 5(\square + 2)$ **3** $2x - 12 = 2(x - \square)$

4 $4m - 16 = 4(m - \square)$ **5** $4t + 12 = \square(t + 3)$ **6** $3n + 18 = \square(n + 6)$

7 $2b - 14 = \square(b - 7)$ **8** $4t - 20 = \square(t - 5)$

Example 10

Factorise

a $5a + 20$ **b** $4x - 12$  **c** $x^2 + 7x$

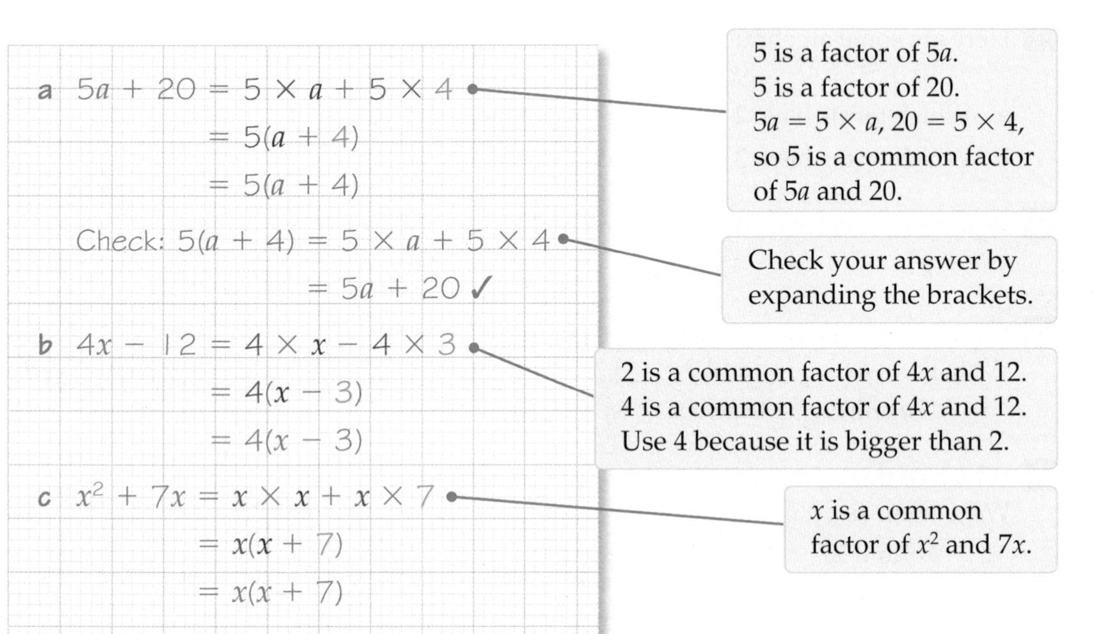

Exercise 11I

1 Factorise these expressions.

a $5p + 20$ **b** $2a + 12$ **c** $3y + 15$
d $7b + 21$ **e** $4q + 12$ **f** $6k + 24$
g $5a + 5$ ($5 = 5 \times 1$) **h** $4g + 8$ **i** $3m + 18$

2 Factorise these expressions.

a $4t - 12$ **b** $3x - 9$ **c** $5n - 20$
d $2b - 8$ **e** $6a - 18$ **f** $7k - 7$
g $4r - 16$ **h** $6g - 12$ **i** $3m - 12$

3 Factorise these expressions.

a $y^2 + 7y$ **b** $x^2 + 5x$ **c** $t^2 + 2t$
d $n^2 + n$ ($n = n \times 1$) **e** $x^2 - 7x$ **f** $z^2 - 2z$
g $p^2 - 8p$ **h** $a^2 - a$ **i** $3m^2 - m$

4 Factorise these expressions.

a $6p + 4$ **(Hint:** $6p = 2 \times 3p$**)**   **b** $4a + 10$ **c** $4t - 6$

d $8m - 12$ **e** $10x + 15$ **f** $6y - 9$

g $4a + 8b$ **h** $10p + 5q$ **i** $7n - 14$

5 Write down the pairs of cards that show equivalent expressions.

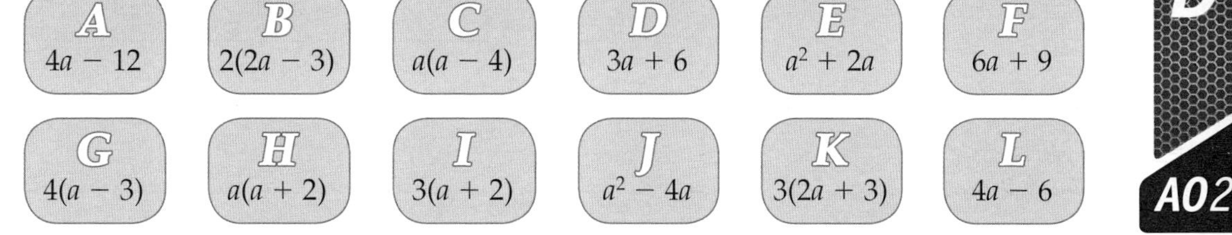

D D AO2

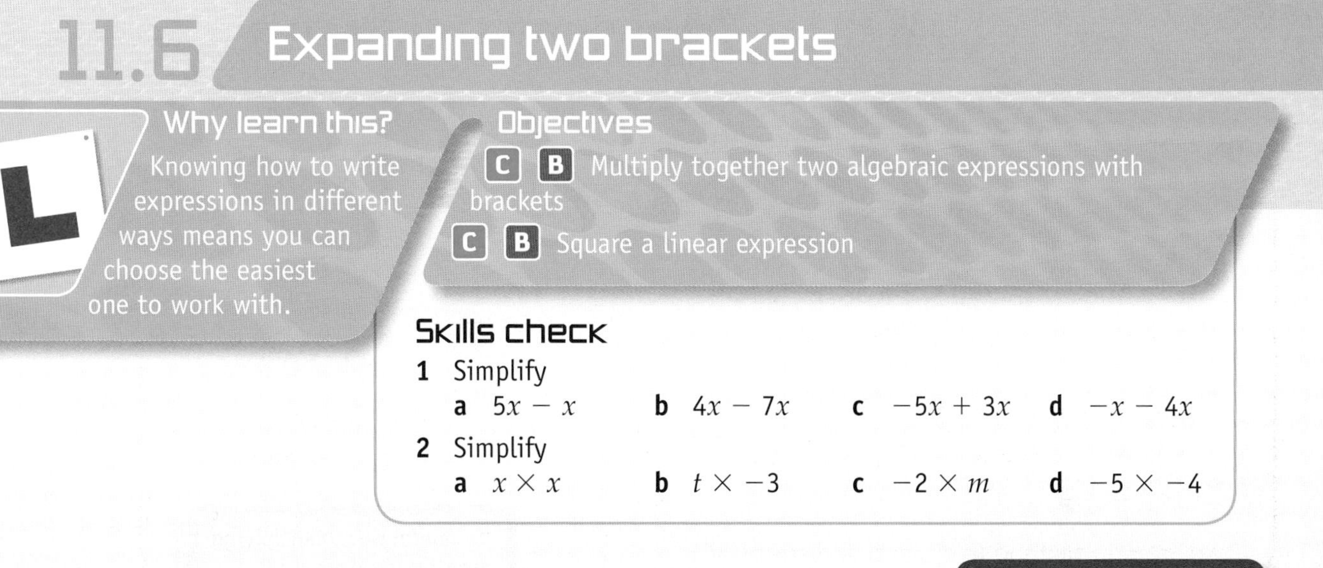

11.6 Expanding two brackets

Why learn this?

Knowing how to write expressions in different ways means you can choose the easiest one to work with.

Objectives

C B Multiply together two algebraic expressions with brackets

C B Square a linear expression

Skills check

1 Simplify

a $5x - x$ **b** $4x - 7x$ **c** $-5x + 3x$ **d** $-x - 4x$

2 Simplify

a $x \times x$ **b** $t \times -3$ **c** $-2 \times m$ **d** -5×-4

Expanding two brackets

For a reminder of the grid method see Section 9.1.

You can use a grid method to multiply two bracketed expressions.

You multiply each term in one bracket by each term in the other bracket.

Example 11

C

Expand and simplify $(x + 2)(x + 5)$.

$(x + 2)(x + 5)$ means $(x + 2) \times (x + 5)$

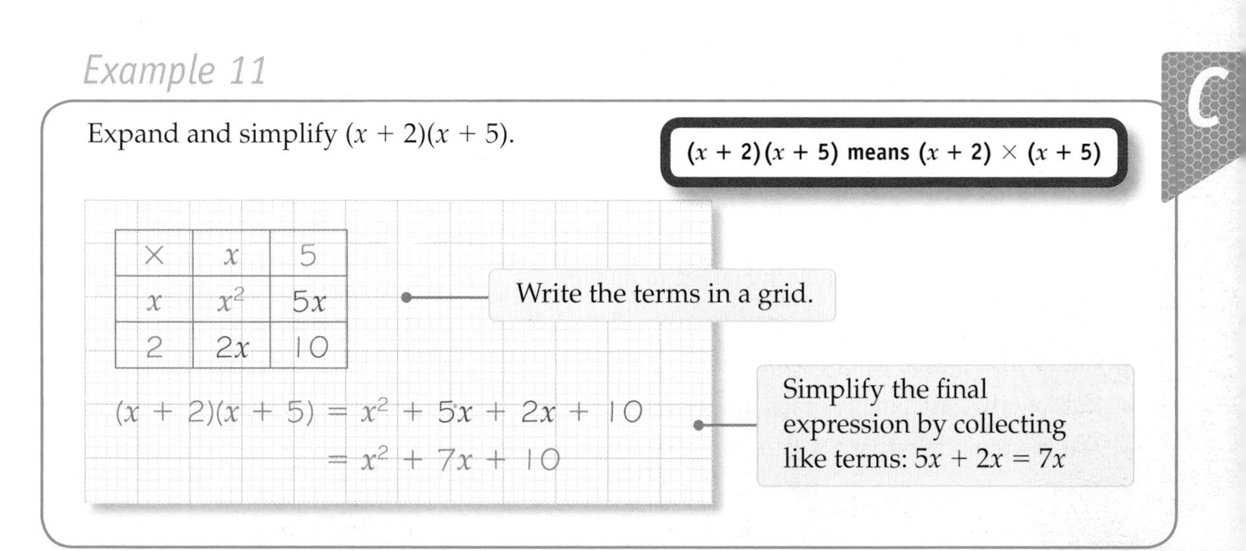

×	x	5
x	x^2	$5x$
2	$2x$	10

Write the terms in a grid.

$(x + 2)(x + 5) = x^2 + 5x + 2x + 10$
$= x^2 + 7x + 10$

Simplify the final expression by collecting like terms: $5x + 2x = 7x$

Example 11 is like working out the area of a rectangle of length $x + 5$ and width $x + 2$.

Total area $= (x + 2)(x + 5)$
$= x^2 + 5x + 2x + 10$
$= x^2 + 7x + 10$

	x	5
x	area $= x \times x = x^2$	area $= x \times 5 = 5x$
2	area $= 2 \times x = 2x$	area $= 2 \times 5 = 10$

Example 12

Expand and simplify $(t + 6)(t - 2)$.

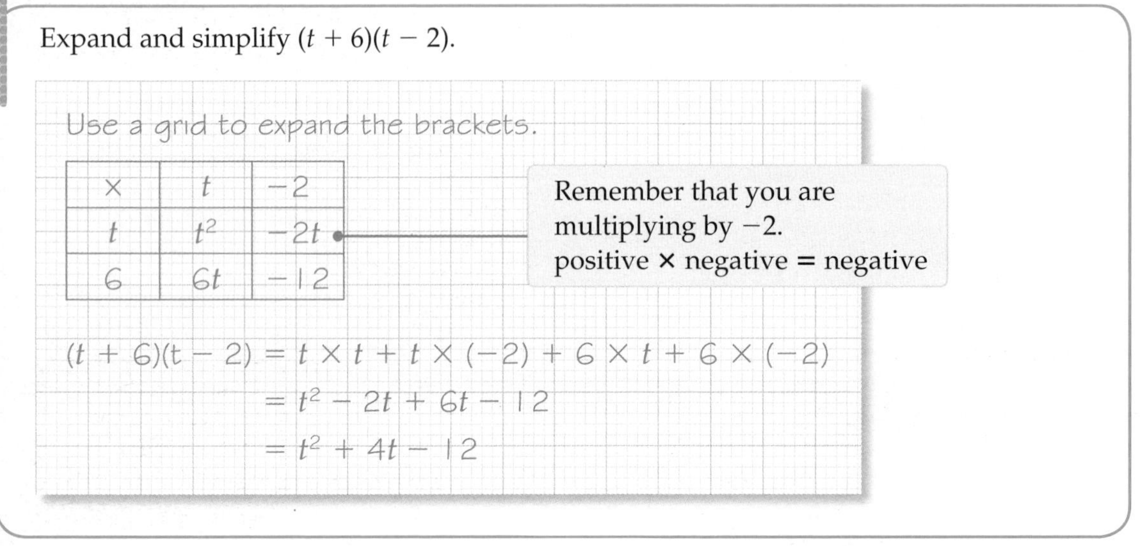

Use a grid to expand the brackets.

$\times$	t	-2
t	t^2	$-2t$
6	$6t$	-12

$(t + 6)(t - 2) = t \times t + t \times (-2) + 6 \times t + 6 \times (-2)$
$= t^2 - 2t + 6t - 12$
$= t^2 + 4t - 12$

Exercise 11J

Use the grid method to expand and simplify these expressions.

Be careful when multiplying negative numbers.

1. $(a + 2)(a + 7)$
2. $(x + 3)(x + 1)$
3. $(x + 5)(x + 5)$
4. $(t + 5)(t - 2)$
5. $(x + 7)(x - 4)$
6. $(n - 5)(n + 8)$
7. $(x - 4)(x + 5)$
8. $(p - 4)(p + 4)$

Look again at Example 12.

$(t + 6)(t - 2)$

- Multiply the **F**irst terms in each bracket to give t^2
- Multiply the **O**utside pair of terms to give $-2t$
- Multiply the **I**nside pair of terms to give $+6t$
- Multiply the **L**ast terms in each bracket to give -12
- Add the terms together $t^2 - 2t + 6t - 12$
- Then simplify $t^2 + 4t - 12$

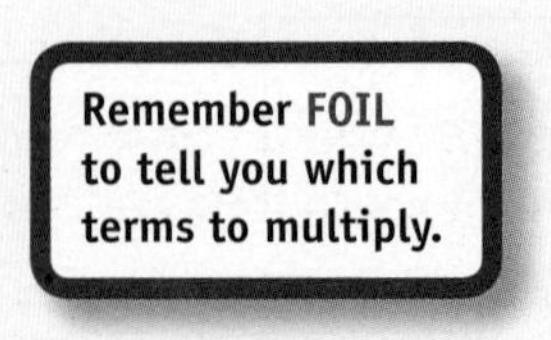

Exercise 11K

1 Use FOIL to expand and simplify these.

a $(y + 5)(y + 3)$ b $(q + 7)(q + 6)$

c $(a + 4)(a - 7)$ d $(m + 7)(m - 8)$

e $(y - 3)(y - 3)$ f $(d + 5)(d - 4)$

g $(x - 3)(x + 3)$ h $(h - 3)(h - 8)$

i $(7 - z)(3 + z)$ j $(x - 9)(x - 4)$

C

2 Expand and simplify these. Use the method you prefer.

a $(x + 4)(x - 4)$ b $(x + 5)(x - 5)$

c $(x + 2)(x - 2)$ d $(x - 11)(x + 11)$

e $(x - 3)(x + 3)$ f $(x - 1)(x + 1)$

g $(x + a)(x - a)$ h $(t + x)(t - x)$

> What happens to the x term when you multiply brackets of the form $(x + a)(x - a)$?

3 Expand and simplify

a $(x + 5)^2 = (x + 5)(x + 5)$

b $(x + 6)^2 = (x + 6)(x + 6)$

c $(x - 3)^2 = (x - 3)(x - 3)$

> Look for a pattern between the terms in the brackets and the final expression.

4 Expand and simplify

a $(x + 1)^2$ b $(x - 4)^2$ c $(x - 5)^2$

d $(x + 7)^2$ e $(x - 8)^2$ f $(3 + x)^2$

g $(2 + x)^2$ h $(5 - x)^2$ i $(x + a)^2$

> Write down the bracket twice and then use FOIL. e.g. $(x + 1)(x + 1)$

5 Copy and complete these.

a $(x + \square)^2 = x^2 + \square x + 36$ b $(x - \square)^2 = x^2 - \square x + 49$

c $(x + \square)^2 = x^2 + 18x + \square$ d $(x - \square)^2 = x^2 - 20x + \square$

C

6 Here is part of a number grid.

1	2	3	4	5	6	7	8	9	10
11	12	13	14	15	16	17	18	19	20
21	22	23	24	25	26	27	28	29	30
31	32	33	34	35	36	37	38	39	40

Maria is investigating blocks of four numbers in the grid. Here is a block of four squares with n in the top left corner.

n	

a Copy this block of four squares and write expressions in terms of n in the other three boxes.

> Use the pattern of numbers in the number grid to help you.

b Multiply together the term in the bottom left box of your block of four and the term in the top right.

c Multiply together the term in the top left box of your block of four (i.e. n) and the term in the bottom right.

d Subtract your answer to part **c** from your answer to part **b**. What do you notice?

AO2

B

Example 13

Expand and simplify $(3x - y)(x - 2y)$.

×	x	$-2y$
$3x$	$3x^2$	$-6xy$
$-y$	$-xy$	$2y^2$

Remember to multiply each term in the first bracket by each term in the second bracket.

$$(3x - y)(x - 2y) = 3x^2 - 6xy - xy + 2y^2$$
$$= 3x^2 - 7xy + 2y^2$$

Be careful when there are negative signs. This is where lots of mistakes are made.
Remember:
+ve × −ve = −ve
−ve × −ve = +ve

Exercise 11L

B

Expand and simplify these expressions.

1. $(3a + 2)(a + 4)$
2. $(5x + 3)(x + 2)$
3. $(2t + 3)(3t + 5)$
4. $(4y + 1)(2y + 7)$
5. $(6x + 5)(2x + 3)$
6. $(4x + 3)(x - 1)$
7. $(2z + 5)(3z - 2)$
8. $(y + 1)(7y - 8)$
9. $(3n - 5)(n + 8)$
10. $(3b - 5)(2b + 1)$
11. $(p - 4)(7p + 3)$
12. $(2z - 3)(3z - 4)$
13. $(5x - 9)(2x - 1)$
14. $(2y - 3)(2y - 3)$
15. $(2 + 3a)(4a + 5)$
16. $(2x + 1)(2x - 1)$
17. $(3y +2)(3y - 2)$
18. $(5n + 4)(5n - 4)$
19. $(3x + 5)(3x - 5)$
20. $(1 + 2x)(1 - 2x)$
21. $(3 + 5x)(3 - 5x)$

Review exercise

E

1. Simplify
   a $4d + 5f + 2f - 3d$ **[2 marks]**
   b $7r - 3p + 4r + 2p$ **[2 marks]**
   c $4x + 2y + x - 5y$ **[2 marks]**

D

2. Eggs are packed in boxes of 6 and in boxes of 12.
   Write down an expression for the total number of eggs in x boxes of 6 and y boxes of 12. **[2 marks]**

3. Expand
   a $6(p + 2)$ **[1 mark]**
   b $t(t - 5)$ **[1 mark]**

4. Factorise
   a $5y + 10$ **[1 mark]**
   b $14a - 7$ **[1 mark]**
   c $m^2 - 6m$ **[1 mark]**
   d $x^2 + x$ **[1 mark]**

5 Expand and simplify

a $3(2x - 5) + 4(x + 3)$ **[2 marks]**

b $4(3x + 2) - 5(2x - 1)$ **[2 marks]**

C

6 Expand and simplify

a $(x + 3)(x + 4)$ **[2 marks]**

b $(y - 5)(y + 2)$ **[2 marks]**

c $(z - 3)(z - 6)$ **[2 marks]**

7 Expand and simplify

a $(a + 3)^2$ **[2 marks]**

b $(b - 5)^2$ **[2 marks]**

8 Show that $(x + 3)(x + 2) - 5x = x^2 + 6$. **[3 marks]**

C AO2

9 Here is part of a number grid.

The shaded shape is called T_{10} because it has the number 10 on the left.
The sum of numbers in T_{10} is 43.

1	2	3	4	5	6	7
8	9	10	11	12	13	14
15	16	17	18	19	20	21
22	23	24	25	26	27	28

a This is T_n.

n

Copy T_n and write expressions for the missing numbers in the empty boxes. **[3 marks]**

b Find the sum of all the numbers in T_n in terms of n.
Give your answer in its simplest form. **[2 marks]**

C AO3

10 Expand and simplify

a $(4x - 3)(2x + 1)$ **[2 marks]**

b $(3n - 5)^2$ **[2 marks]**

B

Chapter summary

In this chapter you learned how to

- simplify algebraic expressions by collecting like terms E
- multiply together two simple algebraic expressions E
- multiply terms in a bracket by a number outside the bracket D
- multiply terms in a bracket by a term that includes a letter D
- recognise factors of algebraic terms D
- simplify algebraic expressions by taking out common factors D
- simplify expressions involving brackets D C
- multiply together two algebraic expressions with brackets C B
- square a linear expression C B

eBay business

Shani lives in Liverpool. She runs a business that imports wooden carvings from Africa. She then sells these to customers via eBay.

Shani sells carvings of elephants and rhinos in three sizes, small (S), medium (M) and large (L). One day she receives three large orders.

Name of customer	Number of elephants			Number of rhinos			Type of delivery
	S	M	L	S	M	L	
Mrs Read	6	3	3	4	2	2	next day
Mr Owen	4	4	0	2	2	0	standard
Mrs Patel	4	3	1	3	2	1	standard

Question bank

1 What is the total cost of the elephants that Mrs Read orders, excluding any delivery charge?

Shani uses two delivery companies. Both companies offer a 'standard' and 'next day' delivery service.

2 Which type of delivery service does Mrs Patel order?

Shani uses two types of boxes to pack the carvings. She always chooses the cheapest delivery company.

Both companies work out the delivery charge by measuring the weight of the package Shani sends.

3 How much does it cost to send a 9 kg package by 'next day' delivery with Liverpool Xpress?

Shani completes a bill for each of her customers. The bill shows the total cost of the carvings and the cost of delivery. Shani says, 'Mrs Read's bill is more than Mr Owen's and Mrs Patel's bills added together.'

4 Is Shani correct? Show working to support your answer.

One day Shani asks her brother to post a package for her. She tells him it contains three identical carvings, but doesn't tell him which animal. Her brother weighs the package and it comes to 950 g. He knows that the packaging weighs 50 g.

5 Solve this equation to work out which animal is in the package:

$$3x + 50 = 950$$

where x is the weight of the mystery animal in grams.

Information bank

Price list

Animal	Weight of carving	Price
Elephant (S)	400 g	£8.50
Rhino (S)	300 g	£6.95
Elephant (M)	1 kg	£12.50
Rhino (M)	900 g	£9.95
Elephant (L)	1.5 kg	£17.50
Rhino (L)	1.3 kg	£13.95

Delivery charges

Liverpool Xpress prices		
Weight (up to)	Standard delivery	Next day delivery
5 kg	£8.70	£12.50
10 kg	£14.40	£24.65
12 kg	£16.80	£25.70
14 kg	£18.55	£28.95
16 kg	£22.95	£30.78

UK Connect prices		
Weight (up to)	Standard delivery	Next day delivery
2 kg	£4.20	£8.25
4 kg	£6.85	£13.60
6 kg	£9.30	£22.00
8 kg	£11.40	£24.30
10 kg	£12.24	£26.50
15 kg	£14.26	£32.20
20 kg	£21.95	£40.54

Types of boxes

A box that can hold up to 9 kg weighs 350 g.

This box measures 30 cm long by 21 cm wide by 25 cm high.

A box that can hold up to 16 kg weighs 450 g.

This box measures 40 cm long by 30 cm wide by 25 cm high.

All boxes containing glass must have a 'Fragile' label.

All the boxes are made from 80% recycled cardboard.

12

Fractions

This chapter is about calculating with fractions.

In Formula 1, the difference between first and second place is incredibly small. As Jenson Button said, 'You're qualifying on pole by a half a tenth of a second.'

Objectives

This chapter will show you how to

- **multiply fractions** E
- **compare fractions** E D
- **add and subtract fractions** E D
- **multiply mixed numbers** D C B
- **divide by a fraction** D C B
- **add and subtract mixed numbers** C
- **work out reciprocals** C

Before you start this chapter

Put your calculator away!

1 Work out

a $\frac{3}{7}+\frac{1}{7}$

b $\frac{4}{9}-\frac{3}{9}$

c $\frac{5}{6}+\frac{2}{6}$

d $\frac{3}{4}+\frac{3}{4}+\frac{3}{4}$

2 Greg starts with one fraction, subtracts another fraction, and gets an answer of $\frac{4}{11}$.
Suggest two fractions that Greg may be using.

3 The diagram shows the total height of a chair. It also shows the height of the seat of the chair from the floor.
What is the height, h m, of the back of the chair?

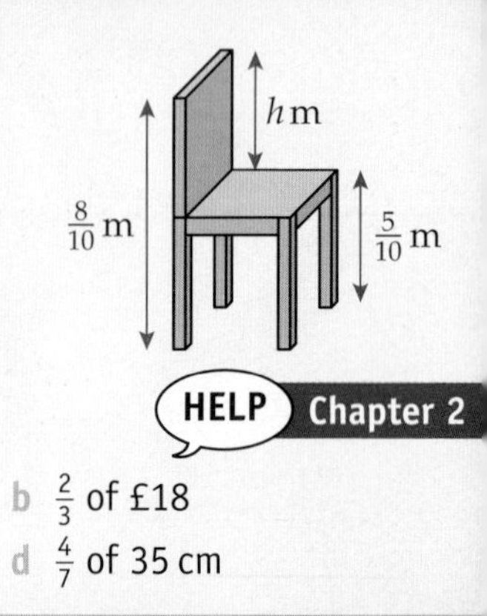

HELP Chapter 2

4 Work out

a $\frac{1}{4}$ of 24 m

b $\frac{2}{3}$ of £18

c $\frac{3}{5}$ of 25 kg

d $\frac{4}{7}$ of 35 cm

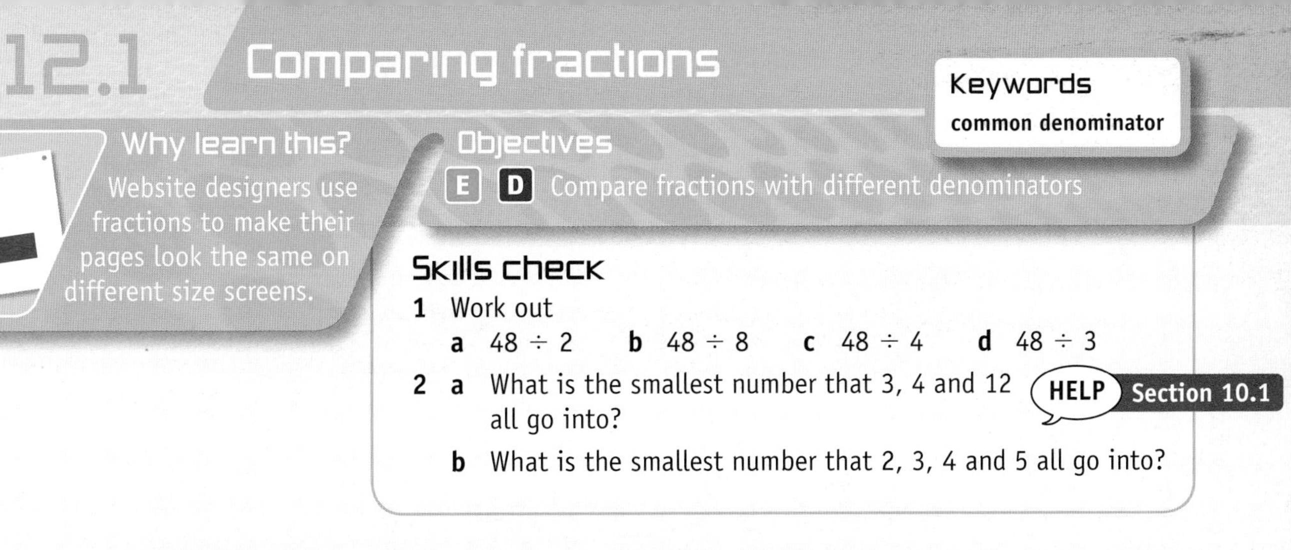

12.1 Comparing fractions

Keywords
common denominator

Why learn this?
Website designers use fractions to make their pages look the same on different size screens.

Objectives
E D Compare fractions with different denominators

Skills check

1 Work out

a $48 \div 2$ b $48 \div 8$ c $48 \div 4$ d $48 \div 3$

2 a What is the smallest number that 3, 4 and 12 all go into?

HELP Section 10.1

b What is the smallest number that 2, 3, 4 and 5 all go into?

Comparing two or more fractions

To compare fractions with different denominators, change them into equivalent fractions with the same denominator. The 'same denominator' is often called the **common denominator.**

Example 1

a Which is larger, $\frac{5}{6}$ or $\frac{7}{9}$?

b Put these fractions in order of size, smallest first. $\frac{2}{3}, \frac{3}{5}, \frac{11}{15}$

a $\frac{5}{6} = \frac{5 \times 3}{6 \times 3} = \frac{15}{18}$ and

$\frac{7}{9} = \frac{7 \times 2}{9 \times 2} = \frac{14}{18}$

$\frac{5}{6}$ is larger because $\frac{15}{18} > \frac{14}{18}$

Start by writing each fraction as an equivalent fraction with the same denominator. In this case use a common denominator of 18. You can then decide which fraction is the larger of the two.

b $\frac{2}{3} = \frac{2 \times 5}{3 \times 5} = \frac{10}{15}$

$\frac{3}{5} = \frac{3 \times 3}{5 \times 3} = \frac{9}{15}$

$\frac{3}{5}, \frac{2}{3}, \frac{11}{15}$

Compare the fractions by writing them as equivalent fractions using a common denominator. The smallest number that 3, 5 and 15 all go into is 15, so use 15 as the common denominator.

Exercise 12A

1 In each of these pairs of fractions, which fraction is the smaller?

a $\frac{3}{4}, \frac{5}{7}$ b $\frac{5}{6}, \frac{7}{8}$ c $\frac{4}{9}, \frac{2}{5}$ d $\frac{5}{12}, \frac{3}{7}$

2 Which of these fractions is closer to $\frac{1}{2}$?

A $\frac{3}{5}$ **B** $\frac{9}{20}$ Give a reason for your answer.

3 David says, '$\frac{4}{5}$ is smaller than $\frac{7}{8}$.'
Is David correct? Give a reason for your answer.

E D E E A02

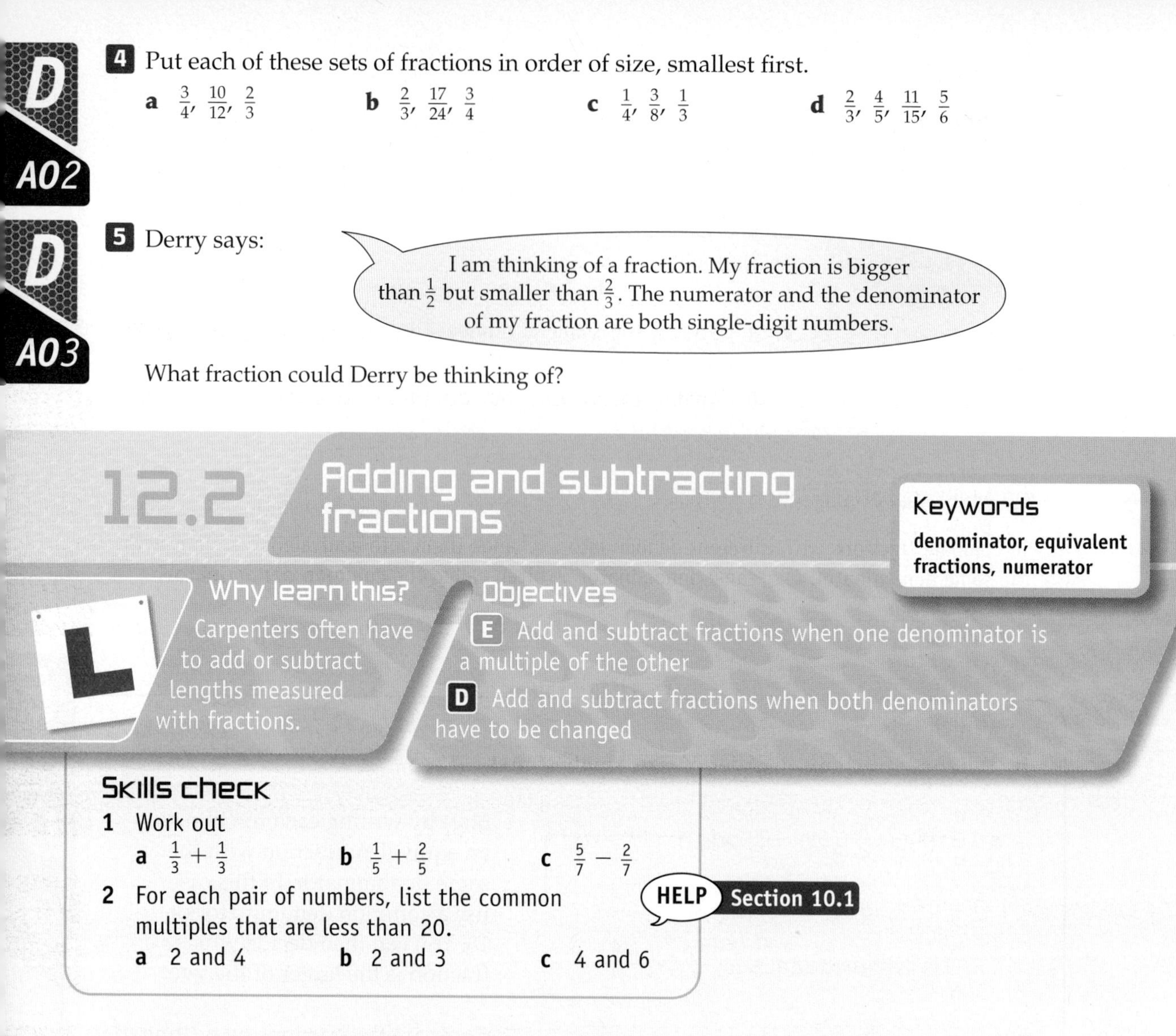

D AO2

4 Put each of these sets of fractions in order of size, smallest first.

a $\frac{3}{4}, \frac{10}{12}, \frac{2}{3}$ **b** $\frac{2}{3}, \frac{17}{24}, \frac{3}{4}$ **c** $\frac{1}{4}, \frac{3}{8}, \frac{1}{3}$ **d** $\frac{2}{3}, \frac{4}{5}, \frac{11}{15}, \frac{5}{6}$

D AO3

5 Derry says:

I am thinking of a fraction. My fraction is bigger than $\frac{1}{2}$ but smaller than $\frac{2}{3}$. The numerator and the denominator of my fraction are both single-digit numbers.

What fraction could Derry be thinking of?

12.2 Adding and subtracting fractions

Keywords
denominator, equivalent fractions, numerator

Why learn this?
Carpenters often have to add or subtract lengths measured with fractions.

Objectives

E Add and subtract fractions when one denominator is a multiple of the other

D Add and subtract fractions when both denominators have to be changed

Skills check

1 Work out

a $\frac{1}{3} + \frac{1}{3}$ **b** $\frac{1}{5} + \frac{2}{5}$ **c** $\frac{5}{7} - \frac{2}{7}$

2 For each pair of numbers, list the common multiples that are less than 20.

a 2 and 4 **b** 2 and 3 **c** 4 and 6

HELP Section 10.1

Adding and subtracting fractions

To add or subtract fractions with different **denominators**, write them as **equivalent fractions** with the same denominator, then add or subtract the **numerators.**

E D

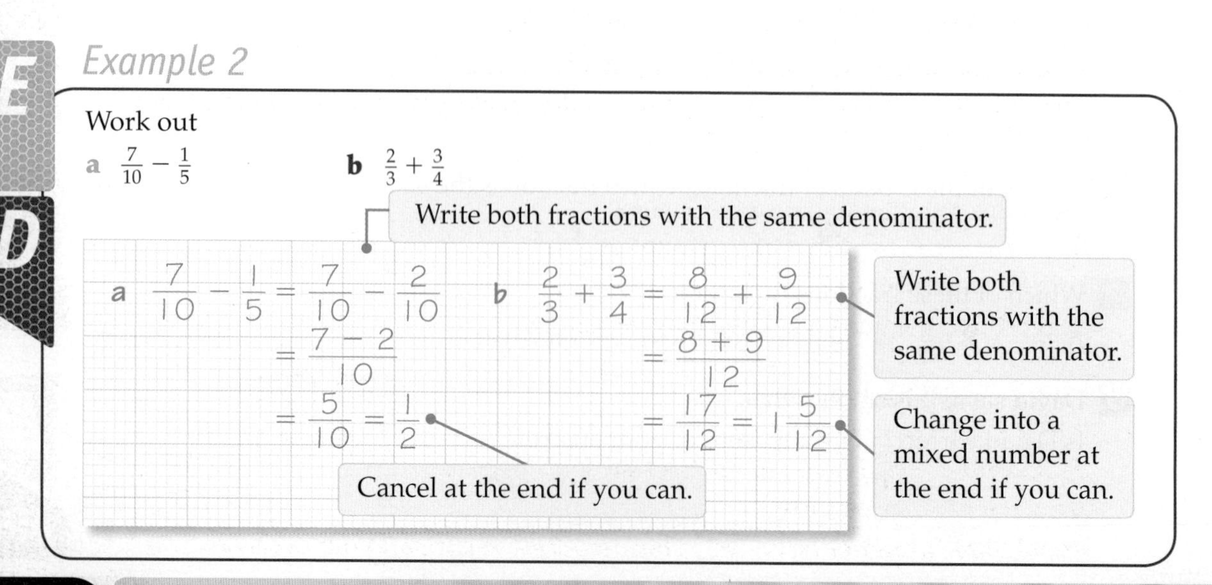

Example 2

Work out

a $\frac{7}{10} - \frac{1}{5}$ **b** $\frac{2}{3} + \frac{3}{4}$

a $\frac{7}{10} - \frac{1}{5} = \frac{7}{10} - \frac{2}{10}$
$= \frac{7-2}{10}$
$= \frac{5}{10} = \frac{1}{2}$

b $\frac{2}{3} + \frac{3}{4} = \frac{8}{12} + \frac{9}{12}$
$= \frac{8+9}{12}$
$= \frac{17}{12} = 1\frac{5}{12}$

Write both fractions with the same denominator.

Cancel at the end if you can.

Change into a mixed number at the end if you can.

Exercise 12B

1 Work out

a $\frac{7}{10} - \frac{2}{5}$ b $\frac{5}{14} + \frac{2}{7}$ c $\frac{1}{3} + \frac{1}{9}$ d $\frac{13}{14} - \frac{6}{7}$

e $\frac{7}{9} - \frac{1}{3}$ f $\frac{5}{12} + \frac{1}{6}$ g $\frac{1}{3} - \frac{2}{15}$ h $\frac{3}{4} - \frac{1}{12}$

E

2 Ros has a bag that contains $\frac{9}{10}$ kg of flour.
She uses $\frac{1}{2}$ kg of the flour to make one loaf of bread.
What is the weight of the flour left in the bag after Ros has made the loaf of bread?

E

3 A bottle contains $\frac{1}{2}$ litre of water. Sue pours $\frac{1}{8}$ litre out of the bottle.
How much water is left in the bottle?

AO2

4 Melita has two fraction cards.
Both the fractions are positive.
One of the fractions is $\frac{3}{5}$.
Melita adds the two fractions and gets an answer of $\frac{8}{15}$.
Could Melita's answer be correct? Show working to support your decision.

$\frac{3}{5}$?

E AO3

5 A hatmaker uses $\frac{1}{8}$ m of red ribbon and $\frac{1}{4}$ m of blue ribbon to decorate one hat.

a What is the total length of ribbon that he uses to decorate one hat?

b The hatmaker decorates 16 identical hats. Ribbon costs 32p per metre.
What is the total cost of the ribbon for the 16 hats?

E AO2

6 Work out

a $\frac{1}{3} - \frac{1}{4}$ b $\frac{2}{5} + \frac{1}{3}$ c $\frac{3}{4} - \frac{1}{5}$ d $\frac{1}{3} - \frac{1}{7}$

e $\frac{3}{5} - \frac{2}{7}$ f $\frac{3}{4} + \frac{3}{5}$ g $\frac{5}{6} + \frac{2}{7}$ h $\frac{1}{4} + \frac{6}{7}$

D

7 In a driving reaction test, one driver reacted in $\frac{1}{8}$ of a second.
Another driver took $\frac{2}{3}$ of a second to react.
What is the difference in their reaction times?

D

8 Clive has two bags of flour.
One bag contains $\frac{1}{8}$ kg and the other contains $\frac{1}{3}$ kg.
Clive needs $\frac{1}{2}$ kg of flour to make one loaf of bread.
Show that Clive has not got enough flour altogether in the two bags to make one loaf of bread.

9 Greg adds together two proper fractions.
Each fraction has a different denominator.
He gets an answer of $1\frac{7}{12}$.
Suggest two fractions that Greg may be using.

AO2

10 Pablo has two fraction cards.
Both the fractions are positive.
Pablo adds the two fractions and gets an answer that simplifies to $\frac{3}{10}$.
Could Pablo's answer be correct? Show working to support your decision.

$\frac{?}{8}$ $\frac{?}{5}$

D AO2

12.3 Adding and subtracting mixed numbers

Keywords
mixed number, improper fraction, lowest terms

Why learn this?
Calculating the distance between two villages may involve adding mixed numbers.

Objectives
C Add and subtract mixed numbers

Skills check

1 Change each mixed number into an improper fraction.
 a $4\frac{1}{2}$ **b** $5\frac{1}{3}$ **c** $7\frac{2}{5}$

2 Change each improper fraction into a mixed number.
 a $\frac{7}{3}$ **b** $\frac{11}{2}$ **c** $\frac{13}{5}$

Adding and subtracting mixed numbers

You can add or subtract **mixed numbers** by changing them into **improper fractions.**
Remember to write your answer in its **lowest terms** and as a mixed number if possible.

C

Example 3

Work out

a $3\frac{1}{3} + 1\frac{3}{4}$ **b** $3\frac{1}{5} - 1\frac{2}{3}$

a $3\frac{1}{3} + 1\frac{3}{4} = \frac{10}{3} + \frac{7}{4}$ — Change to improper fractions.

$= \frac{40}{12} + \frac{21}{12}$

$= \frac{40 + 21}{12}$

$= \frac{61}{12}$ — Write as equivalent fractions with a common denominator, then add.

$= 5\frac{1}{12}$ — Change the answer back to a mixed number.

b $3\frac{1}{5} - 1\frac{2}{3} = \frac{16}{5} - \frac{5}{3}$ — Change to improper fractions.

$= \frac{48}{15} - \frac{25}{15}$

$= \frac{48 - 25}{15}$

$= \frac{23}{15}$ — Write as equivalent fractions with a common denominator, then subtract.

$= 1\frac{8}{15}$ — Change the answer back to a mixed number.

Exercise 12C

1 Work out

a $3\frac{2}{3} + 1\frac{1}{4}$ b $3\frac{7}{10} - 1\frac{3}{40}$ c $4\frac{1}{4} - 1\frac{5}{8}$ d $2\frac{2}{3} + \frac{7}{8}$

e $1\frac{7}{10} - \frac{3}{4}$ f $1\frac{4}{5} - \frac{17}{20}$ g $4\frac{1}{2} + 3\frac{5}{6}$ h $1\frac{1}{12} + 1\frac{1}{8}$

i $3\frac{4}{7} + 2\frac{2}{3}$ j $2\frac{1}{2} - 1\frac{3}{4} + \frac{5}{8}$ k $1\frac{1}{9} + 1\frac{1}{3} + 2\frac{1}{18} - 3\frac{1}{6}$

C

2 Ted buys $1\frac{1}{2}$ kg of apples, $\frac{7}{8}$ kg of oranges and a melon weighing $1\frac{1}{4}$ kg.
He puts all the fruit into one bag.
What is the weight of the fruit in the bag?

C

3 Sandra cycles from her home in Witney to Bampton.
The distance is $6\frac{1}{2}$ miles.
On her way home she cycles $3\frac{1}{5}$ miles and then her bike has a puncture.
She pushes her bike the rest of the way home.
How far does she push her bike?

4 Erin adds together two mixed numbers. The fractions have different denominators.
Erin gets an answer of $3\frac{17}{30}$.
Suggest two different sets of fractions that Erin may have added.

AO2

C

5 Patrick has two mixed-number cards.
Both the mixed numbers are positive.
Patrick adds the two fractions and gets an answer that cancels to $4\frac{1}{4}$.
Could Patrick's answer be correct? Show working to support your decision.

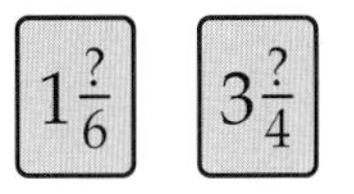

AO3

C

FUNCTIONAL

6 A small lorry is allowed to carry a maximum weight of 7 tonnes.
The lorry is loaded with four pallets of machine parts.
The pallets weigh $1\frac{5}{6}$ tonnes, $1\frac{7}{8}$ tonnes, $1\frac{11}{12}$ tonnes and $1\frac{1}{2}$ tonnes.
Show that the lorry is overloaded.

AO2

12.4 Multiplying fractions

Keywords
cancel

Why learn this?
Recipes often contain fractional amounts like $\frac{1}{2}$ teaspoon. You need to multiply fractions when scaling a recipe up or down.

Objectives

E Multiply a fraction by a fraction

Skills check

1 Work out

a $\frac{1}{4}$ of 12 cm b $\frac{2}{3}$ of 27 kg

c $\frac{4}{5}$ of £30 d $\frac{3}{7}$ of 28 km

2 Write true or false for each of these.

a $1\frac{2}{3} = \frac{5}{3}$ b $\frac{12}{5} = 2\frac{2}{12}$

c $\frac{35}{4} = 8\frac{3}{4}$ d $6\frac{7}{9} = \frac{61}{9}$

Multiplying fractions

To multiply fractions, multiply the numerators together and multiply the denominators together.

Before you multiply, **cancel** common factors if possible.

E

Example 4

Work out

a $\frac{1}{3} \times \frac{5}{6}$

b $\frac{2}{7} \times \frac{21}{22}$

a $\frac{1}{3} \times \frac{5}{6} = \frac{1 \times 5}{3 \times 6} = \frac{5}{18}$ — Multiply the numerators and multiply the denominators.

b $\frac{{}^{1}\cancel{2}}{\cancel{7}_{1}} \times \frac{{}^{3}\cancel{21}}{\cancel{22}_{11}} = \frac{1 \times 3}{1 \times 11} = \frac{3}{11}$ — Cancel common factors first, then multiply the numerators and multiply the denominators.

Exercise 12D

E

1 Work out

a $\frac{1}{3} \times \frac{2}{7}$ **b** $\frac{2}{3} \times \frac{1}{10}$ **c** $\frac{4}{5} \times \frac{2}{7}$ **d** $\frac{3}{4} \times \frac{2}{5}$

e $\frac{3}{5} \times \frac{10}{12}$ **f** $\frac{1}{7} \times \frac{14}{20}$ **g** $\frac{5}{6} \times \frac{7}{8}$ **h** $\frac{2}{7} \times \frac{3}{4}$

2 a Work out

i $\frac{7}{8} \times \frac{1}{2}$ **ii** $\frac{3}{4} \times \frac{3}{4}$

b Which is the larger fraction, $\frac{7}{8} \times \frac{1}{2}$ or $\frac{3}{4} \times \frac{3}{4}$?

3 a Work out $\frac{2}{3} \times \frac{1}{4}$ and $\frac{4}{5} \times \frac{1}{6}$.

Write your answers in their simplest form.

b Write your answers to part **a** as equivalent fractions with the denominator 30.

c Which is the smaller fraction, $\frac{2}{3} \times \frac{1}{4}$ or $\frac{4}{5} \times \frac{1}{6}$?

E AO2

4 Which is the larger fraction, $\frac{11}{28} \times \frac{2}{3}$ or $\frac{1}{2} \times \frac{11}{21}$?

Show working to support your answer.

E AO2

5 At the end of Nadia's birthday party $\frac{3}{4}$ of a pizza is left.

For lunch the next day Nadia eats $\frac{2}{3}$ of the pizza that is left.

The rest of the pizza she gives to her chickens.

What fraction of the whole pizza did Nadia give to her chickens?

12.5 Multiplying mixed numbers

Why learn this?

When decorating, you may need to work out areas involving fractional measurements.

Objectives

D Multiply a whole number by a mixed number

C Multiply a fraction by a mixed number

B Multiply a mixed number by a mixed number

Skills check

1 Work out

a $\frac{3}{4} \times 24$ **b** $15 \times \frac{2}{5}$ **c** $\frac{1}{2} \times \frac{3}{5}$ **d** $\frac{3}{8} \times \frac{4}{9}$

2 Write true or false for each of these.

a $\frac{11}{3} = 3\frac{1}{3}$ **b** $1\frac{2}{5} = \frac{6}{5}$ **c** $\frac{32}{9} = 3\frac{5}{9}$ **d** $\frac{7}{2} \times 2\frac{1}{2}$

Multiplying fractions and mixed numbers

To multiply mixed numbers, change them to improper fractions first.
Then multiply the numerators together and multiply the denominators together.

Before you multiply, cancel common factors if possible.

If your answer is an improper fraction, change it back to a mixed number.

Example 5

Work out

a $3 \times 1\frac{5}{6}$ **b** $\frac{1}{3} \times 1\frac{5}{6}$ **c** $2\frac{1}{3} \times 3\frac{3}{4}$

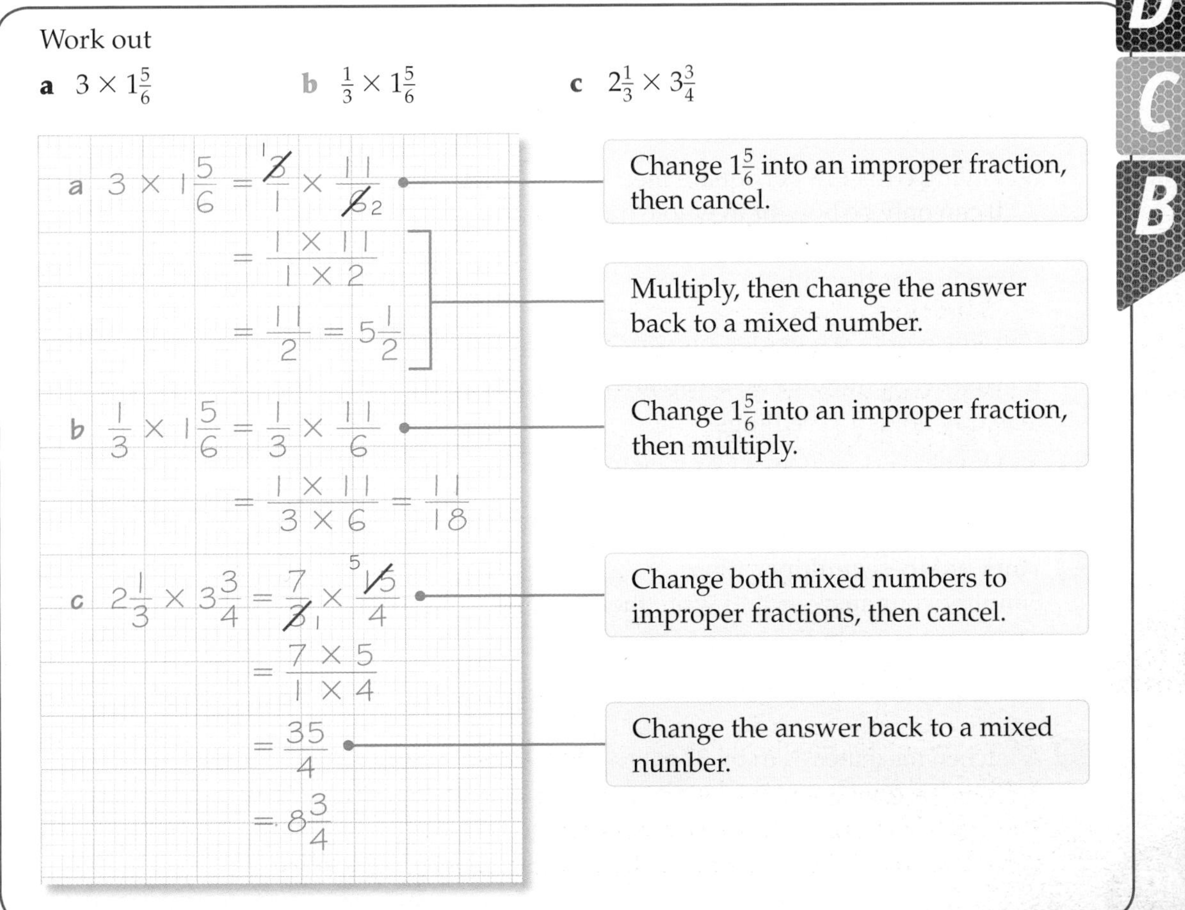

Exercise 12E

D

1 Work out these multiplications.
Give your answers as mixed numbers when possible.

a $5 \times 1\frac{5}{6}$ **b** $1 \times 5\frac{5}{6}$ **c** $2 \times 3\frac{1}{3}$ **d** $3 \times 2\frac{1}{3}$

e $2\frac{1}{3} \times 3$ **f** $2\frac{2}{3} \times 3$ **g** $2\frac{3}{4} \times 3$ **h** $2\frac{4}{5} \times 3$

2 Which is the larger number, $10 \times 1\frac{2}{3}$ or $5 \times 3\frac{1}{3}$?

D

3 A metal rod weighs $4\frac{1}{2}$ kg. Seven of the rods are packed into a crate.

a What is the total weight of the rods?

b The packing crate weighs $3\frac{3}{4}$ kg.
What is the total weight of the rods and crate?

AO2

4 A wooden door weighs $8\frac{3}{5}$ kg. A packing crate weighs $12\frac{1}{2}$ kg.
Twenty of the doors are packed into a crate.
What is the total weight of the doors and crate?

C

5 Work out these multiplications.
Simplify your answers, and write them as mixed numbers when possible.

a $\frac{1}{2} \times 2\frac{1}{2}$ **b** $\frac{1}{3} \times 4\frac{1}{5}$ **c** $\frac{3}{4} \times 1\frac{5}{6}$ **d** $\frac{4}{5} \times 3\frac{1}{3}$

e $6\frac{1}{2} \times \frac{4}{7}$ **f** $5\frac{1}{5} \times \frac{5}{13}$ **g** $3\frac{4}{9} \times \frac{3}{5}$ **h** $8\frac{1}{4} \times \frac{2}{11}$

C

6 It takes $\frac{3}{4}$ of a minute to fill one bucket with water.
How long does it take to fill $10\frac{1}{2}$ buckets with water?

7 Hassan is going to pave the path in his garden.
He has a rectangular path that is $\frac{3}{4}$ m wide and $7\frac{1}{2}$ m long.

FUNCTIONAL

a What is the area of the path?

Area of a rectangle = length × width

b Paving costs £18 per square metre.
It can only be bought in whole numbers of square metres.
What is the smallest amount that Hassan must spend, in order to have enough paving for his path?

8 Caroline works at a garden nursery.
It takes her $4\frac{1}{2}$ minutes to transplant one tray of seedlings.
One tray holds 15 seedlings.

FUNCTIONAL

AO2

a How long does it take Caroline to transplant 1 seedling?

Find $\frac{1}{15}$ of $4\frac{1}{2}$.

b How long does it take Caroline to transplant 200 seedlings?

B

9 Work out these multiplications.
Simplify your answers and write them as mixed numbers.

a $1\frac{1}{2} \times 3\frac{1}{2}$ **b** $2\frac{1}{3} \times 1\frac{2}{5}$ **c** $2\frac{3}{4} \times 1\frac{4}{7}$ **d** $4\frac{3}{5} \times 5\frac{1}{3}$

e $1\frac{1}{6} \times 2\frac{3}{8}$ **f** $2\frac{4}{5} \times 2\frac{5}{9}$ **g** $4\frac{2}{9} \times 3\frac{5}{12}$ **h** $8\frac{1}{2} \times 4\frac{3}{4}$

B

AO2

10 A kitchen measures $4\frac{1}{2}$ m by $3\frac{3}{4}$ m.
What is the floor area of the kitchen?

11 A large bag of flour weighs $1\frac{1}{2}$ kg.
Shani has $6\frac{2}{3}$ bags of flour.
Shani says that altogether she has 10 kg of flour.
Is Shani correct? Show working to support your answer.

B

12 John has a watering can that holds $6\frac{4}{5}$ litres of water.
It takes John 70 seconds ($1\frac{1}{6}$ minutes) to fill the watering can.
On one Monday morning, John fills the watering can $8\frac{1}{2}$ times.

a How much water does John use on this Monday morning?

b How long does John spend filling the watering can on this Monday morning?

AO2

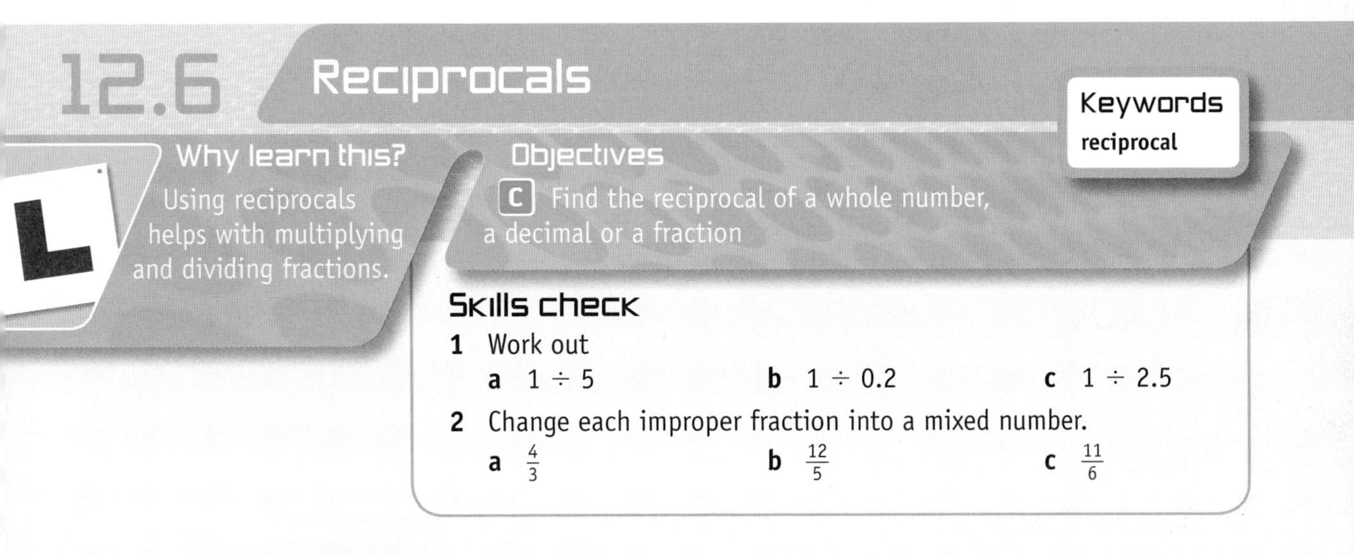

12.6 Reciprocals

Keywords
reciprocal

Why learn this?
Using reciprocals helps with multiplying and dividing fractions.

Objectives
C Find the reciprocal of a whole number, a decimal or a fraction

Skills check

1 Work out

a $1 \div 5$ **b** $1 \div 0.2$ **c** $1 \div 2.5$

2 Change each improper fraction into a mixed number.

a $\frac{4}{3}$ **b** $\frac{12}{5}$ **c** $\frac{11}{6}$

Finding reciprocals

When two numbers can be multiplied together to give the answer 1, then each number is called the **reciprocal** of the other.

The reciprocal of a fraction is found by turning the fraction upside down.
The reciprocal of a number is 1 divided by that number.

Example 6

C

Find the reciprocal of

a 25 **b** 0.4 **c** $\frac{6}{7}$

Give your answers as mixed numbers when possible.

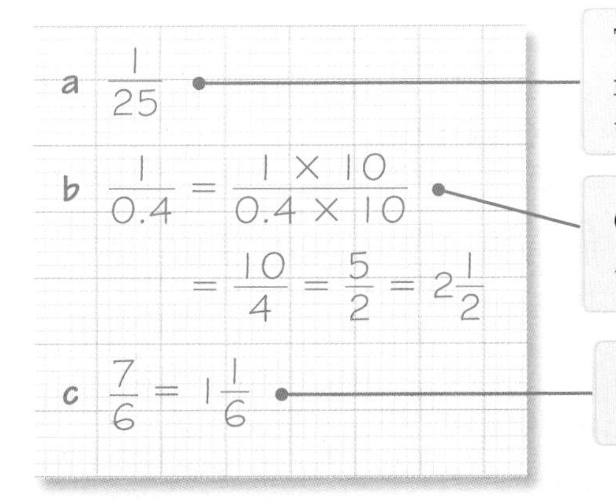

a $\frac{1}{25}$

The reciprocal of a number is 1 divided by that number. Alternatively, think of 25 as $\frac{25}{1}$, and turn it upside down.

b $\frac{1}{0.4} = \frac{1 \times 10}{0.4 \times 10} = \frac{10}{4} = \frac{5}{2} = 2\frac{1}{2}$

Change $\frac{1}{0.4}$ into an improper fraction, then cancel.

c $\frac{7}{6} = 1\frac{1}{6}$

Turn the fraction upside down, then change to a mixed number.

Exercise 12F

C

1 Find the reciprocal of each of these numbers.
Give your answers as fractions, whole numbers or mixed numbers.

a 4 **b** 10 **c** 20 **d** 100 **e** $\frac{1}{2}$ **f** $\frac{1}{5}$ **g** $\frac{2}{3}$ **h** $\frac{3}{10}$

2 Find the reciprocal of each of these numbers.
Give your answers as whole numbers or fractions. Show all your working.

a 0.2 **b** 0.5 **c** 0.04 **d** 1.25

C
A02

3 Multiply each of the numbers in Q1 by its reciprocal.
What do you notice?

12.7 Dividing fractions

Why learn this?

A nurse uses fractions to work out how many $\frac{1}{2}$ m*l* doses she can give from a 5 m*l* vial.

Objectives

D Divide a whole number or a fraction by a fraction

C Divide mixed numbers or fractions by whole numbers

B Divide mixed numbers by mixed numbers

Skills check

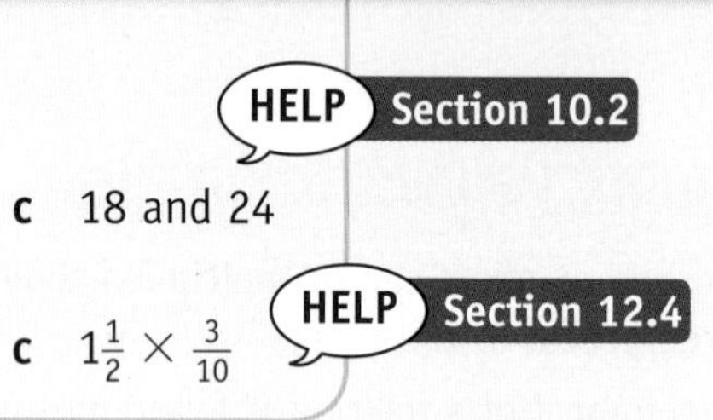

1 List the common factors of

a 4 and 6 **b** 12 and 15 **c** 18 and 24

2 Work out

a $\frac{2}{5} \times \frac{3}{7}$ **b** $\frac{3}{8} \times \frac{2}{9}$ **c** $1\frac{1}{2} \times \frac{3}{10}$

Dividing by a fraction

To divide by a fraction, turn the fraction upside down and multiply.
If the division involves mixed numbers, change them to improper fractions first.

D

Example 7

Work out

a $18 \div \frac{2}{3}$ **b** $\frac{3}{20} \div \frac{9}{40}$

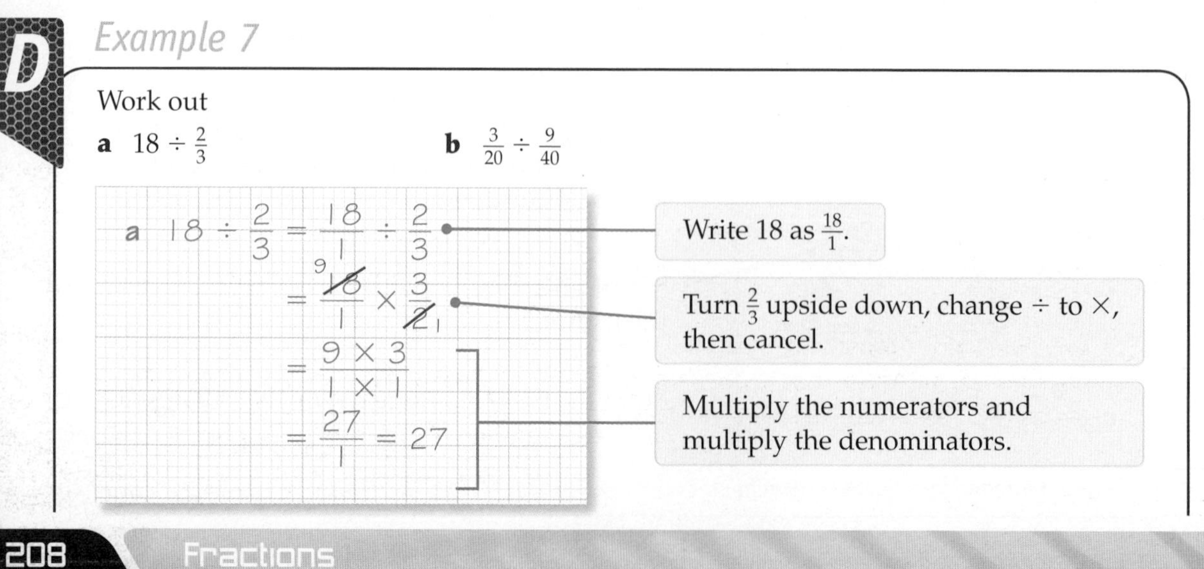

b $\frac{3}{20} \div \frac{9}{40} = \frac{\cancel{3}^{1}}{\cancel{20}_{1}} \times \frac{\cancel{40}^{2}}{\cancel{9}_{3}}$ — Turn $\frac{9}{40}$ upside down, change ÷ to ×, then cancel.

$= \frac{1 \times 2}{1 \times 3}$ — Multiply the numerators and multiply the denominators.

$= \frac{2}{3}$

Example 8

Work out

a $4\frac{1}{2} \div 3$ **b** $1\frac{1}{2} \div 3\frac{4}{5}$

a $4\frac{1}{2} \div 3 = \frac{9}{2} \div \frac{3}{1}$ — Change $4\frac{1}{2}$ into an improper fraction and write 3 as $\frac{3}{1}$.

$= \frac{\cancel{9}^{3}}{2} \times \frac{1}{\cancel{3}_{1}}$ — Turn $\frac{3}{1}$ upside down, change ÷ to ×, then cancel.

$= \frac{3 \times 1}{2 \times 1}$

$= \frac{3}{2} = 1\frac{1}{2}$ — Multiply, then change the improper fraction back to a mixed number.

b $1\frac{1}{2} \div 3\frac{4}{5} = \frac{3}{2} \div \frac{19}{5}$

$= \frac{3}{2} \times \frac{5}{19}$ — Change both mixed numbers into improper fractions, then turn $\frac{19}{5}$ upside down and change ÷ to × .

$= \frac{3 \times 5}{2 \times 19}$ — Multiply the numerators and multiply the denominators.

$= \frac{15}{38}$

Exercise 12G

1 Work out

a $15 \div \frac{1}{2}$ **b** $8 \div \frac{2}{5}$ **c** $12 \div \frac{3}{7}$

d $5 \div \frac{2}{3}$ **e** $6 \div \frac{3}{5}$ **f** $3 \div \frac{2}{9}$

2 This is part of Alvina's homework. Explain where Alvina has gone wrong, and work out the correct answer for her.

$\frac{6}{14} \div 12 = \frac{6}{14} \div \frac{1}{12} = \frac{6}{{}_{7}\cancel{14}} \times \frac{\cancel{12}^{6}}{1}$

$= \frac{6 \times 6}{7 \times 1} = \frac{36}{7} = 5\frac{1}{7}$

C B D D

3 Work out

a $\frac{2}{3} \div \frac{4}{5}$ b $\frac{6}{7} \div \frac{12}{13}$ c $\frac{4}{9} \div \frac{2}{5}$

d $\frac{2}{7} \div \frac{6}{14}$ e $\frac{3}{4} \div \frac{15}{28}$ f $\frac{8}{11} \div \frac{24}{33}$

g $3\frac{3}{4} \div 10$ h $1\frac{2}{7} \div 12$ i $7\frac{1}{5} \div 6$

4 Work out

a $2\frac{1}{4} \div \frac{7}{12}$ b $\frac{3}{4} \div 1\frac{2}{7}$ c $3\frac{2}{3} \div \frac{20}{21}$

5 Adam shares $3\frac{1}{2}$ pizzas between four people.
How much do they each receive?

6 Mathew pours $4\frac{2}{3}$ litres of lemonade into $\frac{1}{4}$ litre glasses.
How many full glasses does he pour?

7 Work out

a $1\frac{1}{3} \div 4\frac{1}{2}$ b $2\frac{3}{5} \div 3\frac{1}{3}$ c $5\frac{4}{7} \div 6\frac{3}{8}$

d $4\frac{1}{4} \div 1\frac{1}{5}$ e $6\frac{3}{5} \div 1\frac{5}{7}$ f $4\frac{1}{2} \div 3\frac{2}{15}$

8 This is part of David's homework.

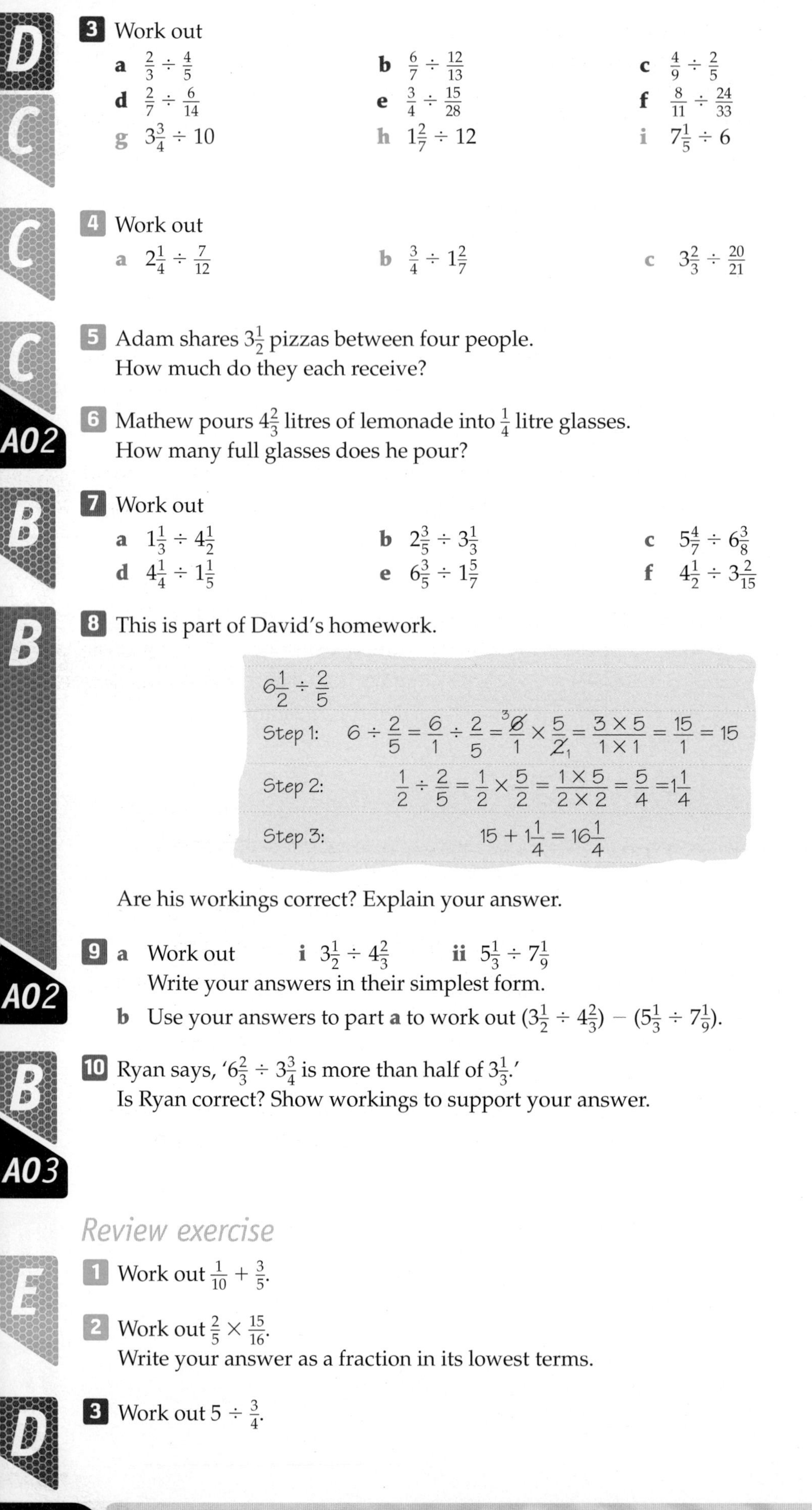

$6\frac{1}{2} \div \frac{2}{5}$

Step 1: $6 \div \frac{2}{5} = \frac{6}{1} \div \frac{2}{5} = \frac{\not{6}^{3}}{1} \times \frac{5}{\not{2}_{1}} = \frac{3 \times 5}{1 \times 1} = \frac{15}{1} = 15$

Step 2: $\frac{1}{2} \div \frac{2}{5} = \frac{1}{2} \times \frac{5}{2} = \frac{1 \times 5}{2 \times 2} = \frac{5}{4} = 1\frac{1}{4}$

Step 3: $15 + 1\frac{1}{4} = 16\frac{1}{4}$

Are his workings correct? Explain your answer.

9 a Work out i $3\frac{1}{2} \div 4\frac{2}{3}$ ii $5\frac{1}{3} \div 7\frac{1}{9}$
Write your answers in their simplest form.

b Use your answers to part **a** to work out $(3\frac{1}{2} \div 4\frac{2}{3}) - (5\frac{1}{3} \div 7\frac{1}{9})$.

10 Ryan says, '$6\frac{2}{3} \div 3\frac{3}{4}$ is more than half of $3\frac{1}{3}$.'
Is Ryan correct? Show workings to support your answer.

Review exercise

1 Work out $\frac{1}{10} + \frac{3}{5}$. **[2 marks]**

2 Work out $\frac{2}{5} \times \frac{15}{16}$.
Write your answer as a fraction in its lowest terms. **[2 marks]**

3 Work out $5 \div \frac{3}{4}$. **[2 marks]**

4 Brian spends $\frac{2}{3}$ of his wages on bills.
He spends $\frac{1}{5}$ of his wages on clothes.
The rest of his wages he saves.
What fraction of his wages does Brian save? **[3 marks]**

D AO2

5 Find the reciprocal of 0.2. Give your answer as a whole number. **[2 marks]**

C

6 Emyr buys $1\frac{1}{4}$ kg of carrots, $\frac{2}{3}$ kg of parsnips and $2\frac{3}{5}$ kg of potatoes.
What is the total weight of the vegetables that Emyr buys? **[3 marks]**

7 It takes $\frac{2}{3}$ of a minute to fill one jug with water.
How long does it take to fill $8\frac{1}{2}$ jugs with water?

a Give your answer in minutes, as a mixed number in its simplest form. **[3 marks]**

b Give your answer in minutes and seconds. **[1 mark]**

C AO2

8 **a** Work out $4\frac{2}{5} \div 8\frac{1}{4}$.
Write your answer in its simplest form. **[3 marks]**

b Use your answer to part **a** to work out the reciprocal of $4\frac{2}{5} \div 8\frac{1}{4}$.
Write your answer as a mixed number in its simplest form. **[2 marks]**

B AO2

Chapter summary

In this chapter you have learned how to

- add and subtract fractions when one denominator is a multiple of the other **E**
- multiply a fraction by a fraction **E**
- compare fractions with different denominators **E** **D**
- add and subtract fractions when both denominators have to be changed **D**
- multiply a whole number by a mixed number **D**
- divide a whole number or a fraction by a fraction **D**
- add and subtract mixed numbers **C**
- multiply a fraction by a mixed number **C**
- divide mixed numbers or fractions by whole numbers **C**
- find the reciprocal of a whole number, a decimal or a fraction **C**
- multiply a mixed number by a mixed number **B**
- divide mixed numbers by mixed numbers **B**

13 Decimals

This chapter is about calculating with decimals.

Your money may be safe if you keep it in a money box, but you don't earn any interest. When you save your money with a bank, the bank will calculate your interest using decimals.

Objectives

This chapter will show you how to

- add and subtract decimal numbers E
- convert decimals to fractions D
- convert fractions to decimals D
- multiply and divide decimal numbers D C
- read and write recurring decimals C
- understand how recurring decimals relate to fractions B

Before you start this chapter

Put your calculator away!

1 Work out

a 82×100

b $82 \div 100$

c 2.65×1000

d $2.65 \div 1000$

2 a What has 120 been multiplied by to give 12 000?

b What has 0.09 been multiplied by to give 90?

3 a What has 657 been divided by to give 6.57?

b What has 0.72 been divided by to give 0.072?

4 Choose any number and put it through this number machine.

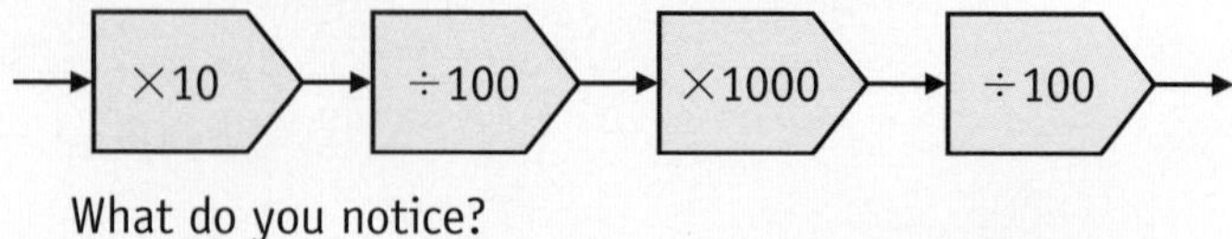

What do you notice?

Why does this happen?

13.1 Adding and subtracting decimals

Keywords
decimal number

Why learn this?
Many calculations with money involve adding and subtracting decimals.

Objectives
E Add and subtract decimal numbers

Skills check
1 Work out 1872 + 35 + 470 + 18.
2 Work out 20 370 − 1605.

Place value diagrams

Here is a reminder of how a place value diagram works for **decimal numbers**.

For the number 234.567

Hundreds	Tens	Units	.	tenths	hundredths	thousandths
100	10	1	.	$\frac{1}{10}$	$\frac{1}{100}$	$\frac{1}{1000}$
2	3	4	.	5	6	7

So the parts of this number are

$$200 + 30 + 4 + \frac{5}{10} + \frac{6}{100} + \frac{7}{1000}$$

The decimal point separates the whole-number part of the number from the fraction part.

Adding and subtracting decimals

To add or subtract whole numbers, you set out the calculation in columns.
The units column is on the right.

```
  1035
    29
+  350
  1414
```

To add or subtract decimals, you need to line up the columns in the same way.
Line up the decimal points, then add or subtract in the usual way.

Example 1

Work out

a 13 + 1.52 + 0.006 + 74.5 **b** 34.5−4.25

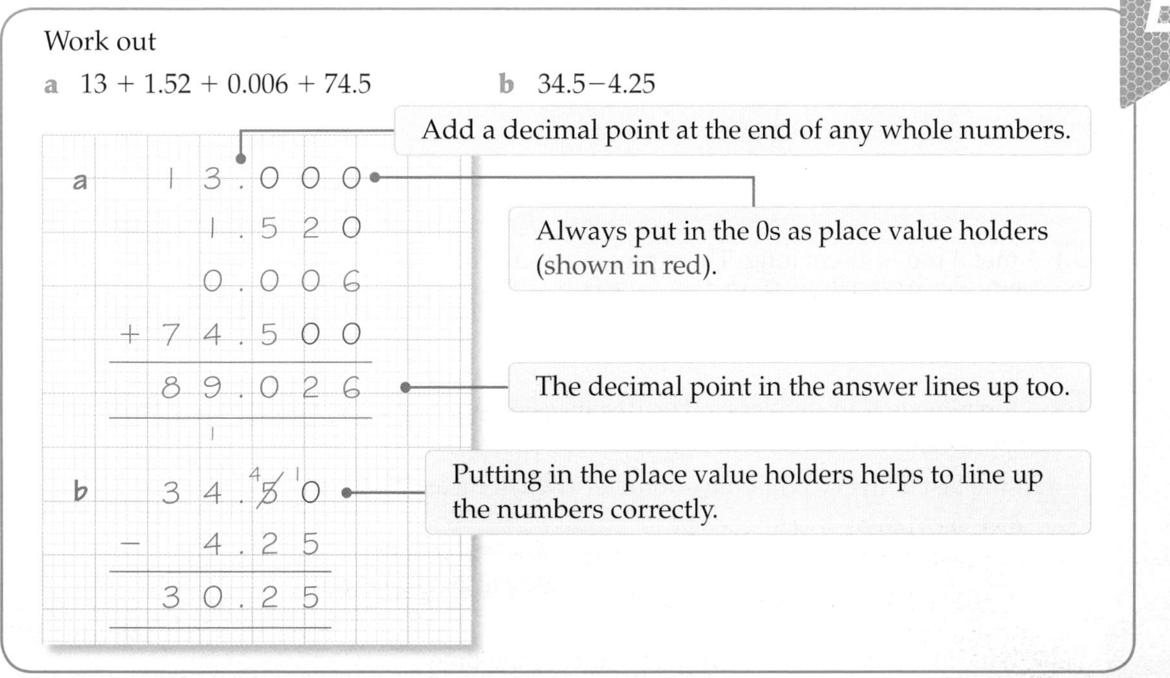

Exercise 13A

E

1 Work out

a
```
   1 2 . 5
     4 . 6
 + 8 6 . 2
 ---------

 ---------
```

b
```
       6 . 8 5
   1 6 2 . 5
 +     0 . 7 1
 -------------

 -------------
```

c
```
   1 0 . 0 0
 -     5 . 6 2
 -----------

 -----------
```

d
```
   3 1 2 . 7 5 3
 -       4 . 6 2 0
 ---------------

 ---------------
```

2 Work out

a 1.25 + 4.6 + 8.92   b 140.25 + 200 + 7.5   c 0.075 + 65 + 3.6

d 88.88 − 4.56   e 58.66 − 4.92   f 49.5 − 0.275

3 Jennie went on a school outing and spent £4.25 on fish and chips, £7.99 on a teddy for her baby brother, £0.99 on an ice cream and £0.72 on a can of drink.

a How much did she spend altogether?

b How much change did she have from a £20 note?

4 At the cinema Paul spent £4.50 on a ticket, £1.35 on popcorn, £0.95 on a drink and £2.79 on a magazine.

a How much did he spend altogether?

b How much change did he have from a £10 note?

5 A carpenter has some lengths of skirting board.
They are 2 m, 4.92 m, 0.75 m and 3.4 m long.
What is the total length of the skirting?

E

6 Here are four numbers.

5.2   10   14.96   19.64

Which two numbers are closest together?

7 A metal rod is 45 cm long. Three pieces are cut from it.
Their lengths are 9.2 cm, 17.5 cm and 11.7 cm.
What length of metal rod is left?

AO2

8 A lorry made four deliveries. The distances between the deliveries were 19.6 miles, 28.3 miles, 17.9 miles and 25.8 miles.
At the last delivery point the odometer (which measures distance) showed:

0 2 5 1 8 4 · 1

What did the odometer show at the start of the day?

13.2 Converting decimals to fractions

Why learn this?

This micrometer measures thicknesses in hundredths of a millimetre. Such measurements are usually written as decimals.

Objectives

D Convert decimals to fractions

Skills check

1 Round each of these numbers to one decimal place.
a 24.27 **b** 9.772 **c** 99.96

2 Round each of these numbers to three significant figures.
a 54.215 **b** 12 345 **c** 0.002 368

Converting decimals to fractions

You can use a place value diagram to convert a decimal to a fraction.

To convert 0.475 to a fraction, look at the place value of the last decimal place.

The value of the '5' is $\frac{5}{1000}$

So $0.475 = \frac{475}{1000} = \frac{95}{200} = \frac{19}{40}$

Cancel by dividing top and bottom by 5, and then dividing by 5 again.

Remember to give your answer in its simplest form.

Example 2

a Convert each of these numbers to a fraction:

i 0.7

ii 0.48

b Convert 2.125 to a mixed number.

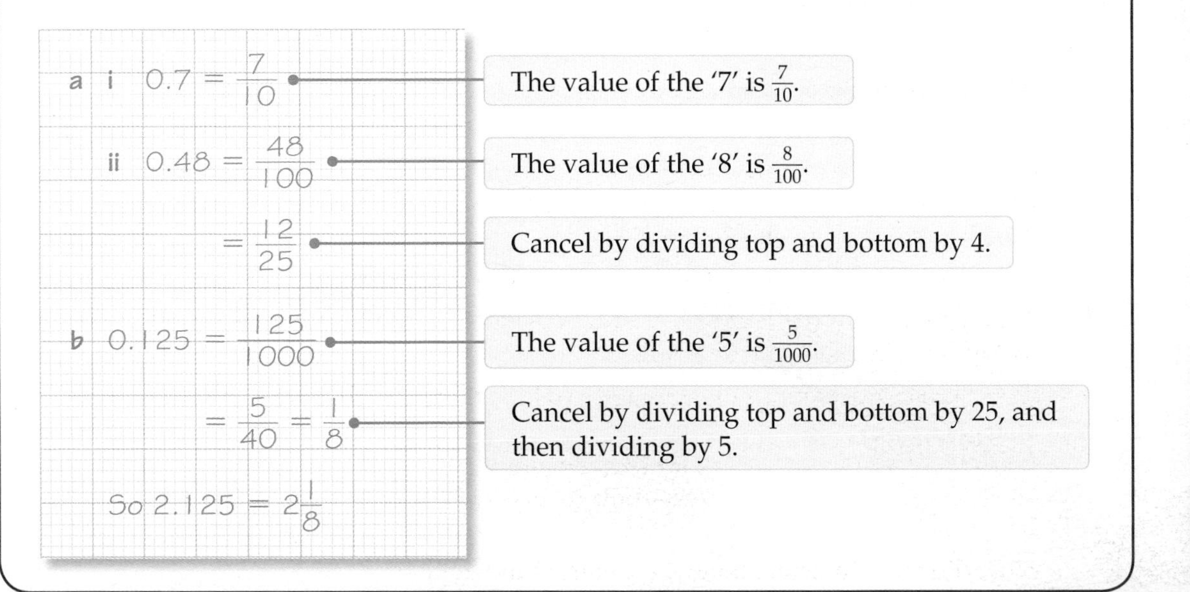

Exercise 13B

D

1 Convert each of these decimals to a fraction in its lowest terms.

a 0.1 **b** 0.5 **c** 0.07 **d** 0.05

e 0.75 **f** 0.28 **g** 0.04 **h** 0.65

i 0.52 **j** 0.79 **k** 0.24 **l** 0.35

2 Convert each of these decimals to a fraction in its lowest terms.

a 0.0001 **b** 0.002 **c** 0.084 **d** 0.009

e 0.035 **f** 0.025 **g** 0.375 **h** 0.425

3 Convert each of these to a mixed number.

a 3.5 **b** 14.8 **c** 5.64 **d** 4.85

4 A decimal fraction of 0.72 of the cats observed in a survey liked Moggymeat best. What fraction of the cats liked Moggymeat best?

5 Michael does a survey and finds that a decimal fraction of 0.44 of the cars in the school car park are diesel. What fraction of the cars are diesel?

6 Jasmine did a survey at her school and found that a decimal fraction of 0.36 of the students took a packed lunch. What fraction of the students took a packed lunch?

13.3 Multiplying and dividing decimals

Why learn this?

L

Most engineers calculate with decimals all the time in their work.

Objectives

D C Multiply and divide decimal numbers

Skills check

1 Work out

a 637×9 **b** 245×3

2 Work out

a $1016 \div 8$ **b** $2298 \div 6$

HELP Section 9.1

Multiplying decimals

You can multiply decimals in the same way as whole numbers.

Example 3

Work out 1.25 × 7.3.

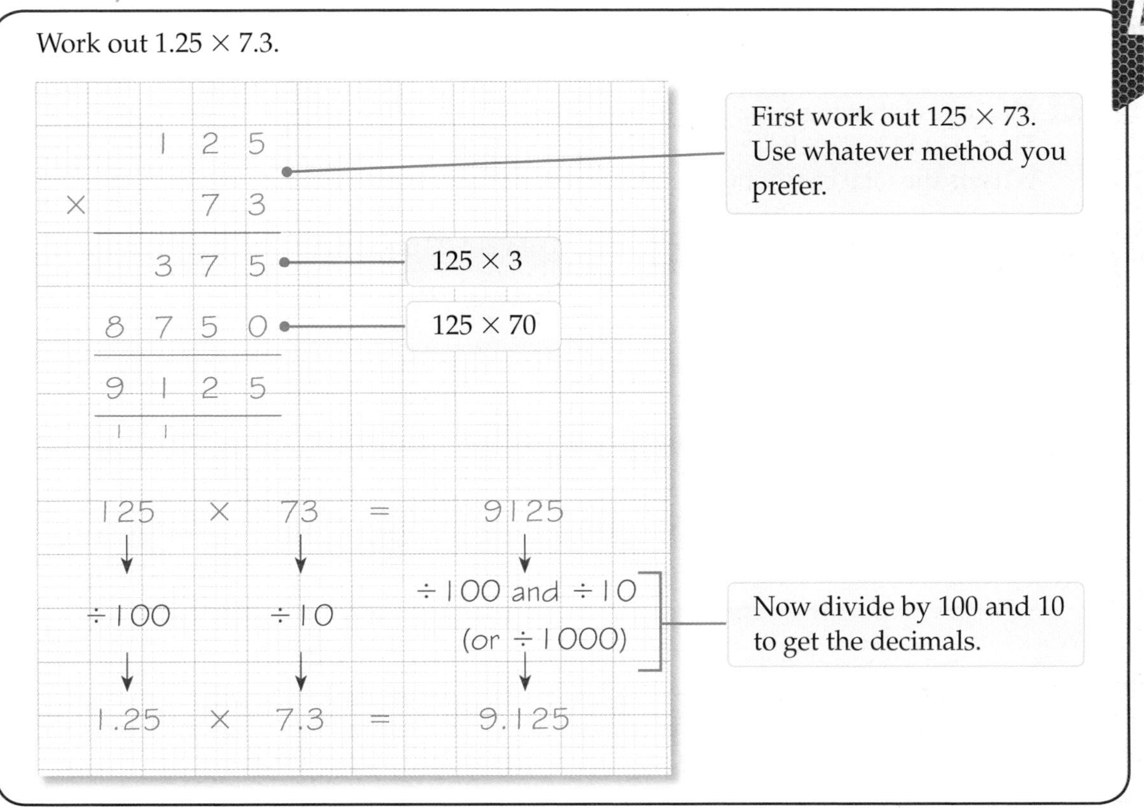

Here is a useful rule for multiplying decimals.

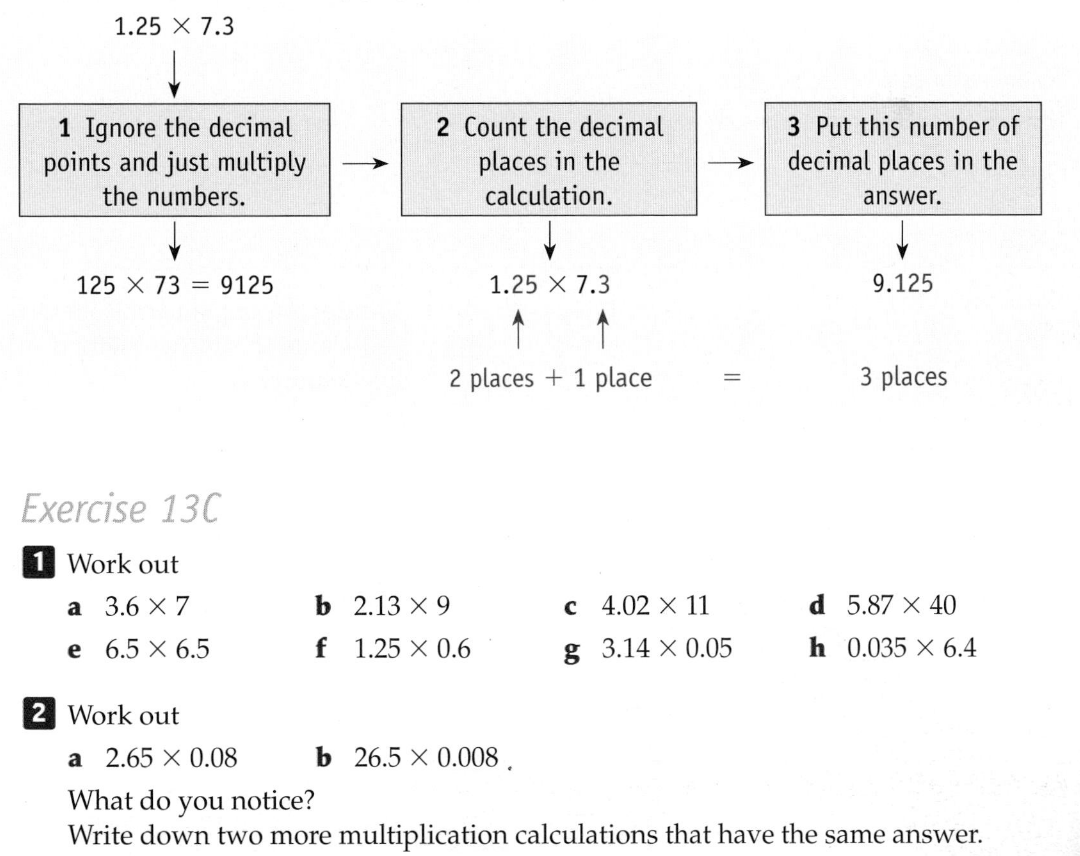

Exercise 13C

D

1 Work out

a 3.6 × 7 **b** 2.13 × 9 **c** 4.02 × 11 **d** 5.87 × 40

e 6.5 × 6.5 **f** 1.25 × 0.6 **g** 3.14 × 0.05 **h** 0.035 × 6.4

D A02

2 Work out

a 2.65 × 0.08 **b** 26.5 × 0.008

What do you notice?

Write down two more multiplication calculations that have the same answer.

3 Henna is making curtains.
She has chosen a material that costs £7.95 per metre. She buys 3.4 metres.
How much will this cost?

4 Mrs Jones is buying a class set of maths text books for her class of 27 students.
They cost £13.99 each.
What is the total cost of buying one for each student, plus one for herself?

A02

5 Eighteen coins are placed in a pile. Each coin is 1.23 mm thick.
What is the height of the pile of coins?

Dividing decimals

You can divide a decimal by a whole number in the usual way.

D

Example 4

Work out 43.41 ÷ 6.

$$6\overline{)4\,3\,.\,4\,{}^{2}1\,{}^{3}0}$$
with answer 7 . 2 3 5 written above

The decimal point in the answer goes over the one below it.

Add zeros as necessary to complete the division.

If you are dividing by a decimal, first write the division as a fraction, and convert the denominator to a whole number.
Then do the division in the usual way.

Example 5

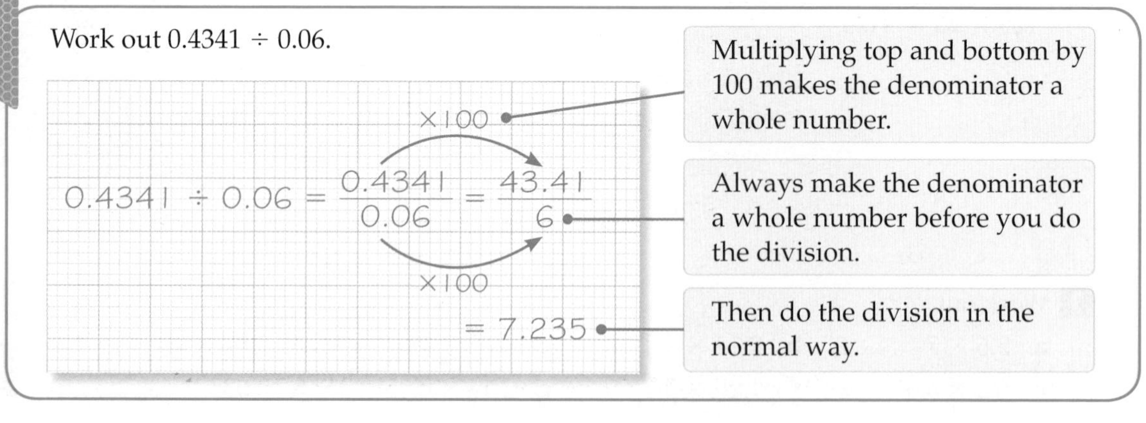

Exercise 13D

C

1 Work out

a 8 ÷ 0.4	**b** 72 ÷ 0.06	**c** 12.78 ÷ 0.3	**d** 0.125 ÷ 0.005
e 3.6 ÷ 0.12	**f** 4.5 ÷ 0.03	**g** 26.72 ÷ 0.008	**h** 0.56 ÷ 0.004

C

2 Work out

a $6.95 \div 0.05$

b $69.5 \div 0.5$

What do you notice?
Write down two more division calculations that have the same answer.

3 George is putting a new fence down the side of his garden.
The fence panels are 0.9 metres long.
The total length of fence is 7.2 metres.

a How many panels does he need to buy?

b How many fence posts does he need?

4 A shelf is 2.1 m long.
Books of thickness 2.8 cm are put on the shelf.
How many of these books will fit on the shelf?

5 A small business signs up for 'BISTEXT', a service that charges 4.8p per SMS text message to customers in the UK.
Their first bill is for £103.20.
How many text messages did they send?

AO2

13.4 Converting fractions to decimals

Keywords
terminate, recurring

Why learn this?
It's easy to compare fractions with different denominators if you convert them to decimals.

Objectives

D Convert fractions to decimals

C Understand recurring decimals

B Understand how recurring decimals relate to fractions

Skills check

1 Cancel each of these fractions to its lowest terms.

a $\frac{8}{10}$ b $\frac{15}{25}$ c $\frac{28}{42}$

d $\frac{26}{39}$ e $\frac{30}{72}$ f $\frac{15}{50}$

Terminating decimals

To convert a fraction to a decimal, divide the numerator by the denominator.

The fraction $\frac{3}{5}$ means $3 \div 5$.

$5\overline{)3.0}$ with quotient 0.6, so $\frac{3}{5} = 0.6$

This decimal ends, or **terminates**.

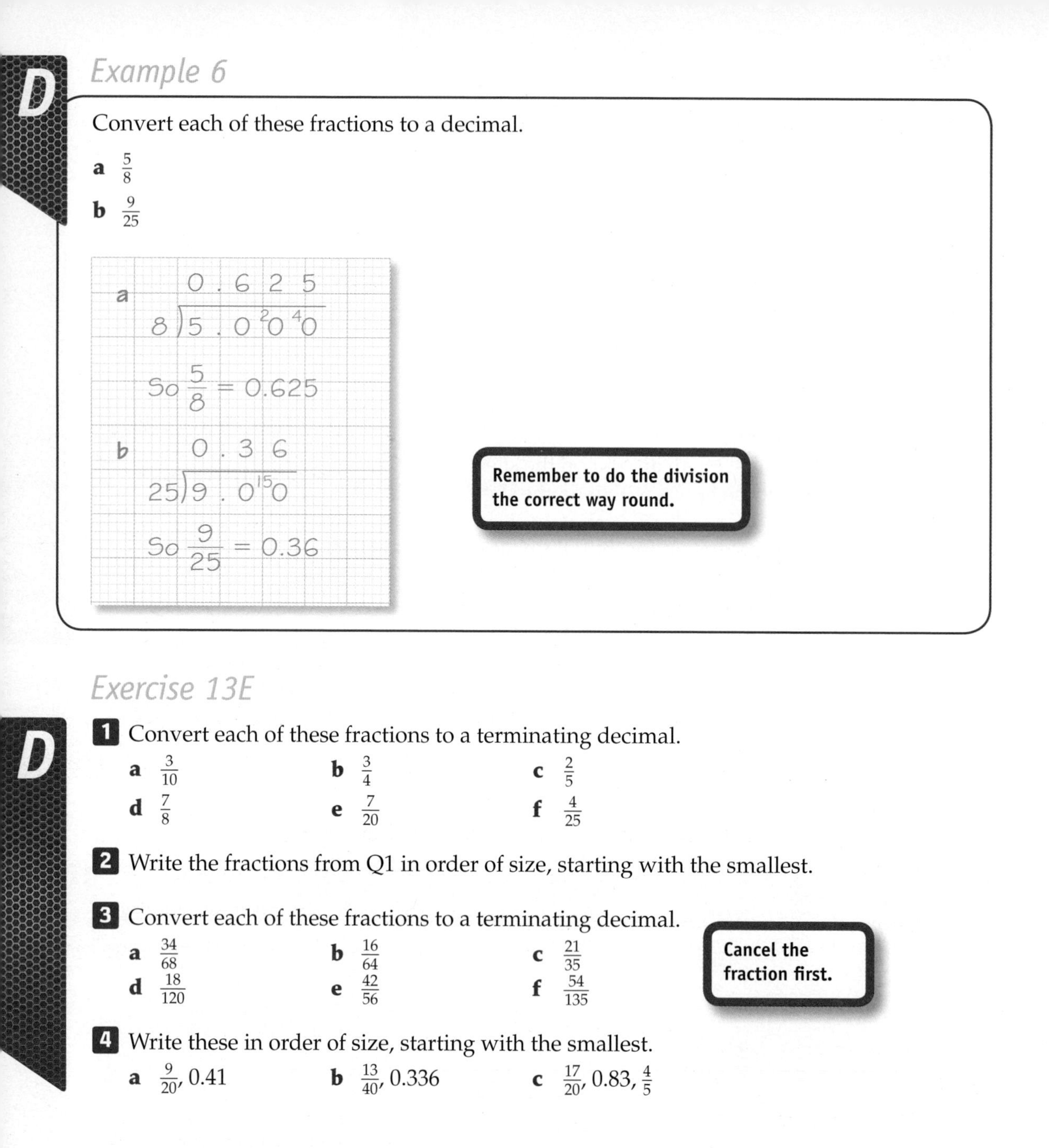

D

Example 6

Convert each of these fractions to a decimal.

a $\frac{5}{8}$

b $\frac{9}{25}$

a

$$8\overline{)5.0^{2}0^{4}0}\quad = 0.625$$

So $\frac{5}{8} = 0.625$

b

$$25\overline{)9.0^{15}0}\quad = 0.36$$

So $\frac{9}{25} = 0.36$

Remember to do the division the correct way round.

Exercise 13E

D

1 Convert each of these fractions to a terminating decimal.

a $\frac{3}{10}$ **b** $\frac{3}{4}$ **c** $\frac{2}{5}$

d $\frac{7}{8}$ **e** $\frac{7}{20}$ **f** $\frac{4}{25}$

2 Write the fractions from Q1 in order of size, starting with the smallest.

3 Convert each of these fractions to a terminating decimal.

a $\frac{34}{68}$ **b** $\frac{16}{64}$ **c** $\frac{21}{35}$

d $\frac{18}{120}$ **e** $\frac{42}{56}$ **f** $\frac{54}{135}$

Cancel the fraction first.

4 Write these in order of size, starting with the smallest.

a $\frac{9}{20}$, 0.41 **b** $\frac{13}{40}$, 0.336 **c** $\frac{17}{20}$, 0.83, $\frac{4}{5}$

Recurring decimals

When you convert some fractions to decimals you get answers that never end.
These are known as **recurring** decimals.

'Recurring dots' are used to make the pattern clear. A dot over a digit shows that it recurs.

$\frac{1}{3} = 0.333\,333\,333\ldots = 0.\dot{3}$

$\frac{7}{12} = 0.583\,333\,333\ldots = 0.58\dot{3}$

In these two examples, a single digit (3) recurs, so there is a dot over the 3.

$\frac{5}{11} = 0.454\,545\,454\ldots = 0.\dot{4}\dot{5}$

$\frac{4}{7} = 0.571\,428\,571\,428\,571\ldots = 0.\dot{5}71\,42\dot{8}$

In these two examples, two or more digits recur. There is a dot over the first and last digits in the sequence that repeats.

Example 7

Convert $\frac{5}{12}$ to a decimal.

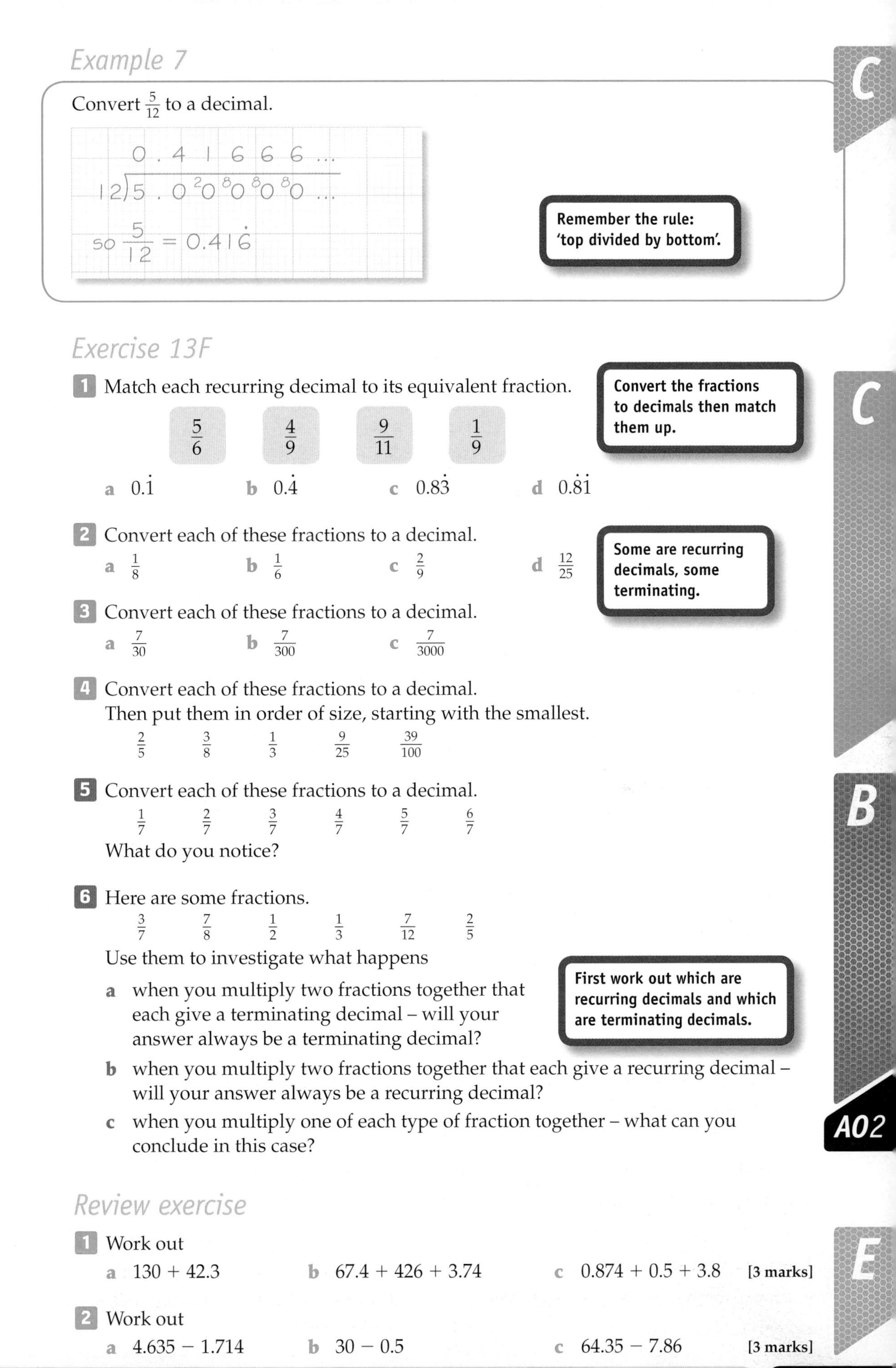

$$\begin{array}{r} 0.41666\ldots \\ 12\overline{)5.0^{2}0^{8}0^{8}0^{8}0\ldots} \end{array}$$

so $\frac{5}{12} = 0.41\dot{6}$

Remember the rule: 'top divided by bottom'.

Exercise 13F

C

1 Match each recurring decimal to its equivalent fraction.

Convert the fractions to decimals then match them up.

$\frac{5}{6}$ $\frac{4}{9}$ $\frac{9}{11}$ $\frac{1}{9}$

a $0.\dot{1}$ **b** $0.\dot{4}$ **c** $0.8\dot{3}$ **d** $0.\dot{8}\dot{1}$

2 Convert each of these fractions to a decimal.

a $\frac{1}{8}$ **b** $\frac{1}{6}$ **c** $\frac{2}{9}$ **d** $\frac{12}{25}$

Some are recurring decimals, some terminating.

3 Convert each of these fractions to a decimal.

a $\frac{7}{30}$ **b** $\frac{7}{300}$ **c** $\frac{7}{3000}$

4 Convert each of these fractions to a decimal.
Then put them in order of size, starting with the smallest.

$\frac{2}{5}$ $\frac{3}{8}$ $\frac{1}{3}$ $\frac{9}{25}$ $\frac{39}{100}$

B

5 Convert each of these fractions to a decimal.

$\frac{1}{7}$ $\frac{2}{7}$ $\frac{3}{7}$ $\frac{4}{7}$ $\frac{5}{7}$ $\frac{6}{7}$

What do you notice?

6 Here are some fractions.

$\frac{3}{7}$ $\frac{7}{8}$ $\frac{1}{2}$ $\frac{1}{3}$ $\frac{7}{12}$ $\frac{2}{5}$

Use them to investigate what happens

a when you multiply two fractions together that each give a terminating decimal – will your answer always be a terminating decimal?

First work out which are recurring decimals and which are terminating decimals.

b when you multiply two fractions together that each give a recurring decimal – will your answer always be a recurring decimal?

c when you multiply one of each type of fraction together – what can you conclude in this case?

AO2

Review exercise

E

1 Work out

a 130 + 42.3 **b** 67.4 + 426 + 3.74 **c** 0.874 + 0.5 + 3.8 **[3 marks]**

2 Work out

a 4.635 − 1.714 **b** 30 − 0.5 **c** 64.35 − 7.86 **[3 marks]**

D

3 Convert each of these decimals to a fraction in its lowest terms.

a 0.78 **[1 mark]**

b 0.95 **[1 mark]**

c 0.024 **[2 marks]**

d 0.625 **[2 marks]**

4 Work out

a 3.4×3.4 **[2 marks]**

b 0.024×62 **[2 marks]**

D

5 Zak went shopping with £50. He bought two T-shirts at £7.99 each, a three-pack of socks for £8.49 and a pair of sunglasses for £22.35.

a How much did he spend altogether? **[2 marks]**

b How much money did he have left? **[2 marks]**

AO2

6 Suzie is making the bridesmaids' dresses for a wedding. She buys 7.6 metres of cotton velvet at £12.95 per metre. How much does this cost her? **[2 marks]**

C

7 Work out

a $37.04 \div 0.8$ **[2 marks]**

b $3.822 \div 0.3$ **[2 marks]**

c $0.075 \div 0.06$ **[2 marks]**

C

8 Ahmed has a lot of empty oil cans that hold 1.2 litres each.
He has an oil drum with 42 litres of oil in it.
How many cans can he fill from the drum? **[2 marks]**

AO2

C

9 Convert each of these fractions to a decimal.

Some are recurring decimals, some terminating.

a $\frac{3}{5}$ **[1 mark]**

b $\frac{63}{1000}$ **[1 mark]**

c $\frac{2}{11}$ **[2 marks]**

d $\frac{11}{12}$ **[2 marks]**

10 Convert these fractions to decimals and put them in order of size, starting with the smallest.

$\frac{2}{3}$ $\frac{3}{5}$ $\frac{4}{7}$ $\frac{5}{8}$ $\frac{7}{10}$ **[3 marks]**

B

11 Convert these factions to recurring decimals.

$\frac{1}{11}$ $\frac{2}{11}$ $\frac{3}{11}$ $\frac{4}{11}$

Deduce a rule for the recurring decimals for 'elevenths'.
Check your rule for another 'elevenths' fraction, e.g. $\frac{8}{11}$. **[4 marks]**

AO3

Chapter summary

In this chapter you have learned how to

- add and subtract decimal numbers E
- convert decimals to fractions D
- convert fractions to decimals D
- multiply and divide decimal numbers D C
- understand recurring decimals C
- understand how recurring decimals relate to fractions B

14 Equations and inequalities

This chapter is about solving equations and inequalities.

For a hot air balloon to rise off the ground, the upthrust (acting upwards) must be greater than gravity (acting downwards).

Objectives

This chapter will show you how to

- solve simple equations by using inverse operations or by transforming both sides in the same way **E**
- solve equations involving brackets **D** **C**
- solve equations where the unknown appears on both sides of the equation **D** **C**
- solve equations which have negative, decimal or fractional solutions **D** **C** **B**
- solve equations involving fractions **C** **B**
- solve simple linear inequalities and represent the solution on a number line **C** **B**
- solve simultaneous linear equations **B**

Before you start this chapter

1 Write down the inverse operations to

a $\times 3$ b $+4$ c $\div 32$ d -7

HELP Chapter 11

2 Expand and simplify

a $2(x + 2)$ b $-3(2x + 3)$ c $2(3 - x) + 4x$ d $2(7 - 3x) - 2x$

3 Factorise

a $7t + 35$ b $64 - 4x$ c $y^2 + 4y$ d $5n^2 - n$

Solving two-step equations

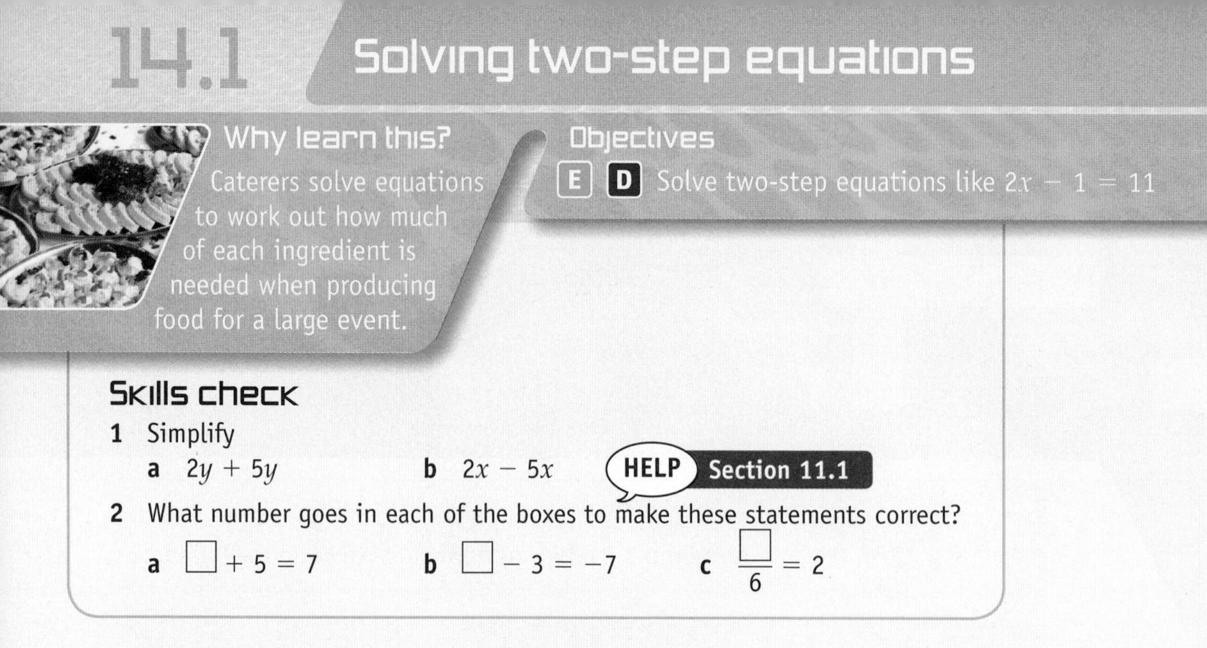

Why learn this?

Caterers solve equations to work out how much of each ingredient is needed when producing food for a large event.

Objectives

E **D** Solve two-step equations like $2x - 1 = 11$

Skills check

1 Simplify

a $2y + 5y$ **b** $2x - 5x$ HELP Section 11.1

2 What number goes in each of the boxes to make these statements correct?

a $\square + 5 = 7$ **b** $\square - 3 = -7$ **c** $\frac{\square}{6} = 2$

Using the balance method

You can think of an equation as a set of balanced scales.

You can use a bag of marbles to represent the unknown letter in the equation.
These bags each contain the same number of marbles. Finding this number is the same as solving the equation.

You will need to use two or more steps to solve some equations.

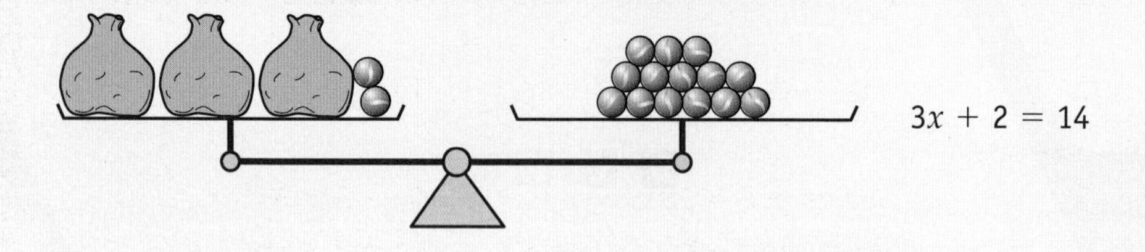

$3x + 2 = 14$

Step 1: Remove 2 marbles from the left-hand side of the scales.
Remove 2 marbles from the right-hand side too, to keep the balance.

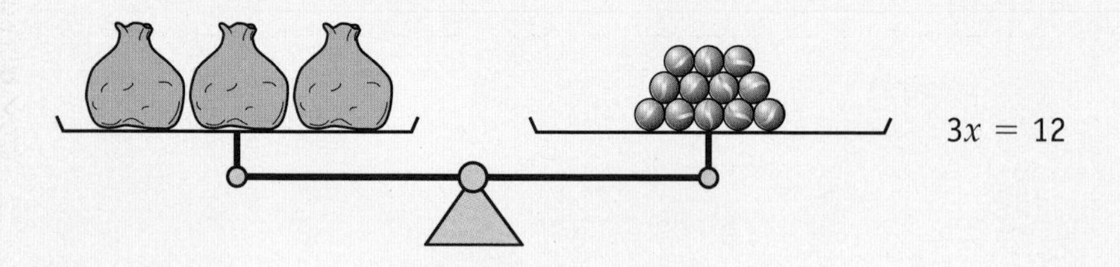

$3x = 12$

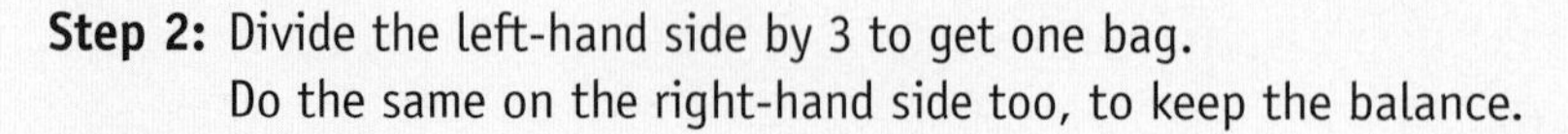

Step 2: Divide the left-hand side by 3 to get one bag.
Do the same on the right-hand side too, to keep the balance.

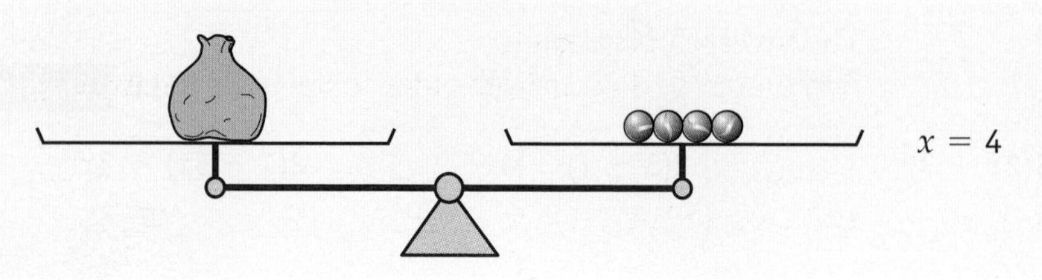

$x = 4$

Example 1

Solve $7x - 3 = 39$.

The expression $7x - 3$ has been formed like this:

$x \rightarrow \boxed{\times 7} \rightarrow \boxed{-3} \rightarrow 7x - 3$

Multiply by 7 then subtract 3.

The inverse is

$x \leftarrow \boxed{\div 7} \leftarrow \boxed{+3} \leftarrow 7x - 3$

Add 3 then divide by 7.

$7x - 3 = 39$

$+3$ on both sides: The inverse operation to -3 is $+3$. Remember to do it to both sides to keep the balance.

$7x = 42$

$\div 7$ on both sides: The inverse operation to $\times 7$ is $\div 7$. Remember to do it to both sides to keep the balance.

$x = 6$

Check: $7 \times 6 - 3 = 42 - 3 = 39$ ✓

Check your answer by substituting it back into the original equation.

Exercise 14A

1 Draw your own set of scales for each equation.
Work out the value of the unknown by working out how many marbles there are in the bag.

a $2x + 1 = 13$ b $4x + 2 = 10$ c $13 = 3x + 1$

2 Solve these equations.

a $3a + 3 = 15$ b $2b + 8 = 18$ c $5c - 5 = 35$

d $4d + 6 = 18$ e $7e + 5 = 68$ f $2g - 2 = -6$

g $5 + 4g = 17$ h $7 - 2x = 3$ i $3f + 6 = -6$

j $19 = 3x - 2$ k $4 = 10 - 2y$ l $15 = 2x + 17$

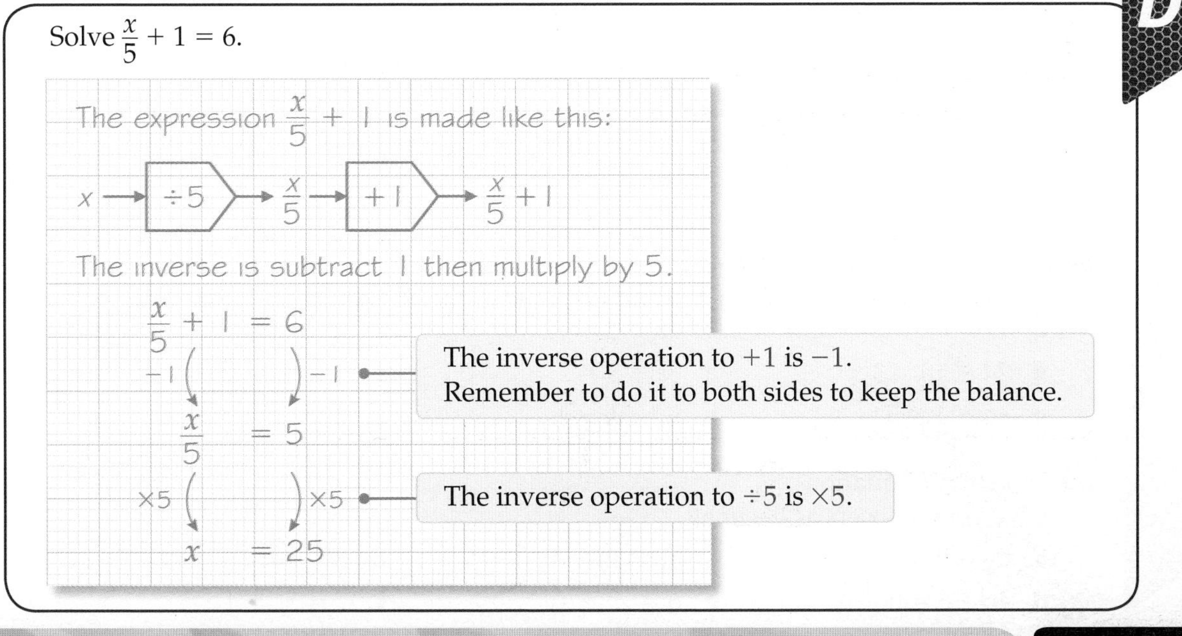

Exercise 14B

D

1 Solve these equations.

a $\frac{a}{3} + 7 = 10$ **b** $\frac{b}{4} - 6 = 1$ **c** $\frac{c}{5} - 2 = 2$

d $7 + \frac{d}{2} = 10$ **e** $12 - \frac{e}{6} = 18$ **f** $\frac{f}{6} + 10 = 4$

g $3 + \frac{g}{10} = 5$ **h** $\frac{h}{7} - 1 = 3$ **i** $20 = \frac{i}{5} + 17$

2 Solve these equations.
Then arrange the letters in order of size (starting with the smallest) to spell the name of a computer game character.

$12a + 10 = -2$ $4m - 1 = -9$ $\frac{o}{5} - 3 = 2$

$12 + \frac{i}{10} = 14$ $\frac{r}{4} + 1 = 1$

Fractional solutions

Solutions to equations are not always whole numbers.

D

Example 3

Solve $5x - 7 = 2$.

Give your answer as a mixed number and as a decimal.

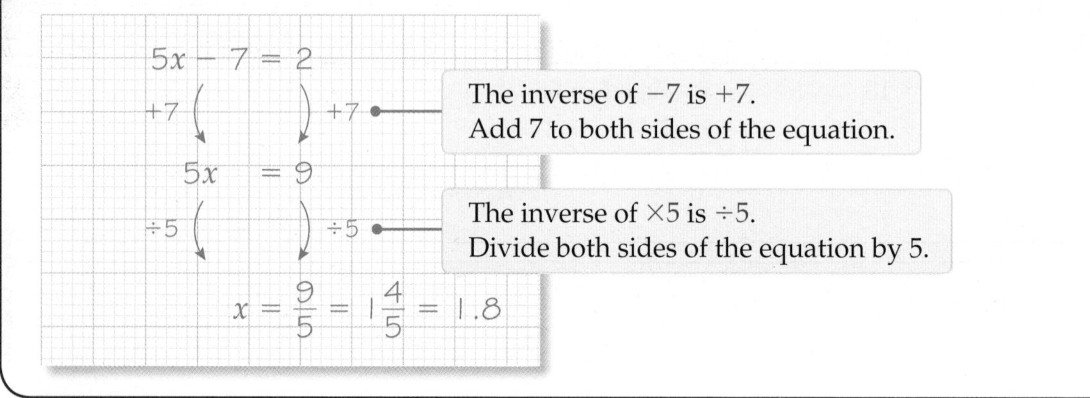

Exercise 14C

D

1 Solve these equations.
Give your answers as mixed numbers.

a $2a = 9$ **b** $7b + 3 = 12$ **c** $3c + 12 = 22$

d $7d - 10 = 3$ **e** $3e - 3 = 2$ **f** $8f - 6 = 6$

g $6g + 1 = 5$ **h** $9 = 3h - 1$ **i** $4i + 1 = -5$

2 Solve these equations.
Give your answers as decimals.

a $4g - 5 = 5$ **b** $10 = 2j + 1$ **c** $8k - 4 = 6$

d $4e - 6 = 0$ **e** $6 = 2a + 1$ **f** $7 = 5c + 1$

Writing and solving equations

Why learn this?

If you have £200 to spend on driving lessons and a driving test, you could write an equation to work out how many lessons you can afford.

Objectives

E D Write and solve equations

Skills check

1 Expand the brackets.

a $2(x + 2)$ b $2(y - 3)$

HELP Section 11.3

2 Solve

a $\frac{x}{3} = 5$ b $x - 3 = 4$ c $2x - 1 = 19$

Writing equations for real-life problems

Many real-life problems can be solved by writing an equation or a set of equations.

When writing equations, decide what your unknown will represent and then write an expression.

Turn this into an equation by looking for expressions or numbers that are equal to one another.

D

AO2

Example 4

The rods have a total length of 14 cm.

Write an equation for the total length of the rods and solve it to find the value of y.

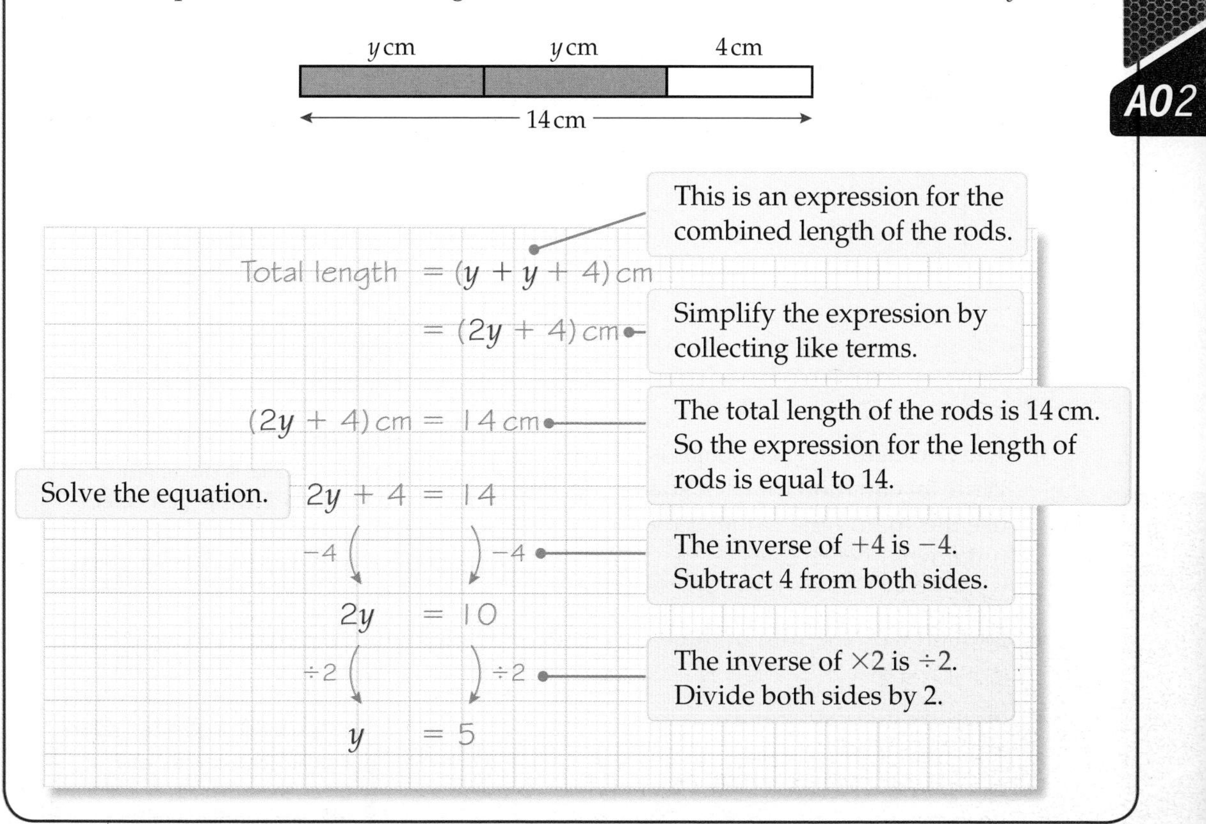

Exercise 14D

E

1 I think of a number and divide it by 7.

a Use the letter n to represent my original number.
Write an expression for my new number in terms of n.

b My answer is 6. Use your answer to part **a** to write an equation.

c Solve your equation to find out what my original number was.

AO2

2 I think of a number, multiply it by 5 and add 3. The answer is 23.
Use an algebraic method to find my original number.

D

3 Write an equation involving the total length of the rods to work out the value of the unknown in each part.

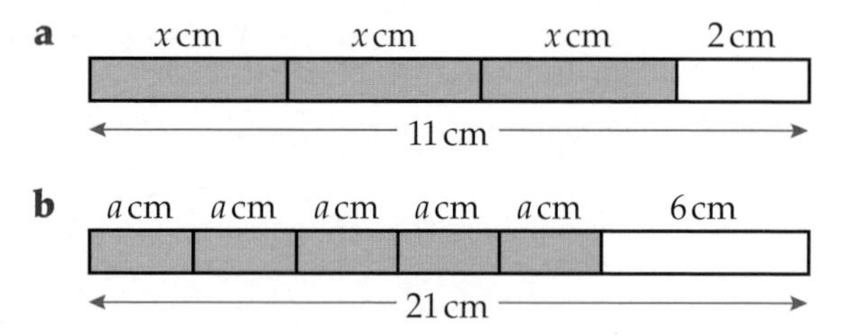

4 Reena is twice the age of her brother Bhavesh.

a Use the letter x for Bhavesh's age.
Write an expression for Reena's age in terms of x.

b Write an expression for the sum of Reena and Bhavesh's ages.

c The sum of their ages is 30.
Write an equation and solve it to find out how old Bhavesh is.

5 **a** Match each shape to an expression for its perimeter.

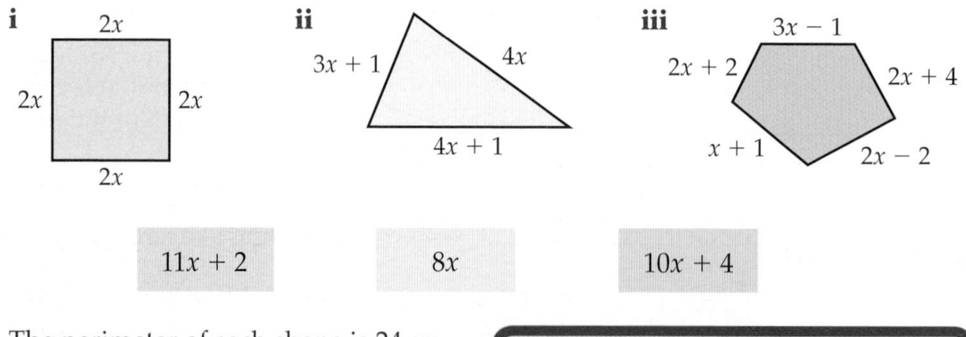

b The perimeter of each shape is 24 cm.
Write an equation for each shape and use this to find the value of the unknown in each case.

> **The perimeter of a shape is the length around its outside. To find the perimeter, add up the lengths of all the sides.**

6 **a** Explain why the sum of three consecutive whole numbers can be written as $x + x + 1 + x + 2$.

b Simplify this expression.

c The sum of three consecutive whole numbers is 306.
Use part **b** to write an equation and solve it to find the value of the first number.

> **Consecutive means 'one after the other'. 3, 4 and 5 are consecutive numbers.**

AO2

Equations with brackets

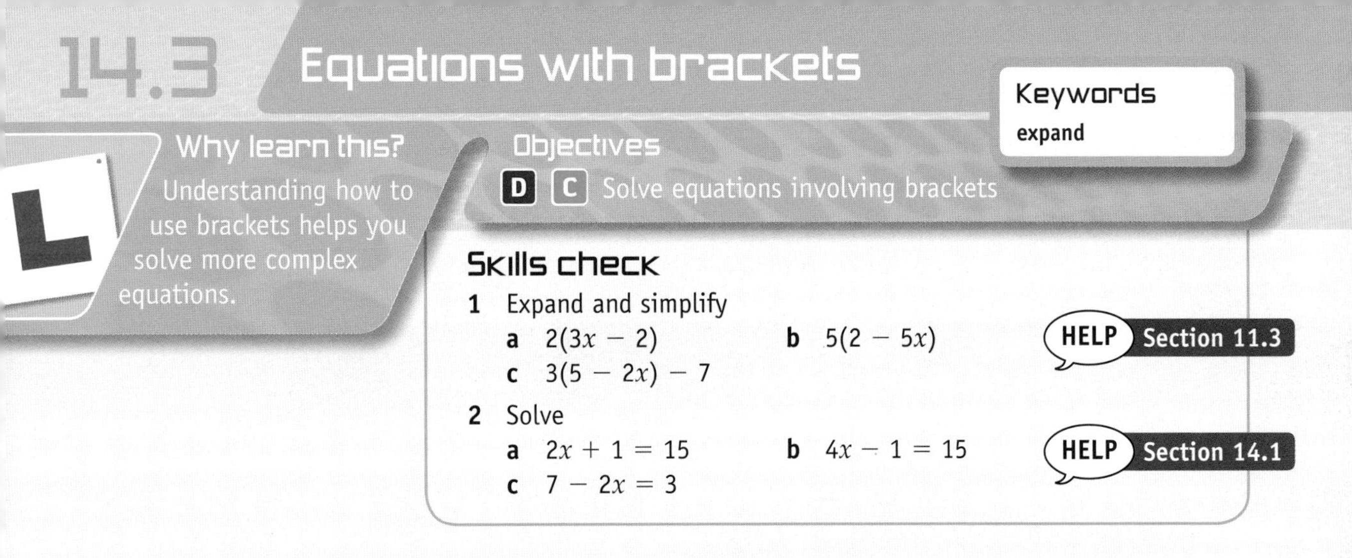

Keywords
expand

Why learn this?
Understanding how to use brackets helps you solve more complex equations.

Objectives
D **C** Solve equations involving brackets

Skills check

1 Expand and simplify
  a $2(3x - 2)$ b $5(2 - 5x)$
  c $3(5 - 2x) - 7$

HELP Section 11.3

2 Solve
  a $2x + 1 = 15$ b $4x - 1 = 15$
  c $7 - 2x = 3$

HELP Section 14.1

Solving equations with brackets

When solving equations involving brackets, **expand** all the brackets first.

Example 5

D

Solve $4(2a - 1) = 12$.

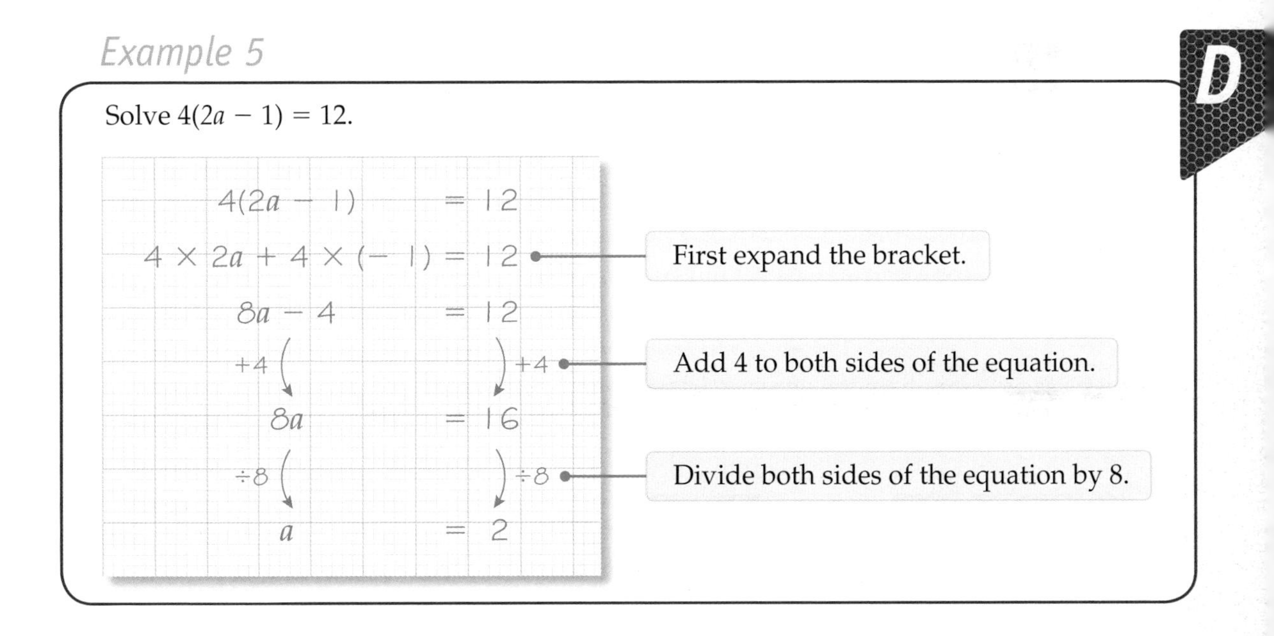

Exercise 14E

Solve these equations.

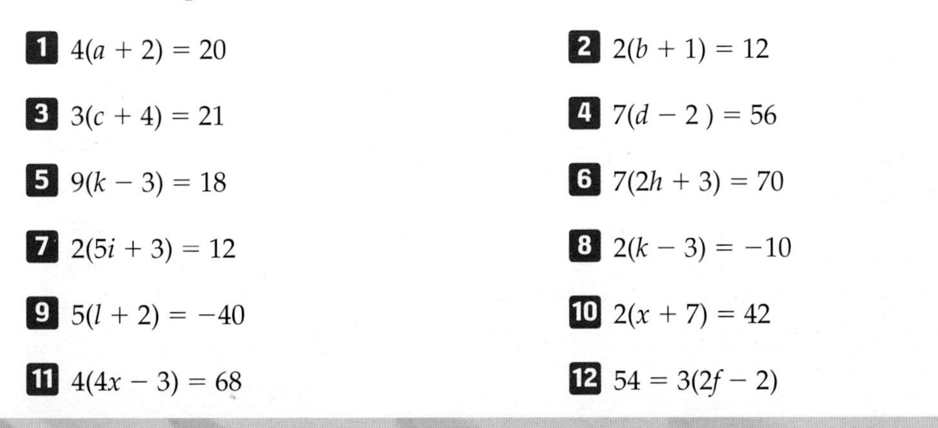

1 $4(a + 2) = 20$

2 $2(b + 1) = 12$

3 $3(c + 4) = 21$

4 $7(d - 2) = 56$

5 $9(k - 3) = 18$

6 $7(2h + 3) = 70$

7 $2(5i + 3) = 12$

8 $2(k - 3) = -10$

9 $5(l + 2) = -40$

10 $2(x + 7) = 42$

11 $4(4x - 3) = 68$

12 $54 = 3(2f - 2)$

D

13 $2(m + 9) + 3 = 17$

14 $8(k + 3) + 3 + 8k = 11$

15 $2(2q + 1) + 3q = 9$

16 $7(n + 11) + n + 9 = 46$

17 $6(2p + 20) + 6p = 30$

18 $3(3r + 1) - 2r = 17$

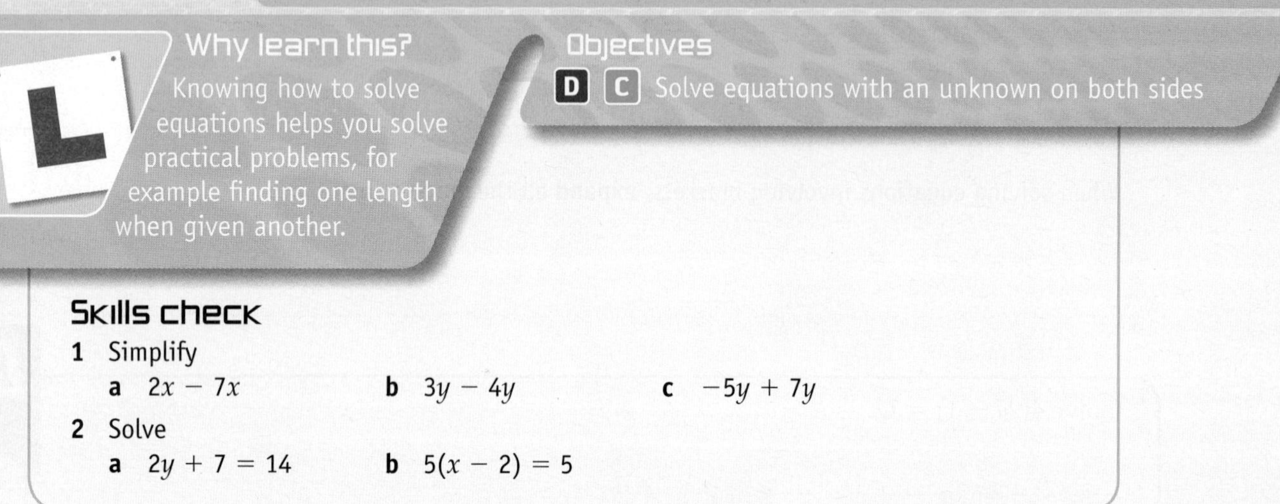

14.4 Equations with an unknown on both sides

Why learn this?

Knowing how to solve equations helps you solve practical problems, for example finding one length when given another.

Objectives

D C Solve equations with an unknown on both sides

Skills check

1 Simplify

a $2x - 7x$ b $3y - 4y$ c $-5y + 7y$

2 Solve

a $2y + 7 = 14$ b $5(x - 2) = 5$

Solving equations with an unknown on both sides

Sometimes equations have the unknown letter on both sides of the equals sign.

With equations like this, you need one additional step.

$8x + 1 = 3x + 6$

Remove 3 bags of marbles from the right-hand side of the scales.
Remember that you need to remove 3 bags of marbles from the left-hand side too, to keep the balance.

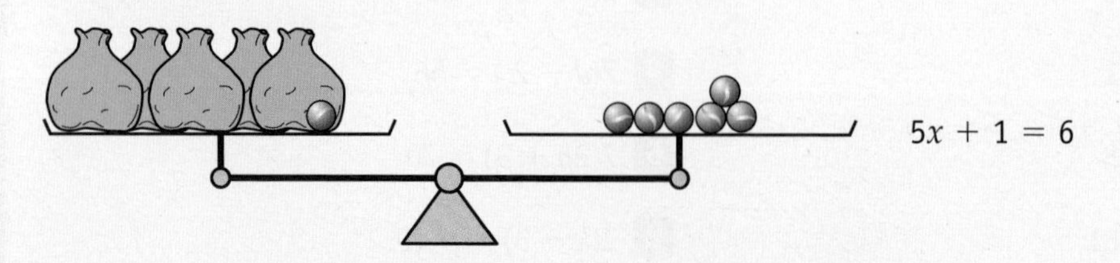

$5x + 1 = 6$

The equation is now like the equations that you have already solved.

In algebra, the extra step was to subtract $3x$ from both sides of the original equation.

Example 6

Solve $7x - 12 = 4x + 3$.

$7x - 12 = 4x + 3$

$-4x$ $-4x$ — Change the equation so that the unknown is on only one side of the equals sign. Take away the smallest number of xs possible. Subtract $4x$ from both sides.

$3x - 12 = 3$ — There is now an unknown on only one side of the equals sign.

$+12$ $+12$ — Add 12 to both sides of the equation.

$3x = 15$

$\div 3$ $\div 3$ — Divide both sides of the equation by 3.

$x = 5$

D

Exercise 14F

1 What algebraic step will change equation 1 into equation 2 in each case?

a $5x + 5 = 3x + 11$ (1) $2x + 5 = 11$ (2)

b $7x - 12 = 4x + 3$ (1) $3x - 12 = 3$ (2)

c $2x - 6 = 3x - 15$ (1) $-6 = x - 15$ (2)

d $10x + 3 = 9 - 7x$ (1) $17x + 3 = 9$ (2)

e $4 - 20x = 30x - 6$ (1) $4 = 50x - 6$ (2)

2 Solve these equations.

a $2a + 3 = a + 8$

b $2b + 1 = b - 1$

c $3c + 3 = 2c + 9$

d $2e - 3 = e + 6$

e $5f - 3 = 2f + 6$

f $8g - 16 = 6g - 4$

g $7i + 11 = 4i + 2$

h $6j - 1 = 2j + 14$

i $3k + 1 = 2k + 6$

j $4n + 2 = 2n + 7$

D

3 This rectangle has a perimeter of 216 cm.

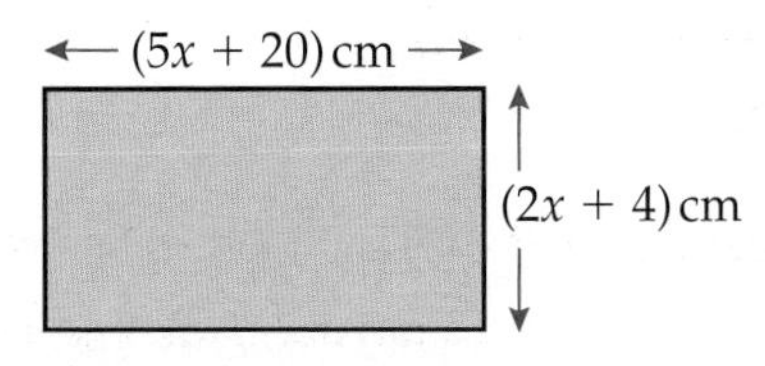

Use an algebraic method to find the length and the width of the rectangle.

C

AO3

D

Example 7

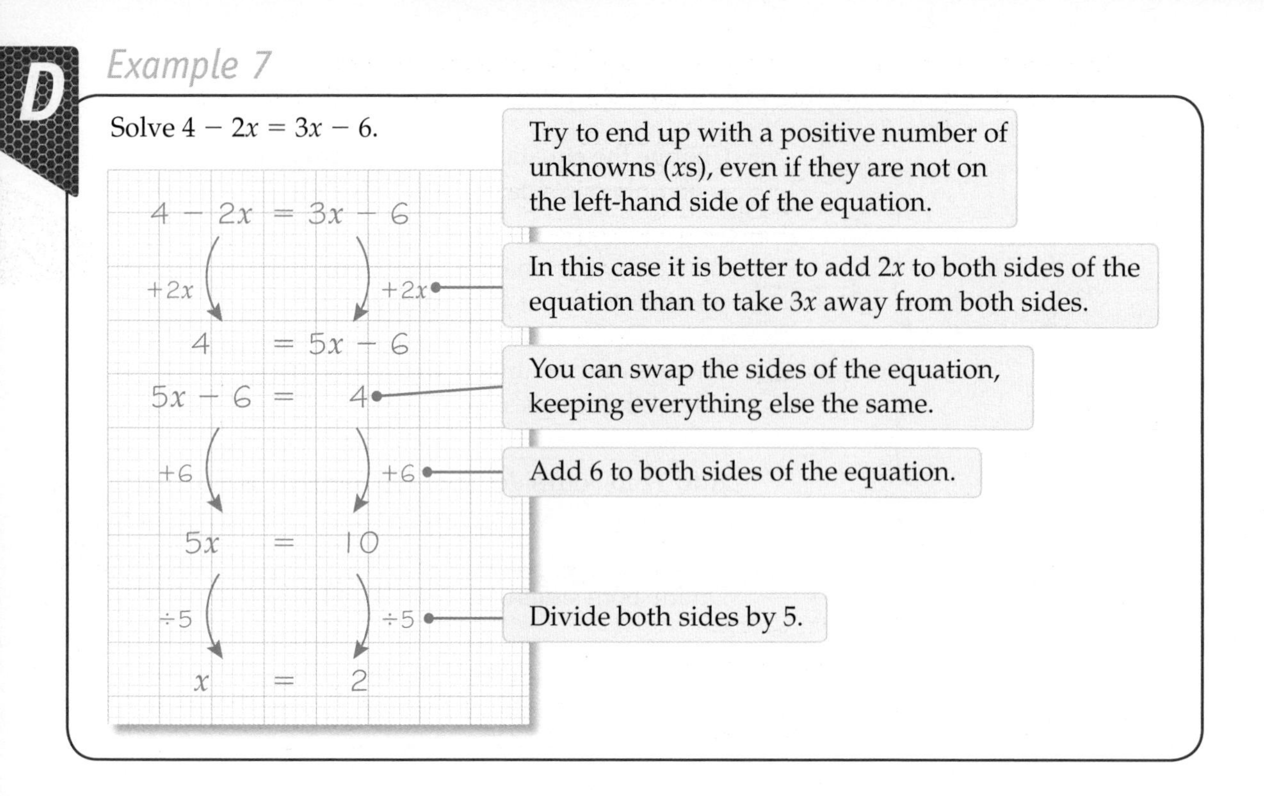

Exercise 14G

D

1 Solve

a $5a + 3 = 17 - 2a$ **b** $2 - 5b = 2b + 16$

c $4 - 20x = 30x - 6$ **d** $7d + 15 = 6 - 2d$

e $2c - 1 = 9 - 3c$ **f** $2z - 6 = 14 - 3z$

g $7 - 3x = 5x - 5$ **h** $6i - 18 = 18 - 2i$

Some solutions may not be whole numbers.

C

2 Parts of the solutions to these equations have been covered in ink.
Rewrite each solution in full.

a $7 - j = 3(5 - j)$
$7 - j =$ ■ $- 3j$
$7 + 2j =$ ■
$2j =$ ■
$j =$ ■

b $10(d - 2) = 6(d + 4)$
$10d$ ■ $= 6d + 24$
■ $- 20 = 24$
■ $= 44$
$d =$ ■

Remember to expand the brackets first.

3 Solve

a $5a + 16 = 3(a + 6)$ **b** $2(2c + 3) = 3c + 12$

c $3(f - 5) = 2(f - 3)$ **d** $-3(2x + 2) = 2(x + 5)$

e $4(1 - 5h) = 6(5h - 1)$ **f** $3(2i - 6) = 14 - 2i$

4 In a hot cross, the sum of the column is equal to the sum of the row.
Write and solve an equation to find the value of the unknown in this hot cross.

	$2e + 3$	
$6e$	$2e + 3$	$2e + 2$
	$2e + 3$	

C

AO2

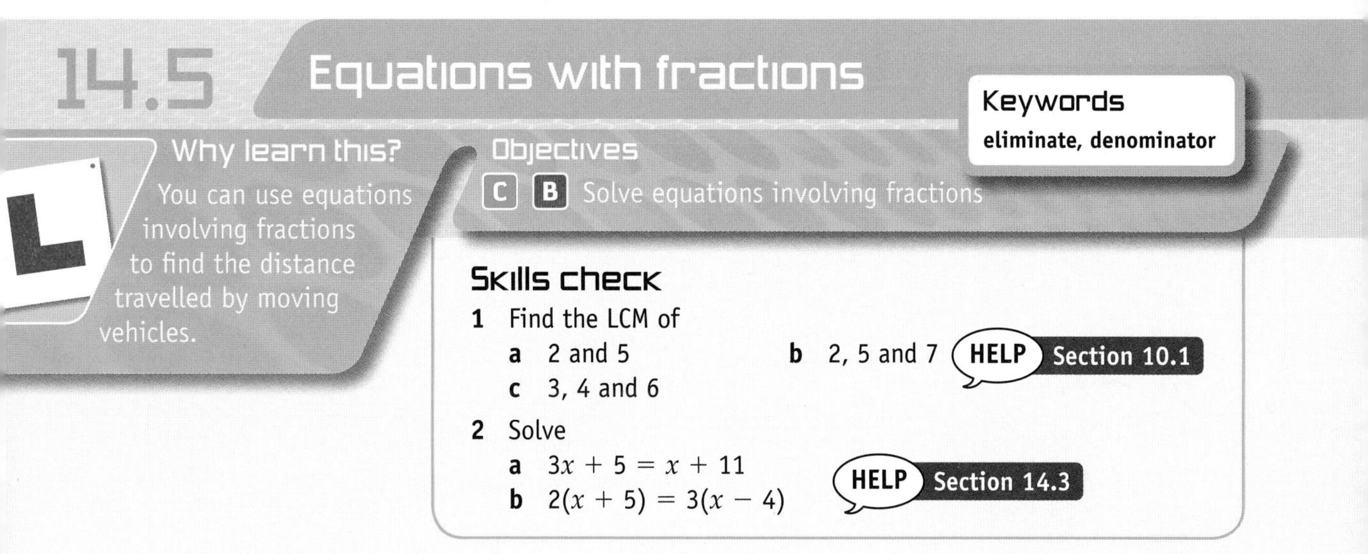

14.5 Equations with fractions

Keywords
eliminate, denominator

Why learn this?
You can use equations involving fractions to find the distance travelled by moving vehicles.

Objectives
C B Solve equations involving fractions

Skills check

1 Find the LCM of
 a 2 and 5
 b 2, 5 and 7
 c 3, 4 and 6

HELP Section 10.1

2 Solve
 a $3x + 5 = x + 11$
 b $2(x + 5) = 3(x - 4)$

HELP Section 14.3

Solving equations with fractions

Some equations involve fractions.

When solving equations with fractions, find a way to **eliminate** the **denominator**.

Example 8

C B

Solve

a $\frac{3x + 10}{4} = 7$ b $\frac{2x}{3} - \frac{x}{4} = 10$ c $\frac{2x}{3} + \frac{x}{2} = 7$

a $\frac{3x + 10}{4} = 7$

$\frac{3x + 10}{4} \times 4 = 7 \times 4$ — Multiply both sides by 4 to eliminate the denominator.

$3x + 10 = 28$

-10 -10 — Subtract 10 from both sides.

$3x = 18$

$\div 3$ $\div 3$ — Divide both sides by 3.

$x = 6$

Check: $\frac{3 \times 6 + 10}{4} = \frac{18 + 10}{4} = \frac{28}{4} = 7$ ✓

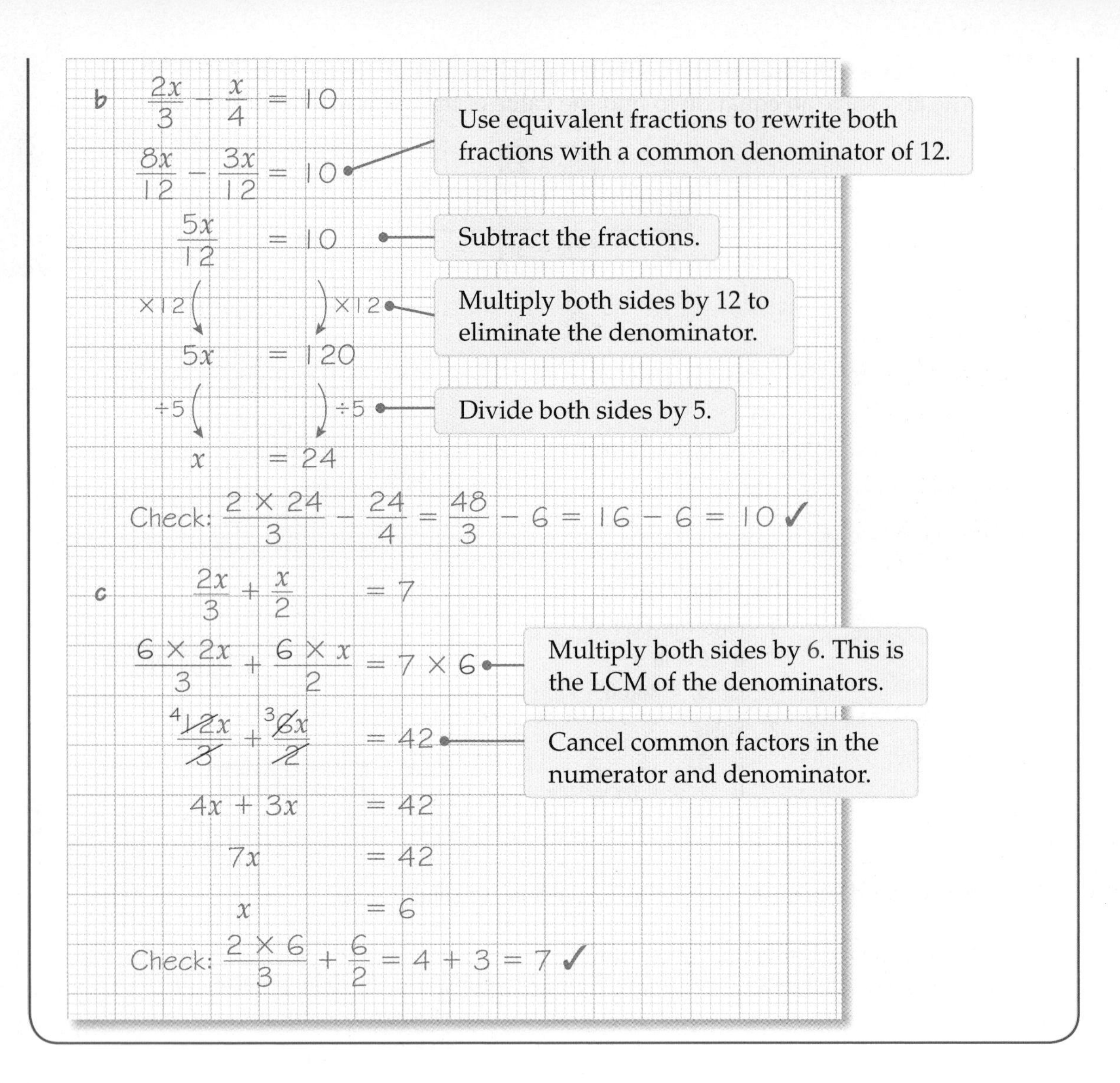

Exercise 14H

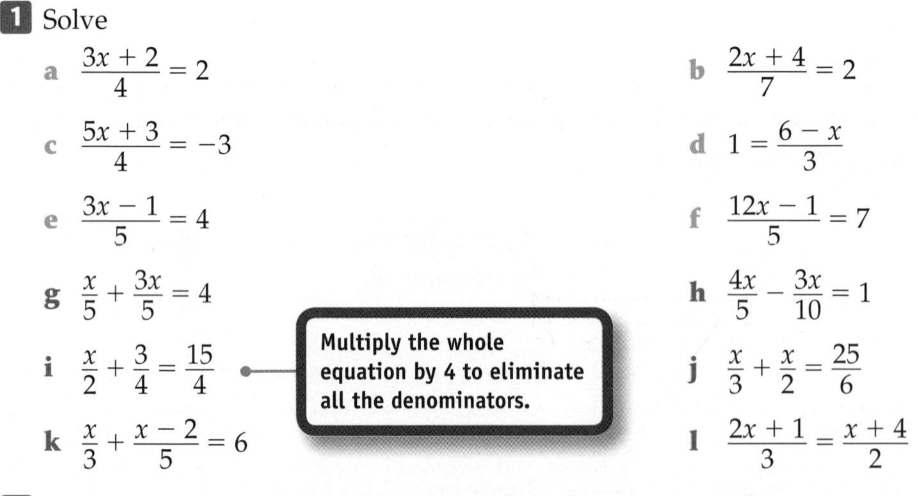

C

1 Solve

a $\frac{3x + 2}{4} = 2$

b $\frac{2x + 4}{7} = 2$

c $\frac{5x + 3}{4} = -3$

d $1 = \frac{6 - x}{3}$

e $\frac{3x - 1}{5} = 4$

f $\frac{12x - 1}{5} = 7$

B

g $\frac{x}{5} + \frac{3x}{5} = 4$

h $\frac{4x}{5} - \frac{3x}{10} = 1$

i $\frac{x}{2} + \frac{3}{4} = \frac{15}{4}$

Multiply the whole equation by 4 to eliminate all the denominators.

j $\frac{x}{3} + \frac{x}{2} = \frac{25}{6}$

k $\frac{x}{3} + \frac{x - 2}{5} = 6$

l $\frac{2x + 1}{3} = \frac{x + 4}{2}$

B

AO2

2 Half of a number added to one third of the number is equal to eight less than the number.

a Write expressions for 'half of a number', 'one third of the number' and 'eight less than the number'.

b Use these expressions to write an equation.

c Solve the equation to find out what the original number is.

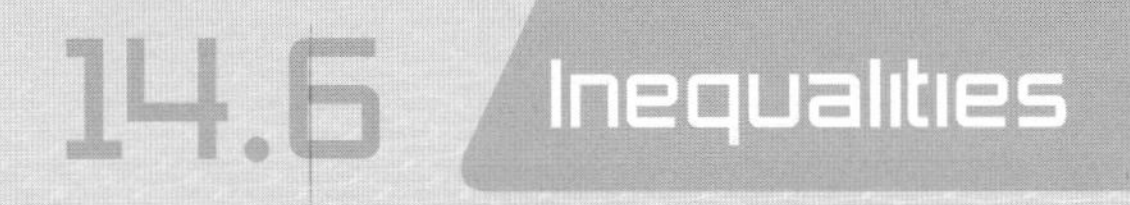

Inequalities

Keywords
inequality

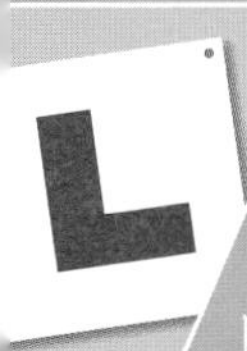

Why learn this?

The inequality $16 \leqslant x < 34$, where x represents age in years, needs to be true for a person to join the UK Armed Forces.

Objectives

E Represent inequalities on a number line

D Write down whole-number values for unknowns in an inequality

C **B** Solve inequalities

Skills check

1 Copy and complete these.
Write $<$ or $>$ between each pair of numbers to make a correct inequality.

a 22 ☐ 13 **b** 9 ☐ 16 **c** 1210 ☐ 1090

2 True or false?

a $3 < 7$ **b** $12 \leqslant 13$ **c** $3.1 \geqslant 2$

Representing inequalities on a number line

$x > 2$ is an **inequality**. It means that x must be greater than 2.

It can be shown on a number line.

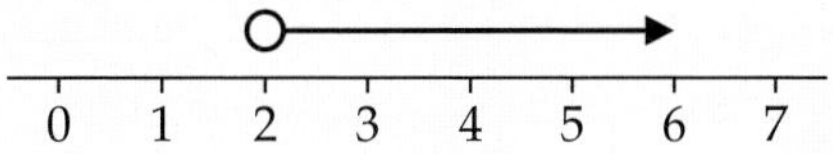

The open circle means that the 2 is *not* included in the solution set.

$x \leqslant 5$ means that x is less than or equal to 5.

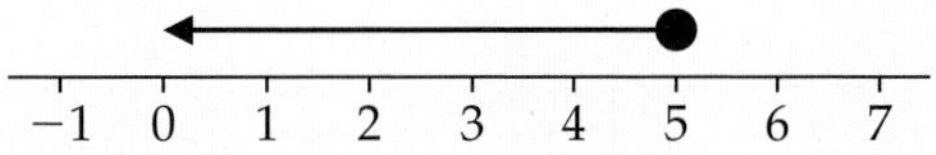

The closed circle shows that the 5 *is* included in the set of solutions.

Example 9

D

Write down all the whole-number values for x in the inequality $-3 < x \leqslant 2$.
Show the inequality on a number line.

x can take the whole-number values −2, −1, 0, 1 or 2.

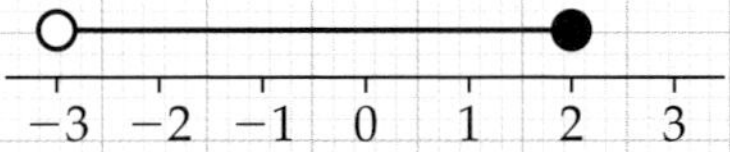

Exercise 14I

E

1 Show each inequality on a number line.

a $x < 12$ **b** $x \geqslant -2$

c $3 < y < 6$ **d** $x \leqslant 2$ or $x > 5$

e $11 \geqslant x$ and $x > 3$ **f** $-3 \leqslant x \leqslant 4$

D

2 Write down all the whole-number values for the unknown in each inequality.

a $3 \leqslant x < 6$ b $-7 < y < -4$ c $-10 \leqslant x < -5$

3 x is an integer. List all the values of x such that

a $2 \leqslant x \leqslant 6$ b $21 \leqslant x < 22$ c $-3 > x \geqslant -4$

Solving inequalities using the balance method

You can use the balance method to solve inequalities.

When you multiply or divide both sides of an inequality by a positive number, the inequality sign remains unchanged.

When you multiply or divide both sides of an inequality by a negative number, you need to change the direction of the sign.

For example, if both sides of the inequality $2 < 3$ are multiplied by -1, the inequality becomes $-2 > -3$. The sign has been reversed to make the inequality true.

C

Example 10

Solve each inequality. Show the solution on a number line.

a $24 < 3x \leqslant 36$ b $4x > 3x - 3$

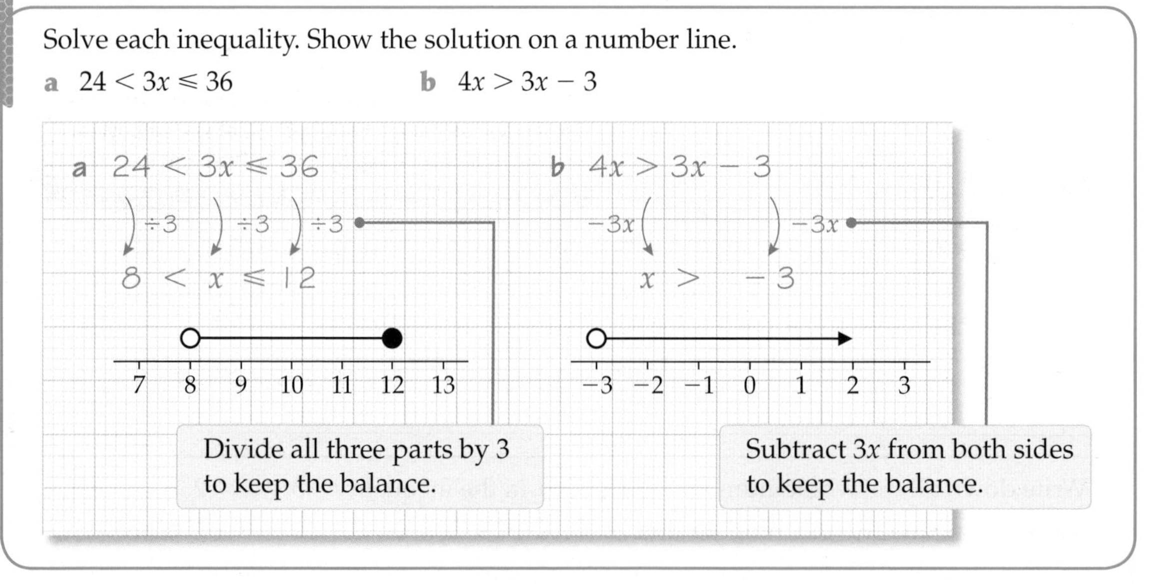

Exercise 14J

C

1 Solve each inequality. Show the solution on a number line.

a $4x < 20$ b $x - 3 \leqslant 5$ c $x + 7 < 10$

d $5x > 30$ e $\frac{x}{2} < 10$ f $x - 3 > -3$

g $6 < 2x \leqslant 18$ h $9 < 3x < 21$ i $2x < 4x + 6$

2 Solve

a $5x - 3 < 7$ b $2x + 1 > 11$ c $\frac{x}{3} + 1 \geqslant 3$

d $3x < 2x + 7$ e $5x > 3x + 10$ f $2x \leqslant 6x + 10$

Example 11

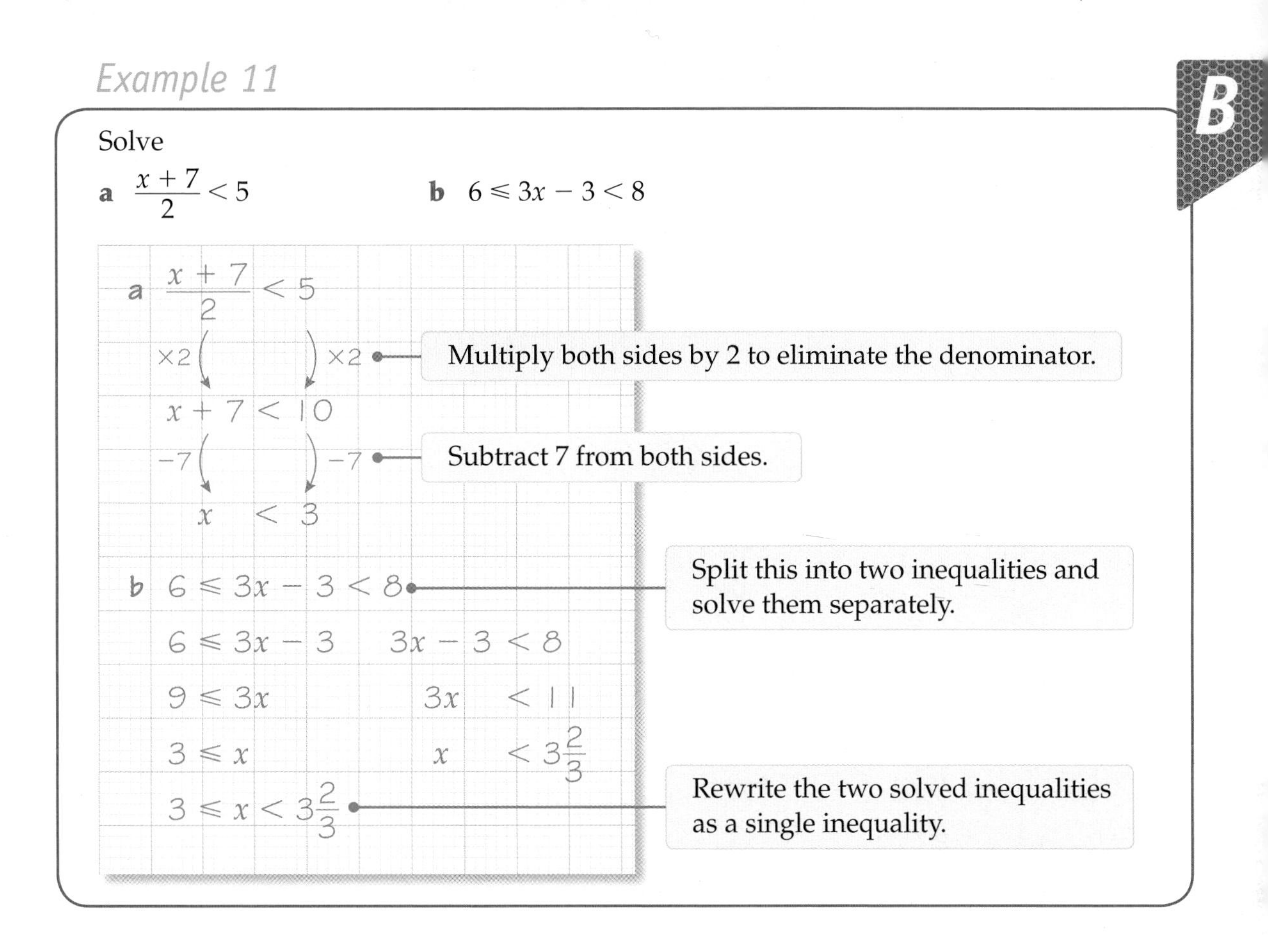

Solve

a $\dfrac{x+7}{2} < 5$ **b** $6 \leqslant 3x - 3 < 8$

a $\dfrac{x+7}{2} < 5$

$\times 2$ both sides — Multiply both sides by 2 to eliminate the denominator.

$x + 7 < 10$

-7 both sides — Subtract 7 from both sides.

$x < 3$

b $6 \leqslant 3x - 3 < 8$ — Split this into two inequalities and solve them separately.

$6 \leqslant 3x - 3$ $3x - 3 < 8$

$9 \leqslant 3x$ $3x < 11$

$3 \leqslant x$ $x < 3\frac{2}{3}$

$3 \leqslant x < 3\frac{2}{3}$ — Rewrite the two solved inequalities as a single inequality.

B

Exercise 14K

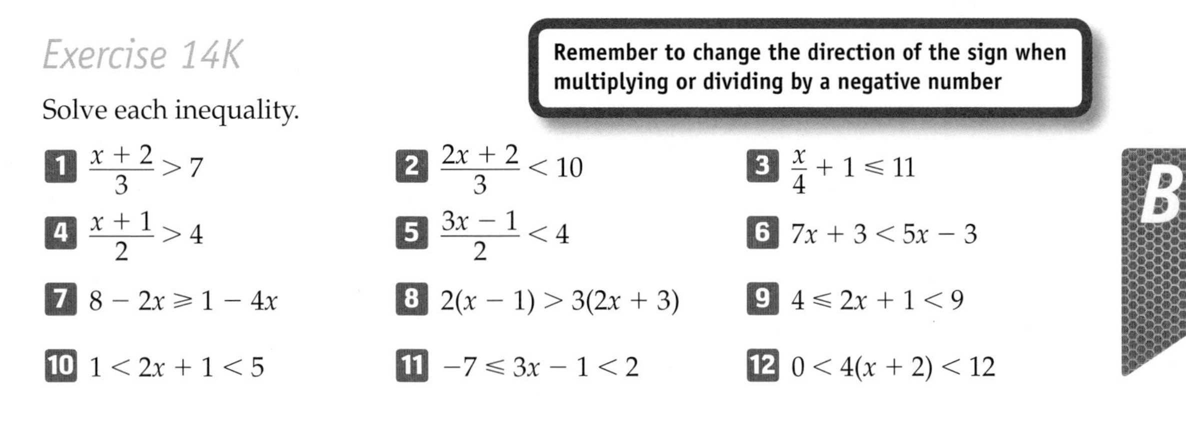

Remember to change the direction of the sign when multiplying or dividing by a negative number

Solve each inequality.

1 $\dfrac{x+2}{3} > 7$ **2** $\dfrac{2x+2}{3} < 10$ **3** $\dfrac{x}{4} + 1 \leqslant 11$

4 $\dfrac{x+1}{2} > 4$ **5** $\dfrac{3x-1}{2} < 4$ **6** $7x + 3 < 5x - 3$

7 $8 - 2x \geqslant 1 - 4x$ **8** $2(x - 1) > 3(2x + 3)$ **9** $4 \leqslant 2x + 1 < 9$

10 $1 < 2x + 1 < 5$ **11** $-7 \leqslant 3x - 1 < 2$ **12** $0 < 4(x + 2) < 12$

B

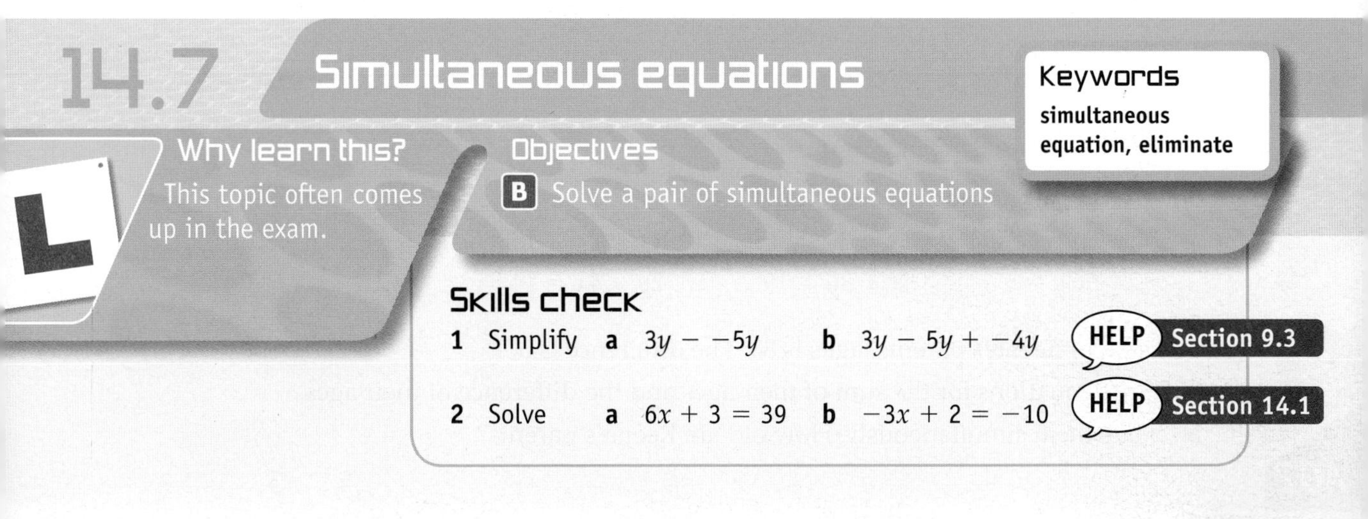

14.7 Simultaneous equations

Keywords
simultaneous equation, eliminate

Why learn this?

This topic often comes up in the exam.

Objectives

B Solve a pair of simultaneous equations

Skills check

1 Simplify **a** $3y - -5y$ **b** $3y - 5y + -4y$ HELP Section 9.3

2 Solve **a** $6x + 3 = 39$ **b** $-3x + 2 = -10$ HELP Section 14.1

Solving simultaneous equations

Simultaneous equations are pairs of equations which have two unknowns and two solutions. Both equations are true at the same time, so they have the same solutions.

To solve simultaneous equations, you need to **eliminate** one of the unknowns using algebraic steps. You should then have a linear equation with one unknown, which you can solve.

B

Example 12

Solve each pair of simultaneous equations.

a $5x + y = 19$
$2x + y = 10$

b $3x - y = 9$
$2x + y = 1$

a
$5x + y = 19$ (1)
$2x + y = 10$ (2)

Start by labelling the equations as (1) and (2) so you can refer to them clearly in your solution.

$3x = 9$

You can eliminate one of the unknowns (y) by subtracting equation 2 from equation 1.

$x = 3$

$5 \times 3 + y = 19$ (1)

Now you need to find the value of y. Substitute the value of x into one of the equations.

$15 + y = 19$
$y = 4$

Check: $2 \times 3 + 4 = 10$ (2)
$10 = 10$ ✓

Check your answer by substituting for both unknowns in the other equation.

b
$3x - y = 9$ (1)
$2x + y = 1$ (2)
$5x = 10$

In part a, equation 2 was subtracted from equation 1. If you did this here, you would end up with $x - 2y = 8$. This would not eliminate an unknown. Instead, *add* the two equations to eliminate y.

$x = 2$

$6 - y = 9$ (1)

Find the value of y by substituting the value of x into equation 1.

$y = -3$

Check: $2 \times 2 + -3 = 1$ (2)
$1 = 1$ ✓

Check your answer by substituting for both unknowns in the other equation.

Exercise 14L

B

1 Solve each pair of simultaneous equations.

a $2x + y = 10$
$-2x + 3y = 6$

b $2x - 3y = -5$
$5x + 3y = 19$

c $4x + y = 13$
$2y - 4x = 14$

d $3x + y = 30$
$-2x + y = -5$

2 The sum of Reena's parents' ages is 83. The difference is 3.

a Write equations for the sum of their ages and the difference of their ages.

b Solve them simultaneously. How old are Reena's parents?

AO2

Multiplying one of the equations

Sometimes you cannot add or subtract the equations to eliminate one of the unknowns because they do not have the same number of either of them.
In these cases, you need to multiply one or both equations so that one of the unknowns appears the same number of times in both equations.

Example 13

B

Solve these simultaneous equations. $2y - x = 6$
$y + 3x = 17$

$2y - x = 6 \quad (1)$
$y + 3x = 17 \quad (2)$

You cannot add or subtract the equations to eliminate one of the unknowns because neither x nor y appears the same number of times in both equations.

$2y + 6x = 34 \quad (3)$ — Instead, multiply equation 2 by 2. Now both equations include $2y$.

$-x - 6x = 6 - 34$ — Subtract equation 3 from equation 1 to eliminate y.

$-7x = -28$ — Solve the equation.

$7x = 28$

$x = 4$

$2y - 4 = 6 \quad (1)$ — Find the value of y by substituting 4 for x in equation 1.

$2y = 10$

$y = 5$

Check: $5 + 3 \times 4 = 17 \quad (2)$
$17 = 17$ ✓ — Check your answer by substituting for both unknowns in the other equation.

Exercise 14M

B

Solve each pair of simultaneous equations.

1 $3x + 2y = 12$
$2x + y = 7$

2 $x + 2y = 8$
$3x + 5y = 19$

3 $2x - 3y = -11$
$9x + y = 23$

4 $-2x - y = 2$
$3x - 4y = -25$

Multiplying both of the equations

Sometimes you need to multiply *both* equations to get the same number of unknowns.

B

Example 14

Solve these simultaneous equations. $3x + 2y = 25$
$-4x + 7y = -14$

$3x + 2y = 25$ (1)
$-4x + 7y = -14$ (2)
$12x + 8y = 100$ (3)
$-12x + 21y = -42$ (4)

Multiply equation 1 by 4 and equation 2 by 3. This means that $12x$ will appear in both equations.

$29y = 58$

Add the equations because the signs of the x terms are different. Then solve to find y.

$y = 2$

$3x + 4 = 25$ (1)

Substitute 2 for y in equation 1 and solve to find x.

$3x = 21$
$x = 7$

Check: $-4 \times 7 + 7 \times 2 = -14$ (2)

Check your answer by substituting for both unknowns in the other equation.

$-28 + 14 = -14$
$-14 = -14$ ✓

Exercise 14N

B

1 Solve each pair of simultaneous equations.

a $2x - 4y = 6$
$x + 15y = 20$

b $2x - 3y = -19$
$5x + 2y = -19$

c $3x - 2y = 8$
$5x + 7y = 3$

d $9x + 2y = 31$
$2x + 3y = 12$

e $2x + 5y = 16$
$3x - 2y = 5$

f $2x + 3y = -5$
$3x + 8y = 3$

2 A music download site makes a total of 1000 sales in a day.
The revenue for the day is £6300.
Albums cost £7.50 and individual songs cost £1.50.

a How many albums were sold?

b On average, an album has 12 songs on it.
Calculate an estimate for the total number of tracks sold.

Review exercise

E

1 Solve

a $\frac{w}{5} = 8$ **[1 mark]**

b $4x - 4 = 6$ **[2 marks]**

E AO2

2 I think of a number. I divide my number by 2 and then add 3.
My answer is 13.
Form an equation and solve it to find the number I was thinking of. **[3 marks]**

3 Solve these equations.

a $4p - 5 = 11$ [2 marks]

b $7t - 8 = 7 + t$ [3 marks]

c $7x - 8 = 3(x + 2)$ [3 marks]

4 a x is an integer.

$0 < x \leq 4$

Write down all the possible values of x. [2 marks]

b x and y are integers.

$0 < x \leq 4$ $\quad y < x$ $\quad x + y < 6$

Write down **two** pairs of values of x and y which satisfy all three inequalities. [2 marks]

D

5 a Solve the inequality $3(x - 1) \leq 6$. [3 marks]

b The inequality $x \leq 3$ is shown on the number line below.

−4 −3 −2 −1 0 1 2 3 4 5 6 x

Copy the number line and draw another inequality on it so that only the following integers satisfy both inequalities.

−2, −1, 0, 1, 2, 3 [1 mark]

6 a Solve the inequality $2x + 2 \leq 0$ [2 marks]

b Write down the inequality shown by the following diagram.

−4 −3 −2 −1 0 1 2 3 4 x

[1 mark]

7 Solve these equations.

a $5x + 19 = 3(x + 8)$ [3 marks]

b $\frac{9 - 2x}{4} = 3$ [3 marks]

C

8 Solve the equation $\frac{3x + 1}{2} - \frac{2x + 3}{3} = 1$. [3 marks]

9 Solve these simultaneous equations. $2x + 3y = 9$

$3x + 2y = 1$ [4 marks]

B

Chapter summary

In this chapter you have learned how to

- represent inequalities on a number line E
- solve two-step equations like $2x - 1 = 11$ E D
- write and solve equations E D
- write down whole-number values for unknowns in an inequality D
- solve equations involving brackets D C
- solve equations with an unknown on both sides D C
- solve equations involving fractions C B
- solve inequalities C B
- solve a pair of simultaneous equations B

15 Indices and formulae

This chapter is about indices and formulae.

The trebuchet is a medieval weapon designed to throw things at or over castle walls. If you know the angle and speed of launch, you can write a formula to work out the range of your weapon.

Objectives

BBC Video

This chapter will show you how to

- substitute numbers to work out the value of simple algebraic expressions E
- substitute numbers into expressions involving brackets and powers E D
- substitute numbers into a variety of formulae E D
- use index laws and notation in algebra E D C
- use algebra to write formulae in different situations E D
- raise a number or variable to the power of 1 or 0 C B
- rearrange a formula to make a different variable the subject of the formula C B
- use index laws for raising a power to another power B

Before you start this chapter

1 Use algebra to write an expression for each of these.

a 6 more than p b 2 taken away from y c p added to q

d 6 lots of d e x divided by 2

HELP Chapter 11

2 Simplify these expressions.

a $r + r + r + r + r$ b $x + x + x + x - x$ c $d + 2d + 4d - 3d$

3 Expand the brackets.

a $3(x + 4)$ b $2(d - 3)$ c $5(x + y - 6)$

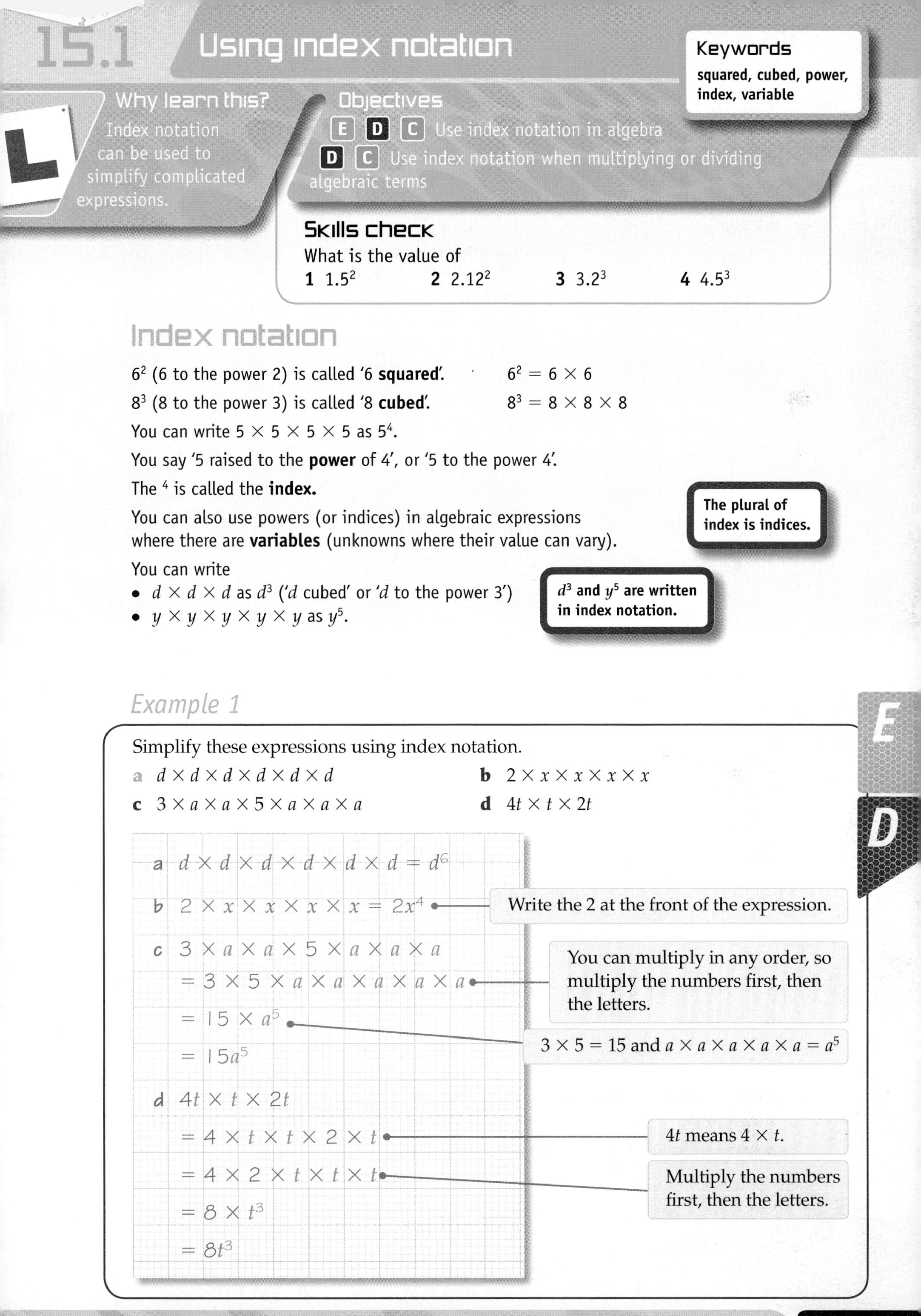

15.1 Using index notation

Keywords
squared, cubed, power, index, variable

Why learn this?
Index notation can be used to simplify complicated expressions.

Objectives

E D C Use index notation in algebra

D C Use index notation when multiplying or dividing algebraic terms

Skills check

What is the value of

1 1.5^2 **2** 2.12^2 **3** 3.2^3 **4** 4.5^3

Index notation

6^2 (6 to the power 2) is called '6 **squared**'. $6^2 = 6 \times 6$

8^3 (8 to the power 3) is called '8 **cubed**'. $8^3 = 8 \times 8 \times 8$

You can write $5 \times 5 \times 5 \times 5$ as 5^4.

You say '5 raised to the **power** of 4', or '5 to the power 4'.

The ^{4} is called the **index.**

You can also use powers (or indices) in algebraic expressions where there are **variables** (unknowns where their value can vary).

The plural of index is indices.

You can write

- $d \times d \times d$ as d^3 ('d cubed' or 'd to the power 3')
- $y \times y \times y \times y \times y$ as y^5.

d^3 and y^5 are written in index notation.

Example 1

Simplify these expressions using index notation.

a $d \times d \times d \times d \times d \times d$ **b** $2 \times x \times x \times x \times x$

c $3 \times a \times a \times 5 \times a \times a \times a$ **d** $4t \times t \times 2t$

a $d \times d \times d \times d \times d \times d = d^6$

b $2 \times x \times x \times x \times x = 2x^4$ — Write the 2 at the front of the expression.

c $3 \times a \times a \times 5 \times a \times a \times a$

$= 3 \times 5 \times a \times a \times a \times a \times a$ — You can multiply in any order, so multiply the numbers first, then the letters.

$= 15 \times a^5$ — $3 \times 5 = 15$ and $a \times a \times a \times a \times a = a^5$

$= 15a^5$

d $4t \times t \times 2t$

$= 4 \times t \times t \times 2 \times t$ — $4t$ means $4 \times t$.

$= 4 \times 2 \times t \times t \times t$ — Multiply the numbers first, then the letters.

$= 8 \times t^3$

$= 8t^3$

Exercise 15A

E

1 Simplify these expressions using index notation.

a $d \times d \times d \times d$ b $a \times a \times a \times a \times a \times a$

c $x \times x \times x$ d $m \times m \times m \times m$

D

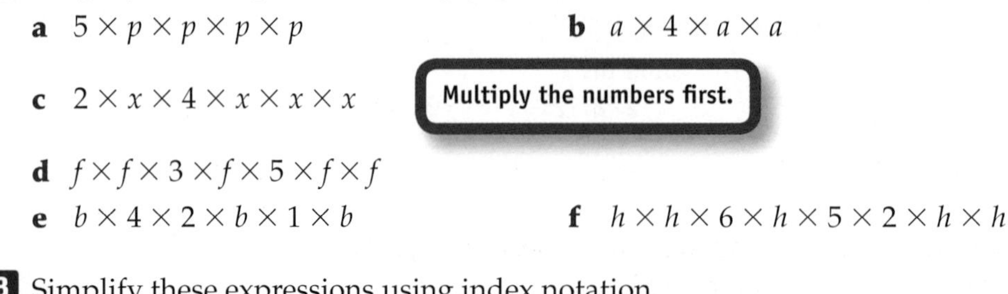

2 Simplify these expressions using index notation.

a $5 \times p \times p \times p \times p$ b $a \times 4 \times a \times a$

c $2 \times x \times 4 \times x \times x \times x$

d $f \times f \times 3 \times f \times 5 \times f \times f$

e $b \times 4 \times 2 \times b \times 1 \times b$ f $h \times h \times 6 \times h \times 5 \times 2 \times h \times h$

3 Simplify these expressions using index notation.

a $2x \times 3x$ b $4y \times y \times 3y$ c $a \times 3a \times 2a \times a$

d $3b \times 2b \times 4b$ e $3x \times 3x \times 3x$ f $6z \times z \times z \times 6z \times z \times z$

Multiplying and dividing with indices

You can also multiply and divide algebraic expressions with indices.

C

Example 2

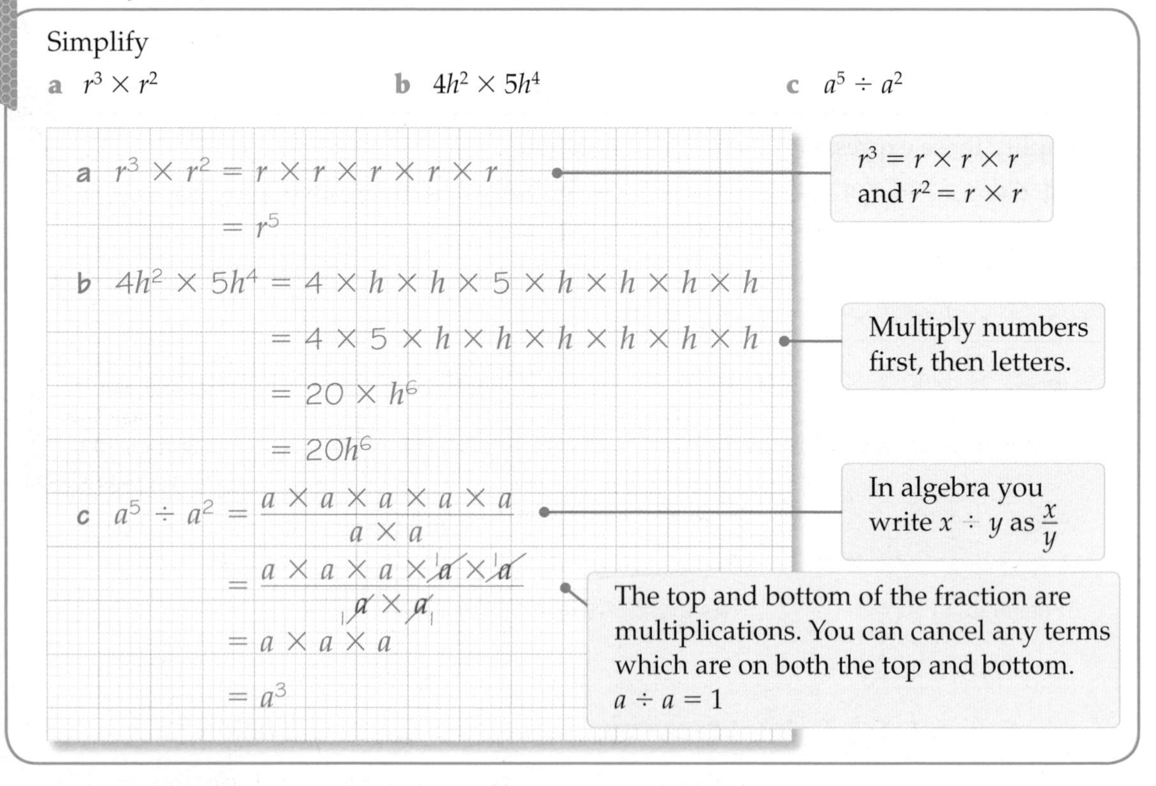

Exercise 15B

C

1 Simplify

a $a^3 \times a^2$ b $b^4 \times b^3$ c $c^2 \times c^4$ d $d^5 \times d^5$

e $x^3 \times x$ f $y^2 \times y$ g $z \times z^4$ h $w \times w^3$

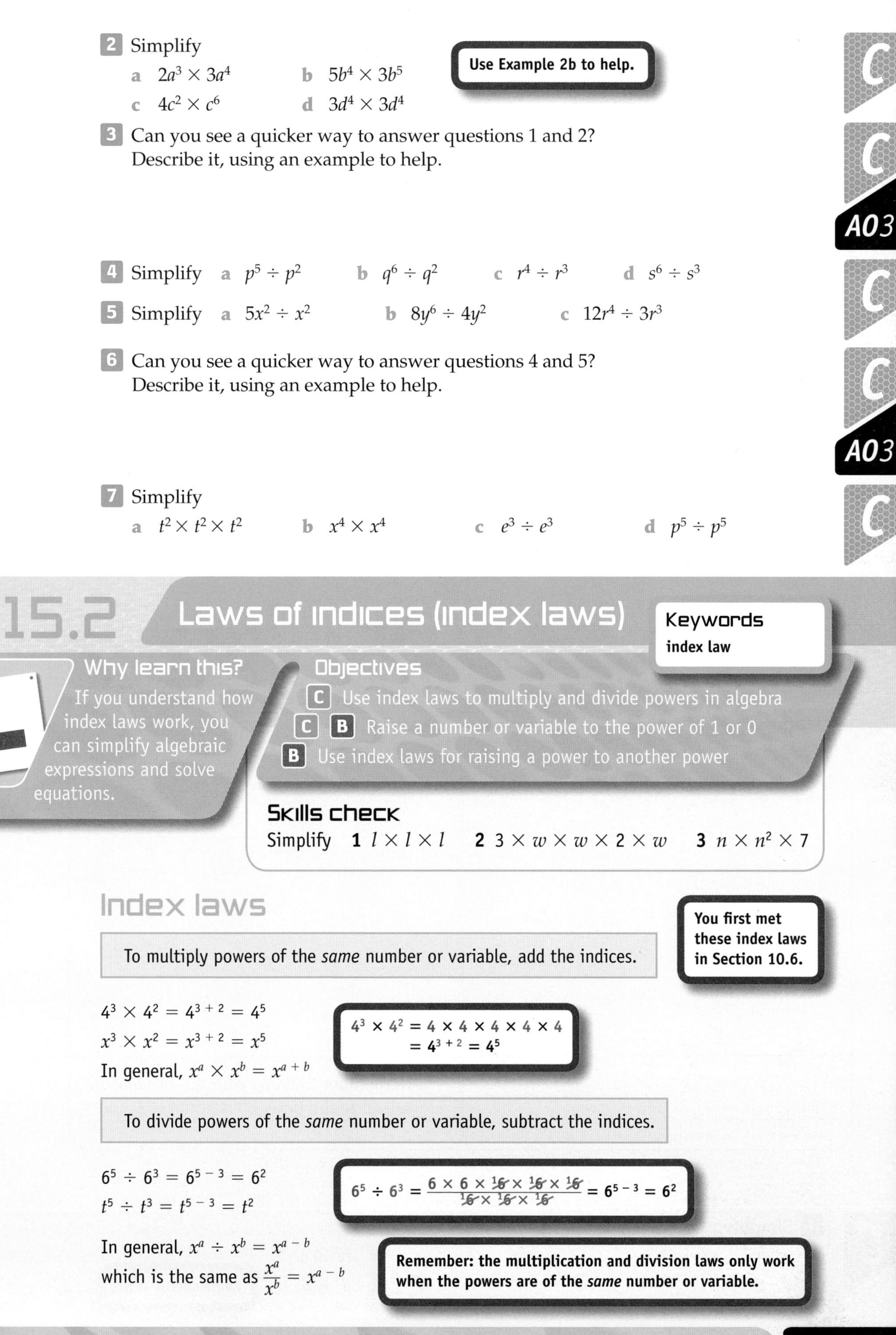

2 Simplify

a $2a^3 \times 3a^4$ b $5b^4 \times 3b^5$

c $4c^2 \times c^6$ d $3d^4 \times 3d^4$

Use Example 2b to help.

C

3 Can you see a quicker way to answer questions 1 and 2?
Describe it, using an example to help.

C AO3

4 Simplify a $p^5 \div p^2$ b $q^6 \div q^2$ c $r^4 \div r^3$ d $s^6 \div s^3$

C

5 Simplify a $5x^2 \div x^2$ b $8y^6 \div 4y^2$ c $12r^4 \div 3r^3$

C

6 Can you see a quicker way to answer questions 4 and 5?
Describe it, using an example to help.

C AO3

7 Simplify

a $t^2 \times t^2 \times t^2$ b $x^4 \times x^4$ c $e^3 \div e^3$ d $p^5 \div p^5$

C

15.2 Laws of indices (index laws)

Keywords
index law

Why learn this?
If you understand how index laws work, you can simplify algebraic expressions and solve equations.

Objectives

C Use index laws to multiply and divide powers in algebra

C B Raise a number or variable to the power of 1 or 0

B Use index laws for raising a power to another power

Skills check

Simplify **1** $l \times l \times l$ **2** $3 \times w \times w \times 2 \times w$ **3** $n \times n^2 \times 7$

Index laws

You first met these index laws in Section 10.6.

To multiply powers of the *same* number or variable, add the indices.

$4^3 \times 4^2 = 4^{3+2} = 4^5$

$x^3 \times x^2 = x^{3+2} = x^5$

In general, $x^a \times x^b = x^{a+b}$

$4^3 \times 4^2 = 4 \times 4 \times 4 \times 4 \times 4$
$= 4^{3+2} = 4^5$

To divide powers of the *same* number or variable, subtract the indices.

$6^5 \div 6^3 = 6^{5-3} = 6^2$

$t^5 \div t^3 = t^{5-3} = t^2$

$6^5 \div 6^3 = \frac{6 \times 6 \times \cancel{6} \times \cancel{6} \times \cancel{6}}{\cancel{6} \times \cancel{6} \times \cancel{6}} = 6^{5-3} = 6^2$

In general, $x^a \div x^b = x^{a-b}$
which is the same as $\frac{x^a}{x^b} = x^{a-b}$

Remember: the multiplication and division laws only work when the powers are of the *same* number or variable.

To raise a power of a number or variable to a further power, multiply the indices.

$(5^2)^3 = 5^{2 \times 3} = 5^6$

$(p^2)^3 = p^{2 \times 3} = p^6$

In general, $(x^a)^b = x^{ab}$

$(5^2)^3 = 5^2 \times 5^2 \times 5^2$
$= (5 \times 5) \times (5 \times 5) \times (5 \times 5) = 5^6$

Any number or variable raised to the power 1 is equal to the number or variable itself.

$3^1 = 3 \qquad x^1 = x$

Any number or variable raised to the power 0 is equal to 1.

$3^0 = 1 \qquad x^0 = 1$

When you divide anything by itself, you always get an answer of 1, so $x^7 \div x^7 = 1$

When you divide powers of the same variable, you subtract the indices, so $x^7 \div x^7 = x^{7-7} = x^0$

So $x^0 = 1$

C

Example 3

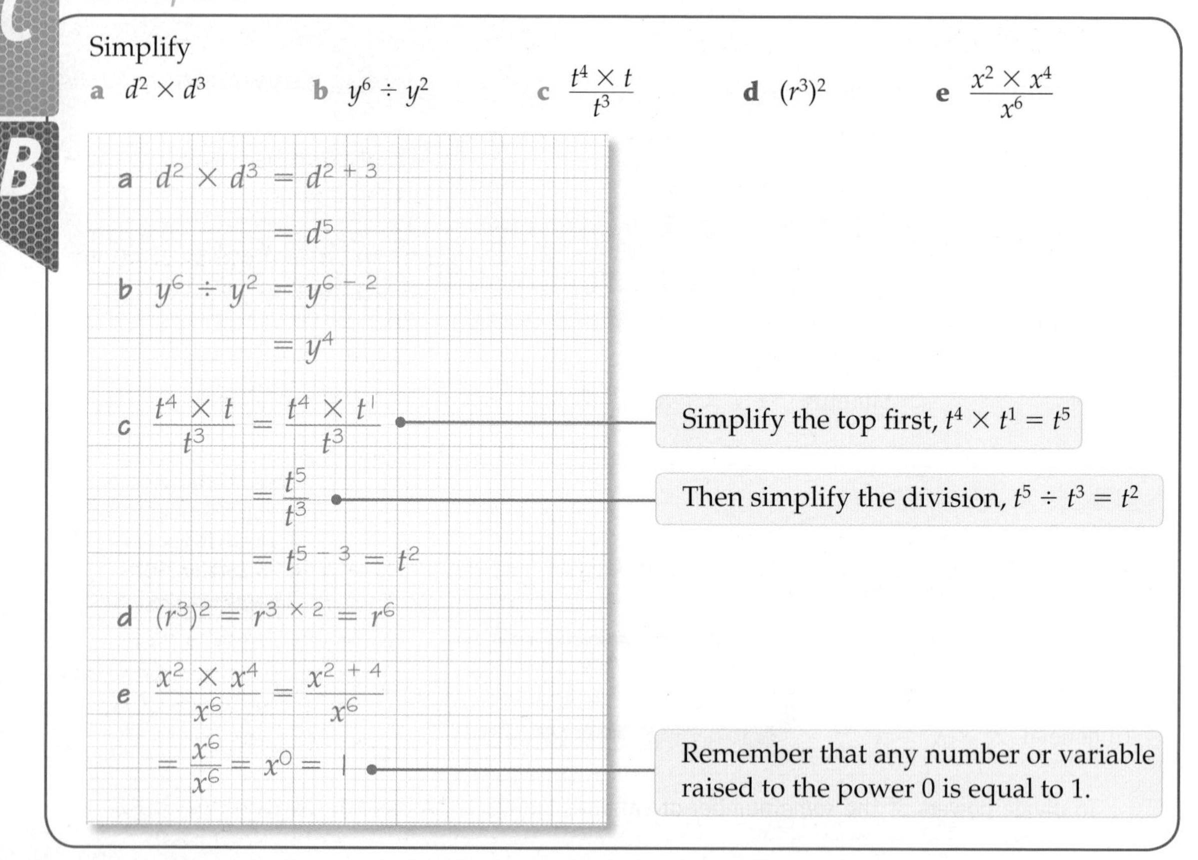

Simplify

a $d^2 \times d^3$ **b** $y^6 \div y^2$ **c** $\frac{t^4 \times t}{t^3}$ **d** $(r^3)^2$ **e** $\frac{x^2 \times x^4}{x^6}$

B

a $d^2 \times d^3 = d^{2+3}$
$= d^5$

b $y^6 \div y^2 = y^{6-2}$
$= y^4$

c $\frac{t^4 \times t}{t^3} = \frac{t^4 \times t^1}{t^3}$ — Simplify the top first, $t^4 \times t^1 = t^5$

$= \frac{t^5}{t^3}$ — Then simplify the division, $t^5 \div t^3 = t^2$

$= t^{5-3} = t^2$

d $(r^3)^2 = r^{3 \times 2} = r^6$

e $\frac{x^2 \times x^4}{x^6} = \frac{x^{2+4}}{x^6}$

$= \frac{x^6}{x^6} = x^0 = 1$ — Remember that any number or variable raised to the power 0 is equal to 1.

Exercise 15C

C

1 Simplify

a $m^3 \times m^2$ **b** $a^4 \times a^3$ **c** $n^2 \times n^4$

d $u^5 \times u^5$ **e** $t^3 \times t^6$ **f** $d^3 \times d$

Remember $d = d^1$

2 Simplify

a $a^5 \div a^2$ b $t^6 \div t^2$ c $e^4 \div e^3$

d $s^6 \div s^3$ e $t^2 \div t^2$ f $k^2 \div k$

3 Simplify

a $\frac{d^4 \times d}{d^3}$ b $\frac{e^2 \times e^2}{e}$ c $\frac{r \times r^3}{r^2}$

d $\frac{a^4 \times a^5}{a^8}$ e $\frac{x^3 \times x^4}{x^7}$ f $\frac{y \times y^4}{y^3 \times y^2}$

4 Simplify

a $h^3 \times h^4 \times h^2$ b $e^4 \times e \times e^3$ c $c^2 \times c^6 \times c$

5 Simplify

a $(p^3)^2$ **b** $(q^2)^4$ **c** $(r^3)^4$ **d** $(f^5)^3$ **e** $(d^4)^3$

C

B

C

B

Example 4

Simplify

a $3a^4 \times 5a^3$ b $12b^5 \div 4b^3$

c $3xy^2 \times 5xy$ d $\frac{10a^3b^2}{2a^2b^2}$

a $3a^4 \times 5a^3 = 3 \times a^4 \times 5 \times a^3$
$= 3 \times 5 \times a^4 \times a^3$ — Multiply numbers first, then the letters.
$= 15 \times a^{4+3}$
$= 15a^7$ — You can write the answer down without the lines of working.

b $12b^5 \div 4b^3 = \frac{12 \times b^5}{4 \times b^3}$ — Divide the numbers first, then the letters.
$= \frac{12}{4} \times \frac{b^5}{b^3}$ — $12 \div 4 = 3$, $b^5 \div b^3 = b^2$
$= 3 \times b^2$
$= 3b^2$

c $3xy^2 \times 5xy = 3 \times x \times y^2 \times 5 \times x \times y$ — Deal with the numbers first, then each of the different letters in turn. Your answer does not need to show the intermediate lines of working.
$= 3 \times 5 \times x \times x \times y^2 \times y$
$= 15 \times x^2 \times y^3$
$= 15x^2y^3$

d $\frac{10a^3b^2}{2a^2b^2} = \frac{10 \times a^3 \times b^2}{2 \times a^2 \times b^2}$ — Deal with the numbers first, then each of the different letters in turn.
$= \frac{10}{2} \times \frac{a^3}{a^2} \times \frac{b^2}{b^2}$ — $\frac{a^3}{a^2} = a^{3-2} = a^1$, $b^2 = b^{2-2} = b^0 = 1$
$= 5 \times a^1 \times 1$
$= 5a$

Exercise 15D

C

1 Simplify

a $2h^3 \times 3h^4$ b $5e^4 \times 3e^5$ c $4g^2 \times g^6$ d $3r \times 3r^4$ e $6e^3 \times 4e$

2 Simplify

a $5x^5 \div x^2$ b $6y^6 \div 2y^2$ c $12r^4 \div 4r^3$ d $2s^7 \div 8s^5$ e $8t^2 \div 4t$

3 Simplify

a $\frac{a^2 \times 3a^2}{a}$ b $\frac{2r \times 3r^3}{r^2}$ c $\frac{4r \times 3r^3}{6r^2}$

d $\frac{4 \times e^3 \times 5e^3}{10e^4}$ e $\frac{b^2 \times 4b^3}{b^5}$ f $\frac{2x^3 \times 6x^4}{x \times 4x^2}$

B

4 Simplify

a $4ab^2 \times 2ab$ b $3f^2g \times 7f^3g^2$ c $5xy^2 \times 4x^3y$

d $6p^4q^3 \times 6p^2q^2$ e $7r^2s \times 4s^5 \times 2r$ f $3c^3 \times 5cd \times 5d^4$

5 Simplify

a $ab^2 \div a$ b $9f^2g \div 3f^2$ c $8xy^2 \div 4x$

d $6p^4q^3 \div 2p^2$ e $12r^2s^3 \div 4s^2$ f $7c^3d^4 \div 7d^3$

6 Simplify

a $\frac{6x^3y^2}{3x^2y^2}$ b $\frac{4j^2k^4}{jk^2}$ c $\frac{6a^2b^3}{6a^2b}$ d $\frac{9m^4n^3}{3mn^3}$ e $\frac{16p^3q^4}{4p^3q^2}$

7 Match each question to an answer from the box.

a $3a^3b \times 2ab^2$ b $a^2 \times 6b \times ab$ c $12a^4b^4 \div 2a^2b$

d $6a^2b^3 \div 3ab$ e $2a^3b^2 \times 6ab^2$ f $8a^4b^3 \div 4a^2b$

$6a^3b^2$	$2ab^2$	$6a^2b^3$	$12a^4b^4$	$2a^2b^2$	$6a^4b^3$

15.3 Writing your own formulae

Keywords
formula

Why learn this?

Organisations often create formulae to use for allocating funding fairly. The money given to run your school may have been decided by a funding formulae.

Objectives

E D Use algebra to write formulae in different situations

Skills check

Write an algebraic expression for

1 d multiplied by 16

2 the total cost of 6 applies at c pence each.

Writing your own formula

A **formula** is a general rule that shows the relationship between quantities.

The quantities in a formula can vary in size. These quantities are called variables.

A formula must contain an equals sign. For example, the formula for the area of a rectangle is $A = lw$, not just the expression lw.

Example 5

Alex buys x packets of sweets.
Each packet of sweets costs 45 pence.
Alex pays with a £5 note.
Write a formula for the change in pence, C, that Alex should receive.

$C = 500 - 45x$

Remember £5 = 500p. The sweets cost 45p per packet so the cost in pence for x packets is $45x$.

D

Exercise 15E

E

1 Nilesh buys y packets of sweets.
Each packet of sweets costs 48 pence.
Write a formula for the total cost in pence, t.

D

2 Apples cost r pence each and bananas cost s pence each.
Sam buys 7 apples and 5 bananas.
Write a formula for the total cost in pence, t, of these fruit.

3 To roast a chicken you allow 45 minutes per kg and then a further 20 minutes.
Write a formula for the time in minutes, t, to roast a chicken that weighs w kg.

4 To roast lamb you allow 30 minutes plus a further 65 minutes per kg.
Write a formula for the time in minutes, t, to roast a joint of lamb that weighs w kg.

5 A rectangle has length $3x + 1$ and width $x + 2$.
Write a formula for the perimeter, p, of this rectangle.

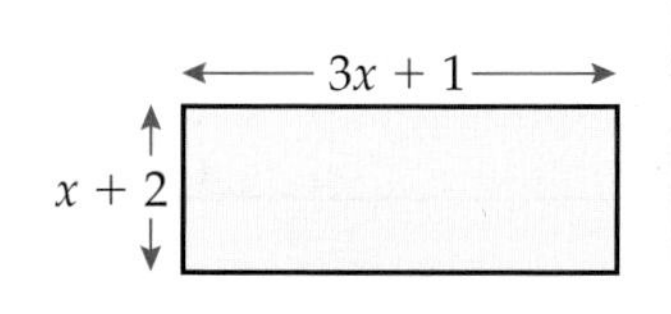

15.4 Substituting into expressions

Keywords
substitute, evaluate

Why learn this?

This topic often comes up in the exam.

Objectives

E Substitute numbers to work out the value of simple algebraic expressions

D **C** Substitute numbers into expressions involving brackets and powers

Skills check

Work out

1 $3 \times 7 - 5$

2 $4 \times 6 + 3 \times (-5)$

3 $3 \times 1.5 + 2 \times (-3)$

4 $5 \times 3^2 - 7$

5 $4^3 - 15$

6 $\dfrac{5 \times (-4)^2}{8}$

Substituting into expressions

You can **substitute** numbers for the variables in an expression. You replace each letter with a number.

This is called **evaluating** the expression.

Use the correct order of operations to do the calculations.

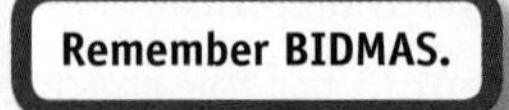

Use the rules for adding, subtracting, multiplying and dividing negative numbers.

E

Example 6

When $x = 5$, $y = \frac{1}{2}$ and $z = -2$, work out the value of these expressions.

a $3x + 4y$ **b** $6x + 2z$ **c** $4x - 6y + 3z$

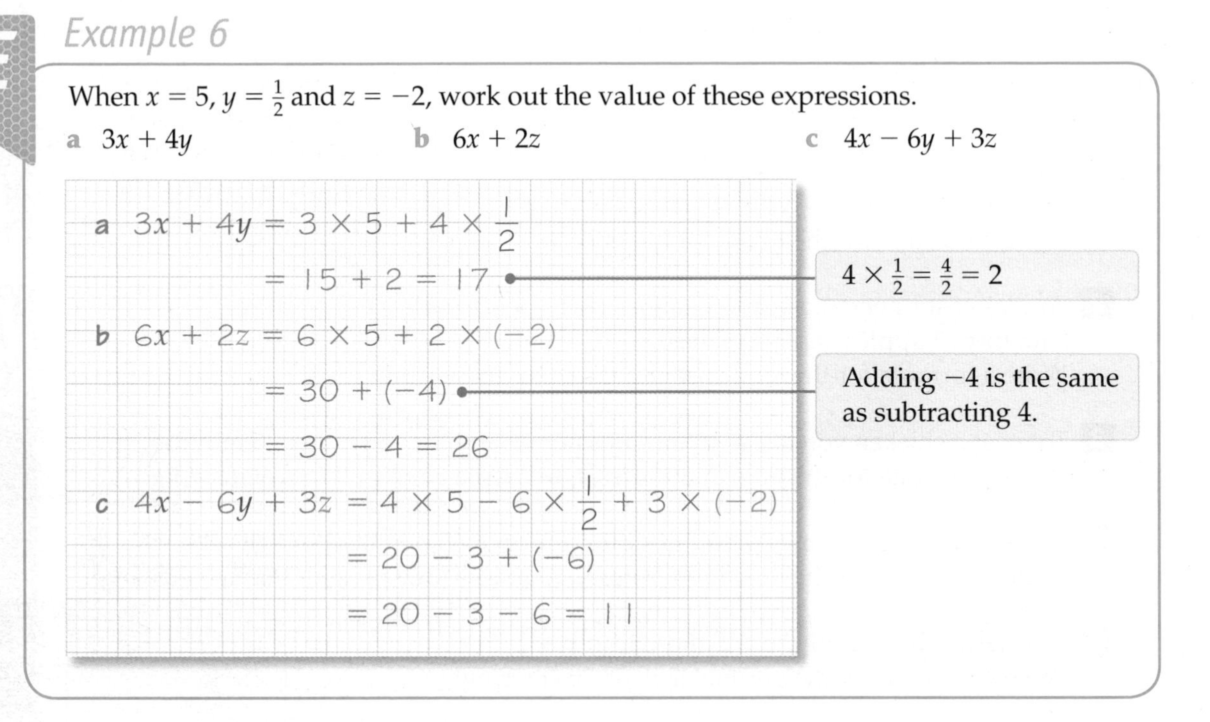

E

Exercise 15F

1 Work out the values of these expressions when $a = 4$, $b = 2$ and $c = -3$.

a $ab + c$ **b** $ac - b$ **c** $ab + ac$

d $bc - 2b$ **e** abc **f** $4ca - 7$

g $8a + 5bc$ **h** $6a - ac$ **i** $5ab - 2bc$

2 When $x = 5$, $y = \frac{1}{2}$ and $z = -2$, work out the value of these expressions.

a $3x + 10y$ **b** $3x + 5y$ **c** $4x + 3z$

d $12y + 2z$ **e** $7x - 8y + 4z$ **f** $4x - 5y + 2z$

3 When $f = 6$, $g = 1.5$ and $h = -3$, work out the value of these expressions.

a $4g + h$ **b** $2f - 5g$ **c** $fg + 8g + 3h$

Substituting into more complicated expressions

You can substitute values into expressions involving brackets and powers (indices).

Example 7

Evaluate these expressions when $a = 5$, $b = 4$ and $c = 3$.

a $\frac{a+3}{2}$

b $\frac{5c+1}{b}$

c $3b^2 - 1$

d $f^3 + f$ when $f = 3$

a $\frac{a+3}{2} = \frac{5+3}{2} = \frac{8}{2} = 4$

$\frac{8}{2} = 8 \div 2$

b $\frac{5c+1}{b} = (5 \times 3 + 1) \div 4$

$= (15 + 1) \div 4 = 16 \div 4 = 4$

c $3b^2 - 1 = 3 \times 4^2 - 1$

$= 3 \times 16 - 1 = 48 - 1 = 47$

Remember the order of operations: indices ($4^2 = 16$), then multiplication (3×16), then subtraction ($48 - 1$).

d $f^3 + f = 3^3 + 3 = 27 + 3 = 30$

Exercise 15G

E

1 When $r = 5$, $s = 4$ and $t = 3$, work out the value of these expressions.

a $\frac{r+3}{2}$

b $\frac{s+5}{3}$

c $\frac{t+7}{2}$

d $4(5s + 1)$

Order of operations: brackets first.

e $t(r + s)$

f $5(2s - 3t)$

D

g $\frac{5t+1}{s}$

h $\frac{4r-2}{t}$

i $\frac{3s+t}{r}$

j $3r^2 + 1$

$3r^2 = 3 \times r^2 = 3 \times r \times r$

k $4t^2 - 6$

l $2s^2 + r$

D

2 When $a = 5$, $b = 1.5$ and $c = -2$, work out the value of these expressions.

a $3a^2 + b$

b $c^2 + a$

c $2a^3 + b$

$2a^3 = 2 \times a^3$

d $10b + 2a^2$

e $4c^2 - a + b$

f $c^3 - b + a$

3 Copy and complete this table.

x	2	3	4	5	6
$x^3 - x$			60		

$4^3 - 4 = 64 - 4 = 60$

4 When $A = 6$, $B = -4$, $C = 3$ and $D = 30$, work out the value of these expressions.

a $D(B + 7)$

b $A(B + 1)$

c $\frac{4B+D}{2}$

d $\frac{2A+3}{C}$

e $A^2 + 2B + C$

f $\frac{A^2+3B}{C}$

C

5 Evaluate the following when $d = 5$, $e = -2$ and $f = 3$.

a $f^2(2d + 3e)$

b $e(d^2 - 2f^2)$

c $\sqrt{d^2 - f^2}$

d $\frac{d+f^2}{d-e}$

e $\frac{\sqrt{4e^2+f^2}}{d}$

f $\frac{\sqrt{2df-3e}}{e}$

15.5 Substituting into formulae

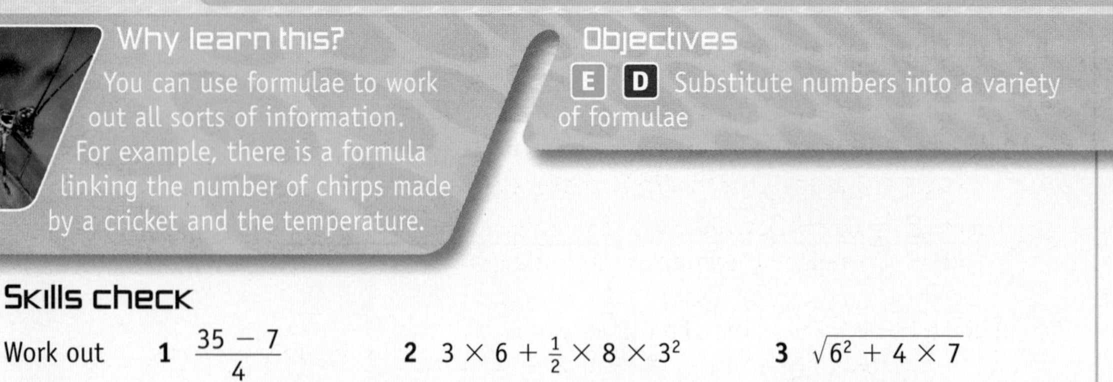

Why learn this?

You can use formulae to work out all sorts of information. For example, there is a formula linking the number of chirps made by a cricket and the temperature.

Objectives

E D Substitute numbers into a variety of formulae

Skills check

Work out **1** $\frac{35 - 7}{4}$ **2** $3 \times 6 + \frac{1}{2} \times 8 \times 3^2$ **3** $\sqrt{6^2 + 4 \times 7}$

Substituting into formulae

You can substitute values into a formula to work out the value of a variable.

E

Example 8

a A formula for working out acceleration is $a = \frac{v - u}{t}$
where v is the final velocity, u is the initial velocity and t is the time taken.
Work out the value of a when $v = 50$, $u = 10$ and $t = 8$.

D

b A formula for working out distance travelled is
$$s = ut + \tfrac{1}{2}at^2$$
where u is the initial velocity, a is the acceleration and t is the time taken.
Work out the value of s when $u = 3$, $a = 8$ and $t = 5$.

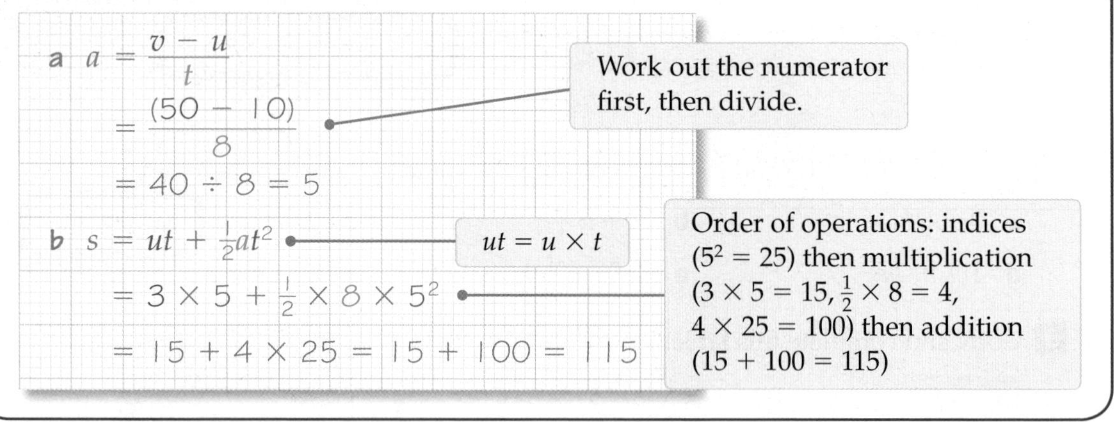

a $a = \frac{v - u}{t}$

$= \frac{(50 - 10)}{8}$ — Work out the numerator first, then divide.

$= 40 \div 8 = 5$

b $s = ut + \frac{1}{2}at^2$ — $ut = u \times t$

$= 3 \times 5 + \frac{1}{2} \times 8 \times 5^2$ — Order of operations: indices ($5^2 = 25$) then multiplication ($3 \times 5 = 15$, $\frac{1}{2} \times 8 = 4$, $4 \times 25 = 100$) then addition ($15 + 100 = 115$)

$= 15 + 4 \times 25 = 15 + 100 = 115$

Exercise 15H

E

1 Use the formula $a = \frac{v - u}{t}$ to work out the value of a when

a $v = 15, u = 3, t = 2$ b $v = 29, u = 5, t = 6$ c $v = 25, u = 7, t = 3$

2 *Talkalot* calculate telephone bills using this formula:
$$C = \frac{n}{10} + 10$$
where C is the total cost (in £) and n is the number of calls.
Work out the telephone bill for

a 80 calls b 160 calls c 350 calls.

3 My water company calculates water bills each quarter using this formula:

$$A = 3V + 5$$

where A is the amount to pay (in £) and V is the volume of water used (in m^3).
In January I used $10\,m^3$, in February I used $11\,m^3$ and in March I used $8\,m^3$.

The first quarter is January, February and March.

a Work out my bill for the first quarter of the year.

b How many m^3 of water would I need to use to make my bill for the quarter more than £100?

E

AO2

4 The formula for the area of a trapezium is

$$A = \frac{1}{2}(a + b)h$$

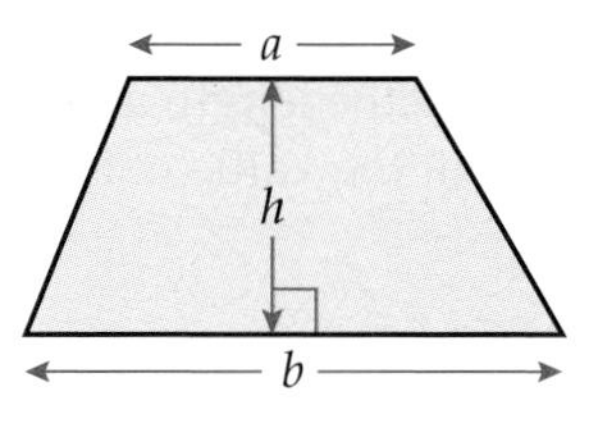

Work out the value of A when

a $a = 6, b = 10, h = 4$

b $a = 9, b = 13, h = 8$

c $a = 6, b = 9, h = 4$

E

5 Use the formula $s = ut + \frac{1}{2}at^2$ to work out the value of s when

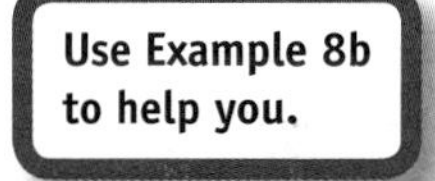

a $u = 3, a = 10, t = 2$ **b** $u = 2.5, a = 5, t = 4$

c $u = -4, a = 8, t = 3$

D

6 A formula for working out the velocity of a car is

$$v = \sqrt{u^2 + 2as}$$

where u is the initial velocity, a is the acceleration and s is the distance travelled.
Work out the value of v when

a $u = 3, a = 4, s = 5$ **b** $u = 9, a = 10, s = 2$ **c** $u = 7, a = 4, s = 15$

7 An approximate formula for the surface area of a cone, including the base, is

$$A = 3r(r + l)$$

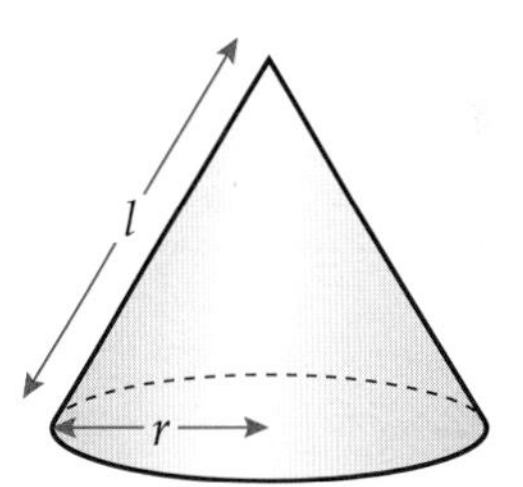

where r is the radius and l is the slant height.
Work out the approximate surface area of a cone with dimensions

a $r = 2\,cm, l = 13\,cm$ **b** $r = 4\,cm, l = 10\,cm$

8 An approximate formula for the total surface area of a cylinder is

$$A = 6r(r + h)$$

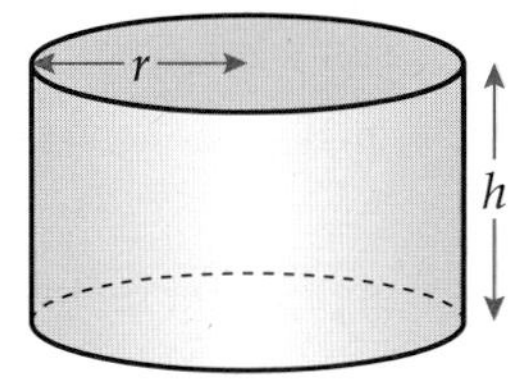

where r is the radius and h is the height.
Work out the approximate surface area of a cylinder with dimensions

a $r = 3\,cm, h = 10\,cm$ **b** $r = 4\,cm, h = 5\,cm$

9 Use the formula $r = \sqrt{\frac{A}{3}}$ to work out the approximate radius, r, of a circle with area $A = 27$.

10 Use the formula $r = \sqrt{\frac{V}{3h}}$ to calculate the approximate radius of a cylinder of height, $h = 2.5\,cm$ and volume, $V = 120\,cm^3$.

Sometimes the value you want is not on the left of the equals sign. Substitute the values you know into the formula. Then solve the equation to find the value of the unknown.

E

D

Example 9

a A formula for working out the perimeter of a regular hexagon is $P = 6x$, where x is the length of each side.
Work out the value of x when $P = 48$.

b The perimeter of a rectangle is given by $P = 2l + 2w$, where l is the length and w is the width.
Work out the value of l when $P = 24$ and $w = 5$.

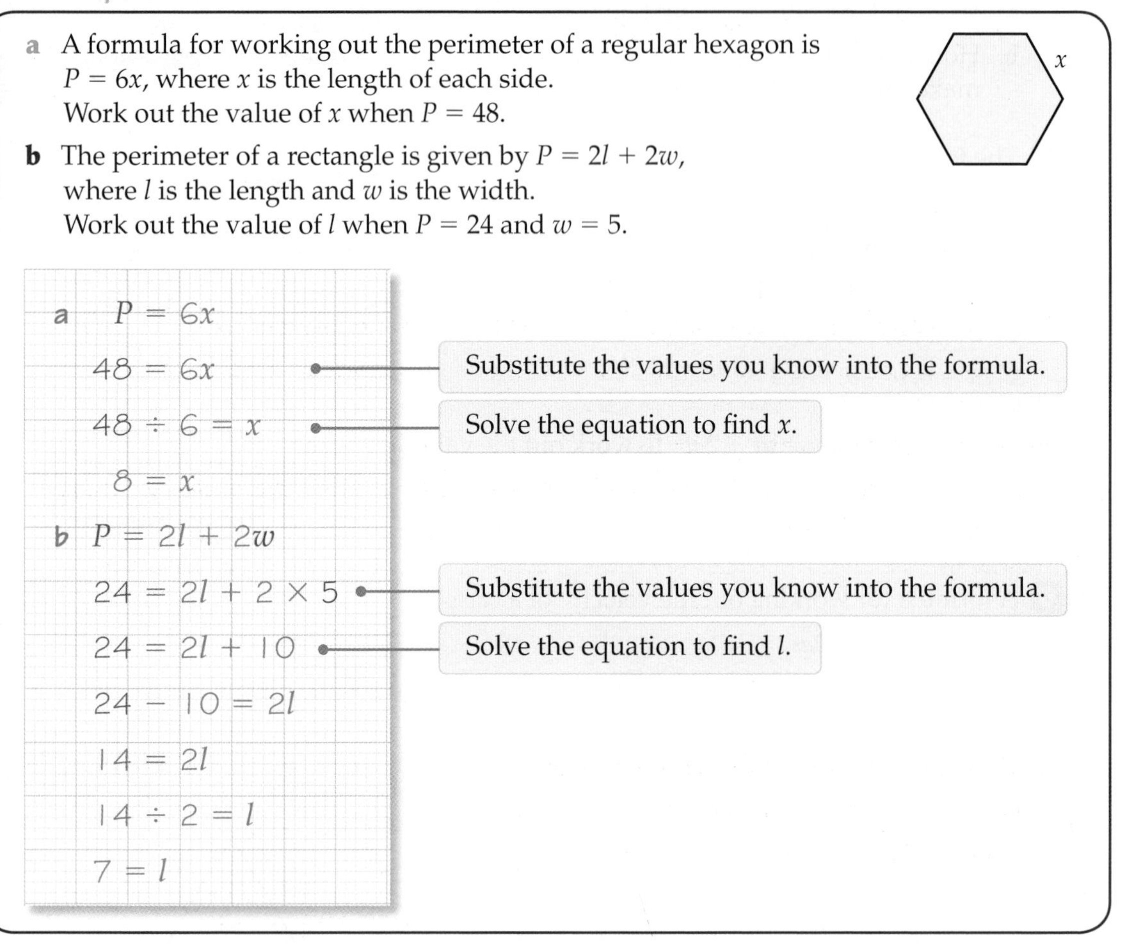

Exercise 15I

E

1 A formula for working out the perimeter of a regular hexagon is $P = 6l$, where l is the length of each side.

Work out the value of l when

a $P = 60$ b $P = 30$ c $P = 120$

D

2 The perimeter of a rectangle is given by $P = 2l + 2w$, where l is the length and w is the width.

Use the formula to find

a l when $P = 18$ and $w = 4$ **b** l when $P = 32$ and $w = 7$
c w when $P = 60$ and $l = 17$ **d** w when $P = 50$ and $l = 13.5$

3 Use the formula $v = u + at$ to find

a u when $v = 30$, $a = 8$ and $t = 3$ **b** u when $v = 47$, $a = 4$ and $t = 9$
c a when $v = 54$, $u = 19$ and $t = 7$ **d** t when $v = 60$, $u = 15$ and $a = 5$
e u when $v = 20$, $a = 7$ and $t = 4$ **f** a when $v = 30$, $u = -12$ and $t = 6$

15.6 Changing the subject of a formula

Why learn this?
In physics, you have to be able to use lots of different formulae confidently.

Objectives
C B Rearrange a formula to make a different variable the subject of the formula

Keywords
subject, rearrange

Skills check

Solve

1 $15 = 9 + x$ **2** $42 = 6t$

3 $23 = 2d + 9$ **4** $11 = \frac{1}{2}m - 5$

Rearranging formulae

In the formula $v = u + at$ the variable v is called the **subject** of the formula.

The subject of a formula is always the letter on its own on one side of the equation. This letter only appears once in the formula.

For example, P is the subject of the formula $P = 2l + 2w$.

You can **rearrange** a formula to make a different variable the subject. This is called 'changing the subject' of the formula.

You can use the same techniques to rearrange a formula as when you solve an equation.

Example 10

C

a Rearrange $d = a + 8$ to make a the subject.

b Rearrange $A = lw$ to make l the subject.

c Make x the subject of the formula $y = 5x - 2$.

a $d = a + 8$ — You need to end up with a on its own on one side.

$d - 8 = a + 8 - 8$ — Subtract 8 from both sides as when solving an equation.

$d - 8 = a$

$a = d - 8$

b $A = lw$ — lw means $l \times w$

$\frac{A}{w} = \frac{lw}{w}$ — l is multiplied by w so divide both sides by w to leave l on its own.

$\frac{A}{w} = l$ or $l = \frac{A}{w}$

c $y = 5x - 2$

$y + 2 = 5x - 2 + 2$ — Add 2 to both sides to get $5x$ on its own.

$y + 2 = 5x$

$\frac{y + 2}{5} = \frac{5x}{5}$ — Now divide both sides by 5 to leave x on its own.

$\frac{y + 2}{5} = x$ or $x = \frac{y + 2}{5}$

Exercise 15J

C

1 Rearrange each of these formulae to make a the subject.

a $c = a + 5$ **b** $k = a - 6$ **c** $w = a - 7$

2 Rearrange each of these formulae to make w the subject.

a $P = 4w$ **b** $A = lw$ **c** $h = kw$

3 Make x the subject of each of these formulae.

a $y = 5x - 6$ **b** $y = 2x + 1$ **c** $y = 6x + 5$

4 Make r the subject of each of these formulae.

a $p = 4r + 2t$ **b** $w = 3r - 2s$ **c** $y = 6r - 5p$

5 Rearrange each of the formulae to make a the subject.

a $b = \frac{1}{2}a + 6$ $\frac{1}{2}a = \frac{a}{2}$ **b** $b = \frac{1}{2}a + 7$ **c** $b = \frac{1}{3}a - 1$

d $b = \frac{1}{4}a - 3$ **e** $b = 2(a + 1)$ **f** $b = 3(a - 5)$

6 A formula used to calculate velocity is

$$v = u + at$$

where u is the initial velocity, a is the acceleration and t is the time taken.

a Rearrange the formula to make a the subject.

b Rearrange the formula to make t the subject.

Example 11

C B

a Rearrange $P = 4g + 2h$ to make g the subject.

b Rearrange $V = \sqrt{w + y}$ to make w the subject.

c Rearrange $p = \frac{q^2}{r} - s$ to make q the subject.

a

$P = 4g + 2h$

$P - 2h = 4g$ — Subtract $2h$ from both sides.

$\frac{P - 2h}{4} = g$ — Divide both sides by 4.

b

$V = \sqrt{w + y}$

$V^2 = w + y$ — Square both sides.

$V^2 - y = w$ — Subtract y from both sides.

c

$p = \frac{q^2}{r} - s$

$p + s = \frac{q^2}{r}$ — Add s to both sides to leave the term with q on its own.

$r(p + s) = q^2$ — Multiply both sides by r.

$\pm\sqrt{r(p + s)} = q$ — Square root both sides.

Exercise 15K

1 Make y the subject of each of these formulae.

a $3x + 2y = 10$ **b** $6x + 3y = 8$ **c** $2x - 3y = 2$

C

2 Rearrange each of these formulae to make x the subject.

a $2(x - y) = y + 5$ **b** $z = \frac{x}{y} - 5$ **c** $3p = \frac{2x}{q} - s$

3 Rearrange each of these formulae to make w the subject.

a $K = \sqrt{w + t}$ **b** $A = \sqrt{w - a}$ **c** $h = 2\sqrt{w} + l$

4 Rearrange each of these formulae to make r the subject.

a $t = \frac{r^2}{g} - m$ **b** $h = \frac{r^2}{4} + 3a$ **c** $V = \frac{1}{3}\pi r^2 h$

5 An approximate formula for the total surface area of a cylinder is

$$A = 6r(r + h)$$

where r is the radius and h is the height.
Rearrange the formula to make h the subject.

6 An approximate formula for the period of a pendulum is

$$T = 6\sqrt{\frac{l}{g}}$$

where T is the time for one complete swing, l is the length and g is a constant.
Rearrange the formula to make l the subject.

B

Review exercise

1 **a** Find the value of $3x + 4y$ when $x = -2$ and $y = 5$. **[2 marks]**

b Find the value of $5r + 4t$ when $r = 6$ and $t = -3$. **[2 marks]**

c Find the value of $g^3 + h^2$ when $g = 2$ and $h = 5$. **[2 marks]**

2 An approximate rule for converting degrees Fahrenheit into degrees Centigrade is

$$C = \frac{F - 30}{2}$$

Use this rule to convert 68°F into °C. **[2 marks]**

3 **a** Use the formula $a = 5b + 2c$ to work out a when $b = 3$ and $c = -4$. **[2 marks]**

b Use the formula $a = 5b + 3c$ to work out c when $a = 16$ and $b = 2$. **[3 marks]**

4 Simplify

a $t \times t \times t \times t \times t$ **[1 mark]**

b $m \times m \times m$ **[1 mark]**

E

5 Simplify

a $3n \times 4n \times n$ **[1 mark]**

b $2s \times 4s \times 5s \times s$ **[1 mark]**

6 Use the formula $s = \frac{1}{2}(u + v)t$ to find t when $s = 27$, $u = 5$ and $v = 13$. **[3 marks]**

D

C

7 Simplify

a $x^3 \times x^6$ **[1 mark]**

b $r^4 \div r^3$ **[1 mark]**

c $\frac{z \times z^6}{z^5}$ **[1 mark]**

d $5n^4 \times 3n^2$ **[1 mark]**

e $10a^4 \div 2a^2$ **[1 mark]**

8 Make r the subject of the formula $p = 3 + 2r$. **[2 marks]**

9 Rearrange $y = 4x - 1$ to make x the subject. **[2 marks]**

C B

10 Make x the subject of

a $w = \frac{1}{2}x - y$ **[2 marks]**

b $w = x^2 + y$ **[2 marks]**

B

11 Simplify

a $(p^2)^5$ **[1 mark]**

b $5a^2b^4 \times 3ab^3$ **[2 marks]**

c $\frac{12a^3b^4}{3a^3b^2}$ **[2 marks]**

12 Rearrange $7(n - p) = 3p + 4$ to make n the subject. **[3 marks]**

Chapter summary

In this chapter you have learned how to

- substitute numbers to work out the value of simple algebraic expressions E
- use algebra to write formulae in different situations E D
- substitute numbers into a variety of formulae E D
- use index notation in algebra E D C
- substitute numbers into expressions involving brackets and powers D C
- use index notation when multiplying or dividing algebraic terms D C
- use index laws to multiply and divide powers in algebra C
- raise a number or variable to the power of 1 or 0 C B
- rearrange a formula to make a different variable the subject of the formula C B
- use index laws for raising a power to another power B

16 Percentages

This chapter is about percentages.

If you don't pay off your credit card bill each month, you will be charged a high percentage interest on the amount you owe.

Objectives

This chapter will show you how to

- perform calculations involving credit E
- perform simple interest calculations E
- calculate a percentage increase or decrease D
- perform calculations involving VAT D
- calculate a percentage profit or loss C
- perform calculations involving a repeated percentage change C
- perform calculations involving finding the original quantity B

Before you start this chapter

Put your calculator away!

1 Work out

a 0.27×100 b $42.5 \div 100$

2 Work out a $5 \div 8$ b $4 \div 25$

3 Cancel $\frac{15}{75}$ to its lowest terms.

4 Convert $\frac{2}{5}$ to a decimal.

5 Convert 0.36 to a fraction.

6 Work out $\frac{17}{200} \times 100$.

P Number skills: fractions, decimals and percentages

1 a Convert 57% to a decimal.
b Convert 0.008 to a percentage.
c Convert 65% to a fraction.
d Convert $\frac{5}{200}$ to a percentage.

2 Write 0.25, $\frac{1}{5}$ and 30% in order of size, starting with the smallest.

3 Write 0.3, $\frac{1}{3}$ and 32% in order of size, starting with the smallest.

4 Ava scored 15 out of 20 in a test. Ben scored 80% in the same test.
Who got more questions correct?

5 Copy and complete this table of equivalent fractions, decimals and percentages.

Percentage	Fraction	Decimal
35%		
	$\frac{1}{20}$	
		0.14

Number skills: using percentages in calculations

1 Work out
a 4% of 56 kg
b 75% of £800
c 6% of 30 litres
d 2% of £6000

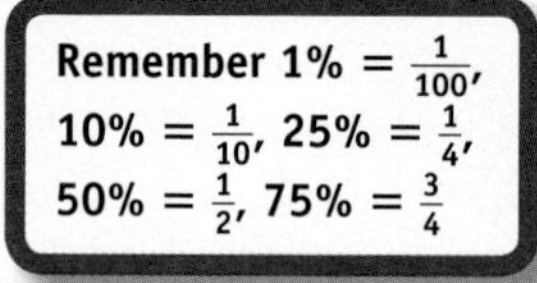

2 Ali earns £27 600 per year and pays 21% of this in income tax.
How much income tax does she pay?

3 In Market Street there are 160 houses. 85% of them have broadband.
How many houses in the street do **not** have broadband?

4 Work out

HELP Section 2.5

a 6 as a percentage of 10
b 13 as a percentage of 20
c 48 as percentage of 60
d 54 as a percentage of 300
e 6 mm as a percentage of 2 cm
f 380 g as a percentage of 0.4 kg

Make sure the units of both quantities are the same.

5 A school has 300 students in Years 10 and 11.
Use the information in the table to work out what percentage of the students are
a Year 10 boys
b Year 11 girls
c Year 10
d boys.

	Boys	Girls
Year 10	72	69
Year 11	75	84

16.1 Percentage increase and decrease

Keywords
original amount, percentage increase, percentage decrease, discount, reduce

Why learn this?

Percentage increase calculations can tell you how much better off you would be if you got a 3% pay rise.

Objectives

D Calculate a percentage increase or decrease

Skills check

1 Convert 83% to a decimal.

HELP Section 2.4

2 Convert 0.38 to a percentage.

3 Convert 45% to a fraction in its simplest form.

4 Convert $\frac{1}{8}$ to a percentage.

Percentage increase and decrease

In Section 2.6 you learned how to work out percentage increase and decrease using these methods.

Method A

1 Work out the actual increase (or decrease).

2 Add to (or subtract from) the **original amount.**

Method B

1 Add the **percentage increase** to 100% (or subtract the **percentage decrease** from 100%).

2 Convert this percentage to a decimal.

3 Multiply it by the original amount.

Here you will practise working out percentage increase and decrease *without* a calculator.

You will probably find Method A more useful.

D

Example 1

Tim used to earn £460 a week. He has had a 4% pay rise.

What does he earn now?

Method A

Pay rise = 4% of £460

$= \frac{4}{100} \times 460$

$= \frac{1840}{100}$

= £18.40

New pay = £460 + £18.40

= £478.40

Method B

Increase = 4%

New pay = 100% + 4% of old pay

= 104% of old pay

= 1.04 × old pay

= 1.04 × 460

= £478.40

$$\begin{array}{r} 460 \\ \times\ 104 \\ \hline 1840 \\ 46000 \\ \hline 47840 \end{array}$$

Divide by 100 to convert a percentage to a decimal.

D

Example 2

The price of a T-shirt is reduced by 20% in a sale. The original price was £15. What is the sale price?

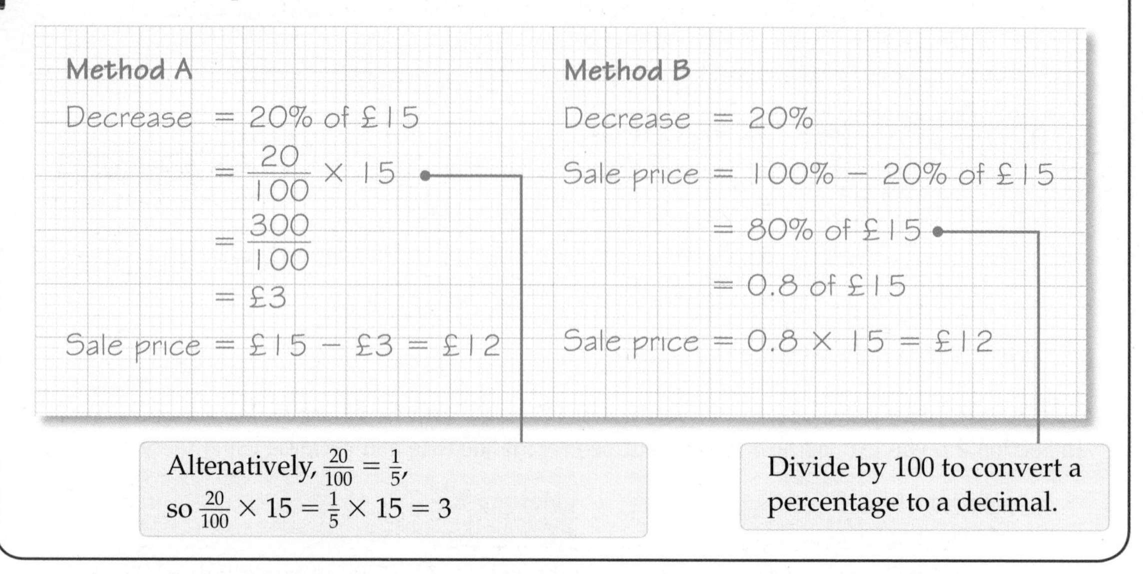

Altenatively, $\frac{20}{100} = \frac{1}{5}$, so $\frac{20}{100} \times 15 = \frac{1}{5} \times 15 = 3$

Divide by 100 to convert a percentage to a decimal.

Exercise 16A

D

1 Increase these quantities by 10%.

a £200 **b** 120 m **c** 4 g **d** 17 litres

2 Decrease these quantities by 5%.

a 20 feet **b** £40 **c** 5 km **d** £7600

3 **a** Increase £95 by 10%. **b** Increase 42 mm by 5%.
c Decrease £160 by 10%. **d** Increase £300 by 7%.
e Decrease 75 m*l* by 8%. **f** Decrease 50 miles by 12%.

4 Sanjeev wants to buy a CD priced at £12.
He gets a 10% discount with his student card.
How much does he have to pay?

A discount is an amount taken off the price. Work out a percentage discount just like a percentage decrease.

5 Nigel is fitting new skirting board around his living room.
He measures the total length he needs as 17 m.
He decides to buy 10% more than this to allow for cutting and wastage.
How much skirting board does he buy?

6 Jan books a flight from London to Rome.
The flight costs £49, but Jan gets a 10% discount as a frequent flier.
How much does she have to pay?

7 Nita earns £285.00 per week. She gets a 4% pay rise.
How much will she earn now?

8 A new car costs £9600. After two years it has lost 40% of its value.
What is it worth now?

9 In February 2009 the price of a barrel of crude oil was $40.
In June 2009 the price was 80% higher than this.
What was the price in June 2009?

10 I pay £1.25 for my bus ticket to work.
The bus company has just announced a 4% increase in fares.
What will my new fare be?

> Work in pence.

D

11 The local cricket club had a membership drive and increased the number of members by 20%.
There used to be 85 members.
How many members are there now?

12 A pair of jeans priced at £35 is reduced by 15% in a sale.
What is the sale price?

> The word *reduced* tells you that this question is about a percentage *decrease*.

13 In April 2008, 175 000 new cars were sold in the UK.
In April 2009, the figure was 25% down on the previous year.
How many cars were sold in April 2009?

D

14 A sum of £200 is increased by 10%, and then the new amount is decreased by 10%.
Will the final amount be greater or less than the original £200?

A02

15 Simon sees three different adverts for the same pressure washer.

Dumbo's DIY	Rock Bottom	*Suit you, Sir*
Pressure washer	**Pressure washer**	**Pressure washer**
Normally £99	**£79**	Normally £120
20% off	*Unbeatable value*	35% off

Which is the best buy?

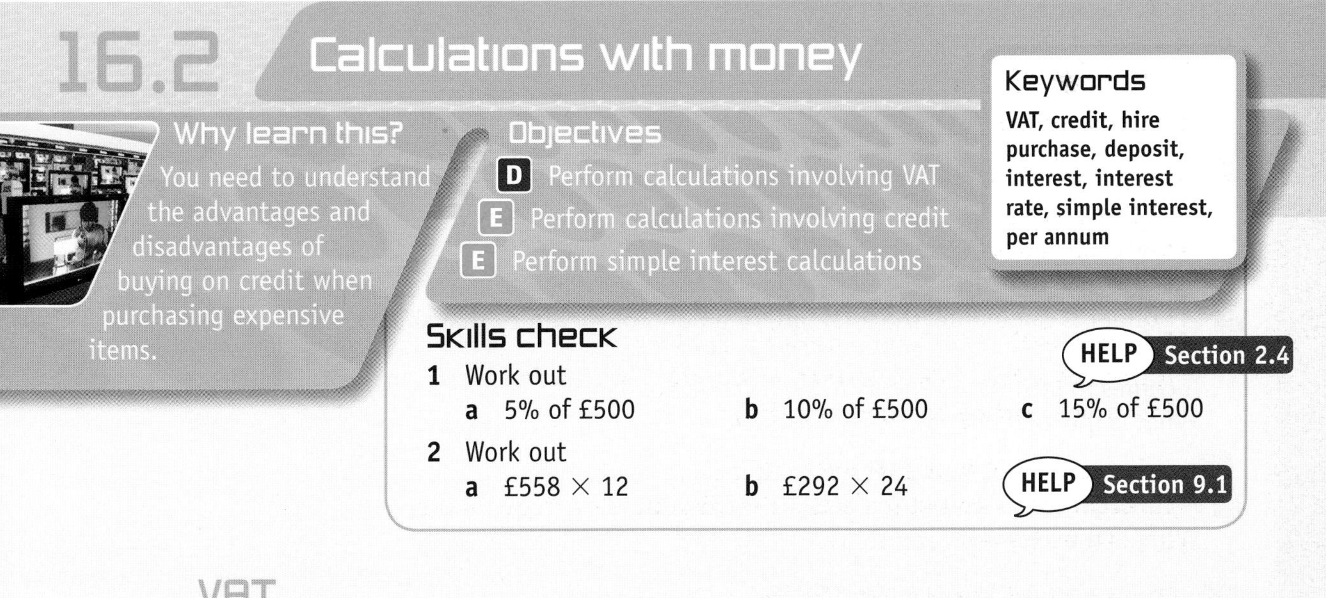

16.2 Calculations with money

Why learn this?
You need to understand the advantages and disadvantages of buying on credit when purchasing expensive items.

Objectives

- **D** Perform calculations involving VAT
- **E** Perform calculations involving credit
- **E** Perform simple interest calculations

Keywords
VAT, credit, hire purchase, deposit, interest, interest rate, simple interest, per annum

Skills check

1. Work out
   a 5% of £500 b 10% of £500 c 15% of £500

HELP Section 2.4

2. Work out
   a £558 × 12 b £292 × 24

HELP Section 9.1

VAT

VAT stands for Value Added Tax.

It is a tax that is added to the price of most items in shops and many other services.

VAT is calculated as a percentage.

In the UK it is generally 17.5%.

> VAT at 17.5% can be worked out by finding 10%
> then 5%
> then 2.5% +
> 17.5%

D

Example 3

A plasma screen television is advertised for sale at £800 + 17.5% VAT.

How much will you have to pay?

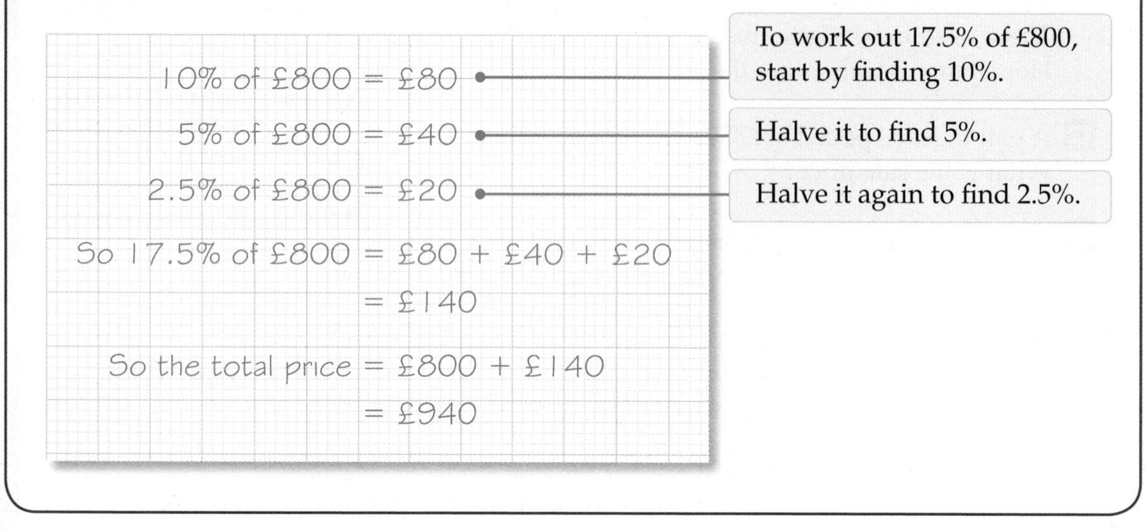

Exercise 16B

D

1 Work out the total price of these items.

a laptop, £400 + 17.5% VAT

b bag, £30 + 17.5% VAT

c mobile phone, £150 + 17.5% VAT

d skateboard, £48 + 17.5% VAT

2 A bill for 500 litres of domestic heating oil comes to £190 + 5% VAT.

a How much VAT will be charged?

b What is the total bill?

VAT for fuel is 5%.

3 A builder is calculating the total bill for a new conservatory that she has recently installed.
It comes to £15 500 + 17.5% VAT.
What is the total bill?

4 A child car seat is advertised at £110 + 5% VAT.
What is the total cost?

5 A car service comes to £280 + 17.5% VAT.
What is the total bill?

6 A meal for four people comes to £84.80 + 17.5% VAT.

a What is the total bill?

b The four people share the bill equally.
How much will each pay?

Credit

If you cannot afford the full price of an expensive item, you might buy it on **credit.**

This is sometimes called **hire purchase.**

When you buy on credit, you pay a **deposit** followed by a number of regular payments, usually monthly.

> **Buying on credit is usually more expensive than paying the cash price.**

Example 4

Mr Smith wants to buy a new five-door diesel saloon. He is offered two options.

Option A: Pay the cash price of £17 900

Option B: Pay a 20% deposit, then 36 monthly payments of £450

a How much is the deposit?

b What is the total of his monthly payments?

c What is the total credit price?

d What is the extra cost of buying on credit rather than paying the cash price?

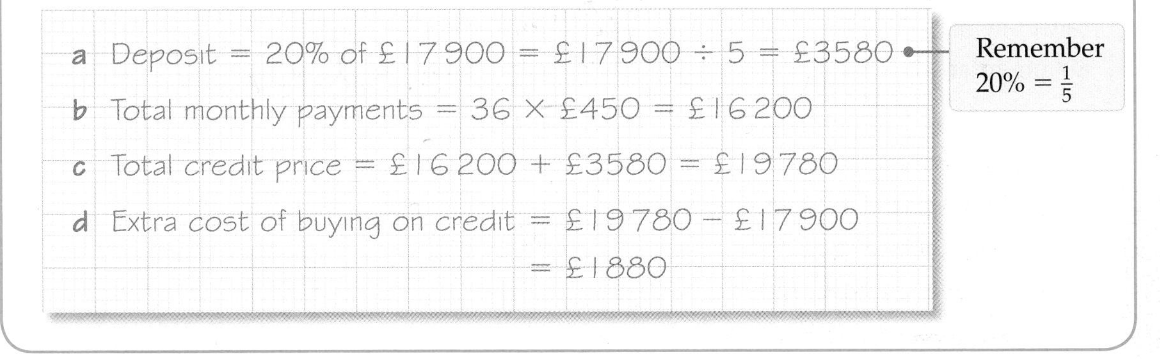

E

Exercise 16C

1 You can buy a motorbike in two ways.

Option A	**Option B**
£3500 cash price	10% deposit and 24 monthly payments of £180

a How much is the deposit?

b What is the total of the monthly payments?

c What is the total cost of buying the motorbike on credit (option B)?

d What is the extra cost of buying on credit (option B) rather than paying the cash price (option A)?

2 Angela wants to buy a washing machine. The cash price is £460.
She decides to buy it on credit by paying a 20% deposit and then 24 monthly payments of £17.

a How much is the deposit?

b What is the total of her monthly payments?

c What is the total credit price?

d How much more does it cost to buy on credit rather than pay the cash price?

E

E

3 Hazel buys a trombone with a cash price of £1380.
The credit terms are a deposit of 20% and 36 monthly payments of £35.

a What is the total credit price?

b How much extra is the credit price compared to the cash price?

E

4 A cycle shop is selling a mountain bike.
You can either pay a cash price of £550, or you can pay a 20% deposit plus 12 monthly instalments of £43.
What is the difference between the cash price and the credit price?

A02

5 The cash price for a black leather sofa is £700.
The credit terms for the same sofa are a 20% deposit plus 12 monthly payments of £56.
What is the difference between the cash price and the credit price?

Calculations involving interest

When you put money into a savings account in a bank or building society, you receive **interest** on your money.

The interest you receive is a percentage of the amount in your account. This percentage is called the **interest rate**.

Simple interest is when you receive the same amount of interest each year.

E

Example 5

Valda has £2300 in a savings account that pays 4% interest **per annum**.

How much simple interest does she receive in five years?

Per annum means 'each year'.

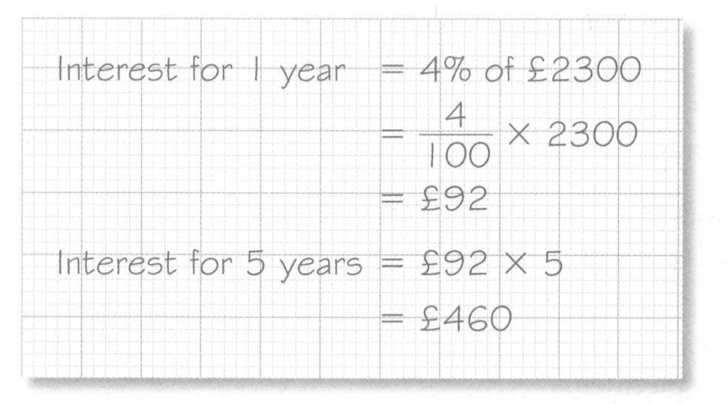

Exercise 16D

E

1 Find the simple interest when

a £600 is invested for three years at a rate of 10% per annum

b £400 is invested for two years at a rate of 5% per annum

c £150 is invested for four years at a rate of 3% per annum

d £1000 is invested for five years at a rate of 2% per annum.

2 Kim takes out a £4000 loan to buy a secondhand car. The interest is 15% per annum.
How much interest will she have to pay after one year?

3 Kylie puts £2000 into a savings account. The interest rate is 2.5% per annum.
Work out the total simple interest after four years.

E

4 George has £3500 to invest.
Grabbitall Bank offers 6% interest for the first year and then 3% simple interest per annum for the following years.
Bonus Buster Bank gives 4% simple interest per annum.
George wants to invest his money for four years.
Which bank will pay him more interest?

E **AO2**

16.3 Percentage profit or loss

Keywords
cost price, selling price, profit, loss, percentage profit, percentage loss, depreciation

Why learn this?

Shop managers use this to compare the profit made on different items.

Objectives

C Calculate a percentage profit or loss

Skills check

1 Work out

a 9 as a percentage of 20 **b** 72 as a percentage of 80

c 294 as a percentage of 300

d 1050 as a percentage of 1750

HELP Section 2.5

Profit and loss

A shopkeeper buys items from a wholesaler at **cost price.**

The shopkeeper sells the items at the **selling price.**

When you make money on the sale of an item, you make a **profit.**

When you lose money on the sale of an item, you make a **loss.**

You can use **percentage profit** (or **loss**) to compare the profitability of items.

$$\text{Percentage profit (or loss)} = \frac{\text{actual profit (or loss)}}{\text{cost price}} \times 100\%$$

where actual profit = selling price − cost price

and actual loss = cost price − selling price

Example 6

C

A DIY store buys 5-litre cans of white emulsion paint for £14.00 each and sells them for £17.50.

What is the store's percentage profit?

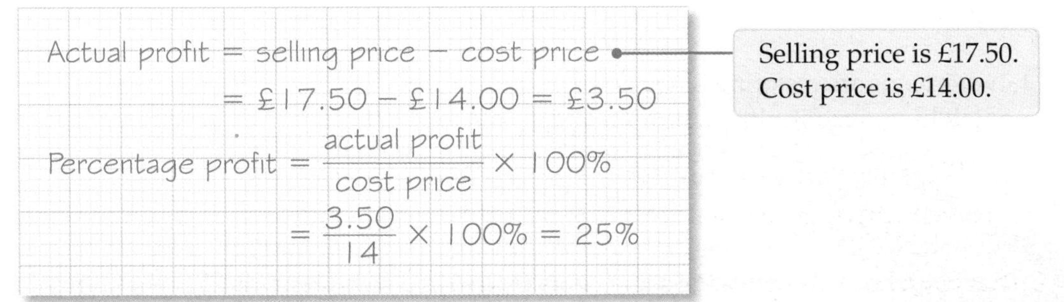

Selling price is £17.50.
Cost price is £14.00.

C

Example 7

Omar bought a car for £7500 and sold it two years later for £4500.

What was the percentage depreciation in the value of the car?

When objects lose value over time, the loss is called **depreciation.**

Actual loss = cost price − selling price
= £7500 − £4500
= £3000

Cost price is £7500.
Selling price is £4500.

$$\text{Percentage depreciation} = \frac{\text{actual loss}}{\text{cost price}} \times 100\%$$
$$= \frac{3000}{7500} \times 100\%$$
$$= \frac{2}{5} \times 100\% = 40\%$$

Exercise 16E

Remember to put the original price on the bottom of the fraction.

C

1 Find the percentage profit or loss for each item.

	Pair of trousers	House	Barbecue set	Car	Television
Cost price	£30	£125 000	£40	£8000	£350
Selling price	£42	£118 750	£60	£6800	£455

2 A collector bought an antique table for £360 and sold it for £420.
What was her percentage profit?

3 Phoebe bought a games console for £160 and later sold it to her friend for £140.
What was her percentage loss?

4 Colin restores lawnmowers.
He bought a petrol mower for £5 and sold it for £45.
What was his percentage profit?

5 Gill bought a clarinet for £600. Later she sold it for £420.
What was her percentage loss?

6 Lucy bought a house for £90 000.
Three years later, she sold it for £92 700.
What was her percentage profit?

7 Mandeep sold his skateboard for £21. He had paid £35 for it.
What was his percentage loss?

C

8 Two friends are restoring old cars.
Sarah buys one for £400, restores it and sells it for £750.
George buys one for £1200, restores it and sells it for £2000.
Which of them makes the bigger percentage profit?

A02

9 Gardens-r-Us buys wellington boots for £15 and sells them for £21.
Yuppies Shoe Shop buys the same boots for £16 and sells them for £22.
Which shop makes the larger percentage profit?

16.4 Repeated percentage change

Keywords
compound interest, multiplier

Why learn this?
Demographers use repeated percentage change to predict population growth.

Objectives
C Perform calculations involving repeated percentage changes

Skills check

1. Jon puts £3000 into a savings account.
   The interest rate is 4% per annum.
   Work out the simple interest after one year.
2. Lucy puts £2500 into a savings account.
   The interest rate is 5% per annum.
   Work out the total simple interest after four years.

HELP Section 16.2

Compound interest

Generally when you invest money, the interest is calculated using **compound interest**.

In Section 8.1 you learned how to work out repeated percentage change using a calculator. Here you will practise these calculations without a calculator.

Compound interest is interest paid on the amount *plus* on the interest already earned.

Example 8

C

Helen puts £1000 into a savings account earning 5% per annum compound interest.
How much does she have after two years?

Year 1

Amount at start of year 1 = £1000

Interest in year 1 = 5% of £1000
= 0.05 × £1000
= £50

Or 5% of £1000 = $\frac{5}{100}$ × £1000 = £50

Year 2

Amount at start of year 2 = £1000 + £50
= £1050

The interest is added to the amount in the account.

Interest in year 2 = 5% of £1050
= 0.05 × £1050
= £52.50

Or 5% of £1050 = $\frac{5}{100}$ × £1050 = £52.50

Amount at end of year 2 = £1050 + £52.50
= £1102.50

You can use 1.05 as a multiplier.
£1000 × 1.05 × 1.05
= £1102.50
See Section 8.1

C

Example 9

It is estimated that a mountain bike loses 10% of its value each year.

A new mountain bike costs £650.

Calculate its value after two years.

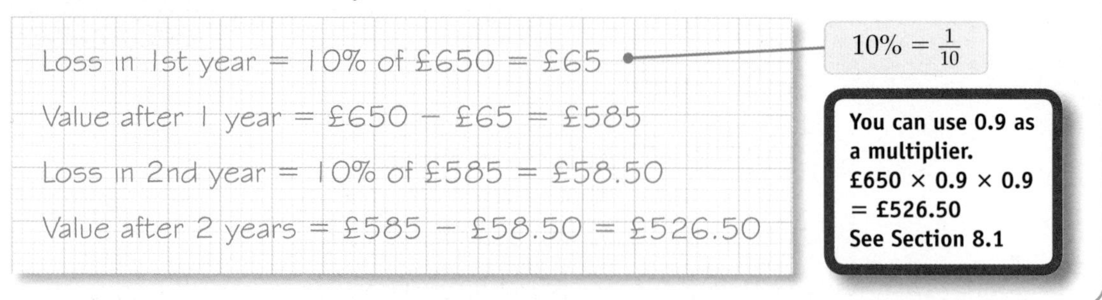

Exercise 16F

C

1 Jenny invests £500 at a rate of 4% per annum compound interest.
How much will she have at the end of two years?

2 Paul invests £300 at a rate of 10% per annum compound interest.
How much will he have after three years?

3 Work out the compound interest on

a £120 invested for two years at a rate of 10% per annum

b £2000 invested for three years at a rate of 5% per annum.

4 The number of rabbits in a field increases at the rate of 20% each month.
If there were 50 rabbits two months ago, how many are there now?

5 The value of a car depreciates by 20% each year.
It was worth £10 000 when it was new three years ago.
How much is it worth now?

'Depreciates' means that it is losing value.

6 There are 5000 whales of a certain species.
Scientists believe that their numbers are reducing by about 10% each year.
Estimate how many whales there will be in two years' time.

7 A colony of bacteria in a Petri dish is increasing at the rate of 50% each hour.
A biologist estimates that there are one million bacteria when she starts her clock.
How many does she estimate there will be three hours later?

8 A baby elephant is getting heavier at the rate of 20% every three months.
It weighs 50 kg now. Estimate what it will weigh in six months' time.

9 The population of a city is 200 000 and is increasing at the rate of 10% per annum.
Estimate the population in two years' time.

10 There is a heat wave, and a reservoir is losing 5% of its water each day.
Today it has about 2 million gallons of water in it.
How much water will it have in two days' time?

C

AO2

11 Hiroshi invests £400 at a rate of 8% per annum compound interest.
Amy invests £380 at a rate of 10% per annum compound interest.
Who has more money after two years?

16.5 Reverse percentages

Keywords
reverse percentage, original quantity

Why learn this?
If you know the sale price of a pair of jeans, you can work backwards and find the original price.

Objectives
B Perform calculations involving finding the original quantity

HELP Section 16.1

Skills check

1 Increase £78 by 10%.
2 Decrease £80 by 6%.
3 Increase 25 mm by 5%.
4 Increase 75 km by 4%.

Finding the original quantity

In Section 8.2 you learned how to work out **reverse percentage** problems using a calculator. Here you will practise these calculations without a calculator.

In reverse percentage problems, you are told the quantity after a percentage increase or decrease and you have to work out the **original quantity**.

Consider this question, for example.

> A dress is reduced by 20% in a sale. The sale price is £36.
> What was the original price?

There are two methods for answering these questions.

Method A

1 Work out what percentage the final quantity represents.
2 Divide by this percentage to find 1%.
3 Multiply by 100 to get the 100% figure.

Method B

1 Work out what percentage the final quantity represents.
2 Divide by 100 to find the multiplier.
3 Divide the final quantity by the multiplier.

Example 10

B

Attendance at a school prom was 30% down on last year's figure.
This year there were 140 students.
How many attended last year?

You do not have last year's figure, which is the original amount.

Method A

100% − 30% = 70%
140 students is the 70% figure.

Work out what percentage the figure you are given represents. '30% down' means this figure represents 70%.

So 1% = 140 ÷ 70
= 2 students

Divide by this percentage to find 1%.

So 100% = 2 × 100
= 200 students

Multiply by 100 to get the 100% figure.

Check: 30% of 200 = 60
200 − 60 = 140 students ✓

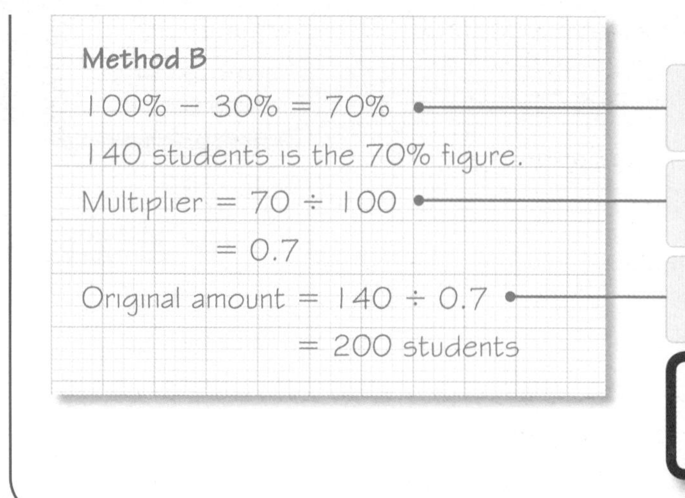

Work out what percentage the figure you are given represents.

Divide by 100 to find the multiplier.

Divide this year's number by the multiplier.

Dividing by the 'multiplier' (see Section 16.4) reverses the process so you can work out the original amount.

B

Example 11

George has had a 10% pay rise.
His new salary is £352 per week.
What was his salary before his pay rise?

Again you do not have the original amount, which is the salary before the rise.

Method A

100% + 10% = 110%

£352 is the 110% figure.

So 1% = £352 ÷ 110
= £3.20

So 100% = £3.20 × 100
= £320

Method B

100% + 10% = 110%

£352 is the 110% figure.

Multiplier = 110 ÷ 100
= 1.1

Original amount = £352 ÷ 1.1
= £320

'10% pay rise' means his new salary represents 110%.

Exercise 16G

B

1 A shop has reduced all its items by 20% in a sale.
Work out the full price of each of these items.

- **a** shirt, sale price £15.60
- **b** jeans, sale price £31.20
- **c** scarf, sale price £16.80
- **d** three pairs of socks, sale price £7.92

2 Helen's car insurance costs £350.
This includes a 60% 'no claims' discount.
How much would the insurance cost without any discount?

3 The value of a car depreciates by 20% each year.
A one-year-old car is worth £7680.
How much was it worth when it was new?

'Depreciates' means that it loses value.

4 A train company has put up all its fares by 10%.
A ticket for a particular journey now costs £28.60.
What did the ticket cost before the price increase?

5 Peter has booked an eight-day package holiday for £1020.
The price includes a 20% surcharge for peak season.
What would the price be without the surcharge?

A 'surcharge' is an additional amount added to the price.

B

6 After a pay rise of 10%, Julia earned £18 975 a year.
What was her salary before the rise?

7 A baby hippopotamus increased its weight by 10% in one week.
It now weighs 58.3 kg.
What did it weigh one week ago?

8 On the island of Mathia, a new 20% tax is placed on petrol.
A litre of petrol now costs 132 Mathia dollars.
What was the price before the new tax?

9 Scientists believe that the population of wildebeest in a game reserve is declining at 10% each year.
They estimate that the present population is about 55 000 animals.

a Estimate what the population will be in one year's time.

b Estimate what the population was one year ago.

B
AO2

10 John bought a trumpet in a sale for £405.
He knew that it had been reduced by 10%.
John said, 'I saved £40.50 by buying the trumpet in the sale.'
Explain why John is wrong.

Review exercise

E

1 A shop is selling a racing bike for the cash price of £1250.
You can also buy it on credit for a 20% deposit plus 24 monthly payments of £50.

a How much is the deposit? **[2 marks]**

b What is the total cost of buying on credit? **[2 marks]**

c How much more does it cost to buy on credit rather than paying cash? **[1 mark]**

2 Beth puts £5000 into a savings account.
The interest rate is 3% per annum.
Work out the interest after one year. **[2 marks]**

D

3 Sarah books a train ticket that would normally cost £45.
She has a rail card that gives her a 20% discount.
How much will she have to pay? **[2 marks]**

4 A football club had 120 members.
After a campaign for new members, the number of members increased by 15%.
How many members are there now? **[2 marks]**

5 A bill for plumbing repairs comes to £360 + 17.5% VAT.
What is the total bill? **[2 marks]**

6 Huw gets £8 pocket money.
He is given an increase of 20%.
How much pocket money does he get now? **[2 marks]**

D

7 Michael gets £6 pocket money.
He is given a 15% increase.
How much more pocket money does he get now? **[2 marks]**

8 In 2008 the population of a town was 84 000.
By 2009 the population had decreased by 3%.
Work out the population of the town in 2009. **[3 marks]**

C

9 Lars bought a house for £120 000.
Three years later, he sold it for £156 000.
What was the percentage profit? **[3 marks]**

10 Will bought a car for £6400 and sold it three years later for £3840.
What was the percentage depreciation in the value of the car? **[3 marks]**

11 The mayor is worried.
The population of rats in his town is increasing at the rate of about 20% each year.
There are about 100 000 rats now.
Estimate how many there will be in three years' time. **[3 marks]**

12 The value of a car depreciates by 20% each year.
It was worth £8000 three years ago, when it was new.
How much is it worth now? **[3 marks]**

B

13 The attendance at the school dance is 20% up on last year.
This year 150 students attended.
How many attended last year? **[2 marks]**

14 The numbers of fish in a lake are increasing at the rate of 20% each year.
There are estimated to be 600 fish in the lake this year.

a Estimate how many there will be in one year's time. **[2 marks]**

b Estimate how many there were one year ago. **[2 marks]**

15 An LCD TV costs £616. It has decreased in price by 23% from its price last year.
What did it cost last year? **[2 marks]**

Chapter summary

In this chapter you have learned how to

- perform calculations involving credit E
- perform simple interest calculations E
- calculate a percentage increase or decrease D
- perform calculations involving VAT D
- calculate a percentage profit or loss C
- perform calculations involving repeated percentage changes C
- perform calculations involving finding the original quantity B

AO2 Example – Algebra

C

This Grade C question challenges you to use algebra in a real-life context – setting up simple equations to represent a problem algebraically, then finding its solution.

Tom and Jerry both cycle from A to B.

Tom's average speed is 20 km/h; Jerry's average speed is 18 km/h.

Tom arrives at B 5 minutes before Jerry.

a Write down expressions for the times taken by Tom and Jerry to complete the journey.

Use x for the distance from A to B.

b Form an equation in x and solve it to work out the distance from A to B.

AO2

What maths should you use?
You need the formula linking time, speed and distance.

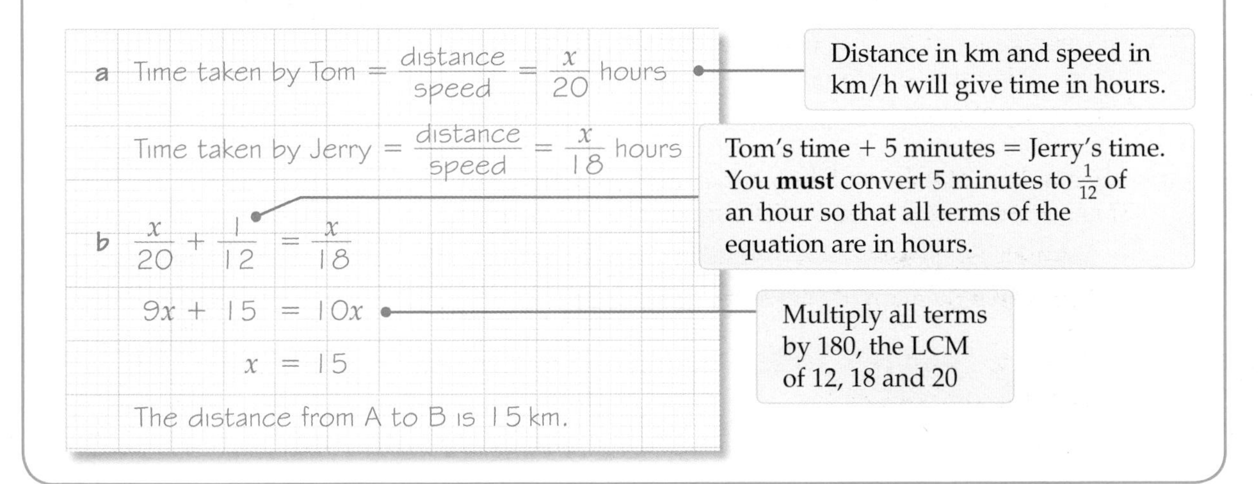

AO3 Question – Algebra

C

Now try this AO3 Grade C question. You have to work it out from scratch.
READ THE QUESTION CAREFULLY.
It's similar to the AO2 example above, so think about where to start.

Simon drove from C to D at an average speed of 42 mph.

On his return journey the traffic was heavier and he averaged only 30 mph.

His return journey took 4 minutes longer than his outward journey.

What is the distance from C to D?

Write down expressions for both journey times.

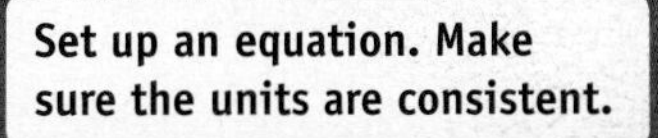

AO3

FUNCTIONAL • FUNCTIONAL •

Camping with wolves

As part of his Duke of Edinburgh award, Carlos has decided to go to the Carpathian Mountains in Slovakia to help with a wolf conservation project. He will be camping while helping with the project.

He books the evening non-stop flight from London Heathrow, on Saturday 15 March. His return flight is on Saturday 22 March.

Question bank

1 What is the cost of his return flight?

Before he leaves, Carlos plans to buy

- the lightest tent he can find
- the cheapest sleeping bag that will be warm enough at night
- £125 worth of euros
- other camping equipment costing a total of £86.

2 Which tent and sleeping bag do you recommend he should buy? Give reasons for your answers.

Carlos says, 'If I round all the prices to the nearest £10, I estimate the total cost of the camping trip is going to be £580.'

3 Is Carlos correct? Show working to support your answer.

Information bank

Return flight prices from London Heathrow (LHR) and London Gatwick (LGW) to Slovakia

Depart	Date	Time	Number of stops	Return flight price
LHR	Sat 15 March	0650	0	£195
LGW	Sat 15 March	1340	0	£215
LHR	Sat 15 March	1400	1	£234
LHR	Sat 15 March	1830	0	£242
LGW	Sat 22 March	0850	1	£199
LHR	Sat 22 March	1350	0	£224

Information bank

Tents

Name of tent	Weight (kg)	Pack dimensions length × width (cm)	Internal height (cm)	Colour	Price (£)
Trekker	2.40	48 × 20	120	Green	199.99
Sport	2.05	53 × 27	95	Red	29.99
Getup	2.05	51 × 14	115	Green	39.99
Shadow	3.15	46 × 32	110	Blue	59.99
Beta	2.35	49 × 12	95	Brown	79.99

Sleeping bags

Name of sleeping bag	Lowest temperature	Pack dimensions length × width (cm)	Price (£)	Customer rating
Arctic square	−12°C	37 × 24	84.99	*
Nordic	−4°C	36 × 19	19.99	****
Greenland	−10°C	45 × 23	54.99	***
Ultralight	0°C	44 × 22	16.99	****
Warm 'n' cosy	−5°C	28 × 17	32.99	*****
Deep down	−24°C	42 × 20	99.99	**

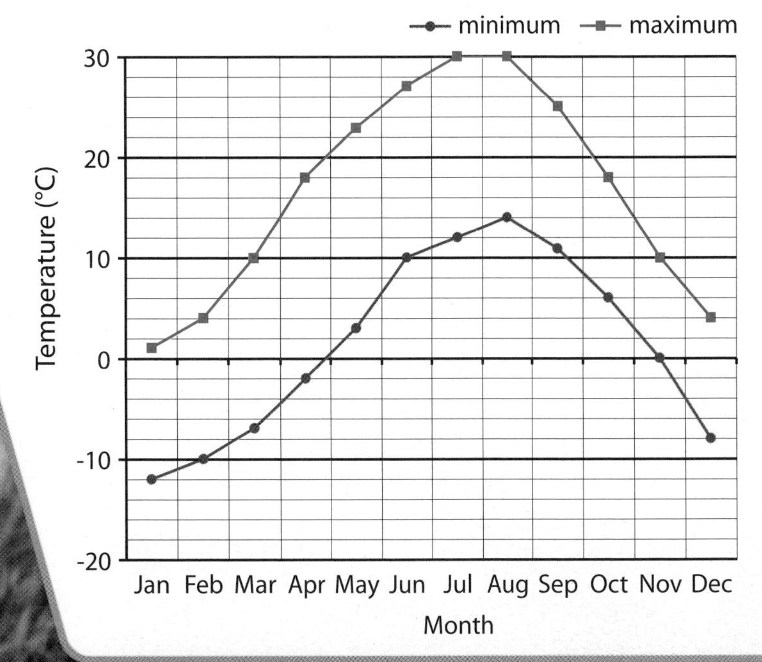

17 Sequences and proof

This chapter is about finding the next terms in a sequence, predicting patterns and justifying them.

Meteorologists look at trends and patterns, along with other data, to predict how fast weather fronts move.

Objectives

This chapter will show you how to

- find the next term in a sequence E
- continue a sequence of diagram patterns and explain how it is formed E
- describe and explain the term-to-term rule E
- continue a sequence by finding differences between consecutive terms E
- find any term of a sequence given a formula for the nth term E D
- find the nth term of a linear or simple quadratic sequence C
- show step-by-step deduction when proving results E D C
- use a counter-example to disprove a statement C

Before you start this chapter

Put your calculator away!

1 Work out the value of each expression when $n = 4$.

a $2n$ b $5n$

c $12n$ d $8n$

HELP Chapter 15

2 What is the value of x^2 when

a $x = 3$ b $x = 5$ c $x = 10$?

HELP Chapter 10

3 Here is a list of numbers.

8 9 20 100 7

Using each of these numbers only once, identify

a an even number b an odd number

c a square number d a power of 2

e a power of 10.

Number patterns

Keywords
sequence, term, consecutive, term-to-term rule

Why learn this?
Shop managers use sales figures for previous months to predict future sales.

Objectives

E D Find the next term in a sequence

E Describe the rule for continuing a sequence

Skills check

1 Here is a sequence of numbers.

3, 6, 9, 12, 15, 18, …

a What is the 4th term of the sequence?
b Write down the next two terms in the sequence.
c Describe in words the rule for finding the next term.

Sequences

A **sequence** is a list of numbers in a given order.

Each number in a sequence is called a **term.**

Consecutive terms are next to each other.

In the sequence

1, 3, 7, 15, 31, …

1 and 3 are consecutive terms.

7 and 15 are consecutive terms.

You can use the pattern in a sequence to predict the next term.

Look at the differences between consecutive terms to work out the **term-to-term rule.**

Example 1

E

a Write down the next three terms of the sequence 14, 10, 6, 2, −2, …

b Write down the term-to-term rule for the sequence.

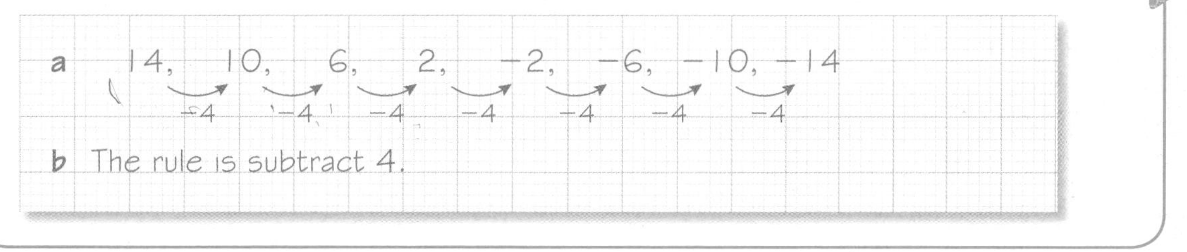

Exercise 17A

E

1 For each sequence
i find the next three terms
ii write down the term-to-term rule.

a −8, −3, 2, 7, … b 5, −1, −7, … c 4, 0, −4, −8, … d 17, 12, 7, 2, …

2 Find the 12th term of each sequence.

a 7, 12, 17, 22, … b −2, −4, −6, −8, … c 23, 17, 11, 5, …

E

3 The first term of a sequence is 15.
The term-to-term rule is subtract 4.

a Write down the first 10 terms of the sequence.

b how many positive terms are there in the sequence?

4 The first term of a sequence is -12. The term-to-term rule is add 5.
How many negative terms are there in the sequence?

D

AO2

5 **a** **i** Write down the first five terms of two different sequences that start 64, 32, …

ii Describe in words the rules for continuing both sequences.

b Here is a sequence with some terms missing.

64, 32, ☐, ☐, -64, …

What is the 10th term of this sequence?

E

Example 2

a Find the next three terms in the sequence 1, 2, 4, 7, 11, 16, ...

b Describe the pattern of differences.

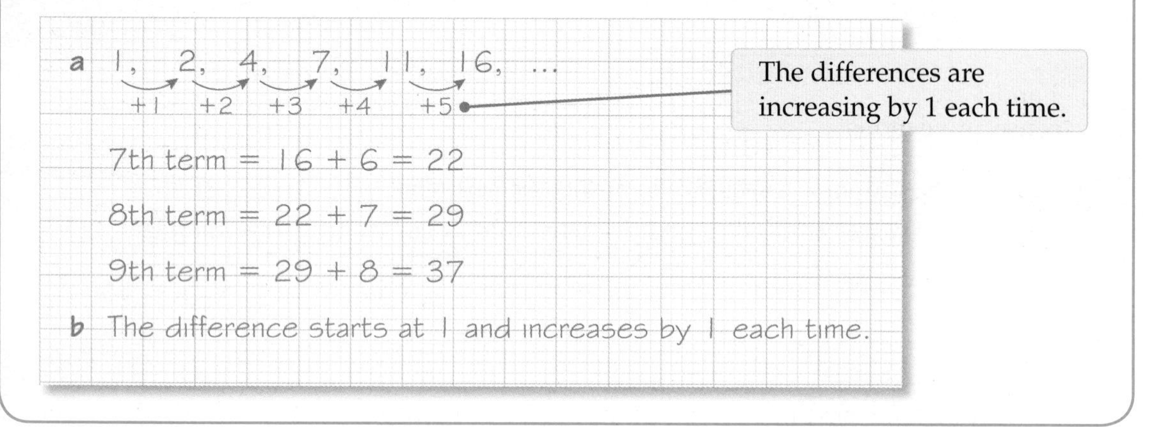

Exercise 17B

E

1 For each sequence

i work out the next two terms

ii describe the pattern of differences.

a 1, 3, 6, 10, …

b 1, 4, 9, 16, 25, …

c 100, 90, 81, 73, …

d 2, 1, -1, -4, …

E

AO2

2 The Fibonacci sequence is a special one.
The first two terms of the sequence are 1 and 1.
The next term is found by adding the previous two terms.
So the 3rd term is $1 + 1 = 2$
The first three terms are 1, 1, 2.

a What is the 4th term?

b Which is the first term that is larger than 20?

c Look at the pattern of differences. What do you notice?

3 The table shows how sales of pay-as-you-go mobile phones at Phones 2U have increased over the last few years.

Year	2006	2007	2008	2009
Sales (to the nearest thousand)	34 000	37 000	41 000	46 000

The company predicts that the pattern will continue.
In which year would sales exceed (be more than) 75 000?

4 The first four terms of a sequence are 50, 45, 39, 32.
What is the value of the first term in the sequence which is less than zero?

5 An engineering firm offers its employees the salary scale in the table.
The salary continues to increase in the same way.
In which year will an employee's salary exceed £18 000?

Year	Salary
Year 1	£10 000
Year 2	£10 500
Year 3	£11 250
Year 4	£12 250

E

AO2

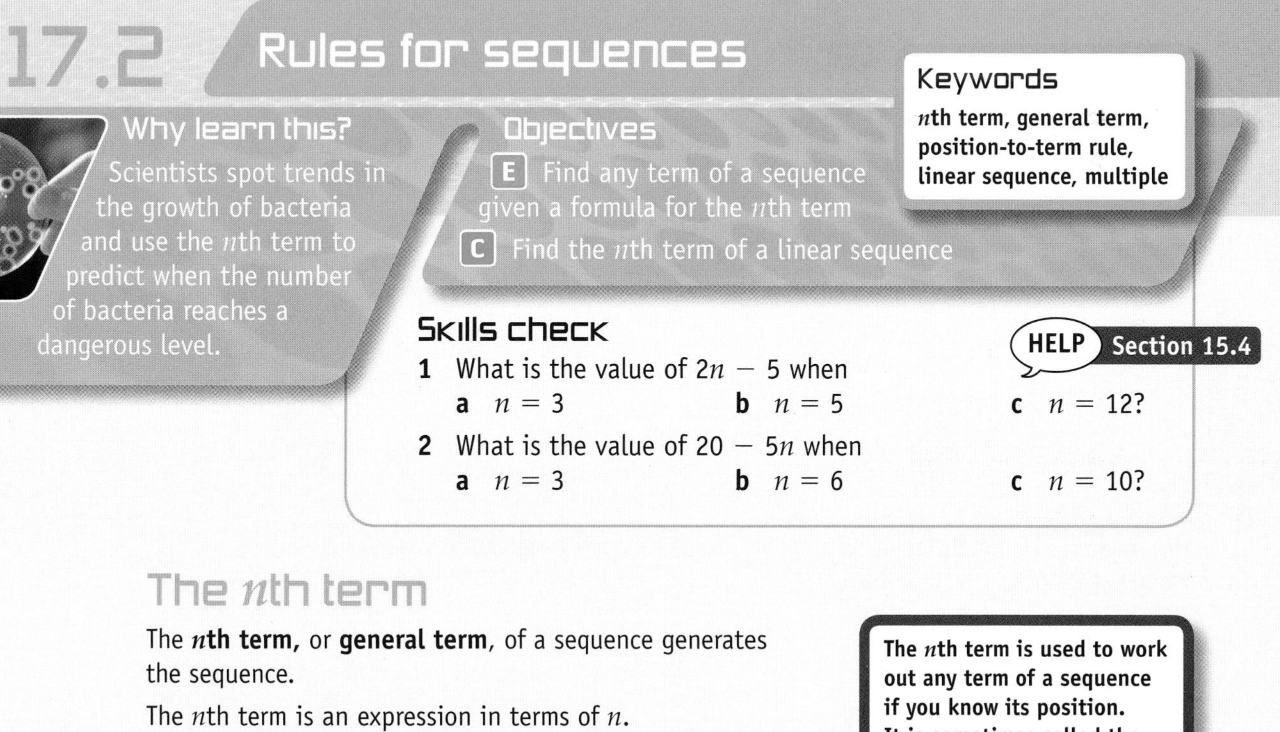

17.2 Rules for sequences

Why learn this?
Scientists spot trends in the growth of bacteria and use the nth term to predict when the number of bacteria reaches a dangerous level.

Objectives
E Find any term of a sequence given a formula for the nth term
C Find the nth term of a linear sequence

Keywords
nth term, general term, position-to-term rule, linear sequence, multiple

Skills check

1 What is the value of $2n - 5$ when
a $n = 3$ b $n = 5$ c $n = 12$?

2 What is the value of $20 - 5n$ when
a $n = 3$ b $n = 6$ c $n = 10$?

HELP Section 15.4

The nth term

The **nth term,** or **general term**, of a sequence generates the sequence.

The nth term is an expression in terms of n.

To find the 1st term, substitute $n = 1$ into the nth term.

To find the 2nd term, substitute $n = 2$, and so on.

The nth term is used to work out any term of a sequence if you know its position. It is sometimes called the position-to-term rule.

Example 3

The nth term of a sequence is $2n - 4$.
What is **a** the 3rd term **b** the 10th term?

a $2 \times 3 - 4 = 6 - 4 = 2$ — Substitute $n = 3$ into the nth term.

b $2 \times 10 - 4 = 20 - 4 = 16$ — Substitute $n = 10$ into the nth term.

E

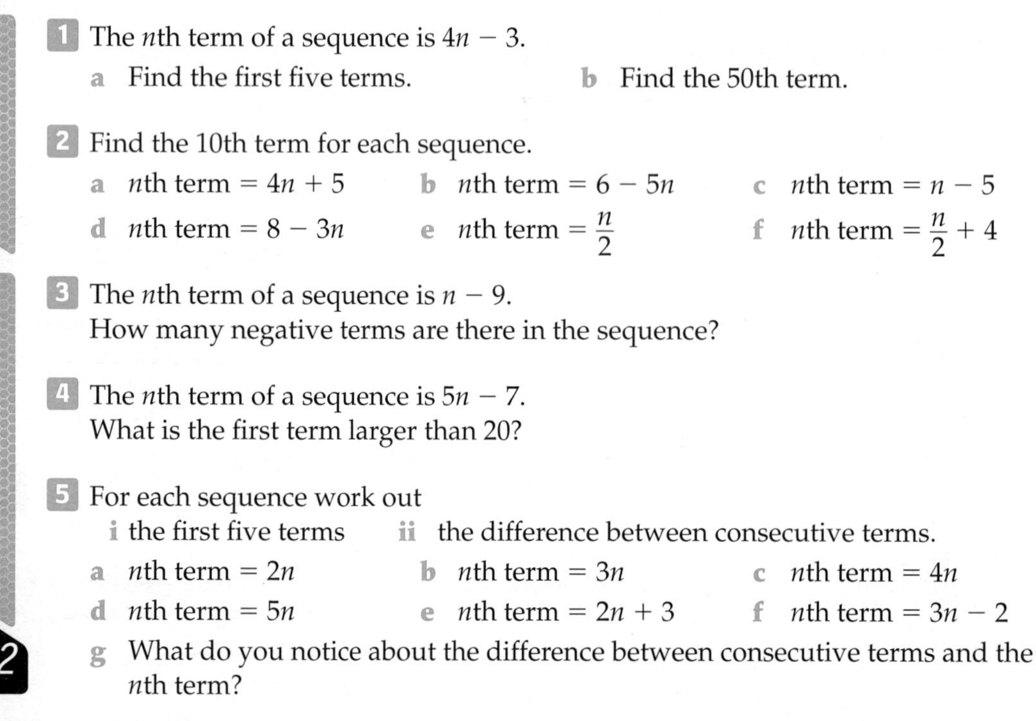

Exercise 17C

E

1 The nth term of a sequence is $4n - 3$.

a Find the first five terms. b Find the 50th term.

2 Find the 10th term for each sequence.

a nth term $= 4n + 5$ b nth term $= 6 - 5n$ c nth term $= n - 5$

d nth term $= 8 - 3n$ e nth term $= \frac{n}{2}$ f nth term $= \frac{n}{2} + 4$

E

3 The nth term of a sequence is $n - 9$.
How many negative terms are there in the sequence?

4 The nth term of a sequence is $5n - 7$.
What is the first term larger than 20?

5 For each sequence work out

i the first five terms ii the difference between consecutive terms.

a nth term $= 2n$ b nth term $= 3n$ c nth term $= 4n$

d nth term $= 5n$ e nth term $= 2n + 3$ f nth term $= 3n - 2$

AO2

g What do you notice about the difference between consecutive terms and the nth term?

Linear sequences

In a **linear sequence**, the differences between consecutive terms are the same.

4, 6, 8, 10, ... is a linear sequence.
+2 +2 +2

1, 4, 9, 16, ... is *not* a linear sequence.
+3 +5 +7

Multiples

The sequence of **multiples** of 3 has nth term (or general term) $3n$.

1st term $= 3 \times 1 = 3$
2nd term $= 3 \times 2 = 6$
3rd term $= 3 \times 3 = 9$
4th term $= 3 \times 4 = 12$

The sequence of multiples of 4 has nth term $4n$.

The sequence of multiples of 5 has nth term $5n$, and so on.

The nth term

A linear sequence is one where the differences between consecutive terms are all the same.

You can find the nth term of a linear sequence by looking at the differences between consecutive terms and comparing the sequence to the multiples of that difference.

Consider the sequence in the table.

	1st term	2nd term	3rd term	4th term
Sequence	1	4	7	10
Multiples of 3	3	6	9	12

The difference between consecutive terms is 3. So compare the sequence to the multiples of 3.

Each term in the sequence is 2 less than the multiples of 3. So the nth term is: $3n - 2$.

Example 4

Find the nth term of each sequence.

a 5, 9, 13, 17, ... **b** 13, 11, 9, 7, ...

a

	1st term	2nd term	3rd term	4th term
Sequence	5	9	13	17
$4n$ (multiples of 4)	4	8	12	16

The difference between consecutive terms is 4.

Compare the sequence to the sequence for $4n$.

nth term is $4n + 1$

Each term is 1 more than the term in the sequence for $4n$.

b

The difference between consecutive terms is -2.

	1st term	2nd term	3rd term	4th term
Sequence	13	11	9	7
$-2n$ (multiples of -2)	-2	-4	-6	-8

Compare the sequence to the sequence for $-2n$.

The nth term is $-2n + 15$ or $15 - 2n$

Each term is 15 more than the term in the sequence for $-2n$.

Exercise 17D

1 Find the nth term of each sequence.

a 3, 6, 9, 12, 15, ...
b 2, 4, 6, 8, 10, ...
c 5, 10, 15, 20, 25, ...
d 100, 200, 300, 400, 500, ...
e 7, 14, 21, 28, 35, ...
f 50, 100, 150, 200, 250, ...

2 The first four terms of a linear sequence are 3, 5, 7, 9.

a Write down the next three terms.
b What is the difference between consecutive terms?
c Copy and complete the table below.

	1st term	2nd term	3rd term	4th term
Sequence	3	5	7	9
Multiples of ☐				

d Use your table to find the nth term of the sequence.

3 Find the nth term of each sequence.

a 5, 9, 13, 17, ...
b 8, 10, 12, 14, ...
c 4, 9, 14, 19, ...
d 9, 20, 31, 43, ...
e 75, 175, 275, 375, ...
f $-5, -10, -15, -20, \ldots$
g 10, 7, 4, 1, ...
h 19, 15, 11, 7, ...
i 77, 67, 57, 47, ...
j 50, 44, 38, 32, ...

C

4 The first four terms of a sequence are 8, 13, 18, 23.

a Find the nth term of the sequence.

b Use the nth term to find the 100th term.

5 A botanist records the length of a plant leaf on five consecutive days. The table shows her measurements.

Day	1	2	3	4	5
Length (mm)	21	23	25	27	29

a Find the nth term of the sequence of lengths.

b What will the length of the leaf be on day 10?

17.3 Using the nth term

Why learn this?

The numbers of pieces of wood in this fence follow a sequence. You can use the nth term to find out how many pieces you would need for 45 panels.

Objectives

C Find the nth term of a linear sequence

C Use the nth term to find terms in a sequence

Skills check

1 Find the nth term of the sequence 2, 6, 10, 14, 18, ...

2 Solve the equation $2n + 4 = 18$ HELP Section 14.1

Using the nth term

You can use the nth term of a sequence to find any term in that sequence.

You can also use the nth term to see if a particular number is in a sequence.

C

Example 5

Here is a sequence.

5, 9, 13, 17, 21, ...

a What is the 50th term of the sequence?

b Is 98 a term in the sequence?

a 5, 9, 13, 17, 21, ...

+4 +4 +4 +4

Term	5	9	13	17	21
$4n$	4	8	12	16	20

The nth term is $4n + 1$

50th term $= 4 \times 50 + 1 = 201$

First find the nth term.

The sequence goes up in 4s, so compare it to the sequence for $4n$.

Each term in the sequence is 1 more than $4n$.

Substitute $n = 50$

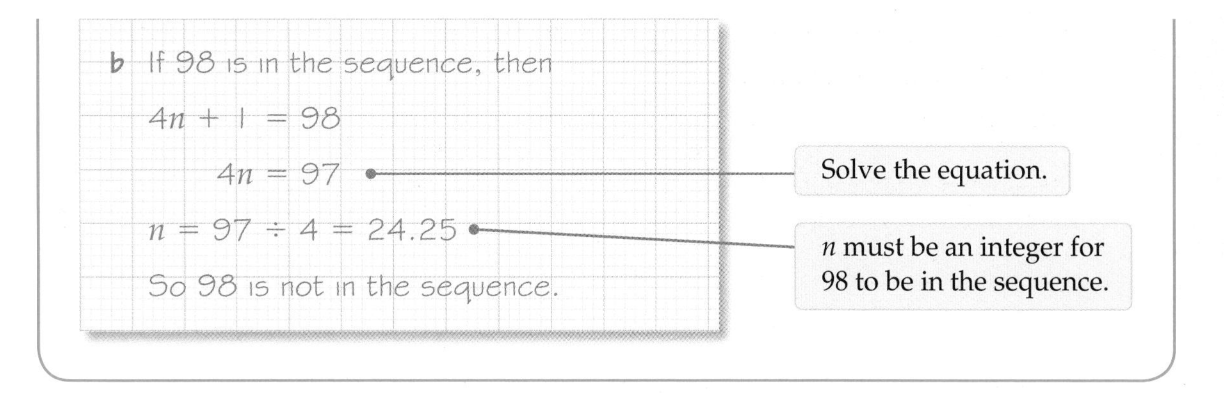

Exercise 17E

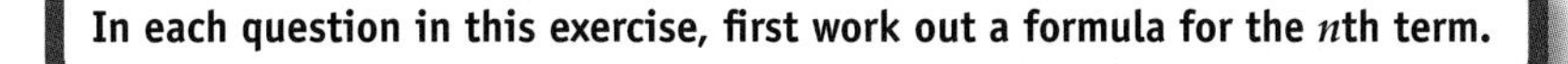

C

1 a Find the 50th term of the sequence 1, 3, 5, 7, …

b Find the 15th term of the sequence 4, 10, 16, 22, …

c Find the 20th term of the sequence 8, 15, 22, 29, …

2 Find the 59th term of the sequence $-8, -6, -4, -2, \ldots$

3 A sequence begins 7, 11, 15, 19, …
One of the terms in the sequence is 47.
Which number term is this?

What is the value of n?

4 Here is a sequence.

8, 17, 26, 35, 44, …

a Is 134 a term in this sequence?

b Is 198 a term in this sequence?

5 Here is a sequence.

19, 23, 27, 31, 35, …

a 75 is a term in this sequence. What number term is it?

b Explain why 52 cannot be a term in this sequence.

AO2

C

6 For each sequence, find the first negative term.

a 112, 101, 90, 79, 68, …

b 47, 41, 35, 29, 23, …

c 220, 208, 196, 184, 172, …

d 58, 51, 44, 37, 30, …

7 For each sequence, find the term closest to 70.

a 4, 11, 18, 25, 32, …

b 5, 8, 11, 14, 17, …

c 23, 32, 41, 50, 59, …

d $-7, -2, 3, 8, 13, \ldots$

AO3

17.4 Sequences of patterns

Keywords
pattern

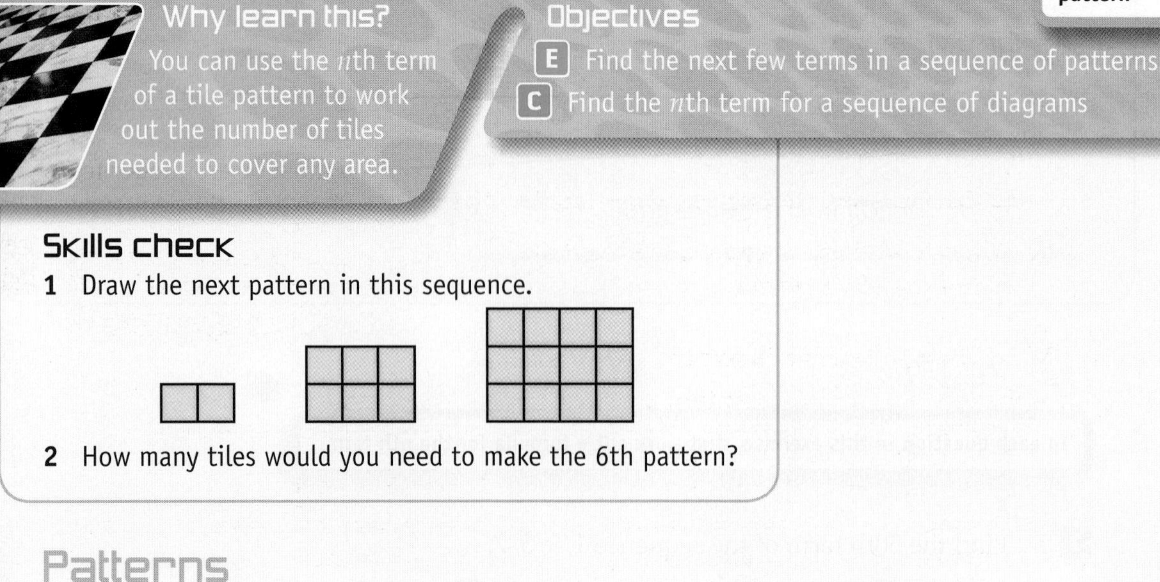

Why learn this?

You can use the nth term of a tile pattern to work out the number of tiles needed to cover any area.

Objectives

E Find the next few terms in a sequence of patterns

C Find the nth term for a sequence of diagrams

Skills check

1 Draw the next pattern in this sequence.

2 How many tiles would you need to make the 6th pattern?

Patterns

Sequences of **patterns** can lead to number sequences.

E

Example 6

AO2

a Draw the next pattern in the sequence.

b How many dots will there be in the 10th pattern of the sequence?

a

Look at how the pattern grows. Each pattern in the sequence has 2 more dots than the one before.

b The 1st pattern has 1 pair of dots.
The 2nd pattern has 2 pairs of dots.
The 3rd pattern has 3 pairs of dots.
So the 10th pattern will have 10 pairs of dots.
$10 \times 2 = 20$ dots

Exercise 17F

E

1 Here is a sequence of dot patterns.

a Without drawing the pattern, work out how many dots are in pattern 5.

b Explain how you worked out your answer to part **a**.

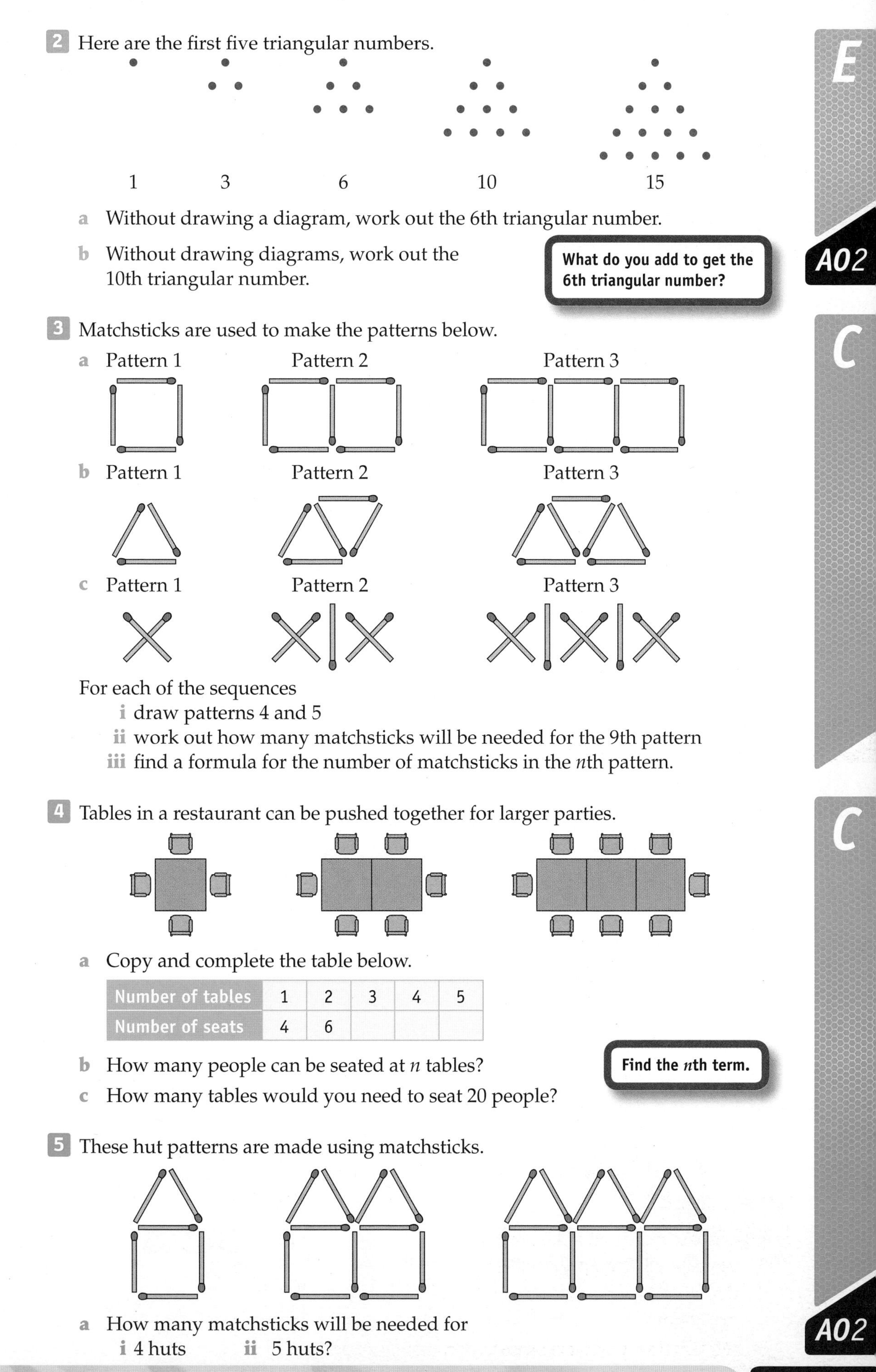

2 Here are the first five triangular numbers.

1 3 6 10 15

a Without drawing a diagram, work out the 6th triangular number.

b Without drawing diagrams, work out the 10th triangular number.

What do you add to get the 6th triangular number?

E

AO2

3 Matchsticks are used to make the patterns below.

a Pattern 1 Pattern 2 Pattern 3

b Pattern 1 Pattern 2 Pattern 3

c Pattern 1 Pattern 2 Pattern 3

For each of the sequences

i draw patterns 4 and 5

ii work out how many matchsticks will be needed for the 9th pattern

iii find a formula for the number of matchsticks in the nth pattern.

C

4 Tables in a restaurant can be pushed together for larger parties.

a Copy and complete the table below.

Number of tables	1	2	3	4	5
Number of seats	4	6			

b How many people can be seated at n tables?

c How many tables would you need to seat 20 people?

Find the nth term.

C

5 These hut patterns are made using matchsticks.

a How many matchsticks will be needed for

i 4 huts ii 5 huts?

AO2

C

b Copy and complete the table below.

Number of huts	1	2	3	4	5	10	15
Number of matchsticks	6	11					

c Work out how many matchsticks will be needed for n huts.

6 I have 90 matchsticks.
I use them to make a row of huts like those in Q5.
If I make as many huts as I can, how many matchsticks will I have left over?

7 Here is a sequence of dot patterns.

a Copy and complete the table below.

Pattern number	Number of dots
1	$1 \times 2 = 2$
2	$2 \times 3 = 6$
3	$3 \times \square = \square$
4	$\square \times \square = \square$
5	$\square \times \square = \square$
10	$10 \times \square = \square$
15	$\square \times \square = \square$
n	$\square \times \square = \square$

AO2

b One of the patterns contains 132 dots. What is its pattern number?

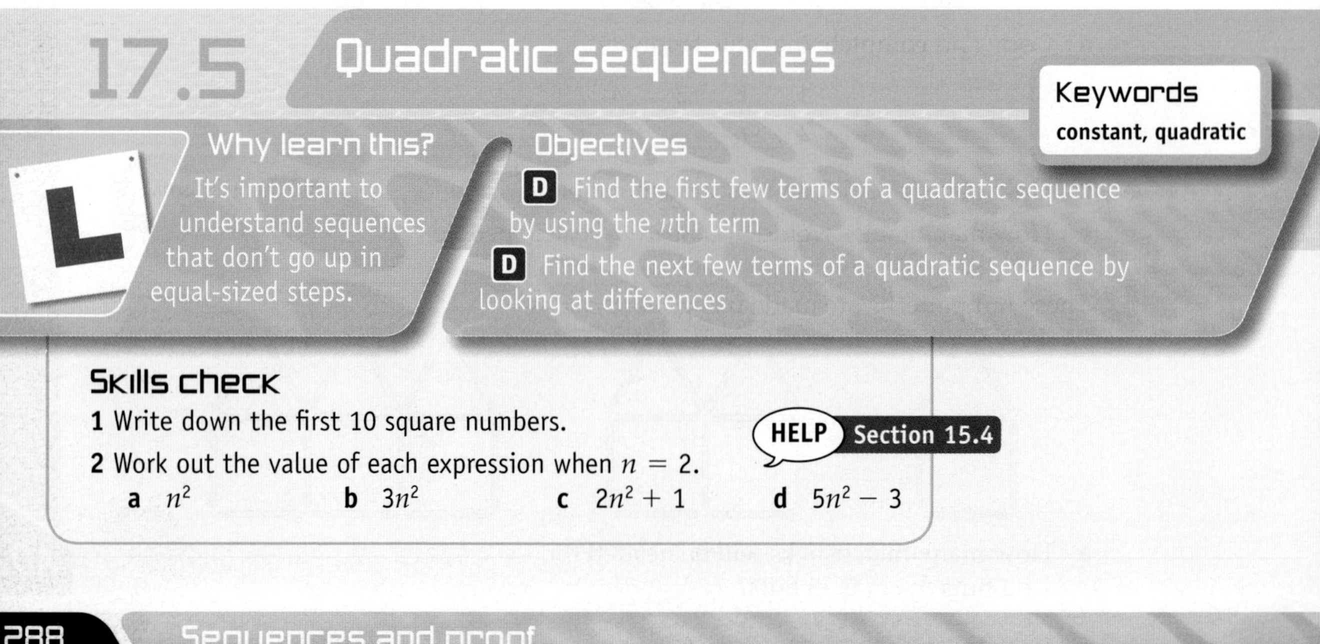

17.5 Quadratic sequences

Keywords
constant, quadratic

Why learn this?
It's important to understand sequences that don't go up in equal-sized steps.

Objectives

D Find the first few terms of a quadratic sequence by using the nth term

D Find the next few terms of a quadratic sequence by looking at differences

Skills check

1 Write down the first 10 square numbers.

2 Work out the value of each expression when $n = 2$.

a n^2 **b** $3n^2$ **c** $2n^2 + 1$ **d** $5n^2 - 3$

HELP Section 15.4

Linear and quadratic sequences

In a linear sequence, the terms go up or down in equal-sized steps.

The difference between consecutive terms is **constant** (the same).

3, 7, 11, 15, ...

+4 +4 +4

This is a linear sequence since the difference between consecutive terms is always 4.

The nth term of a linear sequence includes an n.

For example, $3n + 2$, $4n - 5$, $7 - 2n$ are nth terms of linear sequences.

In a **quadratic** sequence, the difference between consecutive terms is *not* constant.

The nth term of a quadratic sequence includes an n^2.

Example 7

D

AO2

The nth term of a sequence is $n^2 + 3$.

a Work out the first five terms of the sequence.

b Describe the pattern of differences between the terms. Use this to predict the 6th term.

a 1st term: $1^2 + 3 = 1 + 3 = 4$

2nd term: $2^2 + 3 = 4 + 3 = 7$

3rd term: $3^2 + 3 = 9 + 3 = 12$

4th term: $4^2 + 3 = 16 + 3 = 19$

5th term: $5^2 + 3 = 25 + 3 = 28$

4 7 12 19 28

+3 +5 +7 +9

b The differences are increasing odd numbers, so the 6th term is $28 + 11 = 39$

Remember the order of operations: indices before addition.

Write the first five terms as a sequence.

Exercise 17G

D

1 The nth term of a sequence is $n^2 + 2$. Work out

a the 1st term **b** the 2nd term **c** the 3rd term **d** the 10th term.

2 Work out the first five terms of each sequence.

a nth term $= n^2 - 3$ **b** nth term $= n^2 + 6$ **c** nth term $= 3n^2$

d nth term $= \dfrac{n^2}{2}$ **e** nth term $= 4n^2 + 5$

D

3 The nth term of a sequence is n^2. For this sequence, is each statement true or false?

a All the terms are odd. **b** All the terms are even.

c The terms alternate between odd and even.

d All the terms are multiples of 3. **e** All the terms are greater than zero.

4 The nth term of a sequence is $2n^2$.

a Work out the first five terms of the sequence.

b Describe the pattern of differences between the terms. Use this to predict the 6th term.

AO2

D

5 Copy and complete this quadratic sequence.

3, $\square$, $\square$, 15, 23, 33

6 A sequence has nth term $2n^2 - 1$.
Is 72 a number in the sequence? Explain your answer.

7 Jenny, a biologist, is recording the number of colonies of bacteria in a Petri dish after a given number of days.

Number of days (n)	Number of colonies
10	20
20	80
30	180
40	320

She works out that the nth term for calculating the number of colonies is $\frac{n^2}{5}$.

a How many colonies would she expect to see after 50 days?

b How many days was it before the number of colonies was 45?

AO2

8 Iram, another biologist, carries out the same experiment as Jenny (Q7).
She works out an nth term of $n^2 - 90$.
After how many days will the number of colonies in Iram's Petri dish be greater than 100?

17.6 The nth term of a simple quadratic sequence

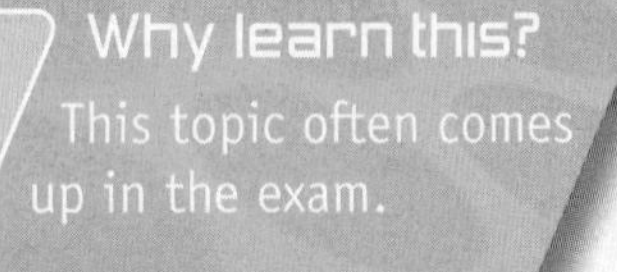

Why learn this?
This topic often comes up in the exam.

Objectives
C Find the nth term of a simple quadratic sequence

Skills check

1 What do you know about the nth term of a quadratic sequence?

2 Work out the first five terms of the sequence with nth term
a $n^2 + 3$ **b** $n^2 - 5$ **c** $n^2 + 8$

HELP Section 17.5

The nth term of a simple quadratic sequence

Here are the first five terms of the quadratic sequence with nth term n^2.

	1		4		9		16		22
Differences		+3		+5		+7		+9	
Second differences			+2		+2		+2		

In a quadratic sequence, the difference between consecutive terms is *not* constant.
But the second difference *is* constant.
If the second difference is 2, the nth term includes n^2.

Example 8

Find the nth term of the sequence 5, 8, 13, 20, 29, …

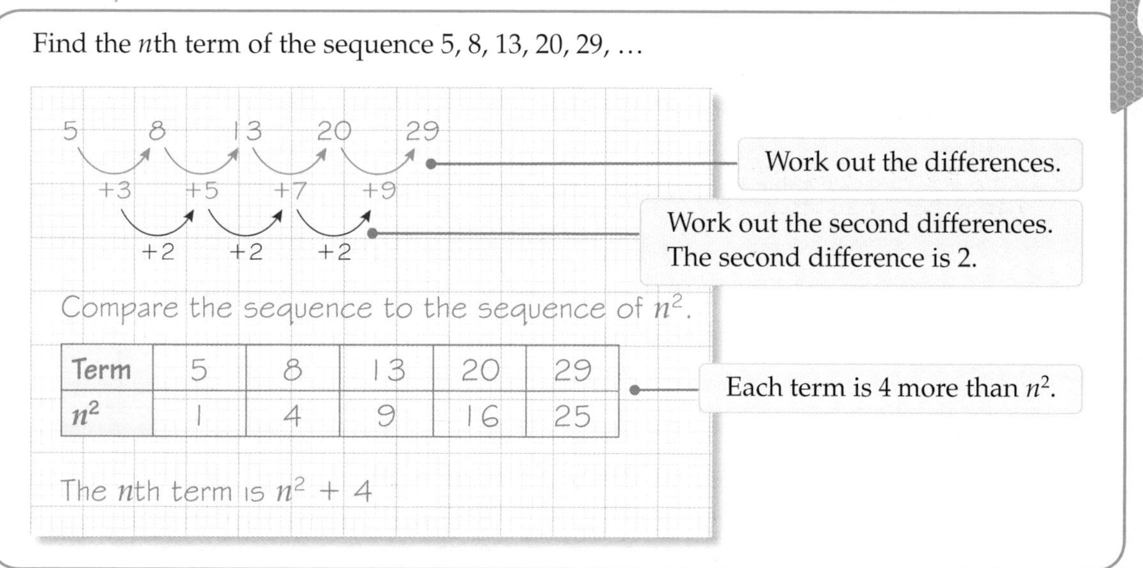

Term	5	8	13	20	29
n^2	1	4	9	16	25

Exercise 17H

Find the nth term and the 10th term of each sequence.

1 4, 7, 12, 19, …

2 −1, 2, 7, 14, …

3 7, 10, 15, 22, …

4 −4, −1, 4, 11, …

5 2, 8, 18, 32, …

6 3, 12, 27, 48, …

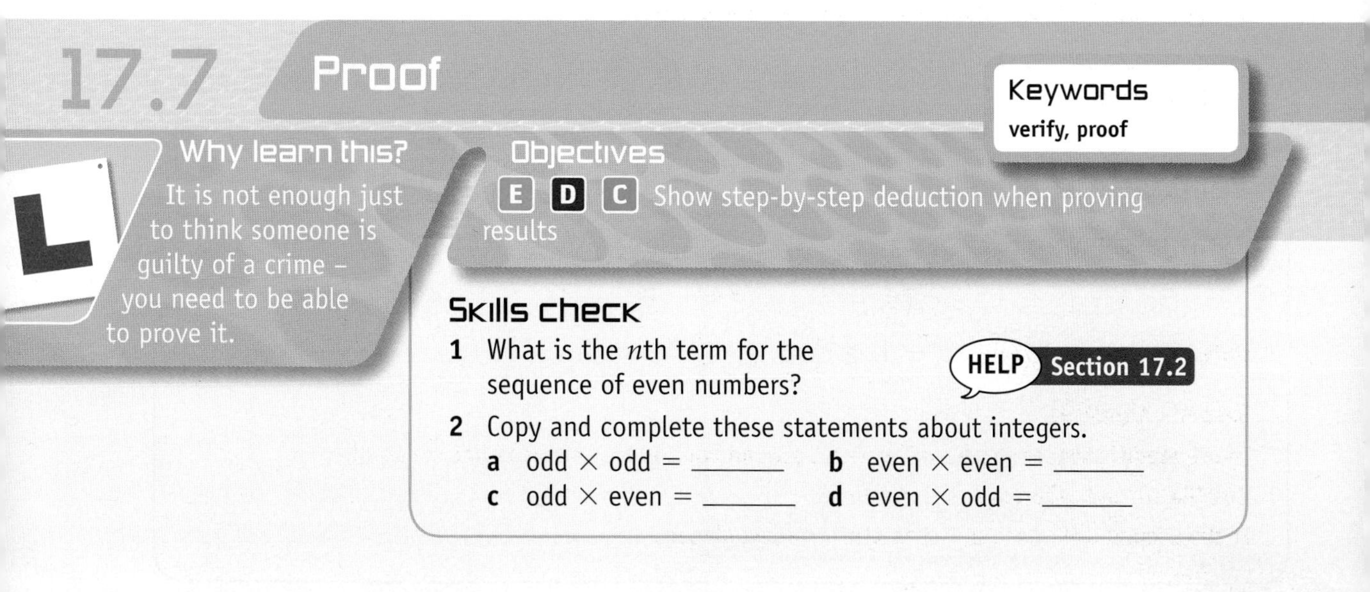

17.7 Proof

Keywords
verify, proof

Why learn this?
It is not enough just to think someone is guilty of a crime – you need to be able to prove it.

Objectives
E D C Show step-by-step deduction when proving results

Skills check

1 What is the nth term for the sequence of even numbers? HELP Section 17.2

2 Copy and complete these statements about integers.

a odd × odd = ______ b even × even = ______

c odd × even = ______ d even × odd = ______

Mathematical proof

Showing that a theory works for *a few* values is called **verifying** a theory. In mathematics, this is not enough.

You need to **prove** the theory by showing that it works for *all* values.

A proof uses logical reasoning to show that something is true.

Proof questions usually ask you to 'prove' or 'show' something.

D AO2

Example 9

The nth term of a sequence is $2n + 1$.

Show that all the terms in the sequence are odd.

Verify means to check the result using numerical values.

1st term $= 2 \times 1 + 1 = 2 + 1 = 3$

2nd term $= 2 \times 2 + 1 = 4 + 1 = 5$

3rd term $= 2 \times 3 + 1 = 6 + 1 = 7$

The first three terms are odd. They go up in 2s and 'odd number' + 2 = odd.

Work out the first few terms of the sequence.

Explain why all the terms will be odd. Use facts you know about odd and even numbers.

Example 10

t represents an integer.

a What type of number (odd or even) is $t + (t + 1)$?

b Explain how you know.

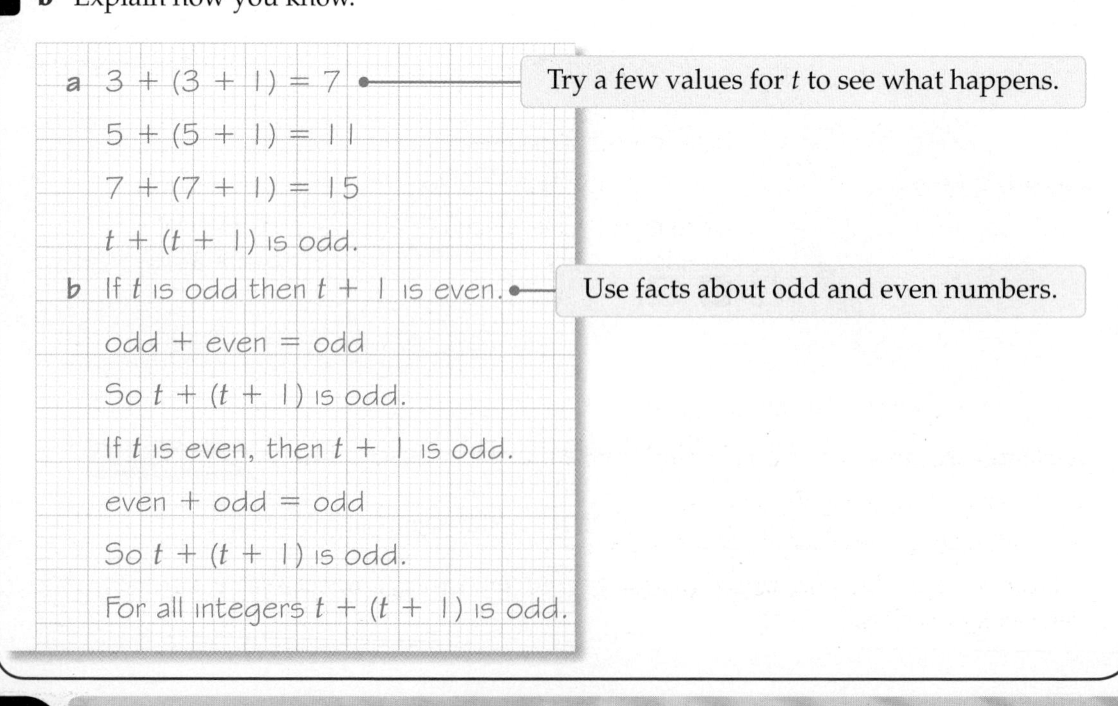

a $3 + (3 + 1) = 7$

$5 + (5 + 1) = 11$

$7 + (7 + 1) = 15$

$t + (t + 1)$ is odd.

b If t is odd then $t + 1$ is even.

odd + even = odd

So $t + (t + 1)$ is odd.

If t is even, then $t + 1$ is odd.

even + odd = odd

So $t + (t + 1)$ is odd.

For all integers $t + (t + 1)$ is odd.

Try a few values for t to see what happens.

Use facts about odd and even numbers.

Exercise 17I

1 I think of a positive integer, multiply it by 2 and take away 1.

- **a** What type of number (odd or even) will I get?
- **b** Explain how you know.

2 Alice is playing a game with her brother Johnny.
She rolls a six-sided dice and doubles the score.
She wins if the answer is even.
Johnny says the game is unfair.
Explain why it is unfair.

3 q is an odd number.
Explain why $q(q + 1)$ is always even.

4 n is a prime number larger than 2.
Explain why $n + 1$ is always even.

5 Show that the product of two consecutive positive integers is always even.

6 x is an odd number.
Explain why $x^2 + 1$ is always even.

7 x is an odd number and y is an even number.
Explain why $(x + y)(x - y)$ is an odd number.

8 Six consecutive positive integers are added.
What type of number (odd or even) will the sum be?
Justify your answer.

17.8 Using counter-examples

Keywords
counter-example

Why learn this?

Generalisations are often made in the media. Finding a counter-example is a good way of disproving statements like these.

Objectives

C Show something is false by using a counter-example

Skills check

1 Here is a list of numbers.

3, 5, 9, 12, 16.5, 17, 20

Write down those which are

- **a** odd numbers
- **b** multiples of 3
- **c** prime numbers.

Counter-examples

A **counter-example** is an example which shows that a statement is false.

For example, here is a statement.

'All teenagers use social networking websites.'

You can disprove this statement if you can find *just one* teenager who doesn't use social networking websites.

Example 11

AO2

The nth term of a sequence is $3n + 1$.
Jeremy says that all the terms in the sequence are even.
Explain why he is wrong.

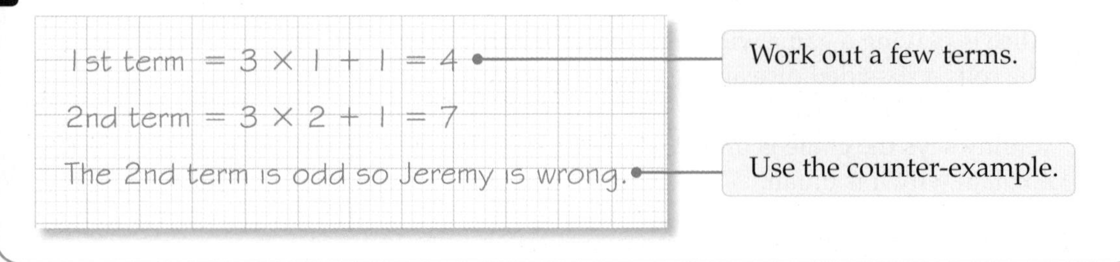

Exercise 17J

C

1 Give a counter-example to show that each of these statements is false.

- **a** Adding a positive number to a negative number always gives a positive result.
- **b** The product of an integer and decimal is always a decimal.
- **c** The sum of two decimals is always a decimal.

2 Rounded to one decimal place, a number is 7.6.
Ben claims that the unrounded number *must* have been larger than 7.55.
Give a counter-example to show that he is wrong.

3 Decide whether each of these statements is true or false.
If they are false, give a counter-example.

- **a** All the terms in the sequence with nth term $n^2 + 1$ are odd.
- **b** If p is odd, then $(p + 1)(p - 1)$ is always even.
- **c** The cube of any number is greater than 0.
- **d** All odd numbers are prime.
- **e** If n is an integer, then $n(n + 1)(n + 2)$ is always even.

C

4 Margaret claims that if x is an integer, x^2 is always even.
Give a counter-example to show that she is wrong.

5 Peta says, 'The difference of two square numbers is always odd.'
Explain why she is wrong.

6 Sam claims, 'The sum of three consecutive numbers is always odd.'
Explain why this is wrong.

7 A website claims, 'If x and y are prime, then $x^2 + y^2$ is an even number.'
Explain why the claim is false.

8 For any number n, the value of $n^2 - n$ is greater than zero.
Explain why this is false.

AO2

9 Dilip says that m^3 is always positive, for any value of m.
Explain why he is wrong.

Review exercise

1 Explain why the sum of two consecutive integers is always odd. **[2 marks]**

E

2 The nth term of a sequence is $2n - 1$.

a Write down the first three terms of the sequence. **[1 mark]**

b Is 12 a term in the sequence? Explain your answer. **[2 marks]**

D

3 Aimee is playing a game with her brother Jon.
He thinks of an integer, she doubles it and subtracts 1.
Aimee wins if the outcome is odd.
Jon says the game is unfair as he cannot win. Explain why this is true. **[2 marks]**

4 Here is a sequence of numbers.

400, 200, □, 50, □, 12.5, …

Write down the two missing numbers in the sequence. **[2 marks]**

5 The nth term of a sequence is $n^2 - 5$.

C

a What is the 8th term of the sequence? **[1 mark]**

b Which term of the sequence is the first to be greater than 150? **[1 mark]**

6 Here is a sequence.

7, 10, 15, 22, 31, …

What is the nth term of the sequence? **[2 marks]**

7 Here is a sequence.

−8, 2, 12, 22, …

What is the nth term of the sequence? **[2 marks]**

8 Here is a pattern is made using tiles.

C

Pattern 1 Pattern 2 Pattern 3

a Copy and complete the table.

Pattern number	Number of tiles
1	1 × 3 = 3
2	2 × □ = □
3	□ × □ = □
4	□ × □ = □
5	□ × □ = □
10	□ × □ = □
n	□ × □ = □

[4 marks]

b Which pattern number contains 80 tiles? **[2 marks]**

AO2

A02

9 n and q are prime numbers.
Adrian says that nq will always be odd.
Find a counter-example to show that he is wrong. **[2 marks]**

10 Rachel says that $m^3 + 1$ cannot be a multiple of 5.
Explain why she is wrong. **[2 marks]**

Chapter summary

In this chapter you have learned how to

- describe the rule for continuing a sequence E
- find the next few terms in a sequence of patterns E
- find any term of a sequence given a formula for the nth term E
- find the next term in a sequence E D
- find the first few terms of a quadratic sequence by using the nth term D
- find the next few terms of a quadratic sequence by looking at differences D
- find the nth term of a linear sequence C
- find the nth term of a simple quadratic sequence C
- find the nth term for a sequence of diagrams C
- use the nth term to find terms in a sequence C
- show step-by-step deduction when proving results E D C
- show something is false by using a counter-example C

18 Linear graphs

BBC Video

This chapter explores coordinates and graphs and their real-life applications.

Video game designers use coordinates and the equations of straight lines to define characters' movement within a game world.

Objectives

This chapter will show you how to

- construct linear functions from real-life problems and plot their corresponding graphs E D
- discuss and interpret graphs modelling real situations E D C
- plot graphs of functions in which y is given explicitly in terms of x, or implicitly (no table or axes given) E D C
- find the coordinates of the mid-point of a line segment D C
- investigate the gradient of parallel lines D C
- understand that the form $y = mx + c$ represents a straight line and that m is the gradient of the line and c is the value of the y-intercept D C B
- interpret simultaneous equations as lines and their common solution as the point of intersection B
- solve linear inequalities in two variables and find the solution set B

Before you start this chapter

Put your calculator away!

1 Write down the value shown by each letter on this scale.

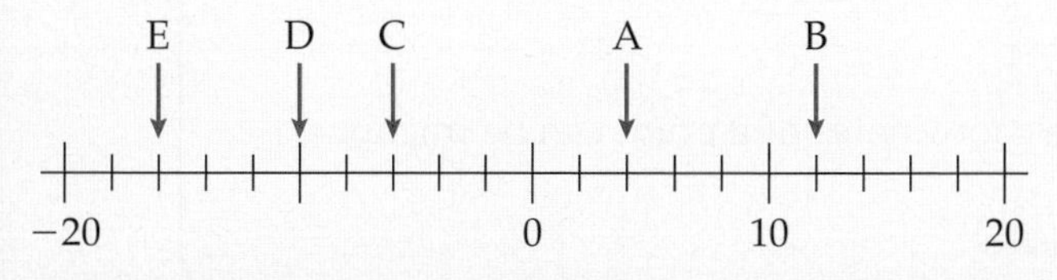

2 Work out the value of each expression when $x = 3$.

a $2x$ b $x + 3$

c $3x - 2$ d $2x + 4$

HELP Chapter 15

3 Work out the value of y when $x = -2$.

a $y = 2x$ b $y = x + 3$

c $y = x - 1$ d $y = 2x - 4$

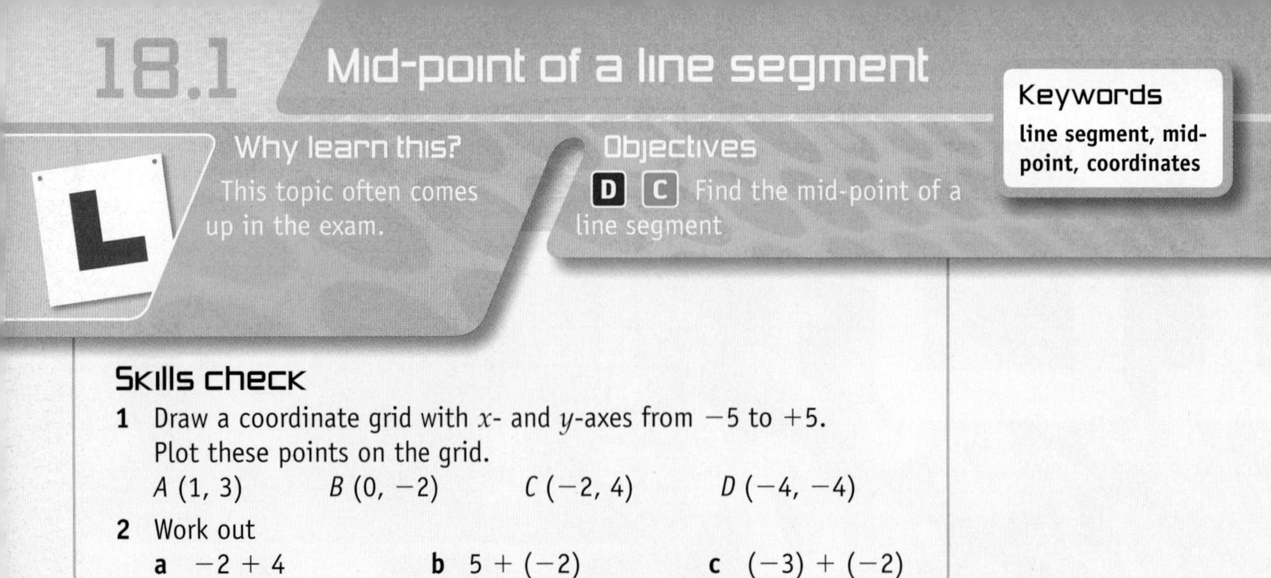

18.1 Mid-point of a line segment

Why learn this?
This topic often comes up in the exam.

Objectives
D **C** Find the mid-point of a line segment

Keywords
line segment, mid-point, coordinates

Skills check

1. Draw a coordinate grid with x- and y-axes from -5 to $+5$.
   Plot these points on the grid.
   A (1, 3) B (0, −2) C (−2, 4) D (−4, −4)
2. Work out
   a $-2 + 4$ **b** $5 + (-2)$ **c** $(-3) + (-2)$

Mid-point of a line segment

A **line segment** is the line between two points.

The **mid-point** of a line segment is exactly half way along the line.

You can work out the coordinates of the mid-point using the **coordinates** of the end points of the line segment.

$$\text{Mid-point } (x, y) = \left(\frac{x_1 + x_2}{2}, \frac{y_1 + y_2}{2}\right)$$

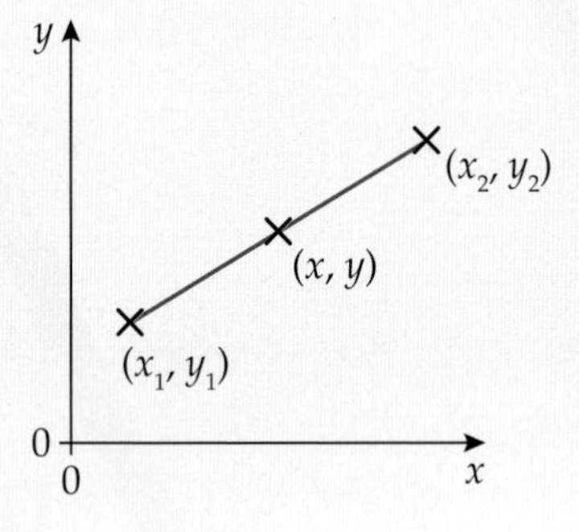

D

Example 1

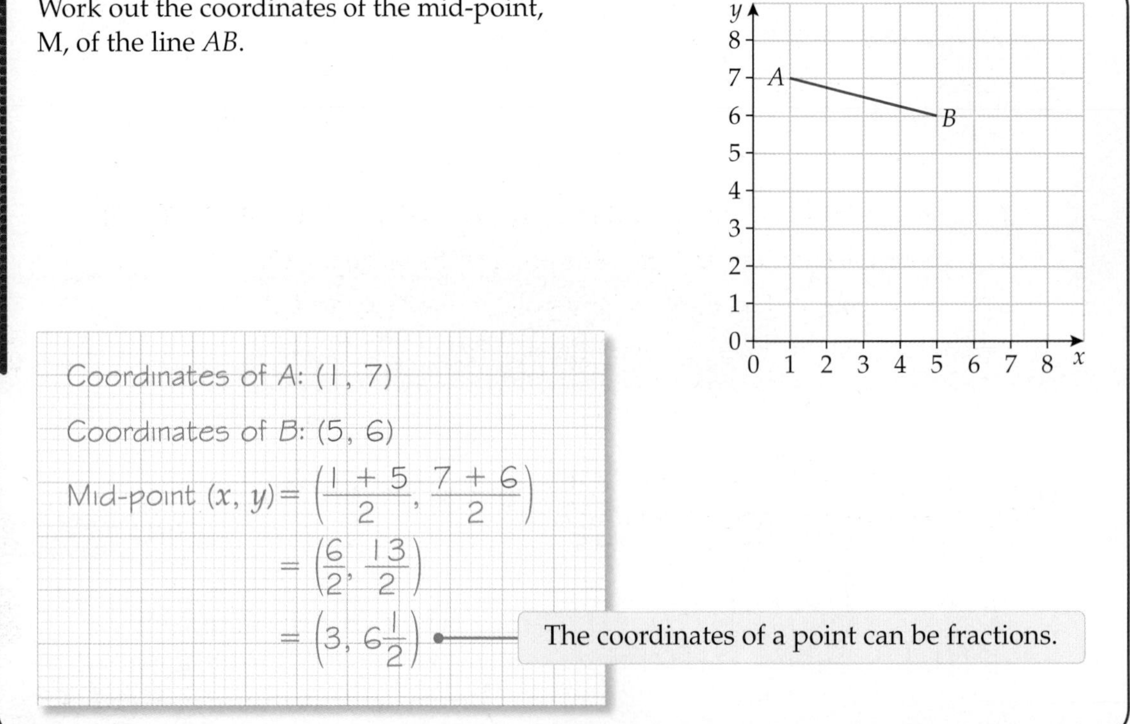

Work out the coordinates of the mid-point, M, of the line AB.

Coordinates of A: (1, 7)

Coordinates of B: (5, 6)

$$\text{Mid-point } (x, y) = \left(\frac{1 + 5}{2}, \frac{7 + 6}{2}\right)$$

$$= \left(\frac{6}{2}, \frac{13}{2}\right)$$

$$= \left(3, 6\frac{1}{2}\right)$$

The coordinates of a point can be fractions.

Exercise 18A

D

1 For each line segment shown on the grid

a write the coordinates of the end-points

b work out the coordinates of the mid-point.

> **You can use the same formula with coordinates that have negative values.**

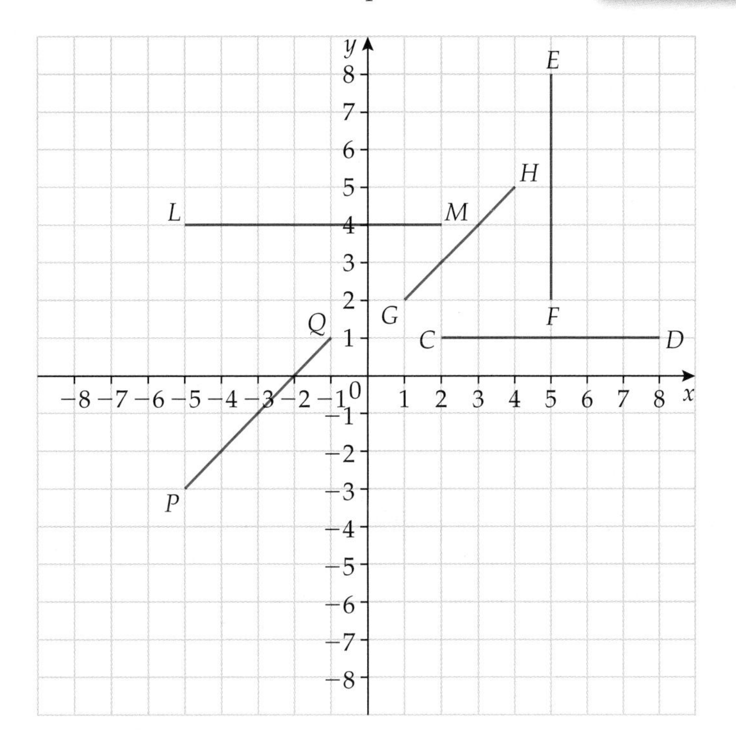

C

2 Work out the mid-points of these line segments.

a $GH : G(1, 1)$ and $H(8, 1)$

b $LM : L(2, 3)$ and $M(5, 9)$

c $ST : S(-2, 2)$ and $T(3, -3)$

d $UV : U(-3, -5)$ and $V(7, -5)$

> **You can draw a sketch to check that your answer is correct.**

C AO2

3 $ABCD$ is a quadrilateral with coordinates $A(-1, 2)$, $B(4, 2)$, $C(4, -3)$, $D(-1, -3)$.

a Work out the coordinates of the mid-point of the diagonal AC.

b Work out the coordinates of the mid-point of the diagonal BD.

c What can you say about the quadrilateral $ABCD$?

4 The coordinates of the mid-point of a line segment PQ are $(1, 2)$.
Work out the coordinates of Q when P is at

a $(1, 4)$ **b** $(-1, 2)$

c $(6, 2)$ **d** $(1, -3)$

Plotting straight-line graphs

Keywords
straight-line graph, parallel, x-axis, y-axis, coordinate pair, linear function

Why learn this?

Straight-line graphs can be used to model and analyse real-life situations such as mobile-phone tariffs.

Objectives

E Recognise straight-line graphs parallel to the x- or y-axis

D Work out coordinates of points of intersection when two graphs cross

D C Plot graphs of linear functions

Skills check

1 Work out the value of each expression when $x = 2$.

a $3x$ b $2x + 1$ c $3x - 4$

2 Work out the value of each expression when $x = -2$.

a $4x$ b $3x + 2$ c $2x - 4$

HELP Section 15.4

Lines parallel to the x- or y-axis

A **straight-line graph parallel** to the **x-axis** has equation $y = b$, where b is a number.

A straight-line graph parallel to the **y-axis** has equation $x = b$, where b is a number.

E

Example 2

Write the equations of the lines A, B and C.

Line A: (0, 2), (2, 2), (−3, 2). — Write down the coordinates of a few points on the line.

The equation of the line is $y = 2$. — All the points on the line have y-coordinate 2.

Line B: The equation of the line is $x = -3$. — All the points on the line have x-coordinate −3.

Line C: The equation of the line is $y = -1$. — All the points on the line have y-coordinate −1.

Exercise 18B

1 Write the equation of each line on this grid that is

a parallel to the x-axis

b parallel to the y-axis.

2 Draw a coordinate grid with both x- and y-axes from -5 to $+5$.
Draw and label these graphs.

a $y = 4$ b $x = 2$ c $y = -2$ d $x = -5$

Equations of other straight lines

$y = 2x + 1$ is a function.

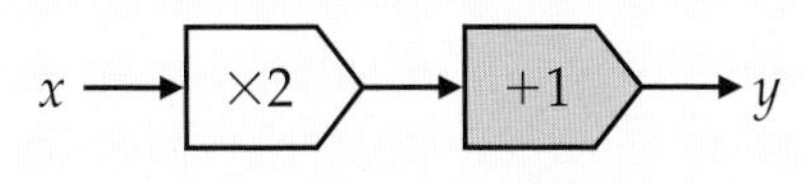

Put in a value of x and you get a value for y.

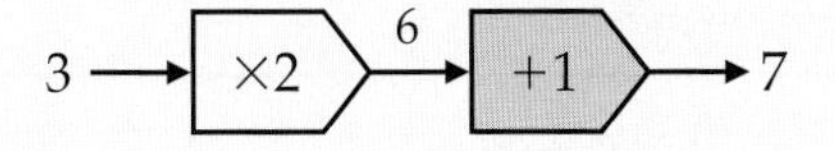

You can draw a graph of the function $y = 2x + 1$ like this:

- Substitute values for x.
- Write the values of x and the corresponding values of y in a table.
- Plot the (x, y) **coordinate pairs** on a grid.
- Join the points with a straight line.

If the graph of the function is a straight line, then it is a linear function.

Example 3

Draw the graph of $y = 2x + 2$ for values of x from -3 to $+2$.

Step 1: Draw a table of values with values of x from -3 to $+2$.

x	-3	-2	-1	0	1	2
y						

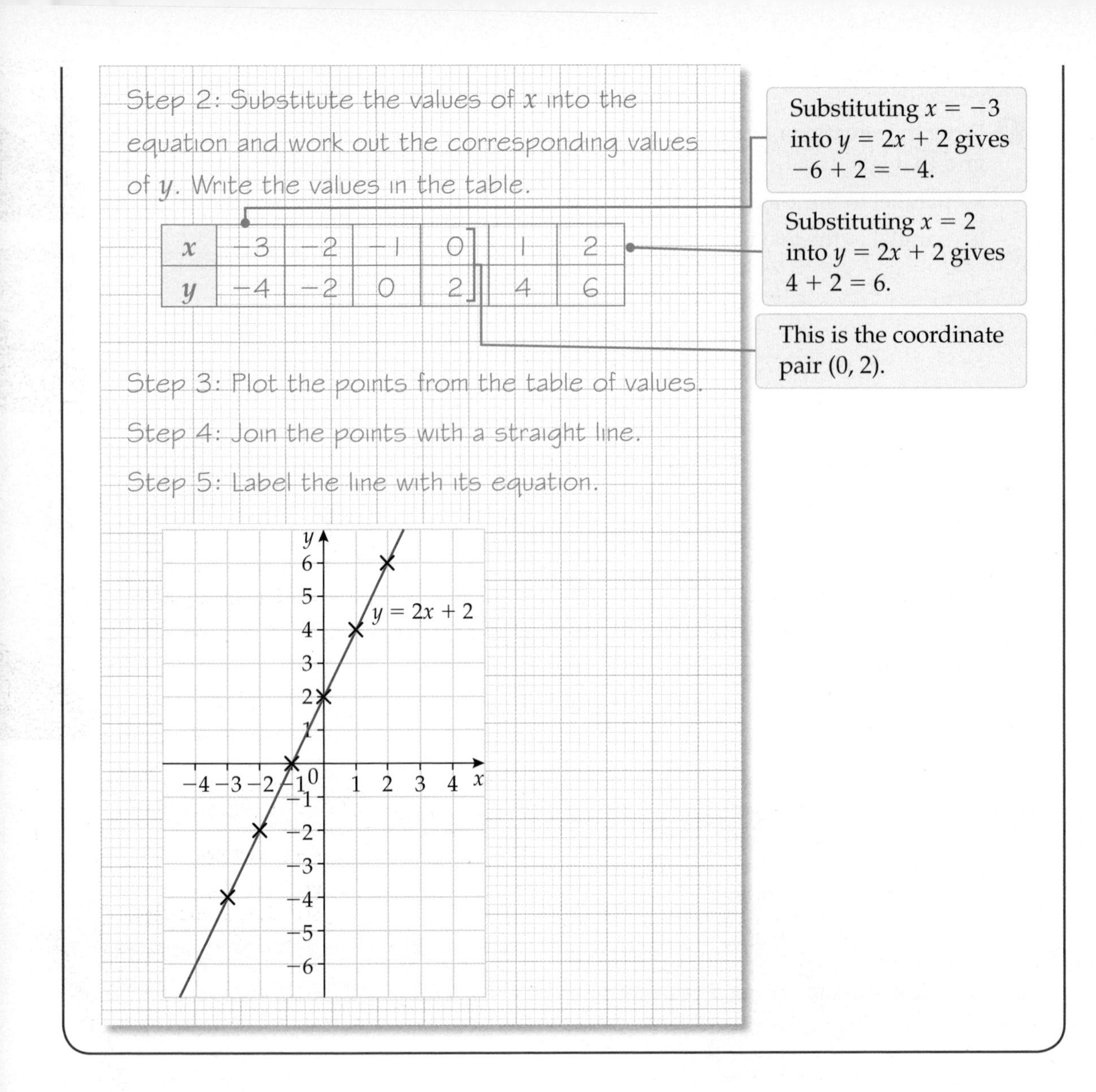

Exercise 18C

D

1 Draw the graph of $y = 2x - 1$ for $-2 \leqslant x \leqslant 2$.

> $-2 \leqslant x \leqslant 2$ means x-values from -2 to $+2$. Draw a coordinate grid with both x- and y-axes from -5 to $+5$.

2 Draw the graph of $y = \frac{1}{2}x - 1$ for $-4 \leqslant x \leqslant 4$.

> Draw x- and y-axes from -4 to $+4$.

D

3 **a** Copy and complete this table of values for $y = 3 - x$.

x	−2	−1	0	1	2
y	5	4			

b Draw the graph of $y = 3 - x$.

c Draw the line $x = 2$ on your graph.

AO2

d A is the point where the two lines cross. Mark the point A and write its coordinates.

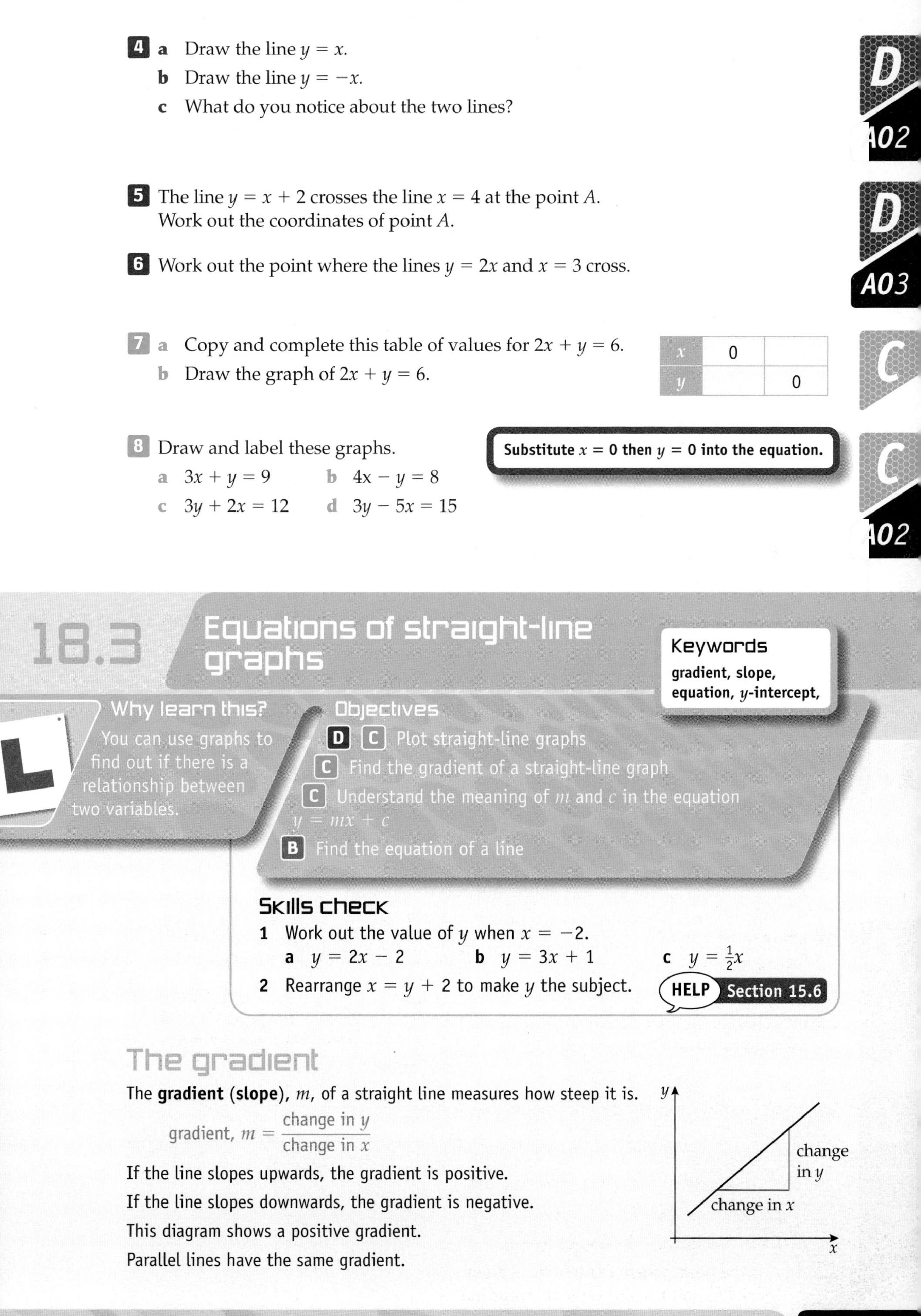

4 a Draw the line $y = x$.
b Draw the line $y = -x$.
c What do you notice about the two lines?

D AO2

5 The line $y = x + 2$ crosses the line $x = 4$ at the point A.
Work out the coordinates of point A.

6 Work out the point where the lines $y = 2x$ and $x = 3$ cross.

D AO3

7 a Copy and complete this table of values for $2x + y = 6$.
b Draw the graph of $2x + y = 6$.

x	0	
y		0

C

8 Draw and label these graphs.
a $3x + y = 9$ b $4x - y = 8$
c $3y + 2x = 12$ d $3y - 5x = 15$

Substitute $x = 0$ then $y = 0$ into the equation.

C AO2

18.3 Equations of straight-line graphs

Keywords
gradient, slope, equation, y-intercept,

Why learn this?
You can use graphs to find out if there is a relationship between two variables.

Objectives
D C Plot straight-line graphs
C Find the gradient of a straight-line graph
C Understand the meaning of m and c in the equation $y = mx + c$
B Find the equation of a line

Skills check
1 Work out the value of y when $x = -2$.
a $y = 2x - 2$ b $y = 3x + 1$ c $y = \frac{1}{2}x$
2 Rearrange $x = y + 2$ to make y the subject.

HELP Section 15.6

The gradient

The **gradient** (**slope**), m, of a straight line measures how steep it is.

$$\text{gradient, } m = \frac{\text{change in } y}{\text{change in } x}$$

If the line slopes upwards, the gradient is positive.
If the line slopes downwards, the gradient is negative.
This diagram shows a positive gradient.
Parallel lines have the same gradient.

C

Example 4

Work out the gradient of these lines.

a $y = 3x - 4$ b $y = -x + 7$

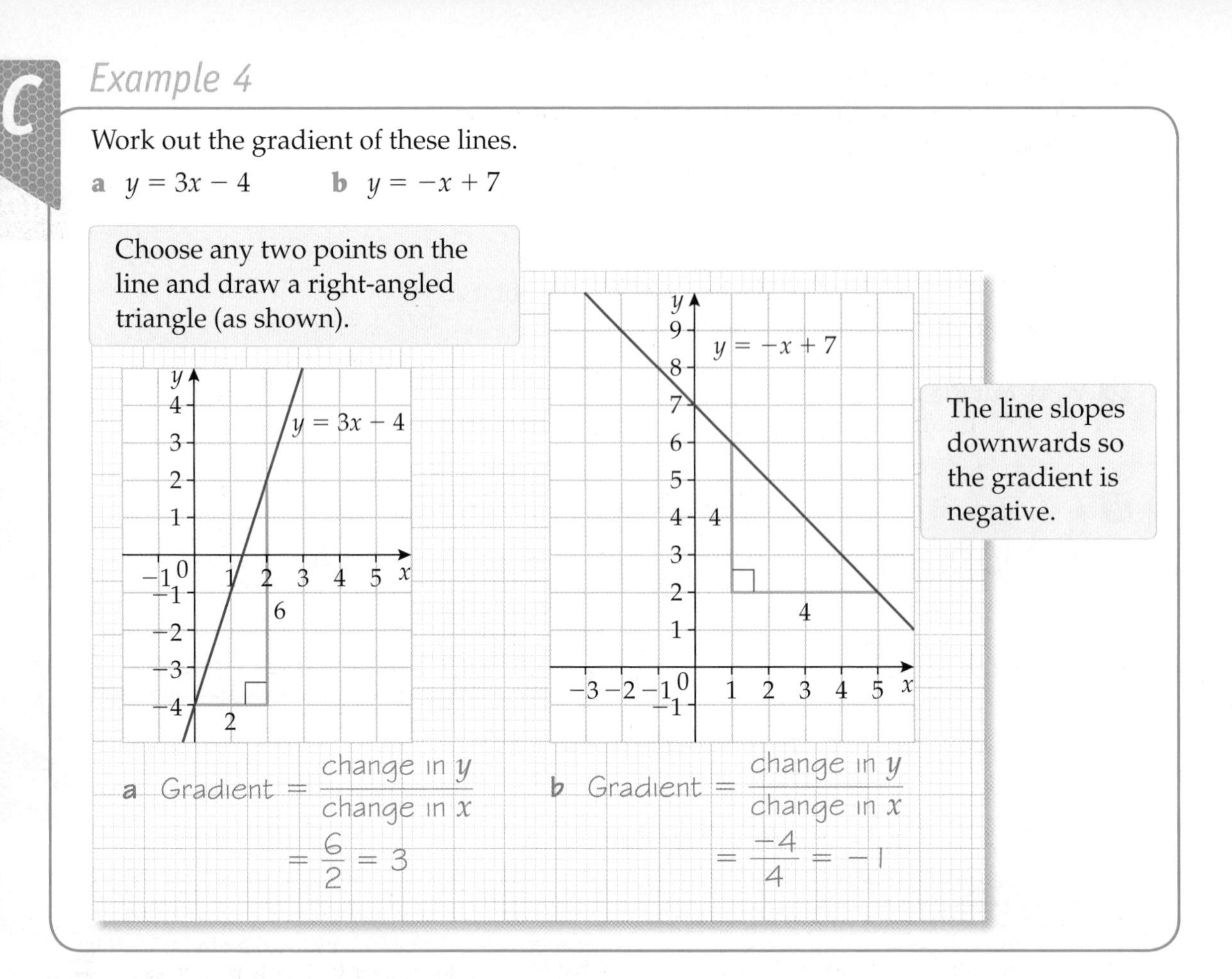

Exercise 18D

D

AO2

1 a Draw and label these graphs.

i $y = 2x + 1$ ii $y = 3x + 1$ iii $y = 4x + 1$

b Which line is the steepest?

c How can you tell which line is the steepest from the equations?

d Draw the graph of $y = 2x + 2$ on your coordinate grid.

e Which lines are parallel to each other?

f How can you tell which lines are parallel from the equations?

Draw a coordinate grid with both x- and y-axes from 0 to +10.

C

2 a Work out the gradients of lines A to D.

b What do you notice about the gradients of lines B and D?

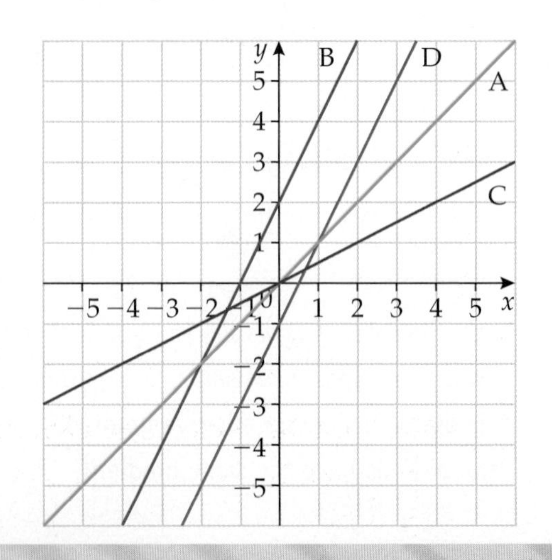

3 Work out the gradient of each line in Q1.
Compare the equation and the gradient for each line.
What do you notice?

C

AO2

4 Write down the gradient of each of these lines.

a $y = 3x + 9$ b $y = 2x + 1$ c $y = 4 + x$
d $y = -\frac{1}{2}x - 6$ e $y = 5 - 2x$ f $2x - 4 = y$

C

5 Which of these lines are parallel to each other?

A	$y = 2x - 3$	B	$y = 3 - 5x$	C	$y = 3x - 1$
D	$y = 4 + 3x$	E	$y = -2x + 3$	F	$2x + 3 = y$

6 Look at the graphs in Q1 again.
Write down where each graph crosses the y-axis. What do you notice?

Equations of straight lines

Straight-line graphs have **equations** of the form

$$y = mx + c$$

The number (m) in front of x is the gradient.

For example, the line with equation $y = 3x - 4$ has gradient 3, and the line with equation $y = -x + 7$ has gradient -1.

Remember that $1x$ is written as x and $-1x$ is written as $-x$.

The value of c is the **y-intercept**.

For example, the line with equation $y = 3x - 4$ crosses the y-axis at $(0, -4)$.

The y-intercept is -4.

Example 5

C

What is the y-intercept value of each of these lines?

a $y = x + 2$ b $y = 2x - 3$ c $y = -x + 2$ d $y = 4 - 2x$

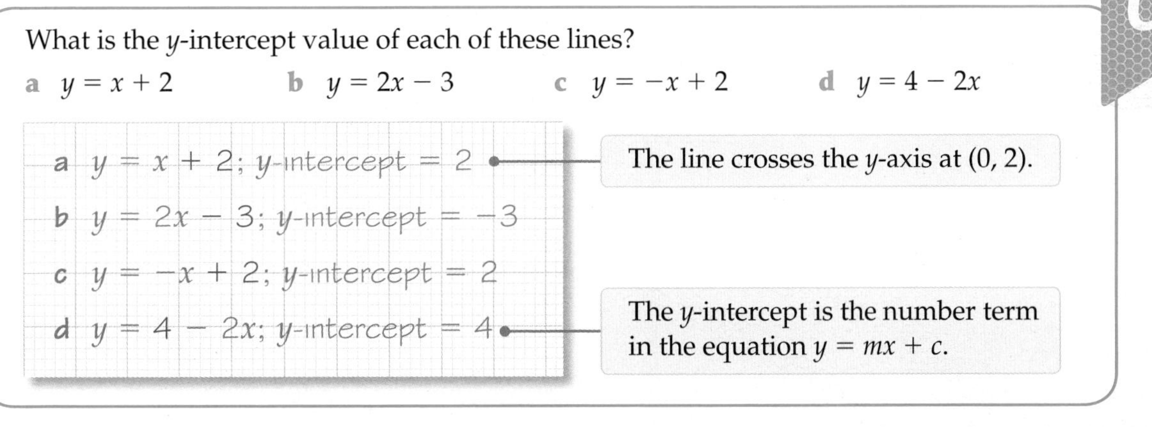

Exercise 18E

C

1 Write down the y-intercepts of these lines.

a $y = 2x + 4$ b $y = -2x + 1$ c $y = x - 4$
d $y = 2x$ e $y = 3 + x$ f $2 + 3x = y$

C

2 Look at these equations of straight lines.

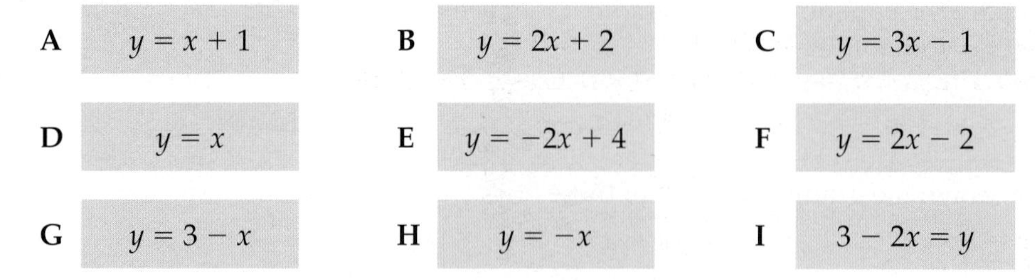

a Write the equation of the line with the steepest gradient.

b Which of the lines are parallel to each other?

c Write the equation of the line that crosses the y-axis at the highest point.

d Write the equation of the line that crosses the y-axis at the lowest point.

Using $y = mx + c$

You can read the gradient and y-intercept of a straight-line graph from its equation in the form $y = mx + c$.

You may have to rearrange the equation first.

C

Example 6

Work out the y-intercepts of these straight lines.

a $2y = x + 4$ **b** $2x - 4y = 6$

a $2y = x + 4$ — The equation needs to be rearranged into the form $y = mx + c$.

$\frac{2y}{2} = \frac{x}{2} + \frac{4}{2}$ — Divide both sides by 2.

$y = \frac{1}{2}x + 2$

The y-intercept is 2. — The line crosses the y-axis at (0, 2).

b $\frac{2x}{4} - \frac{4y}{4} = \frac{6}{4}$ — Divide both sides by 4 to get $1y$.

$\frac{1}{2}x - y = \frac{3}{2}$

$\frac{1}{2}x - y + y = \frac{3}{2} + y$ — Add y to both sides.

$\frac{1}{2}x = \frac{3}{2} + y$

$\frac{1}{2}x - \frac{3}{2} = \frac{3}{2} - \frac{3}{2} + y$ — Subtract $\frac{3}{2}$ from both sides

$\frac{1}{2}x - \frac{3}{2} = y$ — The y can be on the left- or the right-hand side of the equation.

The y-intercept is $-\frac{3}{2}$.

Exercise 18F

1 Write these equations in the form $y = mx + c$.
Then write down the gradient and y-intercept of each line.

a $2y = 2x + 6$ b $2y + 4 = x$
c $y - 2 = -2x$ d $4y = -4x + 8$
e $2y - 7 = -2x$ f $-3x = 4 + y$
g $2y + 3x = 2$ h $4y + 2 + 3x = 0$

C

2 Without plotting these straight lines, identify the ones that are parallel to the line $y = -3x - 2$.

A $y = 3x - 2$ B $y = -3x + 2$
C $y = 2 + 3x$ D $2y = -2 + 3x$
E $x = -3y - 2$ F $3x = -y - 2$

Remember that the equation needs to be in the form $y = mx + c$.

C AO2

Finding the equation of a straight line

You can write the equation of a line in the form $y = mx + c$, where m is the gradient and c is the y-intercept.

The x- and y-coordinates of *any* point on the line satisfy (fit) the equation of the line.

Example 7

B AO2

A line passes through the points $A(4, 0)$ and $B(5, 2)$.
Work out

a the gradient of the line b the equation of the line.

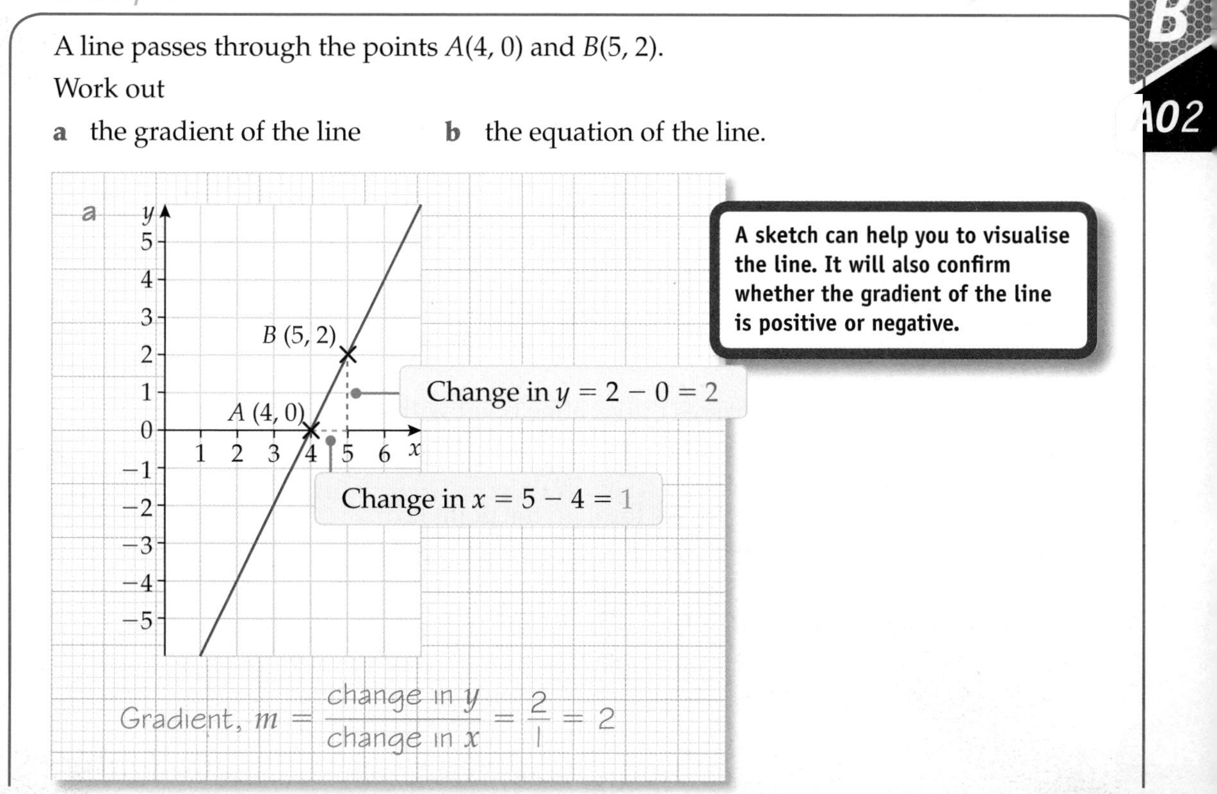

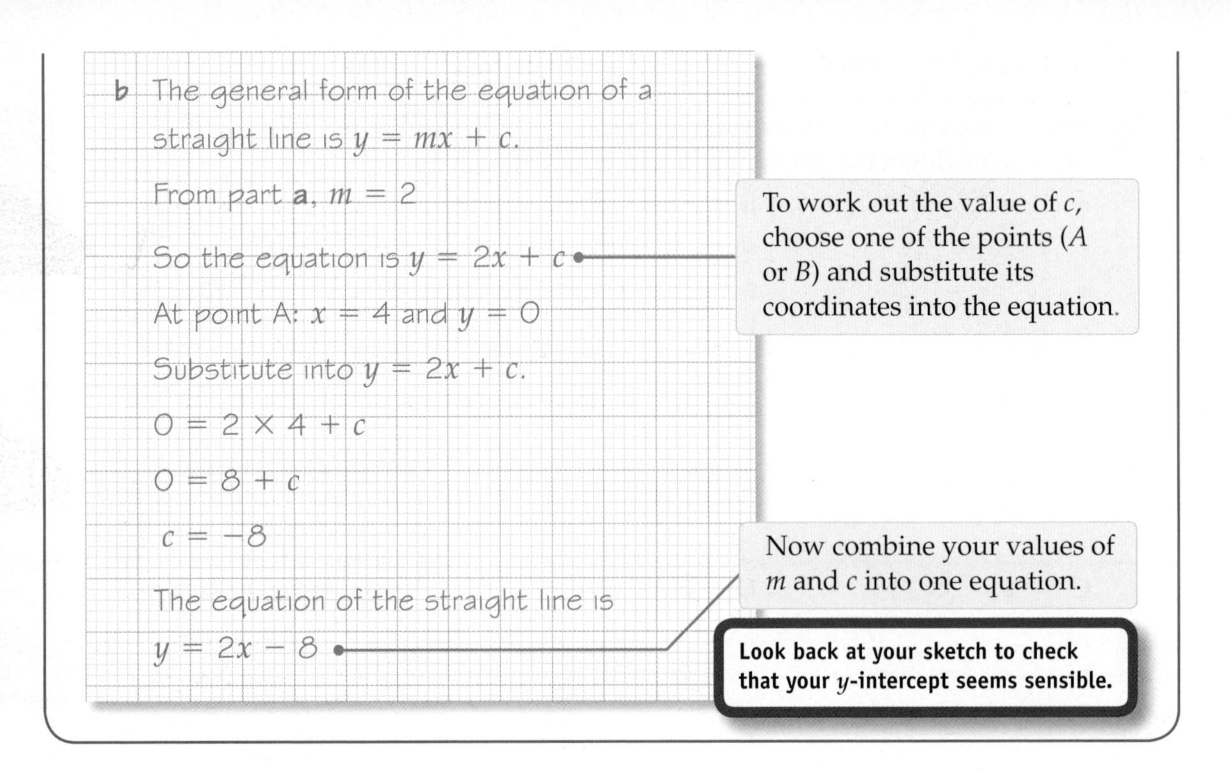

b The general form of the equation of a straight line is $y = mx + c$.

From part **a**, $m = 2$

So the equation is $y = 2x + c$

At point A: $x = 4$ and $y = 0$

Substitute into $y = 2x + c$.

$0 = 2 \times 4 + c$

$0 = 8 + c$

$c = -8$

The equation of the straight line is

$y = 2x - 8$

To work out the value of c, choose one of the points (A or B) and substitute its coordinates into the equation.

Now combine your values of m and c into one equation.

Look back at your sketch to check that your y-intercept seems sensible.

Exercise 18G

B

1 Write down the equation of each line shown on the grid.

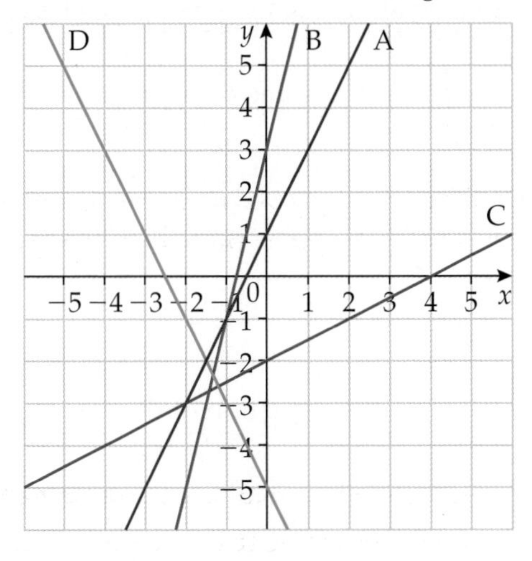

2 Work out the gradient of the line passing through each pair of points.

a (2, 4) and (6, 2)

b (−3, 0) and (0, 3)

3 A line has gradient 3 and passes through the point with coordinates (0, 6).
Work out the equation of the line.

4 A line has gradient −2 and passes through the point (2, 5).
Work out the equation of the line.

B

A02

5 A line passes through the points $A(1, 4)$ and $B(6, 3)$.
Work out

a the gradient of the line

b the equation of the line.

6 A straight line parallel to $y = -4x + 5$ passes through the origin.
Write down the equation of this line.

B

7 A straight line parallel to $y = 3x + 2$ passes through the point $(0, -2)$.
What is the equation of this line?

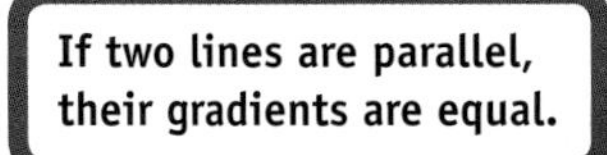

A02

8 Work out the equation of the line that passes through the points $M(4, 1)$ and $N(8, 3)$.

B

A03

18.4 Conversion graphs

Keywords
conversion graph

Why learn this?
Conversion graphs can be used to convert quickly between one currency and another.

Objectives
E Plot and use conversion graphs

Skills check

HELP Section 7.6

1 2 pens cost 42p. How would 4 pens cost?
2 5 bars of chocolate cost £2.50. How much does 1 bar cost?

Conversion graphs

A **conversion graph** converts values from one unit into another, for example amounts of money from pounds sterling (£) to dollars ($) or temperatures from degrees Celsius (°C) to degrees Fahrenheit (°F).

Example 8

E

The conversion rate from pounds to US dollars is approximately £1 = US$1.5.

a Copy and complete this conversion table between pounds and US dollars.

£ (x)	0	2	4	8	10	20
US$ (y)	0	3			15	

b Draw a conversion graph with x-values from 0 to £20 and y-values from 0 to US$40.

c Use your graph to convert
i £12 to US dollars **ii** US$10 to pounds.

a

£ (x)	0	2	4	8	10	20
US$ (y)	0	3	6	12	15	30

This is the coordinate pair (10, 15).

£10 = US$15 so £20 = US$30.

£2 = US$3, so £4 = US$6.

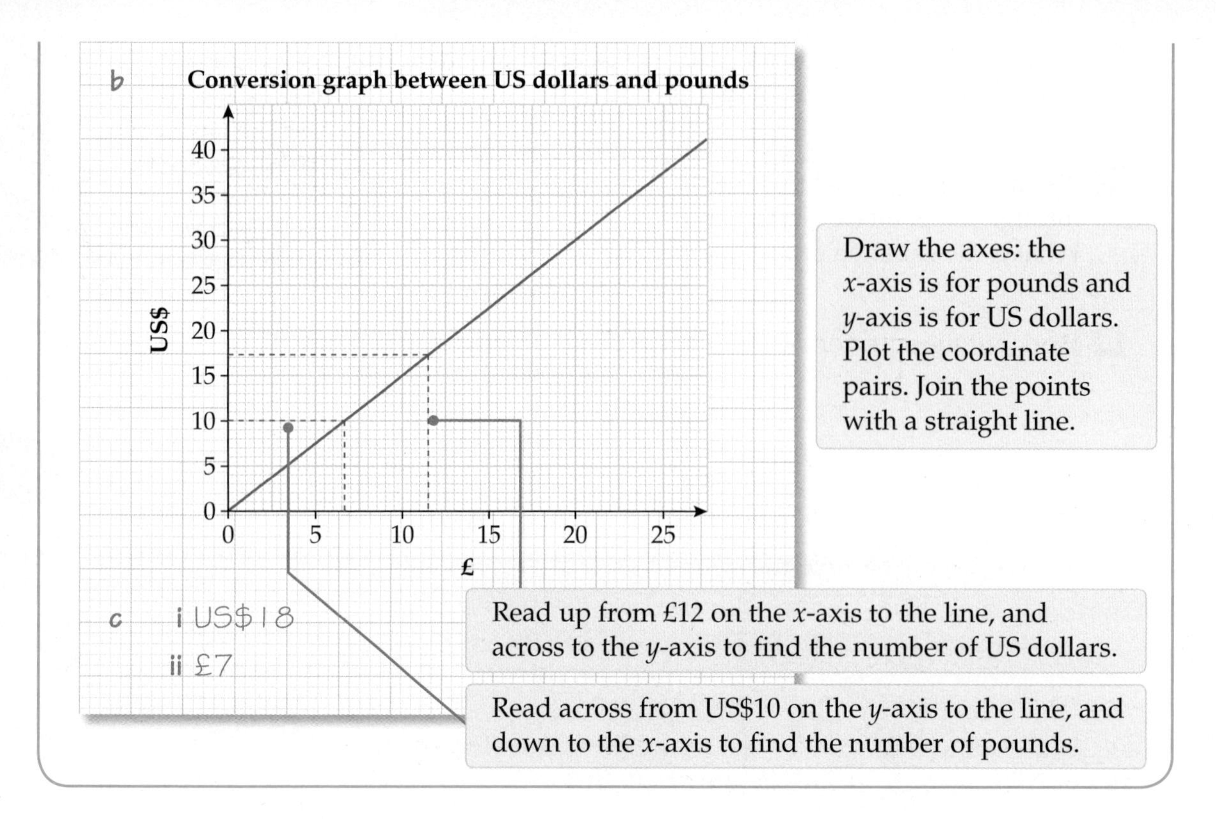

Exercise 18H

E

1 The conversion rate from pounds to Saudi Arabian riyals is approximately £1 = 5.5 riyals.

a Copy and complete this conversion table between pounds and riyals.

£ (x)	0	20	40
Riyals (y)		110	

b Draw a conversion graph with x-values from 0 to £40 and y-values from 0 to 220 riyals.

c Use your graph to convert
i £30 to riyals **ii** 70 riyals to pounds.

d Use your answers to part **c** to work out the value of
i 140 riyals in pounds **ii** £90 in riyals.

2 **a** Draw a coordinate grid with the x-axis going from −100 °C to 100 °C and the y-axis going from −100 °F to 240 °F.

b 60 °C is equivalent to 140 °F and −40 °C is equivalent to −40 °F.
Use this information to draw a conversion graph between degrees Celsius and degrees Fahrenheit.

Only two points are needed to draw a conversion graph but three points provide a good check.

c Use your conversion graph to convert these temperatures to degrees Fahrenheit.
i 20 °C **ii** 50 °C **iii** −20 °C **iv** −60 °C

d Use your conversion graph to convert these temperatures to degrees Celsius.
i 60 °F **ii** 10 °F **iii** −30 °F **iv** −80 °F

e Water freezes at 0 °C. What temperature is this in Fahrenheit?

f Javier wants to go on holiday to a hot country.
In June, the average temperature in Malta is 30 °C and the average temperature in Germany is 70 °F.
Which of these places should Javier choose for his holiday?

AO2

18.5 Real-life graphs

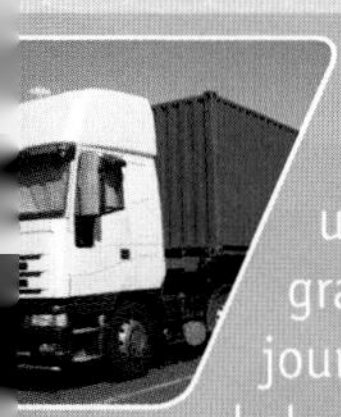

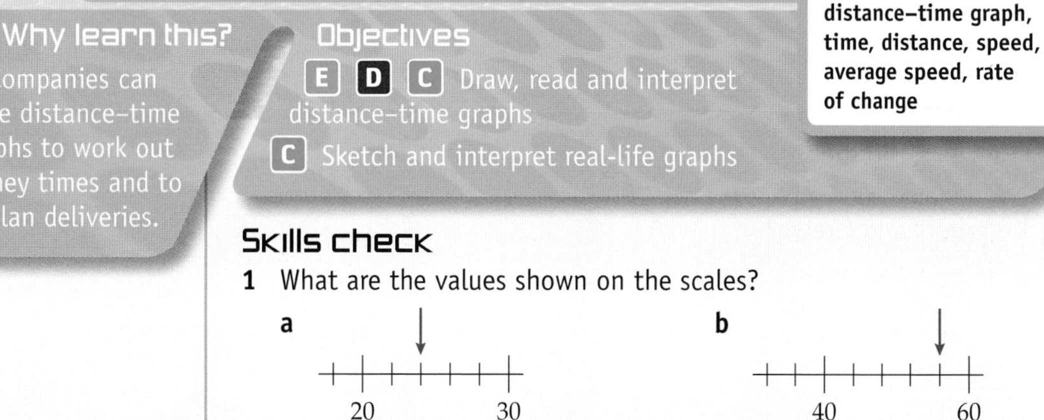

Why learn this?

Companies can use distance–time graphs to work out journey times and to help plan deliveries.

Objectives

E **D** **C** Draw, read and interpret distance–time graphs

C Sketch and interpret real-life graphs

Keywords

distance–time graph, time, distance, speed, average speed, rate of change

Skills check

1 What are the values shown on the scales?

a

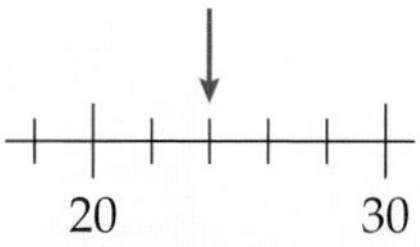

b

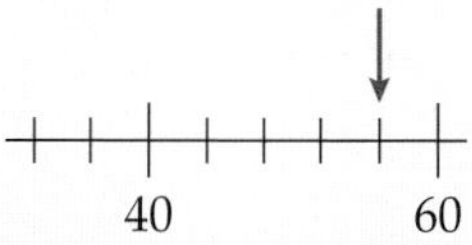

Distance–time graphs

A **distance–time graph** represents a journey.

The x-axis (horizontal) represents the **time** taken.

The y-axis (vertical) represents the **distance** from the starting point.

The gradient represents the **speed** of the journey.

Vikram's sponsored walk

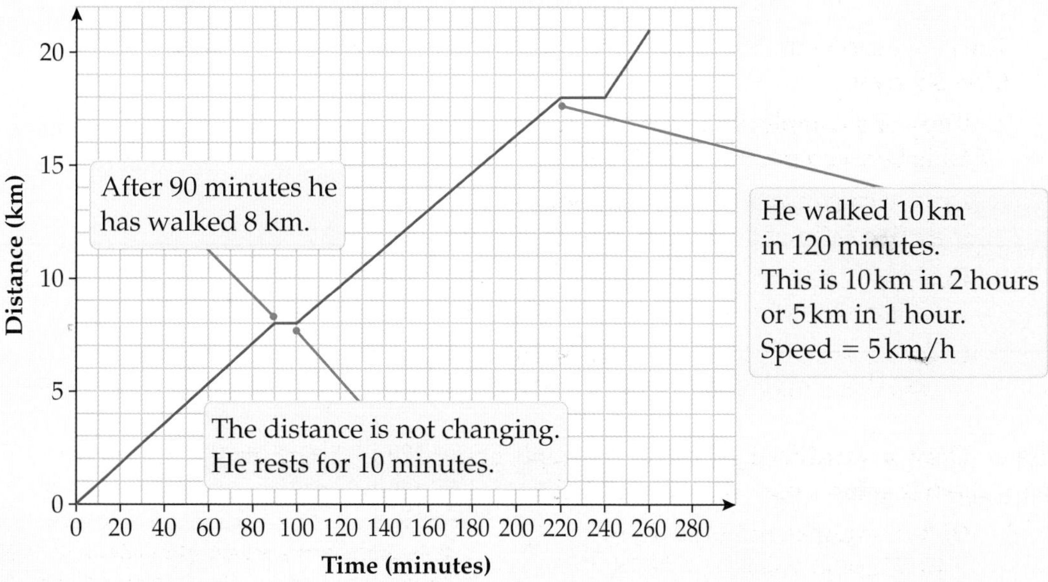

$$\text{Speed} = \frac{\text{distance}}{\text{time}}$$

The units of speed can be metres per second (m/s), kilometres per hour (km/h) or miles per hour (mph).

You can work out the **average speed** of a complete journey using this formula.

$$\text{Average speed} = \frac{\text{total distance}}{\text{total time}}$$

D

Example 9

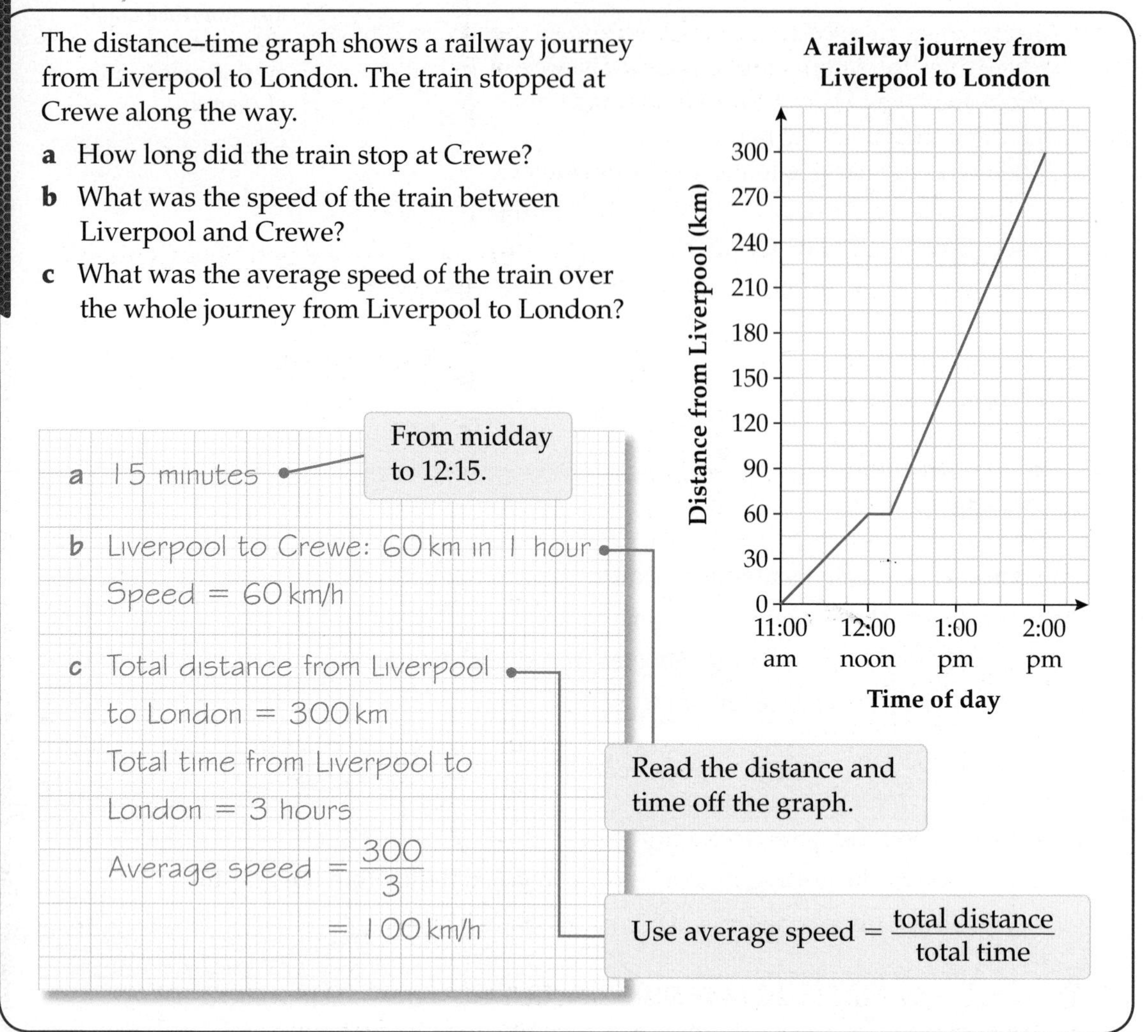

The distance–time graph shows a railway journey from Liverpool to London. The train stopped at Crewe along the way.

a How long did the train stop at Crewe?

b What was the speed of the train between Liverpool and Crewe?

c What was the average speed of the train over the whole journey from Liverpool to London?

a 15 minutes — From midday to 12:15.

b Liverpool to Crewe: 60 km in 1 hour

Speed = 60 km/h

Read the distance and time off the graph.

c Total distance from Liverpool to London = 300 km

Total time from Liverpool to London = 3 hours

Average speed $= \frac{300}{3}$

$= 100$ km/h

Use average speed $= \frac{\text{total distance}}{\text{total time}}$

C

Example 10

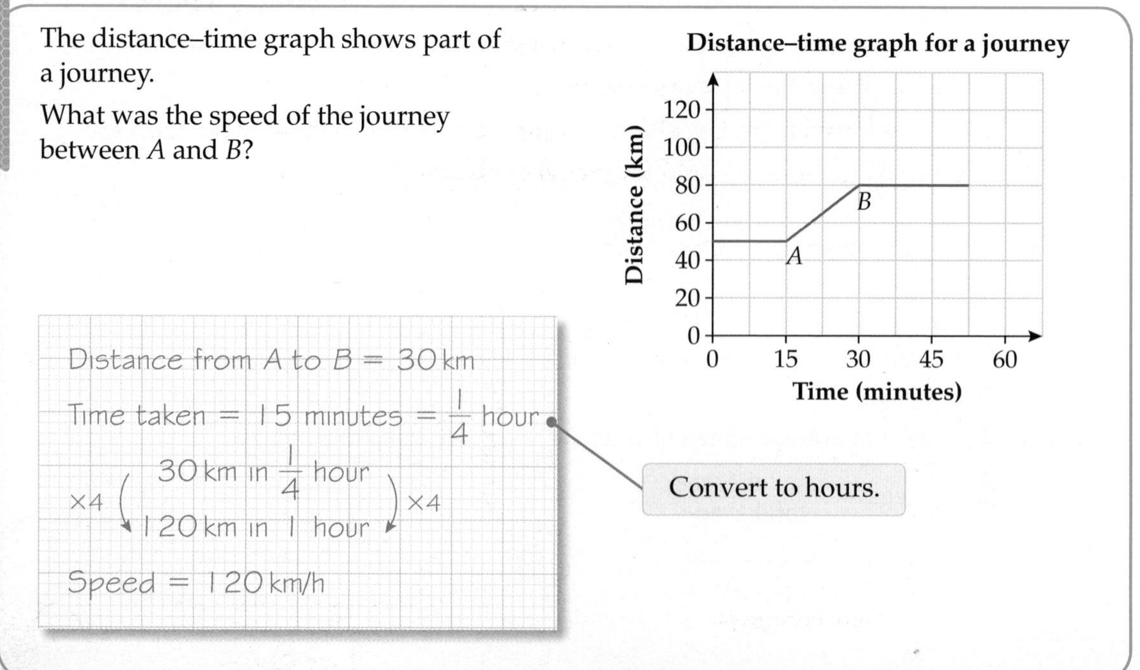

The distance–time graph shows part of a journey.

What was the speed of the journey between A and B?

Distance from A to B = 30 km

Time taken = 15 minutes $= \frac{1}{4}$ hour — Convert to hours.

×4 (30 km in $\frac{1}{4}$ hour) ×4

120 km in 1 hour

Speed = 120 km/h

Exercise 18I

E

1 Jasmina is driving home to visit her parents who live 200 km away.
She drives for 2 hours and covers a distance of 150 km.
Then she takes a break for half an hour.
Jasmina then resumes her journey.
She arrives at her destination $1\frac{1}{2}$ hours later.

a Draw a distance–time graph to show Jasmina's journey.

b Was Jasmina travelling faster during the first or the second part of her journey? Give a reason for your answer.

D

2 On Saturday Yossi went for a bike ride. He stopped twice for breaks before returning home. The graph shows his journey.

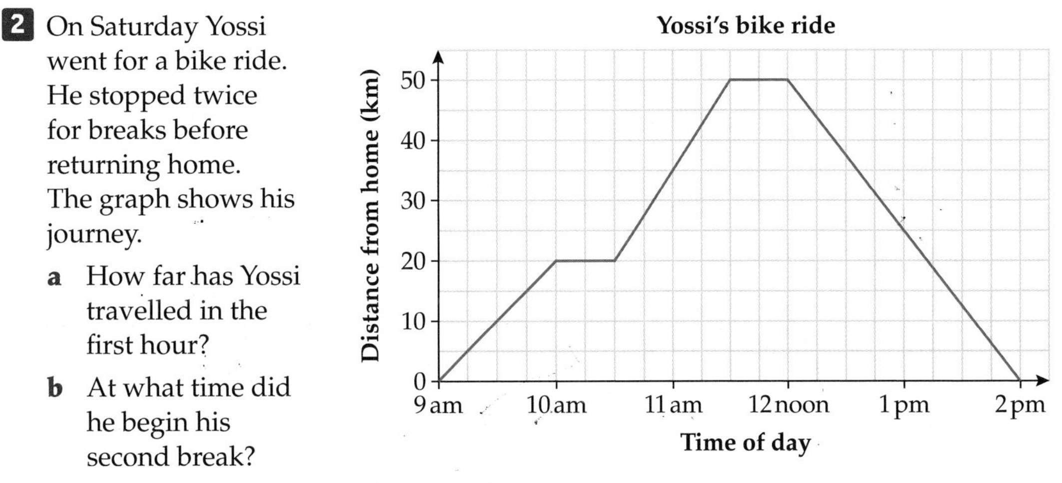

a How far has Yossi travelled in the first hour?

b At what time did he begin his second break?

c How long did Yossi's bike ride take?

d When was he cycling fastest? How can you tell this from the graph?

e Work out his average speed for the whole bike ride.

3 Ryan is travelling to the Lake District on his holidays.
He begins his journey at 1 pm and travels 50 km in the first hour.
Between 2 pm and 3 pm he travels only 25 km owing to heavy traffic.
At 3 pm Ryan stops for a half-hour break.
He reaches the Lake District after a further 2 hours and a further distance of 150 km.

a Draw a distance–time graph for Ryan's journey.

b When did Ryan travel most slowly?

c What speed was Ryan travelling at during the slowest section of the journey?

d What was his average speed for the whole journey?

C

4 Work out the speed of each journey.

a

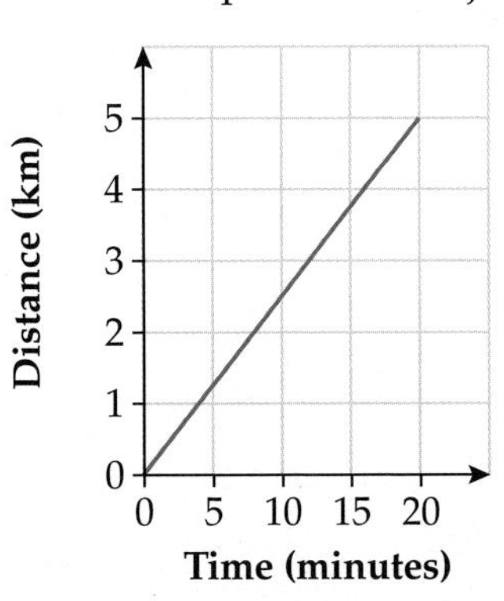

b

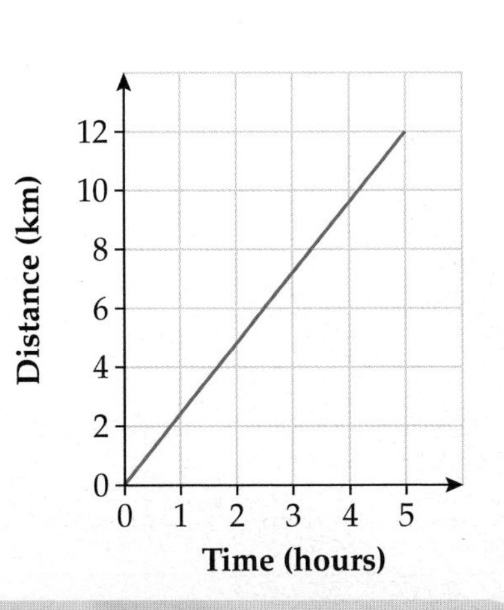

C

5 Llinos is a courier.
She delivers parcels across North Wales from the depot in Llangefni.
She leaves the depot at 8 am.

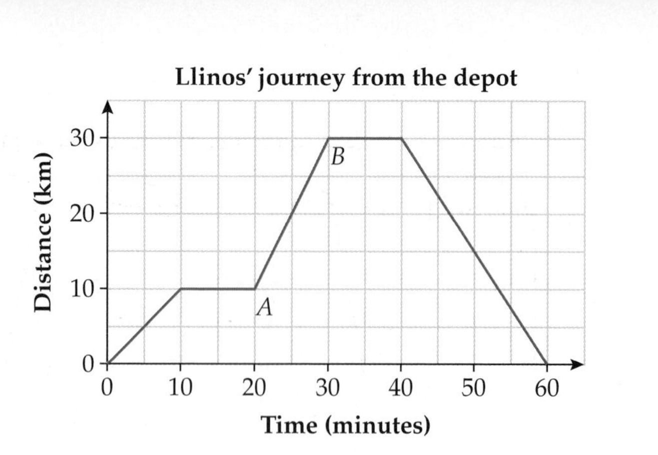

a How far has Llinos travelled in the first 10 minutes?

b At what time does Llinos arrive at her first delivery point?

c Calculate Llinos' speed during section AB.

d At what time does Llinos begin the return journey to the depot?

e Calculate Llinos' average speed for the whole journey.

Other real-life graphs

A straight-line graph shows that the **rate of change** is steady.

A curved graph shows that the rate of change varies.

The steeper the line, the faster the rate of change.

C

Example 11

Water is poured at a constant rate into each of the following containers.

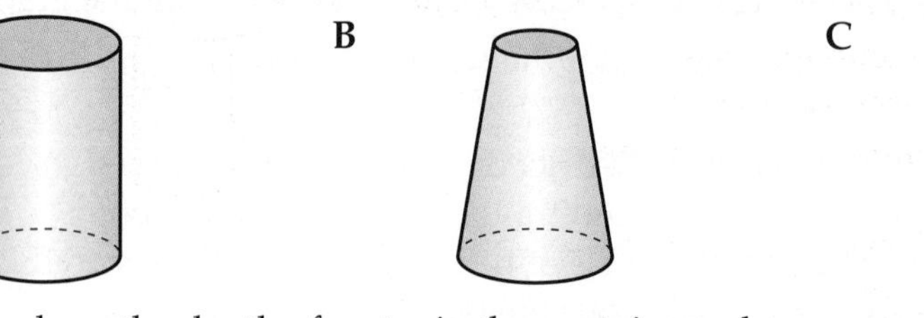

The graphs show how the depth of water in the containers changes over time.

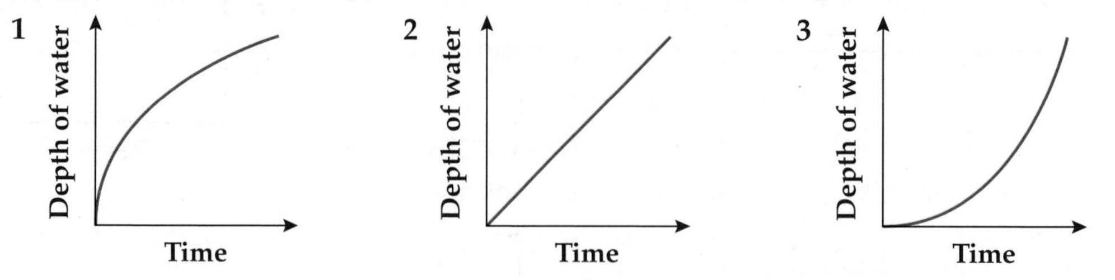

AO3

Match each container to the correct graph. Explain your answers.

Container A matches graph 2: the container has a uniform shape, so it will fill at a constant rate (straight line).

Container B matches graph 3: the container will fill slowly at first (due to its wide bottom) and then more quickly as the container narrows.

Container C matches graph 1: the container will fill quickly at first (due to its narrow bottom) and then more slowly as the container widens.

Exercise 18J

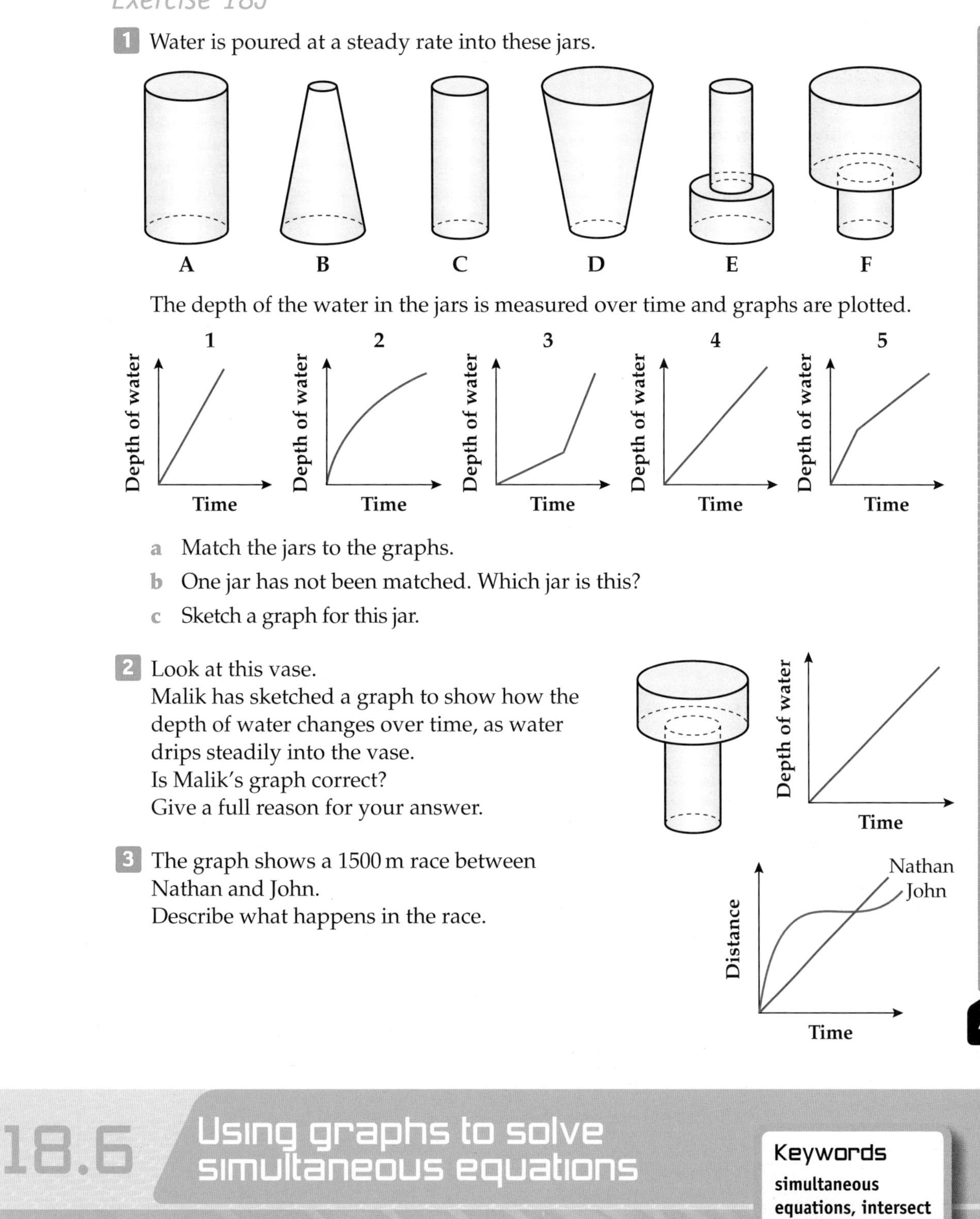

1 Water is poured at a steady rate into these jars.

The depth of the water in the jars is measured over time and graphs are plotted.

a Match the jars to the graphs.

b One jar has not been matched. Which jar is this?

c Sketch a graph for this jar.

2 Look at this vase.
Malik has sketched a graph to show how the depth of water changes over time, as water drips steadily into the vase.
Is Malik's graph correct?
Give a full reason for your answer.

3 The graph shows a 1500 m race between Nathan and John.
Describe what happens in the race.

18.6 Using graphs to solve simultaneous equations

Keywords
simultaneous equations, intersect

Why learn this?

Oil refineries use simultaneous equations to optimise their production of products from crude oil.

Objectives

B Use a graphical method to solve simultaneous equations

Skills check

HELP Section 18.2

1 Draw and label these graphs.

a $x + 3y = 6$ **b** $4x + y = 8$ **c** $3x - y = 9$

Graphs of simultaneous equations

You can solve a pair of **simultaneous equations** by drawing their graphs on the same set of axes.

Solving simultaneous equations algebraically was covered in Section 14.7.

The point on the graph where the two lines **intersect** (cross) is their solution.

Graphical solutions to simultaneous equations are only approximate, as they depend on the accuracy of the drawing.

Example 12

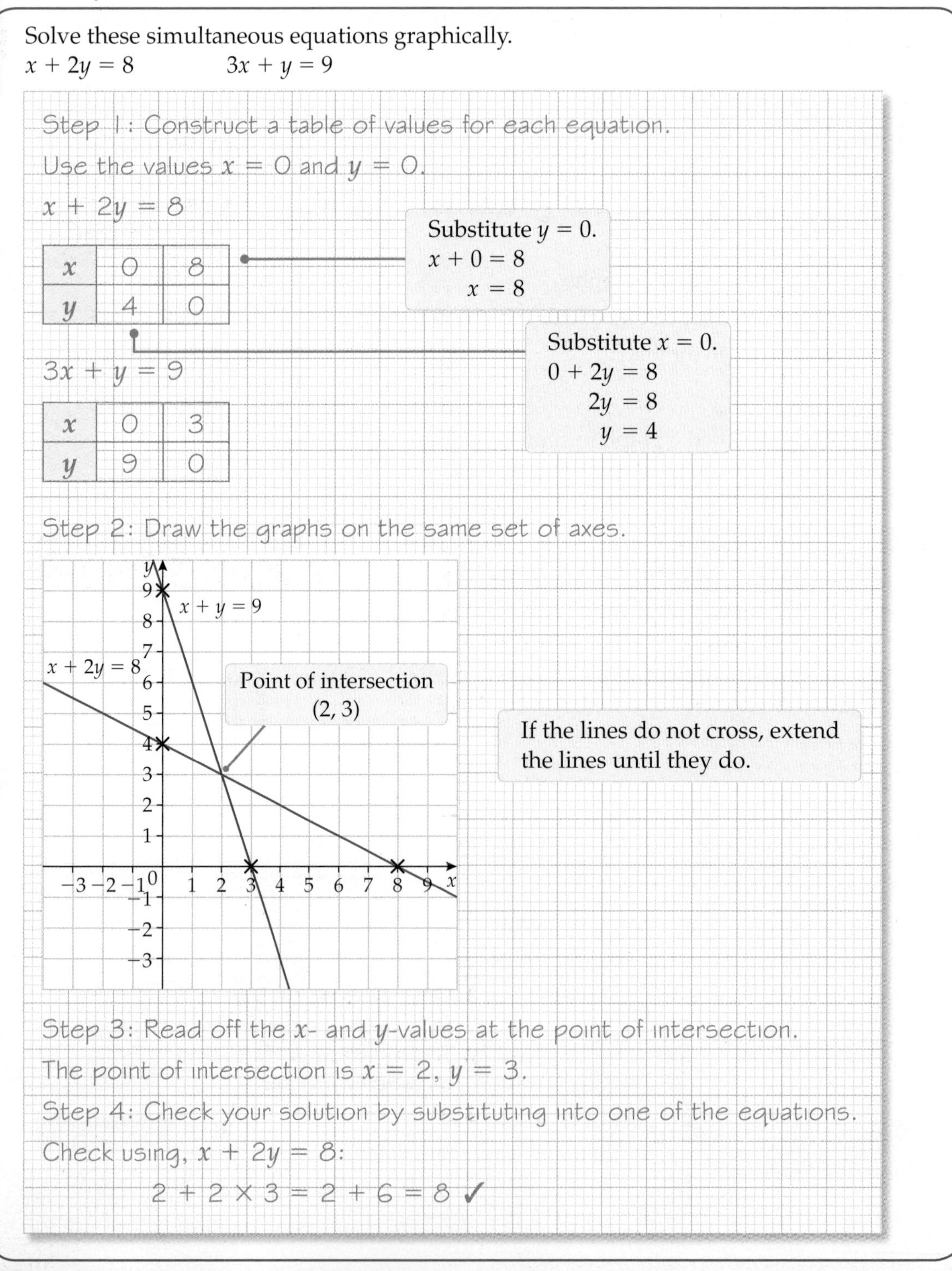

Solve these simultaneous equations graphically.

$x + 2y = 8 \qquad 3x + y = 9$

Step 1: Construct a table of values for each equation.

Use the values $x = 0$ and $y = 0$.

$x + 2y = 8$

x	0	8
y	4	0

Substitute $y = 0$.
$x + 0 = 8$
$x = 8$

Substitute $x = 0$.
$0 + 2y = 8$
$2y = 8$
$y = 4$

$3x + y = 9$

x	0	3
y	9	0

Step 2: Draw the graphs on the same set of axes.

If the lines do not cross, extend the lines until they do.

Step 3: Read off the x- and y-values at the point of intersection.

The point of intersection is $x = 2$, $y = 3$.

Step 4: Check your solution by substituting into one of the equations.

Check using, $x + 2y = 8$:

$2 + 2 \times 3 = 2 + 6 = 8$ ✓

Exercise 18K

1 Solve each pair of simultaneous equations graphically.

a $x + y = 4$
$x + 2y = 3$

b $3x - y = -3$
$x + y = -1$

2 Find a graphical solution to these simultaneous equations. $2x + y = 6$
$x - 2y = 8$

3 Boat A is travelling along the path $-x + 2y = 10$.
Boat B is travelling along the path $y = 2x + 2$.
Use a graphical method to determine whether their paths will cross.
If appropriate, state the coordinates of where they cross.

B

B

AO2

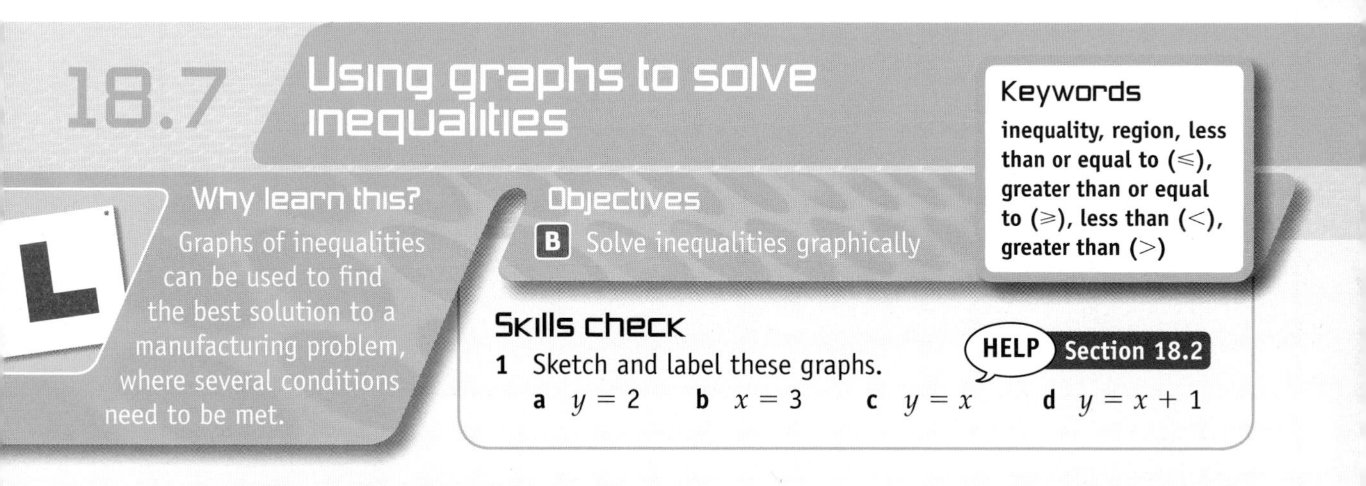

18.7 Using graphs to solve inequalities

Keywords
inequality, region, less than or equal to (≤), greater than or equal to (≥), less than (<), greater than (>)

Why learn this?
Graphs of inequalities can be used to find the best solution to a manufacturing problem, where several conditions need to be met.

Objectives
B Solve inequalities graphically

Skills check

1 Sketch and label these graphs.
a $y = 2$ **b** $x = 3$ **c** $y = x$ **d** $y = x + 1$

HELP Section 18.2

Inequalities

Inequalities can be shown as a shaded **region** on a graph.

HELP To recap inequalities see Section 14.6.

For the inequalities ≤ **(less than or equal to)** and ≥ **(greater than or equal to)**, the boundary is a solid line.
The points on the line *are* included in the region.

For the inequalities < **(less than)** and > **(greater than)**, the boundary is a dashed line.
The points on the line are *not* included in the region.

Example 13

B

Show the regions defined by these inequalities.

a $x \geqslant 3$ **b** $y < 1$ **c** $x + y > 2$

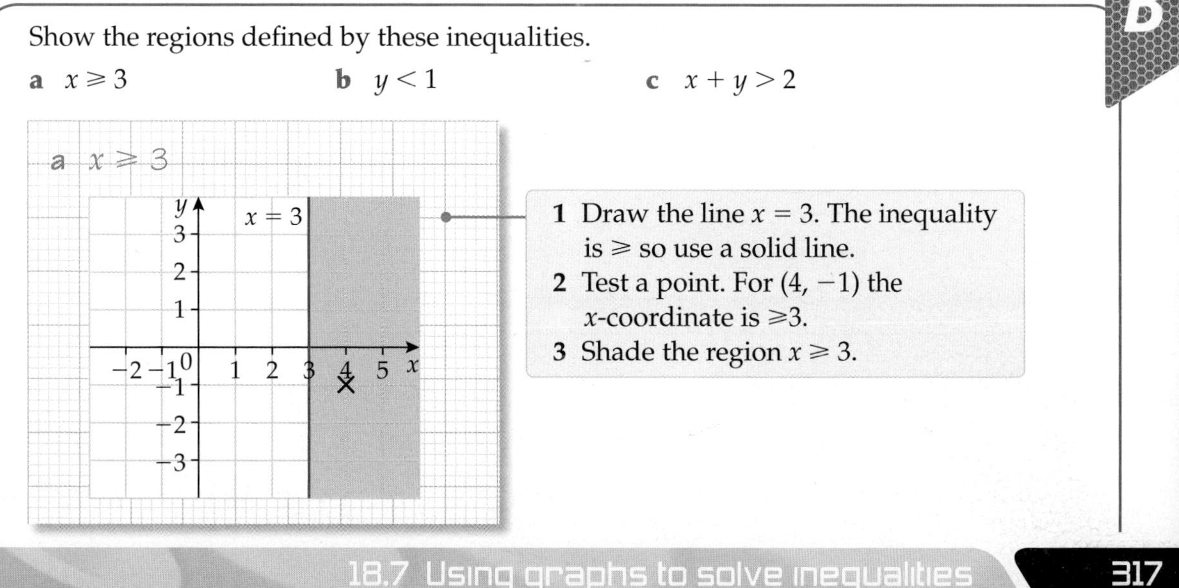

1 Draw the line $x = 3$. The inequality is ≥ so use a solid line.
2 Test a point. For $(4, -1)$ the x-coordinate is ≥3.
3 Shade the region $x \geqslant 3$.

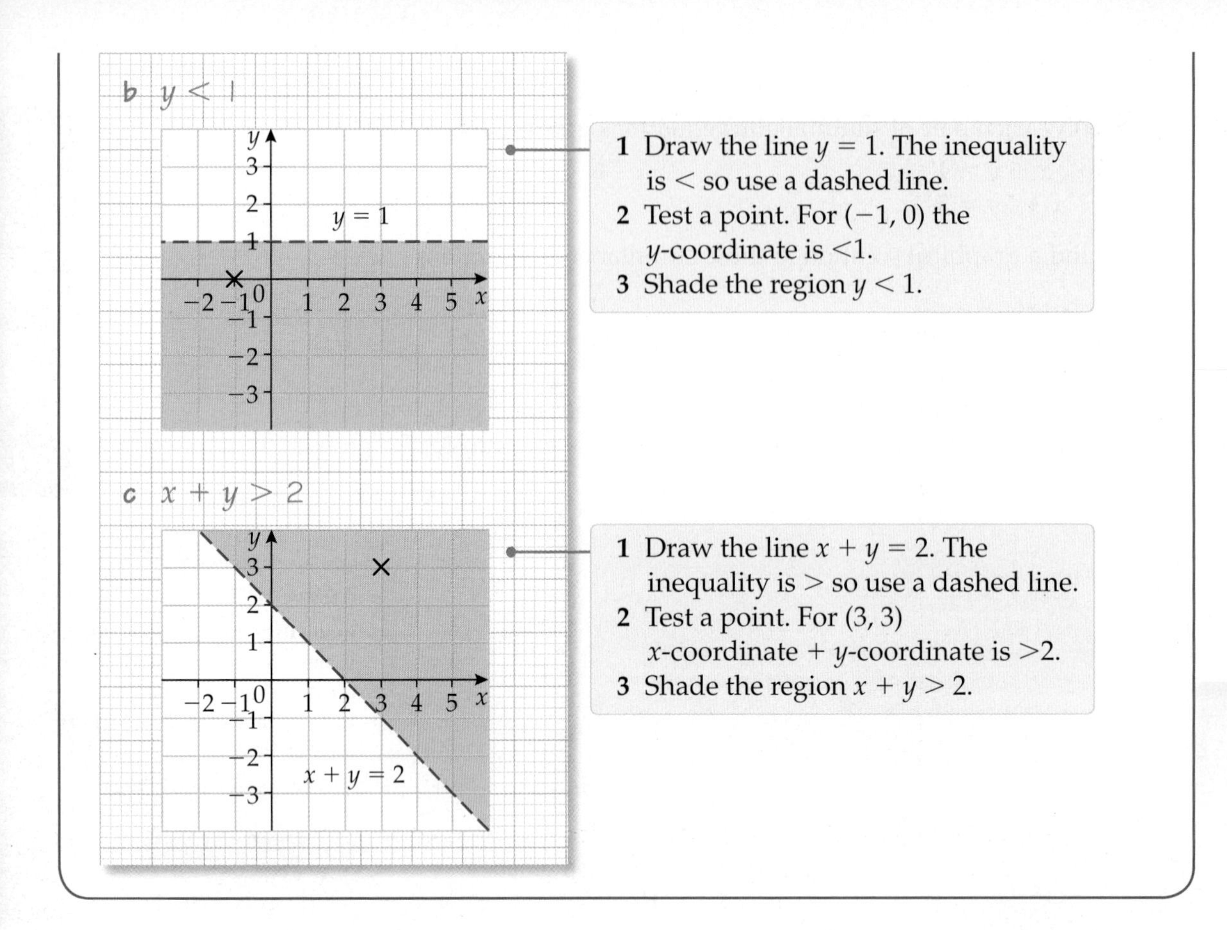

1. Draw the line $y = 1$. The inequality is $<$ so use a dashed line.
2. Test a point. For $(-1, 0)$ the y-coordinate is <1.
3. Shade the region $y < 1$.

1. Draw the line $x + y = 2$. The inequality is $>$ so use a dashed line.
2. Test a point. For $(3, 3)$ x-coordinate + y-coordinate is >2.
3. Shade the region $x + y > 2$.

Example 14

On a grid show the region that is defined by these three inequalities.

$x \geqslant 1$ $\qquad y \leqslant 3$ $\qquad x + y \leqslant 5$

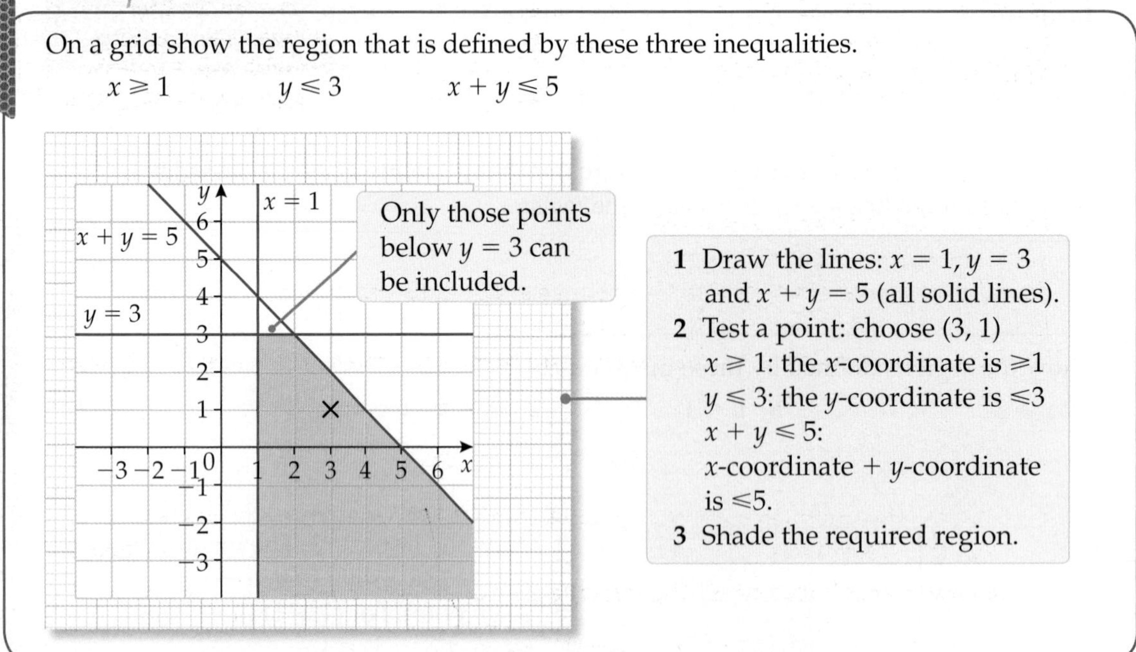

1. Draw the lines: $x = 1$, $y = 3$ and $x + y = 5$ (all solid lines).
2. Test a point: choose $(3, 1)$
   $x \geqslant 1$: the x-coordinate is $\geqslant 1$
   $y \leqslant 3$: the y-coordinate is $\leqslant 3$
   $x + y \leqslant 5$:
   x-coordinate + y-coordinate is $\leqslant 5$.
3. Shade the required region.

Exercise 18L

B

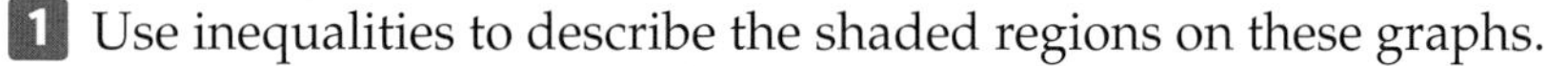

1 Use inequalities to describe the shaded regions on these graphs.

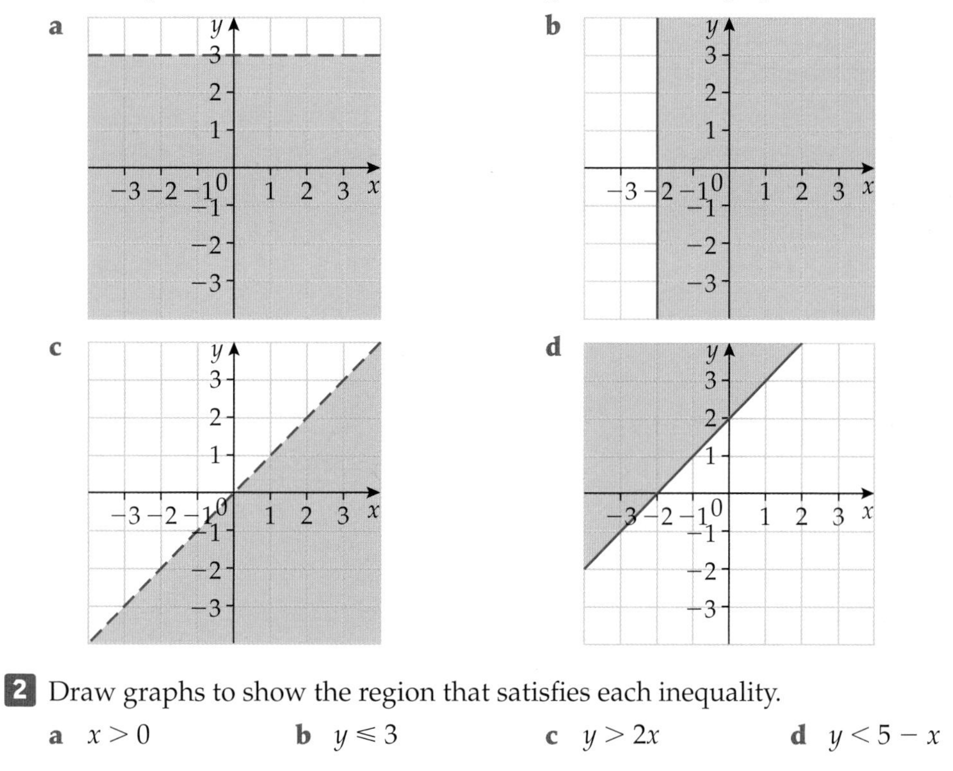

2 Draw graphs to show the region that satisfies each inequality.

a $x > 0$ **b** $y \leqslant 3$ **c** $y > 2x$ **d** $y < 5 - x$

3 Draw graphs to show the region which satisfies each set of inequalities.

a $y > 1$ and $x + y < 5$ **b** $y \leqslant 0$ and $y \geqslant x - 3$

4 Sketch the region defined by these three inequalities.

$x \geqslant 0$ $y < 2x + 1$ $x + y \leqslant 4$

Mark the region with an 'R'.

Draw all the boundary lines before deciding which region to shade.

Review exercise

E

1 Dirk is going to bake a chocolate cake.
The recipe gives weights in ounces.
He needs a conversion graph to help him convert between ounces and grams.
Use the conversion 1 ounce $\approx$ 28 grams.

a Copy and complete this conversion table between ounces and grams. **[1 mark]**

Ounces (x)	0	1	2	4	7
Grams (y)			56		196

b Draw a conversion graph with x-values from 0 to 10 ounces and y-values from 0 to 220 grams. **[2 marks]**

c Use your graph to convert these recipe measurements to grams.

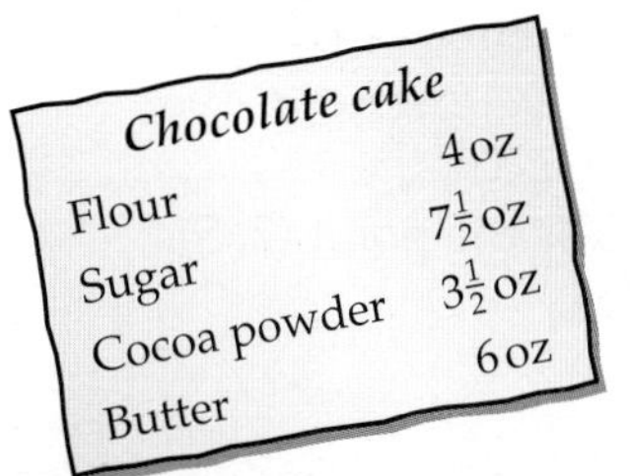

[2 marks]

D

2 Draw the graph of $y = 2x - 3$.

Draw a coordinate grid with both x- and y-axes from -4 to $+5$.

[2 marks]

D

3 **a** Copy and complete this table of values for $y = 2x + 1$. [1 mark]

x	-3	-2	-1	0	1	2	3
y		-3	-1				7

b Draw a coordinate grid with the x-axis from -4 to $+4$ and the y-axis from -8 to $+8$.
Draw the graph of $y = 2x + 1$. [1 mark]

AO2

c Work out the coordinates of the point where the line $y = 2x + 1$ crosses the line $y = -4$. [1 mark]

D

4 The line $y = 3x - 1$ crosses the line $x = -2$ at the point A.
Work out the coordinates of point A. [2 marks]

AO3

C

5 The distance–time graph shows the journey of a train between two stations.
The stations are 6 km apart.

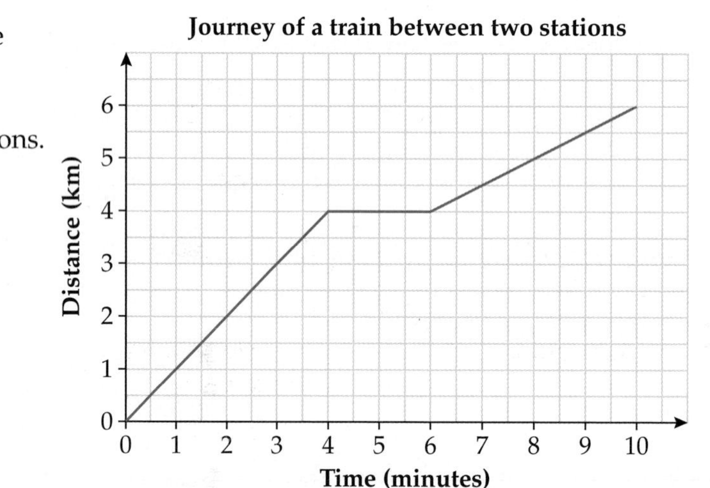

a During the journey the train had to stop at a red signal.
How long was the train stopped? [1 mark]

b What was the average speed of the train for the whole journey?
Give your answer in kilometres per hour. [2 marks]

6 Two rockets are launched.
The path of rocket 1 has the function $x = \frac{1}{2}(y - 4)$.
The path of rocket 2 has the function $x - \frac{1}{2}y - 2 = 0$.

a Rearrange each of these functions into the form $y = mx + c$. [2 marks]

b Draw a graph of each function. [2 marks]

C

7 M is the mid-point of the line LN.
M has coordinates $(4, -1)$. N is the point $(2, 5)$.
Work out the coordinates of L. [2 marks]

AO3

8 Water is poured at a steady rate into this bottle.

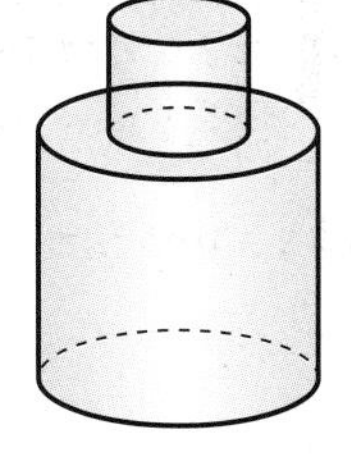

As the bottle is filled, the depth, d, of water in the bottle changes over time, t. Which of the five graphs shows this change? Give a full reason for your choice.

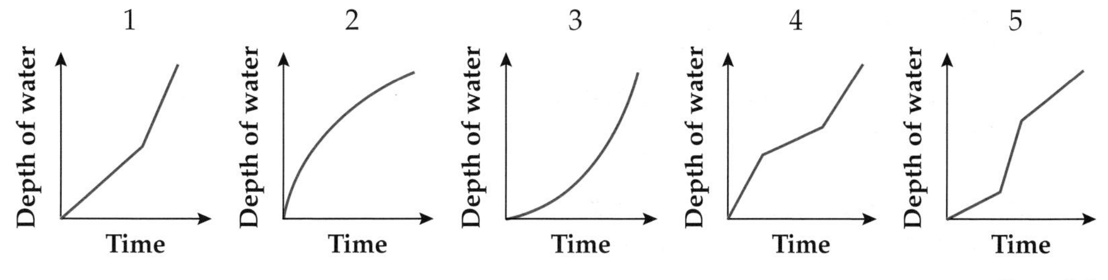

[2 marks]

C AO3

9 Work out the equation of the line that is parallel to the line $y = 3x + 5$ and which passes through the point (2, 9). **[2 marks]**

B AO3

10 Solve this pair of simultaneous equations graphically.

$2x - y = 1$

$3x + 2y = 12$ **[4 marks]**

B

11 Draw x- and y-axes from -6 to $+6$.
Indicate the region defined by these three inequalities.

$x \geqslant -3$ $\quad y < 4$ $\quad y \geqslant x + 2$ **[3 marks]**

Chapter summary

In this chapter you have learned how to

- recognise straight-line graphs parallel to the x- or y-axis **E**
- plot and use conversion graphs **E**
- plot graphs of linear functions **D** **C**
- work out coordinates of points of intersection when two graphs cross **D**
- draw, read and interpret distance–time graphs **E** **D** **C**
- find the mid-point of a line segment **D** **C**
- plot straight-line graphs **D** **C**
- find the gradient of a straight-line graph **C**
- understand the meaning of m and c in the equation $y = mx + c$ **C**
- find the equation of a line **B**
- sketch and interpret real-life graphs **C**
- use a graphical method to solve simultaneous equations **B**
- solve inequalities graphically **B**

19

Quadratic equations

This chapter is about quadratics.

In nature, the growth of a population of rabbits can be modelled by a quadratic equation.

Objectives

This chapter will show you how to

- **factorise quadratic expressions, including the difference of two squares B**
- **solve quadratic equations by rearranging and factorising B**

Before you start this chapter

1 Expand

a $3(m + 4)$ b $5(x^2 - 2)$ c $x(x - 3)$

2 Factorise

a $8x + 4$ b $3y^2 - 2y$ c $6m^3 - m$

3 Expand

a $(x + 5)(x + 2)$ b $(x - 1)(x + 3)$ c $(x - 6)(x - 2)$

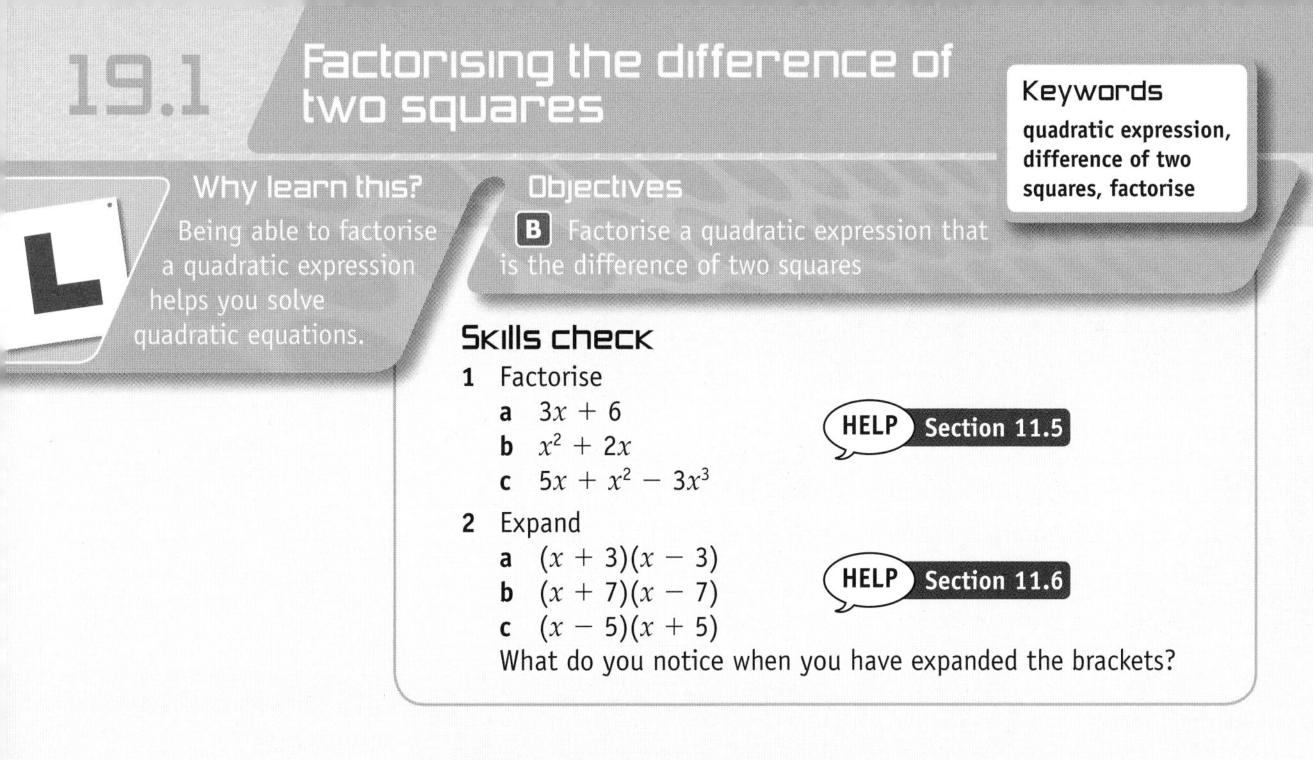

19.1 Factorising the difference of two squares

Keywords
quadratic expression, difference of two squares, factorise

Why learn this?
Being able to factorise a quadratic expression helps you solve quadratic equations.

Objectives
B Factorise a quadratic expression that is the difference of two squares

Skills check

1. Factorise
   a $3x + 6$
   b $x^2 + 2x$
   c $5x + x^2 - 3x^3$

HELP Section 11.5

2. Expand
   a $(x + 3)(x - 3)$
   b $(x + 7)(x - 7)$
   c $(x - 5)(x + 5)$
   What do you notice when you have expanded the brackets?

HELP Section 11.6

The difference of two squares

A **quadratic expression** is an algebraic expression whose highest power of x is x^2.

They are usually of the form $ax^2 + bx + c$, where a, b and c are numbers and $a \neq 0$.

These are all quadratic expressions.

$3x^2 + 2x + 5$ $\qquad$ $x^2 - 3x - 2$ $\qquad$ $4x^2 + 7$ $\qquad$ $12x^2 - 3x$

These expressions all represent one square number subtracted from another.

$x^2 - 9$ $\qquad$ $x^2 - 16$ $\qquad$ $m^2 - 49$

Expressions like this are called the **difference of two squares**.

In general

$$x^2 - b^2 = (x - b)(x + b)$$

Check this by multiplying out. Look back at Section 11.6, where the grid and FOIL methods for expanding brackets are covered.

Example 1

B

Factorise **a** $x^2 - 9$ **b** $m^2 - 36$

Remember that factorising is the inverse of expanding brackets.

a $x^2 - 9 = x^2 - 3^2$ — Write as 'letter squared − number squared'.

$= (x - 3)(x + 3)$ — Use $x^2 - b^2 = (x - b)(x + b)$ with $b = 3$.

b $m^2 - 36 = m^2 - 6^2$ — Use the difference of two squares. $x^2 - b^2 = (x - b)(x + b)$ with $b = 6$.

$= (m - 6)(m + 6)$

Exercise 19A

1 **a** Copy and complete this statement.

$$x^2 - 81 = (x + \square)(x - \square)$$

b Check your answer to part **a** by expanding the brackets.

2 Factorise

a $x^2 - 16$ **b** $x^2 - 25$ **c** $x^2 - 100$ **d** $x^2 - 144$

e $a^2 - 1$ **f** $m^2 - 64$ **g** $t^2 - 121$ **h** $b^2 - 49$

3 Factorise the expression $x^2 - 400$.

4 Joe thinks of a number, squares it and subtracts 16.

a Write down an algebraic expression to illustrate this.

b Factorise your answer to part **a**.

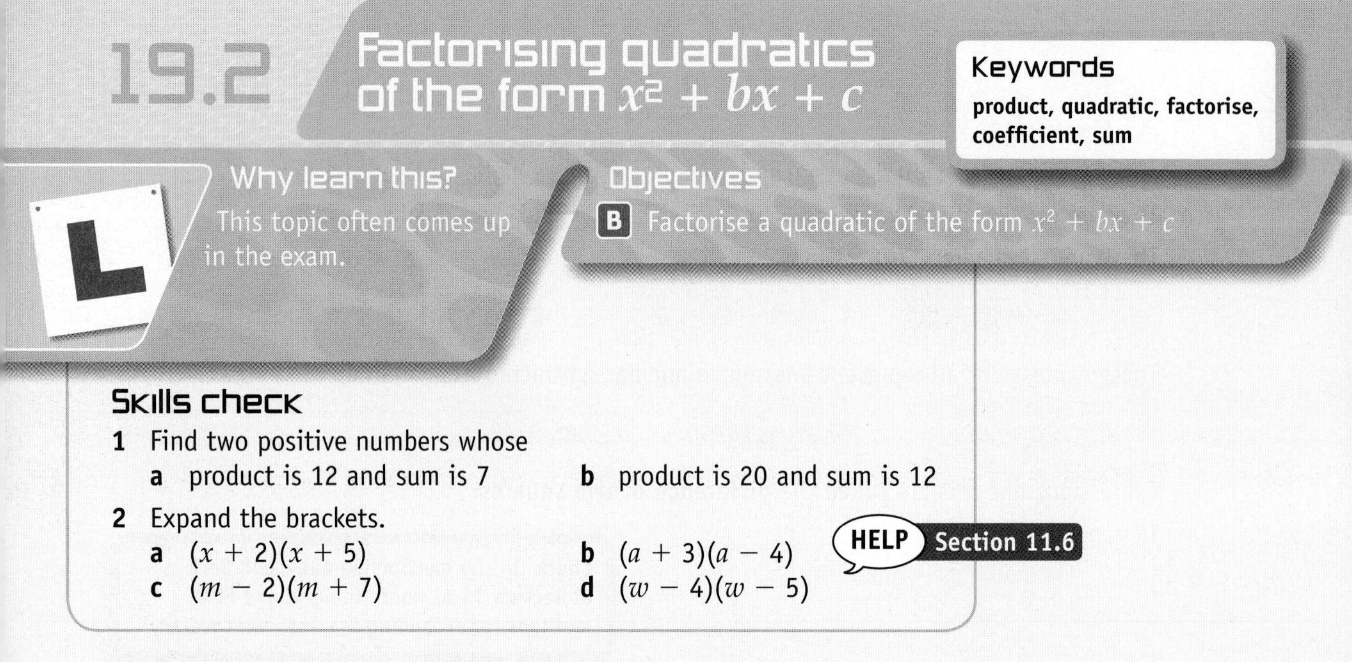

19.2 Factorising quadratics of the form $x^2 + bx + c$

Keywords
product, quadratic, factorise, coefficient, sum

Why learn this?

This topic often comes up in the exam.

Objectives

B Factorise a quadratic of the form $x^2 + bx + c$

Skills check

1 Find two positive numbers whose

a product is 12 and sum is 7 **b** product is 20 and sum is 12

2 Expand the brackets.

a $(x + 2)(x + 5)$ **b** $(a + 3)(a - 4)$

c $(m - 2)(m + 7)$ **d** $(w - 4)(w - 5)$

HELP Section 11.6

Factorising quadratics

Expanding a **product** of two expressions like $(x + 2)$ and $(x + 3)$ gives a **quadratic** expression.

$$(x + 2)(x + 3) = x^2 + 5x + 6$$

Factorising is the inverse of expanding brackets.

To factorise a quadratic expression, you write it as the product of two expressions.

For example

5 is the sum of 2 and 3.

$$x^2 + 5x + 6 = x^2 + 2x + 3x + 6$$
$$= (x + 2)(x + 3)$$

6 is the product of 2 and 3.

The **coefficient** of x is the number multiplying the x.

In general, to factorise the expression $x^2 + bx + c$, you need to find two numbers whose **sum** is b (the coefficient of x) and whose **product** is c.

Example 2

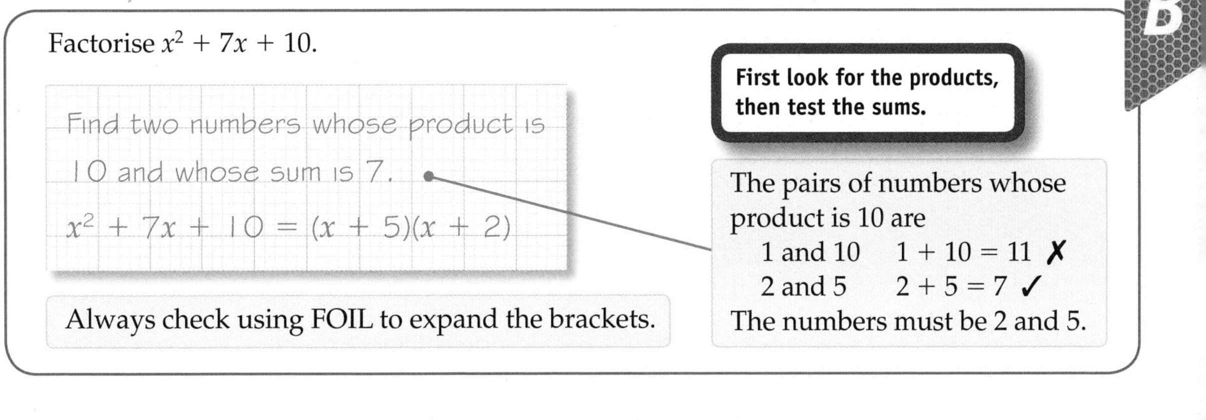

B

Exercise 19B

1 Copy and complete these.

a $x^2 + 14x + 33 = (x + 11)(x + \square)$

b $x^2 + 21x + 110 = (x + 11)(x + \square)$

c $x^2 + 14x + 40 = (x + 10)(x + \square)$

d $x^2 + 10x + 21 = (x + 3)(x + \square)$

2 Factorise each quadratic expression.

a $x^2 + 5x + 6$

b $x^2 + 6x + 8$

c $z^2 + 6z + 5$

d $a^2 + 11a + 10$

e $n^2 + 8n + 15$

f $f^2 + 12f + 36$

g $x^2 + 9x + 20$

h $m^2 + 8m + 12$

i $x^2 + 14x + 24$

j $b^2 + 11b + 30$

B

Example 3

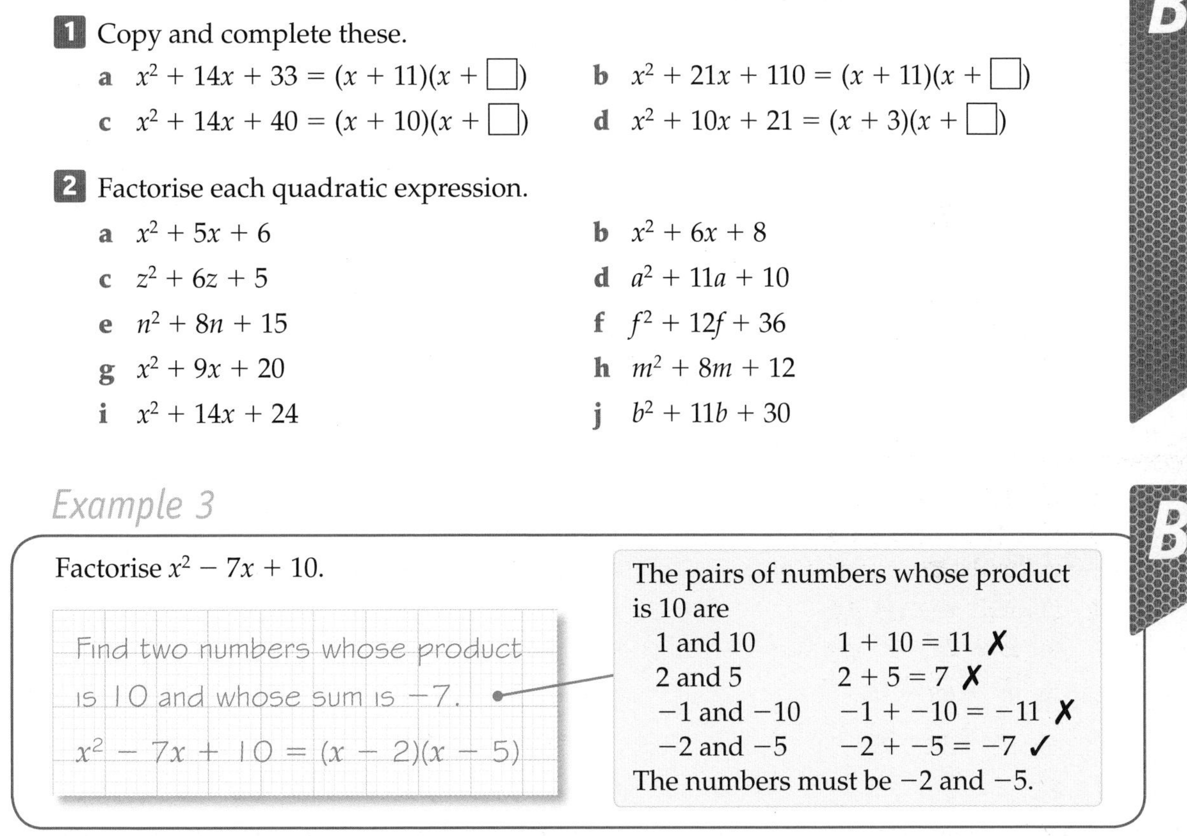

B

Exercise 19C

B

1 Copy and complete these.

a $x^2 - 14x + 24 = (x - 2)(x - \square)$

b $x^2 - 9x + 20 = (x - 5)(x - \square)$

c $x^2 - 25x + 100 = (x - 5)(x \square\square)$

d $x^2 - 14x + 48 = (x - 6)(x \square\square)$

2 Factorise each quadratic expression.

a $x^2 - 5x + 6$

b $x^2 - 9x + 8$

c $z^2 - 7z + 12$

d $a^2 - 9a + 14$

e $n^2 - 10n + 25$

f $f^2 - 8f + 16$

g $x^2 - 13x + 30$

h $b^2 - 11b + 28$

i $x^2 - 12x + 36$

j $p^2 - 10p + 24$

B

Example 4

Factorise **a** $x^2 - 6x - 7$ **b** $x^2 + x - 12$

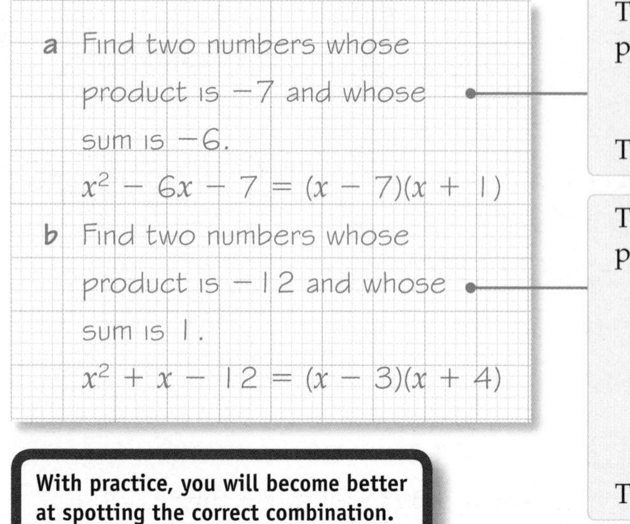

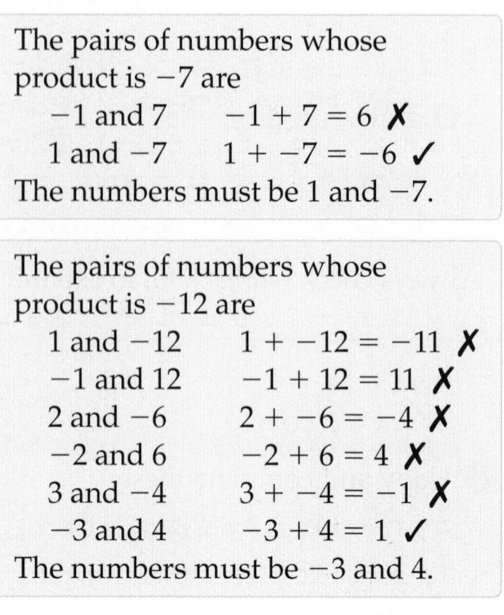

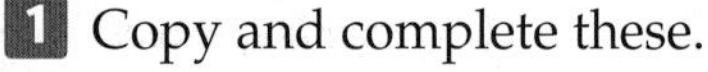

Exercise 19D

B

1 Copy and complete these.

a $x^2 - 13x - 30 = (x + 2)(x - \square)$

b $x^2 + x - 20 = (x - 4)(x \square\square)$

c $x^2 - 2x - 24 = (x + 4)(x \square\square)$

d $x^2 + 3x - 28 = (x + 7)(x \square\square)$

2 Factorise

a $x^2 + 4x - 12$

b $x^2 - x - 20$

c $z^2 - 2z - 15$

d $a^2 + 6a - 7$

e $n^2 + 6n - 16$

f $f^2 - f - 30$

g $m^2 + m - 30$

h $t^2 - 6t - 72$

3 **a** Factorise each expression. Simplify your answers as much as possible.

i $x^2 + 6x + 9$ **ii** $x^2 - 8x + 16$ **iii** $x^2 + 4x + 4$

iv $x^2 - 14x + 49$ **v** $x^2 - 10x + 25$ **vi** $x + 16x + 64$

b What do you notice about all the answers to part **a**?

c Copy and complete these statements, where m and n are numbers.

i $(x + m)^2 = x^2 + \square x + \square$

ii $(x - n)^2 = x^2 - \square x + \square$

B

A02

4 Copy and complete these statements.

a $t^2 + 7t - \square = (t + 10)(t - \square)$

b $m^2 - \square + 15 = (m \square\square)(m - 5)$

c $q^2 - 12q \square\square = (q \square\square)(q - 2)$

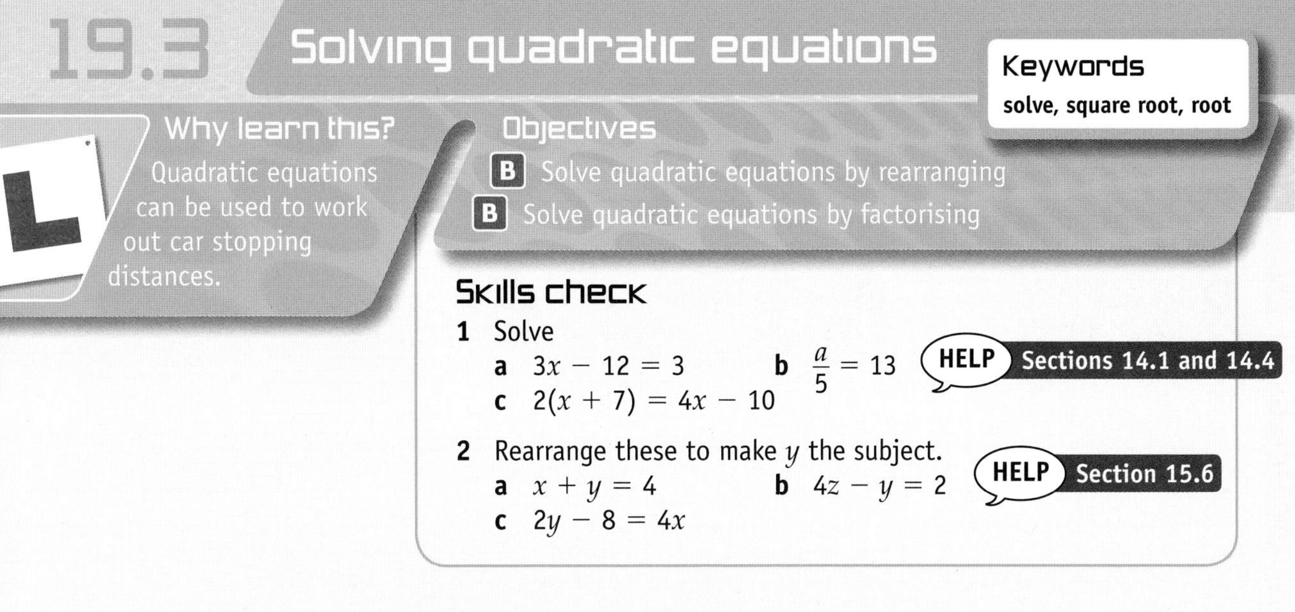

19.3 Solving quadratic equations

Keywords
solve, square root, root

Why learn this?
Quadratic equations can be used to work out car stopping distances.

Objectives
- **B** Solve quadratic equations by rearranging
- **B** Solve quadratic equations by factorising

Skills check

1 Solve
 a $3x - 12 = 3$ b $\frac{a}{5} = 13$
 c $2(x + 7) = 4x - 10$

HELP Sections 14.1 and 14.4

2 Rearrange these to make y the subject.
 a $x + y = 4$ b $4z - y = 2$
 c $2y - 8 = 4x$

HELP Section 15.6

Solving quadratic equations by rearranging

You can **solve** some quadratic equations by rearranging them to make x the subject.

Example 5

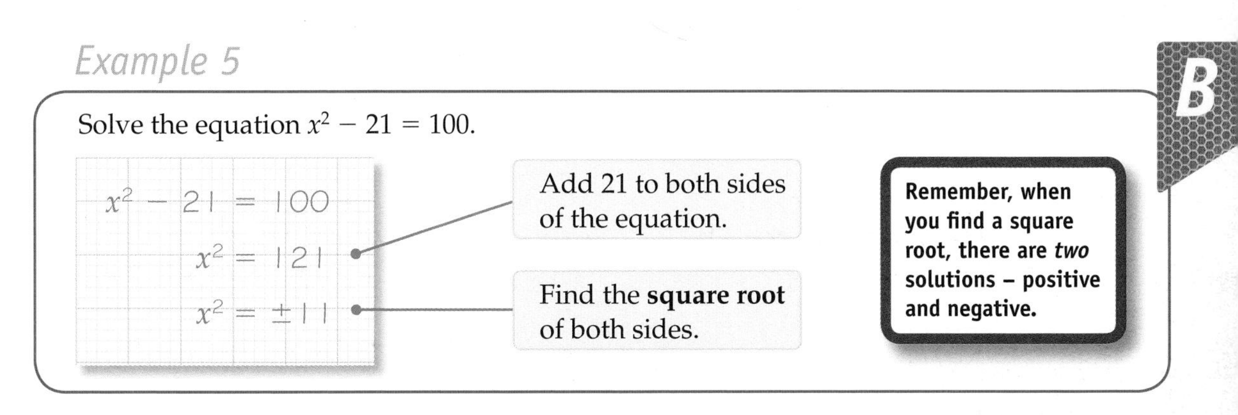

Solve the equation $x^2 - 21 = 100$.

$x^2 - 21 = 100$

$x^2 = 121$ — Add 21 to both sides of the equation.

$x^2 = \pm 11$ — Find the **square root** of both sides.

Remember, when you find a square root, there are *two* solutions – positive and negative.

B

Exercise 19E

Solve these equations.

1 $r^2 = 169$

2 $x^2 + 5 = 14$

3 $18 = 2t^2$

4 $y^2 - 20 = -19$

5 $\frac{m^2}{4} = 6.25$

6 $\frac{p^2}{5} = 20$

B

Example 6

Solve the quadratic equation $3x^2 - 27 = 0$.

$3x^2 - 27 = 0$

$3x^2 = 27$ — Add 27 to both sides of the equation.

$x^2 = 9$ — Divide both sides by 3.

$x = \pm 3$ — Find the square root of both sides.

B

Exercise 19F

Solve these equations.

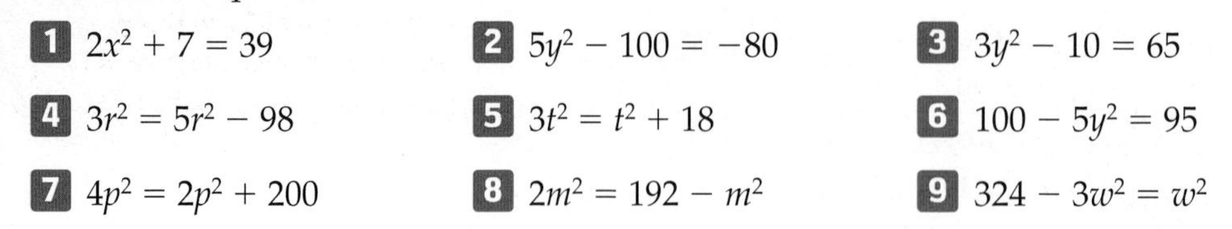

1 $2x^2 + 7 = 39$ **2** $5y^2 - 100 = -80$ **3** $3y^2 - 10 = 65$

4 $3r^2 = 5r^2 - 98$ **5** $3t^2 = t^2 + 18$ **6** $100 - 5y^2 = 95$

7 $4p^2 = 2p^2 + 200$ **8** $2m^2 = 192 - m^2$ **9** $324 - 3w^2 = w^2$

Example 7

Solve the equation $2(x + 3)^2 - 5 = 195$.

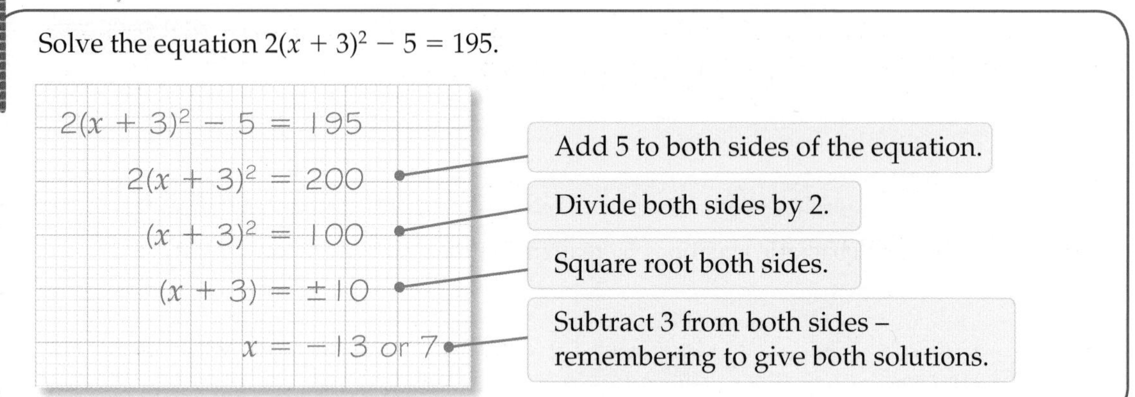

Exercise 19G

1 Solve the equation $4.5 = \frac{(r - 7)^2}{2}$.

2 Solve these equations.

a $(x + 1)^2 - 16 = 20$ **b** $100 = 4t^2 + 36$

c $150 - 3t^2 = 42$ **d** $6 + 3t^2 = 2t^2 + 15$

3 An estimate for the area of a circle is given by the formula
$A = 3 \times r^2$, where r is the radius.
A circle has an area of $75\,\text{cm}^2$.
Work out an estimate for the value of r.

One of the two solutions doesn't make sense!

AO2

4 A field is three times as long as it is wide.

a The field is x metres wide. Write an expression for its length.

b Copy and complete this formula for finding the area:
Area $= x \times \square$

c Simplify your formula in part **b**.

d The area of the field is $1200\,\text{m}^2$. What is the value of x?

e What is the length of the field?

AO3

5 Explain why you cannot find a solution to $x^2 + 20 = 5$.

6 A rectangle has length five times its width.
The area of the rectangle is $845\,\text{mm}^2$.
What is the width of the rectangle?

Solving quadratic equations by factorising

Solving quadratic equations by factorising relies on this fact:

When $a \times b = 0$, a is 0, b is 0 or both are 0.

> **If the product of two things is zero, one of them *must* be zero.**

So if $(x + 2)(x - 4) = 0$,

- either $x + 2 = 0$, which means that $x = -2$
- or $x - 4 = 0$, which means that $x = 4$.

To solve a quadratic equation:

Step 1: Rearrange the equation so that one side is zero.

Step 2: Factorise the quadratic expression.

Step 3: Find the solutions. Usually there are two solutions. However, when the expression factorises to $(x + m)^2 = 0$, there is only one solution.

Example 8

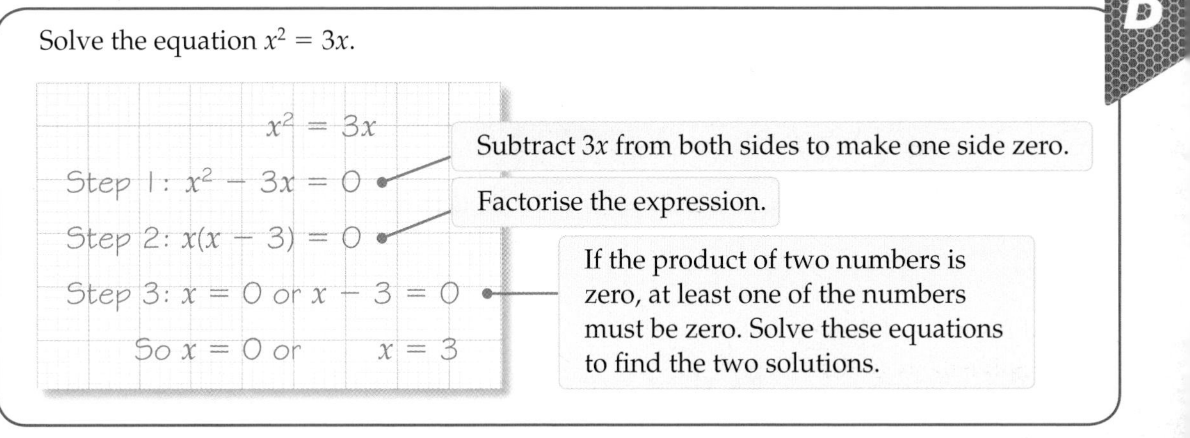

Exercise 19H

1 Find the possible values for x if

a $(x + 3)(x - 12) = 0$ **b** $(x - 8)(x - 9) = 0$

c $(x + 20)(x - 15) = 0$ **d** $(x + 13)(x - 1) = 0$

2 Solve these equations. Leave your answers as fractions where appropriate.

a $x^2 + 7x = 0$ **b** $t^2 - 5t = 0$ **c** $3x^2 + 6x = 0$

d $y^2 = 5y$ **e** $0 = 4w^2 - 12w$ **f** $5y = 20y^2$

g $a - a^2 = 0$ **h** $5t = 30t^2$ **i** $14r = 63r^2$

Example 9

Solve the equation $x^2 + 5x + 6 = 0$.

Step 1: One of the sides is already zero.

Step 2: $(x + 2)(x + 3) = 0$ — Factorise the expression.

Step 3: $x + 2 = 0$ or $x + 3 = 0$

So $x = -2$ or $x = -3$

Exercise 19I

B

Solve these equations.

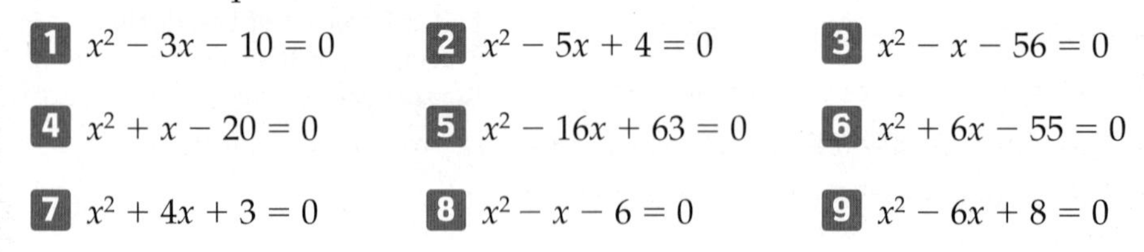

1 $x^2 - 3x - 10 = 0$ **2** $x^2 - 5x + 4 = 0$ **3** $x^2 - x - 56 = 0$

4 $x^2 + x - 20 = 0$ **5** $x^2 - 16x + 63 = 0$ **6** $x^2 + 6x - 55 = 0$

7 $x^2 + 4x + 3 = 0$ **8** $x^2 - x - 6 = 0$ **9** $x^2 - 6x + 8 = 0$

Example 10

B

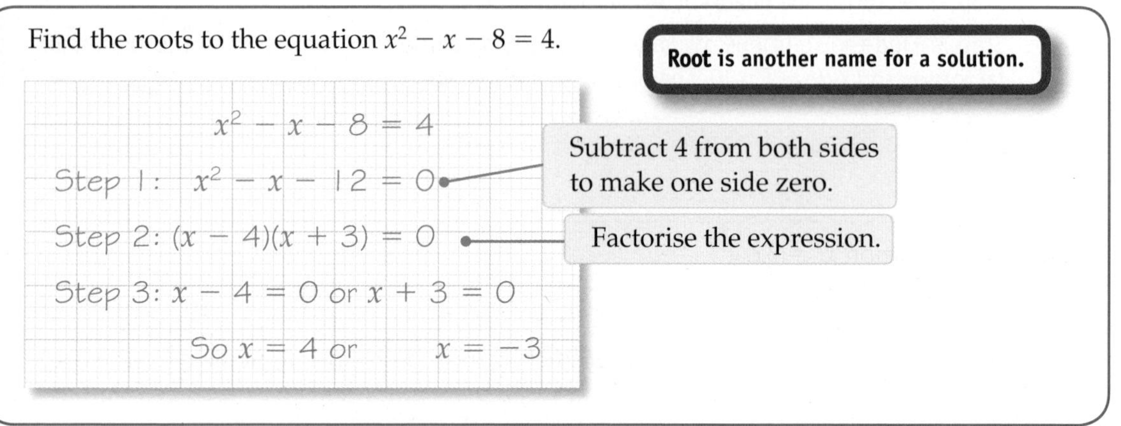

Find the roots to the equation $x^2 - x - 8 = 4$.

Root is another name for a solution.

$x^2 - x - 8 = 4$

Step 1: $x^2 - x - 12 = 0$ — Subtract 4 from both sides to make one side zero.

Step 2: $(x - 4)(x + 3) = 0$ — Factorise the expression.

Step 3: $x - 4 = 0$ or $x + 3 = 0$

So $x = 4$ or $x = -3$

Exercise 19J

B

Find the roots of these equations.

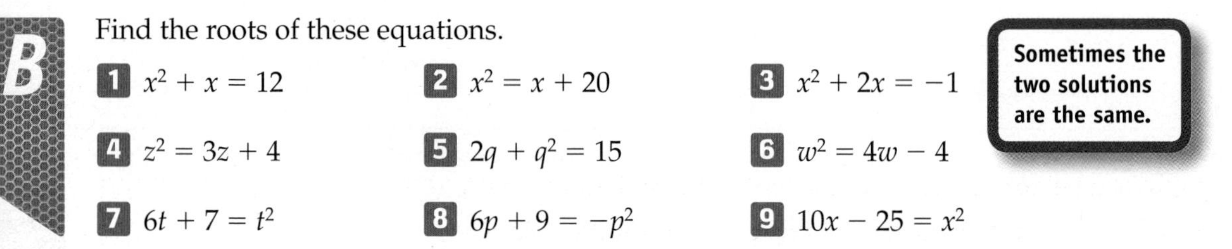

1 $x^2 + x = 12$ **2** $x^2 = x + 20$ **3** $x^2 + 2x = -1$

4 $z^2 = 3z + 4$ **5** $2q + q^2 = 15$ **6** $w^2 = 4w - 4$

7 $6t + 7 = t^2$ **8** $6p + 9 = -p^2$ **9** $10x - 25 = x^2$

Sometimes the two solutions are the same.

19.4 Writing and solving quadratic equations

Why learn this?

The path of a cricket ball can be modelled using a quadratic equation.

Objectives

B Write quadratic equations for problems and then solve them

Skills check

1 x is a mystery number. Write expressions in x for

a 3 times x **b** $\frac{1}{4}$ of x

c 4 times x plus 2 **d** the product of twice x and three times x

Writing and solving quadratic equations

You can solve problems by writing an equation and then solving it.

Example 11

B

AO2

Jane is three years younger than her sister. The product of their ages is 54.

Use x to represent Jane's age.

a Write down an algebraic expression for her sister's age.

b Write down and simplify an algebraic expression for the product of their ages.

c Form and solve an algebraic equation to find the value of x.

d How old is Jane's sister?

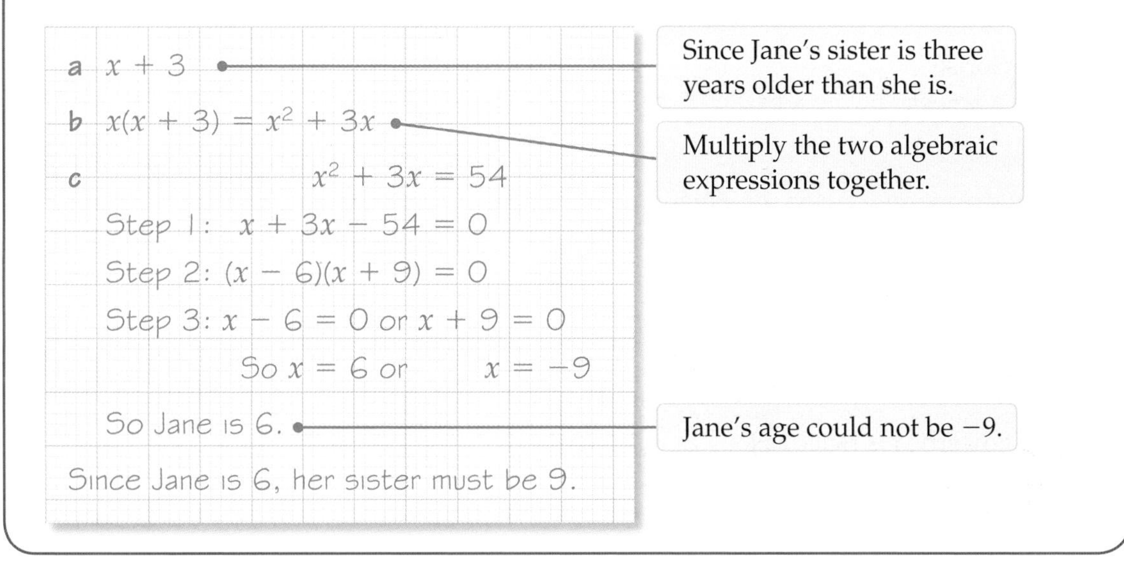

Exercise 19K

B

AO2

1 The height of this rectangle is 3 cm more than its width.

a Write down an algebraic expression for the height of the rectangle.

b Write down an algebraic expression for the area of the rectangle.

c Given that the area of the rectangle is 40 cm^2, form and solve a quadratic equation to work out the value of t.

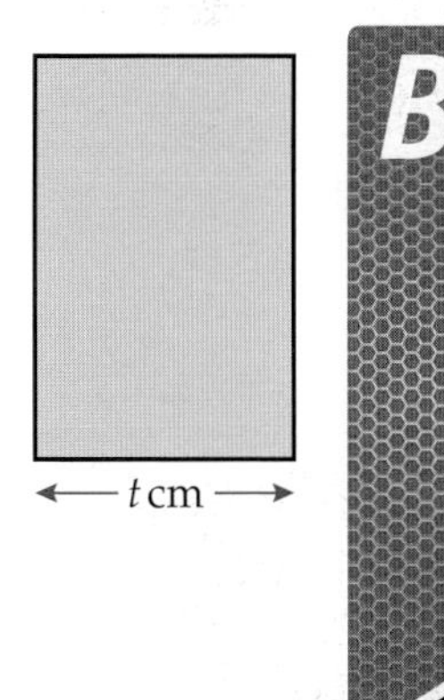

2 A rectangular garden is 4 m longer than it is wide.
Its area is 165 m^2.

a Sketch and label a diagram to show the area.

b Form and solve a quadratic equation to work out the dimensions of the garden.

3 I think of a negative number, square it, and add five times the original number.
My result is 24.
What number did I think of?

4 I think of a positive number.
I square it then subtract six times the number.
The answer is 27.
What was the original number?

Review exercise

B

1 Factorise $x^2 - 2x - 15$. **[2 marks]**

2 Factorise the expression $x^2 - 9$. **[2 marks]**

3 **a** Factorise the expression $x^2 + 6x + 5$.
b Hence solve the equation $x^2 + 6x + 5 = 0$. **[3 marks]**

4 Solve the equation $x^2 + 12 = 7x$. **[3 marks]**

5 Factorise
a $x^2 - 49$ **[2 marks]**
b $x^2 + 2x - 8$ **[2 marks]**

6 By rearranging, solve the equation $\frac{x^2}{3} + 2 = 14$. **[3 marks]**

7 Solve the quadratic equation $x^2 = 11x - 28$. **[3 marks]**

B

8 The product of two algebraic terms is $x^2 - 13x + 36$.
If one of the two terms is $(x - 4)$, what is the other? **[2 marks]**

AO2

9 A flowerbed is twice as long as it is wide.
a The width of the flower bed is t metres.
Write down an algebraic expression for the length. **[1 mark]**
b Write down and simplify an algebraic expression for the area of the flowerbed. **[1 mark]**
c The area is $2.88\,\text{m}^2$. Form and solve an equation to find the value of t. **[2 marks]**

B

AO3

10 I think of a number, square it, then subtract three times the number. The result is 108.
Form and solve an algebraic equation to work out the possible values of the number I thought of. **[4 marks]**

Chapter summary

In this chapter you have learned how to

- factorise a quadratic expression that is the difference of two squares **B**
- factorise a quadratic of the form $x^2 + bx + c$ **B**
- solve quadratic equations by rearranging **B**
- solve quadratic equations by factorising **B**
- write quadratic equations for problems and then solve them **B**

Problem-solving practice: Algebra

Quality of written communication: Some questions on this page are marked with a star ☆. In the exam, this sort of question may earn you some extra marks if you

- use correct and accurate maths notation and vocabulary
- organise your work clearly, showing that you can communicate effectively.

FUNCTIONAL • FUNCTIONAL •

E

1 Dilan is saving his pocket money. This table shows how much he has at the end of each week.

Week 1	Week 2	Week 3	Week 4	Week 5
£13	£18	£23	£28	£33

How many weeks will it take Dilan to save £100? [2]

AO2

D

2 Show that $4(x - 6) - x = 3(x - 8)$ [2]

3 This equilateral triangle and this regular pentagon have the same perimeter.

x

9 cm

The length of one side of the pentagon is 9 cm.
Find the length of one side of the triangle. [2]

4 The perimeter of this triangle is 33 cm.

$x - 1$

$x + 6$

$2x$

Write an equation and solve it to find the value of x. [3]

☆ **5** **a** Factorise $n^2 - n$. [1]

b n is a whole number. Explain why $n^2 - n$ is always an even number. [2]

AO2

C

☆ **6** Look at this sequence: 114, 110, 106, 102, …

a Work out the 50th term of this sequence. [3]

b Is 40 a term in this sequence? Give a reason for your answer. [1]

7 Draw axes from −10 to 10 in both directions. Draw and label these graphs.

a $x + y = 5$ [2]

b $5x + 2y = 15$ [2]

c $x = \frac{1}{2}y + 4$ [2]

d $2x = 6 + 3y$ [2]

AO2

B

8 Copy and complete this statement: $x^2 + \square x - 8 = (x - 2)(x + \square)$ [2]

9 A rectangular field is 10 m longer than it is wide. The area of the field is 75 m^2.

a Write an expression for the area of the field. Use x to represent the width. [1]

b Form an equation in x and solve it. [3]

AO2

Quality of written communication: Some questions on this page are marked with a star ☆. In the exam, this sort of question may earn you some extra marks if you

- use correct and accurate maths notation and vocabulary
- organise your work clearly, showing that you can communicate effectively.

☆ **1** The diagram shows an addition pyramid. Each number is the sum of the two numbers underneath it.

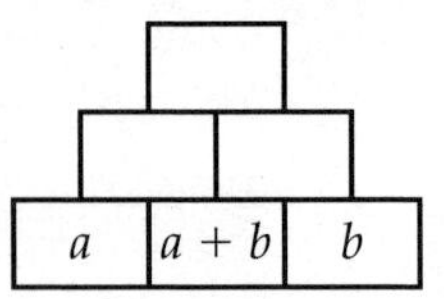

a Copy and complete the pyramid to find expressions for the numbers in each box. [2]

b a and b are whole numbers. Explain why the number in the top box must be a multiple of three. [2]

2 Paul has written down the general term of two different sequences.

Sequence A: nth term $= 3n$
Sequence B: nth term $= 5n - 25$

Work out the smallest value of n for which sequence A is smaller than sequence B. [2]

3 Klara and Lydia are folding programmes for the school play. It takes Klara t seconds to fold one programme. It takes Lydia 8 seconds longer. It takes Lydia 6 minutes to fold 20 programmes. Write an equation and solve it to find the value of t. [3]

4 Karan has started laying square mosaic tiles in his shower. The shower base is a square.

Calculate the area of each mosaic tile. [3]

5 Use a counter example to disprove these statements:

a If n is an even number then $n^2 + 1$ is a prime number. [2]

b If $x < 10$ and $y < 5$ then $xy < 50$. [2]

6 The midpoint of the line segment AB is at (4, 5). The coordinates of A are (−2, 1). Work out the coordinates of B. [2]

7 Find the equation of the straight line which passes through the points (3, −2) and (9, 1). [3]

20

Number skills revisited

BBC Video

This chapter revises essential number skills.

When you go on holiday with friends, you need to be able to work out if you have enough money for accomodation and food, and how to share the bills fairly.

Objectives

This chapter will remind you how to

- understand equivalent fractions
- use ratio notation
- recognise that each terminating decimal is a fraction
- convert simple fractions to percentages and vice versa
- use brackets and the hierarchy of operations
- add, subtract, multiply and divide integers
- use calculators effectively and efficiently; use function keys for squares
- give solutions in the context of the problem to an appropriate degree of accuracy
- round to the nearest integer, to one significant figure and to one, two or three decimal places
- simplify a fraction by cancelling all common factors
- use percentages to compare proportions
- use inverse operations
- understand 'reciprocal' as multiplicative inverse

Number skills: revision exercise 1

1 Copy and complete these equivalent fractions.

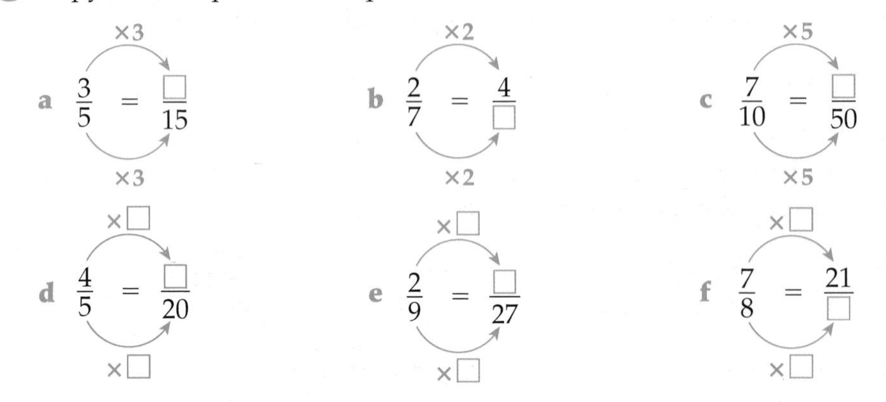

2 Write each fraction in its simplest form.

a $\frac{6}{12}$ b $\frac{10}{15}$ c $\frac{4}{20}$ d $\frac{6}{9}$ e $\frac{20}{24}$

3 Copy and complete the table. Give each fraction in its simplest form.

Fraction	$\frac{1}{10}$		$\frac{1}{2}$		
Decimal		0.25		0.6	
Percentage					75%

4 Convert each of these test scores into a percentage. Give each answer to the nearest whole number.

History	28 out of 32
Geography	54 out of 70
Welsh	32 out of 46
Music	19 out of 24

5 a Write $\frac{9}{16}$ as a decimal. b Write 32% as a decimal.

6 The pie chart shows the results of a survey into the most popular female singers. Estimate

a the percentage of those surveyed that liked Lady Gaga best

b the proportion that liked Rihanna best

c the percentage that did not choose Pixie Lott.

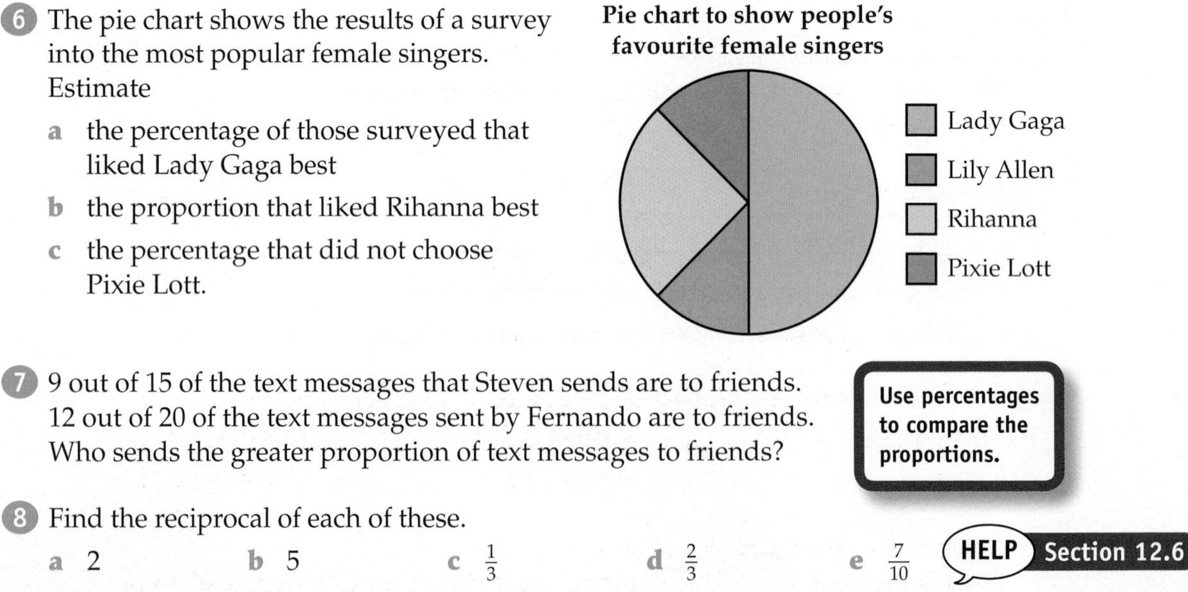

7 9 out of 15 of the text messages that Steven sends are to friends. 12 out of 20 of the text messages sent by Fernando are to friends. Who sends the greater proportion of text messages to friends?

Use percentages to compare the proportions.

8 Find the reciprocal of each of these.

a 2 b 5 c $\frac{1}{3}$ d $\frac{2}{3}$ e $\frac{7}{10}$

HELP Section 12.6

9 a What is the reciprocal of $\frac{3}{4}$?

b Multiply $\frac{3}{4}$ by its reciprocal. What result do you get?

HELP Section 12.6

10 a Work out $\frac{1}{2} \div \frac{3}{2}$.

b Work out $\frac{1}{2} \times \frac{2}{3}$.

c Look at your answers to parts **a** and **b**.
What can you say about these calculations?

11 In a youth group there are 36 girls and 24 boys.
What is the ratio of girls to boys?

HELP Section 7.1

12 A recipe for cake needs 250 g of flour and 100 g of margarine.
What is the ratio of flour to margarine?

13 Work out

a $3 \times 4 - 2$ b $12 \div (3 + 3)$ c $(12 - 6) \times 3$ d $18 - 6 \div 2$

e 3.9×2.5 f 295×0.48 g $43.5 \div 3$ h $12.5 \div 0.25$

14 Work out

a $3 + 4^2$ b $3^2 + 4^2$ c $9^2 - 6^2$

15 Find the sum of 82, 195 and 102.

16 What is 180 minus 63 minus 27?

17 Copy and complete these calculations.

a $37 + \square = 82$ b $\square \times 8 = 232$

c $\square - 45 = 39$ d $380 \div \square = 95$

18 Javier is on holiday and has £20 to spend.
He wants to buy some souvenirs priced at £3.89 each.
How many souvenirs can he buy?

19 Use a calculator to work out $\dfrac{3.7 \times 2.6}{2.3 - 1.1}$.

a Write your full calculator display.

b Write your answer to part **a** to one decimal place.

c Write your answer to part **a** to two decimal places.

20 Round each of these numbers to the degree of accuracy indicated.

a 3.458 (1 d.p.) b 2.473 (2 d.p.) c 83.8154 (3 d.p.)

d 285 (1 s.f.) e 8.24 (1 s.f.) f 19.0956 (1 s.f.)

21 A room measures 12.82 m by 8.9 m.
What is the floor area? Give your answer to an appropriate degree of accuracy.

P Number skills: revision exercise 2

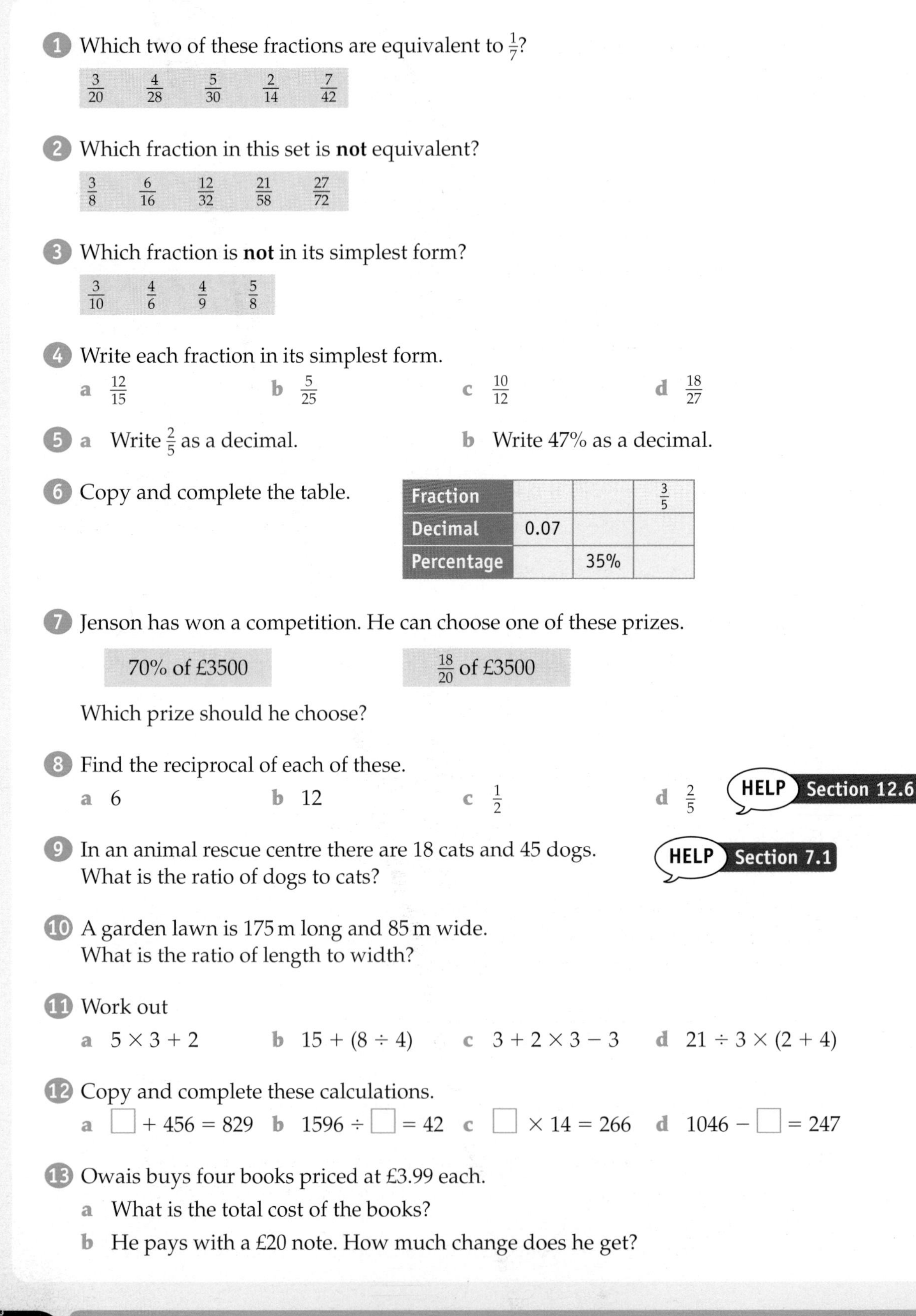

1 Which two of these fractions are equivalent to $\frac{1}{7}$?

$\frac{3}{20}$ $\frac{4}{28}$ $\frac{5}{30}$ $\frac{2}{14}$ $\frac{7}{42}$

2 Which fraction in this set is **not** equivalent?

$\frac{3}{8}$ $\frac{6}{16}$ $\frac{12}{32}$ $\frac{21}{58}$ $\frac{27}{72}$

3 Which fraction is **not** in its simplest form?

$\frac{3}{10}$ $\frac{4}{6}$ $\frac{4}{9}$ $\frac{5}{8}$

4 Write each fraction in its simplest form.

a $\frac{12}{15}$ **b** $\frac{5}{25}$ **c** $\frac{10}{12}$ **d** $\frac{18}{27}$

5 **a** Write $\frac{2}{5}$ as a decimal. **b** Write 47% as a decimal.

6 Copy and complete the table.

Fraction			$\frac{3}{5}$
Decimal	0.07		
Percentage		35%	

7 Jenson has won a competition. He can choose one of these prizes.

70% of £3500 $\frac{18}{20}$ of £3500

Which prize should he choose?

8 Find the reciprocal of each of these.

a 6 **b** 12 **c** $\frac{1}{2}$ **d** $\frac{2}{5}$

HELP Section 12.6

9 In an animal rescue centre there are 18 cats and 45 dogs.
What is the ratio of dogs to cats?

HELP Section 7.1

10 A garden lawn is 175 m long and 85 m wide.
What is the ratio of length to width?

11 Work out

a $5 \times 3 + 2$ **b** $15 + (8 \div 4)$ **c** $3 + 2 \times 3 - 3$ **d** $21 \div 3 \times (2 + 4)$

12 Copy and complete these calculations.

a $\square + 456 = 829$ **b** $1596 \div \square = 42$ **c** $\square \times 14 = 266$ **d** $1046 - \square = 247$

13 Owais buys four books priced at £3.99 each.

a What is the total cost of the books?

b He pays with a £20 note. How much change does he get?

14 What is the product of 18.2 and 3.05?

15 Prize money of £29 805 is to be shared equally between 15 people.
How much does each person receive?

16 Albert got £45 for his birthday.
He wants to buy some T-shirts priced at £6.80 each.

a How many T-shirts can he buy?

b How much money would he have left?

17 Here is a payment plan for a digital LCD TV.
15 monthly payments must be made.

a What is the total cost of the monthly payments?

b What is the total cost of the TV?

TV payment plan	
Deposit	£85.00
Monthly payment	£25.75

HELP Section 16.2

18 Dimitri wants to buy pet insurance for his cat.
How much would he save by making a single annual payment?

Pet insurance	
Monthly payment	£3.99
Annual payment	£46.59

19 What is 3.657 rounded to one decimal place?

20 Use a calculator to work out

a $18 \times (109 - 32)$ **b** $18 \times 109 - 32$ **c** the square of 97.

21 Use a calculator to work out 2.17^2.

a Write down your full calculator display.

b Write your answer to part **a** to two decimal places.

22 Round each of these numbers to one significant figure.

a 278 **b** 65 000 **c** 0.2008 **d** 1.999

23 Mount Everest is 9000 m to one significant figure.
Suzanne thinks the mountain could be about 8755 m high.
Could Suzanne be correct? Explain your answer.

24 A box of sweets contains 50 sweets to one significant figure.
What is the smallest number of sweets that could be in the box?

A 49 sweets **B** 42 sweets **C** 45 sweets **D** 47 sweets

Give a reason for your answer.

25 Round the number in each statement to an appropriate degree of accuracy.

a Nathan is 1.7587 m tall **b** The temperature is 28.345 °C.

21 Angles

BBC Video

This chapter is about angles, which measure the amount of turn.

Olympic divers take care to turn the right amount in the final somersault before straightening out to enter the water.

Objectives

This chapter will show you how to

- understand angle measure using the correct mathematical language E D C
- recall and use properties of angles at a point, angles on a straight line (including right angles), perpendicular lines, and vertically opposite angles E D
- recognise alternate angles and corresponding angles D
- understand and use bearings E D C
- show step-by-step deduction in solving a geometrical problem E D C

Before you start this chapter

1 Describe these turns by giving the fraction of the turn, the angle in degrees and the direction.

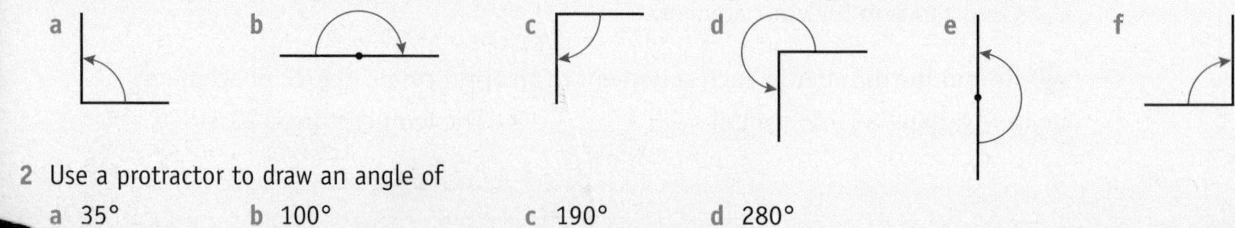

2 Use a protractor to draw an angle of

a 35° b 100° c 190° d 280°

An angle measures a turn. Angles are measured in degrees (°).

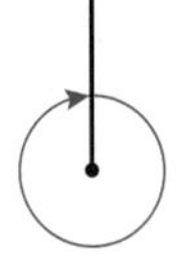
full turn 360°

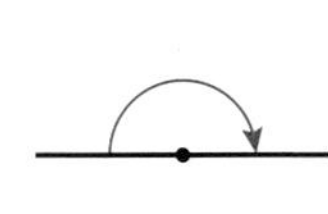
$\frac{1}{2}$ turn 180°

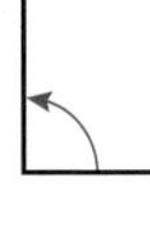
$\frac{1}{4}$ turn 90°

Naming angles

Acute angle		Less than 90°
Right angle		Exactly 90° This symbol means 'right angle'
Obtuse angle		More than 90° but less than 180°
Straight line		Exactly 180°
Reflex angle		More than 180° but less than 360°

Labelling angles

You can use letters on a diagram to label angles.

AB and *BC* are line segments. They meet at *B*.

The arrow is pointing at angle *ABC*.

You can write this angle as $\angle ABC$ or $A\widehat{B}C$ or angle *ABC*.

The point of the angle is at the middle letter.

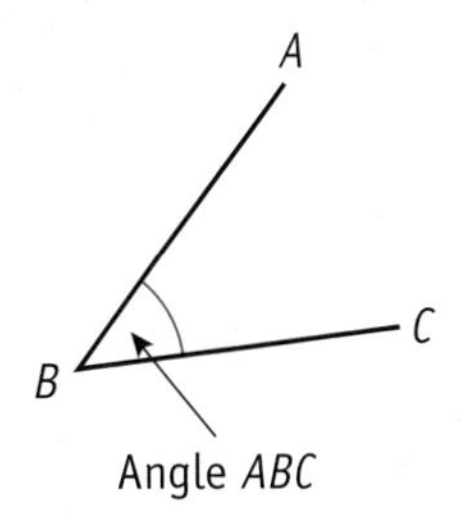

Line and angle facts

Two lines that cross at right angles are called perpendicular.

Angles on a straight line add up to 180°.

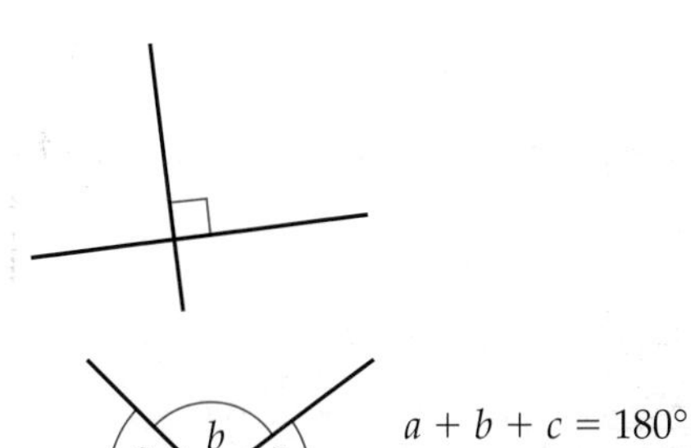

1 Tim faces west.
He makes a half turn.
Where is he facing now?

2 Use a protractor to draw an obtuse angle.
Mark your angle clearly with a curve.

3 Work out the sizes of the angles marked with letters.

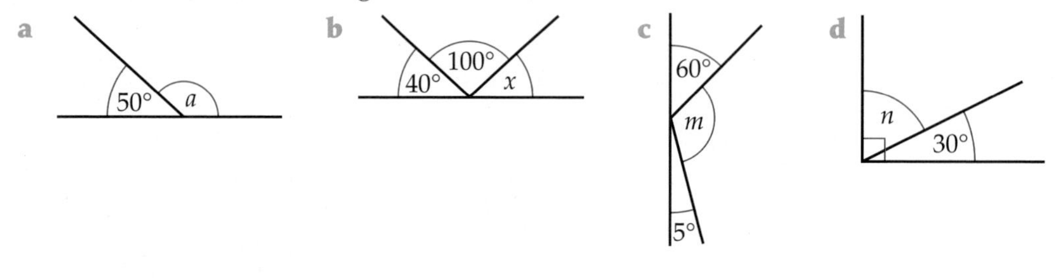

4 The diagram shows a triangle ABC and a point O.
The point O is joined to each vertex (corner) as shown.
Angle x is less than 90°.

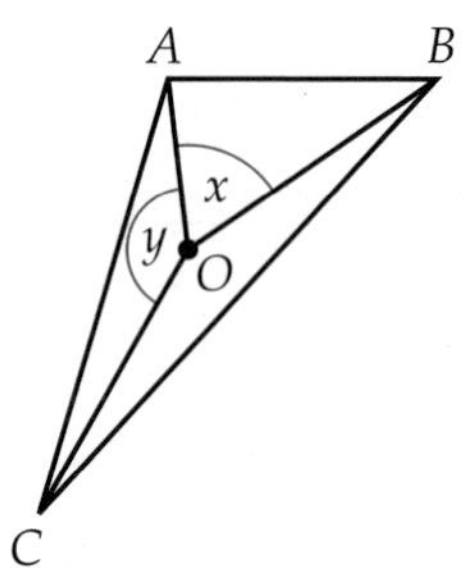

a Write down the name for an angle that is less than 90°.

Angle y is greater than 90°.

b Write down the name for an angle that is greater than 90° and less than 180°.

5 **a** Measure all the angles marked in the diagram.
Use three letters to name each angle when giving your answer.

b Write down the two line segments that are perpendicular to each other.

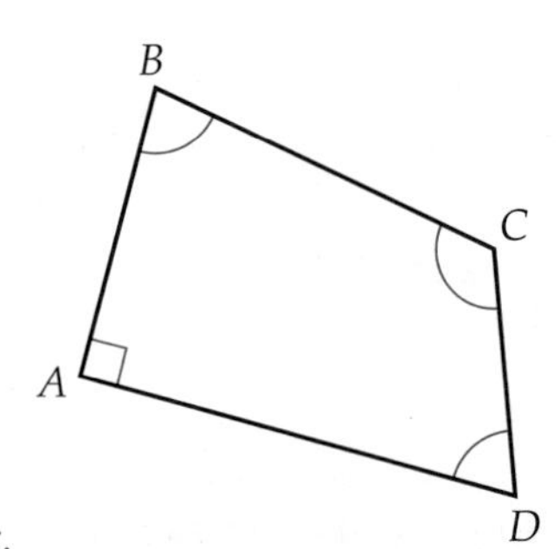

6 Put these angles in order of size, starting with the smallest.

A B C D

E F G H

Angle facts

Keywords
vertically opposite

Why learn this?

Skateboarders describe their moves using angles. A '180' is the same as a half turn.

Objectives

E Calculate angles around a point
E Recognise vertically opposite angles

Skills check

1 Solve each equation to find the value of the letter.
a $x + 30 = 180$ **b** $y + 240 = 360$ **c** $2z + 40 = 180$

2 What is the size of angle x?

70°, x

Angle facts

You need to know these angle facts.

Angles around a point add up to 360°.

$$d + e + f + g + h = 360°$$

All these angles together make a full turn or 360°.

Vertically opposite angles are equal.

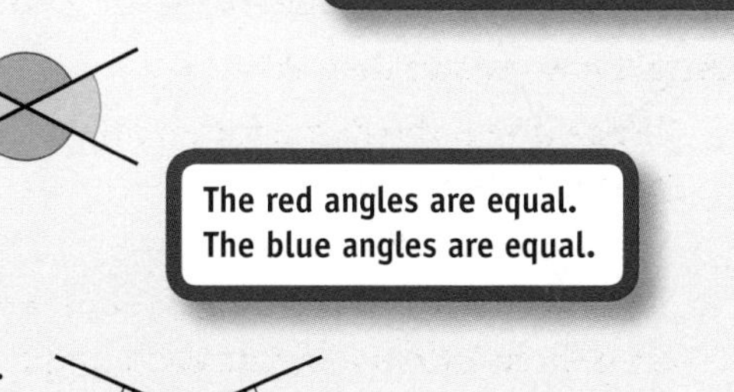

The red angles are equal.
The blue angles are equal.

You can show equal angles using matching arcs.

Example 1

Work out the size of angle b.

120°, b, 140°

$b + 120° + 140° = 360°$ — Select the angle fact to use. Angles around a point add up to 360°.

$b + 260° = 360°$ — Simplify by adding 120° + 140°.

$b = 100°$ — Solve the equation to work out b.

Exercise 21A

1 Work out the sizes of the angles marked with letters.

a

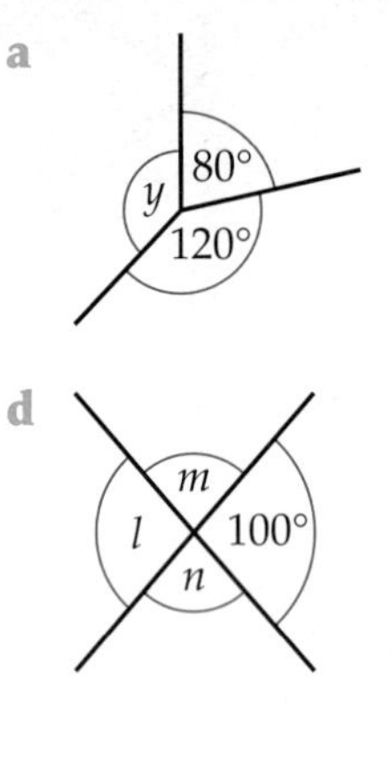

b

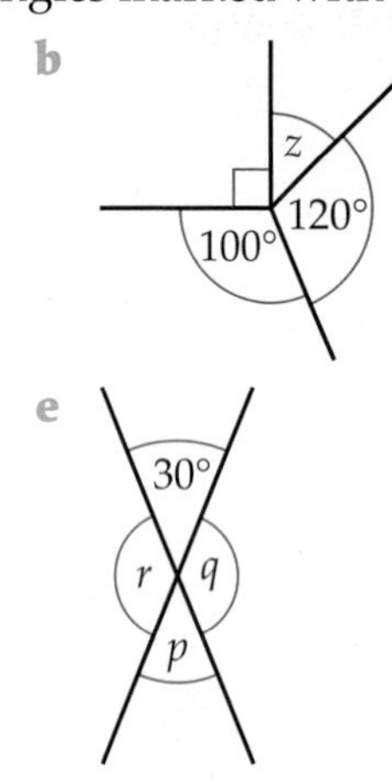

c

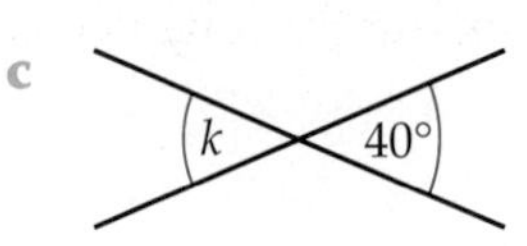

d

e

f

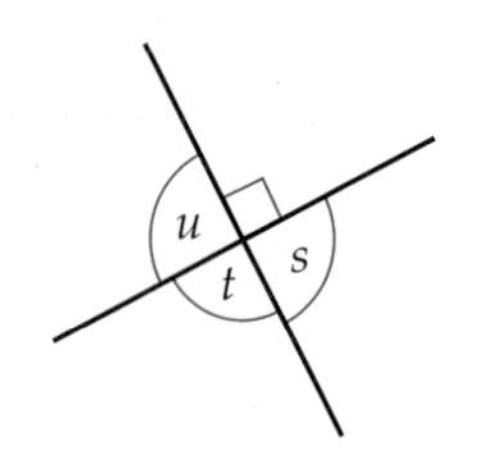

2 Work out the sizes of the angles marked with letters.
The first one has been started for you.

a **b**

c **d**

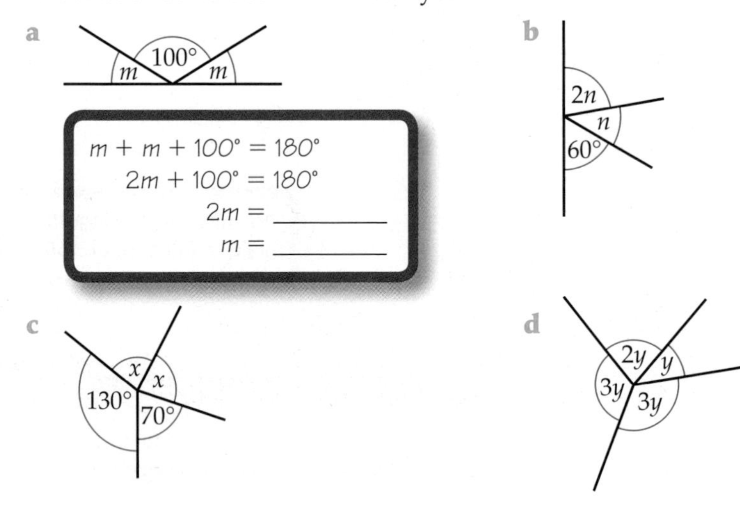

3 Work out the sizes of the angles marked with letters.

a

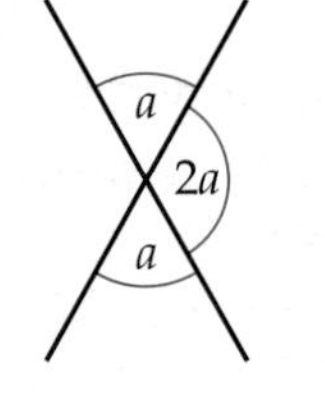

b

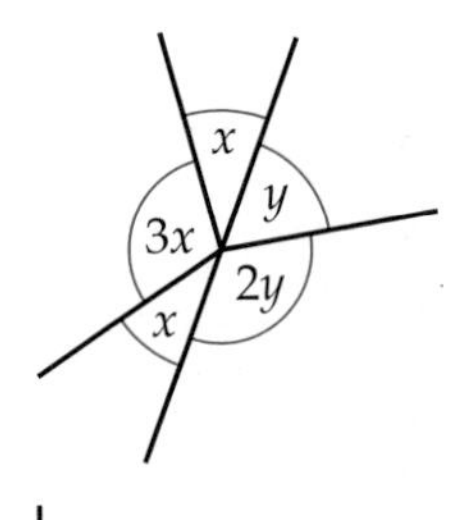

c

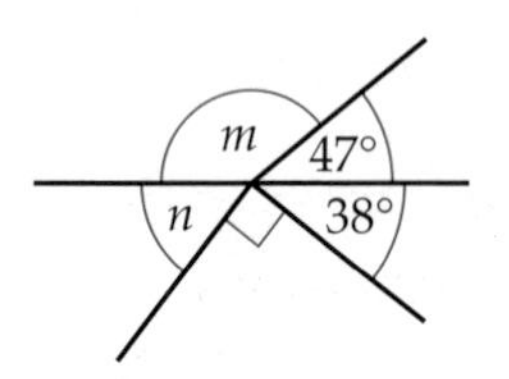

d

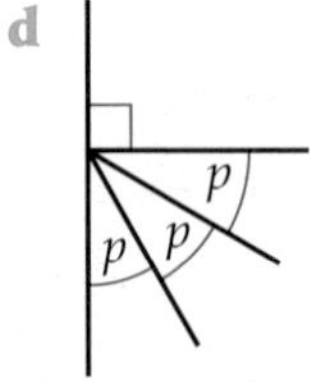

4 The diagram shows the design of a roof truss.
Work out the sizes of angle x and angle y.

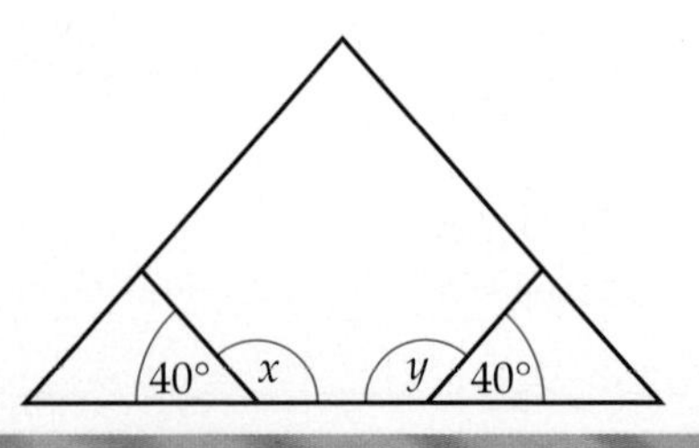

E

E

AO3

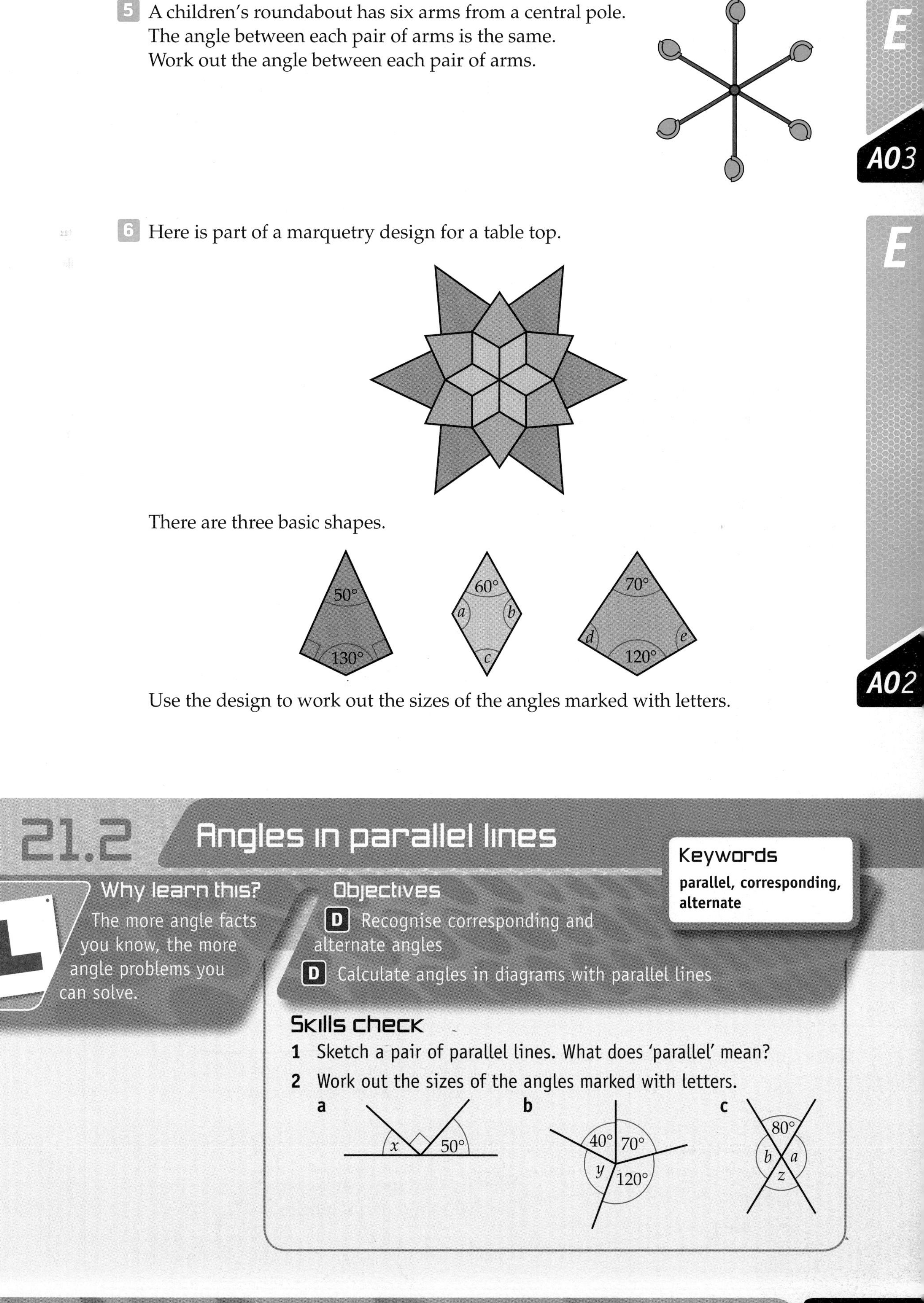

5 A children's roundabout has six arms from a central pole.
The angle between each pair of arms is the same.
Work out the angle between each pair of arms.

E

AO3

6 Here is part of a marquetry design for a table top.

There are three basic shapes.

Use the design to work out the sizes of the angles marked with letters.

E

AO2

21.2 Angles in parallel lines

Why learn this?
The more angle facts you know, the more angle problems you can solve.

Objectives

D Recognise corresponding and alternate angles

D Calculate angles in diagrams with parallel lines

Keywords
parallel, corresponding, alternate

Skills check

1 Sketch a pair of parallel lines. What does 'parallel' mean?

2 Work out the sizes of the angles marked with letters.

a

b

c

Angle facts

Arrowheads are used to show that two lines are **parallel**.

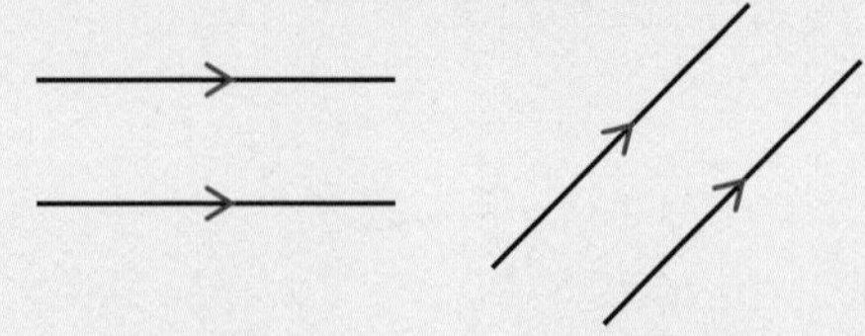

A line crossing two parallel lines creates pairs of equal angles.

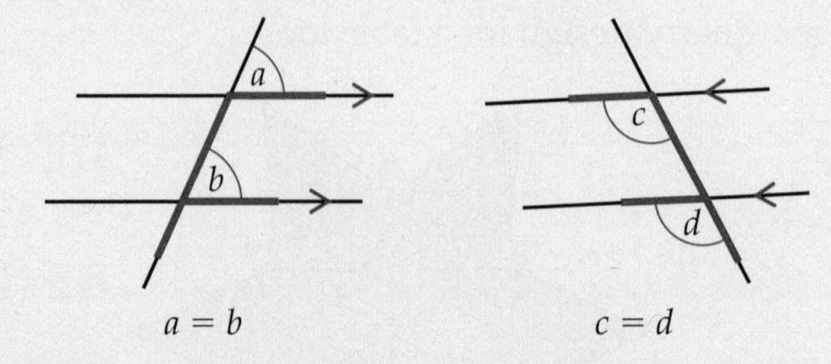

$a = b$ $\quad$ $c = d$

a and b are **corresponding** angles. c and d are also corresponding angles. The lines make an F shape.

Corresponding angles are equal.

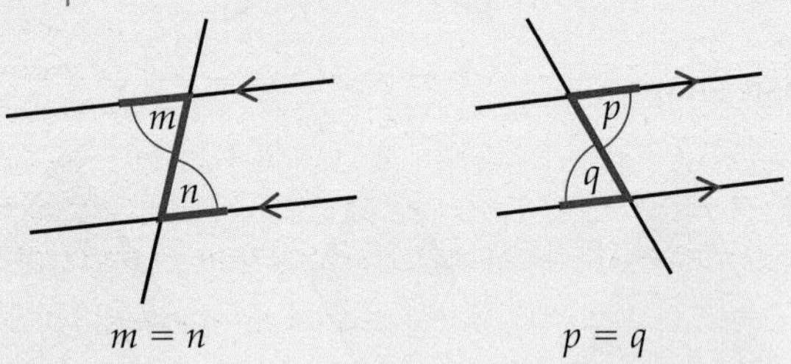

$m = n$ $\quad$ $p = q$

m and n are **alternate** angles. p and q are also alternate angles. The lines make a Z shape.

Alternate angles are equal.

D

Example 2

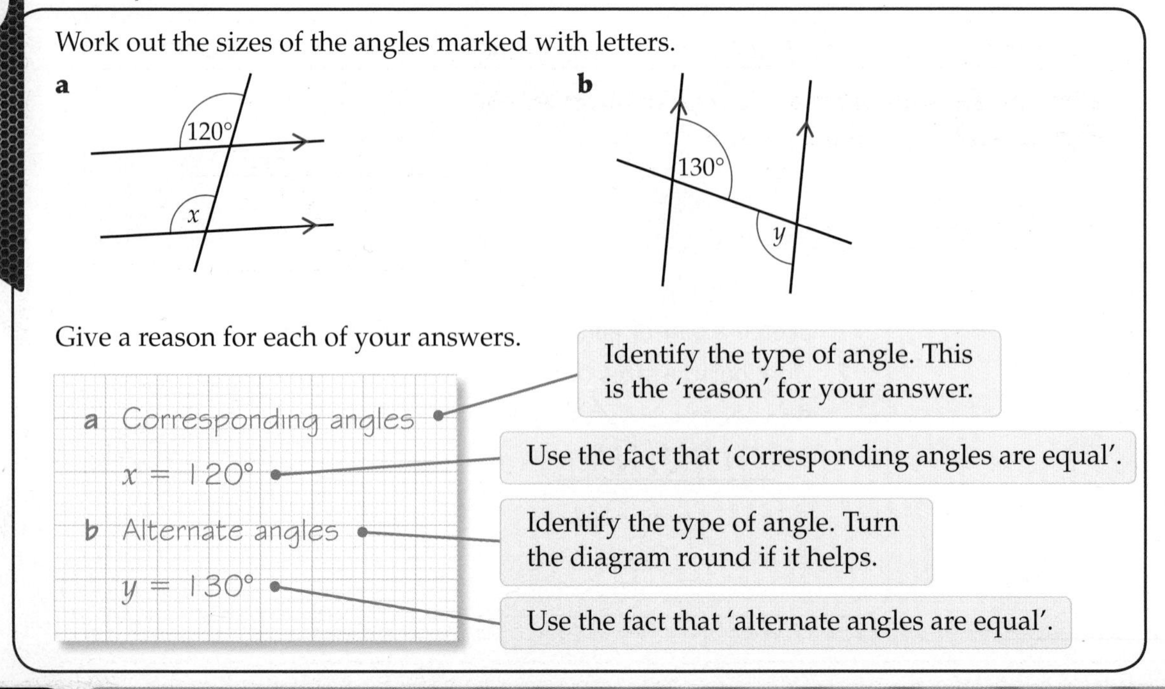

Work out the sizes of the angles marked with letters.

Give a reason for each of your answers.

a Corresponding angles — Identify the type of angle. This is the 'reason' for your answer.

$x = 120°$ — Use the fact that 'corresponding angles are equal'.

b Alternate angles — Identify the type of angle. Turn the diagram round if it helps.

$y = 130°$ — Use the fact that 'alternate angles are equal'.

Exercise 21B

1 Work out the sizes of the angles marked with letters.
Give a reason for your answer each time.

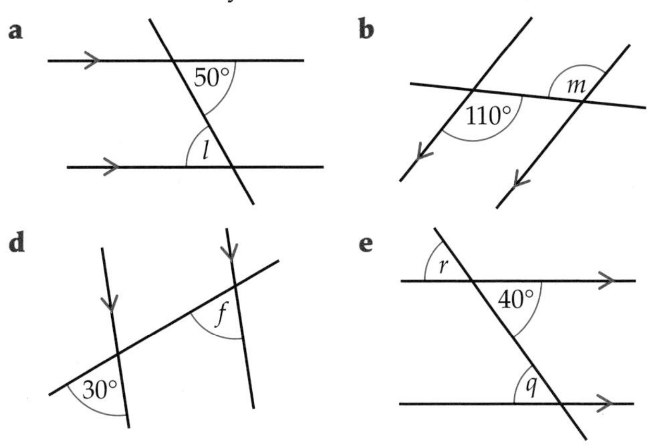

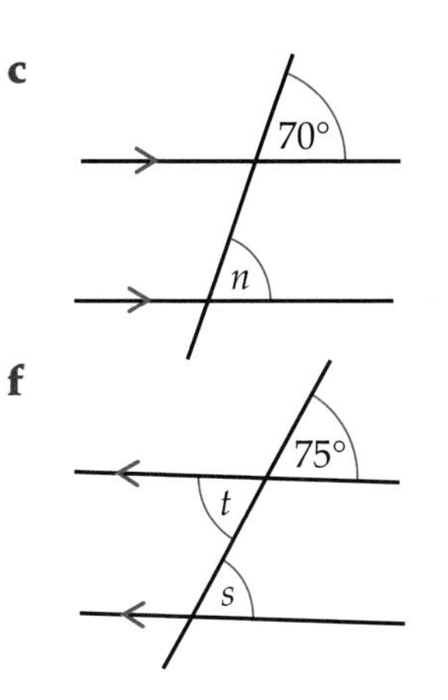

D

2 Alix draws a tessellation of parallelograms.

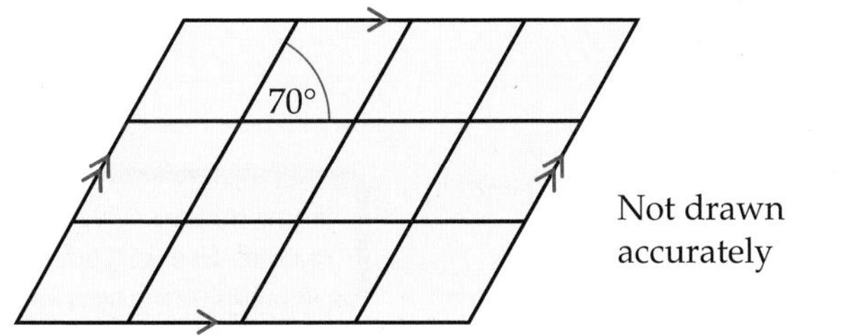

Not drawn accurately

You can't measure the angles because the diagram is not drawn accurately.

a Copy a two-by-two block of four parallelograms from her tessellation.

b Use the angle facts for parallel lines to label all the angles that are 70°.

c How could you use the angle facts you know to work out the other angles?

D

AO2

Using angle facts to solve problems

To solve more complex angle problems, you may need to use more than one angle fact.

Example 3

Work out the sizes of the angles marked with letters.
Give reasons for your answers.

$a + 40° = 180°$ (angles on a straight line)

so $a = 140°$

$b = a = 140°$ (a and b are alternate angles)

$c = b = 140°$ (b and c are vertically opposite angles)

Look for relationships between pairs of angles.

D

Exercise 21C

D

1 Work out the sizes of the angles marked with letters.
Give reasons for your answers.

a

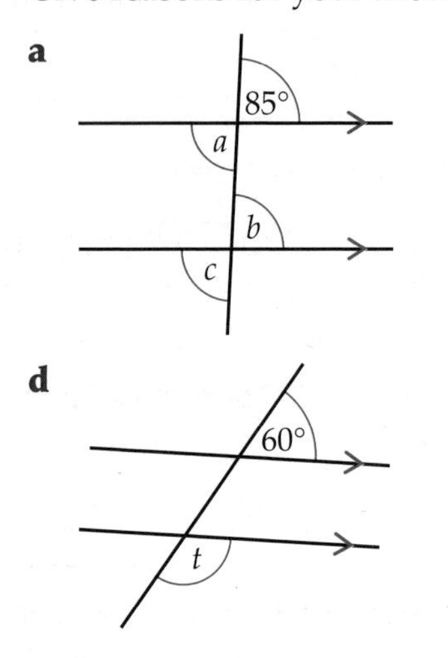

b

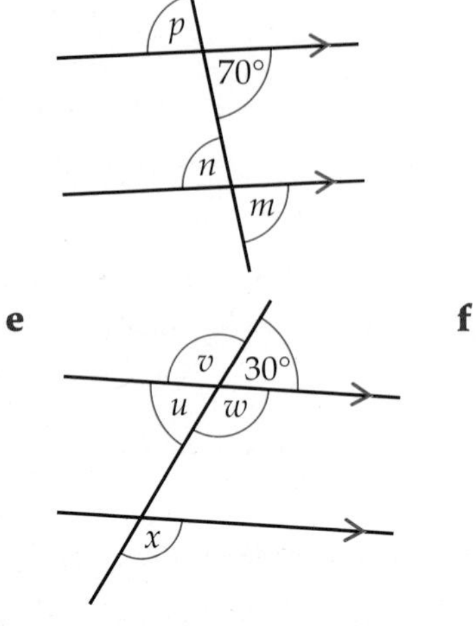

c

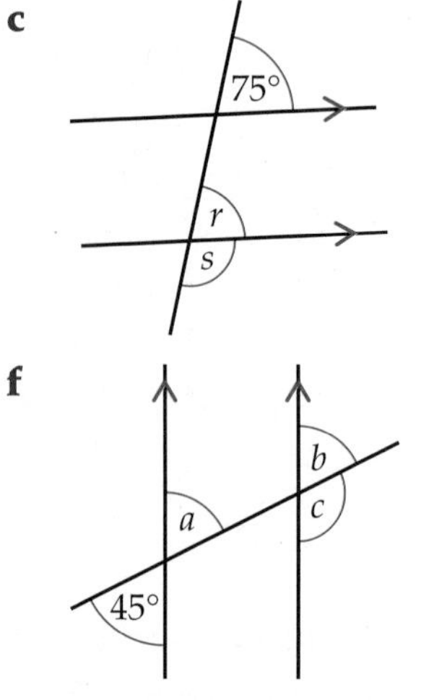

d

60°

t

e

v 30°

u w

x

f

b

a c

45°

D

AO2

2 Show why angle $x = 60°$.

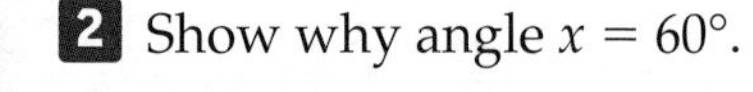

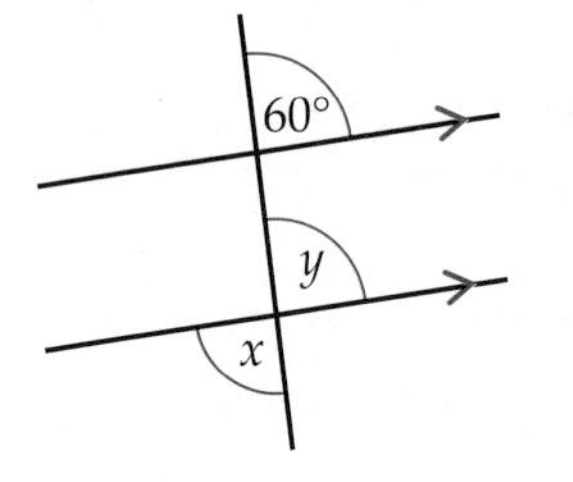

'Show' means 'give a reason'. Use angle facts.

D

3 Show why angle $m = 120°$.

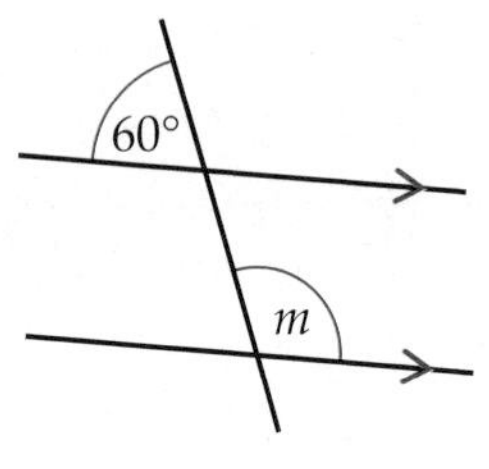

Copy the diagram and label any other angles you need.

4 Nell has drawn a tessellation of triangles.

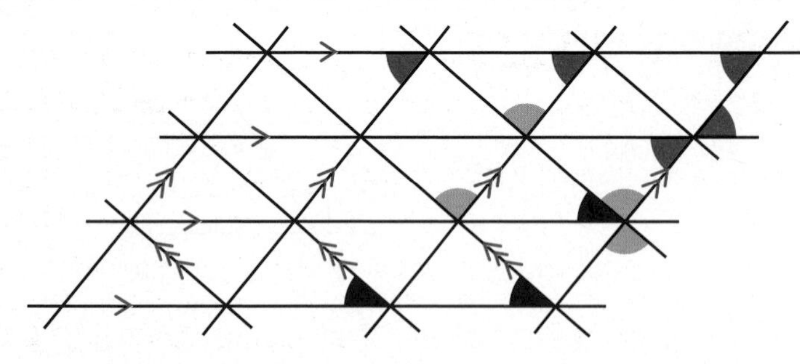

All the horizontal lines are parallel.
The diagonal lines in each set are parallel.
She has used colours to show equal angles.

a Draw a tessellation like this.
Use colours to show equal angles.

b Each red angle equals 50° and each black angle equals 40°.
Work out the size of each blue angle.
Give a reason for your answer.

AO3

5 *PQRS* is a parallelogram.

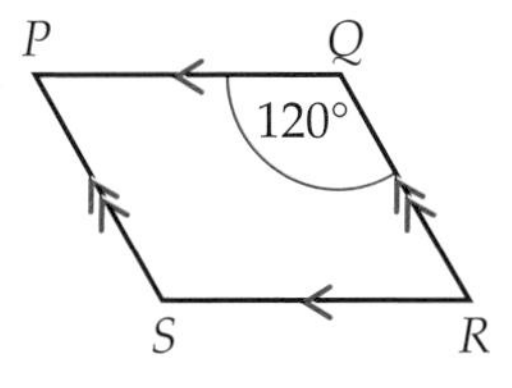

Angle $PQR = 120°$.

a Copy the parallelogram and extend each side as shown.

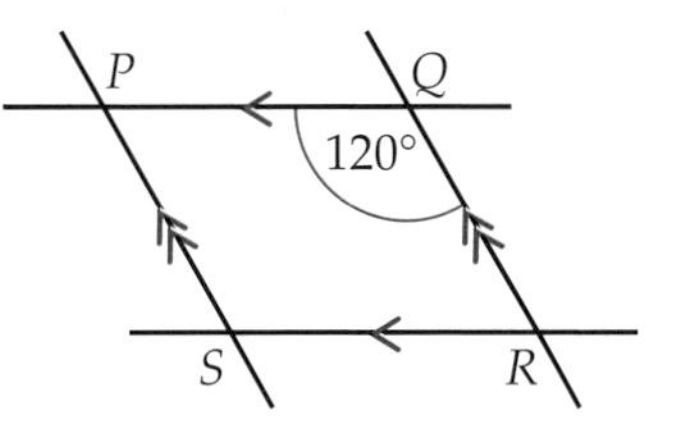

b Use angle facts to work out the sizes of the other angles in the parallelogram.

c What do you notice about the angles in the parallelogram?

d Draw more parallelograms. Does part **c** work for all of them?

D

AO2

6 *ABCD* is a trapezium.

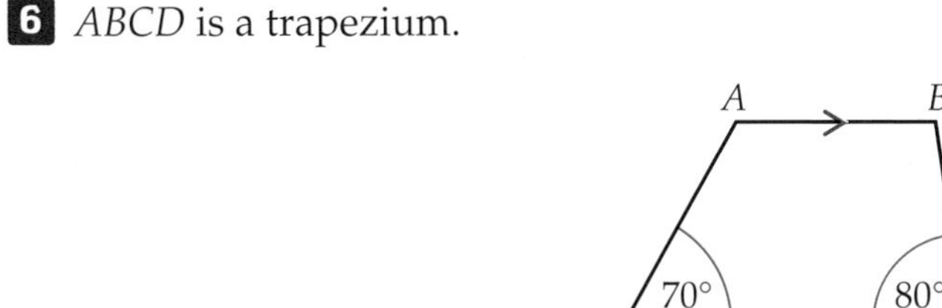

Angle $BCD = 80°$ and angle $CDA = 70°$.
Work out the sizes of the remaining angles.

D

AO3

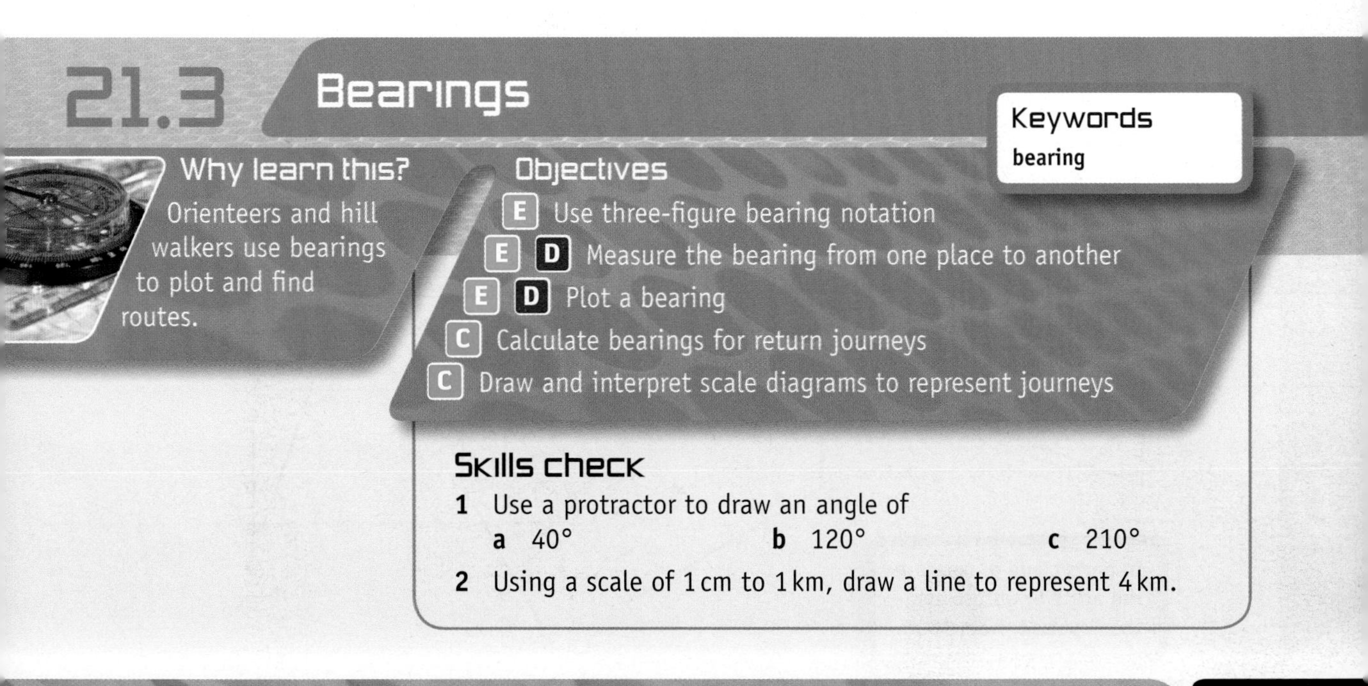

21.3 Bearings

Keywords
bearing

Why learn this?

Orienteers and hill walkers use bearings to plot and find routes.

Objectives

- **E** Use three-figure bearing notation
- **E D** Measure the bearing from one place to another
- **E D** Plot a bearing
- **C** Calculate bearings for return journeys
- **C** Draw and interpret scale diagrams to represent journeys

Skills check

1 Use a protractor to draw an angle of
a 40° **b** 120° **c** 210°

2 Using a scale of 1 cm to 1 km, draw a line to represent 4 km.

Understanding bearings

A **bearing** tells you the direction to travel. It is an angle measured clockwise from north.

A bearing can have any value from 0° to 360°. It is always written with three figures. For a two-digit number, like 72, add a zero in front: 072°.

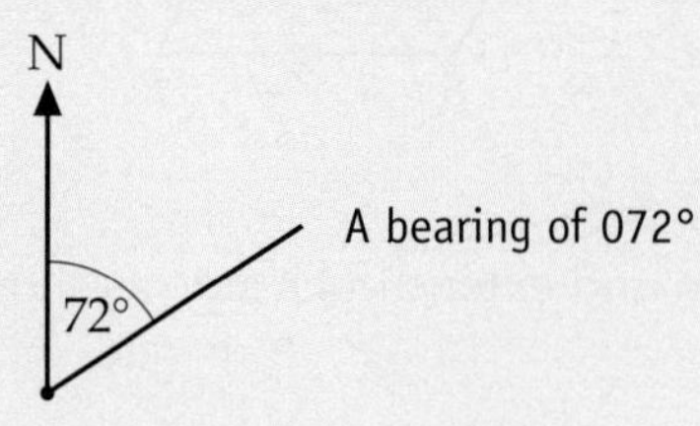

To draw or measure a bearing from a point, draw the north line at that point first, straight up the page.

Example 4

The diagram shows the positions of two towns, Anytown and Smallville.
Measure the bearing of Smallville from Anytown.

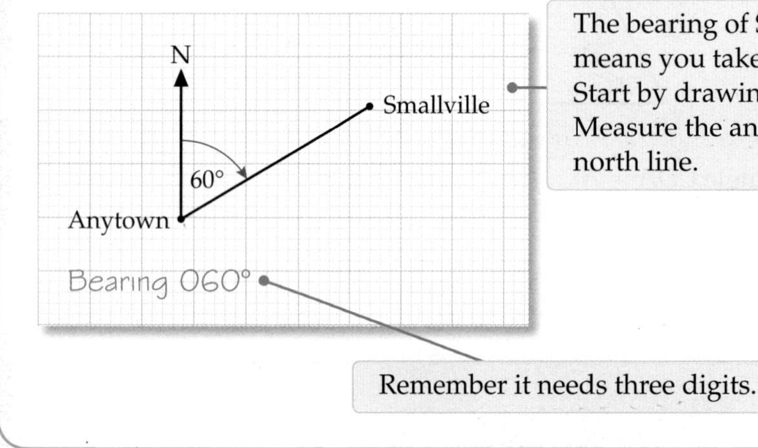

The bearing of Smallville **from** Anytown means you take the bearing **at** Anytown. Start by drawing a north line at Anytown. Measure the angle clockwise from the north line.

Remember it needs three digits.

Exercise 21D

1 For each diagram, measure the bearing of X from Y.

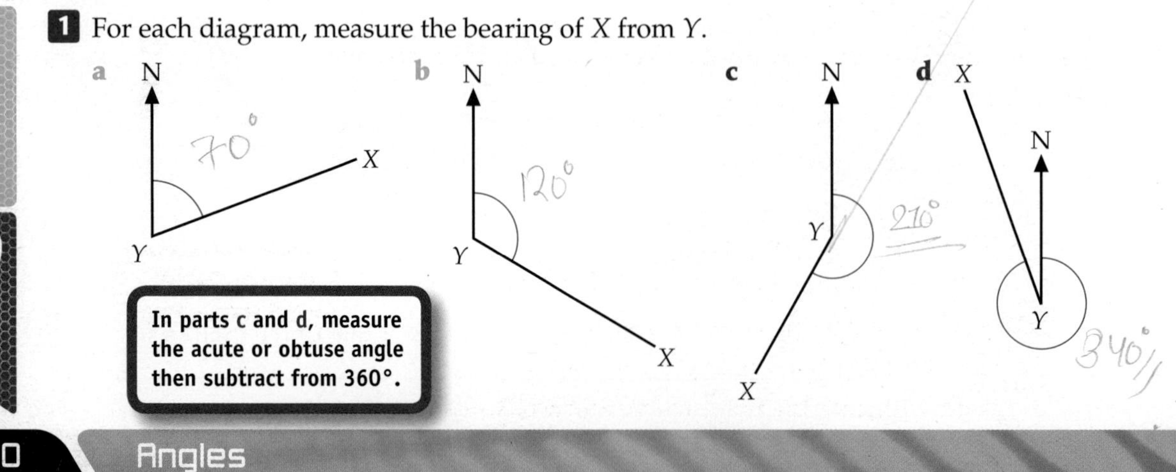

In parts c and d, measure the acute or obtuse angle then subtract from 360°.

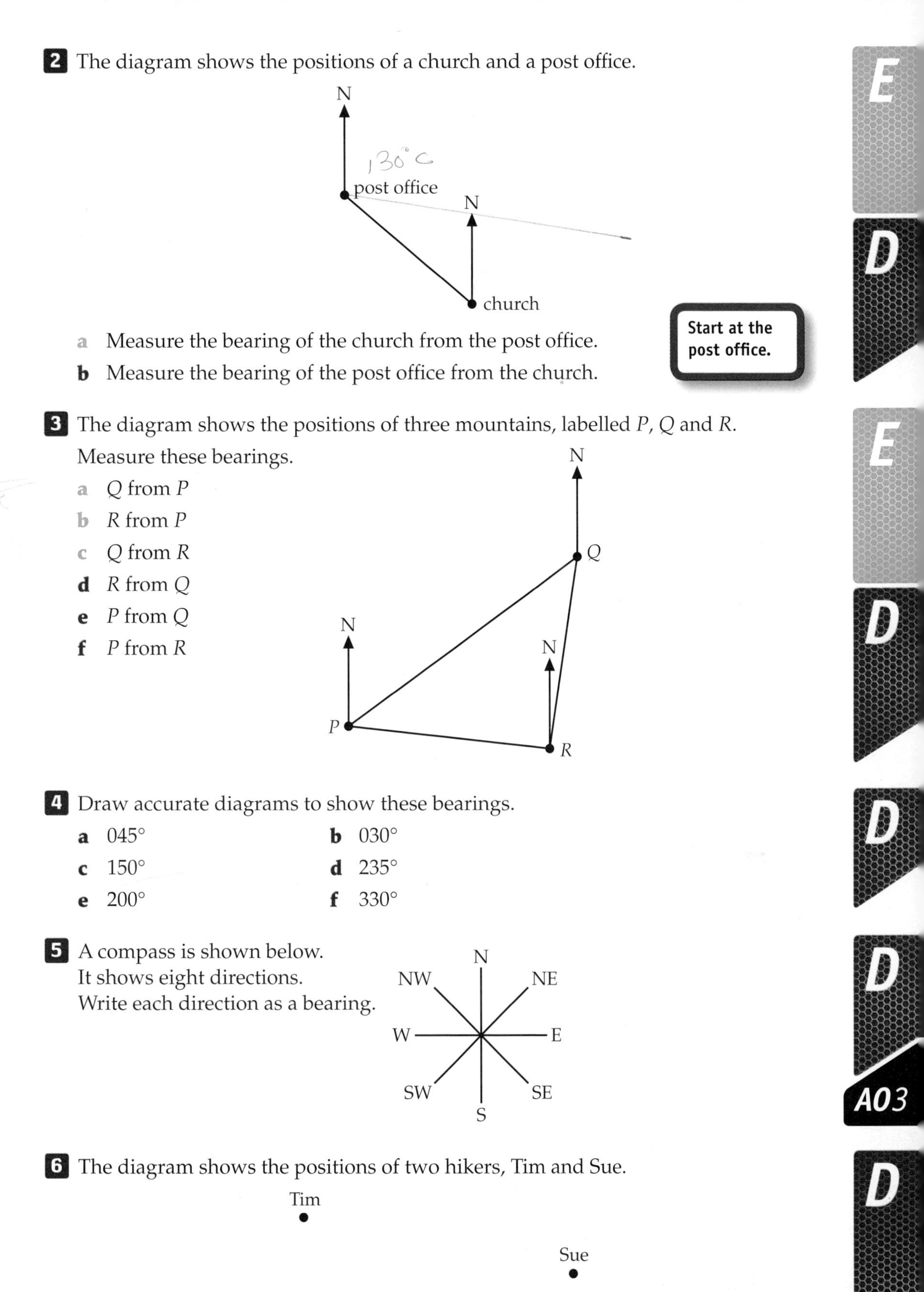

2 The diagram shows the positions of a church and a post office.

a Measure the bearing of the church from the post office.

b Measure the bearing of the post office from the church.

Start at the post office.

3 The diagram shows the positions of three mountains, labelled P, Q and R.
Measure these bearings.

a Q from P

b R from P

c Q from R

d R from Q

e P from Q

f P from R

4 Draw accurate diagrams to show these bearings.

a	045°	**b**	030°
c	150°	**d**	235°
e	200°	**f**	330°

5 A compass is shown below.
It shows eight directions.
Write each direction as a bearing.

6 The diagram shows the positions of two hikers, Tim and Sue.

Tim walks on a bearing of 050°.

a Copy the diagram. Plot and draw Tim's route.

Sue walks on a bearing of 340°.

b Plot and draw Sue's route.

c Mark with an X the point where Sue and Tim's paths cross.

Use a protractor.

7 The diagram shows a ship and a speed boat.

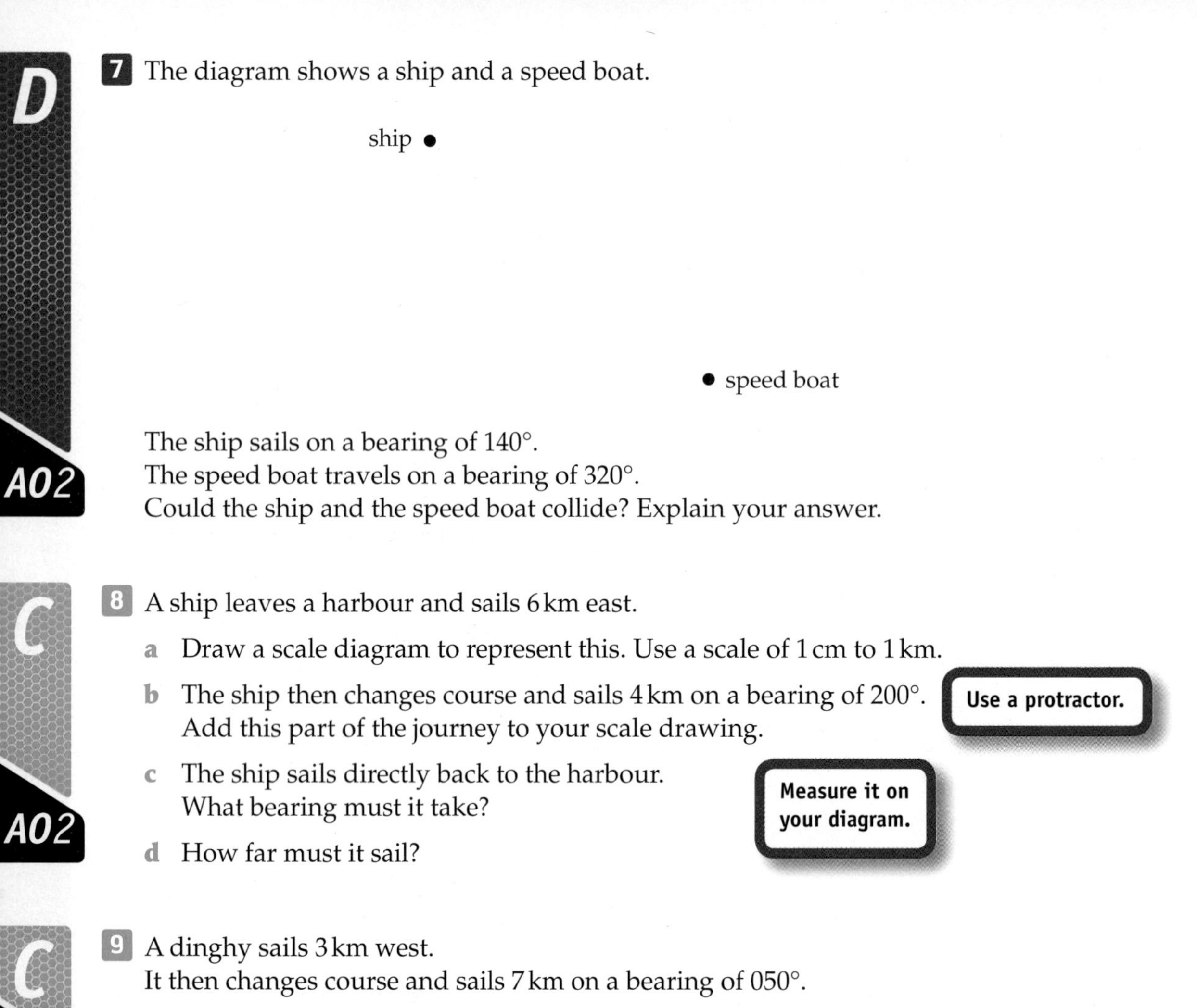

The ship sails on a bearing of 140°.
The speed boat travels on a bearing of 320°.
Could the ship and the speed boat collide? Explain your answer.

C

AO2

8 A ship leaves a harbour and sails 6 km east.

a Draw a scale diagram to represent this. Use a scale of 1 cm to 1 km.

b The ship then changes course and sails 4 km on a bearing of 200°. Add this part of the journey to your scale drawing.

Use a protractor.

c The ship sails directly back to the harbour. What bearing must it take?

Measure it on your diagram.

d How far must it sail?

C

AO3

9 A dinghy sails 3 km west.
It then changes course and sails 7 km on a bearing of 050°.

a How far is the dinghy from its starting point?

b What bearing should it take to return to the starting point?

Parallel lines and bearings

Two north lines are always parallel to each other.

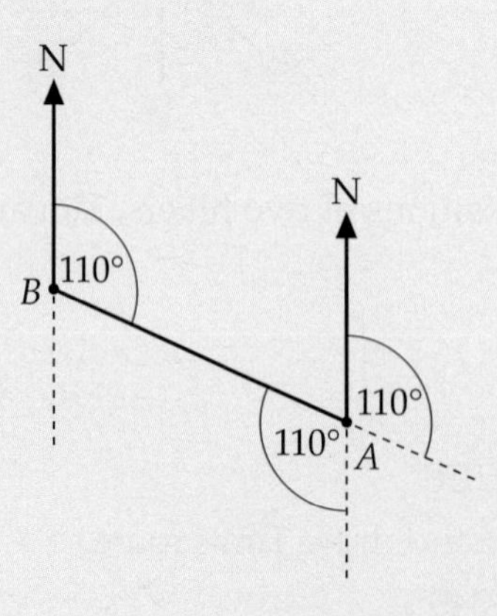

You can use the angle facts for parallel lines to help you work out bearings.

Example 5

a Write down the bearing of A from B.
b Work out the bearing of B from A.

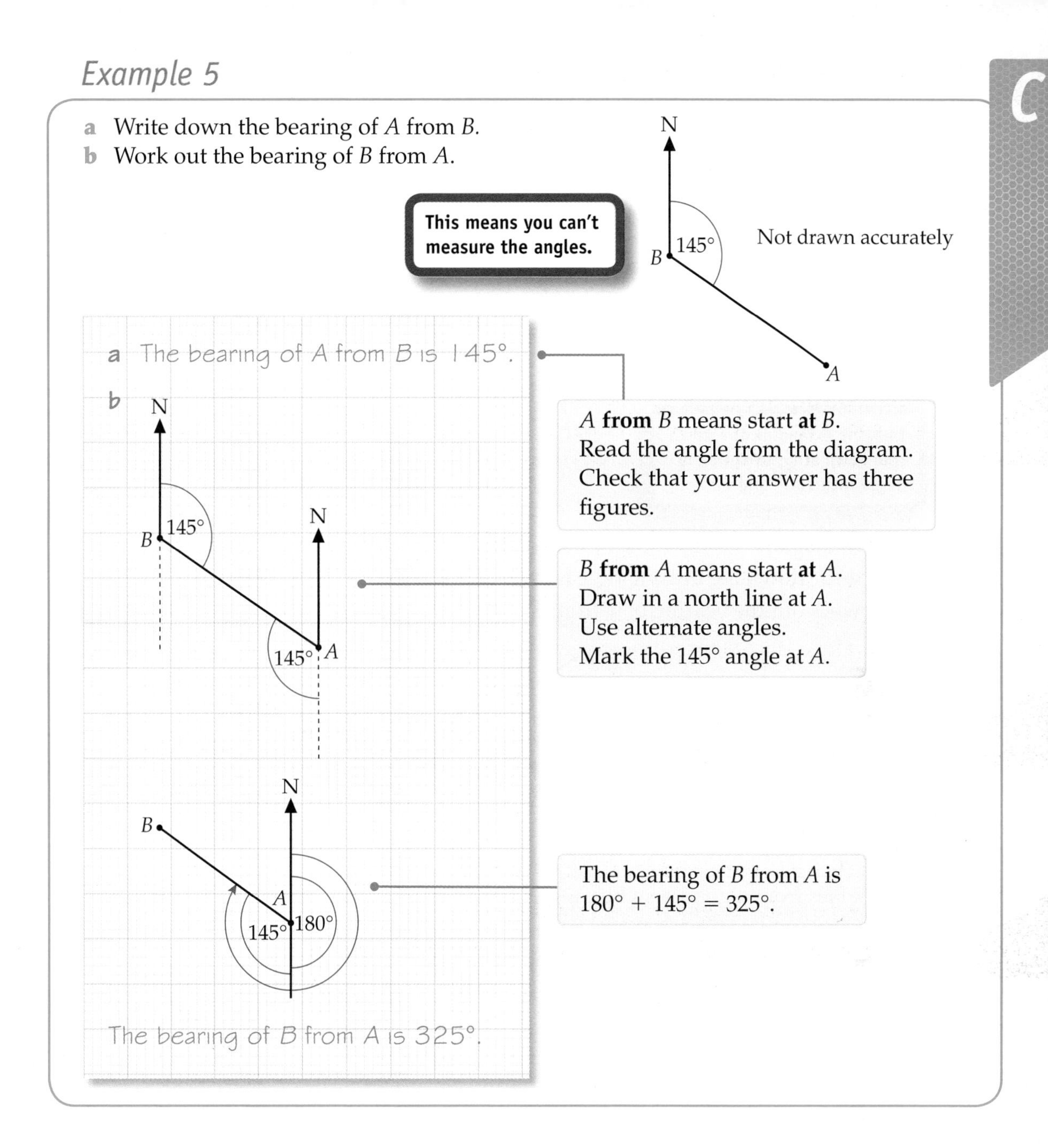

Exercise 21E

1 For each diagram

i write down the bearing of A from B

ii work out the bearing of B from A.

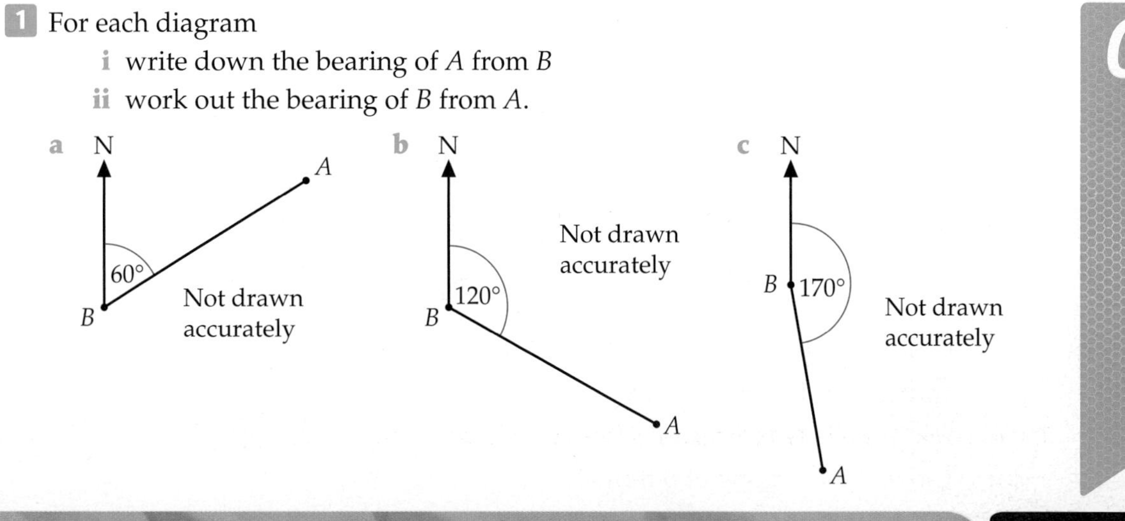

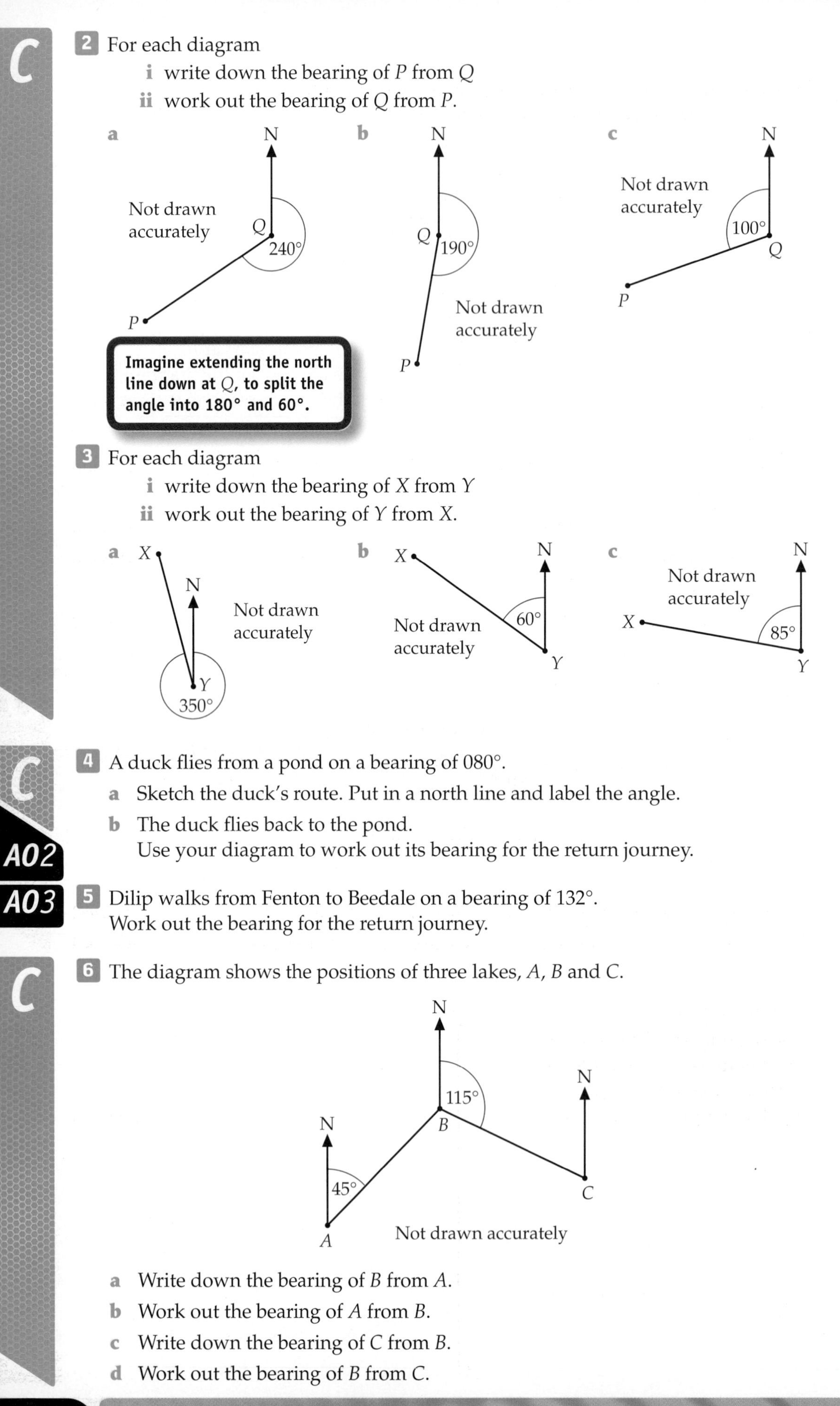

C

2 For each diagram

  i write down the bearing of P from Q

  ii work out the bearing of Q from P.

a

> **Imagine extending the north line down at Q, to split the angle into 180° and 60°.**

b

c

3 For each diagram

  i write down the bearing of X from Y

  ii work out the bearing of Y from X.

a

b

c

C

AO2

AO3

4 A duck flies from a pond on a bearing of 080°.

  a Sketch the duck's route. Put in a north line and label the angle.

  b The duck flies back to the pond.
  Use your diagram to work out its bearing for the return journey.

5 Dilip walks from Fenton to Beedale on a bearing of 132°.
Work out the bearing for the return journey.

C

6 The diagram shows the positions of three lakes, A, B and C.

  a Write down the bearing of B from A.

  b Work out the bearing of A from B.

  c Write down the bearing of C from B.

  d Work out the bearing of B from C.

Review exercise

1 The diagram shows a five-sided shape $ABCDE$ and point O.
The point O is joined to each vertex (corner) as shown.

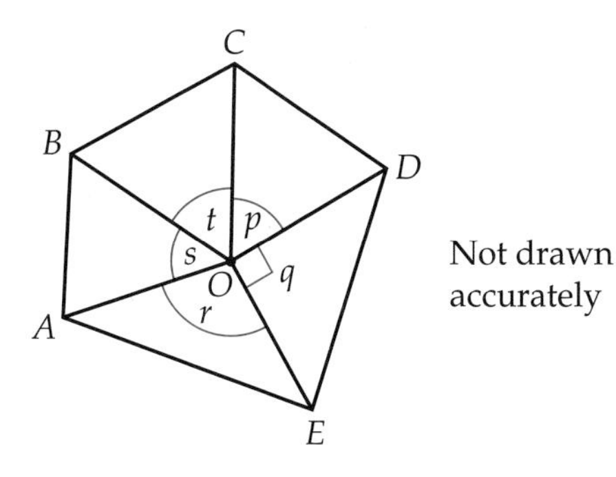

E

a Write down the value of $p + q + r + s + t$. **[1 mark]**

b Angle p is less than $90°$.
Write down the name for this type of angle. **[1 mark]**

c One angle is $90°$ exactly.
Write down the letter of this angle. **[1 mark]**

d What type of angle is angle r? **[1 mark]**

e Angle p + angle $q = 170°$
Work out the size of angle p. **[2 marks]**

D

2 In the diagram, AB is parallel to CD.

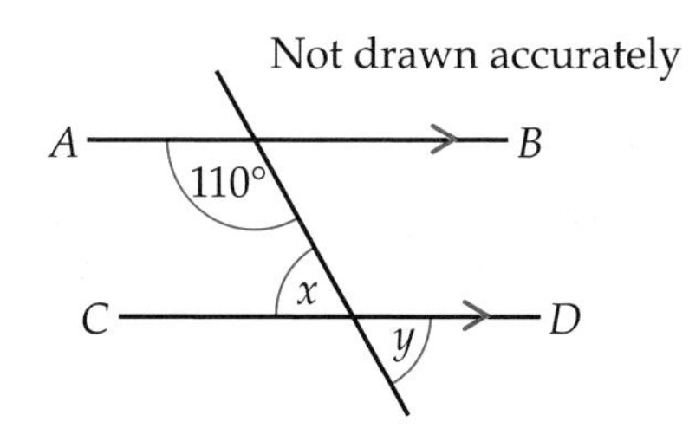

a Work out the size of angle x. Give a reason for your answer. **[2 marks]**

b Write down the size of angle y. Give a reason for your answer. **[2 marks]**

C

3 A is due south of B.
The bearing of C from A is $080°$.
The bearing of C from B is $120°$.
Trace the diagram.
Mark the position of C. **[3 marks]**

N

B

A

AO2

C

4 The diagram shows three points A, B and C.

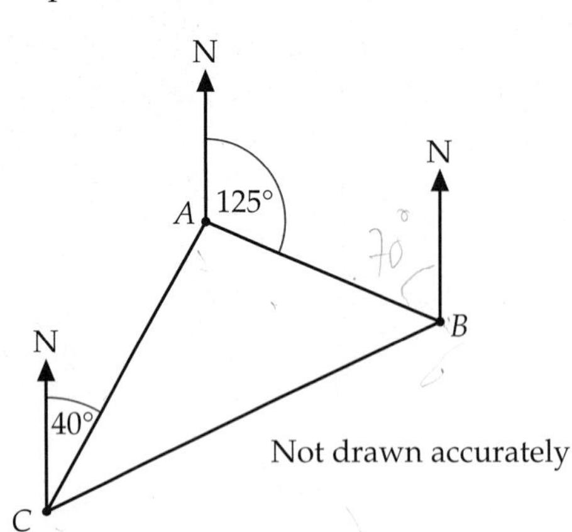

The bearing of B from A is 125°.

a Write down the bearing of A from C. **[1 mark]**

b Work out the bearing of A from B. **[2 marks]**

C

AO3

5 Maya lives 5 km due north of Dan.

a Using a scale of 1 cm to 1 km, draw a diagram to show the positions of Maya's and Dan's houses. **[2 marks]**

b Maya leaves home and walks 5 km due east to the beach.
Add Maya's walk to your diagram. **[2 marks]**

c Dan walks to meet Maya at the beach. He walks in a straight line.
What bearing does he walk on?
How far does he walk? **[2 marks]**

C

AO2

6 An aircraft flies 8 km on a bearing of 060°.
Then it changes direction and flies 6 km on a bearing of 290°.

a Draw a scale diagram to show the aircraft's journey.
Use a scale of 1 km to 1 cm. **[4 marks]**

b Work out the distance and bearing of the return journey to the starting point. **[2 marks]**

Chapter summary

In this chapter you have learned how to

- calculate angles around a point **E**
- recognise vertically opposite angles **E**
- use three-figure bearing notation **E**
- measure the bearing from one place to another **E** **D**
- plot a bearing **E** **D**
- recognise corresponding and alternate angles **D**
- calculate angles in diagrams with parallel lines **D**
- calculate bearings for return journeys **C**
- draw and interpret scale diagrams to represent journeys **C**

22 Measurement 1

This chapter is about converting between metric and imperial units, and using maps and scale drawings.

In 1999 the NASA Mars Climate Orbiter was lost in space because some of its designers forgot to convert between metric and imperial units. The Orbiter was worth $125 million.

Objectives

This chapter will show you how to

- know and use approximate metric equivalents of imperial units E
- solve problems involving time, dates and timetables E
- use and interpret maps and scale drawings E
- recognise that measurements given to the nearest whole unit may be inaccurate by up to one half in either direction C

Before you start this chapter

1 Round 8605.83 to

a the nearest 100   b one decimal place

c the nearest whole number   d three significant figures.

2 Put each of these sets of numbers in order of size, smallest first.

a 8.53   8.9   8.09   10   8.1   b 0.4   0.05   0.9   2   0.71

3 Six cans of fizzy drink cost £2.10. Work out the cost of

a 12 cans   b 60 cans   c 1 can   d 5 cans.

4 Divide 120 m in each of these ratios.

a 2 : 1   b 1 : 3   c 5 : 7   d 3 : 7

HELP Chapter 7

Time and timetables

Why learn this?

Understanding timetables is very important when you are travelling.

Objectives

E Solve problems involving time, dates and timetables

Keywords

12-hour clock, am, pm, midnight, midday, 24-hour clock, hours, minutes, seconds

Skills check

HELP Section 12.4

1 Work out

a $\frac{1}{2} \times \frac{2}{3}$ **b** $\frac{4}{5} \times \frac{1}{4}$ **c** $\frac{2}{15} \times 5$ **d** $\frac{3}{7} \times 4$

2 Work out

a $107 - 15$ **b** $24 - 9$ **c** $219 - 36$ **d** $4020 - 600$

Using timetables

There are two ways of recording times.

- The **12-hour clock** uses **am** for times between **midnight** and **midday** and **pm** for times between midday and midnight.
- The **24-hour clock** counts a whole day from 00 00 to 23 59.

'Midday' is also called 'noon'.

12-hour clock	24-hour clock
7:45 am	07 45
12:00 midday	12 00
1:50 pm	13 50
10:30 pm	22 30
12:00 midnight	00 00

When you are solving problems involving time you need to be very careful when converting between **hours**, **minutes** and **seconds**.

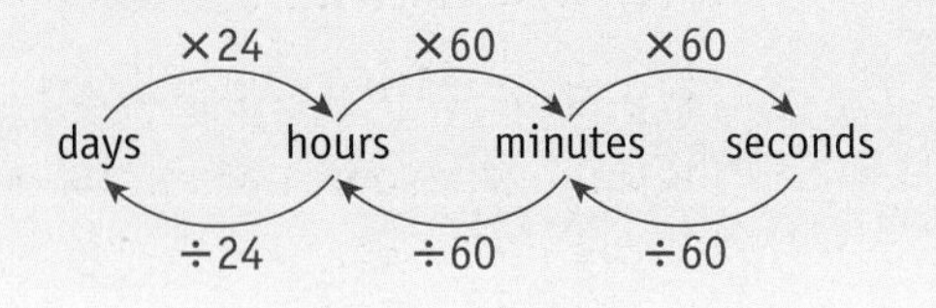

Example 1

This timetable shows bus times in Cardiff.

Cardiff Bay	09 09	then hourly	18 39	18 57	19 27	then at	27	57	until	22 27	23 00
Atlantic Wharf	▼		▼	18 59	19 29		29	59		22 29	23 02
Pier Head Street	▼		▼	19 01	19 31		31	01		22 31	23 04
Bute Street	09 13		18 43	19 06	19 36		36	06		22 36	23 09
Central Station	09 21		18 51	19 12	19 42		42	12		22 42	23 17

a What time does the first bus leave Cardiff Bay?

b How long does this bus take to travel to Central Station?

c I arrive at Atlantic Wharf bus stop at 9:15 pm. What time is the next bus?

a The first bus leaves at 09 09.

b The journey time is 12 minutes.

c The next bus will be at 21 29.

This bus arrives at Central Station at 09 21.

Between 19 29 and 22 29 buses leave from Atlantic Wharf at 29 and 59 minutes past the hour.

Exercise 22A

1 The train journey from Bristol to London should take 1 hour and 53 minutes. Ahmed's train leaves Bristol at 7:18 am.

a What time should Ahmed arrive in London?

b The train is delayed by a quarter of an hour. What is Ahmed's new arrival time?

2 Jeremy begins work at 08 40 and finishes at 16 55. He takes an hour for lunch and has two fifteen-minute breaks.

a Work out the amount of time Jeremy spends working.

b Jeremy is paid £8 per hour. Work out Jeremy's earnings for this day.

3 This is part of an Australian train timetable.

	am	am	am	am	pm	pm	pm	pm	pm
Mount Victoria	9:35	10:05	11:05	12:05	1:05	2:05	3:05	4:05	5:05
Blackheath	9:42	10:12	11:12	12:12	1:12	2:12	3:12	4:12	5:12
Medlow Bath	9:47	10:17	11:17	12:17	1:17	2:17	3:17	4:17	5:17
Katoomba	9:55	10:25	11:25	12:25	1:25	2:25	3:25	4:25	5:25
Leura	9:58	10:28	11:28	12:28	1:28	2:28	3:28	4:28	5:28
Wentworth Falls	10:04	10:34	11:34	12:34	1:34	2:34	3:34	4:34	5:34
Bullaburra	10:10	10:40	11:40	12:40	1:40	2:40	3:40	4:40	5:40

a What time does the last train leave from Katoomba?

b How long is the journey from Blackheath to Wentworth Falls?

c I need to arrive in Katoomba before midday. What train should I catch from Mount Victoria?

d The 1:17 pm train from Medlow Bath is delayed by half an hour. What time will it arrive in Bullaburra?

4 This is a train timetable for journeys from Cleethorpes to Doncaster.

Cleethorpes	08 00	08 45	09 00	09 45	10 00	10 45	11 00	11 45	12 00
Grimsby Town	08 25	09 10	09 25	10 10	10 25	11 10	11 25	12 10	12 25
Habrough	08 50		09 50		10 50		11 50		12 50
Barnetby	09 05		10 05		11 05		12 05		13 05
Scunthorpe	09 35		10 35		11 35		12 35		13 35
Doncaster	10 25	10 25	11 25	11 25	12 25	12 25	13 25	13 25	14 25

A blank space means the train does not stop at that station.

a How long is the journey from Grimsby Town to Scunthorpe?

b I arrive at Grimsby Town at 09 55. How long do I have to wait for the next train to Barnetby?

c The trains that leave Cleethorpes on the hour are stopping services. How much longer does it take to travel from Cleethorpes to Doncaster on a stopping service than on an express service?

d Do you think it is further from Barnetby to Scunthorpe or from Scunthorpe to Doncaster? Give a reason for your answer.

22.2 Converting between metric and imperial units

Keywords
imperial units, convert

Why learn this?

If you are planning a trip to Europe you need to know how to compare distances in kilometres and miles.

Objectives

E Know and use approximate metric equivalents of pounds, feet, miles, pints and gallons

Skills check

1 Use a calculator to work out

a $410 \div 2.2$ **b** $200 \div 1.6$ **c** $46 \div 3$ **d** $14 \times 5 \div 8$

Give your answers correct to two decimal places.

Converting units

You can measure lengths, capacities and masses using **imperial units**.

The imperial units of length are miles, feet and inches.

The imperial units of mass are pounds and ounces.

The imperial units of capacity are gallons and pints.

1 foot = 12 inches
1 pound = 16 ounces
1 gallon = 8 pints

You need to be able to **convert** between metric and imperial units.

To help you do this, you need to remember these approximate conversions.

5 miles ≈ 8 km 1 inch ≈ 2.5 cm

1 kg ≈ 2.2 pounds 1 gallon ≈ 4.5 litres

≈ means 'approximately equal to'.

Example 2

a A dairy cow produces about 30 gallons of milk per week. How many litres is this?

b A marathon is 26 miles. How many kilometres is this?

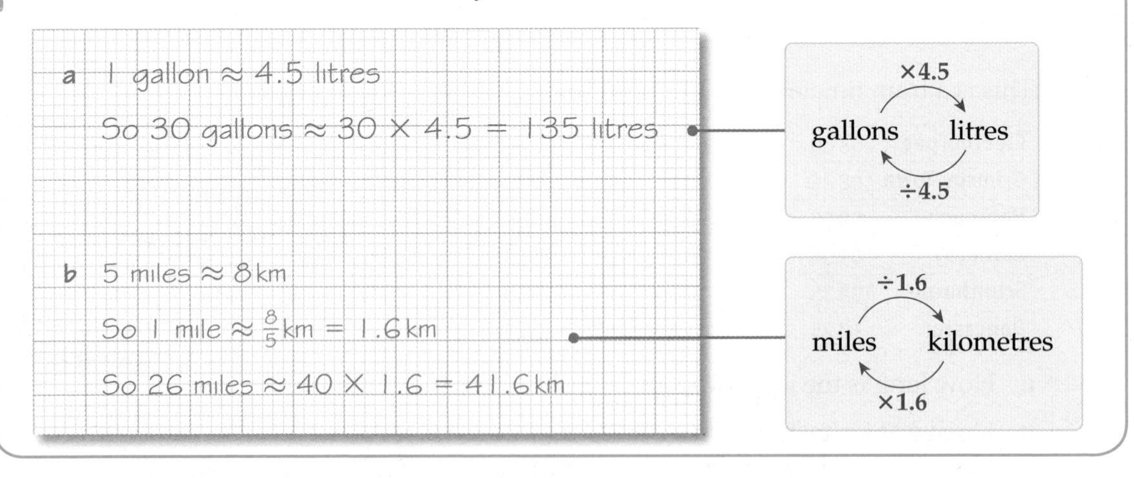

Exercise 22B

1 Convert

a 20 miles into km **b** 20 km into miles **c** 6 kg into pounds

d 9 inches into cm **e** 5 gallons into litres **f** 66 pounds into kg

g 270 miles into km **h** 30 cm into inches **i** 27 litres into gallons

2 The average weight of an emperor penguin is 30 kg.
How many pounds is this?

3 I have a photograph measuring 4 inches by 6 inches.
What are the dimensions of my photograph in centimetres?

4 It is 23 miles from Dover to Calais.
How many kilometres is this?

5 This chart gives the driving distances in km between different cities in Australia.

Adelaide								
1542	Alice Springs							
2063	3012	Brisbane						
3143	2324	1717	Cairns					
3053	1511	3415	2727	Darwin				
728	2270	1674	3054	3781	Melbourne			
2724	3630	4384	5954	4045	3452	Perth		
1420	2644	996	2546	4000	868	4144	Sydney	
2525	2096	1467	374	4556	2857	5728	2494	Townsville

Work out the distance in miles, to the nearest mile, from

a Alice Springs to Brisbane
b Sydney to Adelaide
c Melbourne to Townsville
d Perth to Cairns.

E

6 A caravan water tank holds 12 gallons.
Will it hold 50 litres of water? Explain your answer.

E

7 Christina has 4 pounds of flour.
To make 1 loaf of bread, you need 500 g of flour.
How many loaves of bread can she make?

A02

FUNCTIONAL

8 The petrol tank in a car has a capacity of 14 gallons.

a What is this capacity in litres?
b Petrol costs £1.05 per litre. How much does it cost to fill up this petrol tank?

E

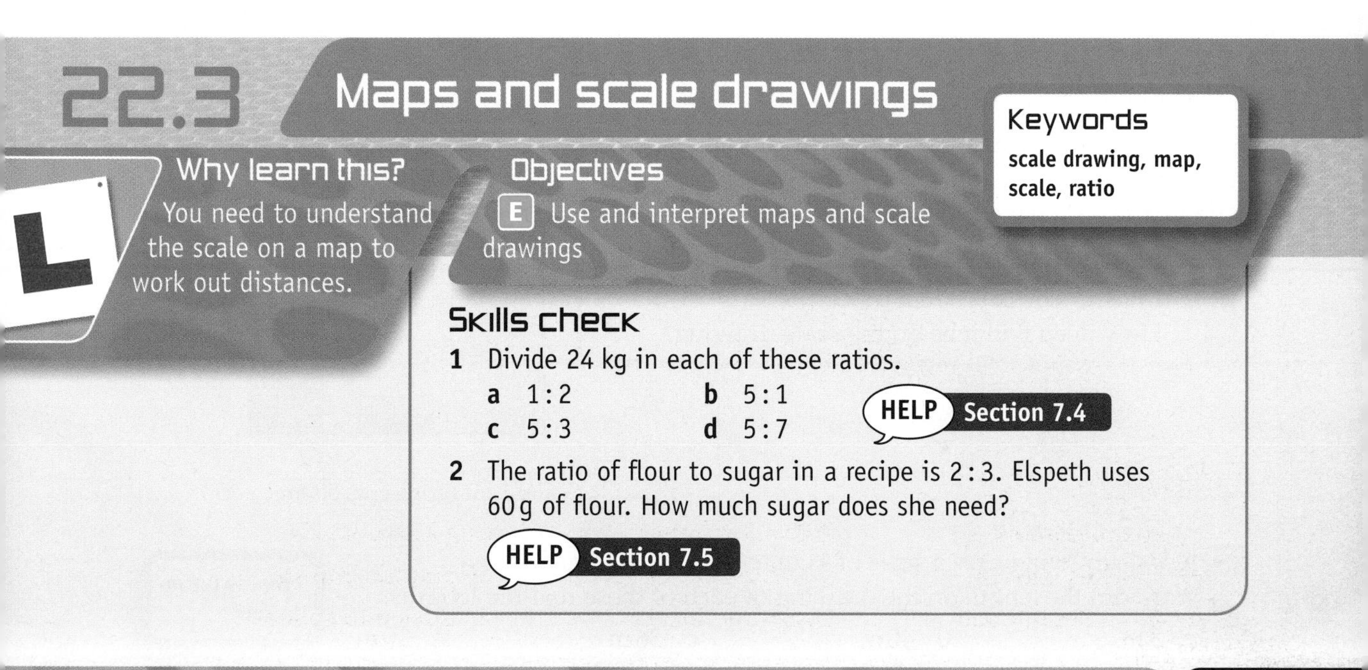

22.3 Maps and scale drawings

Keywords
scale drawing, map, scale, ratio

Why learn this?

You need to understand the scale on a map to work out distances.

Objectives

E Use and interpret maps and scale drawings

Skills check

1 Divide 24 kg in each of these ratios.
a 1 : 2
b 5 : 1
c 5 : 3
d 5 : 7

HELP Section 7.4

2 The ratio of flour to sugar in a recipe is 2 : 3. Elspeth uses 60 g of flour. How much sugar does she need?

HELP Section 7.5

Maps and scale drawings

A **scale drawing** has the same proportions as the object it represents.

A **map** is a type of scale drawing. You can use distances on a map to work out distances in real life.

The **scale** tells you the relationship between lengths on the drawing or map and lengths in real life.

Scales for maps are usually given as **ratios**.

This map has a scale of 1 : 25 000.

There are other ways of writing this ratio:

- 1 cm represents 25 000 cm
- 1 to 25 000

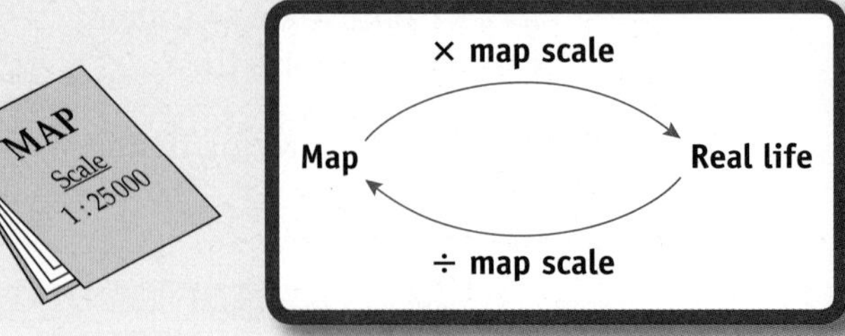

E

Example 3

A map has a scale of 1 : 25 000.

a A stream is 16 cm long on the map. How long is it in real life?

b The distance between two villages is 1.5 km.
How far apart will the two villages be on the map?

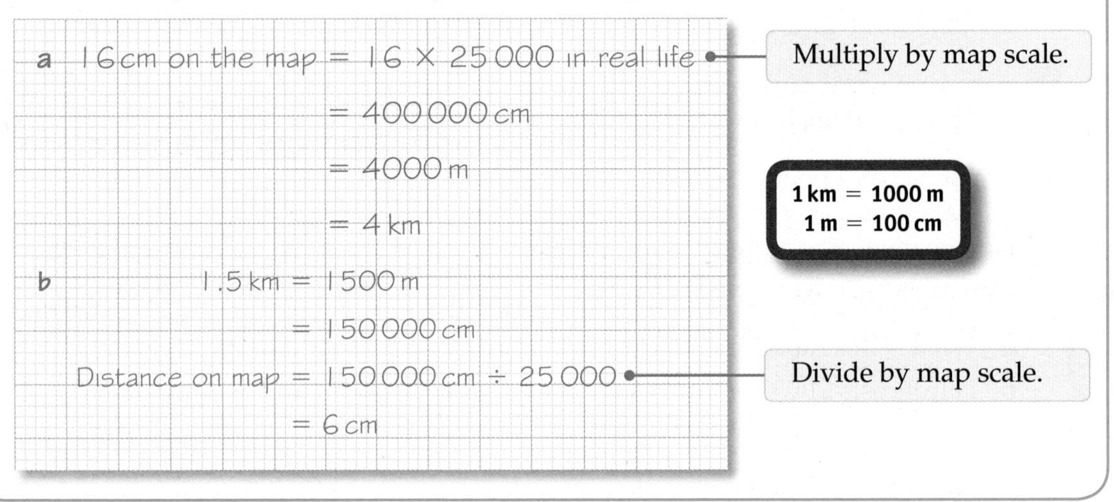

Exercise 22C

E

1 This scale drawing of Martha's bedroom is drawn on cm squared paper.

a Work out the width of Martha's bedroom from A to B.

b Martha's desk is 80 cm deep.
How deep will it be on the scale drawing?

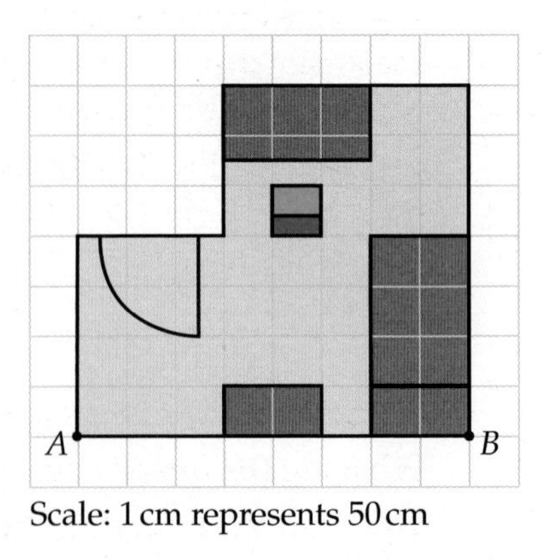

Scale: 1 cm represents 50 cm

2 A scale drawing uses a scale of 1 cm to represent 0.8 m.
Work out the length on the drawing of each of these real-life lengths.

a 8 m　　b 40 m　　c 36 m　　d 10 m

What is 0.8 m in cm?

3 A scale drawing of a new car design uses a scale of 1 to 30.

a On the scale drawing the car is 7 cm wide. How wide is the actual car?

b The car is 4.5 m long. How long is the car in the scale drawing?

4 Ben has made a scale drawing of his house and garden on cm squared paper.

a Work out the width of the house from A to B.

b Work out the length of the garden from B to C.

c Ben wants to build a swimming pool in his garden. The swimming pool is a rectangle 15 m long and 9 m wide. Will it fit in the garden? Give a reason for your answer.

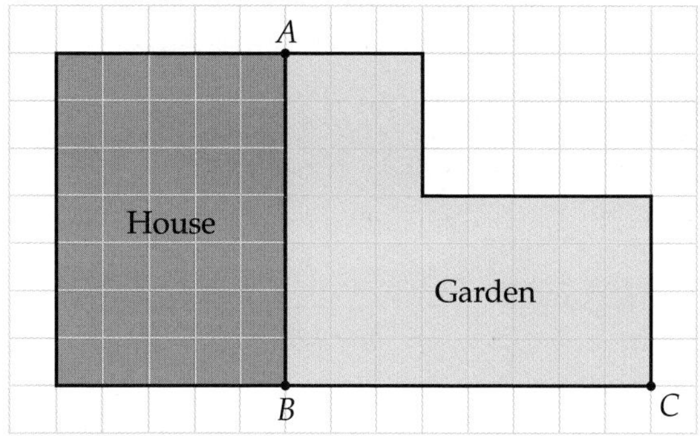

Scale: 1 cm represents 2 m

5 A map uses a scale of 1 : 150 000.

a The distance between two towns on the map is 5 cm.
How far apart are the two towns in real life?

b What length on the map will represent a distance of 24 km in real life?

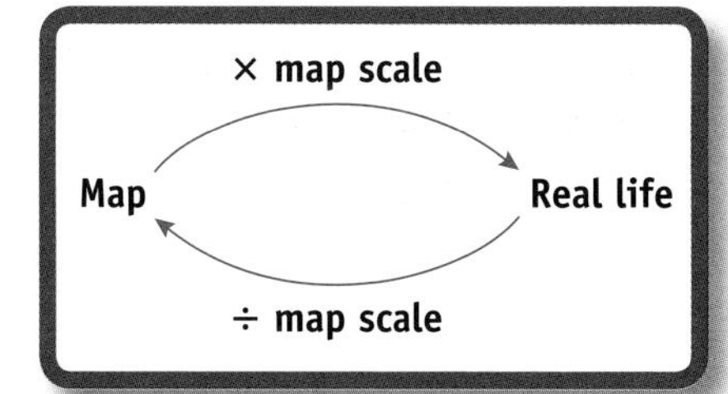

6 Susan is competing in a charity swim on Lake Windermere. The diagram shows a map of the lake with a scale of 1 : 50 000.

a What distance would be represented by 4 cm on the map?

b What length on the map would represent a distance of 700 m?

c Measure the distance from the Steamboat Museum to the checkpoint at Red Nab. Write your answer to the nearest mm.

d Use your answer to part c to work out the actual distance from the Steamboat Museum to the checkpoint at Red Nab.

e The route of the charity swim is marked on the map. Work out the length of the whole swim. Give your answer to the nearest 100 m.

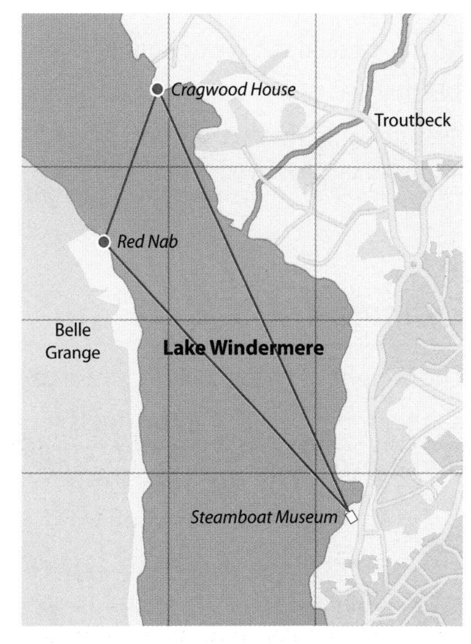

7 On a 1 : 25 000 map, a rectangular field has length 4 cm and width 2.5 cm.

a What are the dimensions of the field in real life?

b Work out the area of the field in real life.

22.4 Accuracy of measurements

Keywords
lower bound, minimum value, upper bound, maximum value

Why learn this?

There are legal bounds of accuracy of a car speedometer. If you are travelling at 60 mph then 60 mph is the lower bound for the display on the speedometer.

Objectives

C Recognise that measurements given to the nearest whole unit may be inaccurate by up to one half unit in either direction

Skills check

1 Work out
  a $8 + 0.5$ b $225 - 0.5$ c $50 - 0.5$ d $6200 + 0.5$

2 Round 350.75 to
  a the nearest hundred b the nearest whole number
  c one decimal place.

Accuracy of measurements

The width of this postcard is 8 cm to the nearest cm.

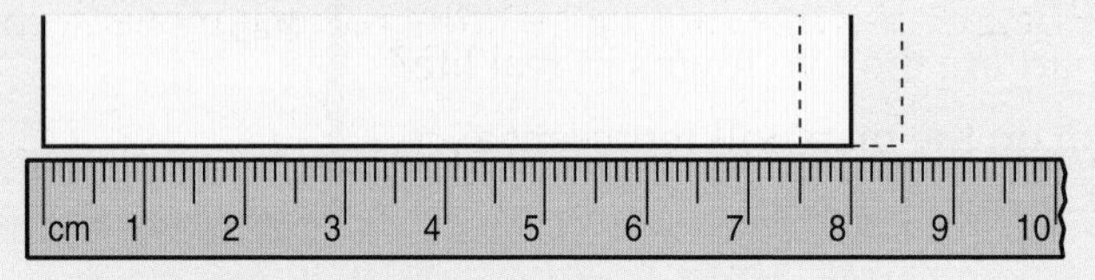

The actual width of the postcard could be anything between 7.5 cm and 8.5 cm.

7.5 cm is called the **lower bound** or **minimum value** for the width.

8.5 cm is called the **upper bound** or **maximum value** for the width.

Measurements that are given to the nearest whole number might be up to half a unit larger or smaller than the given value.

C

Example 4

The weight of a dog is 34 kg, measured to the nearest kg. Work out

a the upper bound for the dog's weight

b the lower bound for the dog's weight.

The weight has been rounded to the nearest kg. The actual weight could be up to 0.5 kg heavier or lighter.

Exercise 22D

C

1 Triona measures her height as 145 cm to the nearest cm. Write down

  a the upper bound for her height

  b the lower bound for her height.

2 The capacity of a drinks can is 330 m*l* to the nearest m*l*. Write down

a the maximum value for the capacity

b the minimum value for the capacity.

3 The temperature in a refrigerator is measured as 4 °C, to the nearest degree. Write down

a the upper bound for the temperature

b the lower bound for the temperature.

4 The lengths on this rectangle are measured to the nearest cm. Work out

a the maximum values for the dimensions of the rectangle

b the maximum value for the area of the rectangle.

3 cm

7 cm

5 The weight of an orange is measured as 73 g to the nearest gram. Charlotte says that the actual weight, w, must be in the range $72.5 \leqslant w \leqslant 73.49$. Is Charlotte right? Give a reason for your answer.

C

C

AO2

Example 5

Aidan records the height of a tree sapling as 1.6 m, correct to one decimal place. Work out

a the upper bound for the height

b the lower bound for the height.

Any value between 1.55 m and 1.65 m would round to 1.6 m.

C

Exercise 22E

1 The capacity of a bathtub is 50 litres, to the nearest 10 litres. Work out

a the maximum value for the capacity of the bathtub

b the minimum value for the capacity of the bathtub.

2 This table shows Olympic gold-medal winning times for the men's 100 m freestyle swimming. Each time is correct to one decimal place.

Beijing 2008	47.2 seconds
Athens 2004	48.1 seconds
Sydney 2000	48.3 seconds

For each time, write down

a the upper bound

b the lower bound.

3 A car weighs 1400 kg, correct to the nearest 100 kg. Write down

a the upper bound for the weight of the car

b the lower bound for the weight of the car.

C

C

AO3

4 The lengths on this rectangle are rounded to one decimal place.
Work out

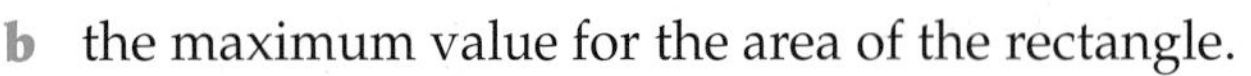

a the minimum value for the area of the rectangle

b the maximum value for the area of the rectangle.

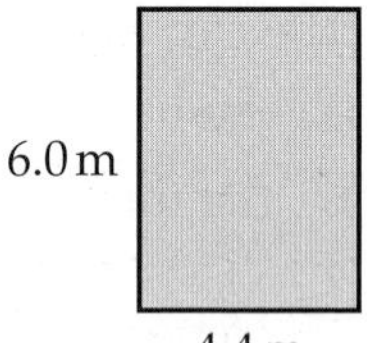

Review exercise

E

1 A rainwater butt contains 220 litres of water. Another contains 48 gallons.
Which one contains the most? **[2 marks]**

2 A car is travelling at 70 miles per hour. What is its speed in kilometres per hour? **[2 marks]**

3 A theme park runs a shuttle bus from the nearest train station. The journey takes 22 minutes. This is a timetable of departures from the train station.

First buses	10 am to 11 am	11 am to 4 pm		4 pm to 5 pm	Last buses
	1005	At these	05		
0920	1020	minutes	25	20	1720
0946	1046	past the	45	50	1750
0952	1055	hour			

a What time is the first bus of the day? **[1 mark]**

b Jodi arrives at the bus stop at 11:30 am. How long does she have to wait for a bus? **[1 mark]**

c What time does the 1055 bus arrive at the theme park? **[1 mark]**

d Paul wants to arrive at the theme park before 11 am.
Which bus should he catch? **[1 mark]**

E

AO2

4 A flag is made in the shape of a trapezium.
The dimensions of the flag are shown.

Construct an accurate scale drawing of the flag. Use a scale of 1 to 20.

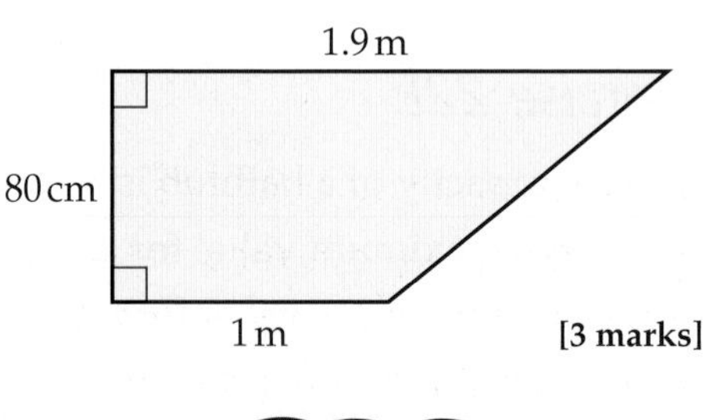

[3 marks]

E

5 Choose a ratio from the cloud to represent each of these scales.

a 1 cm to 5 km

b 1 cm to 50 km

c 1 cm represents 50 m

d 1 cm represents 5 m

1:5 1:500 1:50 000 1:5 000 000 1:500 000 1:50 1:5000

[4 marks]

6 A map uses a scale of 1 : 250 000.

a Work out what distance 1 cm on the map represents in real life.
Give your answer in km. **[2 marks]**

b Layla measures the distance between two towns on the map as 6.1 cm.
How far are these towns apart in real life? **[1 mark]**

c A river is 25 km long. How long will it be on the map? **[1 mark]**

E

7 This is a map of Clapham Common in London.
The scale is 1 to 25 000.
Ivan walks in a straight line from point A by the Long Pond to point B by the recreation ground.

a Measure the distance from point A to point B on the map. **[1 mark]**

b How far did Ivan walk? **[2 marks]**

Megan cycles from point A to point B along Clapham Common South Side and The Avenue.

c Use the map to estimate the distance that Megan cycled. **[2 marks]**

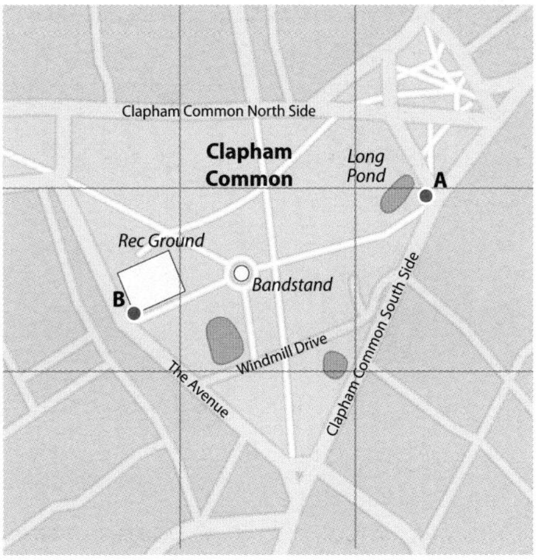

AO2

C

8 The weight of a letter is 52 g to the nearest gram. Work out

a the upper bound for the weight of the letter **[1 mark]**

b the lower bound for the weight of the letter. **[1 mark]**

9 The distance from London to New York is 3500 miles, to the nearest 100 miles. Write down

a the maximum value for the distance **[1 mark]**

b the minimum value for the distance. **[1 mark]**

10 Abi uses a computer program to record her reaction time in an experiment. Her reaction time is 0.31 seconds to two decimal places. Write down

a the maximum value for her reaction time **[1 mark]**

b the minimum value for her reaction time. **[1 mark]**

Chapter summary

In this chapter you have learned how to

- know and use approximate metric equivalents of pounds, feet, miles, pints and gallons **E**
- solve problems involving time, dates and timetables **E**
- use and interpret maps and scale drawings **E**
- recognise that measurements given to the nearest whole unit may be inaccurate by up to one half in either direction **C**

23 Triangles and constructions

This chapter is about working with triangles and constructing triangles.

Triangles are often used in architecture since they are very strong shapes that are difficult to deform.

Objectives

This chapter will show you how to

- **use angle properties of equilateral, isosceles and right-angled triangles** E D
- **show step-by-step deduction in solving a geometrical problem** E D
- **use straight edge and compasses to do standard constructions** E
- **draw triangles using a ruler and protractor given information about their side lengths and angles** D
- **understand congruence** C

Before you start this chapter

1 Which triangle is the odd one out?
Give a reason for your answer.

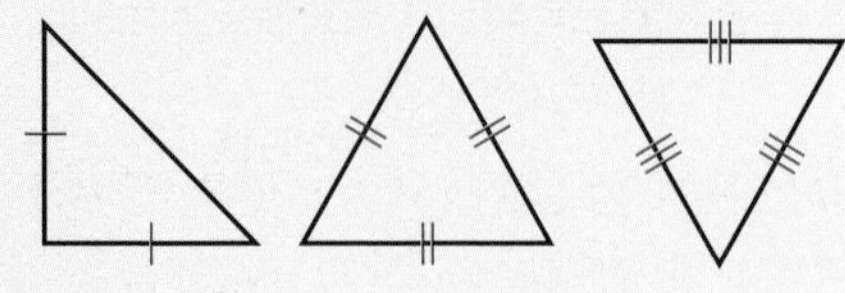

2 Use a protractor to draw an angle of
a 75° b 135°

3 Use a ruler to draw a line of length
a 7 cm b 85 mm

4 The grid shows six shapes A, B, C, D, E and F.
Which shapes are identical to shape A?

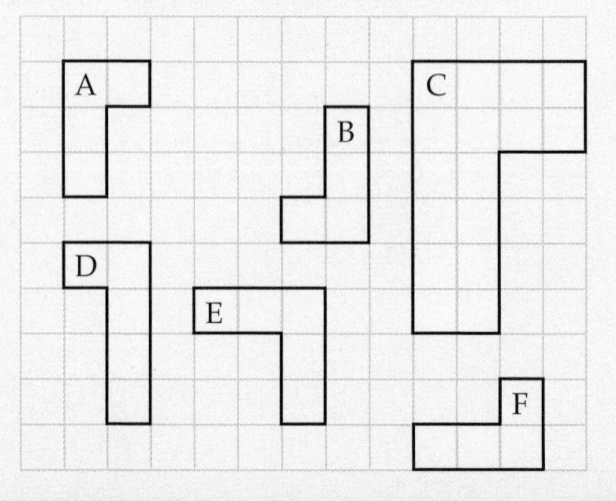

23.1 Interior and exterior angles

Keywords
interior angle, exterior angle

Why learn this?

Carpenters often use angles. If they can't measure the interior angle, they measure the exterior angle instead.

Objectives

E Solve angle problems in triangles

D Solve angle problems in triangles involving algebra

Skills check

1 Work out

a $30 + 45$ **b** $120 + 32$ **c** 2×28

2 Work out

a $180 - 50$ **b** $180 - 124$ **c** $84 \div 2$

Interior and exterior angles of a triangle

Look at this triangle.

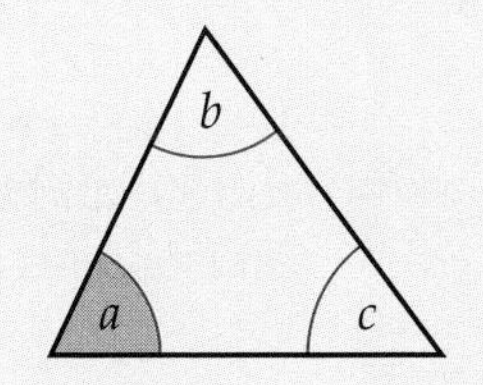

Angles a, b and c are called **interior angles** because they are *inside* the triangle.

If you tear off the three interior angles and place them alongside each other, they fit exactly onto a straight line.

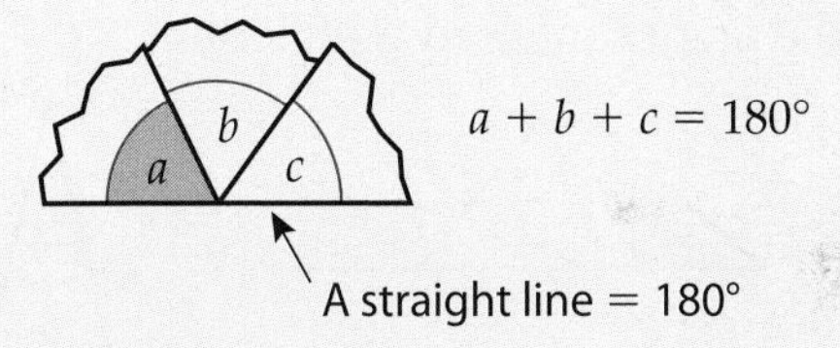

The sum of the angles of a triangle is 180°.

Look at this triangle.

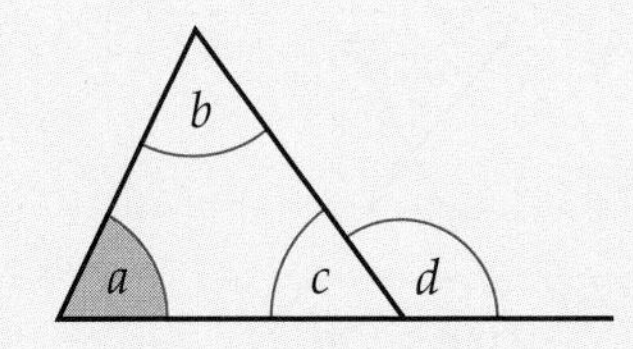

Angle d is called an **exterior angle** because it lies *outside* the triangle on a straight line formed by extending one of the sides of the triangle.

If you tear off the two interior angles a and b, then place them on top of the exterior angle d, they fit exactly.

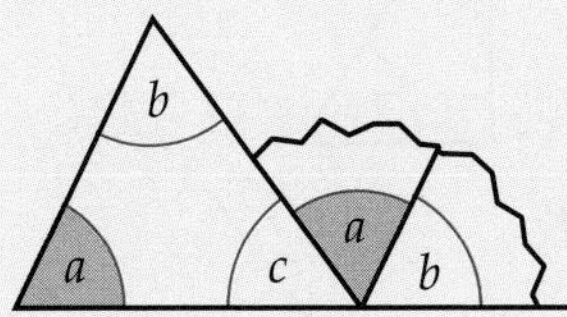

The exterior angle of a triangle is equal to the sum of the two opposite interior angles.

E

Example 1

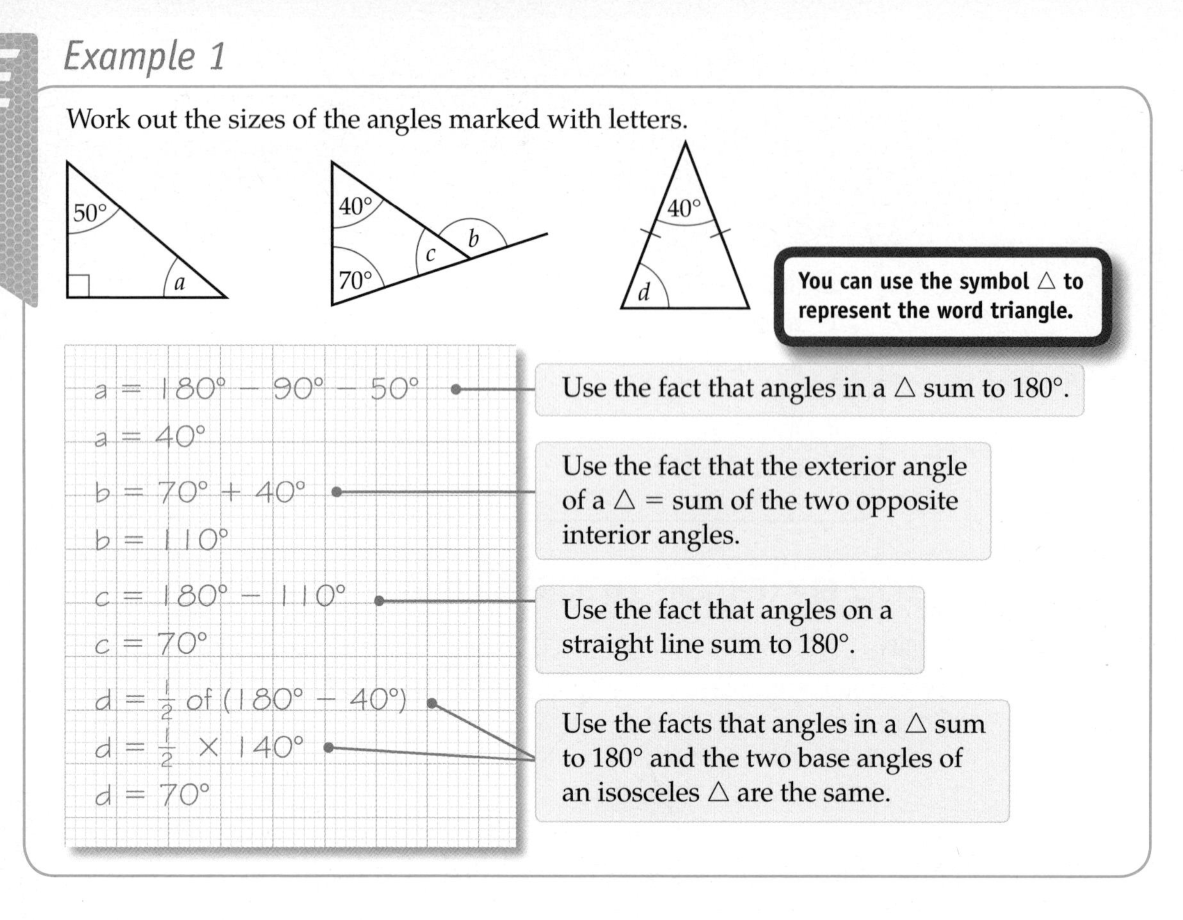

Exercise 23A

E

1 Work out the sizes of the angles marked with letters.

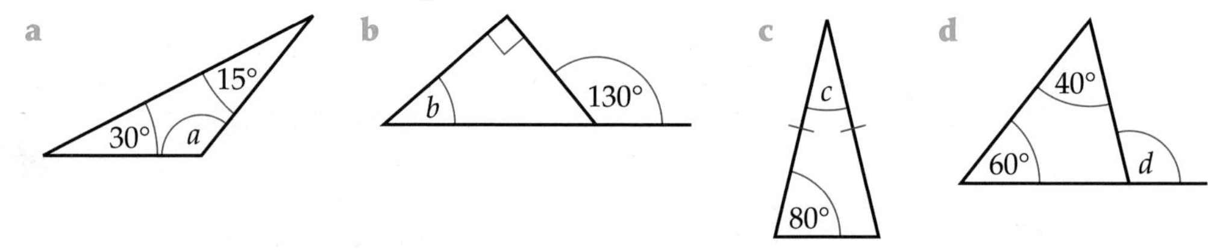

E

AO2

2 Work out the sizes of the angles marked with letters.

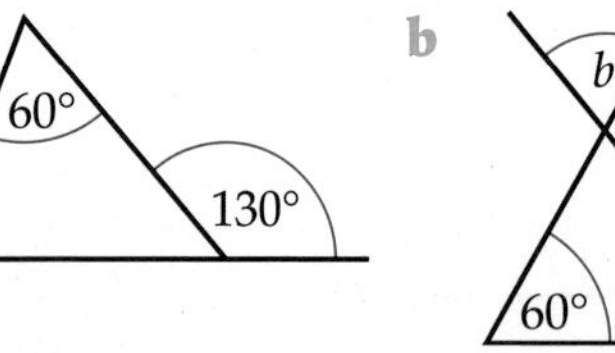

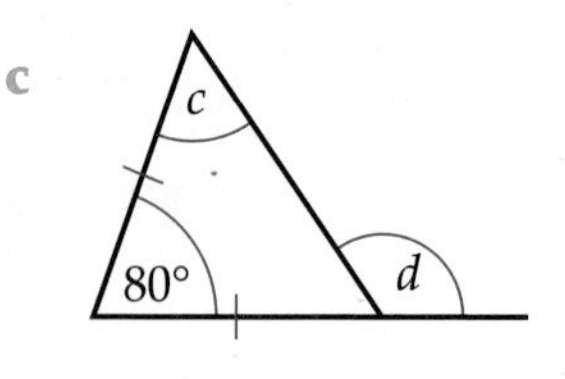

D

AO2

3 Work out the value of x in each diagram.

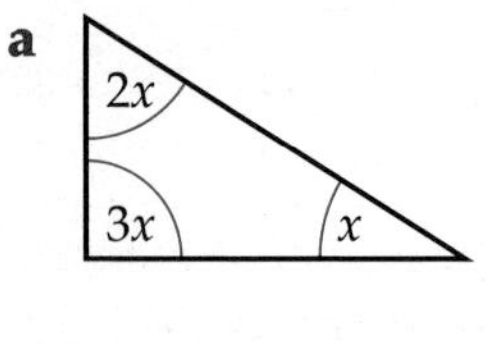

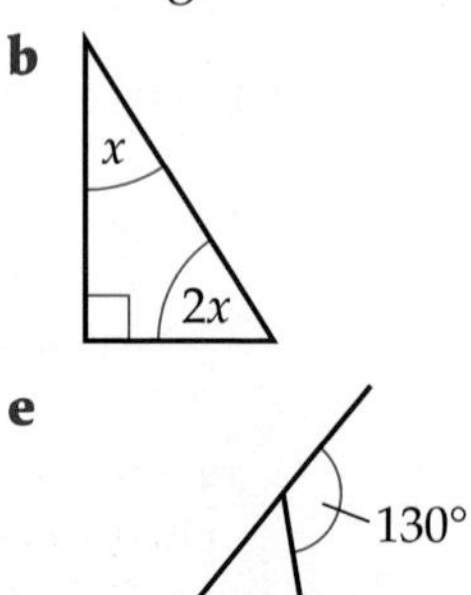

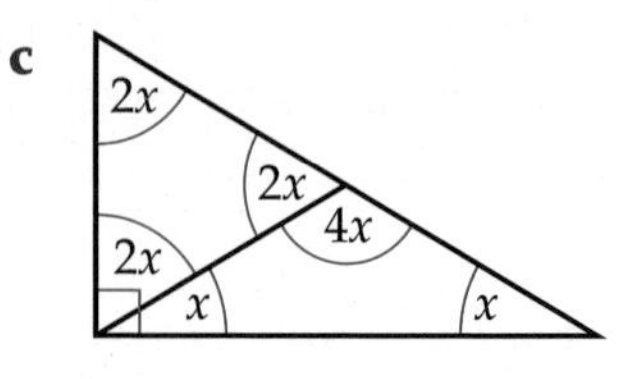

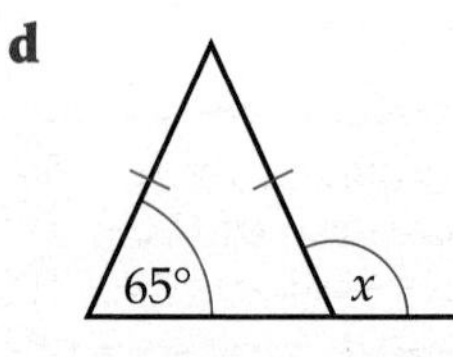

23.2 Constructions

Keywords
construct, arc

Why learn this?
Architects' plans need to be very accurate since a small error on the plans could mean a big mistake in real life.

Objectives

E Draw triangles accurately when given the length of all three sides

D Draw triangles accurately when at least one angle is given

Skills check

1 Use a ruler to draw a line of length 6.7 cm accurately.

2 Use a pair of compasses to draw a circle of radius 45 mm accurately.

Constructing triangles given the lengths of all three sides

The word **construct** means 'draw accurately'.

To construct a triangle when you are given the lengths of all three sides, you need to be able to use a ruler and compasses accurately. Use your compasses to draw an **arc**.

Example 2

Construct triangle ABC when $AB = 8$ cm, $BC = 6$ cm and $AC = 5$ cm.

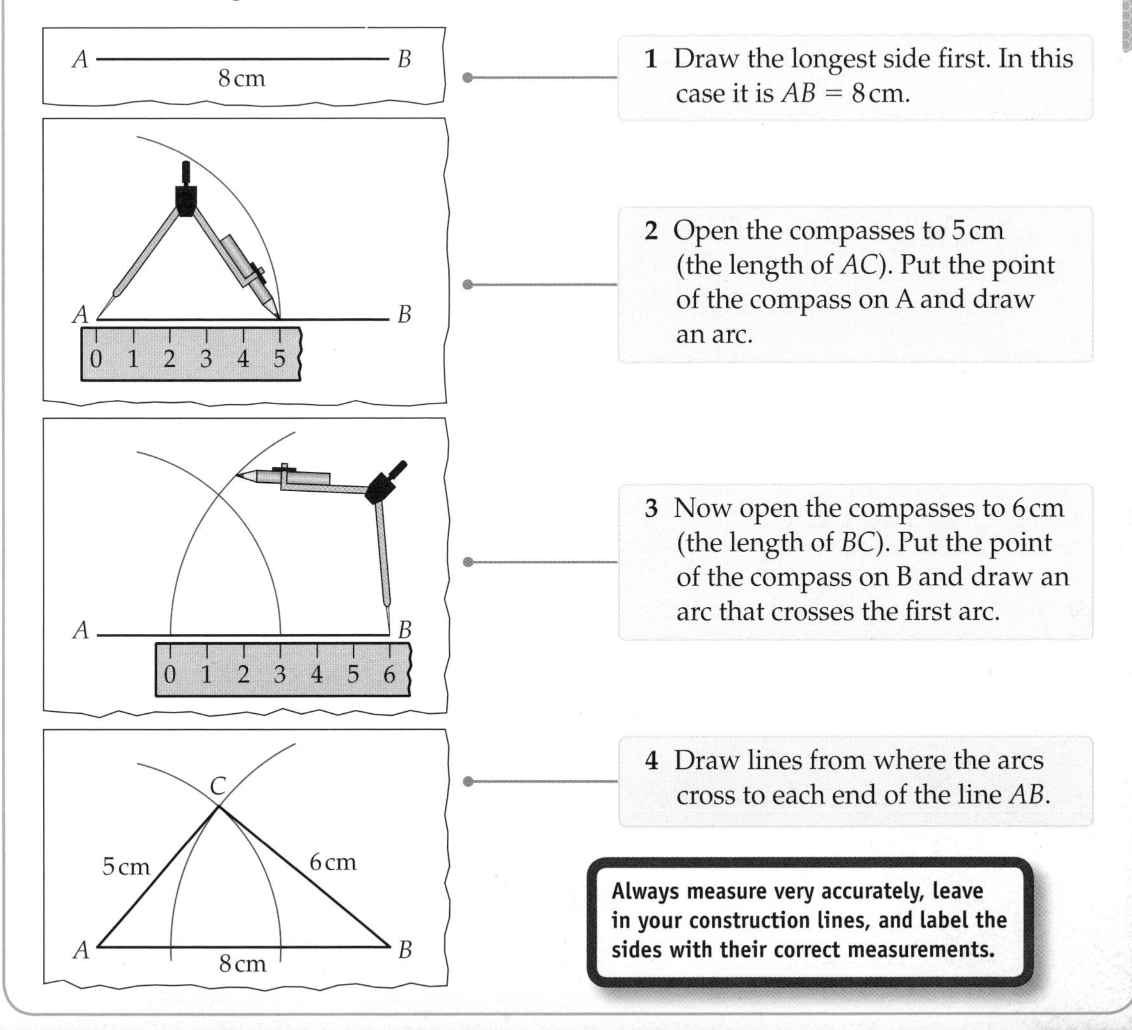

1 Draw the longest side first. In this case it is $AB = 8$ cm.

2 Open the compasses to 5 cm (the length of AC). Put the point of the compass on A and draw an arc.

3 Now open the compasses to 6 cm (the length of BC). Put the point of the compass on B and draw an arc that crosses the first arc.

4 Draw lines from where the arcs cross to each end of the line AB.

Always measure very accurately, leave in your construction lines, and label the sides with their correct measurements.

E

Exercise 23B

1 Construct triangles ABC with the following measurements.

a $AB = 6\,\text{cm}$, $BC = 10\,\text{cm}$ and $AC = 8\,\text{cm}$
Name the type of triangle you have drawn.

b $AB = 7.5\,\text{cm}$, $BC = 6\,\text{cm}$ and $AC = 7.5\,\text{cm}$
Name the type of triangle you have drawn.

c $AB = 55\,\text{mm}$, $BC = 55\,\text{mm}$ and $AC = 55\,\text{mm}$
Name the type of triangle you have drawn.

2 Make an accurate drawing of this shape.

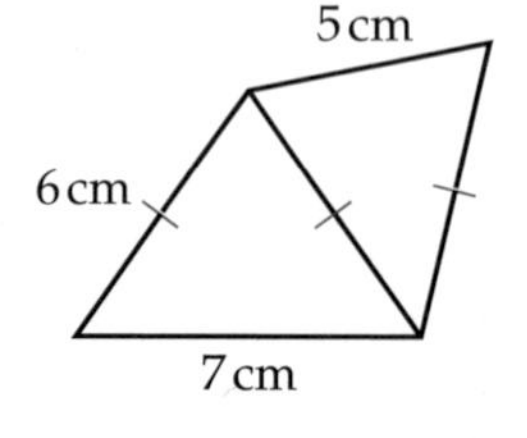

Constructing triangles given SAS and ASA

There are two more types of triangle construction that you need to be able to do.

The first is known as **SAS.** This stands for **s**ide **a**ngle **s**ide.
The second is known as **ASA.** This stands for **a**ngle **s**ide **a**ngle.

D

Example 3

Use a ruler and a protractor to make an accurate drawing of each of these triangles.

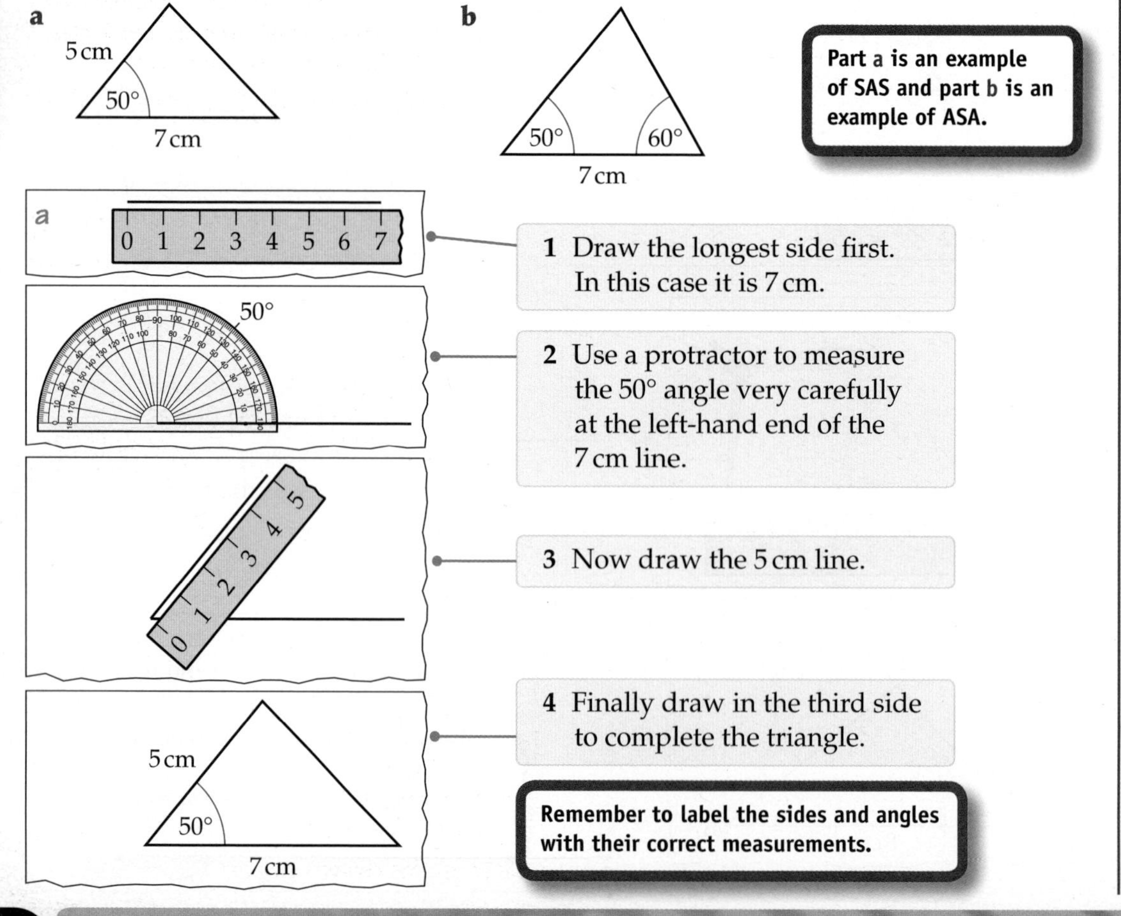

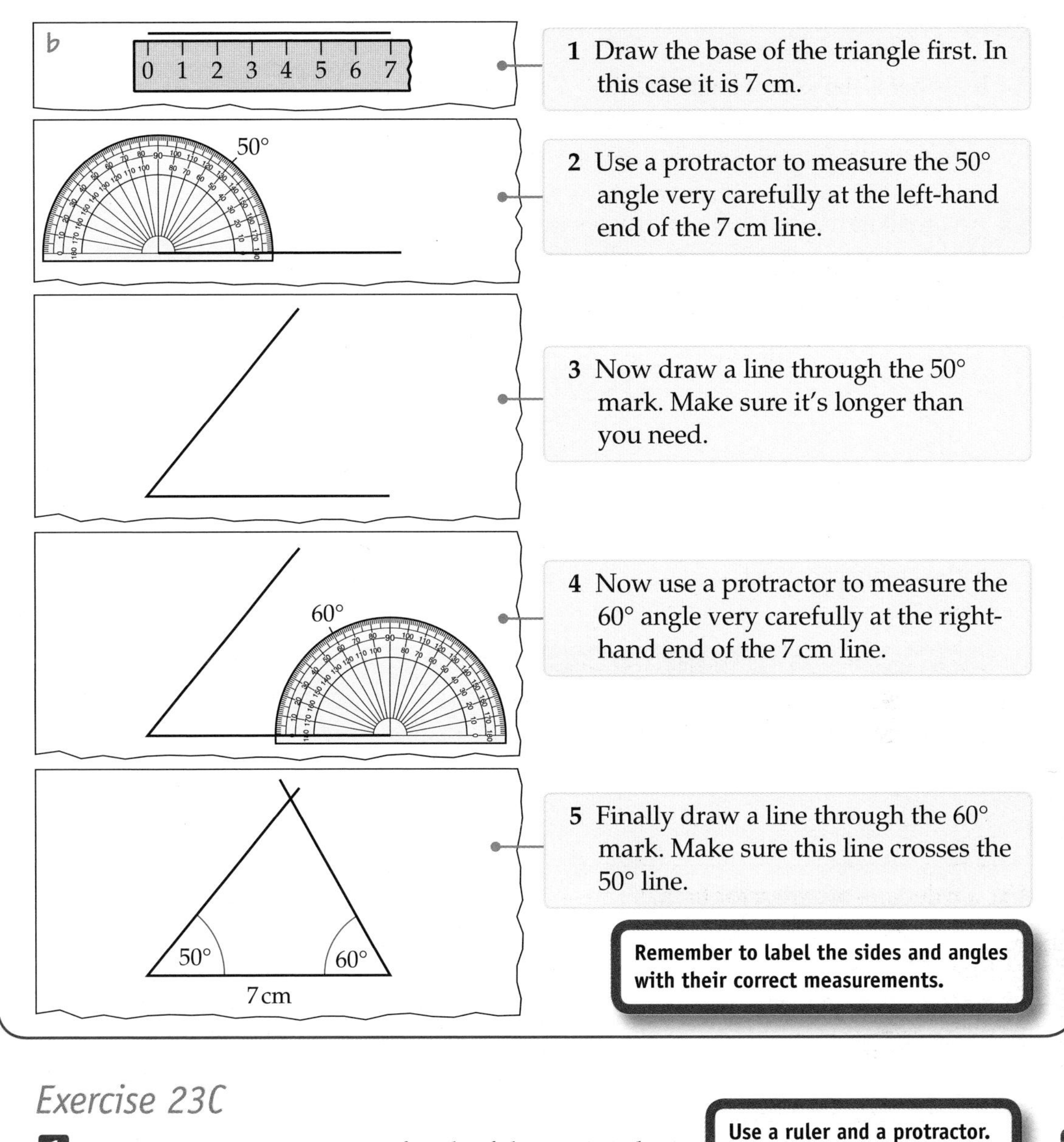

Exercise 23C

Use a ruler and a protractor.

1 **a** Draw an accurate copy of each of these triangles.

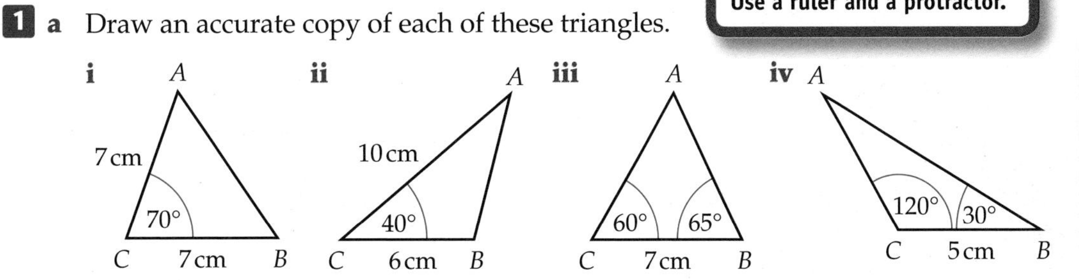

b Write down the length AB for each of the triangles drawn in part **a**.

D

2 **a** Draw an accurate copy of each of these triangles.

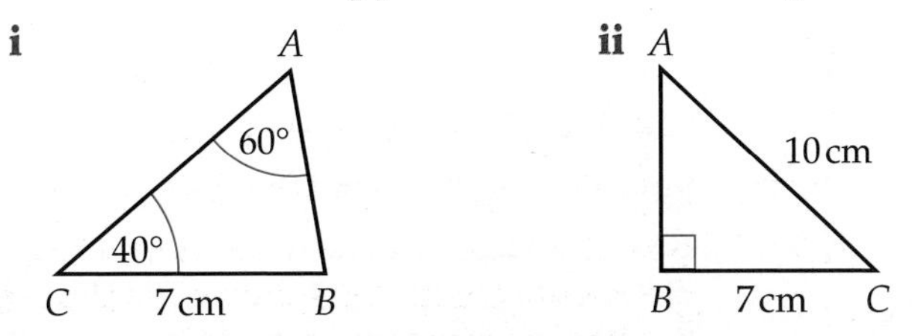

b Write down the length AB for each of the triangles drawn in part **a**.

D

AO2

3 **a** Draw a triangle ABC where $AB = 7$ cm, $BC = 5$ cm and angle $ABC = 60°$.

b What is the size of angle BCA?

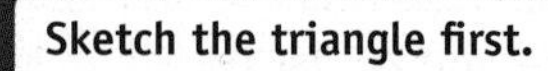

D

AO3

4 An architect has ordered roof trusses to be made for a roof.
He has sent this sketch of the outline of the roof to the manufacturer.

FUNCTIONAL

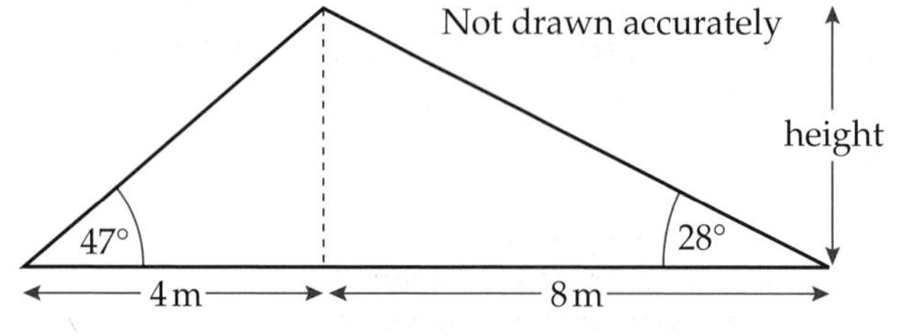

You will need to produce a scale drawing. Use a scale of 1 cm = 1 m.

The architect says that the roof truss should be about 3 m high.
Is the architect correct?

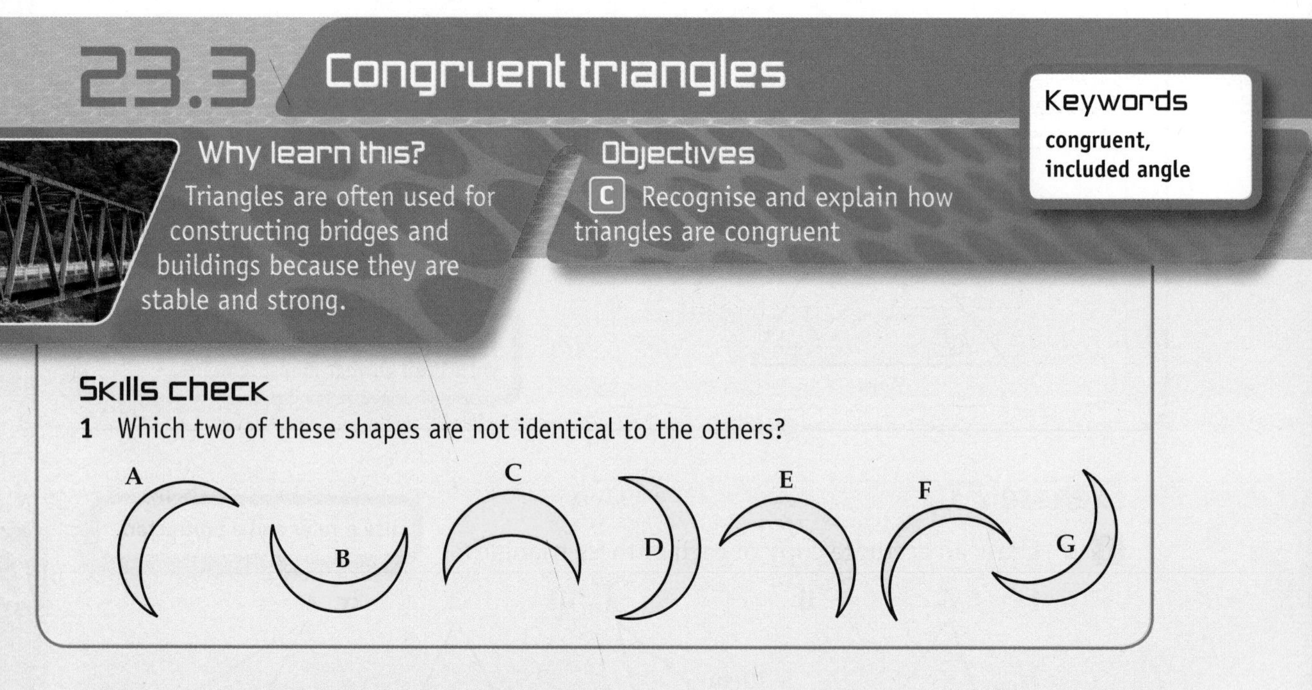

23.3 Congruent triangles

Keywords
congruent, included angle

Why learn this?

Triangles are often used for constructing bridges and buildings because they are stable and strong.

Objectives

C Recognise and explain how triangles are congruent

Skills check

1 Which two of these shapes are not identical to the others?

Showing that triangles are congruent

Congruent means 'exactly the same shape and size', so congruent shapes are identical.
There are four ways to show that triangles are congruent.

1 Using SSS (**s**ide **s**ide **s**ide). If all three sides of one triangle are the same length as another triangle, the triangles are congruent.

2 Using SAS (**s**ide **a**ngle **s**ide). If two sides and the angle between them (the **included angle**) are the same in two triangles, the triangles are congruent.

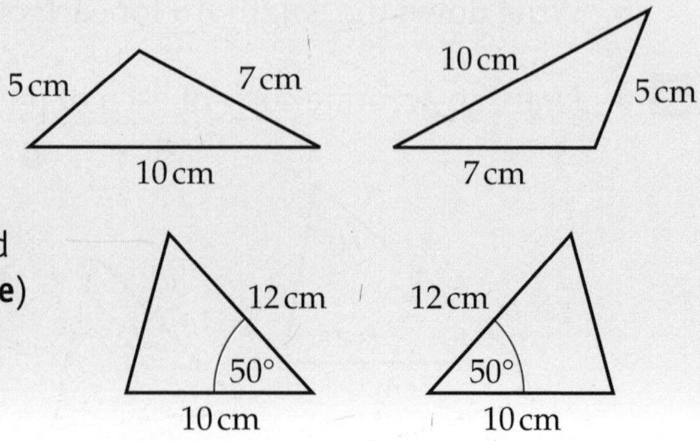

3 Using ASA (**a**ngle **s**ide **a**ngle) or SAA (**s**ide **a**ngle **a**ngle). If one side and two angles of one triangle are the same as the corresponding side and two angles of another triangle, the triangles are congruent.

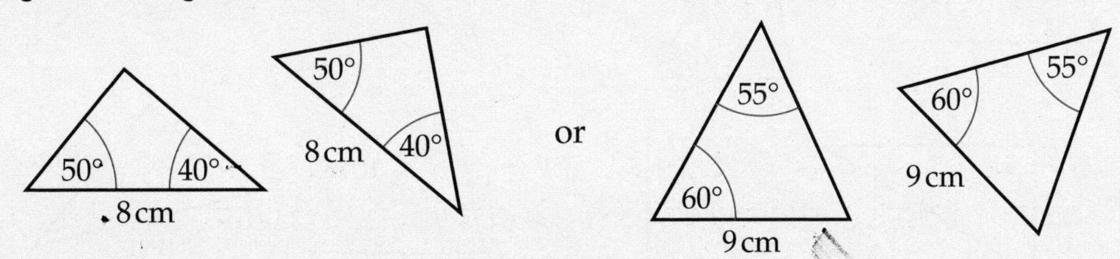

4 Using RHS (**r**ight angle **h**ypotenuse **s**ide). If both triangles have a right angle and the length of the hypotenuse and another side are the same, the triangles are congruent.

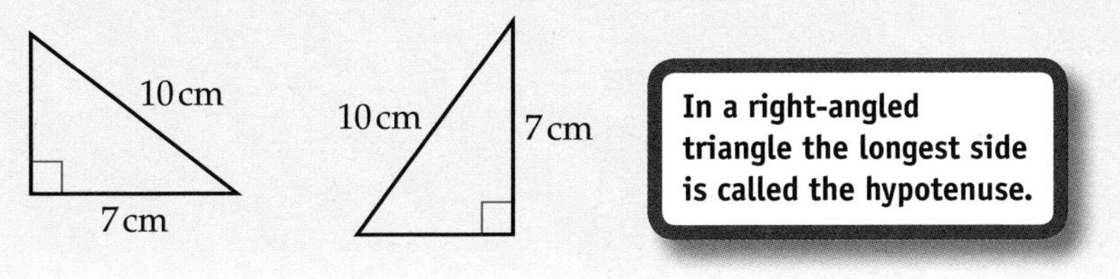

In a right-angled triangle the longest side is called the hypotenuse.

Example 4

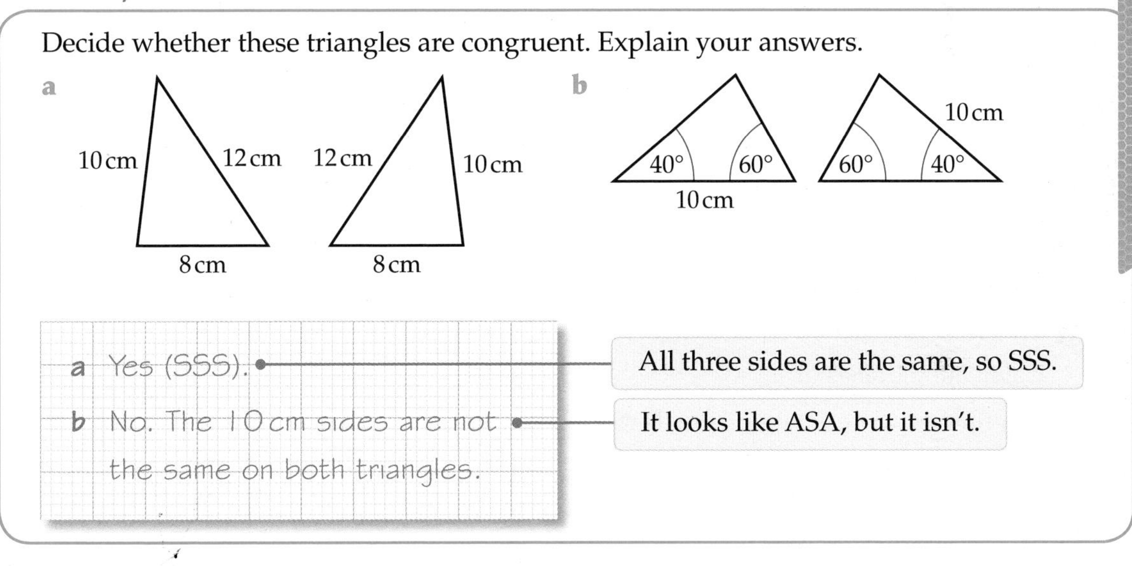

C

Exercise 23D

1 State whether the following pairs of triangles are congruent. If congruent, state which of the four conditions, SSS, SAS, ASA (or SAA) or RHS, is satisfied.

a

b

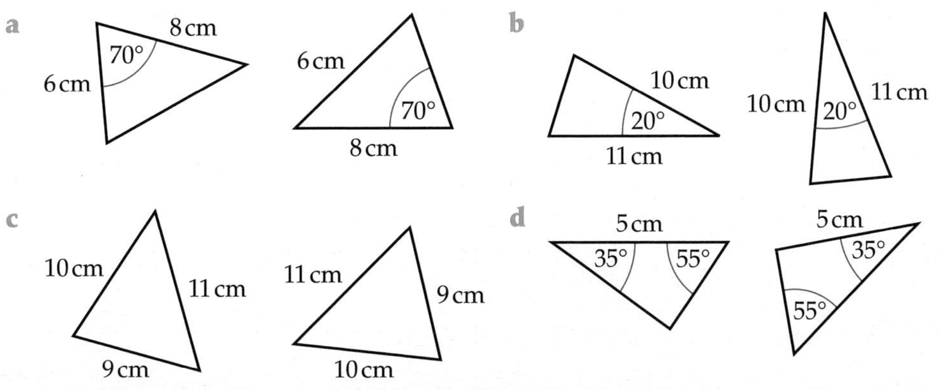

C

C

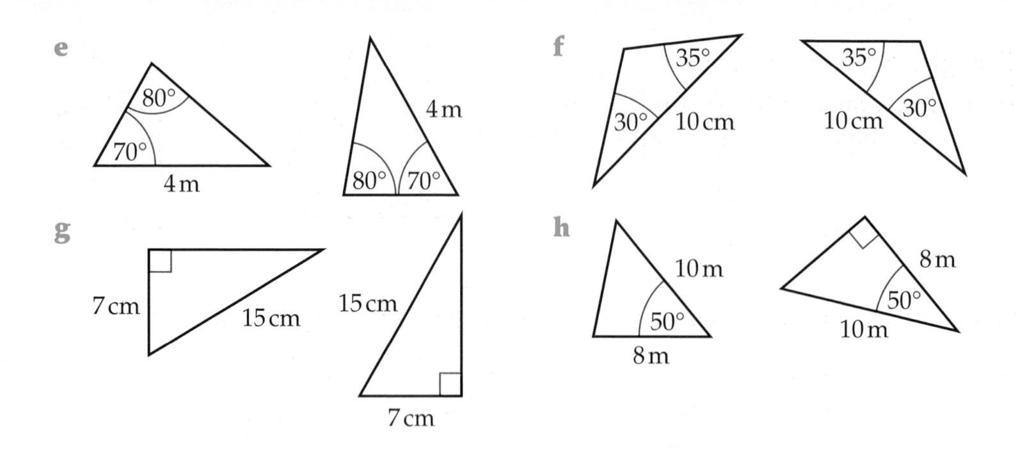

C

2 Explain why these two triangles are congruent.

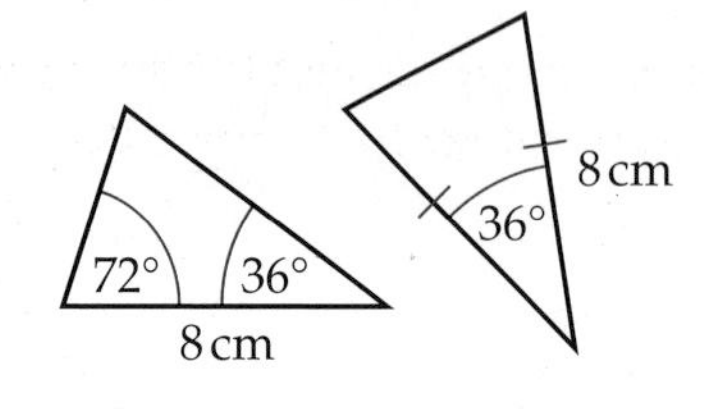

AO2

3 Which triangle, A, B or C, is the odd one out? Explain your answer.

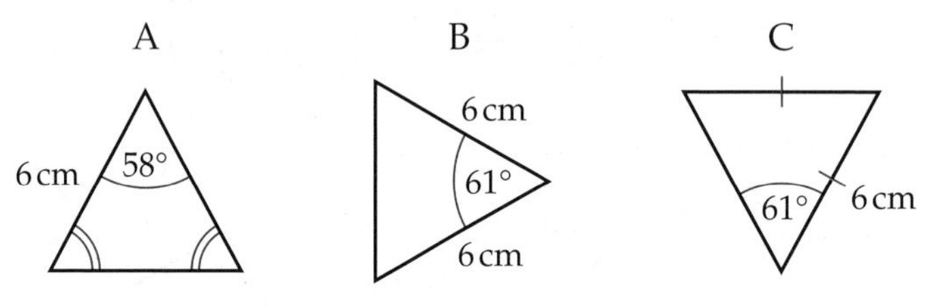

Review exercise

E

1 This triangle has two equal sides.

Calculate the sizes of

a angle x **[2 marks]**

b angle y **[2 marks]**

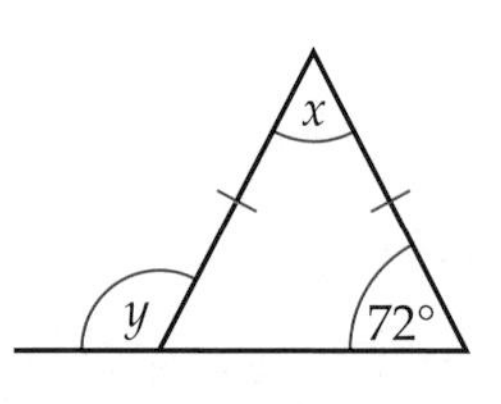

2 **a** Make an accurate drawing of this triangle. **[3 marks]**

b Measure the size of angle ABC. **[1 mark]**

D

3 **a** Make an accurate drawing of this triangle. **[3 marks]**

b Measure the length of AC. **[1 mark]**

4 Tony and Ceri estimate the length of ST in this triangle.

Tony says, 'I think ST is 7 cm long'.

Ceri says, 'I think ST is 8 cm long'.

Who makes the closer estimate to the real length?

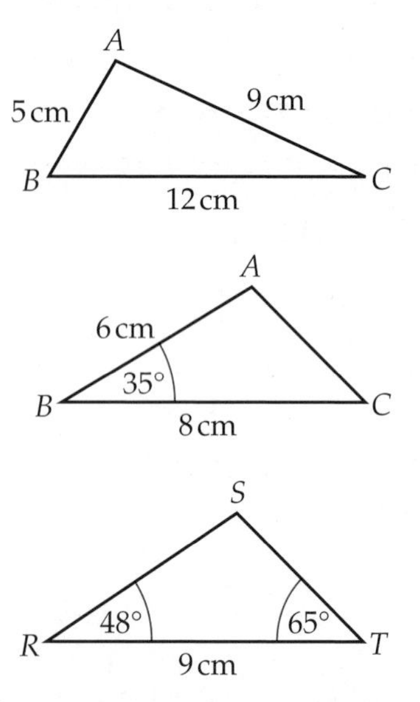

[4 marks]

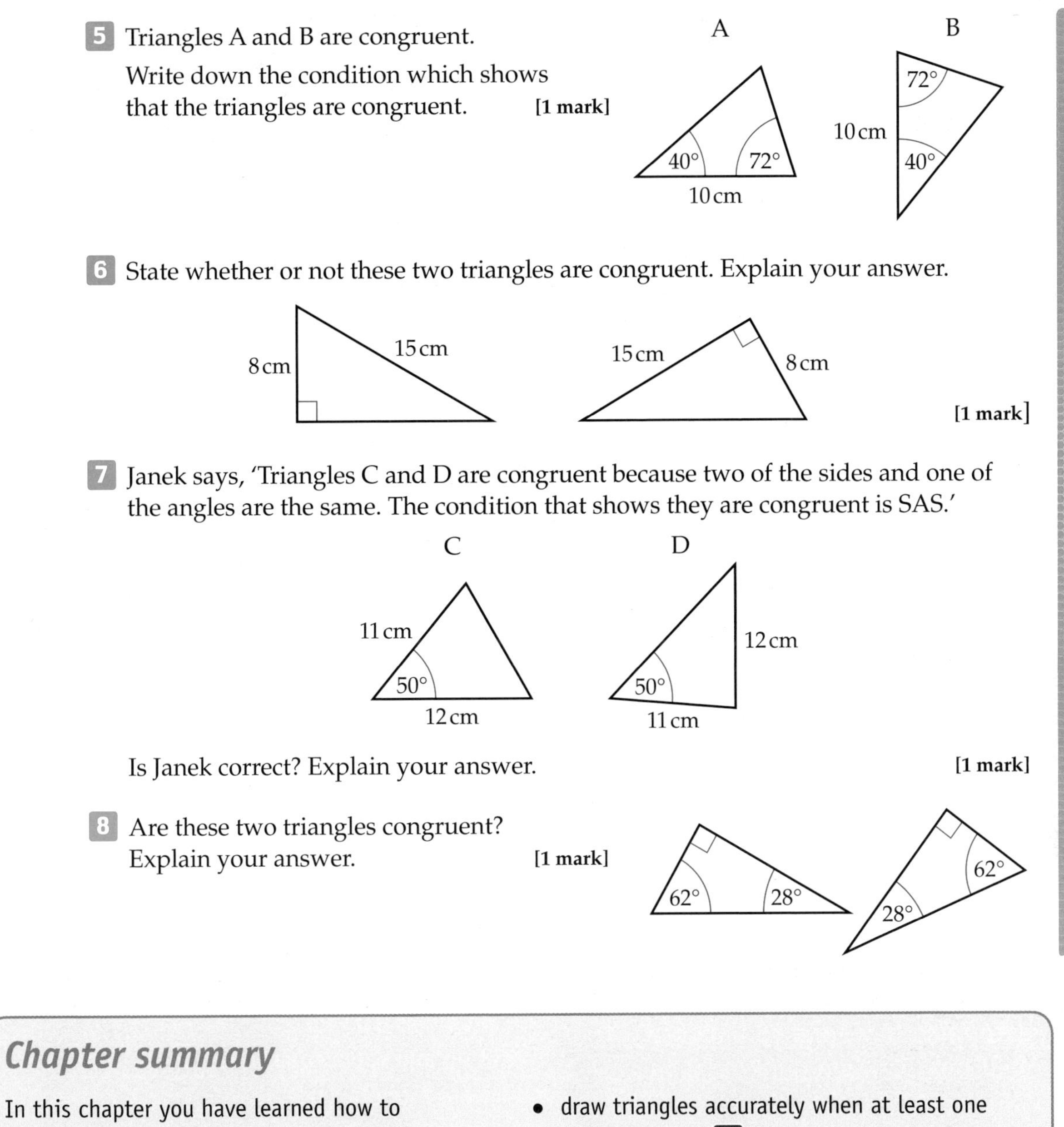

5 Triangles A and B are congruent.
Write down the condition which shows that the triangles are congruent. **[1 mark]**

6 State whether or not these two triangles are congruent. Explain your answer. **[1 mark]**

7 Janek says, 'Triangles C and D are congruent because two of the sides and one of the angles are the same. The condition that shows they are congruent is SAS.'

Is Janek correct? Explain your answer. **[1 mark]**

8 Are these two triangles congruent?
Explain your answer. **[1 mark]**

Chapter summary

In this chapter you have learned how to

- solve angle problems in triangles **E**
- draw triangles accurately when given the length of all three sides **E**
- solve angle problems in triangles involving algebra **D**
- draw triangles accurately when at least one angle is given **D**
- recognise and explain how triangles are congruent **C**

24

Equations, formulae and proof

This chapter is about using formulae and solving equations.

There is even an equation to determine how long to immerse your biscuit when dunking it in your drink!

Objectives

This chapter will show you how to

- write algebraic expressions from word problems **D**
- set up and solve equations **D** **C** **B**
- derive a formula **D**
- substitute into a formula to solve problems **D** **C**
- change the subject of a formula **C** **B**
- prove simple results from geometry **C**
- use trial and improvement to find solutions to equations **C**

Before you start this chapter

1 Work these out.

a $7 + 4 \times 2$ b $22 - (16 - 5)$
c $3 + (2 + 7)^2$ d $22 - \frac{4}{3}$

2 Find the missing numbers.

a $\square + 7 = 21$ b $54 - \square = 30$
c $\square + 4 = -2$ d $6 - \square = 10$

3 Copy and complete the following.

a $\square \times 6 = 42$ b $99 \div \square = 9$
c $\square \times -3 = 9$ d $24 \div \square = -8$

HELP Chapter 11

4 Write down an expression for

a three more than n b six less than y
c four lots of x d one quarter of b.

P

Algebra skills: expressions

1 Write down whether each of these is an expression, an equation or a formula.

a $6x + 1$ **b** $v = u + at$ **c** $\frac{a + b}{2}$

d $E = \frac{1}{2}mv^2$ **e** $2n + 1 = 11$ **f** $2p^2 + pq$

2 Simplify these by collecting like terms.

a $5q + 2q$ **b** $2x + 3y + 5x + y$

c $6e + 2f - e + f + 3 - 9f$ **d** $12s + 10 - 2t - 9s - 8t + 3$

e $2x + x^2 - 5x + 3x^2$ **f** $mn + mn^2 - m + 6mn + 4m$

g $2a + ab - 3a + 4ab + a^2$ **h** $4b^2 + 6b - 2b^2 + 3b^3 - 2b$

3 Find the missing numbers (powers).

a $x \times x \times x \times x \times x = x^{\square}$ **b** $n \times n \times n = n^{\square}$

c $q^2 \times q^5 = q^{\square}$ **d** $k^3 \times k \times k^2 = k^{\square}$

4 Simplify these.

a $8 \times 2x$ **b** $3m \times 2m$ **c** $4x^2 \times 3x^3$

d $3a^3 \times 2a \times 5$ **e** $2a \times 4b$ **f** $6x \times 2y \times x$

g $3p^2 \times q^3 \times 2p \times q$ **h** $4 \times 2m^2 \times n \times 3m \times n^4$ **i** $16x \div 2$

j $8n^2 \div n$ **k** $21h^4 \div 3h$ **l** $100x^5 \div 20x^3$

Algebra skills: brackets

1 Work these out.

a $4 + (13 - 5)$ **b** $30 \div (10 - 4)$ **c** $18 - (3 + 7)$ **d** $(4 + 2) \times (9 - 5)$

2 Multiply out the brackets.

a $5(x + 1)$ **b** $6(2a + 5)$ **c** $3(2 - 3x)$ **d** $10(3m - 5n)$

e $x(4 + x)$ **f** $2a(a + 5)$ **g** $3n(2n - 1)$ **h** $3p(2p - 5q)$

i $-3(x + 1)$ **j** $-2(5x - 3)$ **k** $-a(2b + a)$ **l** $-6m^2(2m - 3)$

3 Multiply out these expressions and simplify by collecting like terms.

a $4(f + 2g) + 2(3f + g)$ **b** $11(2h - 5i) + 7(9i + h)$

c $2(x + 1) - 3(x + 5)$ **d** $6(2j + 2k) + 7(10j - 2k)$

e $5(m - 8n) - 2(6m + n)$ **f** $2a(4 - a) - 3a(2a - 5)$

4 Factorise the following expressions.

a $10m + 8$ **b** $4r - 20$ **c** $26j + 13$ **d** $27h - 81$

e $16 - 4x$ **f** $21 - 60y$ **g** $7x^2 + 2x$ **h** $5m^2 - 9m$

i $y + 11y^2$ **j** $2s^2 + 4s$ **k** $10k + 2k^2$ **l** $4j^3 + 3j^2$

P

Algebra skills: solving equations

1 Solve these equations.

a $2x + 5 = 17$ b $3s - 9 = 18$ c $66 = 8p + 2$

d $12v - 15 = 141$ e $\frac{m}{3} - 4 = 2$ f $20 = \frac{y}{5} + 8$

g $19 = \frac{z}{2} - 11$ h $\frac{d}{9} + 8 = 22$ i $\frac{k}{7} - 7 = 5$

2 Solve these equations.

a $2(x + 3) = 16$ b $3(h - 9) = 33$ c $91 = 7(2k + 7)$

d $8(3g + 7) = 8$ e $\frac{1}{4}(2x - 10) = 3$ f $26 = 13(4x + 10)$

3 Find the value of the unknown in each equation.

a $3x - 8 = x + 12$ b $9y + 7 = 6y + 28$

c $5m - 2 = m$ d $2r + 13 = 5r + 4$

e $5t + 3 = t - 25$ f $8j - 3 = 14j + 33$

g $7(x - 1) = 2x + 28$ h $5z + 19 = 6(2z - 12)$

i $3(1 - 2q) = 4q + 1$ j $6(g + 3) = 2(g + 23)$

k $2(t + 19) = 3(2t - 14)$ l $2(10w + 1) = 16(2 - 5w)$

Algebra skills: formulae

1 If $p = 6$, $q = 3$ and $r = 11$ find the value of

a $3r$ b $2q$ c $\frac{p}{2}$

d $2p + r$ e $3p - q + r$ f $r - 2p + 8q$

2 Substitute $x = 6$ and $y = 9$ into each expression.

a $x^2 + 1$ b $2xy - 3y$ c $3x + 2x^2$

d $3(x + y)$ e $\frac{x + 2y}{3}$ f $(y - x)^3$

3 Use the formula $P = 2(a + b)$ to calculate P when

a $a = 5$ and $b = 8$ b $a = 12$ and $b = 2$

c $a = 0.5$ and $b = 3$ d $a = 120$ and $b = 50$

4 The formula for converting degrees Fahrenheit into degrees Celsius is $C = \frac{5}{9}(F - 32)$.

Work out

a C when F is 41 b C when F is 14

c 77 °F in °C d −4 °F in °C.

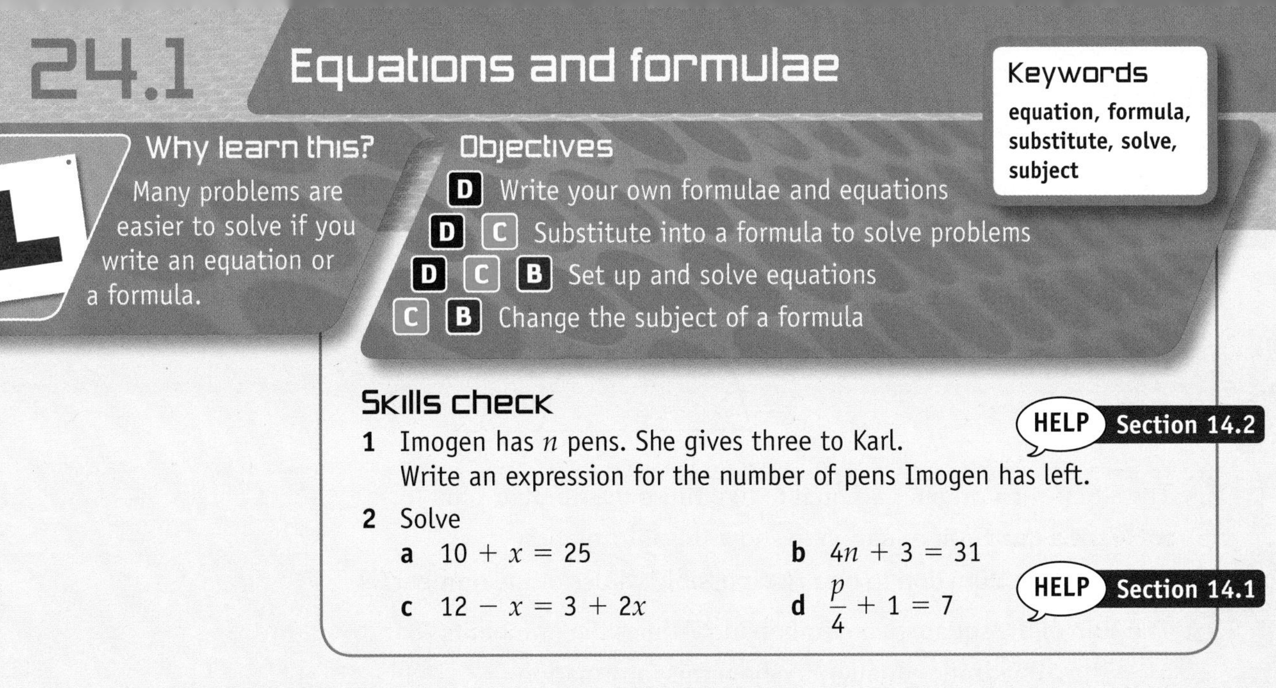

24.1 Equations and formulae

Keywords
equation, formula, substitute, solve, subject

Why learn this?

Many problems are easier to solve if you write an equation or a formula.

Objectives

D Write your own formulae and equations
D C Substitute into a formula to solve problems
D C B Set up and solve equations
C B Change the subject of a formula

Skills check

1 Imogen has n pens. She gives three to Karl.
Write an expression for the number of pens Imogen has left.

HELP Section 14.2

2 Solve

a $10 + x = 25$

b $4n + 3 = 31$

c $12 - x = 3 + 2x$

d $\frac{p}{4} + 1 = 7$

HELP Section 14.1

Writing your own equations

You can solve problems by writing and solving **equations**.

Example 1

D

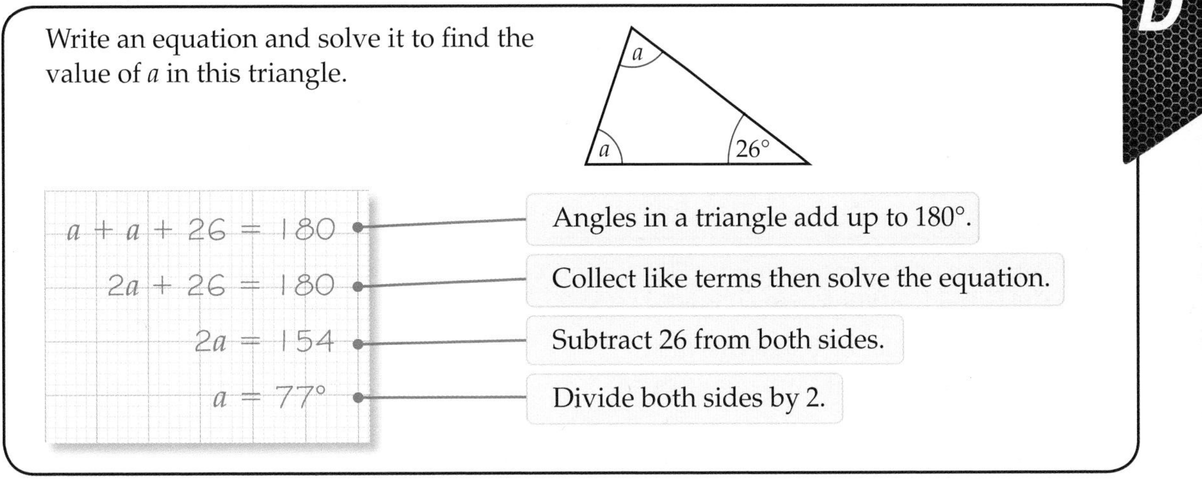

Write an equation and solve it to find the value of a in this triangle.

$a + a + 26 = 180$ — Angles in a triangle add up to 180°.

$2a + 26 = 180$ — Collect like terms then solve the equation.

$2a = 154$ — Subtract 26 from both sides.

$a = 77°$ — Divide both sides by 2.

Exercise 24A

D

1 Work out the value of the letter in each of these diagrams.

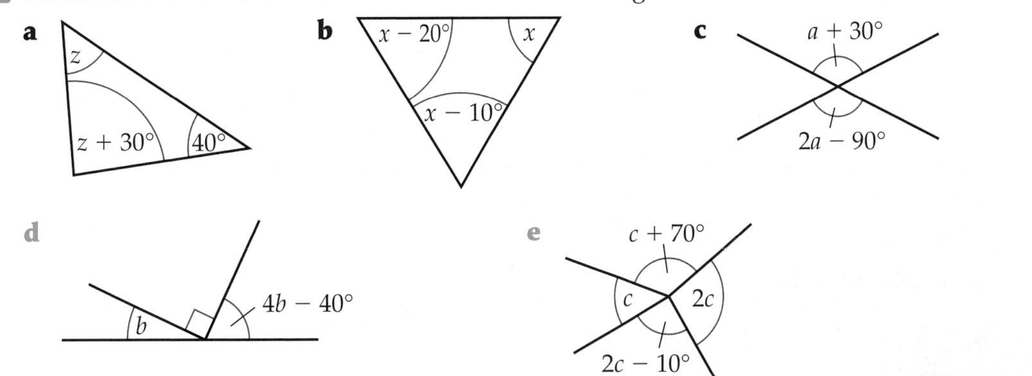

C

2 The lengths of the perimeters of the regular hexagon and the square are the same. The area of the square is 144 cm^2.
Find the length of one side of the hexagon.

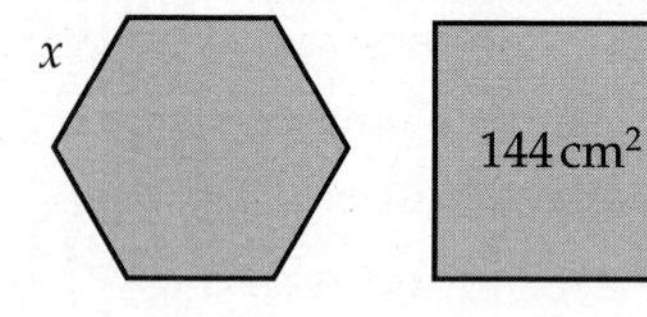

3 Work out the value of x.

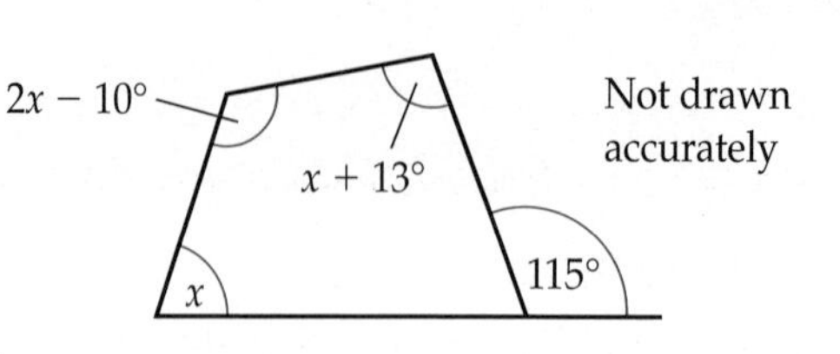

Not drawn accurately

B AO2

4 The square of a number is equal to two times the number plus 15.

a Write a quadratic equation to show this information.

b Solve your equation to find two possible values of the number.

5 The sum of the square of a number and 7 times the number is 78.

a Write a quadratic equation to show this information.

b Solve your equation to find two possible values of the number.

Writing your own formulae

You can use letters or words to write your own **formulae**.

Writing a formula is useful if you need to solve the same type of problem more than once.

Formulae is the plural of formula.

You can **substitute** values into a formulae and **solve** the equation to find the value of an unknown.

D

Example 2

A telephone news service charges 30p a minute plus a connection fee of 80p.
A call lasting x minutes costs C pence.

a Write a formula for C in terms of x.

b Find the length of a call costing £3.50.

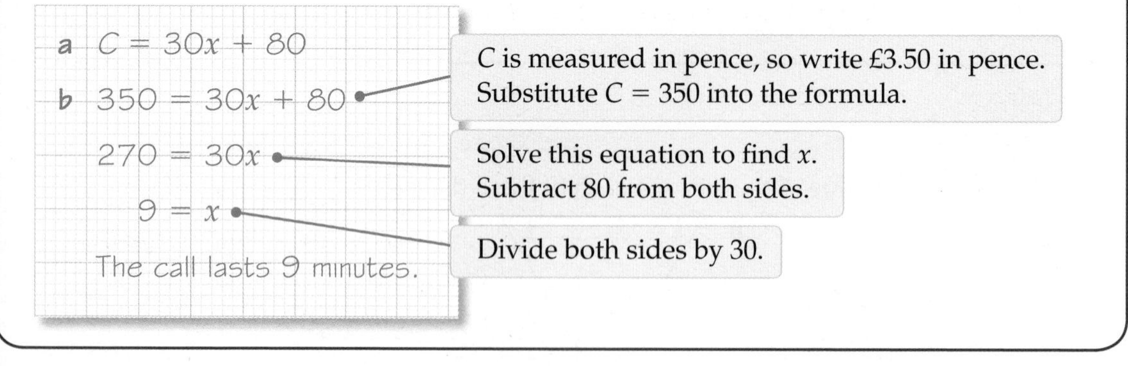

Exercise 24B

D

1 There are n biscuits in a box. The biscuits are shared equally between five friends. Each one gets m biscuits.

a Write a formula for m in terms of n.

b There are 30 biscuits in the box. How many will each person get?

2 A rectangle has length l and width w.

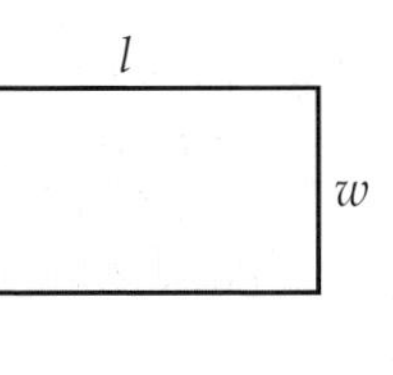

a Write a formula for the perimeter of the rectangle, P, in terms of l and w.

b Find l when $P = 22$ and $w = 8$.

c Find w when $P = 18$ and $l = 4$.

3 The length of a rectangle is 3 more than twice its width, w.

a Write a formula for the perimeter of the rectangle, P.

b Find w when $P = 39$.

4 Apples cost a pence each and bananas cost b pence each. The total cost of 4 apples and 7 bananas is C pence.

a Write a formula for C in terms of a and b.

b Calculate a if $C = 230$ and $b = 18$.

5 Beth has n sweets. She eats three then shares the rest equally amongst x friends. Each friend gets S sweets.

a Write a formula for S in terms of n and x.

b Calculate n when $S = 5$ and $x = 8$.

D

Changing the subject of a formula

This was covered in Section 15.6.

The **subject** of a formula is the letter on its own.
For example, P is the subject of the formula $P = 2l + 2w$.

You can use the rules of algebra to rearrange a formula to make a different letter the subject. This is called changing the subject of the formula.

Changing the subject of a formula is like solving an equation. You need to get a letter on its own on one side of the formula.

Example 3

C

The formula for the perimeter of a rectangle is $P = 2l + 2w$.

a Rearrange this formula to make w the subject.

b Find w when $P = 42$ and $l = 12$.

a $P = 2l + 2w$ — You need to get w on its own on one side of the formula.

$P - 2l = 2w$ — Use the same operations as you would to solve an equation to find w. Subtract $2l$ from both sides.

$\frac{P - 2l}{2} = w$ — Divide both sides by 2.

b $w = \frac{P - 2l}{2}$ — Use the rearranged formula to find w directly.

$w = \frac{42 - 2 \times 12}{2}$ — Substitute $P = 42$ and $l = 12$ into the rearranged formula.

$w = \frac{18}{2} = 9$

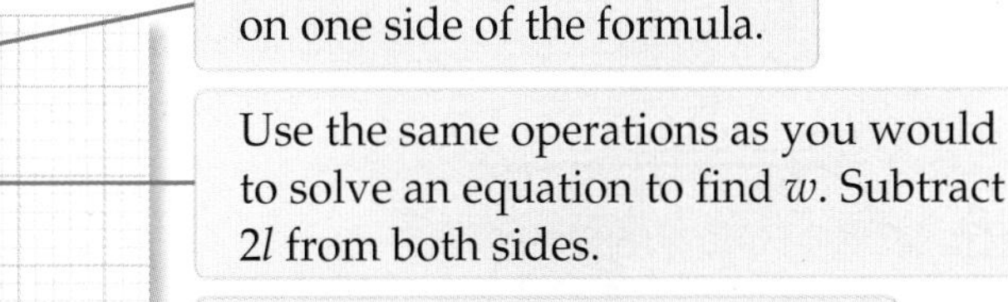

Exercise 24C

C

1 Write down the letter that is the subject of each formula.

a $M = 2(a + b)$ b $ka = R$ c $v = u + at$ d $k = \frac{2B}{a}$

2 Make a the subject of each formulae in Q1.

C

3 The formula for the area of a triangle is

area $= \frac{1}{2} \times$ base $\times$ height

a Rearrange this formula to make 'base' the subject.

b Use your rearranged formula to find the base length (x) of each triangle.

i area = $9\,\text{cm}^2$ ii area = $6\,\text{cm}^2$

iii area = $7.5\,\text{m}^2$ iv area = $6.3\,\text{cm}^2$

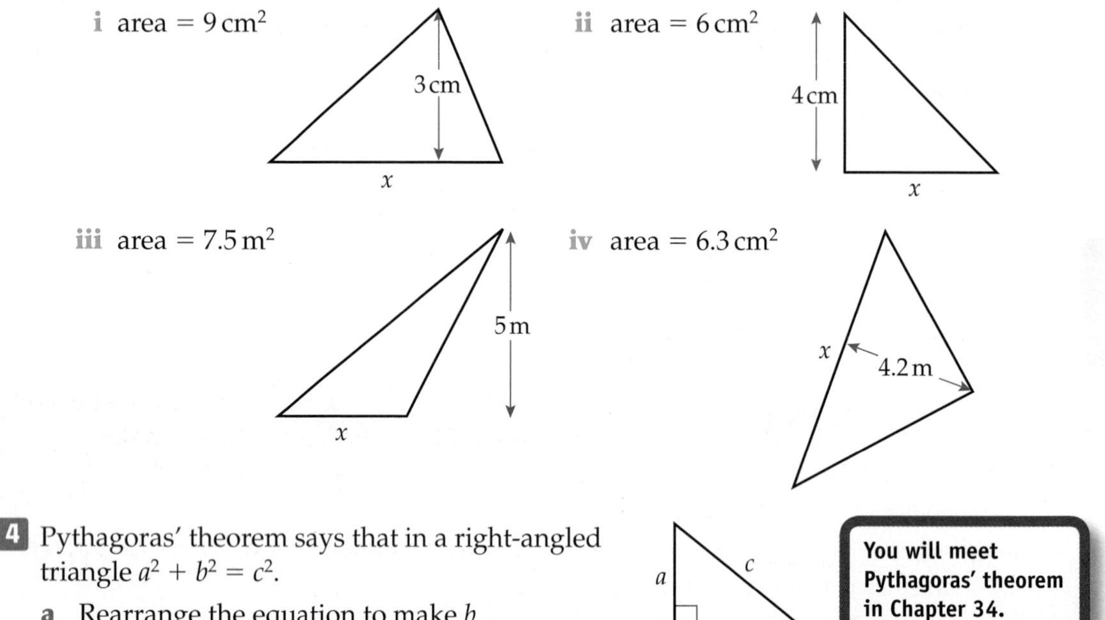

AO2

B

AO2

4 Pythagoras' theorem says that in a right-angled triangle $a^2 + b^2 = c^2$.

a Rearrange the equation to make b the subject.

b Find b when $a = 7$ and $c = 9$. Give your answer to three significant figures.

a c b

You will meet Pythagoras' theorem in Chapter 34.

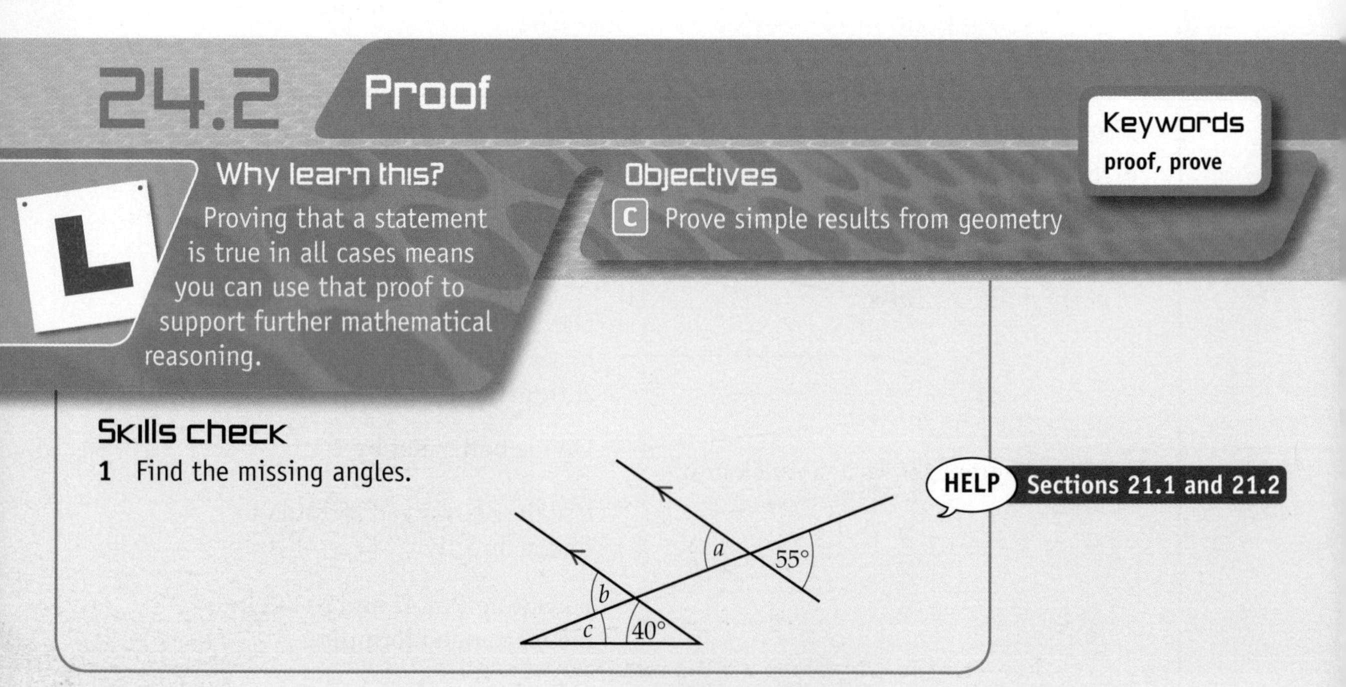

24.2 Proof

Keywords
proof, prove

Why learn this?

Proving that a statement is true in all cases means you can use that proof to support further mathematical reasoning.

Objectives

C Prove simple results from geometry

Skills check

1 Find the missing angles.

HELP Sections 21.1 and 21.2

Proof

A **proof** is a mathematical argument.

To **prove** something you need to explain each step of your working.
You can use these angle facts to prove results in geometry. You need to know the names of each angle fact.

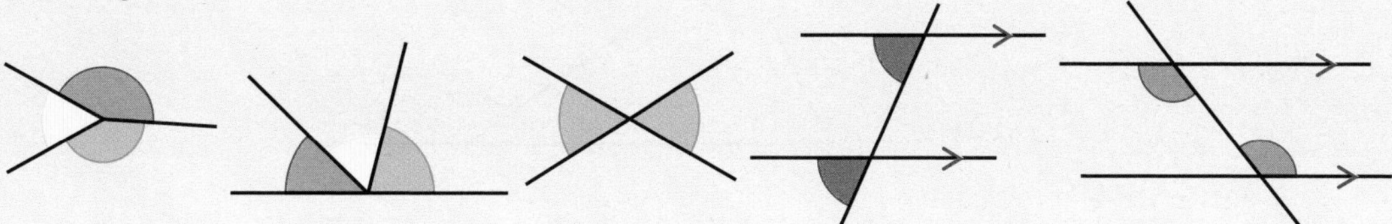

Angles around a point add up to 360°.

Angles on a straight line add up to 180°.

Opposite angles are equal.

Corresponding angles are equal.

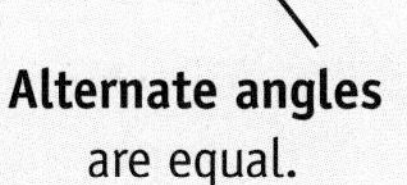

Alternate angles are equal.

Example 4

Prove that the angles in a triangle add up to 180°.

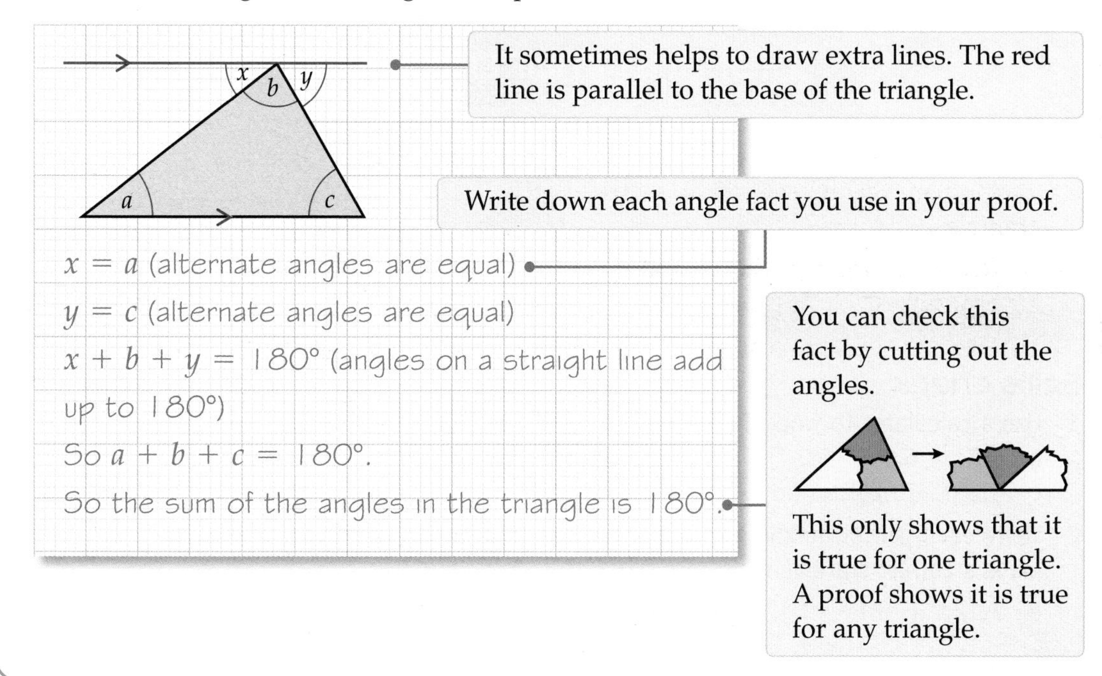

It sometimes helps to draw extra lines. The red line is parallel to the base of the triangle.

Write down each angle fact you use in your proof.

$x = a$ (alternate angles are equal)

$y = c$ (alternate angles are equal)

$x + b + y = 180°$ (angles on a straight line add up to 180°)

So $a + b + c = 180°$.

So the sum of the angles in the triangle is 180°.

You can check this fact by cutting out the angles.

This only shows that it is true for one triangle. A proof shows it is true for any triangle.

Exercise 24D

1 The diagram shows a parallelogram.

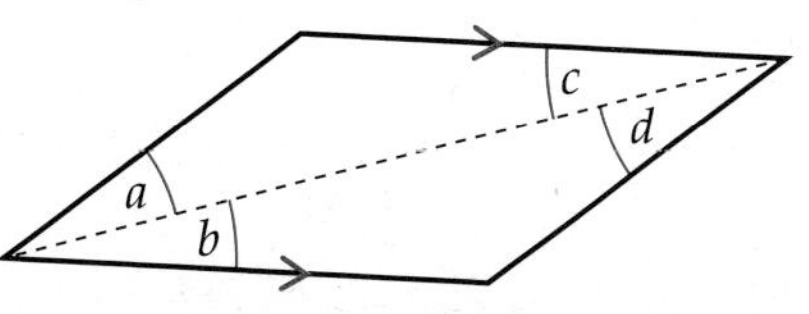

a Give a reason why angle a is the same size as angle d.

b Prove that the opposite angles in a parallelogram are the same size.

C

C

AO2

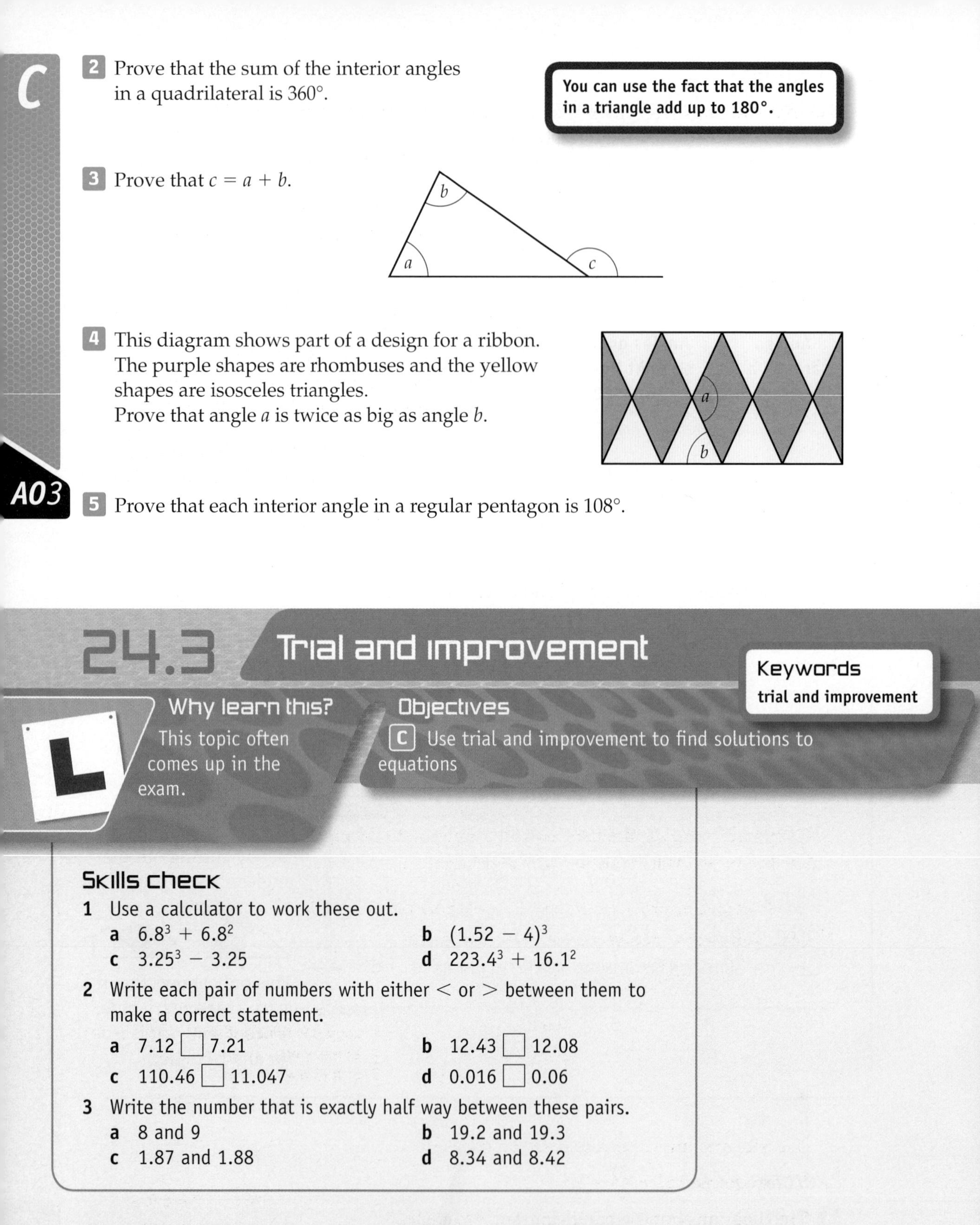

C

2 Prove that the sum of the interior angles in a quadrilateral is 360°.

You can use the fact that the angles in a triangle add up to 180°.

3 Prove that $c = a + b$.

4 This diagram shows part of a design for a ribbon. The purple shapes are rhombuses and the yellow shapes are isosceles triangles.
Prove that angle a is twice as big as angle b.

AO3

5 Prove that each interior angle in a regular pentagon is 108°.

24.3 Trial and improvement

Keywords
trial and improvement

Why learn this?
This topic often comes up in the exam.

Objectives
C Use trial and improvement to find solutions to equations

Skills check

1 Use a calculator to work these out.
 a $6.8^3 + 6.8^2$ b $(1.52 - 4)^3$
 c $3.25^3 - 3.25$ d $223.4^3 + 16.1^2$

2 Write each pair of numbers with either $<$ or $>$ between them to make a correct statement.
 a 7.12 ☐ 7.21 b 12.43 ☐ 12.08
 c 110.46 ☐ 11.047 d 0.016 ☐ 0.06

3 Write the number that is exactly half way between these pairs.
 a 8 and 9 b 19.2 and 19.3
 c 1.87 and 1.88 d 8.34 and 8.42

Trial and improvement

Some equations cannot be solved using algebra. Instead, you can solve them by **trial and improvement.**

Solve an equation by substituting values of x in the equation to see if they give the correct solution. The more values you try, the closer you can get to the solution.

Example 5

Use trial and improvement to find a solution to the equation $2x^3 - x = 70$.
Give your answer to one decimal place.

Draw a table to record your trials.

x	$2x^3 - x$	Comment
3	51	Too low
4	124	Too high
3.5	82.25	Too high
3.3	68.574	Too low
3.4	75.208	Too high
3.35	71.840...	Too high

Choose a starting value for x and use your calculator to work out $2x^3 - x$.

Compare your result with 70.

3 is too low and 4 is too high, so try 3.5.

The solution is between 3.3 and 3.4. To find the answer to one decimal place you need to know which is closer, so try 3.35.

3.35 is too high. This means the solution is closer to 3.3 than 3.4.

The solution is $x = 3.3$ to one decimal place.

C

Exercise 24E

1 Nisha is using trial and improvement to solve the equation $x^3 - 2x = 50$.
Her first two trials are shown in the table.

x	$x^3 - 2x$	Comment
3	21	Too low
5	115	Too high

Copy this table and add as many rows to the table as you need to find the solution.

Copy and complete the table to find a solution to the equation.
Give your answer to one decimal place.

2 Use trial and improvement to find a solution to the equation $\frac{x^3}{2} + x = 300$.
Give your answer correct to one decimal place.

x	$\frac{x^3}{2} + x$	Comment
5	67.5	Too low
10	510	Too high

Copy this table and add as many rows as you need to find the solution.

3 Use trial and improvement to find a solution to the equation $x^3 + \frac{20}{x} = 50$.
Give your answer correct to one decimal place.

x	$x^3 + \frac{20}{x}$	Comment
2	18	Too low
5	129	Too high

4 Use trial and improvement to find a solution to the equation $x^3 + 3x = 100$.
Give your answer correct to one decimal place.

x	$x^3 + 3x$	Comment
5	140	Too high

C

5 The equation $x^3 + \frac{200}{x^2} = 90$ has a solution between 3 and 6.

a Find this solution using trial and improvement. Give your answer correct to one decimal place.

b This equation has another solution between -5 and 0. Find this solution using trial and improvement. Give your answer correct to one decimal place.

6 Michelle is using trial and improvement to solve the equation $x^3 - x = 600$. She says the solution is $x = 8.4$ to one decimal place.

a Show working to explain why Michelle is wrong.

b What is the correct solution?

7 Use trial and improvement to solve the equation $2x^3 + 8x = -90$. Give your answer correct to one decimal place.

8 The volume of water in a barrel is given by the formula $V = \frac{t^3 - t}{10}$.

Use trial and improvement to find the value of t when $V = 30$. Give your answer correct to one decimal place.

C AO2

9 The area of this rectangle is 800 cm².

a Write an equation showing this information.

b Use trial and improvement to find the value of x correct to one decimal place.

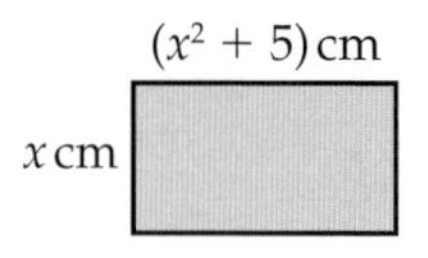

Review exercise

D

1 Work out the value of the letter in each of these triangles.

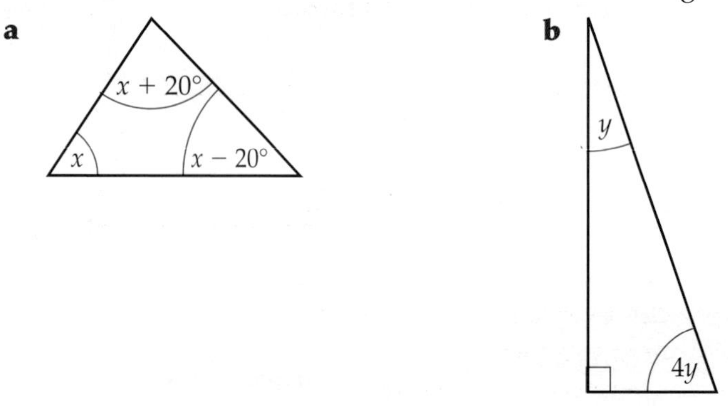

[3 marks each part]

2 Ali is paid £5.50 an hour for normal time and £8 an hour for overtime.

a How much does Ali receive when he works for 7 hours of normal time and 2 hours of overtime? **[1 mark]**

b Construct a formula for working out Ali's pay when he works for p hours of normal time and q hours of overtime. **[2 marks]**

c Use your formula to find Ali's pay when he works for 5 hours of normal time and 6 hours of overtime. **[1 mark]**

C AO3

3 Prove that adjacent angles in a parallelogram add up to 180°. **[4 marks]**

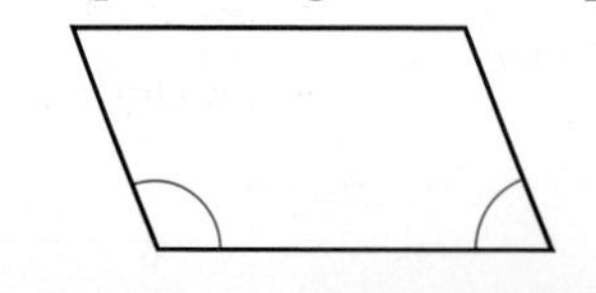

4 Shane has folded over the corner of a rectangular piece of paper and labelled the angles.

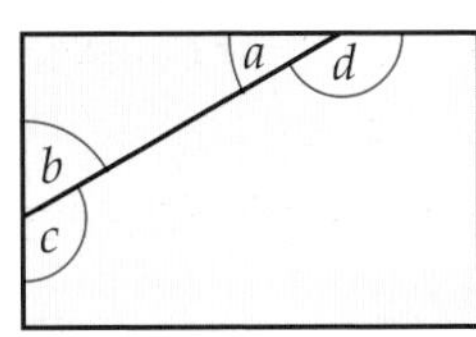

a Prove that $a + b = 90°$. **[2 marks]**

b Prove that $d + c = 270°$. **[4 marks]**

C

AO3

5 Use trial and improvement to find a solution to the equation $2x^3 - 20x = 80$.
Give your answer correct to one decimal place.
You can use a copy of this table to help you.

x	$2x^3 - 20x$	Comment
4	48	Too low

[4 marks]

C

6 Use trial and improvement to solve the equation $x^2 + \sqrt{x + 5} = 50$.
Give your answer correct to one decimal place. **[4 marks]**

7 Alex is using trial and improvement to solve the equation $\frac{x^2}{\sqrt{x}} = 50$.

Her first two trials are shown in this table.
Copy and complete the table to find a solution to the equation.
Give your answer to one decimal place.

x	$\frac{x^2}{\sqrt{x}}$	Comment
10	31.62...	Too low
20	89.44...	Too high

[4 marks]

8 Rearrange the formulae $p = \frac{q + 1}{r} + 5$ to make q the subject. **[2 marks]**

B

9 The area of this trapezium is given by the formula $A = \frac{ah + bh}{2}$.
Rearrange this formula to make b the subject.

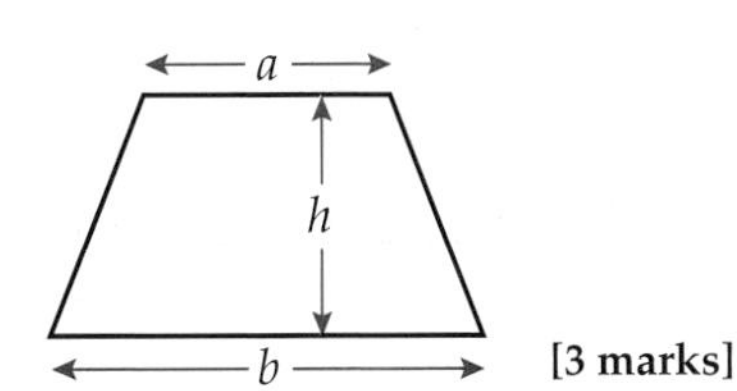

[3 marks]

10 Leanne and Barry have written down a negative number.
Leanne squares the number then subtracts 12.
Barry adds 14 to the number then triples the result.
Both finish with the same answer. What was the mystery number? **[4 marks]**

B

AO3

Chapter summary

In this chapter you have learned how to

- write your own formulae and equations **D**
- substitute into a formula to solve problems **D** **C**
- prove simple results from geometry **C**
- use trial and improvement to find solutions to equations **C**
- set up and solve equations **D** **C** **B**
- change the subject of a formula **C** **B**

25

Quadrilaterals and other polygons

This chapter is about working with quadrilaterals and other polygons.

The Alhambra in southern Spain is one of Europe's most famous examples of Islamic architecture. Many of the walls are covered in tessellating patterns.

Objectives

This chapter will show you how to

- calculate and use the sums of the interior and exterior angles of quadrilaterals, pentagons and hexagons; calculate and use the angles of regular polygons E D C
- show step-by-step deduction in solving a geometrical problem E D C
- recall the essential properties and definitions of special types of quadrilateral, including square, rectangle, parallelogram, trapezium and rhombus E D
- uses axes and coordinates to specify points in all four quadrants; locate points with given coordinates; find the coordinates of points identified by geometrical information; find the coordinates of the mid-point of the line segment *AB*, when points *A* and *B* are plotted on the coordinate axes E

Before you start this chapter

1 Work out the size of the angles marked with letters.

a

b

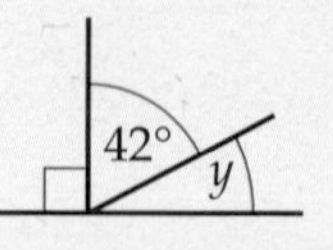

c

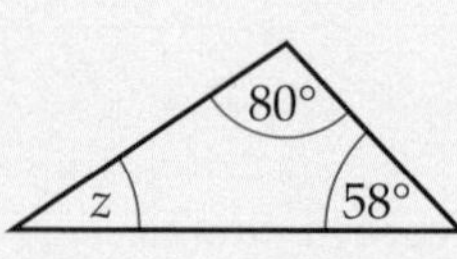

d

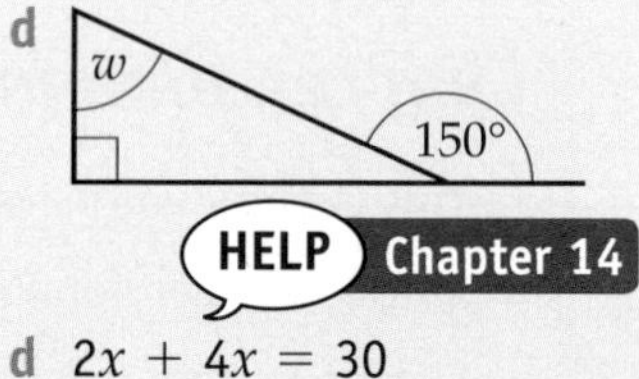

HELP Chapter 14

2 Solve each equation to find the value of the letter.

a $x + 12 = 50$ b $2x = 10$ c $3x + 1 = 16$ d $2x + 4x = 30$

3 Liz has a box of tiles.
Each tile is exactly the same.
Liz uses six of the tiles to make a rectangular pattern.
She wants the pattern to have only one line of symmetry. Draw a pattern that Liz could make.

Quadrilaterals and algebra

Keywords
quadrilateral, diagonal

Why learn this?
Many buildings are made up of rectangles and other quadrilaterals.

Objectives

E Calculate interior angles of quadrilaterals

D Solve angle problems in quadrilaterals involving algebra

Skills check

1 Work out
 a $30 + 60 + 45$
 b $90 + 40 + 50$
 c $120 + 40 + 65$
 d $45 + 85 + 60$

2 True or false?
 a $360 - 160 = 200$
 b $360 - 250 = 90$
 c $360 - 170 = 190$
 d $360 - 145 = 225$

Properties of quadrilaterals

A **quadrilateral** is a 2-D shape bounded by four straight lines.

A **diagonal** is a line joining two opposite corners (vertices).

The diagonal divides the quadrilateral into two triangles.

Diagonal

The 3 yellow angles add up to 180°. The 3 red angles add up to 180°.

The 6 angles from the 2 triangles add up to 360°.

The sum of the interior angles of a quadrilateral is 360°.

Example 1

Work out the size of angle x.

80° 120° 95° x

$180° - 95° = 85°$
(angles on a straight line)

First work out the size of the red angle.

$360° - 85° - 120° - 80° = 75°$
(angle sum of a quadrilateral)

Now work out the size of the blue angle.

$x = 180° - 75° = 105°$
(angles on a straight line)

Finally work out the size of angle x. Remember to give reasons for all of your calculations.

E

E

Exercise 25A

1 Work out the sizes of the angles marked with letters.

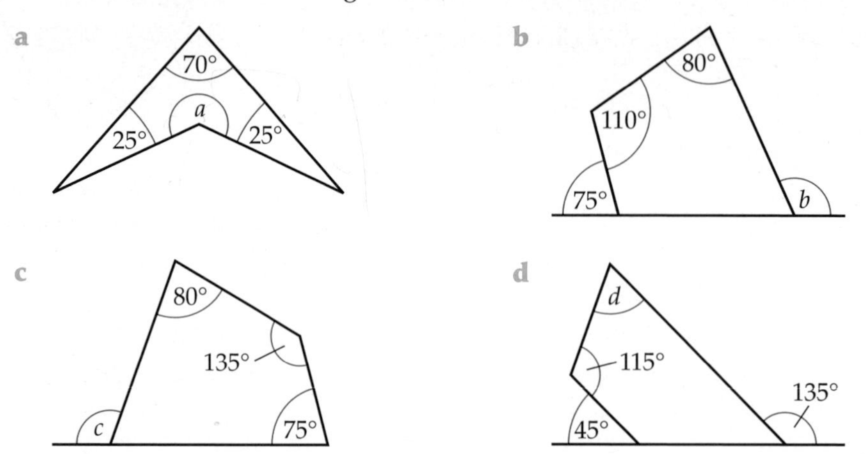

E

2 In this quadrilateral the smallest angle is 25°.

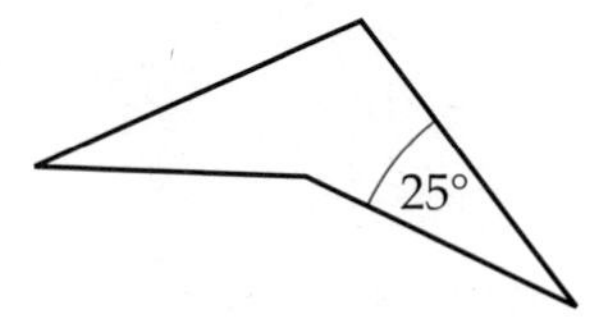

Another angle is twice the size of the smallest angle.
Another angle is three times the size of the smallest angle.
What is the size of the remaining angle?

3 In this quadrilateral the largest angle is 140°.

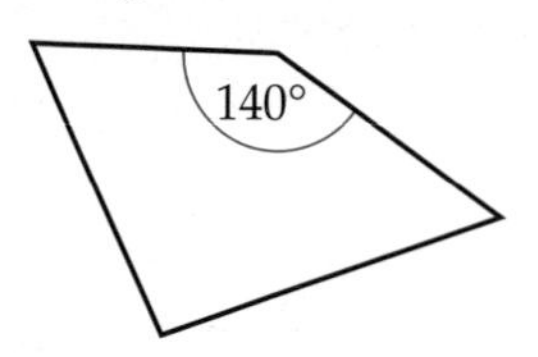

A02

The opposite angle is 35° less than the largest angle.
Another angle is half the size of the largest angle.
What is the size of the remaining angle?

Using algebra to solve problems involving quadrilaterals

Sometimes you may need to use algebra to solve problems involving the angles in a quadrilateral.

You can use the fact that the angle sum of a quadrilateral is 360° to set up an equation. You can then solve the equation and use the answer to work out the solution to the problem.

Example 2

a Work out the size of angle x.

b What is the size of the largest angle in the quadrilateral?

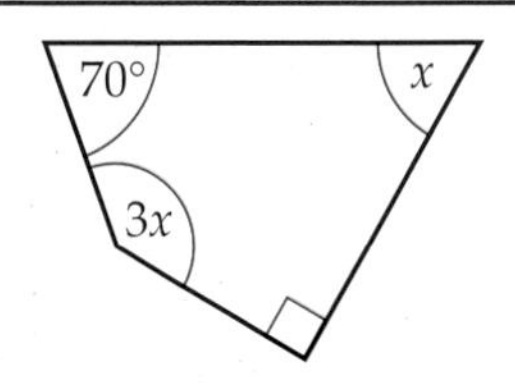

a $70° + x + 90° + 3x = 360°$ — First set up an equation using the angle sum of a quadrilateral.

$160° + 4x = 360°$ — Collect like terms.

$4x = 360° - 160°$ — Solve the equation to find x. Remember to show all stages of your solution.

$4x = 200°$

$x = 200° \div 4$

$x = 50°$

b Angles in the quadrilateral are 70°, 90°, 50° and $3 \times 50° = 150°$. The largest angle is 150°. — Work out the sizes of the angles involving x. Then decide which angle is the largest.

D

Exercise 25B

1 For each of the following quadrilaterals

i form an equation in x

ii solve the equation to find the value of x.

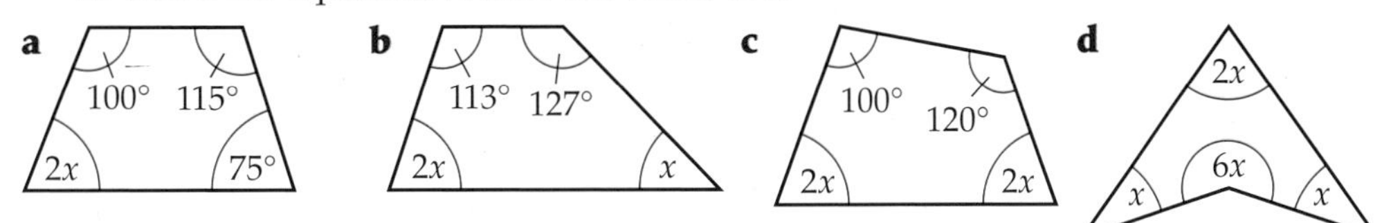

D

2 Work out the value of x.

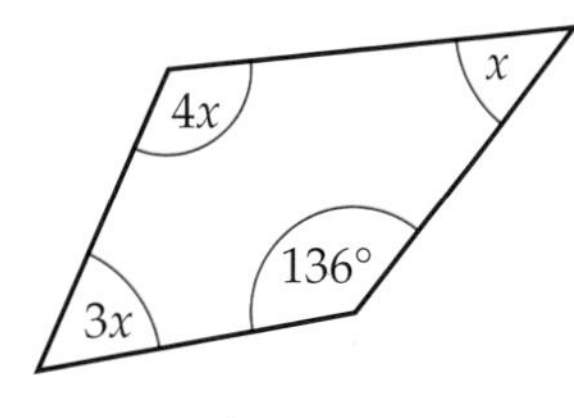

D

3 A quadrilateral has angles of a, $3a$, $3a$ and $5a$.
Work out the size of the largest angle.

AO2

4 Simon says, 'The largest angle in this quadrilateral is four times the smallest.'
Is Simon correct? Show working to support your answer.

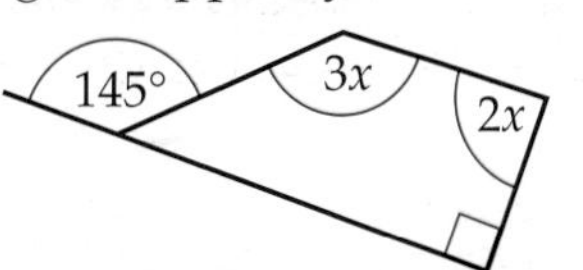

D

AO2

25.2 Special quadrilaterals

Keywords
square, rectangle, rhombus, bisect, parallelogram, trapezium, kite, adjacent

Why learn this?
Kite designers adjust angles in quadrilaterals to design good-looking and more efficient new models.

Objectives

E Make quadrilaterals from two triangles

D Use parallel lines and other angle properties in quadrilaterals

Skills check

1 What is the name given to each of these types of triangle?

a

b

c

d

The names of special quadrilaterals

Some quadrilaterals have special names.

Name	Shape	Properties
Square		• All four sides are the same length • Four right angles
Rectangle		• Opposite sides are equal • Four right angles
Rhombus		• Opposite angles are equal • Opposite sides are parallel • All four sides are the same length • Diagonals **bisect** each other at right angles
Parallelogram		• Opposite angles are equal • Opposite sides are equal and parallel • Diagonals bisect each other **A parallelogram is like a rectangle pushed over at the top.**
Trapezium		• One pair of parallel sides
Kite		• One pair of opposite angles equal • Two pairs of **adjacent** sides equal • One diagonal cuts the other at right angles **A kite is two isosceles triangles joined at the base.**

Example 3

Show how two of these scalene triangles can be joined together to make a parallelogram.

Remember, a parallelogram must have two sets of parallel sides.

Example 4

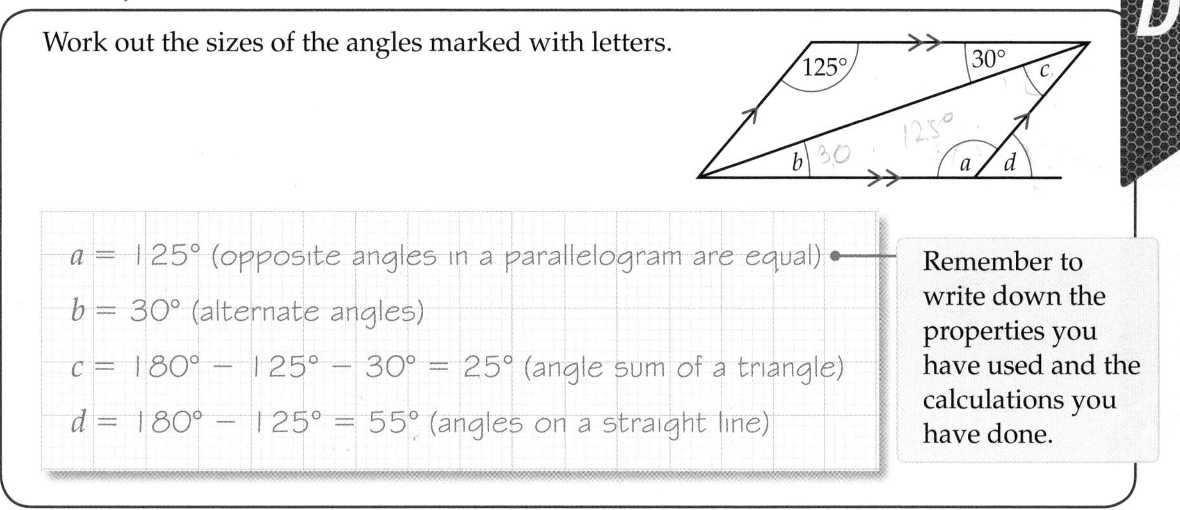

Work out the sizes of the angles marked with letters.

$a = 125°$ (opposite angles in a parallelogram are equal)

$b = 30°$ (alternate angles)

$c = 180° - 125° - 30° = 25°$ (angle sum of a triangle)

$d = 180° - 125° = 55°$ (angles on a straight line)

Remember to write down the properties you have used and the calculations you have done.

Exercise 25C

1 Show how two of these right-angled triangles can be joined together to make

- **a** a rectangle
- **b** a kite
- **c** a parallelogram.

2 Amir said, 'If I join together two identical isosceles triangles, I can make a rhombus.'
Enid said, 'If I join together two identical isosceles triangles, I can make a square.'
Show how Amir and Enid are both correct.

A02

3 Work out the sizes of the angles marked with letters.

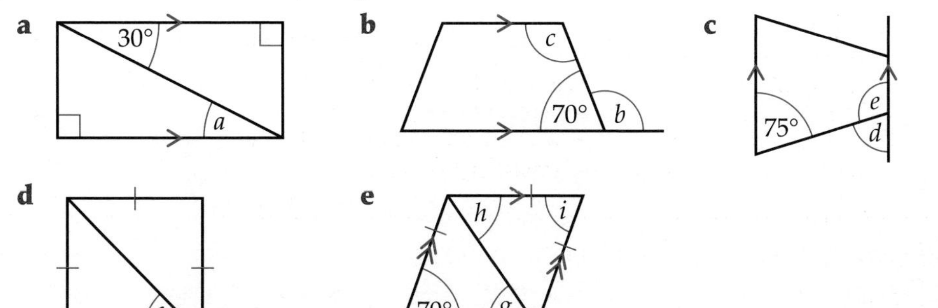

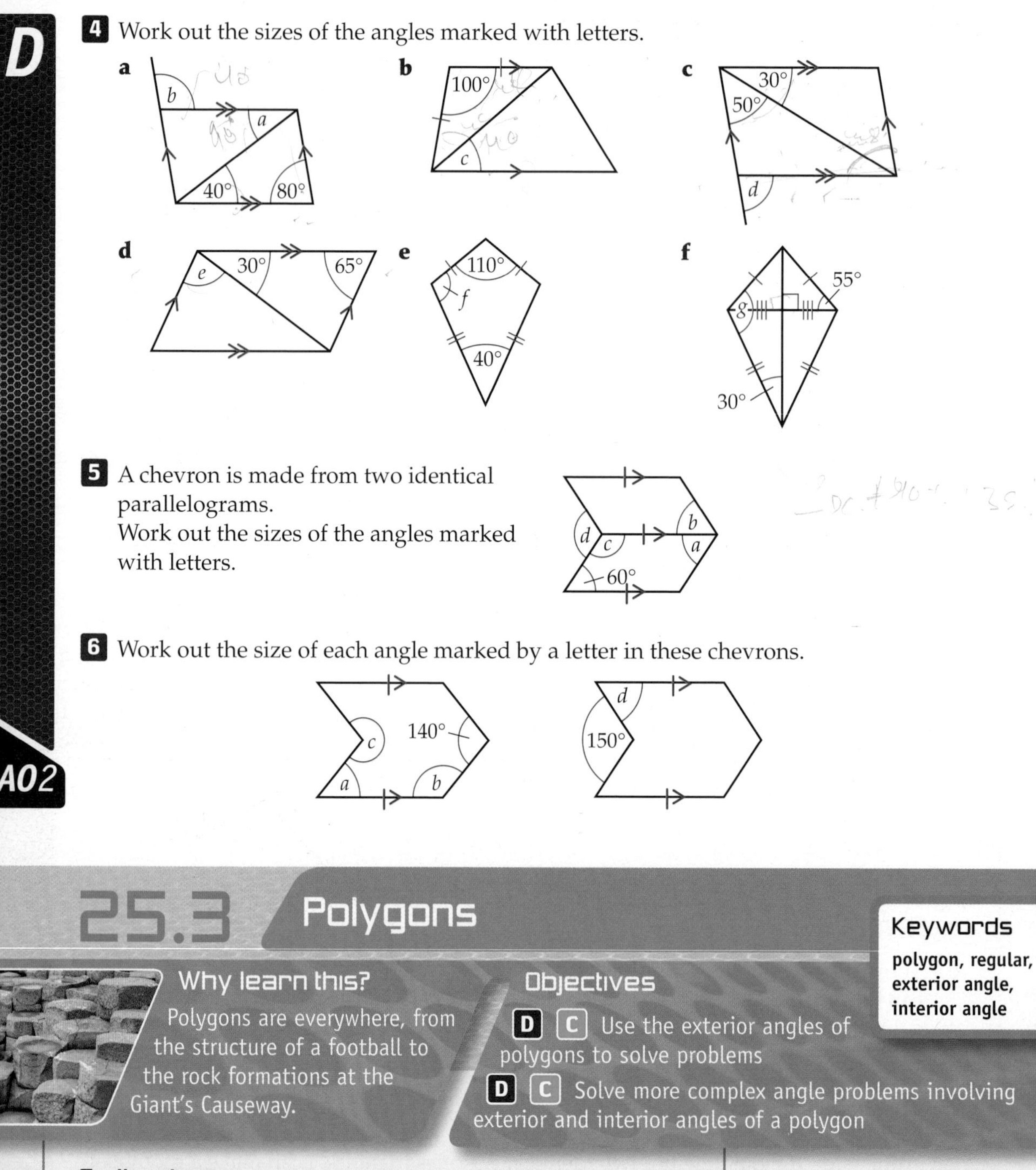

4 Work out the sizes of the angles marked with letters.

a **b** **c** **d** **e** **f**

5 A chevron is made from two identical parallelograms.
Work out the sizes of the angles marked with letters.

6 Work out the size of each angle marked by a letter in these chevrons.

25.3 Polygons

Keywords
polygon, regular, exterior angle, interior angle

Why learn this?

Polygons are everywhere, from the structure of a football to the rock formations at the Giant's Causeway.

Objectives

- **D** **C** Use the exterior angles of polygons to solve problems
- **D** **C** Solve more complex angle problems involving exterior and interior angles of a polygon

Skills check

1. Work out
   a 360 ÷ 4 **b** 360 ÷ 5 **c** 360 ÷ 6
2. Solve each equation to find the value of the letter.
   a $x + 42 = 78$ **b** $20x = 360$

Exterior angles of a polygon

A **polygon** is a 2-D shape bounded by straight lines.

A **regular** polygon has all its sides the same length.
All the angles in a regular polygon are the same size.

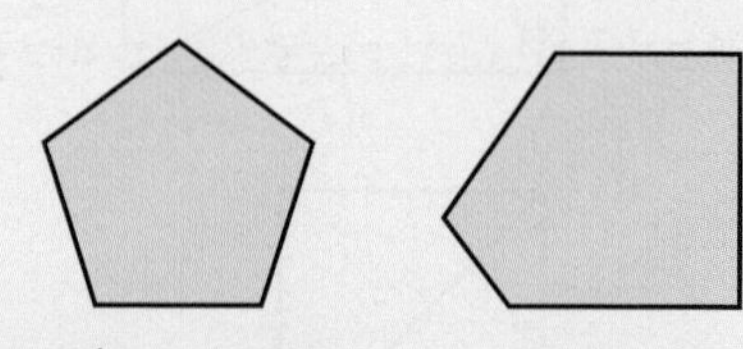

Here are some other polygons you need to know.

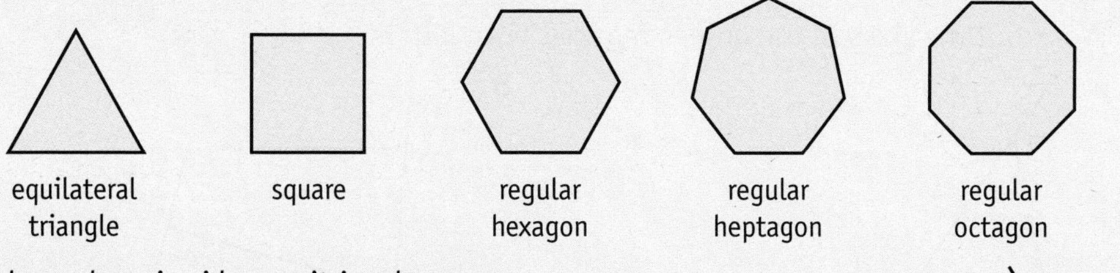

This polygon has six sides, so it is a hexagon.

The sides are not the same length so it is *not* a regular hexagon.
The angles a, b, c, d, e and f are **exterior angles**.

The sum of the exterior angles of any polygon is 360°.

In a regular polygon all the angles are the same size.

You can find the exterior angle of a regular polygon using this formula.

$a + b + c + d + e + f = 360°$

$$\text{Exterior angle of a regular polygon} = \frac{360°}{\text{number of sides}}$$

If you know the exterior angle of a regular polygon and you want to work out how many sides the polygon has, you can use this formula.

$$\text{Number of sides of a regular polygon} = \frac{360°}{\text{exterior angle}}$$

Example 5

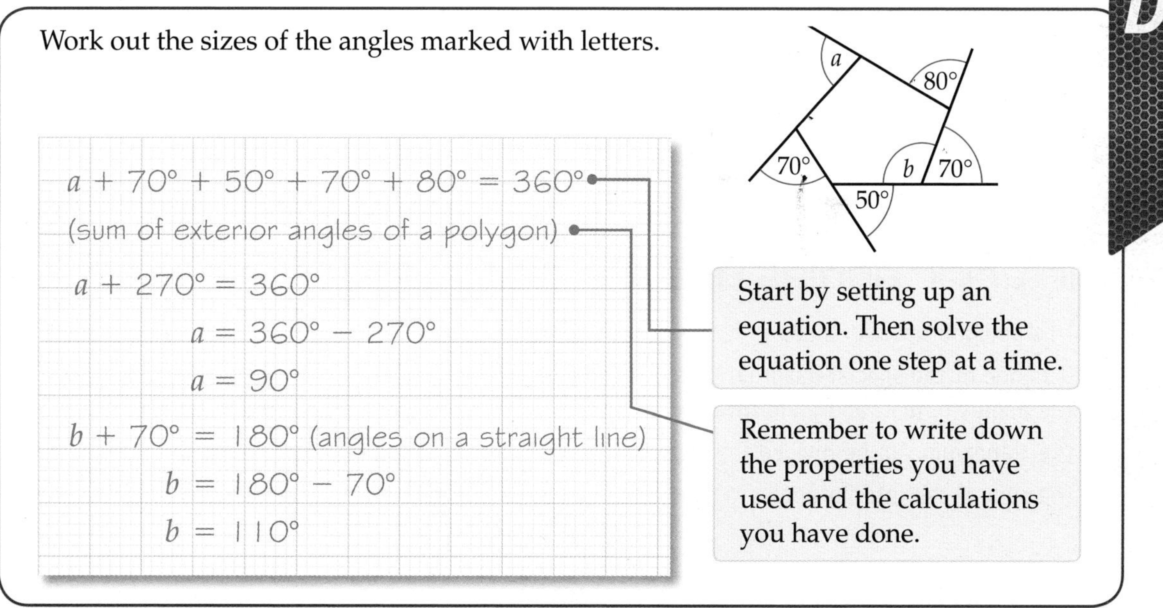

Work out the sizes of the angles marked with letters.

$a + 70° + 50° + 70° + 80° = 360°$
(sum of exterior angles of a polygon)
$a + 270° = 360°$
$a = 360° - 270°$
$a = 90°$

$b + 70° = 180°$ (angles on a straight line)
$b = 180° - 70°$
$b = 110°$

Start by setting up an equation. Then solve the equation one step at a time.

Remember to write down the properties you have used and the calculations you have done.

D

Example 6

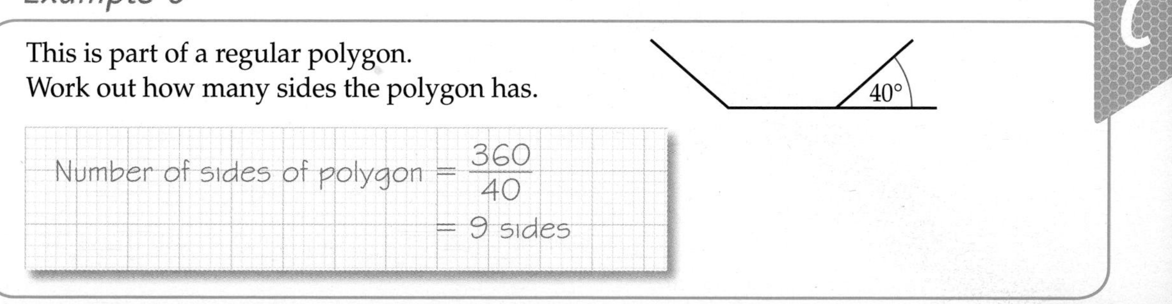

This is part of a regular polygon.
Work out how many sides the polygon has.

Number of sides of polygon $= \frac{360}{40}$
$= 9$ sides

C

Exercise 25D

D

1 Work out the sizes of the angles marked with letters.

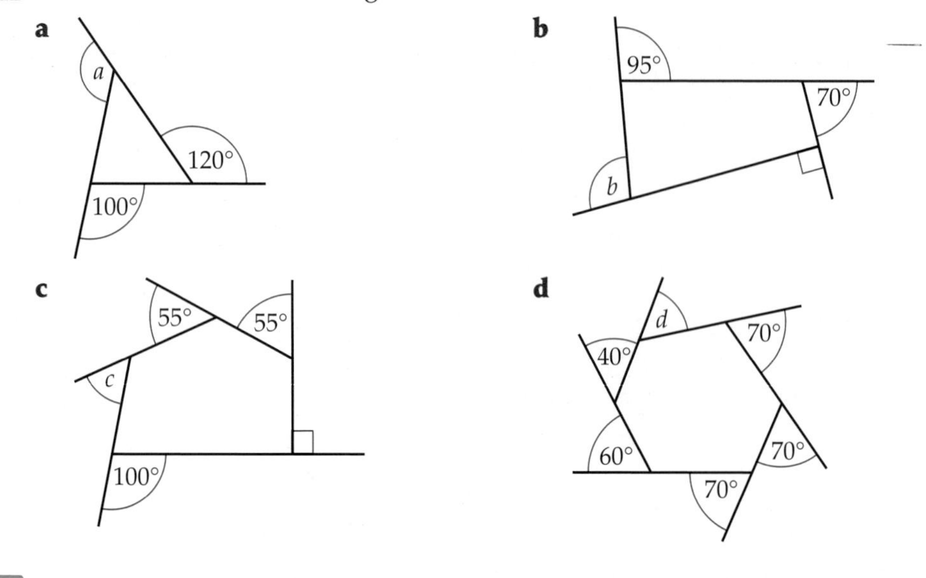

C

2 Work out the size of the exterior angle of a regular polygon with

a 8 sides **b** 10 sides **c** 36 sides.

C

3 Explain why it is not possible for the exterior angle of a regular polygon to be 50°.

4 Two sides of a regular hexagon are extended until they meet.
Work out the sizes of angles x and y.

AO2

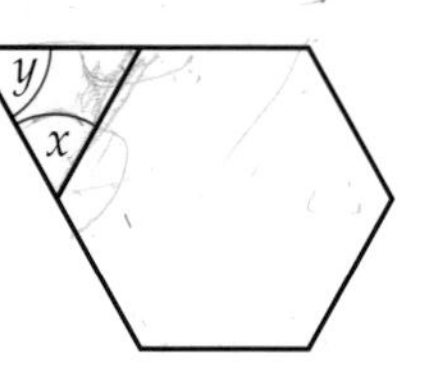

Interior angles of a polygon

This polygon has four sides so it is a quadrilateral.

It is one of the special quadrilaterals, a trapezium.
The angles v, w, x and y are **interior angles**.

The trapezium can be divided into two triangles as shown.

The sum of the interior angles $= 2 \times 180° = 360°$.

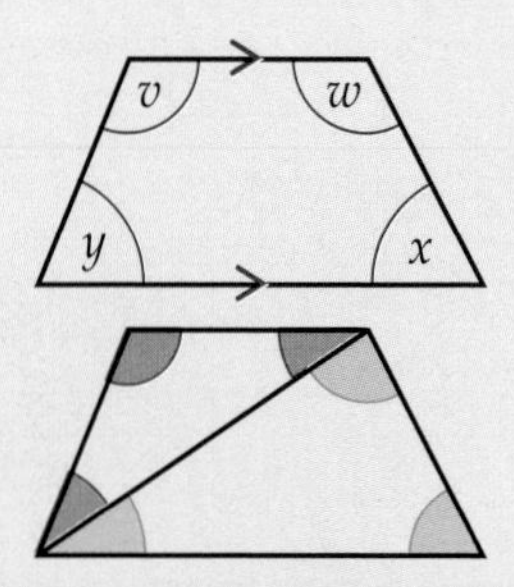

By dividing any polygon into triangles, you can find the sum of the interior angles.

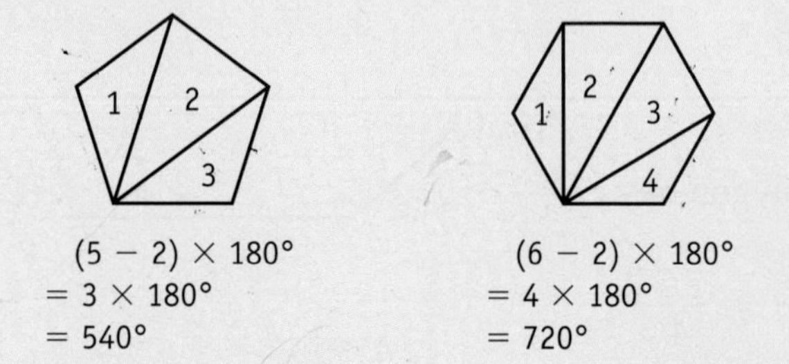

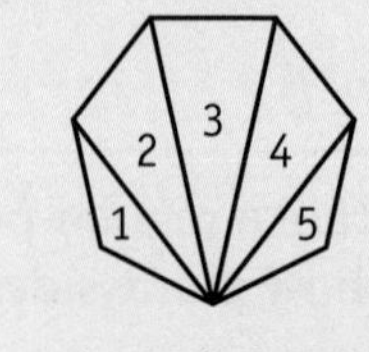

$(5 - 2) \times 180°$
$= 3 \times 180°$
$= 540°$

$(6 - 2) \times 180°$
$= 4 \times 180°$
$= 720°$

$(7 - 2) \times 180°$
$= 5 \times 180°$
$= 900°$

The sum of the interior angles of any polygon is (number of sides $-$ 2) $\times$ 180°.

In a regular polygon, all the angles are the same size.
You can find the interior angle of a regular polygon using this formula.

$$\text{Interior angle of a regular polygon} = 180° - \frac{360°}{\text{number of sides}}$$

If you know the interior angle of a regular polygon and you want to work out how many sides the polygon has, you can use this formula.

$$\text{Number of sides of a regular polygon} = \frac{360°}{(180° - \text{interior angle})}$$

Example 7

D

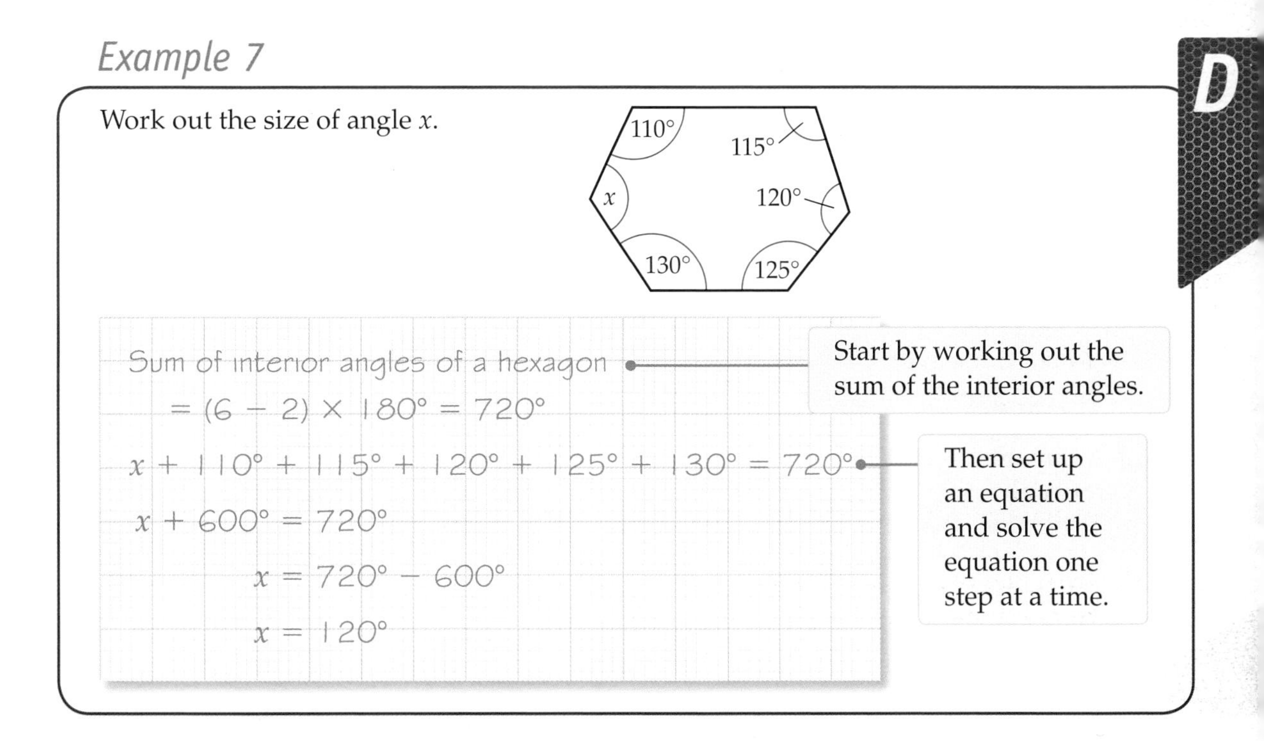

Work out the size of angle x.

Sum of interior angles of a hexagon
$= (6 - 2) \times 180° = 720°$

$x + 110° + 115° + 120° + 125° + 130° = 720°$

$x + 600° = 720°$

$x = 720° - 600°$

$x = 120°$

Start by working out the sum of the interior angles.

Then set up an equation and solve the equation one step at a time.

Example 8

C

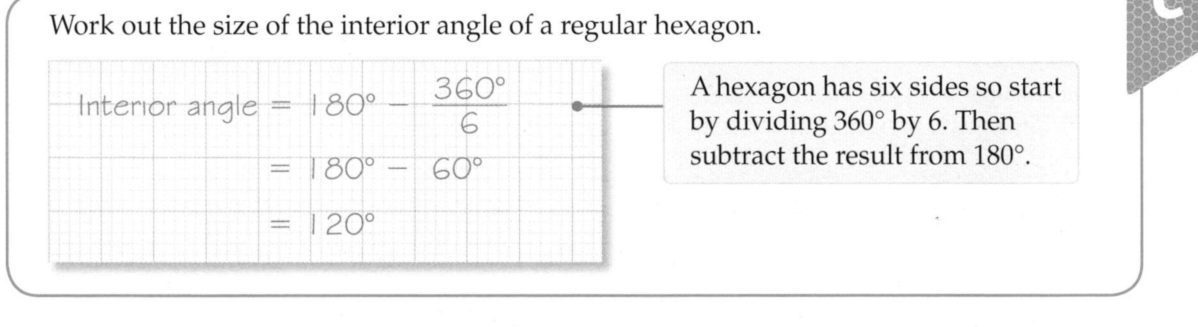

Work out the size of the interior angle of a regular hexagon.

$\text{Interior angle} = 180° - \frac{360°}{6}$

$= 180° - 60°$

$= 120°$

A hexagon has six sides so start by dividing 360° by 6. Then subtract the result from 180°.

Exercise 25E

D

1 Work out the sizes of the angles marked with letters.

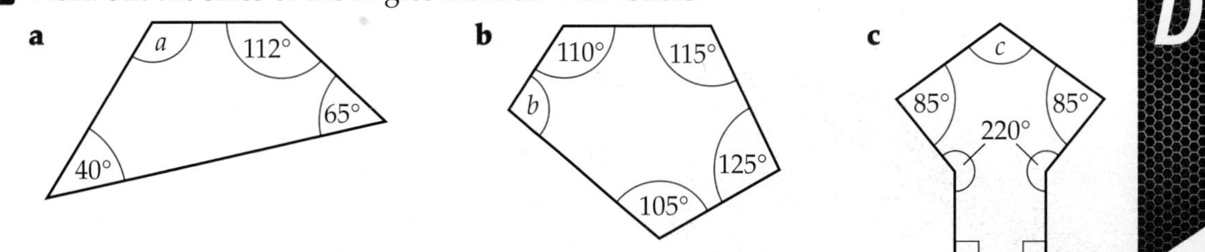

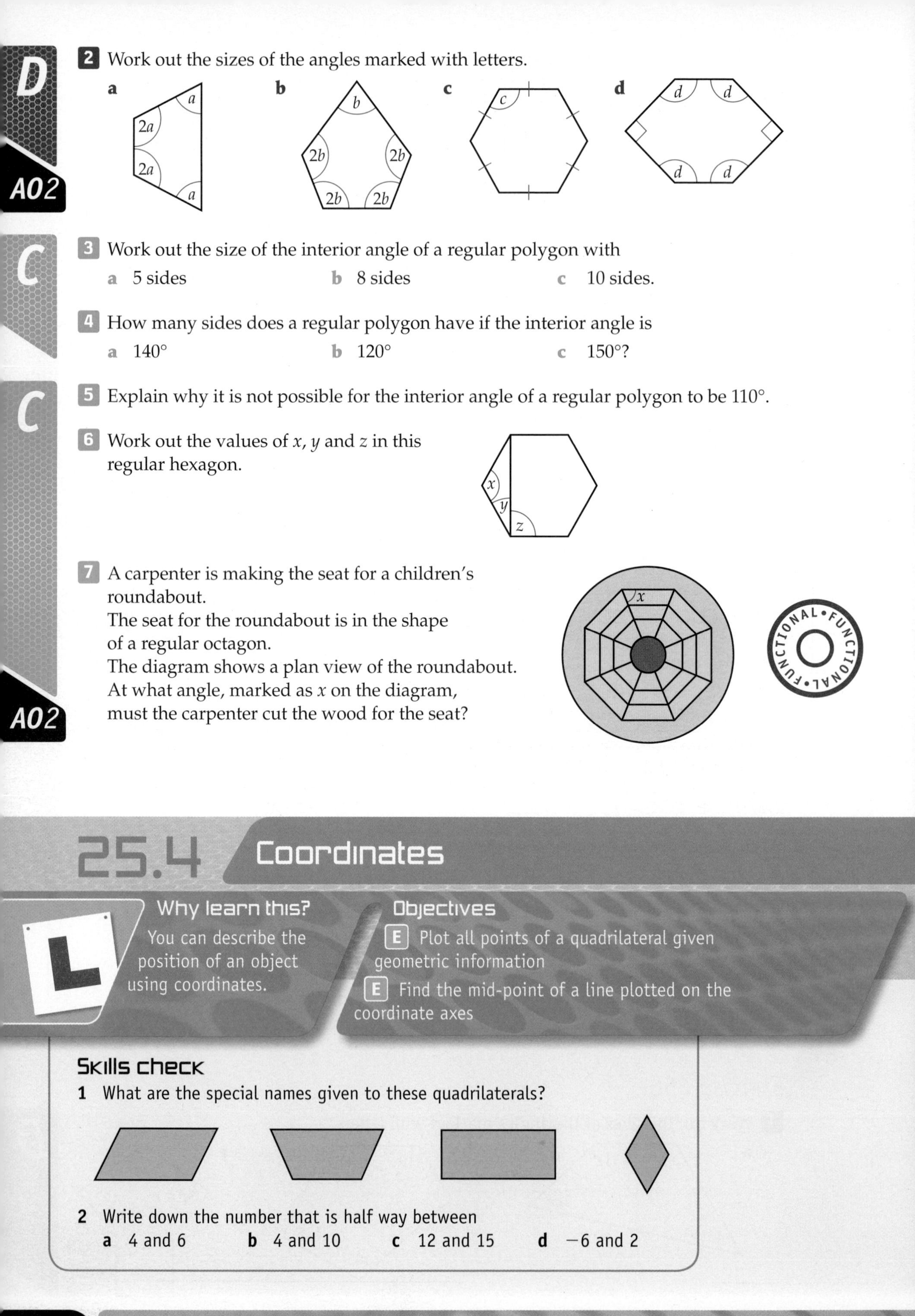

D

2 Work out the sizes of the angles marked with letters.

a **b** **c** **d**

AO2

C

3 Work out the size of the interior angle of a regular polygon with

a 5 sides **b** 8 sides **c** 10 sides.

4 How many sides does a regular polygon have if the interior angle is

a 140° **b** 120° **c** 150°?

C

5 Explain why it is not possible for the interior angle of a regular polygon to be 110°.

6 Work out the values of x, y and z in this regular hexagon.

7 A carpenter is making the seat for a children's roundabout.
The seat for the roundabout is in the shape of a regular octagon.
The diagram shows a plan view of the roundabout.
At what angle, marked as x on the diagram, must the carpenter cut the wood for the seat?

AO2

25.4 Coordinates

Why learn this?

You can describe the position of an object using coordinates.

Objectives

E Plot all points of a quadrilateral given geometric information

E Find the mid-point of a line plotted on the coordinate axes

Skills check

1 What are the special names given to these quadrilaterals?

2 Write down the number that is half way between

a 4 and 6 **b** 4 and 10 **c** 12 and 15 **d** −6 and 2

Plotting geometric shapes

You can plot geometric shapes on a coordinate grid.

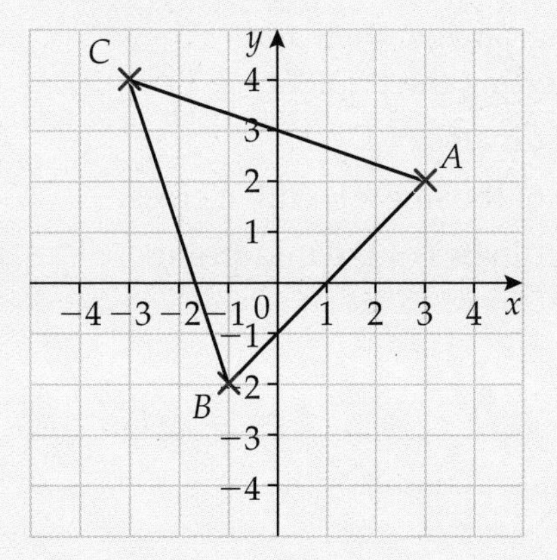

A is the point (3, 2).
B is the point (−1, −2).
C is the point (−3, 4).
Triangle ABC is isosceles.

Example 9

$ABCD$ is a square.

a Complete the square and write down the coordinates of point D.

b Write down the coordinates of the mid-point of the line AB.

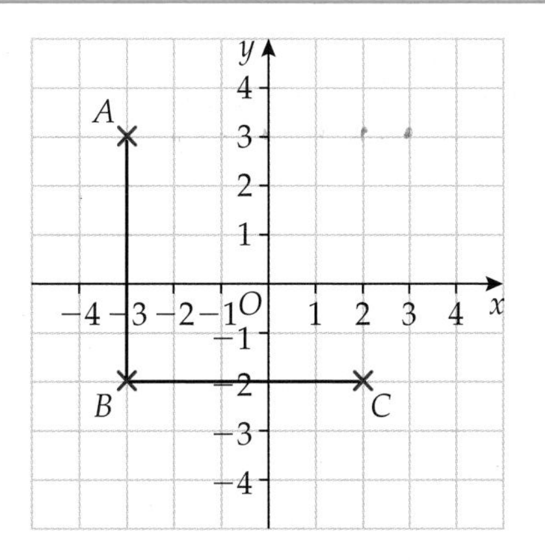

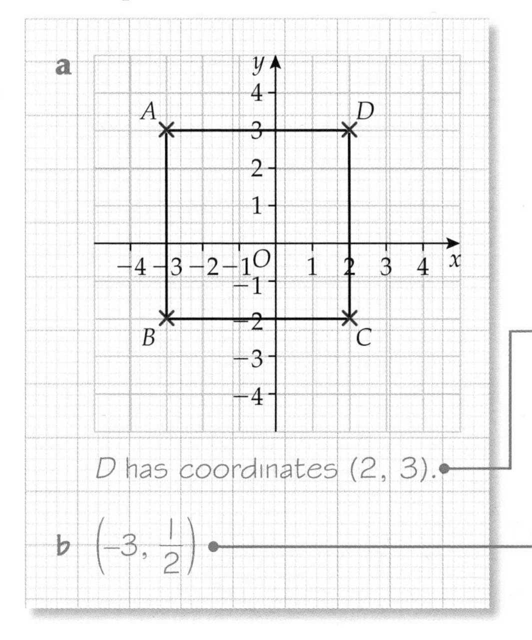

D has coordinates (2, 3).

Remember the x-coordinate is the first number in the bracket and the y-coordinate is the second.

b $\left(-3, \frac{1}{2}\right)$

The mid-point of AB has the same x-coordinate as A and B.
The y-coordinate is half way between 3 and −2.
To find the halfway point add the coordinates together and divide by 2.

$$\frac{(3 + -2)}{2} = \frac{1}{2}$$

E

Exercise 25F

E

1 $ABCD$ is a parallelogram.

a On a copy of the diagram complete the parallelogram by plotting the point D.

b Write down the coordinates of D.

c Mark the mid-point of the line AB with a cross.

d Write down the coordinates of the mid-point of the line AB.

HELP Section 18.1

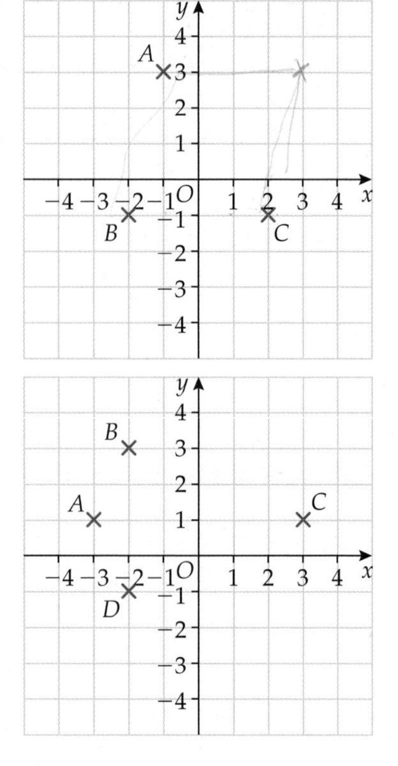

2 a What is the mathematical name of the quadrilateral $ABCD$?

b On a copy of the diagram join A to C.

c Mark the mid-point of AC with a cross and label it E.

d Write down the length of AE.

e Write down the coordinates of E.

E

3 Girish plots the three points on the grid shown.
Girish says, 'If I plot a fourth point at (3, 2) I'll get a rhombus.'
Is Girish correct? Explain your answer.

A02

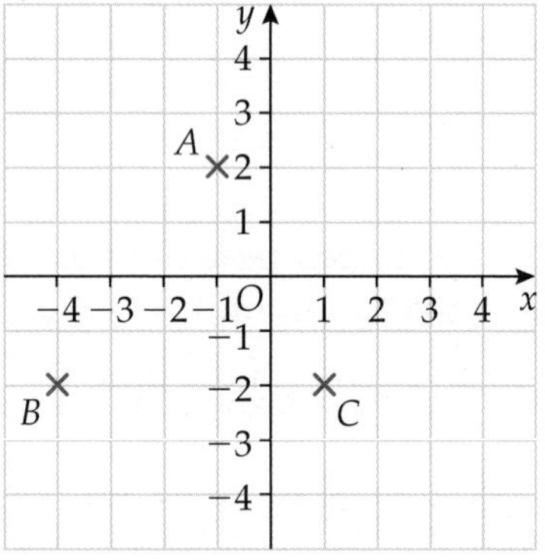

Review exercise

E

1 $ABCD$ is a quadrilateral.
Work out the size of angle a.

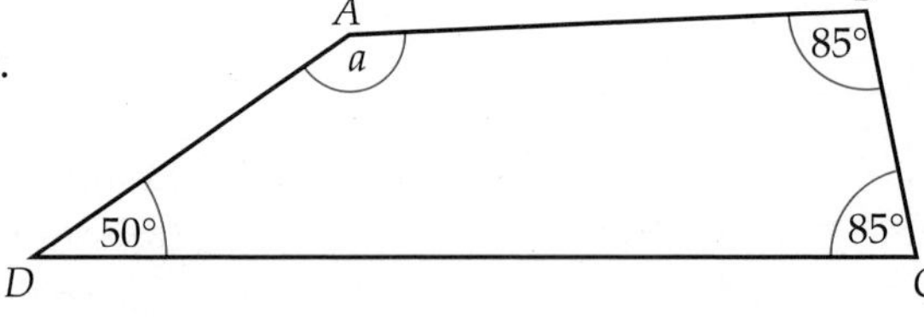

[2 marks]

2 $ABCD$ is a parallelogram.

a On a copy of the diagram complete the parallelogram by plotting point D.

b Write down the coordinates of D.

c Mark the mid-point of the line AB with a cross.

d Write down the coordinates of the mid-point of the line AB.

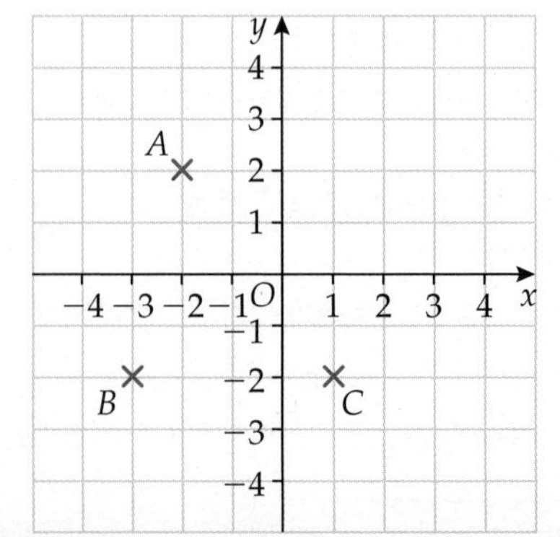

[4 marks]

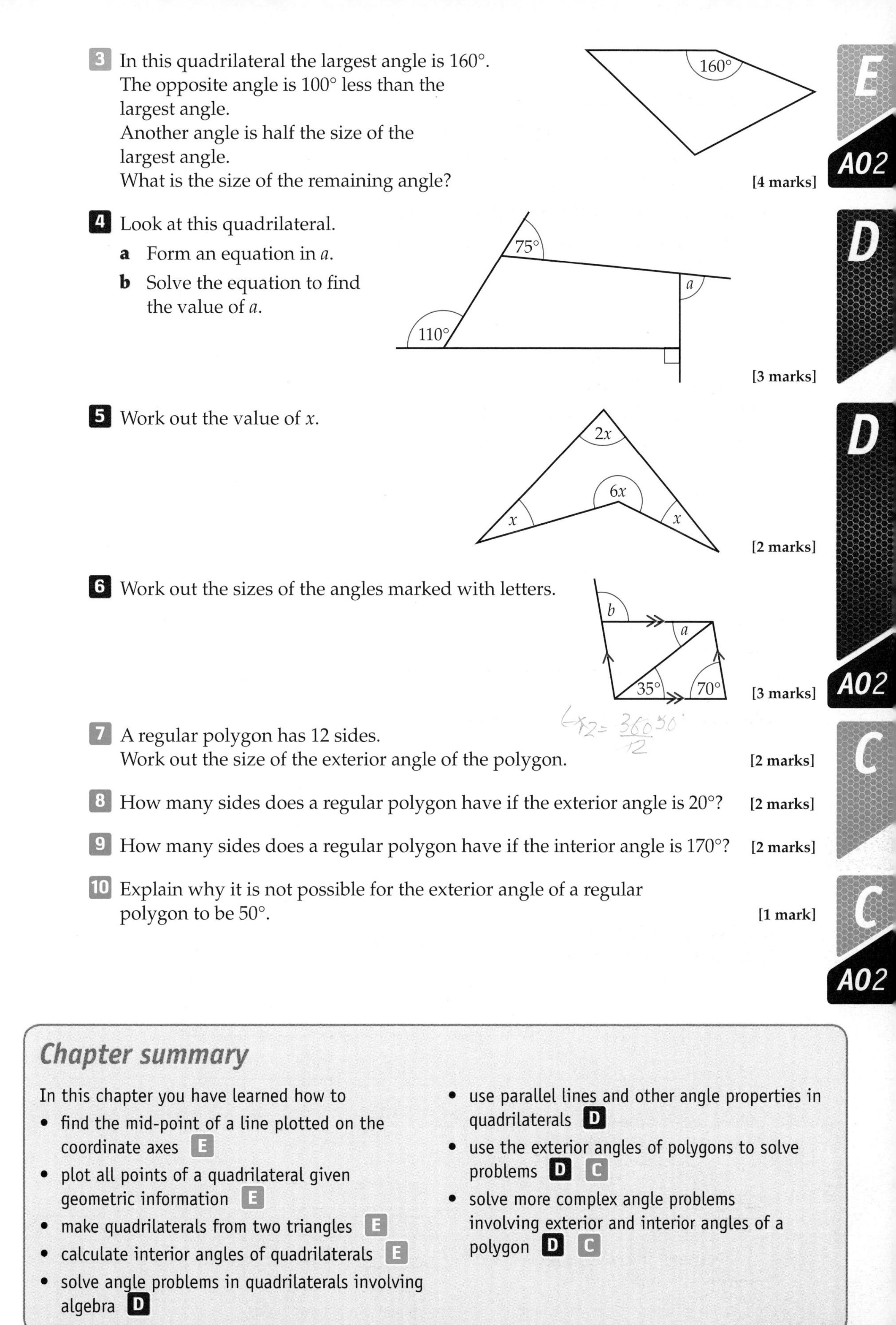

3 In this quadrilateral the largest angle is 160°.
The opposite angle is 100° less than the largest angle.
Another angle is half the size of the largest angle.
What is the size of the remaining angle? **[4 marks]**

4 Look at this quadrilateral.

a Form an equation in a.

b Solve the equation to find the value of a. **[3 marks]**

5 Work out the value of x. **[2 marks]**

6 Work out the sizes of the angles marked with letters. **[3 marks]**

7 A regular polygon has 12 sides.
Work out the size of the exterior angle of the polygon. **[2 marks]**

8 How many sides does a regular polygon have if the exterior angle is 20°? **[2 marks]**

9 How many sides does a regular polygon have if the interior angle is 170°? **[2 marks]**

10 Explain why it is not possible for the exterior angle of a regular polygon to be 50°. **[1 mark]**

Chapter summary

In this chapter you have learned how to

- find the mid-point of a line plotted on the coordinate axes **E**
- plot all points of a quadrilateral given geometric information **E**
- make quadrilaterals from two triangles **E**
- calculate interior angles of quadrilaterals **E**
- solve angle problems in quadrilaterals involving algebra **D**
- use parallel lines and other angle properties in quadrilaterals **D**
- use the exterior angles of polygons to solve problems **D** **C**
- solve more complex angle problems involving exterior and interior angles of a polygon **D** **C**

26 Perimeter, area and volume

This chapter is about calculating perimeters, areas, surface areas and volumes of different objects.

The Sage, Gateshead is a live music venue. To work out the number of stainless steel panels for the curved roof, the architects had to estimate and calculate the areas of complex curved shapes.

Objectives

This chapter will show you how to

- find areas of rectangles, recalling the formula E
- use your knowledge of rectangles, parallelograms and triangles to deduce formulae for the area of a parallelogram, a triangle and a trapezium E D
- recall and use the formulae for the area of a parallelogram and a triangle E D
- calculate perimeters and areas of shapes made from triangles and rectangles E D
- find the surface area of simple shapes using the area formulae for triangles and rectangles D C
- find volumes of cuboids recalling the formula; calculate volumes of right prisms and of shapes made from cubes and cuboids E D C

Before you start this chapter

1 Name these shapes.

a b c d e

2 Sketch three different types of triangle. Mark any equal angles and sides.

26.1 Perimeter and area of simple shapes

Keywords
perimeter, area, parallelogram, perpendicular height, trapezium

Why learn this?
Calculating the area of your roof will help you work out if you can fit in enough solar panels to heat all your hot water.

Objectives
E **D** Find the perimeter and area of rectangles, parallelograms, triangles and trapezia

Skills check
1 Calculate $\frac{1}{2} \times 12 \times 7$.
2 How many centimetres are there in 1.3 metres?

Rectangles, parallelograms and triangles

The **perimeter** of a shape is the sum of the lengths of all its sides.

The **area** of a shape is the amount of space inside it.

Rectangle

This rectangle has length l and width w.

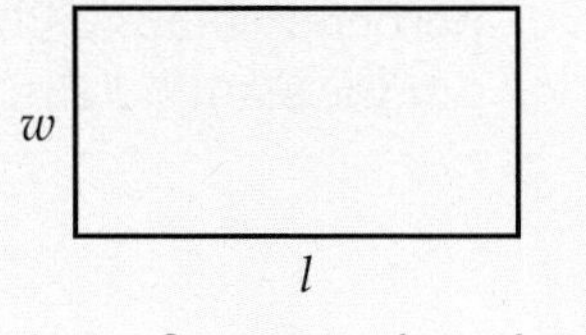

Area of a rectangle = length × width
$= l \times w$

The perimeter of a rectangle $= l + w + l + w$
$= 2l + 2w$

Parallelogram

The base of a **parallelogram** is b and its **perpendicular height** is h.

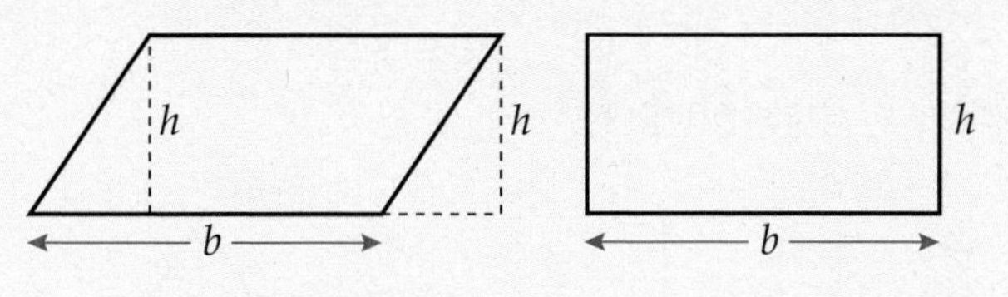

Cutting off a triangle from one end and placing it on the other end turns the parallelogram into a rectangle. The parallelogram and rectangle have equal areas.

Area of a parallelogram = base × perpendicular height
$= b \times h$

You must use the perpendicular height, h, not the slant height.

Triangle

The diagonal in this parallelogram splits it into two identical triangles.

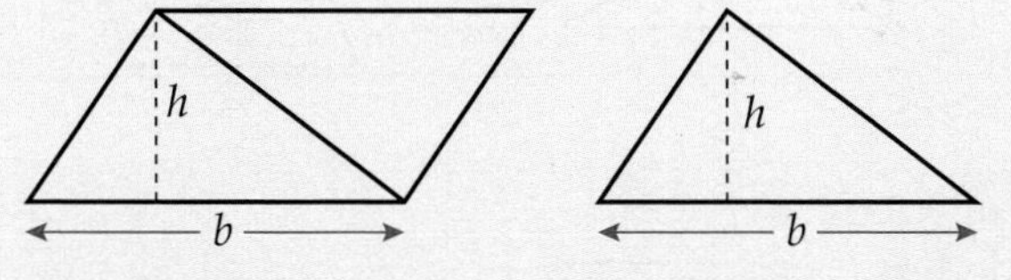

2 triangles = 1 parallelogram

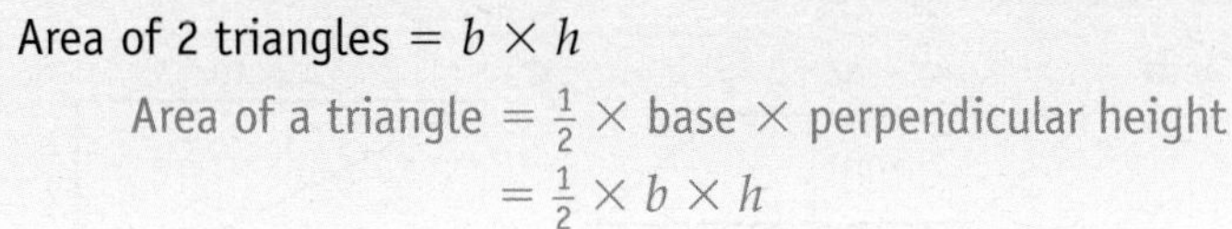

Area of 2 triangles $= b \times h$

Area of a triangle $= \frac{1}{2} \times$ base × perpendicular height
$= \frac{1}{2} \times b \times h$

E D

Example 1

Calculate the perimeter and area of these shapes.

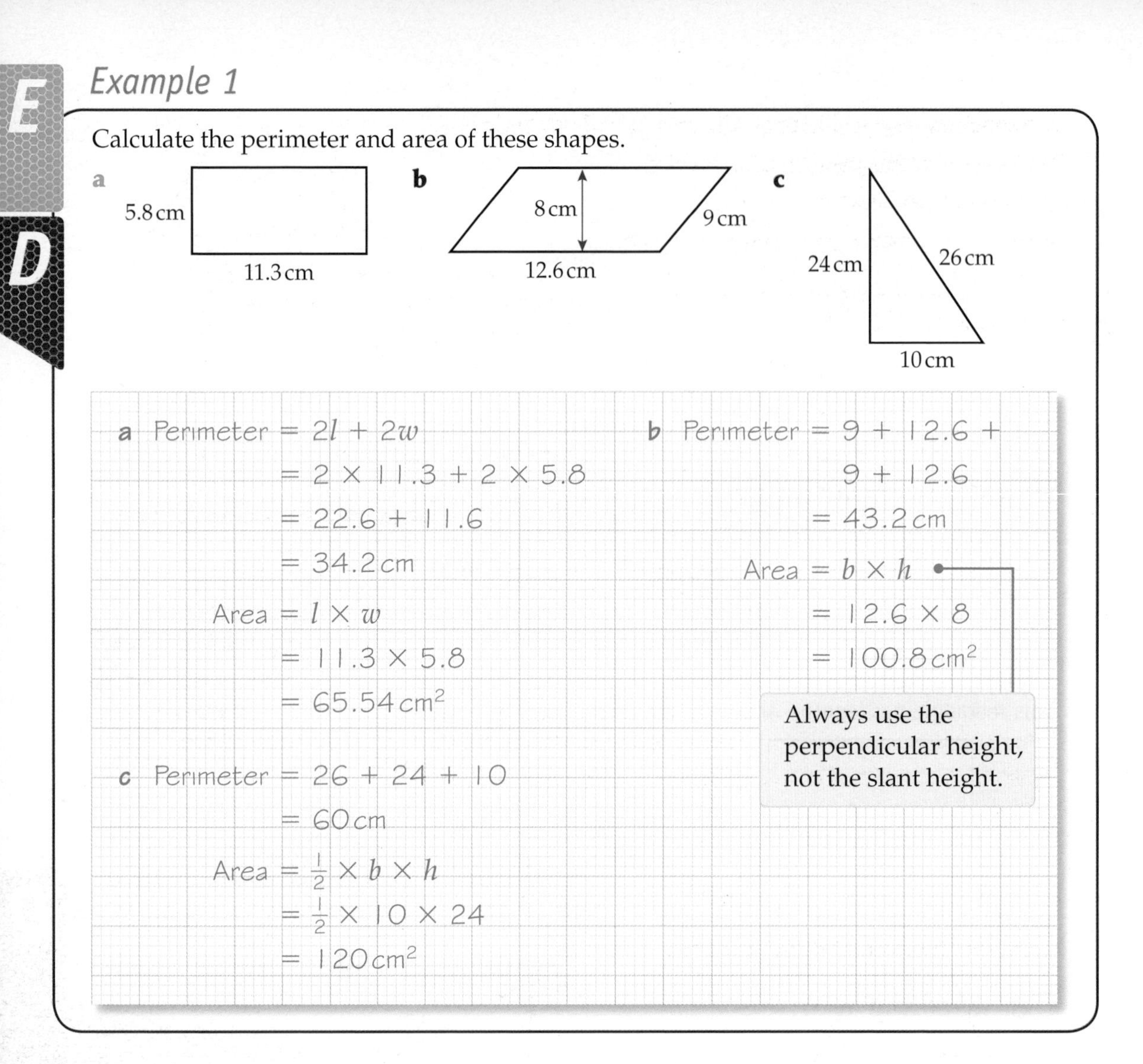

a Perimeter $= 2l + 2w$
$= 2 \times 11.3 + 2 \times 5.8$
$= 22.6 + 11.6$
$= 34.2$ cm

Area $= l \times w$
$= 11.3 \times 5.8$
$= 65.54\text{ cm}^2$

b Perimeter $= 9 + 12.6 + 9 + 12.6$
$= 43.2$ cm

Area $= b \times h$
$= 12.6 \times 8$
$= 100.8\text{ cm}^2$

Always use the perpendicular height, not the slant height.

c Perimeter $= 26 + 24 + 10$
$= 60$ cm

Area $= \frac{1}{2} \times b \times h$
$= \frac{1}{2} \times 10 \times 24$
$= 120\text{ cm}^2$

Exercise 26A

E D

1 Calculate the perimeters and areas of these shapes.
All lengths are in centimetres.

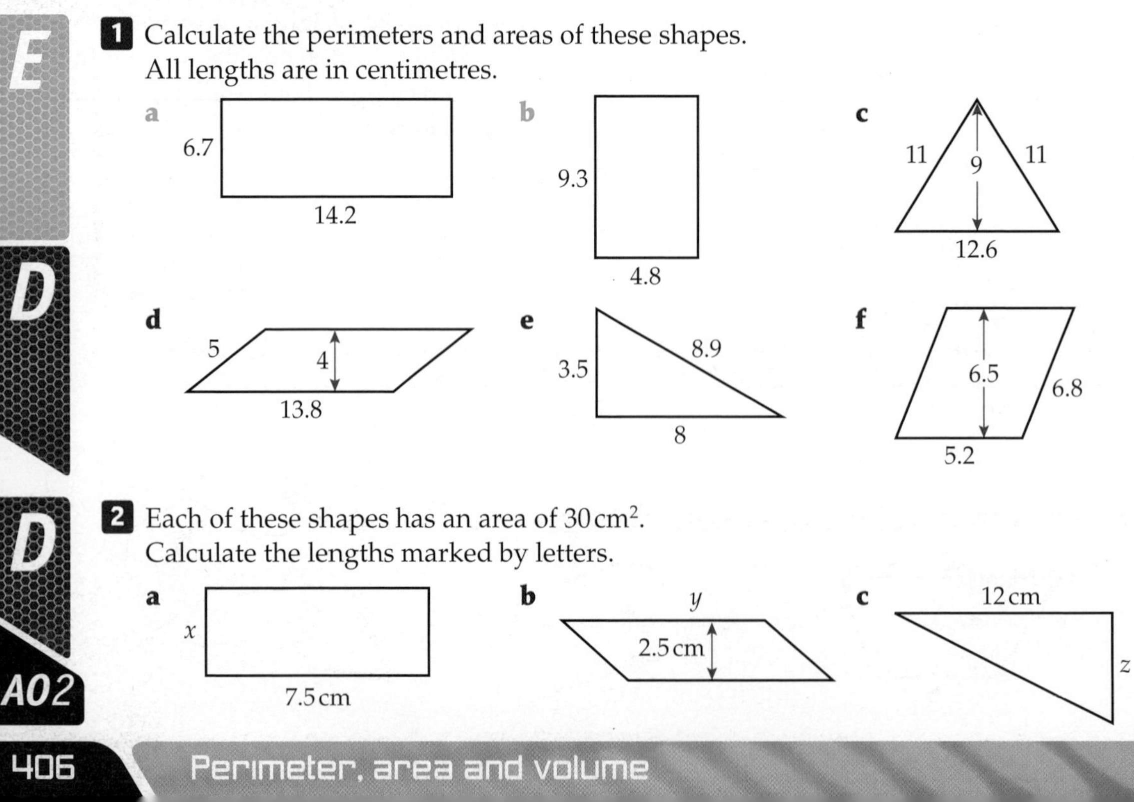

D A02

2 Each of these shapes has an area of 30 cm^2.
Calculate the lengths marked by letters.

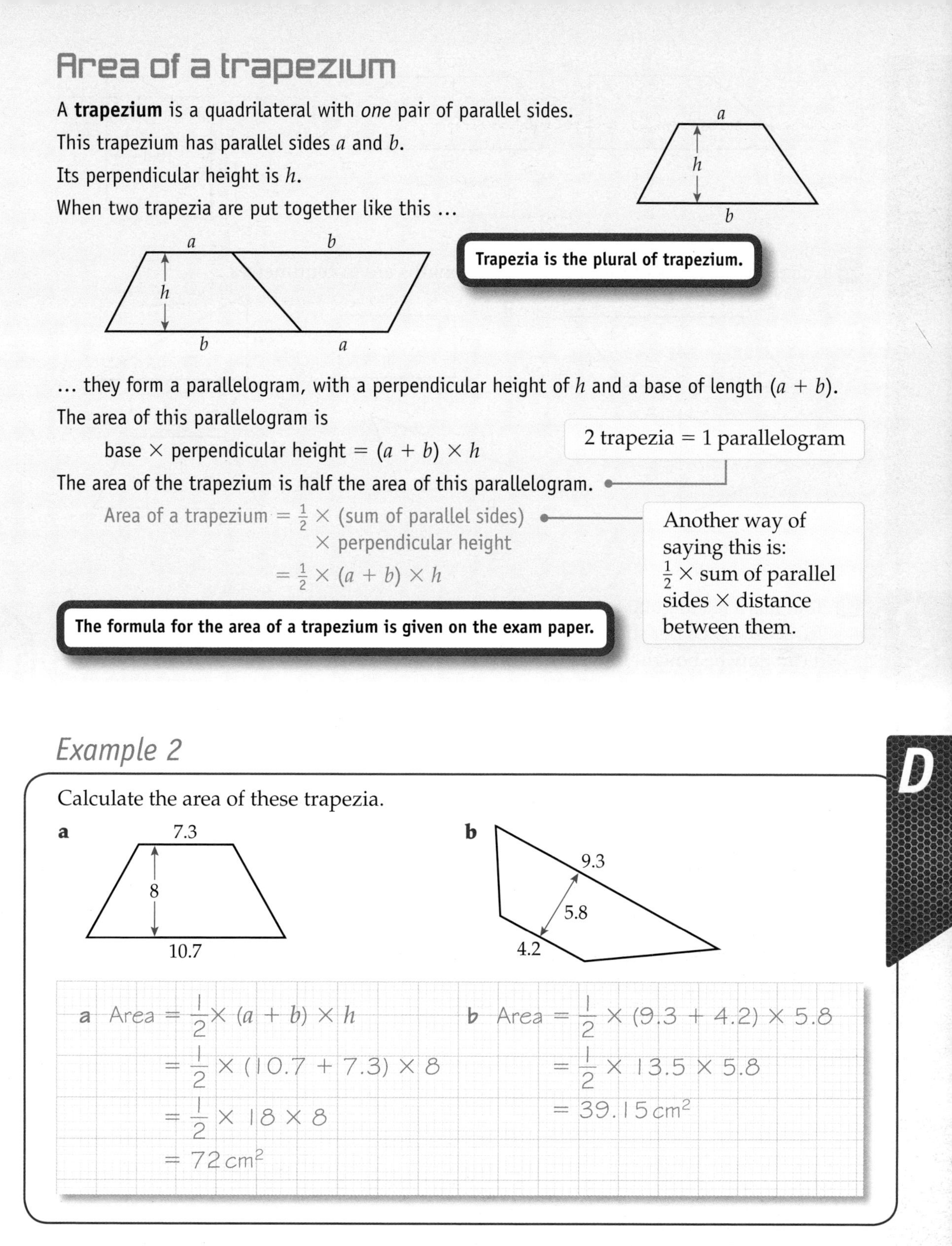

Area of a trapezium

A **trapezium** is a quadrilateral with *one* pair of parallel sides.

This trapezium has parallel sides a and b.

Its perpendicular height is h.

When two trapezia are put together like this ...

Trapezia is the plural of trapezium.

... they form a parallelogram, with a perpendicular height of h and a base of length $(a + b)$.

The area of this parallelogram is

base × perpendicular height $= (a + b) \times h$

2 trapezia = 1 parallelogram

The area of the trapezium is half the area of this parallelogram.

Area of a trapezium $= \frac{1}{2} \times$ (sum of parallel sides) × perpendicular height

$= \frac{1}{2} \times (a + b) \times h$

Another way of saying this is: $\frac{1}{2} \times$ sum of parallel sides × distance between them.

The formula for the area of a trapezium is given on the exam paper.

Example 2

Calculate the area of these trapezia.

a (7.3, 8, 10.7) **b** (9.3, 5.8, 4.2)

a Area $= \frac{1}{2} \times (a + b) \times h$

$= \frac{1}{2} \times (10.7 + 7.3) \times 8$

$= \frac{1}{2} \times 18 \times 8$

$= 72\,\text{cm}^2$

b Area $= \frac{1}{2} \times (9.3 + 4.2) \times 5.8$

$= \frac{1}{2} \times 13.5 \times 5.8$

$= 39.15\,\text{cm}^2$

D

Exercise 26B

1 Calculate the areas of these trapezia. All lengths are in centimetres.

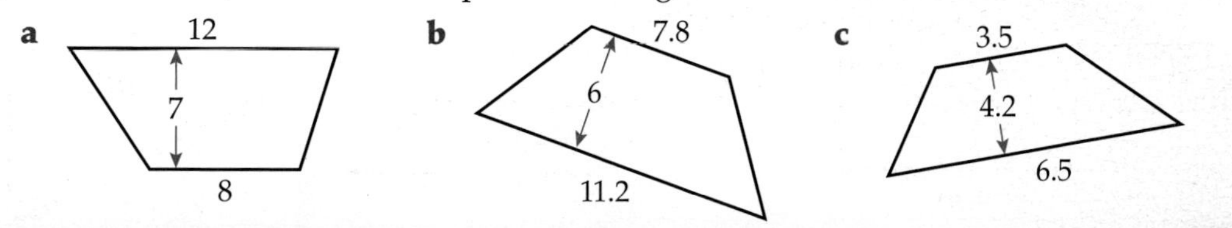

D

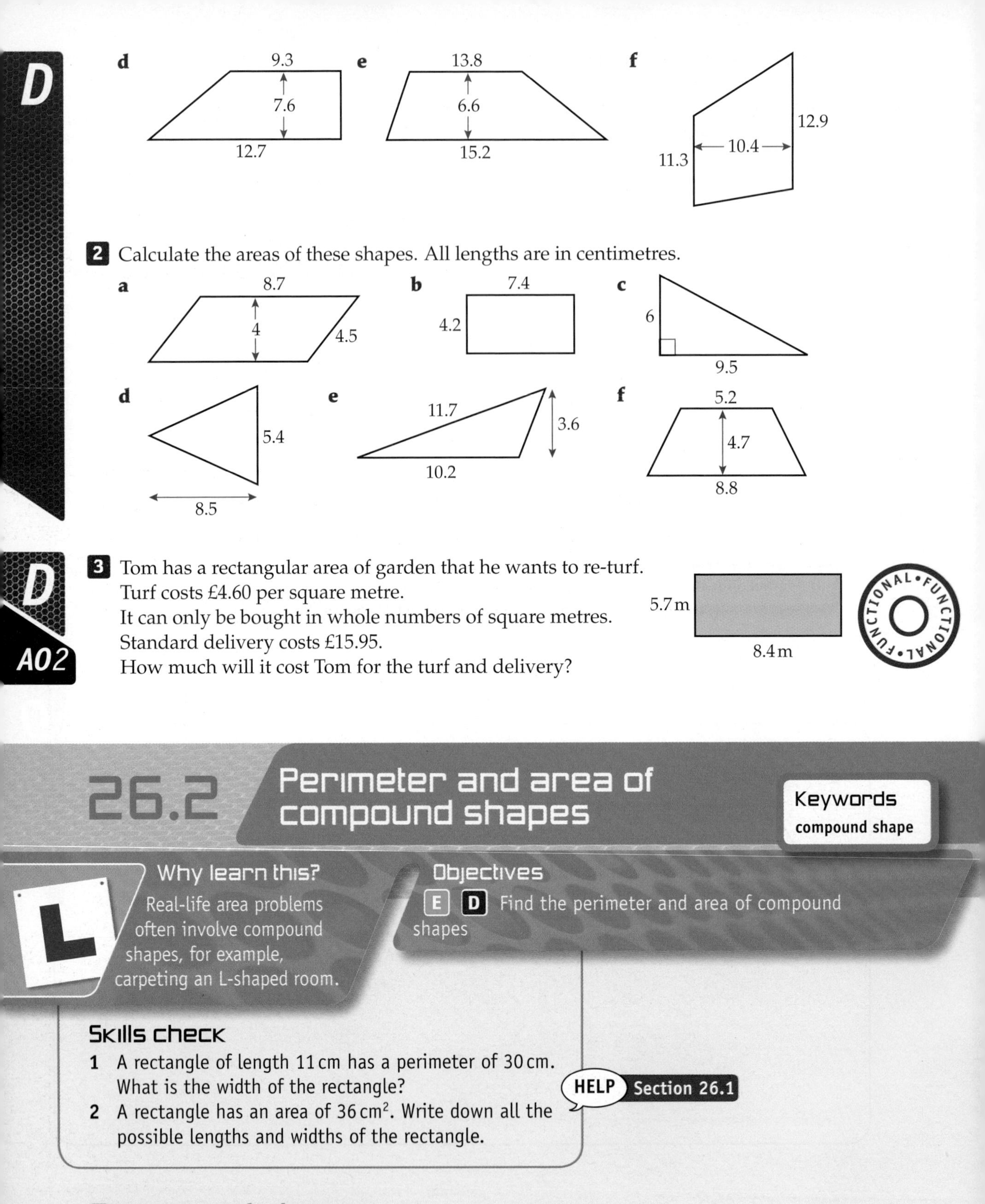

2 Calculate the areas of these shapes. All lengths are in centimetres.

3 Tom has a rectangular area of garden that he wants to re-turf.
Turf costs £4.60 per square metre.
It can only be bought in whole numbers of square metres.
Standard delivery costs £15.95.
How much will it cost Tom for the turf and delivery?

26.2 Perimeter and area of compound shapes

Keywords
compound shape

Why learn this?

Real-life area problems often involve compound shapes, for example, carpeting an L-shaped room.

Objectives

E **D** Find the perimeter and area of compound shapes

Skills check

1. A rectangle of length 11 cm has a perimeter of 30 cm. What is the width of the rectangle?
2. A rectangle has an area of 36 cm². Write down all the possible lengths and widths of the rectangle.

HELP Section 26.1

Compound shapes

A **compound shape** is a shape made up of simple shapes.

To find the area of a compound shape, you split it into simple shapes.

Then use the formulae for areas of simple shapes.

Example 3

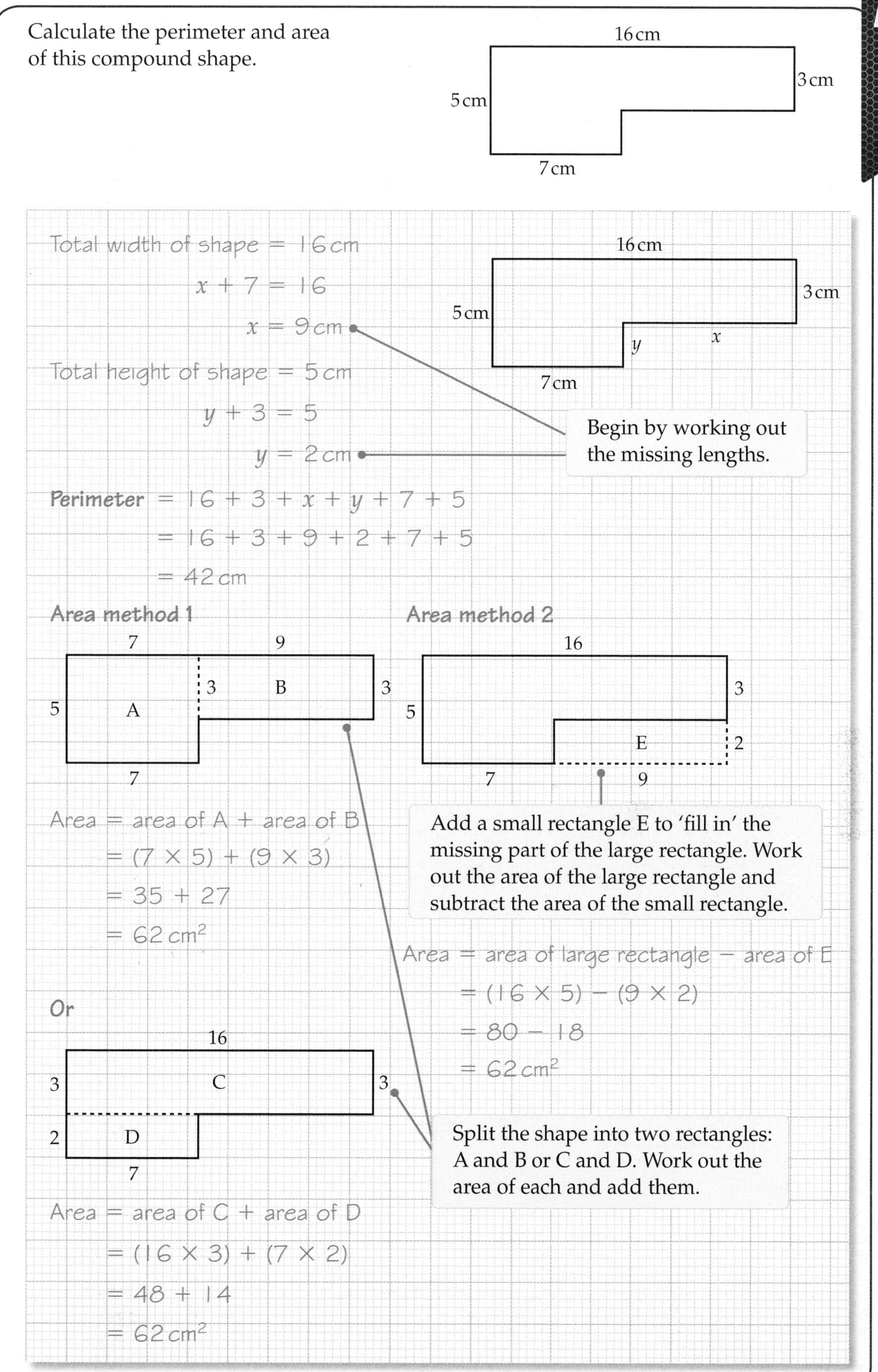

Calculate the perimeter and area of this compound shape.

Total width of shape = 16 cm

$x + 7 = 16$

$x = 9$ cm

Total height of shape = 5 cm

$y + 3 = 5$

$y = 2$ cm

Begin by working out the missing lengths.

Perimeter $= 16 + 3 + x + y + 7 + 5$

$= 16 + 3 + 9 + 2 + 7 + 5$

$= 42$ cm

Area method 1

Area = area of A + area of B

$= (7 \times 5) + (9 \times 3)$

$= 35 + 27$

$= 62\text{ cm}^2$

Or

Area = area of C + area of D

$= (16 \times 3) + (7 \times 2)$

$= 48 + 14$

$= 62\text{ cm}^2$

Split the shape into two rectangles: A and B or C and D. Work out the area of each and add them.

Area method 2

Add a small rectangle E to 'fill in' the missing part of the large rectangle. Work out the area of the large rectangle and subtract the area of the small rectangle.

Area = area of large rectangle − area of E

$= (16 \times 5) - (9 \times 2)$

$= 80 - 18$

$= 62\text{ cm}^2$

D

Example 4

Calculate the area of this shape.

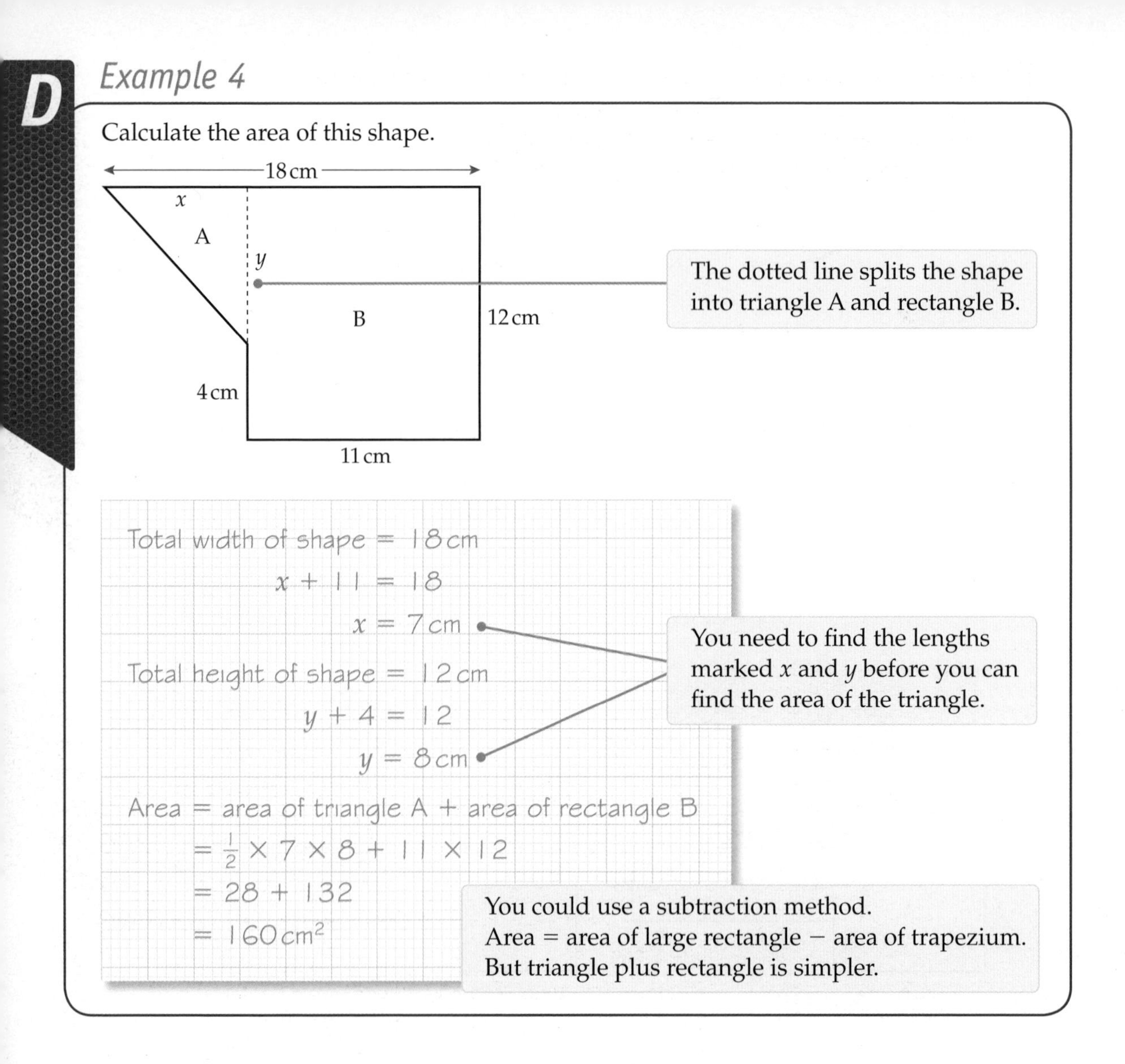

The dotted line splits the shape into triangle A and rectangle B.

Total width of shape = 18 cm

$x + 11 = 18$

$x = 7$ cm

Total height of shape = 12 cm

$y + 4 = 12$

$y = 8$ cm

You need to find the lengths marked x and y before you can find the area of the triangle.

Area = area of triangle A + area of rectangle B

$= \frac{1}{2} \times 7 \times 8 + 11 \times 12$

$= 28 + 132$

$= 160\text{ cm}^2$

You could use a subtraction method.
Area = area of large rectangle − area of trapezium.
But triangle plus rectangle is simpler.

Exercise 26C

D

1 Calculate the perimeter and area of each of these compound shapes.
All lengths are in centimetres.

Choose one of the methods used in Example 3.

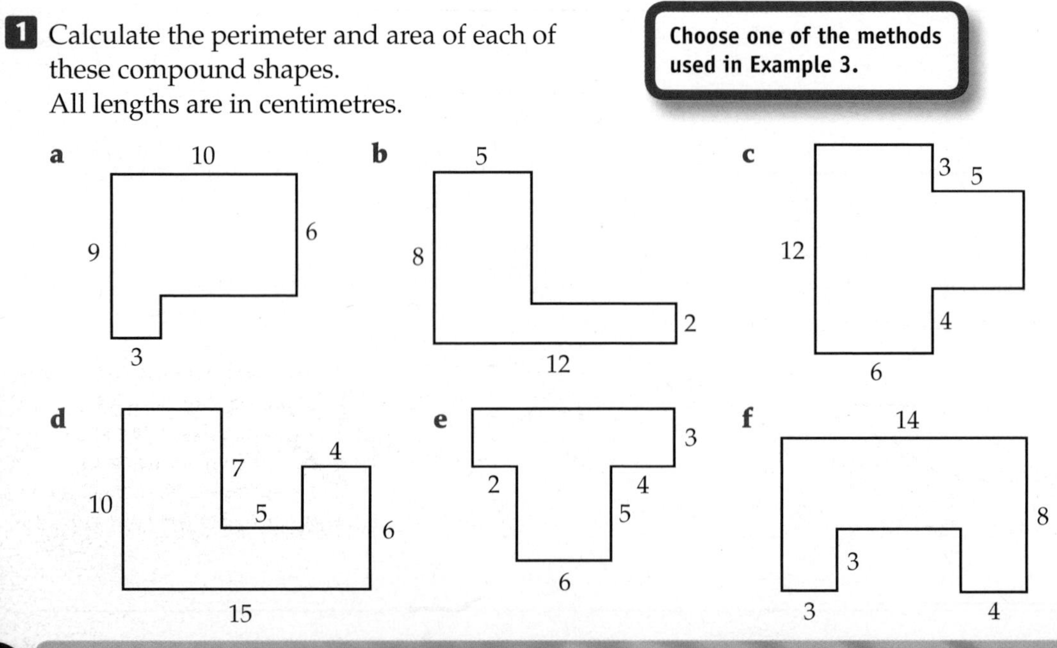

2 Calculate the areas of these shapes. All lengths are in centimetres.

Choose a method similar to that used in Example 4.

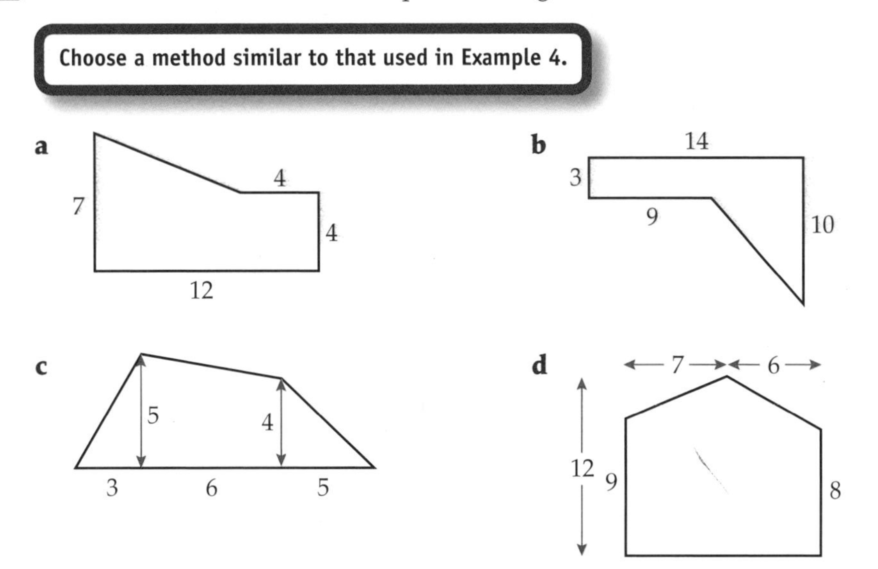

26.3 Volume and surface area of prisms

Keywords
prism, cross-section, cuboid

Why learn this?

To work out how much wrapping paper you need, you could calculate the surface area of the box.

Objectives

Find the volume and surface area of a prism

Skills check

1 What are the areas of these shapes?

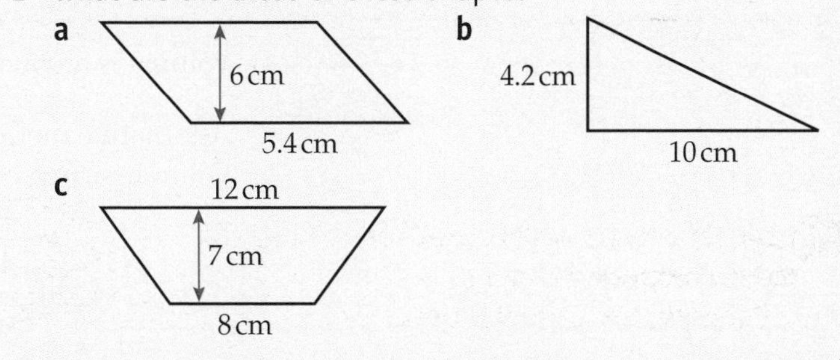

Volume of a prism

A **prism** is a 3-D object whose **cross-section** is the same all through its length.

In these prisms the cross-section is shaded.

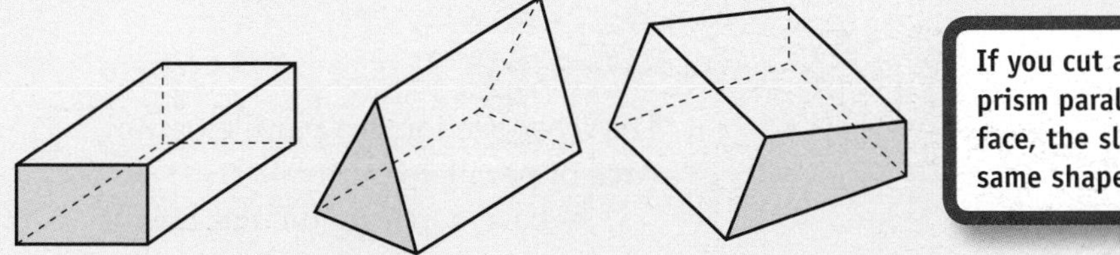

If you cut any 'slice' of a prism parallel to the end-face, the slice will be the same shape as the end-face.

A **cuboid** is a 3-D object. Its cross-section is a rectangle.

To calculate the volume you use the formula

Volume of cuboid = length × width × height

$= l \times w \times h$

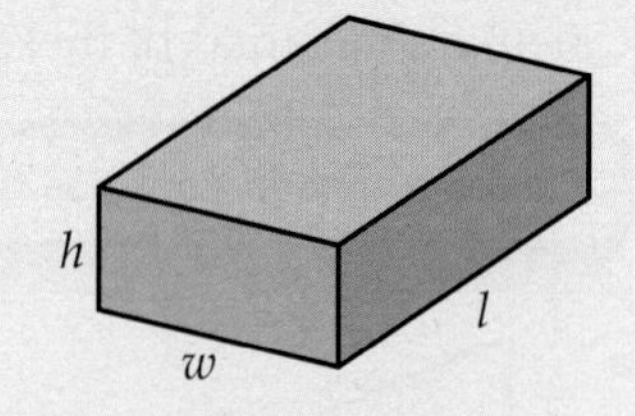

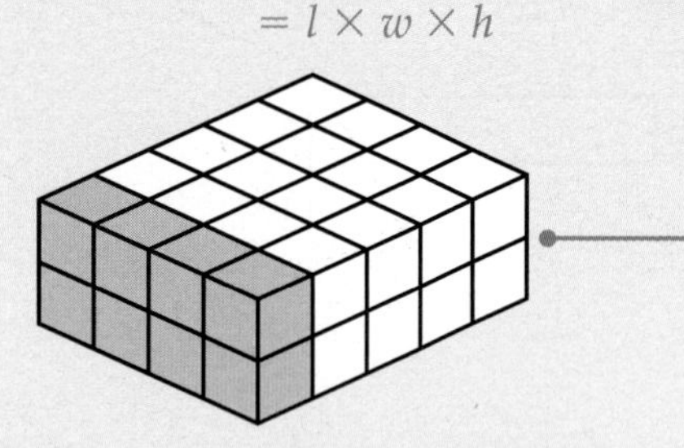

Imagine a cuboid of length 5 cm, width 4 cm and height 2 cm made from 1 cm cubes. The end-face has 4 × 2 cubes. There are 5 'slices' of 4 × 2 cubes so there are 5 × 4 × 2 = 40 cubes in total.

Another way of writing this formula is

Volume of cuboid = area of end face × length

You can use a similar formula to calculate the volume of *any* prism.

Volume of prism = area of cross-section × length

Area of cross-section = area of end-face

This formula is given on the exam paper.

E D C

Example 5

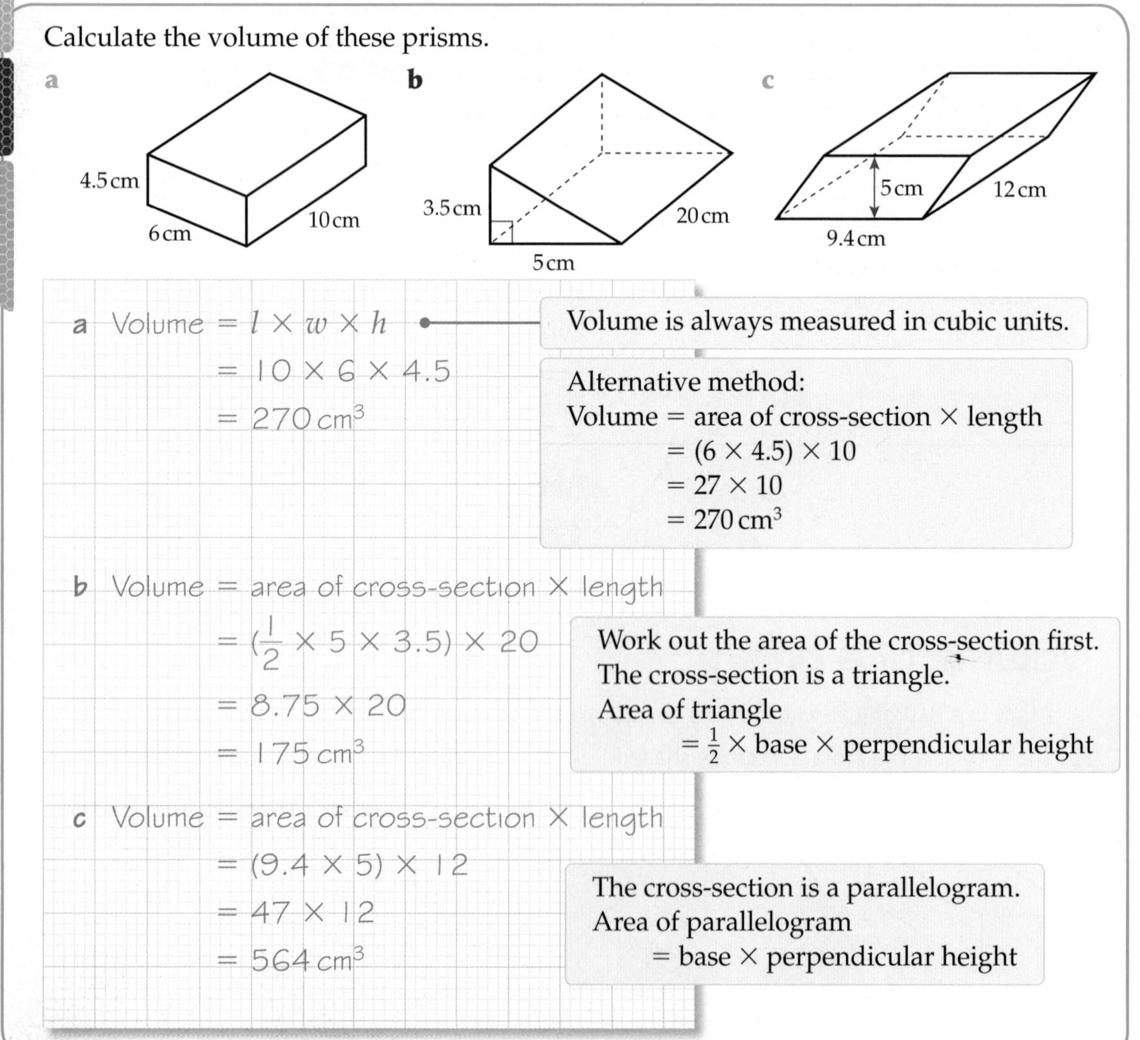

Calculate the volume of these prisms.

a Volume $= l \times w \times h$

$= 10 \times 6 \times 4.5$

$= 270\text{ cm}^3$

Volume is always measured in cubic units.

Alternative method:

Volume = area of cross-section × length

$= (6 \times 4.5) \times 10$

$= 27 \times 10$

$= 270\text{ cm}^3$

b Volume = area of cross-section × length

$= (\frac{1}{2} \times 5 \times 3.5) \times 20$

$= 8.75 \times 20$

$= 175\text{ cm}^3$

Work out the area of the cross-section first. The cross-section is a triangle.

Area of triangle

$= \frac{1}{2} \times$ base × perpendicular height

c Volume = area of cross-section × length

$= (9.4 \times 5) \times 12$

$= 47 \times 12$

$= 564\text{ cm}^3$

The cross-section is a parallelogram.

Area of parallelogram

= base × perpendicular height

Exercise 26D

1 Calculate the volume of these prisms. All lengths are in centimetres.

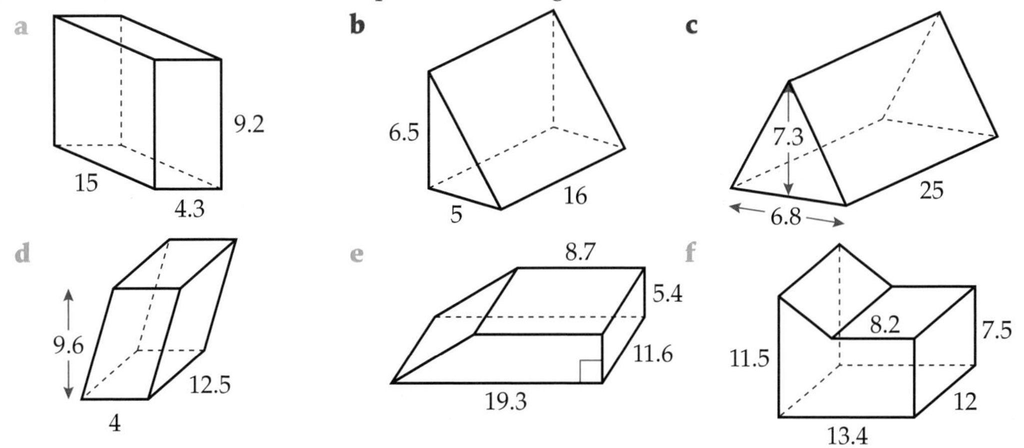

2 Ben has a pond in the shape of a cuboid.
The cuboid is 5.5 m long, 1.8 m wide and 1 m deep.
Ben wants to empty the pond and fill it in with hardcore and pebbles.
Hardcore and pebbles are sold in $1\,m^3$ bags.
A $1\,m^3$ bag of hardcore costs £31.65. A $1\,m^3$ bag of pebbles costs £52.20.
Ben estimates that he will need *two* $1\,m^3$ bags of pebbles for the top surface.
He will fill the rest of the pond with hardcore.
Calculate how much it will cost him to do the work.

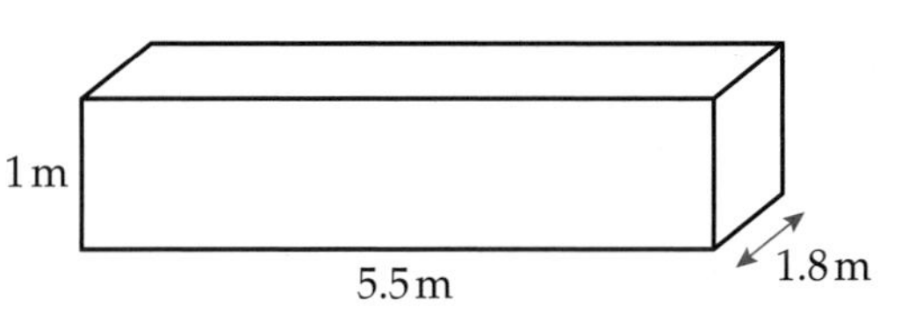

AO3

Surface area of a prism

This is the net of a cuboid.

The surface area of a 3-D object is the sum of the area of all its surfaces.

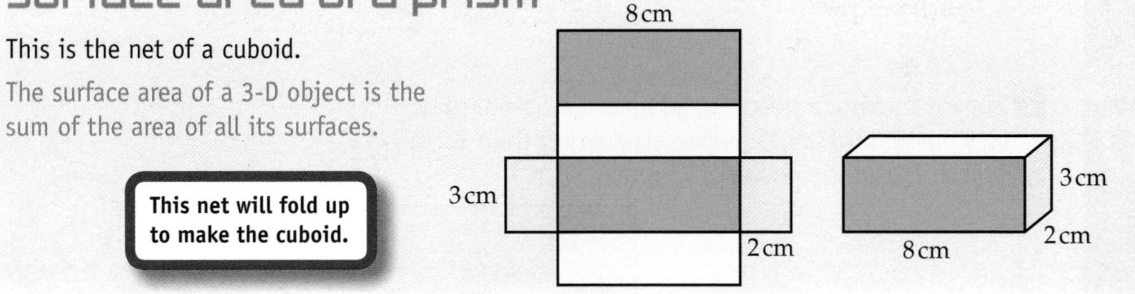

Example 6

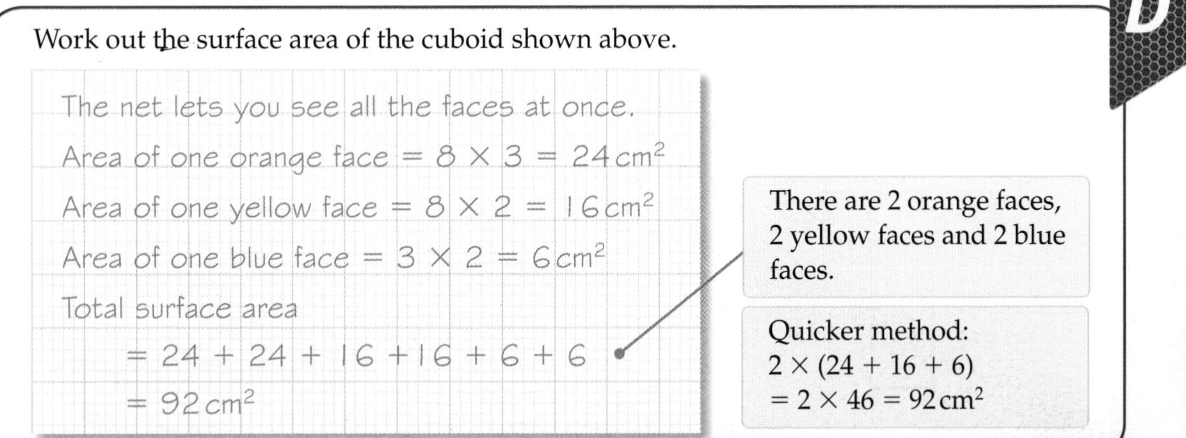

Work out the surface area of the cuboid shown above.

The net lets you see all the faces at once.

Area of one orange face $= 8 \times 3 = 24\,cm^2$

Area of one yellow face $= 8 \times 2 = 16\,cm^2$

Area of one blue face $= 3 \times 2 = 6\,cm^2$

Total surface area

$= 24 + 24 + 16 + 16 + 6 + 6$

$= 92\,cm^2$

There are 2 orange faces, 2 yellow faces and 2 blue faces.

Quicker method:
$2 \times (24 + 16 + 6)$
$= 2 \times 46 = 92\,cm^2$

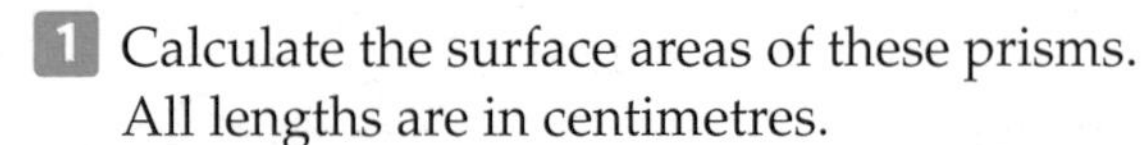

Exercise 26E

D

1 Calculate the surface areas of these prisms.
All lengths are in centimetres.

a

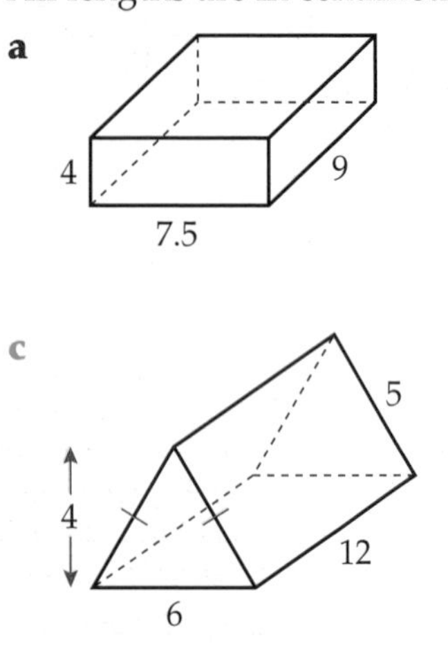

b

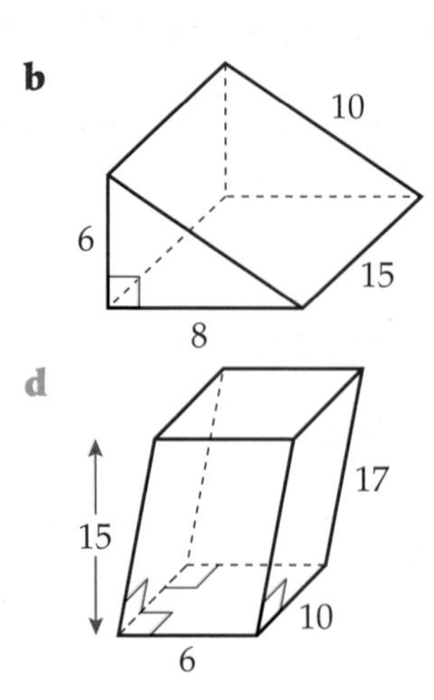

C

c

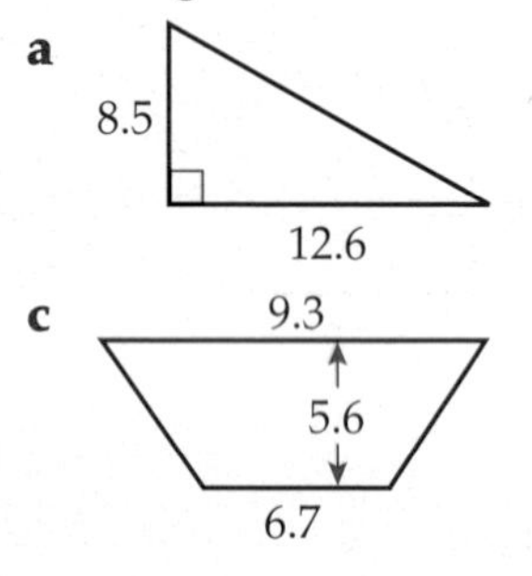

d

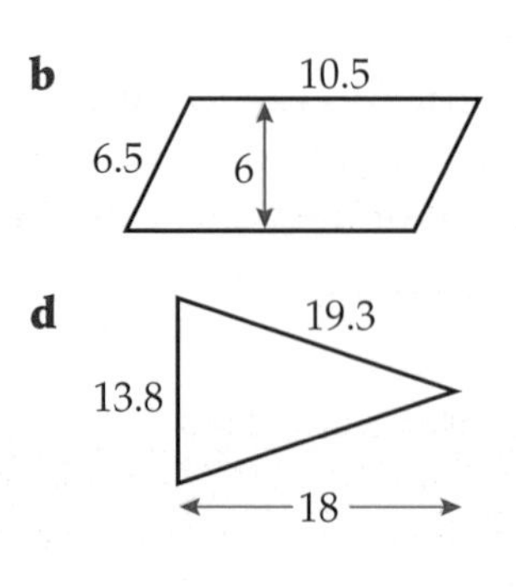

Review exercise

D

1 Calculate the areas of these shapes.
All lengths are in centimetres.

a 8.5, 12.6

b 10.5, 6.5, 6

c 9.3, 5.6, 6.7

d 19.3, 13.8, 18

[2 marks each part]

D

2 Lucy is making a poster to advertise a school play.
Her poster must not have an area larger than $1.2\,m^2$.
The length of the poster is 150 cm.

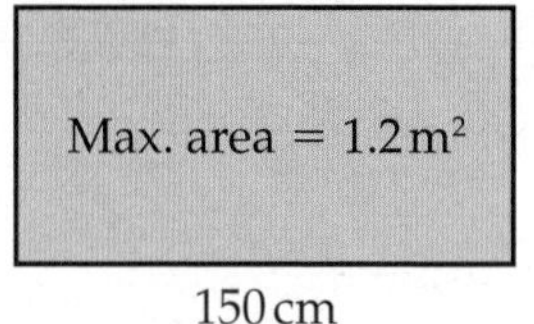

AO2

What is the greatest width the poster can be? **[3 marks]**

D

3 Calculate **i** the area and **ii** the perimeter of these compound shapes.
All lengths are in centimetres.

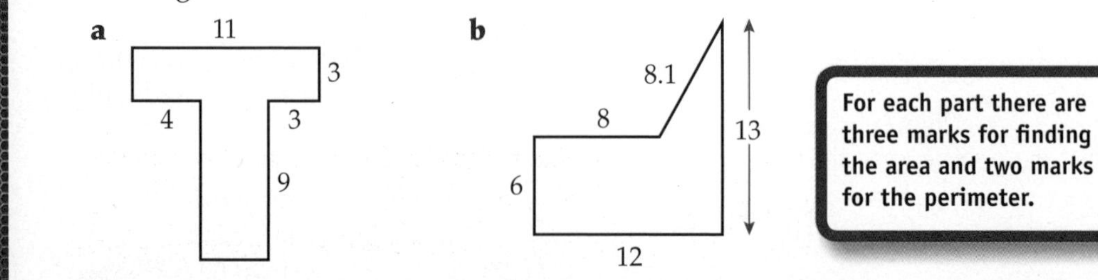

For each part there are three marks for finding the area and two marks for the perimeter.

[5 marks each part]

4 The diagram shows a shaded triangular shape made up from two right-angled triangles, one inside the other.
The larger triangle has a base of 15.6 cm and a height of 12 cm.
The smaller triangle has a base of 6.5 cm and a height of 5 cm.
Calculate the shaded area. **[4 marks]**

12 5 6.5 15.6

D AO2

5 Calculate **i** the volume and **ii** the surface area of these 3-D objects.
All lengths are in centimetres.

a 16 12 10

b 8.6 5 10 7

For each part there are two marks for finding the volume and three marks for the surface area.

[5 marks each part]

D C

6 A garden shed has dimensions as shown in the diagram.
The total height of the shed is 4 metres.

FUNCTIONAL

a Calculate the area of the cross-section of the shed. **[3 marks]**

b Calculate the volume of the shed. **[2 marks]**

c Roofing felt is sold in rolls.
Each roll is 5 m long and 1 m wide and costs £7.33.
What is the cheapest way to cover the roof of the shed?
How much will it cost? **[4 marks]**

2.4 m 4 m 2.8 m 6 m 4.2 m

D C AO2

7 A tent is in the shape of a triangular prism.
The two end faces are isosceles triangles of base 2.8 m and sides 3.8 m.
The two sides and the base of the tent are rectangles.
The length of the tent is 5 m and its perpendicular height is 3.5 m.
Tent material costs £5.60 per square metre.
You can only buy the material in whole numbers of square metres.
How much would it cost to make this tent? **[6 marks]**

FUNCTIONAL

3.5 m 3.8 m 5 m 2.8 m

C AO2

Chapter summary

In this chapter you have learned how to

- find the perimeter and area of rectangles, parallelograms, triangles and trapezia E D
- find the perimeter and area of compound shapes E D
- find the volume and surface area of a prism E D C

27

3-D objects

This chapter is about 3-D objects and the ways in which they can be drawn.

3-D objects that you need to know are cube, cuboid, cylinder, sphere, pyramid, tetrahedron and prism.

Objectives

This chapter will show you how to

- **use 2-D representations of 3-D shapes and analyse 3-D shapes through 2-D projections and cross-sections, including plans and elevations** E D

Before you start this chapter

1 Here are some nets of 3-D objects. Sketch each object.

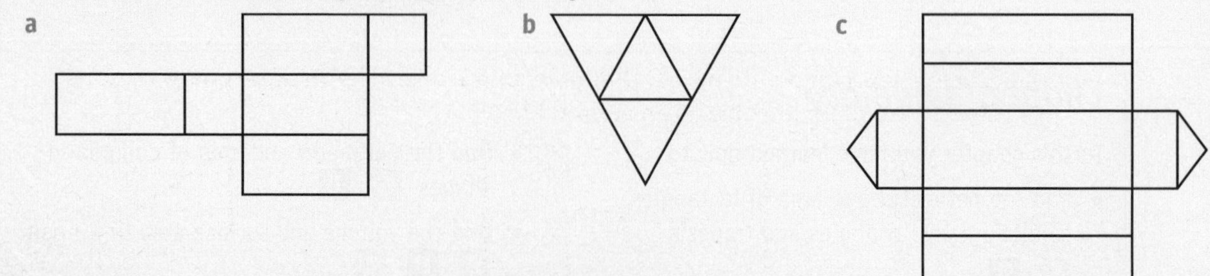

27.1 Drawing 3-D objects

Keywords
cross-section, plan, front elevation, side elevation, plane of symmetry

Why learn this?

Architects produce drawings to show what a new building will look like when it's finished.

Objectives

E Make a drawing of a 3-D object on isometric paper

D Draw plans and elevations of 3-D objects

D Identify planes of symmetry of 3-D objects

Skills check

1 What would each of these 3-D objects look like when viewed from above?

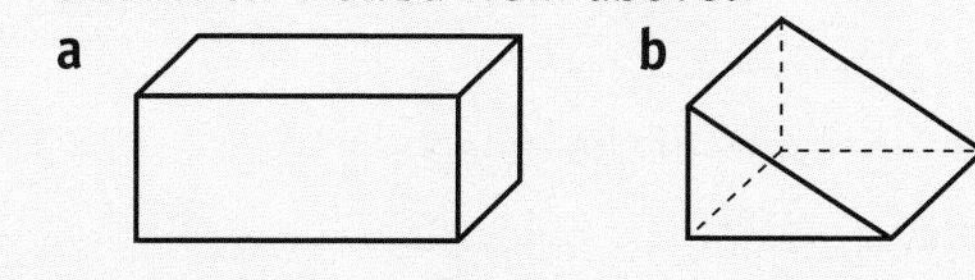

Imagine you are hovering 20 metres in the air, looking down on the objects. What would you see?

Using isometric paper

You can draw 3-D objects on isometric paper.
Draw along the printed lines of the paper.
Vertical lines on the paper represent the vertical lines of the object.
The lines at an angle on the paper represent the horizontal lines on the object.

Isometric paper is a grid made up of triangles. When you use it 3-D objects are viewed at an angle.

Example 1

This is the **cross-section** of a 3-D object.
The object is 3 cm wide.
Draw the 3-D object on isometric paper.

The cross-section is the shape that runs through the whole of the solid.

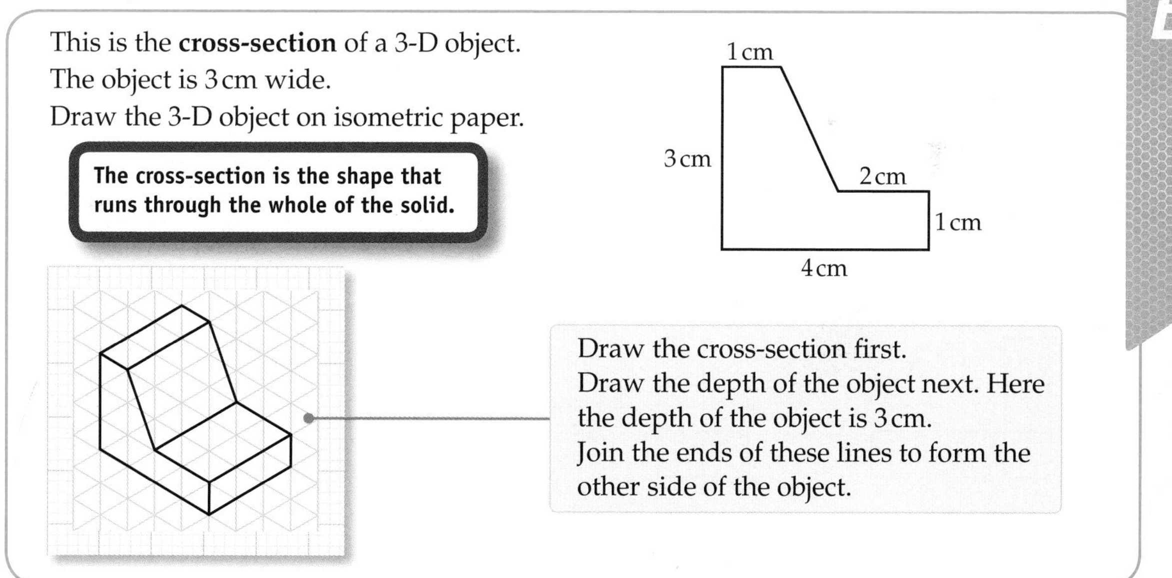

Draw the cross-section first.
Draw the depth of the object next. Here the depth of the object is 3 cm.
Join the ends of these lines to form the other side of the object.

Exercise 27A

1 On isometric paper draw 3-D diagrams of the objects with these cross-sections.
Assume that all the objects are 4 cm wide.

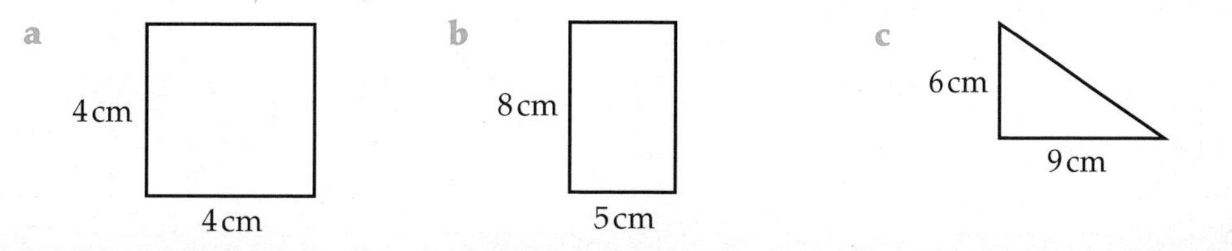

E

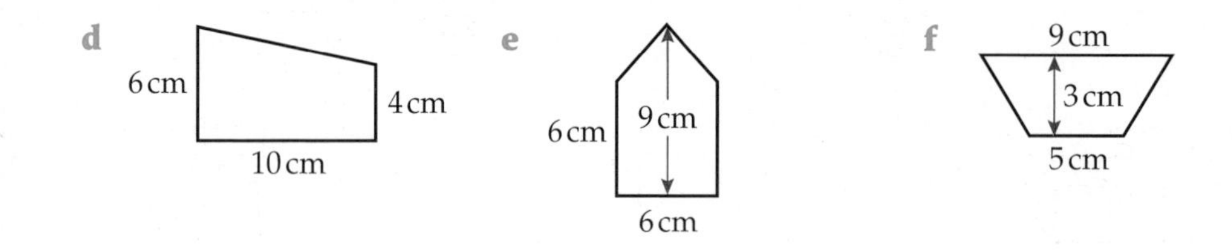

Drawing plans and elevations of a 3-D object

Look at a 3-D object from above, from the front and from the side and think about exactly what you can see and what you can't see.

The **plan** is the view from above the object.

The **front elevation** is the view from the front of the object.

The **side elevation** is the view from the side of the object.

D

Example 2

Draw the plan, the front elevation and the side elevation (from the right-hand side) of this block of seven cubes.

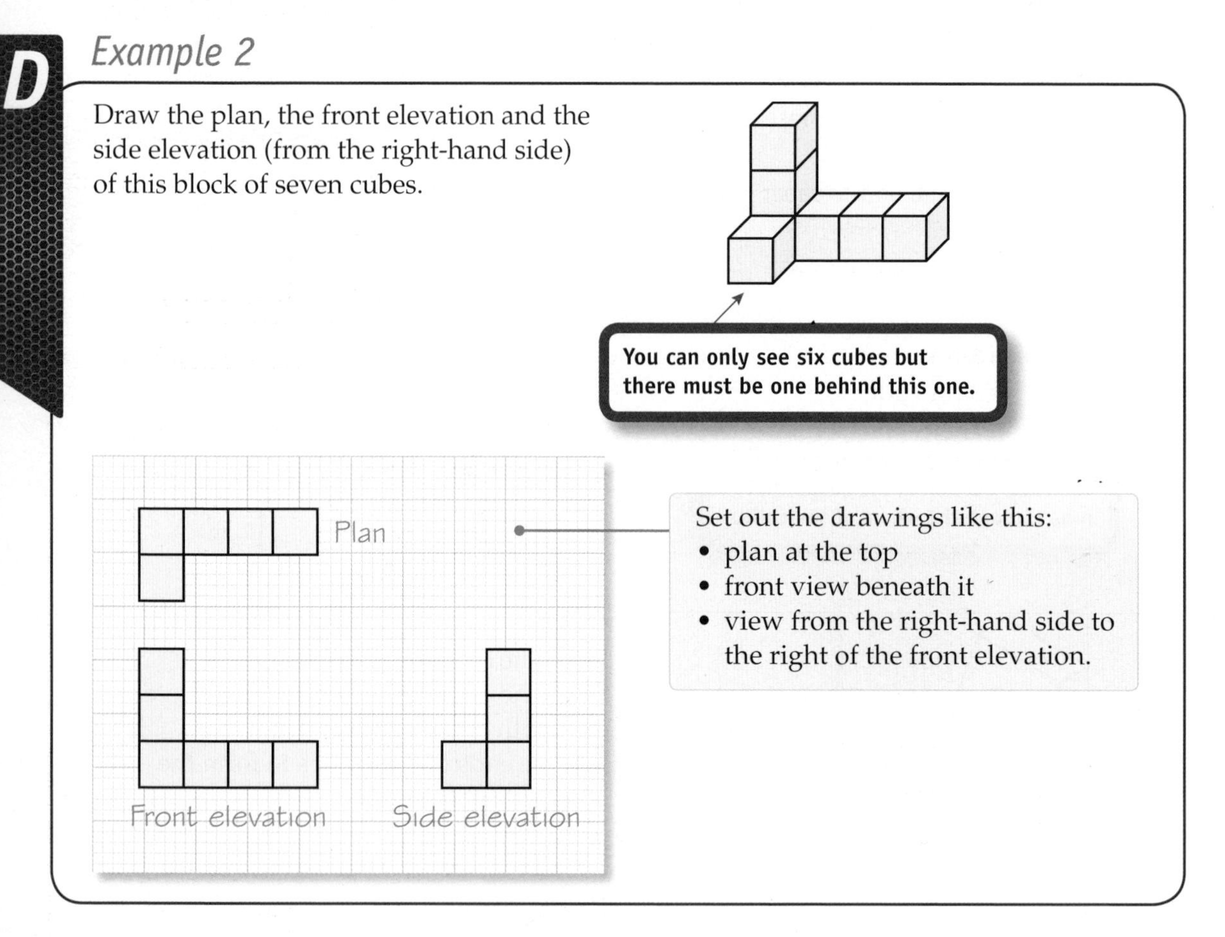

Set out the drawings like this:
- plan at the top
- front view beneath it
- view from the right-hand side to the right of the front elevation.

D

Exercise 27B

For each block of cubes, draw
a a plan
b a front elevation
c a side elevation
(from the right-hand side).

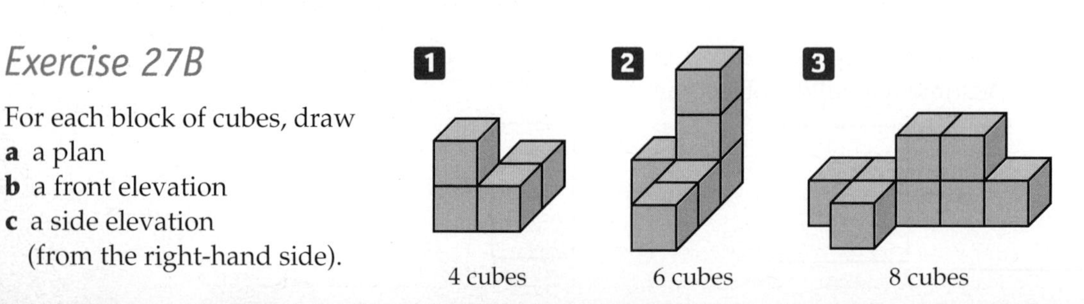

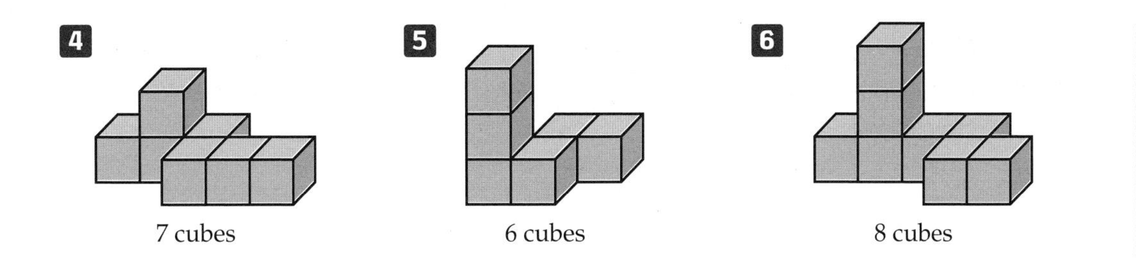

Using plans and elevations

You could be asked to find the volume or the surface area of a block of cubes.

Usually the cubes will have a side of 1 cm so each cube will have a volume of $1\,\text{cm}^3$.
The volume of the object can then be found by counting the cubes.
For seven cubes (Example 2), volume $= 7 \times 1 = 7\,\text{cm}^3$.

Don't forget to look at the object from underneath and from the back.

The surface area is the sum of all the faces you can see if you look at the object from all sides.
The area of each face of a cube of side 1 cm is $1\,\text{cm}^2$.
Surface area of the object = number of faces you can see $\times 1\,\text{cm}^2$.

Example 3

D

Calculate the surface area of this block of seven cubes.
All the cubes have side length 1 cm.

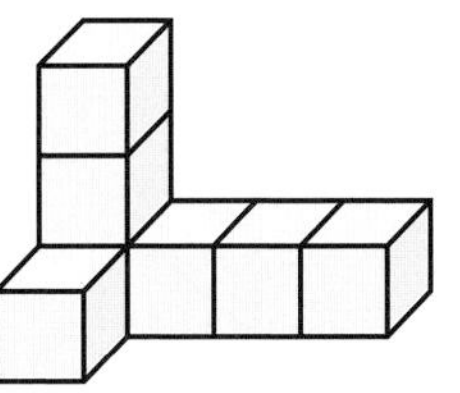

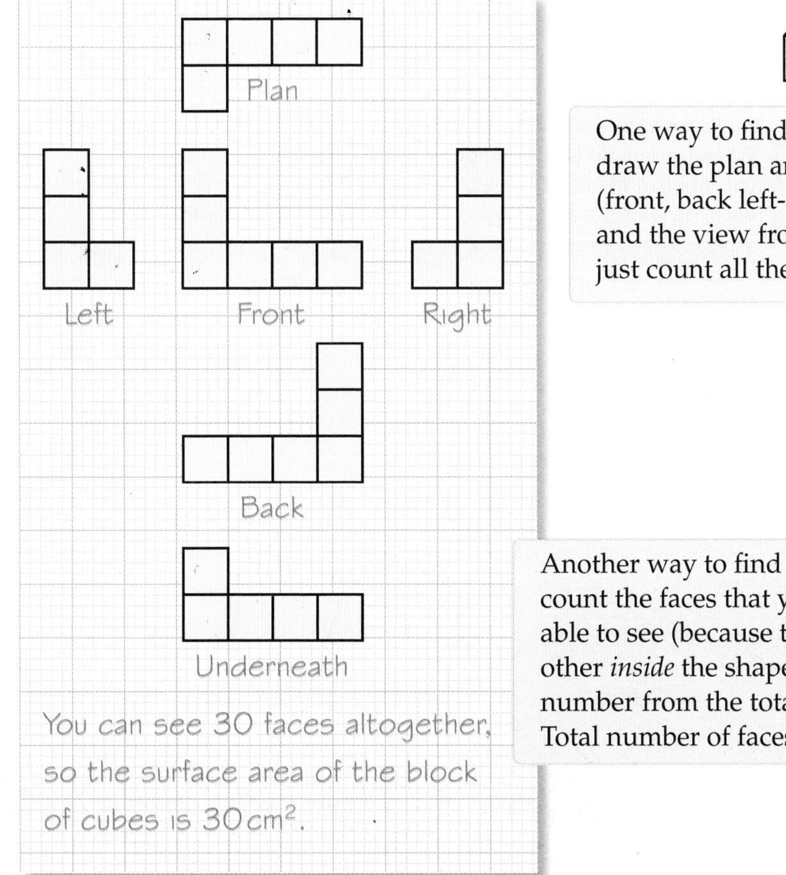

One way to find the surface area is to draw the plan and all the elevations (front, back left- and right-hand sides) and the view from underneath. Then just count all the squares.

Another way to find the surface area is to count the faces that you would never be able to see (because they are touching each other *inside* the shape), then subtract this number from the total number of faces.
Total number of faces = $6 \times$ number of cubes.

Exercise 27C

1 Work out the volume and surface area of each block of cubes in Q1–6 in Exercise 27B. All the cubes have side length 1 cm.

Symmetry in 3-D

Symmetry can exist in 3-D objects as well as in 2-D shapes.

If symmetry exists in 3-D objects, you don't call it a line of symmetry, it is a **plane of symmetry.**

A plane of symmetry divides a 3-D object into two equal halves where one half is the mirror image of the other.

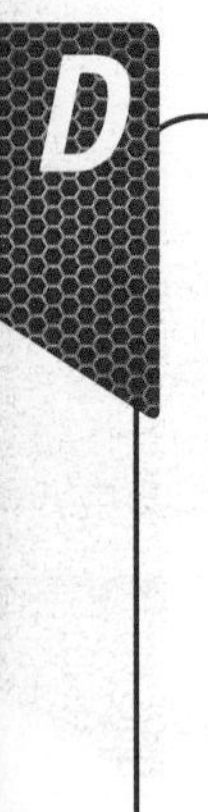

Example 4

Copy this cuboid.

Show all its planes of symmetry.

Draw a separate diagram for each plane of symmetry.

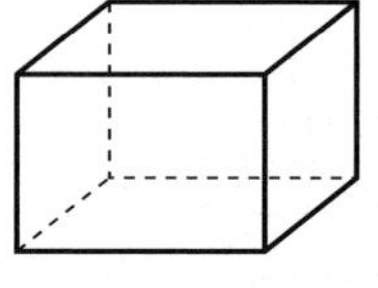

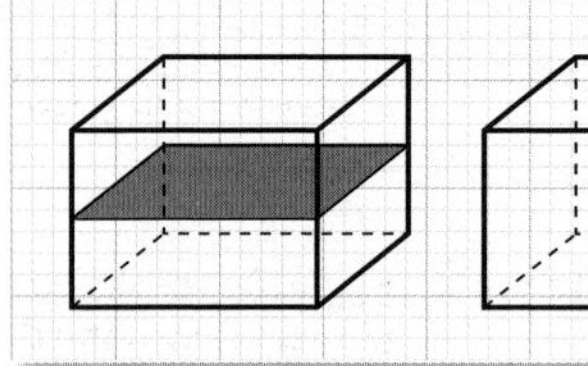

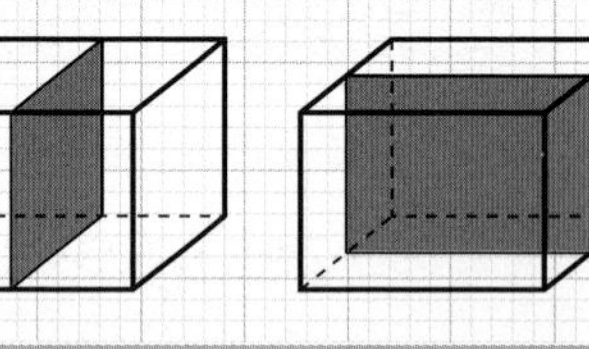

In each diagram the plane of symmetry has been shaded in red.

When the cuboid is cut along the planes marked in red, it will be cut into two identical halves. This is the test you should apply when deciding whether a 3-D object has any planes of symmetry.

Exercise 27D

1 Copy these 3-D objects.
For each object show all its planes of symmetry.
Draw a separate diagram for each plane of symmetry.

a

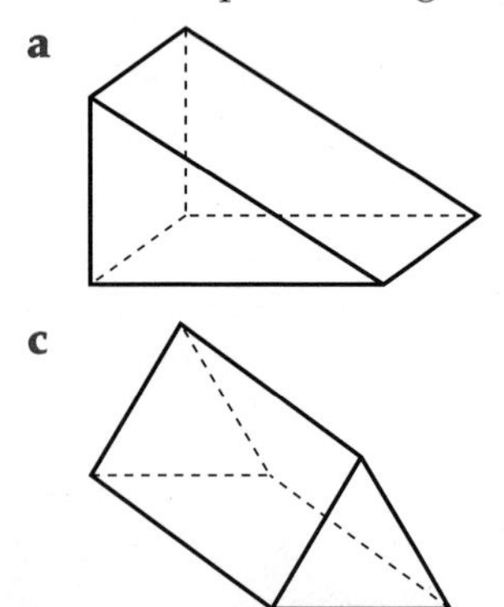

b

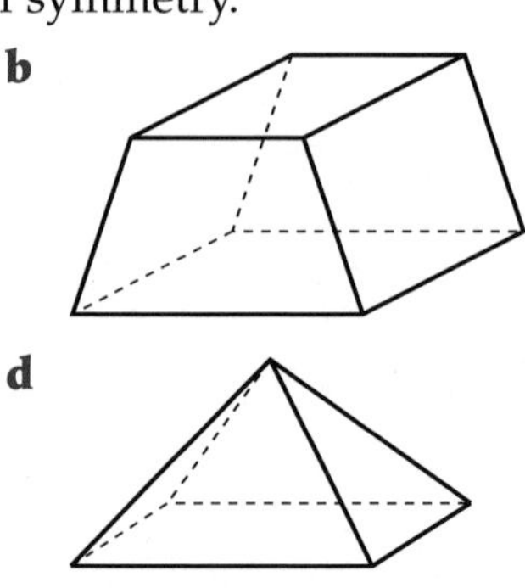

c

d

Review exercise

1 Amira is designing a new building.
She begins by making a scale drawing.
The next step is to make the scale model.
On isometric paper draw a 3-D diagram of the scale model with this cross-section.
Assume the scale model is 4 cm wide. **[3 marks]**

E

2 Copy this 3-D object.
Show all its planes of symmetry.
Draw a separate diagram for each plane of symmetry. **[2 marks]**

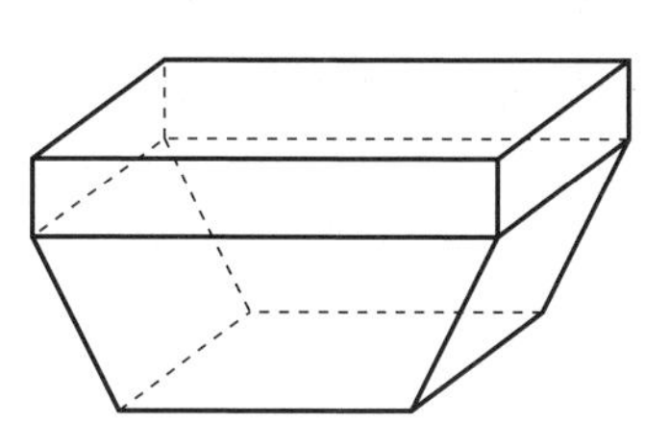

D

3 For each block of cubes, draw **i** a plan view, **ii** a front elevation, and **iii** a side elevation (from the right-hand side).

a

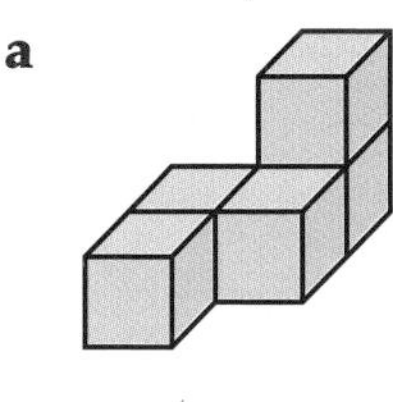

b

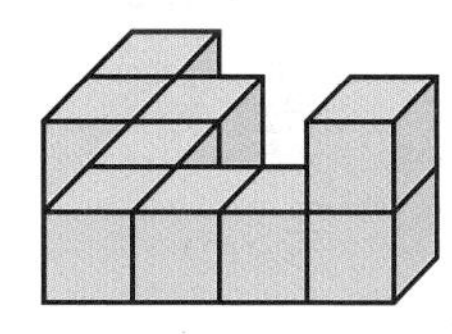

[3 marks each part]

4 Calculate the volume and surface area of each block of cubes in Q3.
All the cubes have side 1 cm. **[3 marks each part]**

Chapter summary

In this chapter you have learned how to

- make a drawing of a 3-D object on isometric paper E
- draw plans and elevations of 3-D objects D
- identify planes of symmetry of 3-D objects D

28 Reflection, translation and rotation

BBC Video

This chapter is about transforming shapes: reflections, translations and rotations.

Architects use reflective materials in buildings for decorative effect. This glass building reflects the other buildings around it and the sky, which is constantly changing.

Objectives

This chapter will show you how to

- understand that reflections are specified by a mirror line, at first using a line parallel to an axis, then a mirror line such as $y = x$ or $y = -x$ E D C
- understand that rotations are specified by a centre, an angle and a direction; rotate a shape about the origin, or any other point; measure the angle of rotation using right angles, simple fractions of a turn or degrees D C
- understand that translations are specified by a distance and a direction (or a vector) D C
- understand and use vector notation for translations C
- transform triangles and other 2-D shapes by translation, rotation and reflection and combinations of these transformations, recognising that these transformations preserve length and angle E D C

Before you start this chapter

1 Copy these letters. Draw in any lines of symmetry.

H L M O P E D Q

2 Match the shapes into congruent pairs.
Which shapes are left over?

HELP Chapter 23

A B C D E F G H

28.1 Reflection on a coordinate grid

Keywords
reflection, object, image, congruent

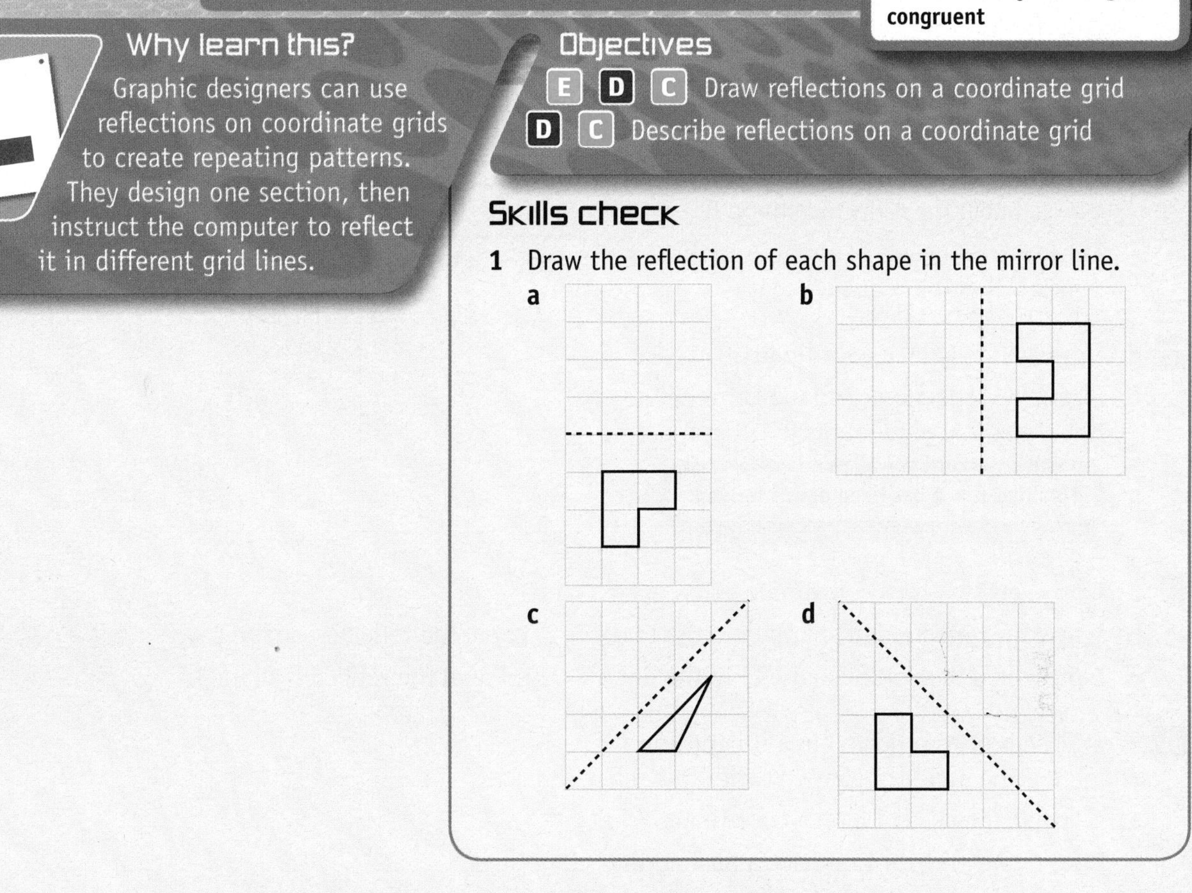

Why learn this?

Graphic designers can use reflections on coordinate grids to create repeating patterns. They design one section, then instruct the computer to reflect it in different grid lines.

Objectives

E D C Draw reflections on a coordinate grid

D C Describe reflections on a coordinate grid

Skills check

1 Draw the reflection of each shape in the mirror line.

a **b** **c** **d**

Reflection on a coordinate grid

To describe a **reflection** on a grid, you need to give the equation of the mirror line.

In a reflection, the **object** and the **image** are the same perpendicular distance from the mirror line, on opposite sides.

When you reflect a shape in a mirror line, the object and the image are **congruent**.

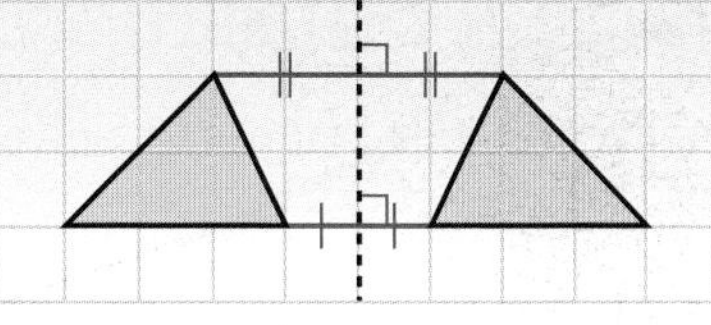

Example 1

Draw the reflection of the shape in the line $x = 1$.

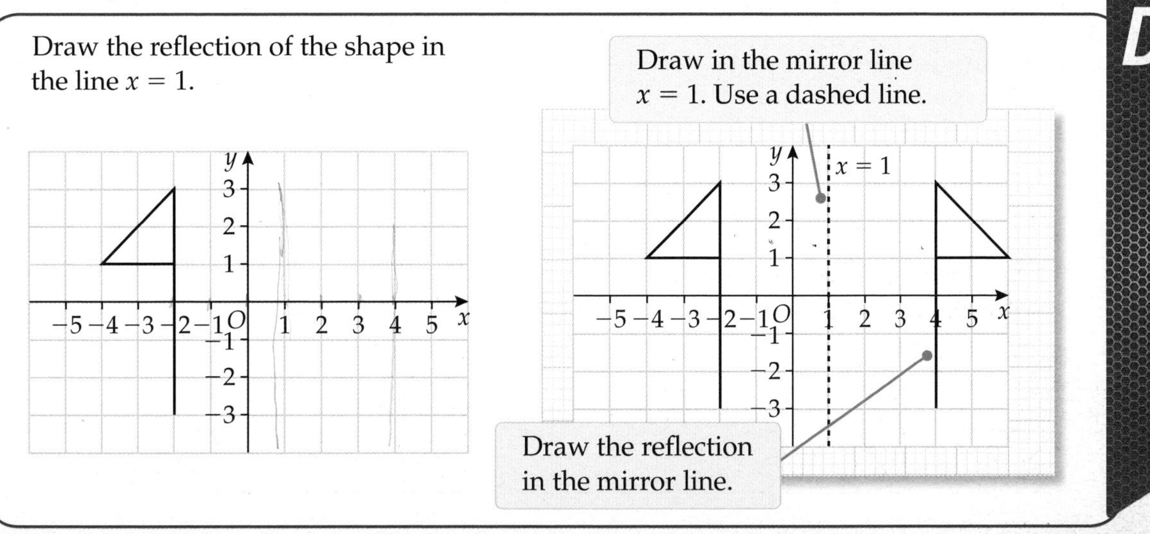

D

Exercise 28A

For Q1–4, copy the shape and draw the reflections according to the instructions in the question.

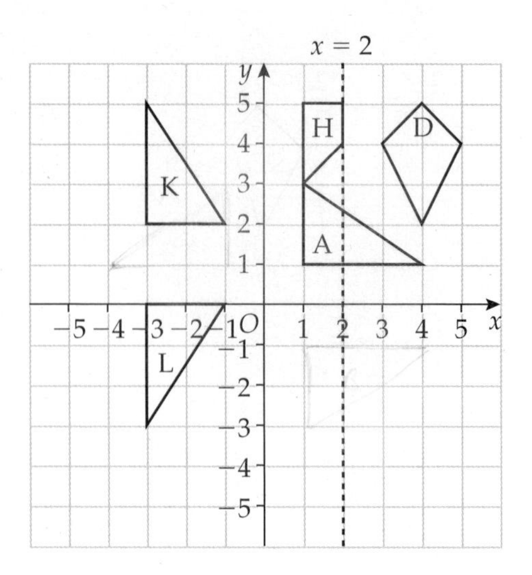

E

1 **a** Copy shape A on a coordinate grid.

b Draw the reflection of shape A in the x-axis. Label the reflected shape B.

c Draw the reflection of shape A in the y-axis. Label the reflected shape C.

D

2 **a** Copy shape D on a coordinate grid.

b Draw the reflection of shape D in the line $x = 2$. Label the reflected shape E.

> **This line $x = 2$ has been drawn for you.**

3 **a** Copy shape H on a coordinate grid.

b Draw the reflection of shape H in the line $y = 2$. Label the reflected shape I.

c Draw the reflection of shape H in the line $x = -1$. Label the reflected shape J.

4 Shape K has been reflected in a mirror line.
Shape L is the image of shape K after this reflection.

a Copy shapes K and L on a coordinate grid.

b Draw in the mirror line with a dashed line.

c Write down the equation of the mirror line.

Describing a reflection

To describe a reflection fully you need to give the equation of the mirror line.

D

Example 2

Shape R is reflected in a mirror line.

Shape S is the image of shape R after this reflection.

Describe this transformation.

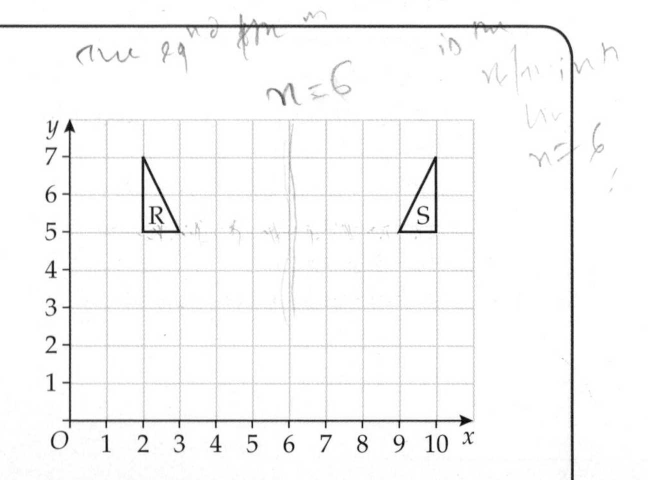

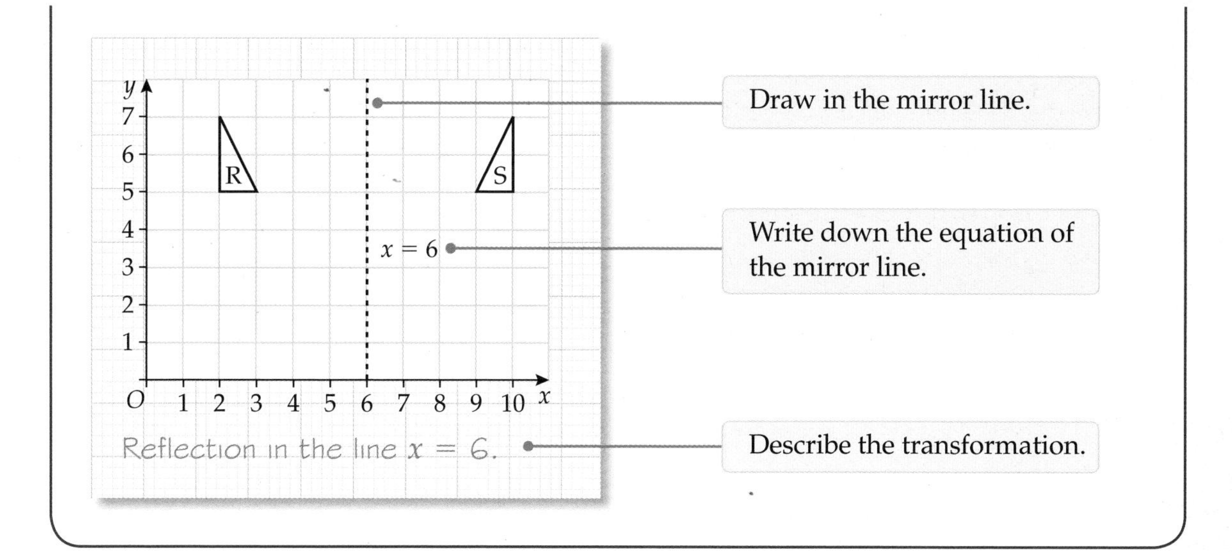

Exercise 28B

D

1 **a** Shape M is reflected in a mirror line.
Shape N is the image of shape M after this reflection.
Describe this transformation.

b Shape N is reflected in a mirror line.
Shape P is the image of shape N after this reflection.
Describe this transformation.

c Shape P is reflected in a mirror line.
Shape Q is the image of shape P after this reflection.
Describe this transformation.

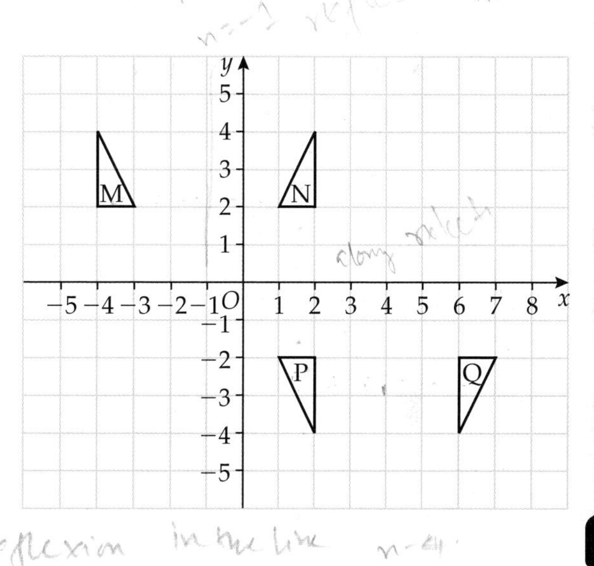

AO2

C

2 **a** Copy the diagram on a coordinate grid.

b Draw in the line $y = x$ with a dashed line.

c Draw the reflection of shape R in the line $y = x$.
Label the reflected shape S.

d Draw the reflection of shape T in the line $y = x$.
Label the reflected shape U.

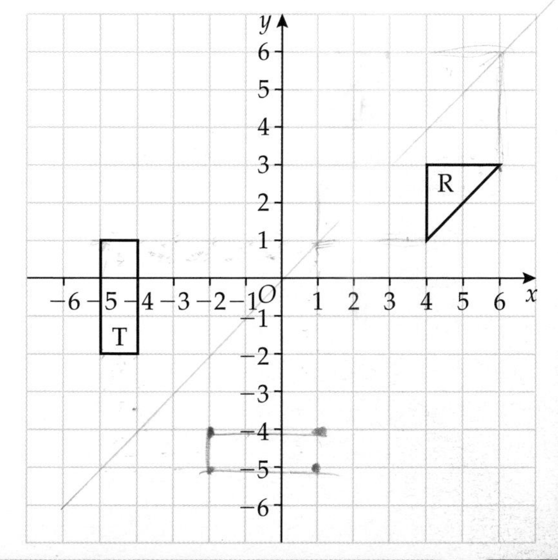

C

3 Describe the transformation that takes

a shape V to shape W

b shape X to shape Y

c shape A to shape B.

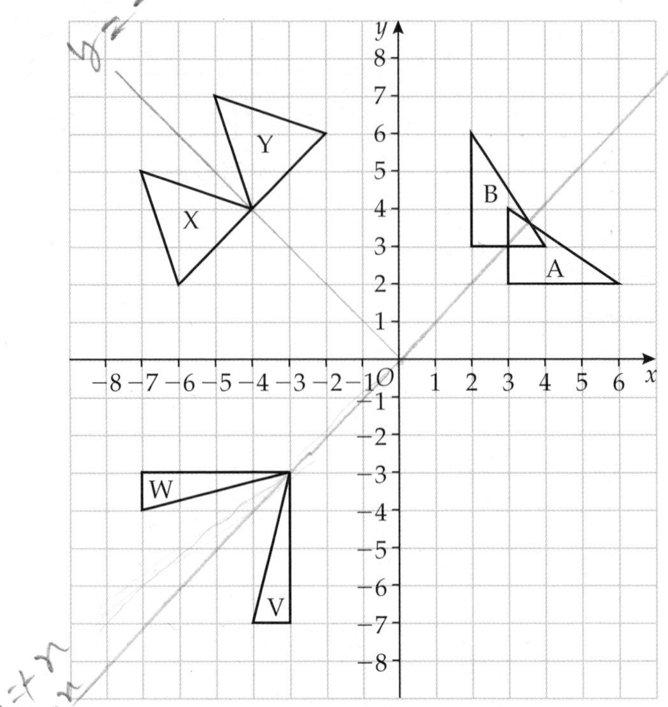

4 This pattern was made by reflecting a shape in a mirror line.
Draw the original shape on a coordinate grid.
Describe the transformation.

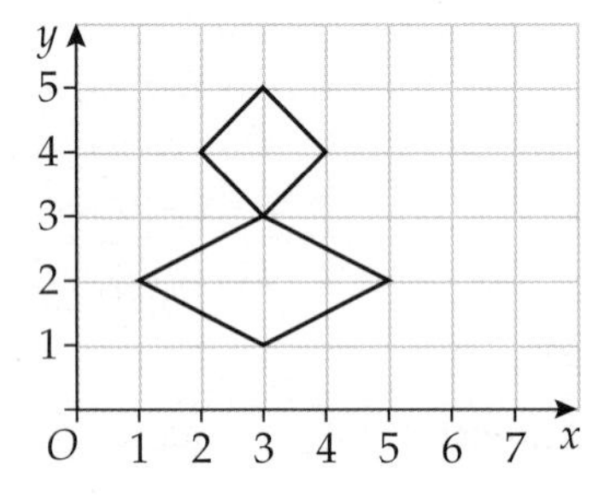

AO2

C

5 Jade starts with this shape.

She transforms the shape to make this pattern.
Describe the transformations she uses.

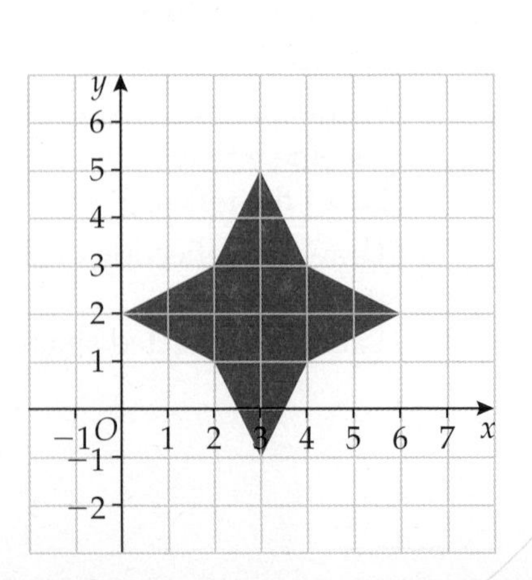

AO3

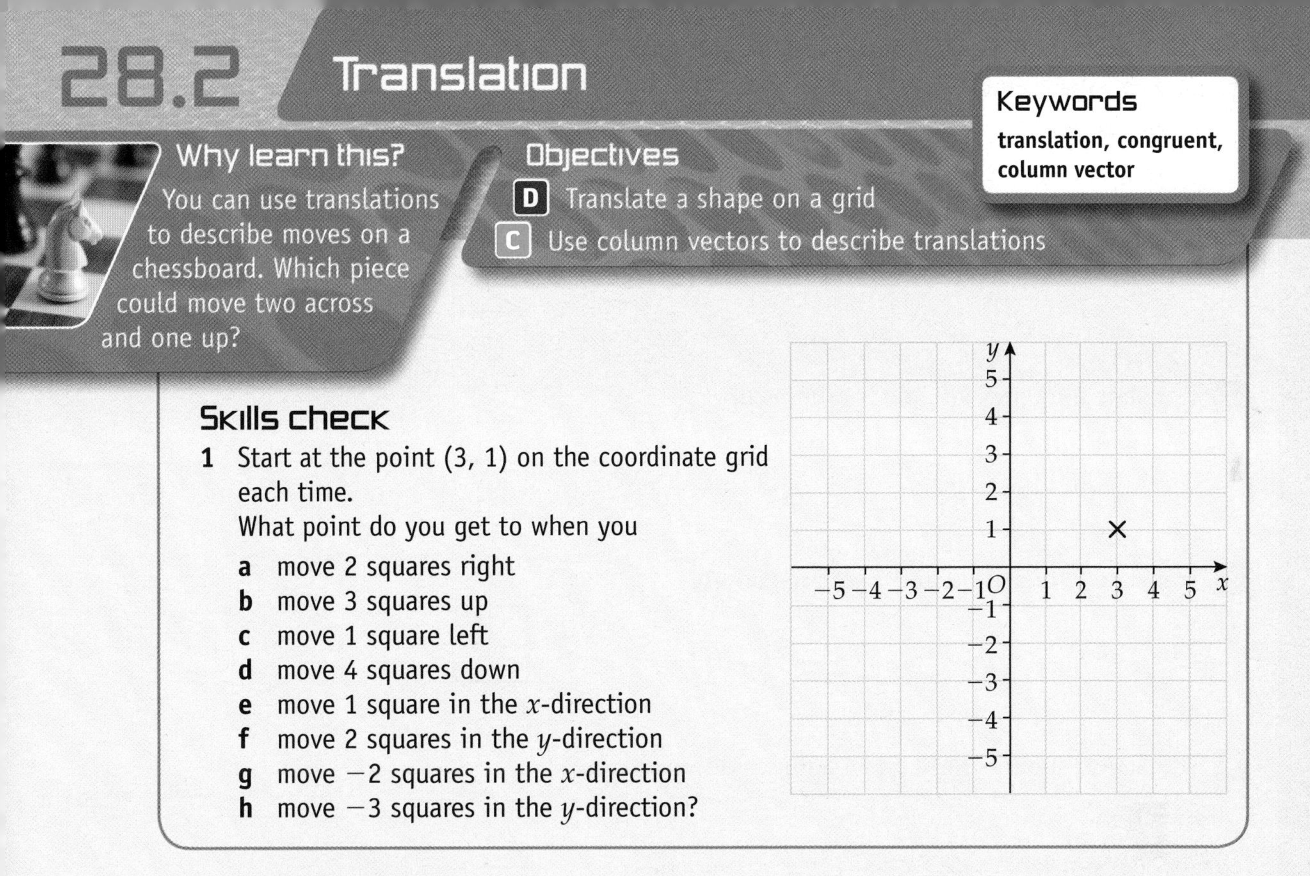

28.2 Translation

Keywords
translation, congruent, column vector

Why learn this?
You can use translations to describe moves on a chessboard. Which piece could move two across and one up?

Objectives

D Translate a shape on a grid
C Use column vectors to describe translations

Skills check

1 Start at the point (3, 1) on the coordinate grid each time.
What point do you get to when you

a move 2 squares right
b move 3 squares up
c move 1 square left
d move 4 squares down
e move 1 square in the x-direction
f move 2 squares in the y-direction
g move -2 squares in the x-direction
h move -3 squares in the y-direction?

Translation

A **translation** slides a shape across a grid. It can slide right or left, and up or down.

In a translation, all points on the shape translate the same number of squares in the same direction.

An object and its image after a translation are **congruent**.

Example 3

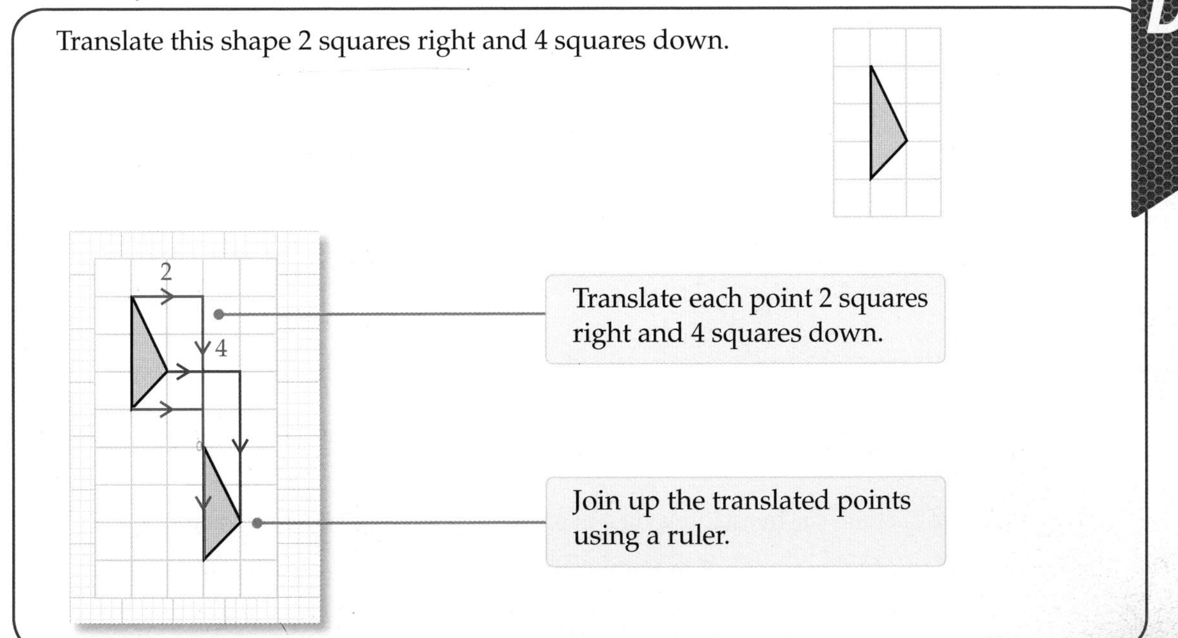

Translate this shape 2 squares right and 4 squares down.

Translate each point 2 squares right and 4 squares down.

Join up the translated points using a ruler.

Exercise 28C

D

For Q1–4, first copy the shape on squared paper.
Then do the translation.

1 Translate 3 squares right and 1 square up.

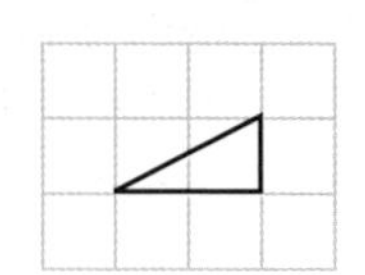

2 Translate 2 squares left and 3 squares up.

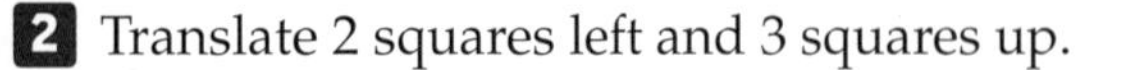

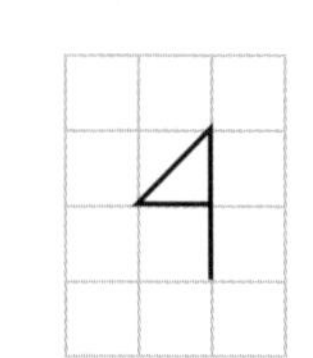

3 Translate 1 square left and 2 squares down.

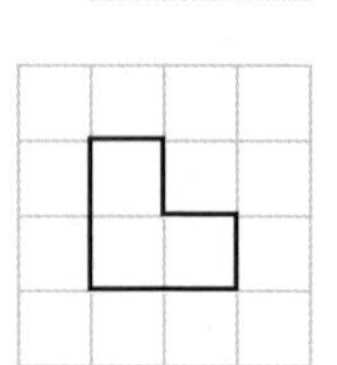

4 Translate 2 squares right and 4 squares down.

D

5 Look back at the translations you have drawn in Q1–4.

a How does the image compare to the object in each one?

b Use a mathematical word to complete this sentence.

In a translation, the object and the image are ____________ .

6 Describe the translation that takes

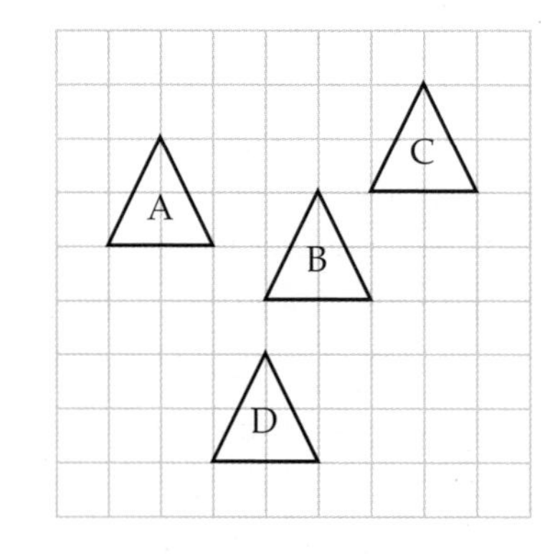

a shape A to shape B

> **Write the number of squares right or left, and the number of squares up or down**

b shape A to shape C

c shape A to shape D

d shape D to shape A.

AO2

e What do you notice about your answers to parts **c** and **d**?

Translations and column vectors

You can give instructions for translations on a coordinate grid using a **column vector.**

The column vector $\begin{pmatrix} 3 \\ 2 \end{pmatrix}$ means move 3 in the x-direction and then move 2 in the y-direction.

Example 4

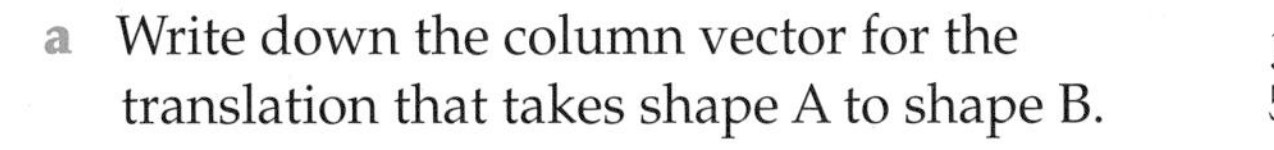

a Write down the column vector for the translation that takes shape A to shape B.

b Write down the column vector for the translation that takes shape B to shape C.

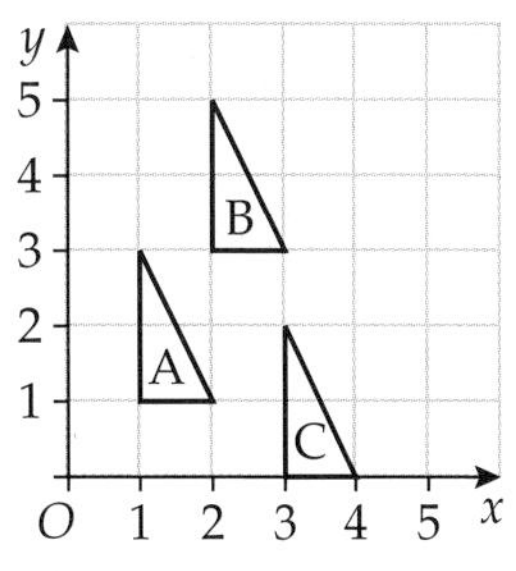

a $\begin{pmatrix} 1 \\ 2 \end{pmatrix}$

Write down the move in the x-direction.

Write down the move in the y-direction.

b $\begin{pmatrix} 1 \\ -3 \end{pmatrix}$

The shape moves 1 in the x-direction and -3 in the y-direction.

C

Exercise 28D

1 Some shapes are translated on this coordinate grid.
Write down the column vector for each translation.

a shape A to shape B

b shape B to shape C

c shape C to shape D

d shape D to shape E

e shape E to shape F

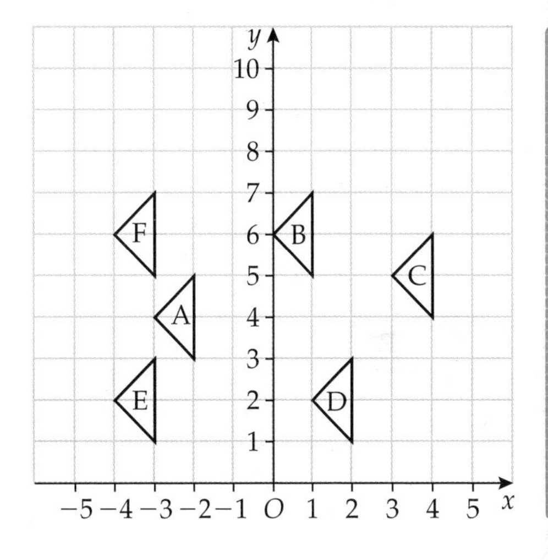

C

2 Look at the coordinate grid in Q1 again.
Write down the column vectors for these translations.

a Shape F to shape C

b Shape C to shape F

c What do you notice about your answers in parts **a** and **b**?

d The translation that takes shape B to shape D is $\begin{pmatrix} 1 \\ -4 \end{pmatrix}$.

Write down the column vector for the translation that takes shape D to shape B.

C

3 Look at the coordinate grid in Q1 again.
Write down the column vector for these translations.

a shape E to shape A

b shape A to shape B

c shape E to shape B

d What do you notice about your answers in parts **a**, **b** and **c**?

AO2

C

4 On a grid, $\begin{pmatrix} 2 \\ -1 \end{pmatrix}$ translates shape A to shape B and $\begin{pmatrix} 3 \\ 2 \end{pmatrix}$ translates shape B to shape C.
Write down the column vector that translates shape A directly to shape C.

AO2

5 On a grid, $\begin{pmatrix} -1 \\ 4 \end{pmatrix}$ translates shape D to shape E and $\begin{pmatrix} 2 \\ -5 \end{pmatrix}$ translates shape E to shape F.
Write down the column vector that translates shape D directly to shape F.

C

6 Sadie starts with this shape on a grid.

She translates it to make this tessellation.

A tessellation is a pattern of repeated shapes, with no gaps in between.

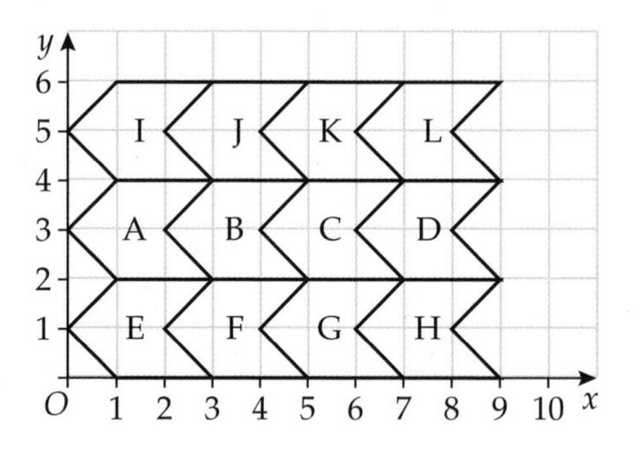

Use column vectors to write instructions for each translation.

Each shape in the tessellation has been labelled to help you.

AO3

28.3 Rotation

Keywords
rotation, centre of rotation

Why learn this?

Engineers need to understand rotation and the forces acting on rotating objects to design safe theme park rides.

Objectives

D **C** Draw the position of a shape after rotation about a centre

D **C** Describe a rotation fully giving the size and direction of turn and the centre of rotation

Skills check

1 How many degrees are there in
   a a full turn **b** a half turn **c** a quarter turn?

2 For each turn, write down the number of degrees and whether the direction is clockwise or anticlockwise.

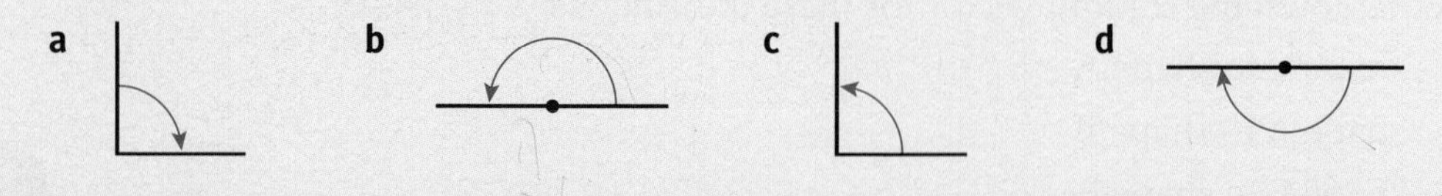

Rotation

A **rotation** turns a shape around a fixed point, called the **centre of rotation**.

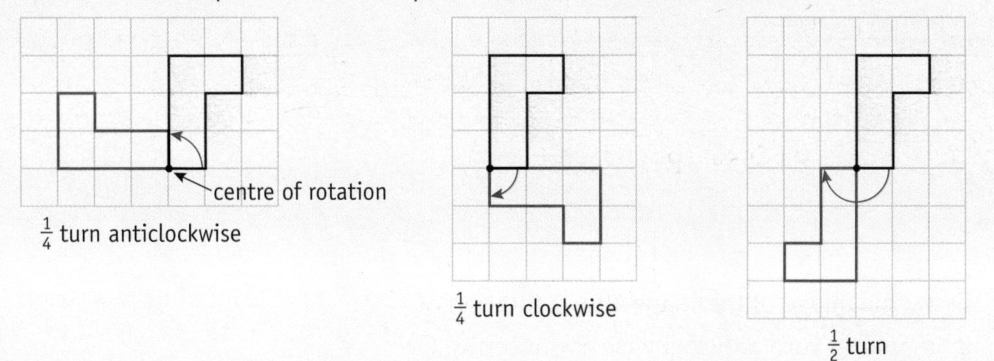

You can use tracing paper to draw a rotation.

Step 1: Trace the shape and the centre of rotation.	**Step 2:** Use a pencil to hold the tracing paper on the centre of rotation. Rotate the tracing.	**Step 3:** Copy the rotated shape from the tracing paper.

In a rotation, the object and the image are congruent.

Example 5

D

Draw the image of this shape after a rotation of a quarter turn anticlockwise about centre C.

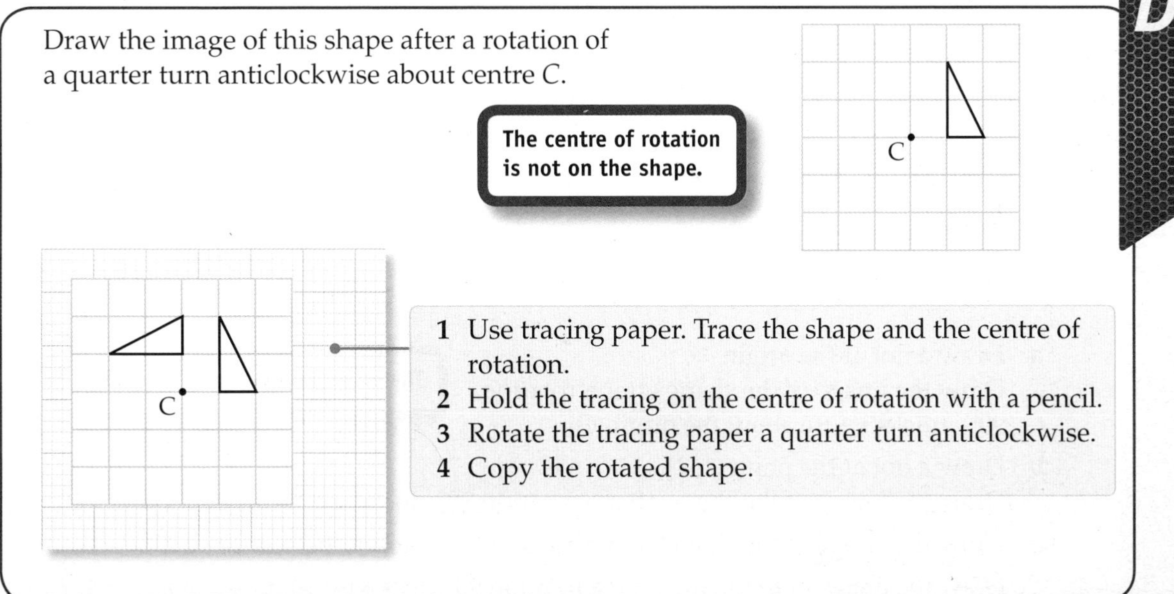

1. Use tracing paper. Trace the shape and the centre of rotation.
2. Hold the tracing on the centre of rotation with a pencil.
3. Rotate the tracing paper a quarter turn anticlockwise.
4. Copy the rotated shape.

Exercise 28E

D

1 For each part of this question, copy this shape and the centre of rotation on squared paper.

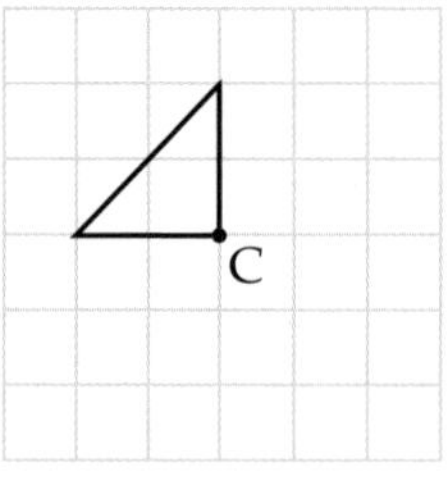

Draw the image of the shape after a rotation of

a a quarter turn anticlockwise about centre C

b 90° clockwise about centre C

c 180° about centre C.

D

2 Copy this shape on squared paper.
Follow the instructions to make a pattern.

- Rotate the shape a quarter turn clockwise about the centre C.
- Draw the image.
- Rotate the image a quarter turn clockwise about the centre C.
- Draw the image.
- Repeat until the pattern is complete.

What is the order of rotational symmetry of the pattern?

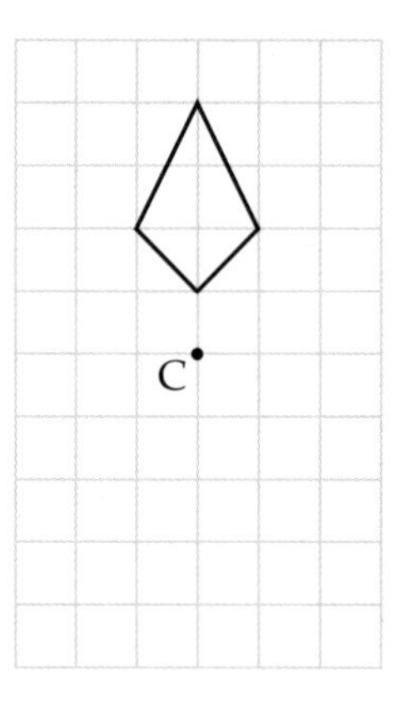

A02

D

3 Copy this grid. For each part of this question, draw the image and label it.

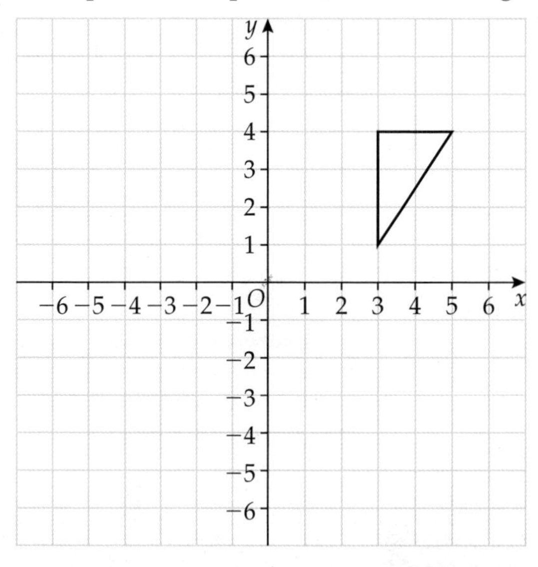

a Draw a dot at the origin.
Draw the image of the shape after a rotation 90° anticlockwise about the origin O.

The origin (O) is the point (0, 0).

C

b Draw a dot at the point (1, 2).
Draw the image of the shape after a rotation 90° clockwise about the point (1, 2).

c Draw the image of the shape after a rotation 180° about the point (−1, 3).

d Draw the image of the shape after a rotation 90° clockwise about the point (−4, 2).

4 Copy this pattern on a coordinate grid. Rotate the pattern about the origin to make a pattern with rotational symmetry of order 4.

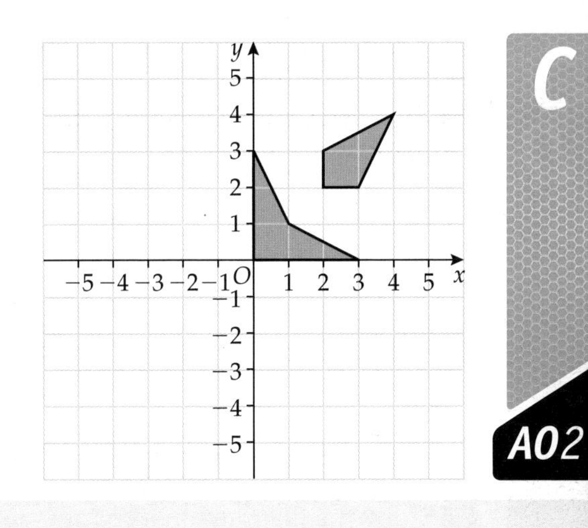

C

AO2

Describing rotations

To describe a rotation fully, you need to give

- the centre of the rotation
- the size of turn
- the direction of turn.

You can use tracing paper to find the centre and the size and direction of turn.

Step 1: Trace the object shape.	**Step 2:** Rotate the tracing until the shape looks like the image, though it might not be in the same place. What size turn have you rotated it? In what direction have you rotated it?	**Step 3:** Put your tracing back over the object shape. Rotate the tracing again, holding a point fixed with your pencil. Repeat for different points, until your tracing ends up on the image shape. That is the centre of rotation.
object	object	object

Example 6

C

Describe fully the rotation that maps shape A onto shape B.

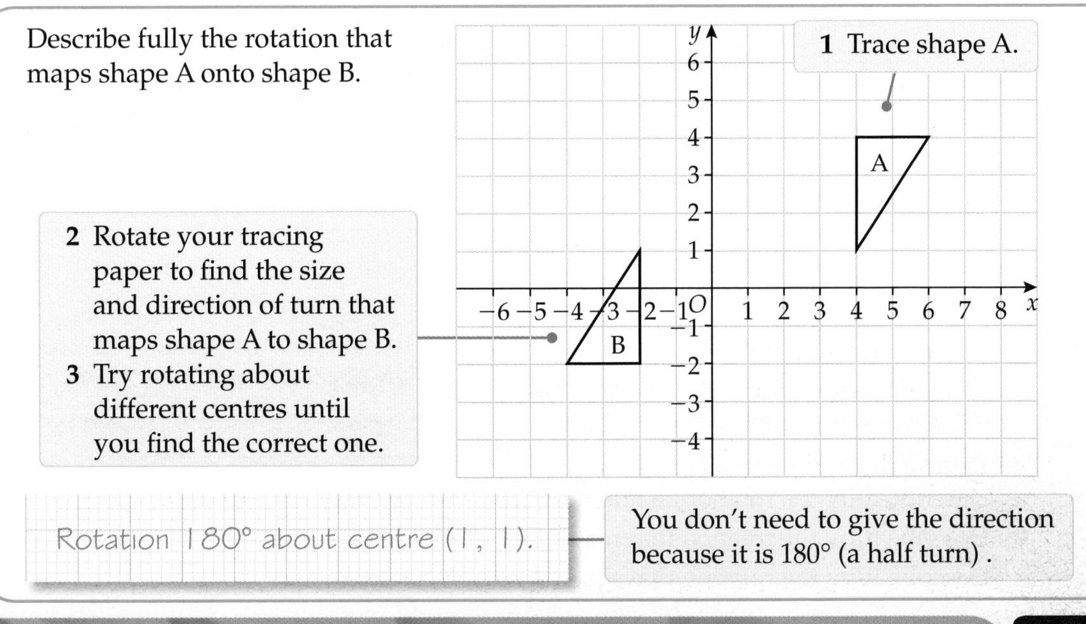

1 Trace shape A.

2 Rotate your tracing paper to find the size and direction of turn that maps shape A to shape B.

3 Try rotating about different centres until you find the correct one.

Rotation 180° about centre (1, 1).

You don't need to give the direction because it is 180° (a half turn) .

Exercise 28F

D

1 Copy this diagram on squared paper.

Shape A rotates onto shape B.

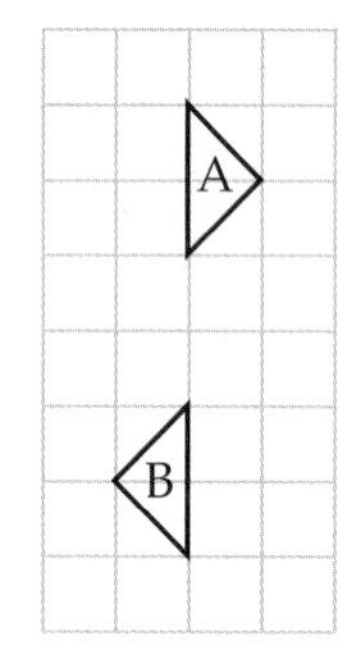

a What size turn is the rotation?

b What direction is the rotation?

c Find the centre of the rotation. Label it C on your diagram.

2 Copy the diagram on a coordinate grid.

Shape X rotates onto shape Y.

a What size turn is the rotation?

b What direction is the rotation?

c Find the centre of the rotation. Label it C on your diagram.

d Write down the coordinates of C.

D

C

3

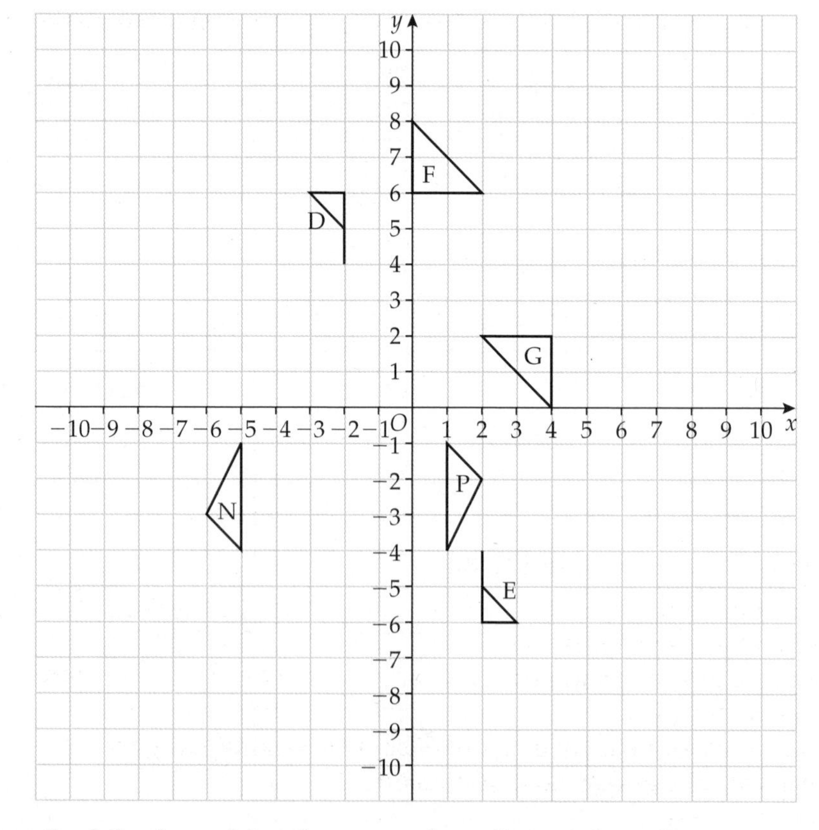

a Describe fully the rotation that maps shape D onto shape E.

b Describe fully the rotation that maps shape F onto shape G.

c Describe fully the rotation that maps shape N onto shape P.

AO2

4 Some of the shapes on this grid are rotations of each other.

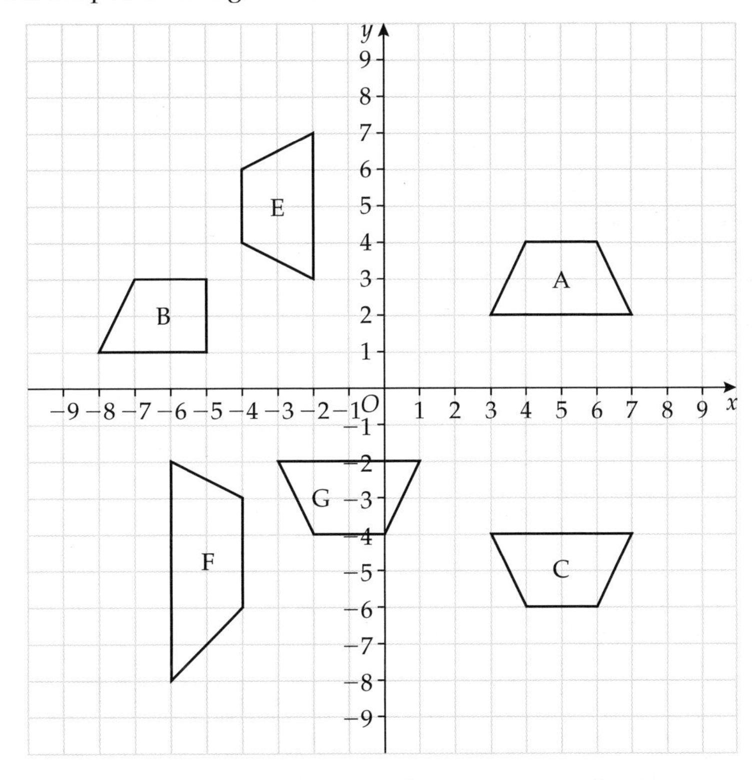

Write down the letters of pairs of shapes that rotate to each other.
Describe fully each rotation.

AO2

5 Describe fully the rotations needed to create a square from this triangle.

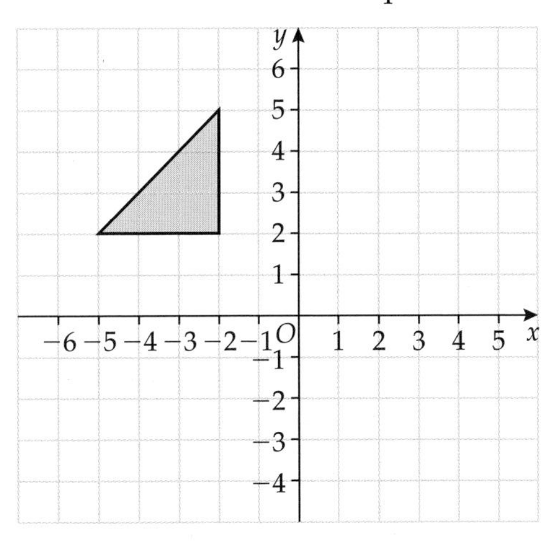

C

6 A roundabout has eight seats like this.

Each seat is fixed to a pole.
The poles fit into the central post.
The seats are placed symmetrically about the central post.

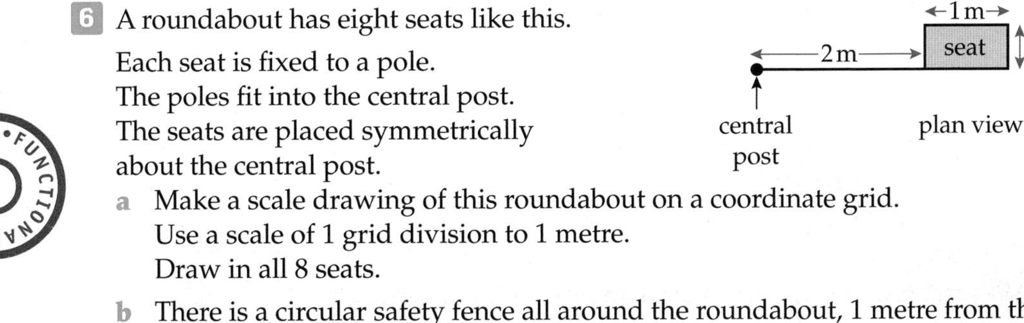

a Make a scale drawing of this roundabout on a coordinate grid.
Use a scale of 1 grid division to 1 metre.
Draw in all 8 seats.

b There is a circular safety fence all around the roundabout, 1 metre from the outside edge of the seats.
Draw the safety fence on your scale drawing.

AO3

Keywords
transformation

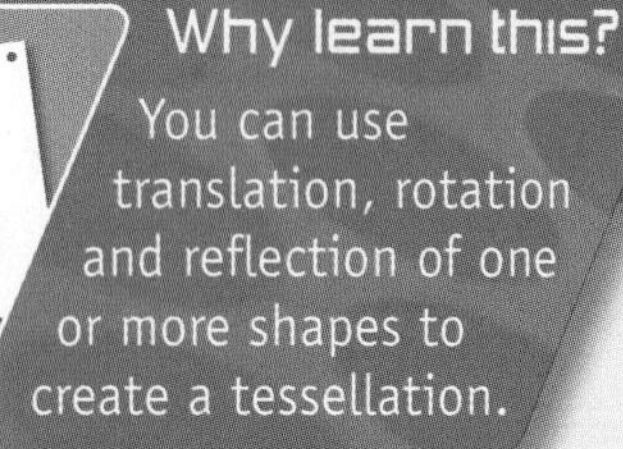

Why learn this?
You can use translation, rotation and reflection of one or more shapes to create a tessellation.

Objectives

- C Transform shapes using more than one transformation
- C Describe combined transformations of shapes on a grid

Skills check

1 Copy this diagram on to a coordinate grid.

- a Draw the image of shape D after reflection in the line $x = 2$. Label it E.
- b Draw the image of shape D after a translation $\begin{pmatrix} 3 \\ 2 \end{pmatrix}$. Label it F.
- c Draw the image of shape D after a rotation 90° anticlockwise about (3, 1). Label it G.

Combined transformations

Reflection, translation and rotation are **transformations**. They transform an object to an image. For these transformations, the object and its image are congruent.

You can combine transformations by doing one, then another.

C

Example 7

a Reflect shape P in the line $x = 1$. Label the image Q.

b Rotate shape Q 90° clockwise around the point (1, 4). Label the image R.

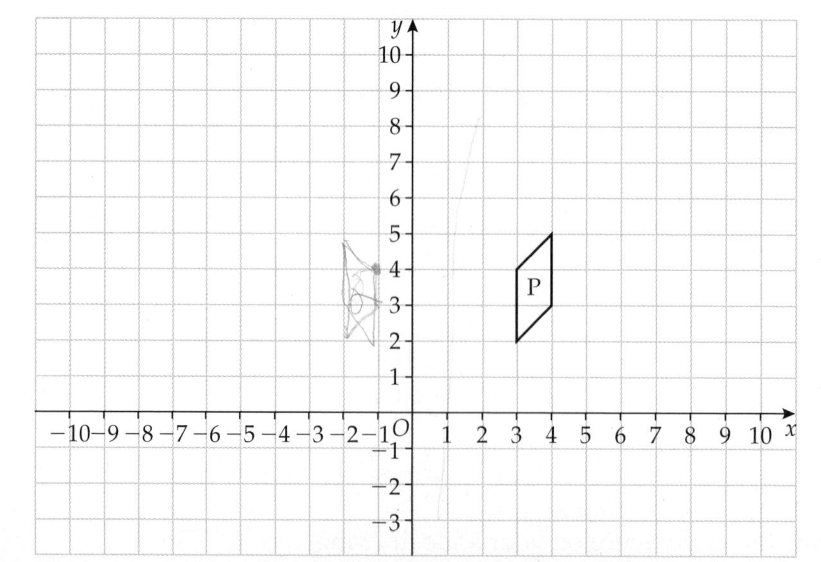

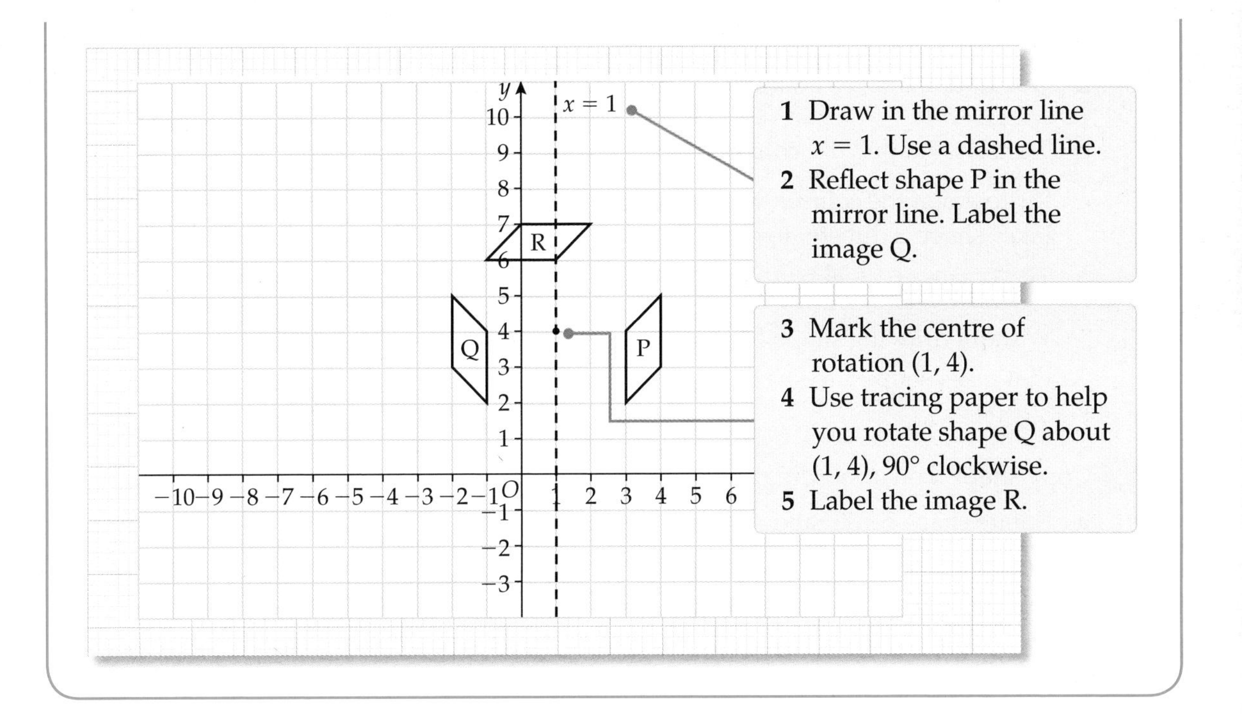

Exercise 28G

Use this diagram for Q1–3 and Q7–8.

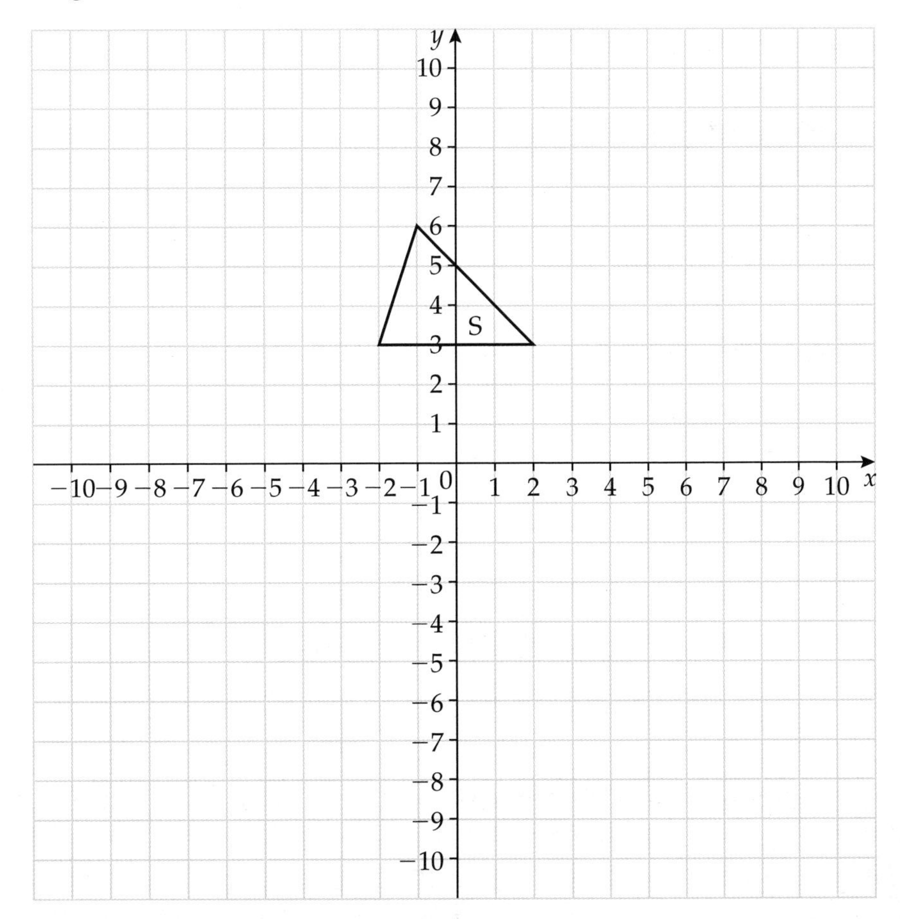

C

1 Copy shape S on a coordinate grid.

a Translate shape S by column vector $\begin{pmatrix} 4 \\ -3 \end{pmatrix}$. Label the image V.

b Translate shape V by column vector $\begin{pmatrix} -2 \\ 1 \end{pmatrix}$. Label the image W.

c Write down the column vector for the transformation that takes shape S directly to shape W.

d The translation $\begin{pmatrix} 4 \\ -2 \end{pmatrix}$ takes shape V to shape X.

Without drawing shape X, write down the column vector that takes shape S to shape X.

2 Copy shape S on a coordinate grid.

a Rotate shape S 180° about (0, 0). Label the image X.

b Rotate shape X 180° about (3, 0). Label the image Y.

c Describe the single transformation that takes shape S to shape Y.

3 Copy shape S on a coordinate grid.

a Reflect shape S in the line $x = 3$. Label the image A.

b Reflect shape A in the line $x = -1$. Label the image B.

c Describe the single transformation that takes shape S to shape B.

C

4 Describe a single transformation that is equivalent to a reflection in the x–axis followed by a reflection in the y–axis.

AO2

5 Describe a single transformation that is equivalent to a reflection in the line $y = x$ followed by a reflection in the line $y = -x$.

C

AO3

6 A rotation of 180° about the origin is the equivalent of two transformations, T1 followed by T2.
Give *two* possible descriptions of transformations T1 and T2.

C

7 Copy shape S on a coordinate grid.

a Translate shape S by $\begin{pmatrix} 4 \\ 0 \end{pmatrix}$. Label the image T.

b Reflect shape T in the line $x = 6$. Label the image U.

c Describe fully the single transformation that takes shape S to shape U.

8 Copy shape S on a coordinate grid.

a Rotate shape S 90° anticlockwise about the point (2, 2). Label the image M.

b Reflect shape M in the line $y = 1$. Label the image N.

9 Draw a right-angled triangle on a coordinate grid.

a Reflect your shape in the line $x = 2$.
Then reflect the image in the line $x = 4$.
Describe the transformation that is equivalent to this double reflection.

b Describe the transformation that you think is equivalent to a reflection in the line $x = 1$ followed by a reflection in the line $x = 3$.
Use your original shape to check if you are correct.

C

A02

Review exercise

D

1 Copy the diagram.

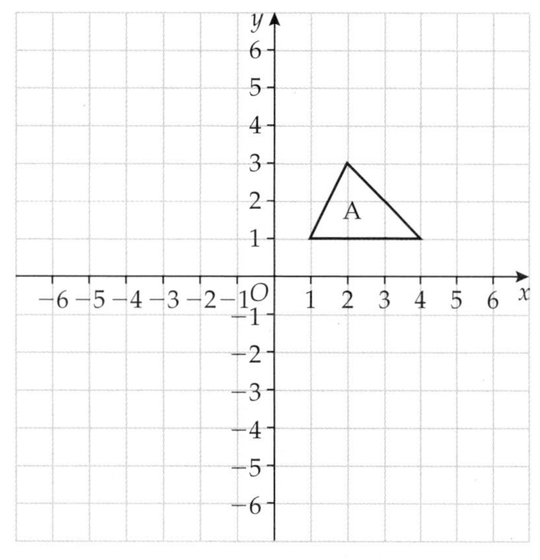

a Rotate shape A 90° anticlockwise about the origin O.
Label the image B. **[2 marks]**

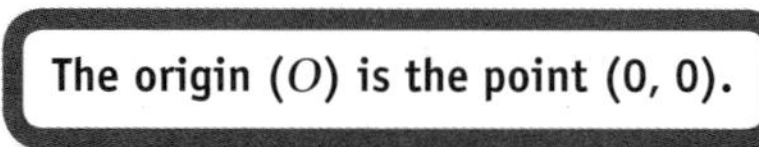

b Rotate shape A 180° about centre $(-1, 2)$.
Label the image C. **[2 marks]**

2 Copy the diagram.

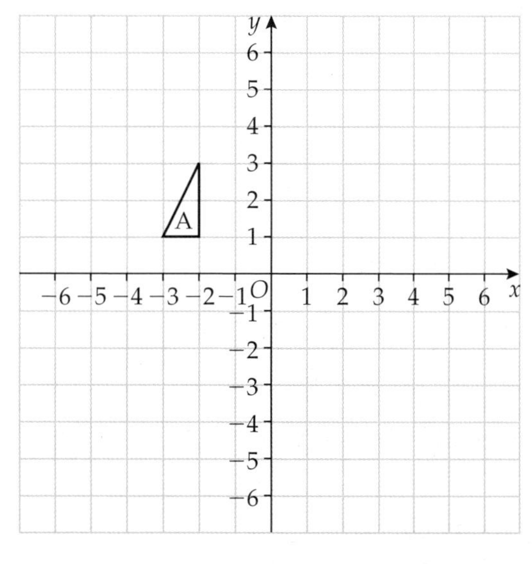

a Reflect triangle A in the line $y = -2$. Label the image B. **[2 marks]**

b Rotate triangle A a quarter of a turn anticlockwise about the origin O.
Label the image C. **[3 marks]**

C

3 The diagram shows two rectangles A and B.

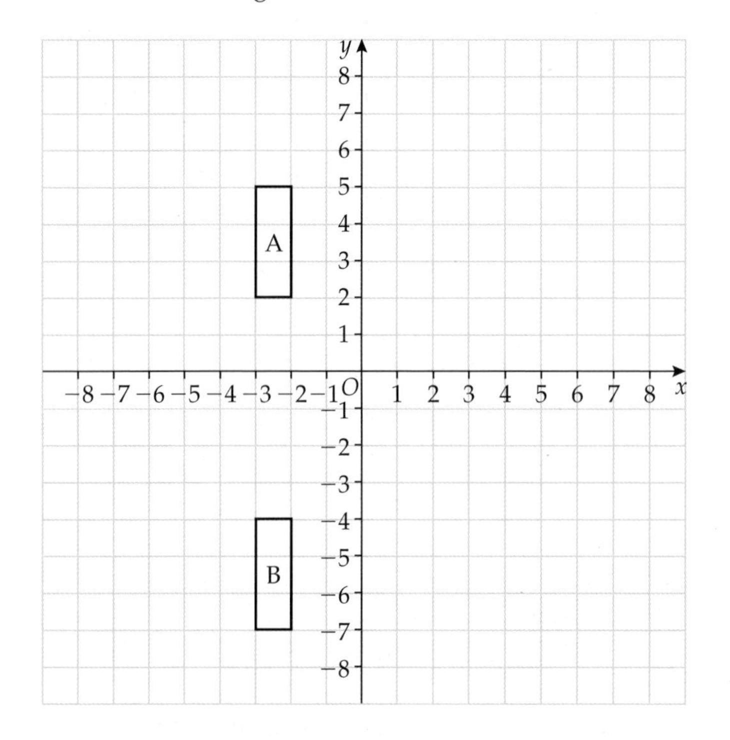

Copy and complete these statements.

a Rectangle B is a reflection of rectangle A in the line $\square = \square$. **[1 mark]**

b Rectangle B is a translation of rectangle A by the vector $\begin{pmatrix} \square \\ \square \end{pmatrix}$ **[1 mark]**

c Rectangle B is a rotation of rectangle A through $\square$ degrees about the point ($\square$, $\square$). **[2 marks]**

4

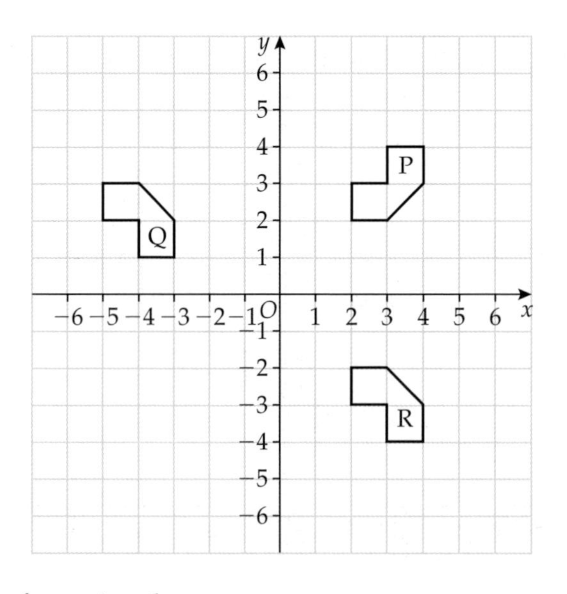

Describe the transformation that maps

a shape P to shape Q. **[1 mark]**

b shape Q to shape R. **[1 mark]**

c shape R to shape P. **[1 mark]**

5 The grid shows several transformations of the shaded triangle.

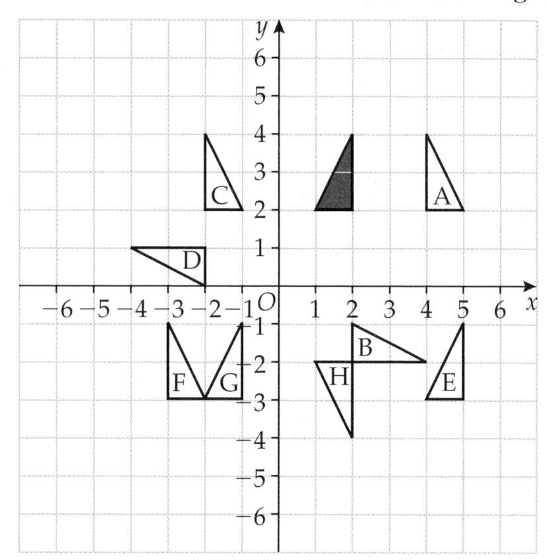

a Write down the letter of the image

i after the shaded triangle is reflected in the line $x = 3$ **[1 mark]**

ii after the shaded triangle is translated by 3 squares to the right and 5 squares down **[1 mark]**

iii after the shaded triangle is rotated 90° clockwise about O. **[1 mark]**

b Describe fully the single transformation which takes triangle F onto triangle G. **[2 marks]**

C

6 Look at the grid in Q5 again.
Which two triangles are a reflection of each other in the line $y = -x$? **[2 marks]**

C

7 Look at the grid in Q5 again.
To go from one of the triangles to another, you reflect in the line $x = 3$ and then reflect in the y-axis.
Write down the letters of the two triangles. **[3 marks]**

AO2

8 Triangle A is transformed onto triangle B by a rotation followed by a reflection.

C

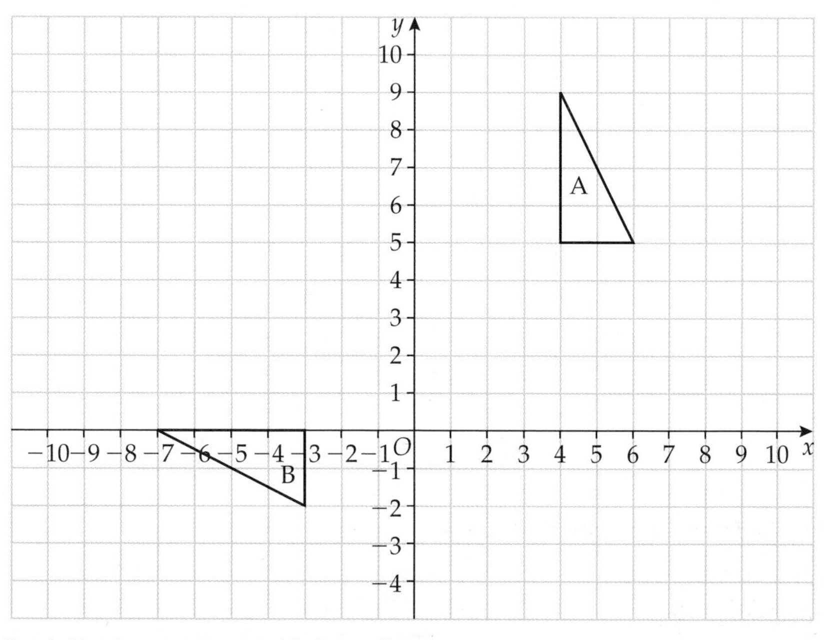

Describe fully the rotation and the reflection. **[3 marks]**

AO3

9 a Investigate reflecting a shape in one line, and then in another.
The two lines must not be parallel.
What type of transformation is equivalent to this double reflection? **[2 marks]**

b Describe how to find the single transformation that is equivalent to two reflections in lines that are not parallel. **[2 marks]**

Chapter summary

In this chapter you have learned how to

- draw reflections on a coordinate grid E D C
- translate a shape on a grid D
- describe reflections on a coordinate grid D C
- draw the position of a shape after rotation about a centre D C
- describe a rotation fully giving the size and direction of turn and the centre of rotation D C
- use column vectors to describe translations C
- transform shapes using more than one transformation C
- describe combined transformations of shapes on a grid C

29 Circles and cylinders

BBC Video

This chapter is all about circles and cylinders.

A drinking straw is cylindrical and so is a tin of baked beans. What is the biggest example of a cylinder you can think of? How about the smallest?

Objectives

This chapter will show you how to

- **find circumferences of circles and areas enclosed by circles, recalling relevant formulae** D C
- **calculate volumes of cylinders** C
- **solve problems involving surface areas and volumes of cylinders** C

Before you start this chapter

1 cm^2, m^3, mm, m*l*, mm^3, m^2, m, cm^3, km, *l*

Which of the units above is a measurement of

a length b area c volume d capacity?

2 Calculate the area of this shape.

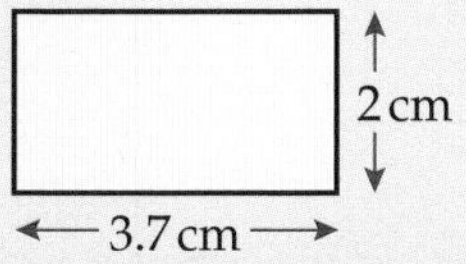

3 Calculate the area of a square with side length 5 cm.

4 Calculate the surface area of a cube with sides of length 3 cm.

5 Calculate the volume of a cuboid with sides of length 2 cm, 3 cm and 4 cm.

HELP Chapter 26

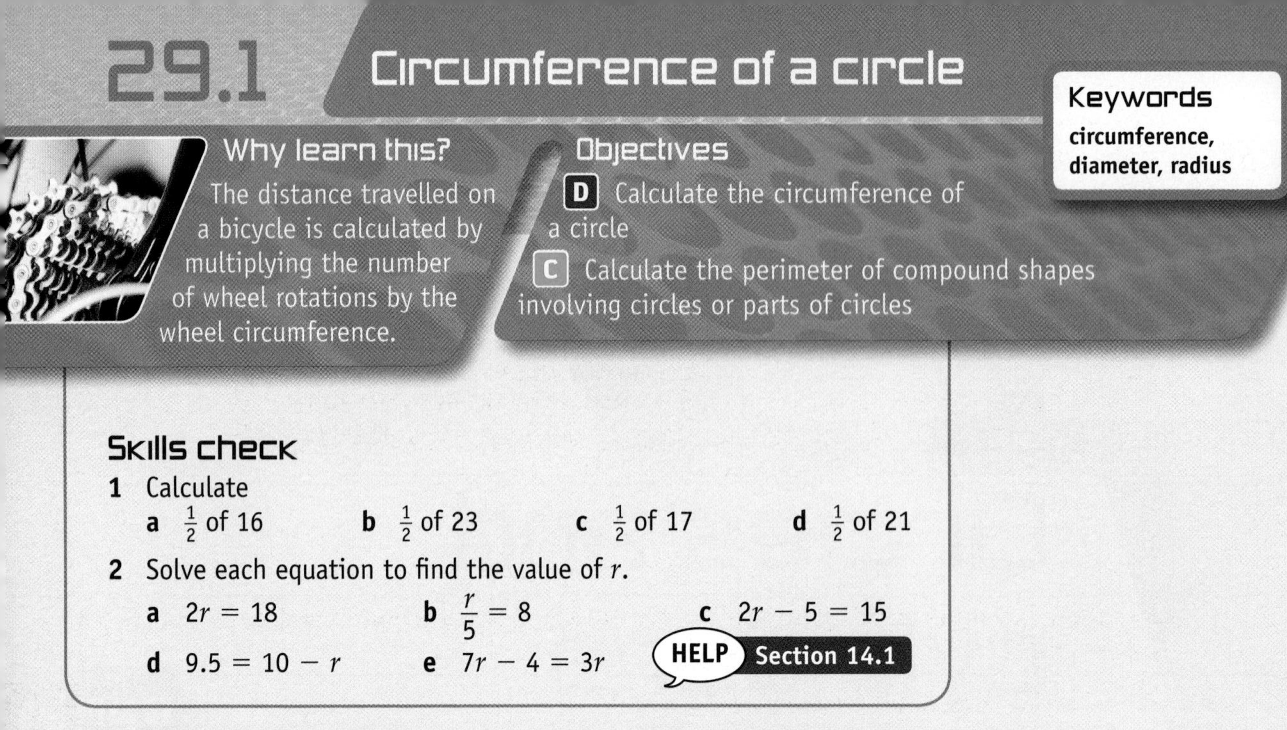

29.1 Circumference of a circle

Keywords
circumference, diameter, radius

Why learn this?

The distance travelled on a bicycle is calculated by multiplying the number of wheel rotations by the wheel circumference.

Objectives

D Calculate the circumference of a circle

C Calculate the perimeter of compound shapes involving circles or parts of circles

Skills check

1 Calculate

a $\frac{1}{2}$ of 16 **b** $\frac{1}{2}$ of 23 **c** $\frac{1}{2}$ of 17 **d** $\frac{1}{2}$ of 21

2 Solve each equation to find the value of r.

a $2r = 18$ **b** $\frac{r}{5} = 8$ **c** $2r - 5 = 15$

d $9.5 = 10 - r$ **e** $7r - 4 = 3r$

HELP Section 14.1

Calculating the circumference of a circle

The **circumference** of a circle is calculated by multiplying the **diameter** by π.

$C = \pi d$, where C = circumference and d = diameter

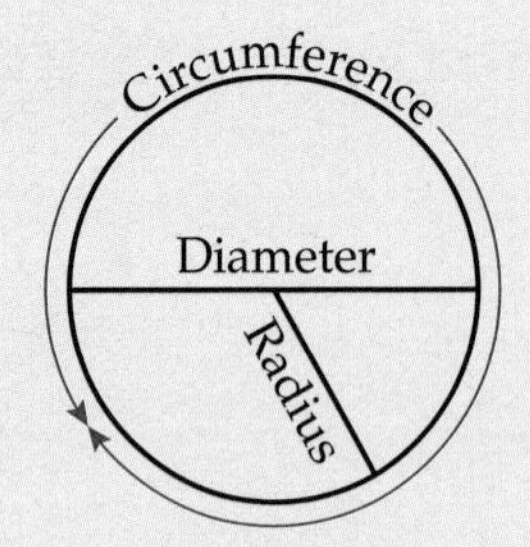

π (pi) has a value of 3.141 592 654....
Your scientific calculator has a π key. In some calculations you use 3.14 or 3.142 as an approximate value for π.

The diameter is twice the length of the **radius.**

$d = 2r$, where r = radius

So the formula can also be written as

$C = \pi \times 2 \times r$

or $C = 2\pi r$

Example 1

Calculate the circumference of a circle with diameter 5 cm.
Give your answer to two significant figures.

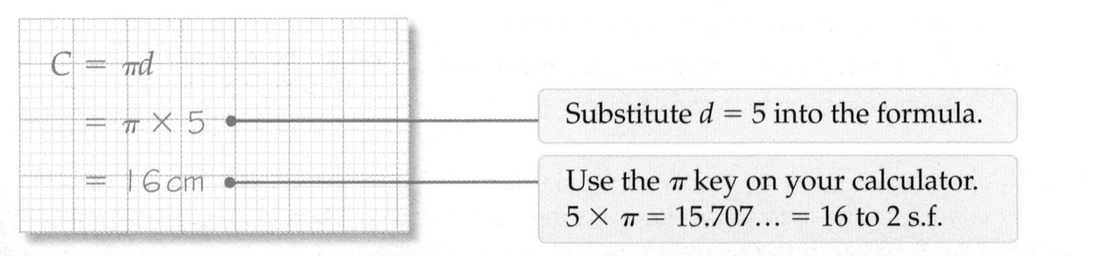

$C = \pi d$

$= \pi \times 5$ — Substitute $d = 5$ into the formula.

$= 16\text{cm}$ — Use the π key on your calculator. $5 \times \pi = 15.707... = 16$ to 2 s.f.

Example 2

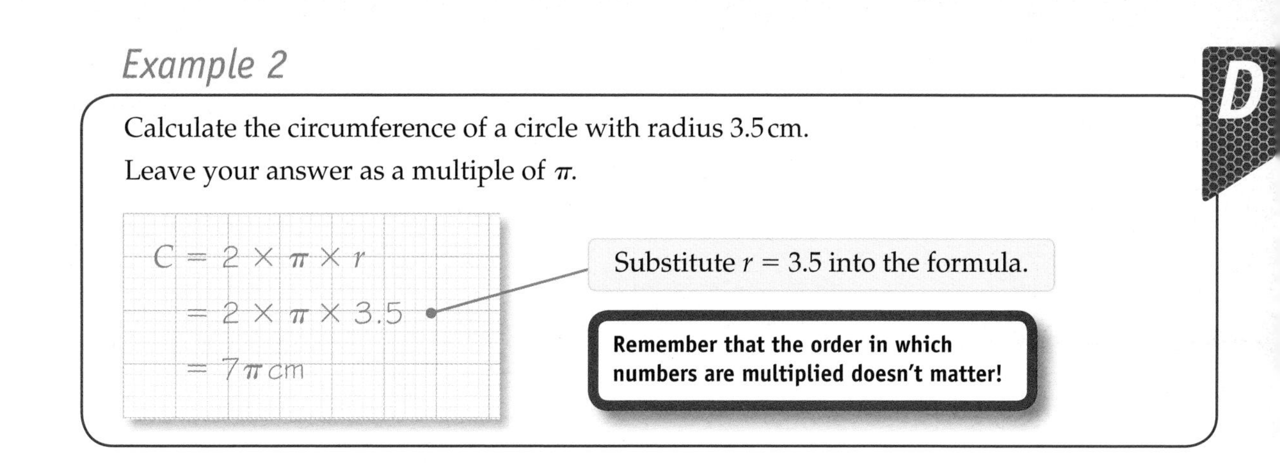

Exercise 29A

Use the π key on your calculator unless you are asked to leave your answer in terms of π.

D

1 Calculate the circumferences of these circles.
Give your answers to one decimal place.

a diameter = 10 cm
b diameter = 9.5 cm
c diameter = 3.2 cm
d radius = 12 cm
e radius = 3.5 cm
f radius = 7.8 cm

2 Calculate the circumferences of these circles.
Leave your answers in terms of π.

a

b

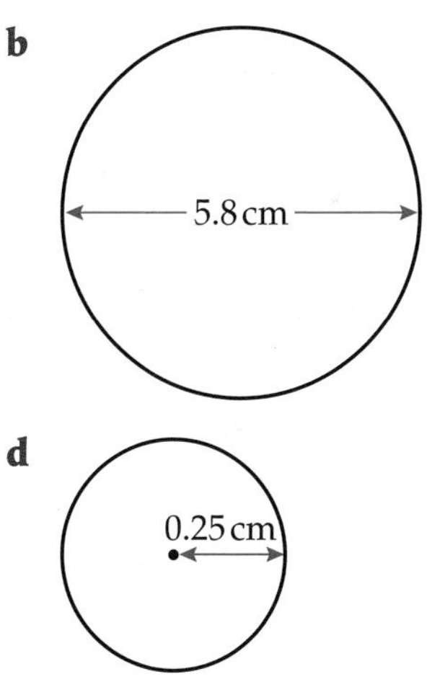

c

d

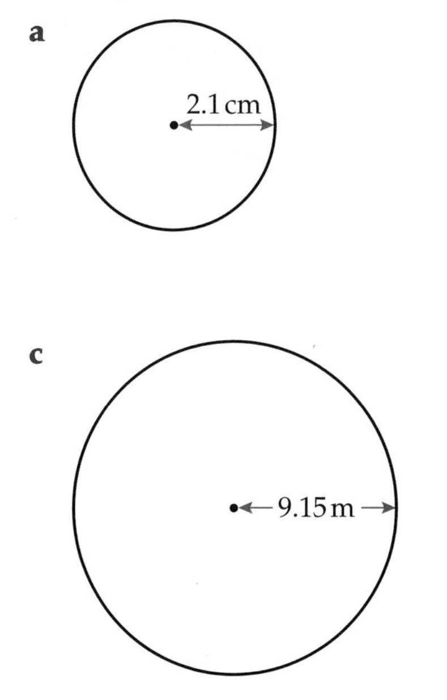

D

3 A bicycle wheel has a diameter of 65 cm.

a Work out the circumference of the wheel.

b How many complete revolutions does the wheel make when the bicycle travels 800 m?

You will need to change the units so that they are the same.

A02

4 A circular box has a strip of ribbon glued round it. There is a 2 cm overlap of ribbon. What is the length of ribbon required if the diameter of the box is 7.5 cm? (Give your answer to the nearest mm.)

Finding the diameter or the radius

You can calculate the diameter of a circle using $C = \pi d$.

You can calculate the radius of a circle using $C = \pi d$ or $C = 2\pi r$.

If you use $C = \pi d$, remember that you need to divide d by 2 to get r.

D

Example 3

A circular flower bed has a circumference of 18.8 m.

Calculate the diameter of the flower bed to one decimal place.

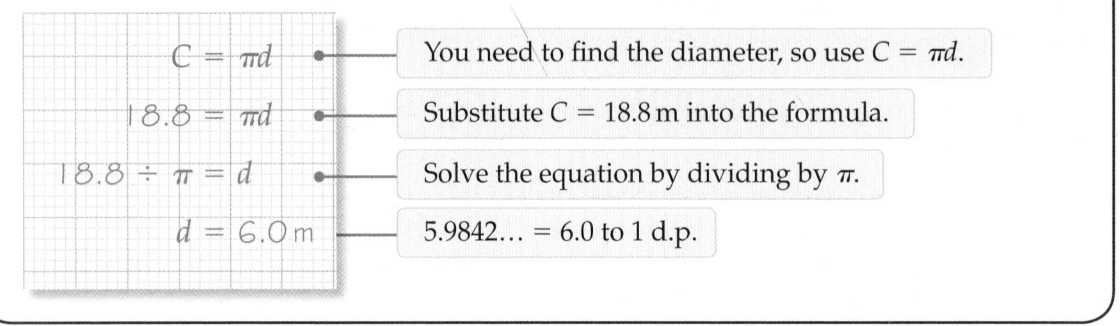

Exercise 29B

D

1 Calculate the diameters of these circles.
Give your answers to three significant figures.

a circumference = 3 m **b** circumference = 25 cm

c circumference = 12.5 mm **d** circumference = 18.7 cm

e circumference = 85.4 m **f** circumference = 105.7 cm

2 **a** The circumference of a circle is 9π cm. What is the diameter?

b The circumference of a circle is 8π cm. What is the radius?

3 The distance around the edge of a circular pond is 10.5 m.
Calculate the radius of the pond to three significant figures.

D

4 A penny-farthing bicycle has a front wheel of radius 90 cm and a back wheel of radius 21 cm.
In a journey the front wheel turned 170 times.

AO2

a How far was the journey? Give your answer to the nearest metre.

b How many complete turns did the rear wheel make on the journey?

Perimeters of shapes involving parts of circles

Some problems involve a shape that is made up of a number of different smaller shapes, such as a rectangle and a semicircle.

A semicircle is half a circle.

Example 4

C

A semicircular pond has a diameter of 6 m.

The curved edge of the pond has a wall running along it.

How long is the inside of the wall?
Give your answer to the nearest centimetre.

6 m

The circumference of a whole circle with diameter 6 m is

$C = \pi \times 6 = 18.849\ldots$ m

The pond is semicircular so the wall is

$18.849\ldots \div 2 = 9.42$ m (to the nearest cm).

Exercise 29C

C

1 Calculate the perimeter of this 'quarter-light' window of radius 52 cm.

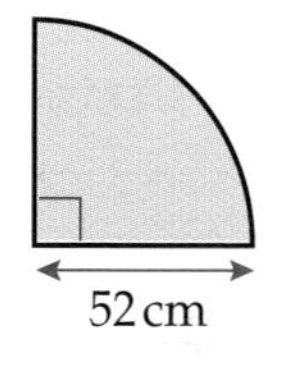

Don't forget to add on the straight edges.

C

2 The diagram shows a running track.
Maria runs round the track three times.
How far has she run?
Give your answer to the nearest metre.

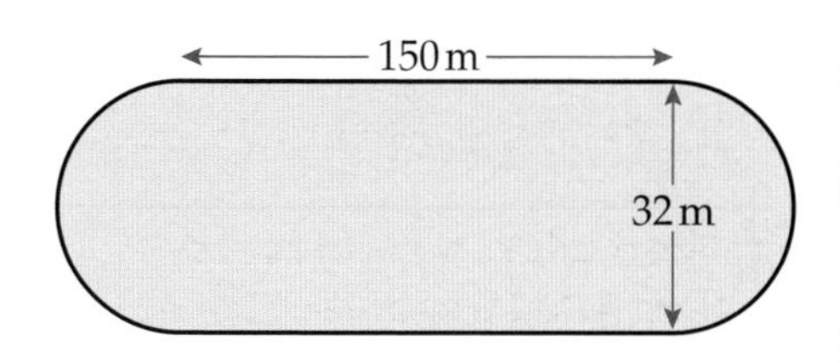

3 Here is a picture of a stained glass window.

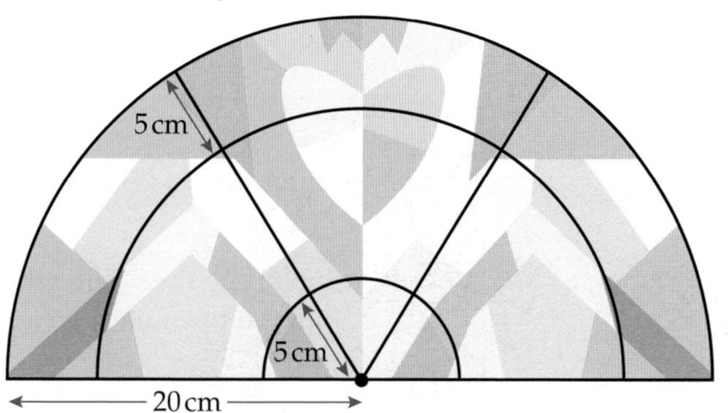

The lead that holds the glass in place is shown in black.
Calculate the length of lead to the nearest centimetre.

AO2

Area of a circle

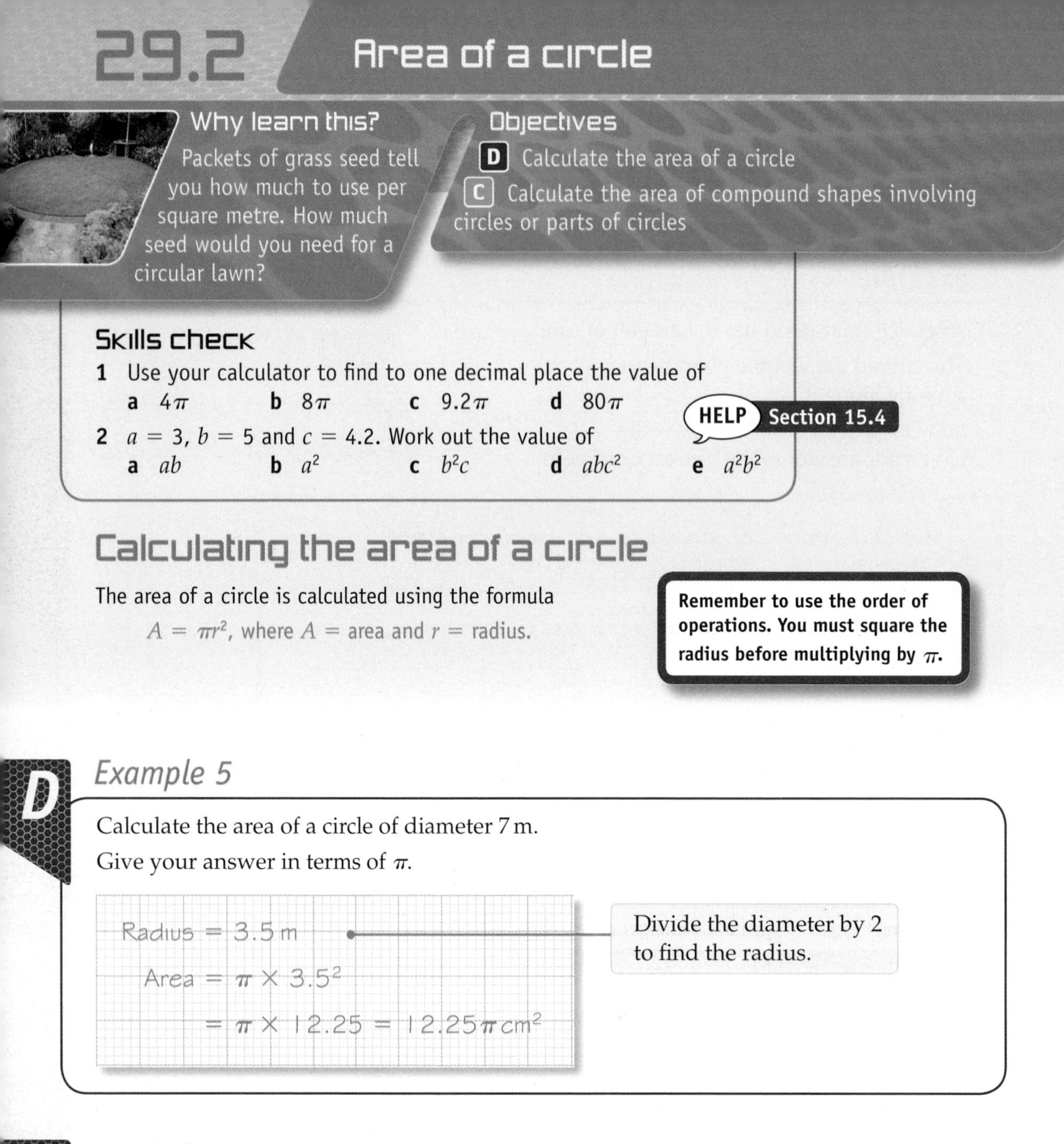

Why learn this?

Packets of grass seed tell you how much to use per square metre. How much seed would you need for a circular lawn?

Objectives

D Calculate the area of a circle

C Calculate the area of compound shapes involving circles or parts of circles

Skills check

1 Use your calculator to find to one decimal place the value of

a 4π **b** 8π **c** 9.2π **d** 80π

2 $a = 3$, $b = 5$ and $c = 4.2$. Work out the value of

a ab **b** a^2 **c** b^2c **d** abc^2 **e** a^2b^2

HELP Section 15.4

Calculating the area of a circle

The area of a circle is calculated using the formula

$A = \pi r^2$, where A = area and r = radius.

Remember to use the order of operations. You must square the radius before multiplying by π.

D

Example 5

Calculate the area of a circle of diameter 7 m.
Give your answer in terms of π.

Radius = 3.5 m

Area = $\pi \times 3.5^2$

= $\pi \times 12.25 = 12.25\pi$ cm²

Divide the diameter by 2 to find the radius.

D

Example 6

Calculate the radius of a circle with area **a** 16π cm² **b** 148 m².

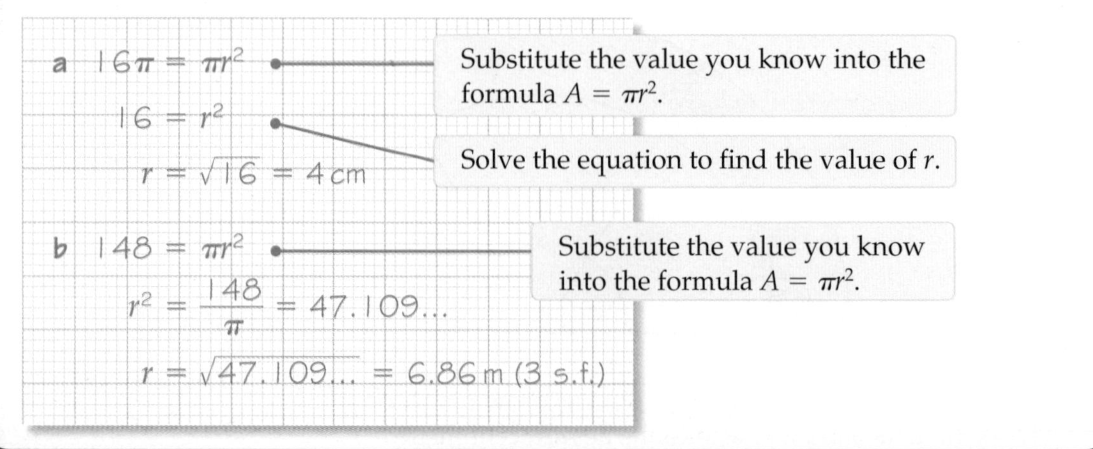

a $16\pi = \pi r^2$

$16 = r^2$

$r = \sqrt{16} = 4$ cm

Substitute the value you know into the formula $A = \pi r^2$.

Solve the equation to find the value of r.

b $148 = \pi r^2$

$r^2 = \frac{148}{\pi} = 47.109...$

$r = \sqrt{47.109...} = 6.86$ m (3 s.f.)

Substitute the value you know into the formula $A = \pi r^2$.

Exercise 29D

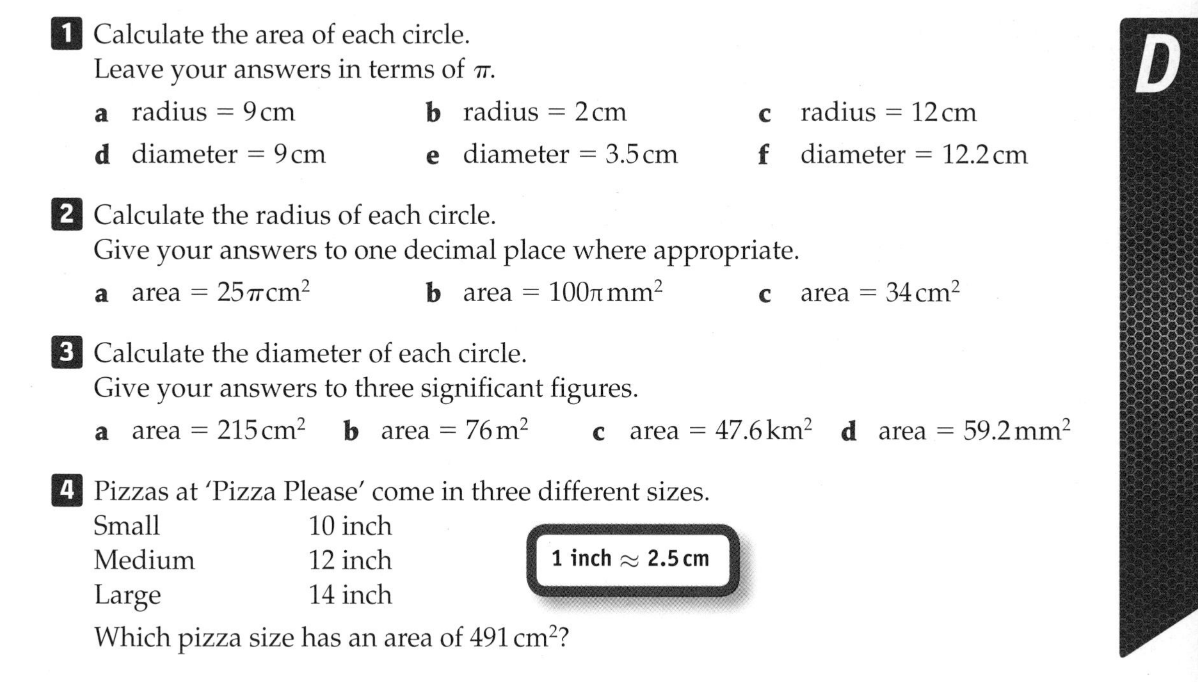

D

1 Calculate the area of each circle.
Leave your answers in terms of π.

a radius = 9 cm **b** radius = 2 cm **c** radius = 12 cm

d diameter = 9 cm **e** diameter = 3.5 cm **f** diameter = 12.2 cm

2 Calculate the radius of each circle.
Give your answers to one decimal place where appropriate.

a area = $25\pi \text{cm}^2$ **b** area = $100\pi \text{mm}^2$ **c** area = 34cm^2

3 Calculate the diameter of each circle.
Give your answers to three significant figures.

a area = 215cm^2 **b** area = 76m^2 **c** area = 47.6km^2 **d** area = 59.2mm^2

4 Pizzas at 'Pizza Please' come in three different sizes.

Small 10 inch
Medium 12 inch
Large 14 inch

1 inch ≈ 2.5 cm

Which pizza size has an area of 491cm^2?

Areas of shapes involving parts of circles

You can calculate the area of compound shapes involving circles or parts of circles.

Example 7

C

A02

FUNCTIONAL

A circular photo frame has a wooden surround as shown.
Calculate the area of the wood to the nearest square centimetre.

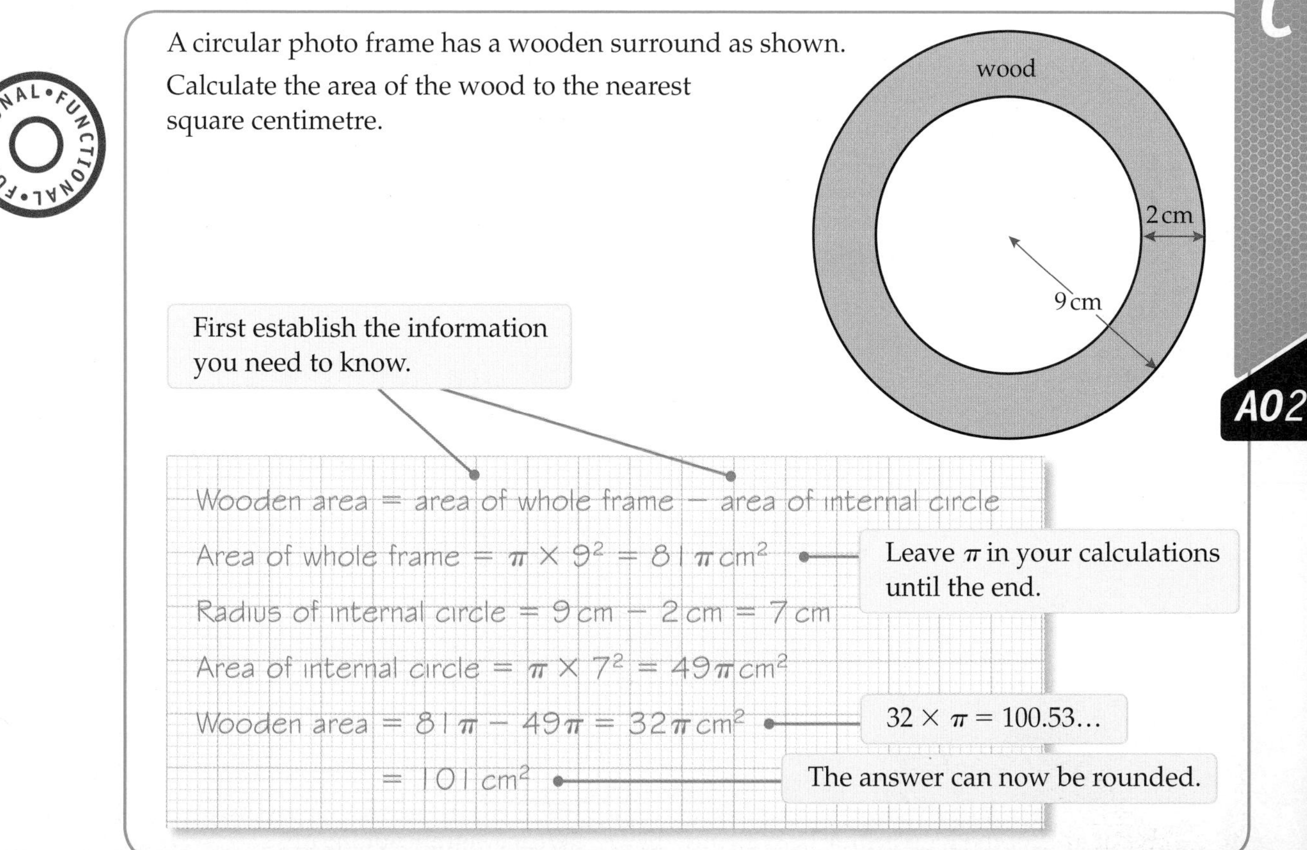

Exercise 29E

C

1 Calculate the area of each shape. Give your answers to two decimal places.

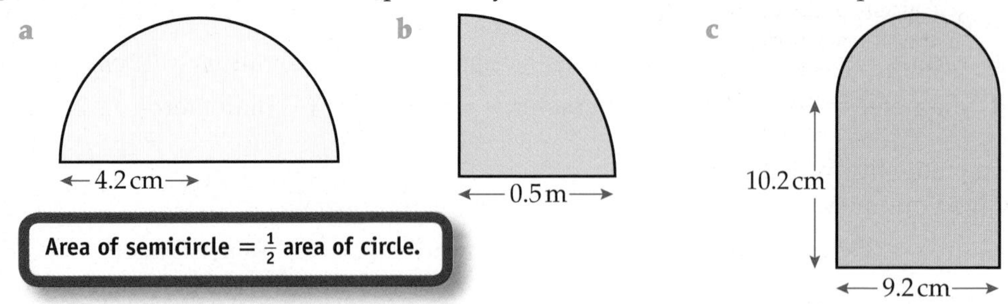

Area of semicircle = $\frac{1}{2}$ area of circle.

AO2

C

2 Calculate the shaded area in each of these diagrams.
Give your answers to one decimal place.

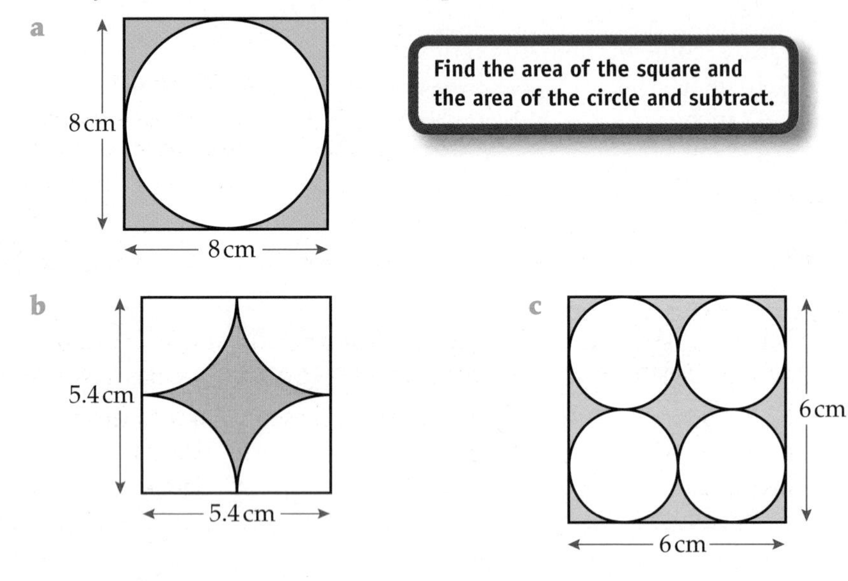

Find the area of the square and the area of the circle and subtract.

AO3

C

3 A circular flower bed has a diameter of 4 m.
The council wants to plant 100 bulbs per square metre.
How many bulbs will be needed for the flower bed?

Give your answer to the nearest whole number.

AO2

C

4 Here is a logo for a sports company.

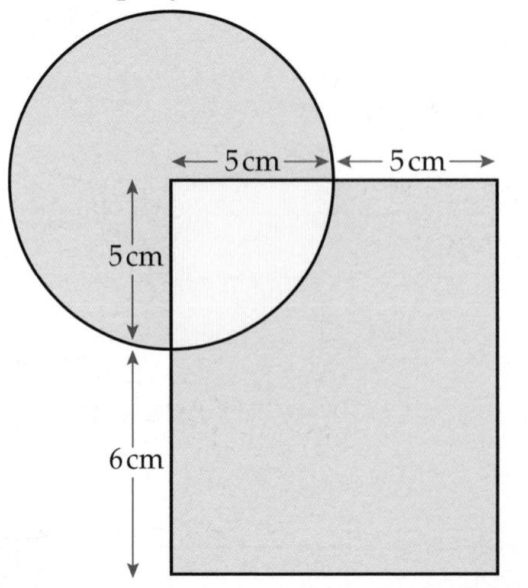

What area of the logo is blue? Give your answer to three significant figures.

AO3

5 Four concentric circles are drawn and shaded.

Concentric circles are two or more circles which have been drawn using the same position for their centres.

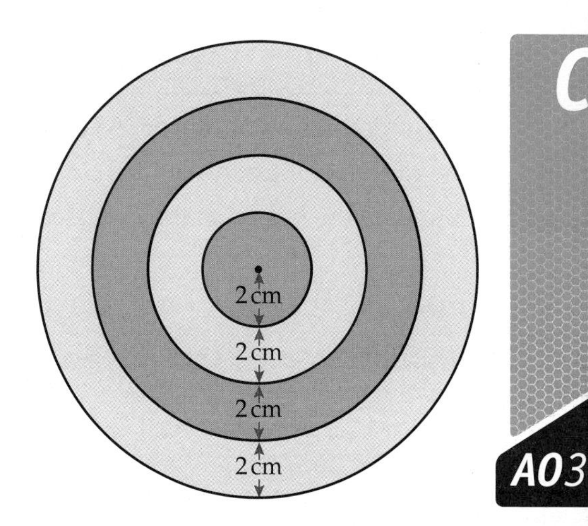

Calculate

a the area that is shaded red

b the area that is shaded blue.

Give each of your answers in terms of π.

C

AO3

29.3 Cylinders

Keywords
cylinder, total surface area, curved surface area

Why learn this?
Some musical instruments use different steel cylinders to create different notes.

Objectives

C Calculate the volume of a cylinder

C Solve problems involving the surface area of cylinders

Skills check

1 Calculate the areas of these circles. Give your answers in terms of π.

a radius = 5.5 cm **b** diameter = 12 cm **c** diameter = 9 cm

2 $1\,l = 1000\,ml$ $1\,ml = 1\,cm^3$. Convert these volumes to litres.

a 2000 cm^3 **b** 1500 cm^3 **c** 300 cm^3

d 425 cm^3 **e** 86 cm^3 **f** 15 cm^3

Cylinders and volume

A **cylinder** is a prism with a circular cross-section.

To calculate the volume of a cylinder, multiply the area of the circular face by the height.

$V = \pi r^2 h$, where V = volume, r = radius and h = height

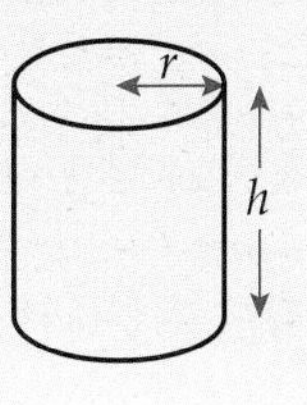

Example 8

C

Calculate the volume of this cylinder.
Leave your answer in terms of π.

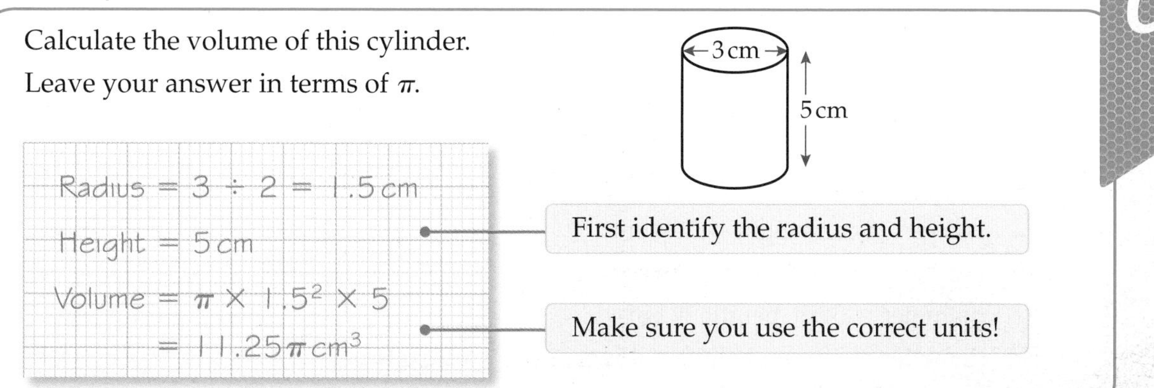

C

AO2

Example 9

A cylinder has a volume of 339 cm^3 to the nearest cm^3.

The radius of the cylinder is 3 cm.

Calculate the height of the cylinder to the nearest centimetre.

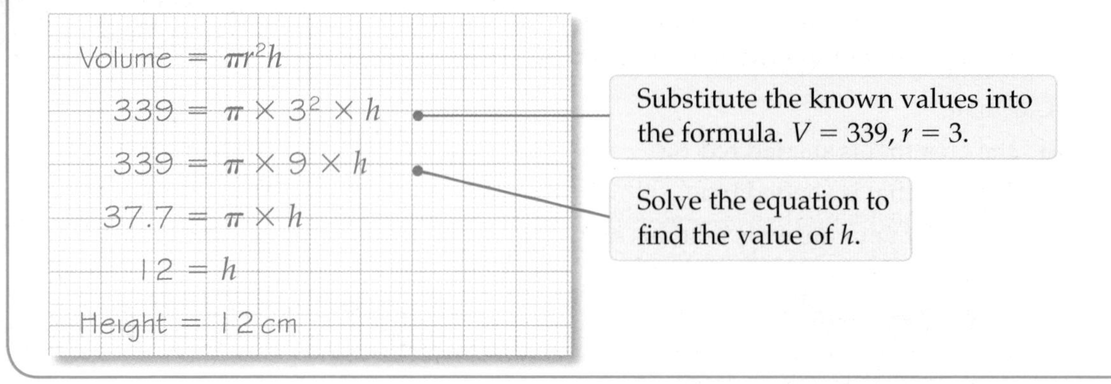

C

AO2

Example 10

A cylinder has a volume of 1130 cm^3.

The height of the cylinder is 17 cm.

Calculate the diameter of the cylinder.

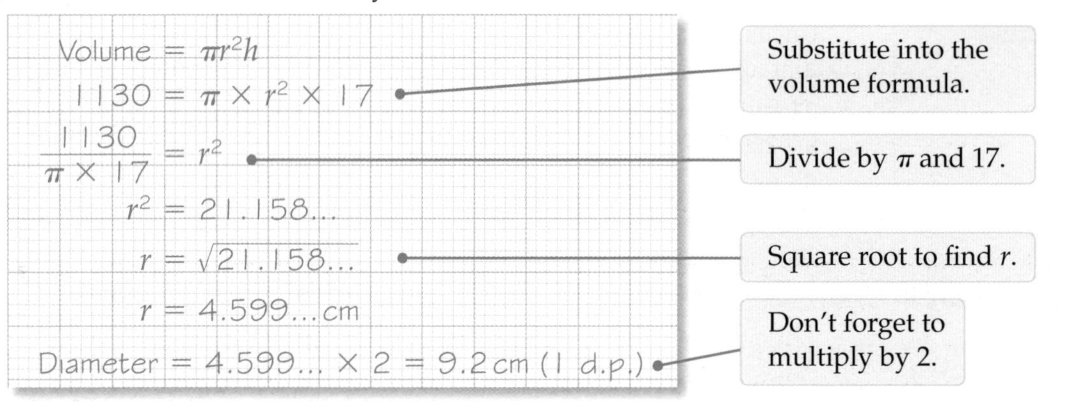

Exercise 29F

C

1 Calculate the volume of each cylinder.
Leave your answers in terms of π.

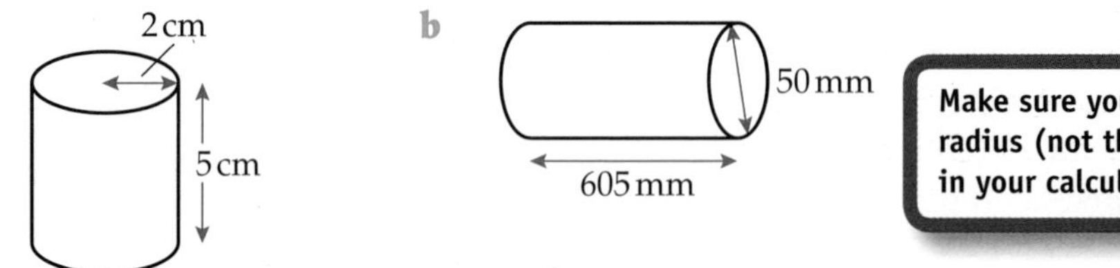

Make sure you use the radius (not the diameter) in your calculations.

2 Calculate the volumes of these cylinders.

a radius = 5 cm, height = 23.4 cm

b radius = 5 mm, height = 1.5 cm

c diameter = 3 cm, height = 1 m

d diameter = 30 mm, height = 45 cm

Give your answers to two decimal places.

Before carrying out the calculation make sure the units are the same!

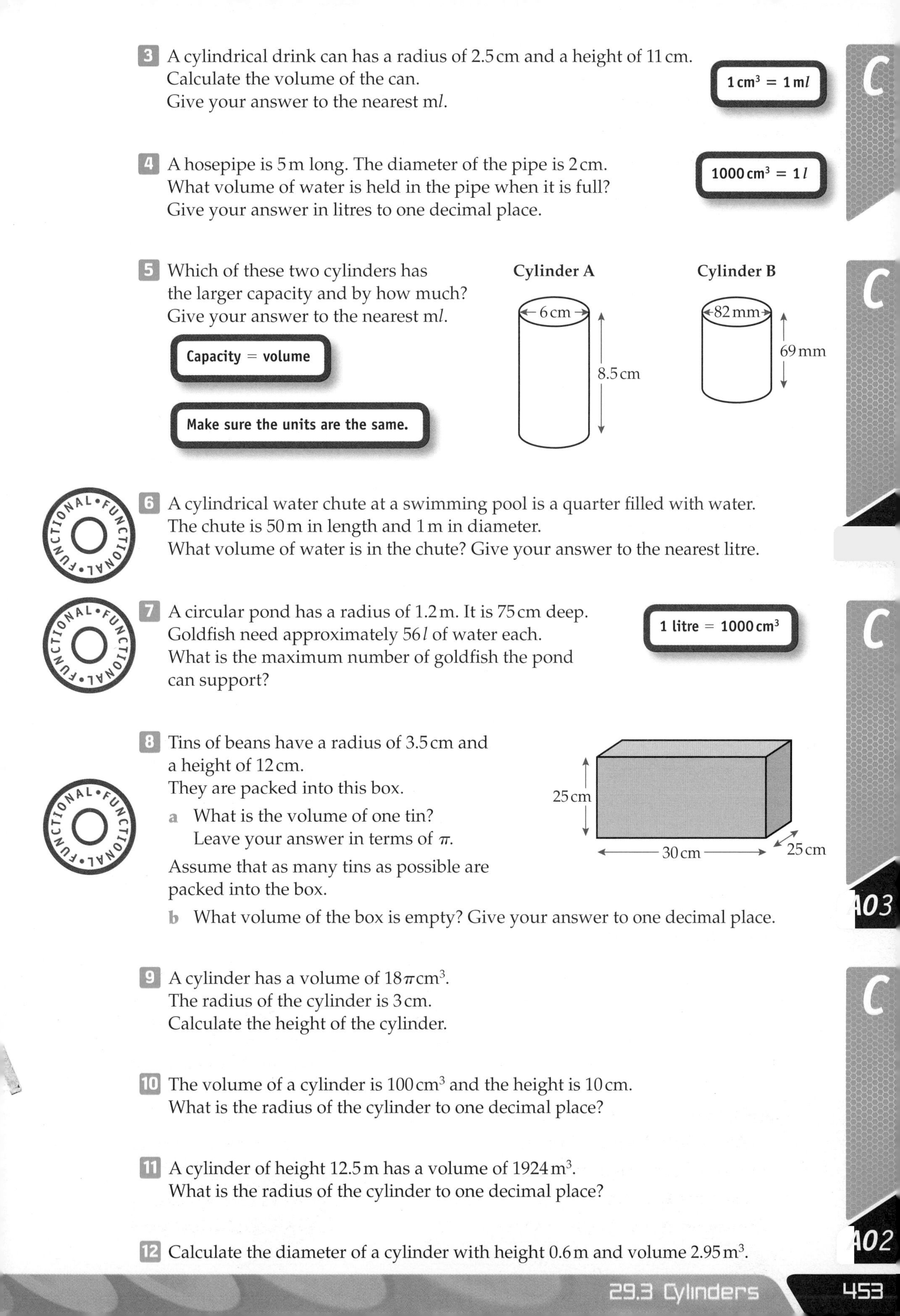

3 A cylindrical drink can has a radius of 2.5 cm and a height of 11 cm.
Calculate the volume of the can.
Give your answer to the nearest m*l*.

$1\text{ cm}^3 = 1\text{ m}l$

4 A hosepipe is 5 m long. The diameter of the pipe is 2 cm.
What volume of water is held in the pipe when it is full?
Give your answer in litres to one decimal place.

$1000\text{ cm}^3 = 1\,l$

5 Which of these two cylinders has the larger capacity and by how much?
Give your answer to the nearest m*l*.

Capacity = volume

Make sure the units are the same.

6 A cylindrical water chute at a swimming pool is a quarter filled with water.
The chute is 50 m in length and 1 m in diameter.
What volume of water is in the chute? Give your answer to the nearest litre.

7 A circular pond has a radius of 1.2 m. It is 75 cm deep.
Goldfish need approximately 56 *l* of water each.
What is the maximum number of goldfish the pond can support?

$1\text{ litre} = 1000\text{ cm}^3$

8 Tins of beans have a radius of 3.5 cm and a height of 12 cm.
They are packed into this box.

a What is the volume of one tin?
Leave your answer in terms of π.

Assume that as many tins as possible are packed into the box.

b What volume of the box is empty? Give your answer to one decimal place.

9 A cylinder has a volume of $18\pi\text{ cm}^3$.
The radius of the cylinder is 3 cm.
Calculate the height of the cylinder.

10 The volume of a cylinder is 100 cm^3 and the height is 10 cm.
What is the radius of the cylinder to one decimal place?

11 A cylinder of height 12.5 m has a volume of 1924 m^3.
What is the radius of the cylinder to one decimal place?

12 Calculate the diameter of a cylinder with height 0.6 m and volume 2.95 m^3.

Calculating the surface area of a cylinder

The surface area of a solid is calculated by finding the total area of all the faces.

The **total surface area** of a cylinder equals the **curved surface area** plus the area of its two circular ends.

To work out the curved surface area of a cylinder, consider its net.

The length of the rectangle is the same as the circumference of the circular top.

cylinder

Net of cylinder

So curved surface area $= 2\pi r \times h$

$= 2\pi rh$

Area of each circular end $= \pi r^2$

So total surface area of cylinder = curved surface area + area of circular ends

$= 2\pi rh + 2\pi r^2$

There are two circular ends.

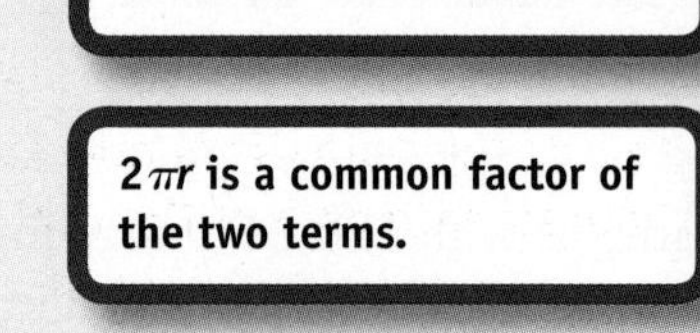

$= 2\pi r(h + r)$

$2\pi r$ is a common factor of the two terms.

The formula for the total surface area of a cylinder is *not* given on the exam paper.

C

Example 11

Calculate the surface area of a cylinder with height 5 cm and radius 3 cm.
Leave your answer in terms of π.

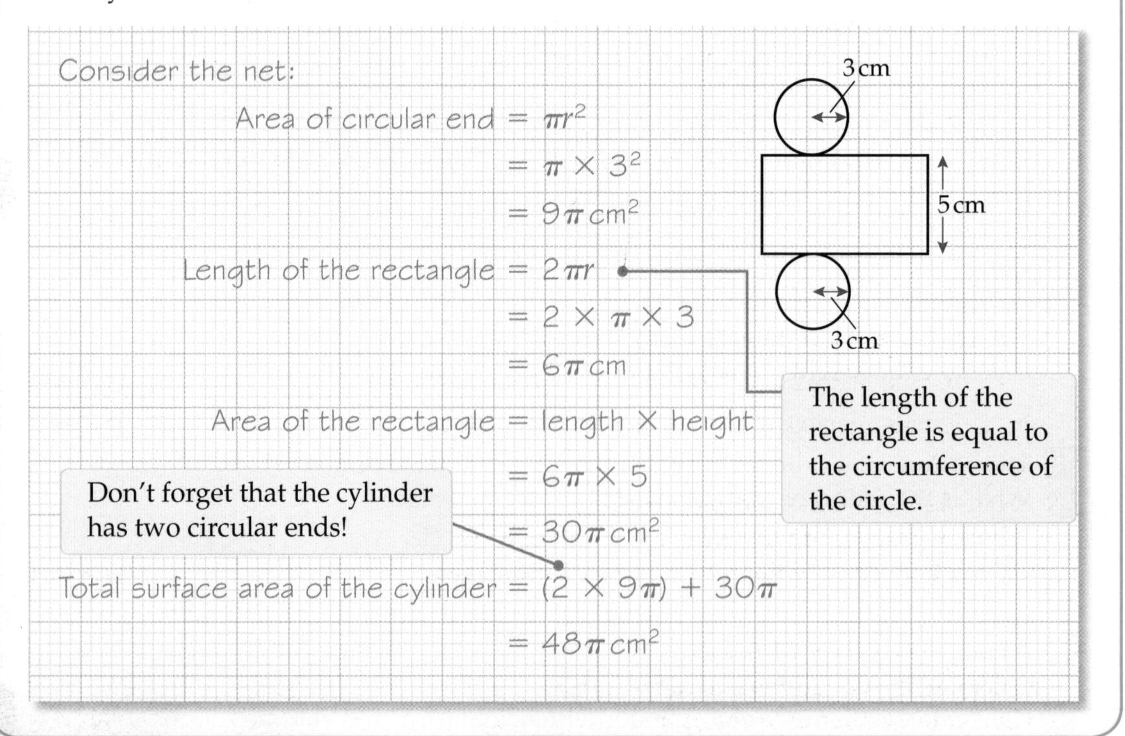

Exercise 29G

1 Calculate the total surface area of this cylinder to the nearest cm^2.

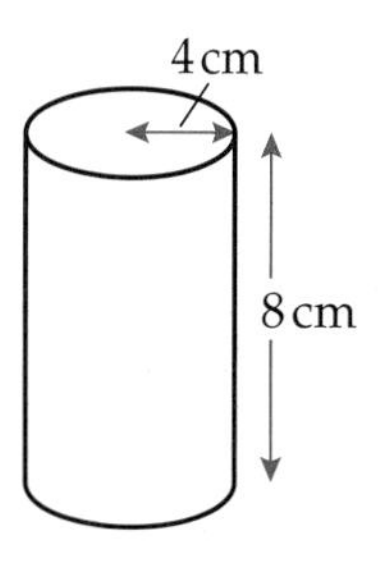

2 Calculate the total surface areas of these cylinders.
Give your answers to one decimal place.

a height = 10 cm, radius = 3 cm

b height = 15 cm, radius = 7 cm

c height = 20 cm, diameter = 12 cm

d height = 13.5 cm, diameter = 7 cm

3 A factory is making cylindrical pots for pens.
The radius of the pot is 4 cm and the height is 10 cm.
What is the surface area of the pot?
Leave your answer in terms of π.

How many circular ends are there?

4 A hollow hosepipe is 3 m long and has a diameter of 2 cm.
Calculate the surface area of the hosepipe.
Give your answer to one decimal place.

5 A tin of beans has a label around it.
The tin is 11 cm tall and has a radius of 4 cm.
The label overlaps by 1 cm.
Calculate the area of paper required to make the label.
Give your answer to three significant figures.

6 A sheet of cardboard has dimensions 150 cm by 100 cm.
What is the maximum number of cylindrical cardboard tubes of radius 3 cm and height 12 cm that can be made from it?

Review exercise

1 A circle has a circumference of 15.7 cm.

a Calculate the radius of the circle to one decimal place. [2 marks]

b Calculate the area of the circle to the nearest cm^2. [3 marks]

2 A circular lawn has a radius of 5 m.
A box of grass seed will cover $10\,m^2$ and costs £5.25.
Calculate the cost of the seed required for the lawn. [3 marks]

3 A circle has an area of $16\pi\,cm^2$.
Calculate the circumference of the circle.
Give your answer in terms of π. [3 marks]

C

4 The minute hand on a clock is 15 cm long.
How far does the tip of the hand move in 3 hours?
Give your answer to the nearest millimetre. **[2 marks]**

5 An athletics track consists of two semicircular ends and two straights.
The straights are 150 m long.
Calculate the area enclosed within the track to one decimal place.

150 m
80 m

[5 marks]

6 A school logo is shown below.

2 cm
3 cm
4 cm

Calculate the shaded area of the logo.
Give your answer to two significant figures. **[3 marks]**

7 A cylindrical paint tin has a height of 20 cm and a radius of 8 cm.
What is the maximum volume of liquid it can hold?
Give your answer to the nearest m*l*. **[4 marks]**

FUNCTIONAL • FUNCTIONAL •

8 A cylindrical cushion needs to be covered.
The diameter is 0.3 m and the thickness of the cushion is 6 cm.

Work in m.

a Calculate the area of fabric required to cover the cushion.
Give your answer to two decimal places. **[4 marks]**

FUNCTIONAL • FUNCTIONAL •

A factory has to cover 300 cushions.
The fabric costs £12.99 per m^2 but can only be bought in whole units.

b What is the cost of the fabric required? **[2 marks]**

AO2

9 A soft drink can is 10 cm tall. It holds 330 m*l* of liquid.
What is the radius of the can?
Give your answer to one decimal place. **[3 marks]**

Chapter summary

In this chapter you have learned how to

- calculate the circumference of a circle **D**
- calculate the area of a circle **D**
- calculate the perimeter of compound shapes involving circles or parts of circles **C**
- calculate the area of compound shapes involving circles or parts of circles **C**
- calculate the volume of a cylinder **C**
- solve problems involving the surface area of cylinders **C**

AO2 Example – Geometry

C

This Grade C question challenges you to use geometry in a real-life context – applying maths you know to solve a problem.

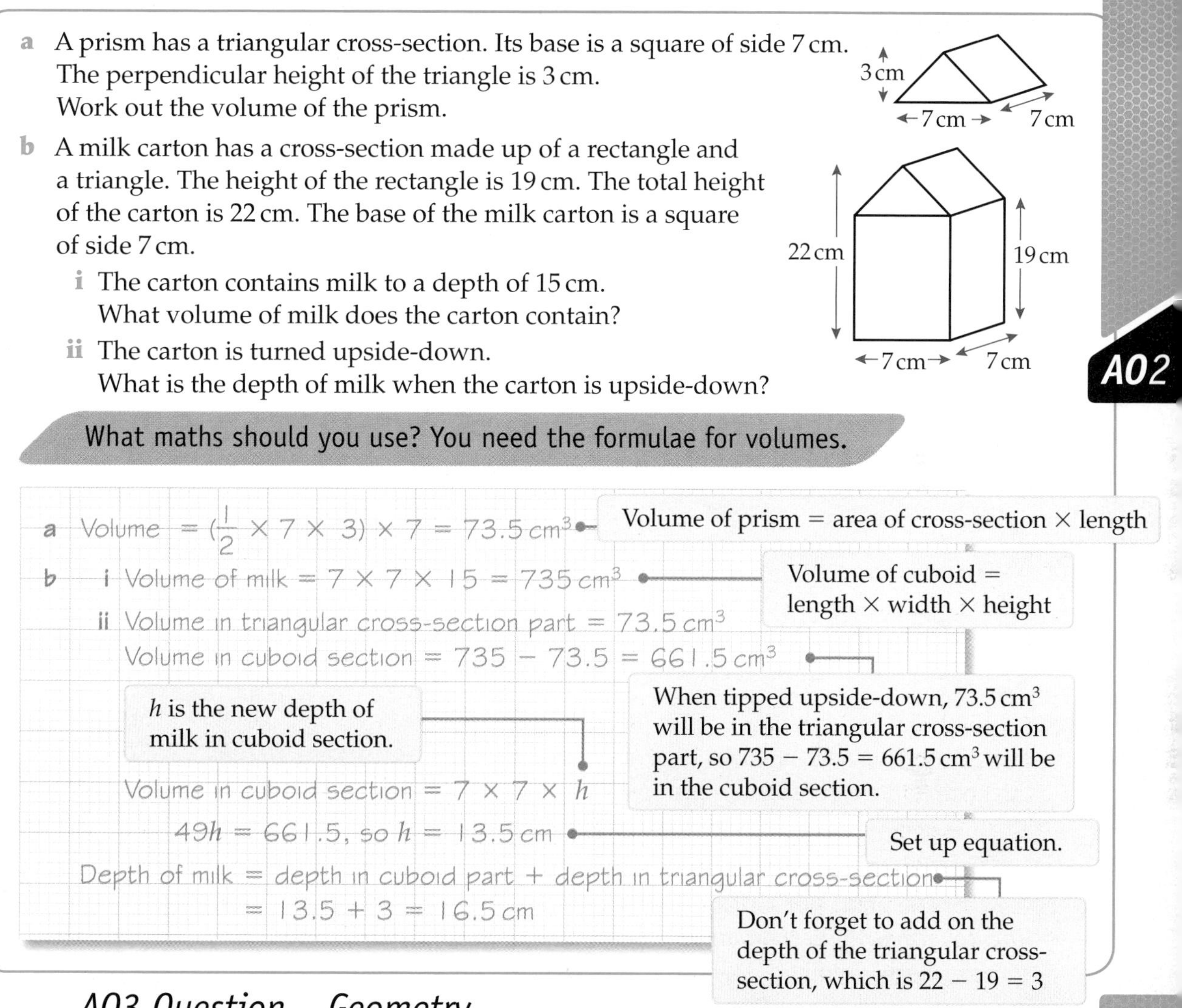

a A prism has a triangular cross-section. Its base is a square of side 7 cm. The perpendicular height of the triangle is 3 cm. Work out the volume of the prism.

b A milk carton has a cross-section made up of a rectangle and a triangle. The height of the rectangle is 19 cm. The total height of the carton is 22 cm. The base of the milk carton is a square of side 7 cm.

 i The carton contains milk to a depth of 15 cm. What volume of milk does the carton contain?

 ii The carton is turned upside-down. What is the depth of milk when the carton is upside-down?

AO2

What maths should you use? You need the formulae for volumes.

a Volume $= (\frac{1}{2} \times 7 \times 3) \times 7 = 73.5\text{ cm}^3$ — Volume of prism = area of cross-section × length

b **i** Volume of milk $= 7 \times 7 \times 15 = 735\text{ cm}^3$ — Volume of cuboid = length × width × height

 ii Volume in triangular cross-section part $= 73.5\text{ cm}^3$

 Volume in cuboid section $= 735 - 73.5 = 661.5\text{ cm}^3$ — When tipped upside-down, 73.5 cm^3 will be in the triangular cross-section part, so $735 - 73.5 = 661.5\text{ cm}^3$ will be in the cuboid section.

 h is the new depth of milk in cuboid section.

 Volume in cuboid section $= 7 \times 7 \times h$

 $49h = 661.5$, so $h = 13.5\text{ cm}$ — Set up equation.

Depth of milk = depth in cuboid part + depth in triangular cross-section

$= 13.5 + 3 = 16.5\text{ cm}$ — Don't forget to add on the depth of the triangular cross-section, which is $22 - 19 = 3$

AO3 Question – Geometry

C

Now try this AO3 Grade C question: You have to work it out from scratch.
READ THE QUESTION CAREFULLY.
It's similar to the AO2 example above, so think about where to start.

A juice carton has a cross-section made up of a rectangle and a trapezium.

First find the volume of the top section of this carton.

The trapezium has parallel sides of 8 cm and 3 cm.
The height of the rectangle is 20 cm.
The total height of the carton is 24 cm.
The base of the juice carton is a rectangle 8 cm by 6 cm.

The carton contains juice to a depth of 16 cm.
The carton is turned upside-down?
What is the depth of juice when the carton is upside-down?

3 cm
24 cm
20 cm
8 cm
6 cm

AO3

Roof garden

Anil moves into a small flat in Manchester. The flat has a roof garden. Anil makes a scale drawing of the new design for his roof garden.

The garden will have wooden fencing on three sides. There will also be five large plant pots, each 1 m high.

Question bank

1 What is the total length of fencing that Anil needs to buy?

2 What is the cost of the circular plant pot that Anil wants?

Anil is going to cover the sides of all the plant pots with a bamboo covering. He is then going to fill the pots with compost, leaving a 10 cm gap at the top.

3 What length of bamboo covering does Anil need to go around one of the square plant pots?

4 How much compost does Anil need for one of the square plant pots?

Anil says, 'The total cost of the pots, bamboo, compost and fencing is just under £1200.'

5 Is Anil correct? Use π on your calculator and show working to support your answer.

Information bank

Haroldston garden centre price list

Plant pots (1 m high)	square 1 m × 1 m square 1.2 m × 1.2 m square 1.4 m × 1.4 m rectangular 1 m × 2 m rectangular 1 m × 2.5 m rectangular 1.2 m × 2.5 m circular pot, radius 0.9 m circular pot, radius 1.1 m circular pot, radius 1.3 m	£49 £59 £69 £49 £69 £89 £99 £119 £139
Fencing	wooden wire	£9 per metre £7 per metre
Bamboo	roll (0.5 m high) roll (1 m high) roll (1.5 m high)	£3.50 per metre £5.50 per metre £7.50 per metre
Compost	200 litre bag	£8.95

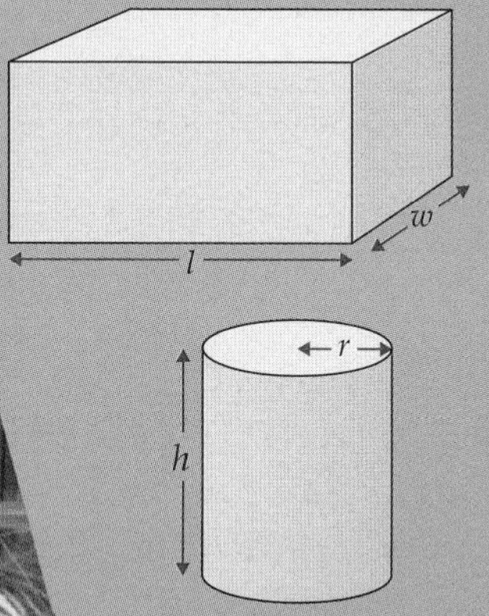

Perimeter of a rectangle $= 2(l + w)$

Area of a rectangle $= lw$

Volume of a cuboid $= lwh$

Perimeter of a circle $= 2\pi r$

Area of a circle $= \pi r^2$

Volume of a cylinder $= \pi r^2 h$

$1\ \text{cm}^3 = 1\ \text{m}l$

$1000\ \text{m}l = 1$ litre

1000 litres $= 1\ \text{m}^3$

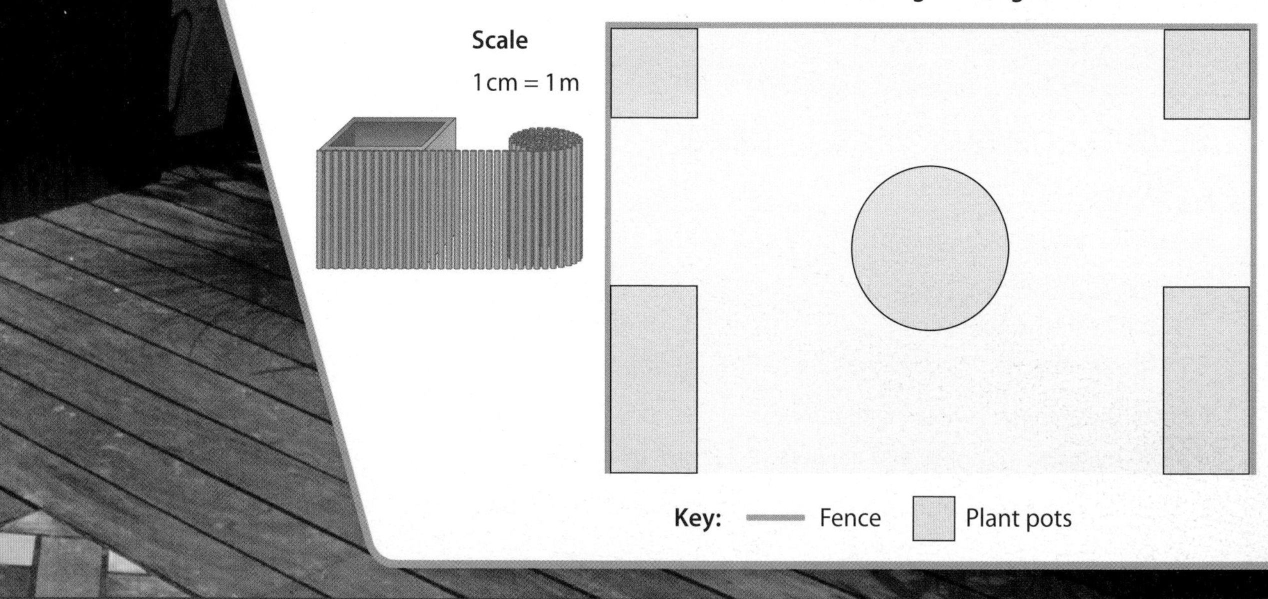

30 Measurement 2

This chapter is about using different units, and calculating speed and density.

Sea water has a higher density than fresh water so it is easier to float in sea water.

Objectives

This chapter will show you how to

- **convert between different units of area** D
- **convert between different units of volume** C
- **calculate average speeds** D
- **make calculations using density** C
- **recognise formulae for length, area or volume by considering dimensions** B

Before you start this chapter

1 Convert
   a 300 m into km b 220 mm into cm c 1.6 m into cm d 0.52 km into cm

2 Round 20 645.98 to
   a the nearest whole number b three significant figures
   c the nearest 10 d one decimal place

3 Write these times in hours and minutes.
   a 80 minutes b 400 minutes c a quarter of an hour d $2\frac{2}{5}$ hours

4 Work out these calculations, rounding your answers to one decimal place.
   a $80 \div 6$ b $4 \div 9$ c $200 \div 42$ d $60 \div 35$

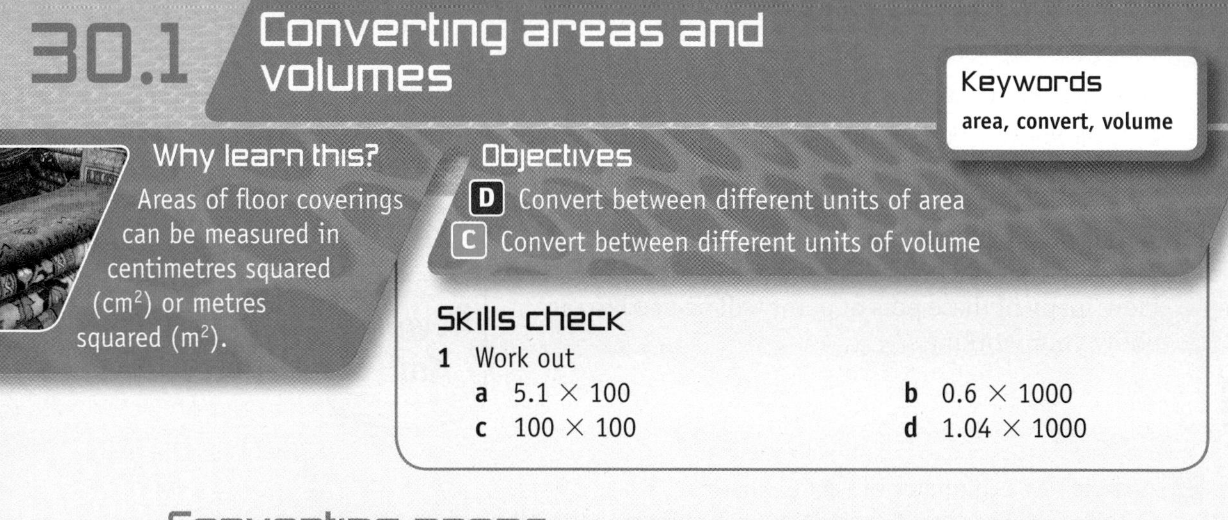

30.1 Converting areas and volumes

Keywords
area, convert, volume

Why learn this?

Areas of floor coverings can be measured in centimetres squared (cm^2) or metres squared (m^2).

Objectives

D Convert between different units of area

C Convert between different units of volume

Skills check

1 Work out

a 5.1×100 **b** 0.6×1000

c 100×100 **d** 1.04×1000

Converting areas

Area is measured in mm^2, cm^2, m^2 or km^2.

These two squares have the same area.

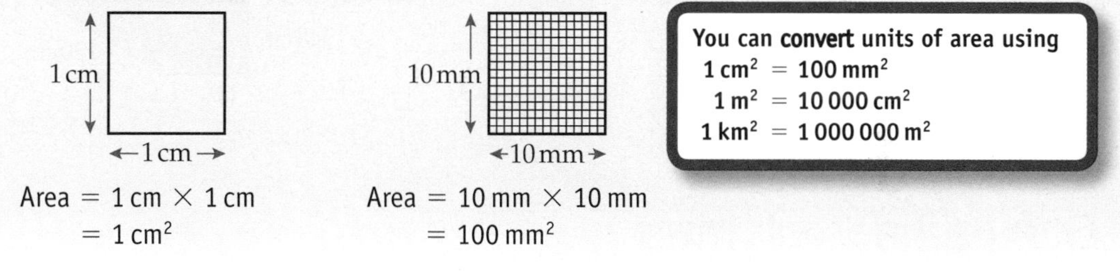

Area = 1 cm × 1 cm
= 1 cm^2

Area = 10 mm × 10 mm
= 100 mm^2

You can convert units of area using
1 cm^2 = 100 mm^2
1 m^2 = 10 000 cm^2
1 km^2 = 1 000 000 m^2

Example 1

Convert

a $4.5\,m^2$ into cm^2 **b** $220\,mm^2$ into cm^2.

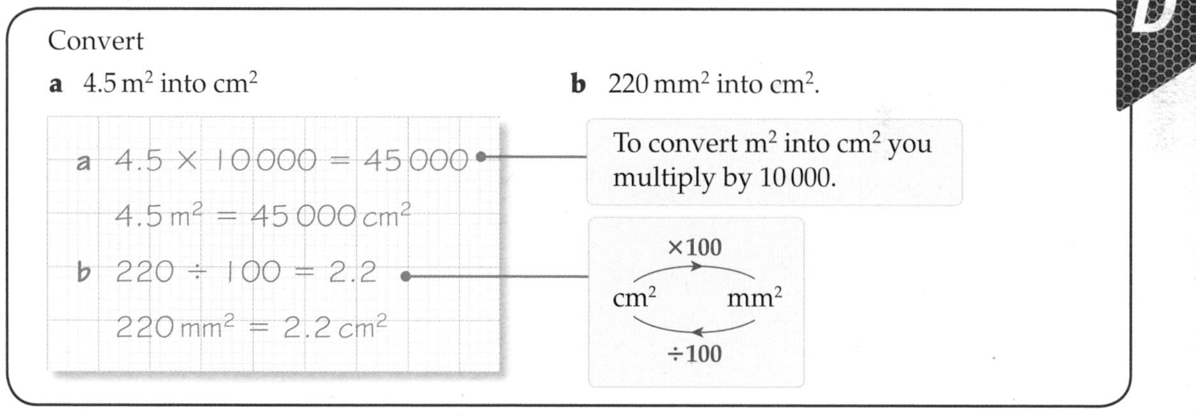

a $4.5 \times 10\,000 = 45\,000$

$4.5\,m^2 = 45\,000\,cm^2$

b $220 \div 100 = 2.2$

$220\,mm^2 = 2.2\,cm^2$

To convert m^2 into cm^2 you multiply by 10 000.

Exercise 30A

1 Convert

a $6\,m^2$ into cm^2 **b** $0.2\,m^2$ into cm^2 **c** $2400\,cm^2$ into m^2 **d** $650\,cm^2$ into m^2

2 Copy and complete these conversions.

a $2.5\,m^2$ = _____ cm^2 **b** $4\,km^2$ = _____ m^2

c $8900\,m^2$ = _____ km^2 **d** $5000\,mm^2$ = _____ cm^2

3 Work out the area of this rectangle

a in cm^2

b in mm^2.

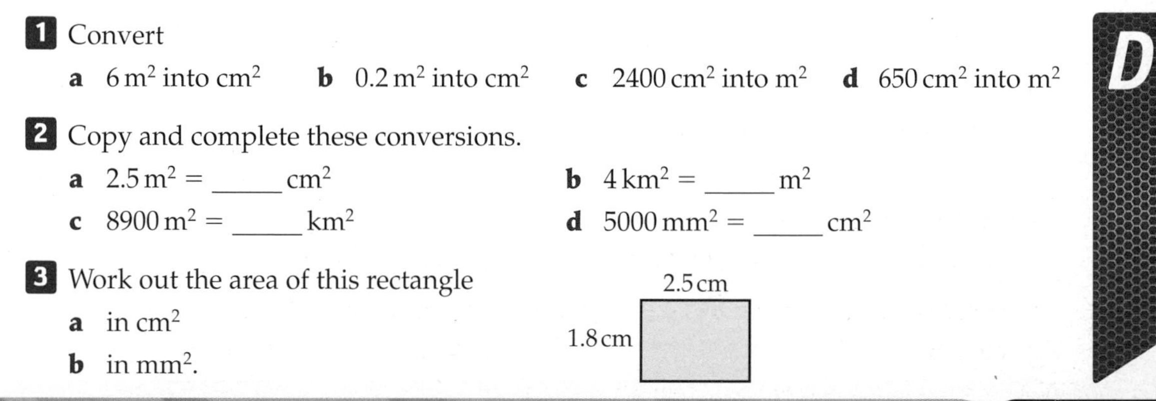

D

4 Claire's living room is a rectangle 850 cm long and 440 cm wide.

a Work out the area of Claire's living room in m^2.

b Oak flooring costs £70 per square metre. How much will it cost Claire to buy enough oak flooring to cover her living room?

5 Karl is painting his bedroom.
The total area he needs to paint is 355 000 cm^2.
How many of these pots of paint will he need to buy?
Show your working.

AO2

D

6 A circle has a diameter of 0.6 m.
Calculate the area of this circle in cm^2, correct to three significant figures.

7 **a** Calculate the surface area of this cuboid in cm^2.

b Convert this area into m^2.

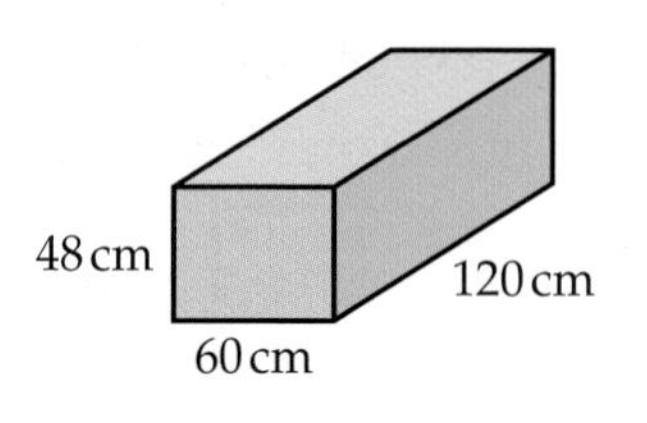

Converting volumes

Volume is measured in mm^3, cm^3, m^3 or km^3.

These two cubes have the same volume.

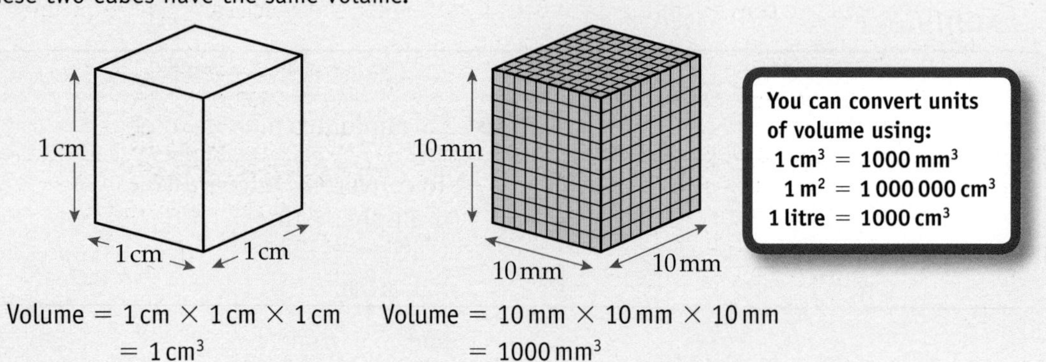

Volume = 1 cm × 1 cm × 1 cm
= 1 cm^3

Volume = 10 mm × 10 mm × 10 mm
= 1000 mm^3

C

Example 2

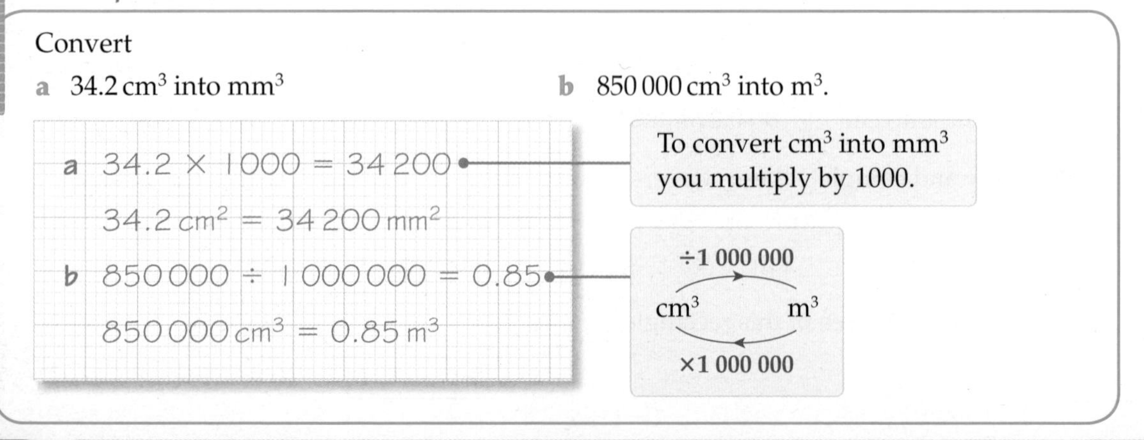

Convert

a 34.2 cm^3 into mm^3

b 850 000 cm^3 into m^3.

a 34.2 × 1000 = 34 200

34.2 cm^2 = 34 200 mm^2

To convert cm^3 into mm^3 you multiply by 1000.

b 850 000 ÷ 1 000 000 = 0.85

850 000 cm^3 = 0.85 m^3

Exercise 30B

1 Convert

a $3.5\,m^3$ into cm^3

b $0.79\,m^3$ into cm^3

c $9200\,mm^3$ into cm^3

d $7\,200\,000\,cm^3$ into m^3.

C

2 Danny says that to convert from m^3 to cm^3 you multiply by 100.

a What mistake has Danny made?

b Copy and complete this sentence.
'To convert from m^3 to cm^3 you multiply by _____.'

C AO2

3 Convert

1 litre = $1000\,cm^3$

a $4300\,cm^3$ into litres

b 0.7 litres into cm^3

c $850\,ml$ into cm^3

d 5700 litres into m^3.

C

4 Work out the volume of this fish tank in

a cm^3

b m^3

c litres.

40 cm, 120 cm, 90 cm

5 A hydroelectric power station uses $2000\,m^3$ of water every second. The reservoir above the power station contains $0.8\,km^3$ of water.

a Convert $0.8\,km^3$ into m^3.

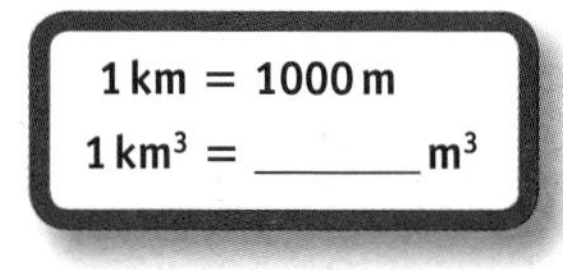

b How long will it take for the power station to use up all the water in the reservoir? (Assume there is no rainfall in the meantime.)
Give your answer to the nearest hour.

C AO3

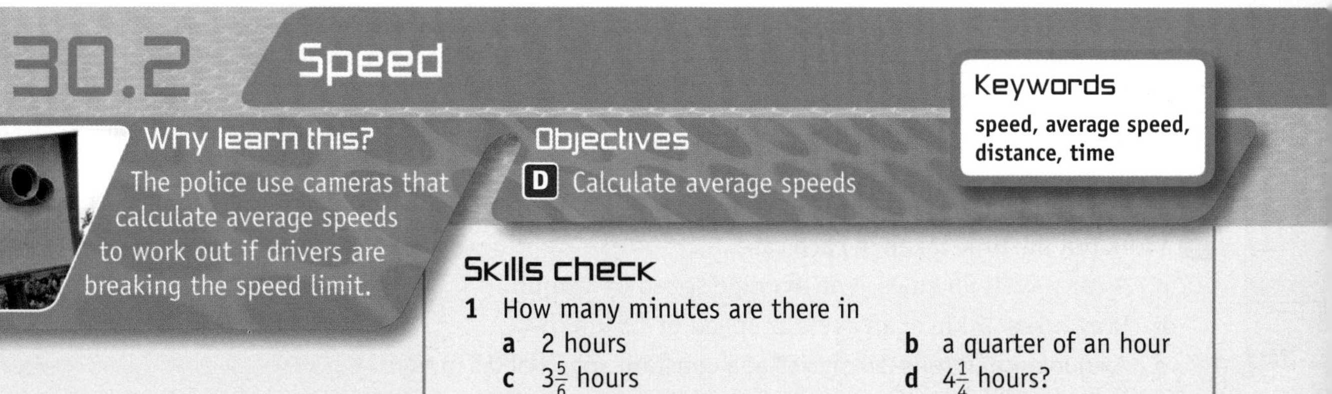

30.2 Speed

Why learn this?

The police use cameras that calculate average speeds to work out if drivers are breaking the speed limit.

Objectives

D Calculate average speeds

Keywords
speed, average speed, distance, time

Skills check

1 How many minutes are there in

a 2 hours

b a quarter of an hour

c $3\frac{5}{6}$ hours

d $4\frac{1}{4}$ hours?

2 Write these times as a fraction of 1 minute.

a 10 seconds

b 30 seconds

c 45 seconds

d 12 seconds

Speed, distance and time

Speed is a measurement of how fast something is travelling.

You can calculate the **average speed** of an object if you know the **distance** travelled and the **time** taken.

$$\text{speed} = \frac{\text{distance}}{\text{time}}$$

$$\text{distance} = \text{speed} \times \text{time}$$

$$\text{time} = \frac{\text{distance}}{\text{speed}}$$

You can use this triangle to remember all three formulae.

- **Cover up the letter you are trying to find.**
- **The position of the other two letters tells you whether to multiply or divide.**

D
S | T

The most common metric units of speed are metres per second (m/s) and kilometres per hour (km/h). The most common imperial unit of speed is miles per hour (mph).

The units given for distance and time will tell you which unit to use for speed.

D

Example 3

An athlete runs 400 m in 52 seconds.
Calculate his average speed, correct to one decimal place.

$\text{speed} = \frac{\text{distance}}{\text{time}}$

$400 \div 52 = 7.7$ (1 d.p.)

The average speed is 7.7 m/s.

Distance is measured in metres and time is measured in seconds, so the unit for speed in this example is metres per second (m/s).

Exercise 30C

D

1 Work out the average speed for each journey.

- **a** A train travels 100 km in 2 hours.
- **b** A swimmer swims 50 m in 40 seconds.
- **c** An aeroplane flies 2400 miles in 4 hours.
- **d** A cyclist rides 22.5 km in $2\frac{1}{2}$ hours.

2 Work out the distance travelled for each journey.

- **a** A car travels at 40 mph for 3 hours.
- **b** A cheetah runs at 16 m/s for 12 seconds.
- **c** A skydiver falls at 90 m/s for 45 seconds.
- **d** A cyclist rides at 12 km/h for half an hour.

You will need to use the formula distance = speed × time.

3 Work out the time taken in each case.

- **a** A car travels 96 miles at an average speed of 32 mph.
- **b** Prav hikes 27 km at an average speed of 4.5 km/h.
- **c** A snooker ball rolls 0.8 metres at a constant speed of 0.5 m/s.

D

AO2

4 The distance from London to Southampton is 80 miles.
Jonathan leaves home at 1:30 pm and arrives in Southampton at 3:15 pm.

- **a** How long does Jonathan's journey take?
- **b** Calculate his average speed in mph. Give your answer to one decimal place.

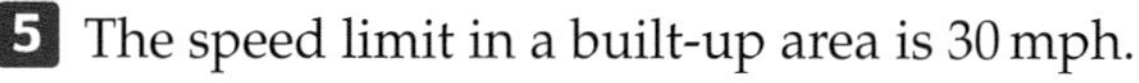

5 The speed limit in a built-up area is 30 mph.

a Convert 30 miles into km.

b Write down this speed limit in km/h.

6 Alison wants to convert 54 km/h into m/s.
Copy and complete this diagram showing each stage of the conversion.

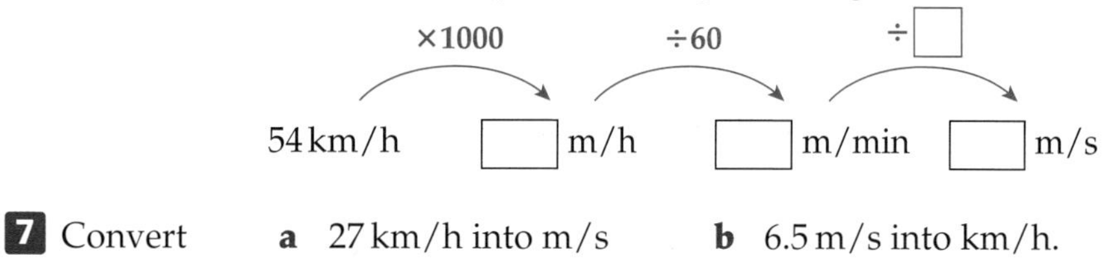

7 Convert **a** 27 km/h into m/s **b** 6.5 m/s into km/h.

8 Ben cycled 8 km in 45 minutes.
He rested for an hour, then cycled a further 10 km in an hour.
Work out his average speed for the whole journey.
Give your answer in km/h, correct to one decimal place.

9 A snail travels at a speed of 0.1 km/h.
Pete's pet tortoise walks 20 metres in 3 minutes.
Which one is faster?

D

AO2

D

AO3

30.3 Density

Keywords
density, volume, mass

Why learn this?
Icebergs float because ice has a lower density than water.

Objectives
C Make calculations using density

Skills check

1 Work out these calculations, giving your answers to one decimal place.
a $40 \div 6$ **b** $8 \div 0.3$ **c** $200 \div 15.4$ **d** $347.1 \div 27$

2 The formula $v = u + at$ is used in physics.
Work out v if
a $u = 4$, $a = 2$ and $t = 10$ **b** $u = 6.5$, $a = 2.5$ and $t = 52$

HELP Section 15.5

Density, mass and volume

Density is the amount of a substance contained in a certain **volume**.

Density is measured in grams per cubic centimetre (g/cm^3) or kilograms per cubic metre (kg/m^3).

You can calculate the density of a material if you know its **mass** and its volume.

A 1 cm^3 block of iron has a mass of 8 grams. The density of iron is 8 g/cm^3.

$$\text{density} = \frac{\text{mass}}{\text{volume}}$$

$$\text{mass} = \text{density} \times \text{volume}$$

$$\text{volume} = \frac{\text{mass}}{\text{density}}$$

You can use this triangle to remember all three formulae.
- **Cover up the letter you are trying to find.**
- **The position of the other two letters tells you whether to multiply or divide.**

M
D | V

C

Example 4

The density of cork is 240 kg/m^3. Work out the mass of 3.4 m^3 of cork.

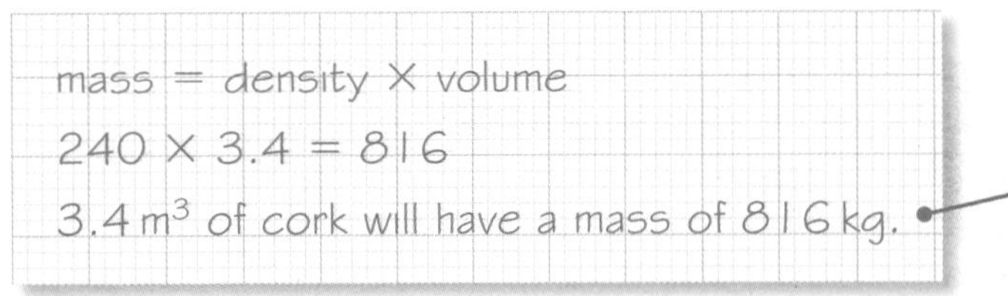

Density is given in kg/m^3 and volume is given in m^3 so the unit for mass in this example is kg.

Exercise 30D

C

1 Andrew records the volume and mass of four different substances for a physics experiment.
Work out the density of each substance.
Give your answers in g/cm^3 correct to one decimal place.

Substance	Volume (cm^3)	Mass (g)
gold	2	38
iron	25	195
lead	18.5	209
aluminium	30	80

2 Use this table to work out the mass of

a 20 cm^3 of platinum

b 14 cm^3 of mercury

c 55.2 cm^3 of copper.

Substance	Density (g/cm^3)
magnesium	1.7
copper	9
mercury	13.6
platinum	21.4

3 Use the table in Q2 to work out the volume of

a 450 g of copper **b** 200 g of magnesium **c** 35 g of platinum.

4 Karl weighs a 20 cm^3 block of silver. It has a mass of 210 g.

a Work out the density of silver.

b Use your answer to part **a** to calculate the volume of a 480 g block of silver.

C

5 Petrol has a density of 0.7 g/cm^3. The fuel tank on a car holds 45 litres of petrol.

a Convert 45 litres into cm^3.

b Work out the mass of the petrol in the tank when it is full.

6 Abi measures the mass of this wooden block as 2 kg.

a Work out the volume of the block.

b Work out the density of the wood in g/cm^3.

c Work out the mass of a 600 cm^3 block of the same wood.

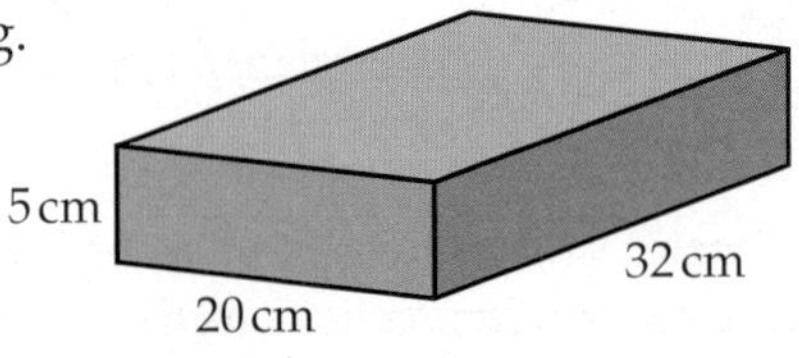

A02

7 This steel beam has a cross-sectional area of 0.15 m^2 and a length of 3.6 m.
It weighs 4200 kg.

Work out the density of the steel used to make the beam in kg/m^3.
Give your answer to the nearest whole number.

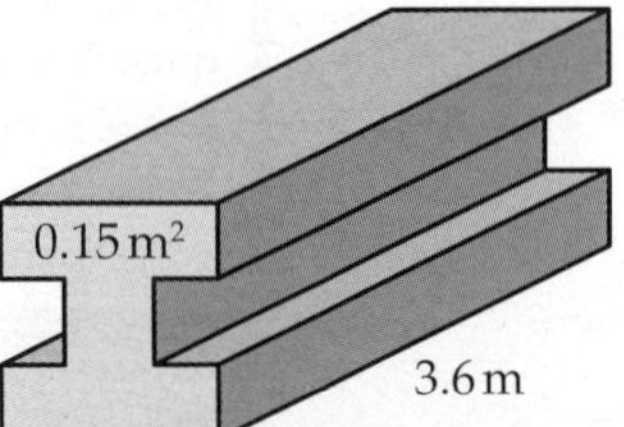

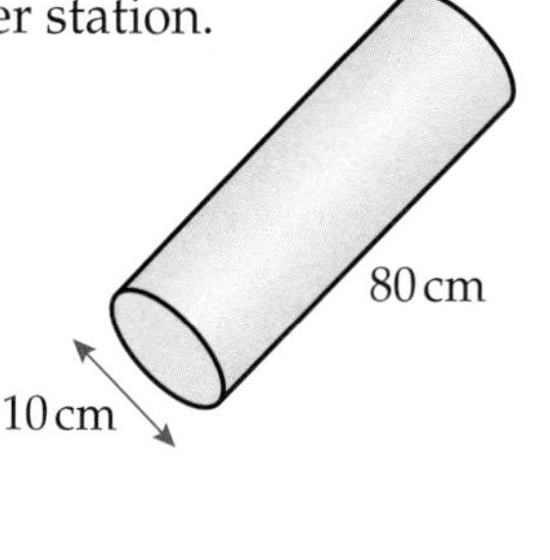

8 The diagram shows a cylindrical fuel rod from a nuclear power station.
The rod is made from uranium with a density of 19 g/cm³.

a Calculate the volume of the rod in terms of π.

b Calculate the mass of the rod in kg. Give your answer to the nearest whole number.

C

9 A silver bracelet is made from 4.5 cm³ of silver.
The density of silver is 10.5 g/cm³.
Silver bracelets are priced by weight at 80p per gram.
How much does the bracelet cost?

AO2

10 An iceberg has a volume of 18 000 m³.
The density of ice is 0.92 g/cm³.
Calculate the mass of the iceberg in tonnes.

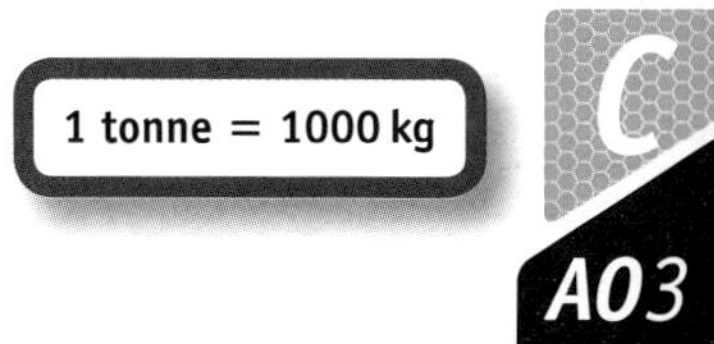

C

AO3

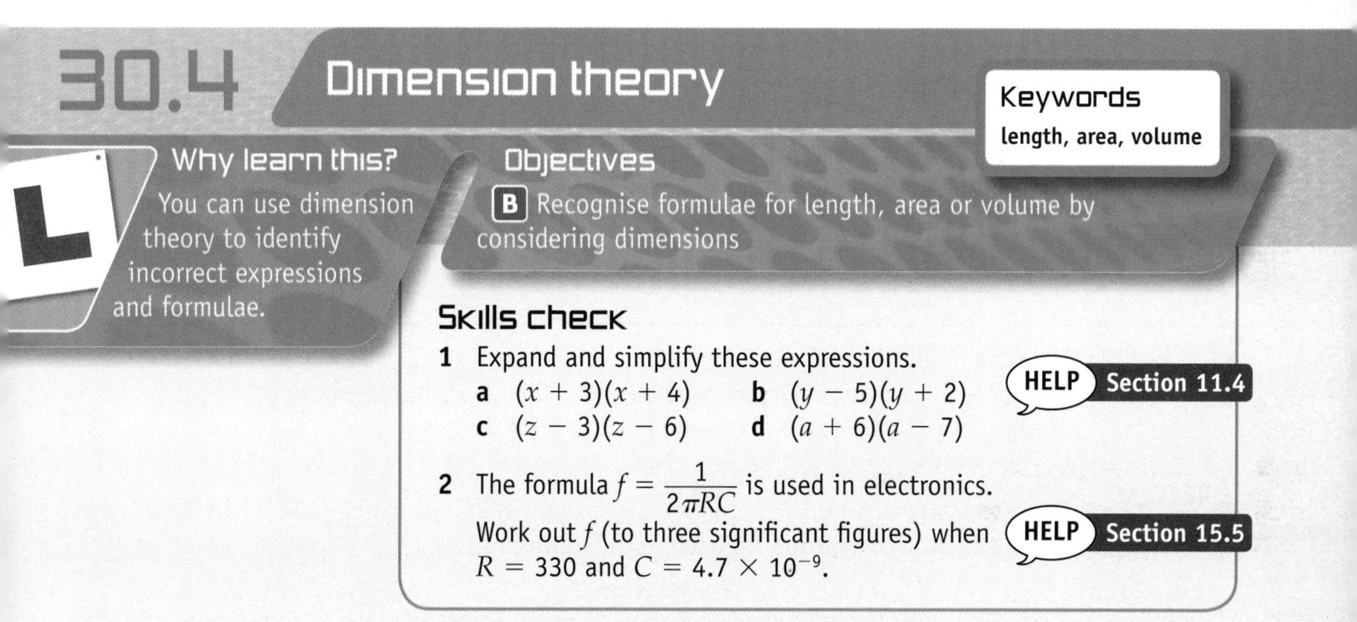

30.4 Dimension theory

Keywords
length, area, volume

Why learn this?
You can use dimension theory to identify incorrect expressions and formulae.

Objectives
B Recognise formulae for length, area or volume by considering dimensions

Skills check

1 Expand and simplify these expressions.
a $(x + 3)(x + 4)$ **b** $(y - 5)(y + 2)$
c $(z - 3)(z - 6)$ **d** $(a + 6)(a - 7)$

HELP Section 11.4

2 The formula $f = \dfrac{1}{2\pi RC}$ is used in electronics.
Work out f (to three significant figures) when $R = 330$ and $C = 4.7 \times 10^{-9}$.

HELP Section 15.5

Length, area and volume

Length is a one-dimensional quantity.

In a formula for a length, each term must contain
- one letter representing a length.

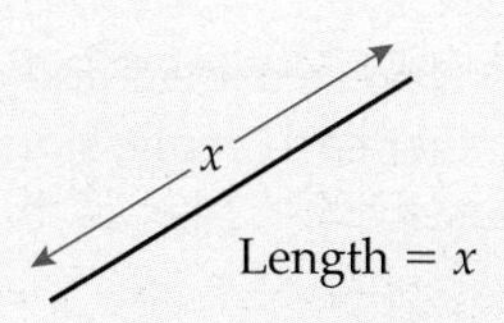

Area is a two-dimensional quantity.

In a formula for an area, each term must contain either
- two lengths multiplied together, or
- one letter representing an area.

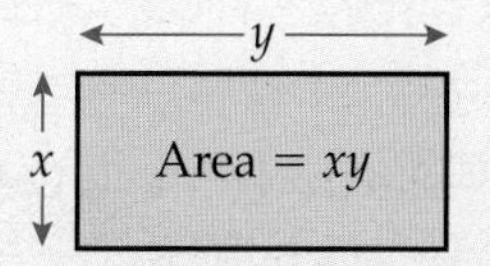

Volume is a three-dimensional quantity.

In a formula for a volume, each term must contain
- three lengths multiplied together, or
- a length multiplied by an area, or
- one letter representing a volume.

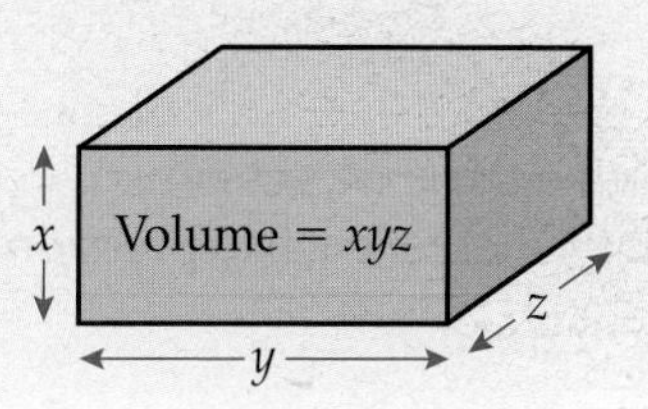

Example 5

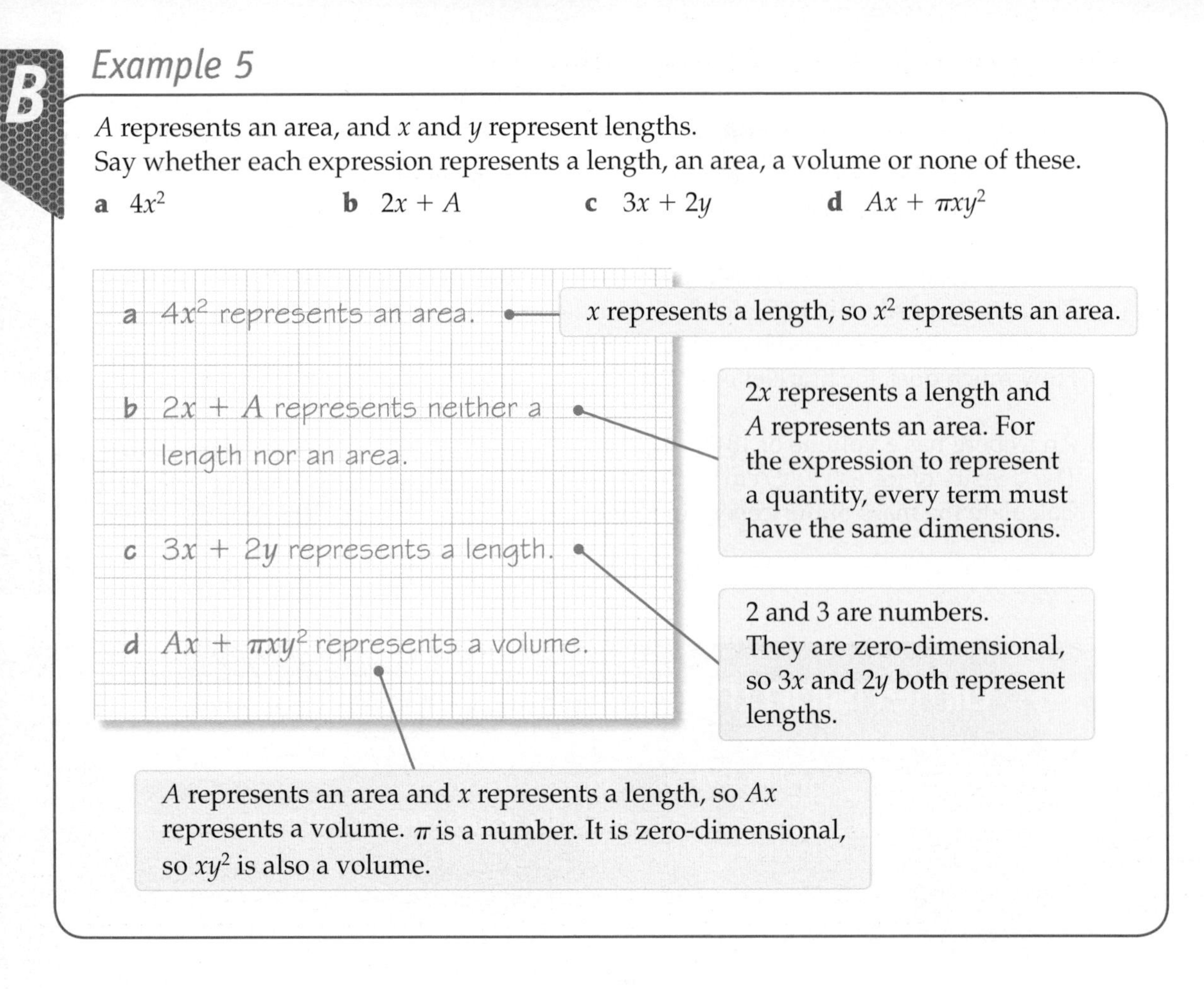

Exercise 30E

1 Write down length, area or volume for each of the following.

a $24\,\text{cm}^3$ **b** $8\,\text{km}$

c $24\,\text{mm}^2$ **d** $42\,\text{m}^3$

2 P and Q represent areas and a, b and c represent lengths.
Say whether each expression represents a length, an area, a volume or none of these.

a $P + Q$ **b** $3a^2b$

c $2ab + 2c^2$ **d** $3x^3 + 2\pi Q$

3 In these expressions, m is measured in cm, n is measured in cm^2 and p is measured in cm^3. Write down the units for each of the following.

a m^2 **b** $2mn$

c $nm + 2p$ **d** $2\pi p - m^3$

4 In the these expressions x, y and z all represent lengths.
Say whether each expression represents a length, an area, a volume or none of these.

a $2x(x + y)$ **b** $3(x + y)$

c $z(2x^2y + z^3)$ **d** $4x^2(y + 2z)$

5 A torus is the mathematical name for a ring.

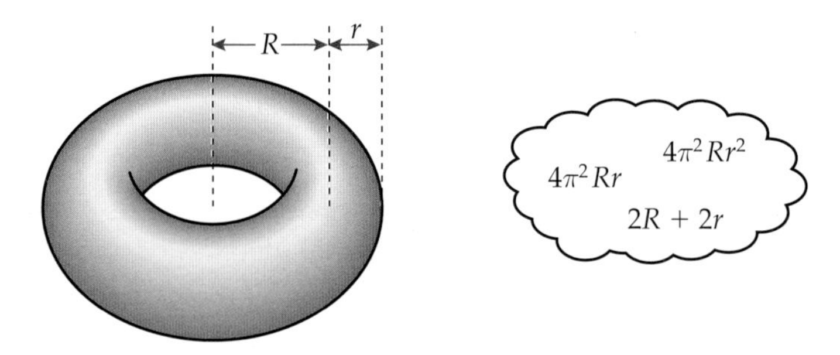

Write down the expression from the cloud which represents

a the diameter of the torus

b the surface area of the torus

c the volume of the torus.

6 Zoe writes down the following formula for the volume of a sphere of radius r.

$$V = \frac{4}{3}\pi r^2$$

Use dimensions to show why Zoe's formula must be wrong.

7 In these expressions a is measured in m, b is measured in m^2 and c is measured in m^3. Write down the units for

a $\frac{ab}{2}$ **b** $\frac{b}{a}$ **c** $\frac{2c}{a^2}$ **d** $\frac{2a(b + a^2)}{b}$

8 Which formula is most likely to be correct for the perimeter of this shape. Give a reason for your answer.

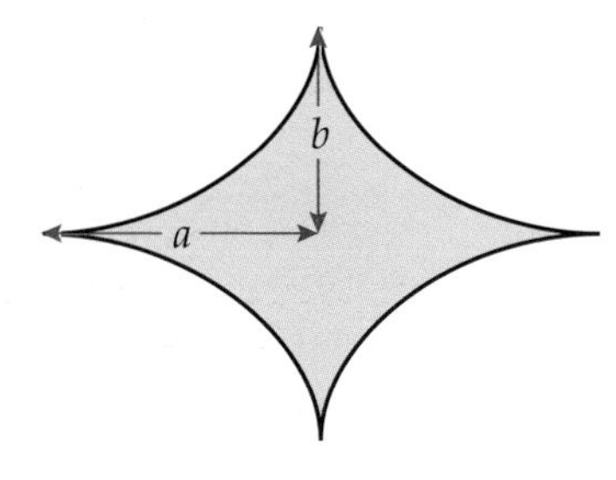

A $P = \frac{2\pi(a^2 + b^2)}{(a + b)^2}$ **B** $P = \frac{4(a^2 + ab + b^2)}{(a + b)}$ **C** $P = \frac{2a(a^2 + b^2)}{a + b}$

AO2

B

9 A garden centre sells plant pots in different sizes. This notice is displayed in the garden centre.

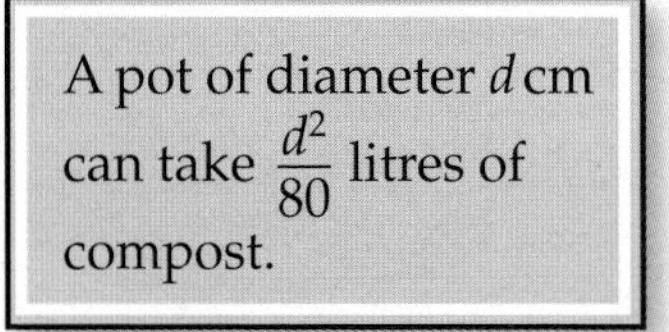

Do you think the notice will provide an accurate guide for all the different sizes of plant pot?
Give a reason for your answer.

AO3

Review exercise

D

1 Convert $280\ \text{cm}^2$ into m^2. **[2 marks]**

2 The course for a cycling race is 56 km long.
The winning cyclist averages 25.8 km/h.
How long did the winning cyclist take to complete the course?
Give your answer in hours correct to one decimal place. **[2 marks]**

3 Nick is travelling at 110 km/h when he passes this road sign.

SPEED
LIMIT
70
MPH

You need to remember the miles to km conversion factor.

AO2

Is Nick breaking the speed limit?
Show working to support your answer. **[2 marks]**

C

4 A company uses foam packing blocks when it transports its products.
The foam has a density of $300\ \text{kg/m}^3$.

a Work out the mass of a $2.6\ \text{m}^3$ block of foam. **[2 marks]**

b Work out the volume of a block of foam with a mass of 120 kg. **[2 marks]**

C

5 Gold costs £22 per gram. The density of gold is $19.3\ \text{g/cm}^3$.

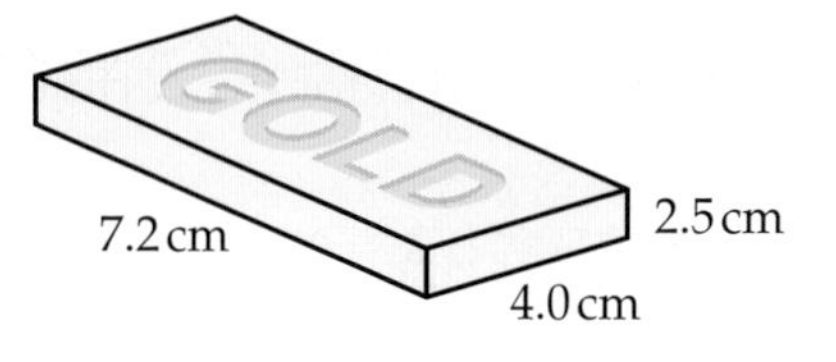

AO3

Find the value of this gold bar to the nearest pound. **[5 marks]**

C

6 The loading bay on a van has a volume of $3.5\ \text{m}^3$.
Write this volume in cm^3. **[2 marks]**

C

7 A cylindrical water butt has a radius of 30 cm and a height of 80 cm.

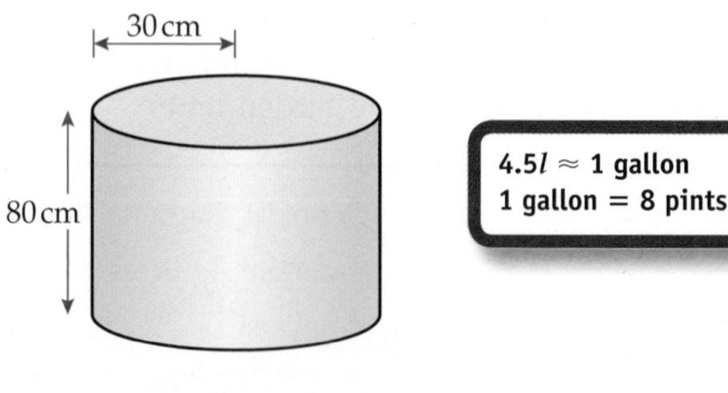

a Work out the volume of the water butt in cm^3 correct to 3 s.f. **[2 marks]**

AO2

b Katherine says the water butt can hold about 400 pints of water.
Show working to demonstrate that Katherine is right. **[3 marks]**

8 This diagram shows a cylindrical jug.

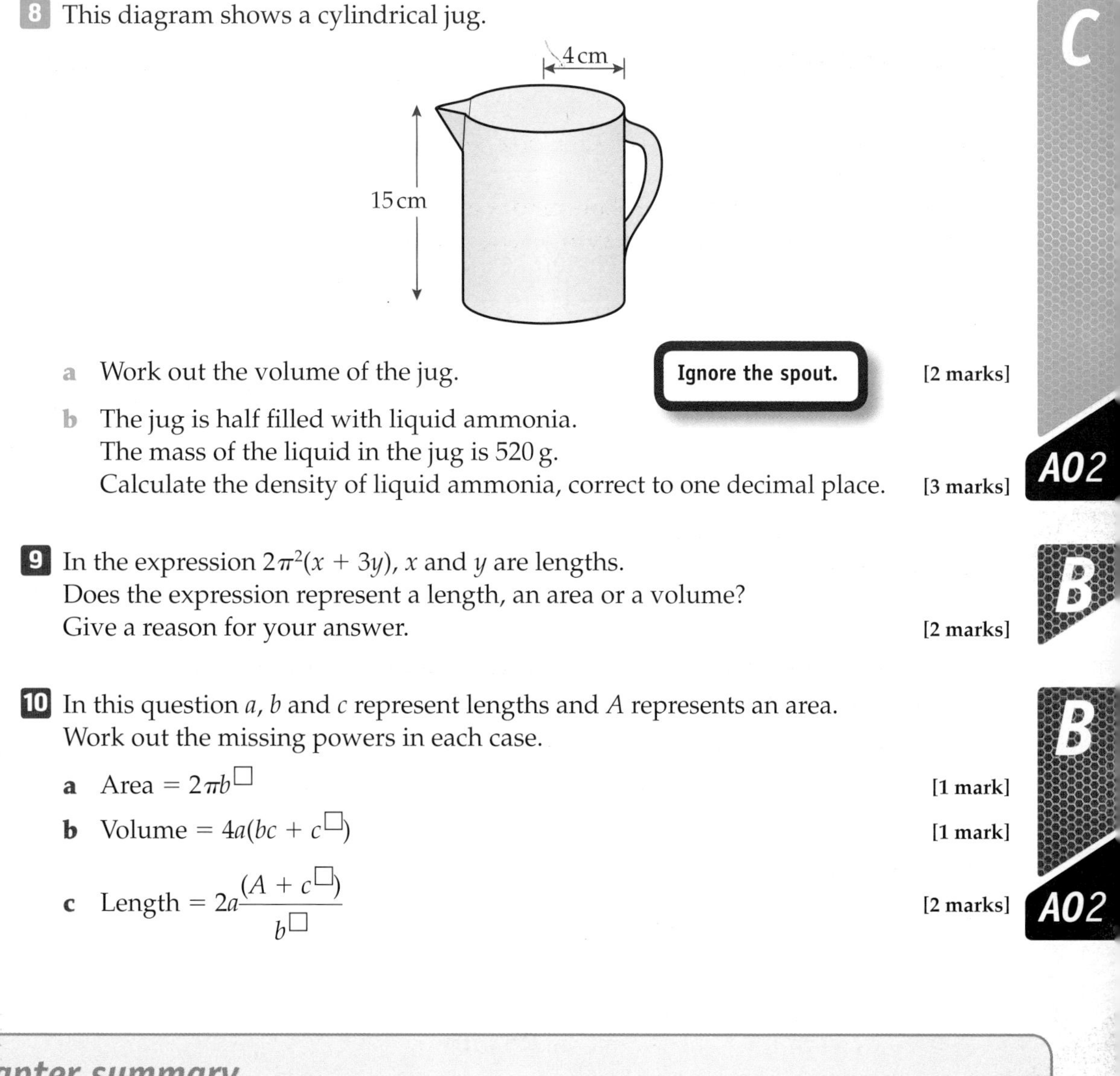

C

a Work out the volume of the jug. **Ignore the spout.** [2 marks]

b The jug is half filled with liquid ammonia.
The mass of the liquid in the jug is 520 g.
Calculate the density of liquid ammonia, correct to one decimal place. [3 marks]

AO2

9 In the expression $2\pi^2(x + 3y)$, x and y are lengths.
Does the expression represent a length, an area or a volume?
Give a reason for your answer. [2 marks]

B

10 In this question a, b and c represent lengths and A represents an area.
Work out the missing powers in each case.

B

a Area $= 2\pi b^{\square}$ [1 mark]

b Volume $= 4a(bc + c^{\square})$ [1 mark]

c Length $= 2a\dfrac{(A + c^{\square})}{b^{\square}}$ [2 marks]

AO2

Chapter summary

In this chapter you have learned how to

- convert between different units of area **D**
- calculate average speeds **D**
- make calculations using density **C**
- convert between different units of volume **C**
- recognise formulae for length, area or volume by considering dimensions **B**

Pembrokeshire cycle ride

Ted is one of the organisers of the Pembrokeshire cycle ride. Cyclists follow one of two routes; the 115 mile route or the 64 mile route. Both routes start and finish at Begelly.

There are feeding stations, where cyclists are given refreshments, on both routes. On the 115 mile route there are feeding stations after Fishguard, Newgale and Castlemartin. On the 64 mile route there are feeding stations after Newgale and Pembroke.

Question bank

1 At which mile marker is the Newgale feeding station on the 64 mile route?

Ted uses the slowest and fastest average speeds of last year's cyclists to work out the times he expects the first and last cyclist to arrive at each feeding station.

2 How long would it take a cyclist travelling at the fastest average speed to reach the Fishguard feeding station on the 115 mile route?

The cyclists can start their route between 7:30 am and 8:30 am.

3 At what time would a cyclist starting at 8:30 am and travelling at the slowest average speed reach the Pembroke feeding station on the 64 mile route?

4 Copy and complete the table to show the times Ted expects the first and last cyclist to arrive at each feeding station.

Altogether 550 cyclists have registered for the event. Two fifths of the cyclists have registered for the 64 mile route, and the rest have registered for the 115 mile route.

At each feeding station, 1 cereal bar, 1 banana and 2 litres of drink are provided for each cyclist.

5 Work out the total number of cereal bars, bananas and litres of drink that are needed at each feeding station.

When Ted orders the refreshments, he orders an extra 20% of each item, so he is certain he has enough.

6 Work out the total number of cereal bars, bananas and litres of drink that Ted needs to order.

115 mile route	
Feeding station	**Mile marker**
Fishguard	35
Newgale	62
Castlemartin	95
64 mile route	
Feeding station	**Mile marker**
Newgale	28
Pembroke	49

	Fastest average speed (mph)	Slowest average speed (mph)
115 mile route	18.4	10.2
64 mile route	18.5	7.3

	Expected time of arrival	
Feeding station	first cyclist	last cyclist
Fishguard		
Newgale		
Castlemartin		
Pembroke		

31

Enlargement and similarity

BBC Video

This chapter is about constructing enlargements of shapes.

You can use a microscope to view enlarged images of very small objects. The photograph shows an electron microscope image of pollen grains – in reality, these are less than a thousandth of a centimetre in diameter.

Objectives

This chapter will show you how to

- enlarge shapes using positive scale factors, including fractional scale factors **D** **C**
- identify the relationship between the scale factor of an enlargement and the ratio of the lengths of any two corresponding line segments and apply this to 2-D shapes **D**
- understand the implications of enlargement for perimeter **D**
- understand that enlargements are described by a centre of enlargement and a scale factor **D**
- understand and use similarity **B**

Before you start this chapter

1 Match each shape to one of the words.

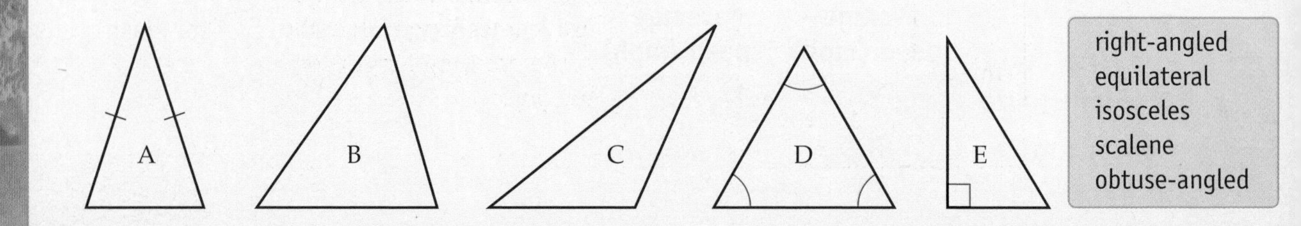

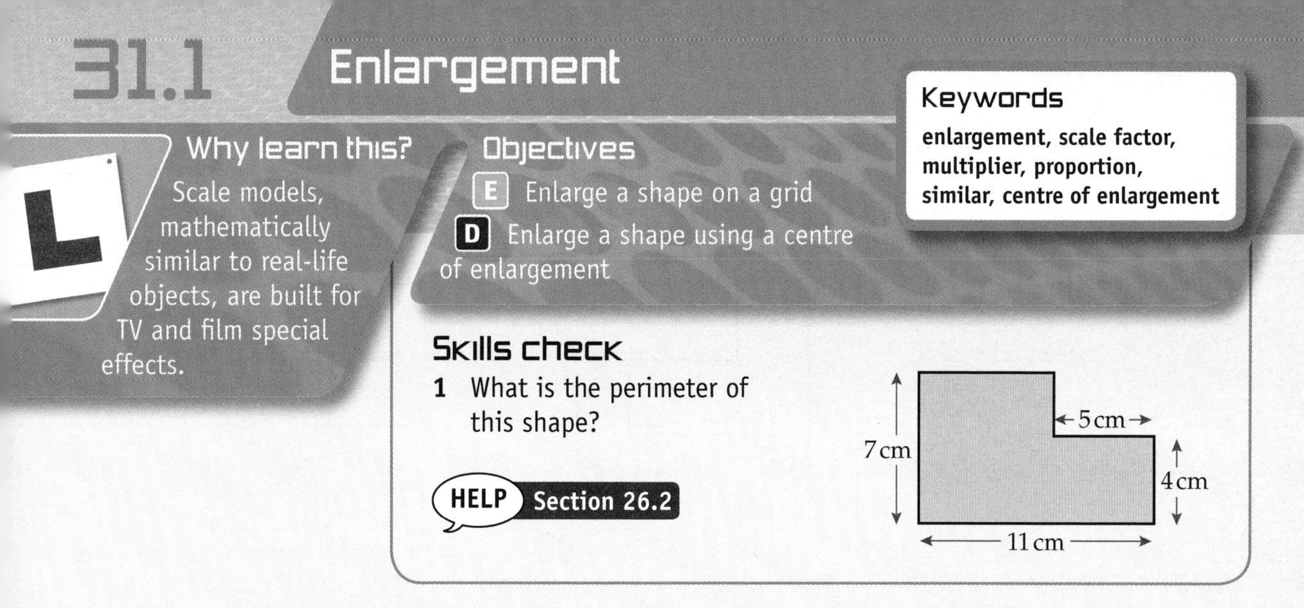

31.1 Enlargement

Why learn this?

Scale models, mathematically similar to real-life objects, are built for TV and film special effects.

Objectives

E Enlarge a shape on a grid

D Enlarge a shape using a centre of enlargement

Keywords

enlargement, scale factor, multiplier, proportion, similar, centre of enlargement

Skills check

1 What is the perimeter of this shape?

HELP Section 26.2

Enlargement

An **enlargement** of an object changes the size of the object but keeps its shape the same.

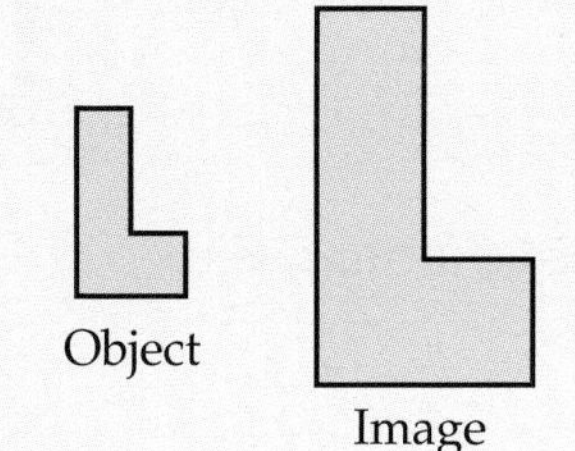

Look at the two L shapes.

The image is an enlargement of the object.

The **scale factor**, or **multiplier**, is 2.

This means that every length on the image is twice as long as that on the object.

In an enlargement all the angles stay the same, but all the lengths are changed in the same **proportion**. The image is **similar** to the object.

Using a grid helps you draw enlargements.

Example 1

E

Enlarge the object by scale factor 3.

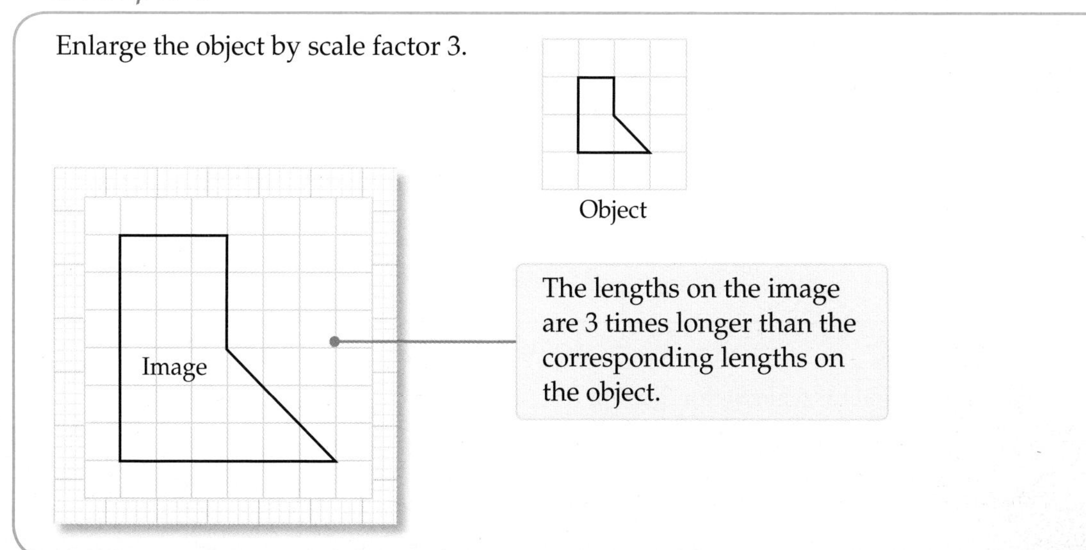

E

Exercise 31A

1 Enlarge each object by the scale factor given.

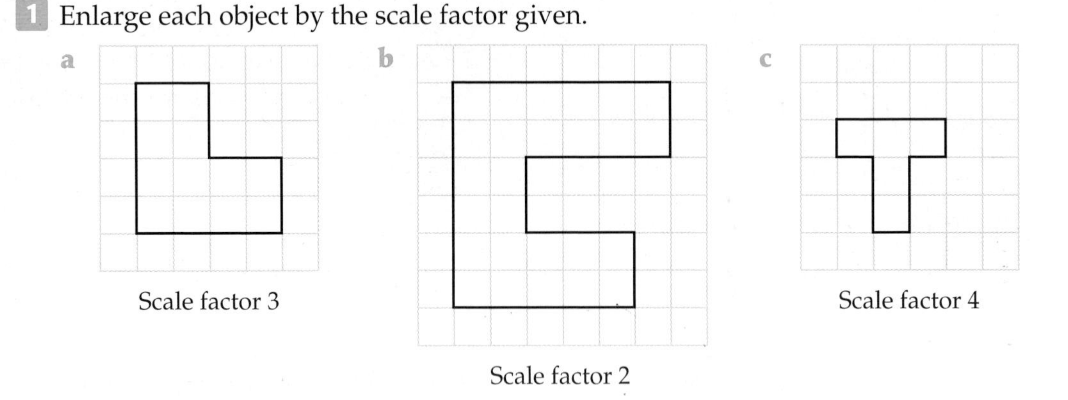

a Scale factor 3 **b** Scale factor 2 **c** Scale factor 4

2 Look at each part of your answer to Q1.

i By counting squares, write down the perimeter of the object and the perimeter of the image.

ii How many times longer is the perimeter of the image than the perimeter of the object?

> **When you do an enlargement, you should find that the perimeter is multiplied by the same scale factor.**

3 Enlarge each object by the scale factor given.

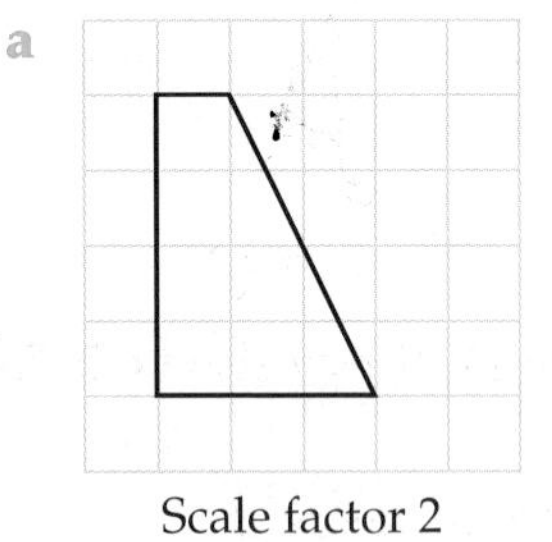

a Scale factor 2

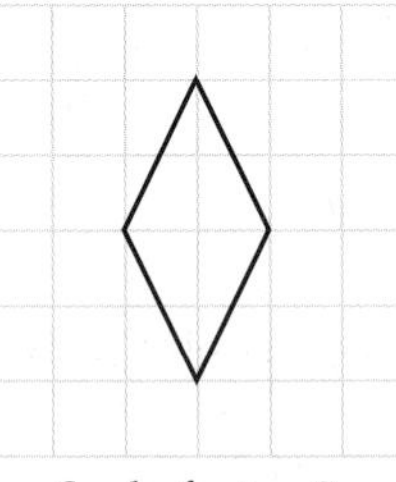

b Scale factor 2

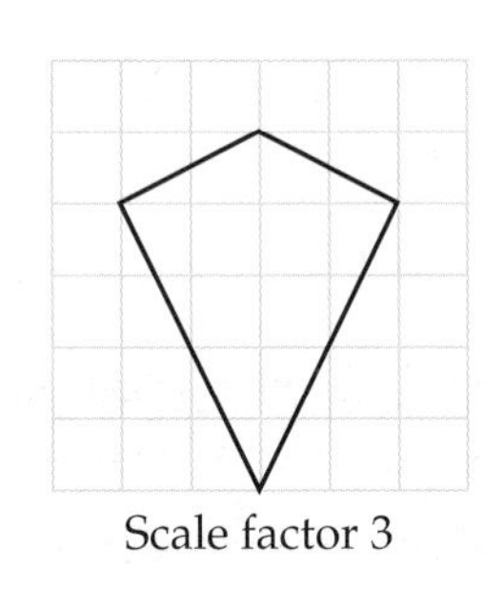

c Scale factor 3

Centre of enlargement

When you enlarge a shape from a **centre of enlargement**, the distances from the centre to each point are multiplied by the scale factor.

D

Example 2

Copy triangle ABC.

Enlarge the triangle by scale factor 2 using the point O as the centre of enlargement.

O •

C

A B

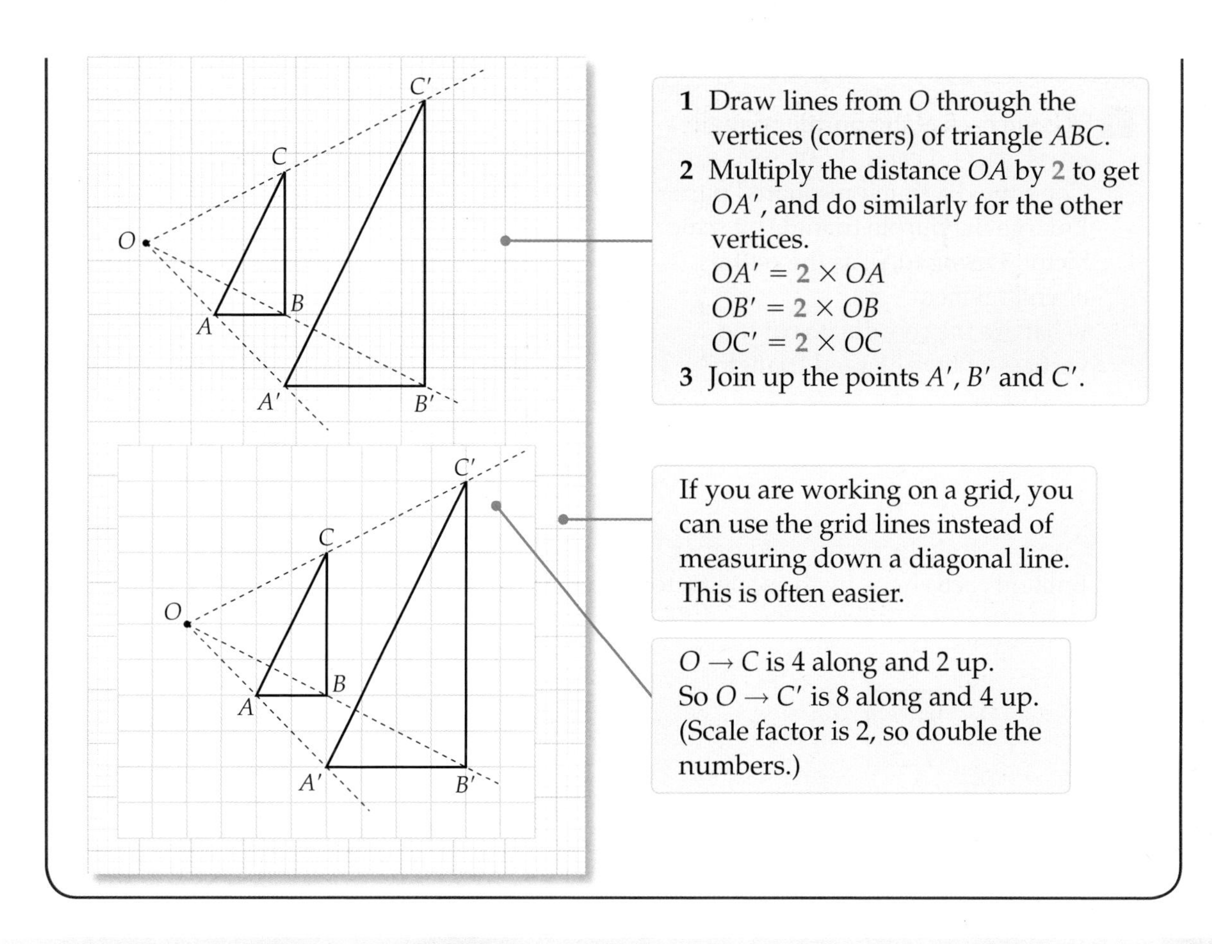

The position of the centre of enlargement

The centre of enlargement may be outside the object (as in Example 2). It may also be inside the object, on an edge or at a corner.

This means that the object and image may overlap, or one may be inside the other.

Example 3

a Enlarge the blue object by scale factor 2 using C as the centre of enlargement.

b Enlarge the blue object by scale factor 2 using O as the centre of enlargement.

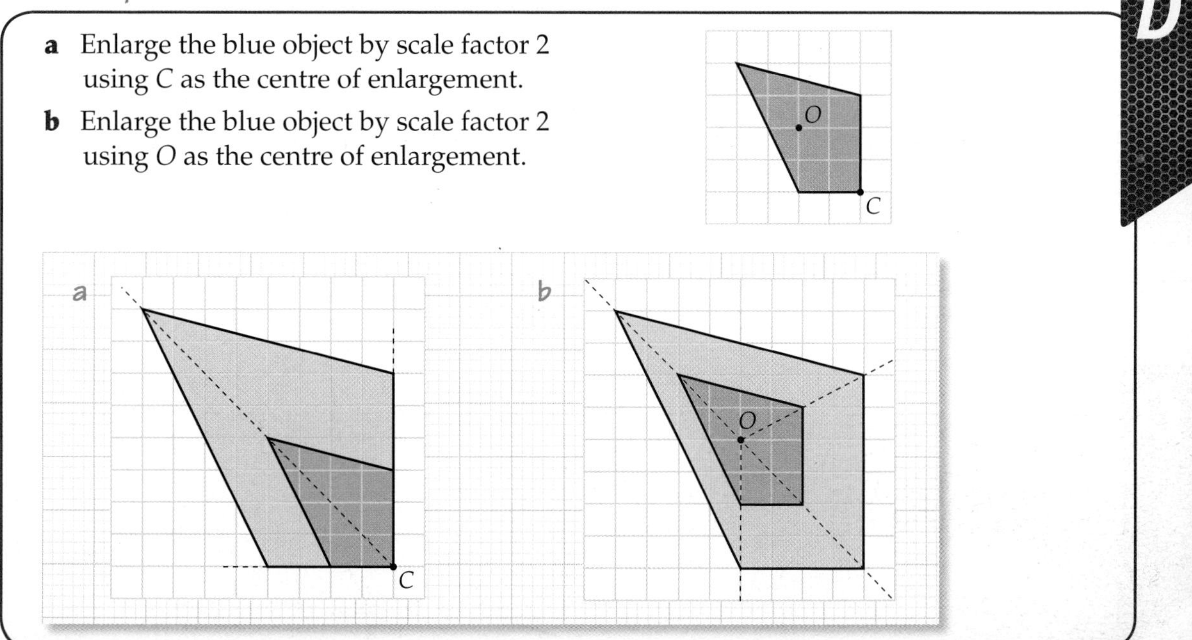

Exercise 31B

D

1 The vertices of the purple triangle are (1, 1), (4, 1) and (1, 3).
Copy the diagram on squared paper.
Enlarge the purple triangle by scale factor 3 using (0, 0) as the centre of enlargement.
What are the coordinates of the vertices of the enlarged triangle?

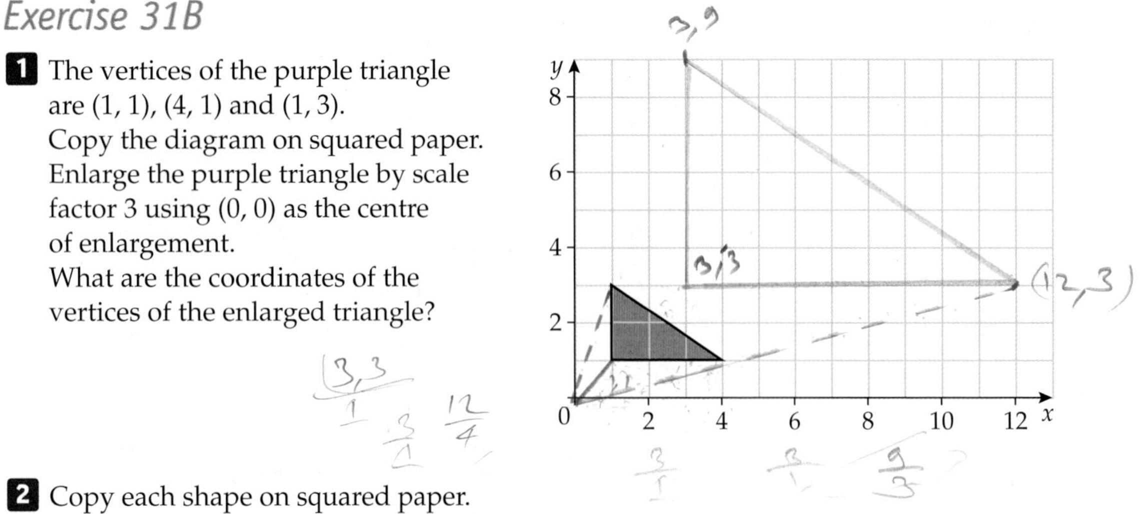

2 Copy each shape on squared paper.
Enlarge each shape by the scale factor given using O as the centre of enlargement.

a

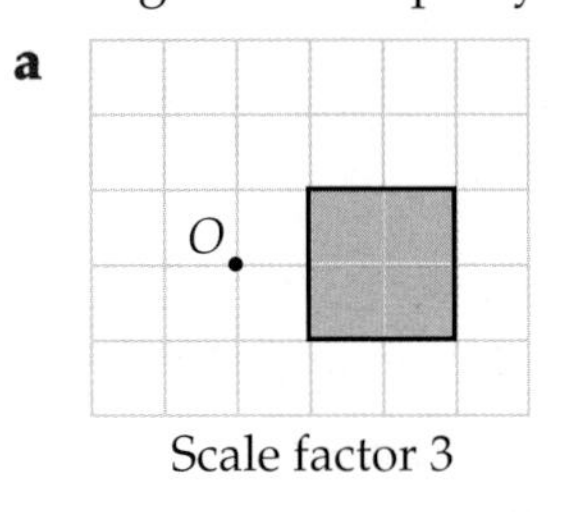

Scale factor 3

b

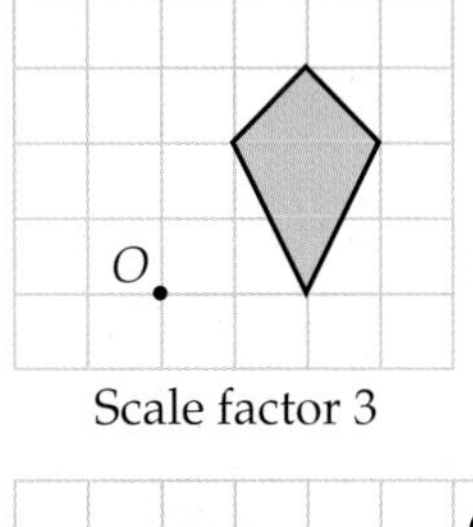

Scale factor 3

c

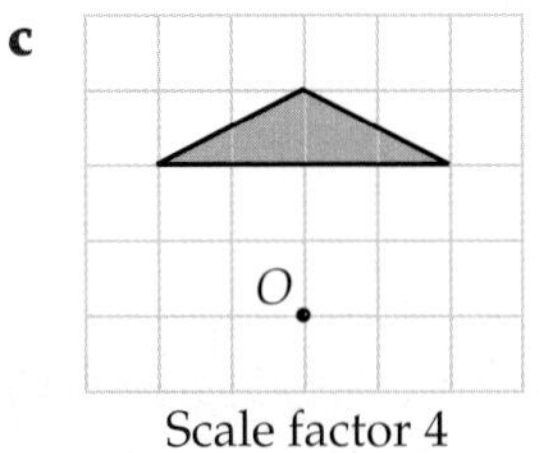

Scale factor 4

d

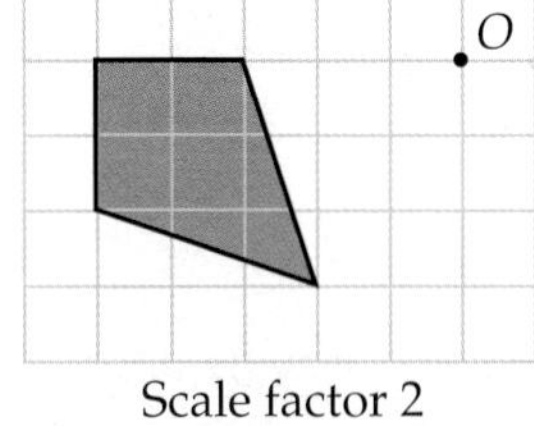

Scale factor 2

3 Copy each shape on squared paper.
Enlarge each one by the scale factor given using O as the centre of enlargement.

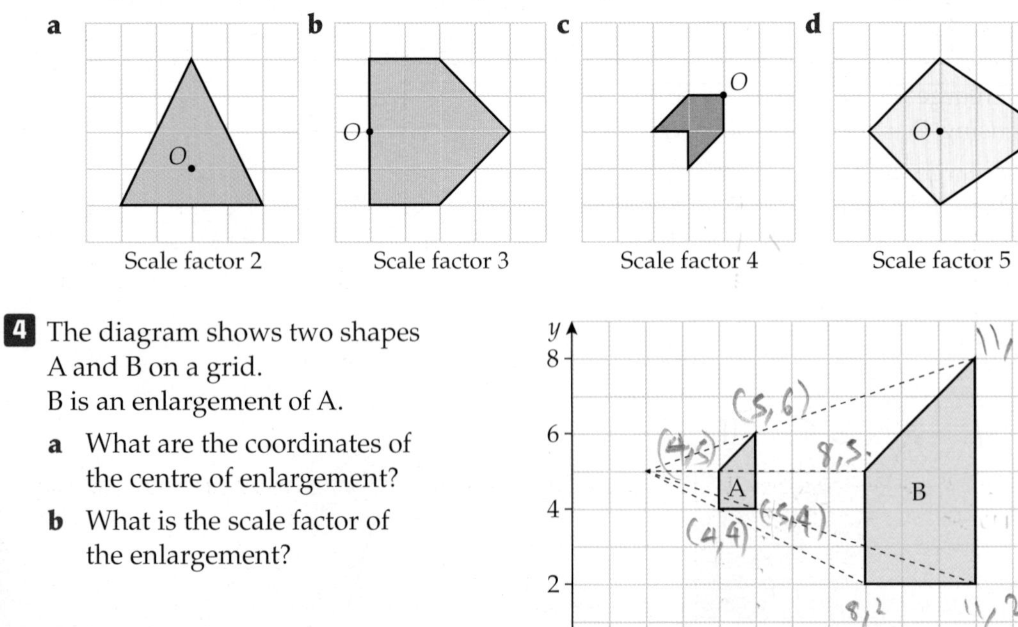

4 The diagram shows two shapes A and B on a grid.
B is an enlargement of A.

a What are the coordinates of the centre of enlargement?

b What is the scale factor of the enlargement?

5 Shape Q is an enlargement of shape P.

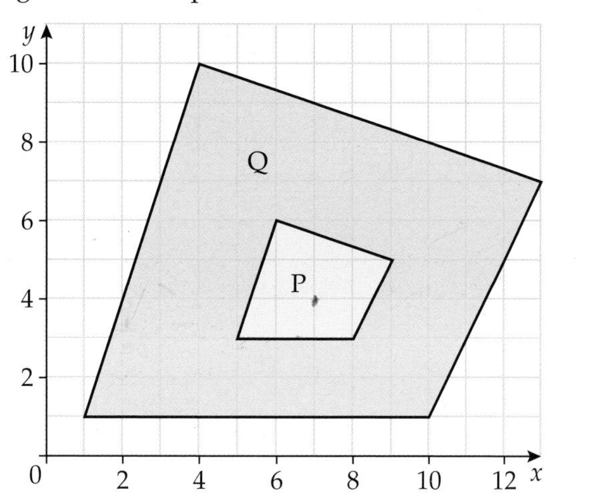

a What is the scale factor of the enlargement?

b What are the coordinates of the centre of enlargement?

6 On squared paper draw a coordinate grid with *x*- and *y*-axes from 0 to 7. Plot a triangle with vertices at (2, 2), (5, 2) and (2, 5). Label the triangle A.

a Draw the image of triangle A after an enlargement by scale factor 2 with (3, 3) as the centre of enlargement. Label the image B.

b What are the coordinates of the vertices of B?

31.2 Enlargements with fractional scale factors

Why learn this?

Surprising but true! A mathematical enlargement can be *smaller* than the original object.

Objectives

C Enlarge a shape using a fractional scale factor

Skills check

1 Work out

a $6 \times \frac{1}{2}$ **b** $9 \times \frac{1}{2}$ **c** $4 \times \frac{1}{3}$

2 Work out

a $-\frac{1}{2} \times 2$ **b** $-\frac{1}{2} \times 5$ **c** $-\frac{1}{3} \times 6$

Fractional scale factors

You can have an enlargement with a scale factor less than 1. In this case, the enlarged image is smaller than the object.

A scale factor greater than 1 gives an image that is larger than the object.

A scale factor less than 1 gives an image that is smaller than the object.

A scale factor of 1 gives an image that is the same size as the object.

C

Example 4

a Enlarge shape A by scale factor $\frac{1}{3}$ using point O as the centre of enlargement. Label the enlarged shape B.

b What enlargement would transform B back to A?

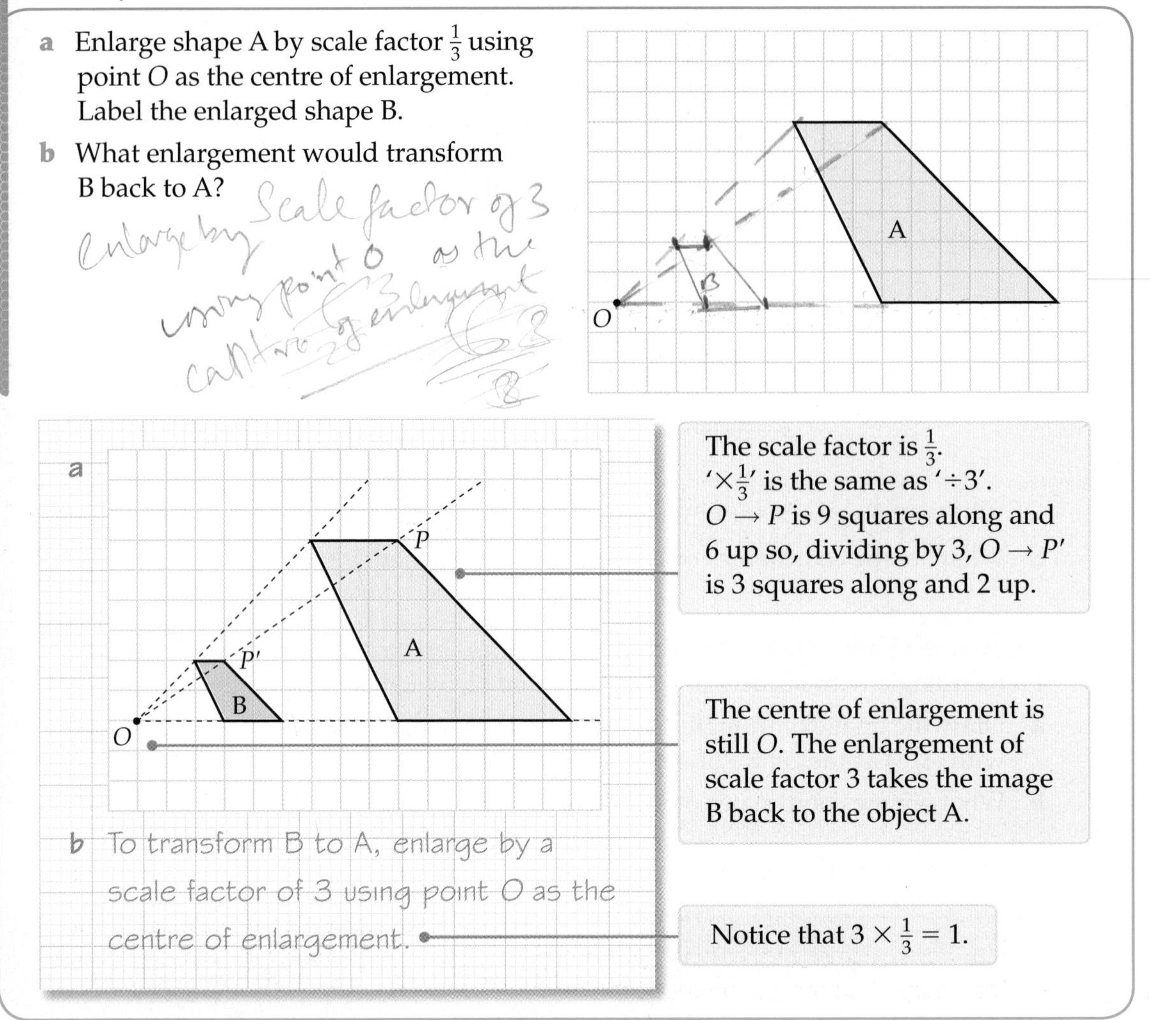

The scale factor is $\frac{1}{3}$.
'$\times\frac{1}{3}$' is the same as '$\div 3$'.
$O \rightarrow P$ is 9 squares along and 6 up so, dividing by 3, $O \rightarrow P'$ is 3 squares along and 2 up.

The centre of enlargement is still O. The enlargement of scale factor 3 takes the image B back to the object A.

b To transform B to A, enlarge by a scale factor of 3 using point O as the centre of enlargement.

Notice that $3 \times \frac{1}{3} = 1$.

Exercise 31C

C

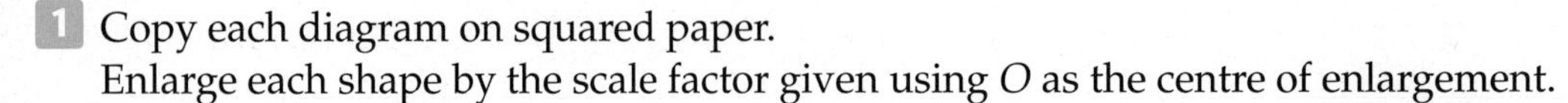

1 Copy each diagram on squared paper.
Enlarge each shape by the scale factor given using O as the centre of enlargement.

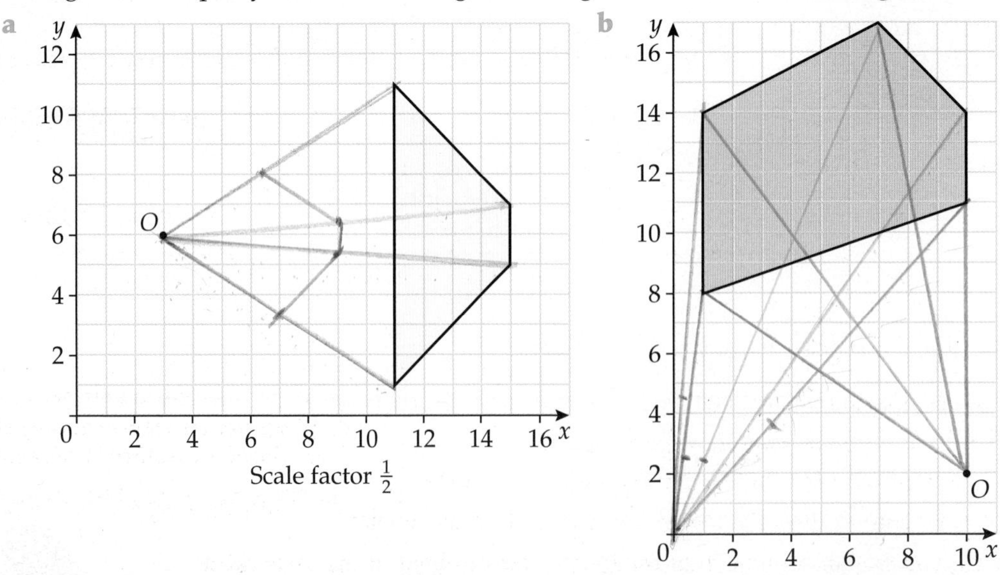

c

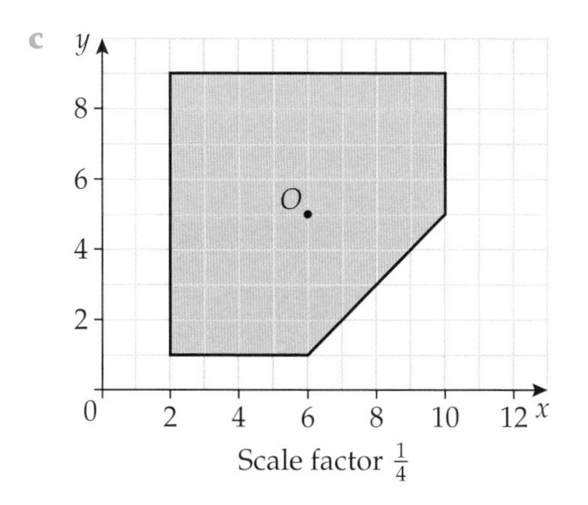

Scale factor $\frac{1}{4}$

d

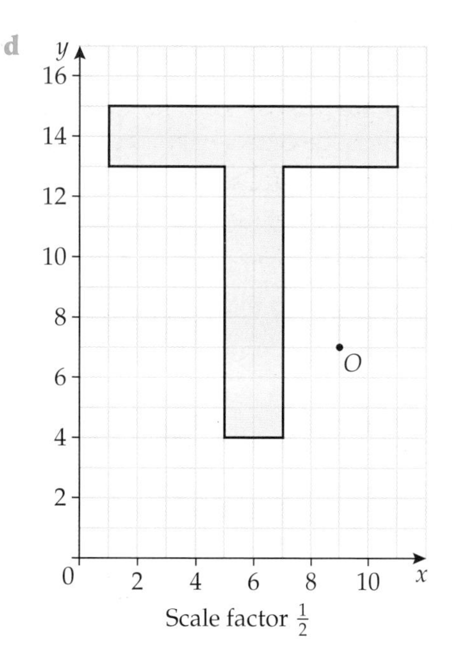

Scale factor $\frac{1}{2}$

2 For each part of Q1, describe the enlargement that would take the image back to the object.

3 Copy the diagram on squared paper.

a Enlarge the blue shape by scale factor $\frac{2}{3}$ with O as the centre of enlargement.

b What enlargement would take the image back to the object?

c What is the perimeter of the blue object?

d What is the perimeter of the image?

e How many times bigger is the perimeter of the image than the perimeter of the object?

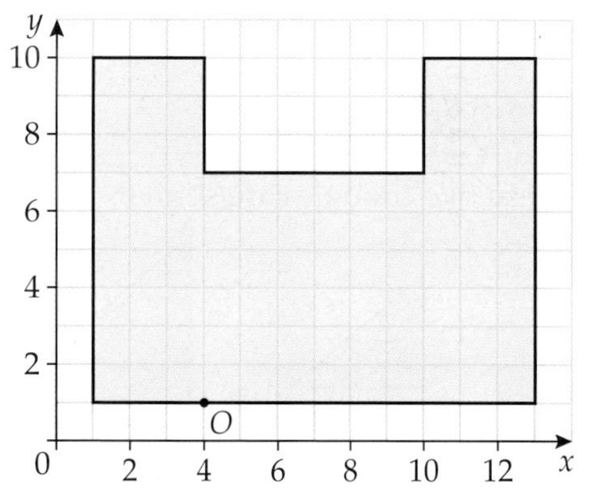

31.3 Similarity

Keywords
similar, ratio

Why learn this?

Engineers use understanding of similarity to calculate lengths on a model.

Objectives

B Understand similarity and the link with enlargement

Skills check

HELP Section 7.1

1 Simplify these ratios.

a 24 : 16	**b** 15 : 9	**c** 35 : 28
d 2.5 : 6	**e** 3.5 : 7	**f** 2.5 : 12.5

Similarity

Two objects are **similar** when they are exactly the same shape but not the same size.

For two shapes to be similar

- the corresponding angles must be equal
- the ratios of corresponding sides must be the same.

An enlargement always produces two shapes that are similar.

When you enlarge an object all the lengths are changed in the same **ratio**.

All squares are mathematically similar.
All circles are mathematically similar.

Here are two triangles. Triangle Q is an enlargement of triangle P.

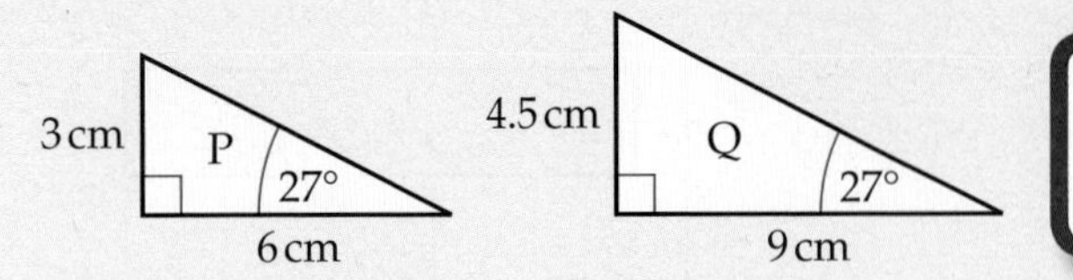

The two triangles are similar shapes. The bases of the triangles are 6 cm and 9 cm.

The ratio of the bases of the triangles is 6 : 9.

This can be simplified to 2 : 3.

Similarly, the heights of the triangles are in the ratio 3 : 4.5.

This can be simplified to 6 : 9, and further simplified to 2 : 3.

The two perimeters are also in the same ratio of 2 : 3.

The corresponding angles in the two triangles are equal.

All the sides should be in the same ratio because Q is an enlargement of P. The scale factor is $\frac{3}{2}$ or $1\frac{1}{2}$. $\frac{3}{2} \rightarrow$ ratio 3 : 2.

B

Example 5

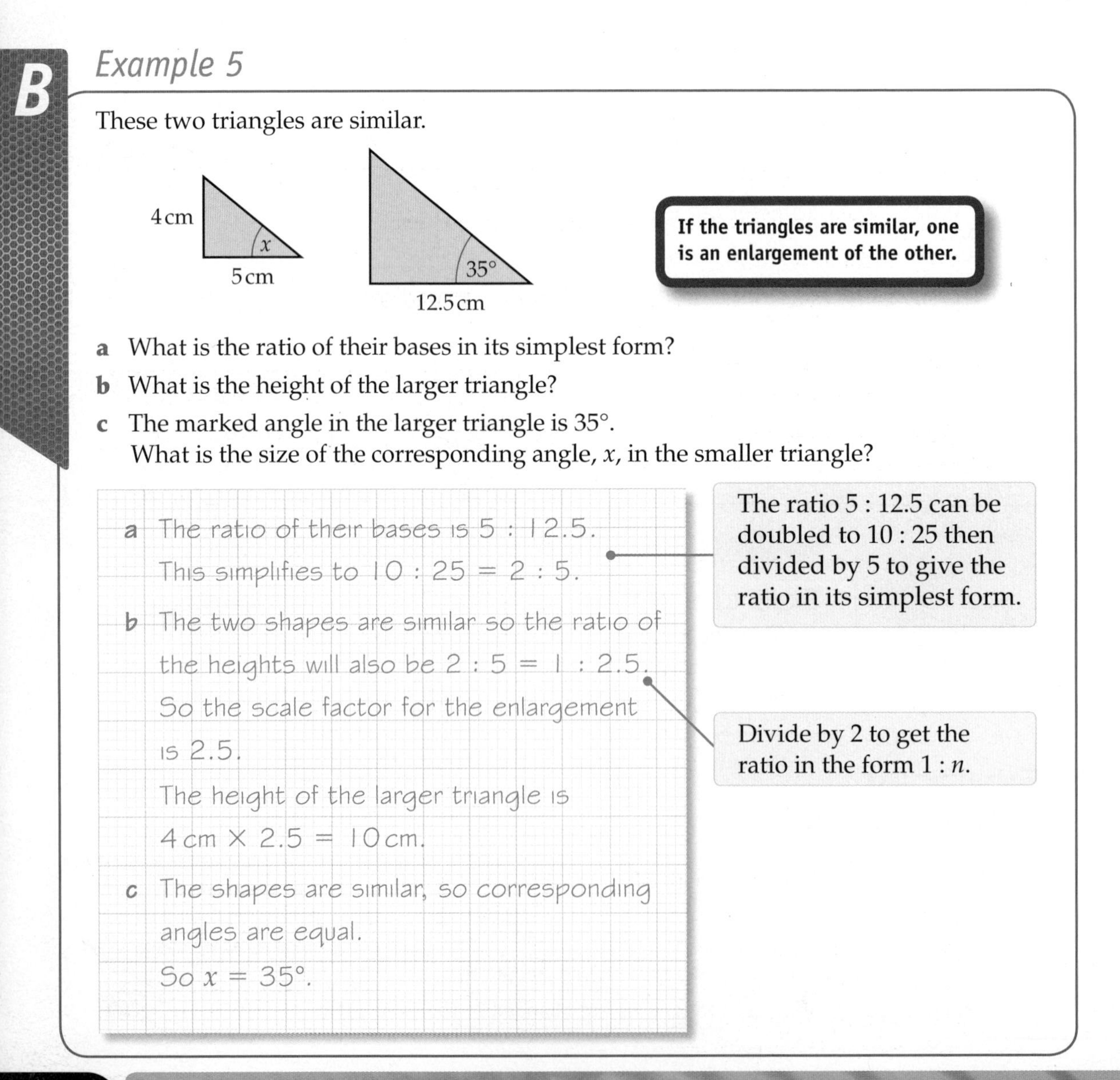

These two triangles are similar.

If the triangles are similar, one is an enlargement of the other.

a What is the ratio of their bases in its simplest form?

b What is the height of the larger triangle?

c The marked angle in the larger triangle is 35°.
What is the size of the corresponding angle, x, in the smaller triangle?

a The ratio of their bases is 5 : 12.5.
This simplifies to 10 : 25 = 2 : 5.

b The two shapes are similar so the ratio of the heights will also be 2 : 5 = 1 : 2.5.
So the scale factor for the enlargement is 2.5.
The height of the larger triangle is
4 cm × 2.5 = 10 cm.

c The shapes are similar, so corresponding angles are equal.
So x = 35°.

The ratio 5 : 12.5 can be doubled to 10 : 25 then divided by 5 to give the ratio in its simplest form.

Divide by 2 to get the ratio in the form 1 : n.

Exercise 31D

1 Triangle B is an enlargement of triangle A.

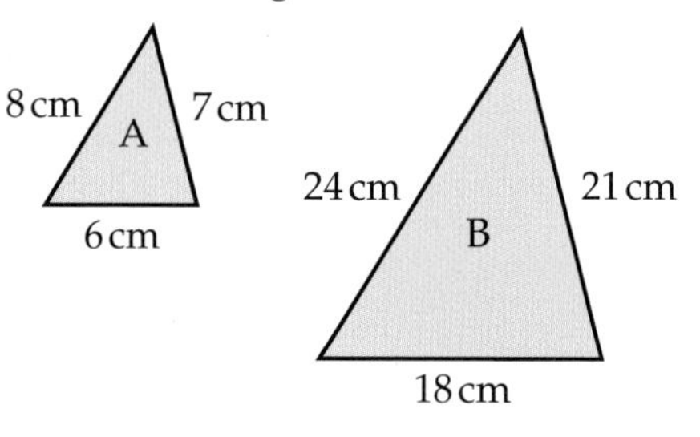

a What is the ratio of the bases of the triangles in its simplest form?

b What is the ratio of the sides of the triangles in its simplest form?

c What is the ratio of the perimeters in its simplest form?

d What is the scale factor of the enlargement that takes A to B?

e The angle at the top of triangle A is 52°.
What is the size of the corresponding angle in triangle B?

2 Triangle T is an enlargement of the isosceles triangle S.

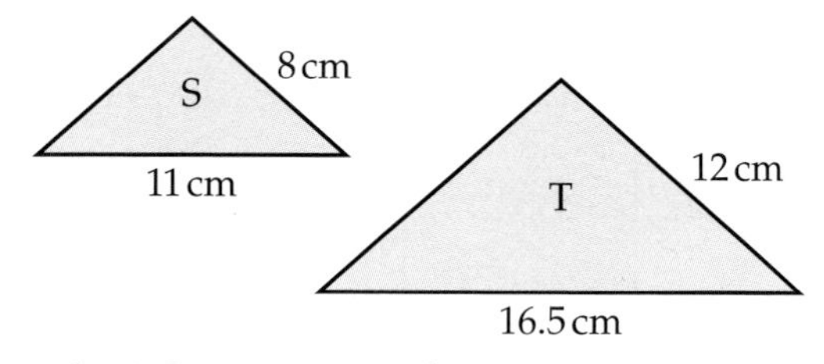

a Is triangle T isosceles? Give a reason for your answer.

b What is the ratio of the bases of the two triangles?

c What is the ratio of the perimeters of the two triangles?

3 Triangles P and Q are similar.

a Find the size of angle *ABC*.

b What is the ratio of side *AC* to side *A′C′*?

c Work out the length of *B′C′*.

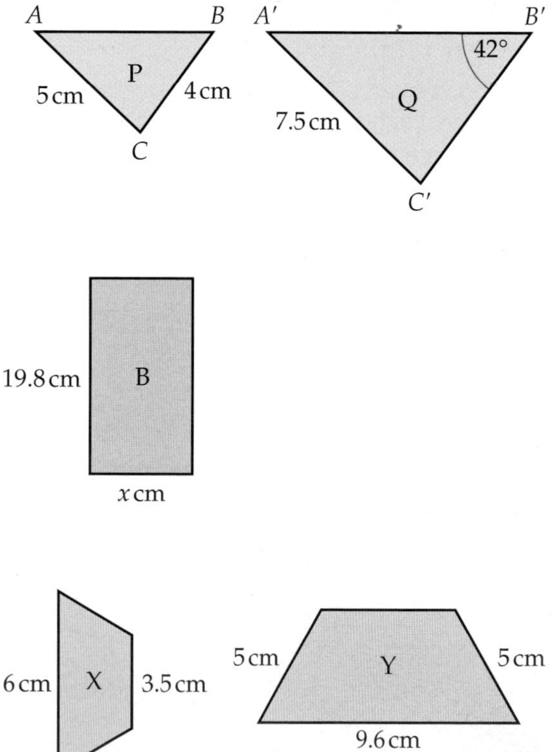

4 Rectangles A and B are similar.

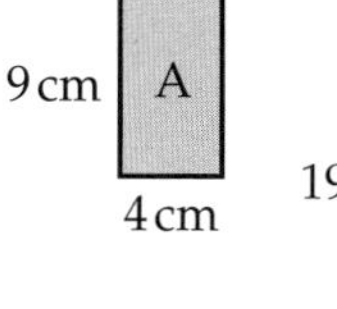

Work out the value of x.

5 Trapezium Y is an enlargement of trapezium X.

a Work out the ratio of the corresponding lengths.

b Calculate the lengths of the unknown sides in both trapezia.

B

AO2

6 Julie says that these two triangles are similar.
Robert says the bigger triangle is an enlargement of the smaller triangle.
Is either of them correct?
Give reasons for your answer.

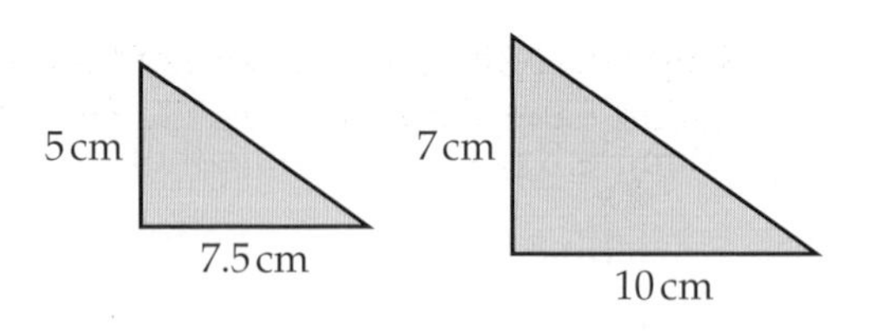

Review exercise

E

1 Enlarge this shape by scale factor 3.

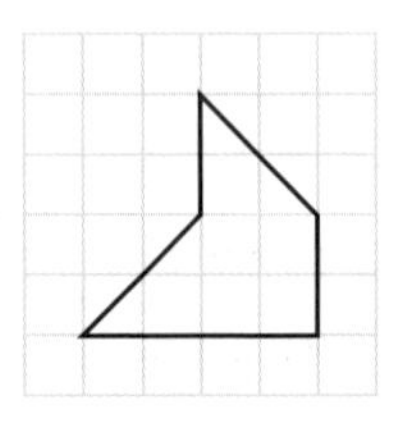

[3 marks]

D

2 Triangle B is an enlargement of triangle A.

a What is the scale factor of the enlargement?

b What are the coordinates of the centre of enlargement?

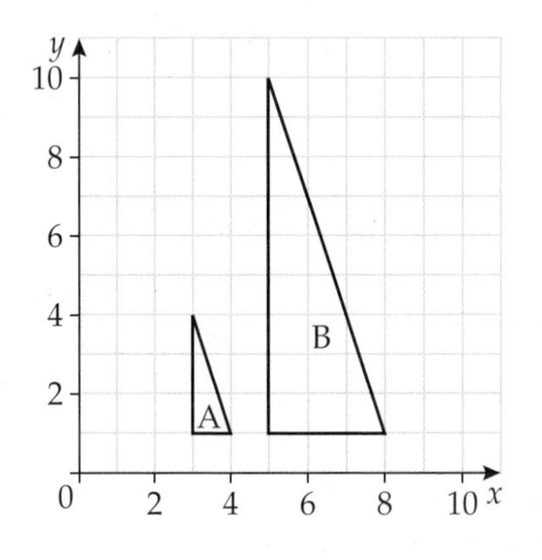

[3 marks]

3 Copy each shape on squared paper.
Enlarge the shape by the scale factor given, using O as the centre of enlargement.

a

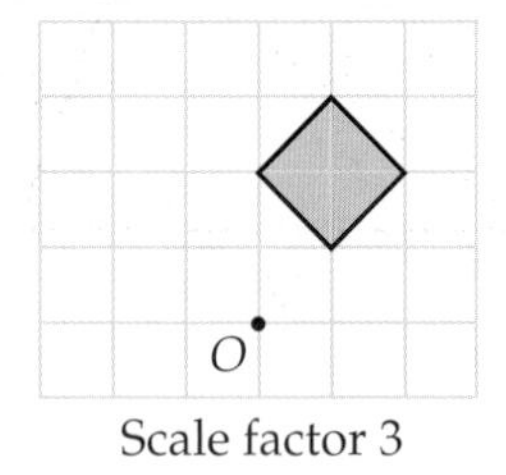

Scale factor 3

b

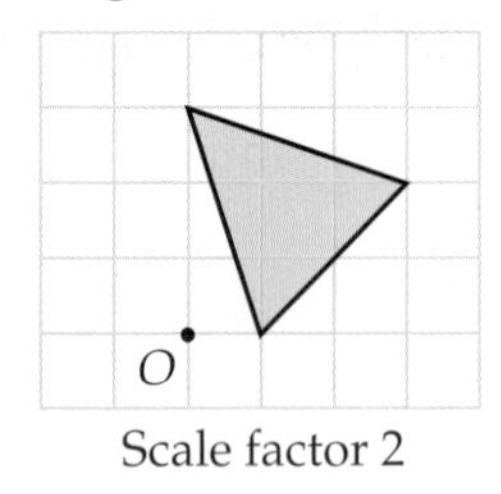

Scale factor 2

[3 marks each part]

4 Copy each shape on squared paper.
Enlarge the shape by the scale factor given from the centre O.

a **b**

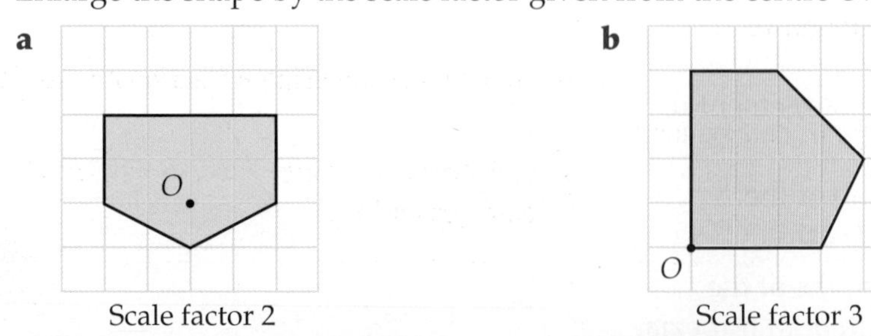

Scale factor 2 Scale factor 3

[3 marks each part]

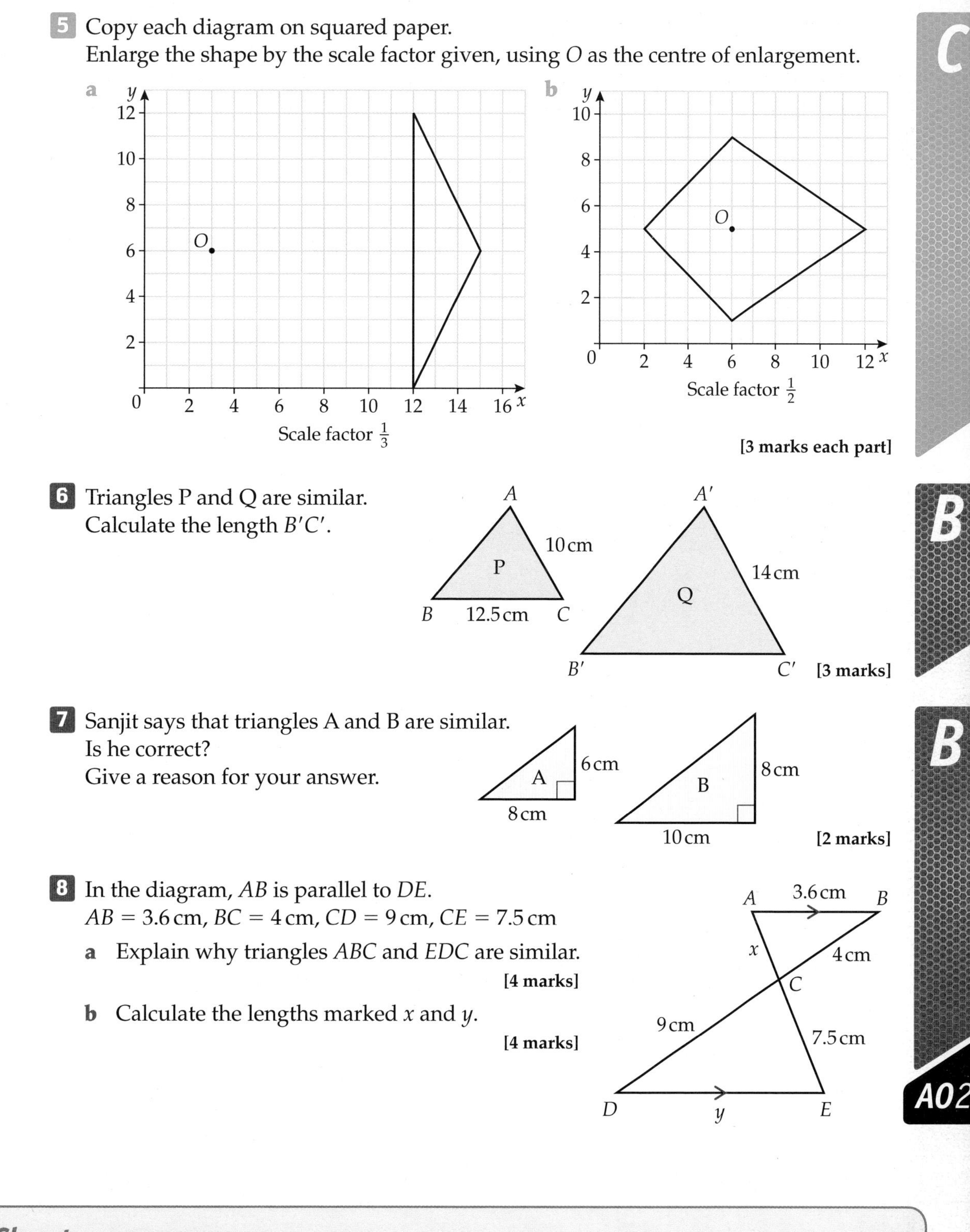

5 Copy each diagram on squared paper.
Enlarge the shape by the scale factor given, using O as the centre of enlargement.

a Scale factor $\frac{1}{3}$

b Scale factor $\frac{1}{2}$

[3 marks each part]

6 Triangles P and Q are similar.
Calculate the length $B'C'$.

[3 marks]

7 Sanjit says that triangles A and B are similar.
Is he correct?
Give a reason for your answer.

[2 marks]

8 In the diagram, AB is parallel to DE.
$AB = 3.6$ cm, $BC = 4$ cm, $CD = 9$ cm, $CE = 7.5$ cm

a Explain why triangles ABC and EDC are similar. **[4 marks]**

b Calculate the lengths marked x and y. **[4 marks]**

Chapter summary

In this chapter you have learned how to

- enlarge a shape on a grid **E**
- enlarge a shape using a centre of enlargement **D**
- enlarge a shape using a fractional scale factor **C**
- understand similarity and the link with enlargement **B**

32 Non-linear graphs

BBC Video

This chapter explores the shapes of different graphs, in particular, curved graphs.

The cable holding up a suspension bridge takes the shape of a curve called a parabola.

Objectives

This chapter will show you how to

- generate points and plot graphs of quadratic functions **D** **C**
- discuss and interpret graphs modelling real situations **D** **C**
- find approximate solutions of a quadratic equation from the graph of the corresponding quadratic function **C**
- find the intersection points of the graphs of a linear function and a quadratic function **C**
- plot graphs of simple cubic functions and use them to find approximate solutions to cubic equations **B**

Before you start this chapter

1 a Complete the table of values for $y = 2x + 3$.

x	-2	-1	0	1	2
y		1			

b Draw the graph of $y = 2x + 3$ for values of x from -2 to $+2$.

HELP Chapter 18

2 Draw a coordinate grid with x- and y-axes from -4 to $+4$. Draw and label these graphs.

a $x = 2$ b $y = 0$

c $x = -3$ d $y = -2$

HELP Chapter 18

3 Find the value of each expression when i $x = 3$ and ii $x = -2$.

a x^2 b $x^2 + 2$

c $x^2 - 5$ d $x^2 + 2x$

HELP Chapter 15

P Algebra skills: quadratic expressions

1 Factorise

a $x^2 - 9$
b $a^2 - 4$
c $t^2 - 1$
d $49 - s^2$
e $9x^2 - 1$
f $25a^2 - 9$
g $16x^2 - 25y^2$
h $12a^2 - 27b^2$
i $36x^2 - 4y^2$

To recap factorising quadratic expressions, see Chapter 19.

2 Factorise

a $x^2 + 4x + 3$
b $a^2 + 9a + 18$
c $r^2 + 7r + 10$
d $d^2 + 2d - 3$
e $x^2 + 2x - 8$
f $b^2 - b - 6$
g $m^2 - 2m - 8$
h $x^2 - 3x - 18$
i $x^2 - 8x + 16$

3 Factorise

a $x^2 + x - 2$
b $m^2 - 4m - 5$
c $t^2 + 4t - 5$
d $x^2 - 2x - 8$
e $15x^2 - 5x$
f $y^2 - 4y + 4$
g $m^2 - 2m - 3$
h $x^2 + 4x - 5$
i $x^2 + 4x - 12$

4 Solve these equations.

a $f^2 = 25$
b $2g^2 = 72$
c $s^2 - 81 = 0$
d $3t^2 - 27 = 0$
e $(x - 3)^2 = 49$
f $4(x + 4)^2 = 36$
g $2(t - 5)^2 - 50 = 0$
h $3(x + 3)^2 - 48 = 0$
i $2(7 + m)^2 - 98 = 0$

5 Solve these equations.

a $m^2 + 5m = 0$
b $3x^2 - 2x = 0$
c $4z^2 + 3z = 0$
d $t^2 - 2t - 15 = 0$
e $x^2 - 6x + 8 = 0$
f $y^2 + 6y - 7 = 0$
g $x^2 + x = 6$
h $x^2 - 5x = 24$
i $k^2 - 5k = 14$

6 A rectangle has length six times its width.
The area of the rectangle is 864 mm^2.
What is the width of the rectangle?

7 I think of a positive number, square it, then subtract five times the number.
The answer is 84.
What number did I think of?

8 By rearranging, solve the equation $\frac{x^2}{5} = 10 + x$.

Graphs of quadratic functions

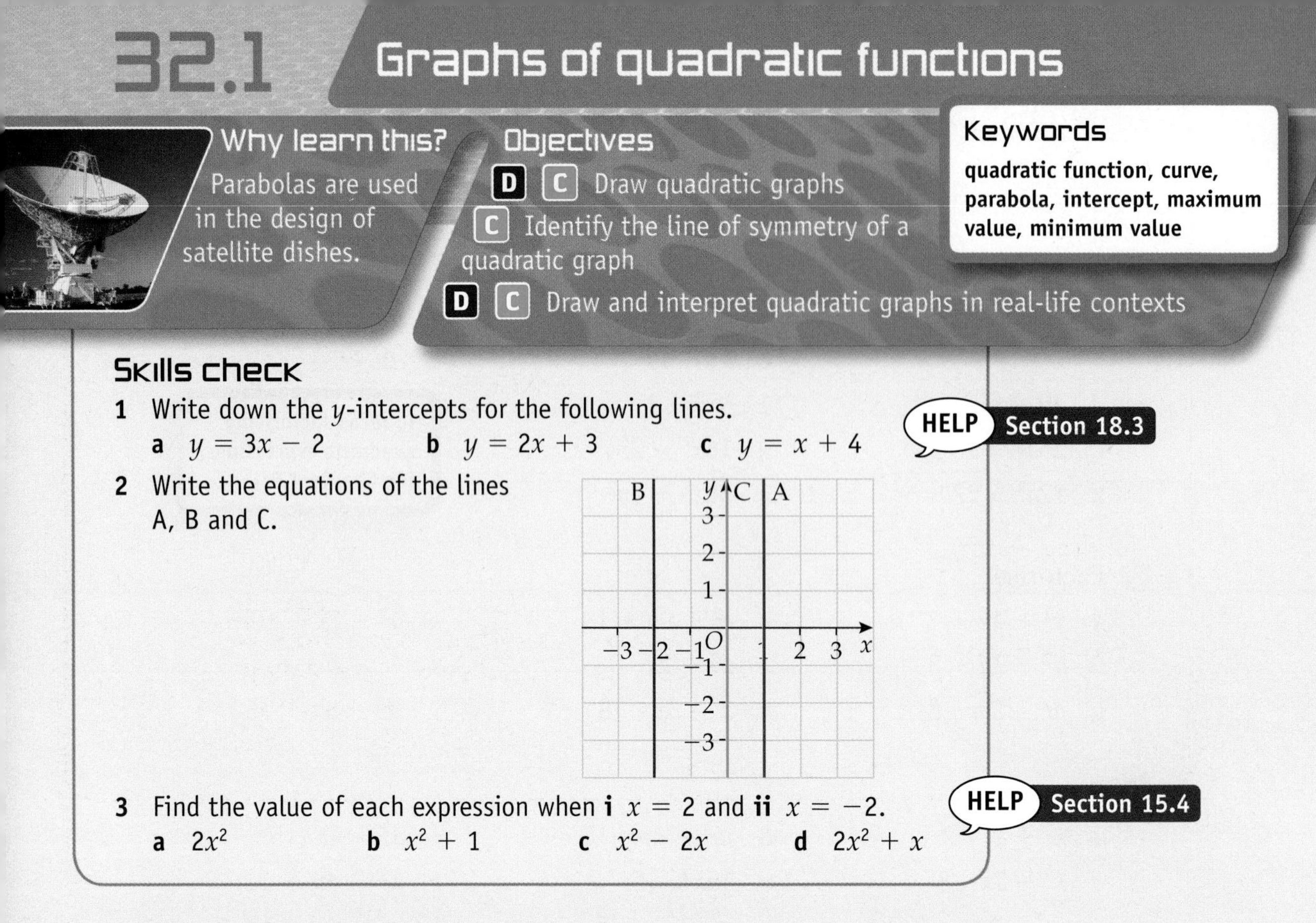

Why learn this?

Parabolas are used in the design of satellite dishes.

Objectives

D C Draw quadratic graphs

C Identify the line of symmetry of a quadratic graph

D C Draw and interpret quadratic graphs in real-life contexts

Keywords

quadratic function, curve, parabola, intercept, maximum value, minimum value

Skills check

1 Write down the y-intercepts for the following lines.

   a $y = 3x - 2$ b $y = 2x + 3$ c $y = x + 4$

HELP Section 18.3

2 Write the equations of the lines A, B and C.

3 Find the value of each expression when **i** $x = 2$ and **ii** $x = -2$.

   a $2x^2$ b $x^2 + 1$ c $x^2 - 2x$ d $2x^2 + x$

HELP Section 15.4

A **quadratic function** has a term in x^2. It may also have a term in x and a number.
It does not have any terms with powers of x higher than 2.
For example, x^2, $x^2 + 2$, $x^2 + x$, $2x^2 - 6x + 7$ are all quadratic functions.

You can plot the graphs of quadratic functions by creating tables of values and plotting the pairs of coordinates.

You used this method to draw graphs of linear functions in Section 18.2.

The graph of a quadratic function is a **curve** called a **parabola**.

All quadratic graphs are U-shaped. The U shape can open upwards (∪) or open downwards (∩).

D

Example 1

a Draw the graph of $y = x^2 - 2$ for values of x from -3 to $+3$.

b Use your graph to estimate the value of y when

   i $x = 2.5$ **ii** $x = -1.5$

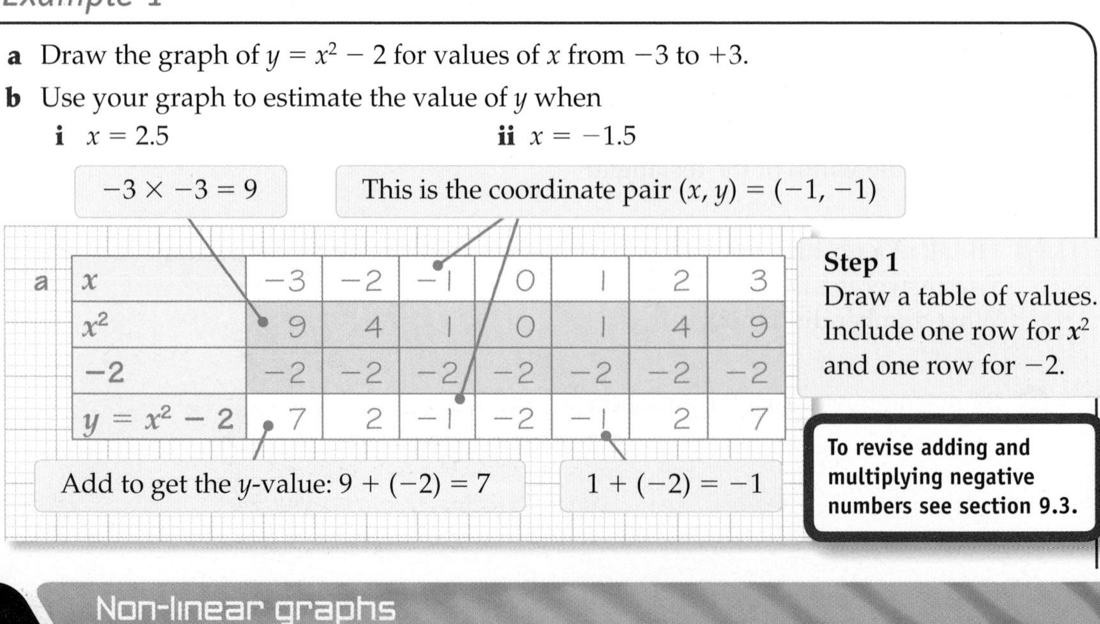

$-3 \times -3 = 9$

This is the coordinate pair $(x, y) = (-1, -1)$

a

x	−3	−2	−1	0	1	2	3
x^2	9	4	1	0	1	4	9
−2	−2	−2	−2	−2	−2	−2	−2
$y = x^2 - 2$	7	2	−1	−2	−1	2	7

Add to get the y-value: $9 + (-2) = 7$

$1 + (-2) = -1$

Step 1
Draw a table of values.
Include one row for x^2 and one row for -2.

To revise adding and multiplying negative numbers see section 9.3.

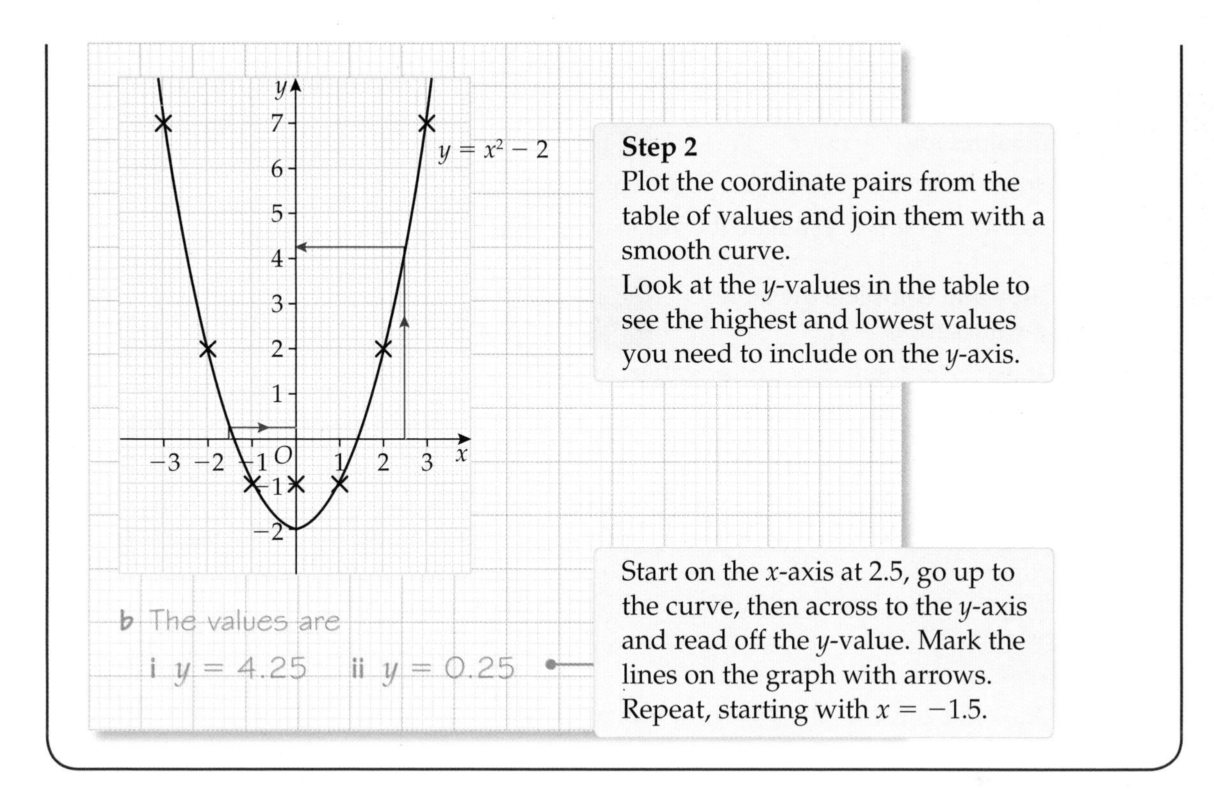

Step 2
Plot the coordinate pairs from the table of values and join them with a smooth curve.
Look at the y-values in the table to see the highest and lowest values you need to include on the y-axis.

Start on the x-axis at 2.5, go up to the curve, then across to the y-axis and read off the y-value. Mark the lines on the graph with arrows.
Repeat, starting with $x = -1.5$.

Example 2

Draw the graph of $y = x^2 - 2x$ for values of x from -2 to $+4$.

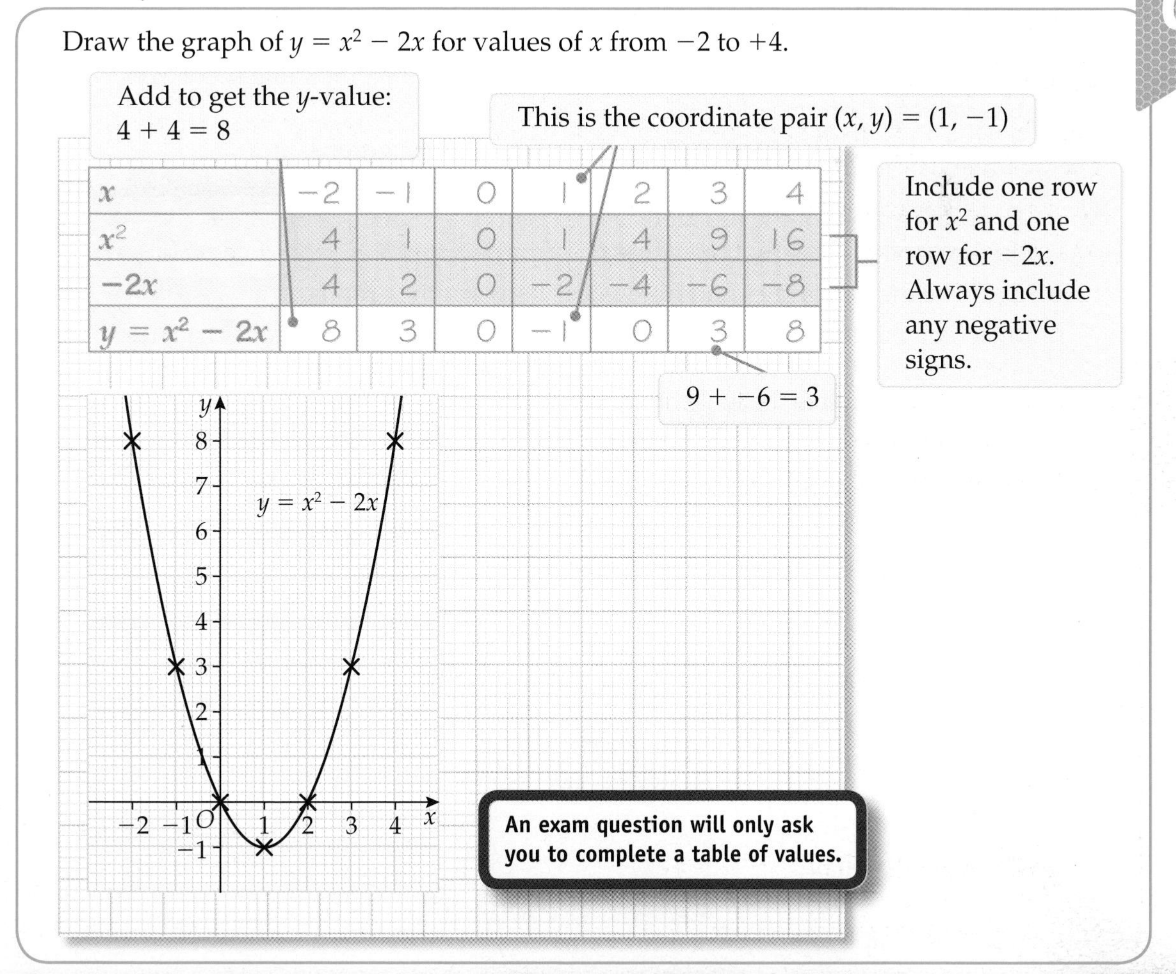

Add to get the y-value: $4 + 4 = 8$

This is the coordinate pair $(x, y) = (1, -1)$

x	−2	−1	0	1	2	3	4
x^2	4	1	0	1	4	9	16
$-2x$	4	2	0	−2	−4	−6	−8
$y = x^2 - 2x$	8	3	0	−1	0	3	8

Include one row for x^2 and one row for $-2x$. Always include any negative signs.

$9 + -6 = 3$

An exam question will only ask you to complete a table of values.

Exercise 32A

D

1 **a** Copy and complete the table of values for $y = x^2$.

x	−3	−2	−1	0	1	2	3
y	9	4				4	

b Draw the graph of $y = x^2$ for values of x from −3 to +3.

c Use your graph to estimate the value of y when x is **i** 1.5 **ii** −2.5

2 **a** Copy and complete the table of values for $y = 3x^2$.

x	−3	−2	−1	0	1	2	3
y	27	12				12	

b Draw the graph of $y = 3x^2$ for values of x from −3 to +3.

3 Describe the similarities and differences between your graphs in Q1 and Q2.

4 **a** Copy and complete the table of values for $y = x^2 + 2$.

x	−4	−3	−2	−1	0	1	2	3	4
x^2	16	9	4		0			9	
+2	+2	+2	+2	+2	+2	+2		+2	
$y = x^2 + 2$	18	11	6					11	

b Draw the graph of $y = x^2 + 2$ for values of x from −4 to +4.

5 **a** Copy and complete the table of values for $y = 5 - x^2$.

x	−3	−2	−1	0	1	2	3
5			5			5	5
$-x^2$			−1			−4	−9
$y = 5 - x^2$			4			1	−4

b Draw the graph of $y = 5 - x^2$.

c What is the highest point on the curve?

6 **a** Draw the graphs of $y = x^2$, $y = x^2 + 3$ and $y = x^2 - 3$ on the same set of axes. Use values of x from −3 to +3.

b Describe the similarities and differences between the graphs.

C

7 **a** Draw the graphs of the functions $y = 2x^2$ and $y = -2x^2$ on the same set of axes. Use values of x from −3 to +3.

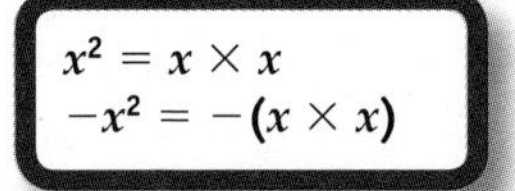

b Describe the similarities and differences between the graphs.

c By looking at the functions, how can you tell if the graph will be ∩-shaped?

8 **a** Copy and complete the table of values for $y = x^2 - 3x$.

x	−2	−1	0	1	2	3	4	5
x^2	4		0	1			16	
$-3x$	6		0	−3			−12	
$y = x^2 - 3x$	10		0	−2			4	

b Draw the graph of $y = x^2 - 3x$ for values of x from −2 to +5.

c Use your graph to find the value of y when $x = 4.5$.

d Use your graph to find the values of x that give a y-value of 5.

9 **a** Copy and complete the table of values for $y = 2x^2 - 4x - 1$.

x	−2	−1	0	1	2	3
$2x^2$	8	2				18
$-4x$	8	4				−12
−1	−1	−1				−1
$y = 2x^2 - 4x - 1$	15	5				5

b Draw the graph of $y = 2x^2 - 4x - 1$.

c Use your graph to find the value of y when $x = -0.5$.

Symmetry in quadratic graphs

All quadratic graphs are symmetrical about a line parallel to the y-axis.

The line of symmetry is always given as '$x =$ '.

For quadratic graphs that open upwards (∪), the line of symmetry passes through the lowest point on the curve. The graph crosses the x-axis at the x-**intercepts**.

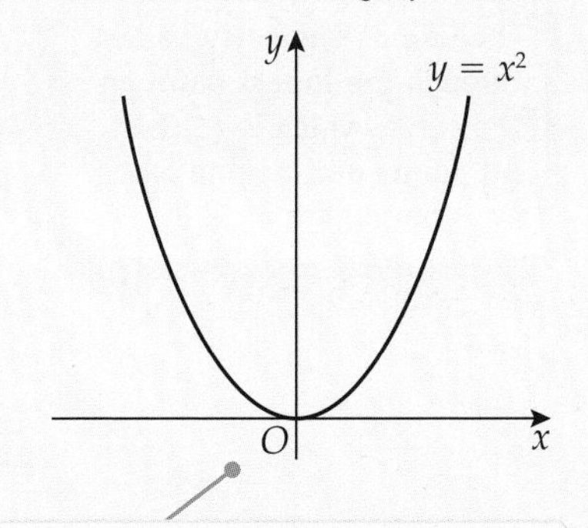

The graph of $y = x^2$ is symmetrical about the y-axis.

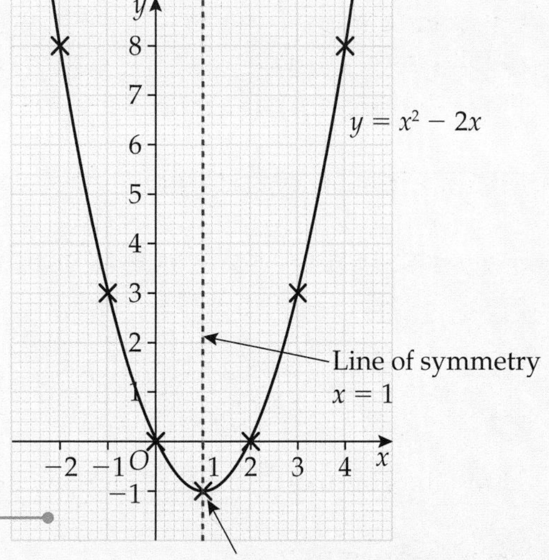

Minimum value of y

The graph of $y = x^2 - 2x$ is symmetrical about the line $x = 1$. The x-intercepts are (0, 0) and(2, 0).

At the point where the curve turns, y has either a **maximum value** or a **minimum value**.

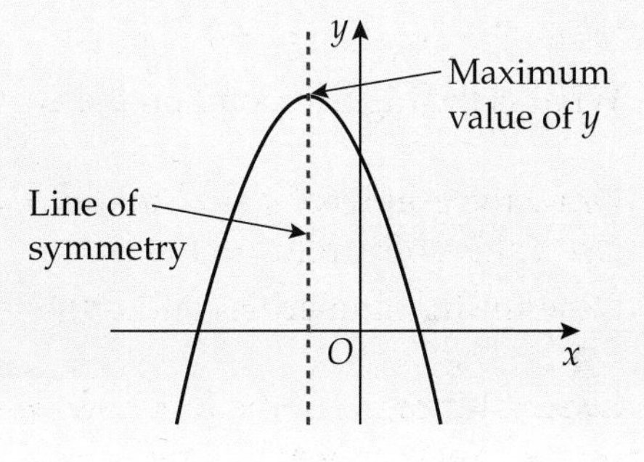

Example 3

a Draw the graph of $y = x^2 - 2x - 1$ for values of x from -2 to $+4$.

b State its line of symmetry.

c Find the minimum value of y.

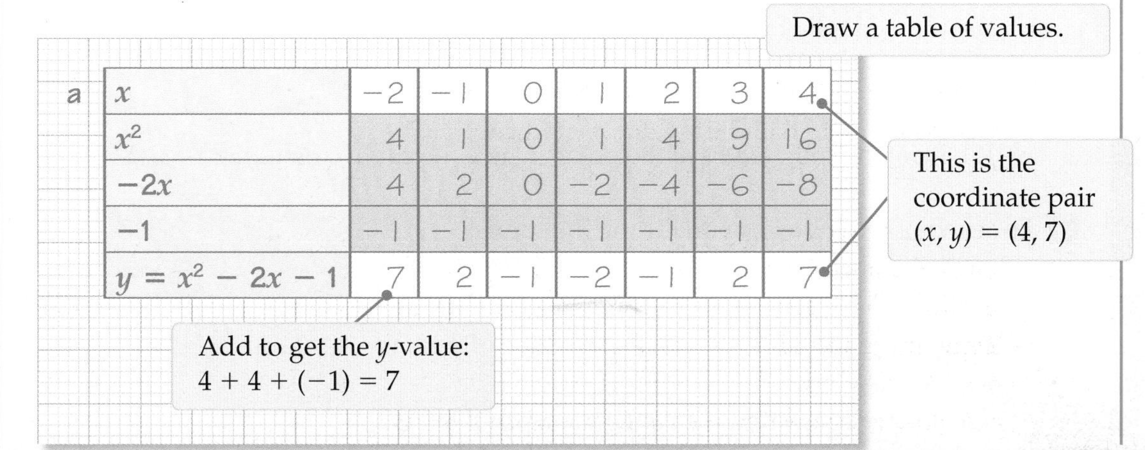

Draw a table of values.

a

x	−2	−1	0	1	2	3	4
x^2	4	1	0	1	4	9	16
$-2x$	4	2	0	−2	−4	−6	−8
-1	−1	−1	−1	−1	−1	−1	−1
$y = x^2 - 2x - 1$	7	2	−1	−2	−1	2	7

This is the coordinate pair $(x, y) = (4, 7)$

Add to get the y-value: $4 + 4 + (-1) = 7$

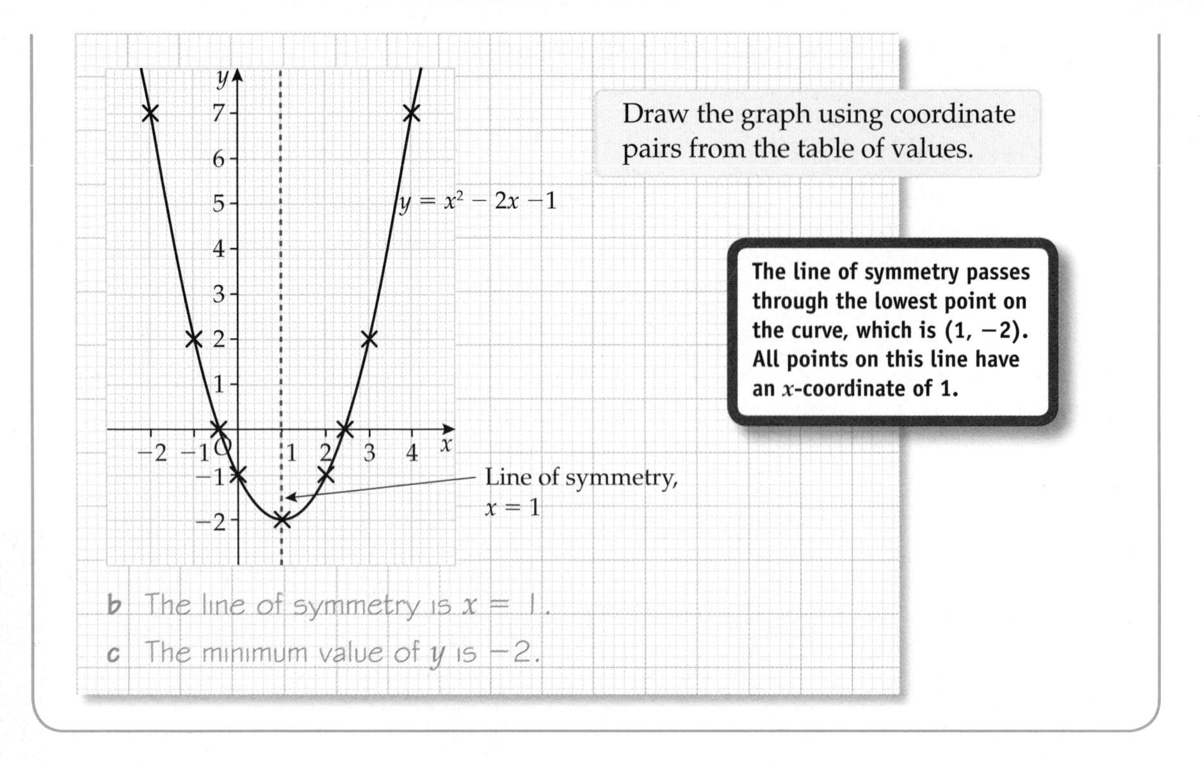

Exercise 32B

C

1 a Copy and complete the table of values for $y = x^2 + x - 2$.

b Draw the graph of $y = x^2 + x - 2$.

c State the line of symmetry.

x	−3	−2	−1	0	1	2
x^2	9	4	1	0	1	4
$+x$	−3	−2		0		
−2	−2	−2	−2	−2	−2	−2
$y = x^2 + x - 2$	4	0		−2		

2 a Copy and complete the table of values for $y = x^2 - 4x - 5$.

x	−2	−1	0	1	2	3	4	5	6
x^2	4			1	4	9	16		
$-4x$	8			−4	−8	−12			
−5	−5			−5	−5	−5	−5		
$y = x^2 - 4x - 5$	7			−8	−9	−8			

b Draw the graph of $y = x^2 - 4x - 5$.

c What are the coordinates of the lowest point?

d State the line of symmetry.

e What are the coordinates of the points where the curve crosses the x-axis?

3 For each of the following quadratic functions

i make a table of values

ii draw the graph

iii state the maximum or minimum value of y

iv state the line of symmetry

v state the coordinates of the x-intercepts of the graph.

a $y = x^2 + 2x - 5$ for values of x from -4 to $+2$
b $y = 2x^2 - 2x + 3$ for values of x from -3 to $+4$
c $y = x^2 - 4x + 4$ for values of x from -1 to $+5$
d $y = -x^2 + 4x - 7$ for values of x from -1 to $+5$

C

Real-life use of quadratic functions

Quadratic graphs can be used to represent real-life situations. For example, a quadratic function describes the path of a rocket or a tennis ball thrown vertically into the air.

Example 4

D

The graph shows the heights reached by two rockets during a test flight.

a What was the maximum height reached by rocket A?

b Estimate how much higher rocket A went than rocket B.

c Estimate the time after the start when the two rockets were at the same height.

d How long was rocket B more than 200 m above the ground?

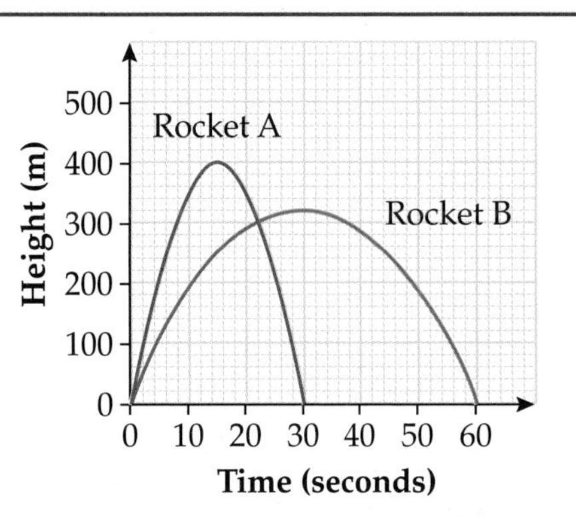

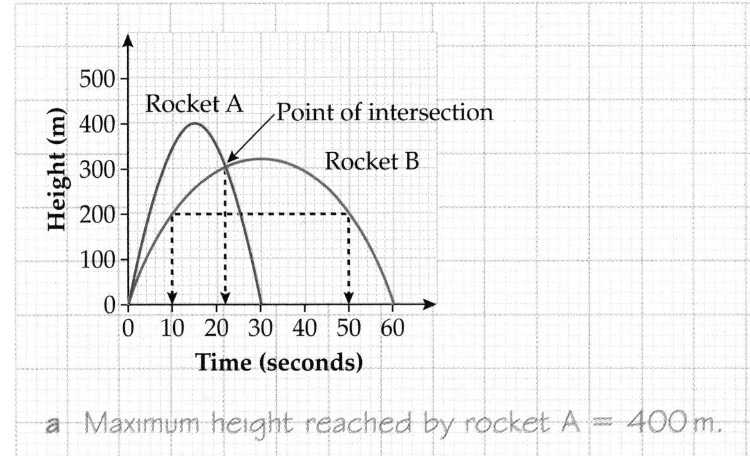

a Maximum height reached by rocket A = 400 m.

b Rocket A = 400 m; rocket B ≈ 320 m.

So rocket A went about 80 m higher than rocket B.

c At about 22 seconds

The point of intersection of the two graphs is the point where the rockets were at the same height. Draw a vertical line down from this point to the x-axis.

d $50 - 10 = 40$ s

Look at the part of the graph for rocket B that is above 200 m. Draw vertical lines down to the x-axis from the point where the rocket goes higher than 200 m (about 10 s) and when it drops below 200 m (about 50 s).

Exercise 32C

D

1 The graph shows the heights obtained by toy rockets when shot vertically upwards.

a What is the maximum height reached by rocket X?

b Which rocket reached its maximum height in the quickest time?

c What was the height of rocket Y after 2.5 seconds?

d How long does it take rocket X to reach a height of 50 m?

e How long after the start does rocket Z return to the ground?

f Were the rockets at the same height at any time after 0 seconds.

g Match each function to the correct rocket.

i $h = 25t - 5t^2$ **ii** $h = 35t - 5t^2$ **iii** $h = 15t - 5t^2$

C

2 Xabi needs to fence off an area of his garden for one of his pets.

$(x - 2)$ m

$(x - 4)$ m

He is investigating different sizes of rectangular enclosure.

The graph shows how the area of the enclosure changes as the value of x varies.

a What is the area of the enclosure when $x = 1$?

b What is the area of the enclosure when $x = 5$?

c Look at your answers to parts **a** and **b**. Which of these values of x would not be possible in real life? Give a reason for your answer.

d The area of the enclosure needs to be greater than 9 m². What is the minimum value of x that Xabi can consider?

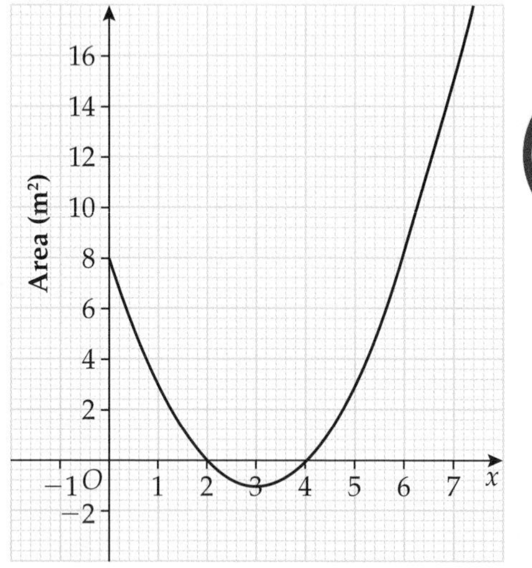

AO2

C

3 A toy rocket is shot vertically upwards.
The height of the rocket above the ground after t seconds is given by the function $h = 30t - 5t^2$ where h is the height above the ground (in metres).

a Copy and complete the table of values for $h = 30t - 5t^2$.

t	0	1	2	3	4	5	6
h	0	25		45			

b Draw the graph of $h = 30t - 5t^2$.

c Use your graph to find the height of the rocket after 1.5 seconds.

d How long does it take the rocket to reach its maximum height?

> **Plot the t-values along the x-axis and the h-values along the y-axis.**

Solving quadratic equations graphically

Keywords
quadratic equation

Why learn this?

You can use quadratic equations to work out car acceleration and stopping distances.

Objectives

C Use a graph to solve quadratic equations

Skills check

1 a Draw the graph of $y = x + 3$ for values of x from -4 to $+1$.

b Draw the line $x = -3$ on your graph.
What are the coordinates of the point where $x = -3$ intersects $y = x + 3$?

2 Draw the graphs of $y = x + 4$ and $y = 2x$ on the same set of axes. Where do the lines cross?

HELP Section 18.2

You can solve a **quadratic equation** by drawing its graph and finding where the graph crosses the x-axis ($y = 0$).

Solving quadratic equations by factorising was covered in Chapter 19.

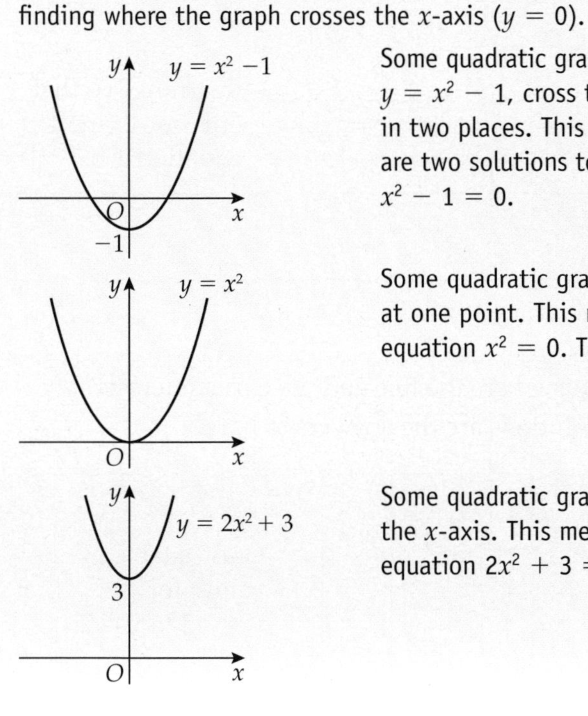

Some quadratic graphs, such as $y = x^2 - 1$, cross the x-axis in two places. This means there are two solutions to the equation $x^2 - 1 = 0$.

Some quadratic graphs, such as $y = x^2$, just touch the x-axis at one point. This means there is only one solution to the equation $x^2 = 0$. The solution is $x = 0$.

Some quadratic graphs, such as $y = 2x^2 + 3$, do not cross the x-axis. This means there are no solutions to the quadratic equation $2x^2 + 3 = 0$.

Example 5

a Draw the graph of $y = x^2 - 4x$ for values of x from -1 to $+5$.

b Find the solutions to the equation $x^2 - 4x = 0$.

c Draw the line $y = 2$ on the same set of axes.
What are the x-coordinates of the points where the line and the curve intersect?

d Write the quadratic equation whose solutions are the answers to part **c**.

a

x	−1	0	1	2	3	4	5
x^2	1	0	1	4	9	16	25
$-4x$	4	0	−4	−8	−12	−16	−20
$y = x^2 - 4x$	5	0	−3	−4	−3	0	5

Draw a table of values for $x = -1$ to $x = 5$.

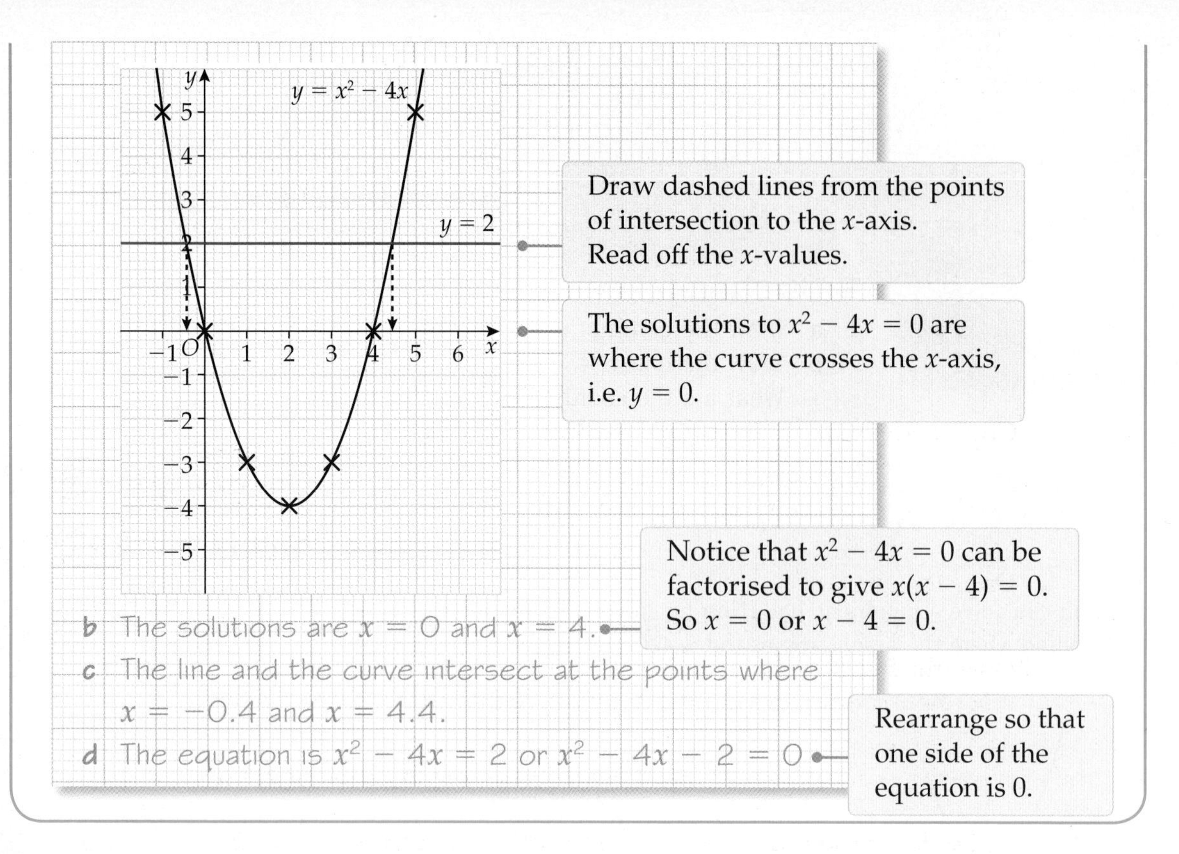

b The solutions are $x = 0$ and $x = 4$.

c The line and the curve intersect at the points where $x = -0.4$ and $x = 4.4$.

d The equation is $x^2 - 4x = 2$ or $x^2 - 4x - 2 = 0$

Example 6

a Draw the graph of $y = x^2 + 2x - 5$ for values of x from -3 to $+2$.

b Draw the line $y = -2$ on the same axes.
What are the x-coordinates of the points where the line and the curve intersect?

c Write the quadratic equation whose solutions are the answers to part **b**.

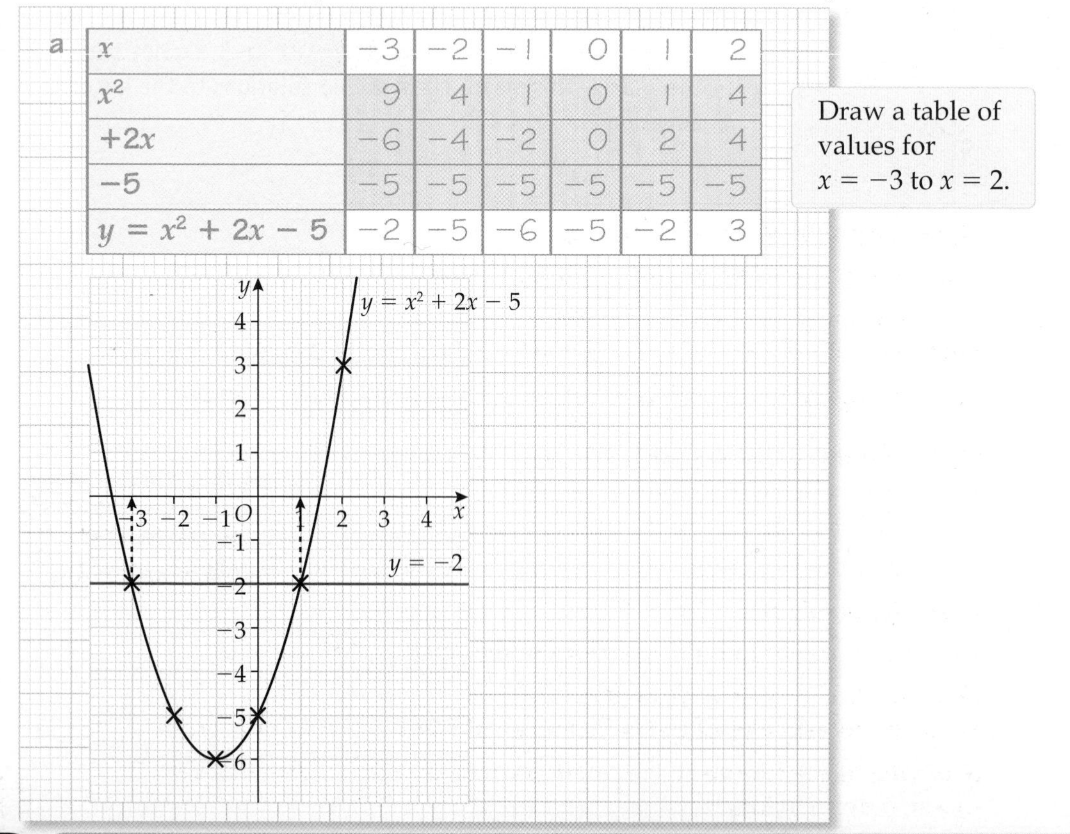

a

x	-3	-2	-1	0	1	2
x^2	9	4	1	0	1	4
$+2x$	-6	-4	-2	0	2	4
-5	-5	-5	-5	-5	-5	-5
$y = x^2 + 2x - 5$	-2	-5	-6	-5	-2	3

b The line and the curve intersect at the points where
$x = -3$ and $x = 1$.

c The equation is $x^2 + 2x - 5 = -2$
or $x^2 + 2x - 3 = 0$

Notice that $x^2 + 2x - 3 = 0$ can be factorised to give $(x - 1)(x + 3) = 0$. So $x - 1 = 0$ or $x + 3 = 0$ and $x = 1$ or $x = -3$.

Exercise 32D

C

1 **a** Copy and complete the table of values for $y = x^2 + 3$.

x	−3	−2	−1	0	1	2	3
y	12		4	3		7	

Expand the table if you find it easier.

b Draw the graph of $y = x^2 + 3$.

c Use your graph to work out how many solutions there are to the equation $x^2 + 3 = 0$.

d Draw the line $y = 7$ on your graph.
Find the x-coordinates of the points where the line $y = 7$ crosses the curve $y = x^2 + 3$.

e Write the quadratic equation whose solutions are the answers to part **d**.

2 **a** Draw the graph of $y = x^2 + x - 3$ for values of x from −3 to +2.

b An approximate solution to the equation $x^2 + x - 3 = 0$ is $x = 1.2$.

i Explain how you can find this solution from the graph.

ii Use the graph to find another solution to the equation.

3 **a** Copy and complete the table of values for $y = x^2 - 4x + 1$.

x	−1	0	1	2	3	4	5
x^2	1		1	4		16	
$-4x$	4		−4	−8		−16	
$+1$	+1		+1	+1		+1	
$y = x^2 - 4x + 1$	6		−2	−3		1	

b Draw the graph of $y = x^2 - 4x + 1$.

c Use your graph to solve the equation $x^2 - 4x + 1 = 0$.

d Draw the line $y = -1$ on your graph.
Find the x-coordinates of the points where the line and the curve intersect.

e Write the quadratic equation whose solutions are the answers to part **d**.

C

4 **a** Copy and complete the table of values for $y = 3x^2 - 6$.

x	−3	−2	−1	0	1	2	3
y		6	−3				21

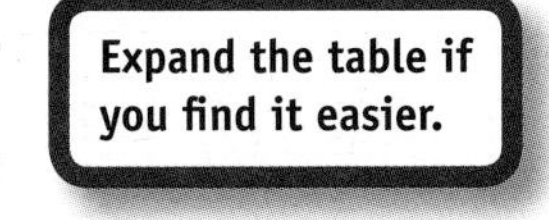

b Draw the graph of $y = 3x^2 - 6$.

c Use your graph to find the solutions to the equation $3x^2 - 6 = 0$.

d Draw the line $y = 10$ on your graph.
Write the coordinates of the points where the line and the curve cross.

e Show that the solutions to the quadratic equation $3x^2 - 16 = 0$ can be found at these points.

AO2

C

5 **a** Copy and complete the table of values for $y = -x^2 - 3x + 4$.

x	-4	-3	-2	-1	0	1
$-x^2$	-16		-4		0	
$-3x$	12		6		0	
$+4$	$+4$		$+4$		$+4$	
$y = -x^2 - 3x + 4$	0		6		4	

b Draw the graph of $y = -x^2 - 3x + 4$.

A02

c Draw the line $y = 6$ on your graph.
Find the solutions to the equation $-x^2 - 3x + 4 = 6$.

C

6 **a** Draw the graph of $y = x^2 - 4x + 4$ for values of x from -1 to $+4$.

b Use your graph to solve these equations.
i $x^2 - 4x + 4 = 0$ **ii** $x^2 - 4x + 4 = 6$

A03

c Can the quadratic equation $x^2 - 4x + 4 = -1$ be solved? Explain your answer.

32.3 Graphs of cubic functions

Keywords
cubic function

Why learn this?
Part of the track of a roller coaster can be described using a cubic function.

Objectives

B Draw cubic graphs

B Use a graph to solve cubic equations

Skills check

HELP Section 15.4

1 Find the value of each expression when **i** $x = 2$ and **ii** $x = -2$.
a x^3 **b** $-x^3$ **c** $2x^3 - 5$ **d** $x^3 - x^2$

A **cubic function** has a term in x^3. It may also have terms in x^2 and x and a number. It does not have any terms with powers of x higher than 3.

For example, x^3, $x^3 + 3x$, $x^3 - x^2 - 2x$ are all cubic functions.

This is the graph of $y = x^3$. This is the graph of $y = -x^3$.

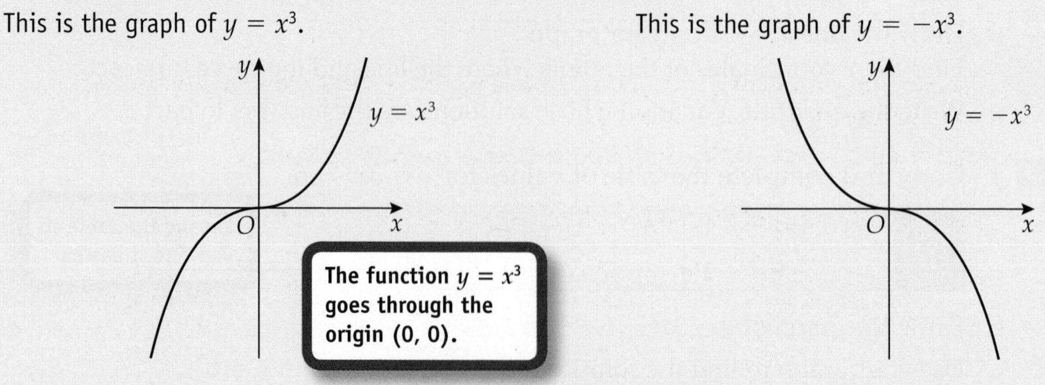

You can solve a cubic equation by drawing its graph and finding where the graph crosses the x-axis.

In general, a cubic function can intersect the x-axis at one, two or three places.

Example 7

a Draw the graph of $y = x^3 - 5x + 2$ for $-3 \leqslant x \leqslant 3$.

b Use the graph to find the solutions to the equation $x^3 - 5x + 2 = 0$.

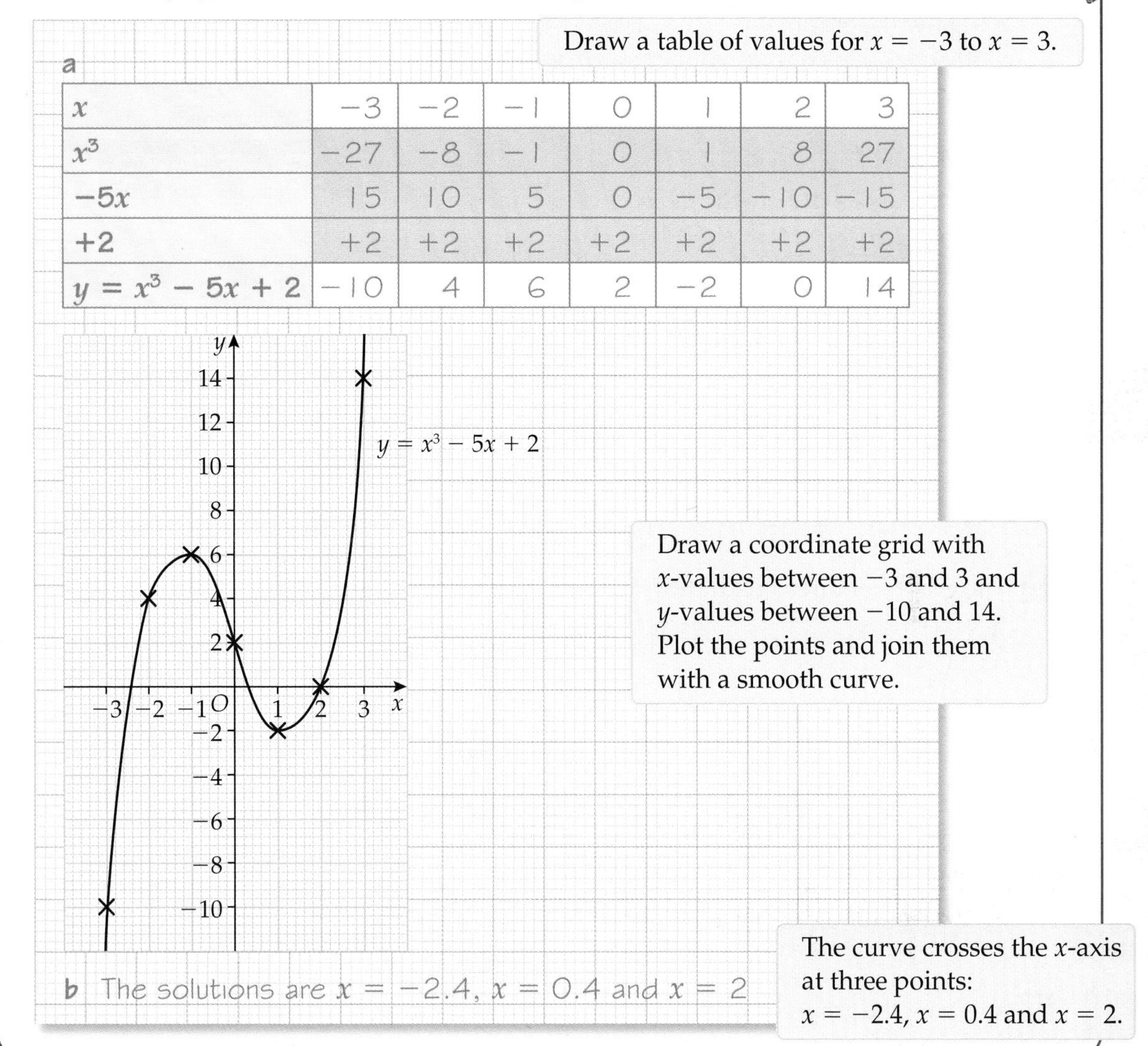

Draw a table of values for $x = -3$ to $x = 3$.

a

x	−3	−2	−1	0	1	2	3
x^3	−27	−8	−1	0	1	8	27
$-5x$	15	10	5	0	−5	−10	−15
$+2$	+2	+2	+2	+2	+2	+2	+2
$y = x^3 - 5x + 2$	−10	4	6	2	−2	0	14

Draw a coordinate grid with x-values between −3 and 3 and y-values between −10 and 14. Plot the points and join them with a smooth curve.

b The solutions are $x = -2.4$, $x = 0.4$ and $x = 2$

The curve crosses the x-axis at three points: $x = -2.4$, $x = 0.4$ and $x = 2$.

B

Exercise 32E

1 Draw the graph of $y = -x^3$ for values of x from −3 to +3.

2 **a** Draw the graphs of $y = x^3$, $y = \frac{1}{2}x^3$ and $y = x^3 + 2$ on the same set of axes. Use values of x from −3 to +3.

b Describe the similarities and differences between the graphs.

3 **a** Copy and complete the table of values for the function $y = x^3 - 2x + 2$.

x	−2.5	−2	−1	0	1	2	2.5
x^3		−8		0	1		15.6
$-2x$		4		0	−2		−5
$+2$		+2		+2	+2		+2
$y = x^3 - 2x + 2$		−2		2	1		12.6

b Draw the graph of $y = x^3 - 2x + 2$ for $-2.5 \leqslant x \leqslant 2.5$.

c Find the solutions to $x^3 - 2x + 2 = 0$.

B

B

4 **a** Draw the graph of $y = x^3 - 2x^2 - 6x$ for $-3 \leqslant x \leqslant 5$.

b Use your graph to find the solutions to $x^3 - 2x^2 - 6x = 0$.

5 **a** Write down the solutions to $(x - 1)(x - 3)(x - 4) = 0$.

b Draw the graph of $y = (x - 1)(x - 3)(x - 4)$ for $0 \leqslant x \leqslant 5$ to confirm your answers in part **a**.

Write a row for each of $(x - 1)$, $(x - 3)$ and $(x - 4)$ and *multiply* your answers.

Review exercise

D

1 **a** Copy and complete the table of values for $y = x^2 - 4$.

x	-3	-2	-1	0	1	2	3
x^2	9			0	1		
-4	-4	-4	-4	-4	-4		
$y = x^2 - 4$	5			-4	-3		

[2 marks]

b Draw the graph of $y = x^2 - 4$ for values of x from -3 to $+3$. **[2 marks]**

c Use your graph to find the value of y when

i $x = 1.5$ **ii** $x = -2.5$. **[2 marks]**

D

2 The graphs of three quadratic functions are shown below.

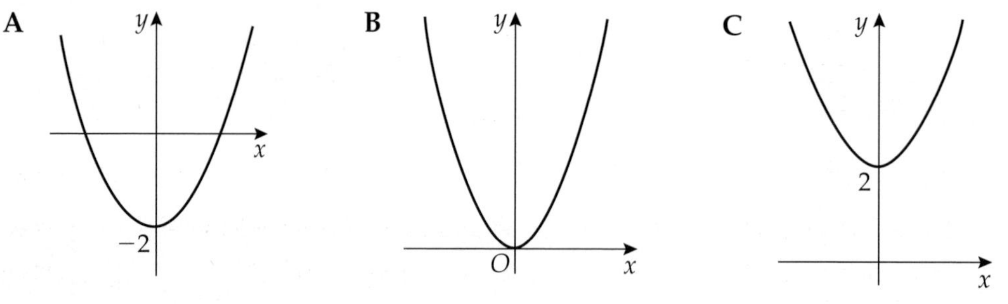

AO2

Match each graph to the correct function.

$y = 2x^2$ $\quad y = x^2 + 2$ $\quad y = x^2 - 2$ **[3 marks]**

C

3 **a** Draw the graph of $y = 2x^2 - 2x + 15$ for values of x from -2 to $+3$. **[4 marks]**

b Use your graph to find

i the x-intercept of the curve

ii the coordinates of the lowest point

iii the line of symmetry. **[3 marks]**

$2x^2 = 2 \times x^2$

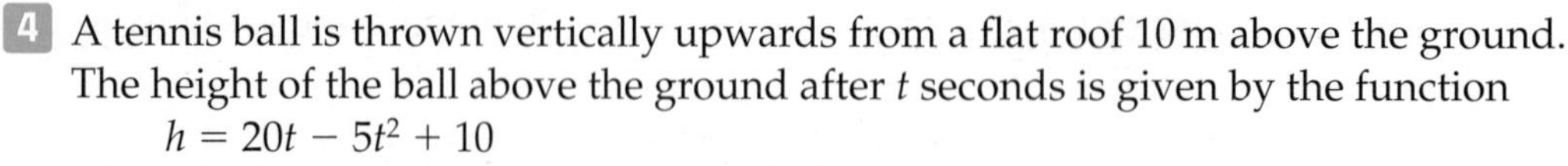

4 A tennis ball is thrown vertically upwards from a flat roof 10 m above the ground. The height of the ball above the ground after t seconds is given by the function

$$h = 20t - 5t^2 + 10$$

where h is the height above the ground (in metres).

a Copy and complete the table of values for $h = 20t - 5t^2 + 10$.

t	0	1	2	3	4	5
h	10	25			10	

Expand the table.

[2 marks]

b Draw the graph of $h = 20t - 5t^2 + 10$. **[2 marks]**

c What is the maximum height reached by the ball? **[1 mark]**

d How long does it take the ball to reach its maximum height? **[1 mark]**

e How long will it take the ball to reach the ground? **[1 mark]**

5 a Copy and complete the table of values for $y = x^2 + 2x - 5$.

x	−4	−3	−2	−1	0	1	2
y	3	−2		−6	−5		

[2 marks]

b Using a grid with y-values from −7 to +3 draw the graph of $y = x^2 + 2x - 5$ for values of x from −4 to +2. [2 marks]

c An approximate solution of the equation $x^2 + 2x - 5 = 0$ is $x = 1.5$.

i Explain how you can find this from the graph.

ii Use the graph to find another solution to this equation. [2 marks]

C

6 Match each graph with one of the following functions.

$y = x^2 + 4$ $y = x + 4$ $y = 4x^2$ $y = 4 - x^2$

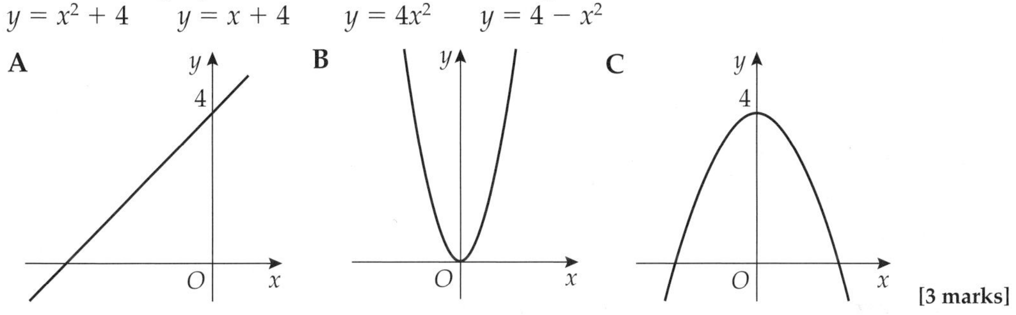

[3 marks]

C

AO2

7 a Copy and complete the table of values for $y = 2x^2 - x - 3$.

x	−3	−2	−1	0	1	2	3	4
y	18			−3	−2	3		

[2 marks]

b Draw the graph of $y = 2x^2 - x - 3$ for values of x from −3 to +4. [2 marks]

c Use your graph to solve the equation $2x^2 - x - 3 = 0$. [2 marks]

C

8 a Copy and complete the table of values for $y = x^2 - 5$.

x	−3	−2	−1	0	1	2	3
y		−1	−4	−5			

[2 marks]

b Draw the graph of $y = x^2 - 5$ for values of x from −3 to +3. [2 marks]

c Use your graph to

i solve the equation $x^2 - 5 = 0$

ii find the minimum value of y. [3 marks]

C

9 a Copy and complete the table of values for $y = x^2 + 2x + 5$.

x	−3	−2	−1	0	1	2
x^2	9		1		1	
$2x$	−6		−2		2	
+5	+5		+5		+5	
$y = x^2 + 2x + 5$	8		4		8	

[2 marks]

b Draw the graph of $y = x^2 + 2x + 5$. [2 marks]

c By drawing a suitable straight line, solve the equation $x^2 + 2x + 5 = 7$. [2 marks]

C

AO2

C

10 **a** Copy and complete the table of values for $y = x^2 - 5x$.

x	-1	0	1	2	3	4	5	6
y		0	-4	-6		-4	0	6

[2 marks]

b Using a grid with y-values from -7 to $+6$ draw the graph of $y = x^2 - 5x$. **[2 marks]**

c Use your graph to find the solutions to the equation $x^2 - 5x = -3$. **[2 marks]**

AO3

d Explain why you cannot find solutions to the equation $x^2 - 5x = -8$. **[1 mark]**

B

11 **a** Copy and complete the table of values for the function $y = x^3 - 2x^2 - 3x$.

x	-2	-1	0	1	2	3	4
x^3	-8			1		27	
$-2x^2$	-8			-2		-18	
$-3x$	6			-3		-9	
$y = x^3 - 2x^2 - 3x$	-10			-4		0	

[2 marks]

b Draw the graph of $y = x^3 - 2x^2 - 3x$ for $-2 \leq x \leq 4$. **[2 marks]**

c Write the x-coordinates of the points where the graph cuts the x-axis. **[2 marks]**

d Write the equation whose solutions are the answers to part **c**. **[1 mark]**

B

12 The graphs of three functions are shown below.

A

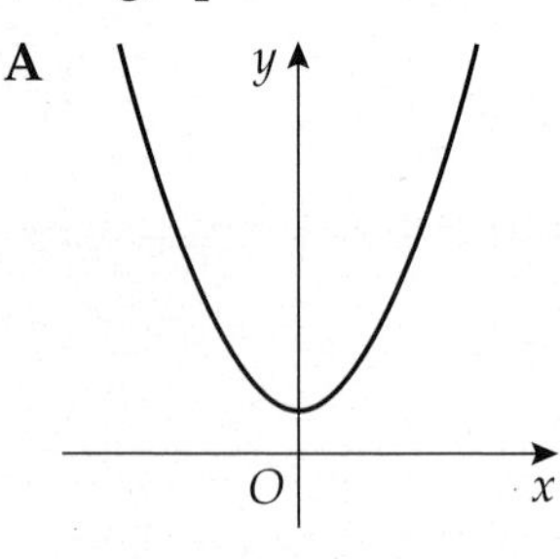

B **C**

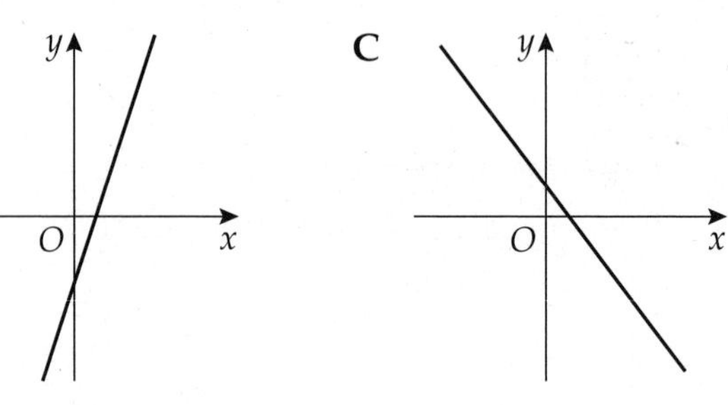

a Match each graph to the correct function.

$y = 3x - 1$ $\quad$ $y = x^2 + 1$ $\quad$ $y = x^3 - 1$ $\quad$ $3x + 2y = 1$ **[3 marks]**

AO2

b Sketch the graph of the function that is not used. **[2 marks]**

Chapter summary

In this chapter you have learned how to

- draw quadratic graphs D C
- identify the line of symmetry of a quadratic graph D C
- draw and interpret quadratic graphs in real-life contexts D C
- use a graph to solve quadratic equations C
- draw cubic graphs B
- use a graph to solve cubic equations B

33

Constructions and loci

BBC Video

This chapter is about doing constructions using only compasses and a straight edge, and solving locus problems.

The path you follow on a fairground ride can be described by a locus. Fairground designers need to think about loci when they decide how close together two rides should be.

Objectives

This chapter will show you how to

- use straight edge and compasses to do standard constructions, including the perpendicular bisector of a line segment, the perpendicular from a point to a line, the perpendicular from a point on a line, and the bisector of an angle C
- find loci to produce shapes and paths C

Before you start this chapter

HELP Chapter 23

1 Using a ruler and compasses, draw these diagrams on squared paper.

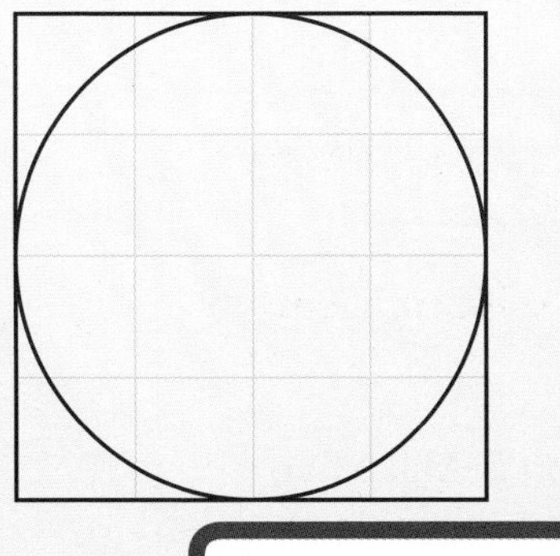

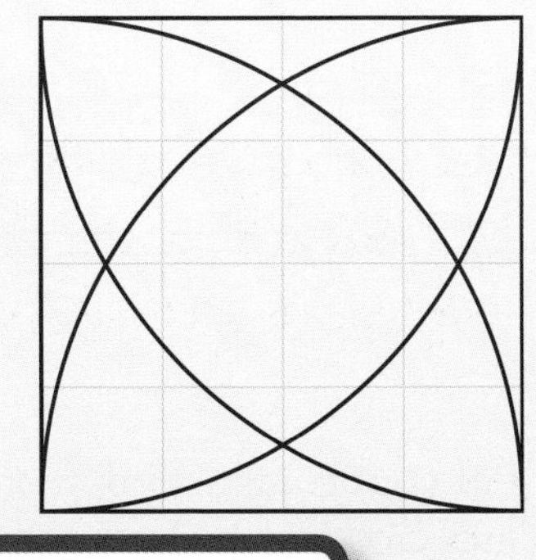

The side of each small square is 1 cm.

2 In triangle *EFG* each side is 6 cm long.
 a What type of triangle is *EFG*?
 b Draw triangle *EFG* accurately.
 c Use a ruler to find the mid-point of *FG*. Label it *M*. Draw the line *EM*.
 d Measure angles *FEM* and *GEM*. What do you notice?
 e Fill in the missing word in this sentence.

 EM is ______________ to *FG*.

33.1 Constructions

Keywords
construction, arc, perpendicular, line segment, perpendicular bisector, bisect, angle bisector

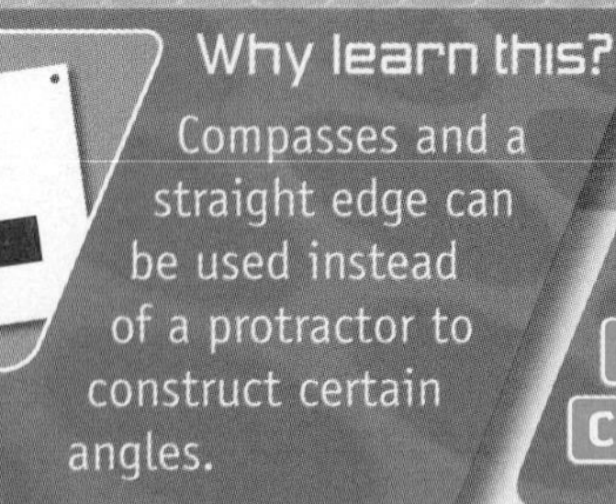

Why learn this?
Compasses and a straight edge can be used instead of a protractor to construct certain angles.

Objectives
- C Construct perpendiculars
- C Construct the perpendicular bisector of a line segment
- C Construct angles of 90° and 60°
- C Construct the bisector of an angle

Skills check

1 Draw a triangle.
Estimate and mark the mid-point of each side of the triangle.
Check how accurate you have been by measuring with a ruler.

Constructions

Constructions are accurate diagrams drawn using only

- compasses

You are not allowed to use a protractor when doing constructions.

- a straight edge.

You may use a ruler as the straight edge.

When doing constructions, you will often draw **arcs** (parts of a circle), rather than complete circles.

The arcs will show how you did the construction. You must not rub them out, even if you think they look untidy.

No arcs, no marks!

Constructing the perpendicular at a point on a line

A line at right angles (90°) to a given line is called a **perpendicular**.

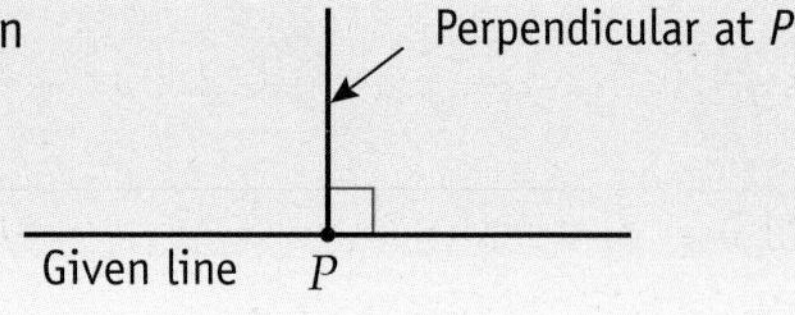

Example 1

P is a point on a line.
Construct a perpendicular at P.

Step 1

1 Put your compass point on P. Make two arcs crossing the line, one either side of P. Label these points A and B.

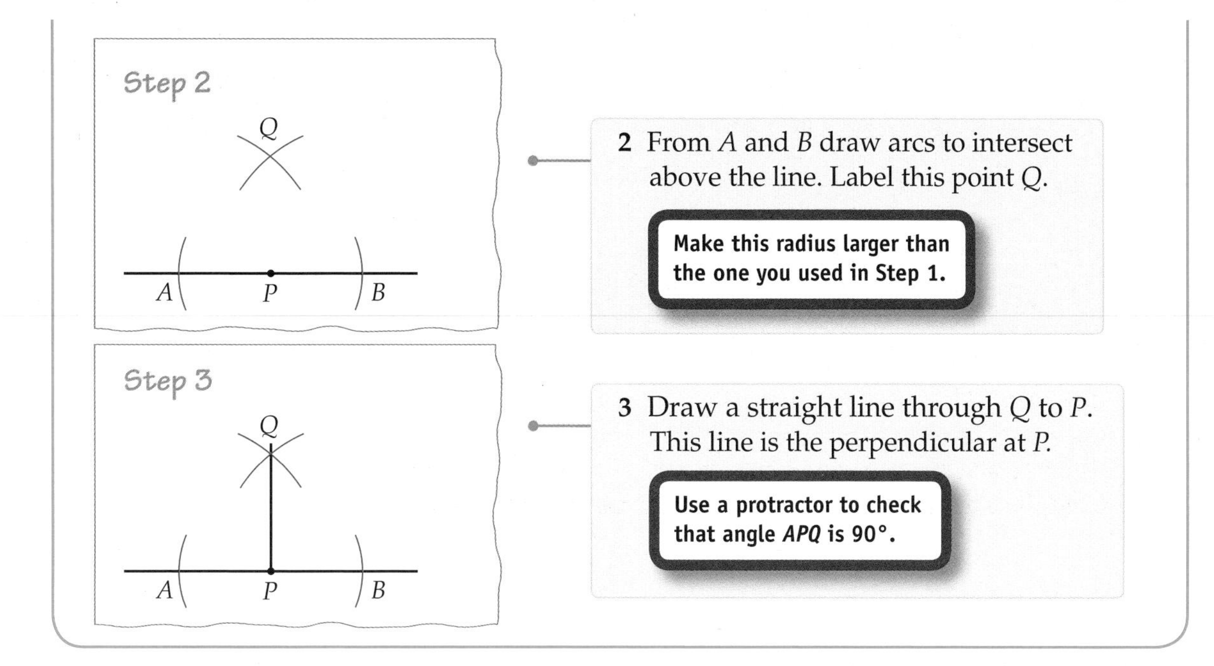

Exercise 33A

1 Draw a line at an angle, as shown, and mark a point P on it.
Construct a perpendicular at point P.

2 Draw a line and mark two points, X and Y, on it.

- **a** Construct a perpendicular at point X.
- **b** On the other side of the line, construct a perpendicular at point Y.

3 Construct a square without using a protractor or measuring with a ruler.

4 Draw a horizontal line and mark three points on it.
Construct a perpendicular at each point.
What is the geometrical relationship between these perpendiculars?

Constructing the perpendicular from a point to a line

When you are asked to find the distance from a point, P, to a line, you will need the shortest distance.

This is the length of the perpendicular from P to the line.

The perpendicular from P to the line is at right angles to it.

Example 2

Construct the perpendicular from point P to the line.

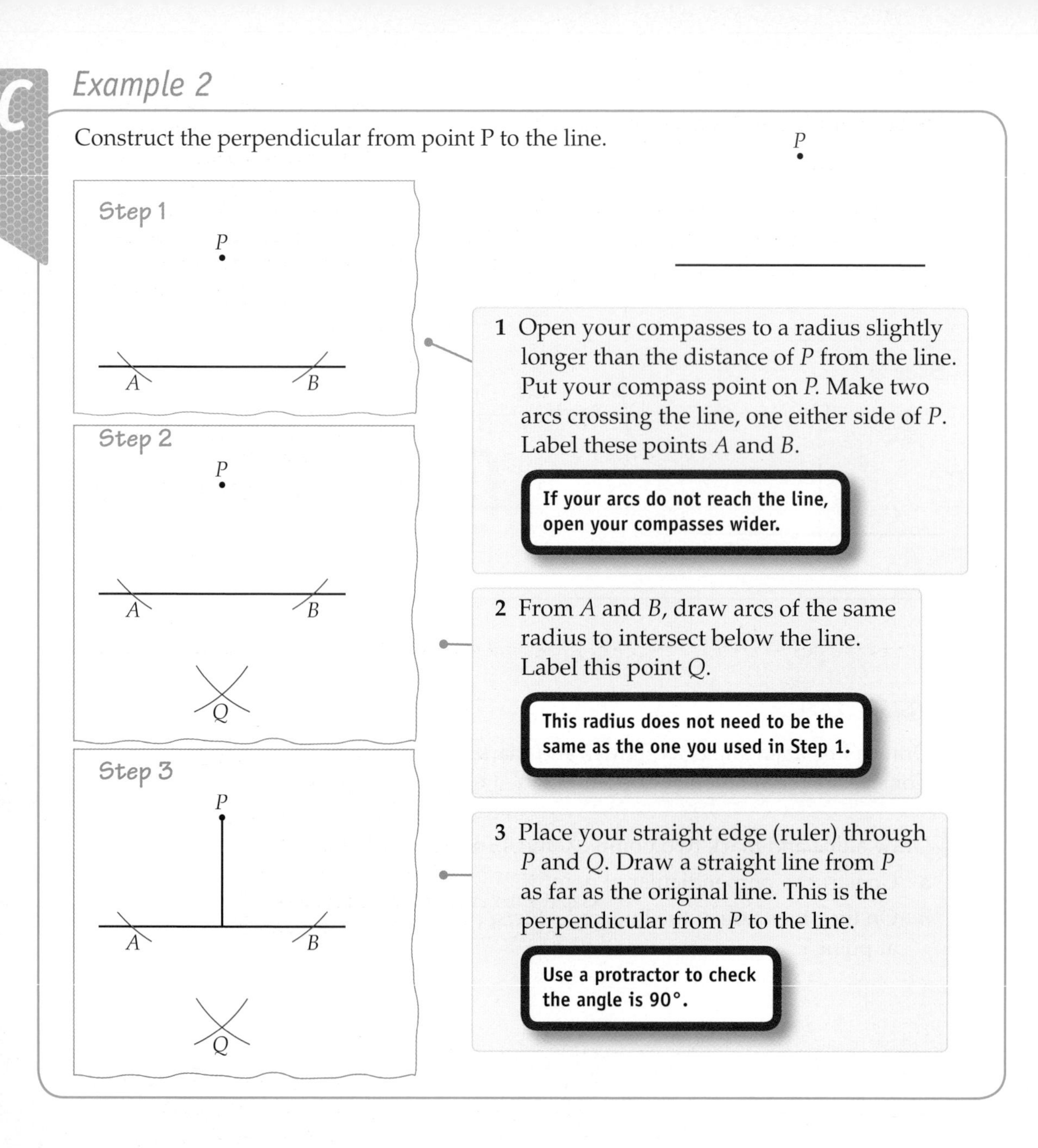

Exercise 33B

1 Copy the diagrams and construct the perpendicular from the point P to the line.

a P

b P

2 Draw a line and mark two points A and B, one either side of the line.
Construct perpendiculars from A and B to the line.

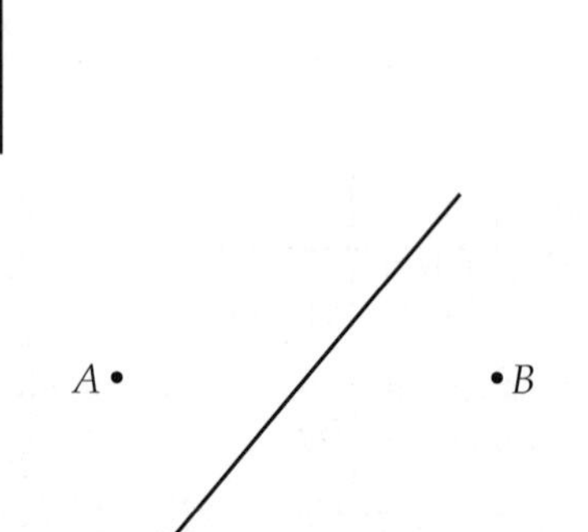

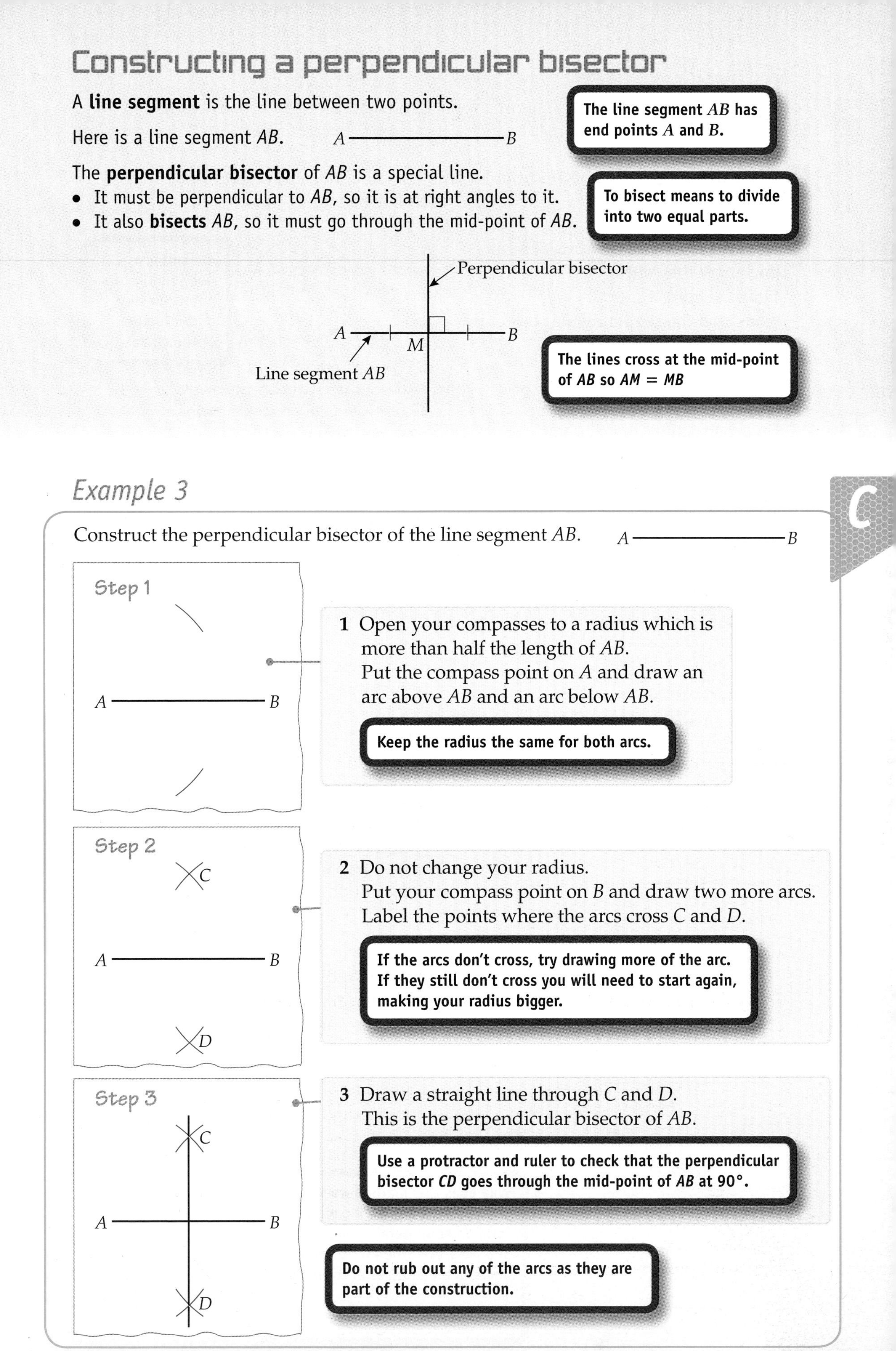

Constructing a perpendicular bisector

A **line segment** is the line between two points.

Here is a line segment AB.

The line segment AB has end points A and B.

The **perpendicular bisector** of AB is a special line.

- It must be perpendicular to AB, so it is at right angles to it.
- It also **bisects** AB, so it must go through the mid-point of AB.

To bisect means to divide into two equal parts.

The lines cross at the mid-point of AB so $AM = MB$

C

Example 3

Construct the perpendicular bisector of the line segment AB.

Step 1

1 Open your compasses to a radius which is more than half the length of AB.
Put the compass point on A and draw an arc above AB and an arc below AB.

Keep the radius the same for both arcs.

Step 2

2 Do not change your radius.
Put your compass point on B and draw two more arcs.
Label the points where the arcs cross C and D.

If the arcs don't cross, try drawing more of the arc. If they still don't cross you will need to start again, making your radius bigger.

Step 3

3 Draw a straight line through C and D.
This is the perpendicular bisector of AB.

Use a protractor and ruler to check that the perpendicular bisector CD goes through the mid-point of AB at 90°.

Do not rub out any of the arcs as they are part of the construction.

Exercise 33C

C

For these questions, use compasses and a straight edge to do the constructions. Then check your accuracy by measuring with a protractor and ruler.

1. Draw three line segments of different lengths. Label the ends A and B. Construct the perpendicular bisector of each line segment AB.

2. Draw a circle with a radius of 5 cm and label the centre O.
   Draw a chord XY.
   Construct the perpendicular bisector of XY.
   What do you notice?

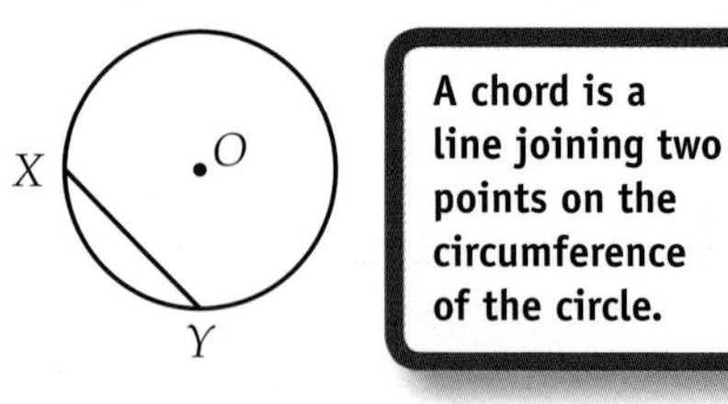

A chord is a line joining two points on the circumference of the circle.

3. Draw a line segment PQ.
   By construction, find the mid-point of PQ.

Constructing angles

Example 1 showed you how to construct an angle of 90°.

Here you will find how to construct an angle of 60°.

Example 4

C

Construct an angle of 60°.

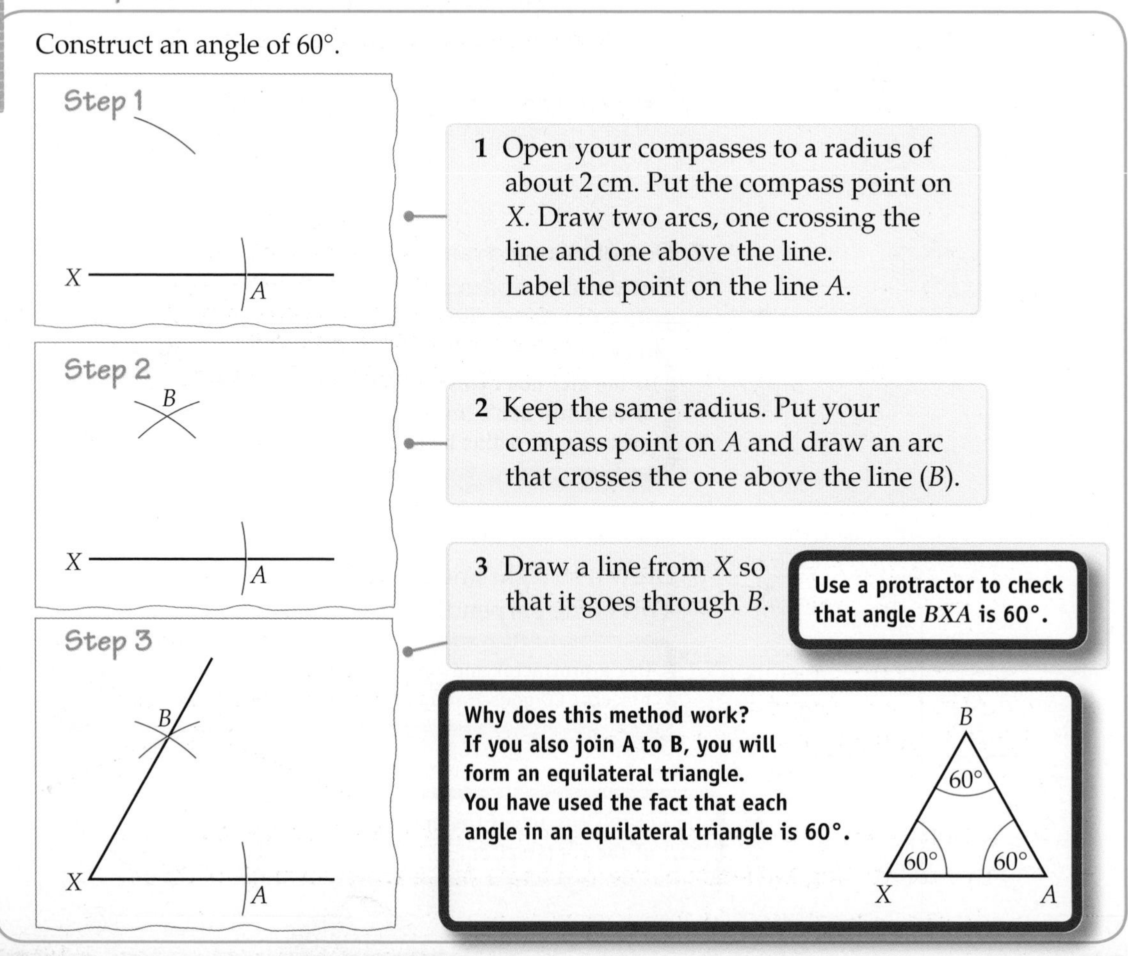

1 Open your compasses to a radius of about 2 cm. Put the compass point on X. Draw two arcs, one crossing the line and one above the line.
Label the point on the line A.

2 Keep the same radius. Put your compass point on A and draw an arc that crosses the one above the line (B).

3 Draw a line from X so that it goes through B.

Use a protractor to check that angle BXA is 60°.

Why does this method work?
If you also join A to B, you will form an equilateral triangle.
You have used the fact that each angle in an equilateral triangle is 60°.

Constructing the bisector of an angle

The bisector of an angle divides an angle into two equal parts.

The line is called an **angle bisector**.

Angle bisector

Example 5

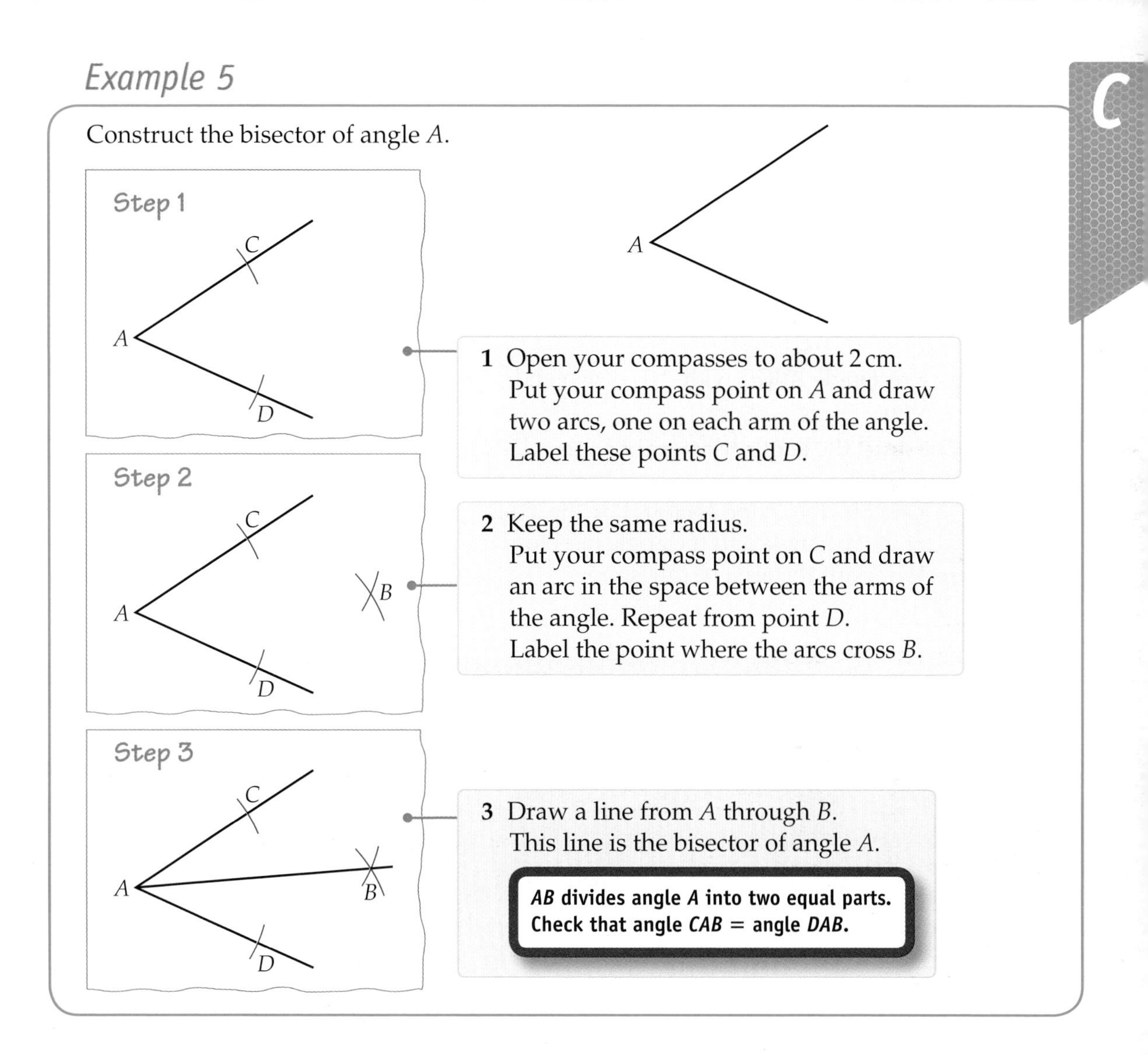

Exercise 33D

1 Draw an acute angle X. Draw an obtuse angle Y.
Construct the angle bisector of each angle.

2 Draw two lines crossing at A.
Construct the angle bisectors of angles x and y.
What do you notice about the angle bisectors?

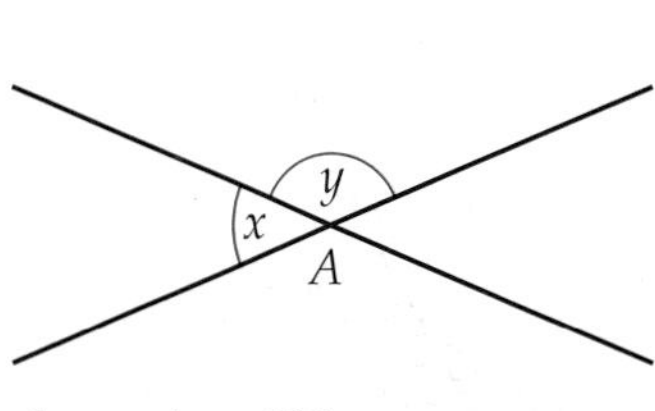

3 **a** Using a ruler and compasses only, construct angle A where $A = 60°$.
b By constructing the angle bisector of angle A, construct an angle of 30°.
c Explain how you would construct an angle of 15°.

4 a Jack says he can draw an angle of 45° without using a protractor. Explain how he can do this.

b Using a ruler and compasses only, construct an angle of 45°.

c State an obtuse angle Jack could construct accurately without using a protractor. Explain how.

5 Sam drew a triangle and then constructed a line. She correctly described the line as both an angle bisector and a perpendicular bisector. Explain how this could happen.

AO2

6 a Draw a circle.

b Construct an angle of 60° at the centre of the circle.

c Now construct a regular hexagon so that it fits *exactly* inside the circle.

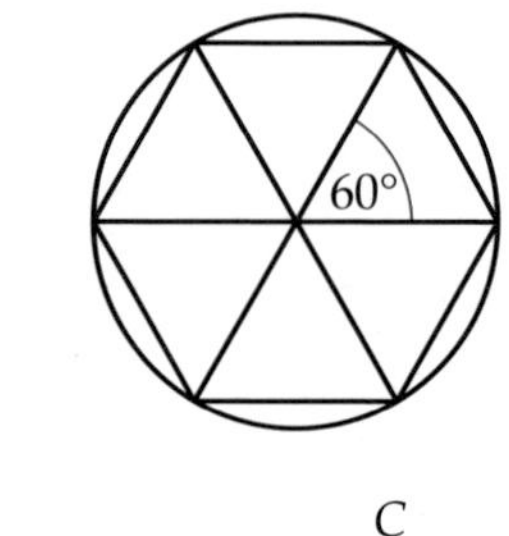

7 Draw a rectangle $ABCD$ with $AB = 8$ cm and $BC = 6$ cm.

a Construct the angle bisectors of angle ADC and angle ABC.

b i Describe the shape enclosed by the sides of the rectangle and the angle bisectors.

ii Work out the area of this shape.

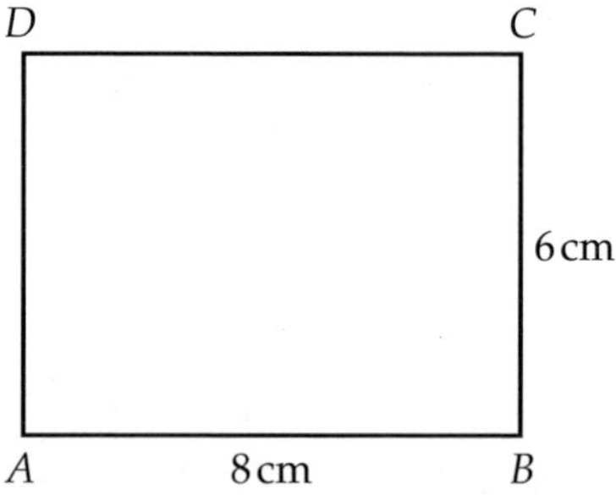

AO2

33.2 Locus

Keywords
locus (loci), equidistant

Why learn this?
Takeaway food outlets often have a free delivery area, which can be described as a locus, e.g. 'anywhere within two miles'.

Objectives
C Construct loci
C Solve locus problems, including the use of bearings

Skills check

1 Sketch the following paths.
  a A ball being dropped vertically to the ground.
  b A ball being thrown from one person to another.
  c The tip of your finger when you write a text message.
  d The centre of a ball as it is rolled along the ground.
  e The tip of a car's windscreen wiper in the rain.

Locus

The paths of many moving objects are unpredictable, such as the trail left by a snail in the garden. Some paths are predictable, such as a roundabout ride at a fair.

A **locus** is a set of points that obey a given rule.

You can use standard constructions to show **loci**.

Example 6

C is a fixed point. Draw the locus of points that are always 1 cm from C.

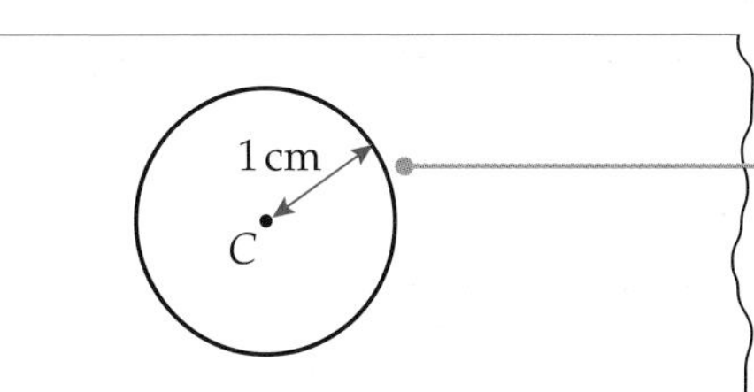

All points on the circle obey this rule. They are 1 cm from C.
The points are **equidistant** (the same distance) from C.

All points *inside* the circle are *less than* 1 cm from C.
All points *outside* the circle are *more than* 1 cm from C.

The locus of points that are the same distance from a fixed point is a circle, with the fixed point as the centre.

C

Example 7

Draw the locus of points that are exactly 1 cm from

a a given line

b a given line segment AB.

a

1 cm Given line 1 cm

1 cm 1 cm

A line has infinite length.

Using a ruler, draw parallel lines 1 cm either side of the given line.

Remember to measure in at least two places.

b

1 cm

A B

The locus is 'racetrack' shape.

The line segment AB has end points A and B.

Using a ruler, mark points 1 cm from different points on AB.
Use your compasses to draw the semicircles at A and B.

Points inside the shape are less than 1 cm away from AB.

The locus of points that are the same distance from a fixed *line* is two parallel lines, one each side of the given line.
The locus of points that are the same distance from a fixed *line segment AB* is a 'racetrack' shape. The shape has two lines parallel to *AB* and two semicircular ends.

C

Exercise 33E

C

1 The line segment XY is 5 cm long.

X ———————— Y
5 cm

Draw the line segment XY.
Draw the locus of points that are exactly 3 cm from XY.

2 Signals from a transmitter can be picked up at any point within a distance of 30 km from it.
Draw a diagram to show where the signal can be picked up.
Use a scale of 1 cm to 10 km.

3 A disc has centre C. It is rolled around the inside of a rectangular frame so that it is always touching the frame.
Draw a rectangle and sketch the locus of C.

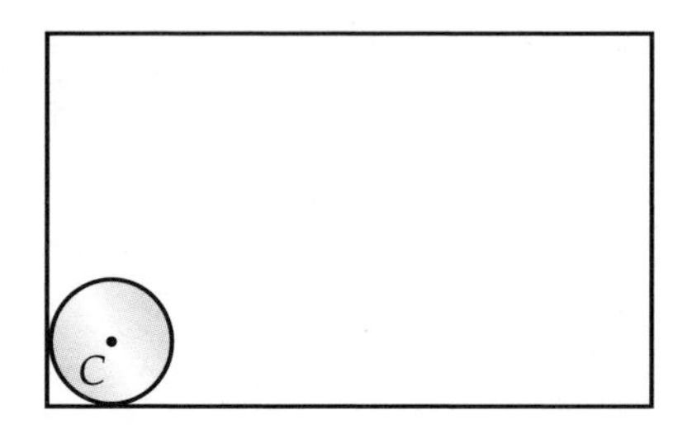

4 A circular pond has a radius of 3 m.
A gardener wants to plant flowers exactly 1 m away from the pond.

a Draw an accurate diagram of the pond. Use a scale of 1 cm to 1 m.

b Show accurately where the flowers could be planted.

5 A disc has centre C.
Sketch the locus of C as the disc is rolled from X to Y.

a

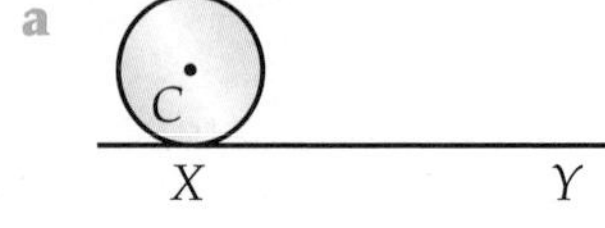

b

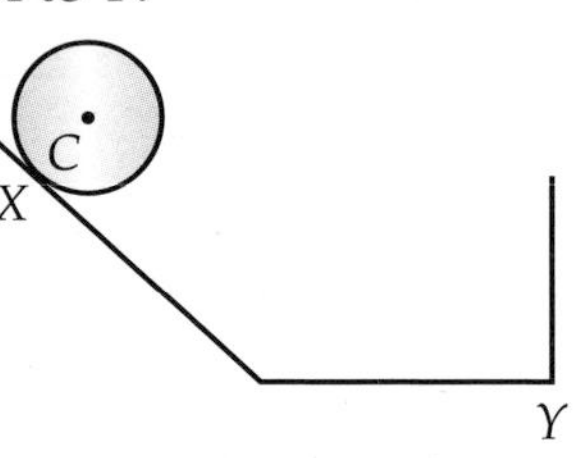

6 A ladder PQ is leaning against a wall.
A bucket is placed on the ladder, making the ladder slide to the ground.

a Sketch the locus of P.

b Sketch the locus of Q.

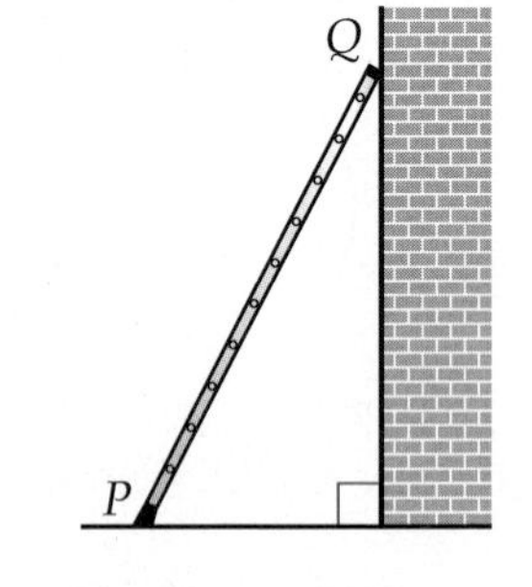

7 A fenced enclosure at a zoo is in the shape of a rectangle 8 m by 6 m.
Visitors are allowed to approach the enclosure, but must stand at least 1 m away from the fence.
Draw an accurate diagram to show where people may not stand.
Use a scale of 1 cm to 1 m.

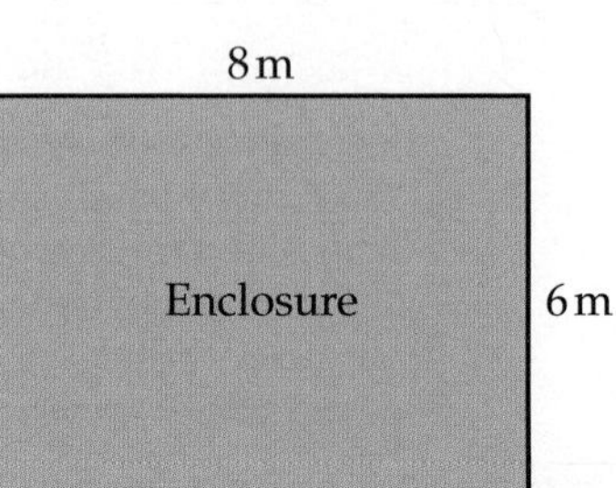

8 A dog is tied to a 3 m length of rope.
At the other end of the rope is a ring.
The ring can slide along a wire, attached to a wall at X and Y.
The wire is 5 m long.
Draw an accurate scale diagram to show the region where the dog can move.
Use a scale of 1 cm to 1 m.

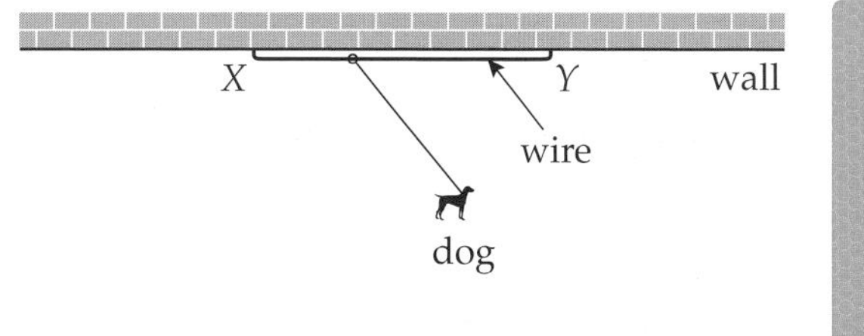

9 Draw the shapes on centimetre squared paper.
For each shape, draw the locus of points that are 1 cm from the shape.

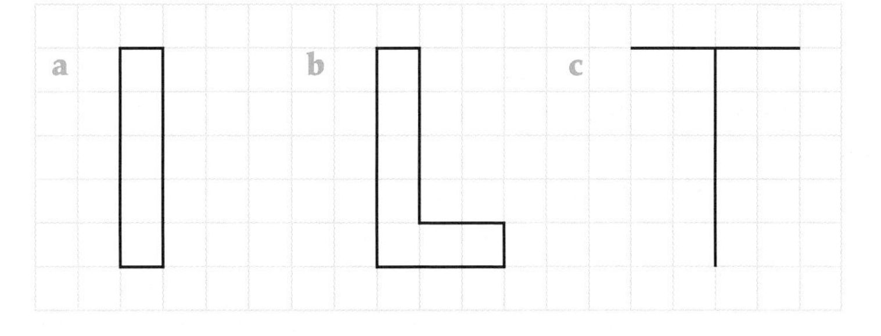

10 A guard dog is tied by a chain to a post at the corner of a shed.
The chain is 3 m long.

a Draw a scale diagram of the shed, using a scale of 2 cm to 1 m.

b Show on the diagram all the possible positions of the dog if the chain is tight.

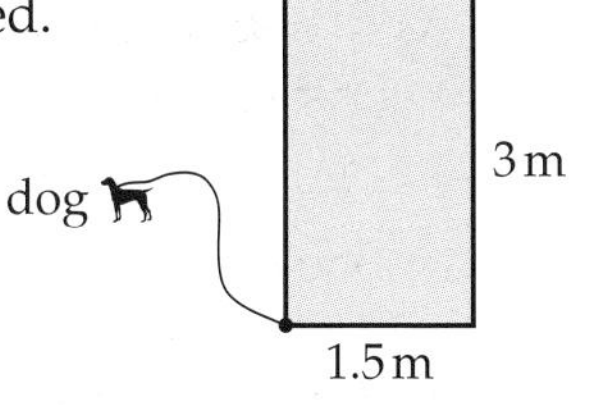

Special loci

The locus of points equidistant from two fixed points is the perpendicular bisector of the line segment joining the two points.

To construct the locus, follow the steps in Example 3.

The locus of points equidistant from two fixed lines is the angle bisector of the angle formed by the lines.

To construct the locus, follow the steps in Example 5.

'Equidistant from' means 'the same distance from'.

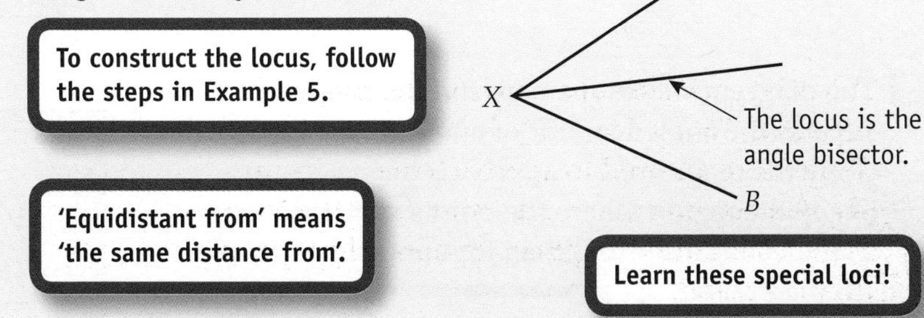

Learn these special loci!

Exercise 33F

C

1 There are two trees in a field. A fence is to be built across the field so that it is the same distance from each tree.

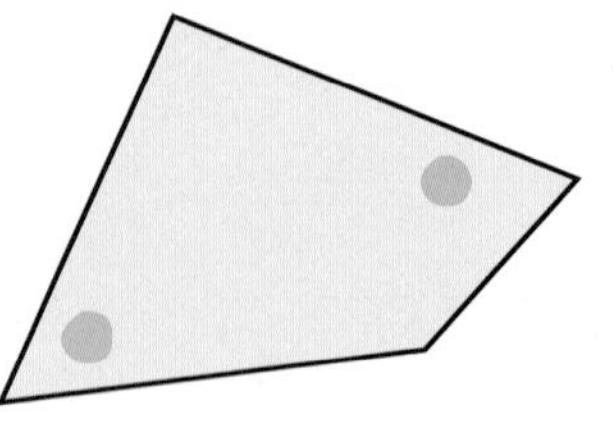

a Copy the diagram and construct the position of the fence.

b Copy and complete this sentence.

'The locus of the fence is ________________.'

2 A drainage pipe is to be laid across a plot of land so that it is the same distance from AD as it is from CD.

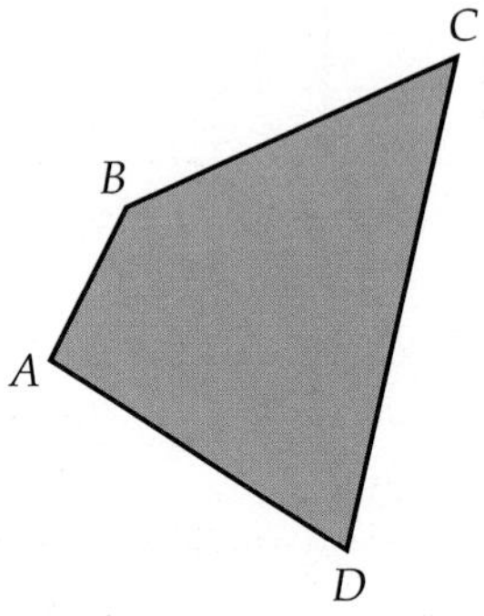

a Copy the diagram and construct the locus of the drainage pipe.

b Copy and complete this sentence.

'The locus of the drainage pipe is ________________.'

3 The diagram shows the perpendicular bisector of AB.
Copy the diagram.
Shade the region inside the triangle where the points are closer to B than to A.

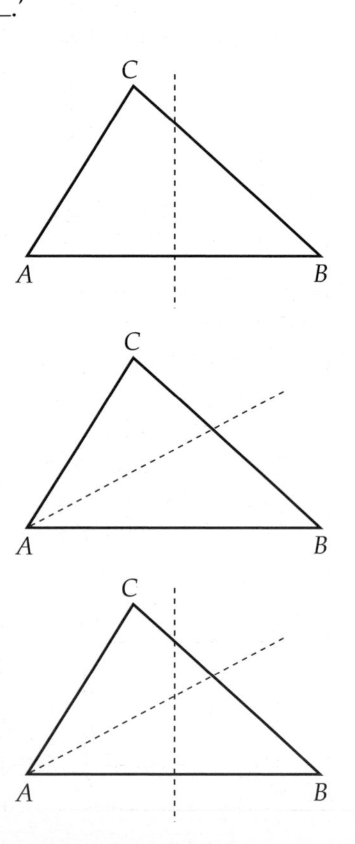

4 The diagram shows the angle bisector of angle A.
Copy the diagram.
Shade the region where the points are closer to AB than to AC.

5 The diagram shows the perpendicular bisector of AB and the angle bisector of angle A.
Copy the diagram.
Shade the region where the points inside the triangle are closer to B than to A and closer to AB than to AC.

Use Q3 and Q4 to help you.

Intersecting loci

Questions often involve intersecting loci. You may have to locate a region satisfied by two or more constraints.

Example 8

Ben and Kate are devising a game for their stall at a fete.

Players will try to locate the buried treasure on this rectangular board.

Ben says the treasure should be closer to *C* than it is to *D*.

Kate says the treasure should be less than 45 cm from *D*.

They decide that both these constraints should be met.

By accurate construction, show the region where they should bury the treasure.

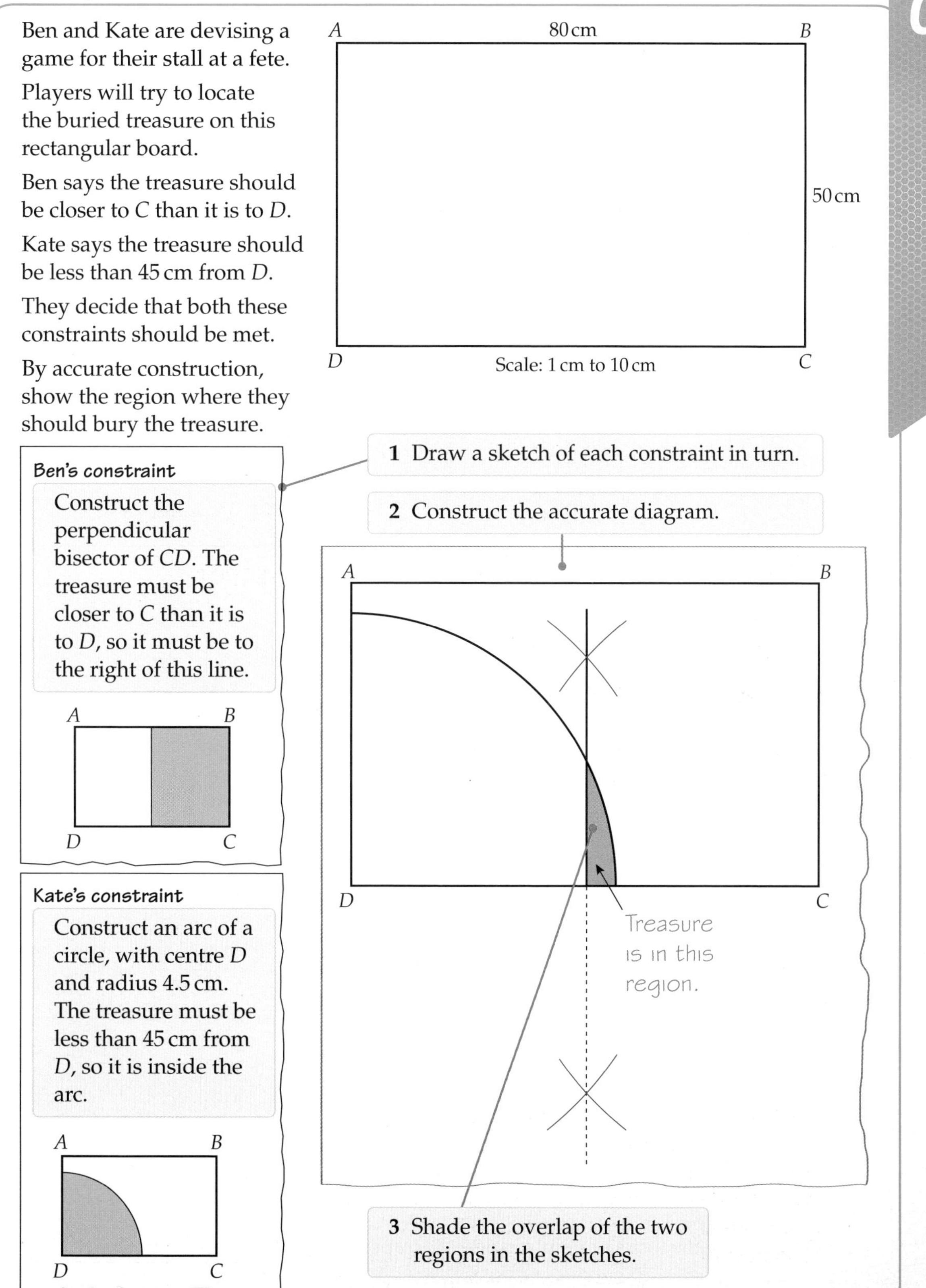

C

Example 9

The positions of a ship S and a lighthouse L are shown in the diagram. The scale is 1 cm to 10 km.

The ship sails on a bearing of 070°.

Show clearly where the ship is

a 10 km from the lighthouse

b less than 10 km from the lighthouse.

HELP Section 21.3

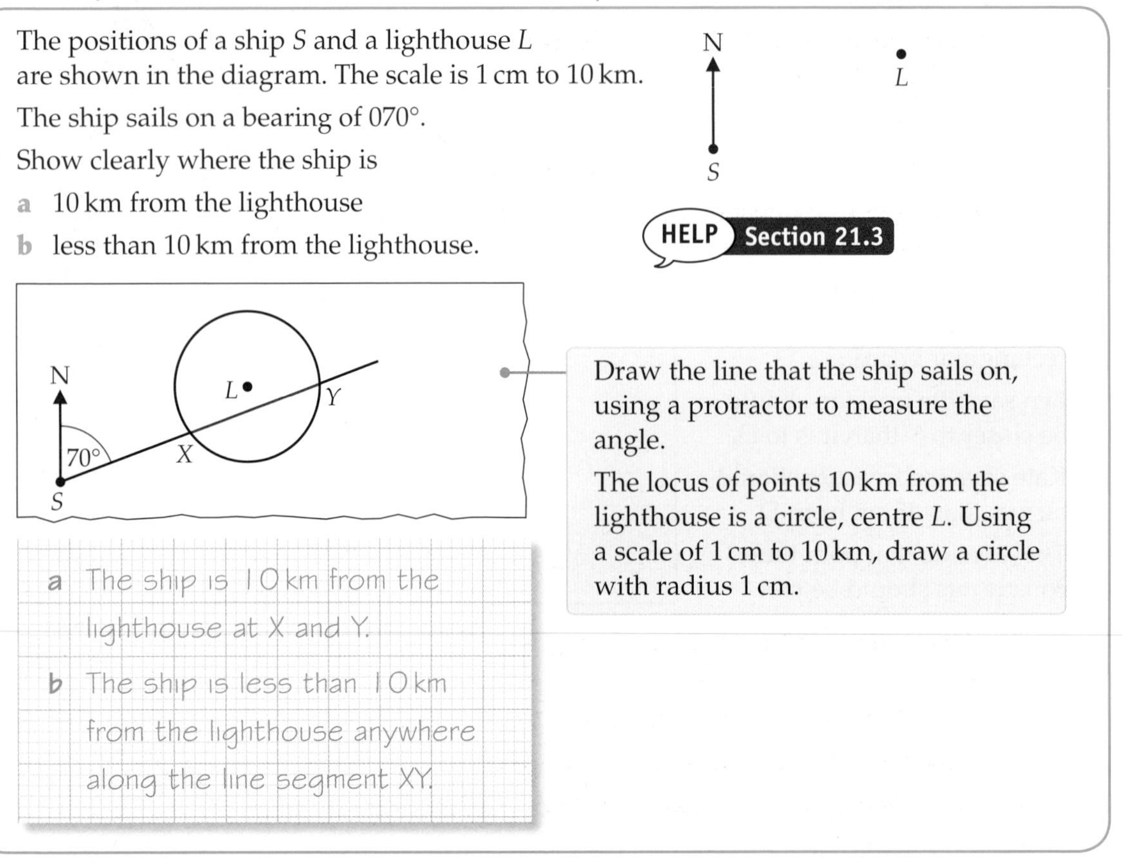

Draw the line that the ship sails on, using a protractor to measure the angle.

The locus of points 10 km from the lighthouse is a circle, centre L. Using a scale of 1 cm to 10 km, draw a circle with radius 1 cm.

a The ship is 10 km from the lighthouse at X and Y.

b The ship is less than 10 km from the lighthouse anywhere along the line segment XY.

Exercise 33G

C

1 A and B are two sprinklers on a lawn, 6 m apart.
Water from sprinkler A can reach anywhere within 3 m of A.
Water from sprinkler B can reach anywhere within 4 m of B.
Draw a scale diagram and show clearly the region of the lawn that will be watered by both sprinklers.

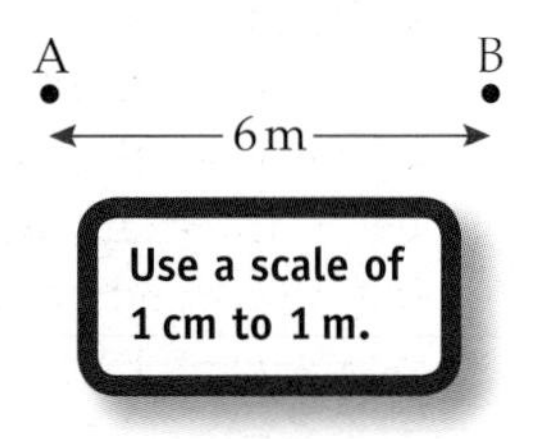

Use a scale of 1 cm to 1 m.

C

2 Make a copy of the trapezium $PQRS$.

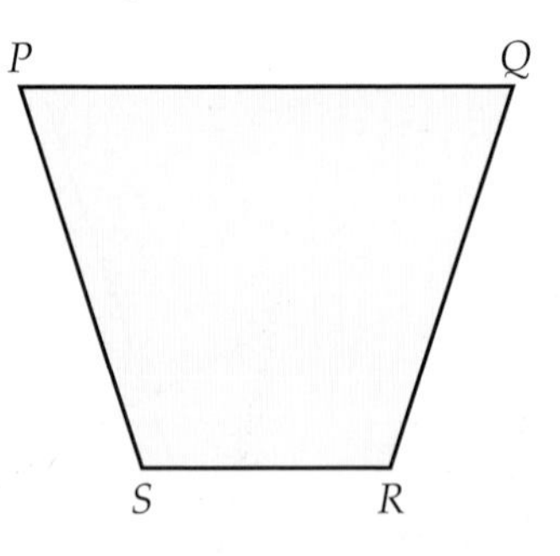

a Construct the locus of points inside the trapezium that are equidistant from QR and RS.

b Construct the locus of points inside the trapezium that are 2 cm from PQ.

c Shade the region inside the trapezium where the points are nearer to QR than to RS **and** more than 2 cm from PQ.

AO2

3 In triangle ABC, $AB = 5$ cm, $AC = 7$ cm and $BC = 6$ cm.

a Construct triangle ABC accurately using a ruler and compasses.

b Construct the perpendicular bisector of AB.

c Construct the locus of points that are 4 cm from C.

d Shade the region where the points are less than 4 cm from C **and** nearer to A than to B.

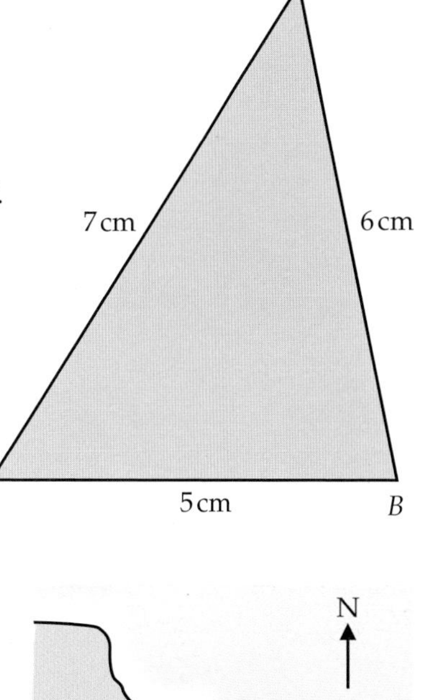

C

4 Ship A is at the port.
Ship B is anchored 20 km away from the port on a bearing of 130°.

a Using a scale of 1 cm to 4 km, draw the positions of the two ships.

Ship A sails out of port on a bearing of 110°.

b Show the path of ship A on your diagram.

c Show clearly on your diagram where ship A is 12 km from ship B.

N
Port A

5 Three towns, A, B and C, are connected by main roads.
$AB = 16$ miles, $AC = 14$ miles and $BC = 12$ miles.
A new leisure centre is to be built so that it is

- inside triangle ABC
- closer to road AC than it is to road AB, and
- between 8 miles and 10 miles from B.

Copy the diagram and construct the region where the leisure centre could be built.
The scale is 1 cm to 2 miles.

C
14 miles
12 miles
A
16 miles
B

C

6 In triangle PQR, $PQ = 6$ cm, $PR = 4$ cm and $QR = 8$ cm.
Construct the diagram accurately.
Shade the region inside the triangle where points are nearer to R than to P and closer to PR than to QR.

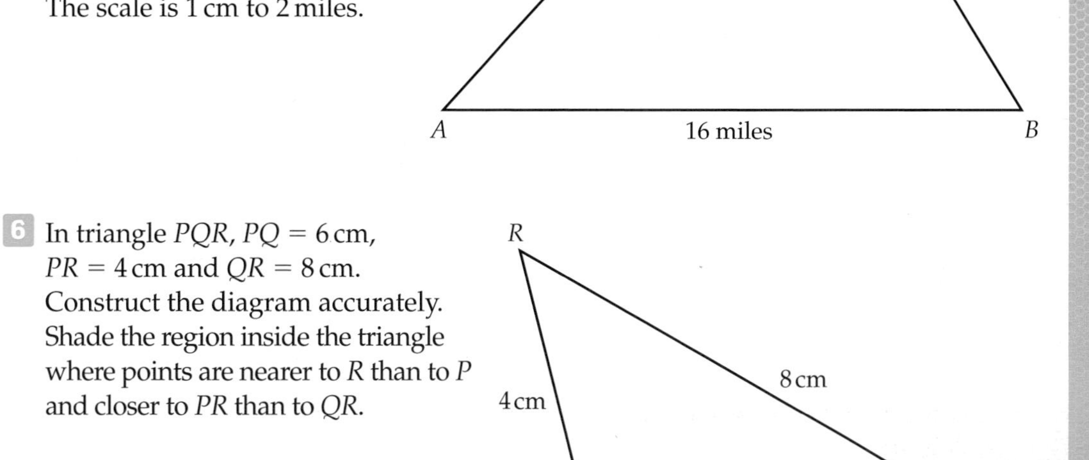

AO3

Review exercise

C

1 Copy the diagram and, using ruler and compasses only, construct the bisector of angle ABC.

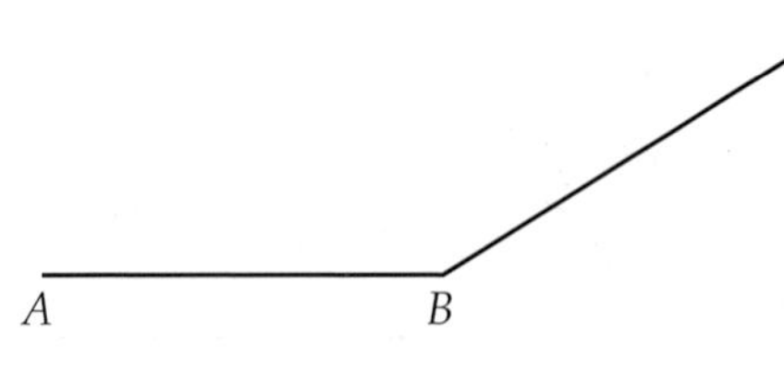

[2 marks]

2 Two radio beacons, X and Y, are 50 km apart.
An aircraft flies so that it is the same distance from each beacon.

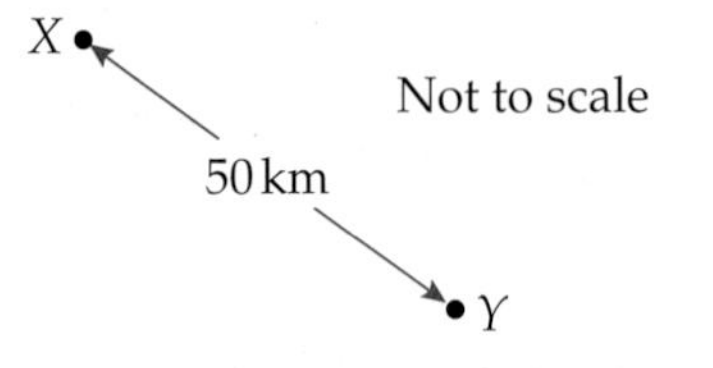

Show the radio beacons and the path of the aircraft on an accurate diagram.
Use a scale of 1 cm to 10 km. **[3 marks]**

3 There are two mobile phone masts on an island.
The mobile phone mast at A has a range of 20 km.
The mobile phone mast at B has a range of 30 km.
Trace the diagram.
Show clearly the area in which the signal from both masts can be received.

[3 marks]

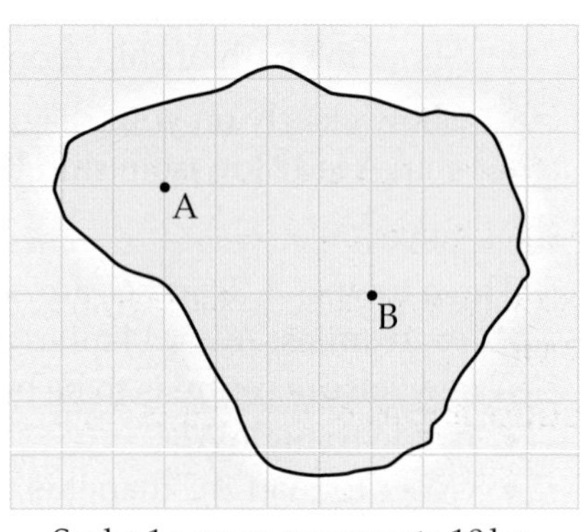

Scale: 1 square represents 10 km

4 Using a ruler and compasses only, construct an angle of 60°.
Show all your construction lines and arcs. **[2 marks]**

5 Copy rectangle $ABCD$.

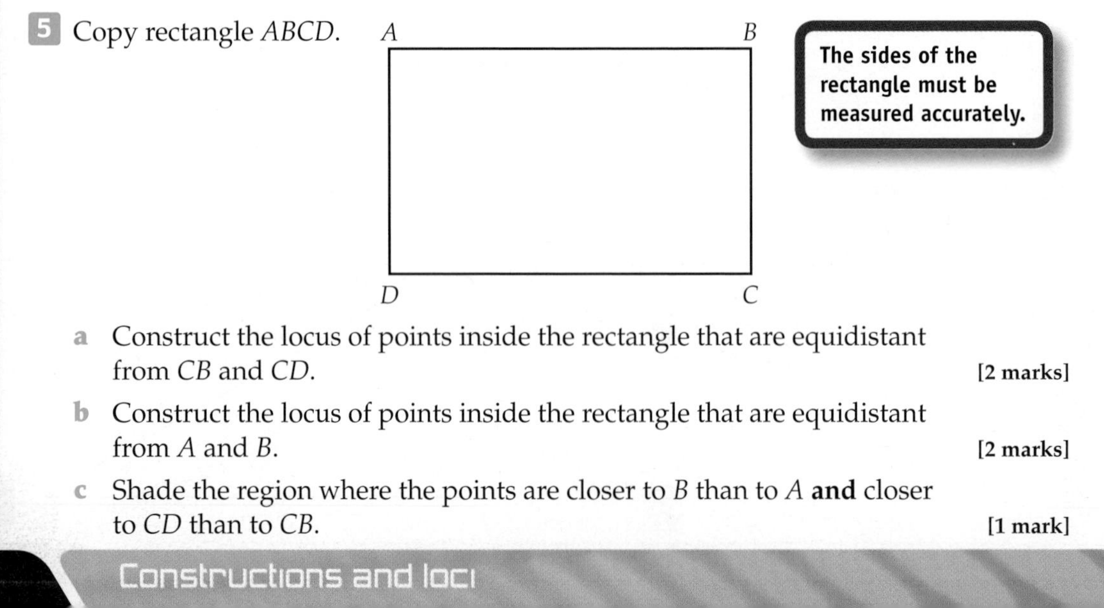

The sides of the rectangle must be measured accurately.

a Construct the locus of points inside the rectangle that are equidistant from CB and CD. **[2 marks]**

b Construct the locus of points inside the rectangle that are equidistant from A and B. **[2 marks]**

c Shade the region where the points are closer to B than to A **and** closer to CD than to CB. **[1 mark]**

6 *AB* and *AC* represent two walls.
A flagpole is to be erected that is equidistant from AB and AC and between 20 m and 50 m from A.
Trace the diagram.

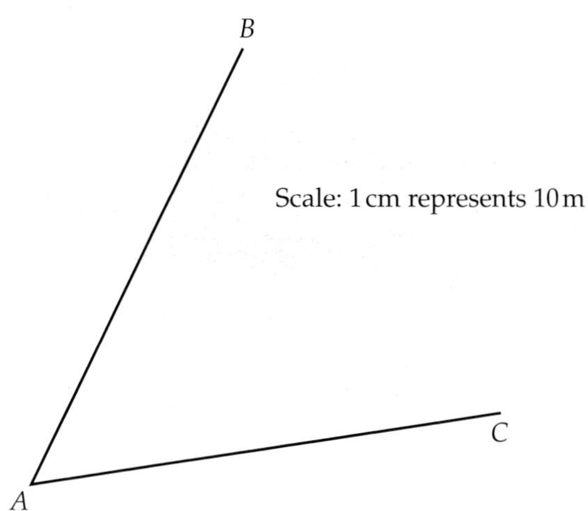

Show clearly all the possible positions of the flagpole. **[3 marks]**

Chapter summary

In this chapter you have learned how to

- construct perpendiculars **C**
- construct the perpendicular bisector of a line segment **C**
- construct angles of 90° and 60° **C**
- construct the bisector of an angle **C**
- construct loci **C**
- solve locus problems, including the use of bearings **C**

34

Pythagoras' theorem

BBC Video

This chapter is about finding unknown lengths in right-angled triangles using Pythagoras' theorem.

When making a stained glass window, you might need to use Pythagoras' theorem to calculate lengths.

Objectives

This chapter will show you how to

- **understand, recall and use Pythagoras' theorem** C
- **calculate the length of line segment *AB*** C

Before you start this chapter

1 Work out

a 5^2 b $\sqrt{64}$ c 7^2 d $\sqrt{100}$

2 Calculate

a 6.3^2 b $\sqrt{640.09}$ c 23.6^2 d $\sqrt{1528.81}$

3 Round these numbers to one decimal place.

a 647.92 b 26.75 c 1.091

4 Round these numbers to two decimal places.

a 16.476 b 23.081 c 106.375

34.1 Pythagoras' theorem

Keywords
Pythagoras' theorem, right-angled triangle, hypotenuse

Why learn this?
This is one of the oldest and best known of all maths theorems!

Objectives
C Understand Pythagoras' theorem

Skills check

1 Work out
 a 9^2 b 4^2 c 6^2 d 10^2

2 Work out
 a $\sqrt{25}$ b $\sqrt{144}$ c $\sqrt{4}$ d $\sqrt{49}$

Pythagoras' theorem

Pythagoras' theorem only applies to **right-angled triangles**.

A right-angled triangle has one angle of 90°.

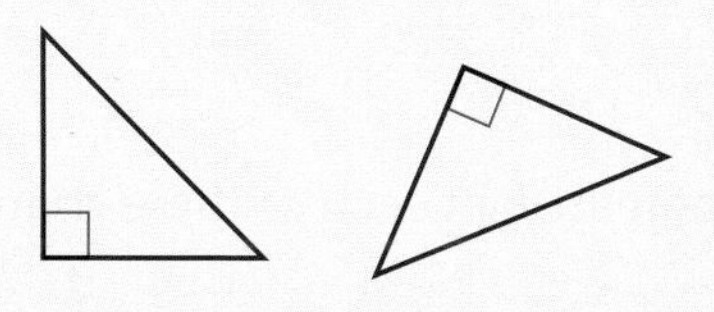

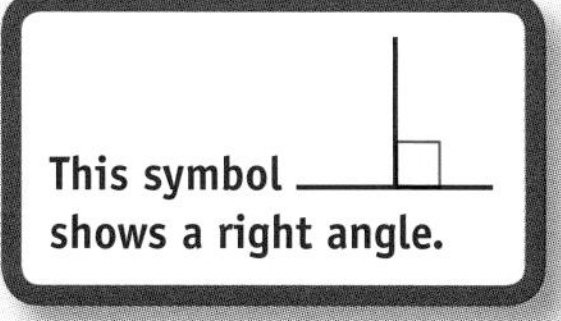

In a right-angled triangle the longest side is called the **hypotenuse**.
The hypotenuse is always opposite the right angle.

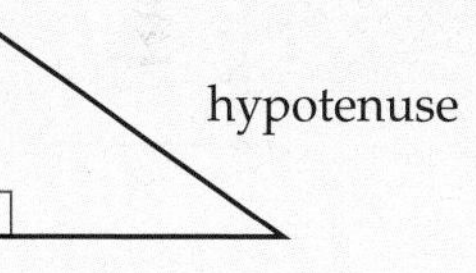

You can construct a square on each side of a right-angled triangle.
This triangle has sides $a = 3$, $b = 4$ and $c = 5$.
If you construct the triangle, you will see that it is a right-angled triangle.

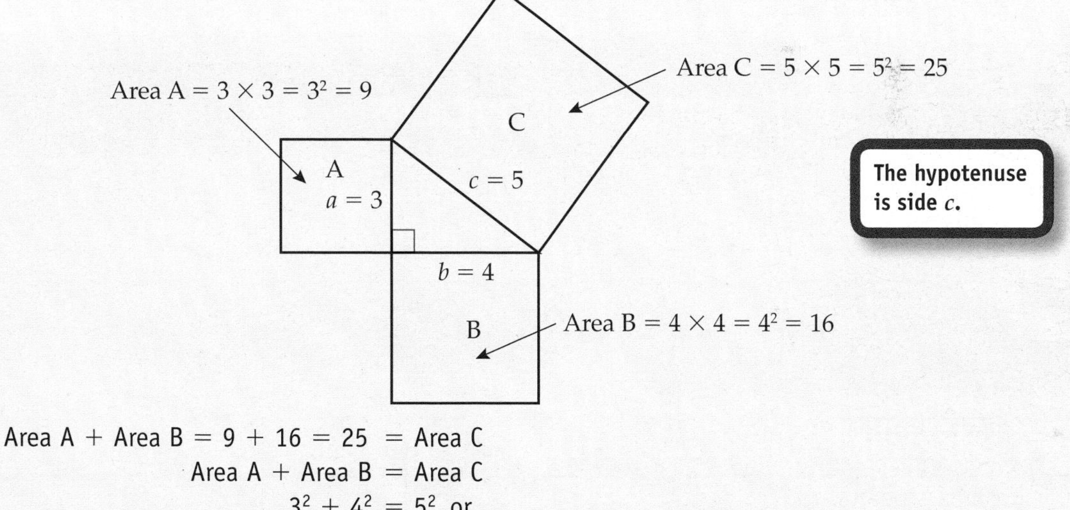

The hypotenuse is side c.

$$\text{Area A} + \text{Area B} = 9 + 16 = 25 = \text{Area C}$$
$$\text{Area A} + \text{Area B} = \text{Area C}$$
$$3^2 + 4^2 = 5^2 \text{ or}$$
$$a^2 + b^2 = c^2$$

This leads to Pythagoras' theorem:

In any right-angled triangle the square of the hypotenuse (c^2) is equal to the sum of the squares on the other two sides ($a^2 + b^2$).

For a right-angled triangle with sides of lengths a, b and c, where c is the hypotenuse, Pythagoras' theorem states that $c^2 = a^2 + b^2$.

You need to learn this formula.

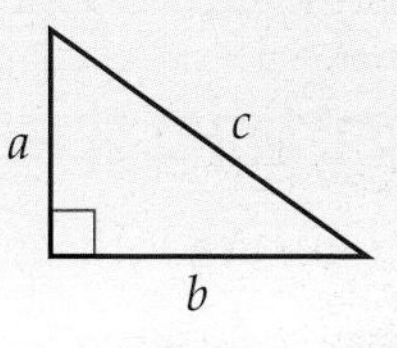

Example 1

a Write the letter that represents the hypotenuse in the triangle.

b Write out the formula using Pythagoras' theorem.

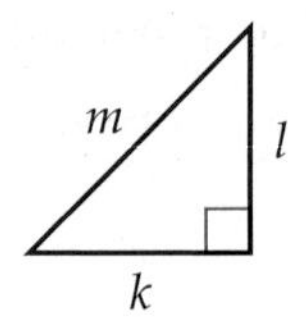

a m is the hypotenuse

The hypotenuse is the longest side and is opposite the right angle.

b $m^2 = l^2 + k^2$

Make the hypotenuse the subject of the formula. The subject appears on its own on one side of the equals sign.

Exercise 34A

1 Write the letter that represents the hypotenuse in each of these triangles.

a

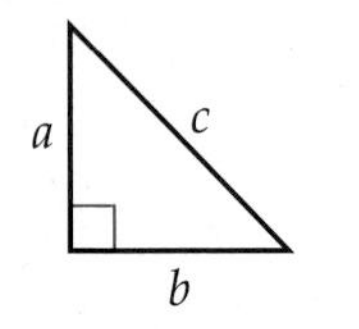

b

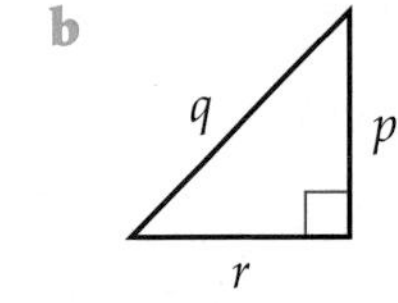

c **d**

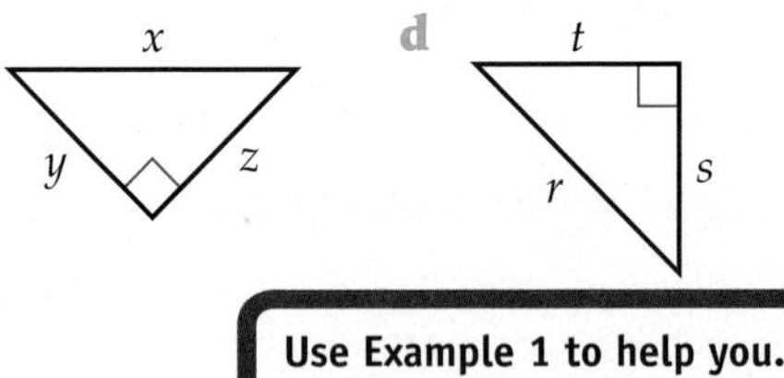

2 For the triangles in Q1, write out the formula using Pythagoras' theorem.
Make the hypotenuse the subject of the formula.

Use Example 1 to help you.

34.2 Finding the hypotenuse

Why learn this?

Being able to use Pythagoras' theorem helps solve a variety of geometrical problems.

Objectives

C Calculate the hypotenuse of a right-angled triangle

C Solve problems using Pythagoras' theorem

Skills check

1 Copy and complete the diagram showing compass points.

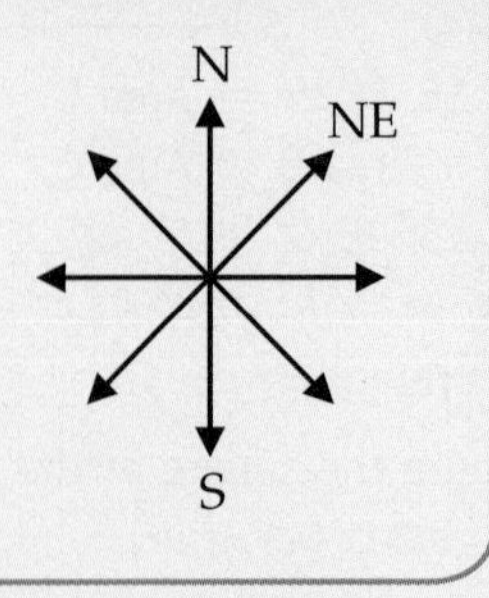

2 Calculate

a $\sqrt{200}$ **b** $\sqrt{625}$

c 7.6^2 **d** 12.1^2

Finding the hypotenuse

You can use Pythagoras' theorem to find the length of the hypotenuse.

Example 2

Calculate the length of the hypotenuse, x, in this right-angled triangle.
Give your answer to one decimal place.

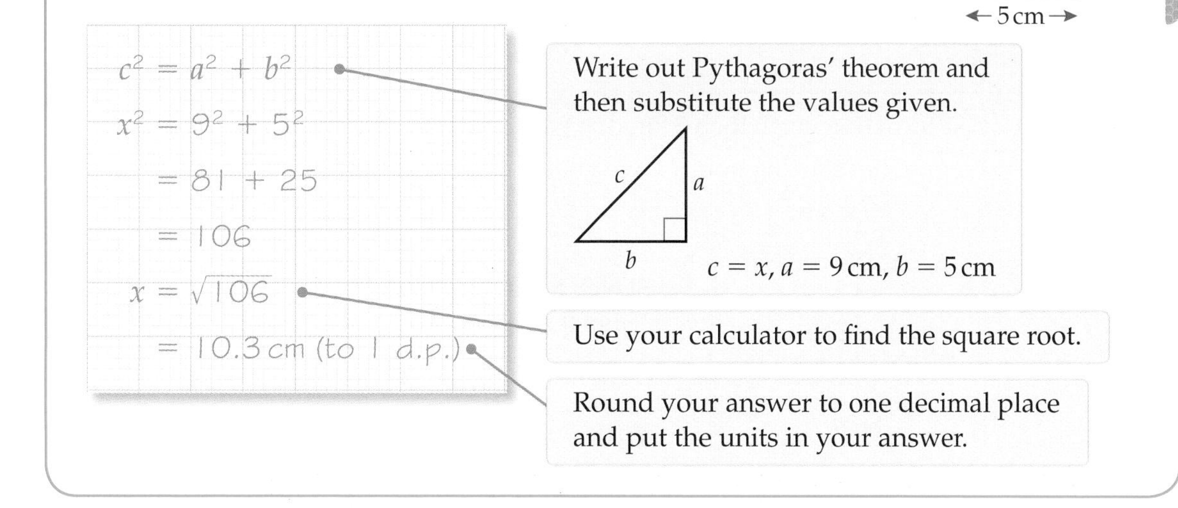

$c^2 = a^2 + b^2$
$x^2 = 9^2 + 5^2$
$= 81 + 25$
$= 106$
$x = \sqrt{106}$
$= 10.3$ cm (to 1 d.p.)

Write out Pythagoras' theorem and then substitute the values given.

$c = x, a = 9$ cm, $b = 5$ cm

Use your calculator to find the square root.

Round your answer to one decimal place and put the units in your answer.

C

Exercise 34B

Calculate the length of the hypotenuse in each triangle.
Give your answer to one decimal place when appropriate.

C

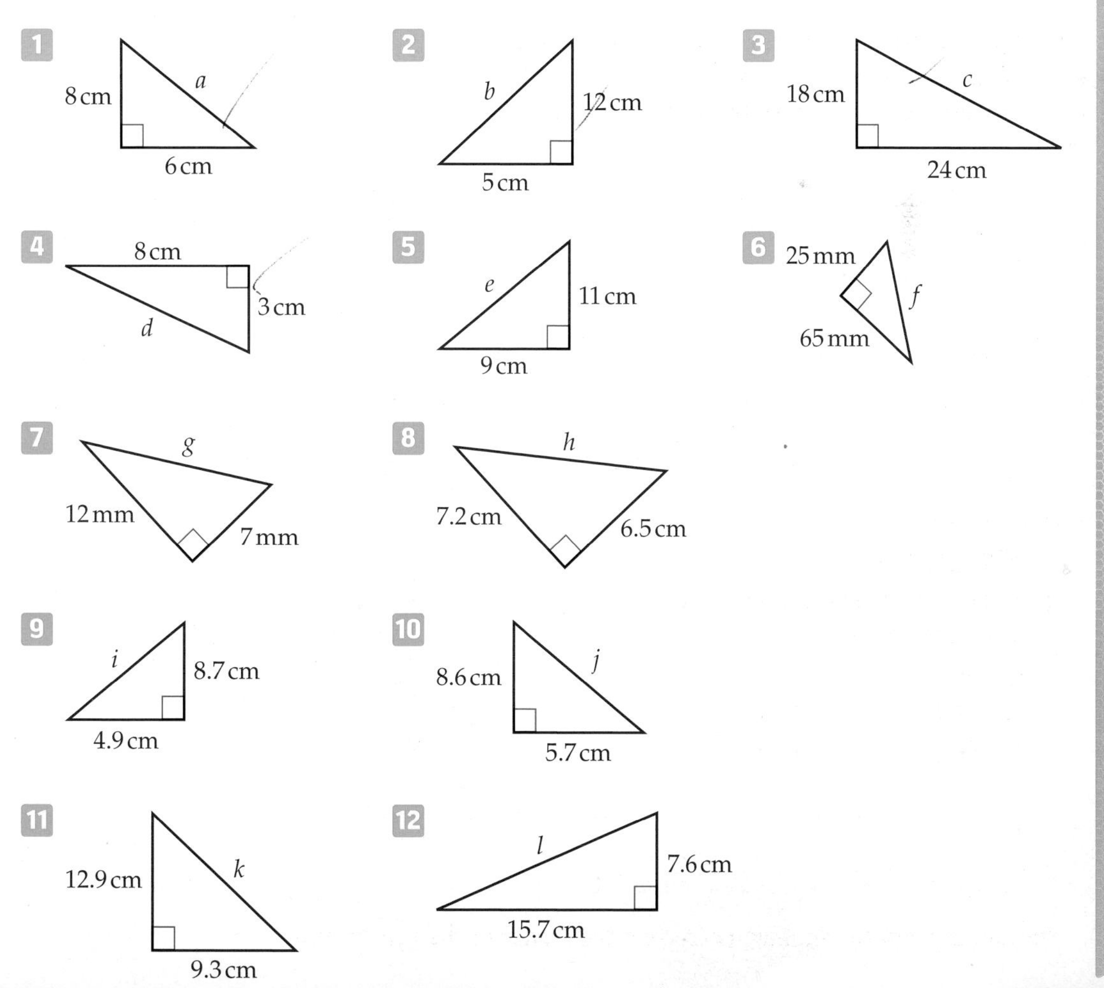

Solving problems using Pythagoras' theorem

You may need to sketch a diagram when solving problems using Pythagoras' theorem.

C

Example 3

Rupinder walks 5 km due north then 3 km due west.
How far is she from her starting point? Give your answer to one decimal place.

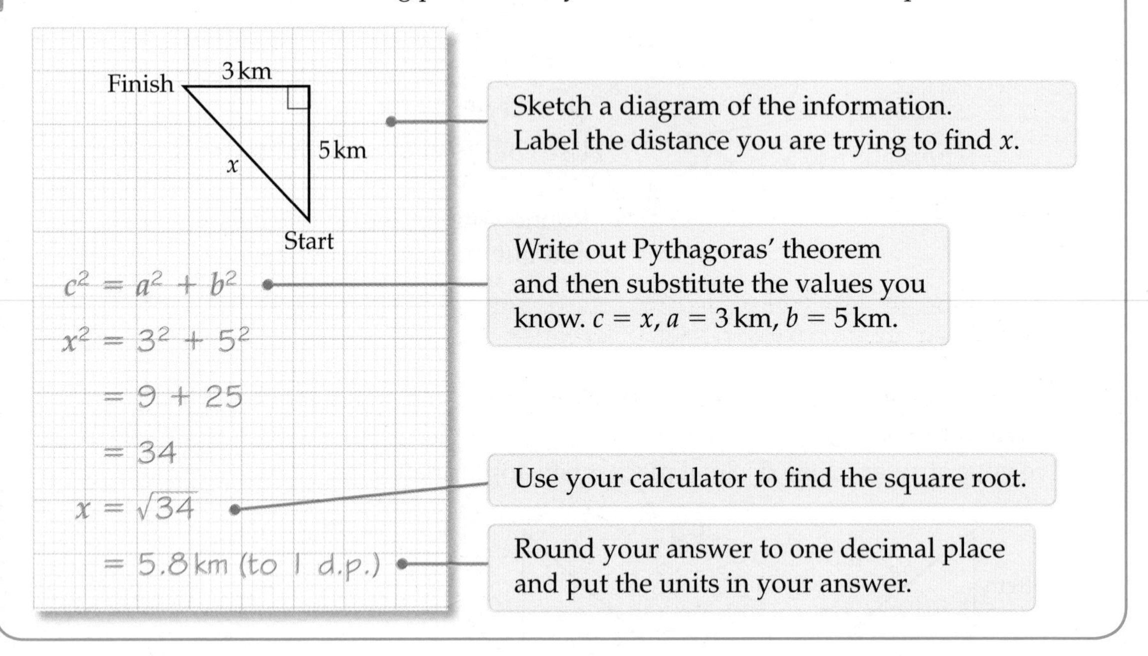

Sketch a diagram of the information. Label the distance you are trying to find x.

Write out Pythagoras' theorem and then substitute the values you know. $c = x$, $a = 3$ km, $b = 5$ km.

Use your calculator to find the square root.

Round your answer to one decimal place and put the units in your answer.

Exercise 34C

C

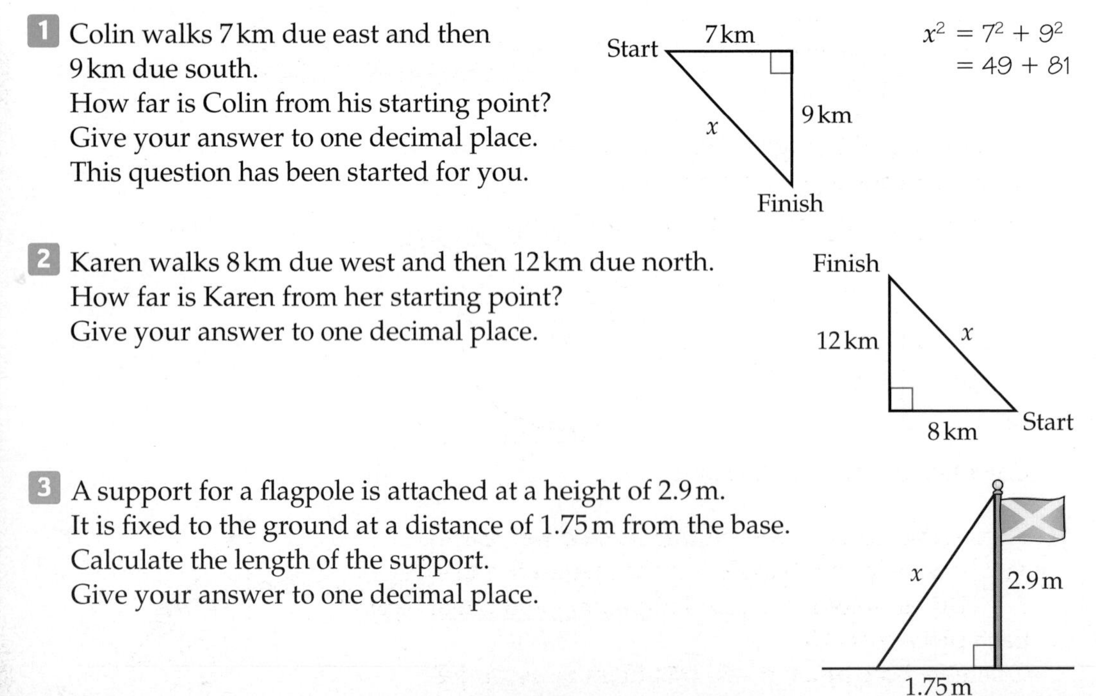

1 Colin walks 7 km due east and then 9 km due south.
How far is Colin from his starting point?
Give your answer to one decimal place.
This question has been started for you.

2 Karen walks 8 km due west and then 12 km due north.
How far is Karen from her starting point?
Give your answer to one decimal place.

3 A support for a flagpole is attached at a height of 2.9 m.
It is fixed to the ground at a distance of 1.75 m from the base.
Calculate the length of the support.
Give your answer to one decimal place.

4 Calculate the length of the diagonal of the rectangle.
Give your answer to one decimal place.

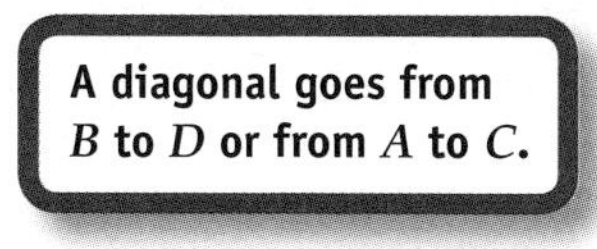
A diagonal goes from B to D or from A to C.

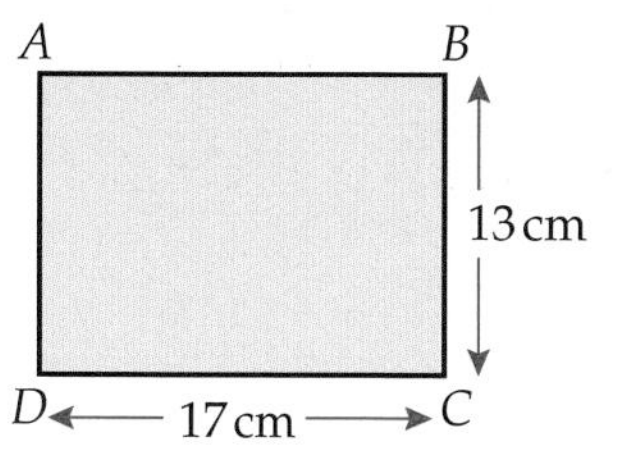

C

5 A ladder is leaning against a wall.
The foot of the ladder is 0.6 m from the base of the wall and it reaches a height of 3 m up the wall.

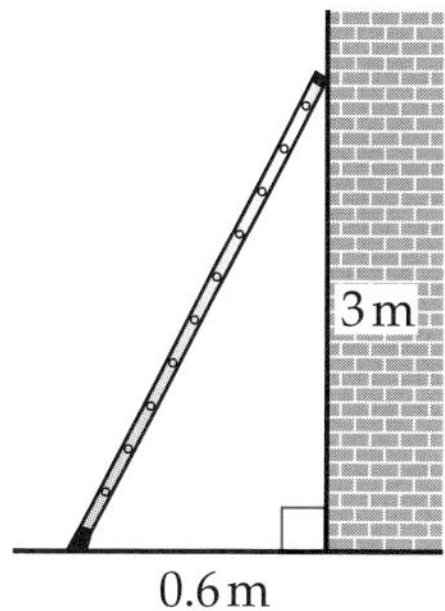

Calculate the length of the ladder.
Give your answer to two decimal places.

6 A right-angled triangle with shorter sides 9 cm and 6 cm is inside a circle with centre O.

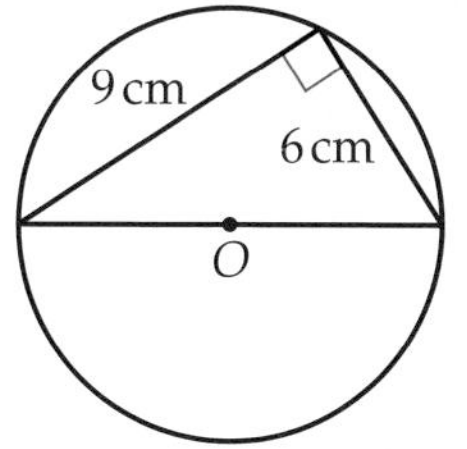

C

a Calculate the length of the diameter of the circle.

b Write down the radius of the circle.

c Calculate the area of the circle.
Give your answer correct to one decimal place.

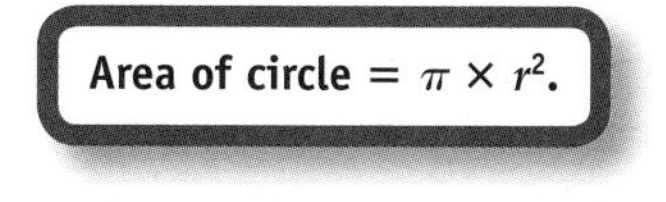
Area of circle $= \pi \times r^2$.

AO2

7 A right-angled triangle with shorter sides 7.5 cm and 5.1 cm is inside a circle with centre O.

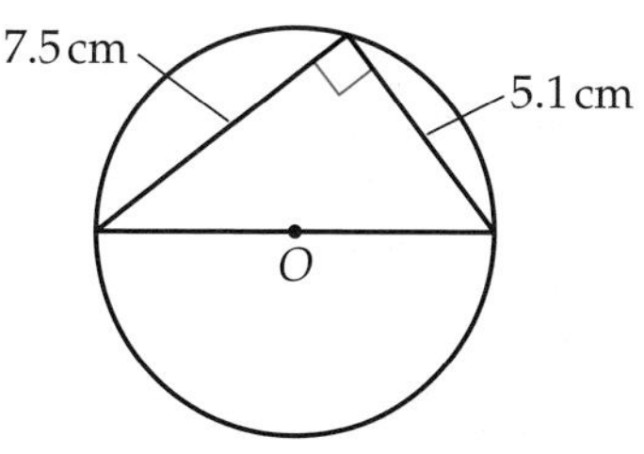

Calculate the area of the circle.

C

8 Ellen wants to put some edging around her lawn.
The lawn is in the shape of a right-angled triangle.
The edging is sold in pieces 110 cm long and 20 cm high.
Each piece costs £5.
Calculate how much it will cost Ellen to do the job.

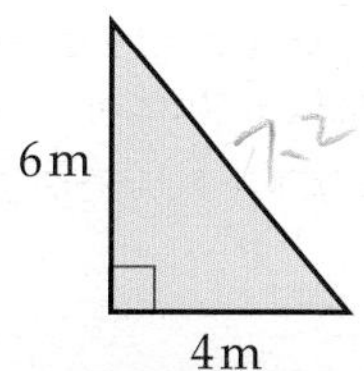

AO3

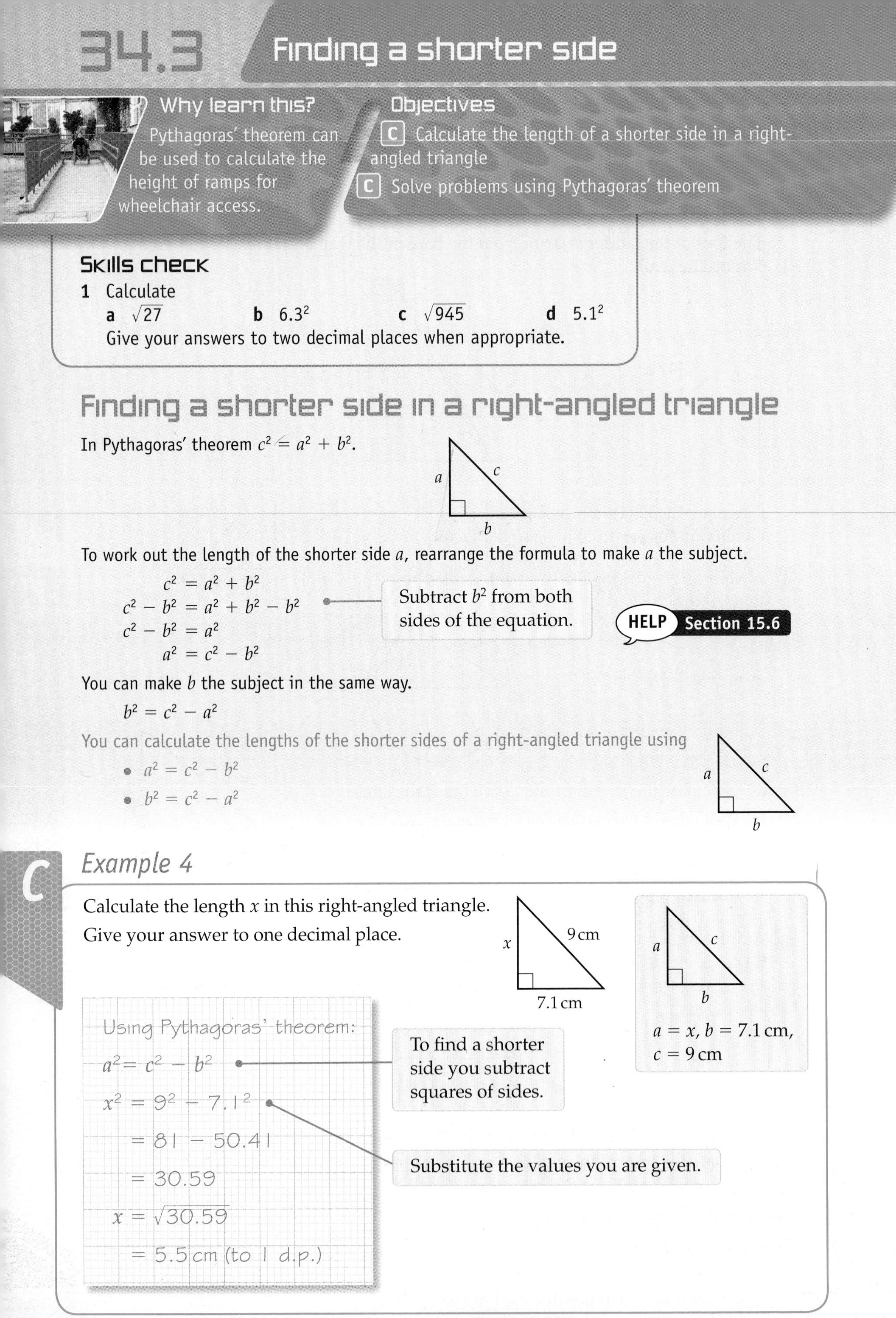

Why learn this?
Pythagoras' theorem can be used to calculate the height of ramps for wheelchair access.

Objectives
C Calculate the length of a shorter side in a right-angled triangle
C Solve problems using Pythagoras' theorem

Skills check

1 Calculate
a $\sqrt{27}$ **b** 6.3^2 **c** $\sqrt{945}$ **d** 5.1^2
Give your answers to two decimal places when appropriate.

Finding a shorter side in a right-angled triangle

In Pythagoras' theorem $c^2 = a^2 + b^2$.

To work out the length of the shorter side a, rearrange the formula to make a the subject.

$$c^2 = a^2 + b^2$$
$$c^2 - b^2 = a^2 + b^2 - b^2$$
$$c^2 - b^2 = a^2$$
$$a^2 = c^2 - b^2$$

Subtract b^2 from both sides of the equation.

HELP Section 15.6

You can make b the subject in the same way.

$$b^2 = c^2 - a^2$$

You can calculate the lengths of the shorter sides of a right-angled triangle using

- $a^2 = c^2 - b^2$
- $b^2 = c^2 - a^2$

C

Example 4

Calculate the length x in this right-angled triangle.
Give your answer to one decimal place.

$a = x$, $b = 7.1$ cm, $c = 9$ cm

Using Pythagoras' theorem:
$a^2 = c^2 - b^2$
$x^2 = 9^2 - 7.1^2$
$= 81 - 50.41$
$= 30.59$
$x = \sqrt{30.59}$
$= 5.5$ cm (to 1 d.p.)

To find a shorter side you subtract squares of sides.

Substitute the values you are given.

Exercise 34D

Calculate the lengths marked with letters in these triangles.
Give your answer to one decimal place when appropriate.
Q1 and Q2 have been started for you.

You subtract to find a shorter side.

C

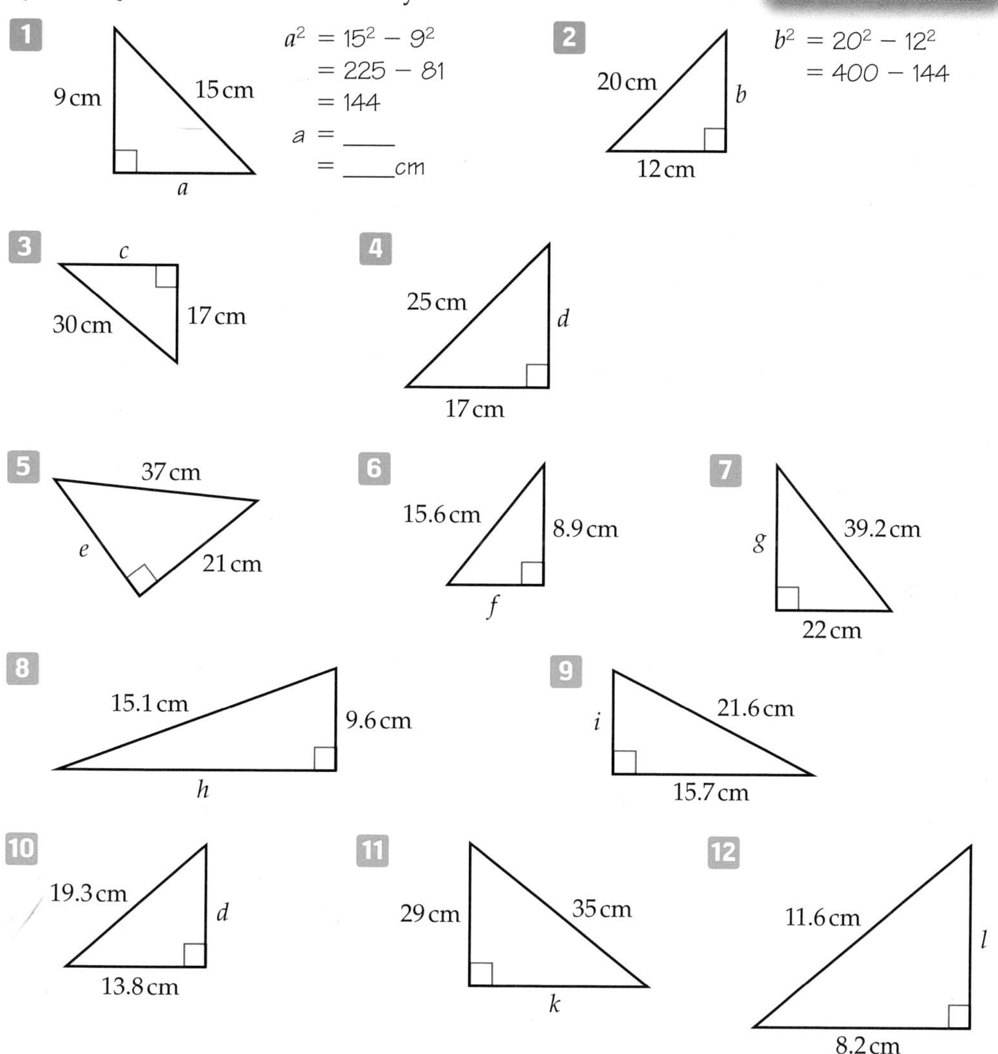

Solving problems using Pythagoras' theorem

Pythagoras' theorem can be used to solve problems which involve finding a shorter side of a right-angled triangle.

Example 5

C

The end-face of a tent is in the shape of an isosceles triangle.
The mid-point of side AC is N. The length of the tent is 3.1 m.

Calculate the volume of the tent.

Give your answer to two decimal places.

B
2.6 m
A
N
C
3.6 m

AO3

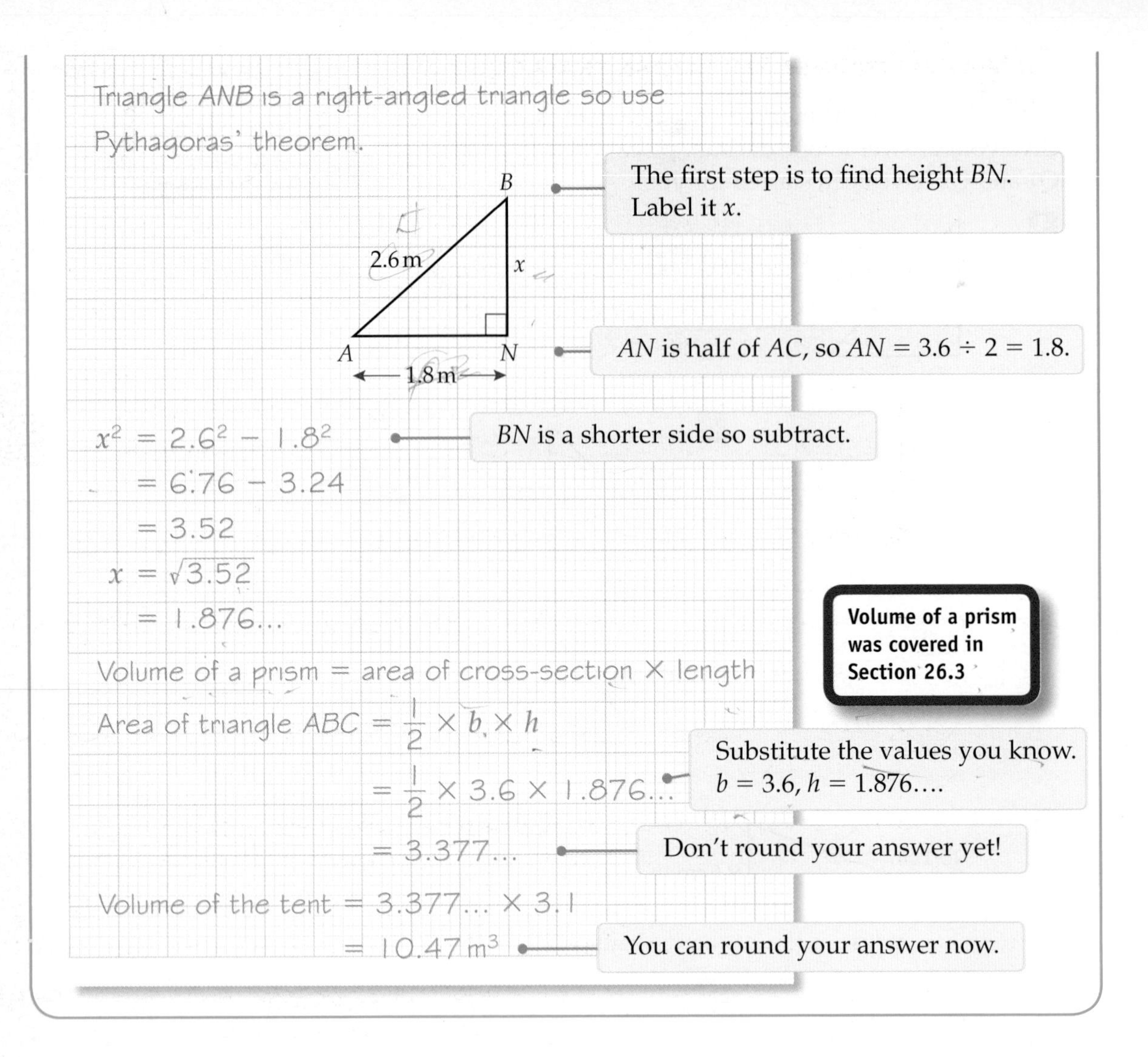

Triangle ANB is a right-angled triangle so use Pythagoras' theorem.

The first step is to find height BN. Label it x.

AN is half of AC, so $AN = 3.6 \div 2 = 1.8$.

$$x^2 = 2.6^2 - 1.8^2$$

BN is a shorter side so subtract.

$$= 6.76 - 3.24$$
$$= 3.52$$
$$x = \sqrt{3.52}$$
$$= 1.876\ldots$$

Volume of a prism = area of cross-section × length

Volume of a prism was covered in Section 26.3

Area of triangle $ABC = \frac{1}{2} \times b \times h$

$= \frac{1}{2} \times 3.6 \times 1.876\ldots$

Substitute the values you know. $b = 3.6$, $h = 1.876\ldots.$

$= 3.377\ldots$

Don't round your answer yet!

Volume of the tent $= 3.377\ldots \times 3.1$

$= 10.47\,\text{m}^3$

You can round your answer now.

Exercise 34E

C

1 A children's slide is 2.9 m long.
The vertical height of the slide above the ground is 1.8 m.
Calculate the horizontal distance between each end of the slide. Give your answer to one decimal place.

The length you are trying to find, x, is a shorter side.

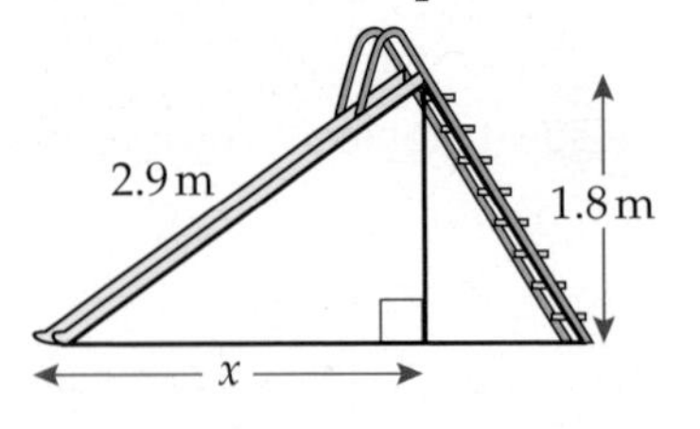

2 A ladder 3.6 metres long is leaning against a wall.
The foot of the ladder is 0.75 m away from the base of the wall.
How far up the wall does the ladder reach?
Give your answer to two decimal places.

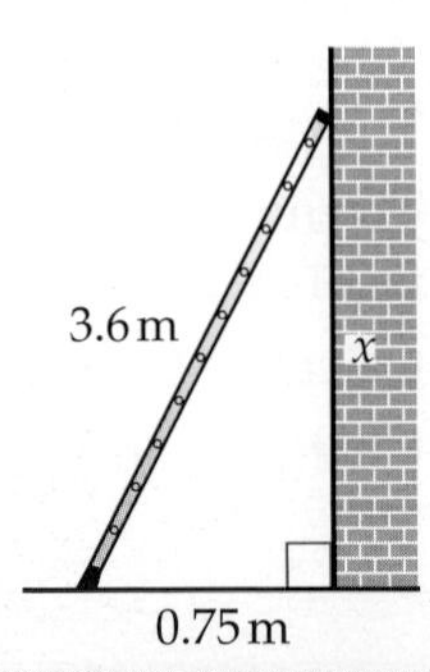

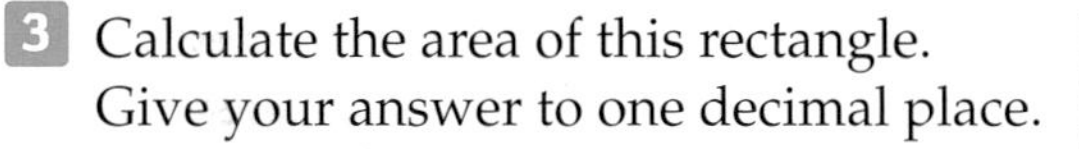

3 Calculate the area of this rectangle.
Give your answer to one decimal place.

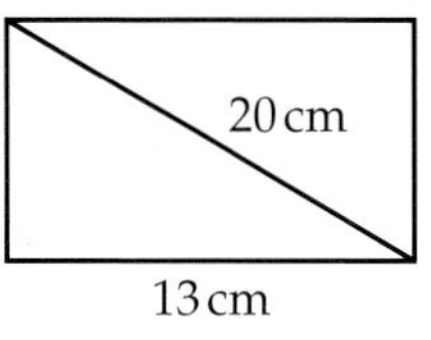

What information do you need to find the area of the rectangle?

C

4 Here is a sketch of the cross-section of a skip.

a Calculate the height of the skip.

b Calculate the area of the cross-section of the skip.

c The skip is 1.7 m long.
Calculate the capacity of the skip in cubic metres.

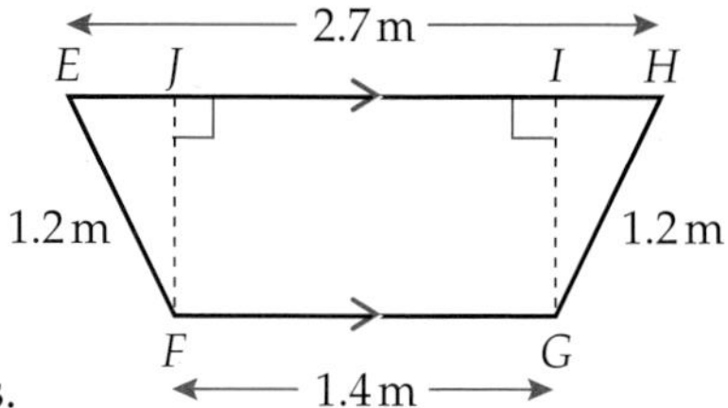

AO2

5 A ship sails 24 km north-east and then 18 km south-east.
How far is the ship from its starting point?

C

6 Here is the sketch of the cross-section of a skip.
The length of the skip is 1.4 m.
Calculate the capacity of the skip in cubic metres.
Give your answer to two decimal places.

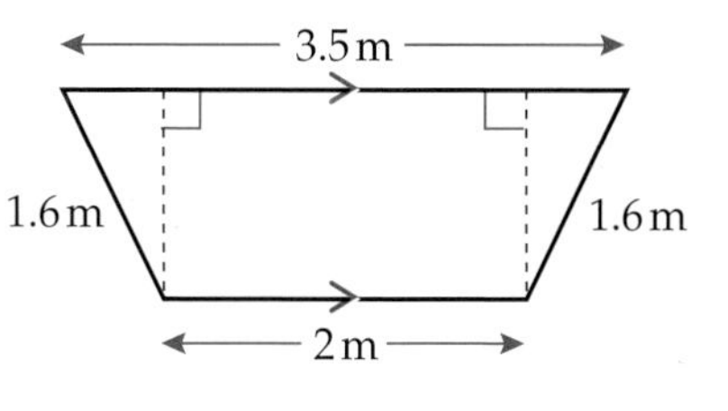

7 A cube is cut through four of its vertices, A, B, C and D, leaving two identical pieces.
The diagram shows one of the pieces.
Calculate the distance AC.

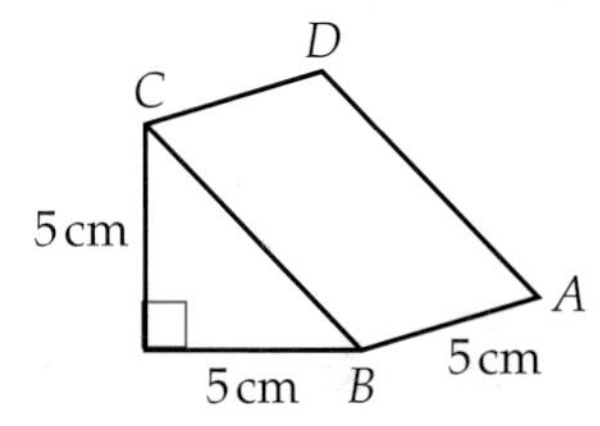

AO3

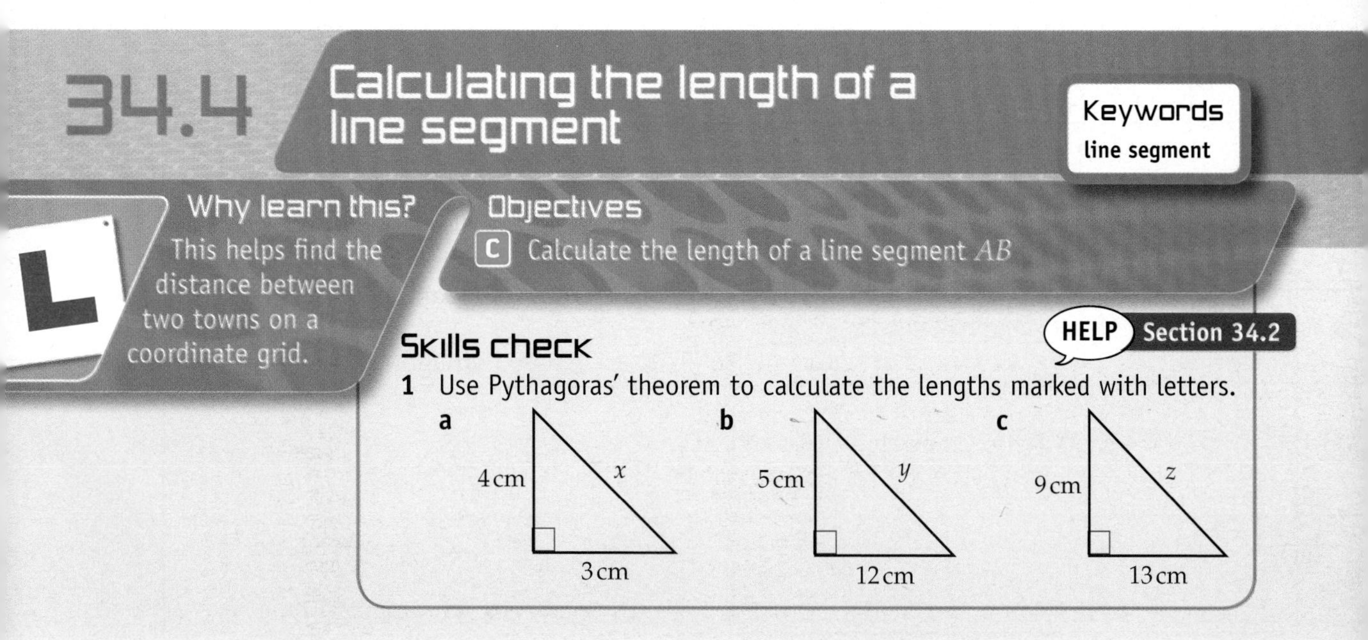

34.4 Calculating the length of a line segment

Keywords
line segment

Why learn this?
This helps find the distance between two towns on a coordinate grid.

Objectives

C Calculate the length of a line segment AB

Skills check

HELP Section 34.2

1 Use Pythagoras' theorem to calculate the lengths marked with letters.

Calculating the length of a line segment

A **line segment** is the line between two points.

The length of a line segment parallel to the x-axis can be found by working out the difference between the x-coordinates of the end-points. You can use a similar method for the length of a line segment parallel to the y-axis.

Pythagoras' theorem can be used to find the length of a sloping line.

C

Example 6

The points A (1, 0) and B (7, 8) are shown.
Calculate the length of AB.
Give your answer to one decimal place.

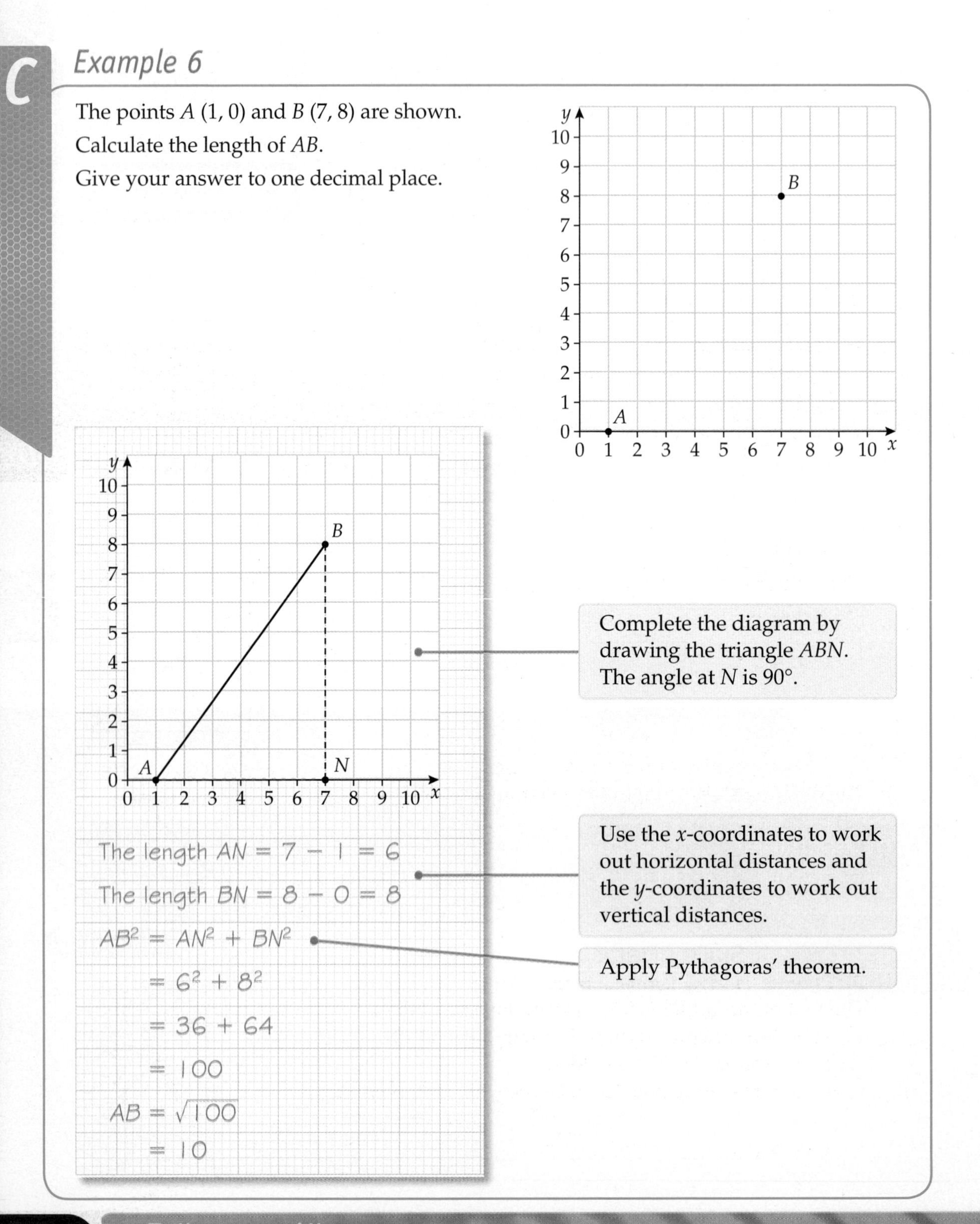

Complete the diagram by drawing the triangle ABN. The angle at N is 90°.

The length $AN = 7 - 1 = 6$

The length $BN = 8 - 0 = 8$

Use the x-coordinates to work out horizontal distances and the y-coordinates to work out vertical distances.

$AB^2 = AN^2 + BN^2$

$= 6^2 + 8^2$

$= 36 + 64$

$= 100$

$AB = \sqrt{100}$

$= 10$

Apply Pythagoras' theorem.

Exercise 34F

1 Calculate the length of each line segment.
Give your answer to one decimal place when appropriate.

a AB

b CD

c EF

d GH

e IJ

f KL

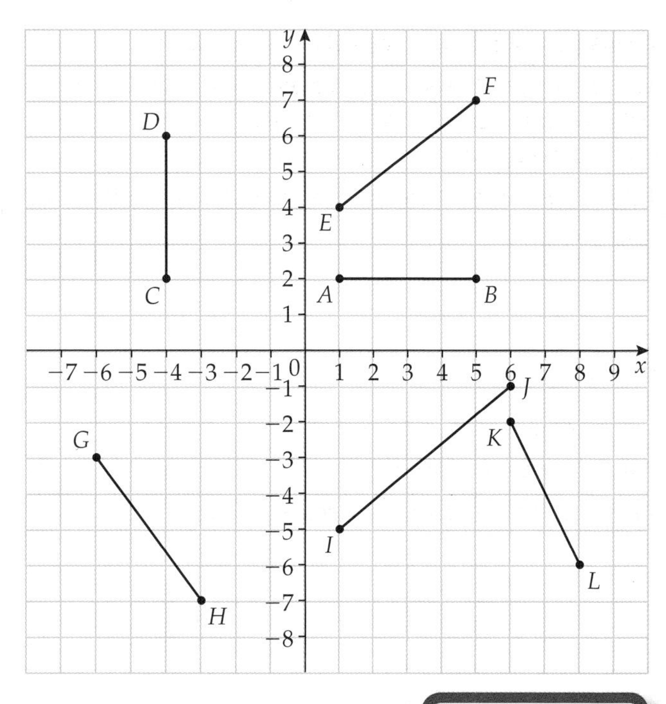

C

2 Calculate the length of each line segment.
Give your answer to one decimal place when appropriate.

a AB: A (2, 2) and B (4, 6)

b CD: C (3, 5) and D (5, 7)

c EF: E (1, 3) and F (5, 7)

d GH: G (−3, 5) and H (5, 6)

e IJ: I (7, 10) and J (−6, 4)

f KL: K (−1, −1) and L (1, 3)

A sketch will help you. Label the ends of the line and draw a right-angled triangle.

Review exercise

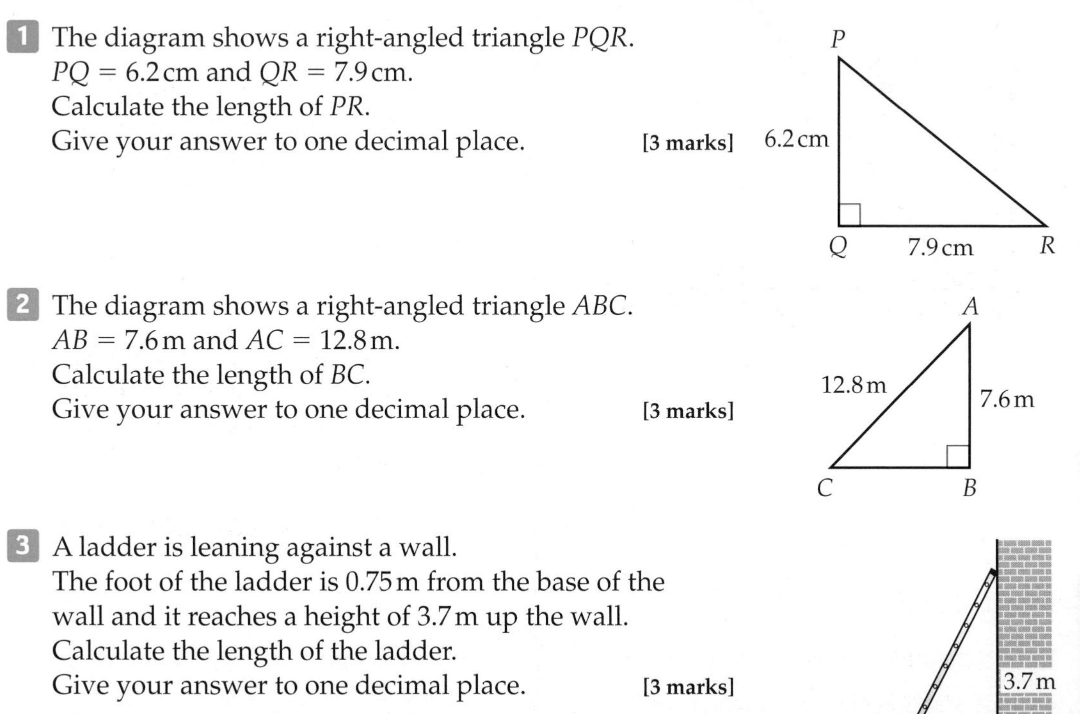

C

1 The diagram shows a right-angled triangle PQR.
$PQ = 6.2$ cm and $QR = 7.9$ cm.
Calculate the length of PR.
Give your answer to one decimal place. **[3 marks]**

2 The diagram shows a right-angled triangle ABC.
$AB = 7.6$ m and $AC = 12.8$ m.
Calculate the length of BC.
Give your answer to one decimal place. **[3 marks]**

3 A ladder is leaning against a wall.
The foot of the ladder is 0.75 m from the base of the wall and it reaches a height of 3.7 m up the wall.
Calculate the length of the ladder.
Give your answer to one decimal place. **[3 marks]**

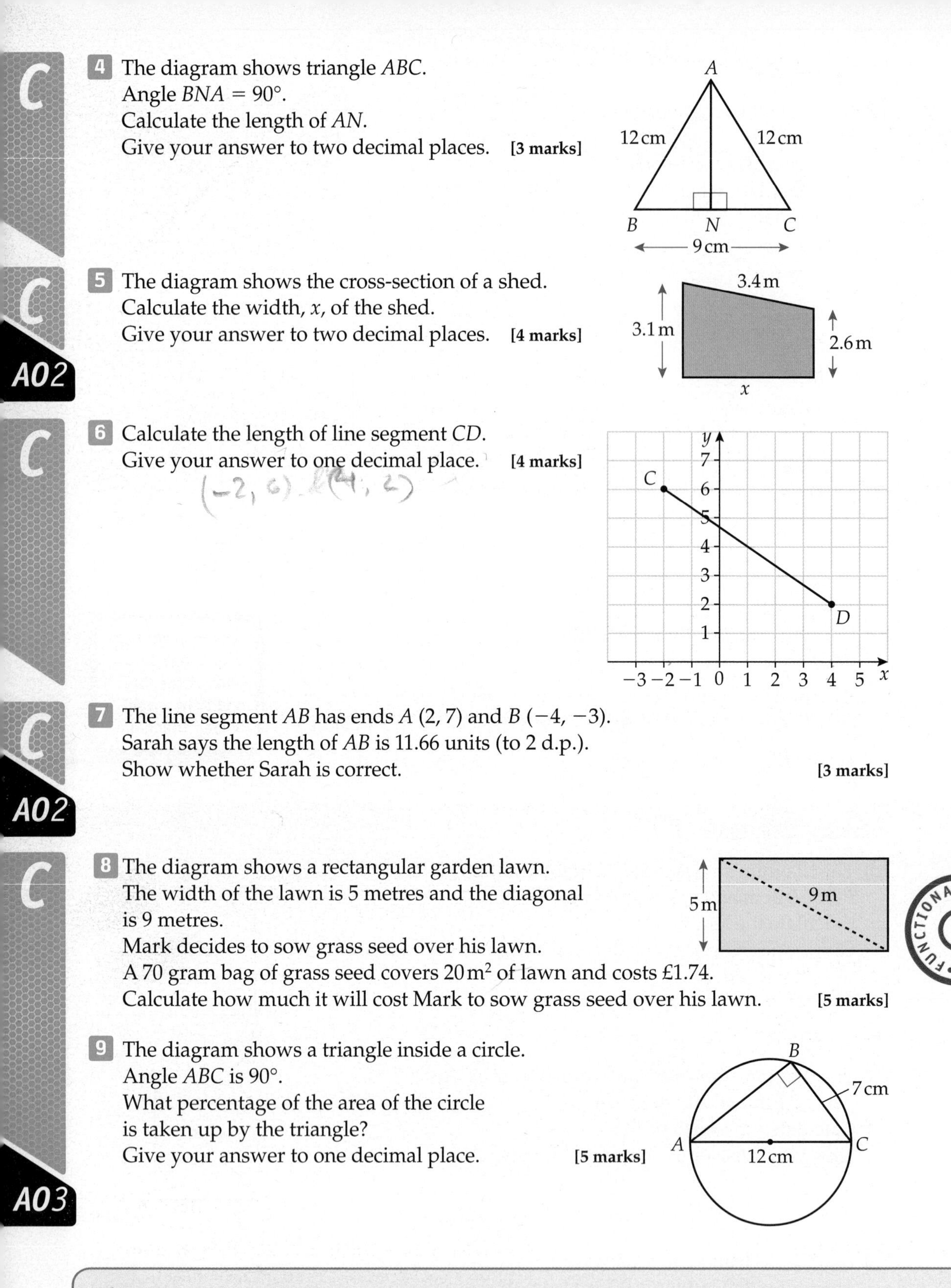

4 The diagram shows triangle ABC.
Angle $BNA = 90°$.
Calculate the length of AN.
Give your answer to two decimal places. **[3 marks]**

5 The diagram shows the cross-section of a shed.
Calculate the width, x, of the shed.
Give your answer to two decimal places. **[4 marks]**

6 Calculate the length of line segment CD.
Give your answer to one decimal place. **[4 marks]**

7 The line segment AB has ends A (2, 7) and B (−4, −3).
Sarah says the length of AB is 11.66 units (to 2 d.p.).
Show whether Sarah is correct. **[3 marks]**

8 The diagram shows a rectangular garden lawn.
The width of the lawn is 5 metres and the diagonal is 9 metres.
Mark decides to sow grass seed over his lawn.
A 70 gram bag of grass seed covers $20\,m^2$ of lawn and costs £1.74.
Calculate how much it will cost Mark to sow grass seed over his lawn. **[5 marks]**

9 The diagram shows a triangle inside a circle.
Angle ABC is 90°.
What percentage of the area of the circle is taken up by the triangle?
Give your answer to one decimal place. **[5 marks]**

Chapter summary

In this chapter you have learned how to

- understand Pythagoras' theorem **C**
- calculate the hypotenuse of a right-angled triangle **C**
- calculate the length of a shorter side in a right-angled triangle **C**
- solve problems using Pythagoras' theorem **C**
- calculate the length of a line segment AB **C**

35 Trigonometry

This chapter is about using trigonometry to find unknown angles and lengths in right-angled triangles.

Car satellite navigation systems use a form of trigonometry to calculate the distance between two places.

Objectives

This chapter will show you how to

- use calculators effectively and efficiently, knowing how to enter complex calculations, including trigonometrical functions **B**
- understand, recall and use trigonometric ratios in right-angled triangles, and use these to solve problems, including those involving bearings **B**

Before you start this chapter

HELP Chapter 34

1 Solve these equations.

a $\frac{x}{2} = 4$

b $\frac{x}{5} = 6$

c $\frac{3x}{4} = 3$

HELP Chapter 14

2 Calculate the length x in the triangles below. Give your answers to two decimal places.

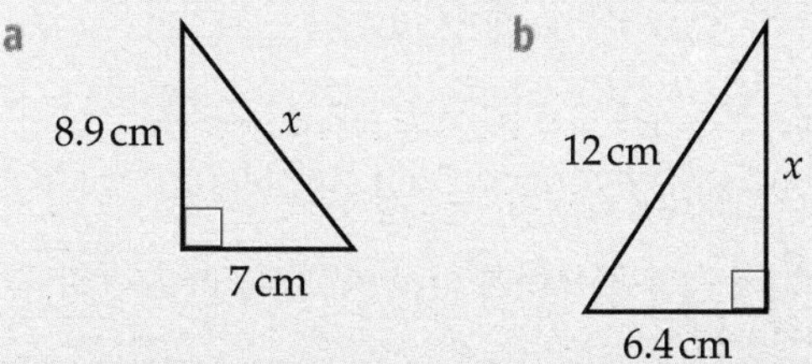

35.1 Trigonometry - the ratios of sine, cosine and tangent

Keywords

trigonometry, sine, cosine, tangent, hypotenuse, opposite, adjacent, inverse

Why learn this?

The trigonometric functions are powerful tools in many areas of maths and science, including acoustics and engineering.

Objectives

B Understand and recall trigonometric ratios in right-angled triangles

B Know how to enter the trigonometric functions on a calculator

Skills check

HELP Section 14.1

1 Solve these equations.

a $\frac{x}{6} = 4$ **b** $\frac{3x}{4} = 6$

2 Round 27.65 to one decimal place.

Sine, cosine and tangent

Trigonometry is concerned with calculating sides and angles in right-angled triangles, and involves three ratios called **sine**, **cosine** and **tangent**.

These ratios are often abbreviated to sin, cos and tan.

Consider this right-angled triangle.

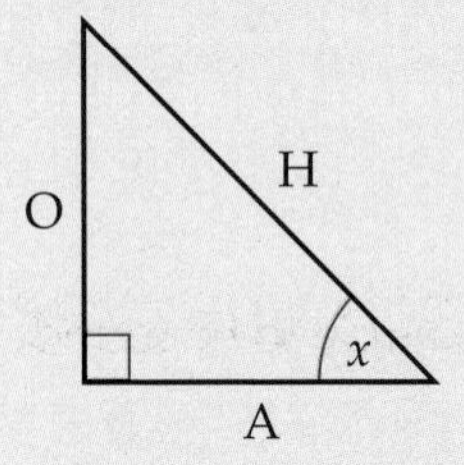

- The side opposite the right angle is called the **hypotenuse** (H).
- The side opposite the angle x is called the **opposite** (O).
- The side next to the angle x is called the **adjacent** (A).

The hypotenuse is always opposite the right angle so it is always the same side. The 'opposite' and 'adjacent' sides switch depending on which angle you are considering.

The ratios for sine, cosine and tangent are defined as follows.

$$\sin x = \frac{\text{opposite}}{\text{hypotenuse}} = \frac{O}{H}$$

$$\cos x = \frac{\text{adjacent}}{\text{hypotenuse}} = \frac{A}{H}$$

$$\tan x = \frac{\text{opposite}}{\text{adjacent}} = \frac{O}{A}$$

You need to learn these formulae for the exam. A useful memory aid is SOHCAHTOA.

Using a calculator

The abbreviated forms (sin, cos and tan) are used on calculator keys.

When using your calculator, make sure it is working in degrees. D or DEG will appear in the display.

To find the sine, cosine or tangent of an angle, press the appropriare key followed by the angle.

For sine 72°, press [sin] [7] [2] [=]

You will sometimes have to work backwards or use the **inverse** function.

The inverse function on your calculator may be labelled [SHIFT] or [2nd F] or [INV].

To find angle x when $\tan x = 0.75$, press [SHIFT] [tan] [0] [·] [7] [5] [=]

Example 1

a Label the sides of the triangle A, O and H, where x is the angle to be found.

b Use a calculator to work out the value of cos 57°.

c Use a calculator to find the value of angle x correct to one decimal place when $\sin x = 0.5784$.

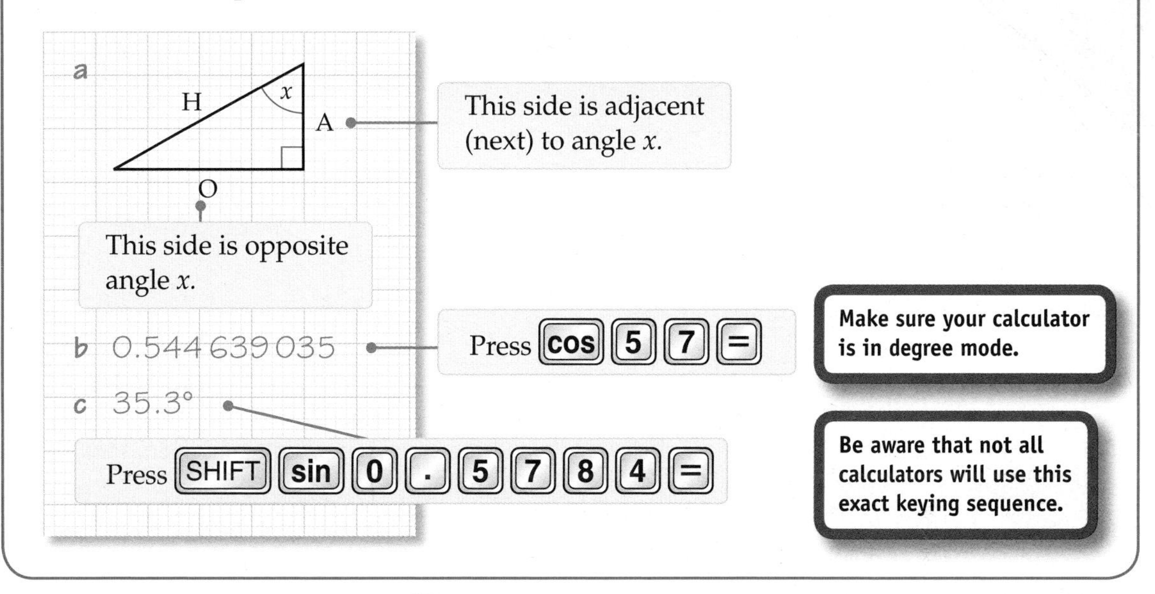

Exercise 35A

1 Draw these triangles and label the sides O, A and H, where x is the angle to be found.

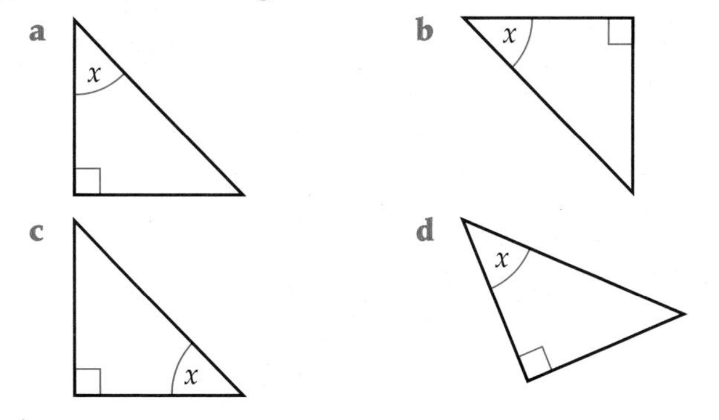

2 Use a calculator to work out the values of the following trigonometric ratios.
Give your answer to five decimal places when appropriate.

a cos 27° **b** sin 21°

c tan 42° **d** sin 69°

e cos 73° **f** tan 19°

g tan 102° **h** sin 106°

3 Use your calculator to find the value of each angle correct to 0.1°.

a $\tan x = 0.4963$ **b** $\sin x = 0.7137$

c $\cos x = 0.2146$ **d** $\cos x = 0.9371$

Correct to 0.1° means give your answer to 1 d.p.

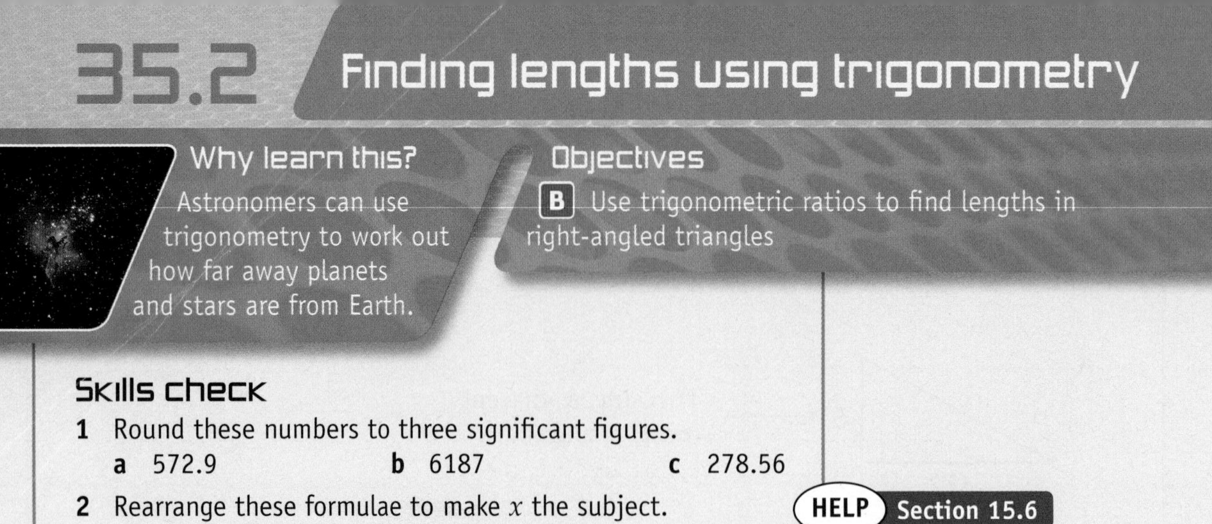

35.2 Finding lengths using trigonometry

Why learn this?

Astronomers can use trigonometry to work out how far away planets and stars are from Earth.

Objectives

B Use trigonometric ratios to find lengths in right-angled triangles

Skills check

1 Round these numbers to three significant figures.

a 572.9 **b** 6187 **c** 278.56

2 Rearrange these formulae to make x the subject.

a $mx = n$ **b** $p = \frac{x}{a}$ **c** $d = \frac{a}{x}$

HELP Section 15.6

Finding lengths

If the length of one side of a right-angled triangle and the size of one angle are given, then the lengths of the remaining sides can be found using trigonometry.

B

Example 2

In triangle ABC, length $AB = 12$ cm, angle $ACB = 90°$ and angle $BAC = 30°$.

Calculate the length of BC.

B
12 cm
30°
A
C

B
(H)
12 cm
(O)
30°
A
(A)
C

$\sin x = \frac{O}{H}$ — Label the sides of the triangle: opposite (O), adjacent (A) and hypotenuse (H). Decide on the ratio (sin, cos or tan) you need to use. You are given H and want to find O so use the sine ratio.

$\sin 30° = \frac{BC}{12}$ — Substitute the values you know. H = 12, angle = 30°

$12 \times \sin 30° = BC$ — Rearrange the equation to make BC the subject.

$BC = 6$ cm — Remember to put in the units.

Example 3

In triangle PQR, angle $PQR = 90°$, angle $QPR = 41°$ and length $PQ = 10$ cm.

Calculate the length of PR.

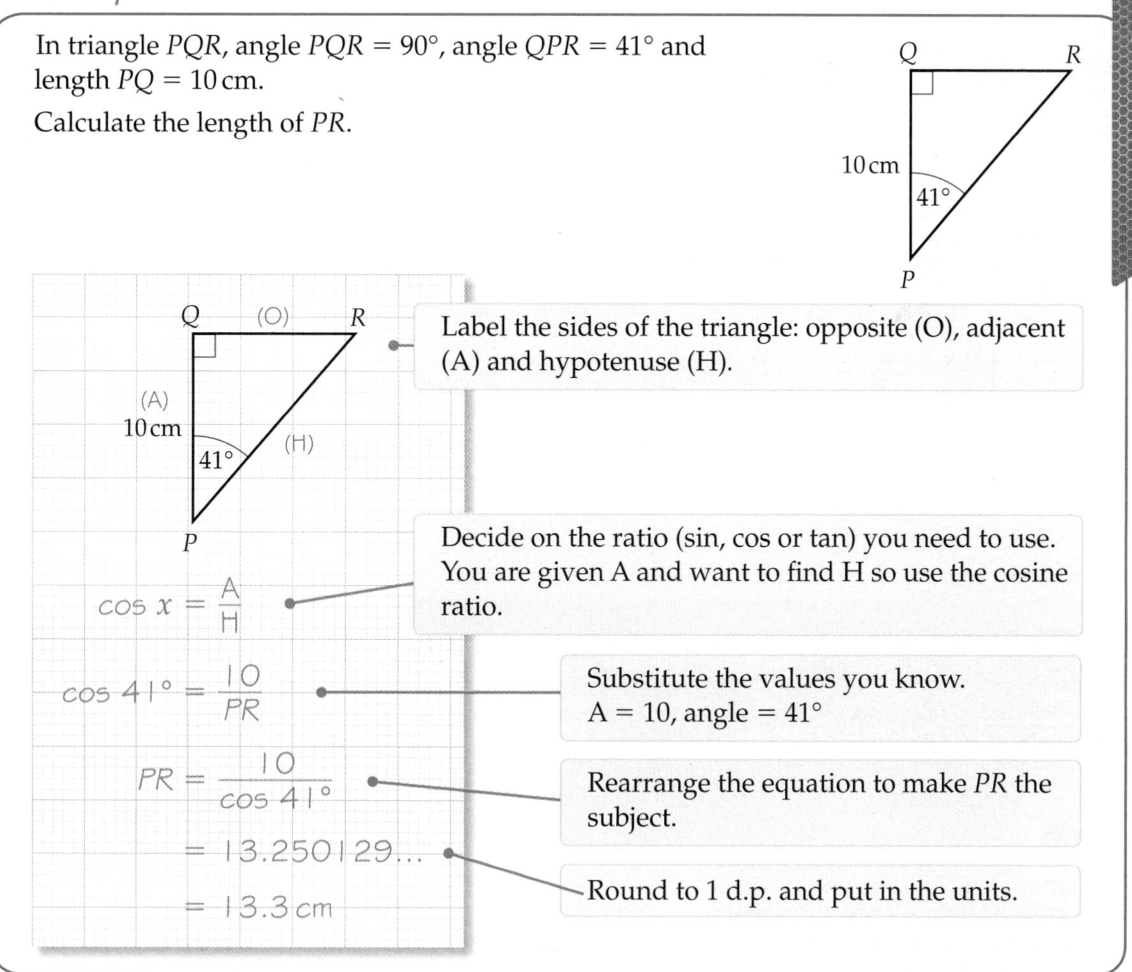

Exercise 35B

1 Calculate the unknown length, x, marked on each diagram. Give your answer to one decimal place when appropriate.

Look back at Examples 2 and 3 to help you.

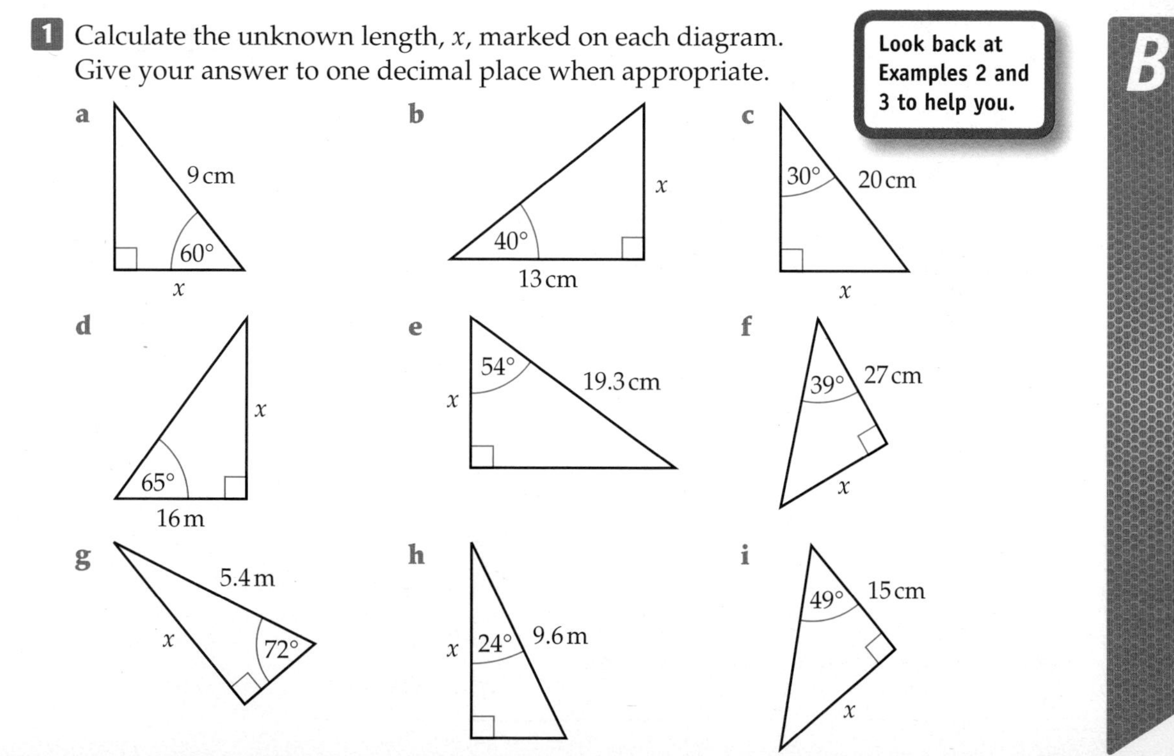

B

2 Calculate the unknown length, x, marked on each diagram.
Give your answer to three significant figures when appropriate.

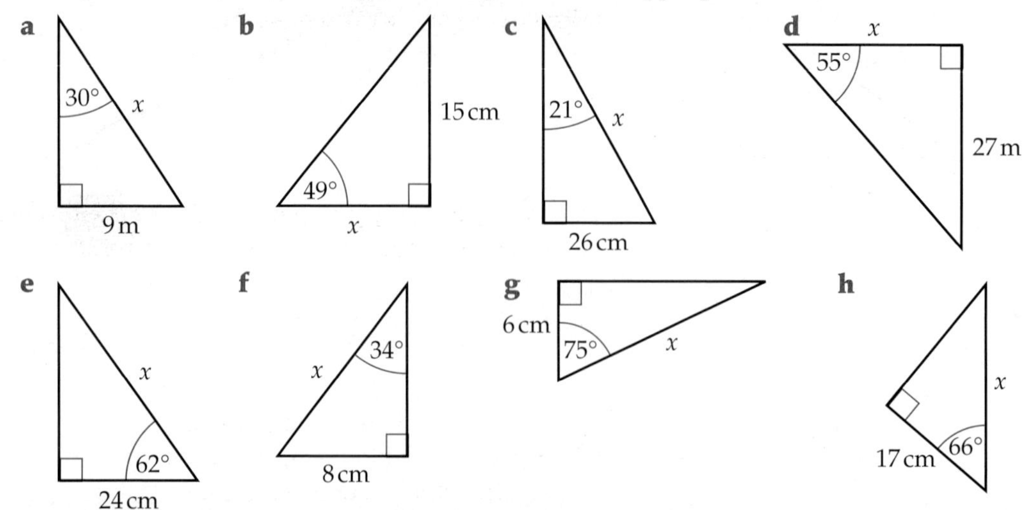

35.3 Finding angles using trigonometry

Why learn this?
Trigonometry is used to manoeuvre the robotic arm on the international space station.

Objectives
B Use trigonometric ratios to find the angles in right-angled triangles

Skills check

1 Round these numbers to one decimal place.
 a 46.52 b 715.85 c 4169.28

2 Convert these fractions to decimals.
 a $\frac{4}{5}$ b $\frac{7}{20}$ c $\frac{3}{25}$

HELP Section 13.4

Finding angles

If the lengths of two sides of a right-angled triangle are given, you will be able to work out an angle using one of the trigonometric ratios.

Example 4

In triangle ABC, calculate the size of angle x.

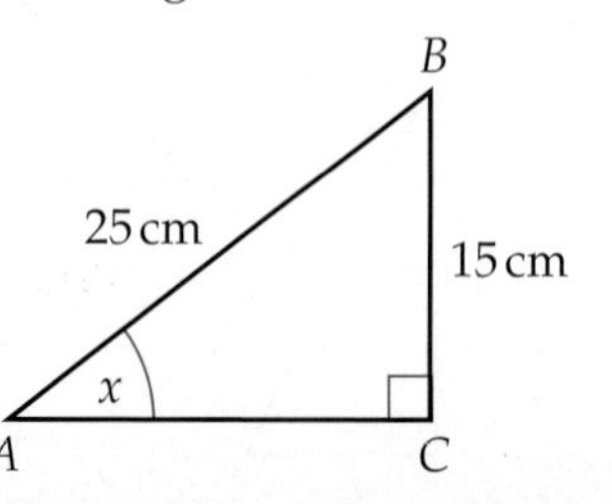

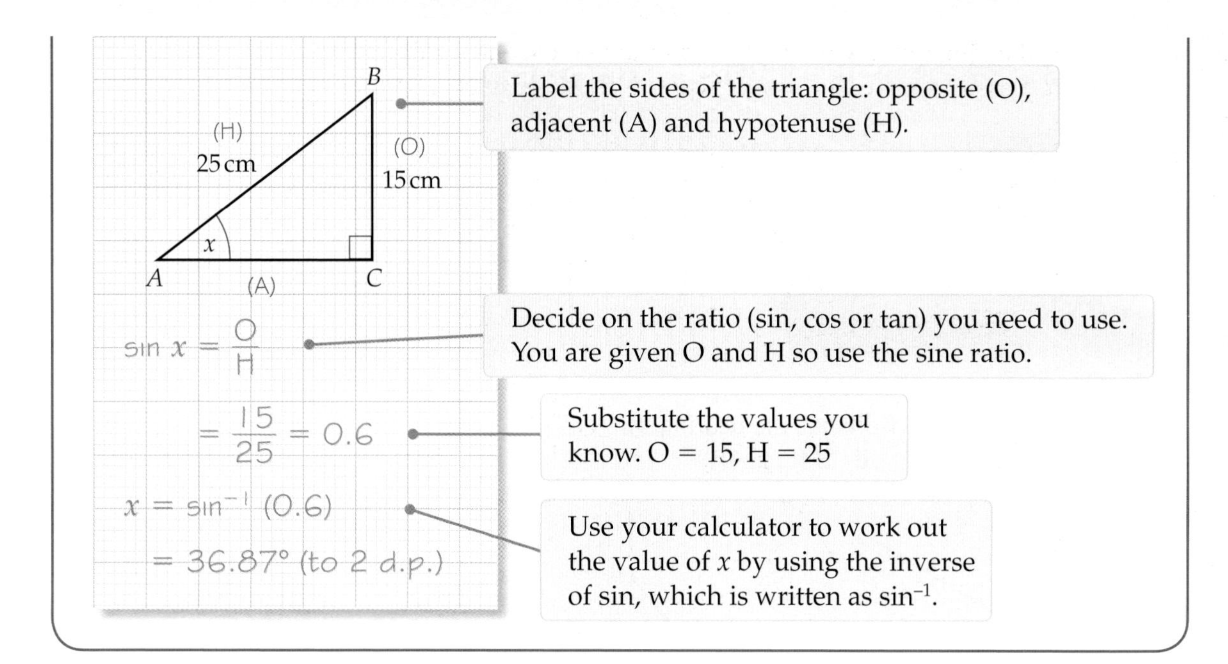

Exercise 35C

Calculate the size of the unknown angle in each of these triangles.
Give your answer to one decimal place when appropriate.

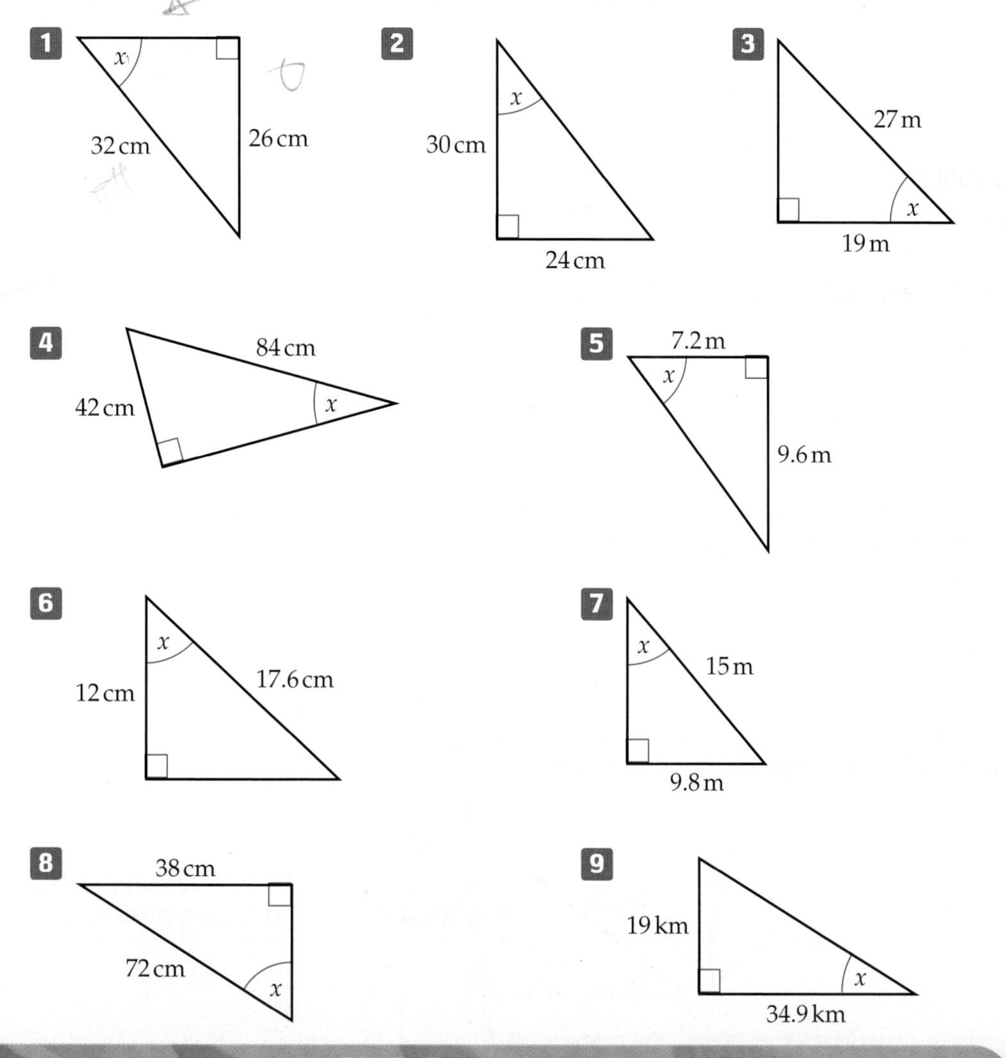

B

35.4 Applications of trigonometry and Pythagoras' theorem

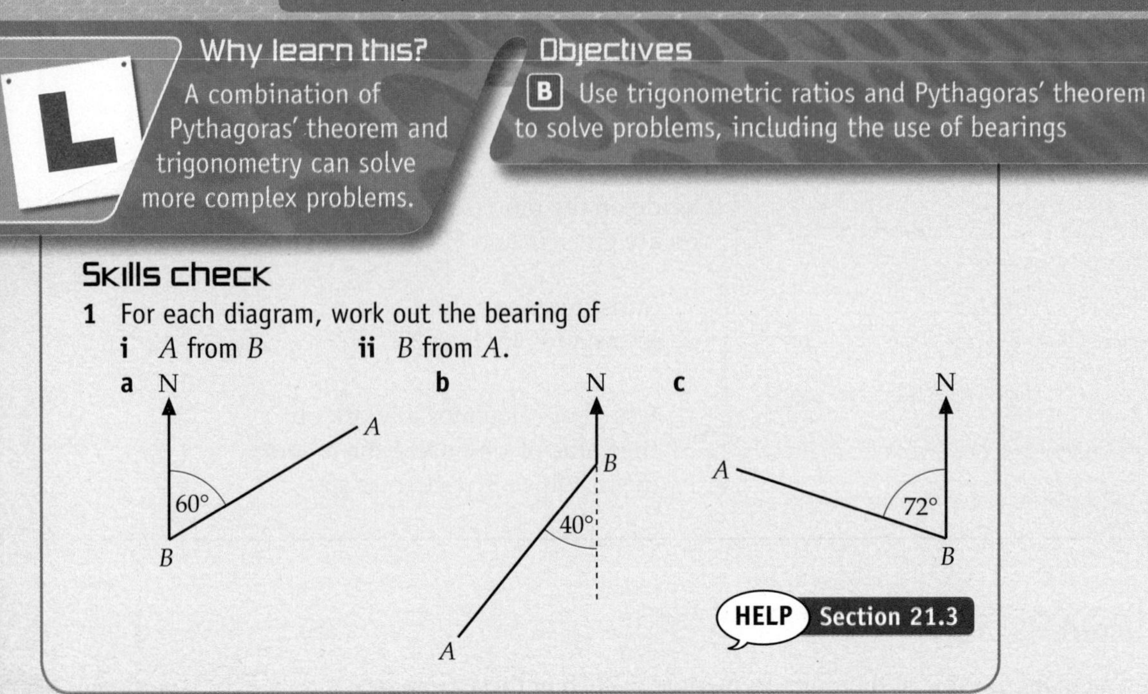

Why learn this?

A combination of Pythagoras' theorem and trigonometry can solve more complex problems.

Objectives

B Use trigonometric ratios and Pythagoras' theorem to solve problems, including the use of bearings

Skills check

1 For each diagram, work out the bearing of

i A from B **ii** B from A.

HELP Section 21.3

Using trigonometry and Pythagoras' theorem to solve problems

The trigonometric ratios and Pythagoras' theorem can be used to solve problems set in a range of real-life contexts.

B

AO2

Example 5

A ship sails on a bearing of 130° for 60 km. How far east has the ship travelled?

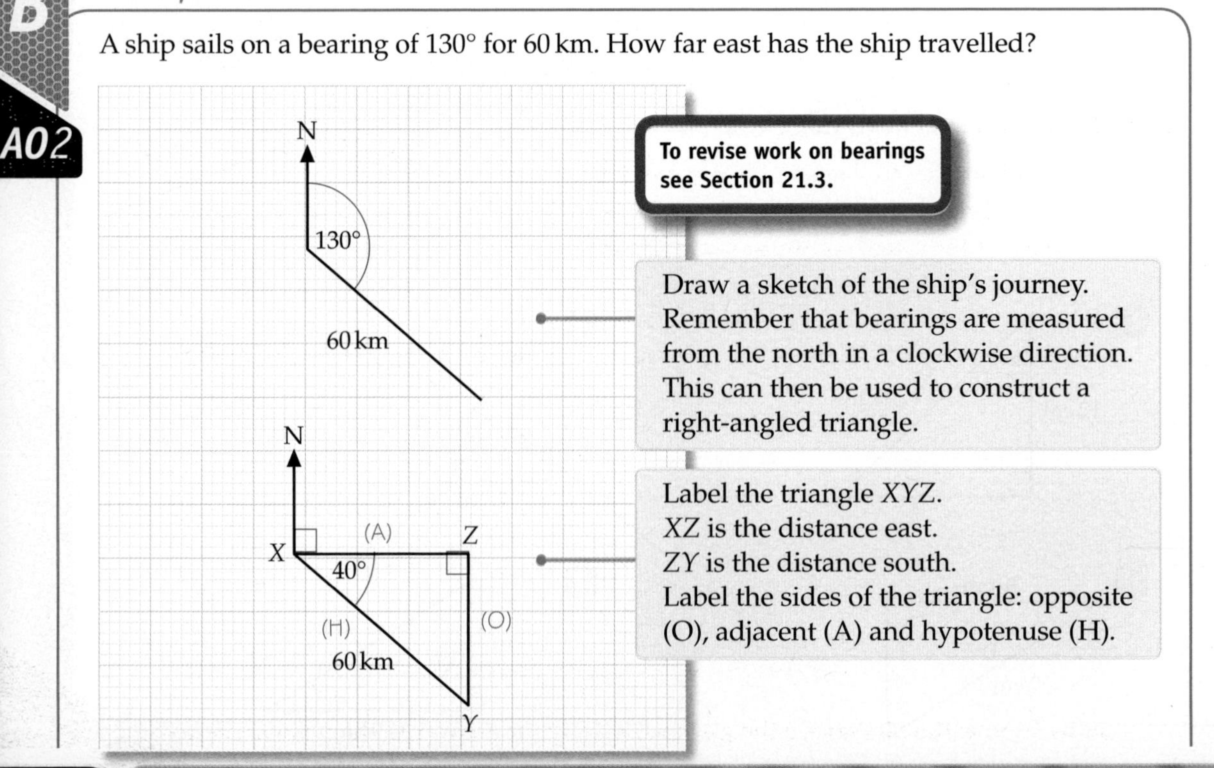

To revise work on bearings see Section 21.3.

Draw a sketch of the ship's journey. Remember that bearings are measured from the north in a clockwise direction. This can then be used to construct a right-angled triangle.

Label the triangle XYZ.
XZ is the distance east.
ZY is the distance south.
Label the sides of the triangle: opposite (O), adjacent (A) and hypotenuse (H).

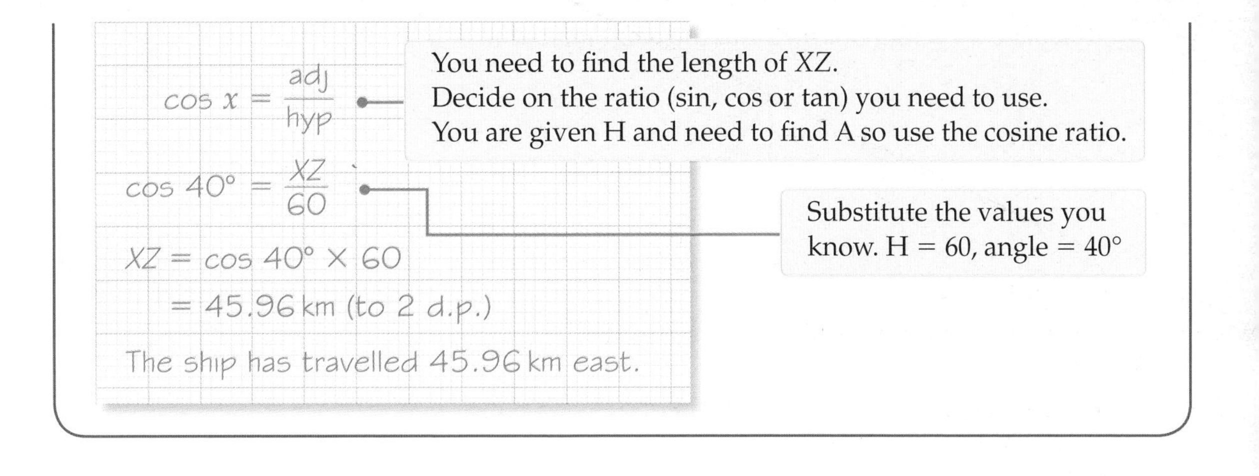

Exercise 35D

1 A ship sails on a bearing of 295° for 80 km.
How far west has the ship travelled?
Give your answer to one decimal place.

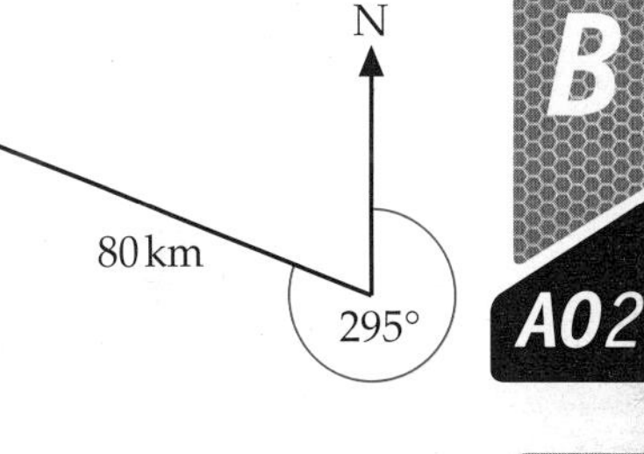

B

AO2

2 A ladder 6 m long rests against a wall.
The foot of the ladder is 2.7 m away from the base of the wall.
What angle does the ladder make with the wall?
Give your answer to one decimal place.

6 m

2.7 m

B

3 The diagram shows a roof truss.

a What angle will the roof truss make with the horizontal? Give your answer to one decimal place.

b What is the length of the sloping strut?

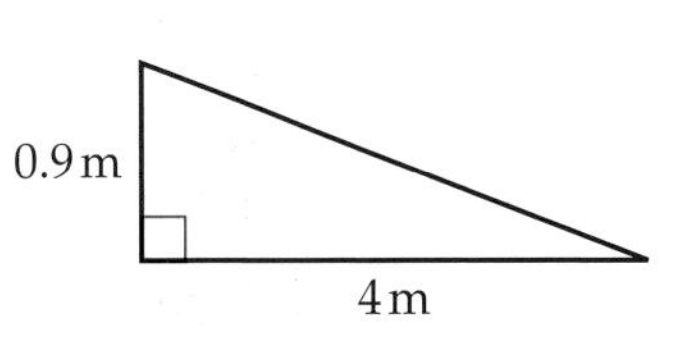

4 A boy is flying a kite on a string 27 m long.
The string is held 1 m above the ground and it makes an angle of 43° with the horizontal.

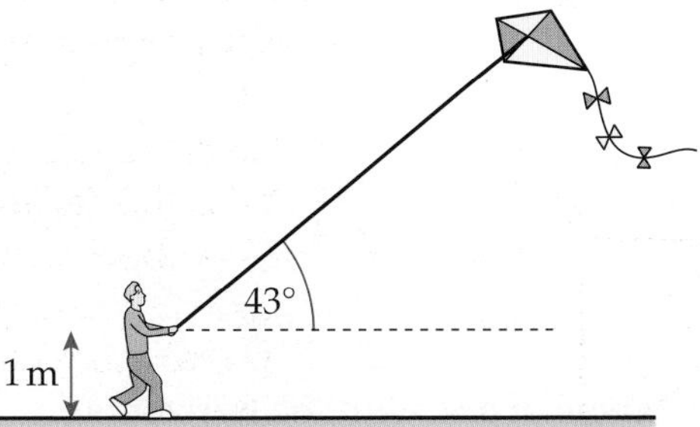

How high is the kite above the ground? Give your answer to the nearest metre.

B

AO2

B

5 Calculate the length marked x in the diagram below.

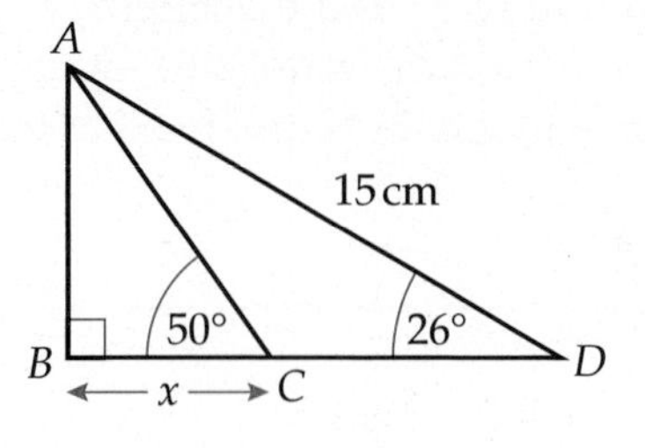

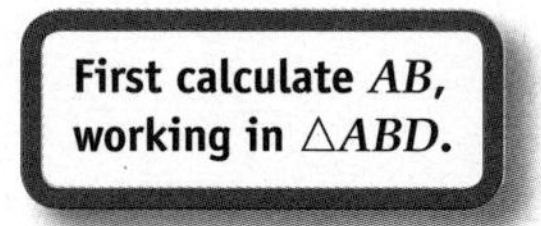

Give your answer to three significant figures.

6 Frith, Soll and Barne are three towns.
Frith is 7.6 km due west of Soll.
Barne is 9.8 km due north of Soll.

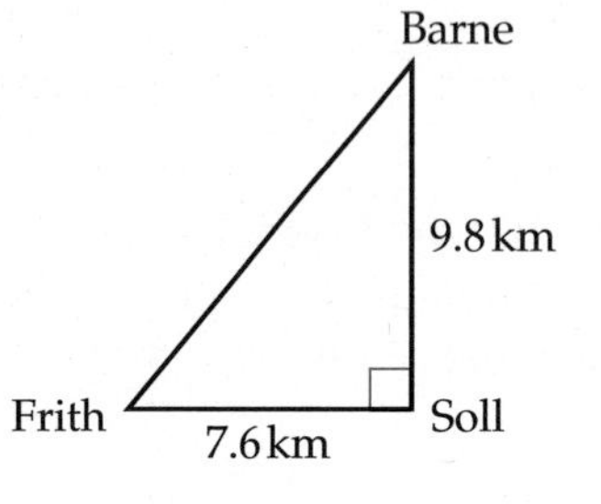

You can use Pythagoras' theorem or trigonometry to calculate the distance.

AO2

Calculate the bearing and distance of Frith from Barne.
Give your answers to three significant figures.

B

7 The diagram shows a right-angled triangle in a circle, centre O.
The diameter of the circle is 15 cm.

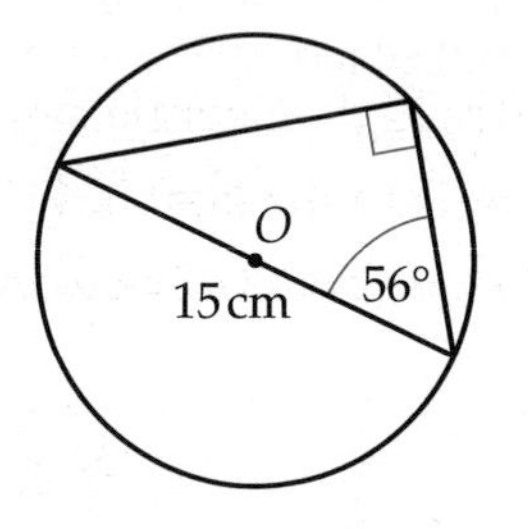

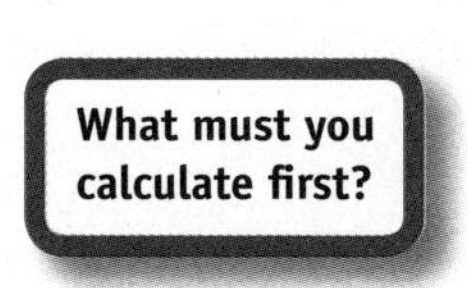

Calculate the area of the triangle.
Give your answer to three significant figures.

8 A wheelchair ramp is to be built to go up four steps.
The steps are 20 cm high and have a depth of 27 cm.

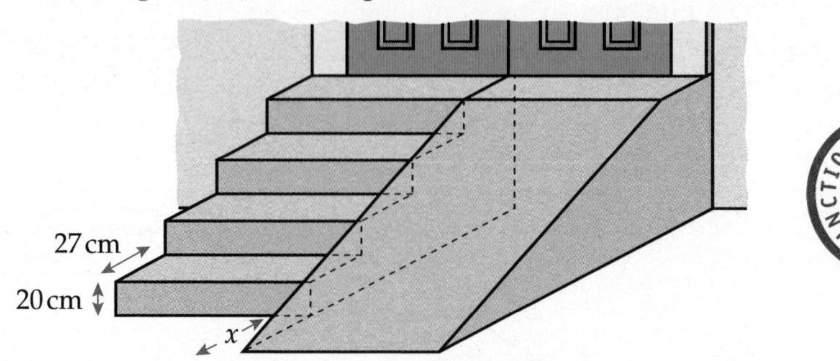

FUNCTIONAL • FUNCTIONAL •

The ramp will have an angle of 4.8° with the horizontal.
Calculate distance x to find out how far away from the steps the ramp should start.
Give your answer to the nearest centimetre.

AO3

35.5 Angles of elevation and depression

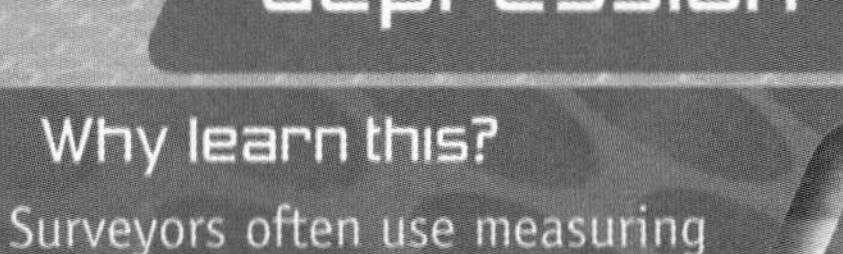

Keywords
angle of elevation, angle of depression

Why learn this?

Surveyors often use measuring instruments which can calculate the angle of elevation of an object from a point.

Objectives

B Solve problems using an angle of elevation or an angle of depression

Skills check

1 Complete the following statements.
   a Angles on a straight line add up to □°.
   b In one full turn there are □°.

Angle of elevation and angle of depression

Problems using Pythagoras' theorem or trigonometry often refer to the **angle of elevation** or the **angle of depression.**

The angle of elevation is the angle measured upwards from the horizontal.

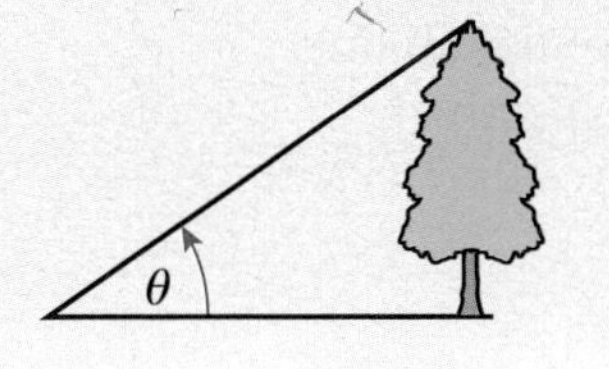

The angle of depression is the angle measured downwards from the horizontal.

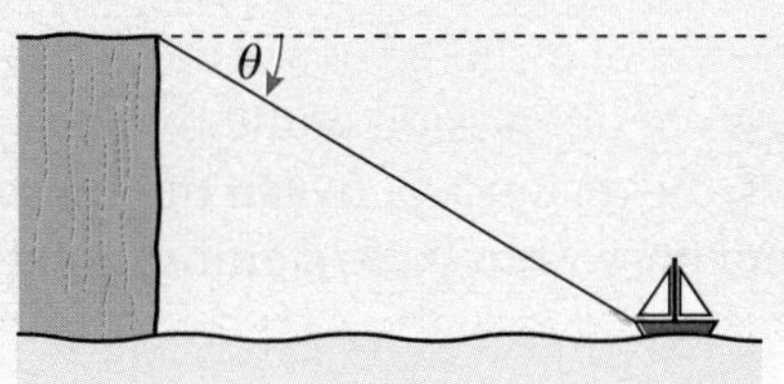

Example 6

B
AO2

From the top of a vertical cliff, 120 m high, Georgia can see a boat out at sea.
The angle of depression from Georgia to the boat is 49°.
How far from the base of the cliff is the boat? Give your answer to one decimal place.

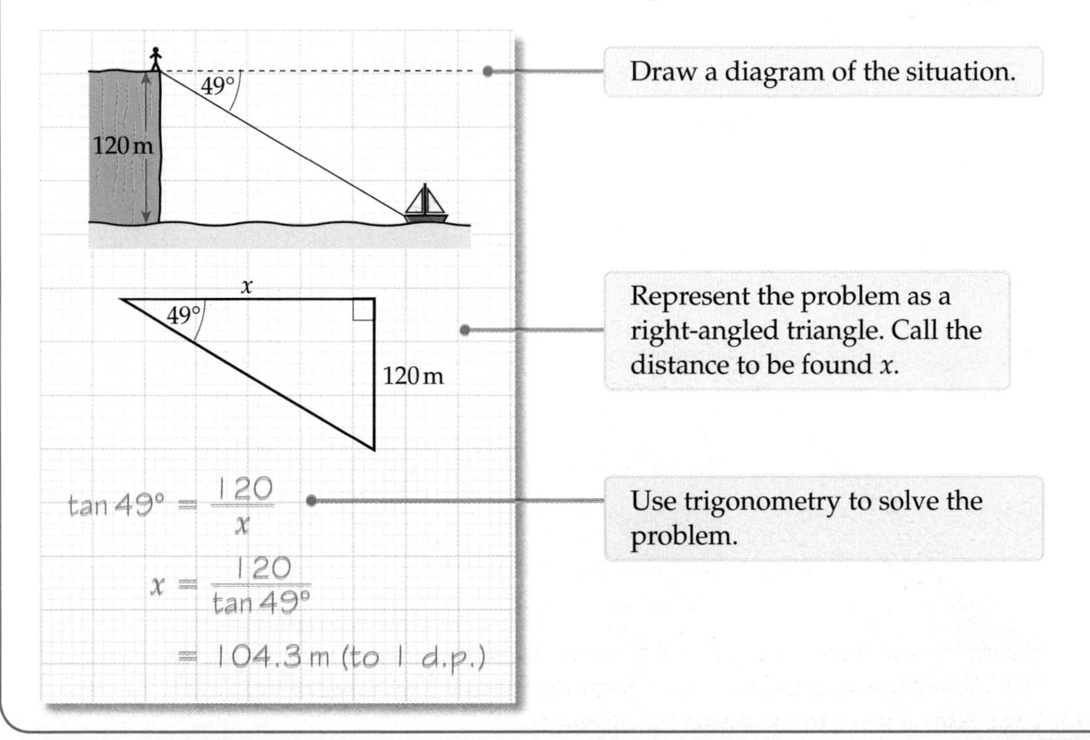

$\tan 49° = \frac{120}{x}$

$x = \frac{120}{\tan 49°}$

$= 104.3$ m (to 1 d.p.)

Exercise 35E

1 William sees an aircraft in the sky. The angle of elevation is 37°.
The aircraft is at a horizontal distance of 31 km from William.
How high is the aircraft? Give your answer to one decimal place.

2 Shamil is standing 1420 m from a wind turbine.
The angle of elevation to the top of the wind turbine is 7.6°.
What is the height of the wind turbine? Give your answer to one decimal place.

3 From the top of a 180 m high vertical cliff, a boat has an angle of depression of 61°.
How far is the boat from the base of the cliff?
Give your answer to one decimal place.

4 A surveyor stands 70 m from a castle.
The angle of elevation to the top of the castle is 55°.
A flagpole stands on top of the castle.
The angle of elevation to the tip of the flagpole is 59°.
Calculate the height of the flagpole. Give your answer to one decimal place.

5 From the top of a cliff 85 m above sea level, two buoys can be seen in line.
The angles of depression of the two buoys are 22° and 14° respectively.
Calculate the distance between the two buoys.
Give your answer to three significant figures.

Review exercise

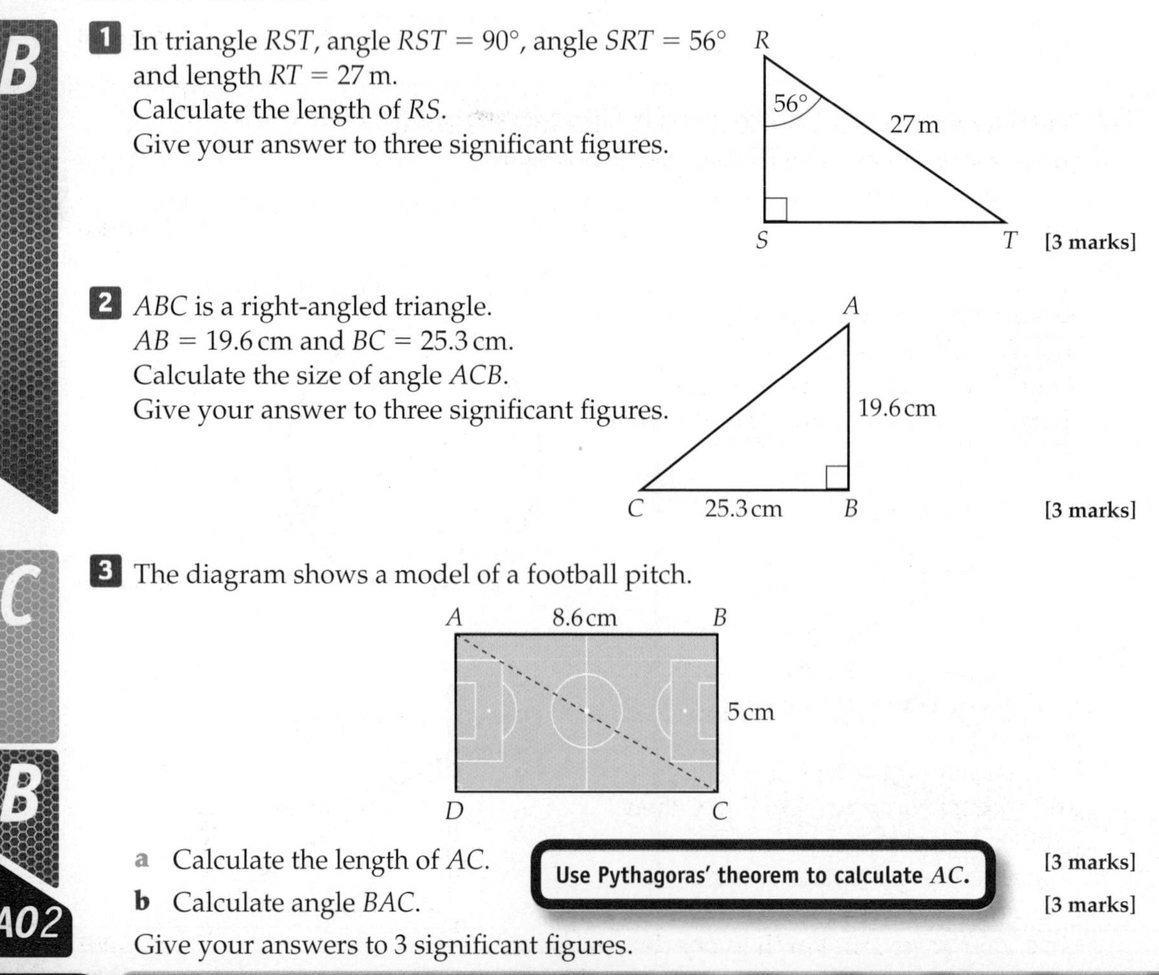

1 In triangle RST, angle $RST = 90°$, angle $SRT = 56°$ and length $RT = 27$ m.
Calculate the length of RS.
Give your answer to three significant figures. **[3 marks]**

2 ABC is a right-angled triangle.
$AB = 19.6$ cm and $BC = 25.3$ cm.
Calculate the size of angle ACB.
Give your answer to three significant figures. **[3 marks]**

3 The diagram shows a model of a football pitch.

a Calculate the length of AC. **[3 marks]**

Use Pythagoras' theorem to calculate AC.

b Calculate angle BAC. **[3 marks]**

Give your answers to 3 significant figures.

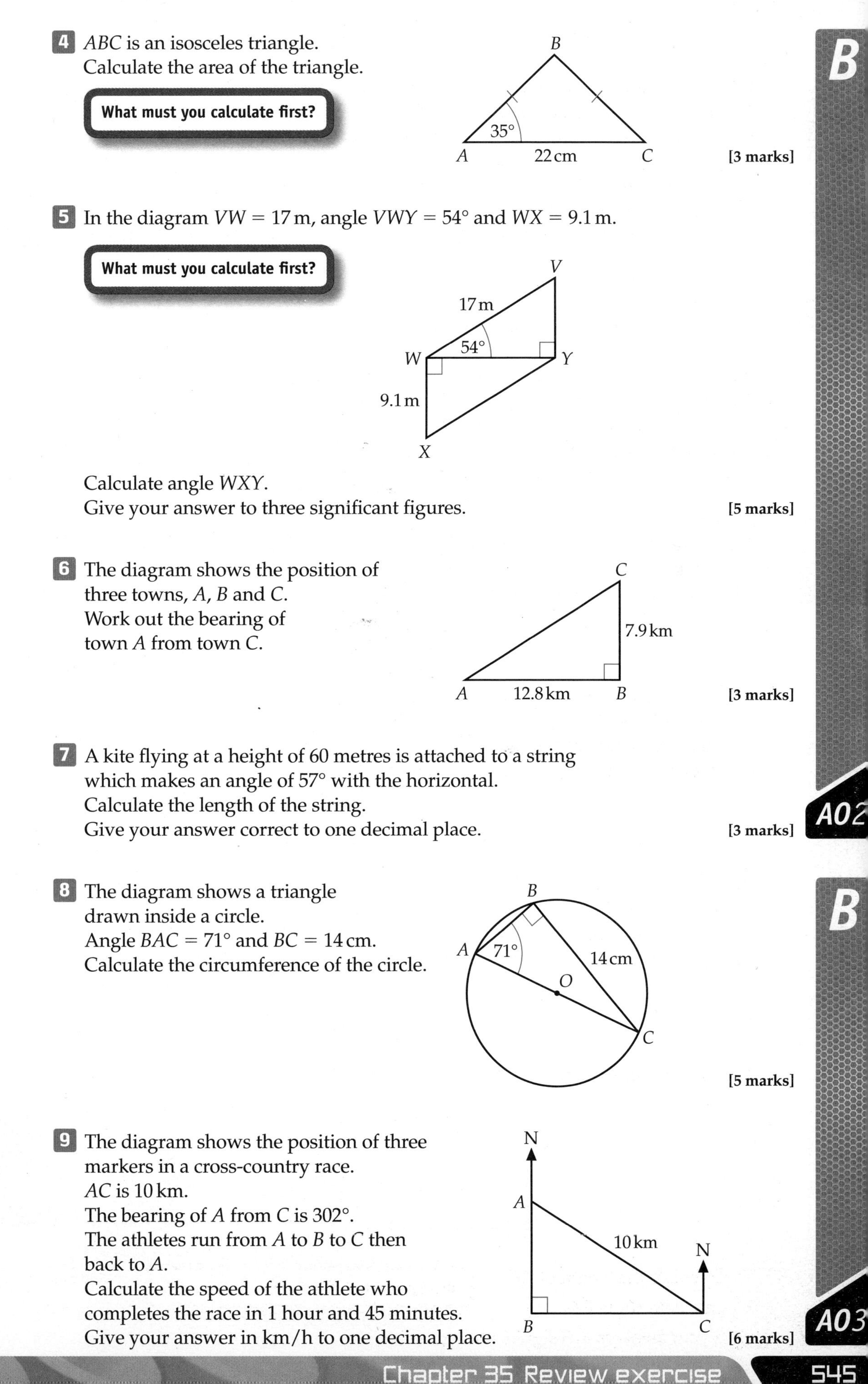

4 ABC is an isosceles triangle.
Calculate the area of the triangle.

What must you calculate first?

B, $35°$, A, $22\,\text{cm}$, C

[3 marks]

5 In the diagram $VW = 17\,\text{m}$, angle $VWY = 54°$ and $WX = 9.1\,\text{m}$.

What must you calculate first?

Calculate angle WXY.
Give your answer to three significant figures.

[5 marks]

6 The diagram shows the position of three towns, A, B and C.
Work out the bearing of town A from town C.

[3 marks]

7 A kite flying at a height of 60 metres is attached to a string which makes an angle of 57° with the horizontal.
Calculate the length of the string.
Give your answer correct to one decimal place.

[3 marks]

8 The diagram shows a triangle drawn inside a circle.
Angle $BAC = 71°$ and $BC = 14\,\text{cm}$.
Calculate the circumference of the circle.

[5 marks]

9 The diagram shows the position of three markers in a cross-country race.
AC is 10 km.
The bearing of A from C is 302°.
The athletes run from A to B to C then back to A.
Calculate the speed of the athlete who completes the race in 1 hour and 45 minutes.
Give your answer in km/h to one decimal place.

[6 marks]

B

AO2

B

AO3

10 A rocket flies 7 km vertically, then 10 km at an angle of 16° to the vertical, then 30 km at an angle of 28° to the vertical.
Calculate its vertical height at the end of the third stage.
Give your answer correct to three significant figures. **[4 marks]**

11 From the top of a sea cliff 40 m above the water, two boats can be seen in line.
The angles of depression of the two boats, at P and Q, are 24° and 9° respectively.
Calculate the distance between the two boats.
Give your answer correct to three significant figures. **[5 marks]**

Chapter summary

In this chapter you have learned how to

- understand and recall trigonometric ratios in right-angled triangles **B**
- enter the trigonometric functions on a calculator **B**
- use trigonometric ratios to find lengths in right-angled triangles **B**
- use trigonometric ratios to find the angles in right-angled triangles **B**
- use trigonometric ratios and Pythagoras' theorem to solve problems, including the use of bearings **B**
- solve problems using an angle of elevation or an angle of depression **B**

36 Circle theorems

This chapter is about circle theorems and their applications.

Circle theorems could be used to predict the path of a rollerblader on a semi-circular ramp.

Objectives

This chapter will show you how to

- use circle properties and theorems, including
  - the tangent and the radius at any point are perpendicular
  - tangents from an external point are equal in length
  - the angle subtended by an arc at the centre of a circle is twice the angle subtended at the circumference
  - the angle in a semi-circle is a right angle
  - angles in the same segment are equal
  - opposite angles of a cyclic quadrilateral are supplementary **B**

Before you start this chapter

1 Draw a circle. Identify and label the following parts.

a radius b diameter c circumference

2 Explain the meaning of these words.

HELP Chapter 33

a bisect b perpendicular

3 What is the sum of the angles

a in a triangle b around a point?

HELP Chapter 21

4 What are these angles called?

a b

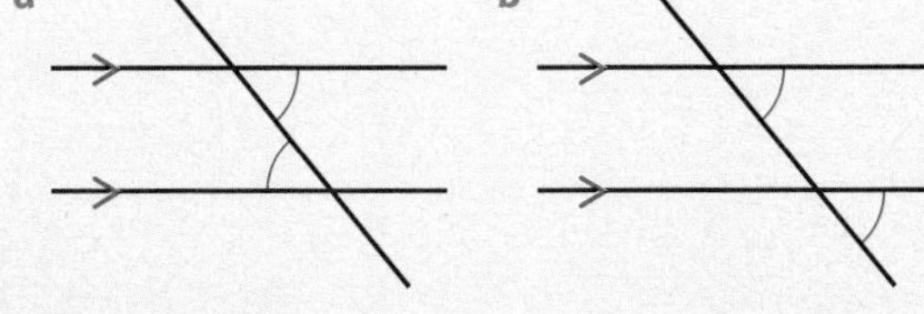

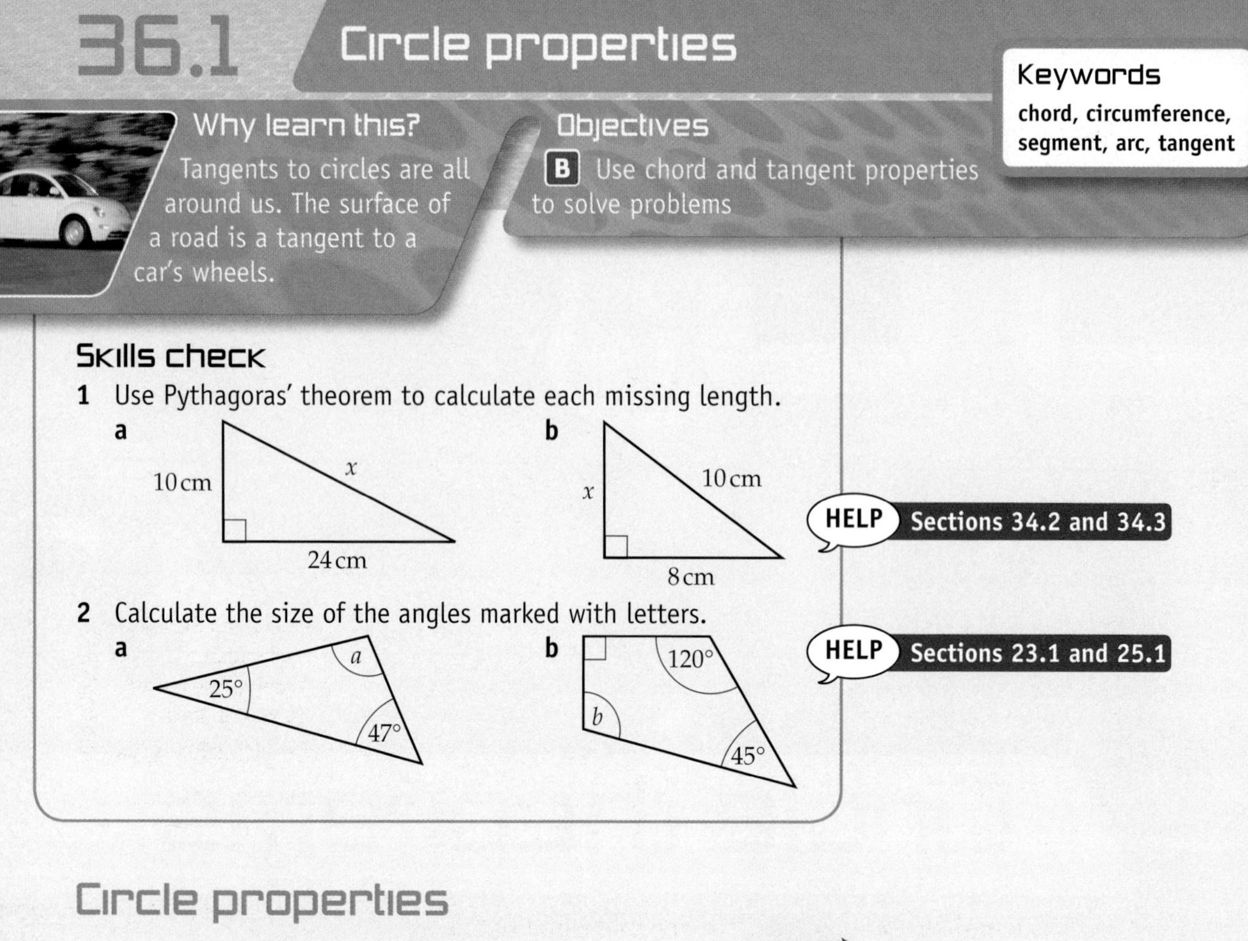

36.1 Circle properties

Why learn this?

Tangents to circles are all around us. The surface of a road is a tangent to a car's wheels.

Objectives

B Use chord and tangent properties to solve problems

Keywords
chord, circumference, segment, arc, tangent

Skills check

1 Use Pythagoras' theorem to calculate each missing length.

a

b

HELP Sections 34.2 and 34.3

2 Calculate the size of the angles marked with letters.

a

b

HELP Sections 23.1 and 25.1

Circle properties

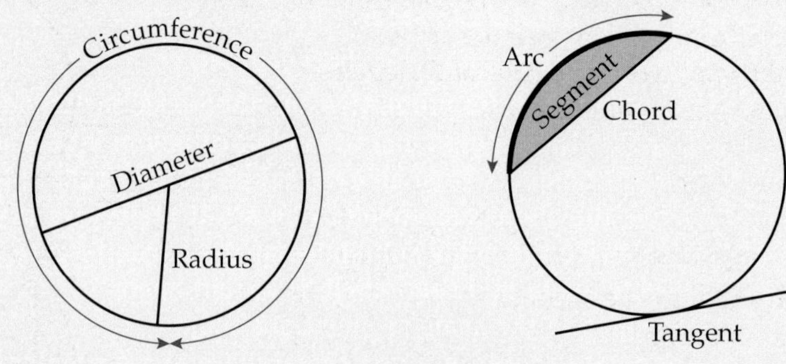

The diagrams above show some of the terms you will use when solving problems involving circles.

A **chord** is a straight line joining two points on the **circumference**.

A chord divides a circle into **segments.** The minor segment is shaded in the diagram.

An **arc** connects two separate points around the circumference of a circle.

A **tangent** to a circle is a straight line that touches the circumference at exactly one place.

Property 1

The perpendicular from the centre of a circle to a chord bisects the chord.

$AM = BM$

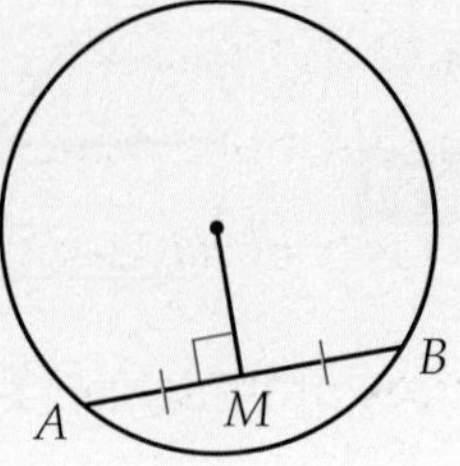

Example 1

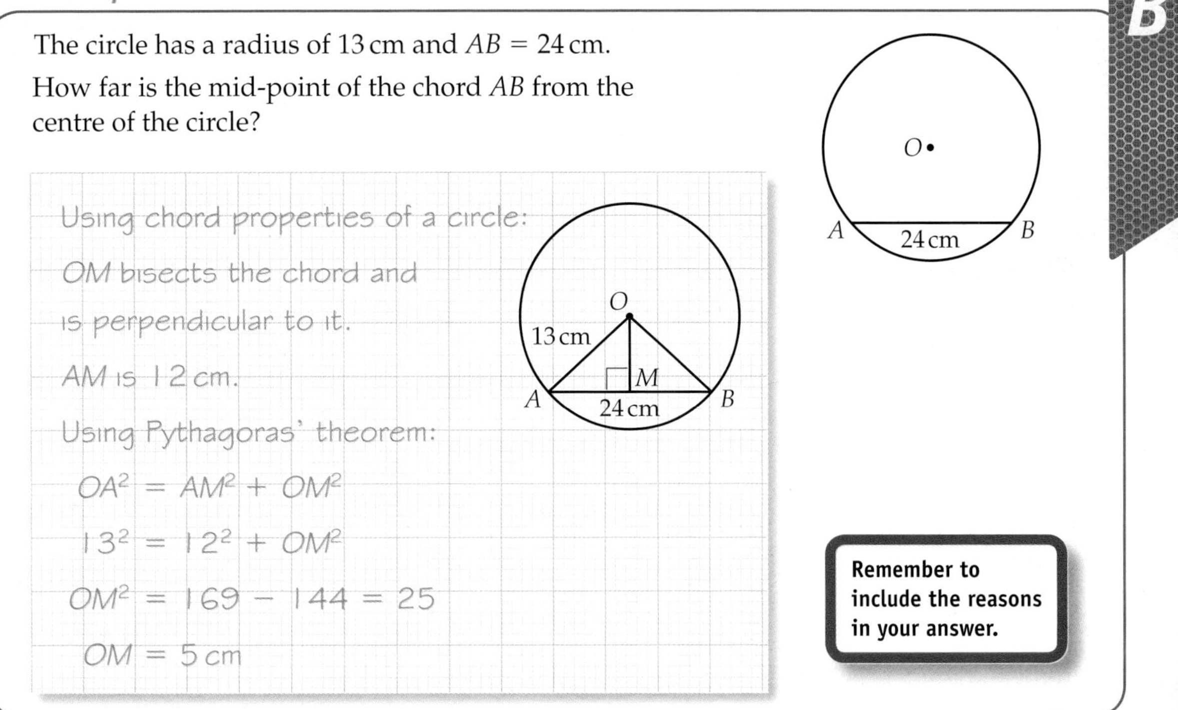

The circle has a radius of 13 cm and $AB = 24$ cm.
How far is the mid-point of the chord AB from the centre of the circle?

Using chord properties of a circle:
OM bisects the chord and is perpendicular to it.
AM is 12 cm.
Using Pythagoras' theorem:

$OA^2 = AM^2 + OM^2$

$13^2 = 12^2 + OM^2$

$OM^2 = 169 - 144 = 25$

$OM = 5$ cm

Remember to include the reasons in your answer.

Exercise 36A

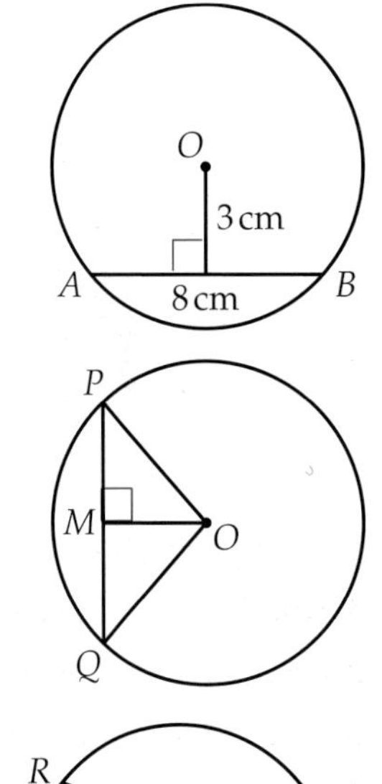

1 The length of chord AB is 8 cm. The mid-point of the chord is 3 cm from the centre of the circle.
Calculate the length of the radius of the circle.
State any circle properties you use.

2 The radius of the circle is 10 cm. $OM = 6$ cm.
Calculate the length of the chord PQ.

3 A circle with centre O has a radius of 7 cm and chords PQ and RS. $PQ = 4.8$ cm and $RS = 9.8$ cm.

- **a** Write down the length of the lines OR and OS and add them to the diagram.
- **b** M is the mid-point of the chord RS.
  Use Pythagoras' theorem to calculate the length of OM.
- **c** N is the mid-point of the chord PQ. Calculate the length ON.
- **d** Which chord is closer to the centre of the circle?

4 PQ and RS are two chords in a circle of radius 6 cm.
$PQ = 10.4$ cm and $RS = 6.6$ cm.
Which chord is closer to the centre of the circle? Explain your answer.

Property 2

Tangents drawn to a circle from an external point are equal in length.

$TA = TB$

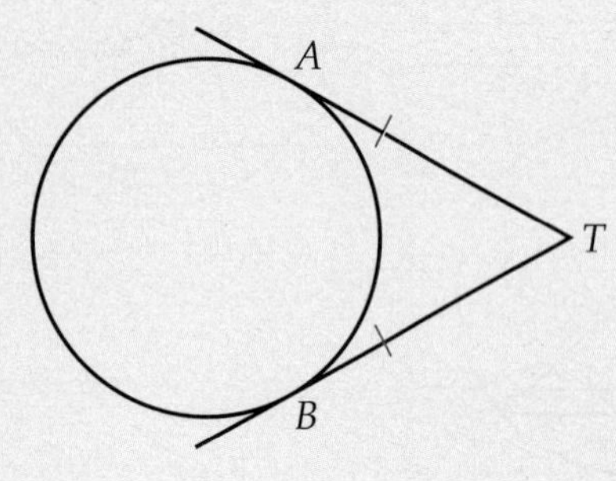

Property 3

The angle between a tangent and a radius is 90 degrees.

Angle $OAT = 90°$

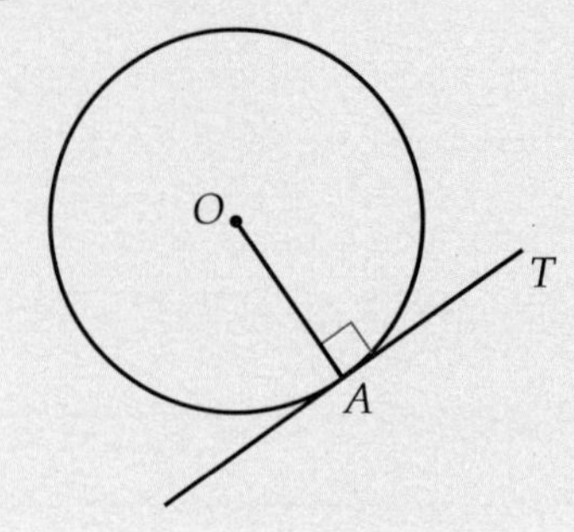

Example 2

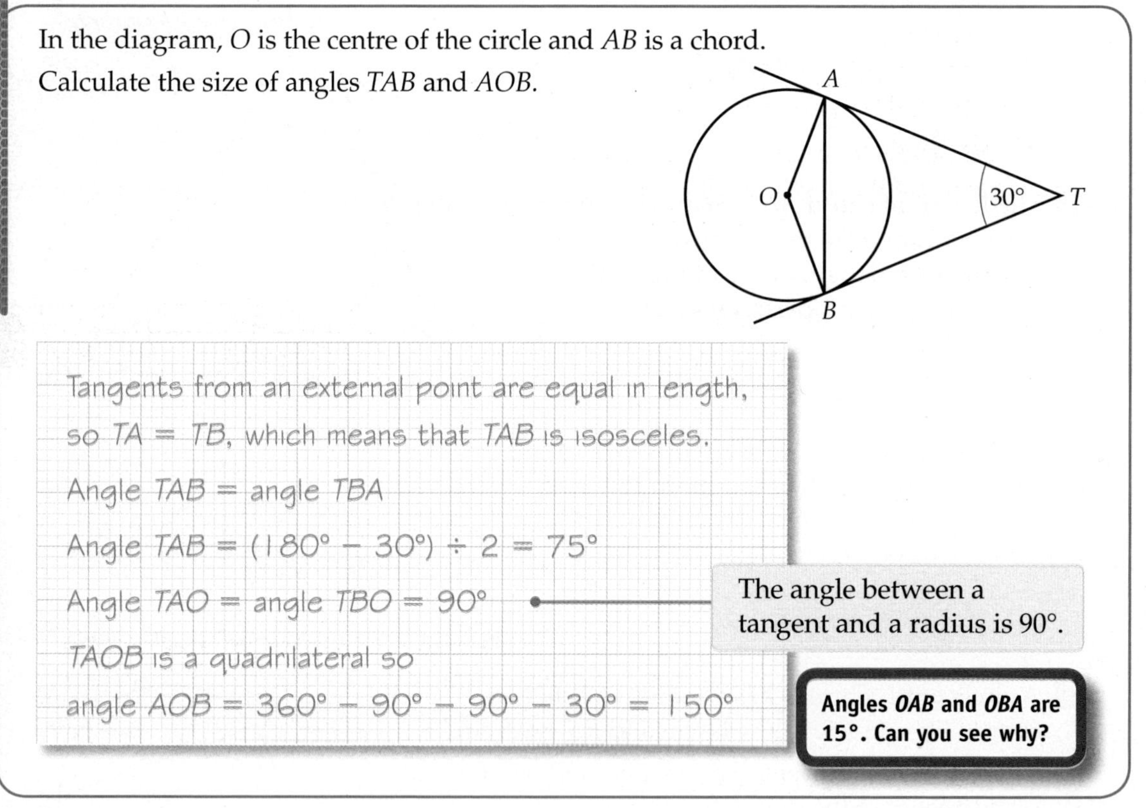

In the diagram, O is the centre of the circle and AB is a chord.
Calculate the size of angles TAB and AOB.

Tangents from an external point are equal in length, so $TA = TB$, which means that TAB is isosceles.

Angle TAB = angle TBA

Angle $TAB = (180° - 30°) \div 2 = 75°$

Angle TAO = angle $TBO = 90°$

The angle between a tangent and a radius is 90°.

$TAOB$ is a quadrilateral so
angle $AOB = 360° - 90° - 90° - 30° = 150°$

Angles *OAB* and *OBA* are 15°. Can you see why?

Exercise 36B

1 Calculate the size of angle x.

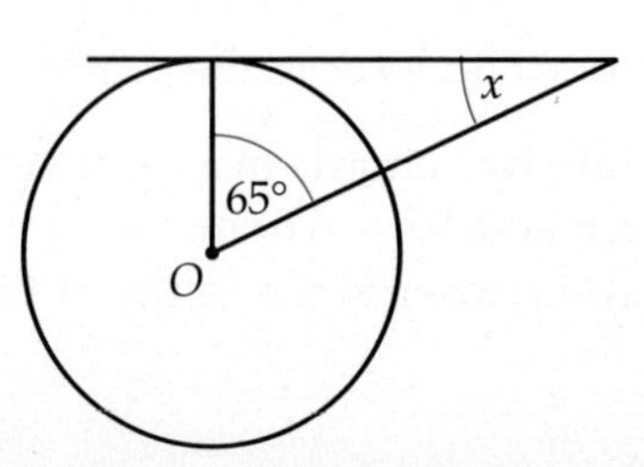

B

2 Calculate the size of angle y.
Explain your answer.

Use Example 2 to help you.

3 Calculate the sizes of angles a and b.
Explain your answer.

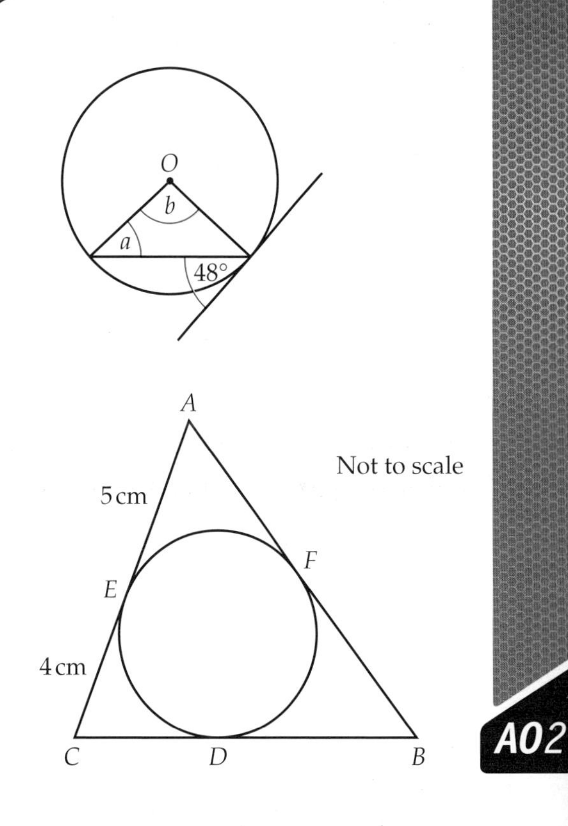

4 The sides of triangle ABC are tangents to the circle.
$AE = 5\,\text{cm}$ and $EC = 4\,\text{cm}$.

a Write down the length of CD.

b The perimeter of the triangle is 34 cm.
Calculate the length of DB.

AO2

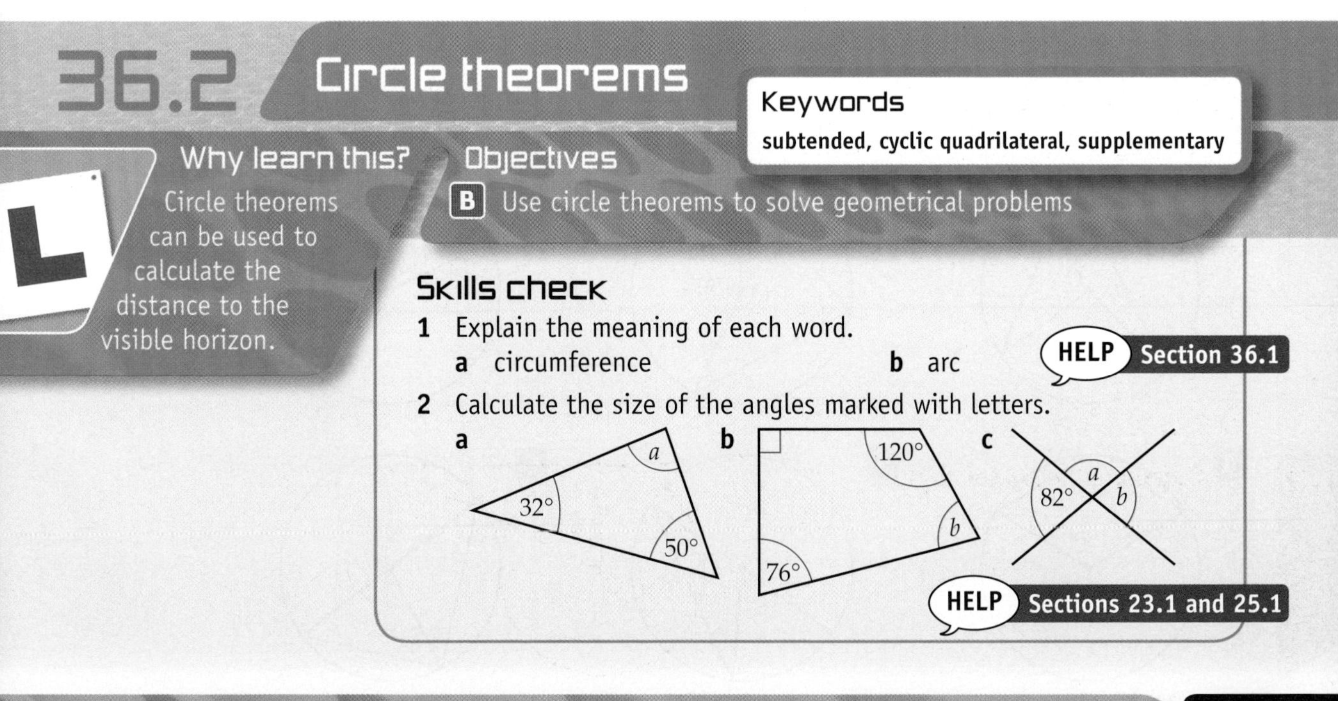

36.2 Circle theorems

Keywords
subtended, cyclic quadrilateral, supplementary

Why learn this?
Circle theorems can be used to calculate the distance to the visible horizon.

Objectives
B Use circle theorems to solve geometrical problems

Skills check

1 Explain the meaning of each word.
 a circumference **b** arc

HELP Section 36.1

2 Calculate the size of the angles marked with letters.
 a **b** **c**

HELP Sections 23.1 and 25.1

Theorem 1

The angle **subtended** by an arc at the centre of a circle is twice the angle that it subtends at the circumference.

'Subtended' means that the angle starts and finishes at the ends of the arc.

Here is a special case of Theorem 1.

The angle in a semicircle is a right angle.

Angle $AOB = 180°$
since AOB is a straight line.

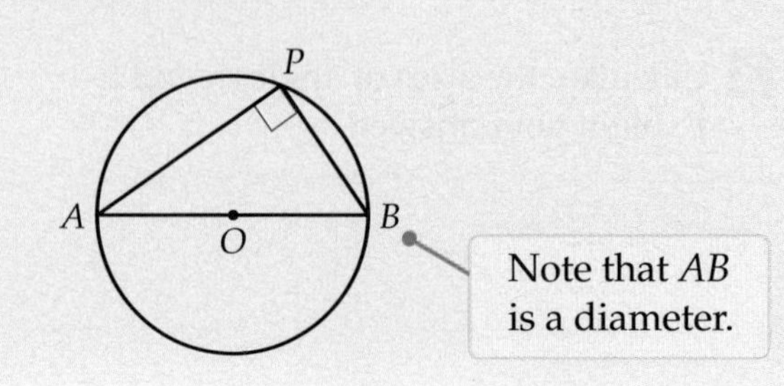

Example 3

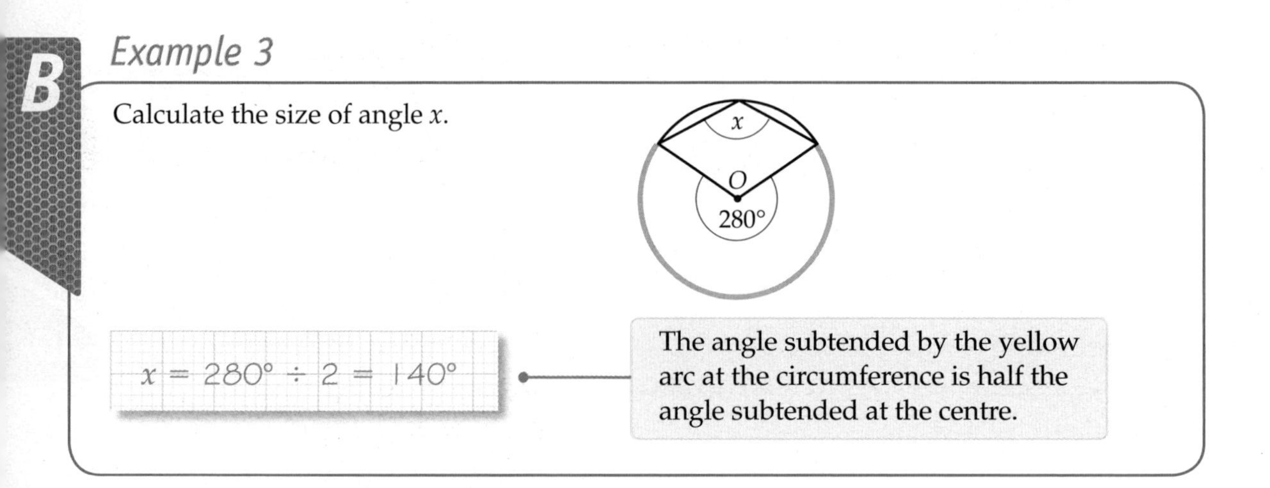

Calculate the size of angle x.

$x = 280° \div 2 = 140°$

The angle subtended by the yellow arc at the circumference is half the angle subtended at the centre.

Exercise 36C

Calculate the sizes of the angles marked with letters. Explain each step in your reasoning.

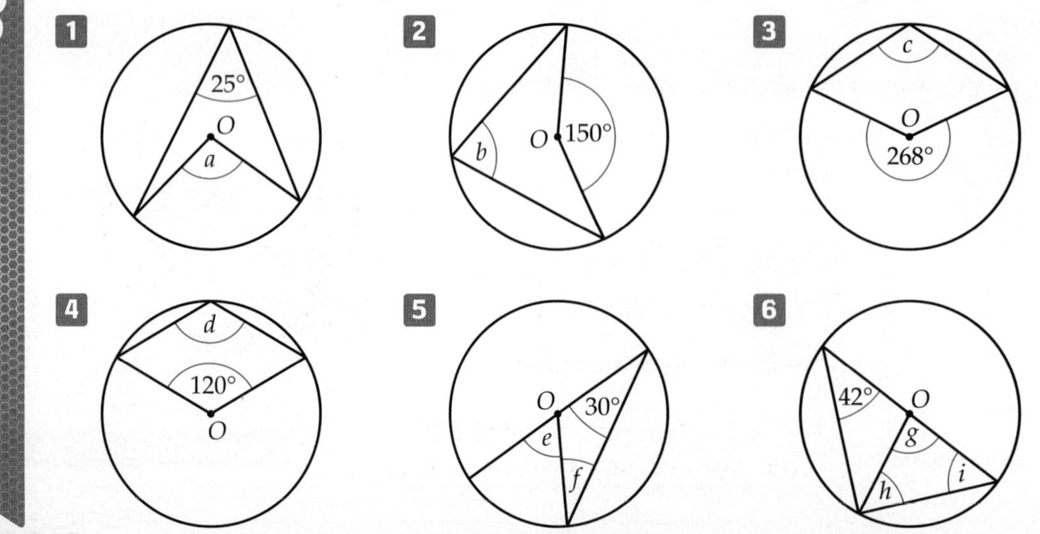

7

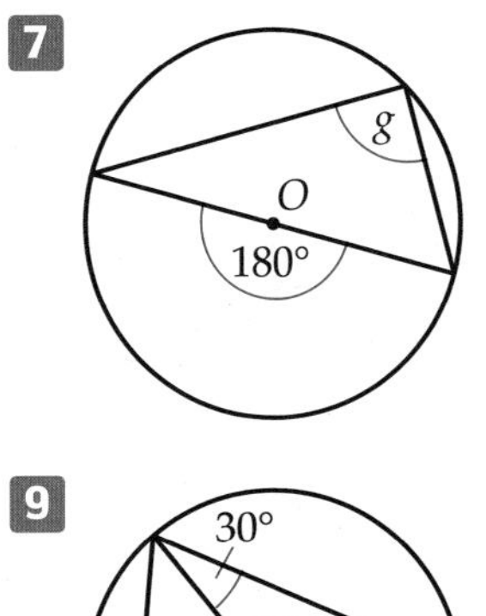

8

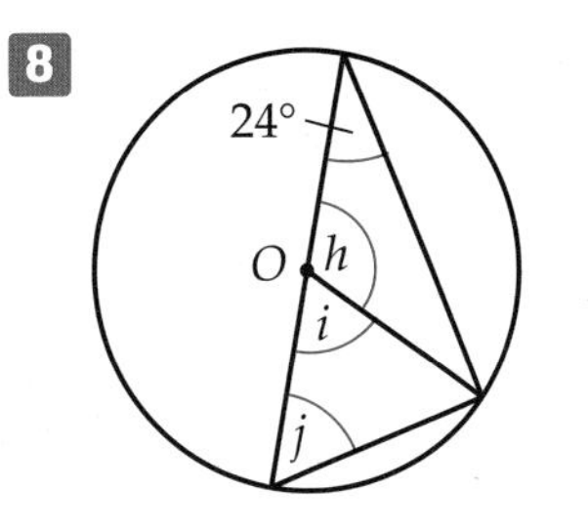

9

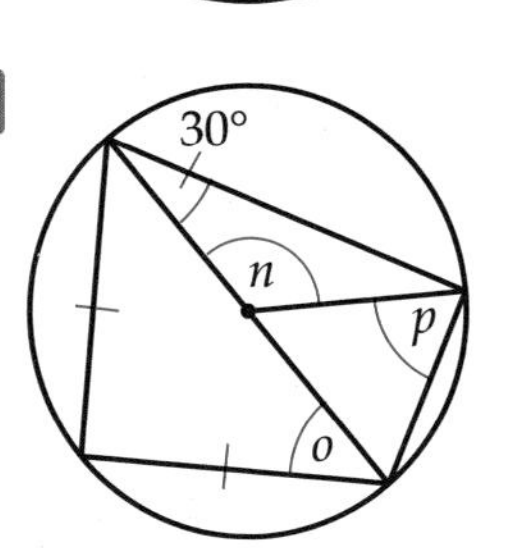

10

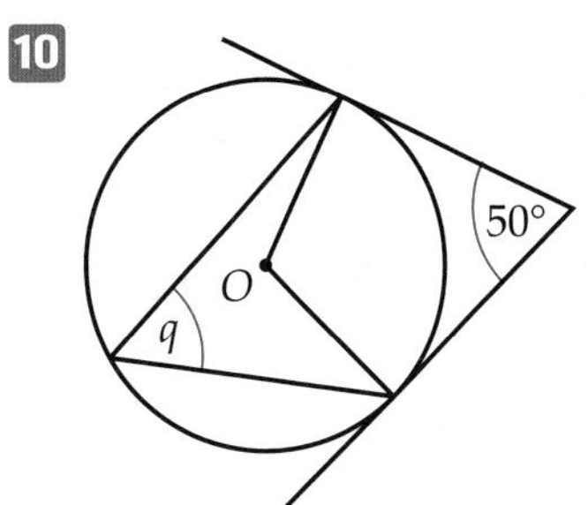

11 A, B and C are points on the circumference of a circle with centre O.
Angle $ABC = 30°$
Explain why triangle OAC is equilateral.

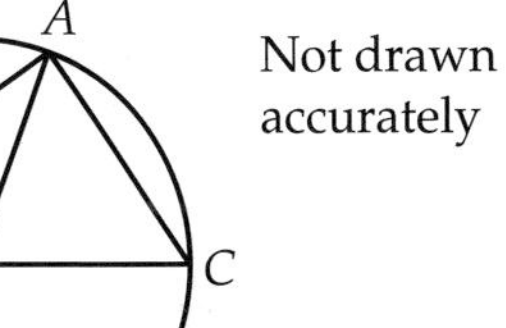

A02

Theorem 2

Angles subtended by the same arc are equal.
All the shaded angles are subtended by arc PQ.

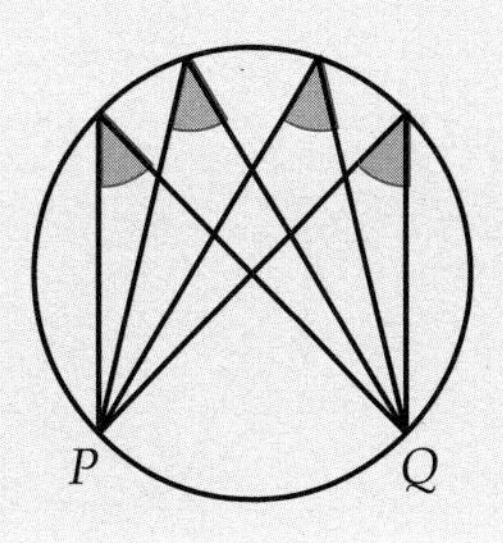

Example 4

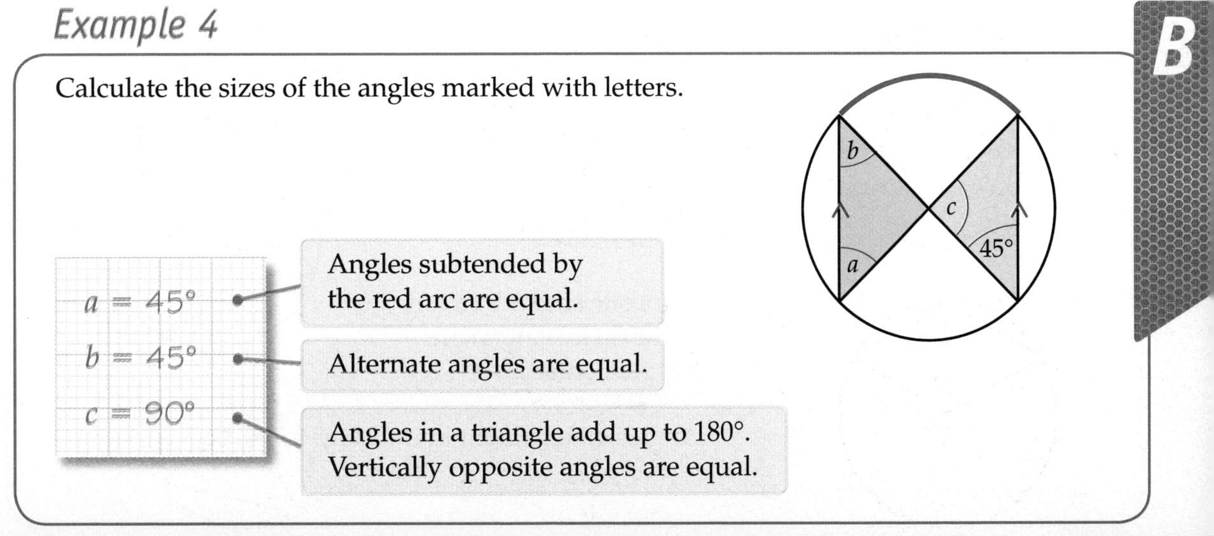

Calculate the sizes of the angles marked with letters.

$a = 45°$ — Angles subtended by the red arc are equal.

$b = 45°$ — Alternate angles are equal.

$c = 90°$ — Angles in a triangle add up to 180°. Vertically opposite angles are equal.

Exercise 36D

B

Calculate the sizes of the angles marked with letters.

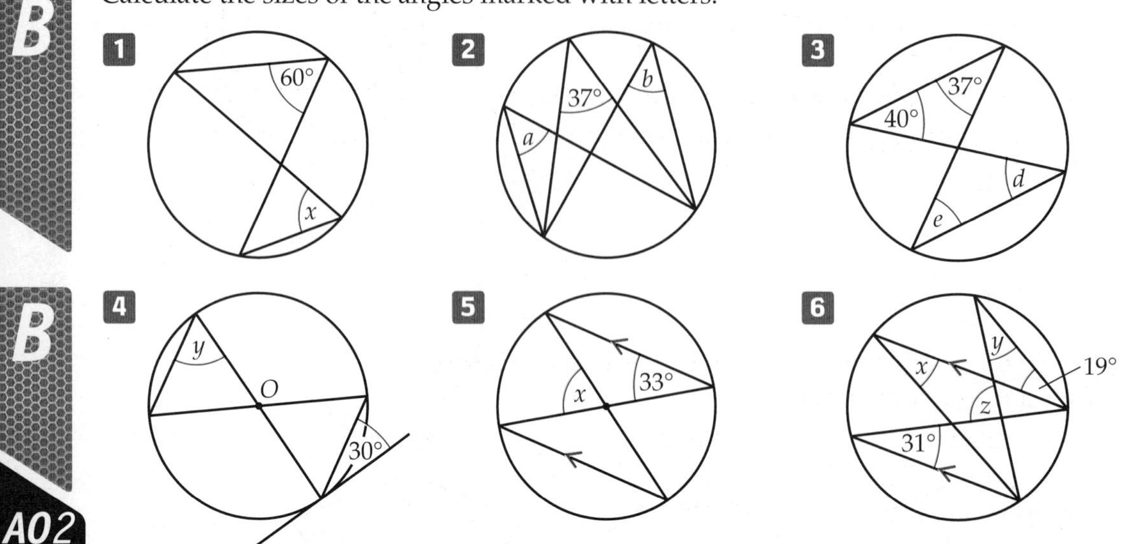

Cyclic quadrilaterals

A **cyclic quadrilateral** is a quadrilateral whose vertices lie on the circumference of a circle.

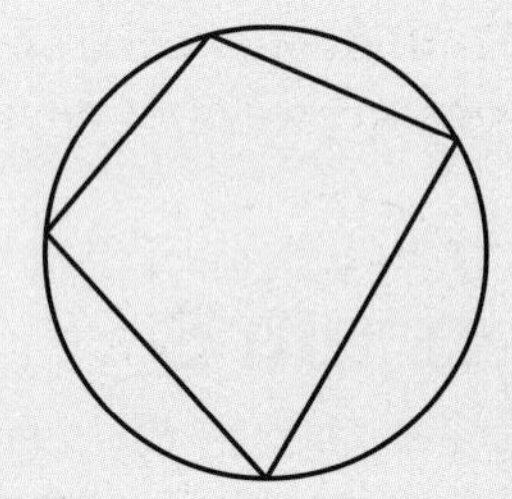

Theorem 3

Opposite angles in a cyclic quadrilateral are **supplementary**.

$a + c = 180°$

$b + d = 180°$

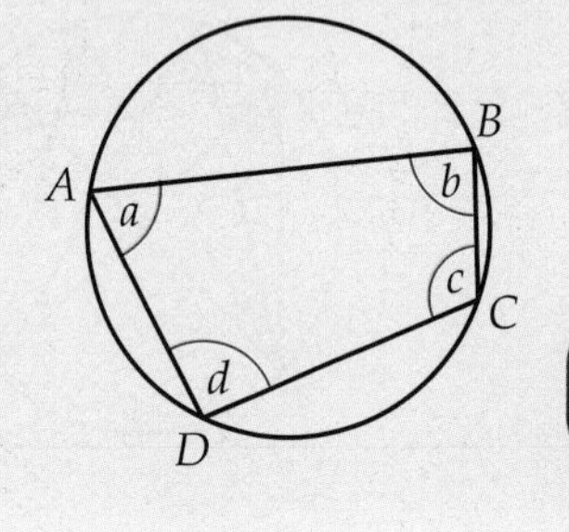

Supplementary angles add up to 180°.

B

Example 5

$ABCD$ is a cyclic quadrilateral.

Calculate the size of the exterior angle x.

B, 95°, C, A, x, D

$CDA = 85°$ — Opposite angles in a cyclic quadrilateral are supplementary.

$x = 180° - 85°$ — Angles on a straight line add up to 180°.

$= 95°$

Theorem 4

Example 5 illustrates another circle theorem.

The exterior angle of a cyclic quadrilateral is equal to the opposite interior angle.

$a + b = 180°$ (opposite angles of cycic quadrilateral)

$b + c = 180°$ (angles on a straight line)

Therefore $a = c$

Exercise 36E

1 Which **one** of the following kites is a cyclic quadrilateral? Give a reason for your answer.

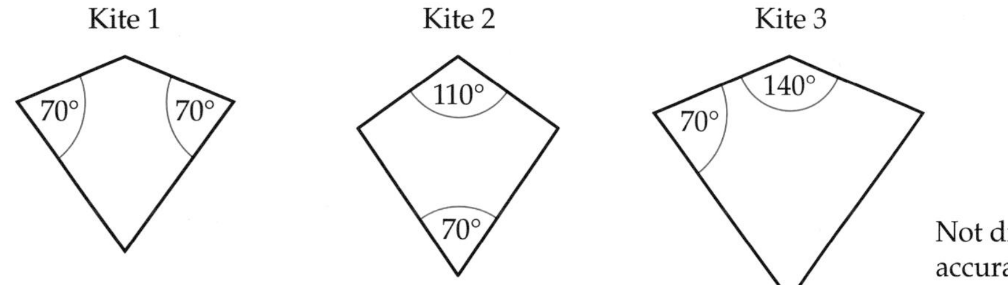

Not drawn accurately

For Q2–10 calculate the sizes of the angles marked with letters. Give reasons for your answers.

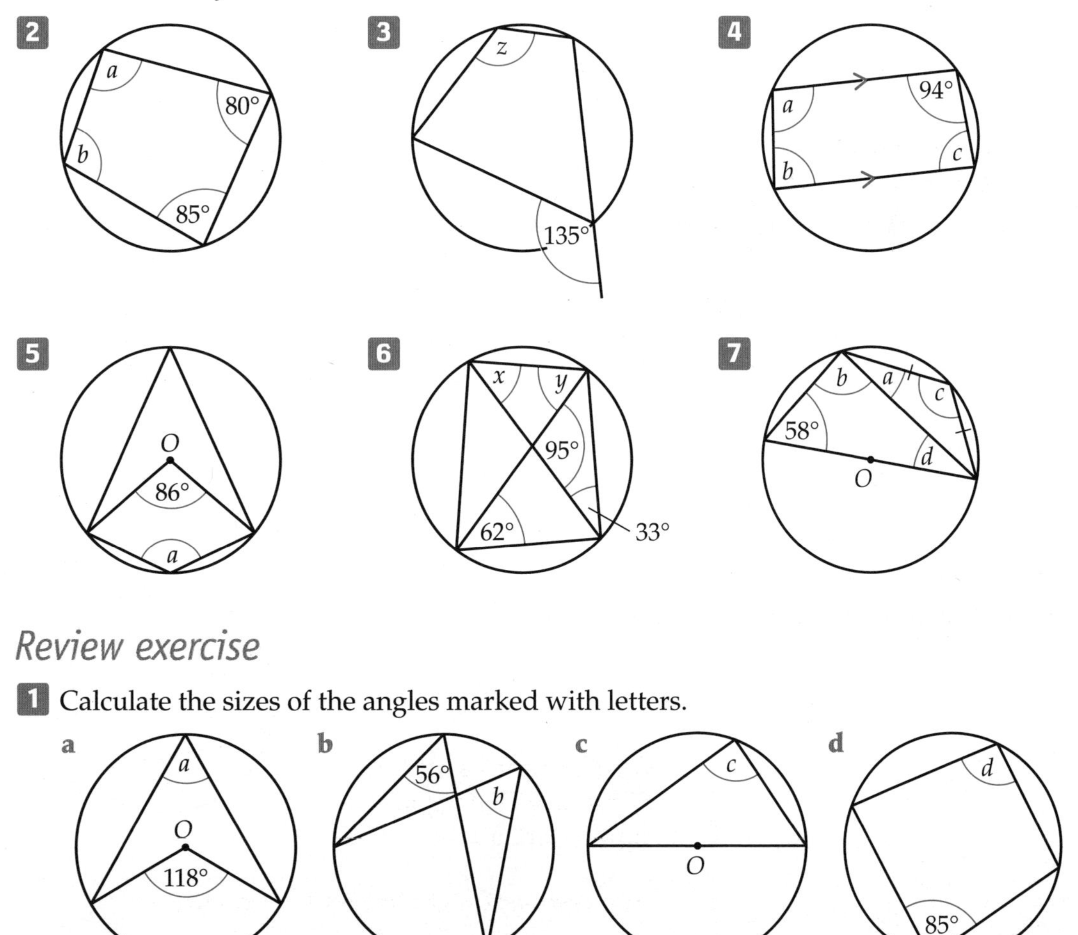

Review exercise

1 Calculate the sizes of the angles marked with letters.

a **b** **c** **d**

[4 marks]

B

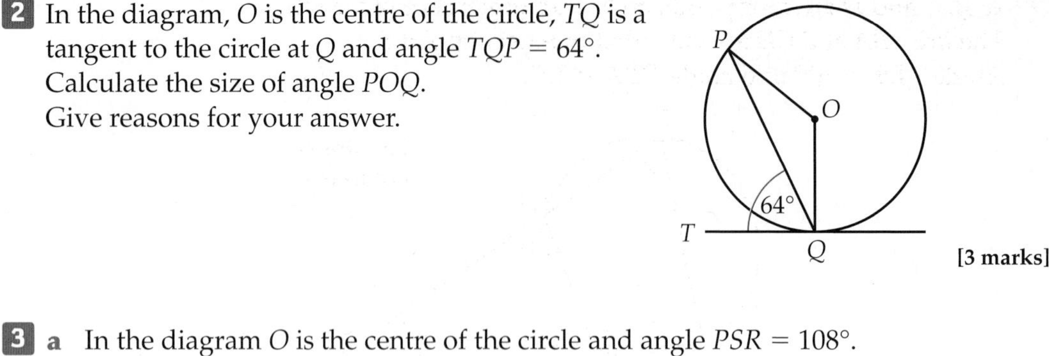

2 In the diagram, O is the centre of the circle, TQ is a tangent to the circle at Q and angle $TQP = 64°$.
Calculate the size of angle POQ.
Give reasons for your answer.

[3 marks]

3 **a** In the diagram O is the centre of the circle and angle $PSR = 108°$.

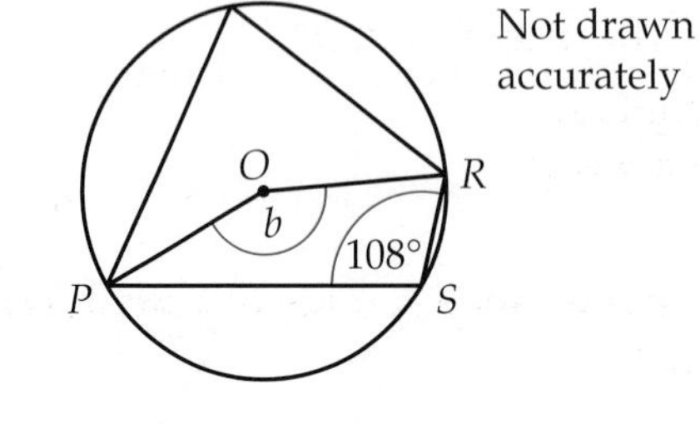

Not drawn accurately

Calculate the size of angle b. **[2 marks]**

b PQ and PR are tangents to the circle centre O.
Angle QPR is 65°.

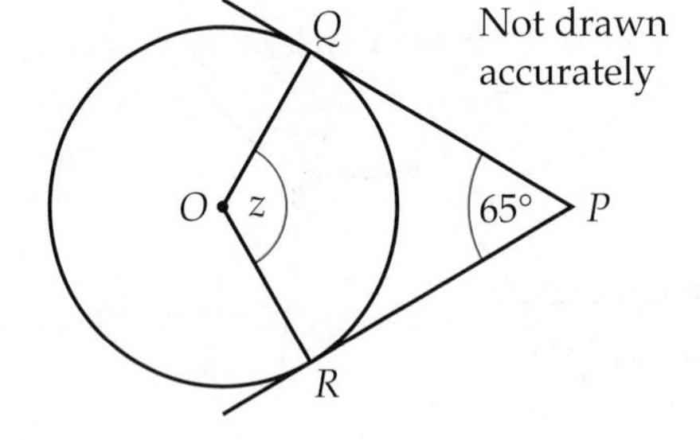

Not drawn accurately

Calculate the size of angle z. **[2 marks]**

4 A, B, C and D are four points on the circumference of a circle.
AC meets BD at X. Angle $ABD = 54°$ and angle $CXD = 85°$.
Calculate the size of angle d.
Give reasons for your answer.

Not drawn accurately

[3 marks]

5 In the diagram, O is the centre of the circle and AC is a diameter.

AO2

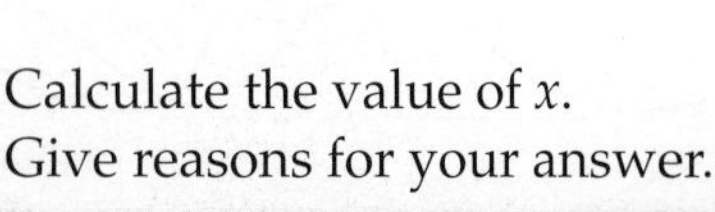

Calculate the value of x.
Give reasons for your answer. **[3 marks]**

6 A, B, C and D are four points on the circumference of a circle.
The lines BA and CD are extended to meet at point E.
Angle $ADE = 67°$ and angle $DEA = 30°$.

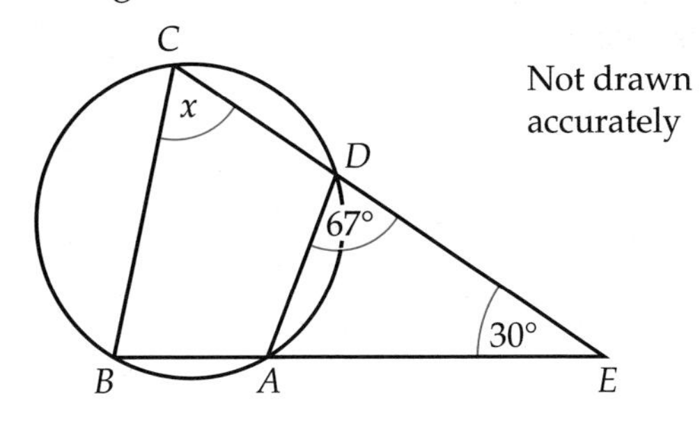

a What is the name of quadrilateral $ABCD$? **[1 mark]**

b Calculate the size of angle x.
Give reasons for your answer. **[3 marks]**

B

AO2

7 A, B, C and D are points on the circumference of a circle, centre O.
AC is a diameter of the circle.
Angle $DCA = 65°$.

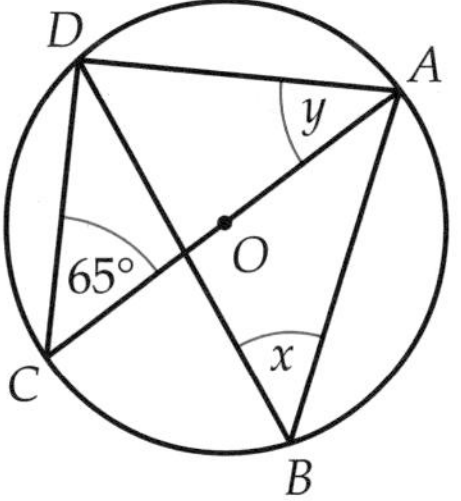

a Write down the size of angle x. **[1 mark]**

b Calculate the size of angle y. **[1 mark]**

B

Chapter summary

In this chapter you have learned how to

- use chord and tangent properties to solve problems **B**
- use circle theorems to solve geometrical problems **B**

AO2 Problem-solving practice: Geometry

Quality of written communication: Some questions on this page are marked with a star ☆. In the exam, this sort of question may earn you some extra marks if you

- use correct and accurate maths notation and vocabulary
- organise your work clearly, showing that you can communicate effectively.

E AO2

1 Arnav is making friendship bracelets. He needs 14 inches of gold thread to make each bracelet. He buys a 1.8 m roll of gold thread. How many friendship bracelets can he make? [3]

D AO2

☆ **2** **a** Use ruler and compasses to accurately construct this triangle. [3]

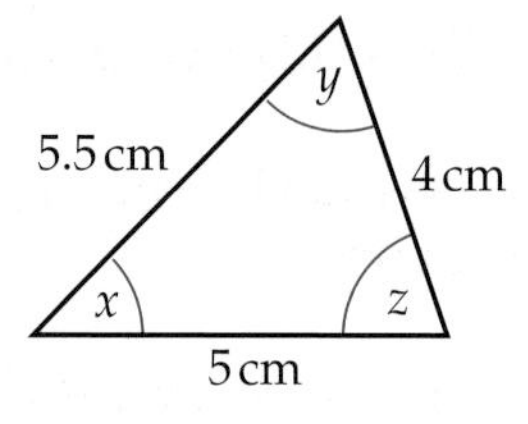

b Measure the size of angles x, y and z. [3]

c Work out $x + y + z$. Comment on your answer. [1]

C AO2

3 The diagram shows a rubber door wedge.

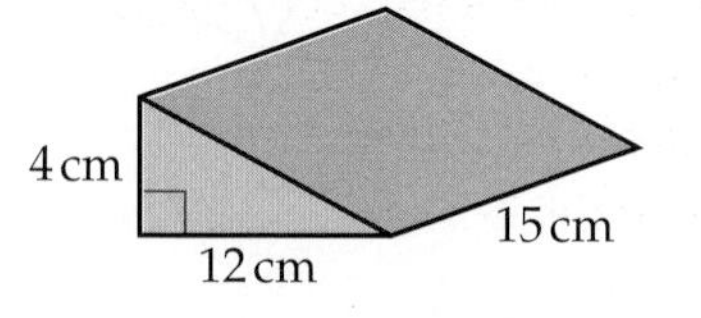

a Work out the volume of the wedge. [3]

b Rubber has a density of $11\ \text{g/cm}^3$. Calculate the mass of the wedge. [2]

4 Calculate the length of the line segment from $A(-3, 4)$ to $B(6, 0)$. Give your answer correct to two decimal places. [3]

C AO3

5 The diagram shows the dimensions of a football pitch. The goalkeeper is standing within 20 m of the line AC. He is also within 30 m of the corner flag at C.

A ← 105 m → B
68 m
C D

a Make an accurate scale drawing of the football pitch using a scale of 1 cm to 10 m. [3]

b Shade the locus of points where the goalkeeper could be standing. [2]

B AO2

6 The diagram shows a triangular course for a yacht race. Point C is due west of point A.

N B
5 km
A 8 km C

Calculate

a the total distance of the race [3]

b the bearing of B from A. [3]

Problem-solving practice: Geometry

Quality of written communication: Some questions on this page are marked with a star ☆. In the exam, this sort of question may earn you some extra marks if you
- use correct and accurate maths notation and vocabulary
- organise your work clearly, showing that you can communicate effectively.

E

1 Dan's car can travel 48 miles on 1 gallon of petrol.

The fuel tank holds 9 gallons. The diagram shows the reading on the fuel gauge.

Dan needs to drive from Norwich to Peterborough. The journey is 100 km.

Will Dan have enough fuel? Show all of your working. [4]

AO3

D

2 The bearing north-north-west is exactly half way between north and north-west. Write north-north-west as a bearing. [1]

☆ **3** Imran's remote controlled car has a maximum speed of 28 km/h. An Olympic sprinter can run 100 m in 9.8 seconds. Imran says his car would beat an Olympic sprinter in a 100 m race. Is he correct? Give a reason for your answer. [3]

4 The diagram shows a triangle.

Amir says that it is possible to construct two different triangles with these measurements. Construct both triangles to prove that Amir is correct. [6]

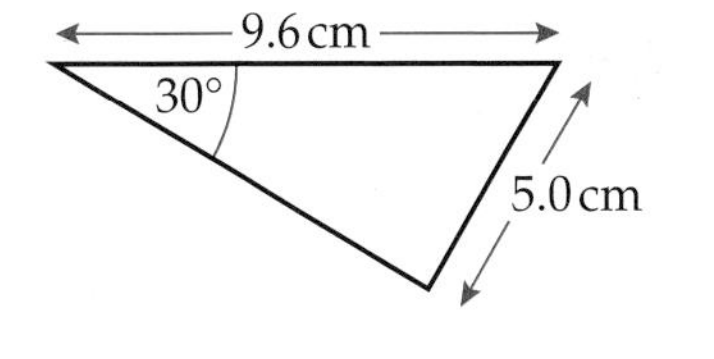

AO3

C

5 The diagram shows the bucket of a wheelbarrow.

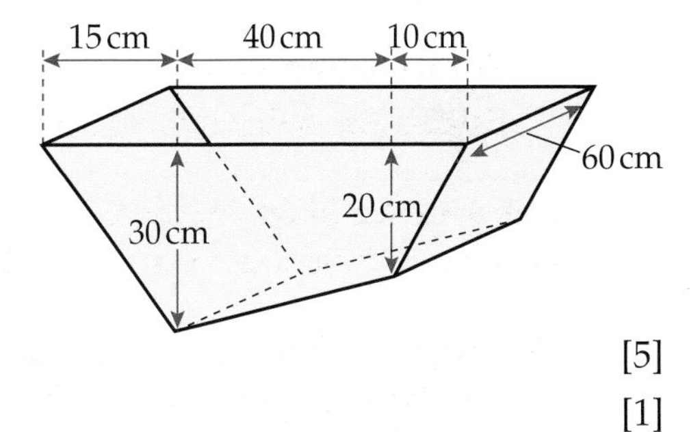

Calculate the capacity of the bucket in

a cm^3 [5]

b litres. [1]

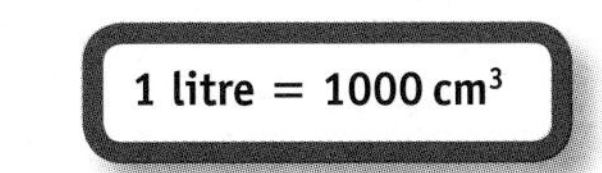

AO3

B

☆ **6** A triangle ABC is drawn inside a circle of radius 4 cm. AC is a diameter of the circle.

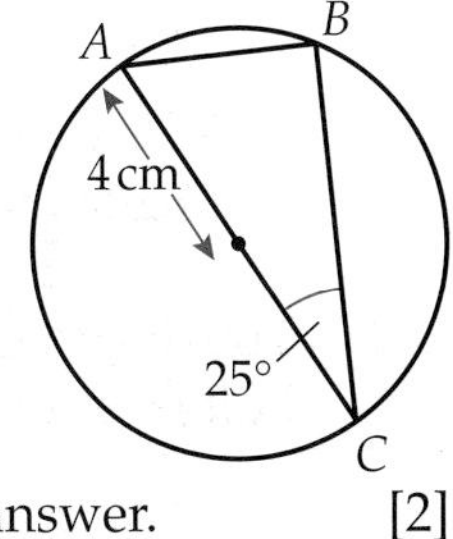

a Write down the size of angle ABC. Give a reason for your answer. [2]

b Calculate the area of the triangle.
Give your answer correct to three significant figures. [4]

Answers

Chapter 1

Exercise 1A

1 a e.g. 'Boys in Year 10 can run faster than girls in Year 10'
b e.g. 'People do more of their shopping at the supermarket than at their local shop'

Exercise 1B

1 a e.g. Need to define what age 'young' and 'old' people are.
b e.g. 'Girls' and 'boys' are very general terms, better to make a more specific statement.

Exercise 1C

1 a Secondary data. Look at official government statistics or 2008 sales figures from car makers/distributors. Need to identify which cars have petrol engines and which have diesel and compare total sales of each.
b Secondary data. Get official weather statistics from the MET office or similar. Need to compare the average number of hours of sunshine received in June for Tenby and Southend. This should be done over a sample of several years.
c Primary data. Conduct a survey of people in your street. Ask people whether they prefer Chinese or Indian takeaways and add up the totals.
d Secondary data. Find official government statistics for the last general election. Compare the total number of people aged 40–50 who voted with those aged 20–30.
e Primary data. Carry out a survey of Year 11 students. Ask them where they'd prefer to go on an end-of-term trip and add up the totals.
f Secondary data. Ask the local cinema for attendance figures over the last 12 months. You could plot a graph of the total figures for each month to see whether attendance is falling.

Exercise 1D

1 a Qualitative, e.g. brown hair
b Quantitative, discrete, e.g. 60
c Quantitative, continuous, e.g. 240 g
d Quantitative, discrete, e.g. 30
e Qualitative, e.g. Dove
f Quantitative, discrete, e.g. 32 inches
g Qualitative, e.g. Fiat
h Quantitative, continuous, e.g. 10.7 m

Exercise 1E

1 a

Colour	Tally	Frequency
red	𝍸 𝍸 |	11
blue	𝍸 ||||	9
pink	𝍸 𝍸 𝍸	15
white	𝍸 𝍸	10
	Total	45

b Pink c 45

2 a

Number of children	Tally	Frequency
1	𝍸 |	6
2	𝍸 |||	8
3	𝍸 |	6
4	𝍸	5
5	||	2
6	|	1
	Total	28

b 2
c 28
d For each row in the frequency table, multiply the 'number of children' by the 'frequency', then add up the total.

3 a

Number of pets	Tally	Frequency
1		10
2		8
3		5
4		5
5		1
6		1
	Total	30

b 1
c For each row in the frequency table, multiply the 'number of pets' by the 'frequency', then add up the total.

Exercise 1F

1 a

Marks	Tally	Frequency
1–5	||	2
6–10	𝍸 ||||	9
11–15	𝍸 𝍸 |	11
16–20	𝍸 |	6
	Total	28

b 31 c 11–15 d Over 50%

2 e.g.

Pocket money received	Tally	Frequency
£5.50–£8.74	𝍸 |	6
£8.75–£11.99	𝍸 |||	8
£12.00–15.24	𝍸 ||||	9
£15.25–18.49	||||	4
	Total	27

3 a 80 b 20 c 45 d $28 \geqslant h < 34$

4 a

Weight, w kg	Tally	Frequency
$45 \leqslant w < 55$	𝍸	5
$55 \leqslant w < 65$	𝍸 |||	8
$65 \leqslant w < 75$	𝍸 ||	7
$75 \leqslant w < 85$	𝍸 |	6
$85 \leqslant w < 95$	||||	4
	Total	30

b 20 c 17 d $55 \leqslant w < 65$

Exercise 1G

1 a

	A*	A	B	C	D	E	F	G	Total
Boys	10	8	12	8	3	9	5	1	**56**
Girls	13	9	10	11	6	8	4	3	**64**
Total	23	**17**	**22**	**19**	9	**17**	**9**	**4**	**120**

b 38 c 15 d Girls e 120

2 a 5 b 22 c 23 d 20%

3 a, b

	Car	Bus	Cycle	Walk	Total
Men	22	1	6	5	34
Women	10	3	1	2	16
Total	32	4	7	7	50

c 7 d 14%

4 a 16 b 21 c 73 d 176

Exercise 1H

1 a i Some people do not like to provide personal details such as their date of birth.
ii This is a leading question, which encourages a particular answer.
iii The responses available overlap.

b i How old are you?
Under 20 years ☐, 20–39 ☐, 40–59 ☐, 60+ ☐
ii Do you think it takes too long to get an appointment to see the doctor?
iii How many times did you visit the doctor last year?
0 ☐, 1–2 ☐, 3–4 ☐, 5–6 ☐, 7+ ☐

2 a It is a leading question. The response categories are poorly chosen, e.g. no option for 'Don't agree'
b It is not a leading question. All possible responses are catered for with no overlap between categories.

3 e.g. For each day, please state whether you'd be able to car-share to and from work, either as the driver or a passenger.

	Able to car-share as: driver	passenger	Unable to car-share
Monday	☐	☐	☐
Tuesday… etc			

4 a 0–1 day ☐, 2–3 days ☐, 4–5 days ☐, 6–7 days ☐
b What types of physical activity do you take part in?
Walking/running ☐, Cycling ☐, Gym ☐, Swimming ☐, Team sports (football, netball etc) ☐, Racquet sports (tennis, squash etc) ☐, Other ☐

5 a, b e.g.

Distance to school, d km	Year 7	Year 8	Year 9	Year 10	Year 11
$0 \leq d < 4$	1	2		1	
$4 \leq d < 8$	2		3	1	2
$8 \leq d < 12$	1	1		1	2
$12 \leq d < 16$	1		1		
$16 \leq d$				1	

Exercise 1I

1 A significant proportion of people will be at work between 9 am and 5 pm and so their views will not be represented. People who work unsociable hours or from home, and people who are pensioners or unemployed may be over-represented.

2 No. Most of the people surveyed are likely to drive and use that particular shopping centre.

3 People who attend a sports centre are likely to do relatively more exercise on average.

4 No. People outside a bus station are likely to travel to work by bus.

5 Not necessarily. The survey will fail to capture people who don't go out in town during the evening and who may spend less on entertainment (or more if they spend a lot on home entertainment).

Review exercise

1 a

Make of car	Tally	Frequency
Ford (F)	卌 卌 |||	13
Vauxhall (V)	卌 |||	8
BMW (B)	卌 卌	10
Audi (A)	卌 ||	7
Kia (K)	||||	4
	Total	42

b Ford c 42

2 e.g. More people go on holiday to America than to Spain.

3 Primary data. Conduct a survey of teenagers in your school. Ask people whether they read *Hi* magazine and/or *Rumours* magazine and add up the totals.

4 a Qualitative, e.g. France
b Qualitative, e.g. red
c Quantitative, discrete, e.g. 130 000
d Quantitative, continuous, e.g. 7 mm

5 a

Shoe size	Tally	Frequency
3–5	卌 卌 ||	12
6–8	卌 卌 卌 卌 ||	22
9–11	卌 卌 卌 卌	20
12–14	卌		6
	Total	60

b 6–8 c By finding the total for each size.

6 a

Height of plants, h	Tally	Frequency
$10 \leq h < 15$	卌 ||	7
$15 \leq h < 20$	卌 ||	7
$20 \leq h < 25$	卌 ||	7
$25 \leq h < 30$	||||	4
$30 \leq h < 35$	卌 |||	8
$35 \leq h < 40$	|||	3
	Total	36

b $30 \leq h < 35$

7 a

	Caravan	B&B	Apartment	Hotel	Total
June	**3**	4	5	14	**26**
July	8	**15**	12	**2**	37
August	11	10	15	**19**	55
September	6	7	**4**	13	30
Total	28	**36**	36	48	**148**

b 36 c 19 d 13 e 148

8 e.g. How many hours (h) do you usually spend doing homework at the weekend?
$0 \leq h < 1$ ☐, $1 \leq h < 2$ ☐, $2 \leq h < 3$ ☐,
$3 \leq h < 4$ ☐, $4 \leq h$ ☐

9 He only asks boys so girls are not represented.
He only asks students in Year 11 so 14 year olds are not represented.

Chapter 2

Exercise 2A

1 a 288 ml b 6 307 c $318\frac{1}{7}$ kg d $12\frac{1}{7}$
e $933\frac{1}{3}$ f 0.004 65

2 a 20 b 1750

3 a $2\frac{2}{5}$ b $416\frac{2}{3}$

4 121

Exercise 2B

1 $\frac{3}{7}$

2 $\frac{8}{28} = \frac{2}{7}$

3 $\frac{4}{52} = \frac{1}{13}$

4 $\frac{16}{30} = \frac{8}{15}$

5 Donna is incorrect. $\frac{5}{5+9} = \frac{5}{14}$

6 a $\frac{2}{7}$ b $\frac{2}{8} = \frac{1}{4}$ c $\frac{4}{20} = \frac{1}{5}$ d $\frac{8}{20} = \frac{2}{5}$
e $\frac{1}{24}$ f $\frac{6}{8} = \frac{3}{4}$

7 a $\frac{20}{100} = \frac{1}{5}$ b $\frac{5}{60} = \frac{1}{12}$ c $\frac{5}{14}$ d $\frac{30}{100} = \frac{3}{10}$
e $\frac{750}{1000} = \frac{3}{4}$ f $\frac{12}{1000} = \frac{3}{250}$

8 a $\frac{175}{500} = \frac{7}{20}$ b $\frac{150}{500} = \frac{3}{10}$ c $\frac{175}{500} = \frac{7}{20}$

Exercise 2C

1 a 9 b 162 c 270

2 a $\frac{8}{35}$ b $\frac{24}{325}$ c $\frac{1}{25}$

3 a $1\frac{7}{18}$ b $\frac{5}{44}$ c $1\frac{271}{420}$

4 a $9\frac{17}{36}$ b $2\frac{7}{8}$ c $18\frac{1}{20}$

5 a 57 600 m b 57.6 km c 36 miles

6 $2\frac{3}{16}$

7 27 min 30 sec

8 None of them. They all give an answer of 21.

9 a $119\frac{7}{22}$ kg
b Answer is almost 120, which is the number of litres. The conversions are all approximations. If they had been exact the answer would have been 120 kg, showing that 1 litre of water weighs 1 kg.

Exercise 2D

1 a £66 b 4.56 m c 100 cm
d 2.025 kg e 2500 f £67.50

2 £437.50
3 £480
4 156
5 2.88 g
6 52
7 £58.90
8 Mike's Bikes by £5.25

Exercise 2E

1 a 10% b 6.25% c $33.\dot{3}\%$
d $16.\dot{6}\%$ e 60% f 12.5%
g 25% h 14.5% i 27.5%
2 5.4% (1 d.p.)
3 62.5%
4 $13.\dot{3}\%$
5 a 27.0% (1 d.p.) b 52.4% (1 d.p.)
6 19.6%
7 16.6%
8 66.0% (1 d.p.)
9 35%
10 8.1% (1 d.p.)

Exercise 2F

1 a 21 b 62.72 c 2822 m*l*
d 326.4 e £145.13 f 815.24 m*l*
2 £436.80
3 329
4 £4816.50
5 £922.35
6 Jane
7 a 80.64 kg b 82.25 kg (2 d.p.)
8 128 960 or 129 000

Exercise 2G

1 DVD: a £22.75 b £152.75
Phone: a £7.00 b £47
2 £148.05
3 £3800 + 17.5% VAT = £4465, so second car is more expensive.
4 £21.62
5 £44.47

Exercise 2H

1 2007: 90.3p 2008: 95p 2009: 104.5p
2 a Down b Gone down by 26%
3 60
4 72p
5 Peter isn't correct. The price of bananas has dropped by $\frac{1}{4}$ so they are $\frac{3}{4}$ of the price they were in 1990.
6 583 333
7 a €1.35 to £1 b 88.1

Review exercise

1 £5850
2 £264
3 51
4 $1\frac{2}{23}$
5 59%
6 Jane £30.10, Peter £18.92
7 £2.63
8 £243.75
9 TVs R US, by £16.60.
10 66.666…%
11 16%
12 17.8%
13 25

Chapter 3

Exercise 3A

1

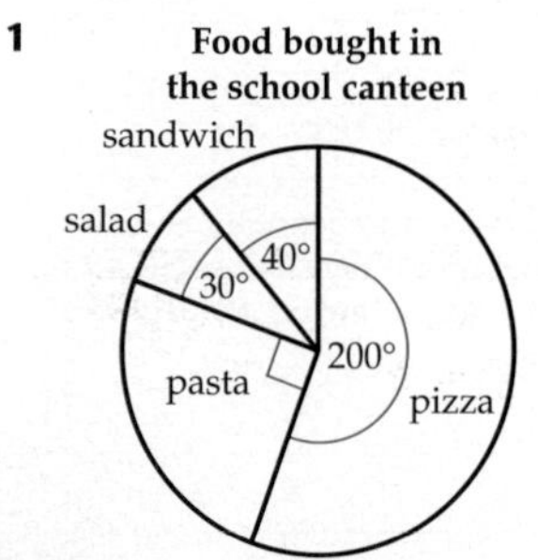

2

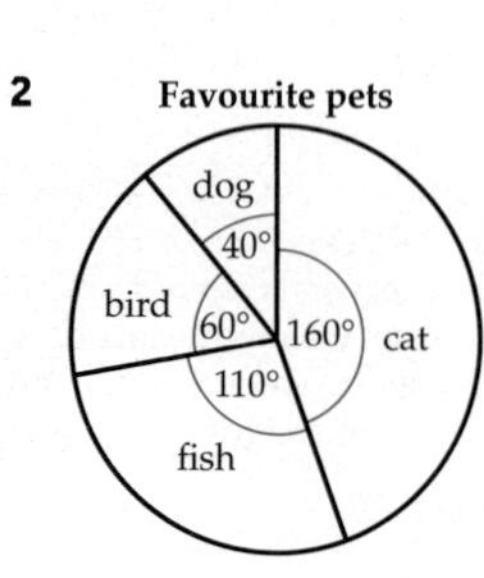

3

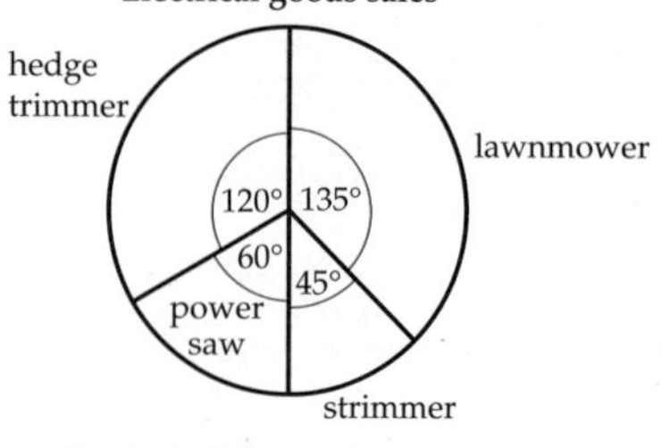

4 a Total of all the angles is less than 360°.
b

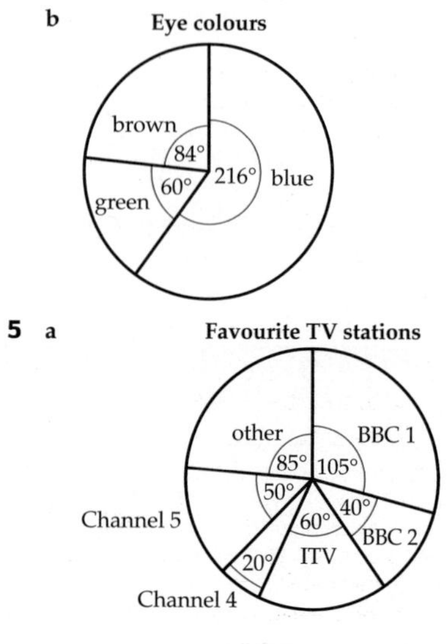

5 a

Exercise 3B

1
0 | 7 8 9 9
1 | 2 2 3 3 3 5 5 6 7 8 9 9 **Key** 1 | 6 means 16°C
2 | 0 0 1 1

2 a 18 CDs
b
2 | 7
3 | 6 6 9
4 | 5 6 7 7 9 **Key** 4 | 5 means 45 minutes
5 | 1 1 3 6 8 8 9
6 | 0 2

Exercise 3C

1 a
2 | 8 8 9 9
3 | 1 2 3 6 7 9 9 **Key** 3 | 6 means 3.6 seconds
4 | 0 1 3 6 7

b 9 girls

2 a
0 | 7 9
1 | 9
2 | 1 3 7 9
3 | 1 2 5 6 **Key** 1 | 9 means 1.9 cm
4 | 2 3 3
5 | 6 7

b 11 seedlings

3 a
2 | 61 93
3 | 62 75 81 **Key** 2 | 93 means £2.93
4 | 63 70 86 91
5 | 23 25 27

b 2 people

Exercise 3D

1 a

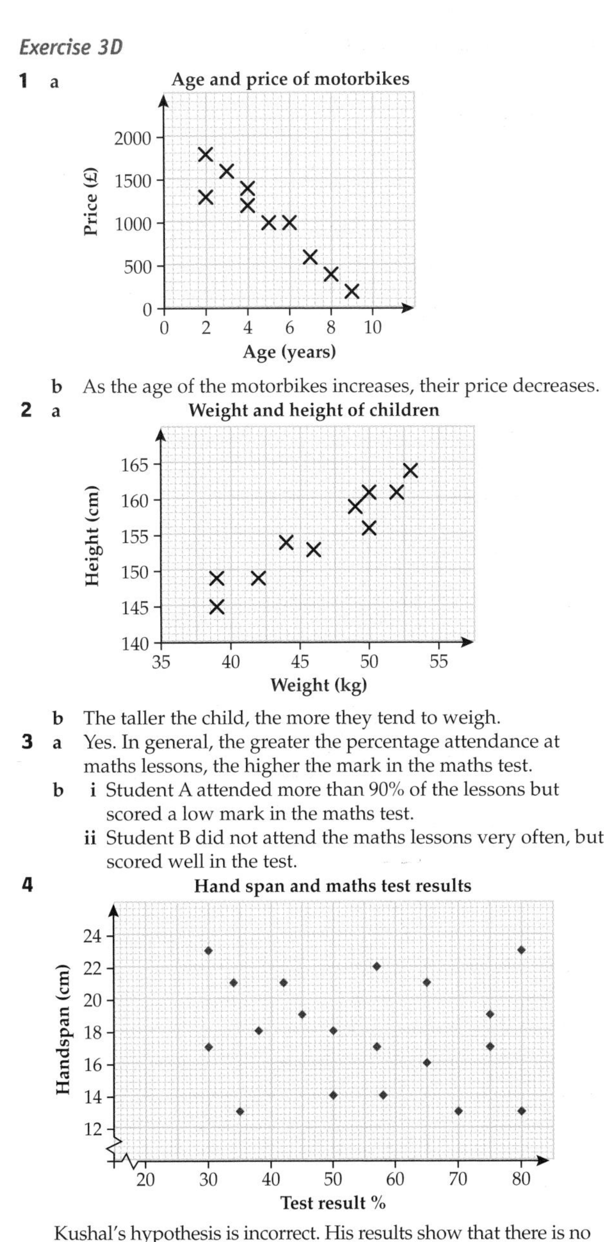

b As the age of the motorbikes increases, their price decreases.

2 a Weight and height of children

b The taller the child, the more they tend to weigh.

3 a Yes. In general, the greater the percentage attendance at maths lessons, the higher the mark in the maths test.

b **i** Student A attended more than 90% of the lessons but scored a low mark in the maths test.

ii Student B did not attend the maths lessons very often, but scored well in the test.

4 Hand span and maths test results

Kushal's hypothesis is incorrect. His results show that there is no connection between the students' test results and their hand spans.

Exercise 3E

1 a Positive **b** B (Arm span)

2 a No correlation **b** Positive correlation

c Negative correlation **d** Positive correlation

3 a and c

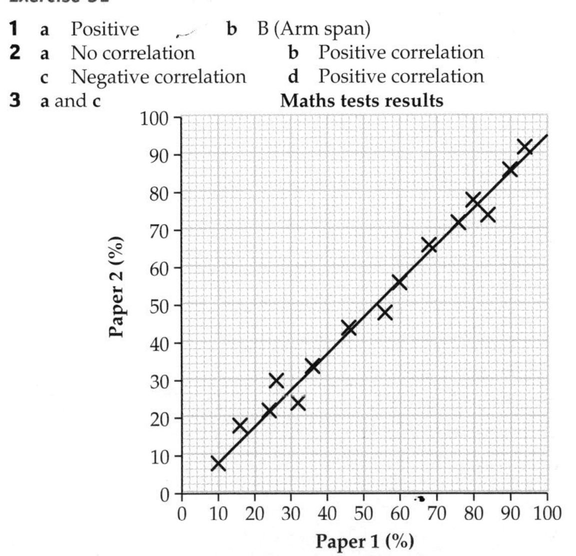

b Positive correlation **d** **i** 60% **ii** 89%

4 a

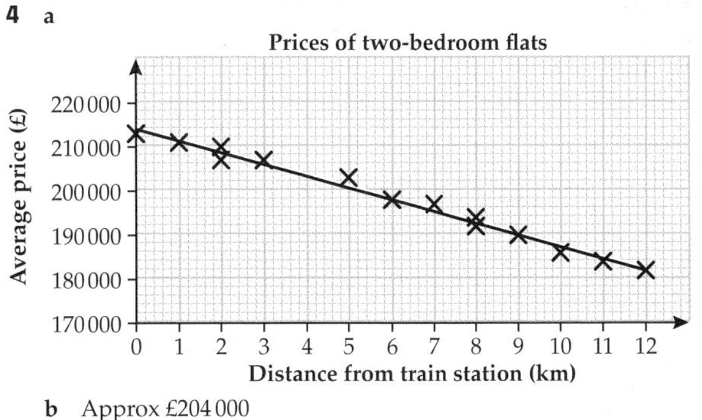

b Approx £204 000

c Negative correlation. The further away from the mainline train station, the cheaper the price of a two-bedroom flat.

d Approx over 7 km

5 a Negative correlation. The older the car, the cheaper the price.

b **i** Approx £5300 **ii** Approx 7 years old

c The line of best fit would indicate a 12-year-old car is worth nothing, which is not correct. It is usually worth at least scrap value.

6 a and b

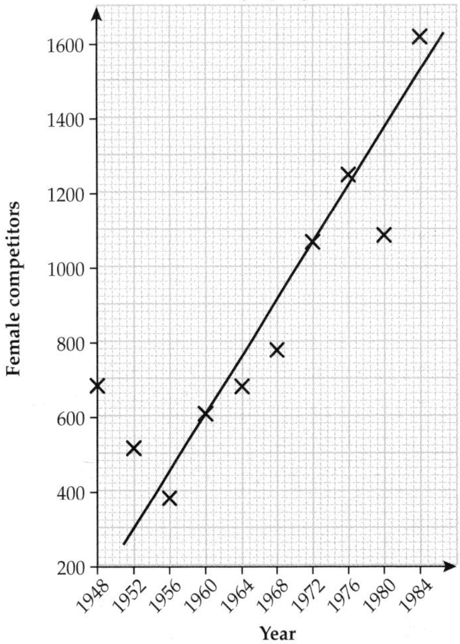

c Weak positive correlation

Exercise 3F

1

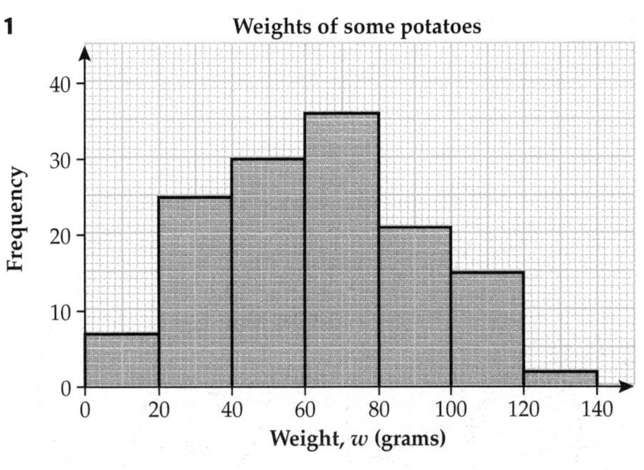

2 a

b It will have varied response. Approximately 30% of the leisure centre users are aged 15–30 – the discounted night club tickets might appeal to them. There would probably be limited response from the remaining 70% of leisure centre users, who are younger or older.

3 a

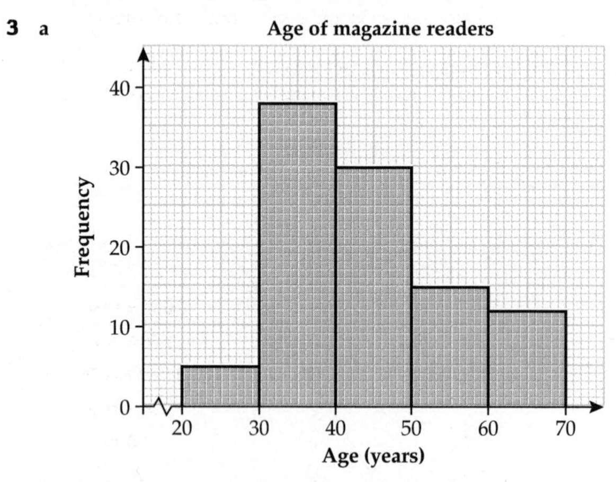

b A large proportion of the magazine's readers are over 30. An article on *High School Musical* would not be appropriate as it is predominantly aimed at a younger audience.

Exercise 3G

1 a Mid-points are: 7.5, 12.5, 17.5, 22.5, 27.5

b

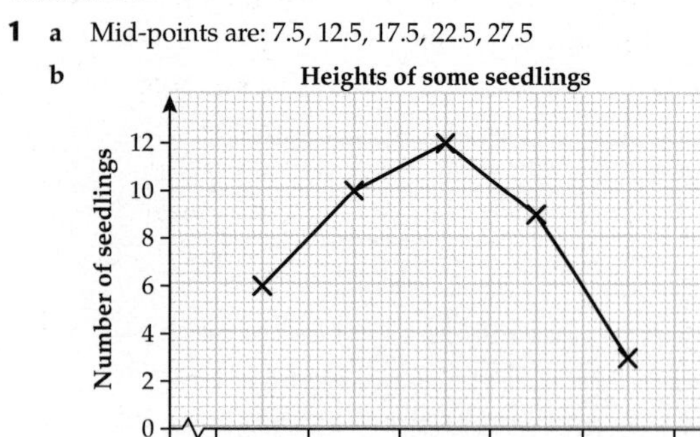

2 a Mid-points are: 35, 45, 55, 65, 75

b

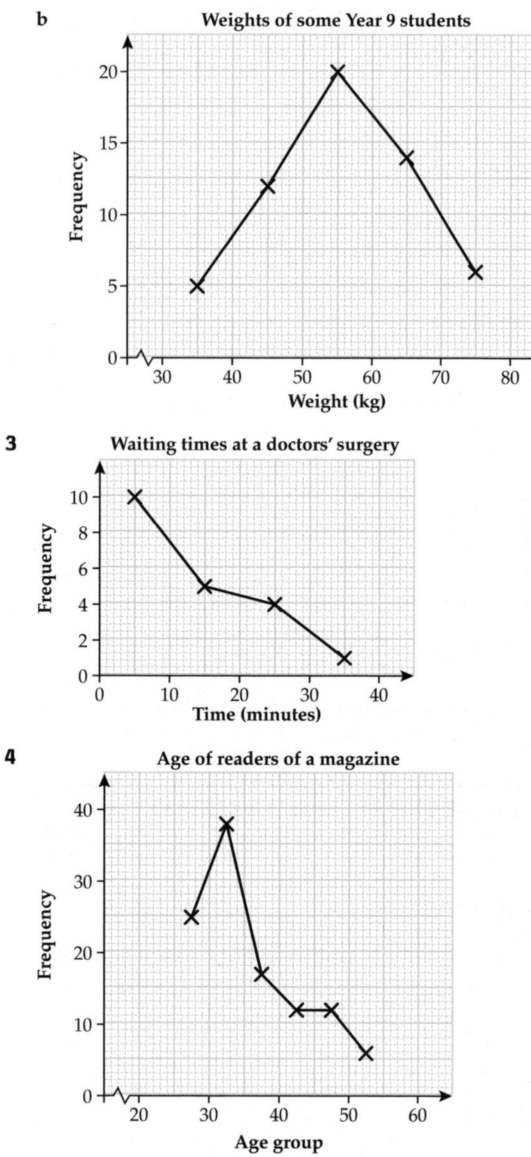

5 The girls are generally taller than the boys. There are more tall girls than tall boys, since there are 4 girls in the 160–165 cm class interval compared to 1 boy. There are also more short boys than short girls, since there are 3 boys in the 135–140 cm class interval, compared to 1 girl.

Review exercise

1 **Sophie's costs at the theme park**

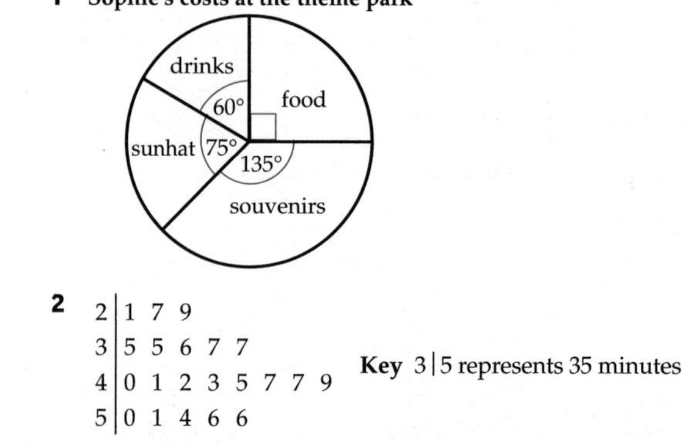

2

2	1 7 9
3	5 5 6 7 7
4	0 1 2 3 5 7 7 9
5	0 1 4 6 6

Key 3 | 5 represents 35 minutes

3 a 17 students

b

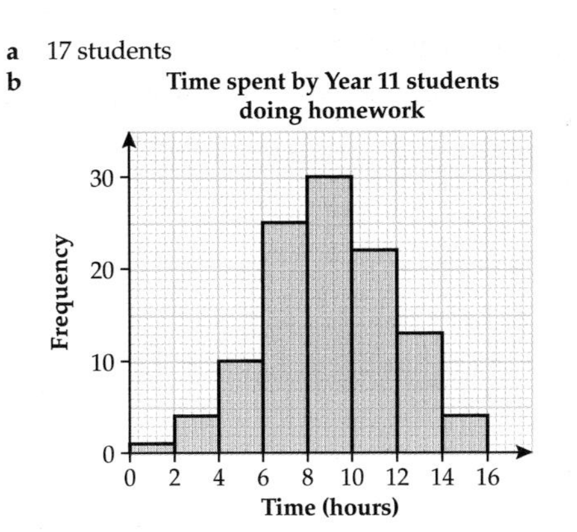

4 a and b

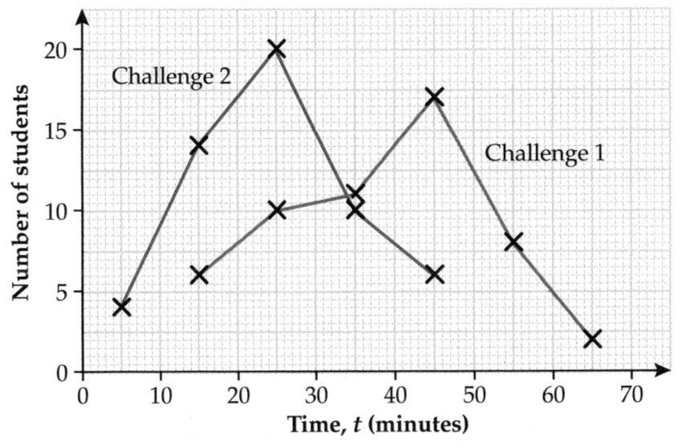

c The students completed Challenge 2 more quickly than Challenge 1. All students had completed Challenge 2 within 50 minutes, whereas for Challenge 1, ten students took 50 minutes or longer.

5 a Positive correlation

b C (Waist measurement)

6 a and c

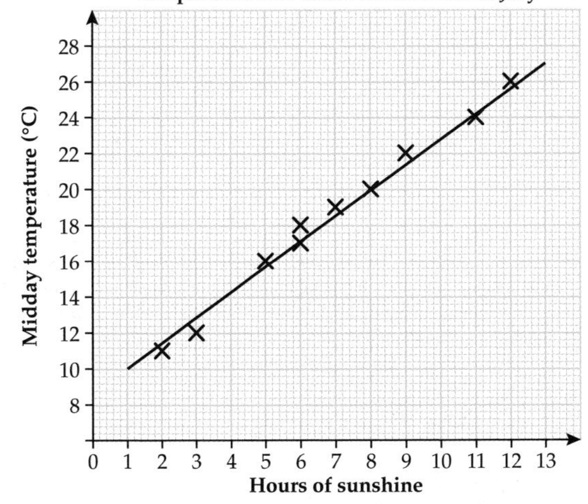

b Strong positive correlation

d 8.6–8.8 hours

e The value is outside the range of the given data.

Chapter 4

Exercise 4A

1 a £250 b £580 c £565

2 a 9 m b 9.5 m c 2.5 m

3 a 1.6 s b 45.0 s c 44.8 s

4 £51.82

5 a 28 b 6

6 2

7 e.g. 5, 6, 7, 8, 9

8 e.g. 6.6, 6.7, 6.9, 7.0

9 4 letters

10 22

11 a 5.0 cm b 8.3 cm

12 a

0	8
1	1 2 2 5 7 8 9
2	0 0 0

Key 1 | 1 means 11 marks

b 12 c 17 d 16

13 a

0	7
1	0 3 5 9 9
2	0 1 2

Key 1 | 0 means 10 strawberries

b 19 c 16

14 95 and 155

15

	Maths	English	Science
William	65%	80%	50%
Laura	35%	95%	65%

Exercise 4B

1 a 4 goals b 1 goal c 1 goal

2 a 4 b 2 c 2

3 a 5 minutes b 3 minutes

c e.g. 75 people surveyed, median is 38th value, median is 4 minutes.

4 a No. Don't know the largest value

b e.g. 33 people surveyed, median is 17th value, median is 2 hours.

Exercise 4C

1 a 25

b

Number of people in a queue	Frequency	Number of people × Frequency
0	4	**0 × 4 = 0**
1	6	**1 × 6 = 6**
2	13	**2 × 13 = 26**
3	2	**3 × 2 = 6**
4	0	**4 × 0 = 0**
Total	**25**	**38**

c 1.52

2

Number of buses	Frequency	Number of buses × Frequency
0	8	**0 × 8 = 0**
1	17	**1 × 17 = 17**
2	12	**2 × 12 = 24**
3	14	**3 × 14 = 42**
4	5	**4 × 5 = 20**
5	4	**5 × 4 = 20**
Total	**60**	**123**

Mean = 2.1

3 a 116 b 214 c 7

d 2.48 e Year 8

Exercise 4D

1 a $55 \leqslant w < 60$ b 20 g c 29

d $55 \leqslant w < 60$

2 a $200 \leqslant t < 300$ b 400 ms c $200 \leqslant t < 300$

3 a 4999 b 3000–3999 c 3000–3999

d i Unchanged

ii New estimated range = 5999

iii Unchanged

e Yes. You don't know the exact data values so you can't calculate the mean exactly.

Exercise 4E

1 a 7.5
b

Time taken (t minutes)	Frequency	Mid-point	Mid-point × frequency
$0 \leqslant h < 5$	3	2.5	$2.5 \times 3 = 7.5$
$5 \leqslant h < 10$	15	**7.5**	**7.5 × 15 = 112.5**
$10 \leqslant h < 15$	8	**12.5**	**12.5 × 8 = 100**
$15 \leqslant h < 20$	2	**17.5**	**17.5 × 2 = 35**
$20 \leqslant h < 25$	5	**22.5**	**22.5 × 5 = 112.5**
Total	**33**		**367.5**

c 11.1 minutes

2

Number of log-ins	Frequency	Mid-point	Mid-point × frequency
0–4	22	**2**	**44**
5–9	31	7	**217**
10–14	17	**12**	**204**
15–19	20	**17**	**340**
20–24	6	**22**	**132**
Total	**96**		**937**

Mean = 9.8 log-ins

3 a No. He does not know the exact data values.
b

Distance thrown, d (metres)	Girls' frequency	Mid-point	Mid-point × frequency (Girls)
$0 \leqslant d < 8$	2	4	8
$8 \leqslant d < 16$	12	12	144
$16 \leqslant d < 24$	5	20	100
$24 \leqslant d < 32$	3	28	84
$32 \leqslant d < 40$	7	36	252
$40 \leqslant d < 48$	0	44	0
Total	29		588

Estimate for girls' mean distance = 20.3 m
c Estimate for whole class mean distance = 19.7 m
d No. The estimate for the girls' mean is greater than the estimated mean for the whole class.

Review exercise

1 a 0.25 seconds b 0.23 seconds c 0.26 seconds
2 a 23 cm b 152 cm c 154 cm
3 e.g. 5, 5, 7, 8, 10
4 a 3 b 0 c 1
d

Number of visits to the doctor	Frequency	Number of visits × frequency
0	18	**0**
1	17	**17**
2	6	**12**
3	3	**9**
4	0	**0**
Total	**44**	**38**

Mean = 0.9 visits

5 a 4 b 4 c 4
d

Number of players	Frequency	Number of players × frequency
2	2	4
3	4	12
4	7	28
5	1	5
6	1	6
Total	15	55

Mean = 3.7 players

6 a 5 cm b $6 \leqslant l < 7$ c $6 \leqslant l < 7$
d

Length (l cm)	Frequency	Mid-point	Mid-point × frequency
$3 \leqslant l < 4$	2	**3.5**	**7**
$4 \leqslant l < 5$	6	**4.5**	**27**
$5 \leqslant l < 6$	11	**5.5**	**60.5**
$6 \leqslant l < 7$	12	**6.5**	**78**
$7 \leqslant l < 8$	9	**7.5**	**67.5**
Total	**40**		**240**

Estimate for mean = 6 cm

7 a 24 b 5–9 c 5–9 d 10.0
e e.g. The data is presented in a grouped frequency table so you do not know the exact data values.
8 a $5 \leqslant t < 10$ b £6.50
c e.g. The data is presented in a grouped frequency table so you do not know the exact data values.
9 a, b e.g.

Fuel consumption, f (mpg)	Frequency	Mid-point	Mid-point × frequency
$10 \leqslant f < 20$	3	15	45
$20 \leqslant f < 30$	0	25	0
$30 \leqslant f < 40$	0	35	0
$40 \leqslant f < 50$	10	45	450
$50 \leqslant f < 60$	13	55	715
$60 \leqslant f < 70$	4	65	260
Total	30		1470

e.g. 49 mpg
c 48.98 mpg
d e.g. The smaller the class intervals the more accurate the estimate.
10 a 14.5 b 15.4
11 2.35

Chapter 5

Exercise 5A

1 a $\frac{1}{10}$ b $\frac{9}{10}$
2 a $\frac{12}{13}$ b $\frac{3}{13}$ c $\frac{10}{13}$
3 0.95
4 a $\frac{97}{100}$ b $\frac{991}{1000}$
5 a $\frac{3}{10}$, 0.3, 30% b $\frac{15}{100} = \frac{3}{20}$, 0.15, 15%
6 a 72% b 1%
7 a 0.7 b 0.75 c 1
d The section numbered 5 isn't on the edge of the spinner, e.g.

1	2
4	3

(with 5 in a central square)

8 The probability of not getting a 6 with this dice is $1 - \frac{1}{3} = \frac{2}{3}$
9 a $\frac{1}{5}$ b 4
10 6

Exercise 5B

1 a $\frac{13}{15}$ b $\frac{9}{15} = \frac{3}{5}$ c $\frac{10}{15} = \frac{2}{3}$ d $\frac{9}{15} = \frac{3}{5}$
e $\frac{8}{15}$ f $\frac{7}{15}$ g $\frac{4}{15}$
2 a 0.55 b 0.2
3 0.3
4 a $\frac{1}{3}$ b 15 red, 12 blue, 9 yellow
5 12

Exercise 5C

1 a 40 b 15 c 7
d i $\frac{15}{40} = \frac{3}{8}$ ii $\frac{2}{40} = \frac{1}{20}$ iii $\frac{10}{40} = \frac{1}{4}$

2 a $\frac{62}{200} = \frac{31}{100}$ b $\frac{58}{200} = \frac{29}{100}$ c $\frac{80}{200} = \frac{2}{5}$

3 a $\frac{6}{50} = \frac{3}{25}$ b $\frac{6}{50} = \frac{3}{25}$ c $\frac{8}{50} = \frac{4}{25}$
d $\frac{27}{50}$ e $\frac{44}{50} = \frac{22}{25}$ f 120 g 520

4 a $\frac{30}{50} = \frac{3}{5}$ b $\frac{3}{24} = \frac{1}{8}$ c $\frac{17}{20}$

5 a $\frac{27}{50}$ b $\frac{28}{50} = \frac{14}{25}$ c $\frac{13}{50}$

6

	Pass	Fail
Male	15	15
Female	16	14

7 a $\frac{22}{30} = \frac{11}{15}$ b $\frac{12}{15} = \frac{4}{5}$ c $\frac{8}{45}$

Exercise 5D

1 250
2 10
3 a 260 b 130 c 40 d 10
4 a 2 b £40 c £20
5 250
6 a 5 b £50 c £80
d No as he is paying out more than he is likely to win.
7 a 15 b 150 c 225 d 0
e 135 f 75
8 £100

Exercise 5E

1 a Relative frequency (blue): 0.75, 0.58, 0.56, 0.62, 0.682, 0.64, 0.631, 0.664, 0.658
Relative frequency (red): 0.25, 0.42, 0.44, 0.38, 0.318, 0.36, 0.369, 0.336, 0.342
b i Approx 0.66 ii Approx 0.34
c Approx 33 blue and 17 red
d Dual line graph showing data from table

2 a $\frac{47}{200} = 0.235$ b 54 050

3 a

Relative frequency	0.05	0.22	0.14	0.16	0.16	0.168

b $0.1\dot{6}$
c No, after 500 throws the relative frequency is very close to the theoretical probability.
d 200

4 a

Score on the dice	1	2	3	4	5	6
Relative frequency	0.175	0.11	0.125	0.135	0.255	0.2

b $0.1\dot{6}$
c No, the dice is biased as the relative frequency of rolling a 5 is much higher than the theoretical probability and the relative frequency of rolling a 2 is much lower.

5 Zoe. She has spun the spinner 200 times, whereas George has only spun the spinner 40 times, so her results are more reliable. Zoe's relative frequencies are all close to the theoretical probabilities of 0.25, so the spinner seems to be fair.

George's results

Relative frequency	0.075	0.35	0.25	0.325

Zoe's results

Relative frequency	0.225	0.265	0.24	0.27

6 a

Relative frequency	0.35	0.15	0.15	0.15	0.2

b 1 as the relative frequency is higher than the other scores
c Spin the spinner more times

Exercise 5F

1 a $\frac{1}{36}$ b $\frac{1}{36}$

2 a $\frac{1}{6}$ b $\frac{1}{16}$ c $\frac{1}{12}$

3 $\frac{1}{40}$

4 a $\frac{1}{1369}$ b $\frac{144}{1369}$

5 a $\frac{1}{30}$ b $\frac{1}{30}$ c $\frac{4}{30} = \frac{2}{15}$
d $\frac{3}{30} = \frac{1}{10}$ e $\frac{6}{30} = \frac{1}{5}$

Exercise 5G

1

1st card	2nd card	Outcomes
R	R	RR
	B	RB
B	R	BR
	B	BB

2

1st flip	2nd flip	Outcomes
H	H	HH
	T	HT
T	H	TH
	T	TT

3

Tom	Marcus	Outcomes
P	P	PP
	F	PF
F	P	FP
	F	FF

4

1st disc	2nd disc	Outcomes
R	R	RR
	B	RB
	Y	RY
B	R	BR
	B	BB
	Y	BY
Y	R	YR
	B	YB
	Y	YY

5

1st breakfast	2nd breakfast	Outcomes
C	C	CC
	T	CT
	P	CP
T	C	TC
	T	TT
	P	TP
P	C	PC
	T	PT
	P	PP

6

1st ball	2nd ball	3rd ball	Outcomes
B	B	B	BBB
		W	BBW
	W	B	BWB
		W	BWW
W	B	B	WBB
		W	WBW
	W	B	WWB
		W	WWW

Exercise 5H

1 a

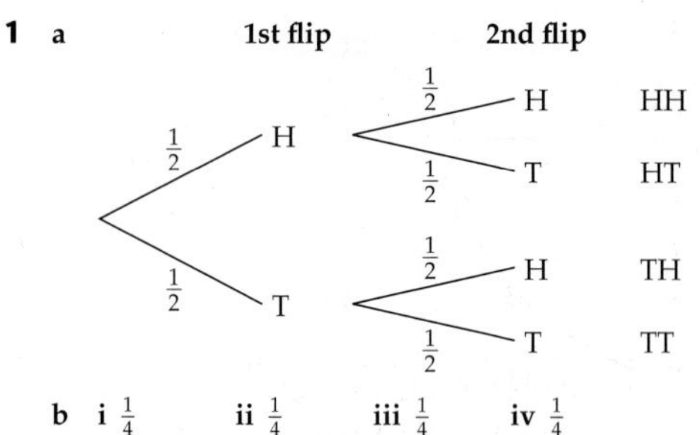

b **i** $\frac{1}{4}$ **ii** $\frac{1}{4}$ **iii** $\frac{1}{4}$ **iv** $\frac{1}{4}$

2 a

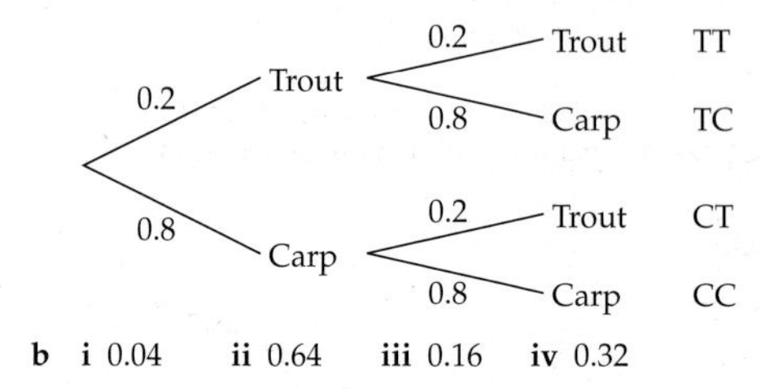

b **i** 0.04 **ii** 0.64 **iii** 0.16 **iv** 0.32

3 a $\frac{4}{52} = \frac{1}{13}$ **b** $\frac{48}{52} = \frac{12}{13}$

c **i** $\frac{1}{169}$ **ii** $\frac{144}{169}$ **iii** $\frac{12}{169}$

4 a

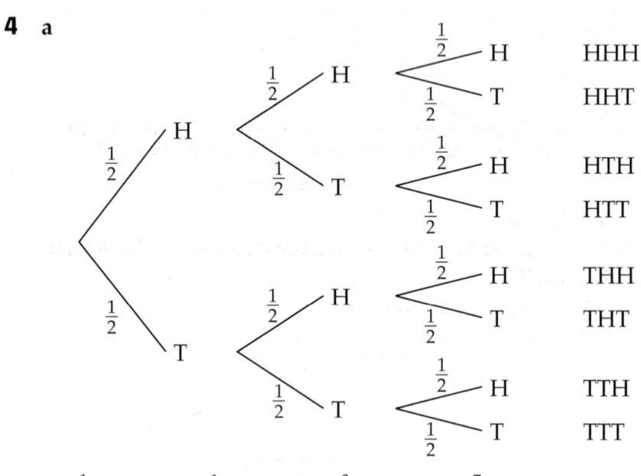

b $\frac{1}{8}$ **c** $\frac{1}{8}$ **d** $\frac{3}{8}$ **e** $\frac{7}{8}$

5 a

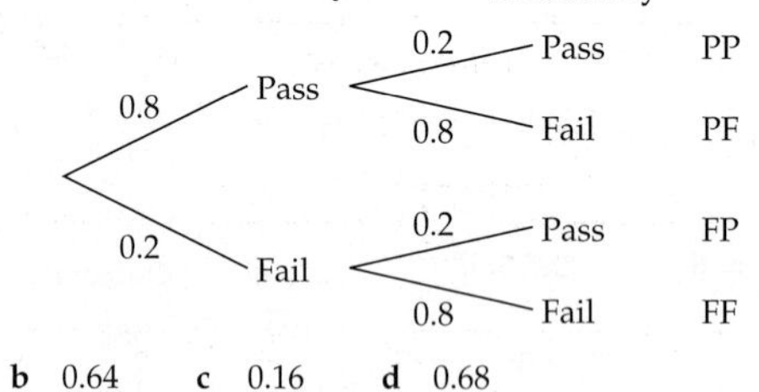

b 0.64 **c** 0.16 **d** 0.68

6 a

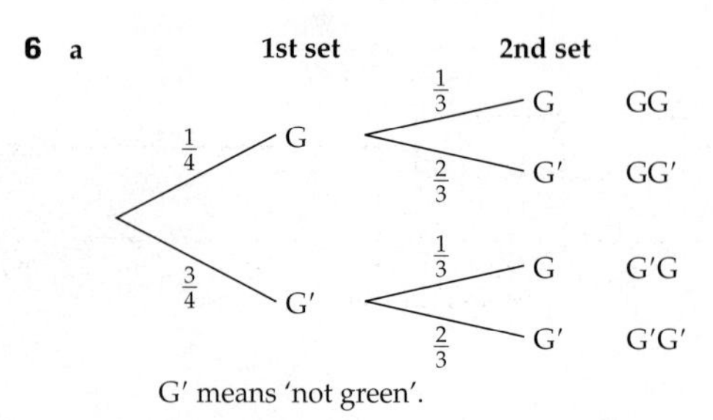

b $\frac{1}{2}$ **c** $\frac{5}{12}$ **d** $\frac{1}{12}$

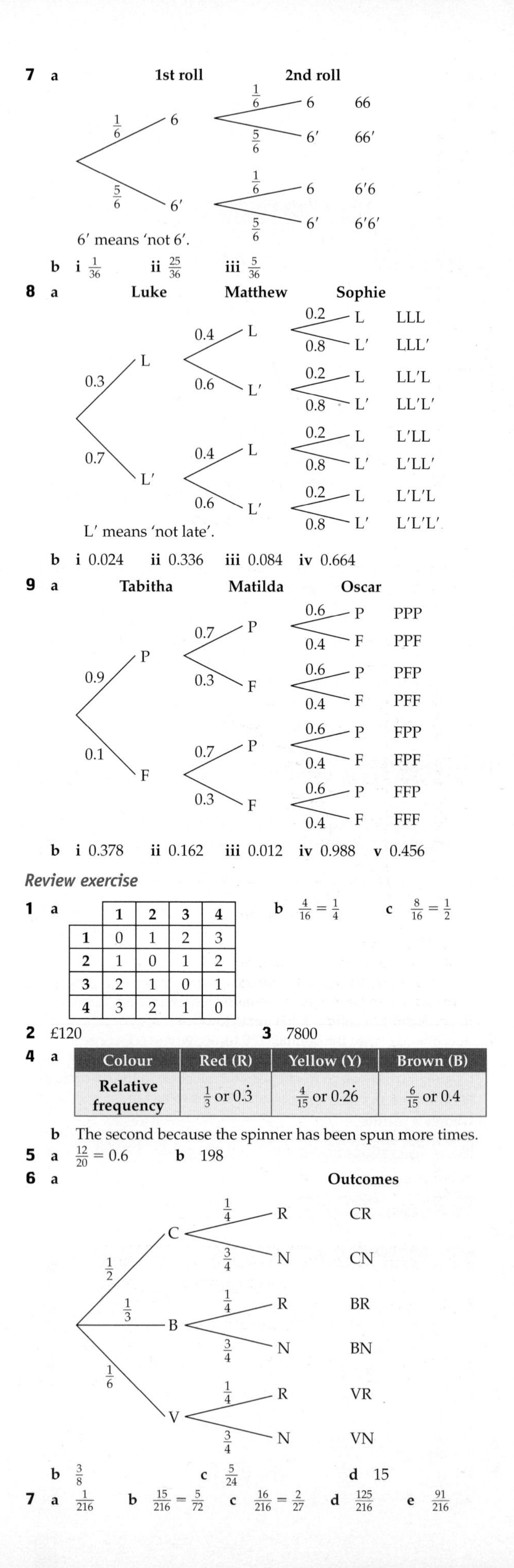

7 a

6′ means 'not 6'.

b **i** $\frac{1}{36}$ **ii** $\frac{25}{36}$ **iii** $\frac{5}{36}$

8 a

L′ means 'not late'.

b **i** 0.024 **ii** 0.336 **iii** 0.084 **iv** 0.664

9 a

b **i** 0.378 **ii** 0.162 **iii** 0.012 **iv** 0.988 **v** 0.456

Review exercise

1 a

	1	2	3	4
1	0	1	2	3
2	1	0	1	2
3	2	1	0	1
4	3	2	1	0

b $\frac{4}{16} = \frac{1}{4}$ **c** $\frac{8}{16} = \frac{1}{2}$

2 £120 **3** 7800

4 a

Colour	Red (R)	Yellow (Y)	Brown (B)
Relative frequency	$\frac{1}{3}$ or $0.\dot{3}$	$\frac{4}{15}$ or $0.2\dot{6}$	$\frac{6}{15}$ or 0.4

b The second because the spinner has been spun more times.

5 a $\frac{12}{20} = 0.6$ **b** 198

6 a

b $\frac{3}{8}$ **c** $\frac{5}{24}$ **d** 15

7 a $\frac{1}{216}$ **b** $\frac{15}{216} = \frac{5}{72}$ **c** $\frac{16}{216} = \frac{2}{27}$ **d** $\frac{125}{216}$ **e** $\frac{91}{216}$

Chapter 6

Exercise 6A

1 a

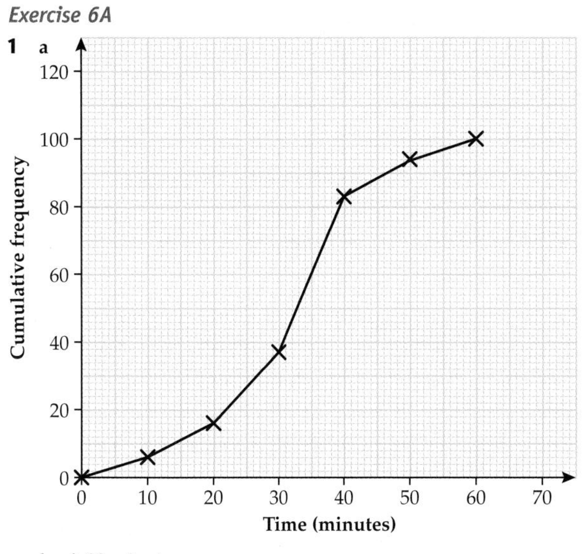

b i 33 minutes

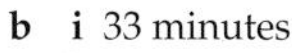

ii Lower quartile = 24 minutes, Upper quartile = 38 minutes

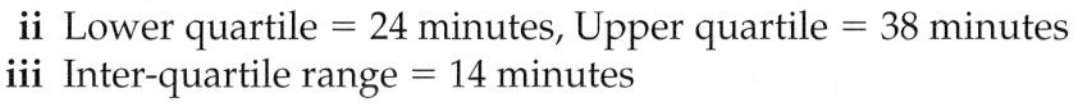

iii Inter-quartile range = 14 minutes

Exercise 6B

1 a

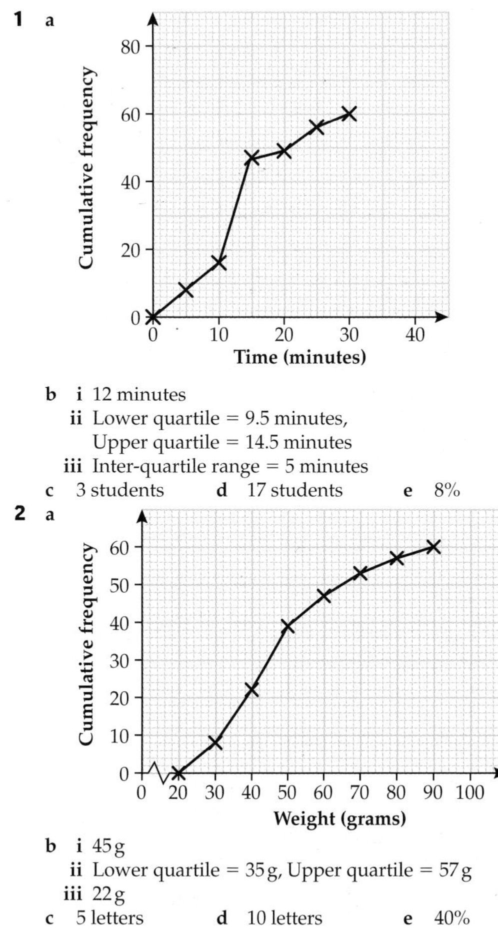

b i 12 minutes
ii Lower quartile = 9.5 minutes,
Upper quartile = 14.5 minutes
iii Inter-quartile range = 5 minutes

c 3 students d 17 students e 8%

2 a

b i 45 g
ii Lower quartile = 35 g, Upper quartile = 57 g
iii 22 g

c 5 letters d 10 letters e 40%

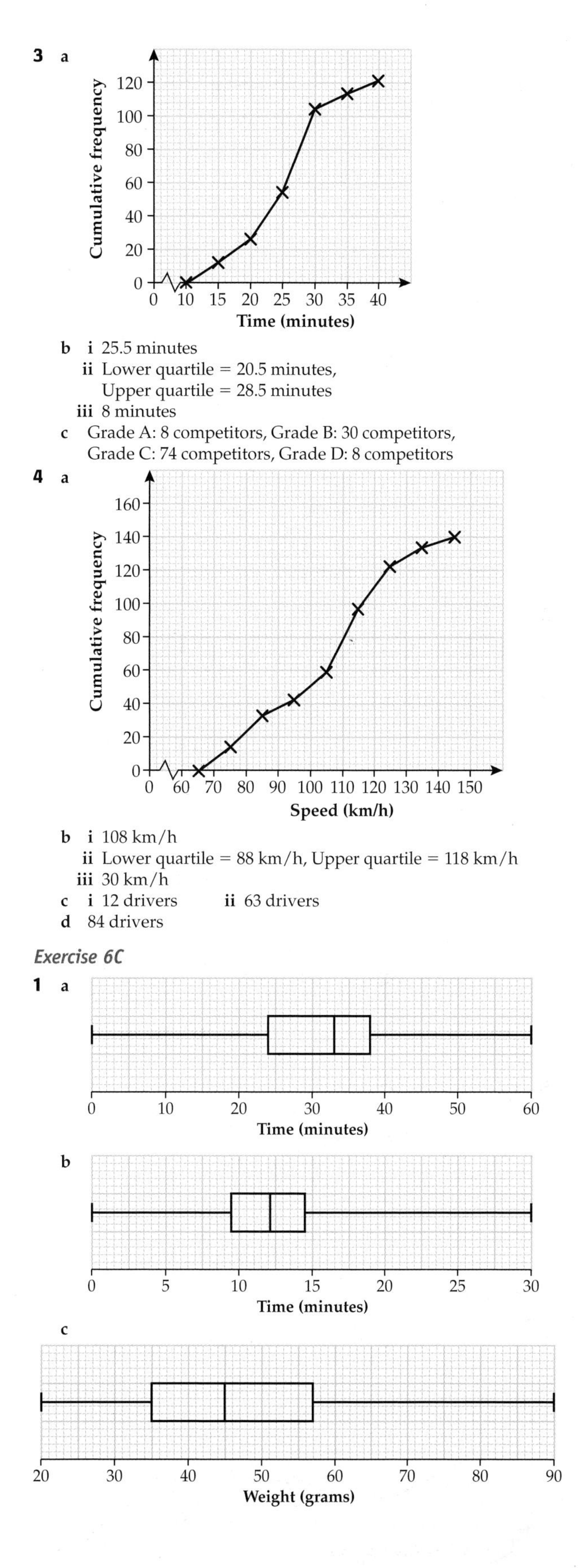

3 a

b i 25.5 minutes
ii Lower quartile = 20.5 minutes,
Upper quartile = 28.5 minutes
iii 8 minutes

c Grade A: 8 competitors, Grade B: 30 competitors,
Grade C: 74 competitors, Grade D: 8 competitors

4 a

b i 108 km/h
ii Lower quartile = 88 km/h, Upper quartile = 118 km/h
iii 30 km/h

c i 12 drivers ii 63 drivers

d 84 drivers

Exercise 6C

1 a

b

c

d

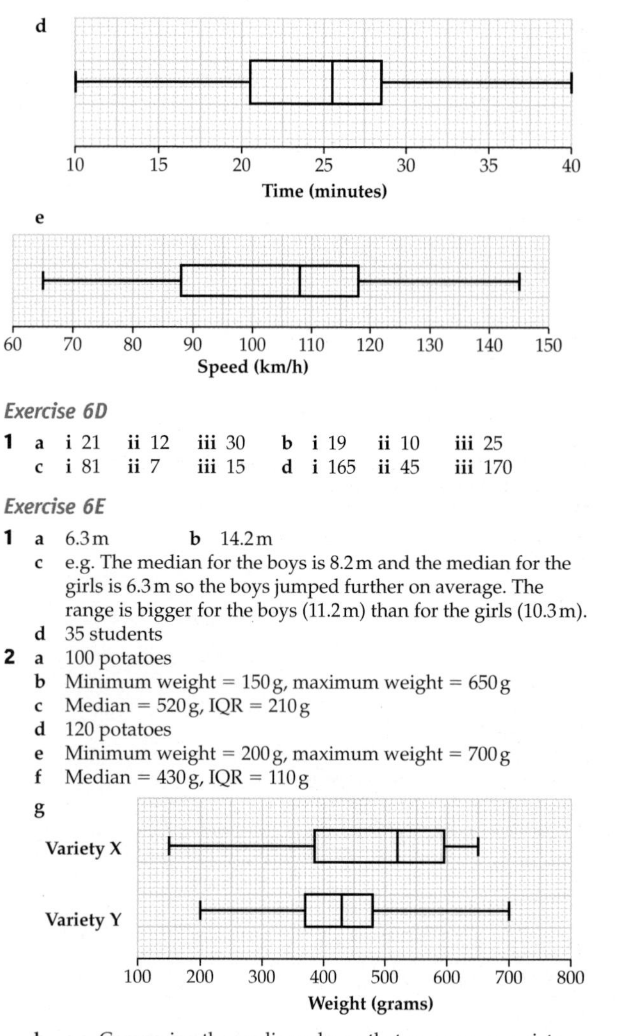

Exercise 6D

1 a i 21 ii 12 iii 30 b i 19 ii 10 iii 25
c i 81 ii 7 iii 15 d i 165 ii 45 iii 170

Exercise 6E

1 a 6.3 m b 14.2 m
c e.g. The median for the boys is 8.2 m and the median for the girls is 6.3 m so the boys jumped further on average. The range is bigger for the boys (11.2 m) than for the girls (10.3 m).
d 35 students

2 a 100 potatoes
b Minimum weight = 150 g, maximum weight = 650 g
c Median = 520 g, IQR = 210 g
d 120 potatoes
e Minimum weight = 200 g, maximum weight = 700 g
f Median = 430 g, IQR = 110 g
g

h e.g. Comparing the medians shows that on average variety X weighs more than variety Y. The range is the same for both varieties, but the IQR is smaller for variety Y. This means that weights of variety Y were more consistent.

Review exercise

1 a

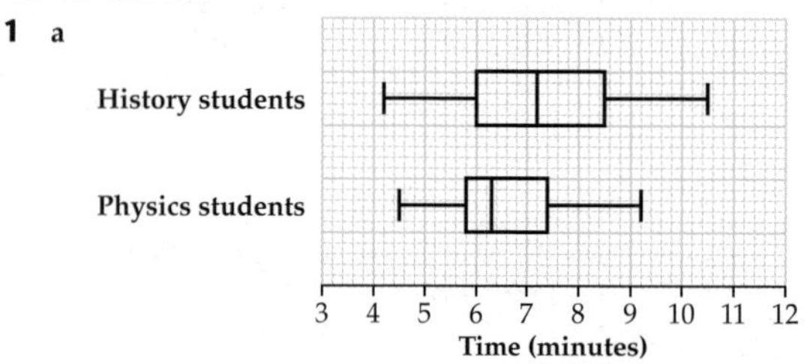

b e.g. Comparing the medians shows that on average the history students took longer than the physics students. The IQR and the range are smaller for the physics students, so their times were more consistent.

2 a e.g. The range and the IQR are smaller for A, so Joe is less likely to have extremes of temperature. The minimum temperature at A is not as cold as the minimum temperature at B.
b e.g. A never has a temperature above 30.5°C. If Joe wants a chance of warmer weather, he should go to B. The median temperature is higher for B, so on average B is warmer.

3 a

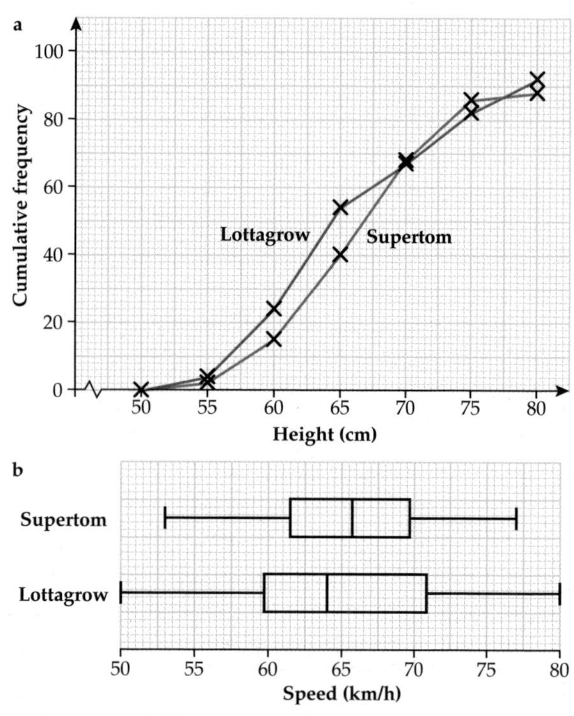

c Lottagrow: Median = 64 cm, IQR = 11 cm
Supertom: Median = 66 cm, IQR = 8 cm
d Supertom. Comparing the medians shows that on average Supertom treated plants grew taller than Lottagrow treated plants. The range and IQR are smaller for Supertom treated plants, so the heights of these plants are more consistent.

Chapter 7

Exercise 7A

1 a 2 : 1 b 1 : 3 c 2 : 3
d 5 : 3 e 2 : 3 f 3 : 4
2 a 2 : 5 b 1 : 2 c 2 : 1
d 25 : 1 e 5 : 1 f 1000 : 1
3 a 1 : 2 b 1 : 6 c 3 : 8
d 2 : 3 e 5 : 7 f 4 : 1
4 B and C
5 A, C and D
6 No. 45 : 50 = 9 : 10
7 5 : 6
8 9 : 2
9 8 : 15
10 5 : 3

Exercise 7B

1 a 1500 g flour, 6 eggs, 900 m*l* milk
b 250 g flour, 1 egg, 150 m*l* milk
c 750 g flour, 3 eggs, 450 m*l* milk
2 a 200 g b 50 g c 350 g
3 a 28 seconds b 33 copies
4 a 12 b 35
5 a 15 kg b 4 kg
6 45 g sugar, 180 m*l* milk, $\frac{3}{4}$ tablespoon custard powder
7 a 175 g onions, 87.5 g butter, 315 g cheddar cheese, 525 m*l* milk
b 7 people

Exercise 7C

1 a $\frac{4}{9}$ b $\frac{5}{9}$
2 a $\frac{2}{7}$ b $\frac{5}{7}$
3 No. The denominator is 5 + 6 = 11. The fraction of pedigree dogs is $\frac{5}{11}$.
4 £900
5 250 g
6 56

Exercise 7D

1 a 3 : 1 b 2 : 1 c 7 : 1 d 1.25 : 1
2 a 1 : 4 b 1 : 3.75 c 1 : 1.71 or 1 : $\frac{12}{7}$
d 1 : 1.5 e 1 : $\frac{1}{3}$ f 1 : $\frac{5}{8}$
3 a 10.90 : 1 b 1.41 : 1
c Necklace A, as for every gram of silver it has 10.9 g of gold
4 1st bar, proportion of copper is $\frac{22}{100} = 0.22$
2nd bar, proportion of copper is $\frac{4}{20} = 0.25$
2nd bar has higher proportion of copper

Exercise 7E

1 a £24, £12 b £9, £3 c 20 litres, 12 litres
d 24 kg, 42 kg e £200, £200, £100 f 100 cm, 60 cm, 90 cm
2 a 3 : 7 b Dan £30 000, Tris £70 000
3 Alix £12 000, Liberty £15 000
4 Pip 50 g, Wilf 100 g, Anna 150 g
5 Red 4 litres, blue 6 litres, white 8 litres
6 Ruby £11 628, Sid £16 279, Fred £22 093

Exercise 7F

1 a 12 b 18 c 27
2 a 6 b 8 c 11
3 a 3 eggs : 150 g flour b 1 egg : 50 g flour
c 600 g d 18 eggs
4 a 88 : 12 b 22 : 3 c 7.5 kg
5 a 16 b 294
6 9 kg

Exercise 7G

1 5
2 14
3 25
4 11 children under 2 years need 4 adults
8 children 2 years old need 2 adults
32 children 3–7 need 4 adults
Total 10 adults

Exercise 7H

1 a £1.92 b 48p c £2.40
2 a £15 b £60
3 a £58.50 b 16
4 £63
5 £19
6 £10.80
7 3 kg
8 3.15 kg
9 £1912.50
10 7.5 bales
11 £7.98

Exercise 7I

1 77p
2 0.25 litres or 250 m*l*
3 a 0.468p b 0.49p
c Large pack as price per gram is lower
4 Small bottle is better buy. Small bottle is 1.75p per m*l*, large bottle is 1.76p per m*l*.
5 Juicy Orange
6 Bottle C

Exercise 7J

1 a 30 euros b US$64 c Aus$58.8
d 69 Polish zloty e 6200 Pakistani rupees
2 a £29.17 b £18.75 c £2.42
d £50 e £86.96 f £81.67
g £40.32 h £255.10 i £217.39
3 a Dale £24.49, Tomas £26.74 b Tomas
4 7000 Pakistani rupees
5 New York. The games console is £30 cheaper than in London (US$399.99 = £249.99).

Exercise 7K

1 75 minutes or 1 hour 15 minutes
2 1.2 days
3 a 9 b 4
4 a 12 hours b 2.4 hours or 2 hours 24 minutes
5 7 people
6 1 hour 47 minutes
7 4 cleaners
8 150 new tiles

Review exercise

1 175 m
2 9 : 11
3 Large packet
4 1200 km
5 30 minutes
6 a 15 b 12
7 £27
8 £1500
9 Al £5000, Deb £3000, Cat £2000
10 1050 mg
11 Camera is £12.50 cheaper in England (€75 = £62.50)
12 a 3.92 litres b 3 tins
13 10.5 hours
14 1667 to nearest whole number
15 77 machinists

Chapter 8

Exercise 8A

1 £146.02
2 £9344.26
3 4% annum for 5 years
4 7 years
5 a £5.20 b £2309.21
6 a £15 571.36 b £1376.64 c £138.95
7 a 12 years b 11 years

Exercise 8B

1 £380
2 £14 000
3 £114.89
4 167 mm
5 6 kg
6 £74.47
7 £4376.25
8 a 40 cm b 40.2 cm

Exercise 8C

1 a 3×10^6 b 7.4×10^3 c 3.2×10^4
d 6.035×10^5 e 1.08×10^2 f 6.8×10^1
g 6.505×10^2 h 9.99×10^1 i 5×10^0
j 2.04×10^0
2 a 1 b 0 c 3 d 4 e 6 f 5

Exercise 8D

1 a 5×10^{-4} b 6×10^{-3} c 4×10^{-1}
d 1.2×10^{-4} e 7.17×10^{-2} f 1.975×10^{-4}
g 9.009×10^{-1} h 1.0003×10^{-3}
2 a 10^{-2} b 10^{-3} c 10^{-5}
d 10^{-8} e 10^{-3} f 10^{-1}

Exercise 8E

1 a 50 000 b 3800 c 0.006
d 7 260 000 000 e 0.084 92 f 4 370 000
g 0.000 100 6 h 6238.7
2 a 148 800 000 b 0.01 m c 0.000 000 000 000 03
3 a 1.23×10^4 b 8×10^6 c 1.7×10^{-1}
d 2.5×10^{-5} e 1.8×10^7 f 1.25×10^{-1}
g 2.16×10^3 h 2×10^1

Exercise 8F

1 a 2.226×10^8 b 4.2×10^1 c 1.4×10^5 d 4×10^1
2 £1.037×10^{10}
3 $1.098066\ldots \times 10^{12}$ km^3
4 1.17×10^{-2} tonnes
5 6.087×10^{12} km^2
6 a 3.78×10^5 cm^2 b 5680 cm = 5.68×10^3 cm

7 2.5×10^{13} miles
8 1.25×10^{-7} cm

Review exercise

1 George
2 1717
3 3 years
4 £58.65
5 a £14 500 b £12 340.43
6 £56.50
7 a 1100 b 1133
8 a 1.26×10^7 b 5.04×10^{13} c 1.44×10^{14}
d 1.62×10^7 e 3.5×10^{-1}
9 1.1×10^{-1} g or 0.11 g

Chapter 9

Number skills: adding and subtracting

1 a 447 b 2462 c 427 d 564
2 £702
3 558
4 a 52 + 48 or 68 + 32 or 58 + 42 b 82 − 32
5 a 9 b 5.86 c 8.92 d 7.56
6 a 2.9 b 2.3 c 2.04 d 1.76
7 £7.82
8 a 5.92 b 0.87 c 21.26 d 6.425
9 a 1.39 b 0.97 c 2.07 d 7.1
10 156.4
11 99 miles
12 a 2 b −4 c −12 d −4
e 4 f −17
13 a −11 b 4 c 10 d −8
e 3 f −7
14 a 8 b 11 c 36

Number skills: multiplying and dividing

1 a 230 b 4500 c 82 000
d 52 600 e 3790
2 1000
3 a 27 b 739 c 4500
d 2 e 325
4 50
5 1380 seats
6 160 kg
7 a 78 b 161 c 720
d 2583 e 3210
8 756
9 a 35 b 46 c 5
d 3.5 e 4.6
10 a 10 b 100 c 10 d 1000
11 2300 boxes
12 8 hundredths
13 a 12 b 26 c 86 d 39
14 £37
15 a 46 boxes b 2 eggs
16 a 5 b 90 c 750
17 a True b False c False d True

Exercise 9A

1 a £986 b No; he needs £36 more.
2 No; he needs £164 more.
3 Yes (Area of land = 12 090 m². Fertiliser bags will cover an area of 12 420 m².)

Exercise 9B

1 17 tins
2 a 42 m² b 15 packs
3 16 packs
4 7 days

Exercise 9C

1 a Estimate B b Estimate C c Estimate B
d Estimate C
2 a 80 b 200 c 20 d 5
3 a 1600 b 1200 c 4 d 30
4 £180
5 a 8 b 5 c 50 d 20 e 100 f 30
6 a 50 b 100 c 15
7 a C. Both wall area and area covered by one litre have been rounded down.
b 100 litres
8 Estimate: $10 \times 5 = 50$ miles of fuel remaining. So Pepe does not have enough fuel.
9 a 200 b 3 c 0.4
10 a 6000 b 15 c 100

Exercise 9D

1 a −4 and 3 b −4 and 3 c −4, 3, 0 and −4, −3, 6
2 a −6 b −9 c −10 d −12 e 8
3 a −5 b −3 c −3 d −4 e 4
4 −16
5 a −6 b 6 c −9 d 6
6 a −3 b −8 c −3 d 10
e 45 f −5
7 −6
8 16, −32, 64
9 3 and −4 or −3 and 4
10 12 and −4
11 a 18 b 24 c −30 d −3
e 1 f −4
12 a 2 b 6 c 6 d −16
13 $a = -3, b = -2, c = 6$

Review exercise

1 a 1470 b 95 c 800 d −5
e −4 f 2
2 a −2 b −15 c 3
3 3, −1.5
4 20
5 a 18 b 10
6 a 792 students
b 31 classes of 25 pupils and 1 class of 17 pupils
7 a 15 coaches b £10 700
c Yes, a profit of £502 (Total cost to youth club £16 250. Money raised by youth club £16 752.)
8 Approximate answer is 15. Daniel is most likely to be correct.
9 6000

Chapter 10

Exercise 10A

1 e.g. 108, 117, 126
2 84
3 No. It is not divisible by 3.
4 a 20, 60, 70, 90 b 21, 35, 70
5 Tom is correct. 128 is a multiple of 8, but 130 is not a multiple of 8.
6 Yes. The number will still be divisible by 8 so it is still a multiple of 8.

Exercise 10B

1 a 6, 12, 18, 24, 30, 36, 42, 48, 54, 60
b 9, 18, 27, 36, 45, 54, 63, 72, 81, 90
c 18
2 a e.g. 6 and 12 b e.g. 20 and 40
c e.g. 10 and 20 d e.g. 9 and 18
3 a 30 b 40 c 10 d 30 e 60 f 60
4 e.g. $6 \times 8 = 48$ but LCM of 6 and 8 is 24.
5 24
6 84 cm

Exercise 10C

1 a 1, 2, 5, 10 b 1, 2, 3, 6, 9, 18
c 1, 2, 3, 4, 6, 8, 12, 24 d 1, 2, 3, 5, 6, 10, 15, 30
2 No. He has missed 1 and 12 from his list.
3 a 6 b 28 c 20 d 8
4 No. Some numbers have an odd number of factors, e.g. the factors of 9 are 1, 3, 9.

Exercise 10D

1 a $3 \times 5 = 15$ b $3 \times 7 = 21$ c $7 \times 9 = 63$ *or* $21 \times 3 = 63$ d $11 \times 11 = 121$ e $3 \times 11 = 33$ f $7 \times 13 = 91$
2 31, 37, 41, 43, 47
3 3 and 13 *or* 5 and 11
4 a 2, 3 b 2, 3, 5 c 2, 5, 7 d 2, 11
5 $13 + 7$ and $17 + 3$

Exercise 10E

1 a 1, 2, 3, 4, 6, 12 b 1, 2, 4, 8 c 4
2 a 5 b 2 c 3 d 4 e 2 f 8
3 e.g. 16 and 24
4 a 12 cm by 12 cm b 30

Exercise 10F

1 a 81 b 125 c 6 d 2
2 a 49 b 100 c 125 d 1000
3 a 6 b 7 c 11 d 8
4 a 5 b 1 c 10 d 3
6

x	1	2	3	4	5	6	7	8
x^2	**1**	**4**	9	**16**	**25**	**36**	**49**	**64**

x	9	10	11	12	13	14	15
x^2	**81**	100	**121**	**144**	**169**	**196**	**225**

6 54 stickers
7 a £405 b 11 paving slabs
8 36 cubes
9 $1 + 49 = 50$ and $25 + 25 = 50$
10 a -7 b -5 c -2 d -3
11 6 and -6
12 a 3 b 10 c 3 d 5
13 a 10 b 4 c 5 d 2
14 a 58 b 98 c 4 d 2
15 96

Exercise 10G

1 a 243 b 256 c 15 625 d 100 000 e 216 f 343
2 a 80 b 192 c 216 d 625 e 24 f 2
3 $1^8, \sqrt{225}, 2^4, 3^3, 2^2 \times 3^2$
4 a 1 b 7^4 c 2744
5 a 1 b 6 c 1 d 17 e 29 f 1
6 a $\frac{1}{4}$ b $\frac{1}{9}$ c $\frac{1}{7}$ d $\frac{1}{4}$ e $\frac{1}{4}$ f $\frac{1}{12}$
7 a i 320 ii 5120
b 80. This is the value of N when $t = 0$.

Exercise 10H

1 a
84
2 42
6 7
2 3

b $2^2 \times 3 \times 7$

2 a $2^2 \times 5$ b $3^2 \times 7$ c 2^6 d $3^2 \times 5$ e $2 \times 5 \times 11$ f 3^4
3 a $2^2 \times 3 \times 13$ b $2^2 \times 3^2 \times 5 \times 11$ c $2^2 \times 5^2 \times 7 \times 11$ d $2^5 \times 19$ e $3^4 \times 5^2$ f $2^2 \times 5 \times 7^2$
4 a e.g.
100
10 10
2 5 2 5

100
2 50
2 25
5 5

b No. The prime factors will always be the same. Student's own example.

Exercise 10I

1 a $2 \times 3^2 \times 5$ b $3 \times 5 \times 11$ c 15
2 a $2 \times 3 \times 7$ b $2 \times 3 \times 5$ c 210
3 a 8 b 8 c 9 d 25 e 16 f 20
4 6
5 a 180 b 96 c 135 d 225 e 240 f 42
6 630
7 504
8 £1
9 a 6
b Using prime factors: $\text{LCM} = 2 \times 3^2 \times 5 = 90$
Using the rule: $\text{LCM} = \frac{18 \times 30}{6} = 90$
c HCF of 16 and 40 is 8.
Using prime factors: $\text{LCM} = 2^4 \times 5 = 80$
Using the rule: $\text{LCM} = \frac{16 \times 40}{8} = 80$
10 7 cards

Exercise 10J

1 a 6^{10} b 4^{10} c 5^4 d 9^6
2 a 3^4 b 6^5 c 4^5 d 5^2
3 a 8^{11} b 2^{13} c 8^9 d 9^5
4 a 125 b 16 c 16 d 343
5 No. To divide powers of the same number you subtract the indices, so $7^{10} \div 7^2 = 7^8$.
6 a 10^8 b 10^6
7 a 5^{-7} b 3^{-1} c 7^{-4} d 9^{-4}
8 a $\frac{1}{16}$ b $\frac{1}{8}$ c $\frac{1}{9}$ d $\frac{1}{16}$
9 a 36 b 27 c 125 d 6

Exercise 10K

1 a 250 b 64 000 c 9 d 12.5 e 5400 f 800
2 a 12 b 0.25 c 0.125 d 125
3 $2^{11} \times 8^2 = 2^{11} \times 8 \times 8 = 2^{11} \times 2^3 \times 2^3 = 2^{17}$

Exercise 10L

1 a 3.4×10^4 b 2.6×10^3 c 7.4×10^2 d 2×10^5 e 6.03×10^6 f 2.26×10^1
2 a 6×10^{-4} b 3.2×10^{-3} c 3.09×10^{-3} d 4.45×10^{-1} e 1×10^{-2} f 9.8×10^{-6}
3 a 20 000 b 6200 c 0.004 54 d 0.207 e 79 000 000 f 0.000 004 551
4 3.2×10^{-2}, 8.66×10^{-1}, 9.31×10^{-1}, 8.04×10^3, 4.8×10^4, 1×10^7
5 5.88×10^7
6 Hydrogen, carbon, iron, tungsten, gold, plutonium
7 2.54×10^{-5} m

Exercise 10M

1 a 8×10^8 b 4.8×10^{10} c 2×10^{10} d 9×10^5 e 5×10^{-5} f 1.86×10^{-6}
2 a 2×10^5 b 4.2×10^3 c 5×10^4 d 2.1×10^{-5} e 2.5×10^{-9} f 7×10^{-5}
3 a 6.7×10^4 b 2.05×10^5 c 1.3×10^4 d 8.25×10^5 e 6.98×10^5 f 9.37×10^4
4 360 g
5 2×10^{-8} g

Review exercise

1 a 18, 21, 24 b 17, 19, 23 c 20, 25
2 a 1, 2, 3, 4, 6, 8, 12, 16, 24, 48 b 2, 3
3 $3 \times 17 = 51$
4 a 125 b 12 c 4
5 6
6 $2^2 \times 3 \times 5^2 \times 11$
7 a 16 b 270
8 a 4^6 b 3^{14} c 6^7
9 20 cm by 20 cm
10 25
11 a 1 b 0.5
12 a 1.86×10^{10} b 1.2×10^{-8}
13 3×10^{-6} kg
14 Yes, the files will all fit. Memory stick holds 1 000 000 kb.
Total size of all files $= 7.1 \times 10^2 + 8.34 \times 10^5 + 9.42 \times 10^4 + 2.41 \times 10^3 = 931\,320$ kb

Chapter 11

Algebra skills: writing simple expressions

1 a $3y$ b $\frac{z}{3}$ c $10n$ d $\frac{12}{x}$ e $\frac{p}{4}$ f $8m$
2 a $x + 2$ b $7x$ c $x - 5$ d $2x$
e $\frac{1}{2}x$ or $\frac{x}{2}$ f $3x - 2$ g $\frac{x}{4} + 5$
3 a £$4x$ b £$6y$ c £$4d$ d £$13t$

Algebra skills: simplifying expressions

1 a $6a$ b $2g$ c $6c$ d $12t$ e $10x$ f $16l$
2 a $3b$ b $5y$ c $4z$ d $6t$ e $4j$ f $2u$
g $3h$ h $5t$ i $2x$

Exercise 11A

1 a $5c + 7d$ b $5m + 4r$ c $7x + 4y$ d $12p + 5$
e $11j + 5$ f $8w + 10$
2 a $3x + 4y$ b $2a + 3b$ c $4k + 3m$ d $8p + 4$
e $2t + 3$ f $z + 10$
3 a $10a + 7b$ b $11q + 9r$ c $2k + 6l$ d $2v - w$
e $y + 3z$ f $c - 13d$
4 a $10x + 6y$ b $4g + h$ c $2a + 8b - c$
d $6j + 4k + 7l$ e $5d + 6e + 2f$ f $x - 5y + 4z$
5 a $9xy + 5x^2$ b $m^2 + 2m$ c $7ab + 4a$
d $8x^2 + 9x$ e $4t^2 + 5$ f $3xy + 6x^2 + 7x$
g $3ab + 7a - 7b$

Exercise 11B

1 a $12a$ b $4x + 4$ c $4x + 2y$ d $10x + 2$

2 a

$6a + 7b$	**$7a$**	$2a + 8b$
$a + 6b$	**$5a + 5b$**	**$9a + 4b$**
$8a + 2b$	**$3a + 10b$**	$4a + 3b$

b

$3a + 2b$	**$4a + 3b$**	$8a - 2b$
$10a - 3b$	$5a + b$	**$5b$**
$2a + 4b$	**$6a - b$**	**$7a$**

c

$7a + b + 2c$	**$6b - 3c$**	**$8a - b + 4c$**
$6a + 3c$	**$5a + 2b + c$**	**$4a + 4b - c$**
$2a + 5b - 2c$	$10a - 2b + 5c$	$3a + 3b$

Exercise 11C

1 a $10k$ b $18b$ c $12x$ d $20a$ e $14h$ f $12m$
g $6ab$ h $12cd$ i $42pq$ j $6gh$ k $5xy$ l $56jk$
m $12t^2$ n $42x^2$ o $30a^2$ p $4n^2$ q $24cd$ r $7x^2$

Exercise 11D

1 a $5p + 30$ b $3a + 15$ c $7k + 14$
d $4m + 36$ e $35 + 5f$ f $16 + 2q$
g $2a + 2b$ h $5x + 5y$ i $8g + 8h + 8i$
j $4u + 4v + 4w$ k $2a + 2b + 2c$ l $9d + 9e + 54$
2 a $2y - 16$ b $3x - 15$ c $6b - 24$
d $7d - 56$ e $14 - 2x$ f $32 - 4n$
g $5a - 5b$ h $2x - 2y$ i $28 + 7p - 7q$
j $8a - 8b + 48$ k $5c - 35 + 5d$ l $3u - 3v + 3w$
3 a $-6k - 8$ b $-6x - 18$
c $-15n - 5$ d $-12t - 20$
e $-12p + 3 = 3 - 12p$ f $-6x + 14 = 14 - 6x$
g $-6x + 18 = 18 - 6x$ h $-10x + 15 = 15 - 10x$
i $-24x + 32 = 32 - 24x$
4 a $6c + 18$ b $12m + 8$ c $20t + 15$
d $24y + 54$ e $12e + 4f$ f $10p + 2q$
g $6a - 3b$ h $18c - 12d$ i $2m - 8n$
j $14x + 7y - 21$ k $18a - 24b + 6c$ l $8u - 20v - 12w$
5 a $2x^2 + 6x + 4$ b $3x^2 + 15x - 18$ c $2a^2 - 2a + 4$
d $4y^2 - 12y - 40$
6 a $5(m + 3) = 5m + 15$ b $4(n + 7) = 4n + 28$
7 A and F, B and K, C and I, D and L, E and G, H and J

Exercise 11E

1 $b^2 + 4b$ **2** $a^2 + 5a$ **3** $k^2 - 6k$
4 $m^2 - 9m$ **5** $2a^2 + 3a$ **6** $4g^2 + g$
7 $2p^2 + pq$ **8** $t^2 + 5tw$ **9** $m^2 + 3mn$
10 $2x^2 - xy$ **11** $4r^2 - rt$ **12** $a^2 - 4ab$
13 $2t^2 + 10t$ **14** $3x^2 - 24x$ **15** $5k^2 + 5kl$
16 $6a^2 + 12a$ **17** $8g^2 + 2gh$ **18** $15p^2 - 10pq$
19 $6xy + 15xz$ **20** $12p^2 + 8pq$

Exercise 11F

1 a $5y + 22$ b $5k + 21$ c $2a + 18$ d $7t - 16$
e $7x + 40$ f $5x - 21$ g $4y - 1$ h $13x + 13$
i $3n + 25$ j $14x - 8$
2 $2(x + 5) + 3x = 2x + 10 + 3x = 5x + 10$
$5(x + 2) = 5x + 10$
3 $6(t - 5) + 6 = 6t - 30 + 6 = 6t - 24$
$6(t - 4) = 6t - 24$
4 a $2b - 5$ b $4m + 2$ c $2k - 14$ d $4p + 14$
e $2g - 8$ f $-4w - 5$
5 $9(x + 1) + 3(x + 2) = 9x + 9 + 3x + 6 = 12x + 15$
$3(4x + 5) = 12x + 15$
6 $2(4p + 1) - 4(p - 3) = 8p + 2 - 4p + 12 = 4p + 14$
$4(p + 3) + 2 = 4p + 12 + 2 = 4p + 14$

Exercise 11G

1 a $10, 2 \times 5$ $4t, 2 \times 2t$ $8x, 2 \times 4x$
b $9, 3 \times 3$ $3y^2, 3 \times y^2$
2 $5x, x \times 5$ $x^2, x \times x$ $wx, x \times w$
3 $12y$ and 6, $3 \times 4y$ and 3×2
$6q$ and 21, $3 \times 2q$ and 3×7
4 $4x^2$ and $2x$, $x \times 4x$ and $x \times 2$
xy and tx, $x \times y$ and $x \times t$
5 a $1, 3$ b $1, 2, 4$ c $1, 2$ d $1, m$ e $1, n$ f $1, 7$

Exercise 11H

1 $3(x + 5)$ **2** $5(a + 2)$ **3** $2(x - 6)$ **4** $4(m - 4)$
5 $4(t + 3)$ **6** $3(n + 6)$ **7** $2(b - 7)$ **8** $4(t - 5)$

Exercise 11I

1 a $5(p + 4)$ b $2(a + 6)$ c $3(y + 5)$
d $7(b + 3)$ e $4(q + 3)$ f $6(k + 4)$
g $5(a + 1)$ h $4(g + 2)$ i $3(m + 6)$
2 a $4(t - 3)$ b $3(x - 3)$ c $5(n - 4)$
d $2(b - 4)$ e $6(a - 3)$ f $7(k - 1)$
g $4(r - 4)$ h $6(g - 2)$ i $3(m - 4)$
3 a $y(y + 7)$ b $x(x + 5)$ c $t(t + 2)$
d $n(n + 1)$ e $x(x - 7)$ f $z(z - 2)$
g $p(p - 8)$ h $a(a - 1)$ i $m(3m - 1)$
4 a $2(3p + 2)$ b $2(2a + 5)$ c $2(2t - 3)$
d $4(2m - 3)$ e $5(2x + 3)$ f $3(2y - 3)$
g $4(a + 2b)$ h $5(2p + q)$ i $7(n - 2)$
5 A and G, B and L, C and J, D and I, E and H, F and K

Exercise 11J

1 $a^2 + 9a + 14$ **2** $x^2 + 4x + 3$ **3** $x^2 + 10x + 25$
4 $t^2 + 3t - 10$ **5** $x^2 + 3x - 28$ **6** $n^2 + 3n - 40$
7 $x^2 + x - 20$ **8** $p^2 - 16$

Exercise 11K

1 a $y^2 + 8y + 15$ b $q^2 + 13q + 42$ c $a^2 - 3a - 28$
d $m^2 - m - 56$ e $y^2 - 6y + 9$ f $d^2 + d - 20$
g $x^2 - 9$ h $h^2 - 11h + 24$ i $21 + 4z - z^2$
j $x^2 - 13x + 36$
2 a $x^2 - 16$ b $x^2 - 25$ c $x^2 - 4$
d $x^2 - 121$ e $x^2 - 9$ f $x^2 - 1$
g $x^2 - 81$ h $x^2 - a^2$ i $t^2 - x^2$
3 a $x^2 + 10x + 25$ b $x^2 + 12x + 36$ c $x^2 - 6x + 9$
4 a $x^2 + 2x + 1$ b $x^2 - 8x + 16$
c $x^2 - 10x + 25$ d $x^2 + 14x + 49$
e $x^2 - 16x + 64$ f $9 + 6x + x^2 = x^2 + 6x + 9$
g $4 + 4x + x^2 = x^2 + 4x + 4$ h $25 - 10x + x^2 = x^2 - 10x + 25$
i $x^2 + 2ax + a^2$
5 a $(x + \mathbf{6})^2 = x^2 + \mathbf{12}x + 36$ b $(x - \mathbf{7})^2 = x^2 - \mathbf{14}x + 49$
c $(x + \mathbf{9})^2 = x^2 + 18x + \mathbf{81}$ d $(x - \mathbf{10})^2 = x^2 - 20x + \mathbf{100}$

6 a

n	$n + 1$
$n + 10$	$n + 11$

b $(n + 1)(n + 10) = n^2 + 11n + 10$

c $n(n + 11) = n^2 + 11n$

d $n^2 + 11n + 10 - n^2 - 11n = 10$

Whatever the value of n, the difference is 10.

Exercise 11L

1 $3a^2 + 14a + 8$ **2** $5x^2 + 13x + 6$ **3** $6t^2 + 19t + 15$

4 $8y^2 + 30y + 7$ **5** $12x^2 + 28x + 15$ **6** $4x^2 - x - 3$

7 $6z^2 + 11z - 10$ **8** $7y^2 - y - 8$ **9** $3n^2 + 19n - 40$

10 $6b^2 - 7b - 5$ **11** $7p^2 - 25p - 12$ **12** $6z^2 - 17z + 12$

13 $10x^2 - 23x + 9$ **14** $4y^2 - 12y + 9$ **15** $12a^2 + 23a + 10$

16 $4x^2 - 1$ **17** $9y^2 - 4$ **18** $25n^2 - 16$

19 $9x^2 - 25$ **20** $1 - 4x^2$ **21** $9 - 25x^2$

Review exercise

1 a $d + 7f$ b $11r - p$ c $5x - 3y$

2 $6x + 12y$

3 a $6p + 12$ b $t^2 - 5t$

4 a $5(y + 2)$ b $7(2a - 1)$ c $m(m - 6)$ d $x(x + 1)$

5 a $10x - 3$ b $2x + 13$

6 a $x^2 + 7x + 12$ b $y^2 - 3y - 10$ c $z^2 - 9z + 18$

7 a $a^2 + 6a + 9$ b $b^2 - 10b + 25$

8 $(x + 3)(x + 2) - 5x = x^2 + 5x + 6 - 5x = x^2 + 6$

9 a

	$n - 6$
n	$n + 1$
	$n + 8$

b $n + n - 6 + n + 1 + n + 8 = 4n + 3$

10 a $8x^2 - 2x - 3$ b $9n^2 - 30n + 25$

Chapter 12

Exercise 12A

1 a $\frac{5}{7}$ b $\frac{5}{6}$ c $\frac{2}{5}$ d $\frac{5}{12}$

2 b $\frac{9}{20}$ is closer to $\frac{10}{20}$ than $\frac{12}{20}$

3 Yes. $\frac{4}{5} = \frac{32}{40}, \frac{7}{8} = \frac{35}{40}$

4 a $\frac{2}{3}, \frac{3}{4}, \frac{10}{12}$ b $\frac{2}{3}, \frac{17}{24}, \frac{3}{4}$ c $\frac{1}{4}, \frac{1}{2}, \frac{3}{8}$ d $\frac{2}{3}, \frac{11}{15}, \frac{4}{5}, \frac{5}{6}$

5 Any of the following: $\frac{5}{9}, \frac{3}{5}, \frac{4}{7}, \frac{5}{8}$

Exercise 12B

1 a $\frac{3}{10}$ b $\frac{9}{14}$ c $\frac{4}{9}$ d $\frac{1}{14}$

e $\frac{4}{9}$ f $\frac{7}{12}$ g $\frac{3}{15} = \frac{1}{5}$ h $\frac{8}{12} = \frac{2}{3}$

2 $\frac{4}{10} = \frac{2}{5}$ kg

3 $\frac{3}{8}$ litre

4 Melita could not be correct. $\frac{3}{5} = \frac{9}{15}, \frac{9}{15} > \frac{8}{15}$

5 a $\frac{3}{8}$ m b £1.92

6 a $\frac{1}{12}$ b $\frac{11}{15}$ c $\frac{11}{20}$ d $\frac{4}{21}$

e $\frac{11}{35}$ f $1\frac{7}{20}$ g $1\frac{5}{42}$ h $1\frac{3}{28}$

7 $\frac{13}{24}$ of a second

8 $\frac{1}{8}$ kg + $\frac{1}{3}$ kg = $\frac{3}{24}$ kg + $\frac{8}{24}$ kg = $\frac{11}{24}$ kg; $\frac{1}{2}$ kg = $\frac{12}{24}$ kg; $\frac{11}{24} < \frac{12}{24}$

9 Any two proper fractions with different denominators that add to give $1\frac{7}{12}$, e.g. $\frac{3}{4} + \frac{5}{6}, \frac{2}{3} + \frac{11}{12}$

10 Pablo could not be correct. $\frac{1}{8} + \frac{1}{5} = \frac{5 + 8}{40} = \frac{13}{40}, \frac{3}{10} = \frac{12}{40}, \frac{13}{40} > \frac{12}{40}$

Exercise 12C

1 a $4\frac{11}{12}$ b $2\frac{5}{8}$ c $2\frac{5}{8}$ d $3\frac{13}{24}$

e $\frac{19}{20}$ f $\frac{19}{20}$ g $8\frac{1}{3}$ h $2\frac{5}{24}$

i $6\frac{5}{21}$ j $1\frac{3}{8}$ k $1\frac{1}{3}$

2 $3\frac{5}{8}$ kg

3 $3\frac{3}{10}$ miles

4 Any two pairs of mixed numbers where the fractions have different denominators that add to give $3\frac{17}{30}$, e.g. $1\frac{1}{15} + 2\frac{1}{2}$, $1\frac{7}{30} + 2\frac{1}{3}$

5 Patrick could not be correct. The smallest possible answer is $1\frac{1}{6} + 3\frac{1}{4} = 4\frac{5}{12}, 4\frac{1}{4} = 4\frac{3}{12}, 4\frac{5}{12} > 4\frac{3}{12}$

6 $1\frac{5}{6} + 1\frac{7}{8} + 1\frac{11}{12} + 1\frac{1}{2} = 7\frac{1}{8}$ tonnes > 7 tonnes

Exercise 12D

1 a $\frac{2}{21}$ b $\frac{1}{15}$ c $\frac{8}{35}$ d $\frac{3}{10}$

e $\frac{1}{2}$ f $\frac{1}{10}$ g $\frac{35}{48}$ h $\frac{3}{14}$

2 a i $\frac{7}{8} \times \frac{1}{2} = \frac{7}{16}$ ii $\frac{3}{4} \times \frac{3}{4} = \frac{9}{16}$ b $\frac{3}{4} \times \frac{3}{4}$

3 a $\frac{1}{6}$ and $\frac{2}{15}$ b $\frac{5}{30}$ and $\frac{4}{30}$ c $\frac{4}{5} \times \frac{1}{6}$

4 Neither as they are both equal to $\frac{11}{42}$

5 $\frac{1}{4}$

Exercise 12E

1 a $9\frac{1}{6}$ b $5\frac{5}{6}$ c $6\frac{2}{3}$ d 7

e 7 f 8 g $8\frac{1}{4}$ h $8\frac{2}{5}$

2 Neither as they are both equal to $16\frac{2}{3}$

3 a $31\frac{1}{2}$ kg b $35\frac{1}{4}$ kg

4 $184\frac{1}{2}$ kg

5 a $1\frac{1}{4}$ b $1\frac{2}{5}$ c $1\frac{3}{8}$ d $2\frac{2}{3}$

e $3\frac{5}{7}$ f 2 g $2\frac{1}{15}$ h $1\frac{1}{2}$

6 $7\frac{7}{8}$ minutes

7 a $5\frac{5}{8}$ m^2 b £108

8 a $\frac{3}{10}$ minute b 60 minutes = 1 hour

9 a $5\frac{1}{4}$ b $3\frac{4}{15}$ c $4\frac{9}{28}$ d $24\frac{8}{15}$

e $2\frac{37}{48}$ f $7\frac{7}{45}$ g $14\frac{23}{54}$ h $40\frac{3}{8}$

10 $16\frac{7}{8}$ m^2

11 Yes, $1\frac{1}{2} \times 6\frac{2}{3} = 10$

12 a $57\frac{4}{5}$ litres

b $9\frac{11}{12}$ minutes, or 9 minutes 55 seconds

Exercise 12F

1 a $\frac{1}{4}$ b $\frac{1}{10}$ c $\frac{1}{20}$ d $\frac{1}{100}$

e 2 f 5 g $1\frac{1}{2}$ h $3\frac{1}{3}$

2 a 5 b 2 c 25 d $\frac{4}{5}$

3 All the answers are 1.

Exercise 12G

1 a 30 b 20 c 28 d $7\frac{1}{2}$ e 10 f $13\frac{1}{2}$

2 She has incorrectly turned the ÷ 12 into ÷ $\frac{1}{12}$

Correct answer $\frac{1}{28}$

3 a $\frac{5}{6}$ b $\frac{13}{14}$ c $1\frac{1}{9}$ d $\frac{2}{3}$ e $1\frac{2}{5}$

f 1 g $\frac{3}{8}$ h $\frac{3}{28}$ i $1\frac{1}{5}$

4 a $3\frac{6}{7}$ b $\frac{7}{12}$ c $3\frac{17}{20}$

5 $\frac{7}{8}$

6 18

7 a $\frac{8}{27}$ b $\frac{39}{50}$ c $\frac{104}{119}$ d $3\frac{13}{24}$ d $3\frac{17}{20}$ f $1\frac{41}{94}$

8 Yes they are, as $6\frac{1}{2} \div \frac{2}{5}$ is the same as $6 \div \frac{2}{5} + \frac{1}{2} \div \frac{2}{5}$

9 a i $\frac{3}{4}$ ii $\frac{3}{4}$ b 0

10 Ryan is correct. $6\frac{2}{3} \div 3\frac{3}{4} = 1\frac{7}{9}$, half of $3\frac{1}{3} = 1\frac{2}{3} = 1\frac{6}{9}, 1\frac{7}{9} > 1\frac{6}{9}$

Review exercise

1 $\frac{7}{10}$ **2** $\frac{3}{8}$ **3** $6\frac{2}{3}$

4 $\frac{2}{15}$ **5** 5 **6** $4\frac{31}{60}$ kg

7 a $5\frac{2}{3}$ minutes b 5 minutes 40 seconds

8 a $\frac{8}{15}$ b $1\frac{7}{8}$

Chapter 13

Exercise 13A

1 a 103.3 b 170.06 c 4.38 d 308.133

2 a 14.77 b 347.75 c 68.675 d 84.32

e 53.74 f 49.225

3 a £13.95 b £6.05

4 a £9.59 b £0.41

5 11.07 m **6** 14.96 and 19.64
7 6.6 cm **8** 025092.5

Exercise 13B

1 a $\frac{1}{10}$ b $\frac{1}{2}$ c $\frac{7}{100}$ d $\frac{1}{20}$
e $\frac{3}{4}$ f $\frac{7}{25}$ g $\frac{1}{25}$ h $\frac{13}{20}$
i $\frac{13}{25}$ j $\frac{79}{100}$ k $\frac{6}{25}$ l $\frac{7}{20}$
2 a $\frac{1}{10000}$ b $\frac{1}{500}$ c $\frac{21}{250}$ d $\frac{9}{1000}$
e $\frac{7}{200}$ f $\frac{1}{40}$ g $\frac{3}{8}$ h $\frac{17}{40}$
3 a $3\frac{1}{2}$ b $14\frac{4}{5}$ c $5\frac{16}{25}$ d $4\frac{17}{20}$
4 $\frac{18}{25}$ **5** $\frac{11}{25}$ **6** $\frac{9}{25}$

Exercise 13C

1 a 25.2 b 19.17 c 44.22 d 234.8
e 42.25 f 0.75 g 0.157 h 0.224
2 a 0.212
b 0.212 Both have the same answer. Student's own examples.
3 £27.03 **4** £391.72 **5** 22.14 mm

Exercise 13D

1 a 20 b 1200 c 42.6 d 25
e 30 f 150 g 3340 h 140
2 a 139
b 139 Both have the same answer. Student's own examples.
3 a 8 panels b 9 posts
4 75 books
5 2150 texts

Exercise 13E

1 a 0.3 b 0.75 c 0.4 d 0.875 e 0.35 f 0.16
2 $\frac{4}{25}, \frac{3}{10}, \frac{7}{20}, \frac{2}{5}, \frac{3}{4}, \frac{7}{8}$
3 a 0.5 b 0.25 c 0.6 d 0.15 e 0.75 f 0.4
4 a 0.41, $\frac{9}{20}$ b $\frac{13}{40}$, 0.336 c $\frac{4}{5}$, 0.83, $\frac{17}{20}$

Exercise 13F

1 a $\frac{1}{9}$ b $\frac{4}{9}$ c $\frac{5}{6}$ d $\frac{9}{11}$
2 a 0.125 b $0.1\dot{6}$ c $0.\dot{2}$ d 0.48
3 a $0.2\dot{3}$ b $0.02\dot{3}$ c $0.002\dot{3}$
4 0.4, 0.375, $0.\dot{3}$, 0.36, 0.39. In order: $\frac{1}{3}, \frac{9}{25}, \frac{3}{8}, \frac{39}{100}, \frac{2}{5}$
5 $0.\dot{1}42\,85\dot{7}$, $0.\dot{2}85\,71\dot{4}$, $0.\dot{4}28\,57\dot{1}$, $0.\dot{5}71\,42\dot{8}$, $0.\dot{7}14\,28\dot{5}$, $0.\dot{8}57\,14\dot{2}$
You get the same sequence of digits, but starting with different digits.
6 a Yes
b No. It can either be terminating or recurring.
c The answer can either be terminating or recurring.

Review exercise

1 a 172.3 b 497.14 c 5.174
2 a 2.921 b 29.5 c 56.49
3 a $\frac{39}{50}$ b $\frac{19}{20}$ c $\frac{3}{125}$ d $\frac{5}{8}$
4 a 11.56 b 1.488
5 a £46.82 b £3.18
6 £98.42
7 a 46.3 b 12.74 c 1.25
8 35 cans
9 a 0.6 b 0.063 c $0.\dot{1}\dot{8}$ d $0.91\dot{6}$
10 $\frac{4}{7} = 0.\dot{5}71\,42\dot{8}$, $\frac{3}{5} = 0.6$, $\frac{5}{8} = 0.625$, $\frac{2}{3} = 0.\dot{6}$, $\frac{7}{10} = 0.7$
11 $0.\dot{0}\dot{9}$, $0.\dot{1}\dot{8}$, $0.\dot{2}\dot{7}$, $0.\dot{3}\dot{6}$.
Recurring digits are the corresponding multiples of 9.
This rule predicts that $\frac{8}{11} = 0.\dot{7}\dot{2}$

Chapter 14

Exercise 14A

1 a $x = 6$ b $x = 2$ c $x = 4$
2 a $a = 4$ b $b = 5$ c $c = 8$ d $d = 3$
e $e = 9$ f $g = -2$ g $g = 3$ h $x = 2$
i $f = -4$ j $x = 7$ k $y = 3$ l $x = -1$

Exercise 14B

1 a $a = 9$ b $b = 28$ c $c = 20$
d $d = 6$ e $e = -36$ f $f = -36$
g $g = 20$ h $h = 28$ i $i = 15$
2 Mario

Exercise 14C

1 a $a = 4\frac{1}{2}$ b $b = 1\frac{2}{7}$ c $c = 3\frac{1}{3}$
d $d = 1\frac{6}{7}$ e $e = 1\frac{2}{3}$ f $f = 1\frac{1}{2}$
g $g = \frac{2}{3}$ h $h = 3\frac{1}{3}$ i $i = -1\frac{1}{2}$
2 a $g = 2.5$ b $j = 4.5$ c $k = 1.25$
d $e = 1.5$ e $a = 2.5$ f $c = 1.2$

Exercise 14D

1 a $\frac{n}{7}$ b $\frac{n}{7} = 6$ c $n = 42$
2 4
3 a $3x + 2 = 11, x = 3$ b $5a + 6 = 21, a = 3$
4 a $2x$ b $3x$ c $x = 10$
5 a i $8x$ ii $11x + 2$ iii $10x + 4$
b i $8x = 24, x = 3$ ii $11x + 2 = 24, x = 2$
iii $10x + 4 = 24, x = 2$
6 a Student's own answer
b $3x + 3$
c $3x + 3 = 306, 3x = 303, x = 101$

Exercise 14E

1 $a = 3$ **2** $b = 5$ **3** $c = 3$
4 $d = 10$ **5** $k = 5$ **6** $h = 3.5$
7 $i = 0.6$ **8** $k = -2$ **9** $l = -10$
10 $x = 14$ **11** $x = 5$ **12** $f = 10$
13 $m = -2$ **14** $k = -1$ **15** $q = 1$
16 $n = -5$ **17** $p = -5$ **18** $r = 2$

Exercise 14F

1 a $-3x$ b $-4x$ c $-2x$ d $+7x$ e $+20x$
2 a $a = 5$ b $b = -2$ c $c = 6$ d $e = 9$ e $f = 3$
f $g = 6$ g $i = -3$ h $j = 3.75$ i $k = 5$ j $n = 2.5$
3 Length 80 cm, width 28 cm

Exercise 14G

1 a $a = 2$ b $b = -2$ c $x = 0.2$ d $d = -1$
e $c = 2$ f $z = 4$ g $x = 1.5$ h $i = 4.5$
2 a $7 - j = \mathbf{15} - 3j$
$7 + 2j = \mathbf{15}$
$2j = \mathbf{8}$
$j = \mathbf{4}$
b $10d - \mathbf{20} = 6d + 24$
$\mathbf{4d} - 20 = 24$
$\mathbf{4d} = 44$
$d = \mathbf{11}$
3 a $a = 1$ b $c = 6$ c $f = 9$ d $x = -2$
e $h = 0.2$ f $i = 4$
4 $6e + 9 = 10e + 5, e = 1$

Exercise 14H

1 a $x = 2$ b $x = 5$ c $x = -3$ d $x = 3$
e $x = 7$ f $x = 3$ g $x = 5$ h $x = 2$
i $x = 6$ j $x = 5$ k $x = 10.5$ l $x = 10$
2 a $\frac{x}{2}, \frac{x}{3}, x - 8$ b $\frac{x}{2} + \frac{x}{3} = x - 8$ c $x = 48$

Exercise 14I

1 a
6 7 8 9 10 11 12
b
−2 −1 0 1 2 3
c
3 4 5 6
d
1 2 3 4 5 6 7

e

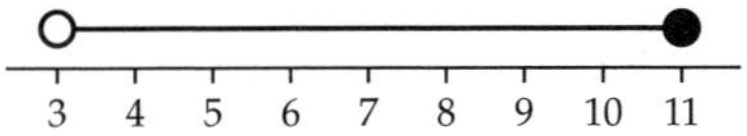

3 4 5 6 7 8 9 10 11

f

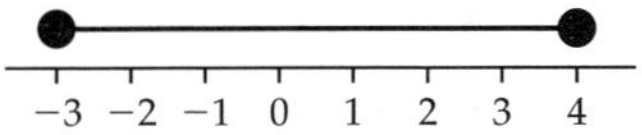

−3 −2 −1 0 1 2 3 4

2 **a** 3, 4, 5 **b** −6, −5 **c** −10, −9, −8, −7, −6
3 **a** 2, 3, 4, 5, 6 **b** 21 **c** −4

Exercise 14J

1 **a** $x < 5$

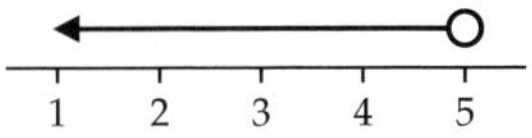

1 2 3 4 5

b $x \leqslant 8$

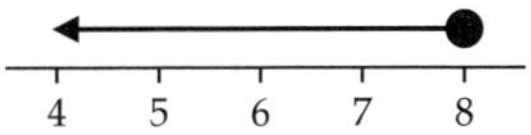

4 5 6 7 8

c $x < 3$

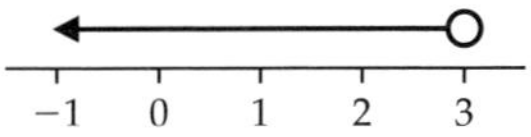

−1 0 1 2 3

d $x > 6$

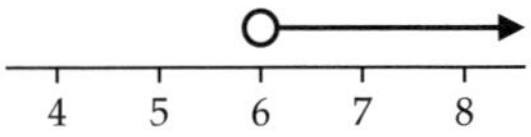

4 5 6 7 8

e $x < 20$

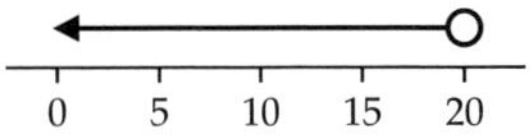

0 5 10 15 20

f $x > 0$

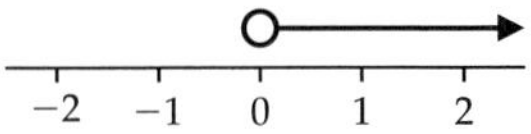

−2 −1 0 1 2

g $3 < x \leqslant 9$

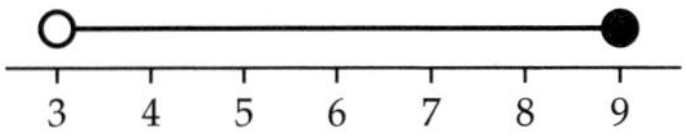

3 4 5 6 7 8 9

h $3 < x < 7$

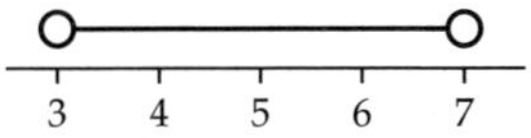

3 4 5 6 7

i $-3 < x$

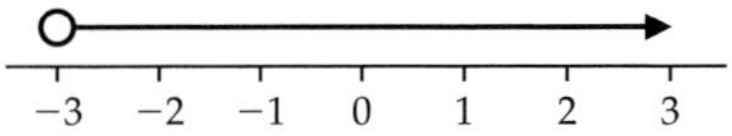

−3 −2 −1 0 1 2 3

2 **a** $x < 2$ **b** $x > 5$ **c** $x \geqslant 6$
d $x < 7$ **e** $x > 5$ **f** $x \geqslant -2.5$

Exercise 14K

1 $x > 19$ **2** $x < 14$ **3** $x \leqslant 40$
4 $x > 7$ **5** $x < 3$ **6** $x < -3$
7 $x \leqslant -3.5$ **8** $-\frac{11}{4} > x$ **9** $\frac{3}{2} \leqslant x < 4$
10 $0 < x < 2$ **11** $-2 \leqslant x < 1$ **12** $-2 < x < 1$

Exercise 14L

1 **a** $x = 3, y = 4$ **b** $x = 2, y = 3$
c $x = 1, y = 9$ **d** $x = 7, y = 9$
2 **a** $x + y = 83, x - y = 3$ **b** 43 and 40

Exercise 14M

1 $x = 2, y = 3$ **2** $x = -2, y = 5$
3 $x = 2, y = 5$ **4** $x = -3, y = 4$

Exercise 14N

1 **a** $x = 5, y = 1$ **b** $x = -5, y = 3$ **c** $x = 2, y = -1$
d $x = 3, y = 2$ **e** $x = 3, y = 2$ **f** $x = -7, y = 3$
2 **a** 800 **b** 9800

Review exercise

1 **a** $w = 40$ **b** $x = 2.5$
2 20
3 **a** $p = 4$ **b** $t = 2\frac{1}{2}$ **c** $x = 3\frac{1}{2}$
4 **a** 1, 2, 3, 4
b Student's own answer, e.g. $x = 3, y = 2$
5 **a** $x \leqslant 3$
b Student's own answer, e.g.

−4 −3 −2 −1 0 1 2 3 4 5 6

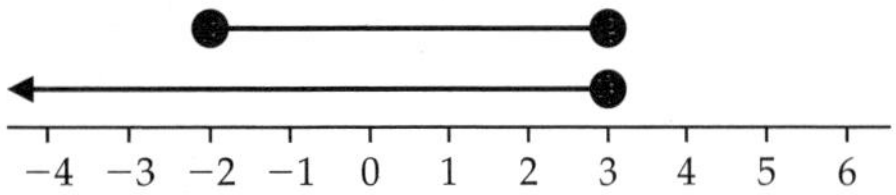

6 **a** $x \leqslant 1$ **b** $x < 2$
7 **a** $x = 2.5$ **b** $x = -1.5$
8 $x = 1.8$
9 $x = -3, y = 5$

Chapter 15

Exercise 15A

1 **a** d^4 **b** a^6 **c** x^3 **d** m^4
2 **a** $5p^4$ **b** $4a^3$ **c** $8x^4$ **d** $15f^5$ **e** $8b^3$ **f** $60h^5$
3 **a** $6x^2$ **b** $12y^3$ **c** $6a^4$ **d** $24b^3$ **e** $27x^3$ **f** $36z^6$

Exercise 15B

1 **a** a^5 **b** b^7 **c** c^6 **d** d^{10} **e** x^4 **f** y^3
g z^5 **h** w^4
2 **a** $6a^7$ **b** $15b^9$ **c** $4c^8$ **d** $9d^8$
3 Add the indices of the variable together, e.g. $a^3 \times a^2 = a^{(3+2)} = a^5$
4 **a** p^3 **b** q^4 **c** r **d** s^3
5 **a** 5 **b** $2y^4$ **c** $4r$
6 Subtract the indices of the variable, e.g. $p^5 \div p^2 = p^{(5-2)} = p^3$
7 **a** t^6 **b** x^8 **c** 1 **d** 1

Exercise 15C

1 **a** m^5 **b** a^7 **c** n^6 **d** u^{10} **e** t^9 **f** d^4
2 **a** a^3 **b** t^4 **c** e **d** s^3 **e** 1 **f** k
3 **a** d^2 **b** e^3 **c** r^2 **d** a **e** 1 **f** 1
4 **a** h^9 **b** e^8 **c** c^9
5 **a** p^6 **b** q^8 **c** r^{12} **d** f^{15} **e** d^{12}

Exercise 15D

1 **a** $6h^7$ **b** $15e^9$ **c** $4g^8$ **d** $9r^5$ **e** $24e^4$
2 **a** $5x^3$ **b** $3y^4$ **c** $3r$ **d** $\frac{1}{4}s^2$ **e** $2t$
3 **a** $3a^3$ **b** $6r^2$ **c** $2r^2$ **d** $2e^2$ **e** 4 **f** $3x^4$
4 **a** $8s^2b^3$ **b** $21f^5g^3$ **c** $20x^4y^3$ **d** $36p^6q^5$ **e** $56r^3s^6$ **f** $75c^4d^5$
5 **a** b^2 **b** $3g$ **c** $2y^2$ **d** $3p^2q^3$ **e** $3r^2s$ **f** c^3d
6 **a** $2x$ **b** $4jk^2$ **c** b^2 **d** $3m^3$ **e** $4q^2$
7 **a** $6a^4b^3$ **b** $6a^3b^2$ **c** $6a^2b^3$ **d** $2ab^2$ **e** $12a^4b^4$ **f** $2a^2b^2$

Exercise 15E

1 $t = 48y$ **2** $t = 7r + 5s$
3 $t = 45w + 20$ **4** $t = 65w + 30$
5 $p = 2(3x + 1 + x + 2) = 8x + 6$

Exercise 15F

1 **a** 5 **b** −14 **c** −4 **d** −10 **e** −24
f −55 **g** 2 **h** 36 **i** 52
2 **a** 20 **b** $17\frac{1}{2}$ **c** 14 **d** 2 **e** 23
f $13\frac{1}{2}$
3 **a** 3 **b** $4\frac{1}{2}$ **c** 12

Exercise 15G

1 **a** 4 **b** 3 **c** 5 **d** 84 **e** 27
f −5 **g** 4 **h** 6 **i** 3 **j** 76
k 30 **l** 37
2 **a** 76.5 **b** 9 **c** 251.5 **d** 65
e 12.5 **f** −4.5

3

x	2	3	4	5	6
$x^3 - x$	6	24	60	120	210

4 **a** 90 **b** −18 **c** 7 **d** 5 **e** 31 **f** 8
5 **a** 36 **b** −14 **c** 4 **d** 2 **e** 1 **f** −3

Exercise 15H

1 a 6 b 4 c 6
2 a £18 b £26 c £45
3 a £92 b 32 litres
4 a 32 b 88 c 30
5 a 26 b 50 c 24
6 a 7 b 11 c 13
7 a $90\,cm^2$ b $168\,cm^2$
8 a $234\,cm^2$ b $216\,cm^2$
9 3
10 4 cm

Exercise 15I

1 a 10 b 5 c 20
2 a 5 b 9 c 13 d 11.5
3 a 6 b 11 c 5 d 9 e −8 f 7

Exercise 15J

1 a $a = c - 5$ b $a = k + 6$ c $a = w + 7$
2 a $w = \frac{P}{4}$ b $w = \frac{A}{l}$ c $w = \frac{h}{k}$
3 a $x = \frac{y+6}{5}$ b $x = \frac{y-1}{2}$ c $x = \frac{y-5}{6}$
4 a $r = \frac{p-2t}{4}$ b $r = \frac{w+2s}{3}$ c $r = \frac{y+5p}{6}$
5 a $a = 2b - 12$ b $a = 2b - 14$ c $a = 3b + 3$ d $a = 4b + 12$ e $a = \frac{b}{2} - 1$ f $a = \frac{b}{3} + 5$
6 a $a = \frac{v-u}{t}$ b $t = \frac{v-u}{a}$

Exercise 15L

1 a $y = \frac{10-3x}{2}$ b $y = \frac{8-6x}{3}$ c $y = \frac{2x-2}{3}$
2 a $x = \frac{3y+5}{2}$ b $x = y(z+5)$ c $x = \frac{q(3p+s)}{2}$
3 a $w = K^2 - t$ b $w = A^2 + a$ c $w = \left(\frac{h-1}{2}\right)^2$
4 a $r = \sqrt{g(t+m)}$ b $r = 2\sqrt{h-3a}$ c $r = \sqrt{\frac{3V}{\pi h}}$
5 $h = \frac{A}{6r} - r$
6 $l = \frac{gT^2}{36}$

Review exercise

1 a 14 b 18 c 33
2 19°C
3 a 7 b 2
4 a t^5 b m^3
5 a $12n^3$ b $40s^4$
6 3
7 a x^9 b r c z^2 d $15n^6$ e $5a^2$
8 $r = \frac{p-3}{2}$
9 $x = \frac{y+1}{4}$
10 a $x = 2(w+y)$ b $x = \sqrt{w-y}$
11 a p^{10} b $15a^3b^7$ c $4b^2$
12 $\frac{10p+4}{7}$

Chapter 16

Number skills: fractions, decimals and percentages

1 a 0.57 b 0.8% c $\frac{13}{20}$ d 2.5%
2 $\frac{1}{5}$, 0.25, 30%
3 0.3, 32%, $\frac{1}{3}$
4 Ben
5

Percentage	Fraction	Decimal
35%	$\frac{7}{20}$	**0.35**
5%	$\frac{1}{20}$	**0.05**
14%	$\frac{7}{50}$	0.14

Number skills: using percentages in calculations

1 a 2.24 kg b £600 c 1.8 litres d £120
2 £5796
3 24
4 a 60% b 65% c 80% d 18% e 30% f 95%
5 24% b 28% c 47% d 49%

Exercise 16A

1 a £220 b 132 m c 4.4 g d 18.7 litres
2 a 19 feet b £38 c 4.75 km d £7220
3 a £104.50 b 44.1 mm c £144 d £321 e 69 *ml* f 44 miles
4 £10.80
5 18.7 m
6 £44.10
7 £296.40
8 £5760
9 $72
10 £1.30
11 102 members
12 £29.75
13 131 250 (rounded to the nearest 1000)
14 Less (£198)
15 *Suit you, Sir* is the best buy. (*Dumbo's DIY* costs £79.20, *Suit you, Sir* costs £78.)

Exercise 16B

1 a £470 b £35.25 c £176.25 d £56.40
2 a £9.50 b £199.50
3 £18 212.50
4 £115.50
5 £329
6 a £99.64 b £24.91

Exercise 16C

1 a £350 b £4320 c £4670 d £1170
2 a £92 b £408 c £500 d £40
3 a £1536 b £156
4 £76
5 £112

Exercise 16D

1 a £180 b £40 c £18 d £100
2 £600
3 £200
4 Bonus Buster pays more interest (£560) than Grabbitall (£525).

Exercise 16E

1 Trousers 40% profit; house 5% profit; barbeque set 50% profit; car 15% loss; TV 30% profit
2 16.7%
3 12.5%
4 800%
5 30%
6 3%
7 40%
8 Sarah makes the bigger percentage profit. (Sarah: 87.5%, George: 66.7%)
9 Gardens-r-Us makes the bigger percentage profit. (Gardens-r-Us: 40%, Yuppies: 37.5%)

Exercise 16F

1 £540.80
2 £399.30
3 a £25.20 b £315.25
4 72

5 £5120 **6** 4050
7 3 375 000 **8** 72 kg
9 242 000 **10** 1 805 000 gallons
11 Hiroshi has more. (Hiroshi: £466.56, Amy: £459.80)

Exercise 16G

1 a £19.50 b £39.00 c £21.00 d £9.90
2 £875
3 £9600
4 £26
5 £850
6 £17 250
7 53 kg
8 110 Mathia dollars
9 a 49 500 b 61 100 (to the nearest 100)
10 Full price of trumpet was £450, so John saved £450 − £405 = £45

Review exercise

1 a £250 b £1450 c £200
2 £150
3 £36
4 138
5 £423
6 £9.60
7 £0.90
8 81 480
9 30%
10 40%
11 172 000
12 £4096
13 125
14 a 720 b 500
15 £800

Chapter 17

Exercise 17A

1 a i 12, 17, 22 ii add 5
b i −13, −19, −25 ii subtract 6
c i −12, −16, −20 ii subtract 4
d i −3, −8, −13 ii subtract 5
2 a 62 b −24 c −43
3 a 15, 11, 7, 3, −1, −5, −9, −13, −17, −21 b 4
4 3
5 a Student's own answer, for example
i 64, 32, 16, 8, 4 or 64, 32, 0, −32, −64
ii divide by 2 or subtract 32
b −224

Exercise 17B

1 a i 15, 21
ii The difference starts at 2 and increases by 1 each time.
b i 36, 49
ii The difference starts at 3 and increases by 2 each time.
c i 64, 58
ii The difference starts at −10 and increases by 1 each time.
d i −8, −13
ii The differences starts at −1 and decreases by 1 each time.
2 a 3 b 21 c It is also the Fibonacci sequence.
3 2013
4 −6
5 Year 8

Exercise 17C

1 a 1, 5, 9, 13, 17 b 197
2 a 45 b −44 c 5 d −22 e 5 f 9
3 8
4 23
5 a i 2, 4, 6, 8, 10 ii 2
b i 3, 6, 9, 12, 15 ii 3
c i 4, 8, 12, 16, 20 ii 4
d i 5, 10, 15, 20, 25 ii 5
e i 5, 7, 9, 11, 13 ii 2
f i 1, 4, 7, 10, 13 ii 3
g The difference between consecutive terms is the coefficient of n.

Exercise 17D

1 a $3n$ b $2n$ c $5n$ d $100n$ e $7n$ f $50n$
2 a 11, 13, 15 b 2
c

	1st term	2nd term	3rd term	4th term
Sequence	3	5	7	9
Multiples of 2	**2**	**4**	**6**	**8**

d $2n + 1$
3 a $4n + 1$ b $2n + 6$ c $5n - 1$
d $11n - 2$ e $100n - 25$ f $-5n$
g $-3n + 13$ h $-4n + 23$ i $-10n + 87$
j $-6n + 56$
4 a $5n + 3$ b 503
5 a $2n + 19$ b 39 mm

Exercise 17E

1 a 99 b 88 c 141
2 108
3 11th term
4 a Yes b No
5 a 15th term b All the terms are odd numbers.
6 a −9 b −1 c −8 d −5
7 a 67 b 71 c 68 d 68

Exercise 17F

1 a 25
b The number of dots needed is the pattern number squared.
2 a 21 b 55
3 a i Pattern 4 Pattern 5
ii 28 iii $3n + 1$
b i Pattern 4 Pattern 5
ii 19 iii $2n + 1$
c i Pattern 4 Pattern 5
ii 26 iii $3n - 1$
4 a

Number of tables	1	2	3	4	5
Number of seats	4	6	**8**	**10**	**12**

b $2n + 2$ c 9
5 a i 21 ii 26
b

Number of huts	1	2	3	4	5	10	15
Number of matchsticks	6	11	**16**	**21**	**26**	**51**	**76**

c $5n + 1$
6 4
7 a

Pattern number	Number of dots
1	1 × 2 = 2
2	2 × 3 = 6
3	3 × 4 = **12**
4	**4 × 5 = 20**
5	**5 × 6 = 30**
10	10 × **11 = 110**
15	**15 × 16 = 240**
n	$n \times (n + 1) = n(n + 1)$

b 11

Exercise 17G

1 a 3 b 6 c 11 d 102

2 a −2, 1, 6, 13, 22 b 7, 10, 15, 22, 31
c 3, 12, 27, 48, 75 d 0.5, 2, 4.5, 8, 12.5
e 9, 21, 41, 69, 104

3 a False b False c True d False e True

4 a 2, 8, 18, 32, 50
b The difference starts at 6 and increases by 4 each time, so the 6th term is 50 + 22 = 72.

5 3, 5, 9, 15, 23, 33

6 No. All the terms in this sequence are odd.

7 a 500 b 15 days

8 14 days

Exercise 17H

1 $n^2 + 3$, 103

2 $n^2 - 2$, 98

3 $n^2 + 6$, 106

4 $n^2 - 5$, 95

5 $2n^2$, 200

6 $3n^2$, 300

Exercise 17I

1 a Odd
b Any number multiplied by 2 is even; even − 1 = odd.

2 Any number doubled will be even.

3 If q is odd then $q + 1$ is even; odd × even = even.

4 All prime number except 2 are odd; odd + 1 = even.

5 For any two consecutive positive integers, one is odd and the other is even; odd × even = even, so the product is always even.

6 If x is odd then x^2 is odd × odd, which is always odd; odd + 1 = even.

7 $x + y$ and $x - y$ are always odd (odd + even = odd and odd − even = odd); odd × odd = odd, so the product is always odd.

8 Odd. In six consecutive integers, there will be three even numbers and three odd numbers;
even + even + even + odd + odd + odd = odd

Exercise 17J

1 a e.g. $-3 + 2 = -1$ b e.g. $12 \times 0.25 = 3$
c e.g. $1.7 + 0.3 = 2$

2 7.55

3 a False, e.g. $n = 1$, $n^2 + 1 = 2$ b True
c False, e.g. $(-1)^3 = -1$ d False, e.g. 9
e True

4 e.g. $x = 1$, $x^2 = 1$ (odd)

5 There is a counter-example, e.g. $9 - 1 = 8$.

6 There is a counter-example, e.g. $3 + 4 + 5 = 12$.

7 There is a counter-example, e.g. $2^2 + 3^2 = 4 + 9 = 13$.

8 There is a counter example, e.g. $n = 1$, $n^2 - n = 1 - 1 = 0$.

9 There is a counter example, e.g. $m = -2$, $m^3 = -8$.

Review exercise

1 If two integers are consecutive then one is odd and one even; odd + even = odd.

2 a 1, 3, 5 b No. All the terms are odd numbers.

3 If you double any number it will be even. Since even − 1 = odd, the outcome is always odd.

4 100 and 25

5 a 59 b 13th

6 $n^2 + 6$

7 $10n - 18$

8

Pattern number	Number of tiles
1	1 × 3 = 3
2	2 × 4 = 8
3	**3 × 5 = 15**
4	**4 × 6 = 24**
5	**5 × 7 = 35**
10	**10 × 12 = 120**
n	$\mathbf{n \times (n + 2) = n(n + 2)}$

b Pattern 8

9 e.g. $n = 2$, $q = 5$, $nq = 10$
(any example where one of the primes is 2)

10 e.g. When $m = 4$, $m^3 + 1 = 64 + 1 = 65$. This is a multiple of 5.

Chapter 18

Exercise 18A

1 a $C(2, 1)$, $D(8, 1)$, $E(5, 8)$, $F(5, 2)$, $G(1, 2)$, $H(4, 5)$, $L(-5, 4)$, $M(2, 4)$, $P(-5, -3)$, $Q(-1, 1)$
b CD: (5, 1); EF: (5, 5); GH: (2.5, 3.5); LM: (−1.5, 4); PQ: (−3, −1)

2 a (4.5, 1) b (3.5, 6) c (0.5, −0.5) d (2, −5)

3 a (1.5, −0.5) b (1.5, −0.5) c $ABCD$ is a square.

4 a (1, 0) b (3, 2) c (−4, 2) d (1, 7)

Exercise 18B

1 a Line B, $y = 3$; Line D, $y = -6$; Line F, $y = 0$
b Line A, $x = -4$; Line C, $x = 2$; Line E, $x = 5$; Line G, $x = 0$

2

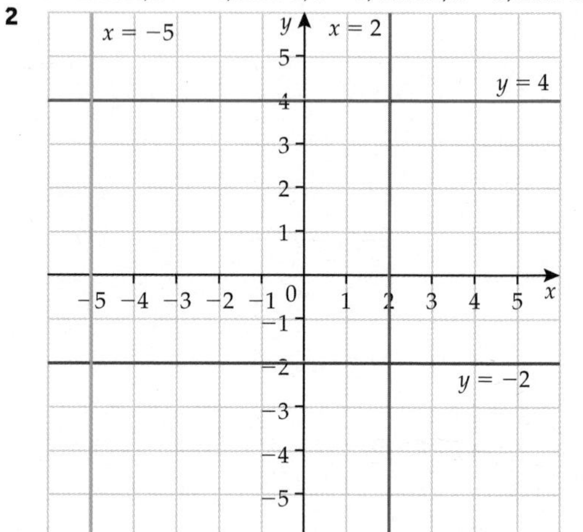

Exercise 18C

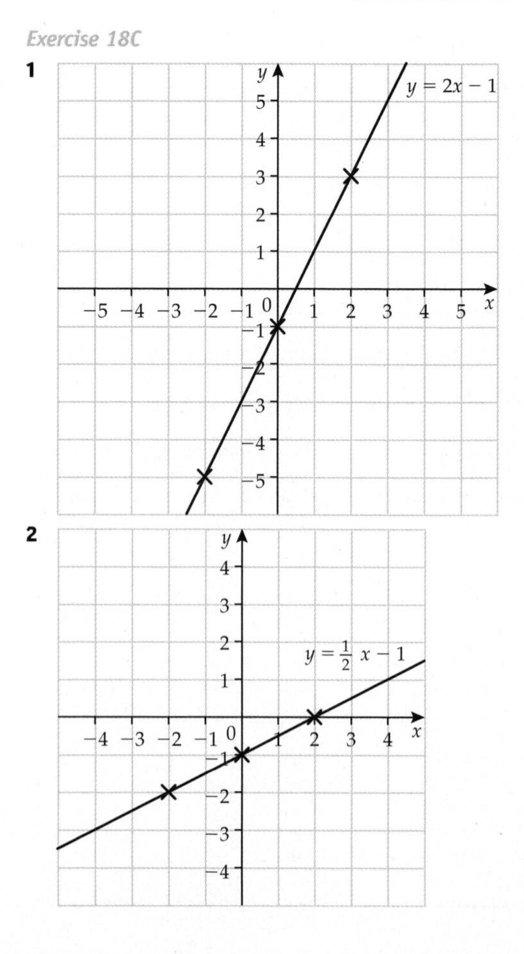

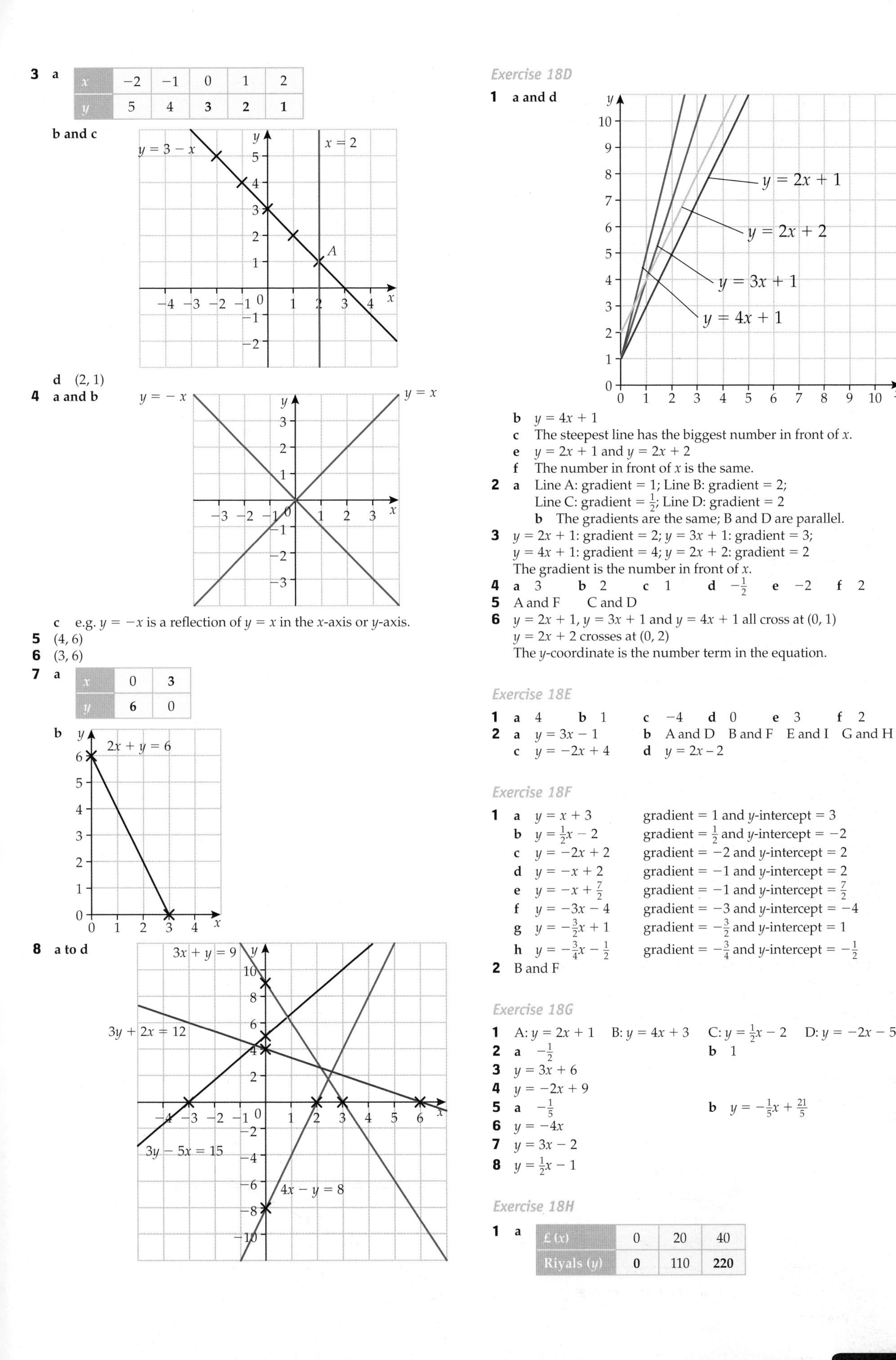

3 **a**

x	−2	−1	0	1	2
y	5	4	**3**	**2**	**1**

b and c

d (2, 1)

4 **a and b**

c e.g. $y = -x$ is a reflection of $y = x$ in the x-axis or y-axis.

5 (4, 6)

6 (3, 6)

7 **a**

x	0	**3**
y	**6**	0

b

8 **a to d**

Exercise 18D

1 **a and d**

b $y = 4x + 1$

c The steepest line has the biggest number in front of x.

e $y = 2x + 1$ and $y = 2x + 2$

f The number in front of x is the same.

2 **a** Line A: gradient = 1; Line B: gradient = 2;
Line C: gradient = $\frac{1}{2}$; Line D: gradient = 2

b The gradients are the same; B and D are parallel.

3 $y = 2x + 1$: gradient = 2; $y = 3x + 1$: gradient = 3;
$y = 4x + 1$: gradient = 4; $y = 2x + 2$: gradient = 2
The gradient is the number in front of x.

4 **a** 3 **b** 2 **c** 1 **d** $-\frac{1}{2}$ **e** −2 **f** 2

5 A and F C and D

6 $y = 2x + 1$, $y = 3x + 1$ and $y = 4x + 1$ all cross at (0, 1)
$y = 2x + 2$ crosses at (0, 2)
The y-coordinate is the number term in the equation.

Exercise 18E

1 **a** 4 **b** 1 **c** −4 **d** 0 **e** 3 **f** 2

2 **a** $y = 3x - 1$ **b** A and D B and F E and I G and H
c $y = -2x + 4$ **d** $y = 2x - 2$

Exercise 18F

1 **a** $y = x + 3$ gradient = 1 and y-intercept = 3
b $y = \frac{1}{2}x - 2$ gradient = $\frac{1}{2}$ and y-intercept = −2
c $y = -2x + 2$ gradient = −2 and y-intercept = 2
d $y = -x + 2$ gradient = −1 and y-intercept = 2
e $y = -x + \frac{7}{2}$ gradient = −1 and y-intercept = $\frac{7}{2}$
f $y = -3x - 4$ gradient = −3 and y-intercept = −4
g $y = -\frac{3}{2}x + 1$ gradient = $-\frac{3}{2}$ and y-intercept = 1
h $y = -\frac{3}{4}x - \frac{1}{2}$ gradient = $-\frac{3}{4}$ and y-intercept = $-\frac{1}{2}$

2 B and F

Exercise 18G

1 A: $y = 2x + 1$ B: $y = 4x + 3$ C: $y = \frac{1}{2}x - 2$ D: $y = -2x - 5$

2 **a** $-\frac{1}{2}$ **b** 1

3 $y = 3x + 6$

4 $y = -2x + 9$

5 **a** $-\frac{1}{5}$ **b** $y = -\frac{1}{5}x + \frac{21}{5}$

6 $y = -4x$

7 $y = 3x - 2$

8 $y = \frac{1}{2}x - 1$

Exercise 18H

1 **a**

£ (x)	0	20	40
Riyals (y)	**0**	110	**220**

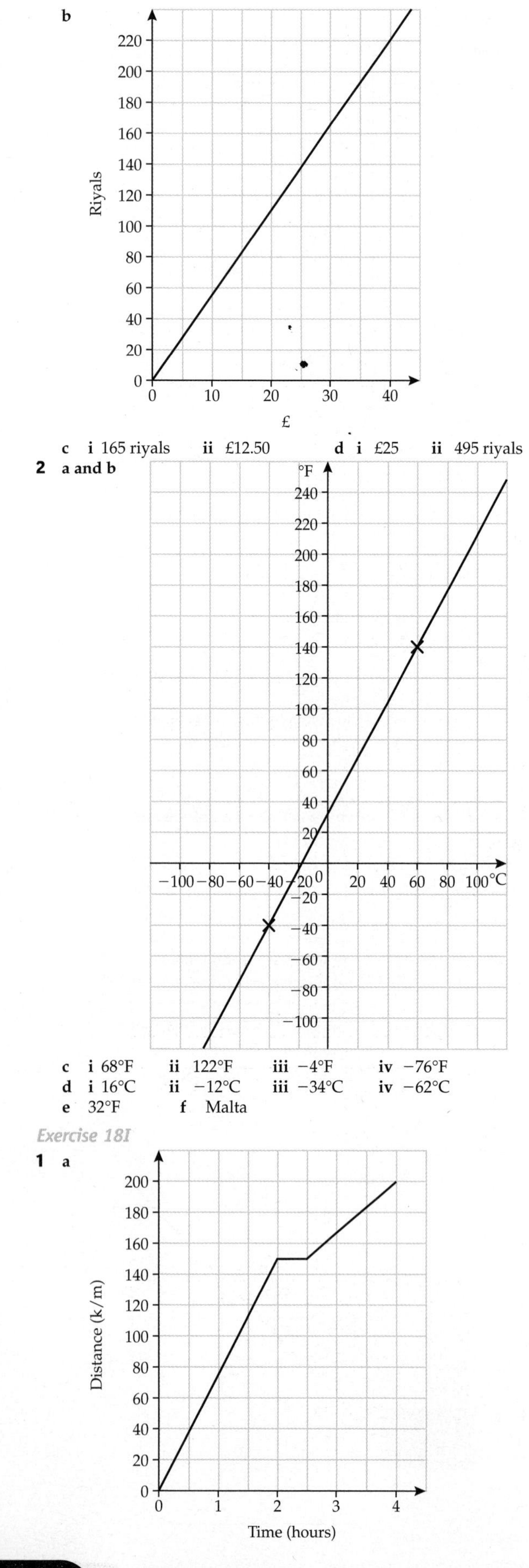

b

c **i** 165 riyals **ii** £12.50 **d** **i** £25 **ii** 495 riyals

2 a and b

c **i** 68°F **ii** 122°F **iii** −4°F **iv** −76°F

d **i** 16°C **ii** −12°C **iii** −34°C **iv** −62°C

e 32°F **f** Malta

Exercise 18I

1 a

b First. The line on the graph is steeper.

2 a 20 km

b 11.30 am

c 5 hours

d Between 10.30 am and 11.30 am.
This is where the line on the graph is steepest.

e 20 km/hour

3 a

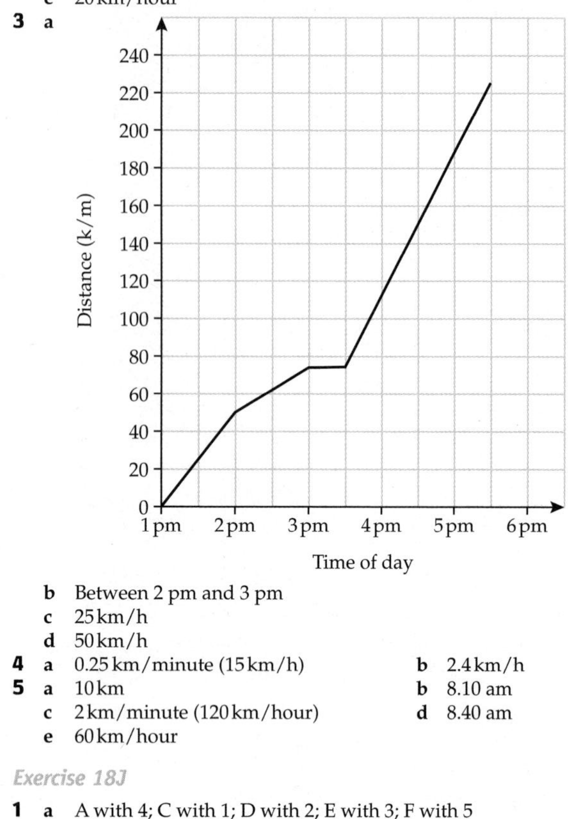

b Between 2 pm and 3 pm

c 25 km/h

d 50 km/h

4 a 0.25 km/minute (15 km/h) b 2.4 km/h

5 a 10 km b 8.10 am

c 2 km/minute (120 km/hour) d 8.40 am

e 60 km/hour

Exercise 18J

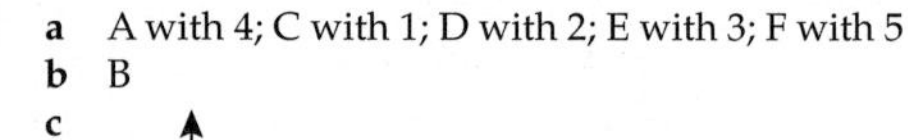

1 a A with 4; C with 1; D with 2; E with 3; F with 5

b B

c

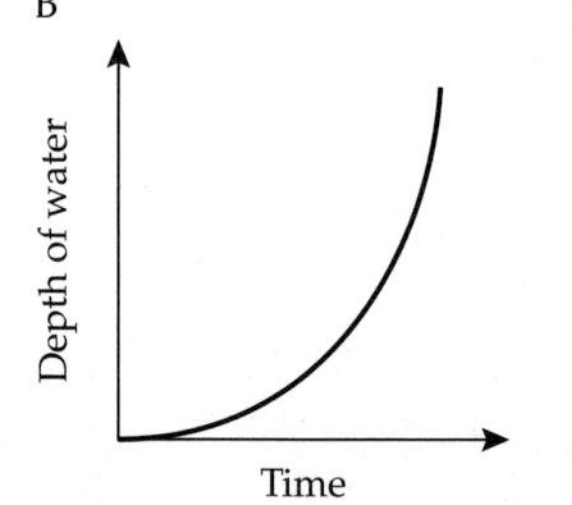

2 No. Malik has only considered the lower part of the vase. The wider part will fill more slowly.

3 e.g. Nathan runs at a constant speed throughout the race. John starts the race quickly but slows down towards the end.

Exercise 18K

1 a

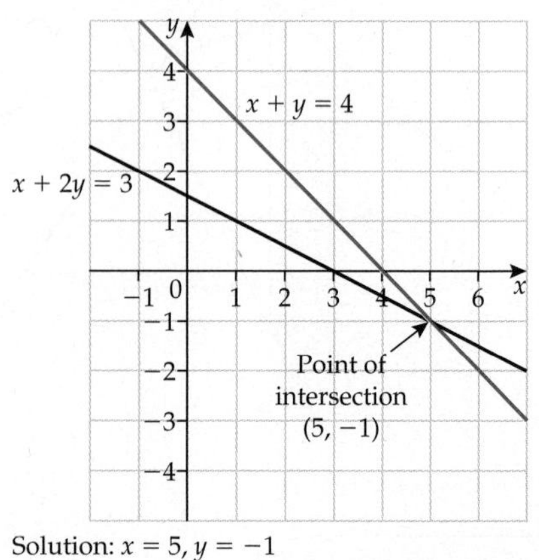

Solution: $x = 5, y = -1$

b

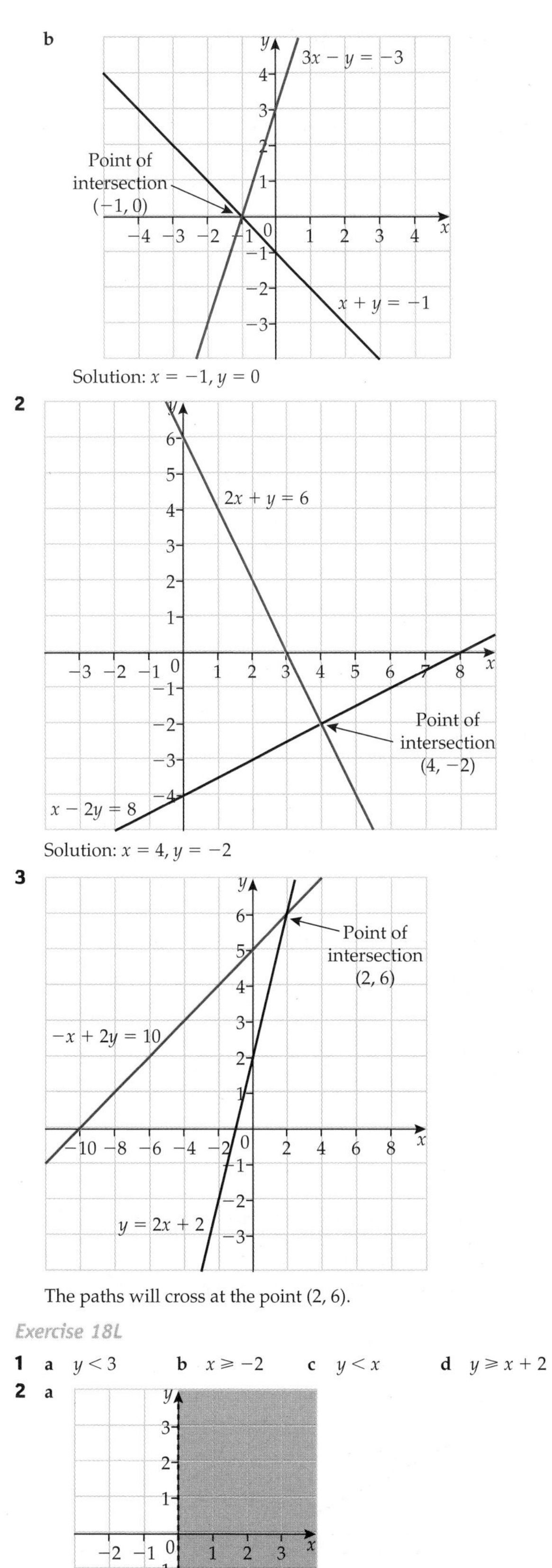

Solution: $x = -1, y = 0$

2

Solution: $x = 4, y = -2$

3

The paths will cross at the point (2, 6).

Exercise 18L

1 a $y < 3$ b $x \geqslant -2$ c $y < x$ d $y \geqslant x + 2$

2 a

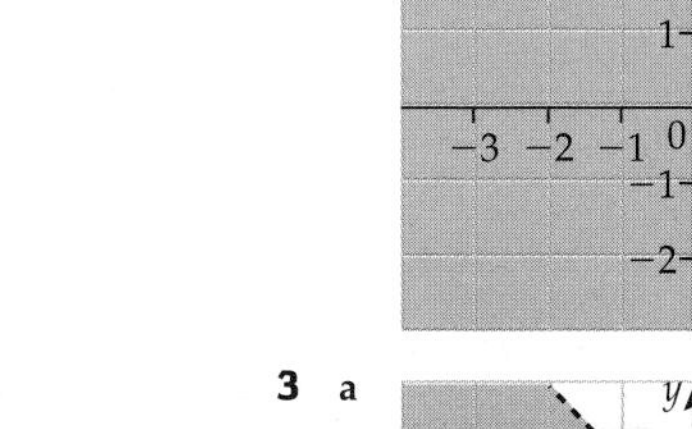

b

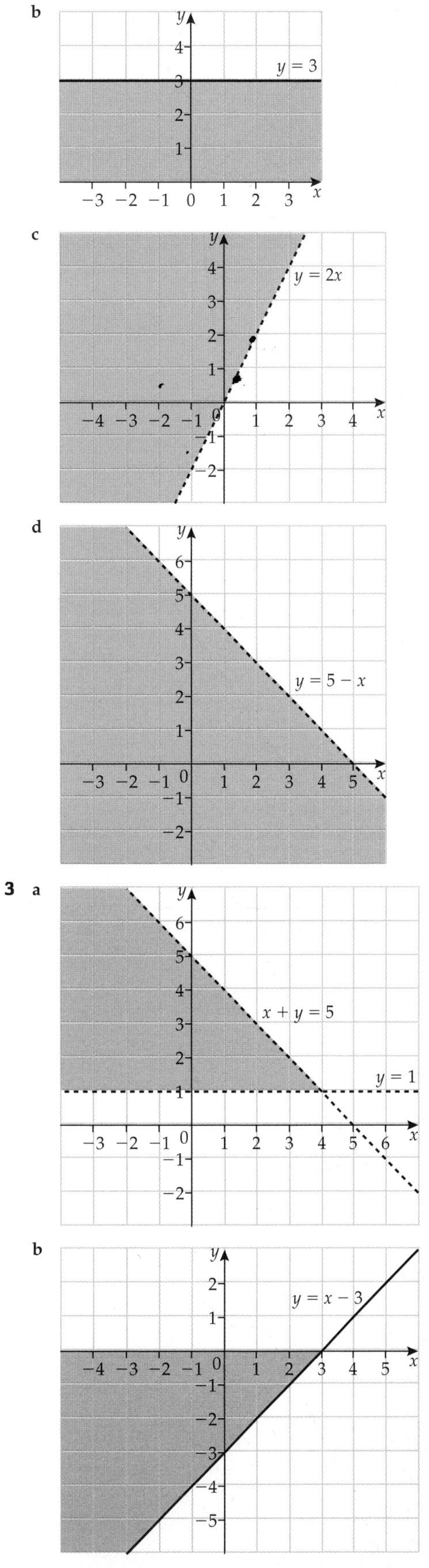

c

d

3 a

b

4

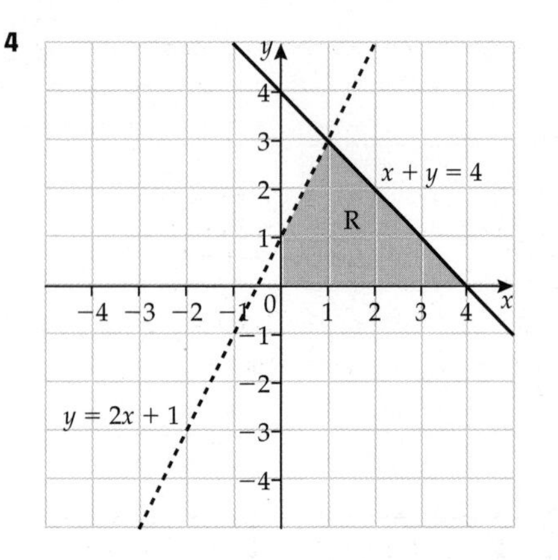

Review exercise

1 a

Ounces (x)	0	1	2	4	7
Grams (y)	**0**	**28**	56	**112**	196

b

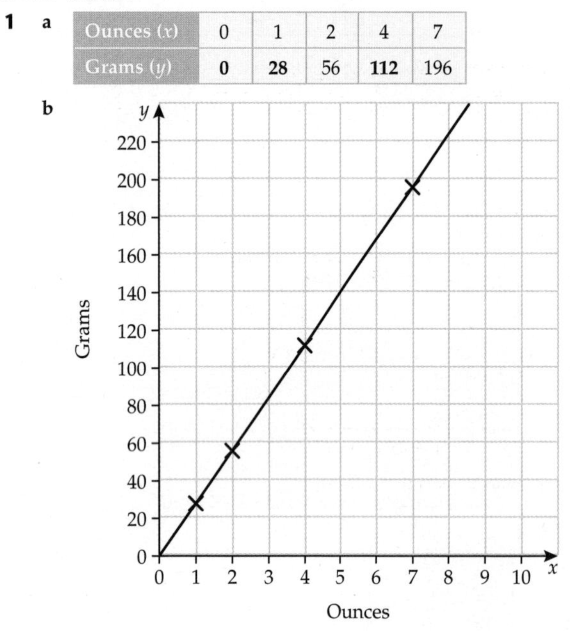

c Flour 112 grams, sugar 210 grams, cocoa powder 100 grams, butter 168 grams

2

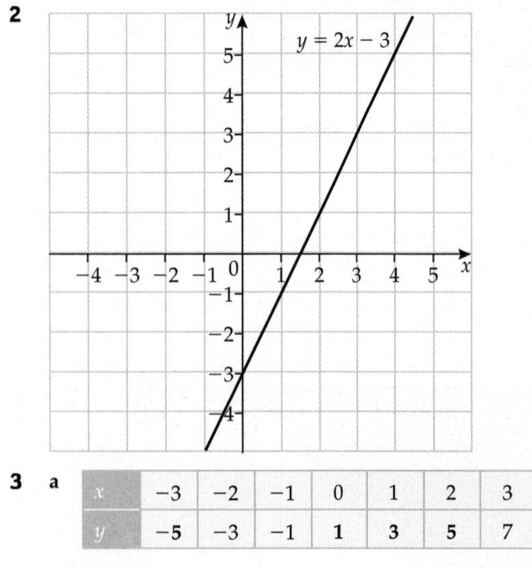

3 a

x	−3	−2	−1	0	1	2	3
y	**−5**	−3	**−1**	**1**	**3**	**5**	7

b

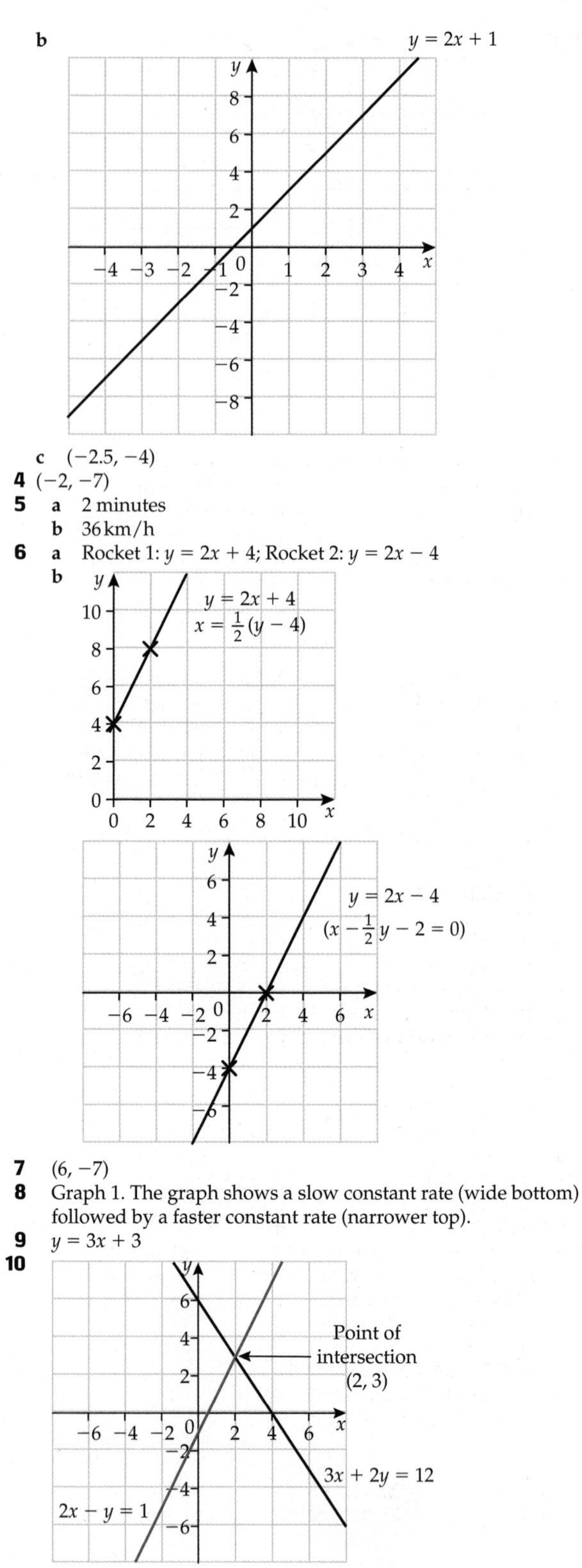

c $(-2.5, -4)$

4 $(-2, -7)$

5 a 2 minutes

b 36 km/h

6 a Rocket 1: $y = 2x + 4$; Rocket 2: $y = 2x - 4$

b

7 $(6, -7)$

8 Graph 1. The graph shows a slow constant rate (wide bottom) followed by a faster constant rate (narrower top).

9 $y = 3x + 3$

10

Solution: $x = 2$, $y = 3$

11

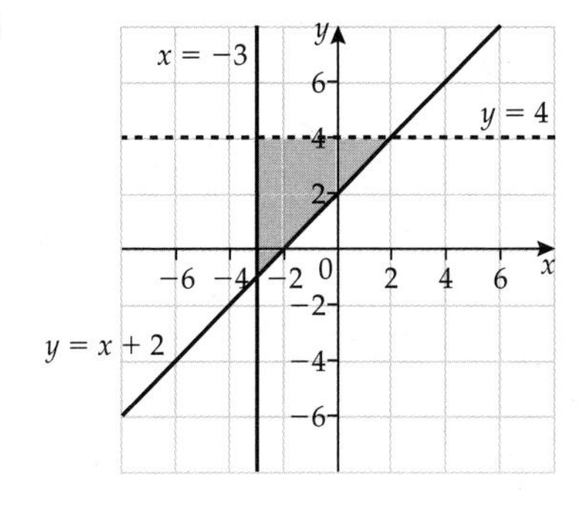

Chapter 19

Exercise 19A

1 a $x^2 - 81 = (x + 9)(x - 9)$ b Student's own answer
2 a $(x - 4)(x + 4)$ b $(x - 5)(x + 5)$
c $(x - 10)(x + 10)$ d $(x - 12)(x + 12)$
e $(a - 1)(a + 1)$ f $(m - 8)(m + 8)$
g $(t - 11)(t + 11)$ h $(b - 7)(b + 7)$
3 $(x - 20)(x + 20)$
4 a $a^2 - 16$ b $(a - 4)(a + 4)$

Exercise 19B

1 a $(x + 11)(x + 3)$ b $(x + 11)(x + 10)$
c $(x + 10)(x + 4)$ d $(x + 3)(x + 7)$
2 a $(x + 2)(x + 3)$ b $(x + 4)(x + 2)$
c $(z + 5)(z + 1)$ d $(a + 10)(a + 1)$
e $(n + 5)(n + 3)$ f $(f + 6)(f + 6) = (f + 6)^2$
g $(x + 5)(x + 4)$ h $(m + 6)(m + 2)$
i $(x + 12)(x + 2)$ j $(b + 5)(b + 6)$

Exercise 19C

1 a $(x - 2)(x - 12)$ b $(x - 5)(x - 4)$
c $(x - 5)(x - 20)$ d $(x - 6)(x - 8)$
2 a $(x - 2)(x - 3)$ b $(x - 8)(x - 1)$
c $(z - 3)(z - 4)$ d $(a - 7)(a - 2)$
e $(n - 5)(n - 5) = (n - 5)^2$ f $(f - 4)(f - 4) = (f - 4)^2$
g $(x - 10)(x - 3)$ h $(b - 7)(b - 4)$
i $(x - 6)(x - 6)$ or $(x - 6)^2$ j $(p - 6)(p - 4)$

Exercise 19D

1 a $(x + 2)(x - 15)$ b $(x - 4)(x + 5)$
c $(x + 4)(x - 6)$ d $(x + 7)(x - 4)$
2 a $(x + 6)(x - 2)$ b $(x - 5)(x + 4)$
c $(z - 5)(z + 3)$ d $(a + 7)(a - 1)$
e $(n + 8)(n - 2)$ f $(f - 6)(f + 5)$
g $(m + 6)(m - 5)$ h $(t - 12)(t + 6)$
3 a i $(x + 3)^2$ ii $(x - 4)^2$ iii $(x + 2)^2$
iv $(x - 7)^2$ iv $(x - 5)^2$ v $(x + 8)^2$
b They are perfect squares.
c $(x + m)^2 = x^2 + 2mx + m^2$
$(x - n)^2 = x^2 - 2nx + n^2$
4 a $t^2 + 7t - 30 = (t + 10)(t - 3)$
b $m^2 - 8m + 15 = (m - 3)(m - 5)$
c $q^2 - 12q + 20 = (q - 10)(q - 2)$

Exercise 19E

1 $r = \pm 13$ **2** $x = \pm 3$ **3** $t = \pm 3$
4 $y = \pm 1$ **5** $m = \pm 5$ **6** $p = \pm 10$

Exercise 19F

1 $x = \pm 4$ **2** $y = \pm 2$ **3** $y = \pm 5$
4 $r = \pm 7$ **5** $t = \pm 3$ **6** $y = \pm 1$
7 $p = \pm 10$ **8** $m = \pm 8$ **9** $w = \pm 9$

Exercise 19G

1 $r = 10$ or 4
2 a $x = 5$ or -7 b $t = \pm 4$
c $t = \pm 6$ d $t = \pm 3$
3 5 cm
4 a $3x$ b Area $= x \times 3x$ c Area $= 3x^2$
d $x = 20$ e 60 m

5 You cannot find the square root of a negative number.
6 13 mm

Exercise 19H

1 a $x = -3$ or $x = 12$ b $x = 8$ or $x = 9$
c $x = -20$ or $x = 15$ d $x = -13$ or $x = 1$
2 a $x = 0$ or $x = -7$ b $t = 0$ or $t = 5$
c $x = 0$ or $x = -2$ d $y = 0$ or $y = 5$
e $w = 0$ or $w = 3$ f $y = 0$ or $y = \frac{1}{4}$
g $a = 0$ or $a = 1$ h $t = 0$ or $t = \frac{1}{6}$
i $r = 0$ or $r = \frac{2}{9}$

Exercise 19I

1 $x = 5$ or $x = -2$ **2** $x = 1$ or $x = 4$ **3** $x = 8$ or $x = -7$
4 $x = -5$ or $x = 4$ **5** $x = 9$ or $x = 7$ **6** $x = -11$ or $x = 5$
7 $x = -3$ or $x = -1$ **8** $x = 3$ or $x = -2$ **9** $x = 4$ or $x = 2$

Exercise 19J

1 $x = -4$ or $x = 3$ **2** $x = 5$ or $x = -4$ **3** $x = -1$
4 $z = 4$ or $z = -1$ **5** $q = -5$ or $q = 3$ **6** $w = 2$
7 $t = 7$ or $t = -1$ **8** $p = -3$ **9** $x = 5$

Exercise 19K

1 a $t + 3$
b $t(t + 3)$
c $t(t + 3) = 40$, $t = -8$ or 5, therefore $t = 5$
2 a

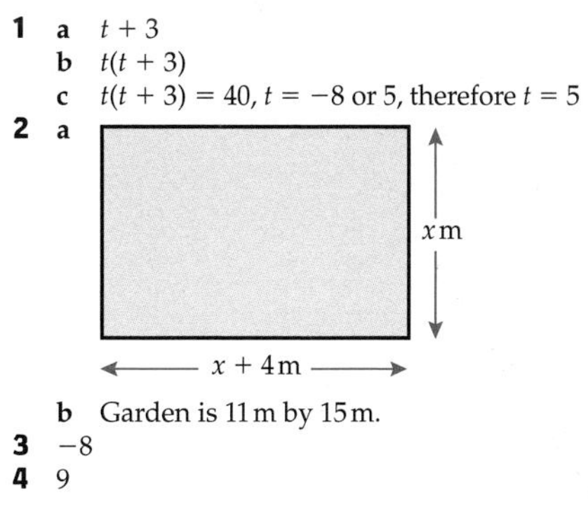

b Garden is 11 m by 15 m.
3 −8
4 9

Review exercise

1 $(x - 5)(x + 3)$
2 $(x - 3)(x + 3)$
3 a $(x + 5)(x + 1)$ b $x = -5$ or -1
4 $x = 3$ or $x = 4$
5 a $(x - 7)(x + 7)$ b $(x + 4)(x - 2)$
6 $x = \pm 6$
7 $x = 7$ or $x = 4$
8 $(x - 9)$
9 a $2t$ b $2t^2$ c $t = 1.2$ m
10 $x = 12$ or $x = -9$

Chapter 20

Number skills: revision exercise 1

1

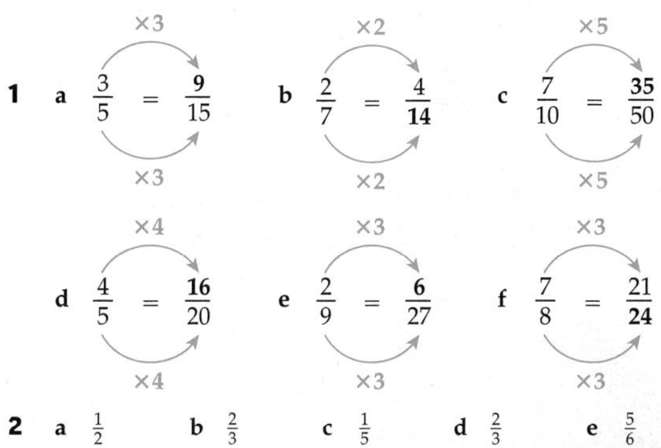

2 a $\frac{1}{2}$ b $\frac{2}{3}$ c $\frac{1}{5}$ d $\frac{2}{3}$ e $\frac{5}{6}$

3 a

Fraction	$\frac{1}{10}$	$\frac{1}{4}$	$\frac{1}{2}$	$\frac{3}{5}$	$\frac{3}{4}$
Decimal	0.1	0.25	0.5	0.6	0.75
Percentage	10%	25%	50%	60%	75%

4 History 88%, Geography 77%, Welsh 70%, Music 79%
5 a 0.5625 b 0.32
6 a 50% b 25% c 87.5%
7 Neither. They both send the same proportion.
8 a $\frac{1}{2}$ b $\frac{1}{5}$ c 3 d $1\frac{1}{2}$ e $1\frac{3}{7}$
9 a $1\frac{1}{3}$ b 1
10 a $\frac{1}{3}$ b $\frac{1}{3}$ c They are the same.
11 3 : 2
12 5 : 2
13 a 10 b 2 c 18 d 15
e 9.75 f 141.6 g 14.5 h 50
14 a 19 b 25 c 45
15 379
16 90
17 a 37 + **45** = 82 b **29** × 8 = 232
c **84** − 45 = 39 d 380 ÷ **4** = 95
18 5
19 a 8.016 666 667 b 8.0 c 8.02
20 a 3.5 b 2.47 c 83.815
d 300 e 8 f 20
21 114.1 m²

Number skills: revision exercise 2

1 $\frac{4}{28}$ and $\frac{2}{14}$
2 $\frac{21}{58}$
3 $\frac{4}{6}$
4 a $\frac{4}{5}$ b $\frac{1}{5}$ c $\frac{5}{6}$ d $\frac{2}{3}$
5 a 0.4 b 0.47
6 a

Fraction	$\frac{7}{100}$	$\frac{7}{20}$	$\frac{3}{5}$
Decimal	0.07	0.35	0.6
Percentage	7%	35%	60%

7 $\frac{18}{20}$ of £3500 = £3150
8 a $\frac{1}{6}$ b $\frac{1}{12}$ c 2 2 $2\frac{1}{2}$
9 5:2
10 35:17
11 a 17 b 17 c 6 d 42
12 a **373** + **456** = 829 b 1596 ÷ **38** = 42
c **19** × **14** = 266 d 1046 − **799** = 247
13 a £15.96 b £4.04
14 55.51
15 £1987
16 a 6 b £4.20
17 a £386.25 b £471.25
18 £1.29
19 3.7
20 a 1386 b 1930 c 9409
21 a 4.7089 b 4.71
22 a 300 b 70 000 c 0.2 d 2
23 Yes. 8755 m to 1 s.f. is 9000 m
24 C 45 sweets. 45 to 1 s.f. is 50 but 42 rounds down to 40.
25 a 1.76 m b 28°C

Chapter 21

Geometry skills: angles

1 East
2 Student's drawing of obtuse angle
3 a $a = 130°$ b $x = 40°$
c $m = 115°$ d $n = 60°$
4 a Acute b Obtuse
5 a Angle *ABC* = 80°, angle *BCD* = 120°, angle *CDA* = 70°, angle DAB = 90°
b *AB* and *AD*
6 D, E, A, B, C, G, H, F

Exercise 21A

1 a $y = 160°$ b $z = 50°$
c $k = 40°$ d $l = 100°, m = n = 80°$
e $p = 30°, r = q = 150°$ f $t = u = s = 90°$
2 a $m = 40°$ b $n = 40°, 2n = 80°$
c $x = 80°$ d $y = 40°, 2y = 80°, 3y = 120°$
3 a $a = 60°, 2a = 120°$
b $y = 60°, 2y = 120°, x = 36°, 3x = 108°$
c $m = 133°, n = 52°$
d $p = 30°$
4 $x = y = 140°$
5 60°
6 $a = b = 120°, c = 60°, d = e = 85°$

Exercise 21B

1 a $l = 50°$, alternate angles
b $m = 110°$, alternate angles
c $n = 70°$, corresponding angles
d $f = 30°$ corresponding angles
e $r = 40°$, vertically opposite angles; $q = 40°$, alternate angles
f $t = 75°$, vertically opposite angles; $s = 75°$, alternate angles *or* $s = 75°$, corresponding angles; $t = 75°$, alternate angles
2 a, b

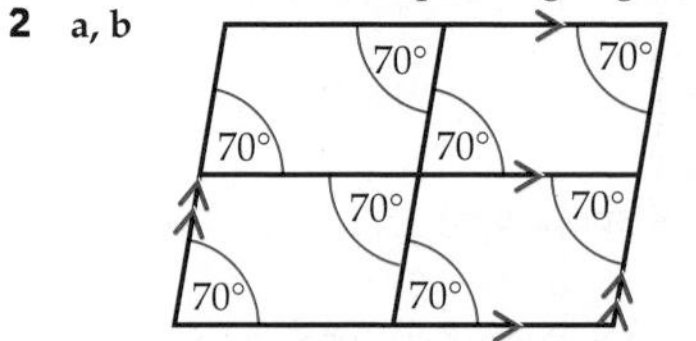

c Use angles on a straight line, vertically opposite, corresponding and alternate angles.

Exercise 21C

1 a $a = 85°$, vertically opposite angles; $b = 85°$, alternate *or* corresponding angles; $c = 85°$, vertically opposite angles
b $p = 70°$, vertically opposite angles; $m = 70°$, corresponding *or* alternate angles; $n = 70°$, vertically opposite *or* corresponding angles
c $r = 75°$, corresponding angles; $s = 105°$, angles on a straight line
d $t = 120°$, angles on a straight line with corresponding angle to 120°
e $u = 30°$, vertically opposite angles; $v = w = 150°$, angles on straight line/vertically opposite; $x = 150°$, corresponding angles
f $a = 45°$, vertically opposite angles; $b = 45°$ corresponding angles; $c = 135°$, angles on a straight line
2 $y = 60°$, corresponding angles; $x = 60°$, vertically opposite angles
3 m is corresponding *or* alternate angle to 120°, found by angles on a straight line
4 a

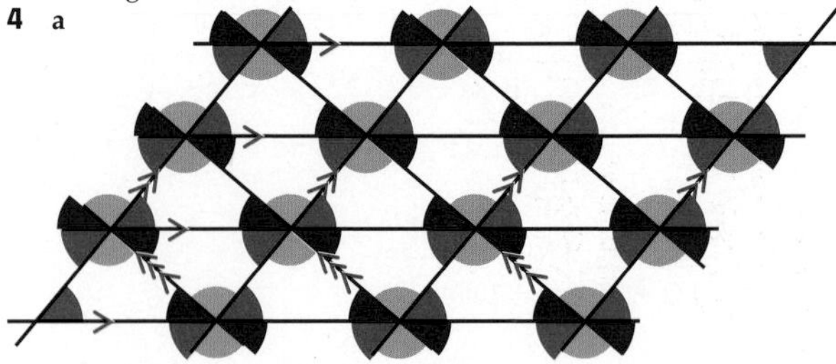

b 90°, angles in a triangle *or* angles on a straight line
5 a Student's drawing
b Angle *PSR* = 120°, angle *QPS* = angle *SRQ* = 60°
c Angles add up to 360°, opposite angles are equal
d Yes
6 Angle *ABC* = 100°, angle *DAB* = 110°

Exercise 21D

1 a 070° b 120°
c 210° d 340°
2 a 130° b 310°
3 a 054° b 095°
c 008° d 188°
e 234° f 275°
4 Student's accurate drawings

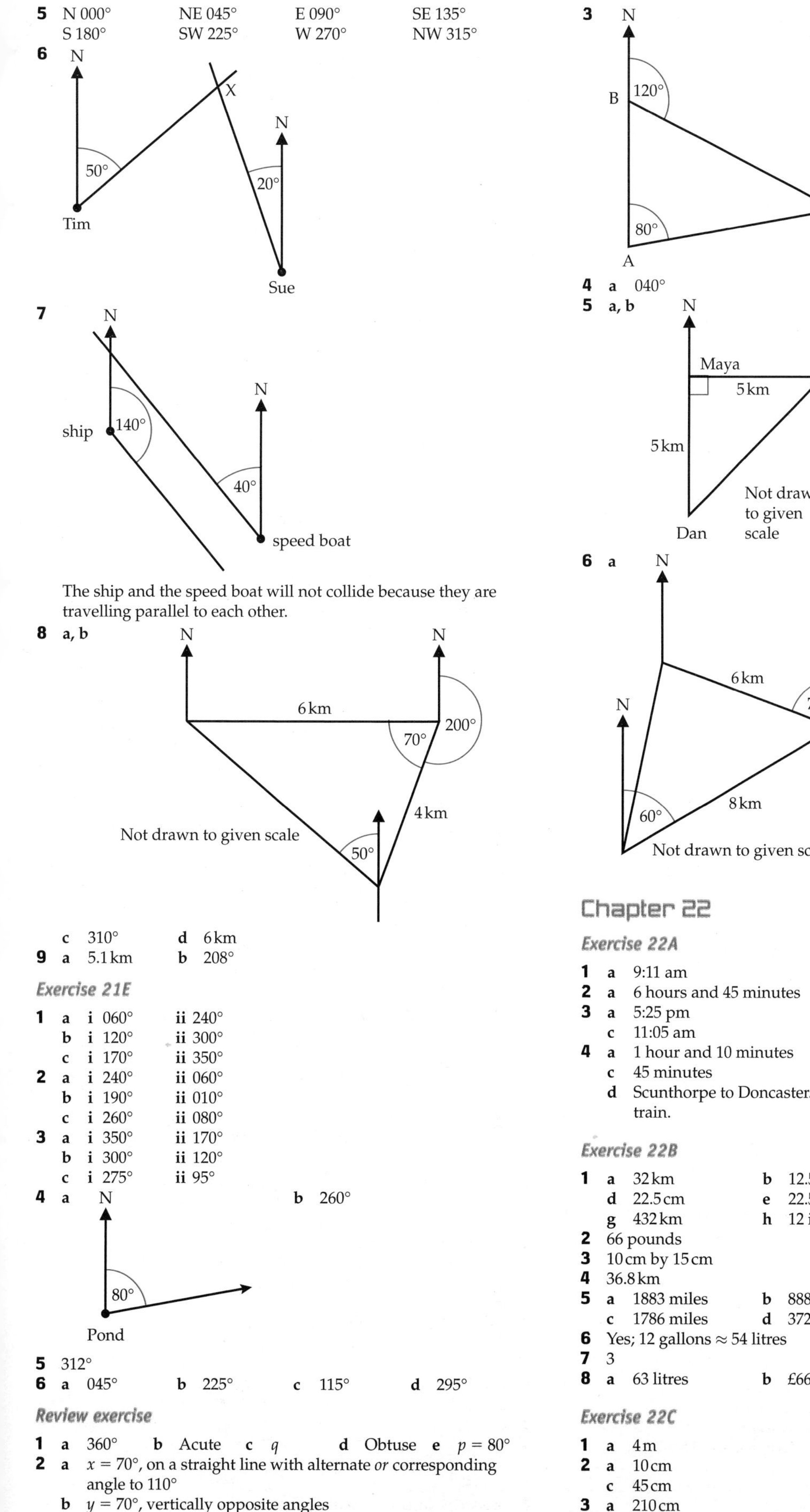

5 N 000° S 180° NE 045° SW 225° E 090° W 270° SE 135° NW 315°

6

7

The ship and the speed boat will not collide because they are travelling parallel to each other.

8 **a, b**

c 310° **d** 6 km

9 **a** 5.1 km **b** 208°

Exercise 21E

1 **a** i 060° ii 240°
b i 120° ii 300°
c i 170° ii 350°

2 **a** i 240° ii 060°
b i 190° ii 010°
c i 260° ii 080°

3 **a** i 350° ii 170°
b i 300° ii 120°
c i 275° ii 95°

4 **a** **b** 260°

5 312°

6 **a** 045° **b** 225° **c** 115° **d** 295°

Review exercise

1 **a** 360° **b** Acute **c** q **d** Obtuse **e** $p = 80°$

2 **a** $x = 70°$, on a straight line with alternate *or* corresponding angle to 110°
b $y = 70°$, vertically opposite angles

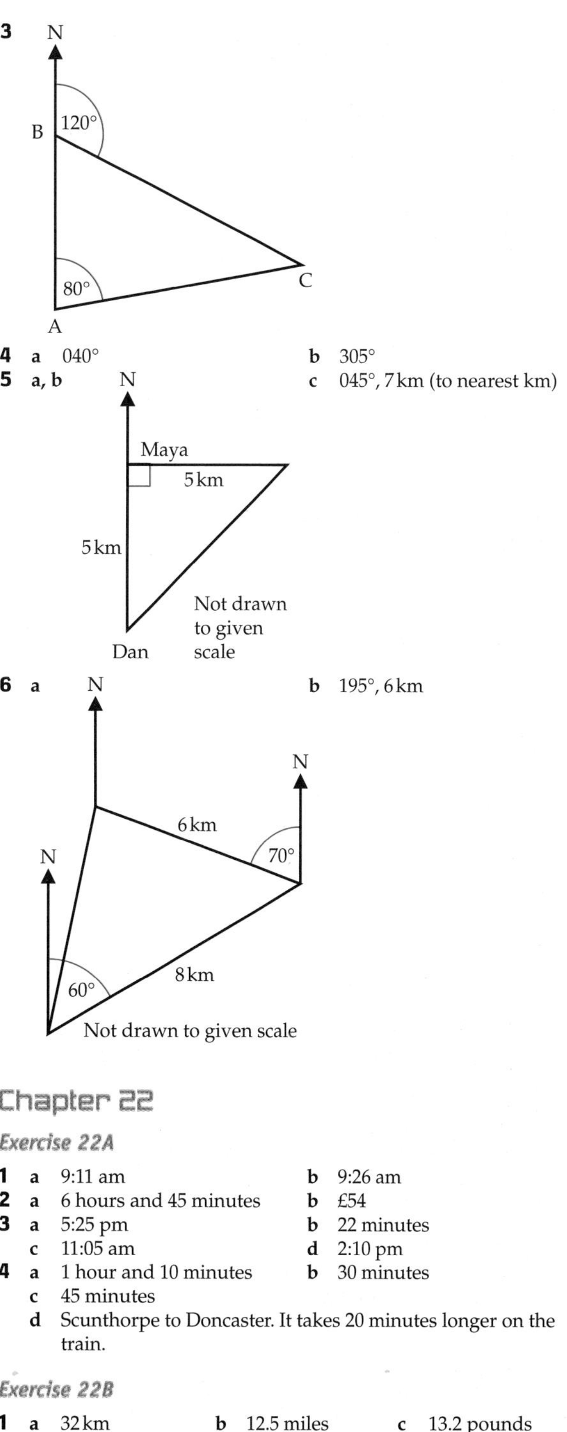

3

4 **a** 040° **b** 305°

5 **a, b** **c** 045°, 7 km (to nearest km)

6 **a** **b** 195°, 6 km

Chapter 22

Exercise 22A

1 **a** 9:11 am **b** 9:26 am

2 **a** 6 hours and 45 minutes **b** £54

3 **a** 5:25 pm **b** 22 minutes
c 11:05 am **d** 2:10 pm

4 **a** 1 hour and 10 minutes **b** 30 minutes
c 45 minutes
d Scunthorpe to Doncaster. It takes 20 minutes longer on the train.

Exercise 22B

1 **a** 32 km **b** 12.5 miles **c** 13.2 pounds
d 22.5 cm **e** 22.5 litres **f** 30 kg
g 432 km **h** 12 inches **i** 6 gallons

2 66 pounds

3 10 cm by 15 cm

4 36.8 km

5 **a** 1883 miles **b** 888 miles
c 1786 miles **d** 3721 miles

6 Yes; 12 gallons ≈ 54 litres

7 3

8 **a** 63 litres **b** £66.15

Exercise 22C

1 **a** 4 m **b** 1.6 cm

2 **a** 10 cm **b** 50 cm
c 45 cm **d** 12.5 cm

3 **a** 210 cm **b** 15 cm

4 a 14 m. b 16 m
c No. The garden is long enough but not wide enough. It needs to be 4.5 cm wide on the drawing.
5 a 7.5 km b 16 cm
6 a 2 km b 1.4 cm
c 48 mm (allow 49 mm) d 2.4 km (allow 2.45 km)
e 6.5 km (allow ±0.2 km)
7 a 625 m by 1000 m b 625 000 m^2 (or 0.625 km^2)

Exercise 22D

1 a 145.5 cm b 144.5 cm
2 a 330.5 ml b 329.5 ml
3 a 4.5°C b 3.5°C
4 a 3.5 cm and 7.5 cm b 26.25 cm
5 No. The mass could be outside this range, e.g. 73.495 g.

Exercise 22E

1 a 55 litres b 45 litres
2 a 47.25 s; 48.15 s; 48.35 s b 47.15 s; 48.05 s; 48.25 s
3 a 1450 kg b 1350 kg
4 a 25.88 m^2 (2 d.p.) b 26.92 m^2 (2 d.p.)

Review exercise

1 220 litres
2 112 km/h
3 a 09 20 b 15 minutes
c 11 17 d 10 20
4 Accurate scale drawing

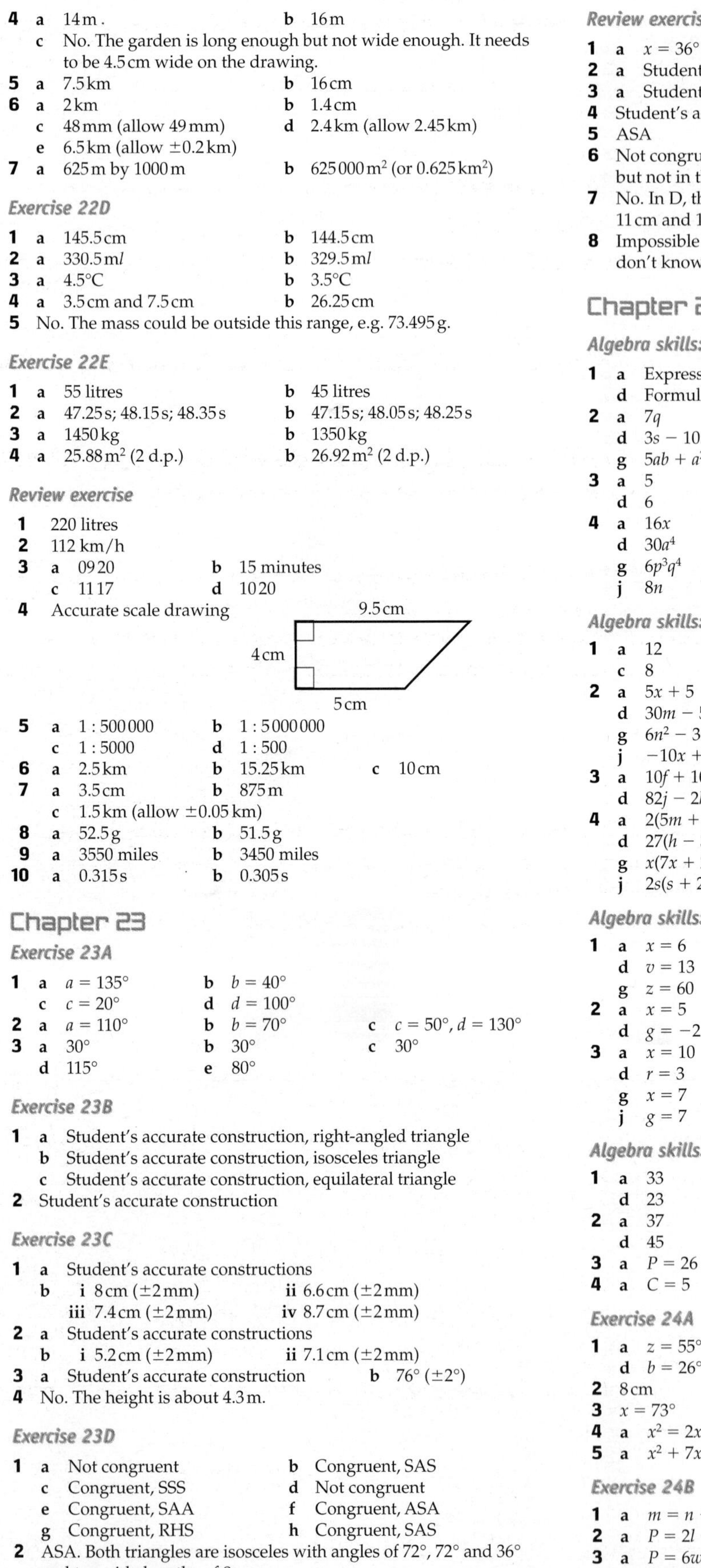

5 a 1 : 500 000 b 1 : 5 000 000
c 1 : 5000 d 1 : 500
6 a 2.5 km b 15.25 km c 10 cm
7 a 3.5 cm b 875 m
c 1.5 km (allow ±0.05 km)
8 a 52.5 g b 51.5 g
9 a 3550 miles b 3450 miles
10 a 0.315 s b 0.305 s

Chapter 23

Exercise 23A

1 a $a = 135°$ b $b = 40°$
c $c = 20°$ d $d = 100°$
2 a $a = 110°$ b $b = 70°$ c $c = 50°, d = 130°$
3 a 30° b 30° c 30°
d 115° e 80°

Exercise 23B

1 a Student's accurate construction, right-angled triangle
b Student's accurate construction, isosceles triangle
c Student's accurate construction, equilateral triangle
2 Student's accurate construction

Exercise 23C

1 a Student's accurate constructions
b i 8 cm (±2 mm) ii 6.6 cm (±2 mm)
iii 7.4 cm (±2 mm) iv 8.7 cm (±2 mm)
2 a Student's accurate constructions
b i 5.2 cm (±2 mm) ii 7.1 cm (±2 mm)
3 a Student's accurate construction b 76° (±2°)
4 No. The height is about 4.3 m.

Exercise 23D

1 a Not congruent b Congruent, SAS
c Congruent, SSS d Not congruent
e Congruent, SAA f Congruent, ASA
g Congruent, RHS h Congruent, SAS
2 ASA. Both triangles are isosceles with angles of 72°, 72° and 36° and two side lengths of 8 cm.
3 B is the odd one out. A and C are SAS as they have two sides of length 6 cm and the included angle of 58°.

Review exercise

1 a $x = 36°$ b $y = 108°$
2 a Student's accurate construction b 43° ±2°
3 a Student's accurate construction b 4.6 cm ±2 mm
4 Student's accurate constructions. ST = 7.3 cm ±2 mm
5 ASA
6 Not congruent. The 15 cm side is the hypotenuse in one triangle but not in the other.
7 No. In D, the 50° angle isn't the included angle between the 11 cm and 12 cm sides.
8 Impossible to tell. All three angles are the same size, but we don't know any of the side lengths.

Chapter 24

Algebra skills: expressions

1 a Expression b Formula c Expression
d Formula e Equation f Expression
2 a $7q$ b $7x + 4y$ c $5e - 6f + 3$
d $3s - 10t + 13$ e $4x^2 - 3x$ f $7mn + 3m + mn^2$
g $5ab + a^2 - a$ h $3b^3 + 2b^2 + 4b$
3 a 5 b 3 c 7
d 6
4 a $16x$ b $6m^2$ c $12x^5$
d $30a^4$ e $8ab$ f $12x^2y$
g $6p^3q^4$ h $24m^3n^5$ i $8x$
j $8n$ k $7h^3$ l $5x^2$

Algebra skills: brackets

1 a 12 b 5
c 8 d 24
2 a $5x + 5$ b $12a + 30$ c $6 - 9x$
d $30m - 50n$ e $4x + x^2$ f $2a^2 + 10a$
g $6n^2 - 3n$ h $6p^2 - 15pq$ i $-3x - 3$
j $-10x + 6$ k $-2ab - a^2$ l $-12m^3 + 18m^2$
3 a $10f + 10g$ b $29h + 8i$ c $-x - 13$
d $82j - 2k$ e $-7m - 42n$ f $-8a^2 + 23a$
4 a $2(5m + 4)$ b $4(r - 5)$ c $13(2j + 1)$
d $27(h - 3)$ e $4(4 - x)$ f $3(7 - 20y)$
g $x(7x + 2)$ h $m(5m - 9)$ i $y(1 + 11y)$
j $2s(s + 2)$ k $2k(5 + k)$ l $j^2(4j + 3)$

Algebra skills: solving equations

1 a $x = 6$ b $s = 9$ c $p = 8$
d $v = 13$ e $m = 18$ f $y = 60$
g $z = 60$ h $d = 126$ i $k = 84$
2 a $x = 5$ b $h = 20$ c $k = 3$
d $g = -2$ e $x = 11$ f $x = -2$
3 a $x = 10$ b $y = 7$ c $m = 0.5$
d $r = 3$ e $t = -7$ f $j = -6$
g $x = 7$ h $z = 13$ i $q = 0.2$
j $g = 7$ k $t = 20$ l $w = 0.3$

Algebra skills: formulae

1 a 33 b 6 c 3
d 23 e 26 f 23
2 a 37 b 81 c 90
d 45 e 8 f 27
3 a $P = 26$ b $P = 28$ c $P = 7$ d $P = 340$
4 a $C = 5$ b $C = -10$ c 25°C d −20°C

Exercise 24A

1 a $z = 55°$ b $x = 70°$ c $a = 120°$
d $b = 26°$ c $c = 50°$
2 8 cm
3 $x = 73°$
4 a $x^2 = 2x + 15$ b $x = 5$ or -3
5 a $x^2 + 7x = 78$ b $x = 6$ or -13

Exercise 24B

1 a $m = n \div 5$ b 6
2 a $P = 2l + 2w$ b $l = 3$ c $w = 5$
3 a $P = 6w + 6$ b $w = 5.5$
4 a $C = 4a + 7b$ b $a = 26$
5 a $S = \frac{n - 3}{x}$ b $n = 43$

Exercise 24C

1 a M b R c v d k

2 a $a = \frac{M}{2} - b$ b $a = \frac{R}{k}$

c $a = \frac{v - u}{t}$ d $a = \frac{2B}{k}$

3 a base $= \frac{2 \times \text{area}}{\text{height}}$

b i $x = 6$ cm ii $x = 3$ cm
iii $x = 3$ m iv $x = 3$ cm

4 a $b = \sqrt{c^2 - a^2}$ b $b = 5.66$ (3 s.f.)

Exercise 24D

1 a Alternate angles are equal.
b e.g. $a = d$ (alternate angles)
$b = c$ (alternate angles)
So $a + b = c + d$

2 e.g

$a + b + c = 180°$ (angles in a triangle)
$d + e + f = 180°$ (angles in a triangle)
So $b + (a + d) + e + (c + f) = 360°$

3 e.g.

$a + b + x = 180°$ (angles in a triangle)
$c + x = 180°$ (angles on a straight line)
So $a + b + x = c + x$
So $a + b = c$

4 e.g.

$y = b$ (base angles in isosceles triangle)
$b + y + x = 180°$ (angles in a triangle)
So $2b + x = 180°$
$a + x = 180°$ (angles on a straight line)
So $2b + x = a + x$
So $2b = a$

5 e.g. Angles add up to 540° (Using sum of interior angles = $180°(n - 2)$ or subdivision into triangles)
All angles equal as regular polygon, so interior angle = $540° \div 5 = 108°$

Exercise 24E

1 $x = 3.9$ (1 d.p.) e.g.

x	$x^3 - 2x$	Comment
3	21	Too low
5	115	Too high
4	56	Too high
3.5	35.875	Too low
3.8	47.272	Too low
3.9	51.519	Too high
3.85	49.3…	Too low

2 $x = 8.4$ (1 d.p.)
3 $x = 3.5$ (1 d.p.)
4 $x = 4.4$ (1 d.p.)
5 a $x = 4.3$ (1 d.p.) b $x = -1.5$
6 a e.g. $8.45^3 - 8.45 = 594 < 600$ and $8.4^3 - 8.4 = 584 < 600$
b $x = 8.5$
7 $x = -3.2$ (1 d.p.)
8 $t = 6.7$ (1 d.p.)
9 a $x(x^2 + 5) = 800$ b $x = 9.1$

Review exercise

1 a $x = 60°$ b $y = 18°$
2 a £54.50 b Pay $= 5.5p + 8q$ c £75.50
3 e.g.

$a = x$ (alternate angles)
$b + x = 180°$ (angles on a straight line)
So $a + b = 180°$

4 a e.g. $a + b + 90° = 180°$ (angles in a triangle)
So $a + b = 90°$
b e.g. $b + c = 180°$ (angles on a straight line)
$a + d = 180°$ (angles on a straight line)
So $a + b + c + d = 360°$
So $90° + c + d = 360°$
So $c + d = 270°$

5 $x = 4.4$ (1 d.p.)
6 $x = 6.8$ (1 d.p.)
7 $x = 13.6$. e.g.

x	$\frac{x^2}{\sqrt{x}}$	Comment
10	31.62…	Too low
20	89.44…	Too high
15	58.09…	Too high
13	46.87…	Too low
13.5	49.60…	Too low
13.6	50.15…	Too high
13.55	49.87…	Too low

8 $q = r(p - 5) - 1$
9 $b = \frac{2A - ah}{h}$
10 $x = -6$

Chapter 25

Exercise 25A

1 a $a = 240°$ b $b = 115°$ c $c = 110°$ d $d = 65°$
2 210°
3 45°

Exercise 25B

1 a i $2x + 290 = 360$ ii $x = 35°$
b i $3x + 240 = 360$ ii $x = 40°$
c i $4x + 220 = 360$ ii $x = 35°$
d i $10x = 360$ ii $x = 36°$
2 $x = 28°$
3 150°
4 He is not correct. $x = 47°$, largest angle $= 3x = 141°$, smallest angle $= 35°$

Exercise 25C

1 a b c

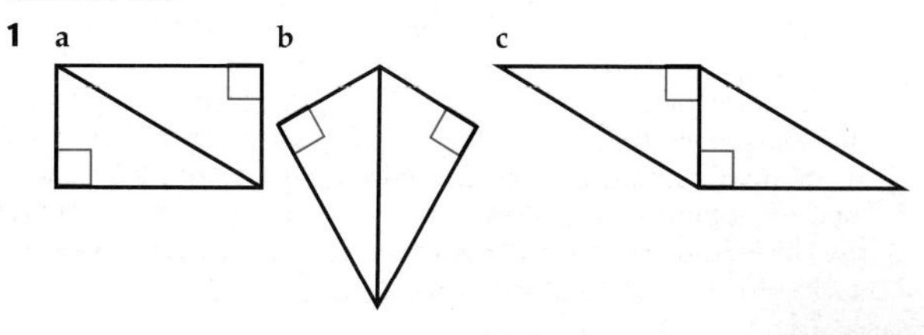

2

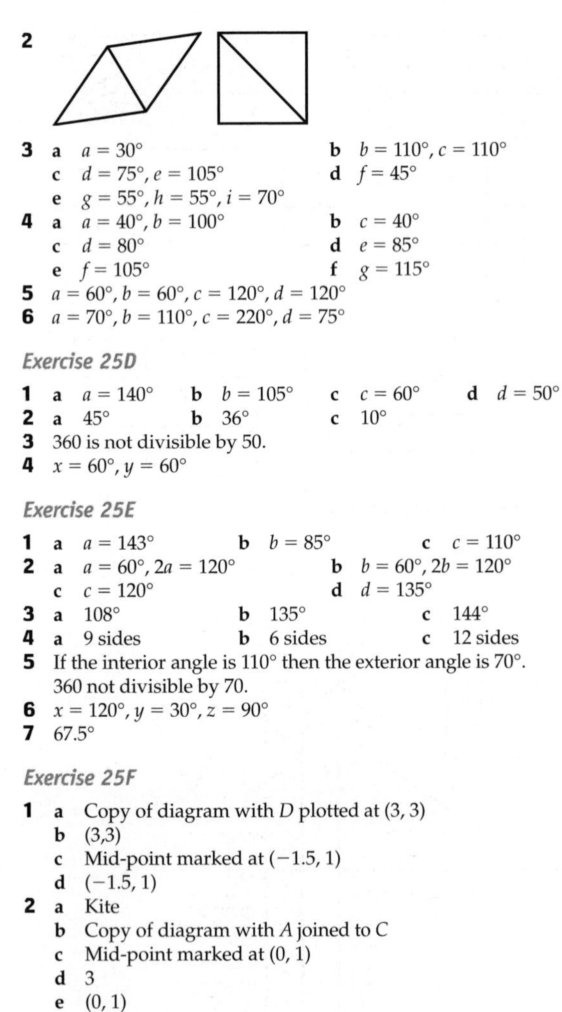

3 **a** $a = 30°$ **b** $b = 110°, c = 110°$
c $d = 75°, e = 105°$ **d** $f = 45°$
e $g = 55°, h = 55°, i = 70°$
4 **a** $a = 40°, b = 100°$ **b** $c = 40°$
c $d = 80°$ **d** $e = 85°$
e $f = 105°$ **f** $g = 115°$
5 $a = 60°, b = 60°, c = 120°, d = 120°$
6 $a = 70°, b = 110°, c = 220°, d = 75°$

Exercise 25D

1 **a** $a = 140°$ **b** $b = 105°$ **c** $c = 60°$ **d** $d = 50°$
2 **a** 45° **b** 36° **c** 10°
3 360 is not divisible by 50.
4 $x = 60°, y = 60°$

Exercise 25E

1 **a** $a = 143°$ **b** $b = 85°$ **c** $c = 110°$
2 **a** $a = 60°, 2a = 120°$ **b** $b = 60°, 2b = 120°$
c $c = 120°$ **d** $d = 135°$
3 **a** 108° **b** 135° **c** 144°
4 **a** 9 sides **b** 6 sides **c** 12 sides
5 If the interior angle is 110° then the exterior angle is 70°. 360 not divisible by 70.
6 $x = 120°, y = 30°, z = 90°$
7 67.5°

Exercise 25F

1 **a** Copy of diagram with D plotted at (3, 3)
b (3,3)
c Mid-point marked at (−1.5, 1)
d (−1.5, 1)
2 **a** Kite
b Copy of diagram with A joined to C
c Mid-point marked at (0, 1)
d 3
e (0, 1)
3 No, BC is 5 units but AD would be 4 units.

Review exercise

1 $a = 140°$
2 **a** Copy of diagram with D plotted at (2, 2)
b (2, 2)
c Mid-point marked at (−2.5, 0) **d** (−2.5, 0)
3 60°
4 **a** $a + 90 + 110 + 75 = 360$ (or equivalent) **b** $a = 85°$
5 $x = 36°$
6 $a = 35°, b = 110°$
7 30°
8 18 sides
9 36 sides
10 50 doesn't divide into 360 exactly.

Chapter 26

Exercise 26A

1 **a** Area = 95.14 cm², perimeter = 41.8 cm
b Area = 44.64 cm², perimeter = 28.2 cm
c Area = 56.7 cm², perimeter = 34.6 cm
d Area = 55.2 cm², perimeter = 37.6 cm
e Area = 14 cm², perimeter = 20.4 cm
f Area = 33.8 cm², perimeter = 24 cm
2 **a** $x = 4$ cm **b** $y = 12$ cm **c** $z = 5$ cm

Exercise 26B

1 **a** 70 cm² **b** 57 cm² **c** 21 cm²
d 83.6 cm² **e** 95.7 cm² **f** 125.84 cm²
2 **a** 34.8 cm² **b** 31.08 cm² **c** 28.5 cm²
d 22.95 cm² **e** 18.36 cm² **f** 32.9 cm²
3 £236.75

Exercise 26C

1 **a** Area = 69 cm², perimeter = 38 cm
b Area = 54 cm², perimeter = 40 cm
c Area = 97 cm², perimeter = 46 cm
d Area = 99 cm², perimeter = 56 cm
e Area = 66 cm², perimeter = 40 cm
f Area = 91 cm², perimeter = 50 cm
2 **a** 60 cm² **b** 59.5 cm² **c** 44.5 cm² **d** 133.5 cm²

Exercise 26D

1 **a** 593.4 cm³ **b** 260 cm³ **c** 620.5 cm³
d 480 cm³ **e** 876.96 cm³ **f** 1330.8 cm³
2 £357.60

Exercise 26E

1 **a** 267 cm² **b** 408 cm² **c** 216 cm² **d** 640 cm²

Review exercise

1 **a** 53.55 cm² **b** 63 cm² **c** 44.8 cm² **d** 124.2 cm²
2 80 cm
3 **a** i 69 cm² ii 46 cm
b i 86 cm² ii 47.1 cm
4 77.35 cm²
5 **a** i 1920 cm³ ii 944 cm²
b i 175 cm³ ii 241 cm²
6 **a** 14.28 m² **b** 85.68 m³
c Use 6 rolls of felt, cost = £43.98
7 £347.20

Chapter 27

Exercise 27A

1

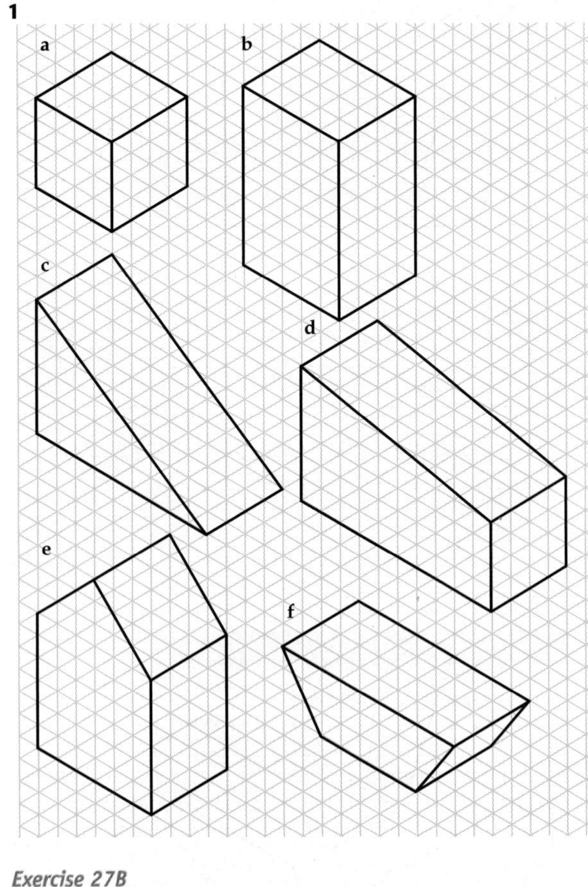

Exercise 27B

1 **a** 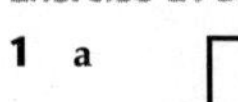 **b** 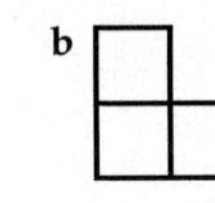 **c** 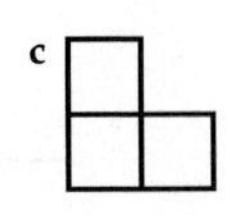

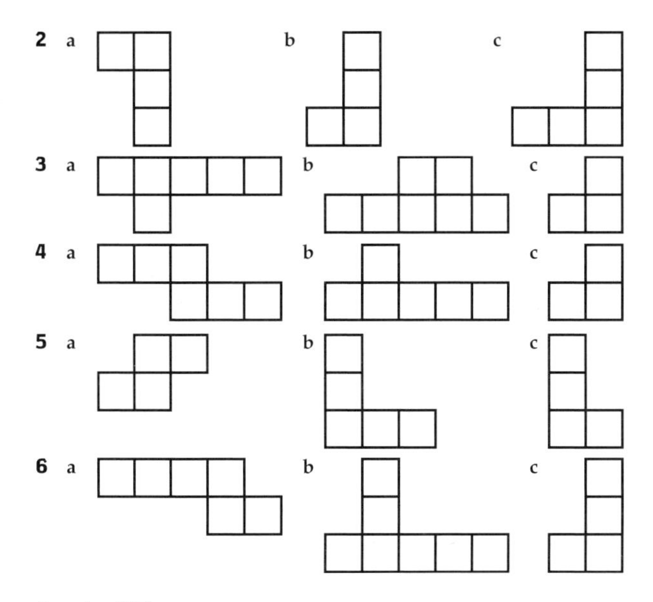

Exercise 27C

1 Volume = $4\,\text{cm}^3$, surface area = $18\,\text{cm}^2$
2 Volume = $6\,\text{cm}^3$, surface area = $26\,\text{cm}^2$
3 Volume = $8\,\text{cm}^3$, surface area = $32\,\text{cm}^2$
4 Volume = $7\,\text{cm}^3$, surface area = $30\,\text{cm}^2$
5 Volume = $6\,\text{cm}^3$, surface area = $26\,\text{cm}^2$
6 Volume = $8\,\text{cm}^3$, surface area = $34\,\text{cm}^2$

Exercise 27D

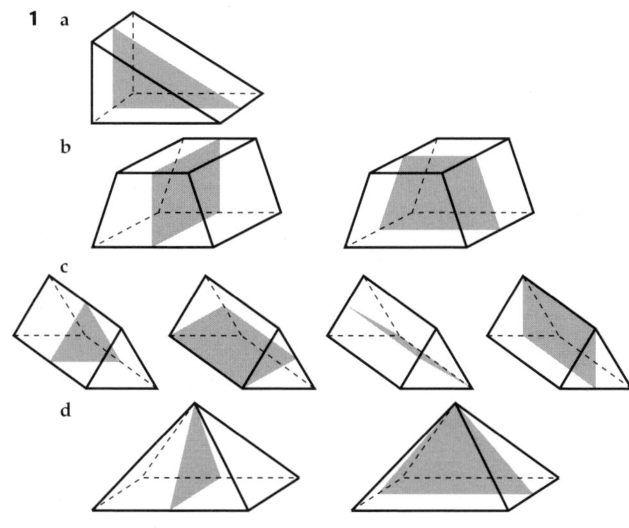

Review exercise

1

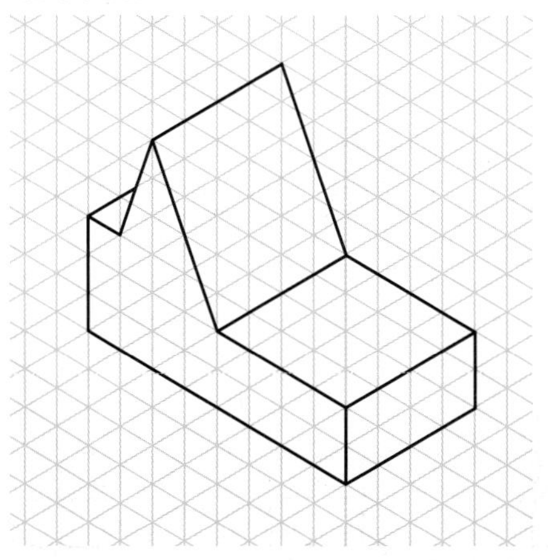

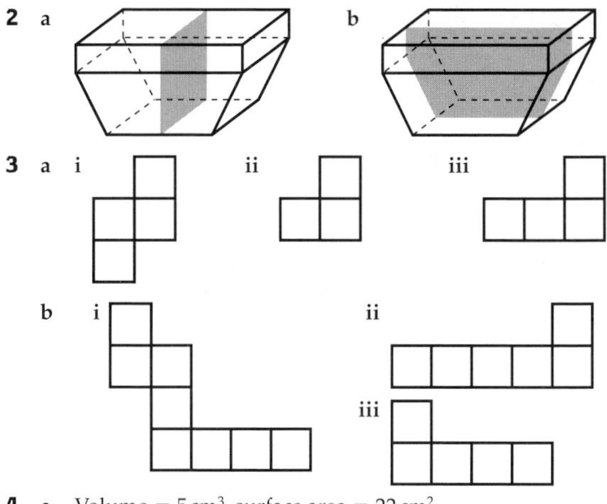

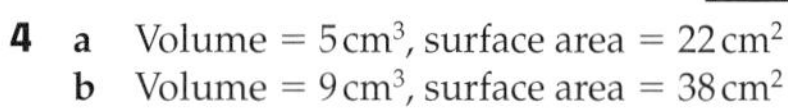

4 a Volume = $5\,\text{cm}^3$, surface area = $22\,\text{cm}^2$
b Volume = $9\,\text{cm}^3$, surface area = $38\,\text{cm}^2$

Chapter 28

Exercise 28A

1-3

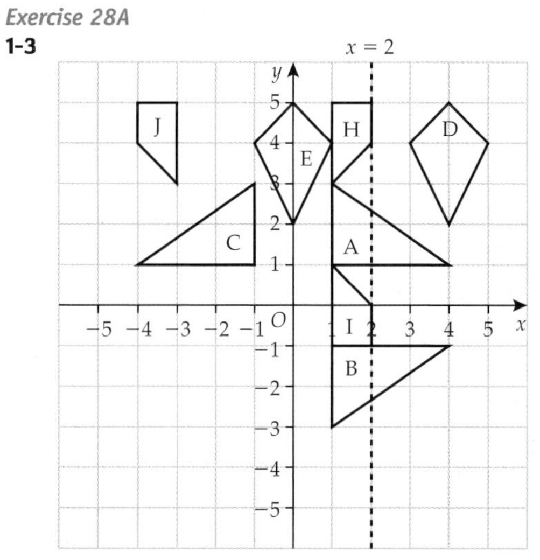

4 a,b Copy of shapes K and L with mirror line drawn at $y = 1$.
c $y = 1$

Exercise 28B

1 a Reflection in the line $x = -1$ b Reflection in the x-axis
c Reflection in the line $x = 4$

2

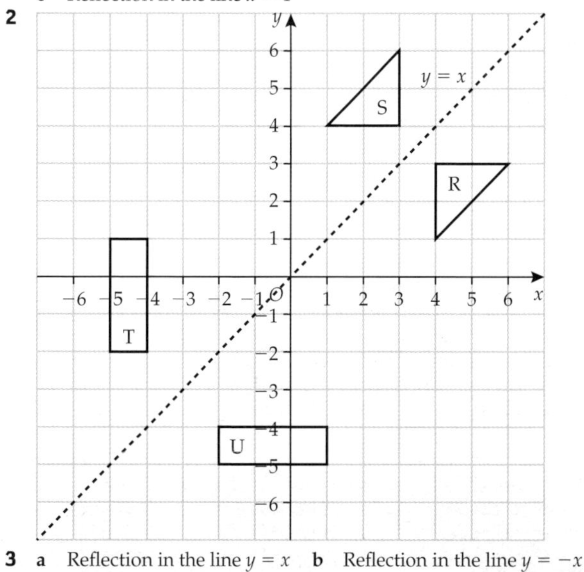

3 a Reflection in the line $y = x$ b Reflection in the line $y = -x$
c Reflection in the line $y = x$

4 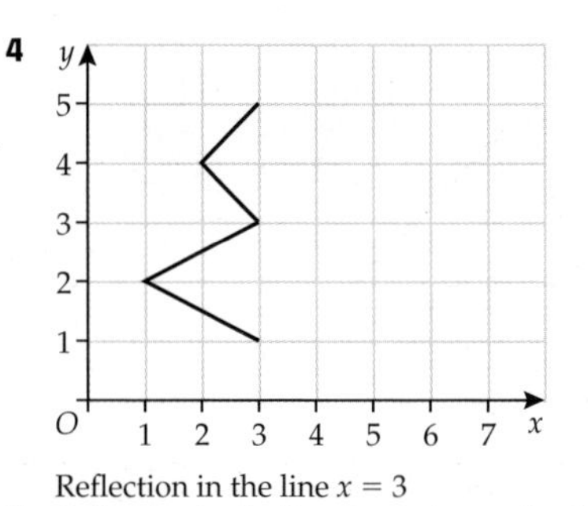

(or the other half of the given shape)

Reflection in the line $x = 3$

5 Reflection in the line $y = 2$, then reflection in the line $x = 3$ (or in the other order)

Exercise 28C

1 **2** **3** **4**

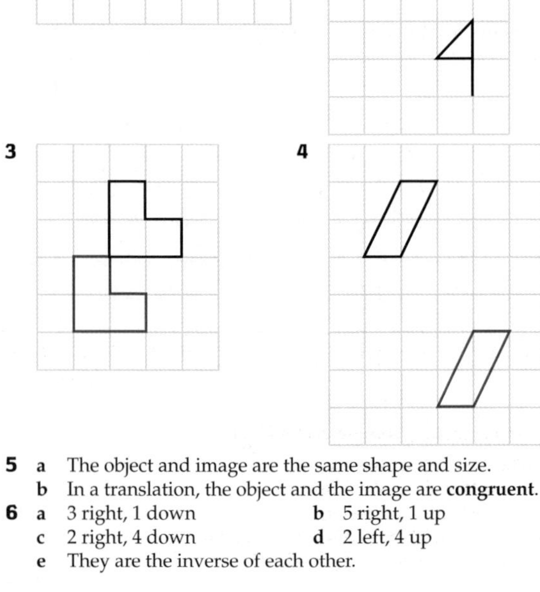

5 **a** The object and image are the same shape and size.

b In a translation, the object and the image are **congruent**.

6 **a** 3 right, 1 down **b** 5 right, 1 up

c 2 right, 4 down **d** 2 left, 4 up

e They are the inverse of each other.

Exercise 28D

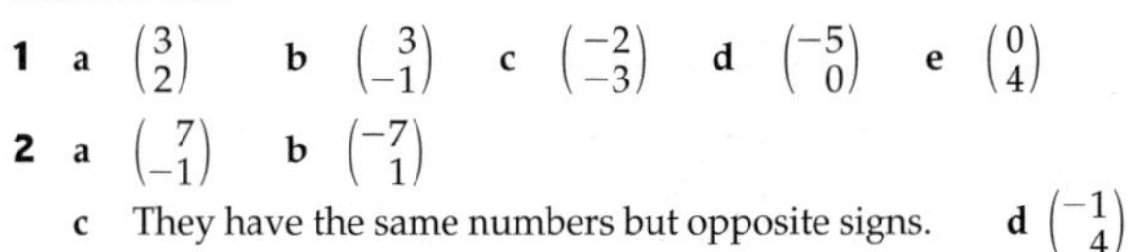

1 **a** $\begin{pmatrix}3\\2\end{pmatrix}$ **b** $\begin{pmatrix}3\\-1\end{pmatrix}$ **c** $\begin{pmatrix}-2\\-3\end{pmatrix}$ **d** $\begin{pmatrix}-5\\0\end{pmatrix}$ **e** $\begin{pmatrix}0\\4\end{pmatrix}$

2 **a** $\begin{pmatrix}7\\-1\end{pmatrix}$ **b** $\begin{pmatrix}-7\\1\end{pmatrix}$

c They have the same numbers but opposite signs. **d** $\begin{pmatrix}-1\\4\end{pmatrix}$

3 **a** $\begin{pmatrix}1\\2\end{pmatrix}$ **b** $\begin{pmatrix}3\\2\end{pmatrix}$ **c** $\begin{pmatrix}4\\4\end{pmatrix}$ **d** $\begin{pmatrix}1\\2\end{pmatrix} + \begin{pmatrix}3\\2\end{pmatrix} = \begin{pmatrix}4\\4\end{pmatrix}$

4 $\begin{pmatrix}5\\1\end{pmatrix}$

5 $\begin{pmatrix}1\\-1\end{pmatrix}$

6 A to B $\begin{pmatrix}2\\0\end{pmatrix}$, A to C $\begin{pmatrix}4\\0\end{pmatrix}$, A to D $\begin{pmatrix}6\\0\end{pmatrix}$, A to E $\begin{pmatrix}0\\-2\end{pmatrix}$,

A to F $\begin{pmatrix}2\\-2\end{pmatrix}$, A to G $\begin{pmatrix}4\\-2\end{pmatrix}$, A to H $\begin{pmatrix}6\\-2\end{pmatrix}$, A to I $\begin{pmatrix}0\\2\end{pmatrix}$,

A to J $\begin{pmatrix}2\\2\end{pmatrix}$, A to K $\begin{pmatrix}4\\2\end{pmatrix}$, A to L $\begin{pmatrix}6\\2\end{pmatrix}$

Exercise 28E

1 **a** **b** **c**

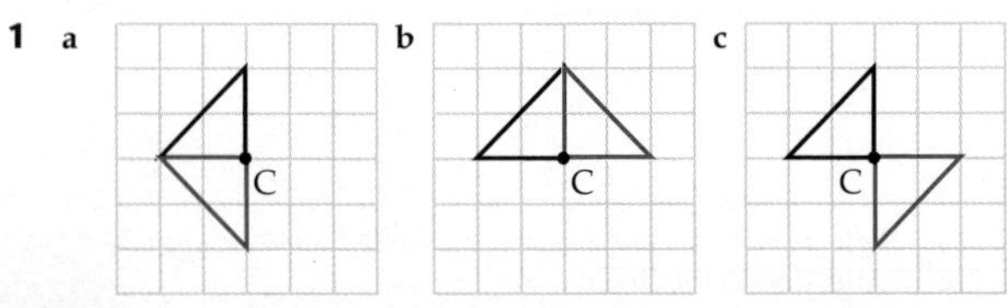

2 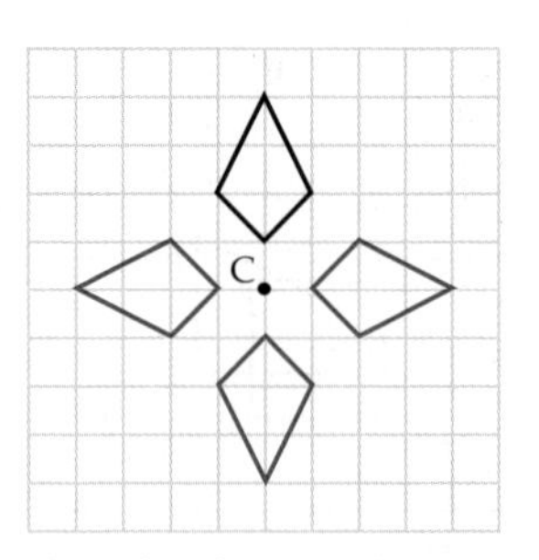

The order of rotational symmetry is 4.

3 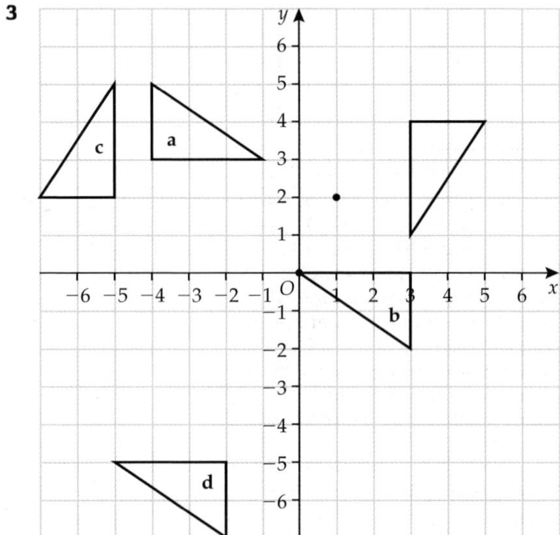

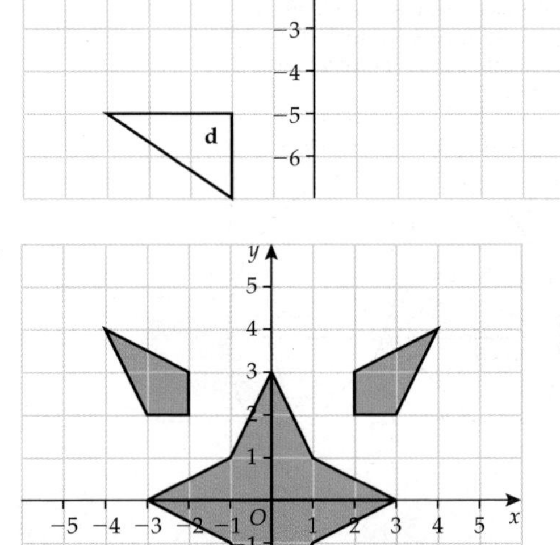

4 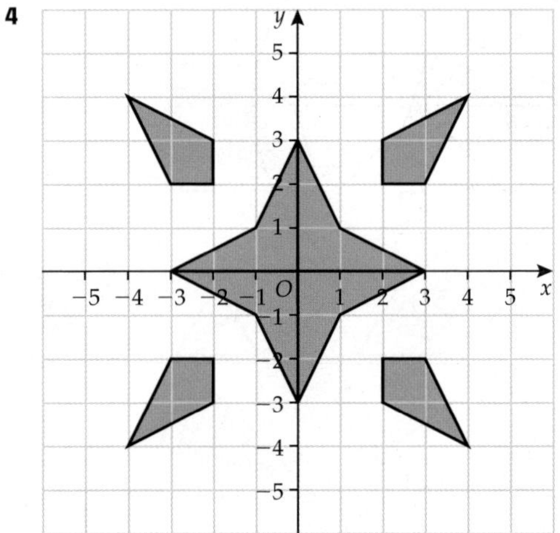

Exercise 28F

1 **a** 180°

b Clockwise or anticlockwise

c

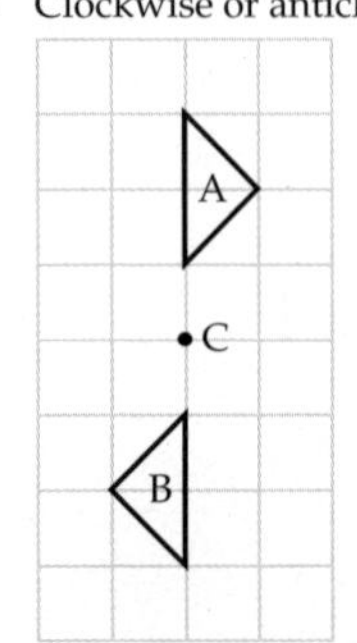

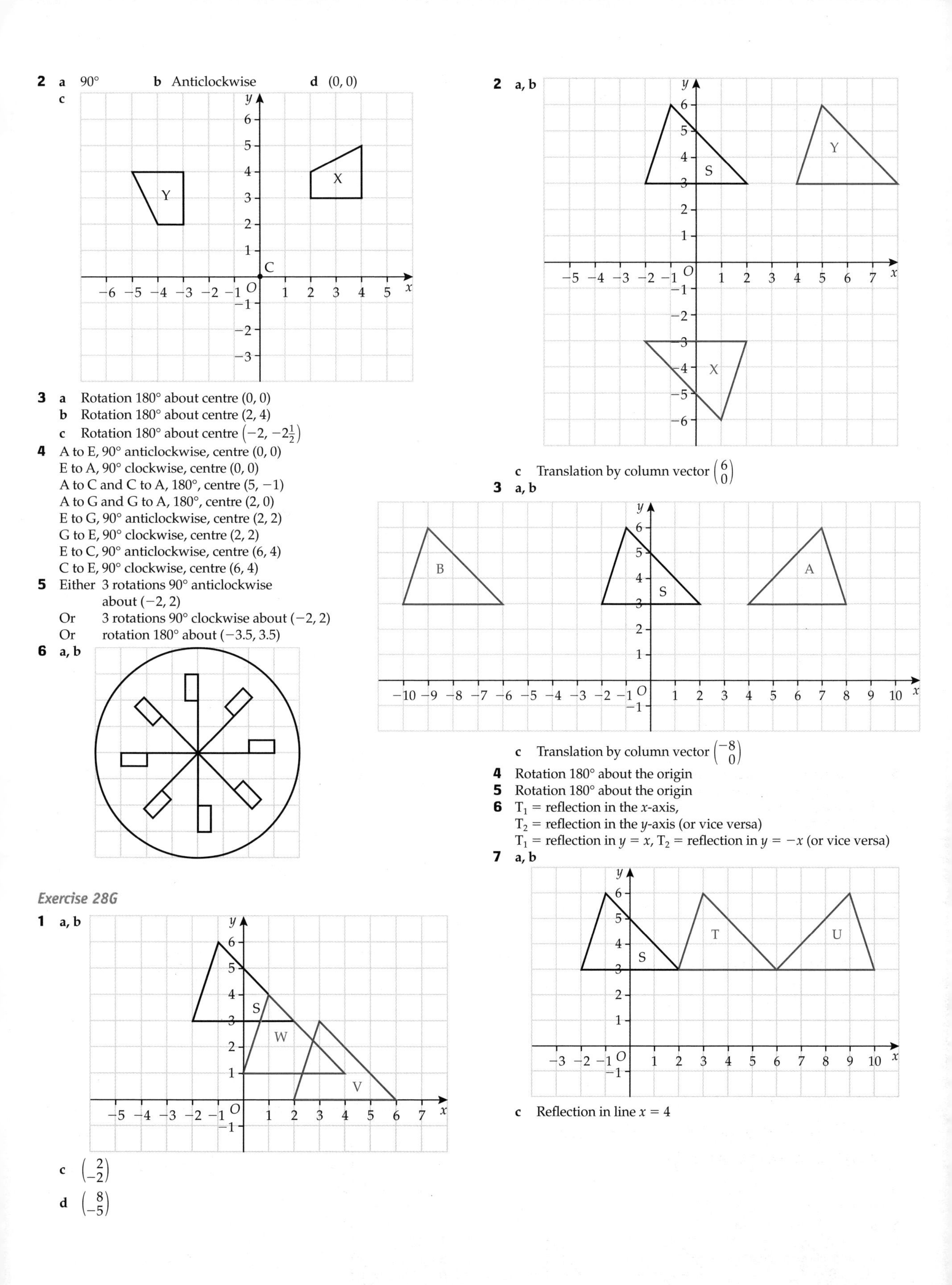

2 **a** 90° **b** Anticlockwise **d** (0, 0)

c

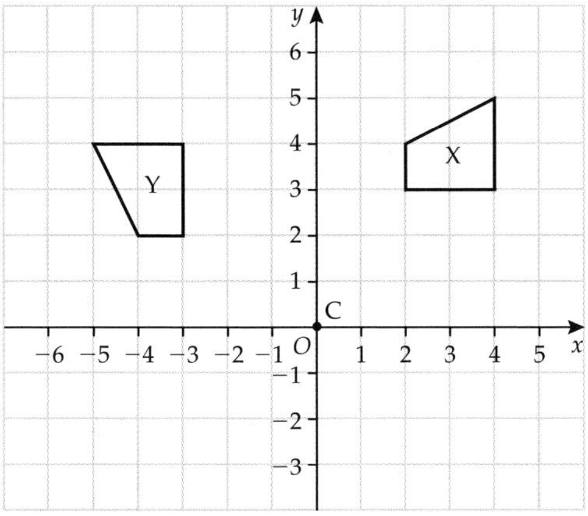

3 **a** Rotation 180° about centre (0, 0)
b Rotation 180° about centre (2, 4)
c Rotation 180° about centre $\left(-2, -2\frac{1}{2}\right)$

4 A to E, 90° anticlockwise, centre (0, 0)
E to A, 90° clockwise, centre (0, 0)
A to C and C to A, 180°, centre (5, −1)
A to G and G to A, 180°, centre (2, 0)
E to G, 90° anticlockwise, centre (2, 2)
G to E, 90° clockwise, centre (2, 2)
E to C, 90° anticlockwise, centre (6, 4)
C to E, 90° clockwise, centre (6, 4)

5 Either 3 rotations 90° anticlockwise about (−2, 2)
Or 3 rotations 90° clockwise about (−2, 2)
Or rotation 180° about (−3.5, 3.5)

6 **a, b**

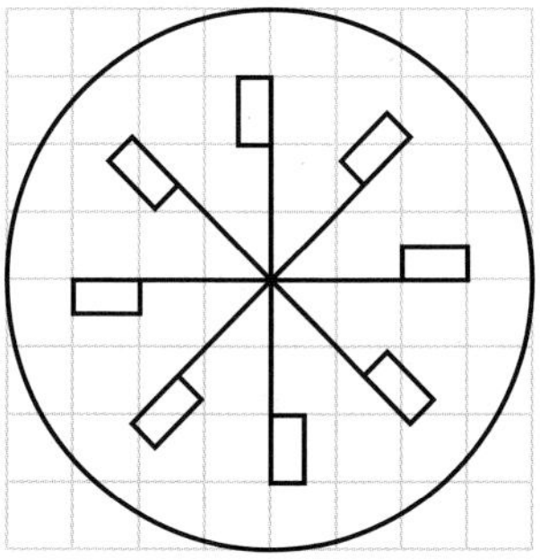

Exercise 28G

1 **a, b**

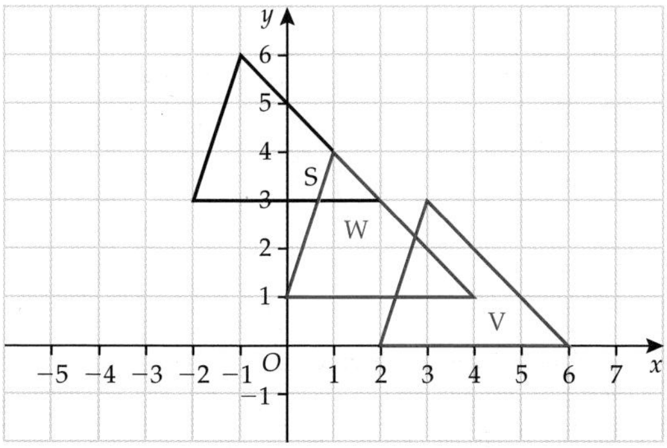

c $\begin{pmatrix}2\\-2\end{pmatrix}$

d $\begin{pmatrix}8\\-5\end{pmatrix}$

2 **a, b**

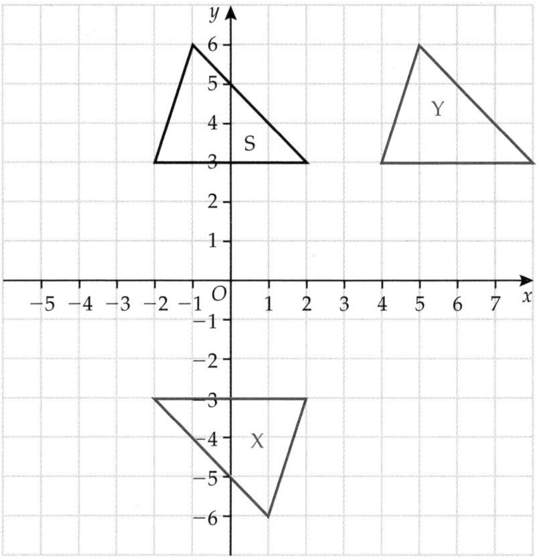

c Translation by column vector $\begin{pmatrix}6\\0\end{pmatrix}$

3 **a, b**

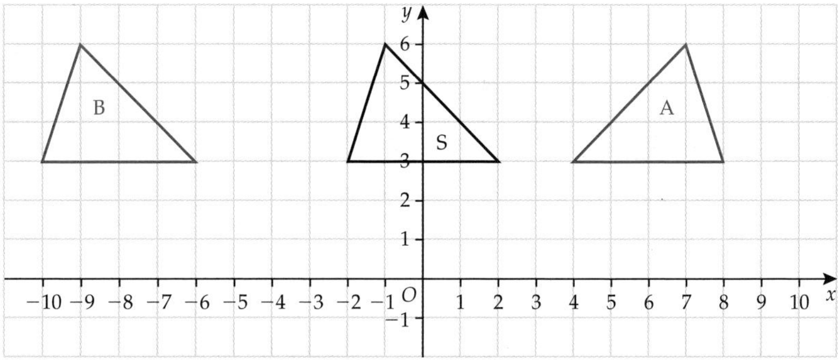

c Translation by column vector $\begin{pmatrix}-8\\0\end{pmatrix}$

4 Rotation 180° about the origin

5 Rotation 180° about the origin

6 T_1 = reflection in the x-axis,
T_2 = reflection in the y-axis (or vice versa)
T_1 = reflection in $y = x$, T_2 = reflection in $y = -x$ (or vice versa)

7 **a, b**

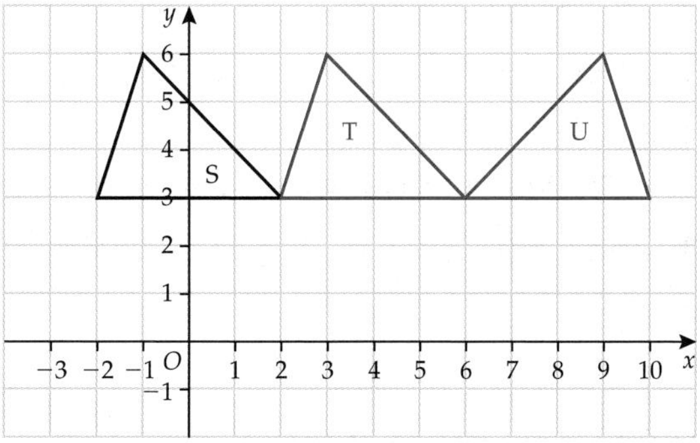

c Reflection in line $x = 4$

8 a, b 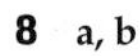

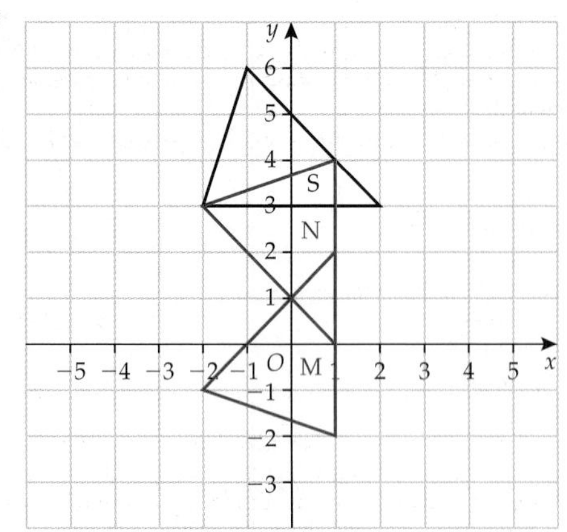

9 a Translation by column vector $\begin{pmatrix}4\\0\end{pmatrix}$

b Translation by column vector $\begin{pmatrix}4\\0\end{pmatrix}$

Review exercise

1 a, b

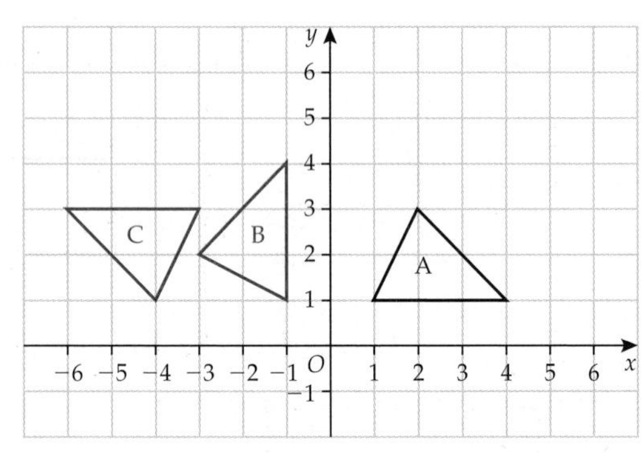

2 a, b

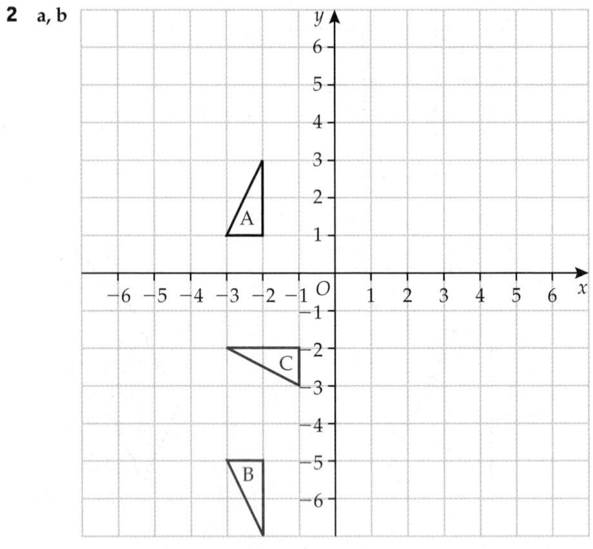

3 a Rectangle B is a reflection of rectangle A in the line $y = -1$

b Rectangle B is a translation of rectangle A by the vector $\begin{pmatrix}0\\-9\end{pmatrix}$

c Rectangle B is a rotation of rectangle A through **180°** about the point **(−2.5, −1)**

4 a Rotation 90° anticlockwise about centre (0, −1)

b Reflection in the line $y = x$

c Reflection in the x-axis

5 a i A ii E iii B

b Reflection in the line $x = -2$

6 H and B

7 E and G

8 Rotation 90° anticlockwise about centre (−1, 3) and reflection in $y = 4$

9 a Rotation

b Two reflections in lines that are not parallel are equivalent to a rotation. The centre of rotation is the point at which the two lines cross. The angle of rotation is equivalent to twice the angle between the lines.

Chapter 29

Exercise 29A

1 a 31.4 cm b 29.8 cm c 10.1 cm
d 75.4 cm e 22.0 cm f 49.0 cm

2 a 4.2π km b 5.8π cm
c 18.3π cm d 0.5π cm

3 a 204 cm b 392 revolutions

4 256 mm

Exercise 29B

1 a 0.955 m b 7.96 cm c 3.98 mm
d 5.95 cm e 27.2 m f 33.6 cm

2 a 9 cm b 4 cm

3 1.67 m

4 a 961 m b 728 turns

Exercise 29C

1 185.7 cm

2 1202 m

3 206 cm

Exercise 29D

1 a 81π cm^2 b 4π cm^2 c 144π cm^2
d 20.25π cm^2 e 3.0625π cm^2 f 37.21π cm^2

2 a 5 cm b 10 mm c 3.3 cm

3 a 16.5 cm b 9.84 m c 7.78 km
d 8.68 mm

4 Small

Exercise 29E

1 a 27.71 cm^2 b 0.20 m^2 c 127.08 cm^2

2 a 13.7 cm^2 b 6.3 cm^2 c 7.7 cm^2

3 1257 bulbs

4 149 cm^2

5 a 24π cm^2 b 40π cm^2

Exercise 29F

1 a 20π cm^3 b $378\,125\pi$ mm^3

2 a 1837.83 cm^3 b 1178.10 mm^3 (or 1.18 cm^3)
c 706.86 cm^3 d 318.09 cm^3 (or 318 086.26 mm^3)

3 216 m*l*

4 1.6 litres

5 Cylinder B by 124 m*l*

6 9817 litres

7 61 goldfish

8 a 147π cm^3 b 7666.5 cm^3

9 2 cm

10 1.8 cm

11 7.0 m

12 2.5 m

Exercise 29G

1 302 cm^2

2 a 245.0 cm^2 b 967.6 cm^2
c 980.2 cm^2 d 373.8 cm^2

3 96π cm^2

4 1885.0 cm^2 (or 0.2 m^2)

5 287 cm^2

6 60

Review exercise

1 a 2.5 cm b 20 cm^2

2 £42

3 8π cm

4 282.7 cm

5 17 026.5 m^2

6 4.3 cm^2

7 4021 m*l*

8 a $0.20\,m^2$ b £779.40
9 3.2 cm

Chapter 30

Exercise 30A

1 a $60\,000\,cm^2$ b $2000\,cm^2$
c $0.24\,m^2$ d $0.065\,m^2$
2 a $2.5\,m^2 = \mathbf{25\,000}\,cm^2$ b $4\,km^2 = \mathbf{4\,000\,000}\,m^2$
c $8900\,m^2 = \mathbf{0.0089}\,km^2$ d $5000\,mm^2 = \mathbf{50}\,cm^2$
3 a $4.5\,cm^2$ b $450\,mm^2$
4 a $37.4\,m^2$ b £2618 (accept £2660)
5 e.g. $355\,000 \div 10\,000 = 35.5$; $35.5 \div 8 = 4.4$ (1 d.p.)
He will need to buy 5 pots.
6 $2830\,cm^2$
7 a $31\,680\,cm^2$ b $3.168\,m^2$

Exercise 30B

1 a $3\,500\,000\,cm^3$ b $790\,000\,cm^3$
c $9.2\,cm^3$ d $7.2\,m^3$
2 a e.g. Danny has converted from m to cm.
b To convert from m^3 to cm^3 you multiply by **1 000 000**.
3 a 4.3 litres b $700\,cm^3$
c $850\,cm^3$ d $5.7\,m^3$
4 a $432\,000\,cm^3$ b $0.432\,m^3$
c 432 litres
5 a $800\,000\,000\,m^3$ b 111 hours

Exercise 30C

1 a 50 km/h b 1.25 m/s
c 600 mph d 9 km/h
2 a 120 miles b 192 m
c 4050 m d 6 km
3 a 3 hours b 6 hours
c 1.6 seconds
4 a 1 hour 45 minutes b 45.7 mph
5 a 48 km b 48 km/h
6

54 km/h —(×1000)→ **54 000** m/h —(÷60)→ **900** m/min —(÷ **60**)→ **15** m/s

7 a 7.5 m/s b 23.4 km/h
8 6.5 km/h
9 Pete's tortoise is quicker.

Exercise 30D

1 Gold $19\,g/cm^3$; iron $7.8\,g/cm^3$; lead $11.3\,g/cm^3$;
aluminium $2.7 = g/cm^3$
2 a 428 g b 190.4 g c 496.8 g
3 a $50\,cm^3$ b $117.6\,cm^3$ c $1.6\,cm^3$
4 a $10.5\,g/cm^3$ b $45.7\,cm^3$
5 a $45\,000\,cm^3$ b 31.5 kg
6 a $3200\,cm^3$ b $0.625\,g/cm^3$ c 375 g
7 $7778\,kg/m^3$
8 a $2000\pi\,cm^3$ b 119 kg
9 £37.80
10 16 560 tonnes

Exercise 30E

1 a Volume b Length
c Area d Volume
2 a Area b Volume
c Area d None of these
3 a cm^2 b cm^3
c cm^3 d cm^3
4 a Area b Length
c None of these d Volume
5 a $2R + 2r$ b $4\pi^2 Rr$
c $4\pi^2 Rr^2$
6 Zoe's formula represents an area so it can't be the correct formula for the volume of a sphere.
7 a m^3 b m c m d m
8 B is most likely to be correct. It is the only expression which could represent a length: the top part of the expression has units of length², and the bottom part of the expression has units of length.
9 No, because the expression cannot represent a volume.

Review exercise

1 $0.028\,m^2$
2 2.2 hours
3 No; $110\,km/h \approx 69$ mph
4 a 780 kg b $0.4\,m^3$
5 £30 571
6 $3\,500\,000\,cm^3$
7 a $226\,000\,cm^3$
b e.g. $226\,000\,cm^3 = 226$ litres; $226 \times 1.8 = 406.8 \approx 400$
8 a $240\pi\,cm^3$ or $754\,cm^3$ (3 s.f.) b $1.4\,g/cm^3$
9 It represents a length. π is dimensionless, and the expression in the bracket is a length plus a length.
10 a $2\pi b^2$ b $4a(bc + c^2)$ c $2a\dfrac{(A + c^2)}{b^2}$

Chapter 31

Exercise 31A

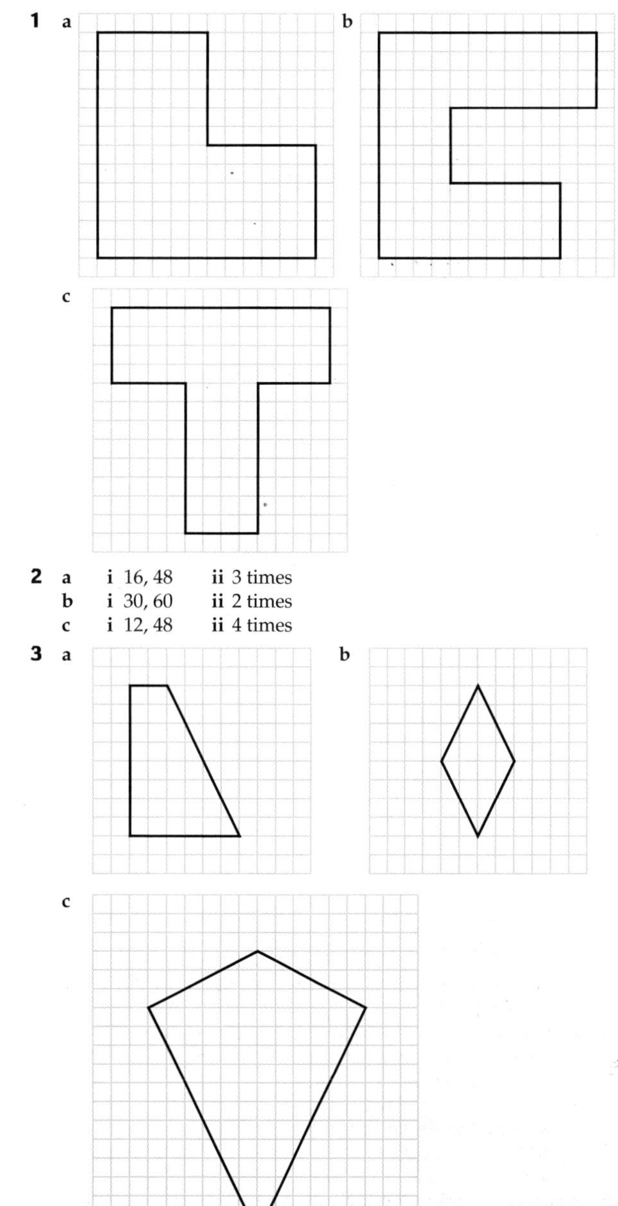

1 a b c

2 a i 16, 48 ii 3 times
b i 30, 60 ii 2 times
c i 12, 48 ii 4 times

3 a b c

Exercise 31B

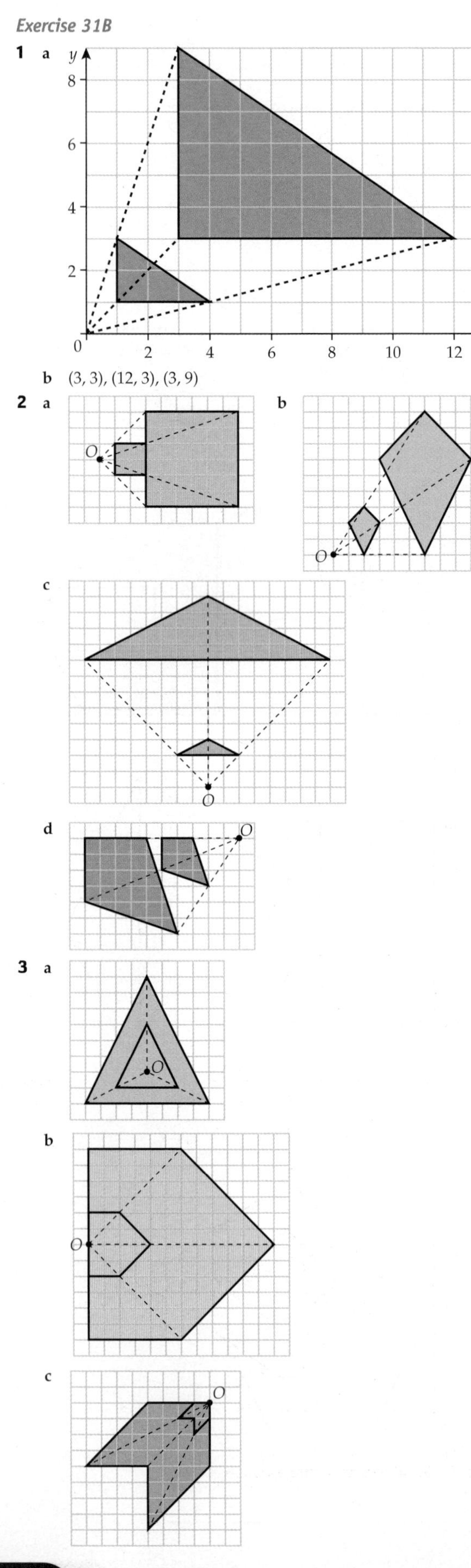

1 a

b (3, 3), (12, 3), (3, 9)

2 a

b

c

d

3 a

b

c

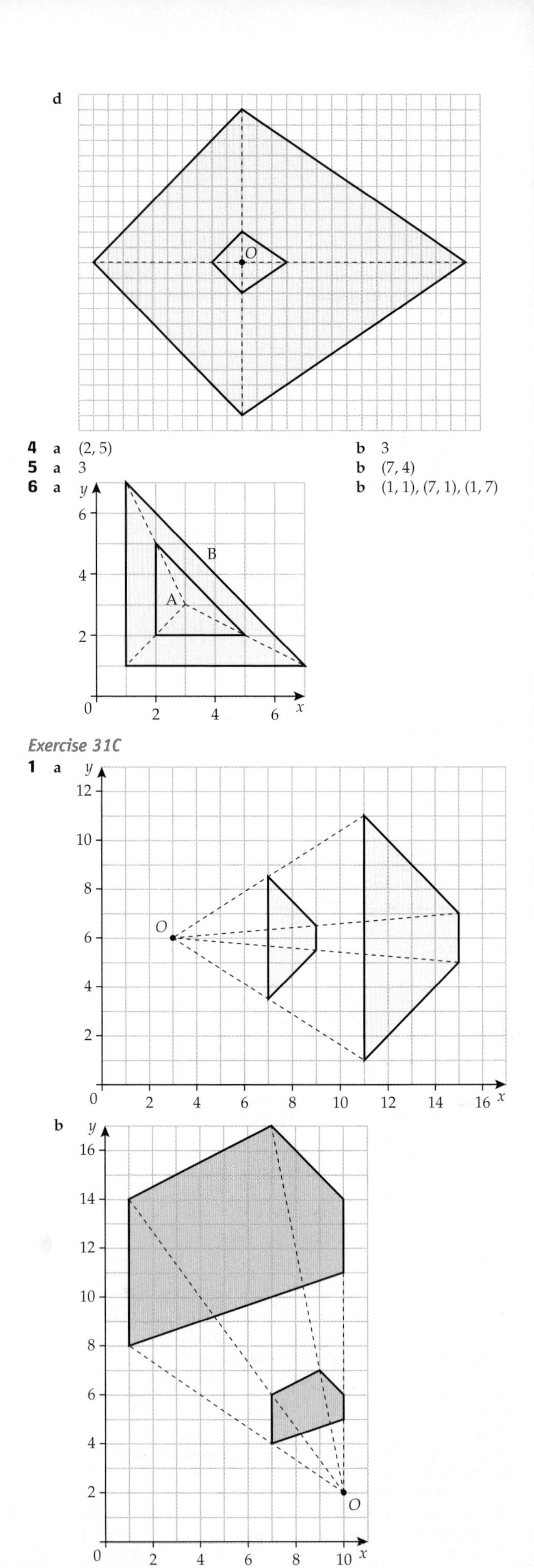

d

4 a (2, 5) b 3

5 a 3 b (7, 4)

6 a

b (1, 1), (7, 1), (1, 7)

Exercise 31C

1 a

b

c

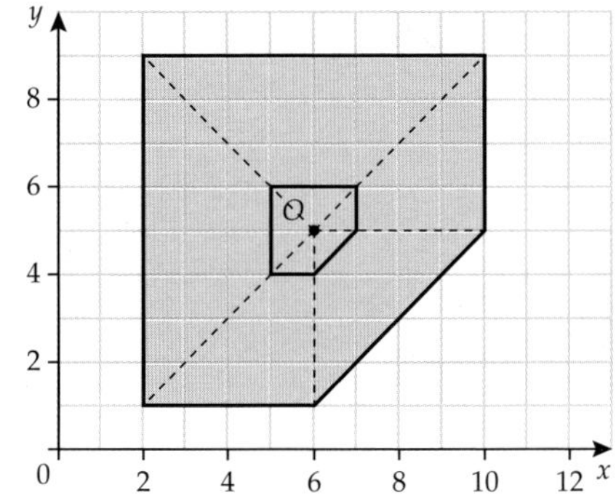

d

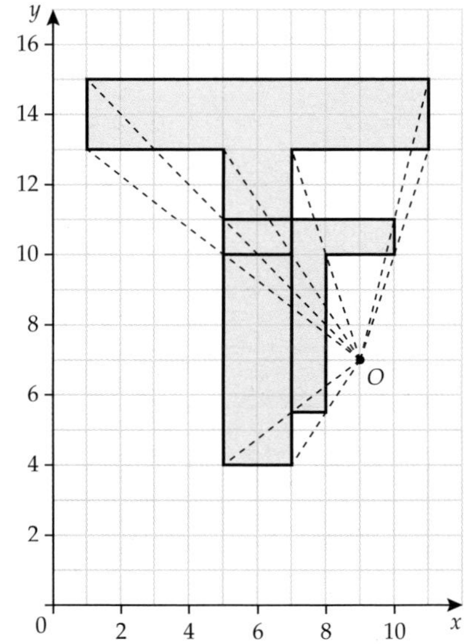

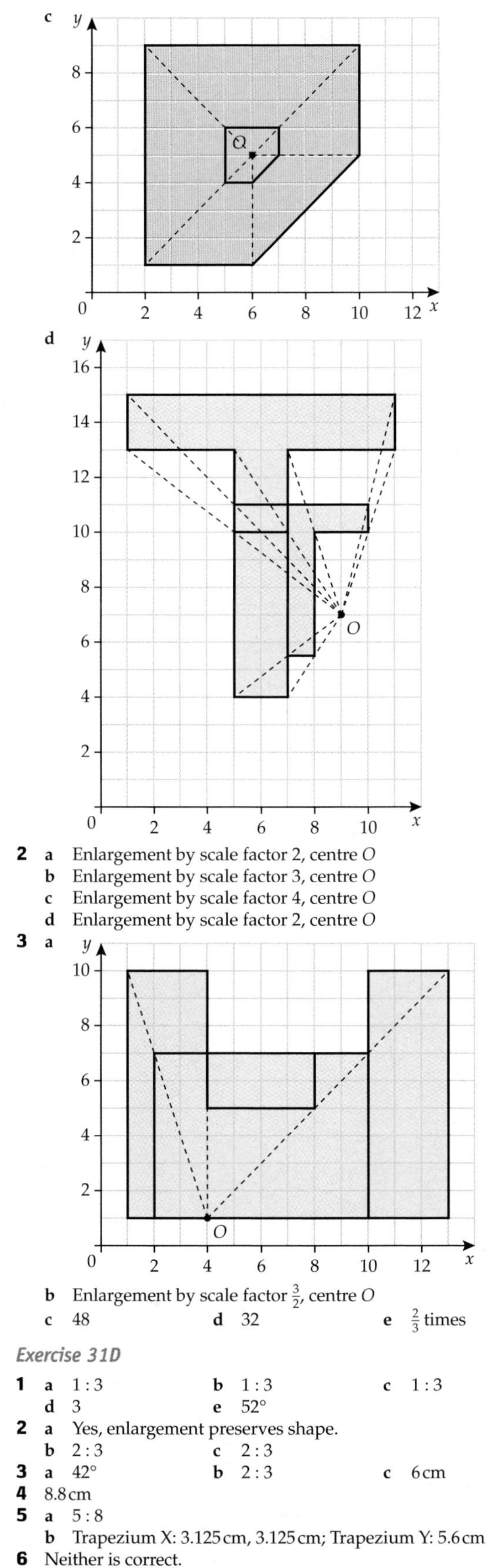

2 **a** Enlargement by scale factor 2, centre *O*
b Enlargement by scale factor 3, centre *O*
c Enlargement by scale factor 4, centre *O*
d Enlargement by scale factor 2, centre *O*

3 **a**

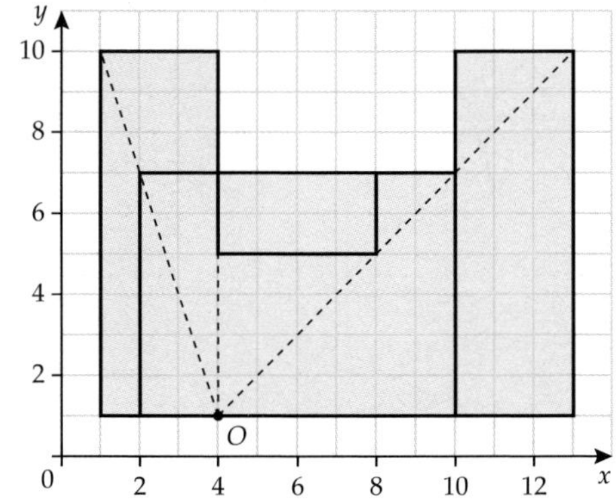

b Enlargement by scale factor $\frac{3}{2}$, centre *O*
c 48 **d** 32 **e** $\frac{2}{3}$ times

Exercise 31D

1 **a** 1 : 3 **b** 1 : 3 **c** 1 : 3
d 3 **e** 52°
2 **a** Yes, enlargement preserves shape.
b 2 : 3 **c** 2 : 3
3 **a** 42° **b** 2 : 3 **c** 6 cm
4 8.8 cm
5 **a** 5 : 8
b Trapezium X: 3.125 cm, 3.125 cm; Trapezium Y: 5.6 cm
6 Neither is correct.
The ratios of corresponding sides are different.

Review exercise

1

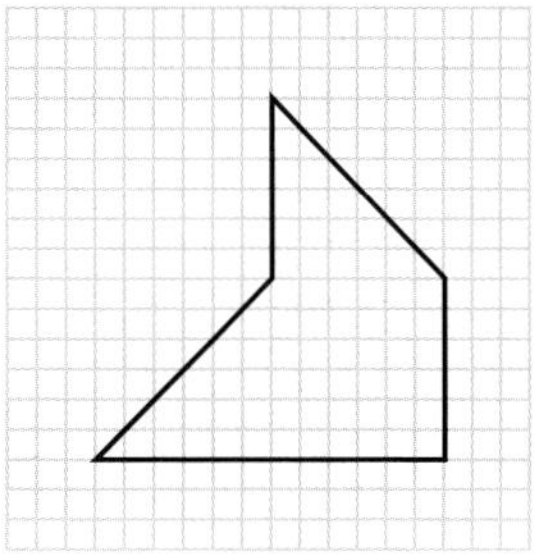

2 **a** 3 **b** (2, 1)

3 **a**

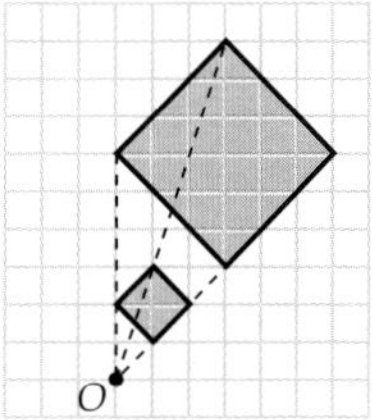

b

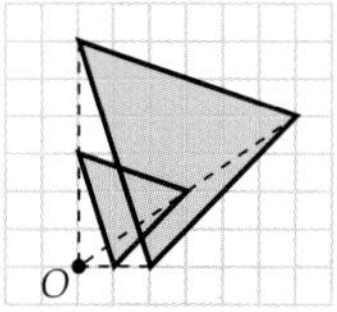

4 **a**

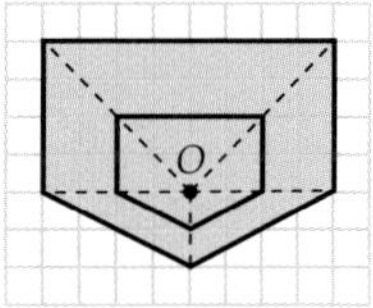

b

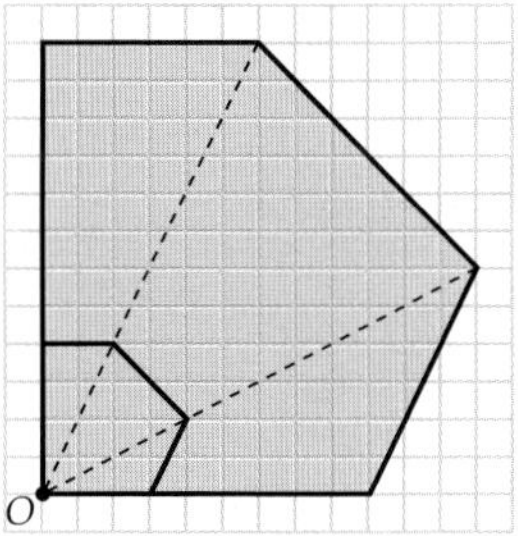

5 **a**

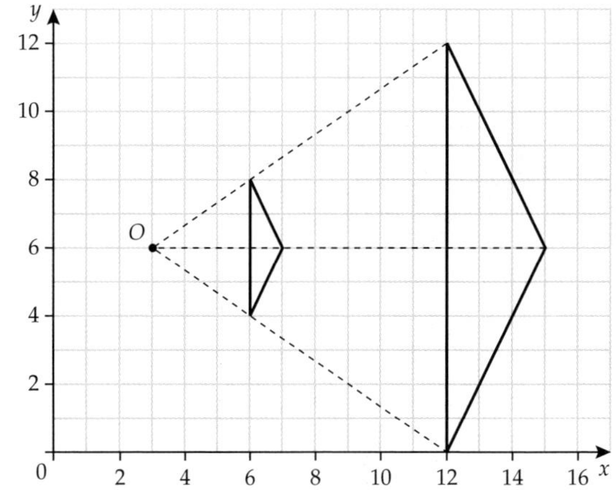

b

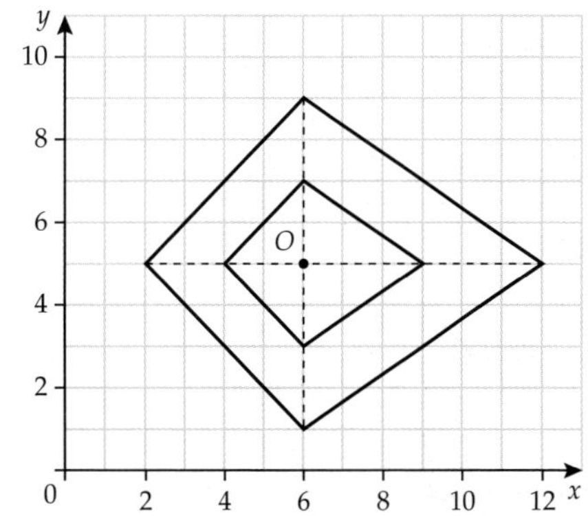

6 17.5 cm
7 No. Ratios of corresponding sides are different.

8 a Angle BAC = angle DEC (alternate angles)
Angle ABC = angle EDC (alternate angles)
Angle ACB = angle ECD (vertically opposite angles)
Corresponding angles are equal so the triangles are similar.
b $x = 3.3$ cm (1 d.p.), $y = 8.1$ cm

Chapter 32

Algebra skills: quadratic expressions

1 a $(x+3)(x-3)$ b $(a+2)(a-2)$
c $(t+1)(t-1)$ d $(7+s)(7-s)$
e $(3x+1)(3x-1)$ f $(5a+3)(5a-3)$
g $(4x+5y)(4x-5y)$ h $3(2a+3b)(2a-3b)$
i $(6x+2y)(6x-2y)$

2 a $(x+1)(x+3)$ b $(a+3)(a+6)$
c $(r+2)(r+5)$ d $(d-1)(d+3)$
e $(x-2)(x+4)$ f $(b+2)(b-3)$
g $(m+2)(m-4)$ h $(x+3)(x-6)$
i $(x-4)(x-4)$ or $(x-4)^2$

3 a $(x-1)(x+2)$ b $(m+1)(m-5)$
c $(t-1)(t+5)$ d $(x+2)(x-4)$
e $5x(3x-1)$ f $(y-2)(y-2)$
g $(m+1)(m-3)$ h $(x-1)(x+5)$
i $(x-2)(x+6)$

4 a $f = \pm 5$ b $g = \pm 6$
c $s = \pm 9$ d $t = \pm 3$
e $e = -4$ or $e = 10$ f $x = -1$ or $x = -7$
g $t = 0$ or $t = 10$ h $x = -7$ or $x = 1$
i $m = -14$ or $m = 0$

5 a $m = -5$ or $m = 0$ b $x = 0$ or $x = \frac{2}{3}$
c $z = -\frac{3}{4}$ or $z = 0$ d $t = -3$ or $t = 5$
e $x = 2$ or $x = 4$ f $y = -7$ or $y = 1$
g $x = -3$ or $x = 2$ h $x = -3$ or $x = 8$
i $k = -2$ or $k = 7$

6 12

7 12

8 $x = 10$ or $x = -5$

Exercise 32A

1 a

x	−3	−2	−1	0	1	2	3
y	9	4	1	0	1	4	9

b

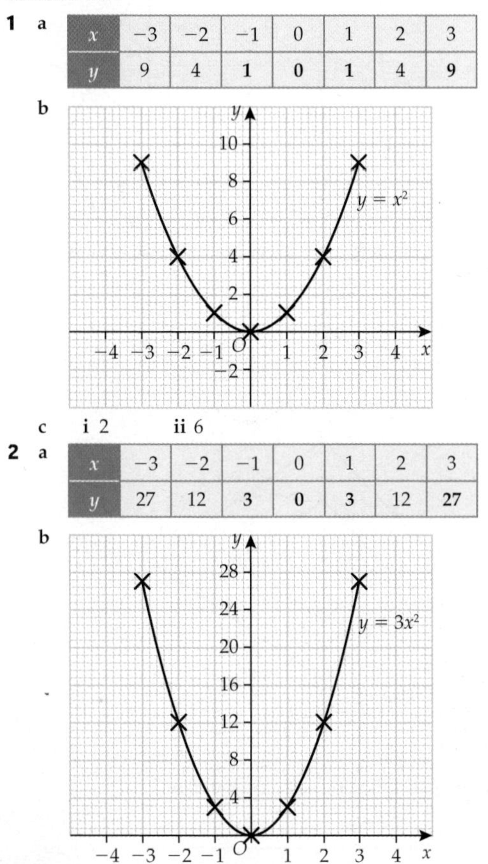

c i 2 ii 6

2 a

x	−3	−2	−1	0	1	2	3
y	27	12	3	0	3	12	27

b

3 Both curves open upwards and are symmetrical about $x = 0$.
(0, 0) is the lowest point of both curves.
y increases at a faster rate in $y = 3x^2$ than in $y = x^2$.

4 a

x	−4	−3	−2	−1	0	1	2	3	4
x^2	16	9	4	1	0	1	4	9	16
+2	+2	+2	+2	+2	+2	+2	+2	+2	+2
$y = x^2 + 2$	18	11	6	3	2	3	6	11	18

b

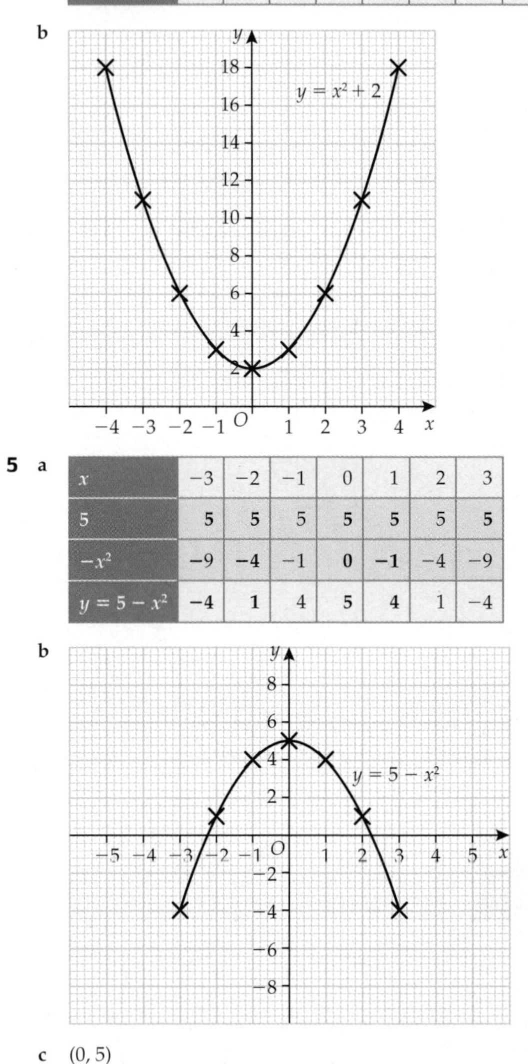

5 a

x	−3	−2	−1	0	1	2	3
5	5	5	5	5	5	5	5
$-x^2$	−9	−4	−1	0	−1	−4	−9
$y = 5 - x^2$	−4	1	4	5	4	1	−4

b

c (0, 5)

6 a

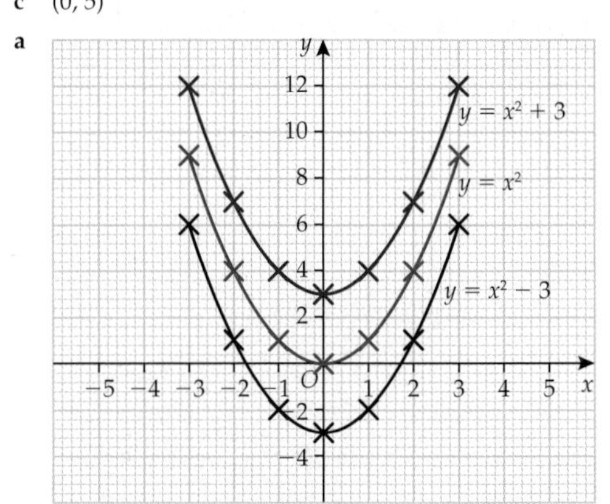

b All curves are the same shape and open upwards. The lowest point of $y = x^2$ is (0, 0), $y = x^2 + 3$ is (0, 3) and $y = x^2 - 3$ is (0, −3).

7 a

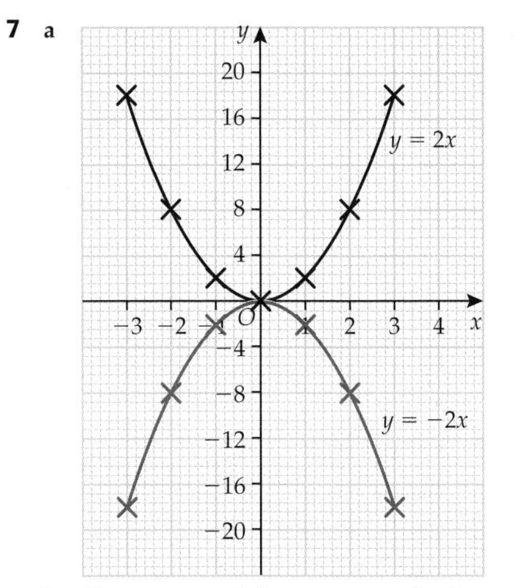

b The curve $y = 2x^2$ opens upwards; the curve $y = -2x^2$ opens downwards. Both are symmetrical about $y = 0$.

c The coefficient of the x^2 term is negative.

8 a

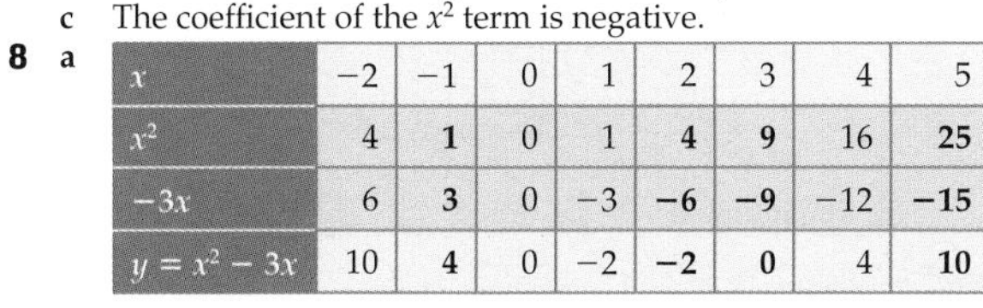

x	−2	−1	0	1	2	3	4	5
x^2	4	**1**	0	1	**4**	**9**	16	**25**
$-3x$	6	**3**	0	−3	**−6**	**−9**	−12	**−15**
$y = x^2 - 3x$	10	**4**	0	−2	**−2**	**0**	4	**10**

b

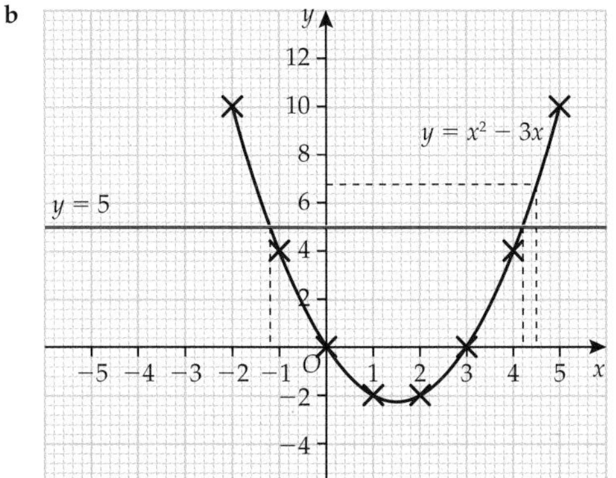

c $y = 6.75$

d $x = -1.2$ and 4.2

9 a

x	−2	−1	0	1	2	3
$2x^2$	8	2	**0**	**2**	**8**	18
$-4x$	8	4	**0**	**−4**	**−8**	−12
-1	−1	−1	**−1**	**−1**	**−1**	−1
$y = 2x^2 - 4x - 1$	10	5	**−1**	**−3**	**−1**	5

b

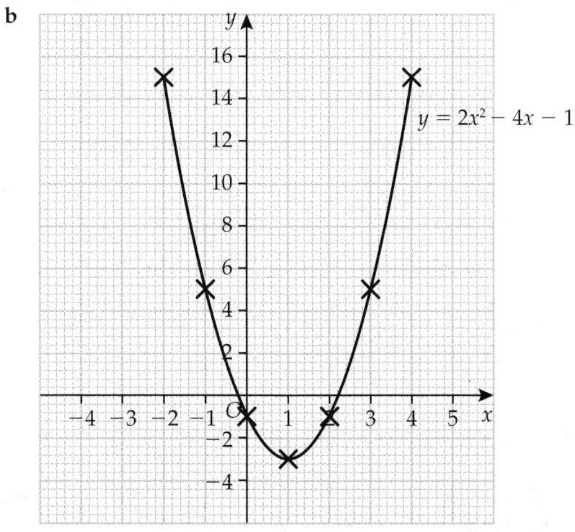

c 1.5

Exercise 32B

1 a

x	−3	−2	−1	0	1	2
x^2	9	4	1	0	1	4
$+x$	−3	−2	**−1**	0	**1**	**2**
-2	−2	−2	−2	−2	−2	−2
$y = x^2 + x - 2$	4	0	**−2**	−2	**0**	**4**

b

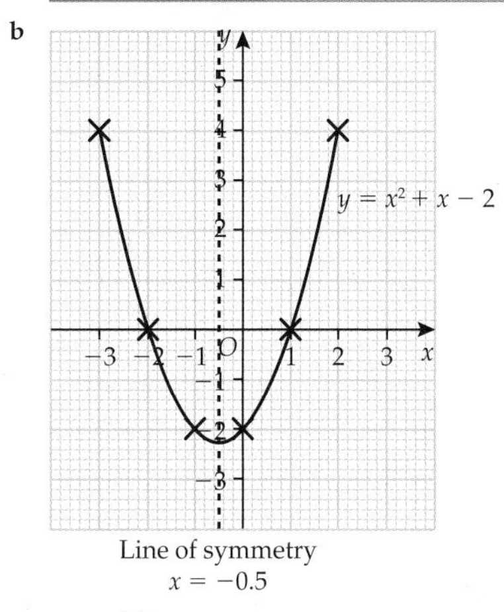

Line of symmetry
$x = -0.5$

c $x = -0.5$

2 a

x	−2	−1	0	1	2	3	4	5	6
x^2	4	**1**	**0**	1	4	9	16	**25**	**36**
$-4x$	8	**4**	**0**	−4	−8	−12	**−16**	**−20**	**−24**
-5	−5	**−5**	**−5**	−5	−5	−5	−5	**−5**	**−5**
$y = x^2 - 4x - 5$	7	**0**	**−5**	−8	−9	−8	**−5**	**0**	**7**

b

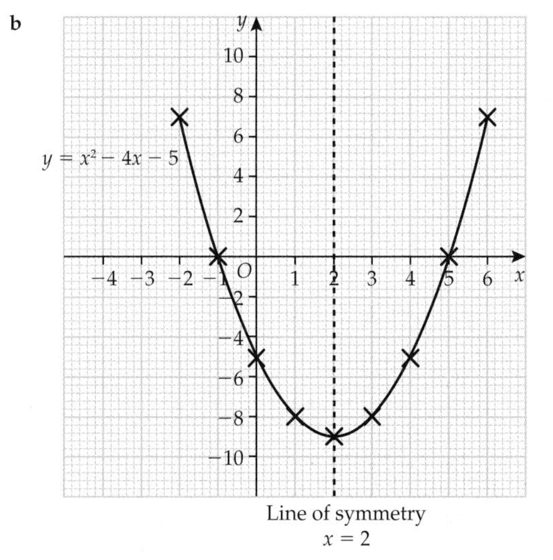

Line of symmetry
$x = 2$

c $(2, -9)$

d $x = 2$

e $(-1, 0)$, $(5, 0)$

3 a i

x	−4	−3	−2	−1	0	1	2
x^2	16	9	4	1	0	1	4
$+2x$	−8	−6	−4	−2	0	2	4
-5	−5	−5	−5	−5	−5	−5	−5
$y = x^2 + 2x - 5$	3	−2	−5	−6	−5	−2	3

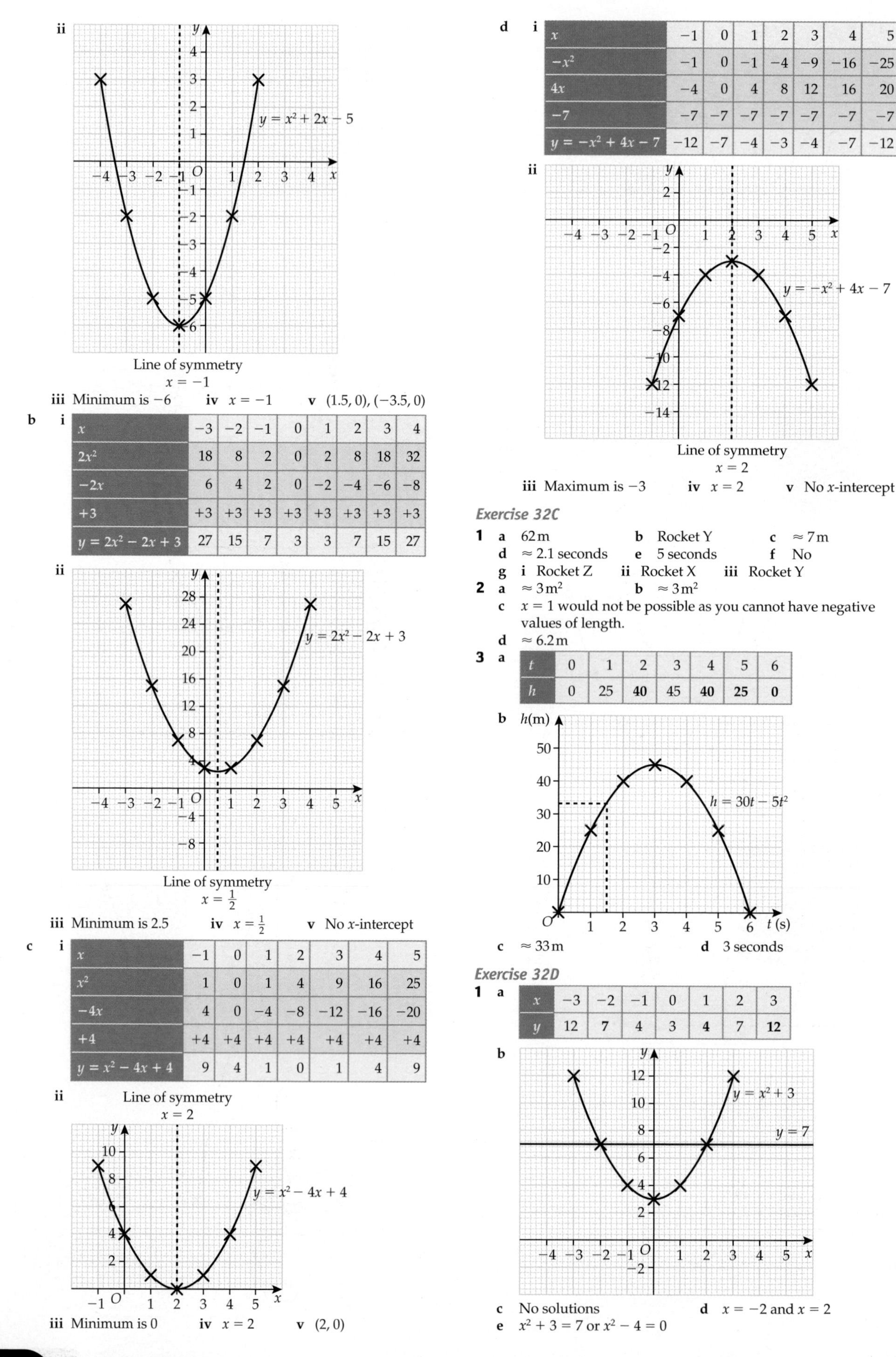

ii

Line of symmetry
$x = -1$

iii Minimum is -6 **iv** $x = -1$ **v** $(1.5, 0), (-3.5, 0)$

b **i**

x	-3	-2	-1	0	1	2	3	4
$2x^2$	18	8	2	0	2	8	18	32
$-2x$	6	4	2	0	-2	-4	-6	-8
$+3$	$+3$	$+3$	$+3$	$+3$	$+3$	$+3$	$+3$	$+3$
$y = 2x^2 - 2x + 3$	27	15	7	3	3	7	15	27

ii

Line of symmetry
$x = \frac{1}{2}$

iii Minimum is 2.5 **iv** $x = \frac{1}{2}$ **v** No x-intercept

c **i**

x	-1	0	1	2	3	4	5
x^2	1	0	1	4	9	16	25
$-4x$	4	0	-4	-8	-12	-16	-20
$+4$	$+4$	$+4$	$+4$	$+4$	$+4$	$+4$	$+4$
$y = x^2 - 4x + 4$	9	4	1	0	1	4	9

ii Line of symmetry
$x = 2$

iii Minimum is 0 **iv** $x = 2$ **v** $(2, 0)$

d **i**

x	-1	0	1	2	3	4	5
$-x^2$	-1	0	-1	-4	-9	-16	-25
$4x$	-4	0	4	8	12	16	20
-7	-7	-7	-7	-7	-7	-7	-7
$y = -x^2 + 4x - 7$	-12	-7	-4	-3	-4	-7	-12

ii

Line of symmetry
$x = 2$

iii Maximum is -3 **iv** $x = 2$ **v** No x-intercept

Exercise 32C

1 **a** 62 m **b** Rocket Y **c** ≈ 7 m
d ≈ 2.1 seconds **e** 5 seconds **f** No
g **i** Rocket Z **ii** Rocket X **iii** Rocket Y

2 **a** $\approx 3\,\text{m}^2$ **b** $\approx 3\,\text{m}^2$
c $x = 1$ would not be possible as you cannot have negative values of length.
d ≈ 6.2 m

3 **a**

t	0	1	2	3	4	5	6
h	0	25	**40**	45	**40**	**25**	**0**

b

c ≈ 33 m **d** 3 seconds

Exercise 32D

1 **a**

x	-3	-2	-1	0	1	2	3
y	12	**7**	4	3	**4**	7	**12**

b

c No solutions **d** $x = -2$ and $x = 2$
e $x^2 + 3 = 7$ or $x^2 - 4 = 0$

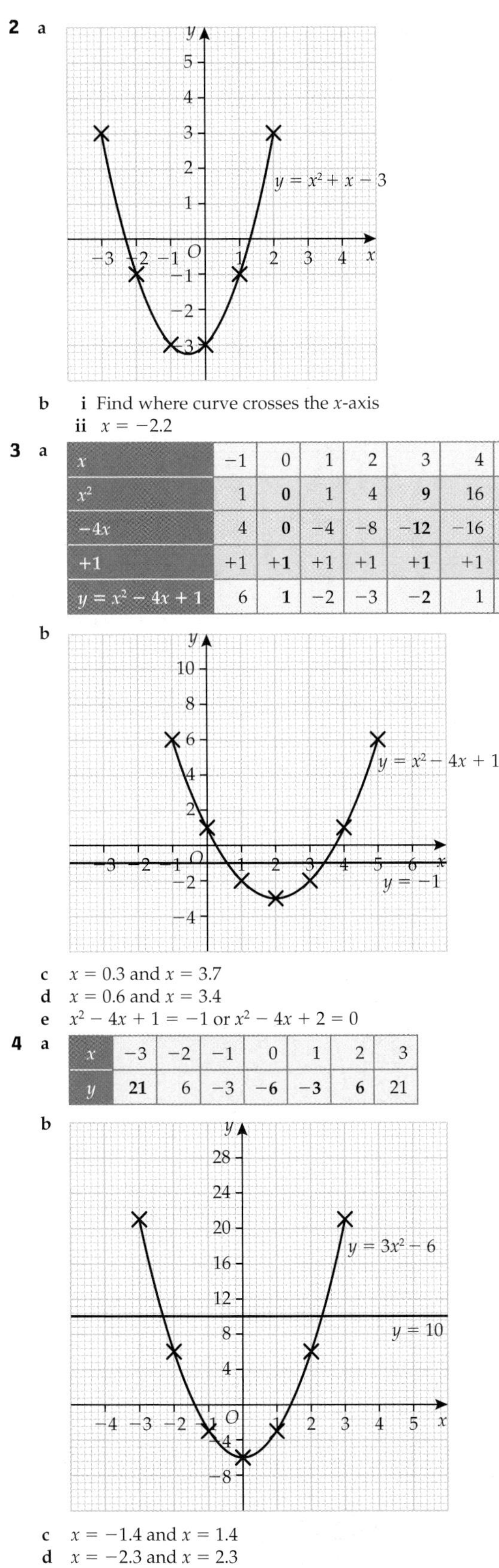

2 a

b i Find where curve crosses the x-axis
ii $x = -2.2$

3 a

x	-1	0	1	2	3	4	5
x^2	1	**0**	1	4	**9**	16	**25**
$-4x$	4	**0**	-4	-8	$\mathbf{-12}$	-16	$\mathbf{-20}$
$+1$	$+1$	$\mathbf{+1}$	$+1$	$+1$	$\mathbf{+1}$	$+1$	$\mathbf{+1}$
$y = x^2 - 4x + 1$	6	**1**	-2	-3	$\mathbf{-2}$	1	**6**

b

c $x = 0.3$ and $x = 3.7$
d $x = 0.6$ and $x = 3.4$
e $x^2 - 4x + 1 = -1$ or $x^2 - 4x + 2 = 0$

4 a

x	-3	-2	-1	0	1	2	3
y	**21**	6	-3	$\mathbf{-6}$	$\mathbf{-3}$	**6**	21

b

c $x = -1.4$ and $x = 1.4$
d $x = -2.3$ and $x = 2.3$
e $3x^2 - 6 = 10$ so $3x^2 - 16 = 0$

5 a

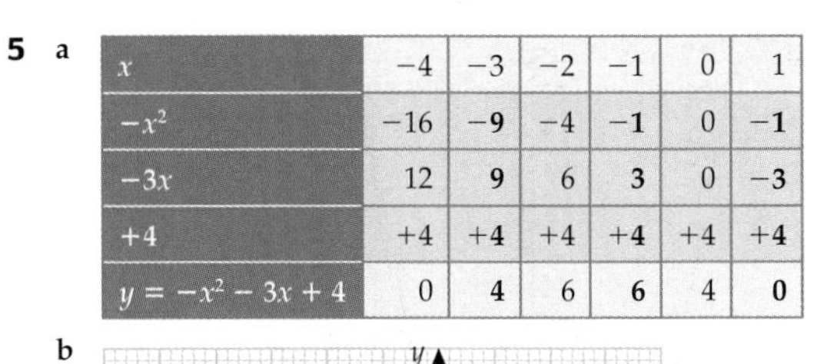

x	-4	-3	-2	-1	0	1
$-x^2$	-16	$\mathbf{-9}$	-4	$\mathbf{-1}$	0	$\mathbf{-1}$
$-3x$	12	**9**	6	**3**	0	$\mathbf{-3}$
$+4$	$+4$	$\mathbf{+4}$	$+4$	$\mathbf{+4}$	$+4$	$\mathbf{+4}$
$y = -x^2 - 3x + 4$	0	**4**	6	**6**	4	**0**

b

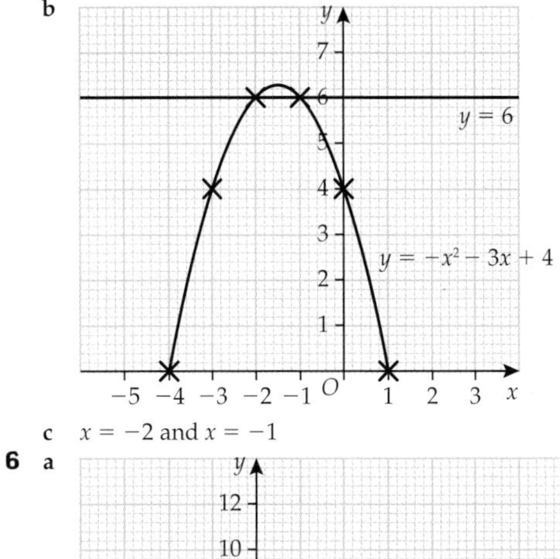

c $x = -2$ and $x = -1$

6 a

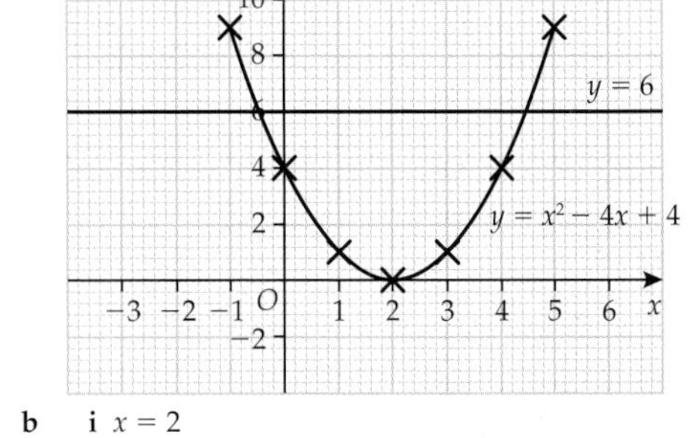

b i $x = 2$
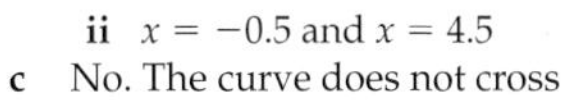
ii $x = -0.5$ and $x = 4.5$
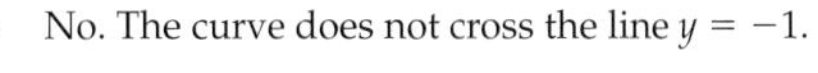
c No. The curve does not cross the line $y = -1$.

Exercise 32E

1

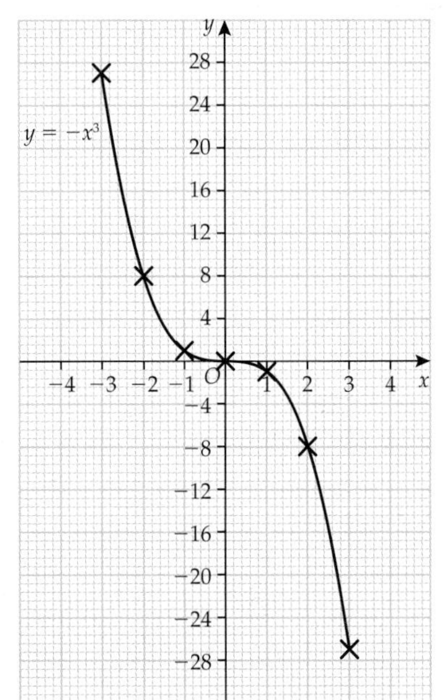

2 a

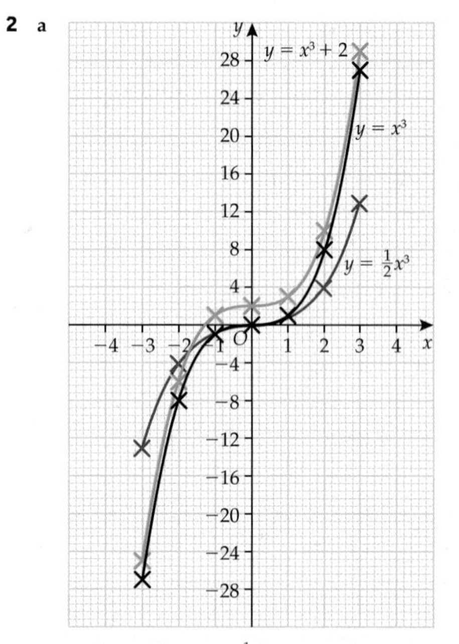

b Both $y = x^3$ and $y = \frac{1}{2}x^3$ pass through the origin whereas $y = x^3 + 2$ intercepts the y-axis at (0, 2). $y = x^3 + 2$ is the same shape as $y = x^3$, moved up 2 units. All have the same general shape but $y = \frac{1}{2}x^3$ lies further away from the y-axis.

3 a

x	−2.5	−2	−1	0	1	2	2.5
x^3	**−15.6**	−8	**−1**	0	1	**8**	15.6
$-2x$	**5**	4	**2**	0	−2	**−4**	−5
$+2$	+2	+2	+2	+2	+2	+2	+2
$y = x^3 - 2x + 2$	**−8.6**	−2	**3**	2	1	**6**	12.6

b

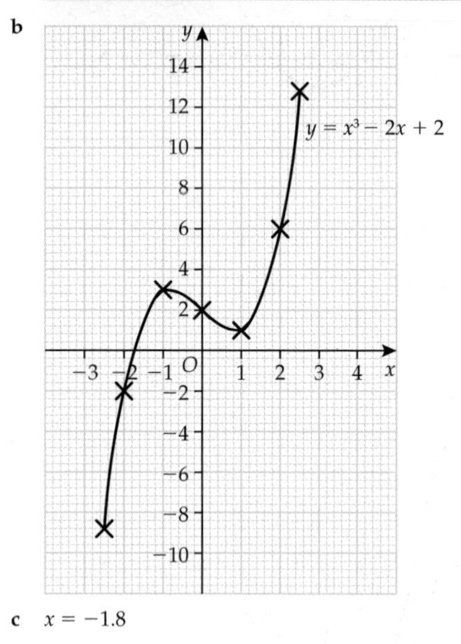

c $x = -1.8$

4 a

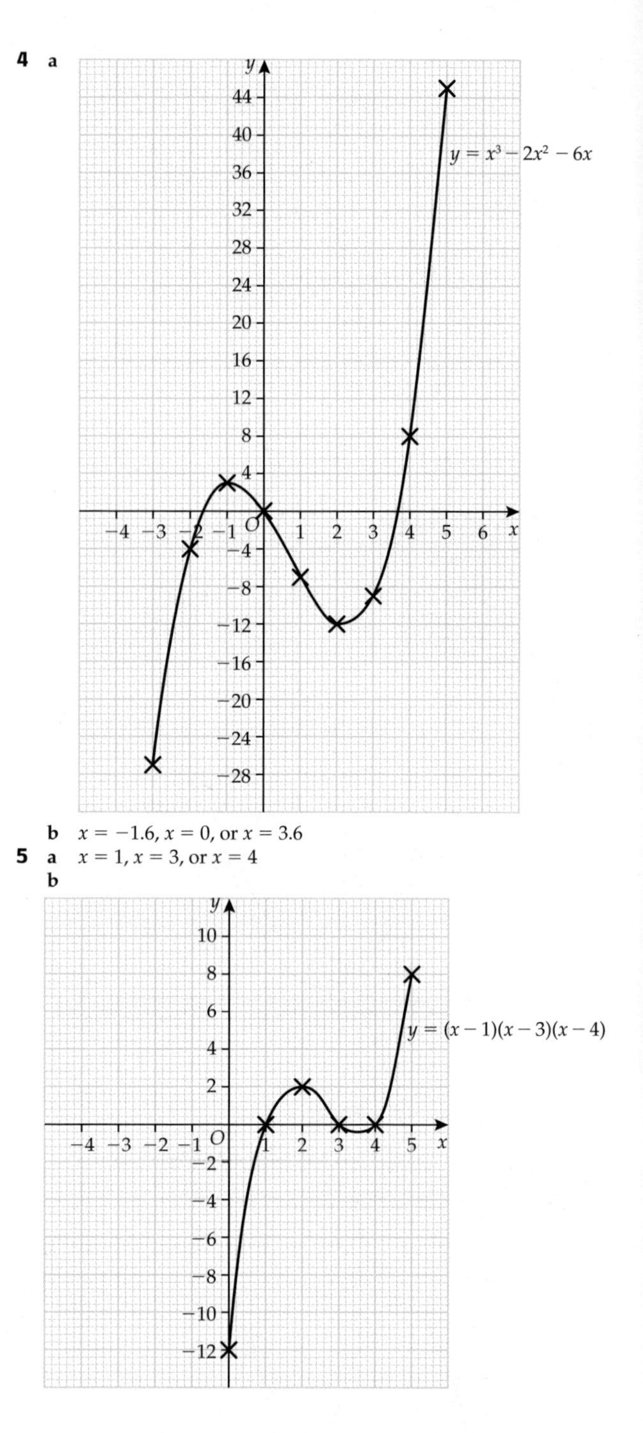

b $x = -1.6$, $x = 0$, or $x = 3.6$

5 a $x = 1$, $x = 3$, or $x = 4$

b

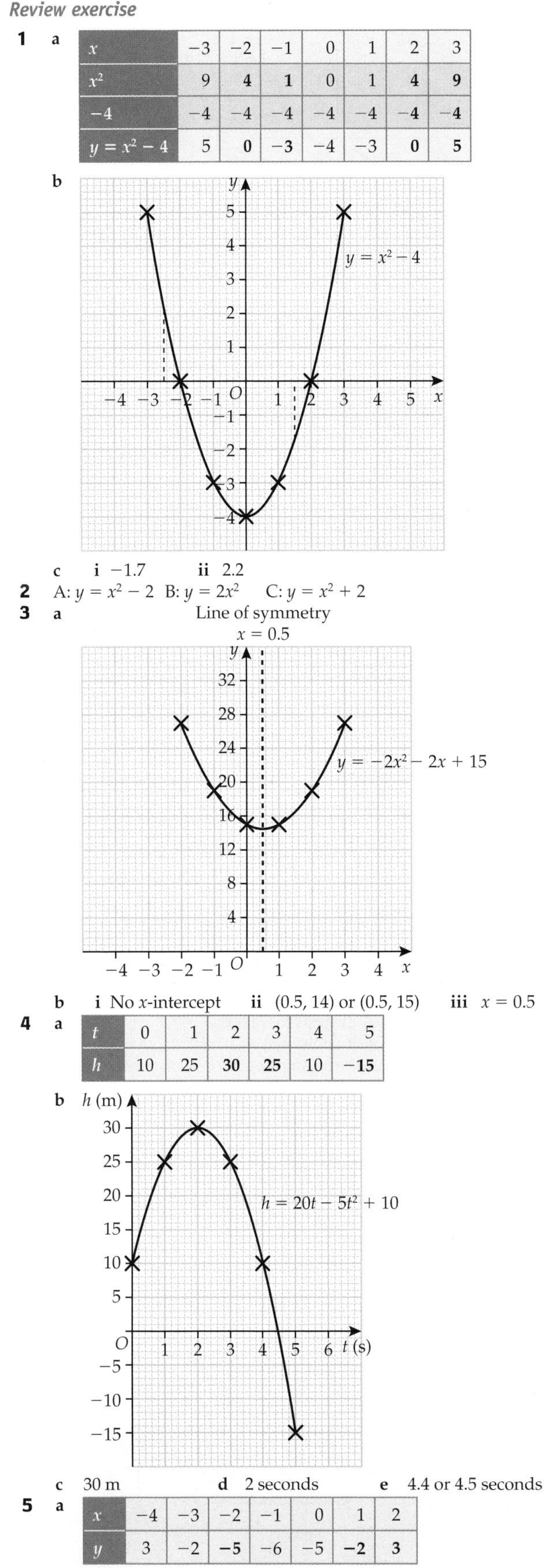

Review exercise

1 a

x	−3	−2	−1	0	1	2	3
x^2	9	**4**	**1**	0	1	**4**	**9**
−4	−4	−4	−4	−4	−4	−4	−4
$y = x^2 - 4$	5	**0**	**−3**	−4	−3	**0**	**5**

b

c i −1.7 ii 2.2

2 A: $y = x^2 - 2$ B: $y = 2x^2$ C: $y = x^2 + 2$

3 a

b i No x-intercept ii (0.5, 14) or (0.5, 15) iii $x = 0.5$

4 a

t	0	1	2	3	4	5
h	10	25	**30**	**25**	10	**−15**

b

c 30 m d 2 seconds e 4.4 or 4.5 seconds

5 a

x	−4	−3	−2	−1	0	1	2
y	3	−2	**−5**	−6	−5	**−2**	**3**

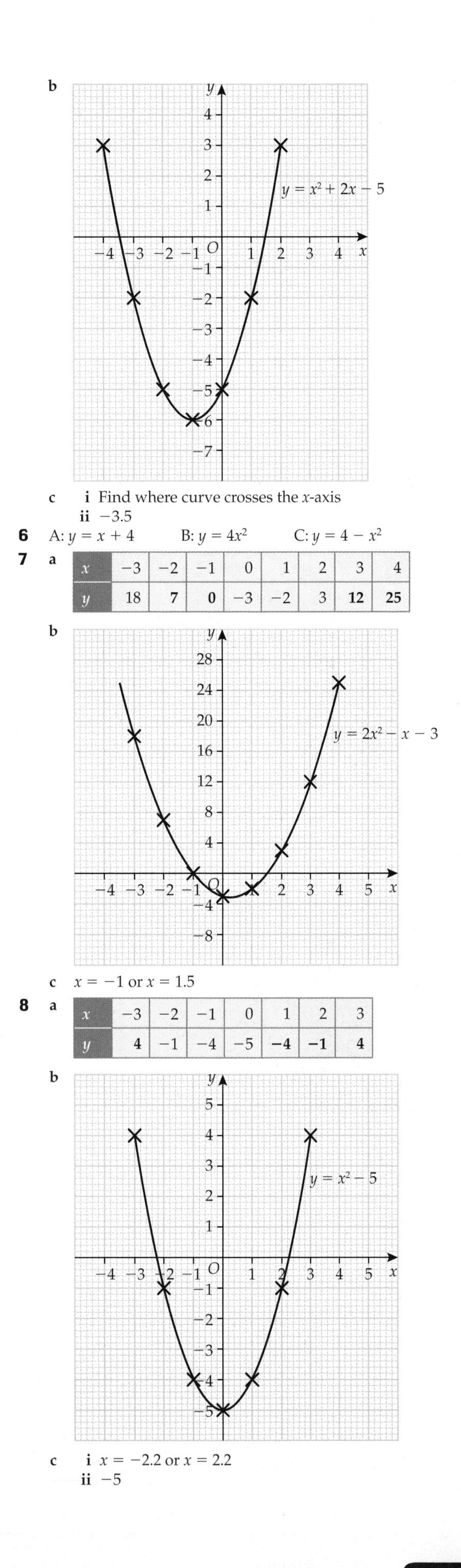

b

c i Find where curve crosses the x-axis
ii −3.5

6 A: $y = x + 4$ B: $y = 4x^2$ C: $y = 4 - x^2$

7 a

x	−3	−2	−1	0	1	2	3	4
y	18	**7**	**0**	−3	−2	3	**12**	**25**

b

c $x = -1$ or $x = 1.5$

8 a

x	−3	−2	−1	0	1	2	3
y	**4**	−1	−4	−5	**−4**	**−1**	**4**

b

c i $x = -2.2$ or $x = 2.2$
ii −5

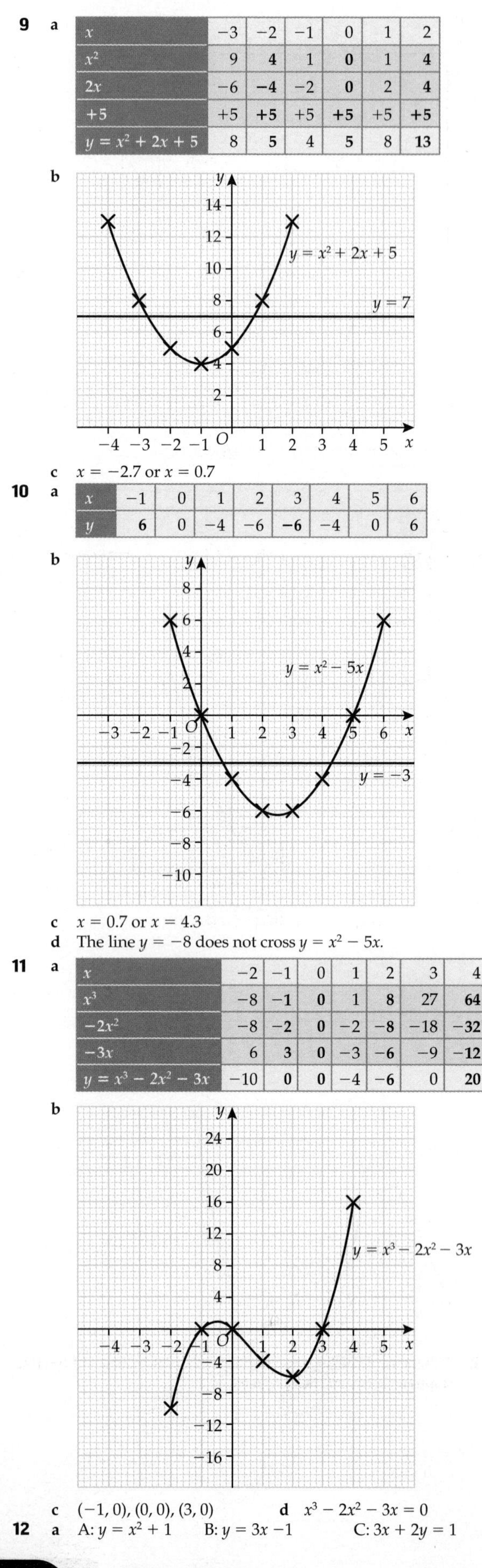

9 a

x	−3	−2	−1	0	1	2
x^2	9	**4**	1	**0**	1	**4**
$2x$	−6	**−4**	−2	**0**	2	**4**
+5	+5	**+5**	+5	**+5**	+5	**+5**
$y = x^2 + 2x + 5$	8	**5**	4	**5**	8	**13**

b

c $x = -2.7$ or $x = 0.7$

10 a

x	−1	0	1	2	3	4	5	6
y	**6**	0	−4	−6	**−6**	−4	0	6

b

c $x = 0.7$ or $x = 4.3$

d The line $y = -8$ does not cross $y = x^2 - 5x$.

11 a

x	−2	−1	0	1	2	3	4
x^3	−8	**−1**	**0**	1	**8**	27	**64**
$-2x^2$	−8	**−2**	**0**	−2	**−8**	−18	**−32**
$-3x$	6	**3**	**0**	−3	**−6**	−9	**−12**
$y = x^3 - 2x^2 - 3x$	−10	**0**	**0**	−4	**−6**	0	**20**

b

c $(-1, 0)$, $(0, 0)$, $(3, 0)$ **d** $x^3 - 2x^2 - 3x = 0$

12 a A: $y = x^2 + 1$ B: $y = 3x - 1$ C: $3x + 2y = 1$

b

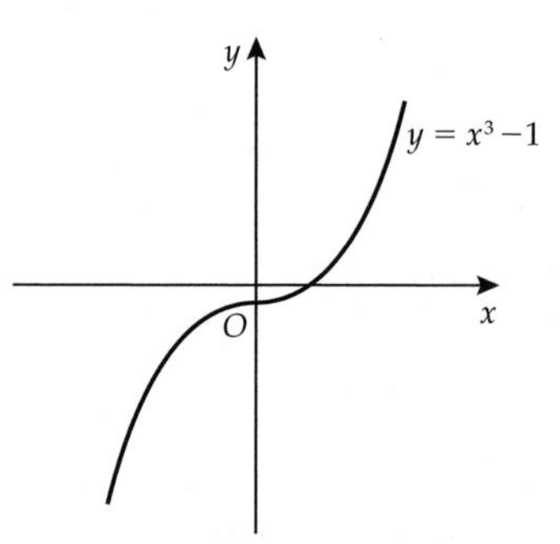

Chapter 33

Some of the diagrams in this chapter are displayed here at a smaller scale than stated in the question.

Exercise 33A

1 Student's accurate construction
2 a,b Student's accurate construction
3 Student's accurate construction, square
4 Student's accurate construction. The lines are parallel to each other.

Exercise 33B

1 a,b Student's accurate constructions
2 Student's accurate construction

Exercise 33C

1 Student's accurate constructions, three different perpendicular bisectors
2 Student's accurate construction. The perpendicular bisector of XY passes through the centre of the circle, O.
3 Student's accurate construction, perpendicular bisector of the line segment PQ. The mid-point of PQ occurs where the perpendicular bisector crosses the line segment.

Exercise 33D

1 Student's accurate constructions
2 Student's accurate construction. The angle bisectors of x and y are at right angles to each other.
3 a Student's accurate construction
b Student's accurate construction
c Construct the angle bisector of 30°
4 a Draw a line, construct the perpendicular to form a right angle, and then construct the angle bisector to create an angle of 45°.
b Student's accurate construction
c e.g. 135°; draw a line, construct the perpendicular to form two back-to-back right angles, and then construct the angle bisector of one of these to create an angle of 45°; 90° + 45° = 135°.
5 If the triangle ABC is an equilateral triangle or an isosceles triangle with angle B = angle C, then the angle bisector of angle A will also be the perpendicular bisector of BC.
6 a, b, c Student's accurate construction
7 a Student's accurate construction
b i Parallelogram
ii $12\,\text{cm}^2$

Exercise 33E

1

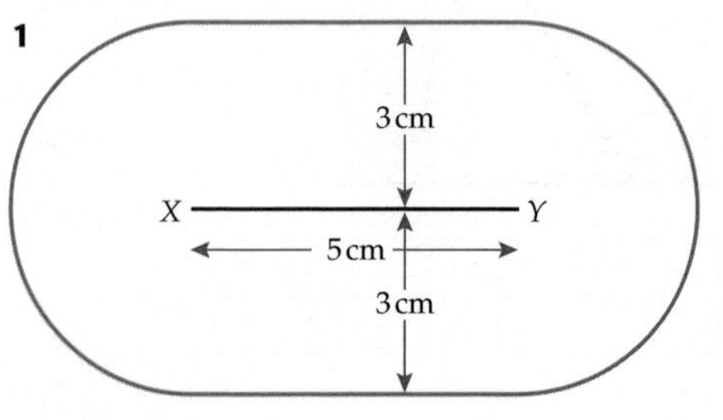

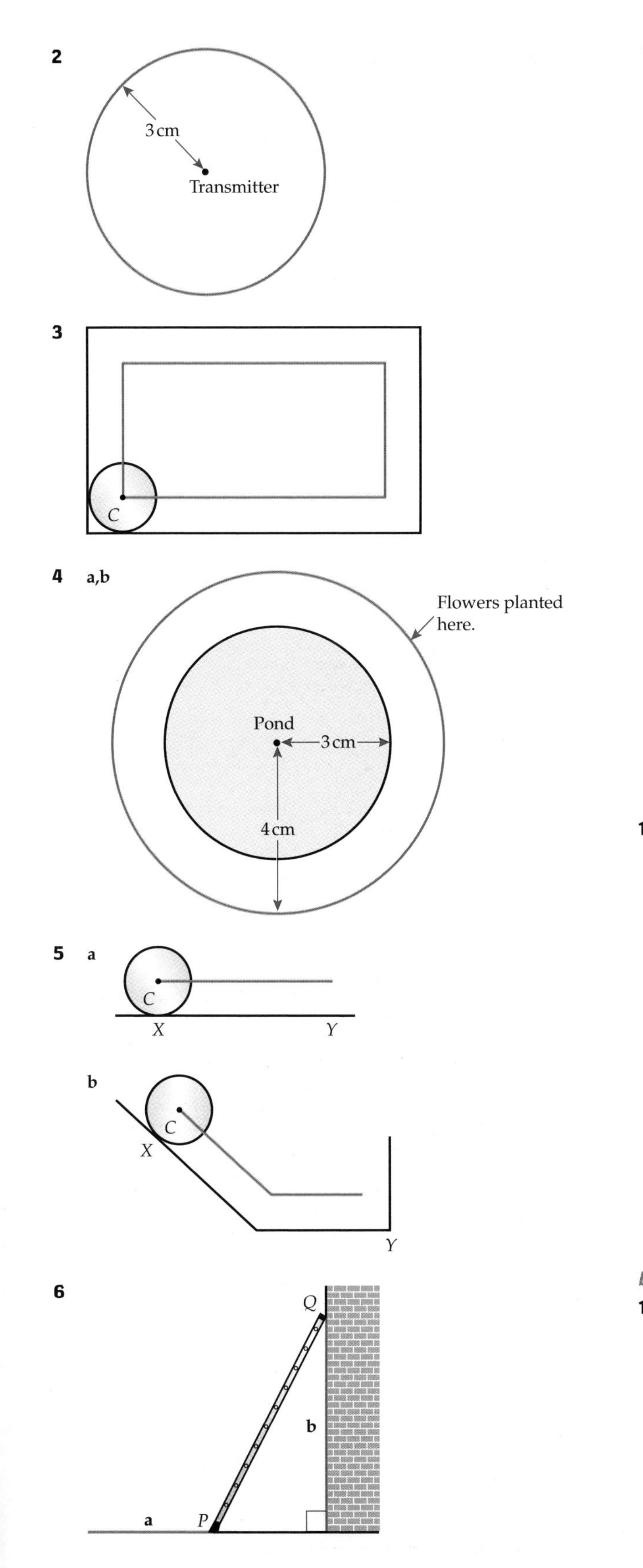

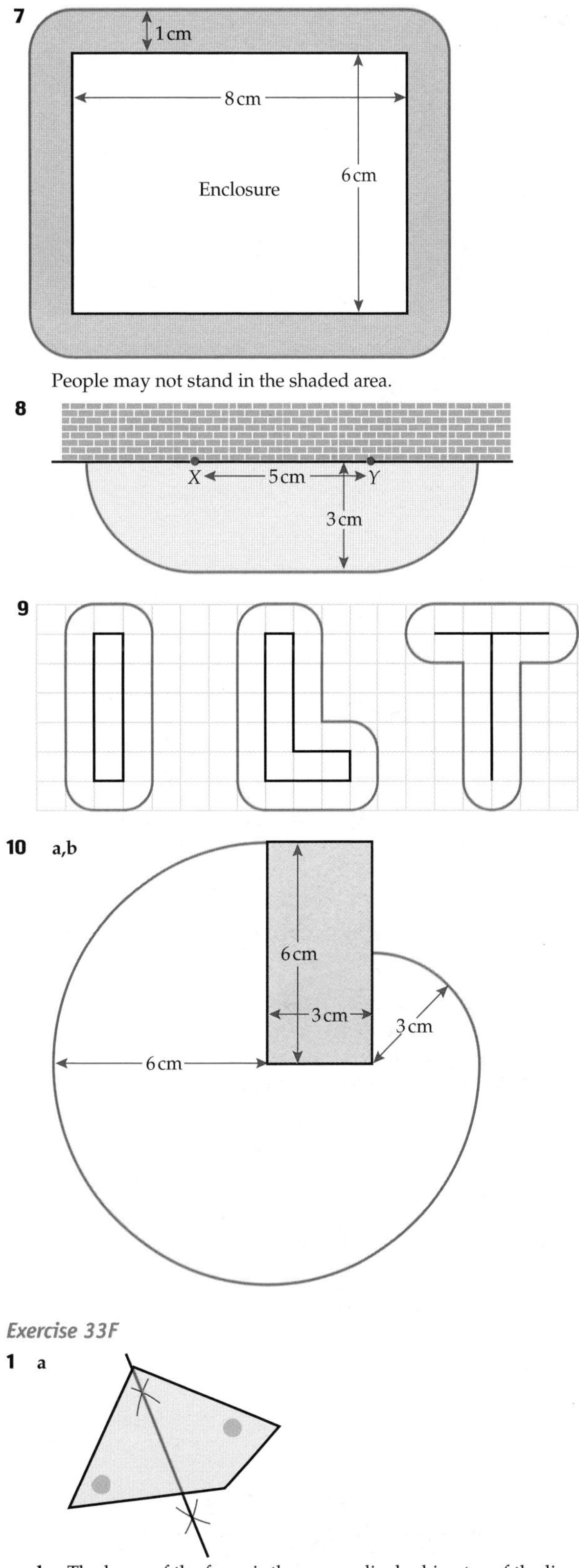

Exercise 33F

1 a

b The locus of the fence is the perpendicular bisector of the line segment between the two trees.

2 a

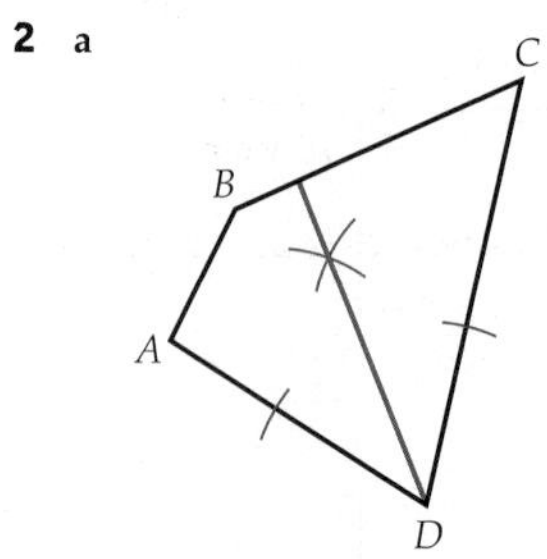

b The locus of the drainage pipe is the angle bisector of angle D.

3

4

5

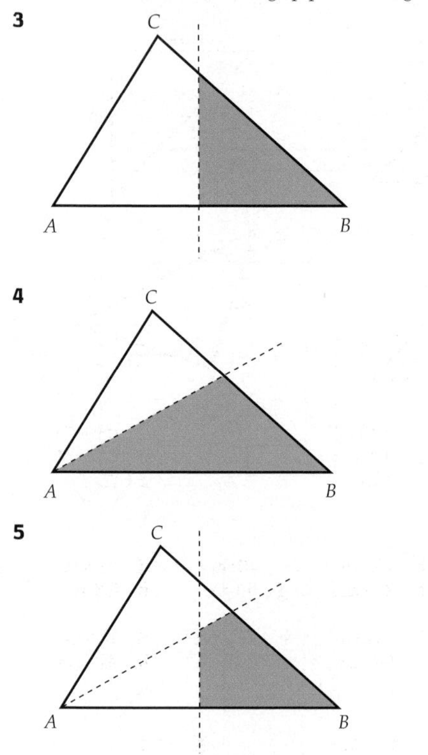

Exercise 33G

1

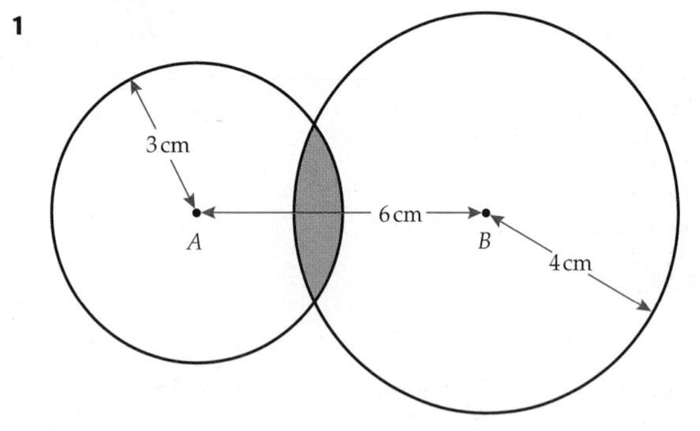

2 a,b,c

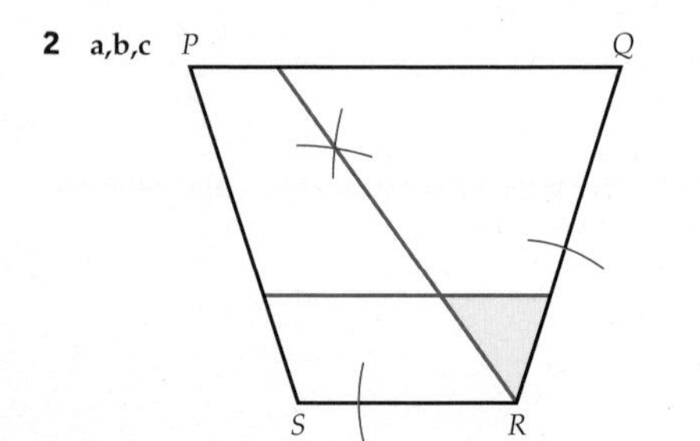

3 a,b,c,d

4 a,b

c Ship A is 12 km away from ship B at points X and Y.

5

6

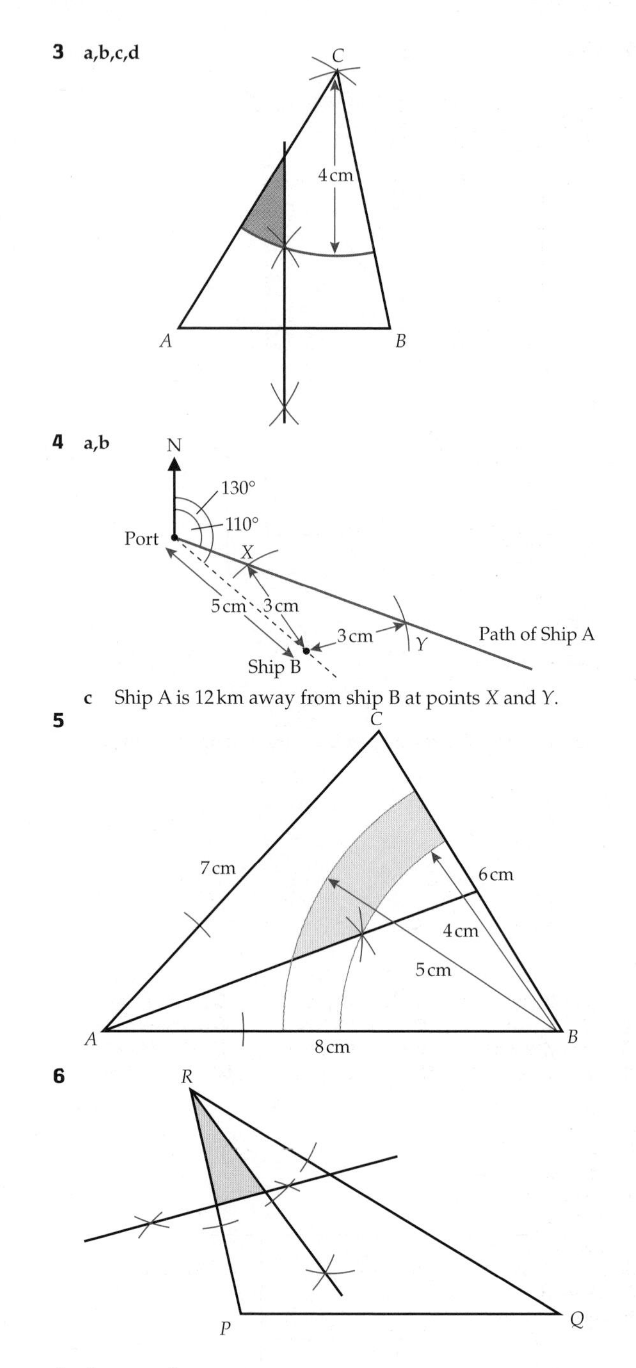

Review exercise

1 Student's accurate construction

2 Student's accurate construction, perpendicular bisector of XY

3

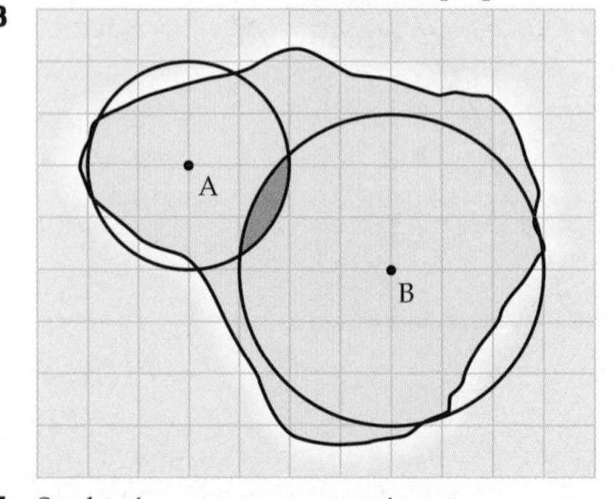

4 Student's accurate construction

5 a,b,c

6

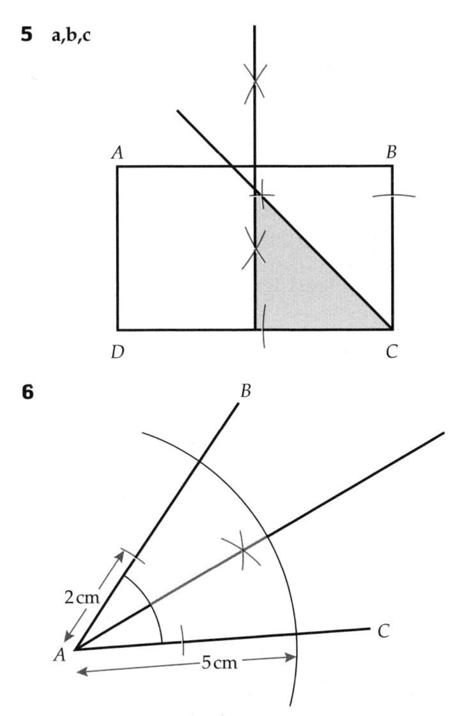

The blue line indicates the possible positions of the flagpole.

Chapter 34

Exercise 34A

1 **a** c **b** q **c** x **d** r
2 **a** $c^2 = a^2 + b^2$ **b** $q^2 = p^2 + r^2$
c $x^2 = y^2 + z^2$ **d** $r^2 = s^2 + t^2$

Exercise 34B

1 $a = 10$ cm **2** $b = 13$ cm **3** $c = 30$ cm
4 $d = 8.5$ cm **5** $e = 14.2$ cm **6** $f = 69.6$ mm
7 $g = 13.9$ mm **8** $h = 9.7$ cm **9** $i = 10.0$ cm
10 $j = 10.3$ cm **11** $k = 15.9$ cm **12** $l = 17.4$ cm

Exercise 34C

1 11.4 km
2 14.4 km
3 3.4 m
4 21.4 cm
5 3.06 m
6 **a** 10.8 cm **b** 5.41 cm **c** 91.9 cm^2
7 64.6 cm^2
8 £80

Exercise 34D

1 $a = 12$ cm **2** $b = 16$ cm **3** $c = 24.7$ cm
4 $d = 18.3$ cm **5** $e = 30.5$ cm **6** $f = 12.8$ cm
7 $g = 32.4$ cm **8** $h = 11.7$ cm **9** $i = 14.8$ cm
10 $j = 13.5$ cm **11** $k = 19.6$ cm **12** $l = 8.2$ cm

Exercise 34E

1 2.3 m
2 3.52 m
3 197.6 cm^2
4 **a** 1.01 m **b** 2.07 m^2 **c** 3.52 m^3
5 30 km
6 5.44 m^3
7 8.66 cm

Exercise 34F

1 **a** 4 units **b** 4 units **c** 5 units
d 5 units **e** 6.4 units **f** 4.5 units
2 **a** 4.5 units **b** 2.8 units **c** 5.7 units
d 8.1 units **e** 14.3 units **f** 4.5 units

Review exercise

1 10.0 cm **2** 10.3 m **3** 3.8 m
4 11.12 cm **5** 3.36 m **6** 7.2 units
7 Sarah is correct.
Horizontal distance = 6 units and vertical distance = 10 units
$AB = \sqrt{6^2 + 10^2} = \sqrt{136} = 11.66$ units (2 d.p.)
8 £3.48
9 30.2%

Chapter 35

Exercise 35A

1 **a** **c**

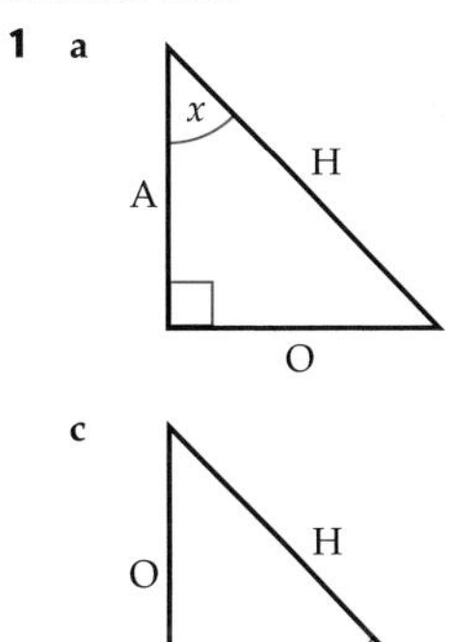

b

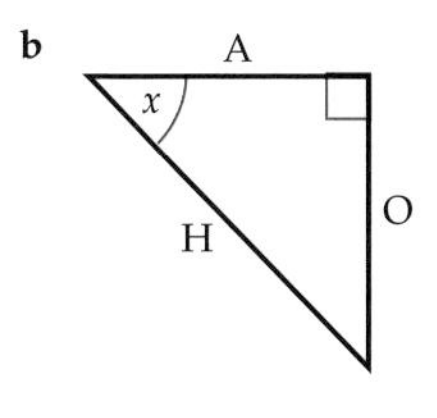

d

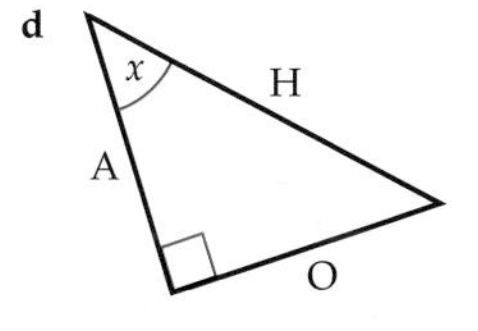

2 **a** 0.89101 **b** 0.35837 **c** 0.90040 **d** 0.93358
e 0.29237 **f** 0.34433 **g** −4.70463 **h** 0.96126
3 **a** 26.4° **b** 45.5° **c** 77.6° **d** 20.4°

Exercise 35B

1 **a** 4.5 cm **b** 10.9 cm **c** 10 cm **d** 34.3 m
e 11.3 cm **f** 21.9 cm **g** 5.1 cm **h** 8.8 m
i 17.3 cm
2 **a** 18 m **b** 13.0 cm **c** 72.3 cm **d** 18.9 m
e 51.1 cm **f** 14.3 cm **g** 23.2 cm **h** 41.8 cm

Exercise 35C

1 54.3° **2** 38.7° **3** 45.3°
4 30° **5** 53.1° **6** 47.0°
7 40.8° **8** 31.9° **9** 28.6°

Exercise 35D

1 72.5 km
2 26.7°
3 **a** 12.7° **b** 4.1 m
4 19 m
5 5.52 cm
6 Bearing 218°, distance 12.4 km
7 52.2 cm^2
8 845 cm

Exercise 35E

1 23.4 km **2** 189.5 m **3** 99.8 m
4 16.5 m **5** 131 m

Review exercise

1 15.1 m
2 37.8°
3 **a** 9.95 cm **b** 30.2°
4 84.7 cm^2
5 47.7°
6 238.3°
7 71.5 m
8 46.5 cm
9 13.6 km/h
10 43.1 km
11 163 m

Chapter 36

Exercise 36A

1 The perpendicular to the chord bisects the chord, therefore AO is the hypotenuse of a triangle with shorter lengths 3 cm and 4 cm. $AO = 5$ cm
2 $PQ = 16$ cm
3 a $OR = OS = 7$ cm b $OM = 5.0$ cm
 c $ON = 6.58$ cm d RS
4 PQ is closer. O to mid-point of $PQ = 3$ cm; O to mid-point of $RS = 5$ cm

Exercise 36B

1 $x = 25°$
2 $y = 120°$
3 $a = 42°, b = 96°$
4 a $CD = 4$ cm b $DB = 8$ cm

Exercise 36C

1 $a = 50°$ 2 $b = 75°$
3 $c = 134°$ 4 $d = 120°$
5 $e = 60°, f = 30°$ 6 $g = 84°, h = 48°, i = 48°$
7 $g = 90°$ 8 $h = 132°, i = 48°, j = 66°$
9 $n = 120°, o = 45°, p = 60°$ 10 $q = 65°$
11 Two sides of the triangle, OA and OC, are the same length (the radius) and angle $AOC = 60°$ (twice the angle subtended by arc AC at the circumference) so the triangle is equilateral.

Exercise 36D

1 $x = 60°$ 2 $a = 37°, b = 37°$
3 $d = 37°, e = 40°$ 4 $y = 60°$
5 $x = 66°$ 6 $x = 31°, y = 31°, z = 81°$

Exercise 36E

1 Kite 2 beacuse it has supplementary opposite angles.
2 $a = 95°, b = 100°$
3 $z = 135°$
4 $a = 94°, b = 86°, c = 86°$
5 $a = 137°$
6 $x = 62°, y = 33°$
7 $a = 29°, b = 90°, c = 122°, d = 32°$

Review exercise

1 a $a = 59°$ b $b = 56°$ c $c = 90°$ d $d = 95°$
2 Angle $OQP = 90° - 64° = 26°$ (tangent properties)
 Angle QPO = angle $OQP = 26°$ (isosceles triangle)
 So angle $POQ = 180 - 26° - 26° = 128°$ (angles in a triangle)
3 a $b = 144°$ b $z = 115°$
4 Angle ACD = angle $ABD = 54°$ (angles subtended by the same arc)
 So $d = 180° - 54° - 85° = 41°$ (angles in a triangle)
5 Angle $ABC = 90°$ (angle in a semicircle is a right angle)
 $180° = 90° + x + 2x$ (angles in a triangle). So $x = 30°$
6 a Cyclic quadrilateral
 b Angle $BCE = 180° - 67° - 30° = 83°$ (angles in a triangle)
 Angle $DCB = 180° - 83° = 97°$ (angles on a straight line)
 $x = 180° - 97° = 83°$ (supplementary angles in cyclic quadrilateral)
7 a $x = 65°$ b $y = 25°$

Functional maths

Missed appointments

1 15 minutes
2 December
3 £10
4 No, total is £26 820.
 Total hygienist fines = 1008 × £10 = £10 080
 Total dentist fines = 1116 × £15 = £16 740
5 Mean = 93, Median = 91, Mode = 88
6 The practice manager is correct since both the mean and median for missed appointments are higher for the dentist than the hygienist. Although the mode is higher for missed hygienist appointments, the mode also happens to be the highest monthly figure so is slightly misleading.

Come dine with Brian

1 100 g
2 380 g
3 1.12 kg rhubarb, 180 g butter, 120 g caster sugar, 120 g self-raising flour, 120 g plain flour, 100 g granulated sugar, 2 eggs (1.6 not possible), 1 orange (0.8 not possible)
4 Brian is not correct. Although he is missing seven ingredients completely from his kitchen, he does not have enough of some of the ingredients so will need to buy more. He needs to buy 11 ingredients in total:
 250 g Stilton cheese, 130 g butter, 1 vegetable stock cube, 90 m*l* white wine, 1 onion, 600 g leeks, 320 g macaroni, 200 m*l* full-fat milk, 40 g chives, 1.12 kg rhubarb, 120 g caster sugar
5 This statement is true if he prepares the crumble while the leek and macaroni bake is cooking, and cooks them together for 15 minutes. If he does this it will take him about 3 hours. If Brian also prepared the leek and macaroni bake while the soup was cooking, he could save even more time.
 If he prepares and cooks the dishes separately, his statement is false – it would take him about 3 hours 45 minutes altogether.
6 e.g. bread rolls to go with the soup, salad or vegetables to go with the leek and macaroni bake, cream, ice cream or custard to go with the crumble, napkins, drinks etc.

eBay business

1 £141
2 Standard
3 £24.65
4 Shani is not correct, Mrs Read's bill = £247.38,
 Mr Owen's bill = £130.04, Mrs Patel's bill = £157.96.
5 Rhino (S)

Camping with wolves

1 £242
2 e.g. Getup tent. It is the joint lightest together with the Sport tent. The Sport tent has a larger pack size and smaller internal height.
 Greenland sleeping bag. The minimum temperature for March is −7°C. The Greenland is designed for a lowest temperature of −10°C so will be warm enough.
 Other answers are acceptable if properly explained.
3 £240 + £40 + £50 + £130 + £90 = £550. Carlos is wrong.
 Answers may vary depending on response to Q2.

Roof garden

1 21 m
2 £119
3 4.8 m
4 1.296 m^3
5 Anil is wrong. Total cost is: £189 (fencing) + £119 (circular pot) + £118 (square pots) + £178 (rectangular pots) + £172.21 (bamboo) + £510.15 (compost) = £1286.36

Pembrokeshire cycle ride

1 28
2 1.9 hours or 1 hour 54 minutes
3 1513
4

	Expected time of arrival	
Feeding station	first cyclist	last cyclist
Fishguard	0924	1156
Newgale	0901	1435
Castlemartin	1240	1749
Pembroke	1009	1513

5

Feeding station	Cereal bars	Bananas	Litres of drink
Fishguard	330	330	660
Newgale	550	550	1100
Castlemartin	330	330	660
Pembroke	220	220	440

6

	Cereal bars	Bananas	Litres of drink
Total	1430	1430	2860
Ted needs to order	1716	1716	3432

Problem-solving

AO3 Statistics

Total number of students = $31 + n$

Total number of minutes

$= (10 \times 12) + (15 \times 3) + (20 \times 11) + (25 \times n) + (30 \times 5) = 535 + 25n$

Therefore

$\frac{535 + 25n}{31 + n} = 19$

$535 + 25n = 19(31 + n)$

$6n = 54$

$n = 9$

AO3 Number

Percentage Jane can afford = $\frac{54}{75} \times 100\% = 72\%$

Percentage of ticket price needed = 28%

4% of ticket price = 5 merits

So 28% of ticket price = $7 \times 4\%$ = 35 merits required

Jane already has 19 merits, so she needs another 16 merits.

AO3 Algebra

Let x = distance from C to D

Time taken from C to D = $\frac{\text{distance}}{\text{speed}} = \frac{x}{42}$ hours

Time taken from D to C = $\frac{\text{distance}}{\text{speed}} = \frac{x}{30}$ hours

$\frac{x}{42} + \frac{1}{15} = \frac{x}{30}$

$5x + 14 = 7x$

$x = 7$ miles

AO3 Geometry

Volume of trapezium prism = $\frac{1}{2}(3 + 8) \times 4 \times 6 = 132\,\text{cm}^3$

Volume of juice = $8 \times 6 \times 16 = 768\,\text{cm}^3$

Volume of juice in cuboid section when carton is upside-down

$= 768 - 132 = 636\,\text{cm}^3$

Let h = depth of juice in cuboid section

$636 = 8 \times 6 \times h$, so $h = 13.25\,\text{cm}$

Total depth = $13.25 + 4 = 17.25\,\text{cm}$

Problem-solving practice

AO2 Statistics

1 a 40 b 18

2 a

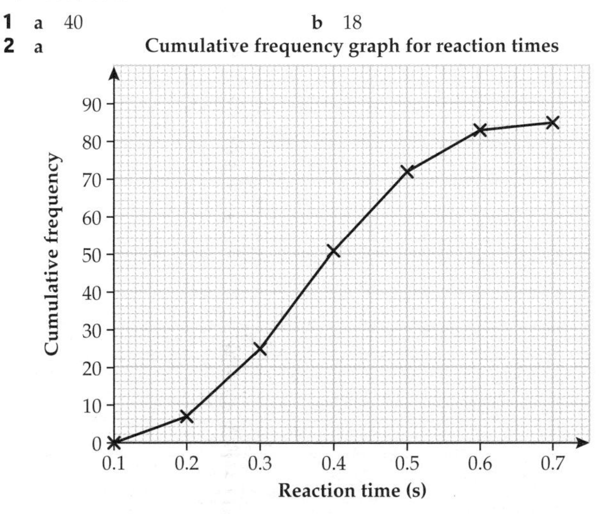

b 16

c

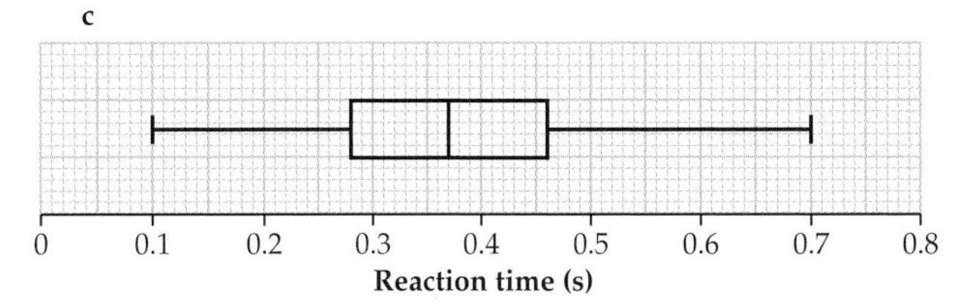
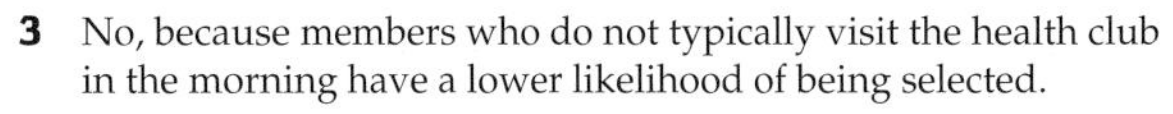

3 No, because members who do not typically visit the health club in the morning have a lower likelihood of being selected.

4 a, b

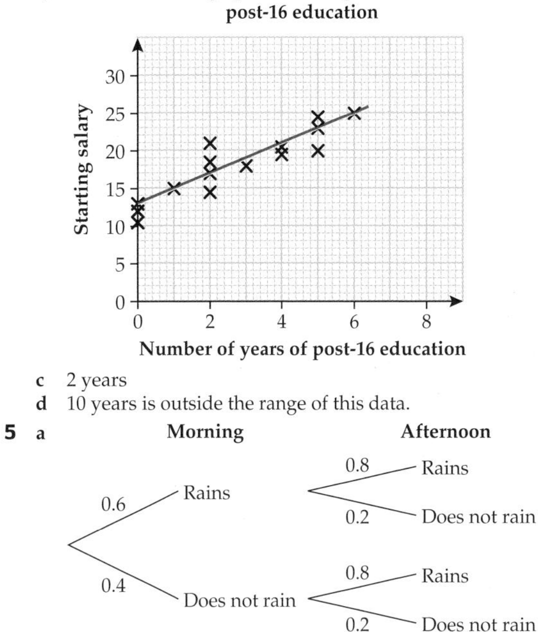

c 2 years

d 10 years is outside the range of this data.

5 a

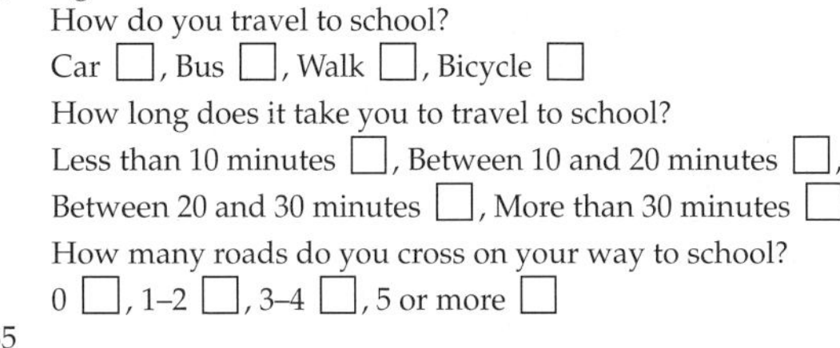

b 0.92 b 0.08

AO3 Statistics

1 e.g. 4, 4, 4, 4, 5.

2 a, b e.g.

How do you travel to school?

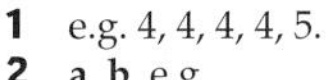

Car □, Bus □, Walk □, Bicycle □

How long does it take you to travel to school?

Less than 10 minutes □, Between 10 and 20 minutes □, Between 20 and 30 minutes □, More than 30 minutes □

How many roads do you cross on your way to school?

0 □, 1–2 □, 3–4 □, 5 or more □

3 35

4 £45

5 a No, Pippa will win a point half the time, but Nisha will only win a point one third of the time.

b Pippa: 75; Nisha: 50

6 a 36

b

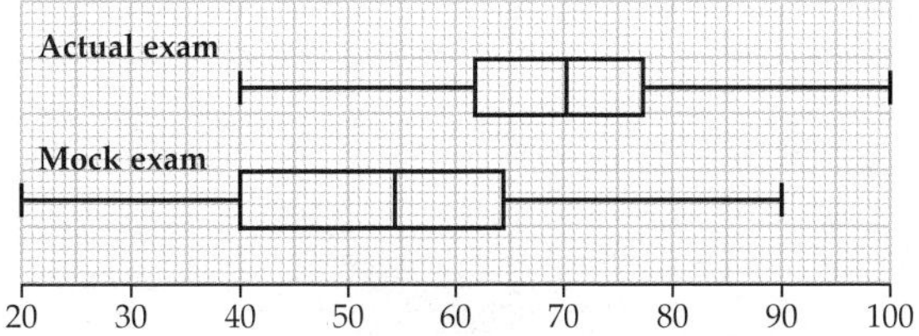

c The scores in the actual exam were better than the scores in the mock exam (higher median and quartiles). The results were more spread out in the mock exam (higher range and interquartile range). The less able students improved more than the more able students (greatest rise in lowest value and lower quartile).

A02 Number

1 55
2 No, the ratio of strawberry yoghurts to vanilla yoghurts is 3 : 5.
3 4 bottles
4 **a** £3.12 **b** £1.24
5 **a** 120 **b** 210 **c** 109.2 pence
6 51 kg
7 2 hours 24 minutes
8 8.4×10^{-5} m

A03 Number

1 858
2 12 hours 45 minutes
3 Business First is cheaper without direct debit (£253.80 per year). Leisure One is cheaper with direct debit (£208 per year instead of £228.42 per year).
4 240 g pack: £12.46 per kg; 400 g pack: £8.73 per kg. 400 g pack is better value.
5 Chloe's
6 62%
7 Theo: £67 500; Deborah: £112 500
8 7

A02 Algebra

1 19
2 LHS $= 4(x - 6) - x = 4x - 24 - x = 3x - 24 = 3(x - 8) =$ RHS
3 $x = 15$ cm
4 $4x + 5 = 33$; $x = 7$
5 **a** $n(n - 1)$
b Either n or $(n - 1)$ must be even, so their product is even.
6 **a** -82
b No. The solution to $118 - 4n = 40$ is $n = 19.5$, which is not a whole number.
7

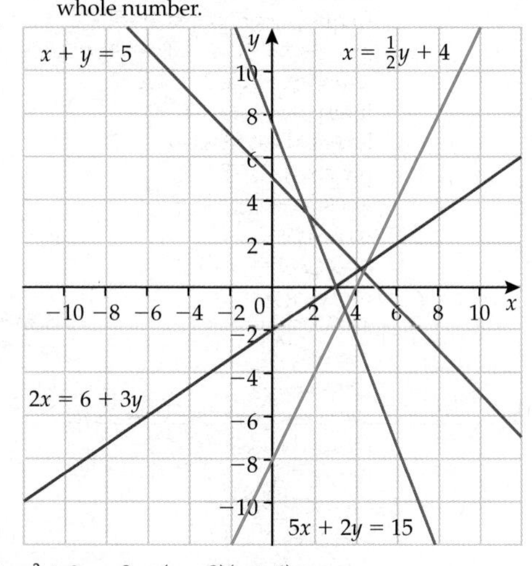

8 $x^2 + 2x - 8 = (x - 2)(x + 4)$
9 **a** $x(x + 10)$
b $x^2 + 10x = 75$; $x = 5$ (disregard solution $x = -15$)

A03 Algebra

1 **a**

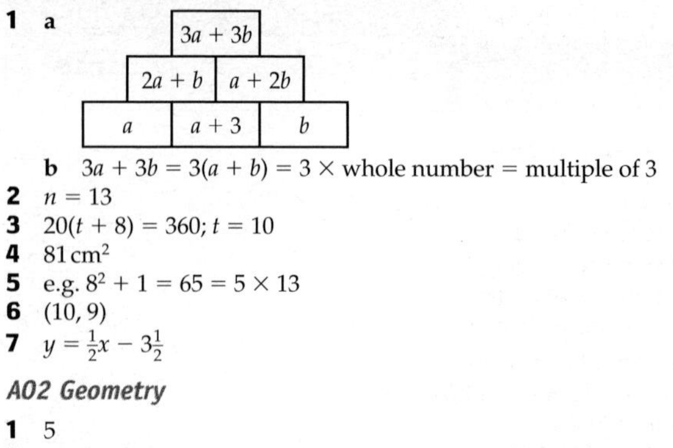

b $3a + 3b = 3(a + b) = 3 \times$ whole number = multiple of 3
2 $n = 13$
3 $20(t + 8) = 360$; $t = 10$
4 81 cm^2
5 e.g. $8^2 + 1 = 65 = 5 \times 13$
6 (10, 9)
7 $y = \frac{1}{2}x - 3\frac{1}{2}$

A02 Geometry

1 5
2 **a** Student's accurate construction

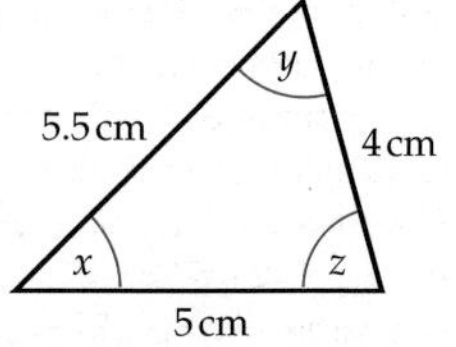

b Actual values (accept ±2°) $x = 44°$; $y = 61°$; $z = 74°$)
c Accept $x + y + z = 180°$ (angles in a triangle add up to 180°) or $x + y + z \neq 180°$ (rounding errors or inaccuracies of drawing/measurement)
3 **a** 360 cm^3 **b** 3.96 kg
4 9.85
5 **a, b** Student's accurate drawing

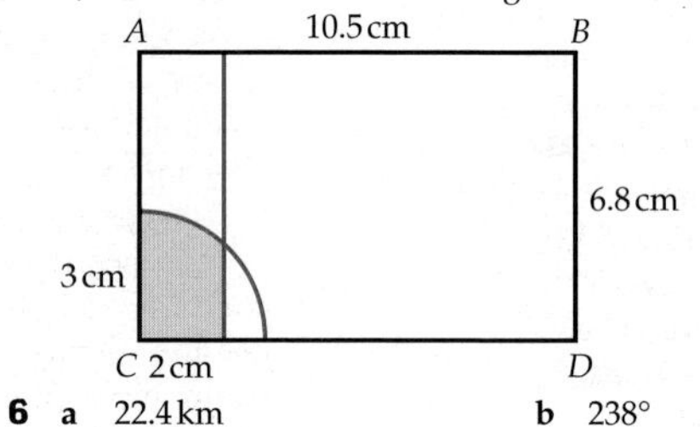

6 **a** 22.4 km **b** 238°

A03 Geometry

1 e.g. 100 km = 62.5 miles
Range of car on full tank $= 48 \times 9 = 432$ miles.
Tank is $\frac{1}{6}$ full so range $= 432 \div 6 = 72$ miles.
Dan has enough fuel.
2 337.5° (accept 338°)
3 No. Average speed of sprinter = 10.2 m/s = 36.7 km/h
4

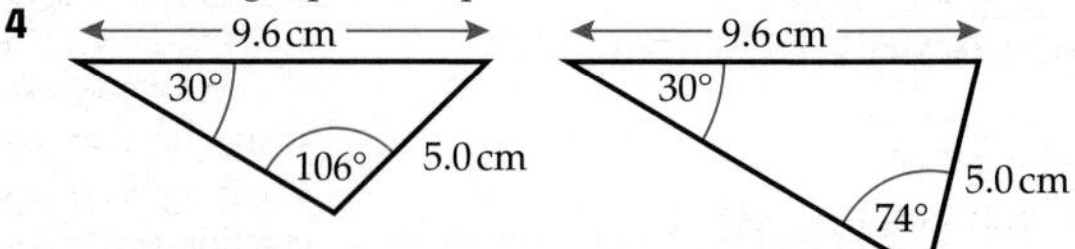

5 **a** 79 500 cm^3 **b** 79.5 *l*
6 **a** 90°; angles in a semicircle **b** 12.3 cm^2

Index

3-D objects 417–21

addition
  decimals 213–14
  fractions 200–3
  number skills 148
  probability 76, 87
  standard form 175–7
adjacent side of triangle 534–5
algebra 179–95
  collecting like terms 181–3
  expanding brackets 184–8, 191–4
  expressions 180, 183–4, 187–91
  factorising expressions 188–91
  FOIL mnemonic 192, 323
  multiplication 183–4
  problem-solving 275, 333–4
  quadrilaterals 391–3
  simplifying expressions 180, 187–8
  skills revision 379–80, 487
  writing expressions 180
alternate angles 345–6, 385
angles 340–56
  alternate 345–6, 385
  bearings 349–54
  circle theorems 550–5
  constructing 60° angle 508
  constructing bisectors 509–10
  corresponding 345–6, 385
  cyclic quadrilaterals 554–5
  elevation/depression 543–4
  parallel lines 345–9, 352–3
  polygons 396–400
  problem-solving 347–9
  proof 384–6
  skills revision 341–2
  supplementary 554
  triangles 369–70, 374, 385
  trigonometry 538–9, 543–4
  useful facts 341, 343–5, 385
approximation 152–3
arcs 504, 548, 552–4
area
  circles 448–51
  compound shapes 408–11
  converting units 461–2
  dimension theory 467–9
  rectangles 186, 206, 405
  triangles 405–6
averages 53–68
  cumulative frequency 96–107
  frequency tables 57–65
  speed 311–14, 463–5

base of index number 31
base of power 166
bearings 349–54, 540–1
best fit lines 43–6
bias 16, 85
bisector construction 507–10
box plots 101–3
brackets
  expansion 184–8, 191–4, 229–30
  skills revision 379
  solving equations 229–30
calculators
  complex calculations 132–42
  fractions 20, 22–4
  percentages 24–5, 28
  standard form 140–1
cancelling 21, 204–5, 207
capacity 453
centre of enlargement 476–9
centre of rotation 430–3
certainty 76
chords of circles 508, 548–9
circle theorems 547–57
  angle in semicircle 552
  angle subtended by arc/
    circumference 552–4
  circumferences 548, 552–3
  cyclic quadrilaterals 554–5
  properties of circles 548–51
  supplementary angles 554
circles 443–51
  area 448–51
  chords 508, 548–9
  circumference 444–7, 548, 552–3
  theorems 547–57
circumference of circle 444–7, 548, 552–3
class intervals
  cumulative frequency 102, 105
  grouped data 8–9, 47–9, 62–5, 97
collecting data 1–18
column vectors 428–30
common denominators 199
common factors
  algebraic expressions 188–90
  fractions 205
  highest 163, 170–1
  ratios 110
common multiples 160–1, 170–1
complex calculations 132–42
  repeated percentage change 133–5
  reverse percentages 135–7
  standard form 138–41
compound interest 133–5, 269–79
concentric circles 451
congruence 374–6, 423–4, 427
consecutive numbers 228
consecutive terms 279
constructions 503–10
  60° angle 508
  angle bisectors 509–10
  perpendicular at point on line 504–5
  perpendicular bisectors 507–8
  perpendicular from point to line
    505–6
  triangles 371–4
continuous data 5–6, 8, 46–50
conversion graphs 309–10
coordinates
  conversion graphs 310
  line segments 298–9
  polygons 400–2
  reflections 423–6
  straight-line graphs 300–2, 304, 307–8
correlation 41, 43–6
corresponding angles 345–6, 385
cosine (cos) 534
counter-examples 293–4
credit/credit cards 259, 265–6
cross-sections 411–12, 417, 451
cube roots 164–6
cubed numbers 243
cubic functions 498–500
cuboids 411–13, 420
cumulative frequency 95–108
  box plots 101–3
  data sets 104–7
  range 96, 98–101, 103–4
currency 124–5, 309–10
cyclic quadrilaterals 554–5
cylinders 451–6

data
  collection 1–18
  comparing sets 104–7
  continuous 5–6, 8, 46–50
  cumulative frequency 95–108
  data handling cycle 2–3, 42, 47
  discrete 5–6, 8, 36, 38
  frequency diagrams 46–8
  frequency polygons 48–50
  grouped 8–10, 46–50, 61–5, 96–101
  handling cycle 2–3, 42, 47
  interpreting 35–52
  lines of best fit 43–6
  pie charts 36–8
  questionnaires 13–15
  recording 11, 14
  representation 35–52
  sampling 15–17
  scatter diagrams 41–2
  sets 104–7
  sources 3–4
  stem-and-leaf diagrams 38–40
  two-way tables 10–12
  types 5–6
decimals 212–22
  addition 213–14
  division 216–19
  fractions 215–16, 219–21, 260
  multiplication 216–19
  percentages 28, 260
  probabilities 74
  ratios 115
  stem-and-leaf diagrams 40
  subtraction 213–14
  see also standard form
denominators 114, 199–200, 233–4
density 465–7
deposits 265–6
depreciation 267–8, 270, 272
depression, angle of 543–4
diagonals 391
diameters of circles 444–6, 548
difference of two squares 323–4
dimension theory 467–9
direct proportion 120, 124

discounts 261–2
discrete data 5–6, 8, 36, 38
distance 311–14, 463–5
division
  decimals 216–19
  divisors 150–1
  fractions 208–10
  indices 244–5
  methods 151–2
  negative numbers 237
  number skills 149–52
  powers 171–4
  proportion 123, 125–6
  ratio 110, 114–17
  repeated subtraction 150–1
  reverse percentages 136
  standard form 175–7
  straight-line graphs 306
  whole numbers 150–2
drawing
  box plots 101–3
  frequency diagrams 46–8, 96–101, 105
  linear graphs 297–321
  representing data 36–48
  scale 361–3
  tree diagrams 88–92

elevation, angle of 543–4
elevations of 3-D objects 418–20
elimination method 233–4, 237–8
enlargement 474–85
  centre of 476–9
  scale factors 475–6, 479–81
  similarity 475, 481–4
equations 224–34, 237–40
  balance method 224–6
  with brackets 229–30
  formulae 381–4
  fractional solutions 226
  containing fractions 233–4
  quadratic 322–32
  real-life problems 227–8
  simultaneous 237–40, 315–17
  skills revision 380
  straight-line graphs 301–9
  trial and improvement 386–8
  two-step 224–6
  unknown on both sides 230–3
  writing 227–8, 381–2
equidistance 511, 513
equivalent fractions 199–200
estimation 62–5, 81–5, 152–4
evaluating expressions 249–50
exchange rates 124–5, 309–10
expanding brackets 184–8, 191–4, 229–30
experimental probability 83
expressions
  algebraic 180, 183–4, 187–91
  evaluation 249–50
  factorising 188–91
  quadratic 487
  skills revision 379

factorising 188–91, 323–6, 329–30
factors 158–78
  algebraic expressions 188–90
  common factors 110, 163, 170–1, 188–90, 205
  factorising 188–91, 323–6, 329–30
  highest common factors 163, 170–1
  prime factors 162–3, 168–71
FOIL mnemonic 192, 323
formulae 248–57
  changing subject 255–7, 383–4
  equations 381–4
  skills revision 380
  substitution 249–54, 382–3
  writing formulae 248–9, 382–3
fractions 19–24, 198–211
  addition 200–3
  using calculator 20, 22–4
  comparisons 199–200
  coordinates 298
  decimals 215–16, 219–21, 260
  division 208–10
  equivalent 199–200
  improper 202–3, 205, 207
  lowest terms 202–3
  mixed numbers 202–3, 205–7
  multiplication 203–10
  one quantity as fraction of another 21–2
  percentages 260
  probability 74
  ratio 110, 113–15
  reciprocals 207–8
  scale factors 479–81
  solving equations 226, 233–4
  subtraction 200–3
frequency
  continuous data 46–8
  cumulative frequency 95–108
  data collection 6–7
  diagrams 46–8
  distributions 7
  expected events 81–2
  polygons 48–50
  relative frequency 82–5
  see also frequency tables
frequency tables 6–10
  averages 57–65
  cumulative frequency 97, 104
  grouped data 8–10, 61–5
  pie charts 36–7
  problem-solving 69

gradient (slope) 303–8
graphs
  conversion 309–10
  cubic functions 498–500
  inequalities 317–19
  linear 297–321
  non-linear 486–502
  quadratic equations 495–8
  quadratic functions 488–94
  real-life 311–15
  simultaneous equations 315–17
  solutions from intersections 316
  straight line 300–9, 314
grouped data
  averages 61–5
  frequency diagrams 46–8, 96–101
  frequency polygons 48–50
  range 61–3
  types of data 8–10

HCF (highest common factor) 163, 170–1
hectares 141
highest common factor (HCF) 163, 170–1
hire purchase 265–6
hypotenuse 375, 521–5, 534–5
hypotheses 2–4, 42

imperial units 24, 360–1
improper fractions 202–3, 205, 207
independent events 86–7, 89
index *see* indices
index numbers, RPI 31–2
indices 166–7, 242–8
  division 244–5
  index laws 171–4, 245–8
  multiplication 244–5
  notation 166–7, 170, 243–5
  standard form 174–7
  see also powers
inequalities 235–7
  balance method 236–7
  graphical solutions 317–19
  number lines 235–6
  standard form 138–9
inflation rates 135
information 3–4
  see also data
integers 138
inter-quartile range 96, 98–101, 103–4
intercepts 305–6, 308
interest 133–5, 266–7, 269–79
inverse operations 225–7
inverse proportion 126–7
investment of money 133–5
isometric paper 417–18

kites 394

LCM (lowest common multiple) 160–1, 170–1
leading questions 14
length 467–9, 536–8
like terms 181–3
likelihood 81, 83
linear functions 301
linear graphs 297–321
  conversion graphs 309–10
  inequalities 317–19
  mid-point of line segment 298–9
  rate of change 314–15
  real-life graphs 311–15
  simultaneous equations 315–17
  straight-line graphs 300–9
linear sequences 282, 289–90
lines of best fit 43–6
loan repayments 133
loci 510–17
  intersecting 515–17
  special 513–14
loss 267–8
lowest common multiple (LCM) 160–1, 170–1

maps 361–3
mass 465–7
maximum values 491
mean 54–7, 59–61, 63–5
measurement 357–67, 460–71
  accuracy 364–6
  area units 461–2
  density 465–7
  dimension theory 467–9

imperial units 24, 360–1
maps 361–3
metric units 24, 360–1
scale drawing 361–3
speed 463–5
timetables 358–9
volume units 462–3
median 54–9
cumulative frequency 96–103, 104–7
frequency tables 57–9, 61–3
metric units 24, 360–1
mid-points
grouped data 48–9, 63–5
line segments 298–9
minimum values 491
minus signs 154–5, 185
mirror lines 423–6
mixed numbers 202–3, 205–7
mode/modal value 54–5, 57–9, 61–3
money 122–4, 133–5, 263–7
multiples 159–61, 170–1, 282
multiplication
algebra 183–4
decimals 216–19
equations 239–40
fractions 203–10
indices 244–5
methods 150–1
negative numbers 192, 194, 237
number skills 149–52
percentages 134, 136
powers 171–4
probability 86
proportion 121, 125
ratio 110–12, 114, 118–19
simultaneous equations 239–40
standard form 175–7
whole numbers 150–2
multipliers 134, 136, 269–72, 475
'multiplying out' brackets 185
see also expanding brackets
mutually exclusive events 76–7

negative numbers 154–6
division 237
multiplication 192, 194, 237
negative powers 139
nets of 3-D objects 413, 417
non-linear graphs 486–502
number lines 235–6
number skills 147–57
addition 148
division 149–52
estimation 152–4
multiplication 149–52
negative numbers 154–6
problem-solving 72, 145–6
revision exercises 335–9
subtraction 148, 150–1
whole numbers 150–2
numerators 200

observation sheets 14–15
opposite angles 385
opposite side of triangle 534–5

parabolas 488
parallel lines
angles 345–9, 352–4
bearings 352–4
linear graphs 300–5, 307, 309
parallelograms 394–5, 405–6
patterns and sequences 286–8
pay calculations 121, 137
percentages 24–32, 133–7, 259–74
decimals 28, 260
fractions 260
increase/decrease 28–30, 135–7, 261–3
index numbers 31–2
money calculations 263–7
one quantity as percentage of another 26–7
original amount 28–9, 135–7, 261–2, 271–3
'percent' definition 24
probability 74–5
problem-solving 72
profit/loss 267–8
repeated change 133–5, 269–70
reverse 135–7, 271–3
scatter diagrams 42
simple 260
use in calculations 260
perimeter 405–11
algebra 182–3
circle parts 447
circumference 444–7
definition 228
perpendiculars 504–8, 548–9
pi (π) 444–51
pie charts 36–8
place value 213
planes of symmetry 420
plans 418–20
polygons 390–402
exterior angles 396–8
frequency 48–50
interior angles 398–400
quadrilaterals 390–6, 400–2, 554–5
population sampling 16
powers 158–78
cube roots 164–6
division 171–4
multiplication 171–4
square roots 164–6
see also indices; standard form
price index 31–2
primary data 3–4, 13
prime factors 162–3, 168–71
prime numbers 162–3
prisms 411–14, 451, 528
probability 73–94
'and' implying multiplication 86
of event happening 81–2
of event not happening 74–5
expected events 81–2
independent events 86–7, 89
mutually exclusive events 76–7
'or' implying addition 76
relative frequency 82–5
tree diagrams 88–92
two-way tables 78–80
problem-solving
algebra 275, 333–4
frequency tables 69
geometry 558–9
number skills 72, 145–6
Pythagoras' theorem 524–5, 527–8, 540–2
statistics 69, 143–4
volume 457
profit 267–8
proof 291–4, 384–6
proportion 120–2, 124–7
best buys 122–4
enlargement 475
exchange rates 124–5
unitary method 121–2, 124
see also ratio
Pythagoras' theorem 384, 520–32
circle theorems 549
hypotenuse 521–5
line segment length 529–31
shorter sides of triangle 526–9
solving problems 524–5, 527–8, 540–2
with trigonometry 540–2

quadratic equations 322–32
difference of two squares 323–4
factorising 323–6, 329–30
general form 324–6
graphical solutions 495–8
rearranging 327–8
solving 327–31
writing equations 330–1
quadratic expressions 487
quadratic functions 488–94
real-life use 493–4
symmetry 491–3
quadratic sequences 288–91
quadrilaterals 390–6
algebra 391–3
coordinates 400–2
cyclic 554–5
properties 391–2
special types 394–6
qualitative data 5–6
quantitative data 5–6
quartiles
inter-quartile range 96, 98–101, 103–4
lower quartile 96–102, 106
upper quartile 96–103
questionnaires 13–15

radius of circle 444–6, 548, 550–1
random sampling 16, 75
range 53–68
cumulative frequency 96, 98–101, 103–4
definition 54
frequency tables 57–9, 61–3
inter-quartile 96, 98–101, 103–4
rate of change 314–15
ratio 109–31
in form $1:n$ 115–16
enlargement 481–2
fractions 110, 113–15
in form $n:1$ 115–16
scale drawing 361–2
sharing in given ratio 116–17
simplifying 110–11
reciprocals 207–8
rectangles 186, 206, 394, 405–6
recurring decimals 220–1
reductions 28–9, 261–3

reflections 423–6, 436–9
regular polygons 396–400
relative frequency 82–5
remainders 150–1
representative samples 16–17
retail prices index (RPI) 31
rhombuses 394
right-angled triangles 375
  see also Pythagoras' theorem
roots of equations 330
roots of numbers 164–6, 327
rotations 430–9
rounding up/down
  number skills 150–1
  proportion 125–6, 128–9
  ratio 119
RPI (retail prices index) 31

sample space diagrams 86
sampling 15–17, 75
savings 133, 212
scale drawing 361–3
scale factors 475–6, 479–81
scatter diagrams 41–6
secondary data 3–4
segments of circles 548
semicircles 447, 450, 552
sequences 278–96
  linear 282, 289–90
  *n*th terms 281–5, 290–1
  patterns 286–8
  proof 291–4
  quadratic 288–91
  rules 281–4
  terms 279–85
significant figures 152–3
similarity 475, 481–4
simple interest 133
simultaneous equations 237–40, 315–17
sine (sin) 534
slope (gradient) 303–8
SOHCAHTOA mnemonic 534
speed 311–14, 463–5
spheres 141
square roots 164–6, 327
squared numbers 164–6, 243
squares 394
standard form 138–41, 158–78
  calculators 140–1
  conversion from decimals 139–40
  index laws 171–4
  numbers greater than 1 138
  numbers less than 1 139
  use of 138, 175–7
stem-and-leaf diagrams 38–40, 54–5
straight-line graphs 300–9, 314
  equations 303–9
  plotting 300–3
substitution methods 249–54, 382–3
subtraction
  decimals 213–14
  fractions 200–3
  number skills 148, 150–1
  standard form 175–7
  straight-line graphs 306
  use in division 150–1
supplementary angles 554
surcharges 273
surface area 411–14, 454–5
surveys 4, 13–17, 48
symmetry 420, 491–3

tally marks 6–7
tangent (tan) 534
tangents to circles 548, 550–1
terminating decimals 219–20
terms in algebra 181–3
terms of sequences 279–85, 290–1
tessellations 430
theoretical probability 83–4
theory verification 291–2
3-D objects 417–21
time
  distance–time graphs 311–14
  proportion 126–7
  speed 463–5
  timetables 358–9
transformations 422–42
  combined 436–9
  reflections 423–6, 436–9
  rotations 430–9
  translations 427–30, 436–9
translations 427–30, 436–9
trapezia 394, 407–8
tree diagrams 88–92
trial and improvement 386–8
triangles 368–77
  area 405–6
  congruence 374–6
  constructions 371–4
  interior/exterior angles 369–70, 385
  perimeter 406
  sum of interior angles proof 385
  see also trigonometry; right-angled triangles
trigonometry 533–46
  finding angles 538–9
  angles elevation/depression 543–4
  using calculator 534–5
  finding lengths 536–8
  with Pythagoras' theorem 540–2
  sin/cos/tan 534
  see also triangles
twelve hour clock 358
twenty-four hour clock 358
two-step equations 224–6
two-way tables 10–12, 14, 78–80

ungrouped frequency tables 57–61
units conversion
  area 461–2
  imperial 24, 360–1
  metric 24, 360–1
  volume 462–3

value added tax *see* VAT
value for money 122–4
VAT (value added tax) 263–4
  excluding VAT 30
  reverse percentages 137
  standard form 141
vectors 428–30
verification of theories 291–2
vertically opposite angles 343
volume 411–14
  converting units 462–3
  cylinders 451–3
  density 465–7
  dimension theory 467–9
  prisms 528
  problem-solving 457
  spheres 141

whiskers, box plots 101
whole numbers 150–2
word problems 118

x-axis 300–1

$y = mx + c$ 305–7
y-axis 300–1
y-intercepts 305–6, 308

zero product 329